Student's t (single mean)

$$t = \frac{\bar{x} - \mu}{s/\sqrt{n}}$$

Student's t (comparing two means)

$$t = \frac{(\bar{x}_1 - \bar{x}_2) - (\mu_1 - \mu_2)}{s\sqrt{\dfrac{1}{n_1} + \dfrac{1}{n_2}}}$$

Pooled Estimate of σ^2

$$s^2 = \frac{\sum\limits_{i=1}^{n_1} (x_{1i} - \bar{x}_1)^2 + \sum\limits_{i=1}^{n_2} (x_{2i} - \bar{x}_2)^2}{n_1 + n_2 - 2}$$

Correlation Coefficient

$$r = \frac{S_{xy}}{\sqrt{S_{xx}S_{yy}}}$$

where

$$S_{yy} = \sum_{i=1}^{n} (y_i - \bar{y})^2 = \sum_{i=1}^{n} y_i^2 - \frac{\left(\sum\limits_{i=1}^{n} y_i\right)^2}{n}$$

$$S_{xx} = \sum_{i=1}^{n} (x_i - \bar{x})^2 = \sum_{i=1}^{n} x_i^2 - \frac{\left(\sum\limits_{i=1}^{n} x_i\right)^2}{n}$$

$$S_{xy} = \sum_{i=1}^{n} (x_i - \bar{x})(y_i - \bar{y})$$

$$= \sum_{i=1}^{n} x_i y_i - \frac{\left(\sum\limits_{i=1}^{n} x_i\right)\left(\sum\limits_{i=1}^{n} y_i\right)}{n}$$

Least Squares Estimators of β_0 and β_1

$$\hat{\beta}_1 = \frac{S_{xy}}{S_{xx}} \text{ and } \hat{\beta}_0 = \bar{y} - \hat{\beta}_1\bar{x}$$

Least Squares Line (single independent variable)

$$\hat{y} = \hat{\beta}_0 + \hat{\beta}_1 x$$

Least Squares Equation (k independent variables)

$$\hat{y}_i = \hat{\beta}_0 + \hat{\beta}_1 x_{1i} + \hat{\beta}_2 x_{2i} + \cdots + \hat{\beta}_k x_{ki}$$

Mann–Whitney U (minimum of U_A and U_B)

$$U_A = n_1 n_2 + \frac{n_1(n_1 + 1)}{2} - T_A$$

$$U_B = n_1 n_2 + \frac{n_2(n_2 + 1)}{2} - T_B$$

Pearson's Chi-square Statistic

$$X^2 = \sum_{i=1}^{k} \frac{(n_i - np_i)^2}{np_i}$$

Spearman's Rank Correlation Coefficient

$$r_s = 1 - \frac{6\sum\limits_{i=1}^{n} d_i^2}{n(n^2 - 1)}$$

STATISTICS
FOR MANAGEMENT
AND ECONOMICS

The Duxbury Series

in Statistics and Decision Sciences

Applications, Basics, and Computing of Exploratory Data Analysis, Velleman and Hoaglin

Applied Regression Analysis and Other Multivariable Methods, Second Edition, Kleinbaum, Kupper, and Muller

Classical and Modern Regression with Applications, Myers

A Course in Business Statistics, Second Edition, Mendenhall

Elementary Statistics for Business, Second Edition, Johnson and Siskin

Elementary Statistics, Fifth Edition, Johnson

Elementary Survey Sampling, Third Edition, Scheaffer, Mendenhall, and Ott

Essential Business Statistics: A Minitab Framework, Bond and Scott

Fundamental Statistics for the Behavioral Sciences, Second Edition, Howell

Fundamentals of Biostatistics, Second Edition, Rosner

Fundamentals of Statistics in the Biological, Medical, and Health Sciences, Runyon

Introduction to Contemporary Statistical Methods, Second Edition, Koopmans

Introduction to Probability and Mathematical Statistics, Bain and Engelhardt

Introduction to Probability and Statistics, Seventh Edition, Mendenhall

An Introduction to Statistical Methods and Data Analysis, Third Edition, Ott

Introductory Statistical Methods: An Integrated Approach Using Minitab, Groeneveld

Introductory Statistics for Management and Economics, Third Edition, Kenkel

Linear Statistical Models: An Applied Approach, Bowerman, O'Connell, and Dickey

Mathematical Statistics with Applications, Third Edition, Mendenhall, Scheaffer, and Wackerly

Minitab Handbook, Second Edition, Ryan, Joiner, and Ryan

Minitab Handbook for Business and Economics, Miller

Operations Research: Applications and Algorithms, Winston

Probability Modeling and Computer Simulation, Matloff

Probability and Statistics for Engineers, Second Edition, Scheaffer and McClave

Probability and Statistics for Modern Engineering, Lapin

Quantitative Forecasting Methods, Farnum and Stanton

Quantitative Models for Management, Second Edition, Davis and McKeown

Statistical Experiments Using BASIC, Dowdy

Statistical Methods for Psychology, Second Edition, Howell

Statistical Thinking for Behavioral Scientists, Hildebrand

Statistical Thinking for Managers, Second Edition, Hildebrand and Ott

Statistics for Business and Economics, Bechtold and Johnson

Statistics for Management and Economics, Sixth Edition, Mendenhall, Reinmuth, and Beaver

Statistics: A Tool for the Social Sciences, Fourth Edition, Ott, Larson, and Mendenhall

Time Series Analysis, Cryer

Time Series Forecasting: Unified Concepts and Computer Implementation, Second Edition, Bowerman and O'Connell

Understanding Statistics, Fourth Edition, Ott and Mendenhall

STATISTICS
FOR MANAGEMENT
AND ECONOMICS

SIXTH EDITION

William Mendenhall
Professor Emeritus, University of Florida

James E. Reinmuth
University of Oregon

Robert Beaver
University of California, Riverside

PWS–KENT Publishing Company
Boston

PWS–KENT
Publishing Company

Library of Congress Cataloging-in-Publication Data

Mendenhall, William.
  Statistics for management and economics.
  Rev. ed. of: Statistics for management and economics/William Mendenhall . . . [et al.]. 5th ed. c1986.
  Includes index.
  1. Social sciences—Statistical methods.
2. Statistics. I. Reinmuth, James E., 1940– . II. Beaver, Robert J. III. Statistics for management and economics. IV. Title.
HA29.S78487 1989   519.5   88-30662
ISBN 0-534-91658-9

Printed in the United States of America.
90 91 92 93 — 10 9 8 7 6 5 4 3

Editor   *Michael Payne*
Production Coordinator   *Pamela Rockwell*
Production   *Technical Texts, Inc.*
Interior Design   *Elise Kaiser*
Cover Design   *Julie Gecha*
Typesetting   *Polyglot Pte. Ltd.*
Manufacturing Coordinator   *Margaret Sullivan Higgins*
Cover Printing   *New England Book Components*
Text Printing and Binding   *Halliday Lithograph/Arcata Graphics*

PREFACE

In the sixth edition of *Statistics for Management and Economics*, we have attempted to revise our text to improve the students' understanding of statistics and its use in managerial decision making. The third through fifth editions featured exercise sets, a majority of which were drawn from real-life settings. These features have been expanded in the sixth edition. The exercise sets that follow each section within a chapter are divided into three types: conceptual-level exercises, which are discussion questions based on the material within the section; skill-level exercises, which are technique exercises designed to give the student experience in applying the necessary methods without reference to any business or economic application; and application exercises, which allow the student to use statistical methods in a practical business or economic setting. Supplementary exercises at the end of each chapter are divided into two types: applications exercises, which are similar to the end-of-section application exercises; and interpretive exercises, in which the student is required to perform analyses and draw conclusions without being "led through" the solution. Once again, exercises are varied according to their primary area of application in management and economics. But, in addition, we have emphasized timely problems currently facing managers— problems resulting from the energy crisis and tax reform, issues related to codetermination in the workplace, and the need for improved productivity and zero quality control in the industrial sector. Furthermore, we have attempted to demonstrate— through certain leading questions within the exercises—that statistics is not an end unto itself but a tool that can be used to assist the manager in becoming a more effective decision maker.

Throughout the preparation of the sixth edition, our primary thrust has been to make our material more relevant and to increase its pedagogical effectiveness. We have revised the exercise sets in order to provide a broad range of practical, relevant exercises, many drawn from current periodicals and journals, that will be meaningful to those preparing for administrative careers in business or government. Also, case studies at the end of most chapters provide students with practical opportunities to investigate the application of statistics to business problems in a broader context than that offered by the exercises. At the same time, we have sought to retain the spirit of the first five editions by presenting a unified, streamlined treatment of the subject of statistics. To this end, statistical inference is offered in the first chapter as the objective of statistics, and the introduction to every subsequent chapter discusses the role of the new material in making inferences and measuring their reliability.

Some specific changes in the sixth edition are the following:

1. In discussions involving univariate data in Chapters 1–10, the response variable is denoted by x instead of y.
2. A new chapter, Chapter 15, on quality control has been added.
3. Two sections introducing the uniform and exponential distributions have been added to Chapter 6.
4. Chapter 8, concerning large-sample statistical inference, has been divided into two parts:
    Part I: Large-Sample Estimation
    Part II: Large-Sample Hypothesis Testing
5. The discussion of probability in Chapter 3 has been expanded to include probability tree diagrams and probability tables. These methods have been used in the explanation of Bayes' Law, Section 3.8.
6. The use of computer packages has been integrated throughout the text; MINITAB output has been updated and added when necessary.
7. Exercises, which have been expanded and updated, have been divided into conceptual, skill, application and interpretive levels.
8. Chapter 8, large-sample statistical inference, has been expanded to include a discussion of the power of a statistical test, along with examples and exercises.
9. The chapters on multiple regression, times series, and forecasting have been improved and expanded.
10. New case studies have been written for Chapters 2, 3, 4, 7, 9, 11, 15, 17, and 18.
11. To demonstrate the breadth of applicability of statistical methodology in management and economics, most exercises in this edition have again been indexed according to their primary area of application. While some exercises are purely instructional, most have been drawn from the functional areas of business and administration. The coding used to indicate the area of application is as follows:

Accounting	Government and public management	
Finance		
	Marketing MKTG	

Economics

Organization and management

Real estate

Transportation

Production and
operations management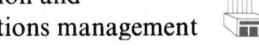

12. We have made many small but, to our mind, significant changes throughout
the text. They make it easier and clearer and also provide more motivation for
the student. Many have originated in student suggestions and in advice given
to us by the users of the past editions.

Consistent with our policy in the previous editions, the sixth edition of *Statistics for
Management and Economics* is designed for a two-term course in statistics. The first nine
chapters provide a natural breaking point for the first-term course, giving a continuous
treatment of probability and statistical inference. The latter half of the text, Chapters
10 through 18, offer an overview of the common methods used in the statistical analysis
of business decision problems. While Chapters 1 through 9 should be covered in order,
Chapters 10 through 18 may be taken in any sequence; however, the chapters on
regression, time series, and forecasting should be covered in order. Since Chapters 10
through 18 require an understanding of the concept of statistical inference, Chapters 1
through 9 must be covered before undertaking a study of the later chapters. As has been
the case in earlier editions, background preparation in college algebra is sufficient to
master the material within the sixth edition.

The authors are grateful to the editorial staff of PWS-KENT, especially to Michael
Payne, for their patience, assistance, and cooperation in the preparation of this edition.
Special thanks are due to Barbara Beaver for reading and editing rough drafts of
manuscript material and for her diligence in preparing the solutions manual for this
and previous editions. Thanks are due to Edward Alban, Savannah State College;
Chung Chen, Washington State University; Robert T. Clemen, University of Oregon;
Carolyn Cuff, University of Pittsburgh; Nicholas R. Farnum, California State
University at Fullerton; John Snyder, Sinclair Community College; Dennis D.
Wackerly, University of Florida; and F. C. Weston, Jr., Colorado State University for
their helpful reviews of the manuscript. We wish to thank authors and organizations
for allowing us to reprint selected material; acknowledgments are made where the
material appears in the text. Finally, we acknowledge the constant encouragement and
assistance of our families in this writing endeavor.

William Mendenhall
James E. Reinmuth
Robert Beaver

CONTENTS

CHAPTER 6

The Normal and Other Continuous Probability Distributions   235

CHAPTER 7

Sampling and Sampling Distributions   275

CHAPTER 8

Large-Sample Statistical Inference   317

CHAPTER 9

Inferences from Small Samples   391

CHAPTER 10

The Analysis of Variance   451

CHAPTER 11

Linear Regression and Correlation   521

CHAPTER 12

Multiple Linear Regression   575

CHAPTER 13

Elements of Time Series Analysis   653

CHAPTER 14

Forecasting Models   697

CHAPTER 15

Quality Control   761

CHAPTER 16

Survey Sampling   813

CHAPTER 17

Analysis of Enumerative Data   871

CHAPTER 18

Nonparametric Statistics   913

CHAPTER 19

Decision Analysis   985

STATISTICS
FOR MANAGEMENT
AND ECONOMICS

TO
THE
READER

What is statistics? The news media and research journals present statistical reports and forecasts concerning politics, the state of the economy, advances in health care, technical advances, business projections, and consumer preferences. Statistics is more than just "a bunch of numbers." In fact, statistics is a field of study that quantifies the information contained in sets of numbers and uses that information to make predictions or decisions.

Chapter 1 offers an introduction to the basic concepts used in statistical reasoning. Chapters 2 through 7 expand on these concepts and provide the basis for the remaining chapters in this text, which deal with the applications of statistics in making business decisions and predictions.

1

WHAT
IS
STATISTICS?

Illustrative Statistical Problems

What is statistics? How does it function? How does it help solve certain business problems? Rather than attempting a definition at this point, let us examine several problems that might come to the attention of the business statistician. From these business problems we can select the essential elements of a statistical problem.

Statistics plays an important role in accounting. For example, auditing the inventory of a large hospital can be very costly and time-consuming. Instead of counting and pricing the thousands of items in stock, the auditor can select a sample of items from the complete inventory. The value of these items actually in stock can be compared with the values shown in the hospital's records, and an estimate can be constructed of the ratio between the total value of supplies on hand to those shown in the hospital records.

Consider the example of an automobile dealership that has complete records of the prices at which its cars have been sold. Understanding and using all this information may be difficult if a complete list of all prices is presented. Some summary of this information—perhaps broken up by model of car, months of the year, or sales representative—would be more easily used.

Another example of the use of statistics is the sampling inspection of purchased items in a manufacturing plant. On the basis of an inspection, each lot of incoming goods must be either accepted or rejected and returned to the supplier. The inspection might involve drawing a sample of 10 items from each lot and recording the number of defectives. The decision to accept or reject the lot would then be based on the number of defective items observed.

Similarly, the production of a manufacturing plant depends on many factors unique to the type of manufacturing plant under consideration. By observing sample data collected on these factors and the production over a period of time, we can construct a prediction equation relating production to the observed factors. Then the plant's future production for a given set of factor conditions can be predicted by substituting values of the factors into the prediction equation. Methods of identifying the important factors needed for the prediction equation as well as a method for assessing the error of prediction will be discussed in subsequent chapters.

Marketing research offers another example of a prediction problem. A representative sample of customers is selected, and each person is asked to give an opinion concerning a manufacturer's product. From the data obtained in this opinion survey, the market analyst must decide whether a sufficient demand exists for the product. If the demand exists, the analyst must select the package design, the best selling price, and the market area. All these questions can be answered from information derived from the sample survey data.

These situations illustrate the need to summarize data and to use those summaries to predict, estimate, and, ultimately, make business decisions. As you will subsequently see, **modern statistics offers a variety of analytical procedures to aid the business statistician in making decisions in the presence of uncertainty.** Of course, we are not implying here that uncertainties exist only within a business context or that modern statistics applies only to business. However, our emphasis in this text is on showing the applications of statistical techniques to business problems. Other areas of application are suggested through the exercises at the end of each chapter.

1.2

The Population and the Sample

The preceding examples are varied in nature and complexity, but each involves summarizing and using the information contained in a set of measurements. In addition, most involve sampling, in which a specified number of items (objects or bits of information)—a *sample*—are drawn from a much larger body of data called the *population*. Note that the word *population* refers to data and not to people. Recording the daily sales for a perishable commodity gives a sample of all possible daily levels of demand (the population) that have occurred in the past or may occur in the future. In the sampling inspection problem we assume that each sample of 10 items is a representative sample of the lot (the population) from which it was selected. The market researcher draws a sample of opinions from the statistical population that represents the entire potential market for her product.

Definition

population

> A **population** is the set representing all observations of interest to the sample collector.

In all the previous examples we were primarily interested in the population. However, in most instances to conduct a **census** of the entire population is impractical or far too costly. So we select a sample that we hope is a small-scale representation of the underlying population (see Figure 1.1). The sample may be of immediate interest, but ultimately, we are interested in describing the population from which the sample is drawn.

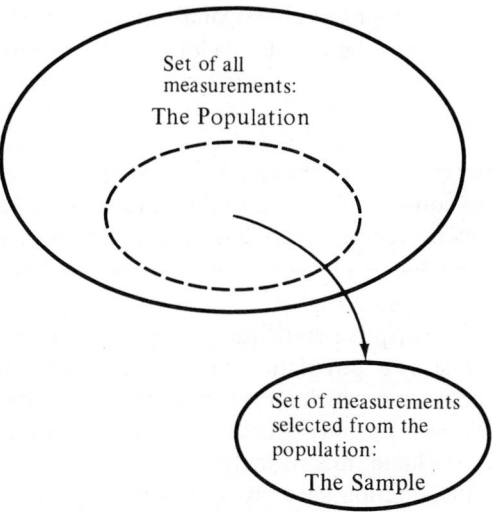

Figure 1.1   Population and Sample

Definition

sample

> A **sample** is a subset of measurements selected from the population of interest.

As used by most people, the word *sample* has two meanings. It can refer to the set of objects on which measurements are to be taken, or it can refer to the measurements themselves. A similar double use can be made of the word *population*.

For example, you read in the newspapers that a Gallup poll was based on a sample of 1,823 people. In this use of the word *sample*, the objects selected for the sample are clearly people. Presumably, each person is interviewed on a particular question, and that person's response represents a single item of data. The collection of data corresponding to the people represents a sample of data.

In a study of sample survey methods we must distinguish between the objects measured and the measurements themselves. To experimenters, the objects measured are called **experimental units**. The sample survey statistician calls them **elements of the sample**.

1.3

Descriptive and Inferential Statistics

Statistical procedures can be classified as belonging to one of two categories: descriptive statistics or inferential statistics. When confronted with a set of measurements, we must first decide how to summarize and quantify the information to best describe and highlight the salient characteristics of the set. Techniques for summarizing information in the form of graphs or charts, and techniques for producing numerical summaries such as averages or percentiles, belong to the category called descriptive statistics.

Definition

descriptive statistics

> **Descriptive statistics** consists of procedures used to summarize the information in a set of measurements and to describe the characteristics of the set.

The procedures of descriptive statistics are employed to describe the measurements in either a sample or a population. Although we usually think in terms of samples, there are situations in which all the items in a population can be measured. Consider a small business with only a limited line of products. When it is time to account for inventory on hand, measuring the entire inventory of the firm may be relatively easy. In such a situation descriptive statistics alone would suffice, since the firm would only want to describe the set of measurements at hand.

Suppose that you cannot measure all the items in a population; instead, you are able to measure only the items in a sample selected from the population. Your objective here would consist not only of describing the sample measurements but also of eventually using the sample information to arrive at general conclusions concerning the population at large.

Definition

inferential statistics

> **Inferential statistics** consists of procedures used to make inferences about population characteristics from information contained in a sample.

The examples cited in the introduction are situations in which inferential statistical techniques would be appropriate.

Before implementing any inferential procedures, we must be able to describe the characteristics of the population as well as the characteristics of the sample. The major portion of this text deals with inferential statistics, but Chapter 2, in which a variety of descriptive techniques are presented, deals mainly with descriptive statistics. In fact, the descriptive measures calculated from sample measurements are the quantities used in making inferences about the corresponding population descriptive measures.

1.4

The Parts of an Inferential Statistical Problem

Some statistical problems end at the descriptive phase. However, most statistical problems are concerned with making inferences about the characteristics of the population, using the information available in a sample drawn from that population. In focusing attention on the inferential aspects of a statistical problem, we can state our objective in the following way.

The Objective of Inferential Statistics

> The **objective of statistics** is to make **inferences** (predictions, decisions) about certain characteristics of the population based on information contained in a sample.

How will we achieve this objective? We will find that every statistical problem involves five parts. By successfully completing each part, we can achieve the objective of statistics.

The first and most important part of a statistical problem is a clear specification of the question to be answered and of the population of data that is related to the question.

The second part of a statistical problem is deciding how the sample will be selected. This part is called the *design of the experiment* or the *sampling procedure*. This second part is important because data cost money and time. In fact, it is not unusual for a business survey to cost $50,000 to $500,000, and the costs of many technological experiments can run into the millions. And what do these experiments and surveys produce? Numbers on a sheet of paper—in brief, information. So planning the experiment is important. Including too many observations in the sample is often costly and wasteful, but including too few observations often results in too little useful information. In addition, the way the sample is selected will often affect the amount of information per observation. Thus, a good sampling design can sometimes reduce the costs of data collection to one-tenth or as little as one-hundredth of the cost of another sampling design.

The third part of a statistical problem involves the analysis of the sample data. No matter how much information the data contain about the practical question, you must use the appropriate method of data analysis to extract the information from the data.

The fourth part of a statistical problem is using the sample data to make an inference about the population. As you will subsequently learn, many different inferential procedures can be employed to make an estimate or decision about some characteristic of a population or to predict the value of some member of the population. For example, 10 different methods might be available to forecast a company's sales, but one procedure might be much more accurate than another. Therefore, you will wish to employ the best inference-making procedure when you use sample data to make an estimate or decision about a population or a prediction about some member of a population.

The final part of a statistical problem identifies what is perhaps the most important contribution of statistics to business decision making. It answers the question, "How good is this inference?" To illustrate, suppose someone conducts a statistical survey for you and estimates that your company's product will gain 34% of the market next year. We hope that you will not be satisfied with this inference but will ask, "How accurate is the estimate?" Of what value is the estimate without a measure of its reliability? Is the estimate accurate to within 1%, 5%, or 20%? Is it reliable enough to be used in setting production goals? As you will subsequently learn, statistical procedures for estimation, decision making, and prediction enable you to calculate a measure of goodness for every inference. Consequently, in a practical inference-making situation, every inference should be accompanied by a measure that tells you how much confidence you can have in the inference.

Parts of a Statistical Problem

1. A clear specification of the question and the population of data related to the question
2. The design of the experiment or the sampling procedure
3. The collection and analysis of data
4. The procedure for making inferences about the population based on sample information
5. The provision of a measure of goodness (reliability) for the inference

The order of presentation of topics in the next several chapters is intended to lay the foundation required for implementing the procedures associated with the various parts of a statistical problem. The estimation and testing procedures presented in Chapters 8 and 9 are expansions of steps 4 and 5 listed in the previous display. The remainder of the text is devoted to a survey of various statistical topics, all of which can be viewed in the context of the general framework presented in Figure 1.2.

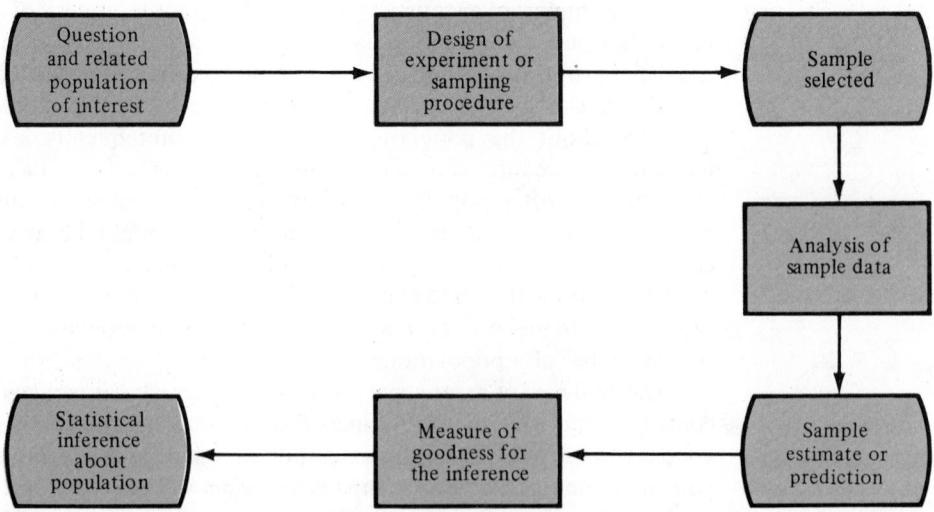

Figure 1.2   The Parts of an Inferential Statistical Problem

1.5

The Statistician and Business Decision Making

How does statistics contribute to business decision making? One of the major contributions of the business statistician is in planning surveys and experiments— buying a specified quantity of information at minimum cost. As an example, a knowledge of market conditions for both raw materials and a company's products is an essential prerequisite to making business decisions. The price of this information, the cost of the survey or experiment to assess market conditions, can be greatly reduced by a good sampling design (part 2 of the statistical problem). A business statistician can also aid in the selection of an appropriate method of data analysis and of a good inference-making procedure (parts 3 and 4 of the statistical problem).

The other major contribution of business statisticians is in providing a measure of goodness (i.e., reliability) to go with each inference. The measure of goodness tells you how much faith you can place in the inference. For example, if you were a businessperson deciding whether to make a capital investment, you would like to know the future course of the economy. A forecast would be helpful, but only if you knew whether the forecast was reliable. An unreliable forecast could lead you to an erroneous and costly decision. Statistical procedures address this problem, providing a measure of goodness for every decision, estimate, or prediction.

1.6

Summary

Statistics is a branch of science concerned with the **design of experiments** or sampling procedures, the **analysis of data**, and the procedures for making inferences about a **population** of measurements from information contained in a **sample**. Statisticians are concerned with developing and using procedures for **design analysis** and for **inference making** that will provide the best inferences for a given expenditure. In addition to making the best inference, they are concerned with providing a quantitative measure of the **goodness** of the inference-making procedure. Also, business statisticians must effectively communicate their experimental conclusions to management.

1.7

Looking Ahead

We have stated the objective of statistics and, we hope, have answered the question, "What is statistics?" The remainder of this text is devoted to the development of the basic concepts involved in statistical methodology. In other words, we wish to explain how statistical techniques actually work and why they work.

Statistics is a theory of information in which applied mathematics plays a major role. Most of the fundamental rules (called *theorems* in mathematics) are developed and based on a knowledge of the calculus or higher mathematics. Since this book is an introductory text, we omit proofs except where they can be easily derived. Where concepts or theorems are intuitively reasonable, we attempt to give an informal explanation. Hence, we attempt to convince you with the aid of examples and intuitive arguments rather than with rigorous mathematical derivations.

As in any other business technique, common sense must be employed when applying statistical methods to the solution of a real problem. Since inferences about a population are made from sample data, the inferences are meaningful only if the sample is selected from the population of interest. For example, if we proposed a study to find the average height of all students at a particular university, we would not select our entire sample from the members of the basketball team. The basketball players should be represented within a sample in approximately the same proportion that they exist within the entire university. Another common misuse of statistics is in making comparisons. The National Safety Council claims that on a holiday weekend there are about twice as many highway deaths as on an ordinary weekend. We might conclude that driving on a holiday weekend is much less safe than driving on an ordinary weekend. But this conclusion is true only if the traffic volume is about the same for all weekends—an unrealistic assumption. It is important that we view all statistics and sets of data with a critical eye and apply common sense and intuition about the problem to our decision format before arriving at a conclusion. The conclusions that apply to a particular problem are unique to that problem and seldom apply to other or related problems.

As you read through the text, refer occasionally to this chapter and review the objective of statistics, its role in business research, and the parts of a statistical problem. Each of the following chapters should, in some way, be directed toward answering the questions posed here. Each is essential to completing the overall picture of the role of statistics in business, management, and economics.

1.8

Communicating Results via Case Studies

Almost all of the remaining chapters close with the presentation of a case study, that is, a problem or situation that can be answered or resolved by using the appropriate statistical concepts and techniques covered in that chapter. In general, the case studies presented are real-life scenarios in which practical problems are solved by following the required steps inherent in the five parts of a statistical problem. Case studies are meant to unify statistical concepts and procedures and to demonstrate how these concepts and procedures are applied in solving a broad spectrum of problems that arise in the areas of business, management, and economics.

Look for the case study at the end of each chapter. In addition to showing you how the chapter material can be used to solve practical problems, case studies provide an opportunity to see how statistical solutions and results are interpreted and communicated to the person or persons with whom the problem originated.

EXERCISES

1.1 What is the distinction between inferential and descriptive statistics?

1.2 Why does business statistics usually involve the use of sample data, instead of the entire population, when one is making business decisions?

1.3 What is the utility of an estimate without an accompanying measure of reliability?

1.4 Within each of the following decision-making situations, identify the population of interest.

    a. A market research firm seeks to explore the market acceptability of a new brand of soft drink.

    b. An investment counselor would like to compare the performance of securities listed on the New York Stock Exchange with those listed on the over-the-counter market.

    c. Concerned with cash shortages, the controller of a chain of department stores is interested in determining the percentage of delinquent credit accounts.

    d. The city manager of a southwestern city seeks to determine whether a capital construction project is likely to be approved in an upcoming municipal election.

    e. Before offering a bid on a large stand of timber, a plywood manufacturer would like to determine the amount (in board feet) of usable timber contained within the stand.

1.5 Refer to Exercise 1.4. Specify how an appropriate sample might be chosen in each decision-making situation. Recall that the sample consists of a set of measurements, not a set of objects.

1.6 The brewing industry has become increasingly competitive in recent years, with national, regional, and local breweries all offering a wide range of products intended to appeal to the varying preferences of the consuming public. Recently, an American brewery discovered an imported beer that could have good U.S. market potential. To determine whether the imported beer would be publicly accepted, the brewer decided to develop a test market—that is, to offer the beer for sale in two or three cities and adjacent suburban areas. The results of this study would be used to reach a decision on whether or not to market the imported beer.

    a. Provide a clear statement of the objective of the test market study.

    b. Define the population of interest related to the objective of the test market study.

    c. Discuss the selection of an appropriate sample for analysis in the test market study.

    d. Why is it advantageous for the brewery to base its new-product decision on sample information instead of information from the entire population?

1.7 A nationwide survey was conducted to determine which issues were of greatest concern among Americans. Each respondent in the survey was randomly selected according to a sampling plan reflecting the proportion of individuals in categories defined by several demographic variables such as age, sex, income, and geographic region. Participants were asked to specify the national problem that caused them the most concern. Some typical responses were poverty, drug abuse, unemployment, and the federal budget deficit.

    a. What is the response that will be measured in this survey?

    b. Define the population of interest to the experimenter.

    c. Describe the sampling procedure used by the experimenter.

    d. What demographic groupings might the experimenter consider as subpopulations within the main population to be studied concerning their response to the survey?

References and Suggested Readings

Careers in Statistics. American Statistical Association and the Institute of Mathematical Statistics, 1979.

Huff, D., and I. Geis. *How to Lie with Statistics*. New York: Norton, 1954.

Markowitz, H. *Portfolio Analysis*. New York: Wiley, 1959.

Reichmann, W. J. *Use and Abuse of Statistics*. London: Methuen, 1961.

Tanur, J. M., et al. *Statistics: A Guide to the Unknown*. 2d ed. San Francisco: Holden-Day, 1978.

TO
THE
READER

Since our objective is to make predictions or inferences about a population based on information contained in a sample from that population, we must first be able to describe a set of measurements in a way that focuses attention on the salient characteristics of the set.

Chapter 2 is concerned with descriptive statistics, which can be divided into two general categories. Graphical descriptive techniques are used to produce a visual description of the data values through the use of charts, graphs, and frequency tabulations. Numerical descriptive techniques are used to summarize and capture the same salient characteristics available through graphical techniques. In fact, inferences about populations are most often phrased in terms of numerical descriptive measures such as the mean and the standard deviation. In Chapter 2 we present descriptive techniques that will serve as a basis for making inferences about population measures based on summary information observed in a sample from that population.

2

DESCRIBING
SETS OF
MEASUREMENTS

2.1

Introduction

The objective of statistics is to make inferences about a large body of data, which we call the population, based on information contained in a sample from that population. A question immediately arises. How do we describe a set of measurements, whether they are from a sample or represent the entire population? If the population were before us, how would we describe this large set of measurements?

Many texts have been devoted to the methods of descriptive statistics—methods used in describing sets of quantitative or qualitative data. These methods can be categorized essentially as *graphical methods* and *numerical methods*. We will restrict our discussion to a few graphical and descriptive measures that are useful not only for descriptive purposes but also for statistical inference. Some common numerical methods found in other texts have been excluded from our discussion to preserve continuity. Such methods have become redundant with the availability of electronic desk calculators and computers and would contribute little, if anything, to the main objective of our study. If you are interested in descriptive statistics not contained in this chapter, refer to the texts in the references at the end of this chapter.

The graphical methods that follow can be applied either to a set of population measurements or to a set of sample measurements without making a specific distinction as to which is involved. Numerical descriptive measures also apply to both population and sample measurements, but different symbols are used to designate whether the measure was obtained from a set of population measurements or from a set of sample measurements. Numerical measures computed from sample measurements are pivotal in making inferences about the corresponding measures in the population.

2.2

Frequency Distributions

When presented with a set of data, most people have difficulty making any sense out of it. One solution to this problem is to present the data in a pictorial or graphical display. One such statistical display, called a frequency histogram, can be explained by example.

Individuals and organizations attempt to maintain investment portfolios that provide for a maximum return at tolerable levels of risk. One measure of the potential return and inherent risk of a security is its price-earnings (P/E) ratio. Generally, securities with a low P/E ratio are preferred to those with a high ratio. The data in Table 2.1 represent the P/E ratios for 25 different common stocks of growth companies listed on the over-the-counter markets.

You can quickly see, by examining Table 2.1, that the largest and smallest P/E ratios are 28.6 and 5.4, respectively. But how are the remaining 23 ratios distributed between the largest and the smallest? To answer this question, we divide the interval into an arbitrary number of subintervals. As a rule of thumb, the number of sub-intervals should be from 5 to 20; the more data available, the more subintervals we will employ. These subintervals, or **classes**, are chosen so that each measurement can fall in *one and only one* subinterval. The measurements are then categorized according to the class into which they fall, and the resulting tabulation is presented graphically in the form of a **frequency histogram**.

classes

frequency histogram

Table 2.1   Price-Earnings Ratios for 25 Common Stocks

20.5	19.5	15.6	24.1	9.9
15.4	12.7	5.4	17.0	28.6
16.9	7.8	23.3	11.8	18.4
13.4	14.3	19.2	9.2	16.8
8.8	22.1	20.8	12.6	15.9

For the P/E ratios of Table 2.1, we choose to use six intervals of equal width. Since the length of the total interval is $(28.6 - 5.4) = 23.2$, a convenient choice is to use six classes of length 4.0. Rather than begin the first interval at the lowest value, 5.4, we choose a more convenient starting value, 5.00, and form the subintervals 5.00 to 8.99, 9.00 to 12.99, 13.00 to 16.99, and so forth. Notice that each measurement can fall in one and only one subinterval. The 25 measurements are now categorized according to the class into which they fall, as shown in Table 2.2.

The six classes are labeled from 1 to 6 for identification purposes. The boundaries for the six classes, along with a tally of the number of measurements falling in each class, are given in the second and third columns of the table. The fourth column gives the number of measurements falling into a particular class, say class i, called the **class frequency** and designated by f_i. The last column of the table presents the fraction or proportion of the total number of measurements falling into each class. We call this

class frequency

Table 2.2   *Relative Frequencies for the 25 Price-Earnings Ratios*

Class, i	Class Boundaries	Tally	Class Frequency, f_i	Class Relative Frequency
1	5.00–8.99	|||	3	3/25
2	9.00–12.99	|||||	5	5/25
3	13.00–16.99	||||| ||	7	7/25
4	17.00–20.99	|||||		6	6/25
5	21.00–24.99	|||	3	3/25
6	25.00–28.99			1	1/25
			25	1

proportion the *class relative frequency.* If we let n represent the total number of measurements—for instance, in our example $n = 25$—then the relative frequency for the ith class is f_i divided by n:

$$\text{relative frequency} = \frac{f_i}{n}$$

The resulting tabulation can be presented graphically in the form of a frequency histogram, as in Figure 2.1. In a frequency histogram, rectangles are constructed over each class interval, their height being proportional to the number of measurements (class frequency) falling into each class interval. When we look at the frequency histogram, we can see at a glance how the price-earnings ratios are distributed over the interval from 5.0 to 29.0.

It is often more convenient to modify the frequency histogram by plotting class relative frequency rather than class frequency. A *relative frequency histogram* is

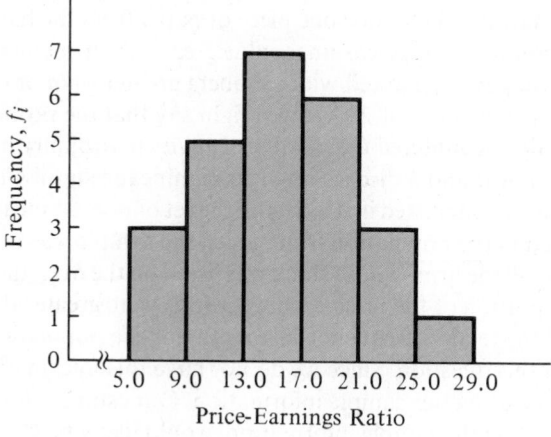

Figure 2.1   *Frequency Histogram*

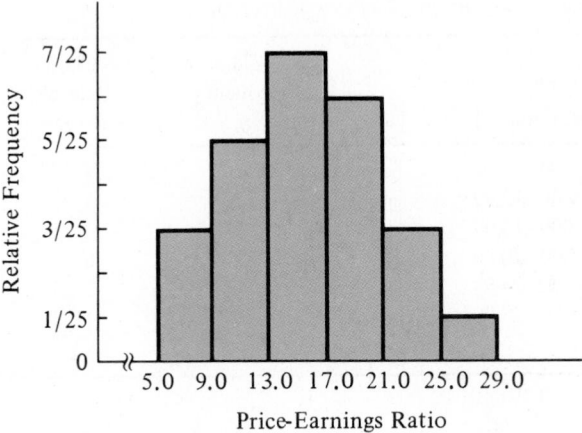

Figure 2.2   Relative Frequency Histogram

presented in Figure 2.2. Statisticians rarely make a distinction between the frequency histogram and the relative frequency histogram and refer to either as a frequency histogram or simply a histogram. If corresponding values of frequency and relative frequency are marked along the vertical axes of the graphs, then the frequency and relative frequency histograms are identical (compare Figures 2.1 and 2.2).

Let us consider the relative frequency histogram for the sample in greater detail. What proportion of the firms had price-earnings ratios equal to 17.0 or greater? Checking the relative frequency histogram, we see that the proportion involves all classes to the right of 17.0. Using Table 2.2, we see that 10 companies had price-earnings ratios greater than or equal to 17.0. Hence, the proportion is 10/25, or 40%. We note that this value is also the percentage of the total area of the histogram in Figures 2.1 and 2.2 that is to the right of 17.0.

Suppose that we write each of the 25 price-earnings ratios on a piece of paper, place them in a hat, and then draw one piece of paper from the hat. What is the chance that this paper contains a price-earnings value greater than or equal to 17.0? Since 10 of the 25 pieces of paper are marked with numbers greater than or equal to 17.0, we say that we have 10 chances out of 25. Or we might say that the probability is 10/25. You have undoubtedly encountered the word *probability* in ordinary conversation; we will defer a definition of it and a discussion of its significance until Chapter 3.

Although we are interested in describing the set of $n = 25$ measurements, we might also be interested in the population from which the sample was drawn, which consists of P/E ratios for all the firms whose stock was listed on the over-the-counter markets in 1988. What proportion of the price-earnings ratios were greater than or equal to 17.0? If we possessed the relative frequency histogram for the population, we could give the exact answer to this question. Since we do not have this information, we are forced to make an *inference* using our sample information. Our estimate for the true population proportion, based on the sample information, would likely be 10/25, or 40%. Without knowledge of the population relative frequency histogram, we would infer that the population histogram is similar to the sample histogram and that approximately 40%

of the price-earnings ratios in the population are greater than or equal to 17.0. Most likely this estimate would differ from the true population percentage. We will examine the magnitude of this error in Chapter 8.

frequency
distribution

The relative frequency histogram is often called a **frequency distribution** because it shows the manner in which the data are distributed along the abscissa (horizontal axis) of the graph. We note that the rectangles constructed above each class are subject to two interpretations. They represent the proportion of observations falling into a given class. Also, if a measurement is drawn from the data, a particular class relative frequency is also the chance or probability that the measurement will fall into that class. The most significant feature of the sample frequency histogram is that it provides information about the population frequency histogram that describes the population. An important point to note here is that different samples from the same population will result in different sample histograms, even when the class boundaries remain fixed. However, we would expect the sample and population frequency histograms to be similar. The degree of resemblance will increase as more and more data are added to the sample. If the sample were enlarged to include the entire population, the sample and population would be the same and the histograms would be identical.

In the preceding discussion we showed how to construct a frequency distribution for the price-earnings ratio data of Table 2.1, and we explained how such a distribution could be interpreted. Before concluding this topic, we summarize the principles one should employ in constructing a frequency distribution for a set of data.

Principles for Constructing a Frequency Distribution

1. **Determine the number of classes.** It is usually best to have from 5 to 20 classes. The larger the amount of data available, the more classes should be employed. If the number of classes is too small, we might be concealing important characteristics of the data by grouping. If we have too many classes, empty classes may result and the distribution will be meaningless. The number of classes should be determined from the amount of data present and the uniformity of the data. A small sample would require fewer classes.

2. **Determine the class width.** As a general rule for finding the class width, divide the difference between the largest and the smallest measurements by the number of classes desired and add enough to the quotient to arrive at a convenient figure for class width. All classes, with the possible exception of the smallest and largest classes, should be of equal width. Classes of equal width allow us to make uniform comparisons of the class frequencies.

3. **Locate the class boundaries.** Begin with the lowest class so that you include the smallest measurement. Then add the remaining classes. Class boundaries should be chosen so that it is impossible for a measurement to fall on a boundary.

To illustrate the use of these principles, suppose we wish to group the 36 incomes of the employees of a small firm into five classes, where the smallest income is $5,500 and the largest is \$29,500. Applying our rule, we have

$$\text{class width} = \frac{\$29,500 - \$5,500}{5} = \$4,800$$

Using \$5,000 as the class width may be more convenient. The class boundaries would then be

$5,000-$9,999
$10,000-$14,999
$15,000-$19,999
$20,000-$24,999
$25,000-$29,999

Here, none of the principles for constructing a frequency distribution is violated. However, if the data were widely dispersed, we might wish to make the lower or upper class open-ended. For instance, if the president's \$100,000 salary were included as the 37th income value, the largest class would be \$25,000 and up. If we were to apply the class width rule, empty classes would occur and would confuse the interpretation of our results.

A histogram for a set of data can be produced using the command HISTOGRAM within the data analysis software package called MINITAB™*, an interactive program package available at most computer facilities. The program automatically chooses classes with conveniently rounded midpoints. If a measurement falls on a class boundary, it is assigned to the class with the larger midpoint. The printout gives the

Table 2.3 MINITAB-Generated Histogram for the Data in Table 2.1

```
MTB>HISTOGRAM C4

MIDDLE OF        NUMBER OF
 INTERVAL       OBSERVATIONS
     6               1        *
     8               2        **
    10               2        **
    12               3        ***
    14               2        **
    16               5        *****
    18               2        **
    20               4        ****
    22               1        *
    24               2        **
    26               0
    28               1        *
```

* MINITAB is the trademark of MINITAB, Inc., 215 Pond Lab., University Park, PA 16802.

midpoints of the class intervals and the number of measurements per interval; it uses a star ($\star$) to represent a measurement in the histogram display. The histogram for the price-earnings ratio data is presented in Table 2.3.

EXERCISES   Conceptual Level

2.1   How do you decide on the number of classes to use in a relative frequency distribution?

 2.2   After the imposition of a voluntary energy conservation program, the manager of a consumer-owned public utility sought to investigate the monthly electric charges to the residential consumers served by the utility. Comment on the effect of using each of the following class interval widths to classify $n = 100$ electric charges for the month of February for the residential consumers, if these charges range from $73 to $584:

$1.00 $10.00 $25.00 $50.00 $100.00 $200.00

 2.3   Three different designations have been proposed for the classification of the hourly wages (in dollars) of cabinetmakers in a certain New England state (see the accompanying table). Criticize the use of each designation.

Designation I	Designation II	Designation III
0–5.00	0–5.50	0–under 5.00
5.00–10.00	5.51–10.00	6.00–under 10.00
10.00–15.00	10.01–15.00	11.00–under 15.00
15.00–20.00	15.01–20.00	16.00–under 20.00
20.00–25.00	Over 20.00	Over 21.00

EXERCISES   Skill Level

2.4   Consider the following set of data:

100.6	84.2	85.2	91.0	85.5	94.7
101.7	87.2	92.7	98.3	73.6	87.6
110.4	105.8	103.7	93.7	89.0	73.1
90.0	68.1	91.6	95.9	94.7	79.1
93.7	79.4	84.5	94.2	88.6	97.8

a.   Construct a relative frequency histogram for these data, using a class width of 5 and beginning at 67.50. (*Hint:* The first class boundaries will be 67.50–72.49.)

b.   What proportion of the observations are less than 87.50?

c.   If one observation is chosen at random from this set of 30, what is the probability that the observation will be greater than or equal to 102.50?

2.5   Consider the following set of data:

1.9	3.1	3.4	4.7	6.9
4.8	.6	3.0	4.9	4.4
7.0	.4	8.3	.1	13.7
7.7	11.1	2.3	1.8	2.8
9.1	15.7	1.7	7.6	1.3

    a.   What is the range of the data?

    b.   How many classes would you use in constructing a relative frequency distribution for the data?

    c.   Define the class boundaries, and produce the relative frequency tabulation for these data.

    d.   Use the results of part c to construct a relative frequency histogram for the data.

2.6   Conduct the following experiment: Toss a coin 10 times and record the number of heads observed. Repeat this process $n = 50$ times, thus providing 50 observations. Construct a relative frequency histogram for these data.

EXERCISES   Applications

2.7   To contend with changing employee values, management has instituted programs in which workers become more involved in making decisions affecting their jobs and improving the quality of work life. After the institution of a quality-of-work-life program in a major manufacturing organization, the general manager recorded the number of employee grievances received for each of 24 consecutive months. The results are as follows:

17	35	30	10	14	3
21	28	24	8	6	7
45	57	19	11	35	18
32	61	17	23	12	14

    a.   Construct a relative frequency histogram for these data.

    b.   Management considers 30 or more grievances in any particular month an indication that some serious problem is affecting employee morale. What fraction of the months provided indication that some serious problem is affecting employee morale?

2.8   A survey of existing home sales in 1987 also listed the median sales prices for existing homes in 53 major metropolitan areas from October to December of 1987, followed by the percentage change from the same period in 1986. The data are given in Table 2.4. Construct a relative frequency histogram for the 53 median sales prices.

2.9   The Federal Trade Commission (FTC) requires automobile manufacturers to substantiate advertised gasoline mileage claims for their vehicles. Many test runs are conducted under a variety of highway, traffic, and driver conditions in order to best simulate actual driving

Table 2.4   Price Data for Exercise 2.8

Metro Area	Price	Change	Metro Area	Price	Change
Akron	$58,100	2.5	Louisville	53,400	2.9
Albany	90,200	18.2	Memphis	75,500	8.0
Albuquerque	79,700	−2.8	Miami	80,300	1.5
Anaheim-SntAna	174,500	14.5	Milwaukee	70,200	1.4
Baltimore	84,400	13.6	Minn.-St. Paul	80,800	4.4
Baton Rouge	65,400	−4.5	Nashville	75,500	5.3
Birmingham	73,800	10.6	New York	180,800	7.9
Boston	178,600	6.4	Oklahoma City	59,400	−5.6
Buffalo	57,600	8.2	Omaha	57,100	−2.9
Chicago	91,500	6.6	Orlando	78,600	9.0
Cincinnati	68,200	n/a	Philadelphia	75,700	−4.9
Cleveland	67,400	3.5	Phoenix	80,300	3.1
Columbus	69,300	8.3	Portland (Ore.)	62,400	−.2
Dallas-Ft.Worth	85,600	−4.7	Providence	127,400	32.2
Denver	86,400	.7	Rochester	72,600	5.1
Des Moines	52,000	−5.5	St. Louis	73,200	1.8
Detroit	65,700	9.3	Salt Lake City	67,800	−1.7
El Paso	58,500	0.0	San Antonio	67,100	−4.4
Ft. Lauderdale	80,200	3.2	San Diego	131,800	10.9
Grand Rapids	53,100	6.8	San Francisco	176,000	6.8
Hartford	163,800	21.7	Syracuse	69,900	5.3
Houston	62,700	−7.2	Tampa-St. Pete	63,300	1.4
Indianapolis	60,600	5.4	Toledo	52,900	−.6
Jacksonville	65,200	4.0	Tulsa	63,000	−.6
Kansas City	68,200	4.8	Washington	122,400	20.0
Las Vegas	76,700	−2.3	West Palm Beach	103,200	13.5
Los Angeles	145,200	11.5	(United States	84,300	5.1)

Source: National Association of Realtors data from The Press Enterprise (Riverside, Calif.), February 17, 1988. Reprinted with permission.

Note: The table lists median sales prices for existing homes in 53 major metropolitan areas from October through December, followed by the percentage change from the same period in 1986.

conditions. In test runs of $n = 50$ cars an automobile manufacturer records the following gasoline mileage ratings for the company's subcompact model:

37.9	39.3	41.8	32.5	44.2
44.2	42.7	36.5	36.4	41.6
45.6	41.0	38.0	43.7	42.0
38.5	37.5	39.8	41.2	38.7
40.0	38.7	43.2	40.5	37.9
41.2	39.5	38.7	33.0	40.1
40.5	41.3	34.9	36.8	39.9
38.7	40.4	41.3	42.7	40.3
43.5	40.5	40.6	45.1	38.6
40.1	40.3	39.6	41.4	42.4

a.   Construct a relative frequency histogram for these data, using five classes of equal width.

b. The automobile manufacturing company has designed this subcompact to achieve a gasoline mileage of at least 40 mpg. Give the fraction of cars in the sample that achieve the company's test standard.

c. Do you feel that it is reasonable for this manufacturer to claim that the subcompact car gets a gasoline mileage of 40 mpg?

2.3

Stem and Leaf Displays

An alternative, but nonetheless straightforward, approach to presenting a set of data was proposed by John Tukey [*Exploratory Data Analysis* (Reading, Mass.: Addison-Wesley, 1977)]. This partly tabular and partly graphical technique is called a **stem and leaf display**. Since a stem and leaf display is best explained by example, we produce a stem and leaf display for the set of personal income data in Table 2.5.

stem and leaf display

A high personal income is indicative of high earning ability as well as high purchasing power. When the personal incomes of individuals are summarized as per capita personal income, this data is indicative of the earnings and purchasing power within the state and may also reflect the kind of economic activity upon which these earnings are based. The data presented in Table 2.5 represent the 1986 per capita personal income (in thousands of dollars) for each of the 50 United States.

In a stem and leaf display the integer part of each number serves as a stem and the

Table 2.5   Per Capita Income for 1986 by State (Thousands of Dollars)

State	Income	State	Income	State	Income	State	Income
AL	9.5	IN	11.1	NE	12.0	SC	9.6
AK	15.0	IA	11.3	NV	12.7	SD	10.7
AZ	11.5	KS	12.2	NH	14.1	TN	10.2
AR	9.2	KY	9.5	NJ	15.3	TX	11.6
CA	14.3	LA	9.8	NM	9.3	UT	9.0
CO	13.1	ME	11.0	NY	14.3	VT	11.0
CT	16.3	MD	13.9	NC	10.5	VA	13.0
DE	12.3	MA	14.6	ND	10.9	WA	12.8
FL	12.3	MI	11.9	OH	11.8	WV	9.0
GA	11.0	MN	12.5	OK	10.4	WI	11.8
HI	12.7	MS	8.2	OR	11.4	WY	11.1
ID	10.1	MO	11.7	PA	12.0		
IL	13.3	MT	10.2	RI	12.9		

Source: Data from U.S. Department of Commerce, Bureau of Economic Analysis, *Survey of Current Business* (1987), Vol. 67, No. 4, pp. 33–36.

Note: Data were computed by using preliminary 1986 Bureau of the Census estimates.

```
 7  |
 8  | 2
 9  | 5 2 5 8 3 6 0 0
10  | 1 2 5 9 4 7 2
11  | 5 0 1 3 0 9 7 8 4 6 0 8 1
12  | 3 3 7 2 5 0 7 0 9 8
13  | 1 3 9 0
14  | 3 6 1 3
15  | 0 3
16  | 3
17  |
```

Figure 2.3   *Stem and Leaf Display of the Data in Table 2.5*
(Leaf Digit Unit = $100)

fractional part as a leaf. For these data the smallest stem value is 8 and the largest is 16. The stem values are recorded in a vertical column as in Figure 2.3, and a vertical line is drawn beside them. For each number in the data set the leaf or fractional part of the number is recorded to the right of the stem. The 1986 per capita personal income data give rise to the stem and leaf display in Figure 2.3. Rather than record the fractional part of the number as a decimal value, we record the value of a leaf digit in units of $100 and note within the display that "leaf digit unit = 100." For example,

$$16 \mid 3 = \$16,300 \quad \text{or} \quad \$16.3 \text{ thousand}$$

The stem and leaf display allows us to pick out several salient characteristics of these data. We immediately see that the per capita incomes (in $1,000 units) range from 8.2 to 16.3, that all but 3 states had per capita personal incomes between 9.0 and 15.0, that exactly 8 states had per capita incomes from 9.0 to 9.9, and that exactly 13 states had incomes from 11.0 to 11.9.

Each stem in the display defines a group or class of incomes, and the number of leaves in a class defines the frequency with which the values in that class occur in the data set. For example, the stem 9 represents a class whose lower limit is 9.0, whose upper limit is 9.9, and whose class frequency is 8. Hence, a stem and leaf display can be thought of as a *frequency distribution* that summarizes the data by providing a set of classes (stems) and the frequency (the number of leaves) with which values are distributed among the classes. If the display in Figure 2.3 is turned so that the vertical line is in a horizontal position, the stem and leaf display looks very similar to a *histogram*, which we discussed in Section 2.2.

There are data sets for which using the first one or two digits as a stem may produce a noninformative stem and leaf display. The following is a set of data in which this is the case. In light of the recent and ongoing energy crisis and its effect on the cost of gasoline, many purchasers of new cars are concerned not only with the purchase price of the vehicle but also with the costs of operating the vehicle after the purchase. Other things being equal, a potential car buyer may opt for a car whose performance in estimated miles per gallon is better than the others. The comparative miles per gallon for city driving for 42 models of cars having four-cyclinder engines and manual transmissions are given in Table 2.6. (The mileages correspond to an alphabetic listing of the cars by model.)

Table 2.6   Comparative Miles per Gallon for City Driving for Autos with Four Cylinders and Manual Transmissions

21	43	30	19	38
24	25	26	26	35
23	23	43	21	37
24	36	25	25	37
39	38	37	39	24
45	49	25	31	24
31	24	26	23	
37	22	19	25	
28	43	37	24	

Constructing a stem and leaf display using only the leading digits as stems gives rise to the stem and leaf display shown in Figure 2.4a. Using only four stem values for these data does not produce an adequate, informative display. One solution is to use each stem value twice and to associate the leaves 0, 1, 2, 3, and 4 with the first value and the leaves 5, 6, 7, 8, and 9 with the second, as in Figure 2.4b. For example, in using the digit 2 twice, we associate the leaves 0, 1, 2, 3, and 4 with the stem "2*" and the leaves 5, 6, 7, 8, and 9 with the stem "2·". If the resulting stem and leaf display does not provide adequate resolution, the next step is to use each stem value five times, associating the leaf values 0 and 1 with "*", 2 and 3 with "t" (*t*wo and *t*hree), 4 and 5 with "f" (*f*our and *f*ive), 6 and 7 with "s" (*s*ix and *s*even), and 8 and 9 with "·", as in Figure 2.4c. Using each stem value twice in Figure 2.4b separates the vehicles into two groups, those whose mileages range from 19 to 29 mpg and those whose mileages range from 30 to 49 mpg. Using each stem value five times in Figure 2.4c further delineates and separates these two groups.

A stem and leaf display can be produced by using the STEM(-AND-LEAF) command in the MINITAB package. When the MINITAB command STEM-AND-LEAF was implemented for the mileage data that were stored in column (C3) of a data array, the stem and leaf plot in Table 2.7 was produced. The first column in this display

(a) 1 | 9 9
     2 | 1 4 3 4 8 5 3 4 2 6 5 5 6 6 1 5 3 5 4 4 4
     3 | 9 1 7 6 8 0 7 7 9 1 8 5 7 7
     4 | 5 3 9 3 3

(b) 1. | 9 9   (c) 1. | 9 9
     2* | 1 4 3 4 3 4 2 1 3 4 4 4   2* | 1 1
      . | 8 5 6 5 5 6 6 6 5 5   t | 3 3 2 3
     3* | 1 0 1   f | 4 4 5 4 5 5 5 5 4 4 4
      . | 9 7 6 8 7 7 9 8 5 7 7   s | 6 6 6
     4* | 3 3 3   . | 8
      . | 5 9   3* | 1 0 1
                                            t |
                                            f | 5
                                            s | 7 6 7 7 7 7
                                             . | 9 8 9 8
                                           4* |
                                            t | 3 3 3
                                            f | 5
                                            s |
                                             . | 9

Figure 2.4   Stem and Leaf Displays of the Data in Table 2.6 (Leaf Digit Unit = 1)

is the number of leaves on that stem or on a stem closer to the nearer end of the display. For example, there are 19 data values less than or equal to 25, while there are 9 data values greater than or equal to 38. For the stem class containing the data value midway from each end, the number in parentheses gives the number of leaves for that stem. The class corresponding to the stem 2S contains this middle value, and there are "(3)" leaves in this stem class. For the computer-produced display in Table 2.7 we find that the leaves are in numerical order, but in Figure 2.4c, which we produced by hand, the leaves are in the same order as the original data set; otherwise, the displays are identical.

A set of data may contain one or more measurements that are either much larger or much smaller than the other measurements in the set. Including these values in a stem and leaf display may produce a display with many stems having no leaves. The stem and leaf display in Table 2.8 shows the distribution of profit after taxes as a percentage of sales for 22 firms. The 2 firms with profit ratios of 12.5% and 15.8% differ for some reason from the remaining 20 firms whose profit ratios range from 2.6% to 8.6%. Hence, the two values 12.5 and 15.8 are not included in the stem and leaf display in Table 2.8 but, instead, are listed below the display in a line labeled "HI." Extremely small values are handled similarly and are displayed in a line labeled "LO."

Table 2.7   MINITAB Stem and Leaf Display of the Data in Table 2.6

```
MTB>STEM-AND-LEAF C3

   STEM-AND-LEAF DISPLAY OF C3
   LEAF DIGIT UNIT  =    1.0000
   1 2 REPRESENTS 12.

    2    1.  99
    4    2*  11
    8    2T  2333
   19    2F  44444455555
  (3)    2S  666
   20    2.  8
   19    3*  011
   16    3T
   16    3F  5
   15    3S  677777
    9    3.  8899
    5    4*
    5    4T  333
    2    4F  5
    1    4S
    1    4.  9
```

Table 2.8   Stem and Leaf Display for Profit After Taxes as a Percentage of Sales for 22 Firms

```
MTB>STEM C1

   STEM-AND-LEAF DISPLAY OF C1
   LEAF DIGIT UNIT = 0.1000
   1 2 REPRESENTS 1.2

    1    2.  6
    7    3*  012244
   10    3.  579
  (3)    4*  004
    9    4.
    9    5*  23
    7    5.  6
    6    6*  24
    4    6.
    4    7*  1
    3    7.
    3    8*
    3    8.  6

        H1   125, 158,
```

If data are to be displayed with more accuracy than we have used, two-digit leaves, separated by commas, may be used in the stem and leaf display. However, increasing the accuracy within the display may obviate the greatest advantage of a stem and leaf display—its simplicity.

EXERCISES   Conceptual Level

2.10   When constructing a stem and leaf display, how do you decide what to use as a stem and what to use as a leaf?

2.11   Why might you need to use the stem values more than once in constructing a stem and leaf display?

EXERCISES   Skill Level

2.12   Use the following set of data:

13.5	13.8	15.2
16.4	15.5	15.8
15.6	14.3	14.4
15.8	14.6	12.9

   a.   Construct a stem and leaf display, using the integer portion of the number as the stem.
   b.   Is there another possible choice of stem and leaf that you might select?
   c.   Use the stem and leaf display to find the smallest observation. The sixth and seventh smallest observations.

2.13   Use the following set of data:

3.4	2.1	2.4	2.8
3.2	3.7	2.5	2.2
2.8	2.6	3.2	3.3
3.3	2.9	2.4	2.5
2.4	3.2	2.3	2.8
2.8	3.1	3.1	3.0

   a.   Construct a stem and leaf display by using the leading digit as the stem.
   b.   Construct a stem and leaf display by using each leading digit twice. Has this technique improved the presentation of the data?
   c.   Construct a stem and leaf display by using each leading digit five times. Which of the three displays that you have constructed is the most informative?

2.14   The MINITAB command STEM was used to generate a stem and leaf display (Table 2.9) for the following data:

6.2	8.3	2.7	.9	6.1	3.8
6.6	10.5	6.1	7.3	3.4	1.7
4.7	8.4	7.1	1.1	.8	2.6
6.8	3.6	5.3	1.4	3.7	6.3
2.3	2.9	3.1	4.8	3.2	1.2

a.   Verify the stem and leaf display shown in Table 2.9.

b.   What is the value of the observation having a stem of 3 and a leaf of 1?

c.   Using the stem and leaf display, find the largest and smallest observations in the data set.

Table 2.9   Stem and Leaf Display for the Data of
            Exercise 2.14

```
MTB>STEM  C1

  STEM-AND-LEAF OF C1     N = 30
  LEAF UNIT = 0.10

    2     0  89
    6     1  1247
   10     2  3679
   (6)    3  124678
   14     4  78
   12     5  3
   11     6  112368
    5     7  13
    3     8  34
    1     9
    1    10  5
```

EXERCISES   Applications

2.15   The data in Table 2.10 are a state-by-state listing of existing home sales in 1987 (in thousands of units) followed by the percentage change from 1986. Construct a stem and leaf display for the percentage change in existing home sales from 1986 to 1987.

2.16   Refer to Exercise 2.15. The MINITAB command STEM was used to produce a stem and leaf display (Table 2.11) for the percentage change in existing home sales from 1986 to 1987. Explain the choice of stem and leaf used in the MINITAB analysis. Does this display differ from the display you constructed in Exercise 2.15?

2.17   Refer to Exercise 2.15. Construct a stem and leaf display for the existing home sales in 1987.

Table 2.10   Sales Data for Exercise 2.15

State	Sales	Change	State	Sales	Change
Alabama	59.7	−1.8	Montana	11.6	−2.5
Alaska	7.7	26.2	Nebraska	23.3	−8.6
Arizona	67.3	−8.8	Nevada	12.2	10.9
Arkansas	48.2	−8.7	New Hampshire	19.2	−5.4
California	507.0	4.6	New Jersey	157.3	−1.4
Colorado	46.7	−1.7	New Mexico	20.0	2.6
Connecticut	52.5	1.0	New York	223.9	3.3
Delaware	13.9	7.8	N. Carolina	140.6	1.0
Florida	184.2	−8.7	North Dakota	11.2	17.9
Georgia	89.0	18.5	Ohio	180.9	.6
Hawaii	11.5	15.0	Oklahoma	44.6	−9.7
Idaho	10.9	−3.5	Oregon	48.5	7.8
Illinois	175.3	−2.2	Pennsylvania	259.5	.5
Indiana	89.8	4.3	Rhode Island	13.4	−.7
Iowa	48.3	4.8	S. Carolina	60.9	5.0
Kansas	54.4	0.0	South Dakota	13.0	−1.5
Kentucky	64.1	9.4	Tennessee	111.2	−.7
Louisiana	40.4	18.8	Texas	224.3	12.8
Maine	35.2	14.3	Utah	14.6	0.0
Maryland	78.5	−9.7	Vermont	13.6	9.7
Massachusetts	100.5	18.8	Virginia	113.0	−13.3
Michigan	169.6	−9.1	Washington	43.6	−13.8
Minnesota	89.4	−7.6	West Virginia	36.6	−7.8
Mississippi	34.2	0.0	Wisconsin	73.3	−4.7
Missouri	88.4	0.0	Wyoming	6.0	−1.6
			(United States	3,885.0	−.3)

Source: National Association of Realtors data from The Press Enterprise (Riverside, Calif.), February 17, 1988. Reprinted with permission.
Note: The data give a state-by-state listing of existing home sales in 1987, followed by the percentage change from 1986. Sales figures are in thousands.

Table 2.11   Stem and Leaf Display for the Data of Table 2.10

```
MTB>STEM C1

  STEM-AND-LEAF DISPLAY OF C1    N = 50
  LEAF UNIT = 1.0

   2    -1  33
  12    -0  9998888775
  25    -0  4322111110000
  25     0  00001123444
  14     0  57799
   9     1  024
   6     1  57888
   1     2
   1     2  6
```

2.4

Other Graphical Methods

statistical table

Data collected from different time periods or geographical areas often are best presented by using statistical tables, charts, or pictograms. A **statistical table** is a classified or subdivided frequency distribution comparing the frequencies or the relative frequencies for samples drawn from two or more different populations. The populations might correspond to different time periods, different geographical areas, different but related firms, different areas within a firm, and so on. Within each sample the classifications must be the same so that we can perform a meaningful cross-analysis of the data.

Table 2.12, an example of a statistical table, shows the breakdown of public expenditures in support of government-sponsored health care programs for the years 1980–1986. Total costs are broken down into those attributable to the Medicaid and Medicare programs for each of the seven years, so we can make meaningful comparisons among the years. Within each year entries list the public expenditures for Medicare, Medicaid, and other programs, as well as the total cost of these programs.

Percentages are also used as entries in statistical tables. Such tables should be examined with care, because comparisons between corresponding entries for different samples can be misleading if the number of measurements differ from sample to sample. For example, two corresponding entries in a table might both show 50%, but the first entry might represent 100 out of a total of 200 and the second might represent 1 out of a total of 2. The first entry (based on a total of 200) is certainly more meaningful than the second (based on a total of 2). Consequently, the two percentages should be visually compared with caution.

Bar charts, line charts, and *pie charts* are designed to serve as visual summaries of data. Usually, line charts are plots of points tracing a firm's profits, sales, or productivity or their change over time. Bar charts and pictograms are pictorial frequency histograms. Many other types of graphical and pictorial methods are useful for the business statistician, but time limits our discussion within this chapter.

bar chart

Figure 2.5 shows how a **bar chart** can be used to display pictorially some of the health care cost data of Table 2.12. Bar charts are not ordinarily as finely subdivided as a classification table, since the extra partitions tend to clutter the appearance of a chart.

Table 2.12   Cost of Federal Government-Sponsored Health Care Services (Billions of Dollars)

Cost	1980	1981	1982	1983	1984	1985	1986
Medicare	35.9	39.1	51.4	46.4	57.5	65.8	70.2
Medicaid	14.5	16.9	18.0	19.0	20.1	22.7	25.0
Other	3.9	4.3	4.1	4.0	4.4	4.3	3.9
Total Costs	54.3	60.3	73.5	69.4	82.0	92.8	99.1

Source: Data from U.S. Office of Management and Budget, *Budget of the U.S. Government, Fiscal Year 1982–1988* (Washington, D.C., 1981–1987).

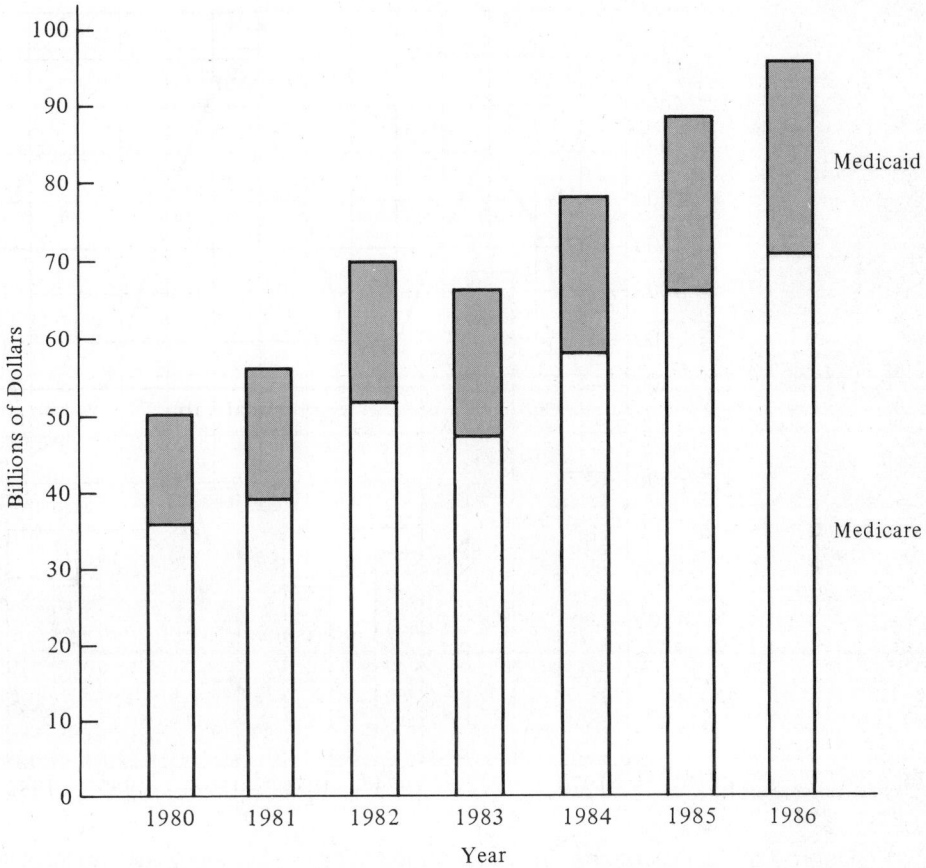

Figure 2.5   Bar Chart Based on the Medicare-Medicaid Data in Table 2.12

The intention is to produce a chart that is easy to read and that provides a quick analysis of the data.

A bar chart could be constructed in other ways. We could have illustrated the same information by drawing three separate rectangles (bars) for each year, showing separately the Medicaid costs, the Medicare costs, and the public expenditures for both health care programs each year. The type of bar chart employed is not important as long as the chart is factual and easy to interpret.

line chart   Figure 2.6 illustrates the use of a bar chart and a **line chart** together. The bars indicate the number of rental units built each year, and the line indicates the number of houses built each year for the years 1970 through 1985. Note the growth in both new rental units and new housing starts over the years 1970 to 1972 and the decline in the prerecession years of 1973 and 1974. Many readers, such as loan companies, commercial banks, and real estate agencies, would be interested not so much in the exact number of rental units and homes built as in the relation of the total of one to the

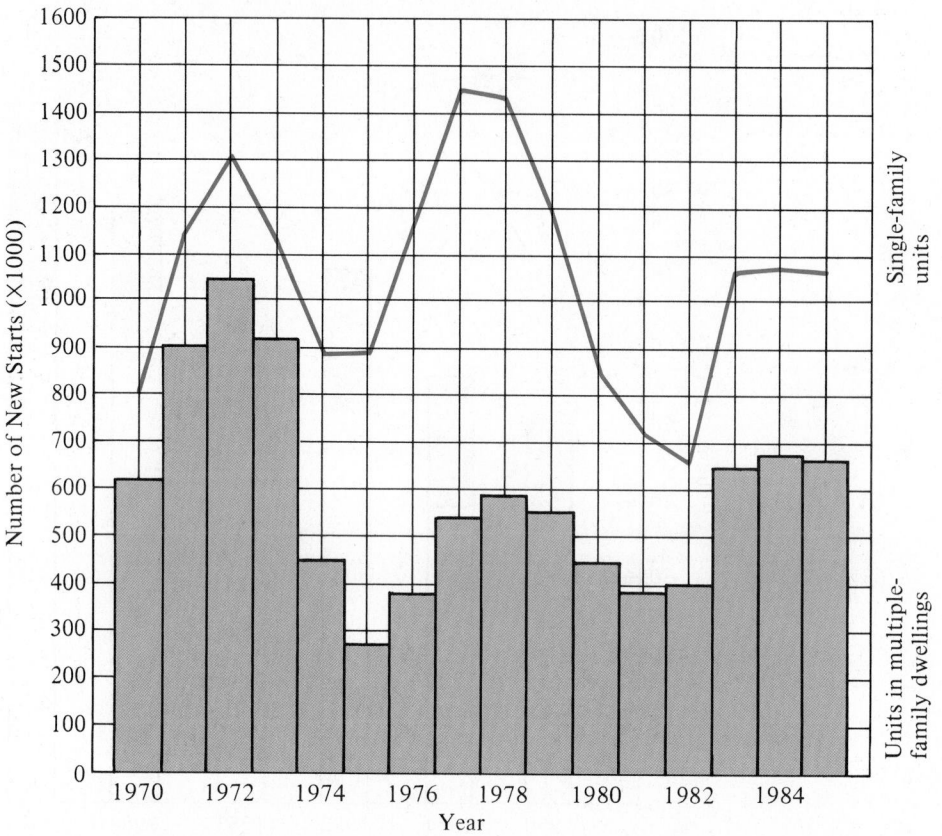

Figure 2.6   Bar Chart and Line Chart Showing New Housing Starts and New Available Rental Units in the United States from 1970 Through 1985

Source: Data from U.S. Department of Commerce, Bureau of Economic Analysis, *Survey of Current Business,* various issues.

total of the other. Contrasting representations like those shown in Figure 2.6 (bar chart and line chart) clearly depict this relation.

pie charts   Bar charts are most useful in representing the total amount of some quantity for each of a given number of years or for each of a group of categories. In contrast, **pie charts** are useful in showing how a single total quantity is apportioned to a group of categories. For example, Figure 2.7 shows the breakdown of the estimated federal budget for the 1988 fiscal year. The pie chart on the left shows where the budget monies will come from. This chart clearly shows that individual income taxes will be the major component (38%); corporation income taxes (11%) and excise taxes (3%) represent only a small portion of the total budget monies. Similar interpretations can be made for the pie chart illustrating where the estimated 1988 budget monies will go. The primary usefulness of the pie chart is that it allows you to see quickly how much of a total is represented by each subdivision of the total.

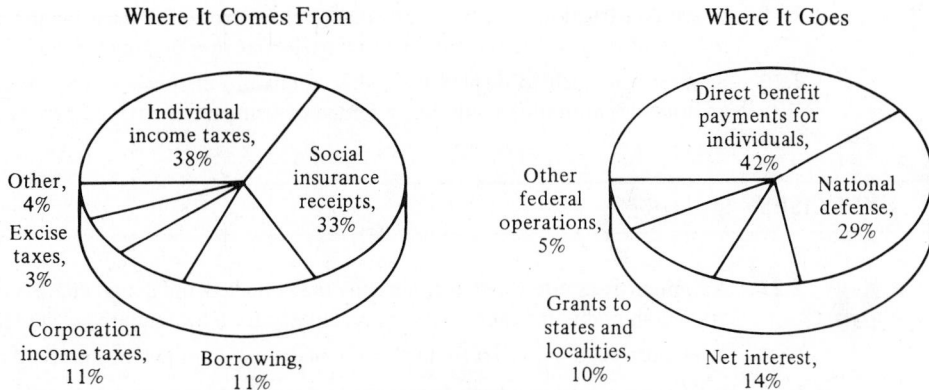

Figure 2.7   Pie Chart Showing the Breakdown of the Estimated Federal Budget for the 1988
Fiscal Year — Where It Comes From and Where It Goes

Source: Data from U.S. Office of Budget and Management, *Budget of the U.S. Government, Fiscal Year 1988*
(Washington, D.C., 1987).

A pie chart is easy to construct if you remember that the total pie contains 360 degrees and that this angle must correspond to 100% of the total represented. For example, consider the pie chart depicting where the estimated 1988 budget monies will come from (Figure 2.7). Let us calculate the angular portion of the pie that should be allocated to borrowing. Since this component represents 11% of the total estimated budget monies, it must be allocated 11% of the 360 degrees of the pie. This angular portion is

$$\frac{11\%}{100\%} \times (360 \text{ degrees}) = 39.6 \text{ degrees}$$

The portions of the pie chart assigned to each of the other cost components are calculated in a similar manner. The total of the pie-shaped portions corresponding to the various components covers 360 degrees of the pie chart.

The main purpose of any chart is to give a quick, easy-to-read-and-interpret pictorial representation of data. The type of chart or graphical presentation used and the format of its construction are incidental to its main purpose. A well-designed graphical presentation can effectively communicate the data's message in a language readily understood by almost everyone.

EXERCISES   Conceptual Level

2.18   In what situations are each of the following graphical methods of presentation most useful?

   a.   Bar charts   **b.**   Line charts   **c.**   Pie charts

2.19   In the construction of a pie chart, describe the procedure used for determining the angular portion of the chart corresponding to a particular proportion of the total.

2.20   Suppose you wish to display the dollar allocation of funds among several categories. What additional information could be presented by using a bar chart rather than a pie chart?

EXERCISES   *Skill Level*

2.21   A company was interested in comparing the budgeted and actual amounts of money spent on a given commodity for each of six consecutive years. The data are given in Table 2.13.

   a.   Construct a bar chart to display the budgeted amounts for the six years.

   b.   Construct a line chart displaying the actual amount superimposed on the bar chart for part a.

Table 2.13   Budget Data for Exercise 2.21

| | Amount (Millions of Dollars) | |
Year	Budgeted	Actual
1	49.1	50.7
2	53.7	54.2
3	59.6	62.5
4	67.4	67.2
5	81.2	79.8
6	78.9	80.1

2.22   Use a pie chart to display the proportions of personnel in the following categories:

Category	Number
Management	4
Clerical	11
Sales	14
Service	21

2.23   A company reports its service calls as follows:

Type	Percentage
New Installation	42
Cleaning Only	22
Adjustment	11
Replacement and Adjustment	25

Construct a pie chart to visually describe this data.

EXERCISES   Applications

2.24   The distribution of employed persons by major occupational groups, based on 1986 averages, is shown in Table 2.14. Construct a pie chart to graphically depict the information.

Table 2.14   Employment Data for Exercise 2.24

Occupational Group	Percentage
Managerial and professional specialty	24.2
Technical, sales, administrative support	31.3
Service occupations	13.4
Precision production, craft, and repair	12.2
Operators, fabricators, laborers	15.7
Farming, fishing, forestry	3.1
	99.9

Source: Data from U.S. Department of Labor, Bureau of Labor Statistics, 1988.

2.25   Table 2.15 lists employment data for the civilian labor force (persons over 16 years of age) in the United States for the years 1980–1986.

a.   Construct a bar chart to represent the total number of people in the civilian work force for the years 1980–1986.

b.   Using the bar chart for part a, construct a bar chart to represent the number of employed and unemployed workers for the period 1980–1986.

c.   Use the chart for part b to discuss changes in the number of unemployed workers for the seven-year period.

Table 2.15   Employment Data for Exercise 2.25

Year	Employed	Unemployed	Unemployment Rate
1980	99,303	7,637	7.1
1981	100,397	8,273	7.6
1982	99,526	10,678	9.7
1983	100,834	10,717	9.6
1984	105,005	8,539	7.5
1985	107,150	8,312	7.2
1986	109,597	8,237	7.0

Source: Data from U.S. Department of Labor, Bureau of Labor Statistics, 1988.

2.26   Refer to Exercise 2.25. Construct a line chart to graphically depict the unemployment rate for the period 1980–1986. Do the results agree with the conclusions drawn in Exercise 2.25?

2.27   On January 8, 1987, the *Wall Street Journal* reported that a total of 11.4 million cars were sold in the United States in 1986. Of these, 71.8% were produced in the United States. The distribution of these domestic sales among U.S. manufacturers was as follows:

Manufacturer	Percentage
General Motors	55.2
Ford	25.2
Chrysler	14.3
Other manufacturers	5.3

Source: Data from U.S. Department of Labor, Bureau of Labor Statistics, 1988.

Construct a pie chart to visually describe these data.

2.5

Cheating with Charts

Although graphical descriptive techniques are very useful for describing data, the figures must be interpreted with care. It is very easy to construct a figure that may lead an unsuspecting reader to the wrong conclusions. For example, one of the simplest methods to lead a reader astray is to shrink or stretch the axes of a graph.

To illustrate, suppose that the number of near collisions between aircraft per month at a major airport is recorded as 13, 14, 14, 15, and 15 for the period January through May. If you want this growth to appear small (perhaps you represent the Civil Aeronautics Administration), you might show the results by using the line chart of Figure 2.8. The growth is apparent, but it does not seem to be very great. If you want the growth to appear large (perhaps you belong to the Citizen's Safety Group), you might construct the graph of the same data shown in Figure 2.9. The vertical axis is stretched and does not include zero. Note the impression of a substantial rise that is indicated by the steeper slope.

Another way to achieve the same effect—to decrease or increase a slope—is to stretch or shrink the horizontal axis. Of course, you are sometimes limited in the amount of shrink or stretch you can apply and still achieve a picture that appears reasonable to the viewer. For example, you could not shrink or stretch the horizontal axes of Figures 2.8 and 2.9 very much because of the limited number of data points ($n = 5$).

Shrinking or stretching axes to increase the slopes in bar graphs, histograms, line charts, or other figures usually catches the hasty reader off guard; the distortions are apparent only if you look closely at the axes. The important point, however, is that increases or decreases in responses are judged large or small depending on the

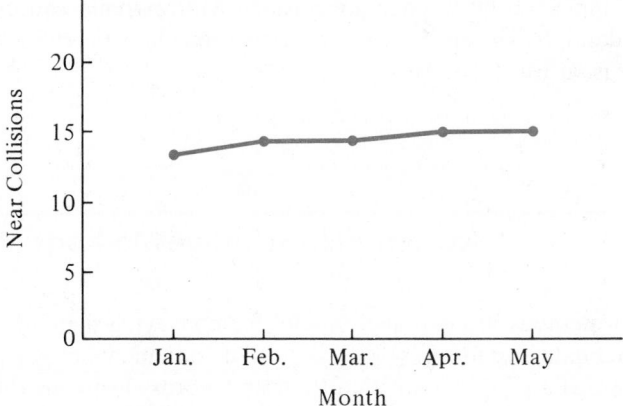

Figure 2.8   Number of Near Collisions per Month

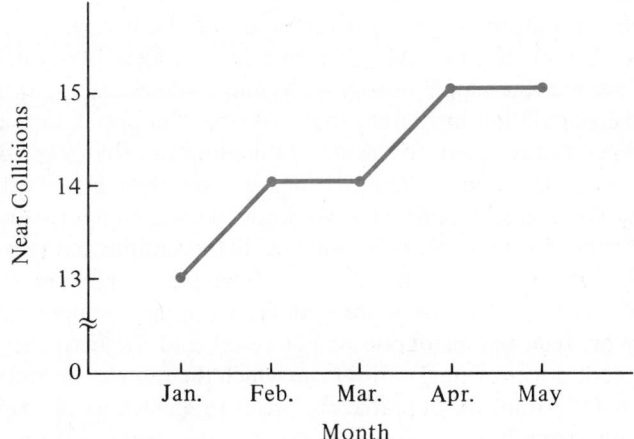

Figure 2.9   Number of Near Collisions per Month

importance of the changes to the observer, not on the slopes shown in graphic representations.

The preceding example provides a simple illustration of how the truth may be distorted, accidentally or purposely, using graphical descriptive methods. To protect yourself, carefully examine graphs and charts and note the scales of measurement. Check to see whether the axes are broken, and ask yourself whether the observed changes in the variable are meaningful from a practical point of view. For example, are you interested in the numerical or the percentage changes in the variable from one condition to another?

To summarize, draw your conclusions with extreme caution. Remember that charts and graphs are sometimes constructed to create a certain illusion and to tell a story that is far from the truth.

2.6

Numerical Descriptive Methods

Graphical methods are extremely useful for conveying a rapid general description of collected data and for presenting data. This statement supports, in many respects, the saying that a picture is worth a thousand words. There are, however, limitations to the use of graphical techniques for describing and analyzing data. For instance, suppose we wish to discuss our data before a group of people and must describe the data verbally. Unable to present the histogram visually, we would be forced to use other descriptive measures that would convey to the listeners a mental picture of the histogram.

A second and not so obvious limitation of the histogram and other graphical techniques is that they are difficult to use for purposes of statistical inference. Presumably, we use the sample histogram to make inferences about the shape and position of the population histogram that describes the population and is unknown to us. Our inference is based on the correct assumption that some degree of similarity exists between the two histograms, but we are then faced with the problem of measuring the degree of similarity. We know when two figures are identical, but this situation is not likely to occur in practice. If the sample and population histograms differ, how can we measure the degree of difference—or, expressing it positively, the degree of similarity? To be more specific, we might wonder about the degree of similarity between the histogram in Figure 2.1 and the frequency histogram for the population of price-earnings ratios from which the sample was drawn. Although these difficulties are not insurmountable, we prefer to seek other descriptive measures that readily lend themselves to use as predictors of the shape of the population frequency distribution.

The limitations of the graphical method of describing data can be overcome by the use of numerical descriptive measures. Thus, we would like to use the sample data to calculate a set of numbers that will convey to the statistician a good mental picture of the frequency distribution and will be useful in making inferences concerning the population.

Definition

parameters
statistics

Numerical descriptive measures computed from population measurements are called **parameters**; those computed from sample measurements are called **statistics**.

2.7

Measures of Central Tendency

measure of
central tendency

In constructing a mental picture of the frequency distribution for a set of measurements, we would most likely envision a histogram similar to that shown in Figure 2.1 for the data on price-earnings ratios. One of the first descriptive measures of interest is a **measure of central tendency**—that is, a measure of the center of the distribution. We note that the price-earnings data ranged from a low of 5.4 to a high of 28.6, the center of the histogram being located in the vicinity of 16.0. Let us now consider some definite rules for locating the center of a distribution of data.

One of the most common and useful measures of central tendency is the arithmetic average of a set of measurements. This measure is also often referred to as the *arithmetic mean*, or simply the *mean*, of a set of measurements. Since we will wish to distinguish between the means for the sample and for the population, we will use the symbol $\bar{x}$ (*x* bar) to represent the sample mean and μ (Greek lowercase letter mu) to represent the mean of the population.

Definition

arithmetic mean

> The **arithmetic mean** of a set of *n* measurements $x_1, x_2, x_3, \ldots, x_n$ is equal to the sum of the measurements divided by *n*.

The procedures for calculating a sample mean and many other statistics are conveniently expressed as formulas. Consequently, we will need a symbol to represent the process of summation. If we denote the *n* quantities that are to be summed as $x_1, x_2, \ldots, x_n$, then their sum is denoted by the symbol

$$\sum_{i=1}^{n} x_i$$

The symbol $\sum_{i=1}^{n}$ ($\sum$ is the Greek capital letter sigma) tells us to sum the elements that appear to the right of the $\sum$ sign, commencing with x_1 (i.e., $i = 1$) and proceeding in order to x_n. Thus,

$$\sum_{i=1}^{3} x_i = x_1 + x_2 + x_3$$

and

$$\sum_{i=1}^{n} x_i = x_1 + x_2 + \cdots + x_n$$

Using this notation, we can express the formula for the sample mean as shown in the following display.

Sample Mean

$$\bar{x} = \frac{\displaystyle\sum_{i=1}^{n} x_i}{n}$$

where n = number of measurements in sample

EXAMPLE 2.1   Find the mean of the measurements 2, 9, 11, 5, 6.

SOLUTION   Substituting the measurements into the formula, we have

$$\bar{x} = \frac{\displaystyle\sum_{i=1}^{n} x_i}{n} = \frac{2 + 9 + 11 + 5 + 6}{5} = 6.6$$

Note that this value, $\bar{x} = 6.6$, falls near the middle of the set of sample measurements, 2, 9, 11, 5, and 6.

◆ ◆

We have seen that $\bar{x}$ is used to locate the center of a set of sample measurements. A more important use of $\bar{x}$ is as an estimator (predictor) of the value of the unknown population mean μ. For example, the mean of the sample given in Table 2.1 is

$$\bar{x} = \frac{\displaystyle\sum_{i=1}^{n} x_i}{n} = \frac{400}{25} = 16.0$$

Note that this value falls approximately in the center of the set of sample measurements. The mean of the entire population of price-earnings ratios is unknown to us, but if we were to estimate its value, our estimate of μ would be 16.0. Although the value of $\bar{x}$ varies from sample to sample, the population mean μ remains the same.

A second measure of central tendency is the *median*, which is the value in the middle position.

Definition

median

The **median** of a set of measurements is the value of x such that at most half the measurements are less than x and at most half are greater than x.

Hence, the median divides a set of measurements into two equal parts. Consistent with this definition, we find the median in the following way.

The Sample Median

> When a set of measurements $x_1, x_2, \ldots, x_n$ are arranged in order of increasing magnitude, the sample median is the middle value of x or the average of the two middle values.

The median is the value in position $.5(n + 1)$. When n is odd, $.5(n + 1)$ will be an integer, and the median is the value of the middle measurement. When n is even, $.5(n + 1)$ will end in .5. In this case the median is, by convention, taken to be the average of the two middle values.

EXAMPLE 2.2   Consider this sample of $n = 5$ measurements:

$$9 \quad 2 \quad 7 \quad 11 \quad 14$$

When arranged in order of increasing magnitude, these measurements are

$$2 \quad 7 \quad 9 \quad 11 \quad 14$$

Since $n = 5$ is odd, the median is in position $.5(5 + 1) = 3$ and equal to 9. Notice that 2/5, or 40%, of the measurements are less than 9 and 40% are greater than 9, consistent with our definition of the median.

◆ ◆

EXAMPLE 2.3   Consider the following set of $n = 6$ measurements, which have been arranged in increasing order of magnitude:

$$2 \quad 6 \quad 7 \quad 9 \quad 11 \quad 14$$

In this case $n = 6$ is even, and $.5(n + 1) = .5(7) = 3.5$. Therefore, the median is the average of the values in the third and fourth positions, given by

$$.5(7 + 9) = 8$$

Notice that 3/6, or 50%, of the measurements are less than 8 and 50% are greater. Even though any number between 7 and 9 would serve the same purpose, we follow convention by taking the median to be the average of the two middle values.

◆ ◆

A third measure sometimes used as a measure of central tendency is the *mode*.

Definition

mode
> The **mode** of a set of n measurements $x_1, x_2, x_3, \ldots, x_n$ is defined to be the value of x occurring with the greatest frequency.

EXAMPLE 2.4   Consider the sample measurements

$$9 \quad 2 \quad 7 \quad 11 \quad 14 \quad 7 \quad 2 \quad 7$$

The value 7 occurs three times, 2 occurs twice, and the others, once each. Thus, 7 is the mode of the sample measurements.

◆ ◆

The mode is not a widely used measure of central tendency, but it is useful in business planning for identifying those products in greatest demand. For example, a shirt or dress manufacturer is interested in the sizes in greatest demand. Similarly, in scheduling the production of a drug, a manufacturer is interested in the drug potency most commonly prescribed by physicians. These measurements are best described by the mode.

The mode, the x value with the greatest frequency, might appear as shown in Figure 2.10. (Note that the relative frequency histogram corresponding to a large quantity of data often will appear, for all practical purposes, as a smooth curve, as shown in Figure 2.10.)

The relationships among the mean (μ), median (Md), and mode (Mo) for a distribution having one mode can be seen by examining Figure 2.11. For a symmetric frequency distribution (one for which values of the variable that are equidistant from the mean occur with equal frequency) such as shown in Figure 2.11a, the values of the mean, median, and mode are identical. If the distribution is skewed to the left (negative skew), the mean, median, and mode are aligned as shown in Figure 2.11b. If skewed to the right (positive skew), the mean, median, and mode appear as shown in Figure 2.11c. For skewed distributions, the median is always between the mean and the mode.

The mean measures the "center of gravity" of a set of data and is consequently influenced by extreme values. This property of the mean can be observed in Figures 2.11b and 2.11c. If the distribution is skewed to the left (Figure 2.11b), the mean shifts to the left of the mode. If the distribution is skewed to the right, the mean shifts to the right of the mode. The greater the skewness, as measured by the preponderance of extreme values lying on one side of the mode, the greater is the shift of the mean in that direction.

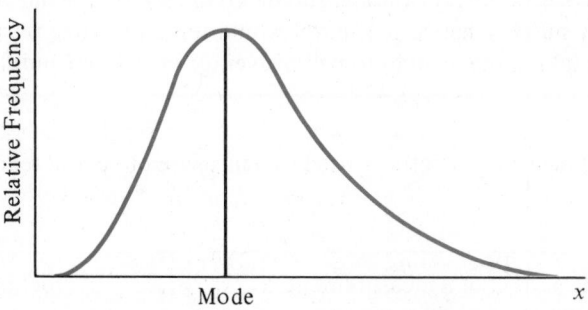

Figure 2.10   Locating the Mode of a Frequency Distribution

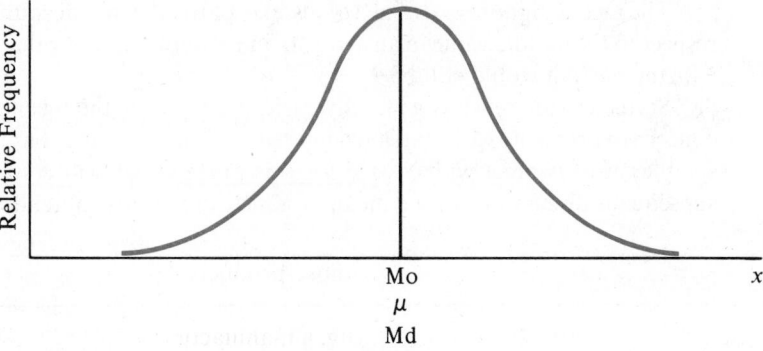

(a) A Symmetric Distribution

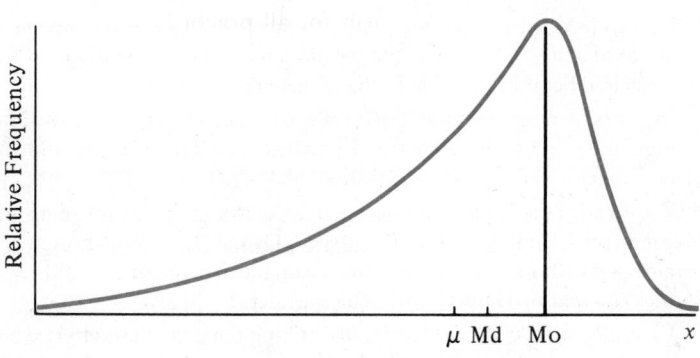

(b) Distribution Skewed to the Left

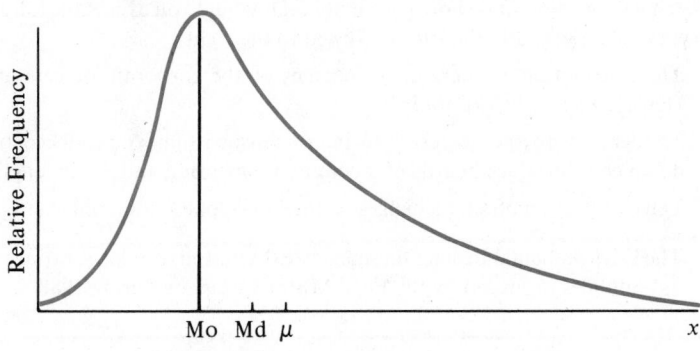

(c) Distribution Skewed to the Right

Figure 2.11   *Relationships Among the Mean (μ), the Median (Md), and the Mode (Mo)*

The median ignores extreme values, except to take into account their location with respect to the middle value in an array. If, in Example 2.2, the value 14 were replaced by 140, the median would still be 9.

Statistical inference is generally easier when using the mean. Because we will be concerned primarily with statistical inference in the following chapters, and because the sample mean is most widely used for this purpose, we will confine our attention in subsequent discussions to the mean as a measure of central tendency.

EXERCISES   Conceptual Level

2.28 Under what conditions are the mean, median, and mode equal?

2.29 Of the mean, median, and mode, which measure is the most sensitive to extreme values? Least sensitive?

2.30 Which measure of central tendency would be most useful in each of the following instances?

a. The production manager for a manufacturer of glass jars is concerned about the proper jar size to manufacture. Would the mean, the median, or the mode of the jar sizes previously purchased be of most value to the manager?

b. The sales manager for a quality furniture manufacturer is interested in selecting the regions most likely to purchase his firm's products. Would he be most interested in the mean or median family income in prospective sales areas?

c. A government economist is interested in determining the prevailing interest rate charged by commercial lending institutions in the Dallas–Fort Worth area for 30-year residential mortgages. Should she examine the mean, the median, or the mode of the mortgage interest rates charged by Dallas–Fort Worth lenders?

d. A security analyst is interested in describing the daily market price change of the common stock of a manufacturing company. Only rarely does the market price of the stock change by more than one point, but occasionally, the price will change by as many as four points in one day. Should the security analyst describe the daily price change of the stock in terms of the mean, the median, or the mode of the daily market price changes?

2.31 Referring to the diagrams shown in Figure 2.11, would you expect the following distributions to be symmetric, skewed to the left, or skewed to the right?

a. The distribution of the annual incomes of the corporate officers and employees of the Hewlett-Packard Corporation

b. The lengths, to the nearest 1/16 in., of finished lumber produced by a milling machine designed to produce boards of a common dimension and 8 ft in length

c. The distribution of daily changes of the Dow-Jones industrial average for the past year

d. The Environmental Protection Agency (EPA) gasoline mileage ratings for gasoline-powered automobiles produced by the Ford Motor Company during 1988

EXERCISES   Skill Level

2.32 Find the mean, the median, and the mode for these observations:

1   3   3   2   6   4   9   0

2.33   Find the mean, the median, and the mode for these observations:

6   4   8   4   1   4   2

2.34   For the following data, compute the mean, the median, and the mode.

24	21	38
31	52	44
64	47	27
43	36	50

2.35   Consider the following set of data:

12	11	16	15	12
13	14	12	11	17
11	13	12	12	15
12	10	14	13	13

   a.   Calculate the mean, the median, and the mode.

   b.   Use the relative values of these three measures to determine whether or not the data are skewed and, if so, in which direction.

   c.   Construct a histogram to confirm your conclusion in part b.

EXERCISES   Applications

2.36   Evidence suggests that during periods of economic recession retail sales on installment credit tend to be smaller than those during periods of economic growth. The manager of a department store selected a sample of $n = 10$ accounts and recorded the following total of installment sales by the ten account holders for a period of recession:

$ 75.00 $ 0.00 $215.00 $ 0.00 $ 97.50
  102.50   68.00   82.50   153.00   148.00

Find the mean, the median, and the mode for the sample of installment sales.

2.37   Refer to Exercise 2.15 and Table 2.10. The state-by-state listing of the percentage change in existing home sales from 1986 to 1987 is shown here in a different format.

−1.8	26.2	−8.8	−8.7	4.6	−1.7
1.0	7.8	−8.7	18.5	15.0	−3.5
−2.2	4.3	4.8	0.0	9.4	18.8
14.3	−9.7	18.8	−9.1	−7.6	0.0
0.0	−2.5	−8.6	10.9	−5.4	−1.4
2.6	3.3	1.0	17.9	.6	−9.7
7.8	.5	−.7	5.0	−1.5	−.7
12.8	0.0	9.7	−13.3	−13.8	−7.8
−4.7	−1.6				

a. Calculate the mean, the median, and the mode for these data.

b. Use the relative values of these three measures to determine whether or not the data are skewed and, if so, in what direction.

c. Confirm the results of part b by constructing a relative frequency histogram for the data.

 2.38 U.S. manufacturers are concerned about trends in manufacturing productivity and labor costs in the United States when compared with trends in competing countries. A study in the *Monthly Labor Review* by Neef and Thomas showed an improvement in productivity and labor costs in the United States for the year 1986. As a part of this study, the authors reported the average annual percent changes in manufacturing productivity for 12 countries for the period 1960–1986. The data are given in Table 2.16. Calculate the mean, the median, and the mode for these data.

Table 2.16   Productivity Data for Exercise 2.38

Country	Output per Hour	Country	Output per Hour
United States	2.8	United Kingdom	3.6
Canada	3.3	Belgium	6.3
Japan	7.9	Denmark	4.6
France	5.2	Netherlands	5.9
Germany	4.6	Norway	3.2
Italy	5.7	Sweden	4.6

Source: A. Neef and J. Thomas, "Trends in Manufacturing Productivity and Labor Costs in the U.S. and Abroad," *Monthly Labor Review*, Vol. 110, No. 12 (December 1987), pp. 25–30. Reprinted with permission.

2.8

Measures of Variability

Once we have located the center of a distribution of data, the next step is to provide a measure of the **variability**, or **dispersion**, of the data. Consider the two distributions shown in Figure 2.12. Both distributions are located with a center at $x = 4$, but there is a vast difference in the variability of the measurements about the mean for the two distributions. The measurements in Figure 2.12a vary from 3 to 5; in Figure 2.12b the measurements vary from 0 to 8.

variability
dispersion

Variation is a very important characteristic of data. For example, if we are manufacturing bolts, excessive variation in the bolt diameter would imply a high percentage of defective product. On the other hand, if we are using an examination to discriminate between good and poor accountants, we would be most unhappy if the examination always produced test grades with little variation, since this would make discrimination very difficult.

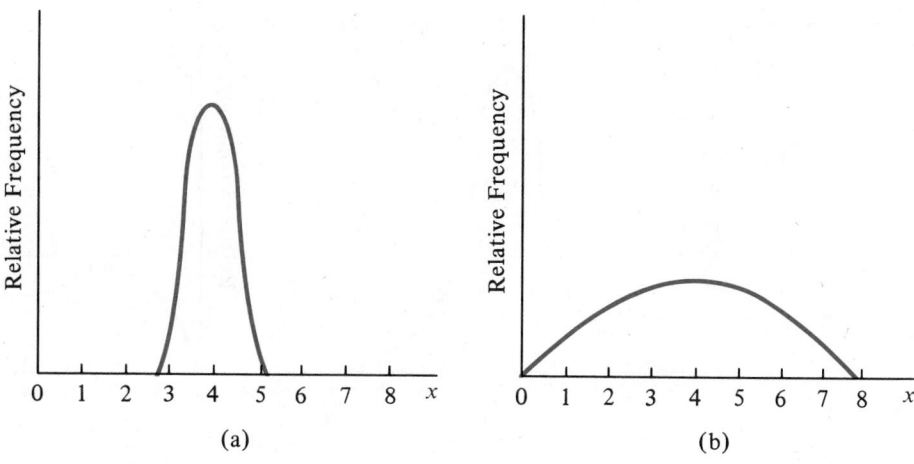

Figure 2.12   Variability or Dispersion of Data

In addition to the practical importance of variation in data, a measure of this characteristic is necessary to the construction of a mental image of the frequency distribution. Numerous measures of variability exist, and we will discuss a few of the most important.

The simplest measure of variation is the *range*.

Definition

range | The **range** of a set of n measurements $x_1, x_2, x_3, \ldots, x_n$ is defined to be the difference between the largest and smallest measurements.

For the price-earnings ratios in Table 2.1, the measurements vary from 5.4 to 28.6. Hence, the range is $(28.6 - 5.4) = 23.2$.

Unfortunately, the range is not completely satisfactory as a measure of variation. Consider the two distributions in Figure 2.13. Both distributions have the same range, but the data of Figure 2.13b are more variable than the data of Figure 2.13a.

To overcome this limitation of the range, we introduce *quartiles* and *percentiles*. Remember that if we specify an interval along the x axis of the histogram, the percentage of area under the histogram lying above the interval is equal to the percentage of the total number of measurements falling into that interval. Since the median is the middle measurement when the data are arranged in order of magnitude, the median would be the value of x such that half the area of the histogram would lie to its left, half to the right. Similarly, we define *quartiles* as values of x that divide the area of the histogram into quarters.

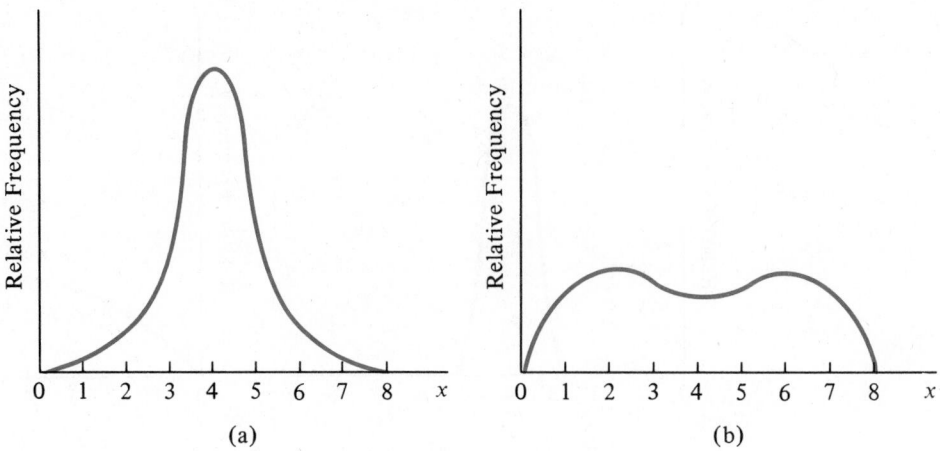

Figure 2.13   Distributions with Equal Ranges and Unequal Variability

Definition

> Let $x_1, x_2, \ldots, x_n$ be a set of n measurements arranged in order of magnitude. The **lower quartile** is a value of x such that at most $1/4$ of the measurements are less than x and at most $3/4$ are greater than x. The **upper quartile** is a value of x such that at most $3/4$ of the measurements are less than x and at most $1/4$ greater than x.

Locating the lower quartile on the histogram in Figure 2.14, we note that $1/4$ of the area lies to the left of the lower quartile, $3/4$ to the right. The upper quartile is the value of x such that $3/4$ of the area lies to the left, $1/4$ to the right.

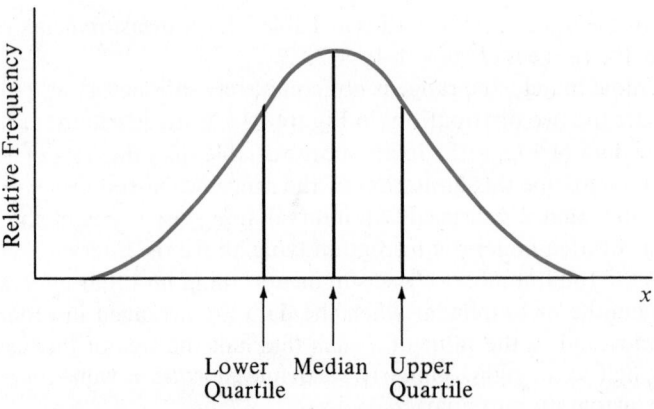

Figure 2.14   Location of Quartiles

For small sets of data the quartiles may fall between two measurements, in which case many numbers satisfy the preceding definition. To avoid this ambiguity, we calculate quartile values in the following way.

Calculating Quartiles

When the measurements $x_1, x_2, \ldots, x_n$ are arranged in order of magnitude, the lower quartile, Q_1, is the value of x in position $.25(n + 1)$, and the upper quartile, Q_3, is the value of x in position $.75(n + 1)$.

When $.25(n + 1)$ and $.75(n + 1)$ are not integers, the quartiles are found by interpolation, using the values in the two adjacent positions.* We demonstrate the procedure with the next example.

EXAMPLE 2.5   Find the median, the lower quartile, and the upper quartile for the following set of $n = 8$ measurements:

$$2 \quad 5 \quad 8 \quad 10 \quad 11 \quad 14 \quad 17 \quad 20$$

SOLUTION   The median is the value in position $.5(n + 1) = .5(9) = 4.5$, so that the median is given by

$$.5(10 + 11) = 10.5$$

The lower quartile is the value in position $.25(n + 1) = .25(9) = 2.25$, and the upper quartile is the value in position $.75(9) = 6.75$. The lower quartile is taken to be the value 1/4 the distance between the second and third ordered measurements, and the upper quartile is taken to be the value 3/4 of the distance between the sixth and seventh ordered measurements. Therefore,

$$Q_1 = 5 + .25(8 - 5) = 5 + .75 = 5.75$$

and

$$Q_3 = 14 + .75(17 - 14) = 14 + 2.25 = 16.25$$

◆ ◆

For some applications, particularly those in which there are large quantities of data, *percentiles* are preferred.

---

* This definition of quartiles is consistent with the one used in the MINITAB package. However, some texts use ordinary rounding when finding quartile positions; others compute a sample quartile as the average of the values in adjacent positions only when $.25(n + 1)$ and $.75(n + 1)$ end in the fraction .5.

Definition

100 *p*th
percentile

Let $x_1, x_2, \ldots, x_n$ be a set of n measurements arranged in order of magnitude. The **100 *p*th percentile** is a value of x such that at most $100p\%$ of the measurements are less than the value of x and at most $100(1 - p)\%$ are greater.

For example, the 90th percentile for a set of data is a value of x that exceeds 90% of the measurements and is less than 10%. Just as in the case of quartiles, 90% of the area of the histogram lies to the left of the 90th percentile.

The rule for finding the value of a percentile is merely an extension of the procedure used in finding the value of the median and the upper and lower quartiles.*

Calculating Percentiles

When the measurements $x_1, x_2, \ldots, x_n$ are arranged in order of increasing magnitude, the 100*p*th percentile is the value in position $p(n + 1)$. If $p(n + 1)$ is not an integer, the 100*p*th percentile is the appropriate interpolated value found by using the measurements in the two adjacent positions.

For example, when $n = 100$, the 20th percentile is the value in position $.2(101) = 20.2$. Since 20.2 is not an integer, the 20th percentile is the value lying 2/10 of the distance between the 20th and 21st ordered measurements in the set.

The range possesses simplicity in that it can be expressed as a single number. Quartiles and percentiles, on the other hand, provide more information about data location and variation, but several numbers must be given to provide an adequate description. Can we find a measure of variability expressible as a single number but more sensitive than the range?

Consider, as an example, the sample measurements 5, 7, 1, 2, 4. We can depict these data graphically, as in Figure 2.15, by showing the measurements as dots falling along the x axis. Figure 2.15 is called a **dot diagram**.

dot diagram

Calculating the mean as the measure of central tendency, we obtain

$$\bar{x} = \frac{\sum_{i=1}^{n} x_i}{n} = \frac{19}{5} = 3.8$$

and we locate $\bar{x}$ on the dot diagram. We can now view variability in terms of distance between each dot (measurement) and the mean $\bar{x}$. If the distances are large, we can say that the data are more variable than if the distances were small. More explicitly, we define the **deviation** of a measurement from its mean to be the quantity $(x_i - \bar{x})$. Note

deviation

---

* Sample percentiles, like sample quartiles, are often computed as the simple average of values in the appropriate adjacent positions.

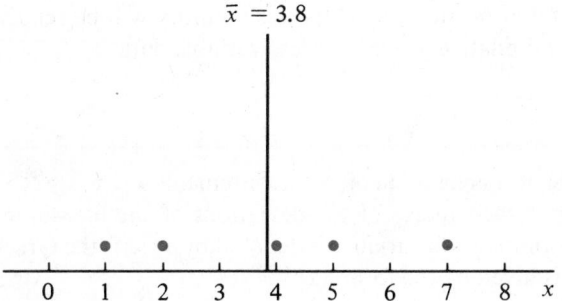

$\bar{x} = 3.8$

Figure 2.15   Dot Diagram

that measurements to the right of the mean produce positive deviations, and those to the left produce negative deviations. The values of x and the deviations for our example are shown in the first and second columns of Table 2.17.

Table 2.17   Computation of $\displaystyle\sum_{i=1}^{n} (x_i - \bar{x})^2$

x_i	$(x_i - \bar{x})$	$(x_i - \bar{x})^2$
5	1.2	1.44
7	3.2	10.24
1	−2.8	7.84
2	−1.8	3.24
4	.2	.04
19	0.0	22.80

If we now agree that deviations contain information on variation, our next step is to construct a formula based on the deviations that will provide a good measure of variation. As a first possibility, we might choose the average of the deviations. Unfortunately, the average will not work, because some of the deviations are positive, some are negative; and the sum is always zero (unless round-off errors have been introduced into the calculations). Note that the deviations in the second column of Table 2.17 sum to zero.

You may have observed an easy solution to this problem. Why not calculate the average of the absolute values* of the deviations? This method has, in fact, been employed as a measure of variability, but it tends to be unsatisfactory for purposes of statistical inference. We prefer to overcome the difficulty caused by the sign of the deviations by working with the sum of their squares,

$$\sum_{i=1}^{n} (x_i - \bar{x})^2$$

---

* The absolute value of a number is its magnitude, ignoring its sign. For example, the absolute value of -2, represented by the symbol $|-2|$, is 2. The absolute value of 2, that is, $|2|$, is 2.

For a fixed number of measurements this quantity will be relatively large for highly variable data and relatively small for less variable data.

Definition

variance of a
population

> The **variance of a population** of N measurements $x_1, x_2, \ldots, x_N$ is defined to be the average of the squares of the deviations of the measurements about their mean μ. The population variance is denoted by σ^2 (σ is the Greek lowercase letter sigma) and is given by the formula
>
> $$\sigma^2 = \frac{1}{N} \sum_{i=1}^{N} (x_i - \mu)^2$$

The letter N is used to denote the number of measurements in the population and n to denote the number of measurements in the sample.

Typically, we do not have all the population measurements available and must be satisfied with sample measurements selected from the population. Thus, we must use the variance of a *sample*, as defined next.

Definition

variance of a
sample

> The **variance of a sample** of n measurements $x_1, x_2, \ldots, x_n$ is defined to be the sum of the squared deviations of the measurements about their mean $\bar{x}$ divided by $(n - 1)$. The sample variance is denoted by s^2 and is given by the formula
>
> $$s^2 = \frac{1}{n-1} \sum_{i=1}^{n} (x_i - \bar{x})^2$$

For example, we may calculate the variance for the set of $n = 5$ sample measurements presented in Table 2.17. The square of the deviation of each measurement is recorded in the third column of Table 2.17. Adding, we obtain

$$\sum_{i=1}^{5} (x_i - \bar{x})^2 = 22.80$$

The sample variance is

$$s^2 = \frac{1}{n-1} \sum_{i=1}^{n} (x_i - \bar{x})^2 = \frac{22.80}{4} = 5.70$$

You may wonder about the apparent inconsistency in the definitions of the population and sample variances. The sample mean $\bar{x}$ is used as an estimator of the population mean μ. Although it was not specifically stated, it is implied that the sample mean provides a good estimate of μ. In the same vein, we might reasonably assume that

$$s'^2 = \frac{1}{n} \sum_{i=1}^{n} (x_i - \bar{x})^2$$

would provide a good estimate of the population variance σ^2, based on a set of sample measurements. However, for small samples (n small) s'^2 tends to underestimate σ^2, and the sample variance s^2 provides better estimates of σ^2 than does s'^2. Note that s^2 and s'^2 differ only in the denominators and that when n is large, s'^2 and s^2 will be approximately equal. In later chapters we will have occasion to use an estimator of the population variance σ^2. **In all our calculations we will use s^2 rather than s'^2 and refer to s^2 as the sample variance.***

At this point you may be understandably disappointed with the practical significance attached to variance as a measure of variability. Large variances imply a large amount of variation, but this statement only permits comparison of several sets of data. When we attempt to say something specific concerning a single set of data, we are at a loss. For example, what can be said about the variability of a set of data with a variance equal to 100? The question cannot be answered with the facts we have. We remedy this situation by introducing a new definition and, in Section 2.9, a theorem and a rule.

Definition

standard deviation

> The **standard deviation** of a set of n sample measurements $x_1, x_2, \ldots, x_n$ is equal to the positive square root of the variance.

The variance is measured in terms of the square of the original units of measurement. If the original measurements are in inches, the variance is expressed in square inches. Taking the square root of the variance, we obtain the standard deviation, which conveniently returns the measure of variability to the original units of measurement.

Sample Standard Deviation

$$s = \sqrt{s^2} = \sqrt{\frac{\sum_{i=1}^{n} (x_i - \bar{x})^2}{n - 1}}$$

The population standard deviation is denoted by σ and is calculated as the positive square root of the population variance σ^2.

Now that we have defined the standard deviation, you might wonder why we bothered to define the variance in the first place. Actually, both the variance and the standard deviation play an important role in statistical inference.

---

* The quantity s^2 is properly referred to as the *estimator* of the population variance.

2.9

On the Practical Significance of the Standard Deviation

We now introduce a useful theorem developed by the Russian mathematician Tchebysheff. Proof of the theorem is not difficult, but we omit it from our discussion.

Tchebysheff's Theorem

Tchebysheff's Theorem

> Given a number k greater than or equal to 1 and a set of n measurements $x_1, x_2, \ldots, x_n$, at least $[1 - (1/k^2)]$ of the measurements lie within k standard deviations of their mean.

Tchebysheff's Theorem applies to any set of measurements and can be used to describe either a sample or a population. We will use the notation appropriate for populations, but you should realize that we could just as easily use the mean and the standard deviation for the sample.

The idea involved in Tchebysheff's Theorem is illustrated in Figure 2.16. An interval is constructed by measuring a distance $k\sigma$ on either side of the mean μ. Note that the theorem is true for any number we choose for k as long as it is greater than or equal to 1. Then, computing the fraction $[1 - (1/k^2)]$, we see that Tchebysheff's Theorem states that at least that fraction of the total number n of measurements lies in the constructed interval.

Let us choose a few numerical values for k and compute $[1 - (1/k^2)]$ (see Table 2.18). When $k = 1$, the theorem states that at least $1 - [1/(1)^2] = 0$ of the measurements lie in the interval from $(\mu - \sigma)$ to $(\mu + \sigma)$, a most unhelpful and uninfor-

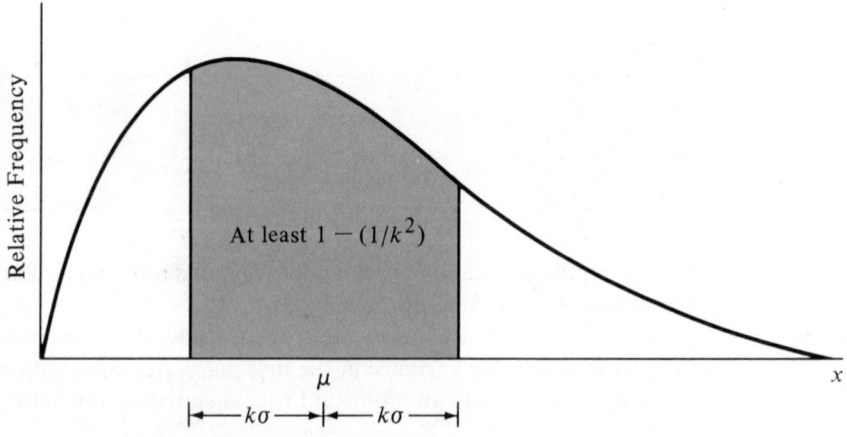

At least $1 - (1/k^2)$

Figure 2.16   Illustrating Tchebysheff's Theorem

Table 2.18   Illustrative Values of $[1 - (1/k^2)]$

k	$1 - (1/k^2)$
1	0
2	3/4
3	8/9

mative result. However, when $k = 2$, we observe that at least $1 - [1/(2)^2] = 3/4$ of the measurements lie in the interval from $(\mu - 2\sigma)$ to $(\mu + 2\sigma)$. At least 8/9 of the measurements lie within 3 standard deviations of the mean, that is, in the interval from $(\mu - 3\sigma)$ to $(\mu + 3\sigma)$. Although $k = 2$ and $k = 3$ are very useful in practice, k need not be an integer. For example, the fraction of measurements falling within $k = 2.5$ standard deviations of the mean is at least $1 - [1/(2.5)^2] = .84$.

To apply Tchebysheff's Theorem to describe sample data, we should use s' rather than s (s', defined in Section 2.8, is a quantity slightly smaller than s) for constructing intervals about the mean. We ignore this minute point because it is of no practical importance. The theorem always holds true when s is used instead of s'. Moreover, s and s' are nearly equal when n is large (and satisfactory for descriptive purposes when n is as small as 10). Finally, note that we are interested primarily in describing populations, not samples. Examples describing small sets of sample measurements are presented solely to demonstrate the use of Tchebysheff's Theorem.

EXAMPLE 2.6   Many major corporations frequently conduct managerial aptitude examinations among their employees in an attempt to identify those who demonstrate management potential. The mean and variance of the managerial aptitude examination scores of $n = 25$ employees of a certain corporation are 75 and 100, respectively. Use Tchebysheff's Theorem to describe the distribution of exam scores.

SOLUTION   We are given $\bar{x} = 75$ and $s^2 = 100$. The standard deviation is $s = \sqrt{100} = 10$. The distribution of measurements is centered about $\bar{x} = 75$, and Tchebysheff's Theorem states the following:

1.   At least 3/4 of the 25 scores lie in the interval $(\bar{x} \pm 2s) = [75 \pm 2(10)]$, or 55 to 95.
2.   At least 8/9 of the scores lie in the interval $(\bar{x} \pm 3s) = [75 \pm 3(10)]$, or 45 to 105.

◆◆

We emphasize the "at least" in Tchebysheff's Theorem because the theorem is very conservative, applying to *any* distribution of measurements. In most situations the fraction of measurements falling into the specified interval will exceed $[1 - (1/k^2)]$. We now state a rule that describes accurately the variability of a particular

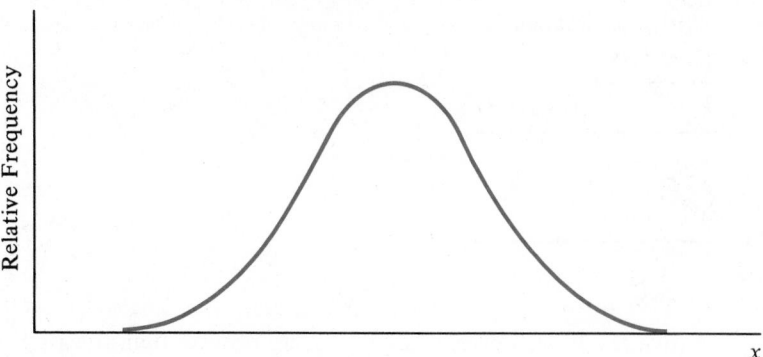

Figure 2.17   Normal (Bell-Shaped) Distribution

bell-shaped distribution and describes reasonably well the variability of other mound-shaped distributions of data. The frequent occurrence of mound-shaped and bell-shaped distributions of data in nature—hence, the applicability of our rule—leads us to call it the Empirical Rule.

Empirical Rule

> Given a distribution of measurements that is approximately bell-shaped (see Figure 2.17), the interval
>
> $(\mu \pm \sigma)$ contains approximately 68% of the measurements
> $(\mu \pm 2\sigma)$ contains approximately 95% of the measurements
> $(\mu \pm 3\sigma)$ contains all or almost all the measurements

The bell-shaped distribution shown in Figure 2.17 is commonly known as the *normal distribution* and will be discussed in detail in Chapter 6. The Empirical Rule applies exactly to data that possess a normal distribution, but it also provides an excellent description of variation for many other types of data.

EXAMPLE 2.7   A time study is conducted to determine the length of time necessary to perform a specified operation in a manufacturing plant. The length of time necessary to complete the operation is measured for each of $n = 40$ workers. The mean and standard deviation are found to be 12.8 and 1.7, respectively. Describe the sample data by using the Empirical Rule.

SOLUTION   To describe the data, we calculate the intervals

$$(\bar{x} \pm s) = (12.8 \pm 1.7) \qquad \text{or} \qquad 11.1 \text{ to } 14.5$$
$$(\bar{x} \pm 2s) = [12.8 \pm 2(1.7)] \qquad \text{or} \qquad 9.4 \text{ to } 16.2$$
$$(\bar{x} \pm 3s) = [12.8 \pm 3(1.7)] \qquad \text{or} \qquad 7.7 \text{ to } 17.9$$

According to the Empirical Rule, we expect approximately 68% of the measurements to fall into the interval from 11.1 to 14.5, approximately 95% to fall into the interval from 9.4 to 16.2, and all or almost all to fall into the interval from 7.7 to 17.9.

If we doubt that the distribution of measurements is mound-shaped or wish for some other reason to be conservative, we can apply Tchebysheff's Theorem and be absolutely certain of our statements. Tchebysheff's Theorem tells us that at least 3/4 of the measurements fall into the interval from 9.4 to 16.2 and at least 8/9 into the interval from 7.7 to 17.9.

Before leaving this topic, we might ask how well the Empirical Rule applies to the price-earnings ratios of Table 2.1. We have already shown that the mean is $\bar{x} = 16.0$, and in Section 2.10 we will show that the standard deviation for the $n = 25$ measurements is $s = 5.6$. The appropriate intervals are calculated and the number of measurements falling into each interval recorded. The results are shown in Table 2.19, with k in the first column and the interval $(\bar{x} \pm ks)$ in the second column, using $\bar{x} = 16.0$ and $s = 5.6$. The frequency, or number of measurements falling into each interval, is given in the third column and the relative frequency in the fourth column. Note that the relative frequency histogram for these data (Figure 2.2) is not bell-shaped. Yet the percentages falling into the intervals $(\bar{x} \pm s)$, $(\bar{x} \pm 2s)$, and $(\bar{x} \pm 3s)$ agree reasonably well with the percentages given by the Empirical Rule.

Table 2.19   Frequency of Measurements Lying Within k
Standard Deviations; Data of Table 2.1

k	Interval, $(\bar{x} \pm ks)$	Frequency in Interval	Relative Frequency
1	10.4–21.6	16	.64
2	4.8–27.2	24	.96
3	−.8–32.8	25	1.00

EXERCISES   Conceptual Level

2.39   When is it appropriate to use s^2 rather than σ^2 as a measure of variation?

2.40   Why is s rather than s^2 often a more useful measure of variability?

2.41   Why is a distinction made between s'^2 and s^2?

2.42   Of what practical importance is Tchebysheff's Theorem?

EXERCISES   Skill Level

2.43   Calculate $\bar{x}$, s^2, and s for the following $n = 5$ measurements:

   2   1   6   4   2

2.44   Calculate $\bar{x}$, s^2, and s for the following $n = 7$ measurements:

   5   3   7   4   2   8   6

2.45   A set of $n = 100$ measurements has a mean of 33 and a standard deviation of 2.5.

   a.   Use Tchebysheff's Theorem to describe the distribution of these measurements.

   b.   If the distribution of the measurements is known to be mound-shaped, use the Empirical Rule to describe the distribution of these measurements.

2.46   Suppose that $\bar{x} = 40$ and $s = 5$ for a set of $n = 50$ observations.

   a.   What can be said about the number of observations in the interval from 30 to 50?

   b.   If the observations have a mound-shaped histogram, what proportion of the observations lie in the interval from 35 to 45? From 30 to 50?

EXERCISES   Applications

 2.47   The yield to maturity of industrial bonds depends on the issuing firm's bond rating and the state of the economy at the time of issue. For the month ending July 1987 the average yield to maturity of industrial bonds issued during the month was 9.92 [*Federal Reserve Bulletin*, Vol. 73, No. 10 (October 1987), p. A24, Table 1.35]. Assume that the distribution of yields on industrial bonds is mound-shaped, with a standard deviation of .11.

   a.   Describe the distribution of industrial bond yields during July 1987.

   b.   During this period the average yield to maturity for long-term U.S. Treasury bonds was 8.70. Approximately what percentage of the industrial bonds exceeded the U.S. Treasury bond average yield during this period?

 2.48   Quality control techniques are used to monitor the quality of a manufacturing process to ensure uniformity and consistency in product output. A quality control engineer for a glass bottle manufacturing company, seeking to establish product quality standards at a time when the manufacturing process was known to be in control, randomly selected $n = 30$ bottles from the manufacturing process and recorded their weights. He found the average weight to be 8.2 oz and the standard deviation to be 1 oz. Describe the distribution of weights when the process is in control.

   a.   Use Tchebysheff's Theorem.

   b.   Use the Empirical Rule. Is the Empirical Rule likely to be applicable in this case? Explain.

   c.   Suppose the quality control engineer had selected a sample of $n = 4$ instead of $n = 30$ bottles from the manufacturing process. Would the Empirical Rule have been suitable for describing the bottle weights? Explain.

2.49   The Elliot Wave system counts, measures, and attempts to predict price movements of stocks based on "waves" or variations in stock prices. Using this system, Elliot Wave followers predicted that the Dow-Jones industrial average (DJIA) would reach 2000 by 1985 ("The Elliot Wave Lift-Off!" *Financial World*, September 15, 1979).

    a.   If records showed that daily recorded values of the DJIA had averaged 820 with a standard deviation of 60 during the decade prior to this prediction, was it likely that the Elliot Wave prediction would come true?

    b.   In 1985 the DJIA reached approximately 1300 points. Although the Elliot Wave prediction was not realized, this value of the DJIA is also very unlikely, using past market activity as a guide. What does this observation suggest regarding the appropriateness of past market activity as a model for future market activity?

2.10

A Short Method for Calculating the Variance

The calculation of the variance and standard deviation of a set of measurements is difficult, regardless of the method employed, but it is particularly tedious if we proceed according to the definition, by calculating each deviation individually as shown in Table 2.17. However, the sum of squares of the deviations can always be obtained by using the alternative formula given in the following display.

Shortcut Formula for Calculating the Sum of Squares of Deviations

$$\sum_{i=1}^{n}(x_i - \bar{x})^2 = \sum_{i=1}^{n}x_i^2 - \frac{\left(\sum_{i=1}^{n}x_i\right)^2}{n}$$

    Table 2.20 presents the individual observations of Table 2.17 in the first column and the squares of these observations in the second column. It follows that

$$\sum_{i=1}^{n}(x_i - \bar{x})^2 = \sum_{i=1}^{n}x_i^2 - \frac{\left(\sum_{i=1}^{n}x_i\right)^2}{n}$$

$$= 95 - \frac{(19)^2}{5}$$

$$= 95 - \frac{361}{5}$$

$$= 95 - 72.2 = 22.8$$

Table 2.20   Table for Simplified Calculations:

$$\sum_{i=1}^{n} (x_i - \bar{x})^2$$

x_i	x_i^2
5	25
7	49
1	1
2	4
4	16
19	95

Substituting this expression for the sum of squares of the deviations into the formula for s^2, we obtain an easy and computationally more precise method for calculating s^2.

Shortcut Formula for Calculating s^2

$$s^2 = \frac{\sum_{i=1}^{n} (x_i - \bar{x})^2}{n-1} = \frac{\sum_{i=1}^{n} x_i^2 - \left[\left(\sum_{i=1}^{n} x_i\right)^2 \Big/ n\right]}{n-1}$$

In the calculation of the variance for a population of N measurements, the sum of squared deviations would be

$$\sum_{i=1}^{N} (x_i - \mu)^2 = \sum_{i=1}^{N} x_i^2 - \frac{\left(\sum_{i=1}^{N} x_i\right)^2}{N}$$

where

$$\mu = \frac{\sum_{i=1}^{N} x_i}{N}$$

The population variance is then found by dividing the sum of squared deviations by N, so that

$$\sigma^2 = \frac{\sum_{i=1}^{N} x_i^2 - \left[\left(\sum_{i=1}^{N} x_i\right)^2 \Big/ N\right]}{N}$$

Because most data sets are sample values selected from a population of measurements, in general we are concerned with calculating the sample mean $\bar{x}$ and the sample variance s^2 and using these pivotal quantities to make inferences about μ and σ^2, the mean and variance in the population sampled.

Most calculators with statistical capabilities have built-in programs that will calculate $\bar{x}$ and s or μ and σ by accumulating $\sum_{i=1}^{n} x_i$ and $\sum_{i=1}^{n} x_i^2$ and then using the shortcut formulas given in this section. The sample standard deviation key is usually marked with s or σ_{n-1}, and the population standard deviation key is usually marked with σ or σ_N. In using any calculator with these built-in functions, be sure that you know which calculation is being carried out by each function key. With some data, like the data in the next example, for which $\bar{x}$, s, and $\sum_{i=1}^{n} (x_i - \bar{x})^2$ are known, verifying which keys are used to calculate $\bar{x}$, s, and σ is a simple exercise.

EXAMPLE 2.8   Calculate $\bar{x}$ and s for the measurements 85, 70, 60, 90, and 81.

SOLUTION   Preliminary calculations are given in Table 2.21. Then we have

$$\bar{x} = \frac{386}{5} = 77.2$$

$$\sum_{i=1}^{n} (x_i - \bar{x})^2 = \sum_{i=1}^{n} x_i^2 - \frac{\left(\sum_{i=1}^{n} x_i\right)^2}{n} = 30{,}386 - \frac{(386)^2}{5}$$

$$= 30{,}386 - 29{,}799.2 = 586.8$$

Thus,

$$s = \sqrt{\frac{\sum_{i=1}^{n} (x_i - \bar{x})^2}{n-1}} = \sqrt{\frac{586.8}{4}} = \sqrt{146.7} = 12.1$$

Table 2.21   *Preliminary Calculations for Example 2.8*

x_i	x_i^2
85	7,225
70	4,900
60	3,600
90	8,100
81	6,561
386	30,386

$\blacklozenge\ \blacklozenge$

Tips on Problem Solving

1. Always use the shortcut formula when calculating $\sum_{i=1}^{n}(x_i - \bar{x})^2$. This procedure will help reduce rounding errors.

2. Be careful about rounding numbers. Carry your calculations of $\sum_{i=1}^{n}(x_i - \bar{x})^2$ to at least six significant figures.

3. After you have calculated the standard deviation s for a set of data, compare its value with the range of the data. The Empirical Rule tells you that approximately 95% of the data should fall in the interval $\bar{x} \pm 2s$; that is, a very approximate value for the range will be $4s$. Consequently, a very rough rule of thumb is that

$$s \approx \frac{\text{range}}{4}$$

This crude check will help you to detect large errors—for example, failure to divide the sum of squares of deviations by $(n - 1)$ or failure to take the square root of s^2.

Many of the numerical descriptive measures that we have discussed are easily found by using the command DESCRIBE in the MINITAB package. As part of its output, this program produces the values of the mean, standard deviation, median, and

Table 2.22   MINITAB Computer Output Using the Command DESCRIBE for (a) the Data in Table 2.6 and (b) the Data in Table 2.1

```
MTB>DESCRIBE C3        MTB>DESCRIBE C4

            C3                      C4
N           42         N            25
MEAN     30.26         MEAN      16.00
MEDIAN   26.00         MEDIAN    15.90
TMEAN    29.97         TMEAN     15.91
STDEV     8.11         STDEV      5.61
SEMEAN    1.25         SEMEAN     1.12
MAX      49.00         MAX       28.60
MIN      19.00         MIN        5.40
Q3       37.00         Q3        20.00
Q1       24.00         Q1        12.20
          (a)                     (b)
```

the lower and upper quartiles as well as the values of some other statistics that we have not discussed. The *trimmed mean* (given as TMEAN) is the mean of the middle 90% of the measurements after excluding the smallest 5% and the largest 5% of the measurements in the set. Unlike the ordinary arithmetic mean, the trimmed mean is not sensitive to extremely large or small values in the data set.

Using the command DESCRIBE to summarize the average city miles per gallon data given in Table 2.6 produced the computer output in Table 2.22a. Table 2.22b displays the equivalent summary for the data in Table 2.1 concerning the price-earnings ratios for common stocks of growth companies listed on over-the-counter markets.

EXERCISES Conceptual Level

2.50 What are the advantages of using the shortcut formula for calculating the variance of a set of measurements?

2.51 Are there any circumstances under which you may not wish to use the shortcut formula for the variance?

EXERCISES Skill Level

2.52 Calculate $\bar{x}$, s^2, and s for the following data:

2 1 3 0 4 2 1 3

Use the shortcut formula to calculate the sum of squares of deviations.

2.53 Calculate $\bar{x}$, s^2, and s for the following data:

2 0 −2 −5 −4 0 −3 0 −1 3

2.54 The MINITAB command DESCRIBE produced the computer output of Table 2.23 for the following data:

35.9	55.4	63.7	48.7	61.6
49.9	32.0	46.2	58.0	49.5
62.5	55.4	54.9	66.9	51.4
43.9	60.0	65.1	56.4	43.1
51.5	58.7	67.6	72.5	57.3

a. Verify the value of $\bar{x}$ and s given in Table 2.23.

b. Calculate the intervals $(\bar{x} \pm ks)$ for $k = 1, 2, 3$, and tabulate the number of observations falling in each of these intervals.

c. Do the results obtained in part b agree with those given by Tchebysheff's Theorem? The Empirical Rule?

Table 2.23   MINITAB Computer Output for Exercise 2.54

```
MTB>DESC C1

                 N      MEAN     MEDIAN      TMEAN     STDEV    SEMEAN
 C1             25     54.72      55.40      54.94      9.78      1.96

               MIN       MAX         Q1         Q3
 C1          32.00     72.50      49.10      62.05
```

EXERCISES   Applications

2.55   Use the data from Exercise 2.38 (reproduced in Table 2.24) to calculate $\bar{x}$, s^2, and s.

Table 2.24   Productivity Data for Exercise 2.55

Country	Output per Hour	Country	Output per Hour
United States	2.8	United Kingdom	3.6
Canada	3.3	Belgium	6.3
Japan	7.9	Denmark	4.6
France	5.2	Netherlands	5.9
Germany	4.6	Norway	3.2
Italy	5.7	Sweden	4.6

Source: A. Neef and J. Thomas, "Trends in Manufacturing Productivity and Labor Costs in the U.S. and Abroad," *Monthly Labor Review*, Vol. 110, No. 12 (December 1987), pp. 25–30. Reprinted with permission.

2.56   Residential properties in a common region often have very different mortgage interest rates, since these rates depend on prevailing economic conditions at the time of the purchase of the property. A sample of 25 properties in a certain residential suburb provided the following mortgage interest rates on conventional loans outstanding on the properties (all listings are percentages):

9.2	10.0	11.5	12.0	14.0
14.8	14.5	13.2	11.4	10.5
12.0	13.0	16.5	15.3	11.0
15.7	12.5	9.5	13.3	13.5
10.5	11.5	10.5	12.1	15.5

a.   Construct a relative frequency histogram for these data.

b.   Calculate the mean, median, and modal interest rates for these properties.

c.   Calculate s^2 and s for these data.

d.   Describe the distribution of interest rates by using Tchebysheff's Theorem.

e.   Describe the distribution of interest rates by using the Empirical Rule. Would you expect the Empirical Rule to be suitable for describing these data? Explain.

f.   Calculate the percentage of measurements lying within one, two, and three standard deviations of the mean. How do these percentages compare with those given in the Empirical Rule? Those that might be expected according to Tchebysheff's Theorem?

g.   Find the 50th percentile of the distribution of interest rates. Find the 90th percentile. Interpret the two values you have computed.

h.   Find the lower quartile and the upper quartile of the distribution of interest rates and interpret these values.

   2.57   The state-by-state listing of the percentage change in existing home sales from 1986 to 1987 (Exercise 2.15 and Table 2.10) is shown here, along with the printout obtained by using the MINITAB command DESCRIBE (Table 2.25).

−1.8	26.2	−8.8	−8.7	4.6	−1.7
1.0	7.8	−8.7	18.5	15.0	−3.5
−2.2	4.3	4.8	0.0	9.4	18.8
14.3	−9.7	18.8	−9.1	−7.6	0.0
0.0	−2.5	−8.6	10.9	−5.4	−1.4
2.6	3.3	1.0	17.9	.6	−9.7
7.8	.5	−.7	5.0	−1.5	−.7
12.8	0.0	9.7	−13.3	−13.8	−7.8
−4.7	−1.6				

a.   Use the printout in Table 2.25 to find $\bar{x}$, s^2, and s.

b.   Use the printout to find the median and the upper and lower quartiles. Interpret these values.

c.   Describe the distribution of measurements by using Tchebysheff's Theorem and the Empirical Rule. Would you expect the Empirical Rule to be suitable for describing these data?

d.   Count the number of observations lying within one, two, and three standard deviations of the mean. How do the percentages falling in these intervals compare with those given in Tchebysheff's Theorem? The Empirical Rule?

Table 2.25   MINITAB Computer Output for Exercise 2.57

```
MTB>DESC C1
```

	N	MEAN	MEDIAN	TMEAN	STDEV	SEMEAN
C1	50	1.64	0.00	1.25	9.25	1.31

	MIN	MAX	Q1	Q3	
C1	−13.80	26.20	−4.87	7.80	

2.11

Estimating the Mean and Variance for
Grouped Data (Optional)

Often, the only data available for analysis are listed in the form of a frequency histogram. Company reports often list data only in terms of class frequencies; governmental and news media sources usually use some type of a bar chart to display pertinent data. In such cases we may not know the exact values of the measurements falling within the class intervals. When this situation occurs, there is no way to compute the exact values of the sample mean and variance.

There is, however, a method for approximating the mean $\bar{x}$ and the variance s^2 when only grouped data are available. **This method is based on the assumption that the midpoint of each class in the grouped frequency classification is approximately equal to the arithmetic mean of the measurements contained within that class.** The midpoint of a particular class i is denoted by the symbol m_i. Now, suppose that the midpoints do actually equal the mean of the measurements within their respective classes. Then for a particular class i, if we multiply m_i by f_i, the frequency within class i, we obtain the total of the measurements within class i. Summing the class totals then gives us the total of all measurements contained within the frequency distribution, and $\bar{x}$ can be found by dividing this sum by the total number of measurements n in the usual manner. Naturally, the accuracy of the approximated mean obtained by using class midpoints depends heavily on the degree to which the midpoints accurately reflect the arithmetic mean of the measurements contained within each respective class. Usually, such approximations are quite reliable, especially when the class frequencies f_i are of sufficient size to guarantee a rather even "coverage" of measurements over each class. For approximating the variance s^2 when only grouped data are available, we follow a procedure that generalizes the shortcut formula for the computation of s^2 introduced in Section 2.10.

Mean and Variance for Grouped Data

If data are grouped according to frequency of occurrence in each of k nonoverlapping classes, then the mean $\bar{x}$ and variance s^2 of the measurements contained within the groups are approximated by

$$\bar{x}_g = \frac{\sum_{i=1}^{k} f_i m_i}{n}$$

$$s_g^2 = \frac{\sum_{i=1}^{k} f_i m_i^2 - \left[\left(\sum_{i=1}^{k} f_i m_i\right)^2 \Big/ n\right]}{n-1}$$

where

m_i = midpoint of class i
f_i = frequency with which measurements occur within class i
$\sum f_i = n$

Table 2.26 Class Frequencies and Class Midpoints for the
25 Price-Earnings Ratios Listed in Table 2.1

Class, i	Class Boundaries	f_i	m_i	$f_i m_i$	$f_i m_i^2$
1	5.00–8.99	3	7	21	147
2	9.00–12.99	5	11	55	605
3	13.00–16.99	7	15	105	1,575
4	17.00–20.99	6	19	114	2,166
5	21.00–24.99	3	23	69	1,587
6	25.00–28.99	1	27	27	729
		25		391	6,809

Table 2.26 summarizes the computations necessary to calculate $\bar{x}_g$ and s_g^2 from the frequency distribution of $n = 25$ price-earnings ratios for 25 common stocks shown in Table 2.1. The computations required by the grouped data formulas given in the display are greatly simplified when the data are organized as in Table 2.26.

Using these formulas, we can approximate the mean from the grouped data as follows:

$$\bar{x}_g = \frac{\sum_{i=1}^{6} f_i m_i}{25} = \frac{391}{25} = 15.64$$

The approximation to the variance of the measurements is found by computing

$$s_g^2 = \frac{\sum_{i=1}^{6} f_i m_i^2 - \left[\left(\sum_{i=1}^{6} f_i m_i\right)^2 \Big/ 25\right]}{24} = \frac{6{,}809 - [(391)^2/25]}{24}$$

$$= \frac{693.76}{24} = 28.91$$

Because the approximation to the variance of the $n = 25$ price-earnings ratios is 28.91, the approximate standard deviation is

$$s_g = \sqrt{28.91} = 5.38$$

In Sections 2.7 and 2.10 we found the actual mean and standard deviation of the ungrouped sample of the $n = 25$ price-earnings ratios to be

$$\bar{x} = 16 \quad \text{and} \quad s = 5.6$$

Thus, the approximations obtained from the grouped frequency distribution of the price-earnings ratios (Table 2.1) appear to be satisfactory approximations to the values of $\bar{x}$ and s calculated from the ungrouped data.

Although it was mentioned in Section 2.2 that the classes should be of equal width, the classes need not be of equal width to apply the grouped data formulas for the mean and variance. All that must be assumed is that the class midpoints are approximately equal to the arithmetic mean of the measurements within the classes. The grouped data procedures are not applicable, however, in the case where one or more of the classes are open-ended (one endpoint is located at either $+\infty$ or $-\infty$). In such cases finding class midpoints for the open-ended classes becomes impossible.

EXERCISES   Conceptual Level

2.58   If a frequency table or histogram constructed from a set of continuous observations is used to calculate a grouped mean, when would the calculated value of the grouped mean be equal to the calculated value of the mean based on the original ungrouped observations?

2.59   A data set consists of the number of accidents occurring per day on a given stretch of freeway during rush hours. Suppose the data were grouped in the following manner:

Number of Accidents	Frequency
0	f_0
1	f_1
2	f_2
$\vdots$	$\vdots$
k	f_k

How would the grouped mean and standard deviation compare with the ungrouped mean and standard deviation?

EXERCISES   Skill Level

2.60   Use the following frequency tabulation to calculate the grouped mean, variance, and standard deviation.

Class Midpoint	Frequency
35	1
40	6
45	10
50	15
55	9
60	5
65	3
70	1

2.61 **a.** Construct a grouped frequency tabulation for the following set of $n = 30$ measurements:

4	7	5	8	4	5
6	2	5	5	5	2
5	5	4	6	6	7
4	7	5	5	6	5
7	2	8	1	7	6

b. Use the frequency distribution in part a to find the mean and standard deviation of these measurements.

EXERCISES   Applications

 2.62   The MINITAB printout in Table 2.27 shows a computer-generated histogram for the $n = 50$ gasoline mileage ratings presented in Exercise 2.9, along with the output generated by the DESCRIBE command.

  a. Use the computing formulas for grouped data to approximate the mean and standard deviation of these data.

Table 2.27   MINITAB Computer Output for Exercise 2.62

```
MTB>HISTOGRAM C1

HISTOGRAM OF C1     N = 50

MIDPOINT    COUNT
      32        1    *
      34        2    **
      36        3    ***
      38       10    **********
      40       15    ***************
      42       12    ************
      44        5    *****
      46        2    **

MTB>DESCRIBE C1
```

	N	MEAN	MEDIAN	TMEAN	STDEV	SEMEAN
C1	50	40.096	40.300	40.216	2.736	0.387

	MIN	MAX	Q1	Q3	
C1	32.500	45.600	38.675	41.650	

b. Compare the value of the grouped mean and standard deviation with the exact values of $\bar{x}$ and s given in the MINITAB printout.

c. Do these data lend themselves to description by the Empirical Rule? Explain.

 2.63 Term life insurance provides a fixed amount of insurance on a person's life for a specified period of time. In most cases the face value of a term life insurance policy is reduced incrementally as the insured increases in age. The data in Table 2.28 represent the term life insurance provided through a corporate insurance plan for the 100 employees of a manufacturing firm.

a. Use the computing formulas for grouped data to approximate the mean and standard deviation of the ages of the employees covered by the insurance policy.

b. Why are the values found in part a approximations to $\bar{x}$ and s and not, in fact, the true values of $\bar{x}$ and s for the data represented in the analysis?

c. Use the computing formulas for grouped data to compute the mean and standard deviation of the amount of insurance carried by the 100 employees. Are these values approximations to $\bar{x}$ and s? Explain.

d. What percentage of the insured employees are 50 years of age or older?

e. What percentage of the term policies carry a face value of $40,000 or less?

Table 2.28   Insurance Data for Exercise 2.63

Age of Employee	Amount of Insurance	Frequency
20–29	$70,000	26
30–39	60,000	33
40–49	40,000	21
50–59	25,000	14
60–69	10,000	6
		100

 2.64 The following frequency distribution gives the amount of property taxes paid during the previous year by each of a sample of 442 residential property owners selected from a city's tax rolls.

Amount Paid (Dollars)	Frequency
Under 200	27
200–399	85
400–599	217
600–799	81
800–999	32

a. Use the computing formulas for grouped data to approximate the mean and standard deviation of the amount of property taxes paid by each of the city's residential property owners.

b. Use the Empirical Rule to approximate the proportion of the city's residential property owners who paid from $312 to $694 in property taxes during the previous year.

2.12

Box Plots for Detecting Outliers (Optional)

We now have available various graphical and numerical techniques that can be used to describe a set of measurements. Stem and leaf displays and frequency histograms show how the measurements are distributed across the scale of measurement and, in effect, provide us with a means of visually assessing various aspects of the data set. Used in combination, numerical measures such as the mean, median, and mode or the median and the quartiles can be used to assess the symmetry or asymmetry of a set of measurements, while the mean and standard deviation can be used to produce intervals within which a known proportion of the measurements are expected to lie. For example, positive skewness is indicated when the mean of a set of measurements is significantly larger than the median. Furthermore, we know that the interval $\bar{x} \pm 2s$ contains at least 75% and more likely 95% of the measurements if the frequency distribution is mound-shaped.

box plot The **box plot** is a graphical display that describes not only the behavior of the measurements in the middle of the distribution but also their behavior at the ends or tails of the distribution. Values that lie very far from the middle of the distribution in either direction are called **outliers**. An outlier may result from transposing digits when recording a measurement, or from incorrectly reading an instrument dial, or perhaps from a malfunctioning piece of equipment, and so on. Even when there are no recording or observational errors, a data set may contain one or more valid measurements which, for one reason or another, differ markedly from the others in the set. These outliers can cause a marked distortion in the values of commonly used numerical measures such as $\bar{x}$ and s. In fact, outliers may themselves contain important information not shared with the other measurements in the set. Therefore, isolating outliers, if they are present, is an important step in any preliminary analysis of a data set. The box plot is designed expressly for this purpose.

hinges A box plot is constructed by using the median and two other measures, known as **hinges**. The median divides the ordered data set into two halves; the hinges are the values in the middle of each half of the data. Hinges are very similar to quartiles and, in effect, serve the same purpose. The actual difference between the value of a hinge and a quartile is quite small and decreases as the number of measurements increases. Nevertheless, we retain the distinction between these two measures and use hinges in constructing a box plot. However, we can think of a hinge as playing the role of a quartile.

Since each hinge is actually a "median" for each half of the data set, the hinges are found in the following way. Define $d(M)$ to be the position of the median, *without* the fraction 1/2 if there is one. Then $d(M)$ gives the "depth" of the median as measured from each end of the ordered measurements. The "depth" of the hinges as measured from each end of the data set is given by

$$d(H) = \frac{d(M) + 1}{2}$$

and the hinges are the values in position $d(H)$ as measured from each end of the data set.

For example, with $n = 10$, the position of the median is $(10 + 1)/2 = 5.5$. Therefore, $d(M) = 5$, and $d(H) = (5 + 1)/2 = 3$. The hinges are then those values in the third position, as measured from each end of the 10 ordered measurements. If $d(H)$ ends in $1/2$, a hinge will be the average of the two adjacent values in the array.

The dispersion of the measurements is now measured in terms of the difference between the hinges, called the **H-spread**, which is approximately equal to $(Q_3 - Q_1)$, the **interquartile range**. A data value will be identified as an outlier depending upon its relative position with respect to boundary points called inner and outer fences. The **inner fences** are defined as follows:

inner fences

$$\text{lower inner fence} = \text{lower hinge} - 1.5(\text{H-spread})$$
$$\text{upper inner fence} = \text{upper hinge} + 1.5(\text{H-spread})$$

outer fences

The **outer fences** are defined as follows:

$$\text{lower outer fence} = \text{lower hinge} - 3(\text{H-spread})$$
$$\text{upper outer fence} = \text{upper hinge} + 3(\text{H-spread})$$

adjacent values

The data values in each tail closest to, but still inside, the inner fences are called the **adjacent values**. Values lying between an inner fence and its neighboring outer fence are termed "outside" and are considered to be mild outliers. Values outside the outer fences are termed "far outside" and are considered to be extreme outliers. A box plot combines all this information in a pictorial display by identifying outliers and plotting their positions with respect to the center of the data set.

We illustrate the procedure by using the data in Table 2.29 which represent the profit after taxes expressed as a percentage of sales for $n = 22$ firms. The stem and leaf display for these data was given in Table 2.8 and is reproduced in Figure 2.18 to simplify the task of finding the median, the hinges, and other values required for the box plot.

With $n = 22$ the median location is $(22 + 1)/2 = 11.5$, so that the median is the average of the eleventh and twelfth ordered values. From the first column in the display we find that the eleventh and twelfth ordered values are in the stem class "4*" and are both equal to 4.0. Therefore, the median is $M = 4.0$. The depth of the hinges is

$$d(H) = \frac{11 + 1}{2} = 6$$

The lower hinge is 3.4, the upper hinge is 6.2, and the H-spread is $(6.2 - 3.4) = 2.8$. The lower and upper inner fences are

$$3.4 - 1.5(2.8) = -.8$$
$$6.2 + 1.5(2.8) = 10.4$$

and the lower and upper outer fences are

$$3.4 - 3(2.8) = -5.0$$
$$6.2 + 3(2.8) = 14.6$$

From the stem and leaf display, we see that 2.6 is the adjacent value closest to, but within, the lower inner fence and that 8.6 is the adjacent value closest to, but within, the upper inner fence.

Table 2.29   Profit After Taxes Expressed as a
             Percentage of Sales for $n = 22$ Firms

5.3	15.8	3.2
4.0	7.1	3.4
12.5	3.7	6.2
3.0	4.4	4.0
3.9	3.5	8.6
6.4	3.4	3.1
5.2	3.2	
2.6	5.6	

```
   1   |  2.   6
   7   |  3*   012244
  10   |  3.   579
  (3)  |  4*   004
   9   |  4.
   9   |  5*   23
   7   |  5.   6
   6   |  6*   24
   4   |  6.
   4   |  7*   1
   3   |  7.
   3   |  8*
   3   |  8.   6
       |       HI 125  158
```

Figure 2.18   Stem and Leaf Display of the
             Data in Table 2.29 (Leaf Digit
             Unit = .1000; 12= 1.2)

The value 12.5, which lies between the upper inner and outer fences, is a mild outlier; the value 15.8, which lies outside the upper outer fence, is an extreme outlier. We now summarize this information in the following display:

Median location	M	11.5	4.0	
Hinge location	H 6	3.4	6.2	2.8 ← H-spread
Adjacent values		2.6	8.6	
Inner fences	f	−.8	10.4	
	(outside)	—	12.5	
Outer fences	F	−5.0	14.6	
	(far outside)	—	15.8	

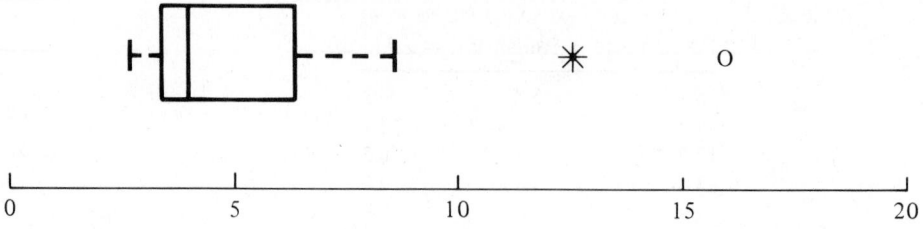

Figure 2.19   Box Plot for the Data in Table 2.29 (Figure 2.18)

Using this summary, we construct a box plot by drawing a box whose ends correspond to the values of the hinges, with a line through the box at the value of the median. Next, a dashed line is drawn from each end of the box to the corresponding adjacent value. Mild outliers and extreme outliers are plotted and prominently labeled by using an asterisk and a capital O. The resulting box plot is displayed in Figure 2.19. The scale of measurement is included in the display so that the values of the median, hinges, and outliers can be read from the box plot.

The box plot emphasizes the fact that the two outliers lie far from the central 50% of the measurements lying between the hinges. The box plot also indicates that these data are positively skewed, since the median is not equally spaced between the hinges but, rather, lies closer to the lower hinge. The lengths of the lines connecting the hinges and adjacent values give further evidence of skewness, the longer line indicating the direction of the skewing.

For these same data the command BOXPLOT in the MINITAB package produced the box plot in Table 2.30. Notice that, except for scale, the box plots are identical and lead to the same conclusions concerning the two observations identified as outliers.

Table 2.30   MINITAB Printout of the Box Plot for the Data in Figure 2.18

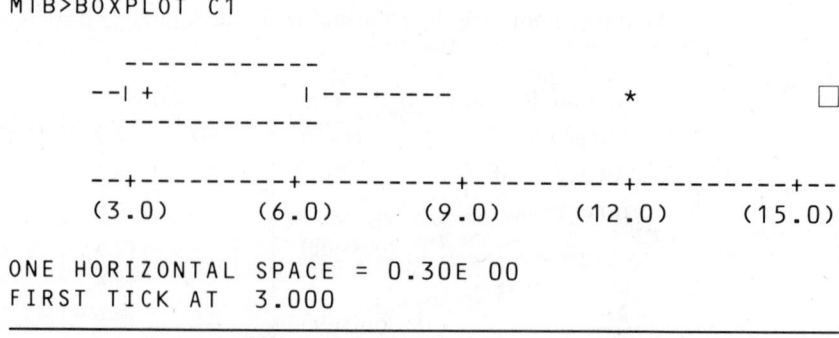

ONE HORIZONTAL SPACE = 0.30E 00
FIRST TICK AT   3.000

EXERCISES   Conceptual Level

2.65   Box plots are designed to aid in detecting outliers. Why is the detection of outliers important?

2.66   Define the following terms in your own words:

   a. $d(M)$ **b.** Hinges   **c.** $d(H)$ **d.** Hinge spread

   e. Inner fences   **f.** Outer fences   **g.** Adjacent values   **h.** Mild outliers

EXERCISES   Skill Level

2.67   Use the following data to construct a box plot, and determine whether or not there are any outliers. If outliers exist, determine whether they are mild or extreme outliers.

13	16	15
12	18	14
8	5	12
27	10	14
21	16	7
13	38	

2.68   Construct a box plot for the following data, and identify any outliers.

60	25	76	58
71	63	98	61
87	80	60	65
47	58	40	56
54	59	57	66
38			

EXERCISES   Applications

 2.69   The state-by-state listing of the percentage change in existing home sales from 1986 to 1987 (Exercise 2.15 and Table 2.10) is shown here. Construct a box plot to identify any outliers. If there are any outliers, how would you explain them?

−1.8	26.2	−8.8	−8.7	4.6	−1.7	7.8	−1.6	−1.5
1.0	7.8	−8.7	18.5	15.0	−3.5	12.8	−.7	−13.8
−2.2	4.3	4.8	0.0	9.4	18.8	−4.7	9.7	−.7
14.3	−9.7	18.8	−9.1	−7.6	0.0	.5	5.0	−7.8
0.0	−2.5	−8.6	10.9	−5.4	−1.4	0.0	−13.3	
2.6	3.3	1.0	17.9	.6	−9.7			

2.70   Refer to Exercise 2.9 in which $n = 50$ gasoline mileage ratings are given. Construct a box plot to identify any outliers.

2.13

Summary

The objective of a statistical study is to make inferences about a characteristic of a population based on information contained in a sample. Since populations are sets of data, we first need to consider ways to phrase an inference about a set of measurements. This latter point constituted the objective of our study in Chapter 2.

Methods for describing sets of measurements fall into one of two categories: graphical methods and numerical methods. The relative frequency histogram and the stem and leaf display are extremely useful graphical methods for characterizing a set of measurements. The box plot is a graphical technique that not only characterizes the symmetry or asymmetry of a data set but also identifies measurements that are deemed to be outliers. Numerical descriptive measures are numbers that attempt to provide a mental image of the frequency distribution. We have restricted the discussion to measures of central tendency and variability, the most useful of which are the mean and standard deviation. Although the mode is not generally a measure of central tendency, its importance in characterizing demand levels should be noted by the business statistician. Although the mean possesses intuitive significance, the standard deviation is meaningful only when used in conjunction with Tchebysheff's Theorem and the Empirical Rule. The objective of sampling is the description of the population from which the sample was obtained. This objective is accomplished by using the sample mean $\bar{x}$ and the sample variance s^2 as estimators of the population mean μ and the variance σ^2.

Many descriptive methods and numerical measures have been presented in this chapter; however, these are only a few of the methods that could have been discussed. Many special computational techniques have been omitted because common use of electronic computers has minimized the importance of special computational formulas. We have chosen to focus on the main objective of modern statistics and of this text—statistical inference.

Supplementary Exercises

Applications

 2.71   Most governmental agencies are authorized to issue tax-exempt bonds (bonds whose interest earnings are tax-exempt to the bondholder) as a means of raising operating revenues at less than the prevailing interest rate. A sampling of 25 tax-exempt revenue bonds listed in an issue of the *Wall Street Journal* provided the following coupon (interest) rates:

6.50	9.25	3.30	7.50	6.00
5.00	9.00	6.25	3.10	7.30
6.38	8.00	9.00	6.00	5.50
5.75	10.25	5.20	9.38	3.75
6.25	6.00	8.25	4.75	9.75

a. Construct a relative frequency histogram for these data.

b. Would you expect the interest rates on tax-exempt bonds to be higher or lower than those assigned to corporate bonds? Explain.

c. Compute $\bar{x}$, s^2, and s.

d. Find the fraction of measurements lying in the interval $\bar{x} \pm s$.

e. Find the fraction of measurements lying in the interval $\bar{x} \pm 2s$. Are these results consistent with Tchebysheff's Theorem?

f. Is the frequency histogram relatively mound-shaped? Does the Empirical Rule adequately describe the variability of the data?

 2.72 Refer to Exercise 2.71 and use the computing formulas for grouped data to approximate the mean and variance of the coupon rates of the listed tax-exempt revenue bonds. How do these approximations compare with $\bar{x}$ and s^2 calculated in Exercise 2.71?

 2.73 Past studies have shown that differing interpretations of such factors as the price range to consider, when to purchase, the choice of neighborhood, home styles, and the availability of mortgage lenders cause the duration of the active search period for home buyers to be quite variable. The following data represent the duration of active search (in weeks) of 25 home buyers in a certain city:

15	17	7	15	20
5	3	19	10	3
11	10	4	8	13
9	15	6	2	8
12	1	2	13	4

a. Construct a relative frequency histogram for these data.

b. What does this graphical description of the data tell you about the lengths of search times for home buyers?

 2.74 Table 2.31 gives the estimated gross proceeds (in millions of dollars) from all new securities publicly offered for cash, for the account of the issuer, with a maturity of more than one year. The securities are classified according to whether they were bonds and notes or preferred or common stocks. Construct a bar graph similar to the graph in Figure 2.5 to visually describe this data.

 2.75 Items that differ significantly from prespecified production limits are classified as defective items that must be reworked or possibly scrapped. A production manager decides to record the difference between the observed measurement, x_1, and the lower specification limit for $n = 10$

Table 2.31   Securities Data for Exercise 2.74

Year	Total	Bonds and Notes	Preferred or Common Stock
1980	65,404	42,502	22,902
1981	64,481	37,293	27,188
1982	73,291	44,719	28,572
1983	102,406	49,485	52,921
1984	85,853	59,483	26,370
1985	127,698	85,828	41,870

Source: Data from U.S. Department of Commerce, Bureau of the Census, *Statistical Abstract of the United States*, 107th ed. (Washington, D.C., 1987).

randomly selected items from a production line. The data are

.105   .090   .112   .005   .070   .085   .164   .095   .100   .045

a. Calculate $\bar{x}$, s^2, and s for these data.

b. Find the range for these data. Find the ratio of the range to s. If we possessed a large number of observations having a mound-shaped distribution, the range would be expected to equal approximately how many standard deviations? (Note that this procedure provides a rough check on the computation of s.)

 2.76 During times of economic recession, dentists and physicians tend to incur special problems with account collection. The number of days since billing 30 accounts receivable of a physician are shown below.

17	57	10	35	26	3
21	11	7	72	5	86
6	20	105	40	14	42
12	32	28	13	19	28
45	8	19	21	38	20

a. Construct a relative frequency histogram for these data.

b. Accounts at least six weeks (42 days) delinquent are considered uncollectible and are turned over to a collection agency. What fraction of the 30 physician's accounts should be turned over to the collection agency?

c. Compute $\bar{x}$, s^2, and s.

d. Find the fraction of measurements lying within $\bar{x} \pm s$.

e. Find the fraction of measurements lying within $\bar{x} \pm 2s$. Are these results consistent with Tchebysheff's Theorem?

2.77 The following data represent the personal consumption expenditures (in billions of dollars) for nondurable goods in the United States for 1986. Construct a pie chart depicting the percentage of nondurable consumer expenditure in the United States in 1986 for each major expense category.

Category	Expenditure
Food and tobacco	532.0
Clothing, accessories, jewelry	209.1
Transportation	365.3
Housing and household operation	779.9
Medical care	257.8
Recreation	198.0
Other	328.2

Source: Data from U.S. Department of Labor, Bureau of Labor Statistics, 1988.

2.78 Give appropriate class boundaries for classifying each of the following sets of measurements into 10 classes.

a. Number of weeks of ownership of an automatic washer before first necessary major repair. Data are available from 964 purchasers, with times before repair ranging from 0 weeks to 113 weeks.

b. The annual amount spent on life insurance premiums by the 583 employees of a manufacturing company. The premium values range from $0 to $934.

c. The annual salaries of all 1,280 employees of a large pulp and paper processing firm. The salaries, including those of the company president and other executives, range from $14,850 to $250,000.

d. The number of customers entering a local discount department store during each of the past 100 days. The data range from 92 arrivals to 471 arrivals per day.

e. The monthly average interest rate charged by a commercial bank to its prime customers since January 1988. Since that time, the monthly average interest rate has fluctuated in the range 7.5% to 12.04%.

2.79 The rule of thumb for estimating s from the range discussed in Exercise 2.75 can be extended to smaller amounts of data obtained in sampling from a bell-shaped distribution. Thus, the calculated value for s should not be very different from the range divided by the appropriate ratio found in the following table:

Number of Measurements	5	10	25
Expected Ratio of Range to s	2.5	3	4

For the following data (from Exercise 2.56), estimate s as suggested. Compare this estimate with the calculated s.

9.2	10.0	11.5	12.0	14.0
14.8	14.5	13.2	11.4	10.5
12.0	13.0	16.5	15.3	11.0
15.7	12.5	9.5	13.3	13.5
10.5	11.5	10.5	12.1	15.5

2.80 Differences in promotional spending among firms reflect differences in how sales respond to communications efforts. For example, pharmaceutical and breakfast food manufacturers generally spend a much greater amount on promotion than do, say, textile firms or manufacturers of primary metals. A survey of 10 pharmaceutical companies reveals the following expenditures on product promotion as a percentage of gross sales revenue:

21   18   25   26   25   20   24   19   28   22

    a. Observe the data and guess the value for s by use of the range approximation method.

    b. Calculate $\bar{x}$ and s and compare with the range approximation of part a.

2.81 Estimate the standard deviation for the underlying distribution of measurements from each of the following data summaries. (Assume that the data in each case are obtained from a bell-shaped distribution.)

    a. A survey of the annual incomes of ten nonunion truck drivers shows a range from \$23,160 to \$34,580.

    b. The personnel officer for a major manufacturing organization reports that for the past year the daily absenteeism rate in her company has ranged from 3 to 27.

    c. Over a two-week period (ten market days) the daily change in the Dow-Jones industrial average ranged from $-14\ 3/4$ on Tuesday of the first week to $+6\ 7/8$ on Thursday of the second week.

    d. The annual advertising and promotional budgets of 20 residential real estate firms that operate in the San Francisco area were recorded. The least expenditure on advertising and promotion among this group was \$29,450; the greatest, \$133,000.

2.82 The U.S. Food and Drug Administration (FDA) specifies that restaurants must maintain a temperature of at least 180°F in the water used to wash dishes. A certain brand of commercial dishwasher is examined and found to provide an average dishwater temperature of 185°F, with a standard deviation of 5°F. If the distribution of dishwater temperatures produced by the dishwasher is bell-shaped (approximately normal), what portion of time would the dishwasher be in compliance with FDA standards?

2.83 An aircraft manufacturer purchases a precision assembly for aircraft engines in large lots from a subcontractor. Over time, experience suggests that the assemblies average 5.1 mm in diameter, with a standard deviation of .1 mm.

    a. If the distribution of diameters of the precision assemblies is bell-shaped (approximately normal), what fraction of each lot received from the subcontractor would possess diameters falling in the interval from 4.9 to 5.3 mm?

    b. Suppose that the aircraft manufacturer specifies that the assembly diameters must be at least 4.8 mm and not greater than 5.2 mm to be usable in their aircraft engines. What fraction of each lot received from the subcontractor would fail to meet specifications?

2.84 Television commercials on a certain television station average 35 seconds in length of airtime, with a standard deviation of 5 seconds. If the distribution of airtime for commercials on the television station is mound-shaped, approximately what fraction of total commercials would last from 30 to 40 seconds?

2.85 Recent studies have shown that television commercials lasting longer than 40 seconds are excessively long and ineffective in communicating the intended message. What percentage (approximately) of the television commercials of the television station described in Exercise 2.84 could be considered excessively long?

2.86 Results of experiments report that advertisers who use electronic speech compression to speed up the audio presentation of their messages often enhance their impact on the public (P. LaBarbera and J. MacLachlan, "Time-Compressed Speech in Radio Advertising," *Journal of Marketing*, January 1979). Suppose an advertiser uses electronic speech compression to effectively reduce by 25% the length of a 40-second television commercial. Estimate the fraction of commercials of the television station described in Exercise 2.84 that are shorter in length than the speech-compressed advertisement. If these experimental results are valid, what policy implications do they suggest for product advertisers? For television stations or networks?

2.87 The issue of property tax relief is of vital concern to all municipal administrators. Suppose administrators from the city discussed in Exercise 2.64 have decided to offer property tax relief to those residential property owners whose past year's property taxes were above the 84th percentile. Use the Empirical Rule to approximate the minimum amount of property taxes necessary to qualify for tax relief in the city.

2.88 Yakima, Washington, is located in the center of a very rich valley known for its production of apples, pears, and cherries. The daily low temperatures in Yakima during the month of May averaged 50°F, with a standard deviation of 18°F. The month of May is a very critical month in the development of these fruits on the trees. If, during this critical development period, the temperature drops below 32°F, the young fruit can be damaged and an entire year's crop wiped out. Using the Empirical Rule, find the approximate percentage of the days in May that are likely to be cold enough to place the Yakima Valley fruit crops in danger.

2.89 The consumption of caffeine has become an increasingly important concern to consumers, since research has suggested that caffeine might be associated with birth defects, cancer, cardio-vascular disease, and other health problems. Federal FDA scientists are conducting studies to assess the possible association between caffeine and various health problems. However, the studies have produced inconclusive or inconsistent results, and according to Chris Lecos ["Caffeine Jitters: Some Safety Questions Remain," *FDA Consumer*, Vol. 21, No. 10 (December 1987–January 1988), pp. 24–26], "most of the evidence does not support a causal role" for caffeine. The caffeine contents for various beverages are listed in Table 2.32.

a. Describe the distribution of caffeine content for drip-brewed coffee, using Tchebysheff's Theorem.

Table 2.32   Data on Caffeine Content (Milligrams) of Beverages
for Exercise 2.89

Item	Average	Standard Deviation
Brewed coffee (5-oz cup)		
Drip	115	40
Percolator	80	35
Brewed tea (5-oz cup)		
U.S. major brands	40	18
Imported brands	60	21
Soft drinks (6-oz serving)		
Regular cola	19	2
Cherry cola	20	1

Source: Data adapted from Chris W. Lecos, "Caffeine Jitters: Some Safety Questions Remain," *FDA Consumer*, Vol. 21, No. 10 (December 1987–January 1988), pp. 24–26.

b. Compare the distribution of caffeine for brewed tea made with U.S. major brands and brewed tea made with imported brands.

c. Does there appear to be a difference in the distribution of caffeine for regular and cherry cola?

Supplementary Exercises

Interpretive

2.90 Concerned with the impact of a quality-of-work-life (QWL) program on industrial productivity, the manager of a company that manufactures laser scanner devices was interested in comparing the output of two manufacturing facilities of equal size and potential. In facility A the QWL program has been fully implemented; in facility B it has not. The data in Table 2.33 provide the output of the two manufacturing facilities for 12 consecutive months. Use the appropriate graphical and numerical descriptive techniques to describe these data. The objective of the presentation is to determine whether or not the QWL program has been effective in increasing productivity.

Table 2.33   Output Data for Exercise 2.90

Month	Facility A	Facility B	Month	Facility A	Facility B
January	216	218	July	239	217
February	212	209	August	243	205
March	222	214	September	250	200
April	247	202	October	254	219
May	231	209	November	235	220
June	252	213	December	246	216

2.91 An accountant for a large retail store is interested in estimating the average accounts receivable balance (in dollars) for the store's 10,000 credit customers. The frequency distribution shown in the accompanying table was constructed from a sample of $n = 100$ accounts selected at random from the store's accounts receivable files. Notice the shape of the frequency distribution. Use appropriate graphical and numerical descriptive techniques to provide an adequate description of the data.

Account Balance	Frequency
$ 0 to under 50	10
50 to under 100	15
100 to under 150	40
150 to under 200	22
200 to under 250	13
250 and above	0
	100

2.92 In the first half of 1987, 927 corporate mergers and acquisitions were announced, a strong decrease from the 1,527 mergers announced during the same period in 1986. The data that follow are the dollar amounts (in billions of dollars) recorded for the 32 mergers or acquisitions involving at least $2.5 billion during the period 1981 to mid-1987.

2.5	3.5	13.3	7.4
3.6	7.4	5.6	3.8
6.2	4.0	2.7	4.2
4.3	2.6	3.4	2.7
3.8	2.7	7.5	4.9
2.7	10.1	5.0	5.7
5.2	3.0	5.2	4.9
4.8	5.6	6.2	3.6

Source: Data from U.S. Department of Labor, Bureau of Labor Statistics, 1988.

Use appropriate graphical and numerical techniques to summarize these data. Do there appear to be outliers in the data set? (The two largest dollar amounts involve the Chevron acquisition of Gulf Oil and the Texaco acquisition of Getty Oil.) If any outliers are statistically identified, do your numerical descriptive measures exhibit a significant change in value if these outliers are not used in their calculation? Would it be reasonable to use the Empirical Rule in describing the data? Why or why not?

Case Study

State Lotteries Produce Millionaires Weekly!

◆

Many states across America have instituted lotteries of various sorts as revenue raisers that require no increases in taxes but whose income can nonetheless be appropriated to specific state budget line items. LOTTO 6/49, the California lottery instituted by means of a ballot initiative, began as a fund-raising venture to compete with neighboring state lotteries, but with the proviso that a fixed percentage of the net lottery income was to be used to augment funding for kindergarten through 12th-grade education in California. The success of the lottery may be counterproductive for education in California if the California legislature sets the K–12 education appropriations by taking into account the monies received from the weekly lottery.

To play LOTTO 6/49, a player selects six numbers plus a seventh or "bonus" number, each between 1 and 49. The winning numbers are randomly drawn one at a time by machine from a container whose contents are Ping-Pong balls numbered 01 through 49. The balls are drawn without replacing any of the balls previously drawn. Any person selecting all six of the first six numbers drawn (this excludes the bonus number, drawn seventh) shares equally in the first-place lottery prize, which has always totaled $1 million or more each week. The data in Table 2.34 provides the actual numbers (six numbers plus the bonus number) drawn each week beginning October 18, 1986, through October 10, 1987, in addition to the number of winners and the amount of the prizes in each of the winning categories.

1. Use the most effective and appropriate tabulation technique to convey the frequency with which the six numbers are drawn in the California LOTTO 6/49 game. (*Hint:* A stem and leaf diagram displays the stem frequencies and provides an effective histogram display of the data.) If the numbers are drawn without bias, all stems in a stem and leaf display should appear with approximately equal frequencies. On the basis of your tabulation, what would you conclude about the fairness of the draws? Would your conclusion change if the 54 bonus numbers were also included in your tabulation?

2. From the results of the number of first-place winners and the amount divided among those first-place winners, what is the "average" amount won by a person selecting all six LOTTO numbers correctly? (If there is no winner for any given week, the winning amount is carried forward to next week's prize, while being increased by a given percentage of the new week's sales.)

3. How would you describe or summarize the distribution of amounts won by those winners correctly selecting five of the first six numbers drawn, plus the bonus number?

Table 2.34   LOTTO 6/49 Prize Data for the Case Study

Draw	Winning Numbers/ Bonus	6 of 6	5 of 6 Plus Bonus	5 of 6	4 of 6	3 of 6 ($5)	Total Prize Winnings
#1 (10/18/1986)	36–09–47–39	0	1	52	4,451	90,495	$ 1,797,417
Sales: 6,361,266	04–33 B/25	($2,417,281)	($679,065)	($6,728)	($71)		
#2 (10/25)	35–21–32–38	0	3	57	5,464	95,628	2,031,922
Sales: 7,342,850	08–36 B/47	($5,211,470)	($261,283)	($7,085)	($67)		
#3 (11/01)	47–30–32–20	0	2	36	3,201	79,725	1,945,745
Sales: 7,308,961	43–44 B/23	($7,738,557)	($390,115)	($11,166)	($114)		

Table 2.34 (Continued)

Draw	Winning Numbers/ Bonus	6 of 6	5 of 6 Plus Bonus	5 of 6	4 of 6	3 of 6 ($5)	Total Prize Winnings
#4 (11/08)	45–07–05–32	1	0	255	10,090	181,001	$ 12,616,491
Sales: 8,058,774	21–12 B/13	($10,874,786)	($860,274)	($1,738)	($39)		
#5 (11/15)	41–48–47–05	0	0	52	3,476	97,927	1,351,267
Sales: 8,208,725	04–18 B/42	($4,621,142)	($876,281)	($8,682)	($118)		
#6 (11/22)	11–27–47–16	0	0	164	10,433	121,747	1,464,162
Sales: 8,157,522	26–21 B/08	($9,035,209)	($870,815)	($2,735)	($39)		
#7 (11/29)	30–19–10–26	0	4	213	10,048	161,601	2,662,876
Sales: 8,797,073	23–02 B/22	($13,771,626)	($234,771)	($2,271)	($43)		
#8 (12/06)	18–15–28–22	1	3	149	10,810	179,975	21,013,873
Sales: 10,305,518	49–24 B/30	($17,939,020)	($366,704)	($3,804)	($47)		
#9 (12/13)	03–13–47–17	0	0	131	9,048	178,042	1,821,406
Sales: 8,870,170	41–19 B/16	($3,188,437)	($946,890)	($3,724)	($49)		
#10 (12/20)	03–05–22–36	1	1	92	7,735	177,291	11,073,927
Sales: 8,943,748	37–33 B/49	($8,300,000)	($954,745)	($5,346)	($57)		
#11 (12/27)	07–01–33–21	0	4	122	9,352	179,085	2,779,335
Sales: 8,930,392	38–42 B/16	($3,208,478)	($238,329)	($4,025)	($47)		
#12 (01/03)	41–14–04–30	0	0	120	6,817	130,094	1,627,746
Sales: 9,342,149	27–37 B/01	($6,565,677)	($997,274)	($4,281)	($68)		
#13 (01/10)	33–34–18–40	0	6	104	7,204	159,450	2,842,286
Sales: 9,659,395	04–21 B/06	($11,685,117)	($171,856)	($5,108)	($67)		
#14 (01/17)	13–20–28–35	4	1	312	10,746	187,371	19,147,753
Sales: 10,841,842	36–21 B/31	($3,980,000)	($1,157,366)	($1,911)	($50)		
#15 (01/24)	48–13–10–21	0	0	151	8,053	173,357	1,889,186
Sales: 9,804,604	08–35 B/12	($3,468,154)	($1,046,641)	($3,571)	($60)		
#16 (01/31)	25–14–24–22	1	2	128	9,351	183,268	12,327,518
Sales: 10,177,805	39–08 B/31	($9,260,000)	($543,240)	($4,373)	($54)		
#17 (02/07)	16–40–26–47	0	2	165	8,163	157,331	2,881,709
Sales: 9,924,644	09–36 B/12	($3,502,333)	($529,727)	($3,308)	($60)		
#18 (02/14)	25–45–41–26	0	1	51	4,730	121,108	2,736,409
Sales: 10,074,246	22–39 B/32	($7,100,002)	($1,075,425)	($10,864)	($106)		
#19 (02/21)	15–24–21–03	0	3	226	11,911	204,774	3,274,168
Sales: 10,672,531	27–34 B/02	($10,882,242)	($379,764)	($2,597)	($44)		
#20 (02/28)	37–06–40–20	2	2	186	9,097	151,256	19,220,569
Sales: 12,209,185	49–46 B/33	($7,940,000)	($651,665)	($3,610)	($67)		
#21 (03/07)	02–28–18–09	0	7	444	19,127	277,778	3,596,533
Sales: 10,456,545	22–04 B/35	($3,700,000)	($159,462)	($1,295)	($27)		
#22 (03/14)	45–22–01–12	0	3	217	10,656	189,940	3,253,290
Sales: 10,881,970	33–49 B/17	($7,654,326)	($387,216)	($2,758)	($51)		
#23 (03/21)	04–28–18–01	2	4	284	14,155	269,024	15,839,088
Sales: 11,424,459	13–27 B/20	($6,040,000)	($304,890)	($2,212)	($40)		
#24 (03/28)	16–40–38–24	0	1	143	8,272	156,417	2,992,252
Sales: 10,442,561	14–10 B/43	($3,814,021)	($1,114,743)	($4,016)	($63)		
#25 (04/04)	23–03–24–02	0	1	207	12,640	238,470	3,460,161
Sales: 10,738,741	08–31 B/43	($7,830,981)	($1,146,360)	($2,853)	($42)		
#26 (04/11)	17–21–49–39	1	8	120	9,917	188,342	15,780,150

(continues)

Table 2.34 (Continued)

Draw	Winning Numbers/ Bonus	6 of 6	5 of 6 Plus Bonus	5 of 6	4 of 6	3 of 6 ($5)	Total Prize Winnings
Sales: 11,147,627	20–33 B/13	($12,480,000)	($148,751)	($5,109)	($56)		
#27 (04/18)	26–47–31–35	1	0	109	6,045	130,881	$ 5,786,961
Sales: 10,414,030	02–15 B/01	($4,040,000)	($1,111,697)	($5,254)	($86)		
#28 (04/25)	41–23–42–03	2	2	191	11,355	213,213	9,702,440
Sales: 10,597,706	11–08 B/04	($3,200,000)	($565,652)	($3,051)	($46)		
#29 (05/02)	15–02–42–01	0	2	140	9,197	188,420	3,209,930
Sales: 10,723,419	23–17 B/30	($4,133,653)	($572,362)	($4,212)	($58)		
#30 (05/09)	31–23–30–15	0	4	208	11,935	203,551	3,275,323
Sales: 10,710,763	33–12 B/29	($8,200,000)	($285,843)	($2,832)	($44)		
#31 (05/16)	03–34–42–09	1	14	357	16,707	279,485	17,624,171
Sales: 12,418,268	01–18 B/25	($13,600,000)	($94,689)	($1,913)	($37)		
#32 (05/23)	39–12–37–18	2	5	313	12,937	209,385	7,632,947
Sales: 10,731,638	06–29 B/34	($2,160,000)	($229,120)	($1,885)	($41)		
#33 (05/30)	03–23–13–08	0	11	376	24,675	317,259	3,813,253
Sales: 10,565,512	36–19 B/09	($4,198,661)	($102,533)	($1,545)	($21)		
#34 (06/06)	26–05–09–25	2	12	550	25,271	352,916	13,069,764
Sales: 11,559,524	13–03 B/20	($4,440,000)	($102,831)	($1,155)	($22)		
#35 (06/13)	47–39–11–29	0	3	115	7,282	151,322	3,119,207
Sales: 11,185,311	48–01 B/33	($4,433,455)	($398,010)	($5,349)	($76)		
#36 (06/20)	37–15–43–41	0	6	115	7,977	179,790	3,353,132
Sales: 11,622,185	13–05 B/28	($8,839,482)	($206,778)	($5,558)	($72)		
#37 (06/27)	12–28–29–10	2	6	307	15,076	276,983	18,751,170
Sales: 13,650,773	08–26 B/22	($7,240,000)	($242,870)	($2,445)	($45)		
#38 (07/04)	07–24–33–06	3	5	460	23,312	363,336	8,653,303
Sales: 11,357,935	04–21 B/37	($1,480,000)	($242,491)	($1,358)	($24)		
#39 (07/11)	16–48–13–20	2	3	96	7,615	173,090	7,790,390
Sales: 11,384,713	11–26 B/41	($2,260,000)	($405,106)	($6,522)	($74)		
#40 (07/18)	36–14–41–40	0	0	95	6,071	129,996	1,827,057
Sales: 11,246,828	25–26 B/21	($4,300,000)	($1,200,598)	($6,511)	($92)		
#41 (07/25)	03–46–48–31	1	5	73	6,965	168,993	15,074,931
Sales: 12,062,103	26–05 B/18	($11,680,000)	($257,525)	($9,087)	($86)		
#42 (08/01)	02–31–49–12	1	2	145	7,770	160,670	7,640,259
Sales: 11,161,175	41–16 B/03	($4,480,000)	($595,727)	($4,233)	($71)		
#43 (08/08)	35–16–33–20	0	5	123	7,929	167,353	3,190,486
Sales: 11,120,723	24–36 B/12	($4,388,329)	($237,427)	($4,972)	($70)		
#44 (08/15)	19–06–45–11	1	7	261	14,534	258,486	13,488,307
Sales: 12,837,257	03–20 B/23	($9,480,000)	($195,768)	($2,705)	($44)		
#45 (08/22)	18–32–37–26	2	16	237	13,807	223,543	8,399,918
Sales: 11,761,493	28–12 B/09	($2,400,000)	($78,471)	($2,729)	($42)		
#46 (08/29)	35–03–06–01	0	2	146	9,849	210,854	3,497,662
Sales: 11,574,622	42–31 B/24	($4,534,859)	($617,795)	($4,360)	($58)		
#47 (09/05)	32–16–13–47	1	6	210	11,694	220,053	13,891,877
Sales: 12,438,992	05–42 B/39	($10,160,000)	($221,310)	($3,257)	($53)		
#48 (09/12)	23–49–08–03	2	3	171	10,775	221,097	8,338,963
Sales: 11,514,676	25–48 B/19	($2,400,000)	($409,730)	($3,703)	($53)		

Table 2.34 (Continued)

Draw	Winning Numbers/ Bonus	6 of 6	5 of 6 Plus Bonus	5 of 6	4 of 6	3 of 6 ($5)	Total Prize Winnings
#49 (09/19)	31–45–02–20	0	2	125	8,545	175,251	$ 3,274,318
Sales: 11,339,288	19–22 B/01	($4,649,108)	($605,234)	($4,989)	($66)		
#50 (09/26/87)	41–42–24–27	1	1	279	12,065	196,523	13,345,533
Sales: 11,943,488	13–35 B/14	($9,840,000)	($1,274,967)	($2,354)	($49)		
#51 (09/30/87)	26–01–10–35	0	0	86	4,807	92,164	977,705
Sales: 4,923,249	47–21 B/02	($2,084,759)	($525,556)	($3,148)	($51)		
#52 (10/03/87)	31–18–04–25	0	6	158	7,967	153,736	2,937,508
Sales: 10,256,542	38–35 B/28	($7,478,874)	($182,480)	($3,570)	($64)		
#53 (10/07/87)	07–32–44–23	1	0	178	8,907	140,459	11,977,409
Sales: 6,848,849	36–09 B/42	($10,560,000)	($731,114)	($2,116)	($38)		
#54 (10/10/87)	45–40–22–36	0	1	65	4,943	120,033	2,757,681
Sales: 10,191,028	23–35 B/14	($5,830,631)	($1,087,892)	($8,623)	($103)		
		38	188	9,870	550,959	10,009,339	$387,828,844

Total sales: $557,225,393   Total winnings: $387,828,844   Total winners: 10,570,394

References and Suggested Readings

Cangelosi, V. E., P. H. Taylor, and P. F. Rice. *Basic Statistics: A Real World Approach*, 3d ed. St. Paul: West, 1983.

Koopmans, L. H. *An Introduction to Contemporary Statistics*, 2d ed. Boston: PWS-KENT, 1987.

Miller, R. B. *Minitab Handbook for Business and Economics*. Boston: PWS-KENT, 1988.

Ryan, B. F., B. L. Joiner, and T. A. Ryan. *MINITAB Handbook*, 2d ed. Boston: PWS-KENT, 1985.

Tukey, J. W. *Exploratory Data Analysis*. Reading, Mass.: Addison-Wesley, 1977.

Velleman, P. F., and D. C. Hoaglin. *Applications, Basics, and Computing of Exploratory Data Analysis*. Boston: PWS-KENT, 1981.

TO
THE
READER

The descriptive techniques of Chapter 2 not only provide us with a means of describing the salient characteristics of a data set but also provide us with a way of phrasing an inference about a population in terms of its parameters. How do we actually *make* an inference about a population based on sample information and assess the reliability of the inference? This step is accomplished through use of the ideas and concepts of probability.

The sampling distribution of a statistic, which plays a crucial part in making an inference, describes the probabilistic behavior of the statistic in repeated sampling. In Chapter 3 we deal with the basic concepts of the theory of probability and provide the framework for inferential statistical techniques presented in Chapter 8 and succeeding chapters.

3

PROBABILITY

3.1

Introduction

As stated in Chapter 1, the objective of statistics is to make inferences about a population based on information contained in a sample. Since the sample provides only partial information about the population, we require a mechanism that enables us to accomplish this objective. Probability is such a mechanism; it enables us to use the partial information contained in a set of sample data to infer the nature of the larger set of data, the population. How we use probability to make inferences is best illustrated by considering an example.

A manufacturer, in comparing two packages (call them A and B) designed for his product, feels that package A is superior and will be preferred over package B by consumers. To test the manufacturer's contention that package A is preferred to package B, 20 consumers are randomly selected, presented with both packages, and asked to select the design they prefer. If all 20 consumers indicate a preference for design B, what would you conclude about the manufacturer's contention?

If the contention that consumers prefer design A is true, the proportion of all potential purchasers of the product who prefer design A must be greater than 1/2, and we would expect that about the same proportion would be observed in the sample of 20 responses. Instead, none of the 20 consumers preferred design A, a result that is highly improbable if at least half of all consumers prefer design A. What can we conclude? Either we conclude that we have observed a very improbable sample, or we conclude that our original contention was false and that less than half of all consumers favor design A. Implicit in this decision process is a reliance on the notion of probability—in particular, the probability of the observed sample results.

In the preceding discussion we found that the sample results were so extremely contrary to the original contention that a decision to reject the contention could be made quite readily. Suppose, however, that the sample of 20 consumer responses indicated that 3 favored design *A* and 17 favored *B*. Would we still conclude that the sample is so improbable that we would reject the original contention? Suppose that 8 favored *A* and 12 favored *B*. What would we then conclude about the original contention? To answer these questions, we need to know "how improbable" a particular sample result is. In other words, we need to find the probability of obtaining a sample as extreme as that observed, given that the original contention is true. Having determined this probability, we are in a position to decide whether the contention is reasonable or should be rejected as untrue. Thus, probability provides the necessary mechanism for making inferences about the population on the basis of sample evidence.

3.2

Two Teaching Objectives

This chapter can be used to achieve varying degrees of understanding of the theory and application of statistics to the solution of business problems. Only a knowledge of the basic concepts and the theory of probability is needed for an understanding of the statistical concepts that commence in Chapter 5 and are developed in the succeeding chapters. Only very basic problem-solving abilities are needed for this objective.

Students seeking only a basic understanding of probability as needed for the discussion of statistical topics in later chapters should cover Sections 3.3 through 3.7 and the Conceptual Level and Skill Level Exercises following those sections. Instructors and students seeking to develop skill in solving probability problems should devote greater time to the optional examples and to the Applications and Supplementary Exercises. Section 3.8 ("Bayes's Law"), Section 3.9 ("Counting Sample Points"), and Section 3.10 ("Random Variables") provide additional material to assist in solving more complicated probability problems.

3.3

The Sample Space

Data are obtained either by observation of uncontrolled events in nature or by controlled experimentation in the laboratory. To simplify our terminology, we seek a word that will apply to either method of data collection, and hence, we define the term *experiment*.

Definition

experiment

An **experiment** is the process by which an observation (or measurement) is obtained.

Note that the observation need not produce a numerical value. Here are some typical examples of experiments:

1. Recording the income of a factory worker
2. Interviewing a buyer to determine brand preference for a particular product
3. Recording the price of a security at a particular time
4. Inspecting an assembly line to determine whether more than the allowable number of defectives is being produced
5. Recording the type and size of policy sold by an insurance salesperson

Definition

population
sample

A **population** is the set of all possible observations that could be generated by the repetition of an experiment. A **sample** consists of a set of observations selected from the population.

For example, we might be interested in the length of life of television tubes produced in a plant during the month of June. Testing a single tube until it fails and measuring the length of its life represents a single experiment. Repetition of the experiment for all tubes produced during this period generates the entire population. A sample consists of the lengths of life measured for certain tubes selected from the population.

events   Each experiment may result in one or more outcomes, which we will call **events** and denote by capital letters. Consider the following experiment.

EXAMPLE 3.1   Toss a die and observe the number appearing on the upper face. Some events would be the following:

Event A: observe an odd number.
Event B: observe a number less than 4.
Event E_1:   observe a 1.
Event E_2:   observe a 2.
Event E_3:   observe a 3.
Event E_4:   observe a 4.
Event E_5:   observe a 5.
Event E_6:   observe a 6.

◆ ◆

There is a distinct difference between events A and B and events $E_1, E_2, E_3, E_4, E_5,$ and E_6. Event A will occur if event $E_1, E_3,$ or E_5 occurs—that is, if we observe a 1, 3, or 5. Thus, A could be *decomposed* into a collection of simpler events, namely, $E_1, E_3,$ and E_5. Likewise, event B will occur if $E_1, E_2,$ or E_3 occurs and could be viewed as a collection of smaller or simpler events. In contrast, we note that it is impossible to decompose events $E_1, E_2, E_3, \ldots, E_6$. Events $E_1, E_2, \ldots, E_6$ are called *simple events* and A and B are *compound events.*

Events $E_1, E_2, \ldots, E_6$ represent a complete listing of all simple events associated with Example 3.1. An interesting property of simple events is readily apparent. **An experiment will result in one and only one of the simple events.**

Definition

simple event

> An event that cannot be decomposed is called a **simple event**. Simple events are denoted by the symbol E with a subscript.

For instance, if a die is tossed, we will observe a 1, 2, 3, 4, 5, or 6, but we cannot possibly observe more than one of the simple events at the same time. Hence, a list of simple events provides a breakdown of all possible outcomes of the experiment.

EXAMPLE 3.2   Toss a coin. The simple events are as follows:

Event E_1:   observe a head.

Event E_2:   observe a tail.

EXAMPLE 3.3

tree diagram

Toss two coins and record the outcome. We can construct a visual model for this experiment by using a device called a tree diagram. In a **tree diagram**, each successive branching of the tree corresponds to a step necessary to generate the possible outcomes of an experiment. A tree diagram for this example is given in Figure 3.1. The resulting simple events are shown in Table 3.1.

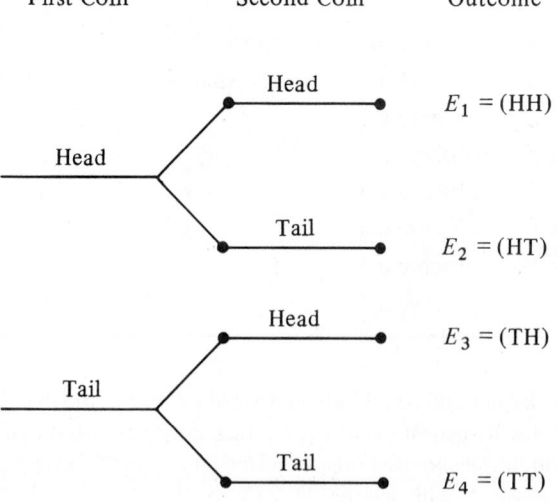

Figure 3.1   Tree Diagram for Example 3.3

Table 3.1   Simple Events for Tossing
Two Coins

Event	Coin 1	Coin 2
E_1	Head	Head
E_2	Head	Tail
E_3	Tail	Head
E_4	Tail	Tail

sample point

Venn diagram

Using the outcomes of the experiment in Example 3.3, we now create a correspondence between the simple events and a set of points. To each simple event we assign a point, called a **sample point**. Thus, the symbol E_1 will now be associated with either simple event E_1 or its corresponding sample point. The resulting diagram—the visual model—is called a **Venn diagram**.

Example 3.1 may be viewed symbolically in terms of the Venn diagram shown in Figure 3.2. Six sample points are shown, corresponding to the six possible simple events enumerated in Example 3.1. Likewise, a Venn diagram for the two-coin-toss experiment of Example 3.3 represents an experiment that has four sample points.

Definition

sample space

> The set of all sample points for an experiment is called the **sample space** and is represented by the symbol S. We say that S is the totality of all sample points.

What is an event in terms of the sample points? Recall that event A in Example 3.1 occurs if any one of the simple events E_1, E_3, or E_5 occurs. That is, we observe event A, an odd number, if we observe a 1, 3, or 5. Event B, a number less than 4, occurs if E_1, E_2, or E_3 occurs. Thus, if an event will occur only when one of a particular set of sample points occurs, the event is as clearly defined as if we had presented a verbal description

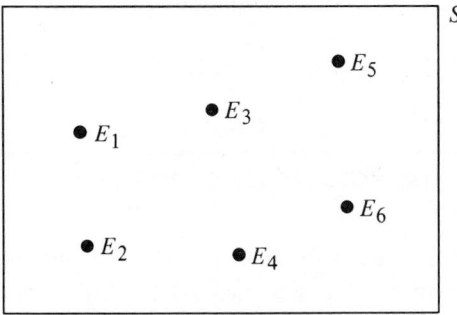

Figure 3.2   Venn Diagram for Die Tossing

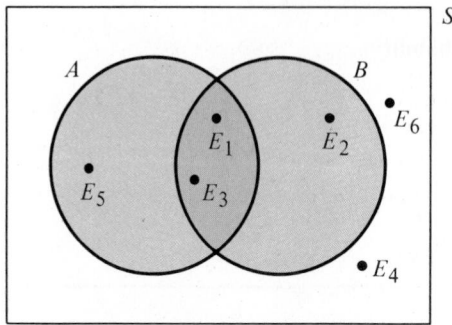

Figure 3.3   Events A and B for Die Tossing

of it. The event "observe E_1, E_3, or E_5" is obviously the same as the event "observe an odd number." Both represent event A.

To decide whether a sample point is included in a particular event, check to see whether the occurrence of the sample point implies the occurrence of the event. If it does, that sample point is included in the event. For example, in the die-tossing experiment the sample point E_1 is in the event A, "observe an odd number," because if E_1 occurs, then A will occur.

Definition

event

> An **event** is a specific collection of sample points.

Keep in mind that the preceding discussion refers to the outcome of a single experiment and that the performance of the experiment will result in the occurrence of one and only one sample point. A particular event will occur if any sample point in the event occurs.

An event can be represented on a Venn diagram by encircling the sample points in that event. Events A and B for the die-tossing problem are shown in Figure 3.3. Note that points E_1 and E_3 are in both events A and B and that both A and B occur if either E_1 or E_3 occurs.

3.4

The Probability of an Event

A population is the set of observations that could be obtained by repeating an experiment an infinite number of times. Some fraction of the experiments will result in E_1, another fraction in E_2, and so on. From a practical point of view, we think of the fraction of the population resulting in an event A as the probability of A. Putting it another way, **if an experiment is repeated a large number N of times and the event A is observed n_A times, the probability of A is approximately**

$$P(A) \approx \frac{n_A}{N}$$

As N becomes infinitely large, the fraction n_A/N approaches $P(A)$. This practical interpretation of the meaning of probability, a view held by most nonstatisticians, is called the **relative frequency concept of probability**.

relative frequency concept of probability

In practice, the composition of the population is rarely known, and hence, the desired probabilities for various events are unknown. We ignore this aspect of the problem mathematically and take the probabilities as given, thus providing a model for a real population and a basis for constructing a theory of probability. For instance, we would assume that for a large population of die tosses (Example 3.1), the numbers 1, 2, 3, 4, 5, and 6 should appear with approximately the same relative frequency and therefore that

$$P(E_1) = P(E_2) = \cdots = P(E_6) = 1/6$$

That is, we assume that the die is perfectly balanced. Is there such a thing as a perfectly balanced die? Probably not, but we are inclined to think that the probability of the sample points is so near 1/6 that our assumption is quite valid for practical purposes and provides a good model for die tossing.

We complete our model for the population by adding the following conditions.

Conditions for Probability Model

To each point in the sample space we assign a number called the probability of E_i, denoted by the symbol $P(E_i)$, such that

1. $0 \le P(E_i) \le 1$, for all i

2. $\sum_S P(E_i) = 1$

where the symbol $\sum_S$ means to sum the sample point probabilities over all sample points in S.

The two requirements placed on the probabilities of the sample points are necessary in order that the model conform to our relative frequency concept of probability. Thus, we require that a probability be greater than or equal to 0 and less than or equal to 1 and that the sum of the probabilities over the entire sample space S be equal to 1. Furthermore, from a practical point of view, we would choose the $P(E_i)$ in a realistic way so that they would agree with the observed relative frequency of occurrence of the sample points.

Keeping in mind that a particular event is a specific collection of sample points, we can now state a simple rule for finding the probability of any event.

Definition

probability of
an event A

> The **probability of an event** A is equal to the sum of the probabilities of the sample points in A.

This definition agrees with the relative frequency concept of probability.

EXAMPLE 3.4   Calculate the probability of the event A for the die-tossing experiment of Example 3.1.

SOLUTION   Event A, "observe an odd number," includes the sample points E_1, E_3, and E_5. Hence,

$$P(A) = P(E_1) + P(E_3) + P(E_5) = (1/6) + (1/6) + (1/6) = 1/2$$

EXAMPLE 3.5   Calculate the probability of observing exactly one head in a toss of two coins.

SOLUTION   Construct the sample space, letting H represent a head and T a tail. See Table 3.2.

Table 3.2   Sample Space and Probabilities for Example 3.5

Event	First Coin	Second Coin	$P(E_i)$
E_1	H	H	1/4
E_2	H	T	1/4
E_3	T	H	1/4
E_4	T	T	1/4

It seems reasonable to assign a probability of 1/4 to each of the sample points. We are interested in

Event A: observe exactly one head.

Since the sample points E_2 and E_3 are in A,

$$P(A) = P(E_2) + P(E_3) = (1/4) + (1/4) = 1/2$$

EXAMPLE 3.6   The personnel director of a company plans to hire two salespeople from a total of four applicants. Suppose she is completely incapable of correctly ranking the applicants according to their ability and, in effect, selects them at random.

    a.   What is the probability that she selects the two best candidates?
    b.   What is the probability that she selects at least one of the two best candidates?

SOLUTION   The experiment consists of selecting two applicants from the four available. Suppose that applicants vary in ability, and let 1, 2, 3, and 4 denote the applicants, with 1 and 2 representing the best and second best. Then the sample points in S are as shown in Table 3.3.

Table 3.3   Sample Points and Probabilities for Example 3.6

Sample Point	Pair Selected	Probability
E_1	1, 2	1/6
E_2	1, 3	1/6
E_3	1, 4	1/6
E_4	2, 3	1/6
E_5	2, 4	1/6
E_6	3, 4	1/6

Since we would expect each of the six pairs to occur with approximately the same relative frequency in many repetitions of the experiment, the probability assigned to each sample point is 1/6.

a.   Define event A as "the personnel director selects the two best applicants." Since A will occur only if E_1 occurs, we have

$$P(A) = P(E_1) = 1/6$$

b.   Define event B as "the personnel director selects at least one of the two best applicants." Since B will occur if $E_1, E_2, E_3, E_4,$ or E_5 occurs, we have

$$P(B) = P(E_1) + P(E_2) + P(E_3) + P(E_4) + P(E_5)$$
$$= (1/6) + (1/6) + (1/6) + (1/6) + (1/6) = 5/6$$

Since the personnel director's selection ability should be far better than a random selection, $P(A)$ and $P(B)$ should be less than the actual probabilities of selecting the "two best" and "at least one of the two best." A good personnel director should be able to make these selections with much higher probabilities.

The sample space and the underlying probabilistic model provide us with a simple, logical, and direct method for calculating the probability of an event or, if you like, the probability of a sample drawn from a theoretical population. A practical limitation to this (or any other) method for calculating probabilities is that the initial assignment of probabilities to the sample points is usually a subjective procedure. Consequently, a valid solution will require that the probability assignments provide a good measure of the likelihood of occurrence of the sample points.

A second limitation that specifically applies to the sample point procedure will be evident when you attempt to apply it to the solution of some probability problems. Listing the sample points can be a difficult problem in itself, and you must be certain that none has been omitted. Since the total number of sample points in S may be quite large, having rules to simplify the counting of sample points is very convenient. Although not necessary for an understanding of the basic concepts of probability, counting rules are included in Section 3.9 (optional). If you wish to develop an ability to solve complex probability problems by using the sample point approach, you should move directly to Section 3.9.

Tips on Problem Solving

To calculate the probability of an event by using the sample point approach, follow these guidelines.

1. Use the following steps for calculating the probability of an event by summing the probabilities of the sample points.
    a. Define the experiment.
    b. Draw a tree diagram, or identify a typical simple event. List the simple events associated with the experiment and test each to make certain that it cannot be decomposed. This list defines the sample space S.
    c. Assign reasonable probabilities to the sample points in S, making certain that $0 \le P(E_i) \le 1$ and $\sum_S P(E_i) = 1$.
    d. Define the event of interest, A, as a specific collection of sample points. (A sample point is in A if A occurs when the sample point occurs. Test *all* sample points in S to locate those in A.)
    e. Find $P(A)$ by summing the probabilities of sample points in A.
2. When the sample points are equiprobable, the sum of the probabilities of the sample points in A, step e, can be acquired by counting the points in A and multiplying by the probability associated with the individual sample points.
3. Calculating the probability of an event by using the five-step procedure described in part 1 is systematic and will lead to the correct solution if all the steps are correctly followed. Major sources of error are the following:
    a. Failing to define the experiment clearly (step a)
    b. Failing to specify simple events (step b)
    c. Failing to list all the simple events
    d. Failing to assign valid probabilities to the sample points

---

EXERCISES   Conceptual Level

3.1   If $E_1, E_2, \ldots, E_N$ are the simple events corresponding to an experiment with N outcomes, the assignment of probabilities, $P(E_i)$, should be done according to the following conditions:

a. _____ $\le P(E_i) \le$ _____   **b.** $\sum_S P(E_i) =$ _____

3.2   If an event A contains the three sample points corresponding to simple events E_1, E_2, and E_3, how is $P(A)$ found?

3.3   An experiment consists of randomly tossing three coins and observing the upper face on each coin. Identify the following as either simple or compound events.

   a.   Observe two heads.

   b.   Observe a head on the first two coins only.

   c.   Observe three heads.

   d.   Observe at least two heads.

EXERCISES   Skill Level

3.4   An experiment involves tossing a single six-sided die. Specify the sample points in the following events:

     S: the sample space.

     A: observe a 4.

     B: observe an even number.

     C: observe a number less than 3.

Assuming the die is balanced, calculate the probabilities of events A, B, and C by summing the probabilities of the appropriate sample points.

3.5   Three fair coins are tossed and the upper face of each coin recorded.

   a.   Construct a tree diagram for this experiment. How many sample points are in the sample space? Use the tree diagram to list the sample points.

   b.   If the coins are fair, what is an appropriate assignment of $P(E_i)$ for the simple events in this experiment?

Use the assignment of probabilities in part b to answer the following questions.

   c.   What is the probability of getting exactly three heads? No heads?

   d.   What is the probability of getting exactly two heads?

   e.   What is the probability of getting at least one head?

 3.6   An investor has decided to choose two of a group of three utility stocks.

   a.   Define the experiment.

   b.   Identify the stocks as 1, 2, and 3, and list the sample points in S.

   c.   Suppose that the stocks differ in growth potential, with stock 1 possessing the least growth potential and stock 3 the greatest growth potential. Identify the sample points in the following events:

     A: the investor chooses the two stocks with the greatest growth potential.

     B: the investor chooses stock 3.

     C: the investor chooses stock 1.

     D: the investor chooses at least one of the two stocks 2 or 3.

   d.   Suppose that the investor has no prior knowledge of the growth potentials of the stocks and, consequently, chooses the two stocks in a random manner (any pair of two is chosen with equal probability). State the probability that will be assigned to each sample point in the sample space.

e. Find the probability of observing events A, B, C, and D.

 3.7 A wine taster is required to taste and rank three varieties of wine, A, B, and C, according to the taster's preference.

a. Define the experiment.

b. Use a tree diagram to identify the sample points in S.

c. If the taster had no ability to distinguish a difference in taste among the three wines, what is the probability that the taster will rank wine variety A as best? As the least desirable?

EXERCISES   Applications

 3.8 Many companies in the rapidly growing high-technology field are seeking to diversify their manufacturing base by locating branch facilities in small but progressive communities. In this way they can develop new sources of labor and improve their rate of retention of key management personnel. Suppose a preliminary analysis by an electronics firm has identified five communities that meet organizational objectives for site location. Of these five communities, two are located in the state of Colorado and three are not. The president of the company plans to select two communities at random for in-depth discussions concerning the location of a manufacturing facility. We are interested in the geographic location of each chosen community.

a. Define the experiment.

b. List the sample points in S.

c. If all pairs of communities have an equal chance of selection, what is the probability that the two Colorado communities will be chosen by the president?

d. What is the probability that the communities chosen by the president will consist of at least one from Colorado? None from Colorado?

 3.9 Product innovation is the lifeblood of modern industry and provides the base from which a firm may redefine its market share relative to competition. Unfortunately, product innovation is very risky, with only a small portion of new products introduced into the market realizing success. A snack food manufacturer will choose for market introduction three from among seven new snack food items that have been created by the firm's new-product research staff. Unknown to the manufacturer, only two of the seven new snack food items will ever realize success in the marketplace. Suppose the manufacturer randomly chooses three of the new snack food items for market introduction without prior pilot market testing.

a. Define the experiment.

b. List the sample points in S.

c. If all possible groups of three snack food items have an equal chance to be chosen for market introduction, what is the probability that the two potentially successful products are selected?

d. What does this experiment suggest regarding the advisability of pilot market testing as a formal part of new-product research?

 3.10 In 1958 the U.S. Congress provided for the establishment of the Small Business Investment Companies (SBIC) to induce private investors to lend money to small business firms. Two small businesses in a certain metropolitan area containing four SBIC offices each seek to borrow money to supplement their working capital. Each small business randomly selects one of the four SBIC offices at which to apply for a loan, and the result of their choices is observed.

a. Define the experiment.

b. Use a tree diagram to identify the sample points in S.

c. Find the probability that both small businesses apply for a loan at the same SBIC office.

d. Find the probability that the two small businesses each choose a different SBIC office.

3.5

Compound Events

Many events of interest in practical situations are compound events that require enumeration of a large number of sample points. However, there is an alternative approach to calculating the probability of events. This method eliminates the necessity of listing the sample points and is therefore simpler and less time-consuming. It is based on the classification of events, event relations, and two probability laws that will be discussed in Sections 3.6 and 3.7.

Compound events, as the name suggests, are formed by a composition of two or more events. Composition takes place in one of two ways, as a *union* or an *intersection* of events, or as a combination of the two.

Definition

union of A and B

> Let A and B be two events in a sample space S. The **union of A and B** is the event containing all sample points in A or B or both. We denote the union of A and B by the symbol $A \cup B$.

A union is the event that **either** event A or event B occurs **or** both A and B occur. For instance, in Example 3.1 we had

$$\text{Event } A: \quad E_1, E_3, E_5$$
$$\text{Event } B: \quad E_1, E_2, E_3$$

The union $A \cup B$ is the collection of points E_1, E_2, E_3, and E_5. The union is shown diagrammatically as the shaded area in Figure 3.4.

Definition

intersection of
A and B

> Let A and B be two events in a sample space S. The **intersection of A and B** is the event composed of all sample points that are in both A and B. An intersection of events A and B is represented by the symbol AB. (Many authors use $A \cap B$.)

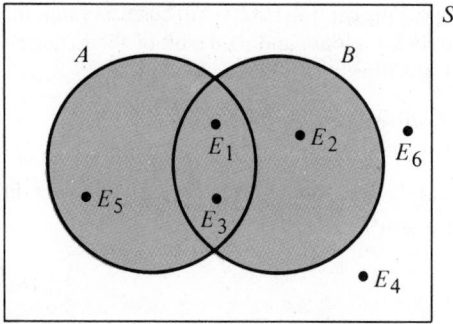

Figure 3.4   Event $A \cup B$ in Example 3.1; $A \cup B$ Is
Represented by the Shaded Area

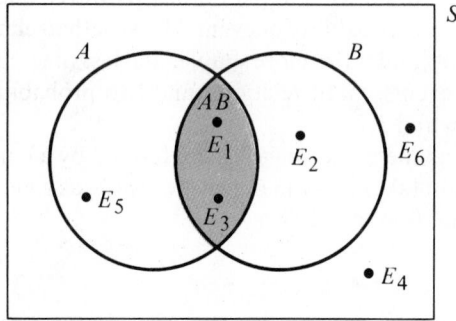

Figure 3.5   Event AB in Example 3.1; AB Is
Represented by the Shaded Area

The intersection AB is the event that **both** A and B occur and would appear in a Venn diagram as the overlapping area between A and B. The intersection AB for Example 3.1 is the event consisting of points E_1 and E_3. If either E_1 or E_3 occurs, both A and B occur. The intersection is shown diagrammatically as the shaded area in Figure 3.5.

EXAMPLE 3.7   Refer to the experiment of Example 3.3, where two coins are tossed, and define events A and B as follows:

Event A: at least one head
Event B: at least one tail

Define events A, B, AB, and $A \cup B$ as collections of sample points.

SOLUTION From Table 3.2, the sample points for this experiment are

E_1: HH (head on first coin, head on second)
E_2: HT
E_3: TH
E_4: TT

Event A occurs if simple event E_1, E_2, or E_3 occurs. Therefore, we can define event A as follows:

Event A: E_1, E_2, E_3

The other events can be defined similarly:

Event B: E_2, E_3, E_4
Event AB: E_2, E_3
Event $(A \cup B)$: E_1, E_2, E_3, E_4

Note that $A \cup B = S$, which is the sample space and thus is certain to occur.

EXERCISES Conceptual Level

3.11 Using a Venn diagram, shade the area corresponding to the following compound events.
  a. $A \cup B$ **b.** AB **c.** $A \cup B \cup C$ **d.** $AB \cup AC$

3.12 Use the union or intersection notation to describe the following compound events.
  a. Events A and B both occur.
  b. Either events A or B or both A and B occur.
  c. Either events A and C both occur, events B and C both occur, or A, B and C all occur.

EXERCISES Skill Level

3.13 Consider the following experiment: Roll two fair six-sided dice and observe the number of dots on the upper faces of each of the two dice.
  a. List the sample points in S.
  b. List the sample points for the following events:

  A: observe a sum of 2.
  B: observe a sum of 7.
  C: observe a total sum less than 7.
  D: observe both A and C.
  E: observe both B and C.
  F: observe either A or C or both.

c. Calculate the probabilities of events $A, B, C, D, E,$ and F by summing the probabilities of the appropriate sample points.

3.14 An experiment can result in 1 of 10 simple events $E_1, E_2, \ldots, E_{10}$, which are equally likely; the events $A, B,$ and C are defined as follows:

Event	Simple Events
A	E_1, E_2, E_3, E_4
B	E_3, E_4, E_5, E_6, E_7
C	E_6, E_7, E_8

a. List the sample points in the following compound events:

$A \cup B$ $B \cup C$
AB $A \cup B \cup C$
AC

b. Calculate the probabilities associated with each of the events in part a by summing the probabilities of the appropriate sample points.

EXERCISES Applications

3.15 A manufacturer of computing machinery has indicated that the monthly demand for the firm's small minicomputer ranges from one through seven. List the sample points in S. Then list the sample points in the following events:

A: two minicomputers are sold during a given month.
B: less than four are sold.
C: no more than five are sold.
D: at least three are sold.
AB
$A \cup B$

If we can assume that one demand level is as likely to occur as any other demand level, calculate the probabilities of events $A, B, C, D, AB,$ and $A \cup B$ by summing the probabilities of the appropriate sample points.

3.16 The Energy Siting Council of a western state consists of four people, two of whom are in favor of nuclear power generation, two opposed. The governor must select two members from the council to represent their state at a regional energy-planning conference. The two representatives are chosen at random by the governor; list the sample points contained in the following events:

S: the sample space.
A: exactly one pro–nuclear power council member is chosen as a representative.
B: exactly two pro–nuclear power council members are chosen.
C: both anti–nuclear power council members are chosen.

Find the probabilities of events $A, B, C, AB, AC,$ and $A \cup B$.

3.17 An investor has the option of investing in two of four recommended securities. Unknown to the investor, only two of the securities will show a substantial profit within the next five years. Suppose the investor selects the two securities at random from among the four that have been recommended. List the sample points in S. Then list the sample points in the following events:

A: at least one of the profitable securities is selected.

B: at least one of the unprofitable securities is selected.

Construct a diagram like the one in Figure 3.4 to depict the event $A \cup B$. Then construct a diagram like the one in Figure 3.5 to depict AB. Calculate the probabilities of events A, B, AB, and $A \cup B$.

3.18 An article in the *New York Times* (February 19, 1988) reported that approximately 70% of the 32,500 workers at a Ford Motor Company plant in Britain had voted to end an 11-day strike and to accept a pay increase of at least 14% over two years. Five workers are chosen at random from those who work at the plant and are asked whether or not they had voted to end the strike.

a. List the sample points in the sample space S.

b. List the sample points in the following events:

A: none of the 5 had voted to end the strike.

B: 2 of the 5 had voted to end the strike.

C: less than 3 had voted to end the strike.

D: $A \cup B$.

E: AB.

F: BC.

G: $A \cup C$.

3.6

Event Relations

In this section we will define three relationships among events that play an important role in finding the probability of an event. The first of these relationships concerns the concept of *complementary events*.

Definition

complement of an
event A

> The **complement of an event** A is the collection of all sample points that are in the sample space S but not in A. The complement of A is denoted by the symbol $\bar{A}$.

In a Venn diagram complementary events appear as shown in Figure 3.6. Because we know that

$$\sum_S P(E_i) = 1 \qquad \text{and} \qquad \sum_{E_i \text{ in } A} P(E_i) + \sum_{E_i \text{ in } \bar{A}} P(E_i) = 1$$

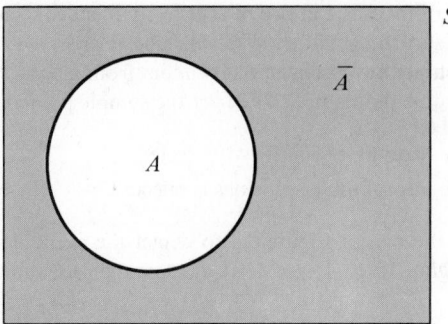

Figure 3.6   Complementary Events

then

$$P(A) + P(\bar{A}) = 1$$

When $P(\bar{A})$ is known or easily calculated, the probability of event A is found to be

$$P(A) = 1 - P(\bar{A})$$

If the probability of occurrence of one event depends on whether a second event has or has not occurred, the two events are said to be *dependent*. For example, suppose that one experiment consists of observing the weather on a specific day. Let A be the event "observe rain" and B be the event "observe an overcast sky." Events A and B are related, because the probability $P(A)$ of rain is not the same as the probability of rain given prior information that the day is cloudy. The probability of A, $P(A)$, is the proportion of the entire population of observations that result in rain. Now let us look only at the subpopulation of observations resulting in B, a cloudy day, and the fraction of them that result in A. This fraction, called the *conditional probability* of A given B, may equal $P(A)$, but we would expect the chance of rain, given that the day is cloudy, to be larger. The conditional probability of A, given that B has occurred, is denoted as

$$P(A \mid B)$$

where the vertical bar in the parentheses is read "given" and the events indicated to the right of the bar are the events that have occurred.

For example, consider a toss of a fair die. Let B be the outcome for which the number is less than 4 (that is, a 1, 2, or 3), and let A be the outcome for which the number is odd. If the event B has occurred, and the outcome is a 1, 2, or 3, then the chance that the number is odd is 2/3. We have just described the conditional probability of A given B.

We define the conditional probabilities of B given A and A given B as shown in the display.

Definition

conditional
probabilities

The **conditional probability of B**, given that A has occurred, is

$$P(B \mid A) = \frac{P(AB)}{P(A)} \qquad \text{if } P(A) \neq 0$$

The **conditional probability of A**, given that B has occurred, is

$$P(A \mid B) = \frac{P(AB)}{P(B)} \qquad \text{if } P(B) \neq 0$$

We can show that this definition of conditional probability is consistent with the relative frequency concept of probability by assigning probabilities to the events in our weather example. The event A denotes rain on a given day; B denotes a cloudy day. Now suppose that 10% of all days are rainy and cloudy [that is, $P(AB) = .10$] and 30% of all days are cloudy [$P(B) = .30$].

The situation we have just described is graphically portrayed in Figure 3.7. Each sample point in event B, denoted by the large circular area, is associated with a single cloudy day. Since 30% of all days will be cloudy, we can regard this area as .3. One-third (10%/30%) of these cloudy days will be rainy. These days are included in the shaded area, event AB. Notice that in this situation the events A and AB are identical. Hence, if a single day is selected from the set of all days representing the population, what is the probability that we will select a rainy day, given that we know the day is cloudy? That is, what is $P(A \mid B)$?

Because we already know that the day is cloudy, we know that the sample point to be selected must fall in event B (Figure 3.7). One-third of these days will result in rain. Hence, the probability that we will select a rainy day is

$$P(A \mid B) = 1/3$$

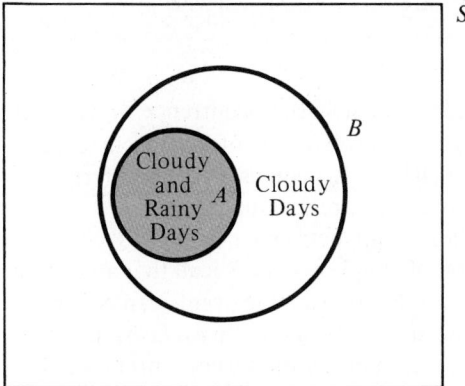

Figure 3.7   Events A and B

You can see that this result agrees with our definition for $P(A\,|\,B)$. That is,

$$P(A\,|\,B) = \frac{P(AB)}{P(B)} = \frac{.10}{.30} = \frac{1}{3}$$

EXAMPLE 3.8   Calculate $P(A\,|\,B)$ for the die-tossing experiment described in Example 3.1.

SOLUTION   Events A and B are defined as follows:

Event A: observe an odd number (E_1, E_3, E_5).
Event B: observe a number less than $4(E_1, E_2, E_3)$.

Given that B has occurred, we are concerned only with sample points E_1, E_2, and E_3, which occur with equal frequency. Of these, E_1 and E_3 imply event A. Hence, as we have shown earlier,

$$P(A\,|\,B) = 2/3$$

We could also obtain $P(A\,|\,B)$ by using the relationship

$$P(A\,|\,B) = \frac{P(AB)}{P(B)} = \frac{1/3}{1/2} = \frac{2}{3}$$

Note that $P(A\,|\,B) = 2/3$ while $P(A) = 1/2$. Because the probability of event A changes if we know that event B has occurred, events A and B are dependent.

Definition

independent events

> Two events A and B are said to be **independent** if either
>
> $$P(A\,|\,B) = P(A) \qquad \text{or} \qquad P(B\,|\,A) = P(B)$$
>
> Otherwise, the events are said to be **dependent**.

dependent events

Two events are independent if the occurrence or nonoccurrence of one of the events does not change the probability of the occurrence of the other event. If $P(A\,|\,B) = P(A)$, then $P(B\,|\,A)$ also equals $P(B)$. Similarly, if $P(A\,|\,B)$ and $P(A)$ are unequal, then $P(B\,|\,A)$ and $P(B)$ are unequal.

A third useful relationship between events was observed but not specifically defined in our discussion of simple events. Recall that an experiment can result in one and only one simple event. No two simple events can occur at exactly the same time. Two events A and B are said to be *mutually exclusive* if when one occurs, the other cannot occur, and hence, the intersection AB contains no sample points. It then follows that $P(AB) = 0$.

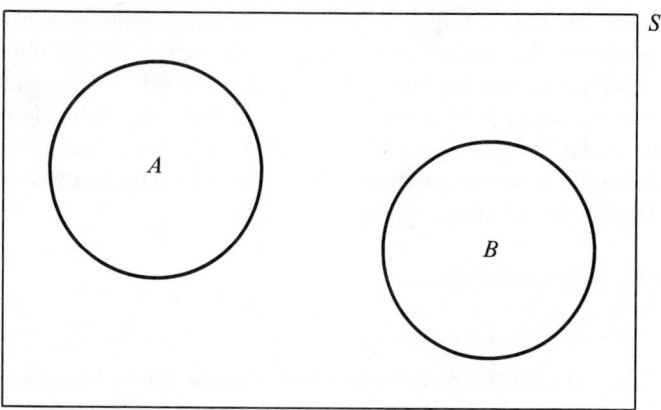

Figure 3.8   Mutually Exclusive Events

Definition

mutually exclusive
events

> Two events A and B are **mutually exclusive** if the event AB contains no sample points.

Mutually exclusive events have no overlapping area in a Venn diagram (see Figure 3.8).

EXAMPLE 3.9   Are events A and B given in Example 3.1 mutually exclusive? Are they complementary? Independent?

SOLUTION   The events are

$$\text{Event } A: \quad E_1, E_3, E_5$$
$$\text{Event } B: \quad E_1, E_2, E_3$$

The event AB is the set of sample points in both A and B. Because AB includes the points E_1 and E_3, A and B are not mutually exclusive. They are not complementary, because B does not contain all points in S that are not in A.

To determine whether A and B are independent, we must check to see whether $P(A \mid B) = P(A)$. From Example 3.8, $P(A \mid B) = 2/3$ and $P(A) = 1/2$. Therefore,

$$P(A \mid B) \neq P(A)$$

and, by definition, events A and B are dependent.

EXAMPLE 3.10   Experience has shown that a particular union-management contract negotiation has led to a contract settlement within a two-week period 50% of the time, that the union strike fund has been adequate to support a strike 60% of the time, and that both of these conditions have been satisfied 30% of the time. What is the probability of a contract settlement given that the union strike fund is not adequate to support a strike? Is settlement of a contract within a two-week period dependent on whether the union strike fund is not adequate to support a strike?

SOLUTION   Define the following events:

A: a contract settlement is reached within a two-week period.
$\bar{A}$:   a contract settlement is not reached within two weeks.
B: the union strike fund is adequate to support a strike.
$\bar{B}$:   the union fund is not adequate to support a strike.

probability table   The probability of interest, $P(A\,|\,\bar{B})$, can be found by using a two-way probability table. The entries in the body of a **probability table** are the four intersection probabilities corresponding to the events AB, $A\bar{B}$, $\bar{A}B$, and $\bar{A}\bar{B}$. Furthermore, the marginal row or column sums are the unconditional probabilities $P(A)$, $P(\bar{A})$, $P(B)$, and $P(\bar{B})$.

We are given that $P(A) = .50$, $P(B) = .60$, and $P(AB) = .30$. These values comprise three of the eight entries, shown in color in the probability table that follows.

	B	$\bar{B}$	
A	.30	.20	.50
$\bar{A}$	.30	.20	.50
	.60	.40	

Since $P(A) + P(\bar{A}) = 1$, the marginal probability $P(\bar{A}) = .50$. Similarly, since $P(B) + P(\bar{B}) = 1$, $P(\bar{B}) = .40$. The remaining entries are found in a similar manner. The required probability, $P(A\,|\,\bar{B})$, is now found as

$$P(A\,|\,\bar{B}) = \frac{P(A\bar{B})}{P(\bar{B})} = \frac{.20}{.40} = .5$$

Since $P(A\,|\,\bar{B}) = P(A) = .5$, by definition, the events A and $\bar{B}$ are independent.

EXERCISES   Conceptual Level

3.19   What does it mean to say that two events are mutually exclusive?

3.20   Explain in your own words what is meant by the statement "Events A and B are independent."

EXERCISES Skill Level

3.21 An experiment can result in one of eight equally likely simple events $E_1, \ldots, E_8$. The events A, B, and C are defined as follows:

A: $\{E_1, E_3, E_4, E_5\}$
B: $\{E_1, E_2, E_4, E_6\}$
C: $\{E_6, E_7, E_8\}$

a. List the sample points in $\bar{A}$, AB, AC, and BC.

b. Find the probabilities associated with the events A, B, C, $\bar{A}$, AB, AC, and BC.

c. Are A and B mutually exclusive? Why or why not?

d. Are A and C mutually exclusive? Why or why not?

e. Find $P(A \mid B)$ and $P(A \mid C)$. Are the events A and B independent? Are A and C independent? Why or why not?

3.22 In Exercise 3.6 an investor randomly selected two utility stocks from a group of three. We defined the following events:

A: the investor chooses the two stocks with the greatest growth potential.
B: the investor chooses stock 3.
C: the investor chooses stock 1.
D: the investor chooses at least one of the two stocks 2 or 3.

a. Find the complement of event A.

b. Find the complement of $B \cup C$.

c. Find $P(A \mid B)$.

d. Find $P(B \mid A)$.

e. Are events A and B independent events? Justify your answer.

f. Are events A and B mutually exclusive events? Justify your answer.

g. Are events C and D mutually exclusive events? Justify your answer.

h. Are events B and D independent events? Justify your answer.

3.23 The accompanying probability table gives the intersection probabilities for four events A, B, C, and D. Using the definitions in Section 3.6, find $P(A)$, $P(B)$, $P(C)$, $P(D)$, $P(C \mid A)$, $P(D \mid A)$, and $P(C \mid B)$.

	A	B
C	.06	.31
D	.55	.08

3.24 Refer to Exercise 3.23.

a. Find $P(\bar{B})$.

b. Find $P(AB)$.

c. Find $P(A \cup B)$.

d.   Are events *B* and *C* independent events? Justify your answer.

e.   Are events *B* and *C* mutually exclusive events? Justify your answer.

f.   Are events *C* and *D* independent events? Justify your answer.

g.   Are events *C* and *D* mutually exclusive events? Justify your answer.

EXERCISES   Applications

3.25   Test marketing involves the duplication of a planned national marketing campaign in a limited geographic area [D. Tull and D. Hawkins, *Marketing Research: Measurement and Method* (New York: Macmillan, 1980)]. The test market of a new home sewing machine was conducted among 75 retail outlets, 50 of which were department stores affiliated with a major department store chain. In 40 of the test market locations the retailer offered customers a free one-year service contract with the purchase of a new sewing machine. Twenty-five of the 50 department stores offered buyers a free one-year service contract. This information is summarized in the table that follows.

Contract	Type of Store		
	D	*D̄*	
F	25		40
F̄			
	50	25	75

A customer buys one of the new sewing machines from a participating test market retailer.

a.   What is the probability that she did not receive a free one-year service contract with the purchase of her sewing machine?

b.   If you know that the customer received a free one-year service contract with the purchase of her sewing machine, what is the probability that she purchased the machine from a department store?

3.26   Oil-drilling contractors typically conduct a seismic sounding of a parcel before deciding whether or not to drill. The detection of a closed structure in the terrain below the site is a hopeful sign, but the detection of no structure usually implies a lower probability of a successful drilling. The accompanying table gives the results of a very large number of seismic soundings on sites where drilling was actually conducted. Table entries are the approximate probabilities of occurrence of the four combinations of "structure type" and "drilling outcome." Calculate the following probabilities and then describe each probability in terms of the foregoing problem.

$P(A_1)$ $P(\bar{A}_1)$ $P(A_1 | B_1)$ $P(A_2 | B_1)$ $P(A_1 | B_2)$ $P(B_2 | A_2)$ $P(A_2 | A_1)$

	B_1, No Structure	B_2, Closed Structure
A_1, unproductive well	.40	.10
A_2, productive well	.15	.35

3.27 A survey of consumers in a particular community showed that 10% were dissatisfied with appliance repair jobs done in their homes. Fifty percent of the complaints dealt with company A, and company A does 40% of the appliance repair jobs in the town.

  a. What is the probability that you will obtain an unsatisfactory appliance repair job, given that company A does the job?

  b. What is the probability that you will obtain a satisfactory repair job, given that company A does the job?

3.7

Two Probability Laws and Their Use

A second approach that can be used to find event probabilities is based on the composition of events, event relations, and two probability laws. These laws can be stated without proof, because they are consistent with our model and with reality. The first law of probability is called the *Multiplicative Law of Probability* and follows directly from the definition of conditional probability. It provides a formula for calculating the probability of an intersection of two events.

Multiplicative Law of Probability

> Given two events A and B, the probability of the intersection AB is
>
> $$P(AB) = P(A)P(B \mid A)$$
> $$ = P(B)P(A \mid B)$$
>
> If A and B are independent, $P(AB) = P(A)P(B)$.

EXAMPLE 3.11 When receiving a shipment of goods from a supplier, the buyer typically conducts an inspection of the quality of the goods received. A discount store has received a lot of 100 portable television sets from a manufacturer. Unknown to the management of the discount store, 10 of the 100 television sets are defective. If 2 sets are randomly selected from the lot of 100 and then subjected to an exhaustive quality inspection, what is the probability that both will be defective?

SOLUTION Define the following events:

  Event A: the first set is defective.
  Event B: the second set is defective.

Then AB is the event that both are defective, and

$$P(AB) = P(A)P(B \mid A)$$

$P(A) = .10$ since there are 10 defectives in the lot of 100. However, $P(B|A) = 9/99$, because after the first is selected and found to be defective, there are 99 sets remaining, of which 9 are defective. Thus,

$$P(AB) = P(A)P(B|A) = (10/100)(9/99) = 1/110$$

The second law of probability, called the *Additive Law*, applies to unions.

Additive Law of Probability

> The probability of the union $A \cup B$ is
>
> $$P(A \cup B) = P(A) + P(B) - P(AB)$$
>
> If A and B are mutually exclusive, then
>
> $$P(AB) = 0 \quad \text{and} \quad P(A \cup B) = P(A) + P(B)$$

The Additive Law conforms to reality and our model. The sum $P(A) + P(B)$ contains the sum of the probabilities of all sample points in $A \cup B$ but includes a double counting of the probabilities of all points in the intersection AB. Subtracting $P(AB)$ gives the correct result.

EXAMPLE 3.12   A recent issue of the *Wall Street Journal* reported that 40% of its subscribers regularly read *Time*, 32% read *U.S. News and World Report*, and 11% read both weekly news-magazines. Define events A and B as follows:

Event A: a *WSJ* subscriber reads *Time*.
Event B: a *WSJ* subscriber reads *U.S. News*.

Find the probability of the events A, B, AB, and $A \cup B$.

SOLUTION   The experiment consists of selecting a single *WSJ* subscriber and recording the magazines he or she reads regularly. The magazine preferences of each subscriber represent a single sample point in the experiment, and the sample points are equiprobable (equally likely to occur). Then, because 40% of the *WSJ* subscribers read *Time*, 40% of the sample points in S are in the event A and

$$P(A) = .4$$

Similarly,

$$P(B) = .32$$

Since 11% of the *WSJ* subscribers read both magazines, we have

$$P(AB) = .11$$

Then

$$P(A \cup B) = P(A) + P(B) - P(AB) = .40 + .32 - .11 = .61$$

◆ ◆

The use of the probability laws for calculating the probability of a compound event is a less direct approach than the listing of sample points, and it requires a bit of experience and ingenuity. The approach involves expressing the event of interest as a union or intersection (or combination of both) of two or more events whose probabilities are known or easily calculated. This step can sometimes be done in more than one way. The trick is to find the right combination, a task that requires considerable creativity in some cases. The usefulness of event relations is now apparent. If the event of interest is expressed as a union of mutually exclusive events, the probabilities of the intersection need not be known. If they are independent, we need not know the conditional probabilities to calculate the probability of an intersection.

EXAMPLE 3.13   In spite of the demand for their services, competition among oil-drilling contractors has increased. As a result, drilling contractors are more cost-conscious and have tended to limit the number of holes they drill in any parcel of land. Suppose a drilling contractor has sufficient resources to drill up to two holes on a given parcel but will stop drilling with his first success. Given that the probability that he is successful on any given drill hole is .2, find the probability that the drilling contractor locates a productive well. Assume that the drilling outcomes are independent from one drill hole to another.

SOLUTION   Let us list the events that, when they occur, satisfy the objective of the problem.

A: a productive well is found within two trials.
S_1:   a productive well is found on the first trial.
S_2:   a productive well is found on the second trial.
F_1:   a dry well is found on the first trial.
F_2:   a dry well is found on the second trial.

A tree diagram depicting how the outcomes of the experiment might occur is given in Figure 3.9. The probabilities listed along the branches are the conditional probabilities associated with this step on the tree, given that the previous steps have occurred. Therefore, probabilities associated with the experimental outcomes are found as the product of the probabilities along the path leading to that outcome.

We see that event A occurs if

1.   he strikes oil on the first trial, S_1, or
2.   he hits a dry well on the first trial and strikes oil on the second, $F_1 S_2$.

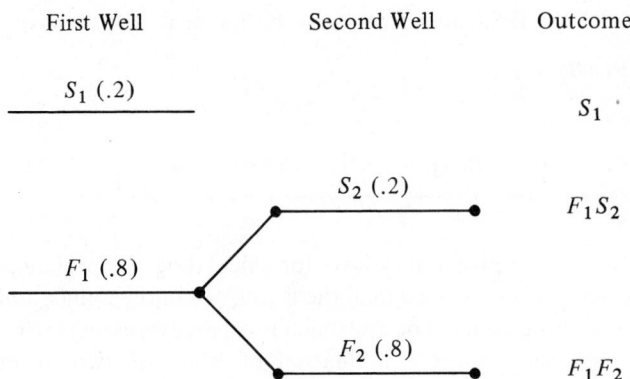

Figure 3.9   Tree Diagram for Example 3.13

Thus,

$$A = S_1 \cup F_1 S_2$$

Since the events S_1 and $F_1 S_2$ are mutually exclusive, we have

$$P(A) = P(S_1 \cup F_1 S_2) = P(S_1) + P(F_1 S_2)$$

We know that

$$P(S_1) = P(S_2) = .2 \quad \text{and} \quad P(F_1) = P(F_2) = .8$$

and the outcomes "success" or "dry" at any trial are independent of the outcome at any other trial. Applying the Multiplicative Law, we have

$$P(F_1 S_2) = P(F_1)P(S_2) = (.8)(.2) = .16$$

Substituting, we find

$$P(A) = P(S_1) + P(F_1 S_2) = .2 + .16 = .36$$

Alternatively, since $\bar{A} = F_1 F_2$, and $P(A) = 1 - P(\bar{A})$, then

$$P(A) = 1 - P(F_1 F_2) = 1 - .64 = .36.$$

◆ ◆

Probability problems can be solved by using either the sample point approach or the event composition approach. However, for any given problem one approach is usually easier than the other. Practice in problem solving, using tree diagrams or probability tables when appropriate, will help you to select the correct method for a given situation.

Tips on Problem Solving

To calculate the probability of an event by using the event composition approach, follow these guidelines.

1. Use the following steps for calculating the probability of an event:
   a.   Define the experiment.
   b.   Clearly visualize the nature of the sample points. Identify a few to clarify your thinking. Use a tree diagram when appropriate.
   c.   Write an equation expressing the event of interest, say A, as a composition of two or more events, using either or both of the two forms of composition (unions and intersections). Make certain that the event implied by the composition and event A represent the same set of sample points. A probability table may be helpful here.
   d.   Apply the Additive and Multiplicative Laws of Probability to step c and find $P(A)$.

2. Be careful with step c. You often can form many compositions that will be equivalent to event A. The trick is to form a composition in which all the probabilities appearing in step d will be known. Thus, you must visualize the results of step d for any composition and select the one for which the component probabilities are known.

3. Always write letters to represent events described in an exercise. Then write the probabilities that are given and assign them to events. Identify the probability that is requested in the exercise. This step may help you to arrive at the appropriate event composition.

EXERCISES   Conceptual Level

3.28   How are the probability laws for $P(A \cup B)$ and $P(AB)$ modified under the following conditions?

   a. A and B are mutually exclusive.

   b. A and B are independent.

3.29   Use the laws of probability to justify your answers to the following questions.

   a.   If $P(A \cup B) = .6$, $P(A) = .2$, and $P(B) = .4$, are A and B mutually exclusive? Independent?

   b.   If $P(A \cup B) = .65$, $P(A) = .3$, and $P(B) = .5$, are A and B mutually exclusive? Independent?

   c.   If $P(A \cup B) = .70$, $P(A) = .4$, and $P(B) = .5$, are A and B mutually exclusive? Independent?

EXERCISES   Skill Level

3.30   An experiment consists of tossing two fair coins. Define the following events:

   A: observe two heads.
   B: observe two tails.
   C: observe one head and one tail.

   a. Find $P(AB)$.   **b.** Find $P(C \mid A)$.   **c.** Find $P(AC)$.
   d. Find $P(A \cup B)$.   **e.** Find $P(A \cup C)$.

 3.31   If an investor randomly selected two utility stocks from a group of three, the probability of selecting the two with the greatest growth potentials is 1/3. Suppose that we view the selection as a two-step procedure.

   a. What is the probability that the first stock chosen will be one of the two stocks that have the greatest growth potentials? Label as event A the event associated with this probability.

   b. Given that event A has occurred, what is the probability that the second stock chosen will be the one with the greatest growth potential of the two remaining stocks?

   c. What is the probability of selecting the two stocks with the greatest growth potentials?

 3.32   To examine lending practices, a federal bank examiner will audit the financial records of two branch banks chosen at random from among the five branches of a branch banking system. Unknown to the examiner, two of the five branches are currently not in compliance with federal standards.

   a. What is the probability that the federal examiner chooses neither noncompliant branch for examination?

   b. What is the probability that the examiner chooses both noncompliant branches for examination?

   c. What is the probability that the examiner chooses either of the noncompliant branches (but not both) for examination?

   d. What is the probability that the examiner chooses at least one of the noncompliant branches for examination?

 3.33   In recent years affirmative action commitments by industrial organizations have led to an increase in the number of women in executive positions. Suppose a company has vacancies that are to be filled by randomly selecting two people from a list of five candidates. On the list are two women and three men, all of whom have worked for the company for a long period of time.

   a. What is the probability that at least one of the women will be selected?

   b. What is the probability that neither woman will be selected?

EXERCISES   Applications

 3.34   A quality control engineer has been asked to examine a complex electronic system that is currently operating out of control. His examination will involve testing six switching mechanisms within the system and identifying the one that is faulty. Assume that one and only one switching mechanism is faulty. If the switching mechanisms are randomly chosen for examination by the engineer but excluded from further consideration if found to be satisfactory, what is the probability that the faulty switching mechanism will be discovered during the second examination? During the third examination? Before the fourth examination?

 3.35 Refer to Exercise 3.34. Construct a tree diagram to depict the possible outcomes of the experiment. Use the diagram to verify the calculation of the probabilities in Exercise 3.34.

 3.36 A manufacturer of a sound movie camera reports that customer complaints are one of two types: those they regard as "defective products" and those they classify as "unsatisfactory products." Unsatisfactory products are those that are not defective but do not meet consumer expectations. Currently, evidence suggests that 2% of the cameras are found to be defective by consumers; 5% are judged to be unsatisfactory. If a camera is found to be defective, the probability of a complaint by a consumer is .9, and the probability of a complaint due to an unsatisfactory camera is .5. The probability of a complaint, given that a camera is good, is zero. Given that a customer issues a complaint about a camera, what is the probability that the complaint is due to the product being judged unsatisfactory by the consumer? Would you then suggest that unsatisfactory products be covered under the product warranty?

 3.37 As long as the future cannot be known with complete certainty, there can be no such thing as risk-free trading. Usually, commodity traders seek to pyramid their winnings, pooling winnings after each successful trade for a predetermined number of trades. But if one trade is lost, the investor loses all his money. Assume that the probability of winning any trade is independent of the outcome of any other commodity trade. If an investor is engaged in a five-trade commodity run, find the probability that she avoids bankruptcy if the probability of winning any one trade is .9. Find the probability that she avoids bankruptcy if the probability of winning an individual trade is .6. How large would the probability of winning a single trade have to be for the investor to have more than a 50–50 chance of winning a five-trade commodity pyramid?

 3.38 Buyers of large quantities of goods from a supplier often use sampling inspection schemes to judge the incoming quality of a supply. The supply of goods is accepted or rejected on the basis of results observed by administering tests to a few sample items selected from the supply. Suppose we know that an inspector for a food-processing firm has accepted 98% of all good shipments and has incorrectly rejected 2% of good shipments. In addition, the inspector accepts 94% of all shipments, and it is known that 5% of all shipments are of inferior quality.

     a. Find the probability that a shipment is rejected.

     b. Find the probability that a shipment is good.

     c. Find the probability that a shipment is good and that it is accepted.

     d. Find the probability that a shipment is of inferior quality and that it is accepted. [*Hint:* Use a probability table.]

     e. Find the probability that a shipment is accepted, given that it is of inferior quality.

     f. Find the probability that a shipment is rejected, given that it is good.

 3.39 If a buyer rejects a supplier's shipment of goods on the basis of evidence gained from sampling inspection, the supplier, if he truly believes the shipment to be of acceptable quality, may insist that the buyer perform a second sampling inspection survey. Refer to Exercise 3.38. Assume that the probability that a shipment is accepted is the same for an initial and a second inspection. Find the probability that a shipment of goods is accepted on the basis of a second sample inspection, given that it was rejected on an initial survey.

 3.40 Quality-of-work-life programs are becoming more frequently instituted by management in order to increase productivity and worker satisfaction in the workplace. In a survey of 74 firms manufacturing consumer goods, 23 firms offer a formal mechanism for worker input to management decision processes (codetermination), 10 have developed flexible work schedules for their employees (flextime), while 6 engage in both codetermination and flextime. One firm from among those surveyed is chosen for further analysis of its employee relations programs. (*Hint:* Use a probability table to answer the following questions.)

a. What is the probability the chosen firm has *both* a codetermination and a flextime program?

b. What is the probability the chosen firm has *neither* a codetermination nor a flextime program?

c. If you know that the firm has a flextime program for its employees, what is the probability it also engages in codetermination?

 3.41 Experience and knowledge of the field are the most important factors considered by personnel directors in a search for managers. A survey of personnel directors from *Fortune* 500 companies gave the percentage of personnel directors listing each of five important factors in the search for a manager (directors were allowed to list more than one factor as important). Use the percentages in the accompanying table as approximate probabilities.

Factor	Percentage
Experience/knowledge of field	60
Technical skills	22
Education	22
Ability to manage	21
Communication skills	12

Source: Data from *USA Today*, February 19, 1988.

a. If a personnel director is chosen at random, what is the probability that he will cite technical skills as an important factor in the search for a manager?

b. If a personnel director is chosen at random, what is the probability that he will cite experience and knowledge of the field as an important factor in the search for a manager?

c. What is the probability that the personnel director does not consider the ability to manage to be an important factor?

d. If two personnel directors are chosen at random, what is the probability that neither will mention ability to manage as an important factor?

e. Are the events listed in the table mutually exclusive? Why or why not?

 3.42 A company that operates as an open shop (not all employees are union members) has initiated a procedure to allow for worker input to company decisions. To allow for such input, the company president will select 4 workers, from a list of 10 suggested by the plant foreman, to join management on a discussion panel. Suppose that the 10 names provided by the foreman consist of 5 union members and 5 workers who are not union members and that the president chooses 4 from among them randomly and individually.

a. What is the probability that the first selection by the president is a union member?

b. What is the probability that the first union member is chosen on the president's third selection?

c. What is the probability that at least 2 of those chosen for the panel are union members?

d. What is the probability that the panel will contain 2 union members and 2 nonunion members?

e. What is the probability that the panel will contain no union members? If it does not, do you have reason to suspect that the president has not chosen membership for the panel in a random manner?

3.43   At the beginning of each month a company decides whether to spend $1000 or $2000 on advertising for that month. Monthly decisions are independent of one another, and the two spending options are equiprobable.

    a.   What is the probability that in three consecutive months a total of more than $4000 is spent on advertising?

    b.   What is the probability that $1000 is spent in each of the three months? [Note: The probability of the intersection of three events A, B, and C is given by the formula

$$P(ABC) = P(A)P(B \mid A)P(C \mid AB)$$

    If A, B, and C are independent events, then we have $P(ABC) = P(A)P(B)P(C)$.]

3.44   Refer to Exercise 3.18 in which the results of a strike vote at a Ford Motor Company plant in Britain were reported. Suppose that five workers are chosen at random from those who work at the plant and that the probability that a single worker has voted to end the strike is .7. Find the probabilities associated with the following events.

    A: none of the five have voted to end the strike.

    B: two of the five have voted to end the strike.

    C: less than three have voted to end the strike.

    D: $A \cup B$.

    E: AB.

    F: BC.

    G: $A \cup C$.

3.8

Bayes's Law

Conditional probabilities allow us to update probabilities by using information as it becomes available. For example, we may know a priori, or before the fact, that all of the items used in a production process are provided by three suppliers, and that supplier S_1 provides 20% of these items, supplier S_2 provides 30%, and supplier S_3 provides the remaining 50%. Therefore, if no other information is available and an item is randomly selected from the production process, the probability that the item came from supplier S_1 is .20, from supplier S_2 is .30, and from supplier S_3 is .50.

From past performance, the percentages of defective items supplied by these three suppliers are .05, .02, and .01, respectively. If an item randomly selected from the production process is found to be defective, what is the probability that the item was supplied by S_1? By S_2? By S_3? That is, how does this information change the *prior* probabilities that the item came from one of the three suppliers?

If D is the event that a defective item was observed, then any one of the three suppliers may have provided the item. We wish to update the *prior* probabilities, $P(S_i)$, using conditional probabilities in the form $P(S_i \mid D)$. The relevant information is given in the tree diagram in Figure 3.10. The second-step probabilities are conditional, given the first step. For example, the probability of a defective item, given supplier S_1, is $P(D \mid S_1) = .05$.

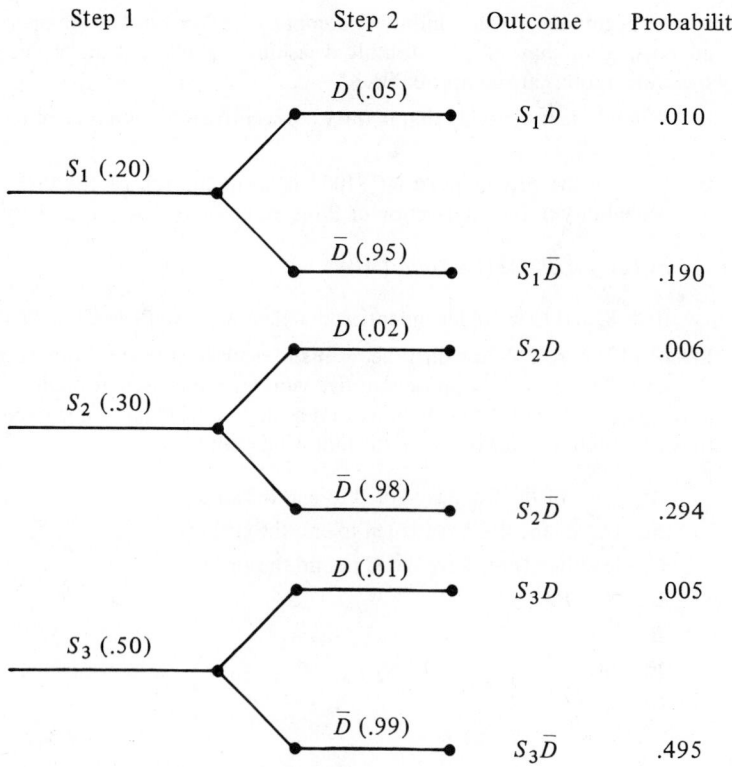

Figure 3.10   Tree Diagram Illustrating Bayes's Law

Suppose we wish to find the *posterior* probabilities that have been updated with the sample information. For example, to find $P(S_1 | D)$, we have

$$P(S_1 | D) = \frac{P(S_1 D)}{P(D)}$$

From the tree diagram in Figure 3.10, $P(D)$ is the sum of the three outcomes resulting in D. Hence,

$$P(D) = P(S_1 D) + P(S_2 D) + P(S_3 D) = P(S_1)P(D|S_1) + P(S_2)P(D|S_2)$$
$$+ P(S_3)P(D|S_3) = .010 + .006 + .005 = .021$$

Since $P(S_1 D) = P(S_1)P(D|S_1) = .20(.05) = .010$, then

$$P(S_1 | D) = \frac{.010}{.021} = .476$$

Furthermore, the two remaining posterior probabilities are

$$P(S_2 | D) = \frac{.006}{.021} = .286 \quad \text{and} \quad P(S_3 | D) = \frac{.005}{.021} = .238$$

By use of the *prior* probabilities, an item was most likely to have been supplied by supplier S_3. By use of the *posterior* probabilities, the sampled item was most likely to have come from supplier S_1.

The method for calculating the posterior probabilities is generally known as *Bayes's Law*, named for the English mathematician Thomas Bayes.

Bayes's Law

Let $S_1, S_2, \ldots, S_k$ represent the k mutually exclusive, only possible states of nature with prior probabilities $P(S_1), P(S_2), \ldots, P(S_k)$. If an event A occurs, the posterior probability of S_i given A is the conditional probability

$$P(S_i \mid A) = \frac{P(S_i)P(A \mid S_i)}{\sum_{j=1}^{k} P(S_j)P(A \mid S_j)}$$

for $i = 1, 2, \ldots, k$.

EXAMPLE 3.14   A department store is considering adopting a new credit management policy in an attempt to reduce the number of credit customers defaulting on their payments. The credit manager has suggested that in the future credit should be discontinued to any customer who has twice been a week or more late with his monthly installment payment. She supports her claim by noting that past credit records show that 90% of all those defaulting on their payments were late with at least two monthly payments.

Suppose from our own investigation we have found that 2% of all credit customers actually default on their payments and that 45% of those who have not defaulted have had at least two late monthly payments. Find the probability that a customer with two or more late payments will actually default on his payments, and, in light of this probability, criticize the credit manager's credit plan.

SOLUTION   Let the events L and D be defined as follows:

Event L: a credit customer is two or more weeks late with at least two monthly payments.

Event D: a credit customer defaults on his payments.

and let $\bar{D}$ denote the complement of event D. We seek the conditional probability

$$P(D \mid L) = \frac{P(DL)}{P(L)} = \frac{P(L \mid D)P(D)}{P(L \mid D)P(D) + P(L \mid \bar{D})P(\bar{D})}$$

From the information given in the problem description, we find that

$$P(D \mid L) = \frac{(.90)(.02)}{(.90)(.02) + (.45)(.98)} = \frac{.0180}{.0180 + .4410} = .0392$$

Therefore, if the credit manager's plan is adopted, the probability is only about .04, or the chances are only about 1 in 25, that a customer who loses his credit privileges would actually have defaulted on his payments. Unless management would consider it worthwhile to detect one prospective defaulter at the expense of losing 24 good credit customers, the credit manager's plan would be a poor business policy.

Further applications of Bayes's Law will be explored in detail in Chapter 19.

EXERCISES   Conceptual Level

3.45 **a.** Use a Venn diagram to verify that for two events A and B, $A = AB \cup A\bar{B}$.

b. Use the Additive Law of Probability to simplify $P(A) = P(AB \cup A\bar{B})$.

c. Use parts a and b to show how to find the Bayes's Law formulation for $P(B \mid A)$.

EXERCISES   Skill Level

3.46 Given that $P(B) = .9$, $P(A \mid \bar{B}) = .7$, and $P(A \mid B) = .2$, find $P(A)$ and $P(B \mid A)$.

3.47 One jar contains two white and two red balls, and a second jar contains one white and four red balls. The first jar can be chosen with probability .6 and the second with probability .4.

a. If a jar is chosen and one ball randomly selected from the jar, what is the probability that the ball chosen is white?

b. If the ball selected is white, what is the probability that it was selected from the first jar?

EXERCISES   Applications

3.48 An assembler of electric fans uses motors from two sources. Company A supplies 75% of the motors and company B supplies the other 25% of the motors. Suppose it is known that 5% of the motors supplied by company A are defective, and 3% of the motors supplied by company B are defective. An assembled fan is found to have a defective motor. What is the probability that this motor was supplied by company B?

3.49 Refer to Exercise 3.38. Find the probability that a shipment is of acceptable quality, given that it has been accepted by the inspector.

3.50 Sampling inspection plans, similar to the application discussed in Exercise 3.38, are also used by a manufacturer of goods to monitor the ongoing quality of the manufactured items. In this way she hopes to detect items of inferior quality and remove them from the lot before a shipment is sent to a buyer. Suppose that in a certain manufacturing plant, as items come to the end of a production line, an inspector chooses those items that are to go through a complete inspection. Ten percent of all items produced are defective, 60% of all defective items go through a complete inspection, and 20% of all good items go through a complete inspection. Given that an item is completely inspected, what is the probability that it is defective?

3.51 Some gasoline distributors mix 9 parts of gasoline with 1 part of ethyl alcohol (distilled from grain) and market the product as gasohol. Suppose that in a certain midwestern city there are two automobile service stations (stations A and B), both of which sell gasohol in addition to gasoline. However, because of the promotion of gasohol's benefits by station A, 40% of all its fuel sales are gasohol, whereas only 10% of the fuel sales for station B are gasohol sales. Assume that the two stations conduct an equal volume of business and are, in general, equally attractive to customers. If you know that a customer purchased a tank of gasohol from a service station within the city, what is the probability the purchase was made from station A? Does this exercise suggest a possible means of examining promotional effectiveness? Explain.

3.52 One of the purposes of an audit is to detect the presence of material errors, procedural errors, or errors in judgment in the recording of accounting information. Suppose a CPA firm has been retained to conduct an audit of the accounting practices of a firm in which accounts are processed by two divisions, a wholesale division and a retail division. The auditor knows that 70% of all accounts are retail accounts. Furthermore, the auditor knows that 10% of all retail accounts and 20% of all wholesale accounts contain some type of accounting error. If the auditor observes an accounting error in an account, find the probability that the account was processed by the firm's retail division.

MKTG **3.53** Refer to Exercise 3.27. Suppose that there are two other appliance repair companies in town. Company A does 40% of the repair business, company B does 45%, and company C does 15%. The dissatisfaction rates, based on prior performance, for the three companies are .5, .4, and .6, respectively. If an appliance repair job is completed and the customer is not satisfied with the job, what is the probability that company B was responsible for the repair?

3.9

Counting Sample Points (Optional)

The preceding sections cover the basic concepts of probability and provide a background to help you understand the role probability plays in making inferences. To illustrate the concepts, we used only examples and exercises for which the total number of sample points in the sample space was small. This approach allowed you to list the sample points in S, to identify the sample points in the event of interest, and then to calculate its probability. Although these examples and exercises were adequate for our learning objective, most real-life problems involve many more sample points. Consequently, we include this optional section for the student who wishes to improve his or her problem-solving ability.

Suppose you are interested in the probability of an event A, and you know that the sample points in S are equiprobable. Then

$$P(A) = \frac{n_A}{N}$$

where

n_A = the number of sample points in A
N = the number of sample points in S

Often, we can use counting rules to find the values of n_A and N and thereby eliminate the necessity of listing the sample points in S.

mn Rule   The first counting rule is known as the ***mn* Rule** and gives the number of pairs of objects that can be formed when selecting one object from each of two different groups.

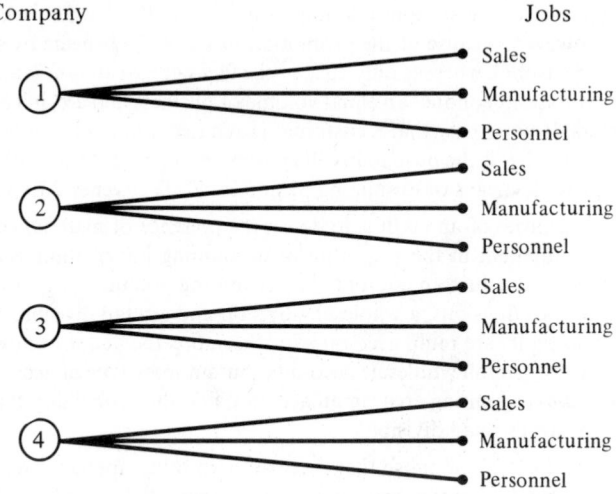

Figure 3.11   Company-Job Combinations

For example, suppose that four companies have job openings in each of three areas: sales, manufacturing, and personnel. How many job opportunities are available to you? You can see that you have two sets of objects: companies (four) and types of jobs (three). Therefore, as shown in Figure 3.11, there are three jobs for each of the four companies, or $(4)(3) = 12$ possible company-job combinations.

mn Rule

> With m elements $a_1, a_2, \ldots, a_m$ and n elements $b_1, b_2, \ldots, b_n$, it is possible to form mn pairs that contain one element from each group.

EXAMPLE 3.15   Two dice are tossed. How many sample points are associated with the experiment?

SOLUTION   The first die can fall in one of six ways, that is, $m = 6$. Likewise, the second die can fall in $n = 6$ ways. The total number N of sample points is

$$N = mn = 6(6) = 36$$

Remember that the *mn* Rule gives the number of pairs you can form in selecting one object from each of two groups. The rule can be extended to apply to triplets formed by selecting one object from each of three groups, quadruplets formed by selecting one object from each of four groups, and so on. The application to triplets is shown in Figure 3.12. If you have m elements in the first group, n in the second, and t in the third, the total number of triplets that you can form, taking one object from each group, is equal to mnt, the number of branchings shown in Figure 3.12.

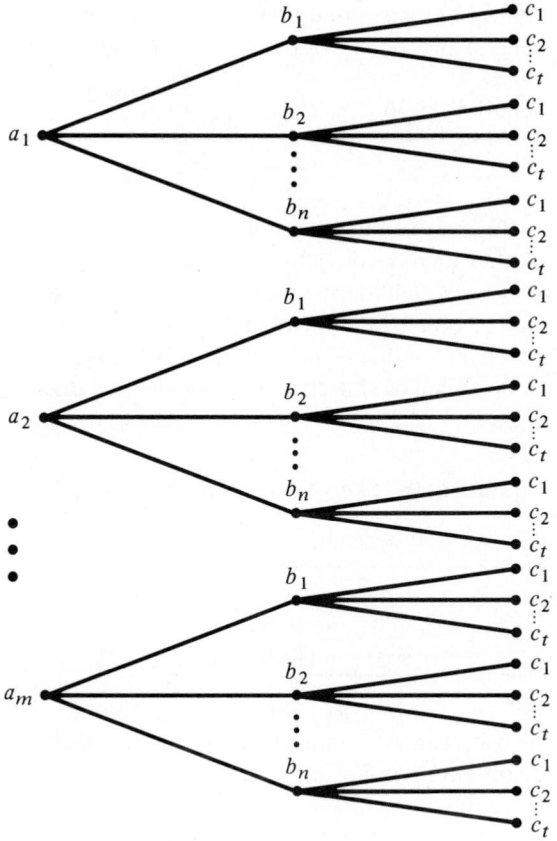

Figure 3.12   Forming *mnt* Triplets

EXAMPLE 3.16   How many sample points are in the sample space when three coins are tossed?

SOLUTION   Each coin can land in one of two ways. Hence,

$$N = 2(2)(2) = 8$$

EXAMPLE 3.17   A truck driver can take three routes in going from city A to city B, four from city B to city C, and three from city C to city D. If, in going from A to D, he must proceed A to B to C to D, how many possible A-to-D routes are available to him?

SOLUTION   Let

$$m = \text{the number of routes from } A \text{ to } B = 3$$
$$n = \text{the number of routes from } B \text{ to } C = 4$$
$$t = \text{the number of routes from } C \text{ to } D = 3$$

Then the total number of ways that you can construct a complete route, taking one subroute from each of the three groups $(A$ to $B)$, $(B$ to $C)$, and $(C$ to $D)$, is

$$mnt = (3)(4)(3) = 36$$

A second useful mathematical result involves orderings, or *permutations*. For instance, suppose that we have three books, b_1, b_2, and b_3. In how many ways can the books be arranged on a shelf, taking them two at a time? We enumerate the ways in Table 3.4, listing all combinations of two in the first column and a reordering of each in the second column. The number of permutations is 6, a result easily obtained from the mn Rule. The first book can be chosen in $m = 3$ ways; and once it is selected, the second book can be chosen in $n = 2$ ways. The result is $mn = 6$.

Table 3.4   Permutations of Three Books Taken Two at a Time

Combinations of Two	Reordering of Combinations
$b_1 b_2$	$b_2 b_1$
$b_1 b_3$	$b_3 b_1$
$b_2 b_3$	$b_3 b_2$

In how many ways can we arrange three books on a shelf using all three at a time? Enumerating, we obtain a total of 6:

$$b_1 b_2 b_3 \qquad b_2 b_1 b_3 \qquad b_3 b_1 b_2$$
$$b_1 b_3 b_2 \qquad b_2 b_3 b_1 \qquad b_3 b_2 b_1$$

This result again could be obtained easily by the extension of the mn Rule. The first book can be chosen and placed in $m = 3$ ways. After choosing the first, we can choose the second in $n = 2$ ways and, finally, the third in $t = 1$ way. Hence, the total number of ways is

$$N = mnt = 3 \cdot 2 \cdot 1 = 6$$

Definition

permutation

> An ordered arrangement of r distinct objects is called a **permutation**. The number of ways of ordering n distinct (different) objects taken r at a time is designated by the symbol P_r^n.

The number of permutations of n objects taken r at a time is equivalent to finding the number of ways of filling r positions with n distinct objects. The counting rule for finding P_r^n follows.

Counting Rule for Permutations

> The number of ways that you can arrange n distinct objects taking them r at a time is
>
> $$P_r^n = n(n-1)(n-2)\cdots(n-r+1) = \frac{n!}{(n-r)!}$$
>
> where
> $$n! = n(n-1)(n-2)\cdots(3)(2)(1) \text{ and } 0! = 1$$

EXAMPLE 3.18   Three lottery tickets are drawn from a total of 50. Assume that order is of importance. How many sample points are associated with the experiment?

SOLUTION   Since $n = 50$ and $r = 3$, the total number of sample points is

$$P_3^{50} = \frac{50!}{47!} = 50(49)(48) = 117,600$$

◆ ◆

EXAMPLE 3.19   A piece of equipment is composed of five parts, which may be assembled in any order. A test is conducted to determine the length of time necessary for each order of assembly. If each order is tested once, how many tests must be conducted?

SOLUTION   Since $n = 5$ and $r = 5$, the total number of tests is

$$P_5^5 = \frac{5!}{0!} = 5(4)(3)(2)(1) = 120$$

◆ ◆

The enumeration of the permutations of books in the previous discussion was performed in a systematic manner, first writing the combinations of n books taken r at a time and then writing the rearrangements of each combination. In many situations ordering is unimportant because we are interested solely in the number of possible combinations. For instance, suppose that an experiment involves the selection of a committee of 5 people from a total of 20 candidates. The simple events associated with this experiment correspond to the different combinations of people selected from the group of 20. How many simple events (different combinations) are associated with this experiment? Since order in a single selection is unimportant, permutations are not necessary. We are interested only in the number of *combinations* of $n = 20$ things taken $r = 5$ at a time.

Definition

number of
combinations

> The **number of combinations** of n objects taken r at a time is denoted by the symbol C_r^n. [*Note:* Some authors prefer the symbol $\binom{n}{r}$.]

The number of combinations of n objects taken r at a time can be found by using the following counting rule.

Counting Rule for Combinations

The number of different combinations that can be formed from n distinct objects taken r at a time is

$$C_r^n = \frac{n!}{r!(n-r)!}$$

EXAMPLE 3.20 A radio tube may be purchased from five suppliers. In how many ways can three suppliers be chosen from the five?

SOLUTION
$$C_3^5 = \frac{5!}{3!2!} = \frac{(5)(4)}{2} = 10$$

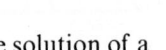

The following example illustrates the use of the counting rules in the solution of a probability problem.

EXAMPLE 3.21 Five manufacturers, of varying but unknown quality, produce a certain type of electronic tube. If we were to select three manufacturers at random, what is the chance that the selection would contain exactly two of the best three?

SOLUTION Without enumerating the sample points, we can see that each point, representing a selection of three manufacturers, would be assigned equal probability. If there are N points in S, then each point receives probability

$$P(E_i) = \frac{1}{N}$$

Let n_A be the number of points in which two of the best three manufacturers are selected. Then the probability of including two of the best three manufacturers in a selection of three is

$$P = \frac{n_A}{N}$$

We will use the counting rules to find n_A and N.

Since order within a selection is unimportant and is unrecorded, each selection is a combination, and hence,

$$N = C_3^5 = \frac{5!}{3!2!} = 10$$

Determining n_A is more difficult, but it can be obtained by using the *mn* Rule. Let *a* be the number of ways of selecting exactly two from the best three manufacturers. Then

$$a = C_2^3 = \frac{3!}{2!1!} = 3$$

Let *b* be the number of ways of choosing the remaining manufacturer from the two poorest, or

$$b = C_1^2 = \frac{2!}{1!1!} = 2$$

Then the total number of ways of choosing two of the best three in a selection of three manufacturers is $n_A = ab = 6$.

Hence, the probability *P* is

$$P = \frac{n_A}{N} = \frac{6}{10}$$

◆◆

Tips on Problem Solving

Many students have difficulty deciding which (if any) of the three counting rules to apply in a given problem. The following tips may help.

1. Look at the problem and note whether a simple event is formed in one of the following ways:
    a. Selecting elements from each of *two (or more)* sets (a situation that suggests the use of the *mn* Rule)
    b. Selecting *r* elements from a *single* set of *n* elements (a situation that suggests the use of either combinations or permutations)

2. If the situation is 1b, you must decide whether you should use combinations or permutations. If every different ordering of the elements in the group of *r* leads to a different simple event, then use permutations. If ordering does not produce a new simple event, then use combinations. Your diagnostic thought process should perform the checks indicated in the decision tree shown in Figure 3.13.

3. If you have difficulty visualizing the appropriate counting rule (or rules) to use for a problem involving a large number of sample points, construct a miniature version of the problem so that you can manually count them. This step may help you to see how the more complex version can be solved.

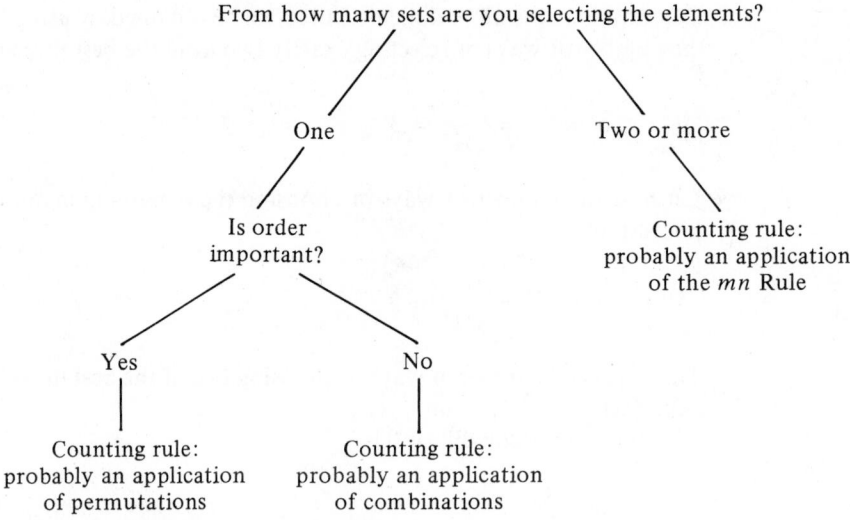

Figure 3.13   *Deciding Which Counting Rule to Use*

EXERCISES   Conceptual Level

3.54   What is the basic difference between a permutation and a combination of *r* items selected from *n*?

3.55   Can a permutation count be used to find the number of possible pairs, with one element from set *A* and the other from set *B*? If not, what is the proper counting rule to use?

EXERCISES   Skill Level

3.56   Evaluate:

    a. $5!$ **b.** P_3^5 **c.** C_3^5 **d.** P_4^4 **e.** C_6^6

3.57   Set *A* contains 10 elements, set *B* contains 5 elements, and set *C* contains 8 elements.

    a.   How many pairs can be formed by using 1 element from set *A* and 1 element from set *B*?

    b.   How many pairs can be formed by using 1 element from set *A* and 1 element from set *C*?

    c.   How many triplets can be formed by using 1 element from each of the sets *A*, *B*, and *C*?

3.58   A set contains the five letters *A, B, C, D* and *E*.

    a.   How many distinct pairs of letters can be chosen if the order of the letters is unimportant?

    b.   How many distinct pairs of letters can be chosen if the order of the letters is important?

EXERCISES   Applications

Check each of the following exercises against the diagnostic clues given in the Tips on Problem Solving.

3.59   In most states automobile license numbers involve a specific arrangement of three letters and three numbers, for instance, HLY – 282. How many different license plates are possible if the same letters and numbers can be repeated on an individual plate? Use the *mn* Rule.

3.60   The marketing director for a manufacturer of consumer products must choose an appropriate test market to examine the market acceptability of a new product. Four variety store chains are chosen, each of which has a retail outlet in each of seven separate cities. Suppose that the test market will involve the selection of a single retail outlet. Use the *mn* Rule to give the number of possible city-store combinations available for consideration as test markets for the company's new product.

3.61   A combination lock will open when the right combination of three digits is selected. Each digit can take a value from 0 to 9. If a particular combination of digits represents a sample point, how many characteristics are used to identify a sample point? How many sets are they selected from? Use the *mn* Rule to give the number of possible combinations of digits.

3.62   The personnel director for a business organization has identified 10 individuals as qualified candidates for 3 managerial training positions her firm seeks to fill. Use the Combinations Rule to give the number of different combinations of the 10 individuals who could be chosen for the 3 positions.

3.63   The president, vice president, secretary, and treasurer are to be selected from a group of 10 candidates. Use the Permutations Rule to give the number of ways the positions may be filled.

3.64   A tennis team has 10 members, who may be assigned to play one of the six singles matches (the singles matches are ordered as No. 1 singles, No. 2 singles, etc.). Use the Permutations Rule to find how many different player-position assignments are available to the coach.

3.65   Five secretaries are selected from a group of 25 to form a secretarial pool. Use the Combinations Rule to give the total number of different pools that could be formed.

3.66   An experiment consists in assigning eight workmen to eight different jobs. In how many different ways can the eight men be assigned to the eight jobs?

3.67   Refer to Exercise 3.66. Suppose there are only four different jobs available for the eight men. In how many different ways can four men be selected from the eight and assigned to the four jobs?

3.68   A study is conducted to determine the attitudes of nurses in a hospital toward various administrative procedures that are currently employed. If a sample of 10 nurses is selected from a total of 75, how many different samples can be selected? (Note that order within a sample is unimportant.)

3.69   A traveler can take one of three airlines to San Francisco, and each airline operates four daily nonstop flights. If the selection of a particular airline-flight combination represents a sample point, how many characteristics (elements) are used to identify a sample point (see Tips on Problem Solving, item 1)? How many sets are they selected from? Use the *mn* Rule to give the number of airline-flight combinations available to the traveler. Construct a diagram similar to Figure 3.11 to identify the airline-flight combinations.

3.70   If 5 cards are to be selected, one after the other, in sequence from a 52-card deck, each card being replaced in the deck before the next draw, how many different selections are possible?

3.71   Refer to Exercise 3.70. Suppose that the 5 cards are drawn from the 52-card deck simultaneously and without replacement. How many different hands could be selected?

3.72   Four men and three women have applied for the two vacancies on a city council finance committee. How many different combinations could be selected from the applicants for the committee?

3.10

Random Variables

In Section 3.3 we defined an experiment as the process by which an observation (or measurement) is obtained. Although the observations resulting from an experiment are not always numerically valued, most experiments produce data that are either quantitative or can be quantified by assigning numbers to represent categories. Thus, we are particularly interested in experiments that result in numerically valued outcomes.

Suppose that the variable measured in an experiment is denoted by the symbol x. The variable x is called a *random variable* if the value that x assumes is a chance or random event.

Definition

random variable

> A variable x is a **random variable** if the value that x assumes, corresponding to the outcome of an experiment, is a chance or random event.

For example, consider a sample of 20 consumers who are asked whether they prefer package design A or B. The number of consumers who indicate that they prefer design A may be regarded as a variable x that assumes any of the values $0, 1, 2, \ldots, 20$. The variable x is a random variable because the value that x assumes is an outcome that cannot be predicted with absolute certainty in advance of the experiment; that is, the particular value that x takes is a chance or random event.

You may recall from our earlier discussion of this sampling problem (Section 3.1) that the probabilities assumed by the random variable x are needed to make an inference about the preferences of all possible purchasers of the manufacturer's product. These probabilities, one associated with each value of x, comprise the *probability distribution* for x. Probability distributions are the subject of Chapter 4.

3.11

Summary

The theories of probability and statistics are concerned with samples drawn from a population. A probabilist assumes that the population is known and calculates the probability of observing a particular sample. A statistician assumes that the sample is known and, with the aid of probability, attempts to describe the frequency distribution of the population, which is unknown. Chapter 3 is concerned with the construction of a model for the repetition of an experiment. Using this model, we can find the probabilities of events associated with the experiment by one of two methods:

1. The summation of the probabilities of the sample points in the event of interest
2. The joint use of event composition (compound events) and the laws of probability

What use will be made of our ability to assess the probability of an event? Keep in mind the objective of statistics. We wish to make inferences about a population based on information contained in a sample. To this end, the experiment is the selection of a sample from a population and the event of interest is the particular sample outcome that we observe. From these sample results we wish to infer the nature of the population. The methods we will employ to accomplish this inference require that we know the probability of obtaining the observed sample.

Although we have not really explained how probability is used to make inferences, we gave an example in Section 3.1 of how it was used to decide which package design was preferred by consumers. Rereading Section 3.1 may help you to understand the role that probability plays in making inferences.

Supplementary Exercises

Applications

3.73   In a study of consumer behavior a customer is given the option of choosing one of five brands of a common household product, two of which are generically labeled.

   a.   If 2 customers are chosen for the experiment, how many different simple events are associated with the experiment?

   b.   If the five brands are actually equally attractive to the 2 customers, what is the probability that both customers choose one of the two generically labeled products?

   c.   Suppose 10 customers are involved in the study and each chooses one of the generic products. Do you think this result would be evidence of distinct consumer preference for the generically labeled products or simply a chance occurrence in the selection from among equally attractive items?

3.74   Corporations must periodically decide whether or not to call in outstanding bonds and replace them with a less costly issue. At a time when prevailing interest rates are low, two corporations decide to exercise their call provision on outstanding bonds. Three investment banking firms have each submitted bids to underwrite each corporation's bond issue. One of the three firms will be selected to underwrite each of the two bond issues, and because the underwriting firm is selected according to the lowest bid, one firm can possibly "win" both bond issues. Assume that none of the bids are identical.

   a.   List the sample points for the experiment.

b. Let A be the event that investment banking firm 1 wins at least one of the underwriting offers. List the sample points in A.

c. If the chance of any of the three investment banking firms submitting the lowest bid on an underwriting offer is the same for all three firms, find $P(A)$.

 3.75   A retailer sells two styles of high-priced high-fidelity consoles that experience indicates are equal in demand. (Fifty percent of all potential customers prefer style 1, 50% prefer style 2.) If she stocks four of each, what is the probability that the first four customers seeking a console all purchase the same style?

a. Define the experiment.

b. List the sample points.

c. Define the event A of interest as a specific collection of sample points.

d. Assign probabilities to the sample points and find $P(A)$.

 3.76   The owner of a small business seeks a business development loan to provide capital for feasibility studies on a new solar-heating device. Seven possible funding sources are available locally, five commercial banks and two private sources. Suppose the business owner randomly selects four possible funding sources to request a loan.

a. What is the probability that all the chosen sources are commercial banks?

b. What is the probability that no more than one private source is involved among the sources contacted?

c. What is the probability that at least one private source is chosen?

 3.77   Lack of product distinctiveness often contributes to the high failure rate of new products. A major publisher of popular books estimated from experience that the probability that a new publication becomes a commercial success is 10% and that the success of one publication is independent of the success of another. Suppose that during a given year the publisher introduces five new books to the bookstores of the country.

a. What is the probability that none are a commercial success?

b. What is the probability that at least one of the new books is a commercial success?

c. What is the probability that no more than two of the five are commercial successes?

 3.78   If the probability that daily demand for a product exceeds two units is 9/10, what is the probability that the daily demand is less than three units? (Assume that the demand will be only for whole number units of the product.)

 3.79   Through a certain stockbroker, 100 stocks are available. A classification of these stocks follows.

	Type		
Exchange	Preferred	Not Preferred	
NYSE	20	40	60
Other	10	30	40
	30	70	100

A man goes to the broker and buys a stock from the New York Stock Exchange (NYSE), unaware that some stocks are preferred. Given that he bought a stock from the NYSE, what is the probability that it is preferred?

3.80   A large municipal bank chain has ordered six microcomputer systems to distribute among its six branches in a certain city. Unknown to the purchaser, three of the six systems are defective. Before installing the systems, an agent of the bank selects two of the six systems from the shipment, thoroughly tests them, and then classifies each as either defective or nondefective.

   a.   List the sample points for this experiment.

   b.   Let A be the event that the selection includes no defectives and hence that the entire shipment is considered acceptable. List the sample points in A.

   c.   Assign probabilities to the sample points and find $P(A)$.

3.81   As a result of airline deregulation and increasing competition, airlines have been forced to cancel many low-capacity and duplicative flights. An airline company at present has four daily flights from New York to London. The first two flights are in the morning; the last two are in the afternoon. If the airline randomly selects two flights to be canceled, what is the probability that a morning and an afternoon flight will still be available?

3.82   In a recent survey of 1,700 companies, 49% of the firms reported that they perform extensive studies of advertising effectiveness, 61% conduct short-term sales forecasts, and 38% undertake both activities. Suppose a single firm is selected from the 1,700 reporting firms. Define the events A and B as follows:

   Event A: the firm studies advertising effectiveness.

   Event B: the firm conducts short-term sales forecasts.

   Construct a probability table for these events. Find $P(A)$, $P(B)$, $P(AB)$, $P(A \cup B)$, and $P(A\,|\,B)$.

3.83   The study referred to in Exercise 3.82 also found that 64% of the firms undertake competitive product studies. Suppose one firm is selected at random from among those reported in the survey. Define the event C as "the firm undertakes competitive product studies."

   a.   Find the probability that it does not undertake competitive product studies.

   b.   Find the probability that it studies advertising effectiveness and undertakes competitive product studies but does no short-term sales forecasting. Assume that event C is independent of events A and B.

   c.   Find the probability that it undertakes all three market research activities. Assume that event C is independent of events A and B.

3.84   A small advertising firm consists of three men and one woman. The firm has two clients who are particularly difficult to deal with. To decide who sees the first client, one person is randomly selected from the four. The same procedure is followed for the second client.

   a.   Find the probability that both clients are served by the same person from the advertising firm.

   b.   Find the probability that both clients are served by men.

   c.   Find the probability that both the events described in parts a and b occur simultaneously.

3.85   Higher oil prices and increased demand for fuel-efficient personal transportation have had a dramatic impact on many major industries in the United States. One industry that has been seriously affected is the recreational vehicle industry. A manufacturer of recreational vehicles must decide which two assembly plants from among five to close due to depressed demand for the firm's products. Unknown to the manufacturer, two of the five assembly plants are located in communities that have the capacity to absorb displaced employees resulting from a plant closure. Suppose the manufacturer randomly selects two assembly plants for closure.

   a.   Define the experiment.

b. List the sample points.

c. Find the probability that neither plant selected for closure is in a community capable of absorbing the displaced employees.

d. Find the probability that both plants chosen are in the communities where other jobs are available.

e. Find the probability that at least one plant chosen for closure is in a community where other jobs are available.

3.86 Corporate bonds are rated A^+, A, B^+, B, or C, depending on the stability of the issuing firm. A novice bond buyer is unaware of the difference in corporate ratings and thus selects two firms at random from five. If each of the five firms has a different rating, what is the probability that she does not buy any bonds rated C? What is the probability that she buys only bonds A^+ or A?

3.87 A state highway department has contracted for the delivery of sand, gravel, and cement at a construction site. Because of other work commitments and labor force problems, contracting firms cannot always deliver items on the agreed delivery date. From past evidence, the probabilities that sand, gravel, and cement will be delivered on the promised delivery dates by the contracting firms are .3, .6, and .8, respectively. Assume that the delivery or nondelivery of one material is independent of another.

a. Find the probability that all three materials will be delivered on time.

b. Find the probability that none of the three materials will be delivered on the promised delivery date.

c. Find the probability that at least one of the materials will be delivered on time.

3.88 According to the National Association of Accountants (NAA), a survey of 3,200 financial executives revealed that during the October 19, 1987, stock market "meltdown," 35% of the company comptrollers cut expenses, 14% cut investments, 13% reduced staff, 9% considered mergers and acquisitions, and 5% used stock buybacks. Only 19% of company comptrollers took no action at all [*The Press Enterprise* (Riverside, Calif.), February 20, 1988]. Suppose two of the surveyed comptrollers were selected at random.

a. What is the probability that both cut expenses?

b. What is the probability that one reduced staff and the other used stock buybacks?

c. What is the probability that both did nothing?

d. What is the probability that at least one comptroller did nothing?

3.89 In a study conducted by the Federal Aviation Administration (FAA) to evaluate the effectiveness of airport security, FAA agents tried to smuggle 2,419 phony guns and other weapons onto airplanes at 28 major U.S. airports in late 1986. Overall, 20% of these weapons went *undetected*; of the 28 airports, considered separately, the best detection rate was 99% and the worst was 34% (*Wall Street Journal*, June 19, 1987). In a major metropolitan area, airport A handles approximately 60% of the total number of air passengers, airport B handles approximately 25%, and airport C handles the remaining 15%. Suppose airports A, B, and C were rated as having 90%, 85%, and 80% security detection rates, respectively.

a. What is the overall airport security detection rate for the whole metropolitan area?

b. If a passenger is found to be carrying a weapon while at an airport security checkpoint, what is the probability that the incident occurred at airport A? At airport B?

3.90 The accompanying table shows the location of all 200 stores in a trade association by city type and geographic region. One store is selected at random from the trade association to be used in testing the sales appeal of a new product. Calculate the following probabilities and then describe each probability in terms of the problem: $P(A_1)$, $P(B_3)$, $P(A_1 B_4)$, $P(B_1 | A_3)$, $P(A_2 \cup B_3)$, $P(B_1 \cup B_4)$, and $P(B_2 B_4)$.

| City Type | Geographic Region | | | |
	East, B_1	South, B_2	Midwest, B_3	Far West, B_4
Large, A_1	35	10	25	25
Small, A_2	15	10	15	15
Suburb, A_3	25	5	10	10

3.91 During periods of peak unemployment those affected most are the young and members of minority groups. The May 26, 1980, issue of *Time* reported that during the peak of the 1980 recession, 16.2% of all Americans between the ages of 18 and 21 were unemployed; the figure was 30% for black Americans in this age range. Recent census data suggest that one out of five Americans is black. Suppose the *Time* study involved discussions of employment prospects with two individuals in the 18-to-21 age range.

a. What is the probability that both were employed?

b. What is the probability that the first one chosen was an unemployed black American?

c. What is the probability that the first one chosen was neither employed nor a black American?

d. What is the probability that the first was an unemployed black American and the second was an employed black American?

3.92 A company is going to temporarily hire two secretaries from a secretarial pool. One hundred secretaries are available. Twenty type under 40 words per minute, 50 type from 40 to 60 words per minute, and the rest type over 60 words per minute. Assume the two secretaries are randomly chosen from the secretarial pool.

a. Find the probability that both secretaries type over 60 words per minute.

b. Find the probability that the first types under 40 words per minute and the second types from 40 to 60 words per minute.

c. Find the probability that the first types from 40 to 60 words per minute and the second types under 40 words per minute.

d. Find the probability that one types under 40 words per minute and the other from 40 to 60 words per minute.

3.93 A housewife is asked to rank four brands (*A, B, C, D*) of common household cleaner according to her preference, No. 1 being the cleaner she prefers best and so on. Suppose the housewife really has no preference among the four brands, and hence, any ordering is equally likely to occur.

a. What is the probability that brand *A* is ranked first?

b. What is the probability that *C* is first and *D* is second in the rankings?

c. What is the probability that *A* is ranked either first or second?

3.94 In a recent annual report on environmental quality in New England, the regional office of the U.S. Environmental Protection Agency (EPA) noted that only 30% of the region's solid-waste disposal facilities meet EPA standards. If this percentage is still accurate and if three New England communities are selected at random, what is the probability that the solid-waste disposal facilities in two of the communities fail to meet EPA standards? At least two fail to meet EPA standards? (Assume that each New England community has a single solid-waste disposal facility and that the ability of a facility in one community to meet EPA standards is independent of the ability of a facility in any other community.)

3.95 A Justice Department study estimates "that employee theft accounts for almost 70% of retailers' losses to all kinds of crime, which amounts to $11.8 billion in 1980, the last year for which an estimate is available" (*Wall Street Journal*, September 17, 1987). Drug use and weakening morality are cited as major causes of employee theft, since Sunday hours and later store closings attract a younger work force. Retail stores are fighting back by offering rewards to employees who report thieving co-workers. Nonetheless, it is estimated that only one of every seven employees who knows of a theft by co-workers will report it.

a. If these figures are correct, what is the probability that two employees who independently are aware of an employee theft both report the theft?

b. What is the probability that at least one reports the theft?

 3.96 To gain funding approval for capital construction projects, most public agencies are required to obtain special bonding authority by a majority approval from the electorate. Typically, however, the same project (at a reduced scale) must be presented to the voters more than once before a final decision is received from the voters. Suppose a municipal government intends to request bonding authority for a new city hall and has decided to submit this matter to the voters a maximum of three times (each time reducing the size of the request). Past evidence suggests that chances are .6 that capital construction bonding authority will be granted in any single election. What is the probability that the city hall construction measure will be approved by the electorate? What is the probability it will not be approved?

 3.97 A shipping firm keeps two vehicles ready for local deliveries. Because of the demand on their time and the chance of mechanical failure, the probability that a specific vehicle will be available when needed is .8. The availability of one vehicle is independent of the other.

a. In the event of two large orders, what is the probability that both vehicles will be available?

b. What is the probability that neither will be available?

c. If one service call is placed, what is the probability that the delivery will be made (i.e., what is the probability that a vehicle will be available)?

 3.98 A certain article is visually inspected successively by two inspectors. When a defective article comes through, the probability that it gets by the first inspector is .1. Of those that do get past the first inspector, the second inspector will miss 4 out of 10. What fraction of the defectives will get by both inspectors?

 3.99 A company is going to buy two new desks for its office. The purchaser of the desks randomly chooses two from a total of five available. Although the purchaser does not know it, one desk has a major defect, two have minor defects, and two are in perfect condition.

a. What is the probability that both desks that are purchased are in perfect condition?

b. What is the probability that the desk with the major defect is purchased?

 3.100 A department store is going to run a sale on a particular item for one, two, or three days. The probability that the store chooses to run the sale for one day is .2, for two days is .3, and for three days is .5. The probabilities of selling all the items in stock if the sale is held one, two, or three days are .1, .7, or .9, respectively. If the store conducts a sale, what is the probability that all items in stock are sold by the store during the sale?

3.101 Five homes and three stores have applied for telephones and only one three-party line is available. Parties are assigned to the available three-party line at random. Let the events A and B be defined as follows:

A: the line contains two homes and one store.

B: the line contains at least one home.

a. Find $P(A)$.   **b.** Find $P(B)$.

3.102 A man takes either a bus or the subway to work, with probabilities .3 and .7, respectively. When he takes the bus, he is late on 30% of the days. When he takes the subway, he is late on 20% of the days. If the man is late for work on a particular day, what is the probability that he took the bus?

 3.103 A manufacturer has two machines that produce a certain product. Machine 1 produces 45% of the product, and machine 2 produces 55%. Machine 1 produces 10% defective items, and machine 2 produces 8% defective items. If a defective item is observed, what is the probability that it was produced by machine 2?

 3.104 An electronic fuse is produced by three production lines in a manufacturing operation. The fuses are costly, are quite reliable, and are shipped to suppliers in 100-unit lots. Because testing is destructive, most buyers of the fuses test only a small number before deciding to accept or reject lots of incoming fuses. All three production lines produce fuses at the same rate and normally produce only 2% defective fuses, which are randomly dispersed in the output. Unfortunately, production line 1 suffered mechanical difficulty and produced 5% defectives during the month of March. This situation became known to the manufacturer after the fuses had been shipped. A customer received a lot produced in March and tested three fuses. One failed. What is the probability that the lot came from one of the two other lines?

 3.105 Each of two appliance stores (stores 1 and 2) sell each of two brands of washing machines (brands A and B). They are the only stores selling these machines. The probability that someone shops at store 1 is 3/4. The probability that someone shopping at store 1 buys brand A is 1/3. The probability that someone shopping at store 2 buys brand A is 1/4. Given that someone bought a brand A washing machine, what is the probability that it was purchased at store 1?

 3.106 Studies indicate that more American families owe installment debt than ever before and that the amount of this debt is rising. In a large western city the probability table classifying families according to income and installment debt is as follows:

Income	Installment Debt		
	None	Less Than $1,000	Exceeding $1,000
Less than $25,000	.25	.12	.04
$25,000–$35,000	.03	.18	.14
More than $35,000	.01	.05	.18

If a certain family from this city has an income within the range of $25,000–$35,000, what is the probability that the family owes installment debt exceeding $1,000?

 3.107 Many real estate professionals are pushing hard for present-value accounting, whereby investment properties can be valued at their current market value instead of their historical cost and, therefore, provide greater annual depreciation write-off. Accountants are generally mixed in their opinion on this subject. In a survey of 15 accountants, 10 favored sanctioning present-value accounting procedures, 6 favored a shortening of the allowable depreciation period, and 3 recommended no change in accounting law relating to asset depreciation. Suppose one of the accountants involved in the survey is selected at random.

a. What is the probability that she favors present-value accounting?

b. What is the probability that she favors either present-value accounting procedures or a shortening of the allowable asset depreciation period?

c. What is the probability that she favors both present-value accounting procedures and a shortening of the allowable asset depreciation period?

3.108 A firm that manufactures precision metal components uses a special machine responsible for both the molding and finishing of the components. The machine is controlled by two electrical relay systems, one for the molding process and one for the finishing process. These two relay systems are connected in a series so that the machine will work only if both electrical relay systems are in proper working order. Suppose the probability that a single relay system will operate satisfactorily is .99, and suppose that whether one will work is independent of whether the other will work. What is the probability that neither of the electrical relay systems will operate satisfactorily? That exactly one of the relay systems will work? That both of the relay systems will work? What is the probability that the special machine will operate satisfactorily?

3.109 Refer to Exercise 3.108. Now assume that to increase the chances that the special machine will operate satisfactorily, a worker installs a second pair of electrical relay systems, connected in series and designed to control, respectively, the molding and finishing processes. The special machine will now operate satisfactorily if one or the other of the pairs of electrical relay systems operates satisfactorily. What is the probability that neither of the pairs of relay systems works? That exactly one of the pairs of relay systems works? That both pairs of relay systems work? That the special machine will operate satisfactorily?

3.110 The following actual case occurred in Gainesville, Florida, in 1976. The eight-member Human Relations Advisory Board considered the complaint of a woman who claimed discrimination, based on her sex, on the part of a local surveying company. The board, composed of five women and three men, voted 5–3 in favor of the plaintiff, the five women voting in favor of the plaintiff, the three men against. The attorney representing the company appealed the board's decision by claiming sex bias on the part of the board members. If the vote in favor of the plaintiff was 5–3 and the board members were not sex-biased, what is the probability that the vote would split along sex lines (five women for, three men against)?

3.111 The executives of a major multinational corporation with four foreign subsidiaries are considering whether or not to insure against losses from war, currency inconvertibility, and expropriation. Actuarial estimates suggest that for any given year the probability is .05 that a foreign subsidiary will be affected by war, .09 that one will experience serious currency inconvertibility problems, and .03 that a foreign subsidiary will be expropriated by the host country. You may assume that the various types of losses are independent and that losses in one country do not affect the stability of subsidiaries in other countries.

a. What is the probability that the parent multinational corporation does not experience losses from war, currency inconvertibility, or expropriation among any of its subsidiaries in a given year?

b. What is the probability that it experiences at least one such loss among its four subsidiaries in a given year?

c. What is the probability that at least two of the foreign subsidiaries are affected by currency inconvertibility, but by no other losses, during a given year?

3.112 On the average, 2% of the 1-lb packages of processed candy contain slightly less than 1 lb of candy. The package weights are independent.

a. What is the probability that the fourth package selected is the first package with weight less than 1 lb?

b. What is the probability that no more than four packages are selected before an underweight package is found?

c. How many packages must be sampled before the probability of finding one or more underweight packages is greater than or equal to .5?

3.113 Many strategies are used by insurance companies in an attempt to ensure continued growth. In particular, insurance companies are using repositioning techniques, such as joint ventures, corporate divestiture, demutualization, and mergers and acquisitions. In a survey of 52 insurance company executives, 88% said that their companies had initiated, or were initiating, some sort of repositioning. As part of this survey conducted by Ernst and Whinney, the merger and acquisition activity of the 52 companies were classified according to type of company (mutual or stock) and whether or not their organizations had either merged with or acquired another company during the last one, two or five years. The data are as follows:

Merger/Acquisition Activity During the Preceding	Mutual Companies, A	Stock Companies, B
One year, C	8	9
Two years, D	4	0
Five years, E	3	6
None during this period, F	14	8

Source: Data from "Insurance Company Growth Strategies Surveyed," *Journal of Accountancy,* January, 1988, p. 136.

One insurance company executive is chosen at random from among the 52 surveyed. Using the event designations given in the table, find the following probabilities.

a. $P(A)$ **b.** $P(D)$ **c.** $P(A \cup D)$ **d.** $P(A \mid D)$

e. $P(F \mid B)$ **f.** $P(F \mid \bar{B})$ **g.** $P(B)$ **h.** $P(A \cup C \cup D)$

i. Are the events B and D independent? Mutually exclusive? Why or why not?

Supplementary Exercises

Interpretive

3.114 In an actual case witnesses to the robbery of a Los Angeles supermarket claimed that the crime was committed by a young man with black hair and a beard who made his getaway in a blue Mustang convertible driven by a blond-haired woman. By examining automobile registrations of owners of blue Mustang convertibles, police found a man who fit the description offered by witnesses. He and his blond wife were then arrested. At their trial an expert witness for the prosecution claimed that the chance of a man having black hair is .20; the likelihood of a man having a beard is .10; the probability that a car is a blue Mustang convertible is (at best) .001; and the probability that a woman has blond hair is .25. He then multiplied these separate probabilities to arrive at the probability of .000005 that the combination of circumstances, as described by witnesses, could have resulted by chance. How would you evaluate the judgment provided by the prosecution's expert witness?

3.115 Each state delegates responsibility to its State Board of Public Accountancy for the administration of an annual CPA examination. The examination is in five parts, with two parts on accounting problems and one each on auditing, business law, and theory of accounts. To be certified, an applicant must pass all five parts. The State Board of Public Accountancy for the state of New York reports that past history shows the percentages given in the accompanying table for applicants passing each part on their first attempt. If success on one part of the examination is independent of success on any other part, what is the probability that an applicant becomes certified on the first attempt to pass the CPA examination of the state of New

York? Do you believe that this probability is a reasonable approximation to reality when you consider that a modest proportion of applicants pass all parts of the examination on their first attempt? Comment on the assumption of independence between parts of the test.

	Accounting 1	Accounting 2	Auditing	Business law	Theory of accounts
Pass percentage	.30	.30	.37	.35	.27

Source: Data from R. G. Allen, CPA, Executive Secretary, State Board of Public Accountancy, Albany, New York.

Case Study

Screening Tests

Screening tests, which used to be associated primarily with medical diagnostic tests, are now finding application in a variety of fields. For example, camera technology and automatic test equipment (ATE) are now routinely used for inspecting parts in high-volume production processes. Other familiar applications include corporate drug testing of employees, home pregnancy tests, inexpensive tests for salmonella contamination in chickens, tests for lead poisoning of water supplies, and AIDS tests (D.G. Savage, "Corporate World Takes to Drug Tests," *Los Angeles Times*, July 27, 1985).

Very few screening tests are perfect. There is always the risk that the test will not catch all defective parts, diseased persons, or contaminated products (i.e., a "false negative" result). At the same time there is the risk that good parts, healthy people, or safe foods could be classified as being defective, sick, or unsafe (a "false positive"). The consequences of a false positive in the case of a corporate drug test or AIDS test can be devastating to the individuals involved. A false negative would have equally important, but different, consequences.

To evaluate the effectiveness and consequences of using a screening test, we have to estimate the probabilities of getting false negatives and false positives. The following event notation will simplify this process. (*Note:* The terminology refers to disease screening, but the concepts apply to any screening test.)

T: the test indicates that the person has the disease (positive result).

$\bar{T}$: the test indicates that the person does not have the disease (negative result).

D: the person actually does have the disease.

$\bar{D}$: the person actually does not have the disease.

Then

$$P(T \mid D) = \text{``sensitivity'' of the test}$$
$$= P(\text{positive test} \mid \text{person actually has disease})$$
$$P(\bar{T} \mid \bar{D}) = \text{``specificity'' of the test}$$
$$= P(\text{negative test} \mid \text{person is free of the disease})$$
$$P(\bar{D} \mid T) = P(\text{false positive})$$
$$P(D \mid \bar{T}) = P(\text{false negative})$$

In 1987 the sensitivity and specificity of the HIV test for AIDS were reported to be 98% and 99%, respectively ("MacNeil/Lehrer Newshour," Segment entitled "Lax Labs," August 12, 1987). The number of known AIDS cases in the United States in 1987 was 45,000 ("Aids Diary," *Discover*, January 1988, p. 38).

1. Using 242,200,000 as an estimate of the 1987 U.S. population, calculate the false positive and false negative rates for this screening test. What are the implications of the magnitude of the false positive rate?

2. Low-incidence diseases, such as AIDS, result in very high false positive rates. Can you suggest a way to improve (reduce) the false positive rate of a screening test?

3. Rephrase all of the probabilities and events above in terms of manufacturing inspection equipment, where one is testing for the presence of defective parts (instead of diseased persons).

4. In the case of inspection equipment, describe how you would determine the sensitivity and specificity of the inspection device.

References and Suggested Readings

Feller, W. *An Introduction to Probability Theory and Its Applications.* Vol. 1, 3d ed. New York: Wiley, 1968.

Mendenhall, W., R. L. Scheaffer, and D. D. Wackerly. *Mathematical Statistics with Applications.* 3d ed. Boston: PWS-KENT Publishing Co., 1986.

Meyer, P. L. *Introductory Probability and Statistical Applications.* 2d ed. Reading, Mass.: Addison-Wesley, 1970.

Riordan, J. *An Introduction to Combinatorial Analysis.* Princeton, N.J.: Princeton University Press, 1980.

Scheaffer, R. L., and W. Mendenhall. *Introduction to Probability: Theory and Applications.* Boston: PWS-KENT Publishing Co., 1975.

TO
THE
READER

What part does the theory of probability play in inferential statistics? In Chapter 3 we calculated the probability associated with an event that depended on a random or chance mechanism. In most cases of interest to a business researcher, an event results in a numerical outcome such as the number of defectives in a random sample of n production items, the number of sales contracts finalized on a given day, or the increase in the price of a stock during a one-week period. If the observed value of a variable depends upon the chance outcome of an experiment, the variable is called a *random variable*.

In Chapter 4 we make a distinction between two types of random variables and then proceed either to develop or simply to describe the properties associated with these two types of random variables. In Chapters 5 and 6 we examine several random variables that arise in many business applications.

4

RANDOM VARIABLES
AND PROBABILITY
DISTRIBUTIONS

Random Variables: How They
Relate to Statistical Inference

In Section 3.3 we defined an *experiment* as the process by which an observation is obtained. The inference that we wish to make determines not only the nature of the experiment but also the manner in which the observation is recorded. In general, inferences focus on certain *characteristics* or *variables* associated with the elements of the population. The variables being measured in an experiment fall into two categories: *quantitative* or *qualitative* variables. Variables such as the closing price of a stock on the New York Stock Exchange, the monthly sales of a retail outlet, or the crude oil production of an oil well in barrels per day are examples of quantitative variables. On the other hand, variables such as the make or model of an automobile, the classification of a production item as "acceptable" or "not acceptable," or the rating of a bond issue are examples of qualitative variables.

In the recording of an observation on a qualitative variable, the classification categories are often assigned numbers to simplify the encoding of the observation. For example, for the toss of a coin, the observation is a "head" or a "tail." However, when we are interested in the number of heads, the recorded measurement is a "1" or a "0." In this way observations on qualitative variables usually give rise to measurements reflecting the number of individuals or elements in each of the defined categories.

In general, most experiments give rise to a numerical measurement that varies from sample point to sample point in a random manner. In Section 3.10 we defined a random variable in the following way.

Definition

random variable

> A variable x is a **random variable** if the value that it assumes, corresponding to the outcome of an experiment, is a chance or random event.

Observing the number of defects on a randomly selected piece of new furniture, selecting a college applicant at random and observing the person's SAT score, or selecting a random sample of microchips from an incoming lot and recording the number of defective microchips are examples of experiments giving rise to random numerical events.

population

The **population** associated with the experiment is the set of all possible observations that could be generated by the experiment. We never actually measure each member of the population, but we can certainly conceive of doing so. In lieu of observing the population, we wish to obtain a small set of these measurements, called

sample

the **sample**, and we wish to use the information in the sample to **describe or make inferences about the population.**

A measurement obtained from an experiment results in a specific value of the random variable of interest and represents a measurement drawn from a population. How can a single measurement or a larger sample of, say, n measurements be used to make inferences about the population of interest? With the consumer preference study of Section 3.1 in mind, we might **calculate the probability of observing the sample value for various populations from which this sample value might have come. Furthermore, we might infer that this sample came from the population that produces the** *highest* **probability of observing that outcome.** This procedure is the basis for one of the important methods for the statistical estimation of population parameters that can be shown to provide "good" inferences in many situations.*

We defer further discussion of inference until Chapter 5. At this point it is sufficient to note that this **procedure requires a knowledge of the probability associated with each**

probability distribution

value of the random variable. In short, we need to know the **probability distribution** for the random variable, a distribution that represents the theoretical frequency histogram for the population of numerical measurements. The theory of probability presented in Chapter 3 provides the mechanism for calculating these probabilities for some random variables.

4.2

Classification of Random Variables

Random variables are classified into two types: *discrete* or *continuous*.

Definition

discrete random
variable

> A **discrete random variable** is one that can assume a countable number of values.

---

* This widely used procedure for finding estimators was proposed by the late Sir Ronald A. Fisher and is known as the *method of maximum likelihood.*

Countable means that you can associate the values that the random variable x can assume with the integers 1, 2, 3, 4, . . . (that is, you can count them). The number of values that x can assume can be finite or can be infinite (because you can conceive of the count as proceeding and never ending).

The number x of errors that a mechanic can make in an assembly operation is a discrete random variable because x can assume a finite number of values, 0, 1, 2, 3, . . . , N, where N is the total number of steps in the assembly operation. The number of years x until a corporation achieves \$1 billion in assets could conceivably be infinite, but nevertheless, x is still a discrete random variable. In the latter example the number of values that x can assume could be infinitely many since we can imagine a count that could be continued indefinitely. Other examples of discrete random variables are the following:

1. The number of automobiles sold per month
2. The number of accidents in a particular manufacturing plant for a given week
3. The number of customers waiting at a supermarket checkout counter
4. The number of television tubes produced in a given hour

Definition

continuous random
variable

> A **continuous random variable** is one that can assume the infinitely many values corresponding to the points on a line interval.

The word *continuous*, an adjective, means "proceeding without interruption." It, in itself, provides the key for identifying continuous random variables. Look for a measurement with a set of values that form points on a line with no interruptions or intervening spaces between them.* For example, suppose that we measure the distance between a supplier and a buyer, and we represent this distance as the random variable x. Even if the finest measuring instrument were used, the values of x could only be given to the decimal accuracy of the measuring instrument. Nonetheless, in building probability models for variables that are measurements, we theoretically view x as one of the infinitely many points on the line interval between the supplier and the buyer. In general, we consider measurement variables to be continuous random variables whose possible values are those associated with the continuum of points on a line interval. Other examples of continuous random variables are the following:

1. The length of time to complete an assembly operation in a manufacturing plant
2. The amount of petroleum pumped per hour from a well
3. The amount, in milligrams, of carbon monoxide in a cubic foot of air
4. The amount of energy produced by a utility company in a given day

---

* Actually, a continuous random variable can be defined over more than one interval, but this tends to be a theoretical consideration. For the cases encountered in applied business statistics, a continuous random variable will be defined over a single interval on a line.

The distinction between discrete and continuous random variables is an important one, because different probability models are required for each. The probabilities associated with each value of a discrete random variable can be assigned so that the probabilities sum to 1. This assignment is not possible with continuous random variables. Accordingly, we will consider the probability distributions for discrete and continuous random variables separately in Sections 4.3 and 4.4, respectively.

EXERCISES   Conceptual Level

 4.1   Identify the following as discrete or continuous random variables.

   a.   The number of loan applications received by a commercial bank during a particular week

   b.   The length of time for an employee to complete a certain task while observed in a time and motion study

   c.   The number of employee grievances received by the personnel officer of an aircraft manufacturing facility

   d.   The rate of interest paid by the federal government on 90-day Treasury bills

   e.   The wellhead price of natural gas to a utility company

4.2   Identify the following as discrete or continuous random variables.

   a.   The length of life of a light bulb observed during a life-testing experiment

   b.   The gross revenue earned by a supermarket during a given day

   c.   The market value of a publicly listed security on a particular day

   d.   The tensile breaking strength, in pounds per square inch, of 1-in.-diameter steel cable

   e.   The daily demand for electrical power by the residential users in Redding, California

4.3   Identify the following as discrete or continuous random variables.

   a.   The number of claims received by an insurance company during a day

   b.   The dollar volume of unemployment compensation disbursed during a week by a state relief assistance office

   c.   The length of time for a customer to make a product selection decision when faced with several alternative choices

   d.   The daily fuel consumption by a fleet of buses serving a large southwestern city

   e.   The impurities content in beer produced by a national brewery

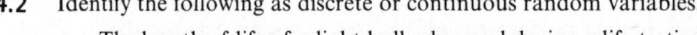

4.3

Probability Distributions for Discrete Random Variables

The **probability distribution** for a discrete random variable can be represented by a formula, a table, or a graph that gives the probability associated with each value of the random variable. Since each value of the random variable x is a numerical event, we can apply the methods of Chapter 3 to obtain $p(x)$, the probability of observing the value x. Since one and only one value of x is assigned to each sample point in the sample space, and $p(x)$ is the sum of the probabilities of all sample points giving rise to the same value of x, then

$$0 \le p(x) \le 1$$

Furthermore, since the values of x represent mutually exclusive events, the sum of the probabilities associated with the values of x is equal to 1. That is,*

$$\sum_x p(x) = 1$$

EXAMPLE 4.1   Consider an experiment that consists of tossing two coins, and let x be the number of heads observed. The sample points for this experiment with their respective probabilities are given in Table 4.1. Find the probability distribution for x.

Table 4.1   Sample Points and Their Probabilities for Example 4.1

Sample Point	Coin 1	Coin 2	$P(E_i)$	x
E_1	H	H	1/4	2
E_2	H	T	1/4	1
E_3	T	H	1/4	1
E_4	T	T	1/4	0

SOLUTION   We assign the value $x = 2$ to point E_1, $x = 1$ to point E_2, and so on. The probability of each value of x may be calculated by adding the probabilities of the sample points in that numerical event. The numerical event $x = 0$ contains one sample point, E_4; $x = 1$ contains two sample points, E_2 and E_3; and $x = 2$ contains one point, E_1. The values of x, with the respective probabilities, are given in Table 4.2. Observe that

$$\sum_{x=0}^{2} p(x) = 1$$

Table 4.2   Probability Distribution for the Number of Heads When Two Coins Are Tossed

x	Sample Points in x	$p(x)$
0	E_4	1/4
1	E_2, E_3	1/2
2	E_1	1/4

$$\sum_{x=0}^{2} p(x) = 1$$

The probability distribution for x is shown graphically in Figure 4.1.

---

* Note that the *probability distribution* of a random variable x is denoted by the symbol $p(x)$ (i.e., lowercase letter p is used). In contrast, the *probability of an event E* is denoted by the symbol $P(E)$ (i.e., capital letter P). We will make this distinction throughout the text.

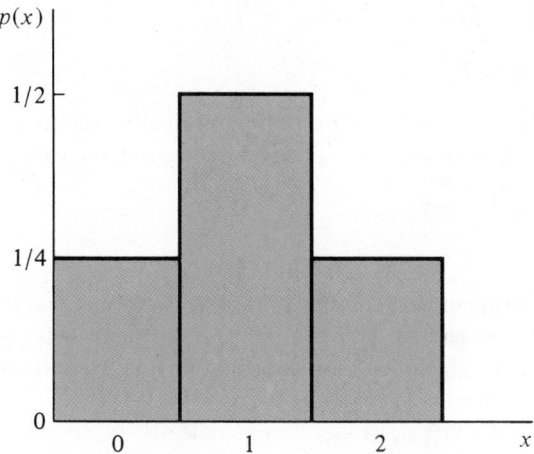

Figure 4.1   Probability Histogram Showing $p(x)$ for Example 4.1

EXAMPLE 4.2   A retail store has received a shipment of six small appliances, two of which are defective. A customer randomly selects and purchases two of the appliances as gifts. Find the probability distribution for x, the number of defective appliances among the two purchased.

SOLUTION   Let D_1 and D_2 represent the two defective appliances and let N_1, N_2, N_3, and N_4 represent the four nondefective appliances. The experiment, which consists of randomly choosing two of the six appliances, results in 15 sample points that are equally probable. These 15 sample points are listed in Table 4.3 together with the corresponding values of x. From Table 4.3 we have

$$
\begin{aligned}
p(0) &= P(x = 0) \\
&= P(E_{10}) + P(E_{11}) + P(E_{12}) + P(E_{13}) + P(E_{14}) + P(E_{15}) = 6/15 \\
p(1) &= P(x = 1) \\
&= P(E_2) + P(E_3) + P(E_4) + P(E_5) + P(E_6) + P(E_7) \\
&\quad + P(E_8) + P(E_9) \\
&= 8/15 \\
p(2) &= P(x = 2) = P(E_1) = 1/15
\end{aligned}
$$

Notice that

$$
\sum_{x=0}^{2} p(x) = 1
$$

and that all the probabilities are nonnegative. The probability distribution for x is shown in Table 4.4.

Table 4.3   Sample Points and Their Probabilities for Example 4.2

Sample Point	$P(E_i)$	x	Sample Point	$P(E_i)$	x
$E_1: D_1 D_2$	1/15	2	$E_9: D_2 N_4$	1/15	1
$E_2: D_1 N_1$	1/15	1	$E_{10}: N_1 N_2$	1/15	0
$E_3: D_1 N_2$	1/15	1	$E_{11}: N_1 N_3$	1/15	0
$E_4: D_1 N_3$	1/15	1	$E_{12}: N_1 N_4$	1/15	0
$E_5: D_1 N_4$	1/15	1	$E_{13}: N_2 N_3$	1/15	0
$E_6: D_2 N_1$	1/15	1	$E_{14}: N_2 N_4$	1/15	0
$E_7: D_2 N_2$	1/15	1	$E_{15}: N_3 N_4$	1/15	0
$E_8: D_2 N_3$	1/15	1			

Table 4.4   Probability Distribution for x, the Number of
Defective Appliances in Example 4.2

x	$p(x)$
0	6/15
1	8/15
2	1/15

To illustrate how we use the probability distribution for a discrete random variable to make inferences, let us consider again the problem (introduced in Section 3.1) of deciding whether package design A is preferred to design B. The experiment that was designed to answer this question consists of randomly selecting 20 consumers and asking each which package design he or she prefers. Once we have recorded the number x of consumers in the sample favoring package design A, a question immediately arises. How small should this number be before we conclude that package design B is preferred to design A? To answer this question, we need to find the probability of observing certain sample results.

This problem is analogous in many ways to a coin-tossing experiment. If the fraction of all consumers favoring package design A is at least 1/2 (let us assume that it is exactly 1/2), then observing the response of a randomly selected consumer is analogous to tossing a fair coin. Likewise, the random sampling of 20 consumer's responses is analogous to the tossing of 20 fair coins (an extension of the two-coin-toss experiment considered earlier). Thus, the random variable x, the number of consumers favoring package design A, corresponds to the number of heads in a toss of 20 coins. For both experiments x can assume any of the values 0, 1, 2, . . . , 20, with associated probabilities $p(0), p(1), p(2), . . . , p(20)$ given by the same probability distribution. Although the exact probability distribution for this experiment will be discussed in Chapter 5, we present its graph here in Figure 4.2.

From the graph of the probability distribution for x, we observe that the largest probability is that associated with the value $x = 10$—which is to be expected if half of

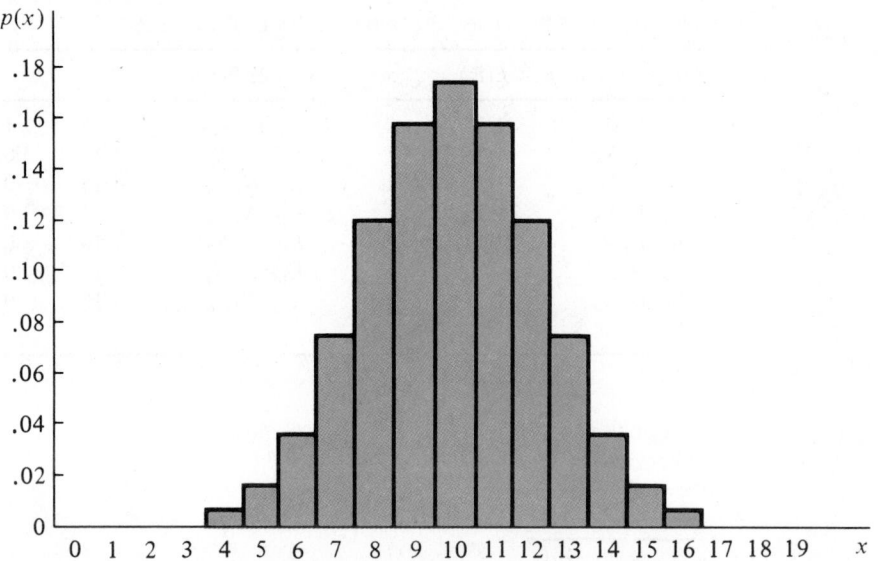

Figure 4.2   Probability Distribution for the Number of Consumers Favoring Package Design *A*

all 20 consumers favor package design *A*. Moreover, we note that it is highly improbable that as few as *x* = 3 or even *x* = 5 consumers favor design *A* in the sample if the fraction in the population of all consumers is 1/2. If *x* = 0, the result is so extremely improbable that we would conclude that the assumption is false. Thus, if in our sample *x* = 0 or some very small value, we would infer that in the population of responses from consumers purchasing the manufacturer's product, less than 1/2 favor package design *A*.

Tips on Problem Solving

To find the probability distribution for a discrete random variable, construct a table listing each value that the random variable *x* can assume. Then calculate *p*(*x*) for each value of *x*.

EXERCISES   Conceptual Level

4.4   Define in your own words the concept of a probability distribution.

4.5   What two restrictions do we place on the probabilities *p*(*x*) associated with a particular probability distribution?

4.6 Determine whether or not the following are valid probability distributions. If they are not, indicate which of the two restrictions given in Exercise 4.5 has been violated.

a.

x	$p(x)$
-1	.2
0	.6
2	.1
2.5	.1

b.

x	$p(x)$
0	$-.2$
1	.2
2	.8

c.

x	$p(x)$
-1	.1
0	.3
1	.5
2	.2

EXERCISES *Skill Level*

4.7 A random variable x has the following probability distribution:

x	$p(x)$
1	.05
2	.20
3	.05
4	.45
5	.25

a. Verify that x has a valid probability distribution.

b. Find the probability that x is greater than 3.

c. Find the probability that x is less than or equal to 2.

d. Find the probability that x is an odd number.

e. Graph the probability distribution for x.

4.8 An investor randomly selected two utility stocks from a group of three. Of these three stocks, two possess higher growth potential than the third. Let x be the number of stocks with the higher growth potential that are contained in the selection.

a. Find the probability that $x = 0, 1, 2$.

b. Graph $p(x)$.

4.9 In Exercise 3.3 we described an experiment in which three fair coins were tossed and the upper face was observed. Let x be the number of heads observed.

a. List the simple events in the experiment, and indicate the value that x assumes for each simple event.

b. Find the probability distribution for x. Graph this probability distribution.

c. Calculate $P(x \geq 2)$.

d. Calculate $P(x \leq 2)$.

e. Find the probability of observing at least one head.

EXERCISES Applications

 4.10 Studies show that in a lower-income area of a certain city, only 20% of shoppers in food stores read unit-pricing labels before making a purchase decision. If $n = 2$ shoppers enter a food store in this lower-income area, what is the probability distribution for x, the number of shoppers who take advantage of the unit-pricing labels? [*Hint:* Let F indicate the event that a shopper does not read the labels. Then $p(0) = P$(both shoppers do not read the labels) $= P(F)P(F) = (.8)^2 = .64$.] Construct a probability histogram for x.

 4.11 Simulate the experiment described in Exercise 4.10 by using five marbles (you could use cards), four of one color and one of another. Draw one marble at random from the five, replace the marble, and repeat the process. Count the number x of times the different-colored marble appears in the two draws and record this observed value of x. The number x will correspond to the number of shoppers in a sample of two (corresponding to the two draws) who take advantage of the pricing labels. Repeat the dual drawing process 100 times and obtain 100 observed values of x. Construct a relative frequency histogram for the data and compare it with the population probability distribution you constructed in Exercise 4.10.

 4.12 Consumers are increasingly interested in becoming knowledgeable in the field of investment planning. A survey conducted in late 1987, but prior to the October market upheaval, concerned the sources of financial advice most frequently used by consumers for investment information and information regarding the impact of the 1986 tax reform policies on investments. The survey found that, at that time, consumers expected the bull market to continue for at least a year and were eager to explore savings and investment opportunities; however, three-fourths of the people surveyed felt confused about the variety of financial products and services available to them (*Journal of Accountancy*, January 1988, p. 24). Suppose that this survey is representative of all consumer attitudes and that three consumers are chosen at random. Let x be the number of consumers who felt confused about the variety of financial products and services available, where $x = 0, 1, 2$, or 3. Find the probability distribution for x, and express your results graphically as a probability histogram.

 4.13 To save time and money but still provide safeguards on the quality of incoming goods, buyers of goods often inspect a portion of the shipment and judge the quality of the entire shipment on the basis of the observed quality of the sample. Suppose a buyer has received a shipment of five photocopy machines, two of which are defective. Two photocopy machines are selected at random from among the five and tested. Let x be the number of defective machines observed, where $x = 0, 1$, or 2. Find the probability distribution for x, and express your results graphically as a probability histogram.

 4.14 Simulate the experiment described in Exercise 4.13 by using five marbles (you could use cards), three of one color and two of another, to represent, respectively, the three good and the two defective photocopy machines. Place the marbles in a hat, mix, draw two, and record x, the number of defectives observed. Replace the marbles and repeat the process until a total of $n = 100$ observations on x have been recorded. Construct a relative frequency histogram for this sample and compare it with the population probability distribution you constructed in Exercise 4.13.

 4.15 Although synthetic fuels offer an opportunity to greatly expand our domestic energy supply, they present environmental problems that worry many government officials. Recent evidence suggests that Environmental Protection Agency (EPA) officials are now evenly divided on the issue of relaxing environmental standards so that businesses may switch to synthetic fuels as their primary source of energy. Suppose the EPA randomly and independently assigns four of its officials to investigate the plans of several New England firms who wish to switch from natural to synthetic fuels. Let x represent the number of EPA officials in the investigative group who support relaxing environmental standards for firms switching to synthetic fuels.

a. Find $p(0)$, $p(1)$, $p(2)$, $p(3)$, and $p(4)$.

b. Give a general formula $p(x)$.

4.4

Probability Distributions for Continuous Random Variables

As indicated in Section 4.2, continuous random variables can assume the infinitely many values corresponding to points on a line interval. However, we cannot assign a positive probability to each of these infinitely many points and still have the probabilities sum to 1, as for discrete random variables. Therefore, a different approach is used to generate the probability distribution for a continuous random variable. The approach that we adopt uses the concept of a relative frequency histogram, such as that for the 25 price-earnings ratios in Figure 2.2. Recall that the width of the class interval in Figure 2.2 was determined in accordance with the number of measurements involved. If more and more measurements are obtained, we might reduce the width of the class interval. The outline of the histogram would change slightly, for the most part becoming less and less irregular. When the number of measurements becomes very large and the intervals very small, the relative frequency histogram would appear, for all practical purposes, as a smooth curve, as shown in Figure 4.3.

The relative frequency associated with a particular class in the population is the fraction of measurements in the population falling in that interval and also is the probability of drawing a measurement in that class. If the total area under the relative frequency histogram is adjusted to equal 1, areas under the frequency curve correspond to probabilities. In fact, this correspondence was the basis for the application of the Empirical Rule in Chapter 2, which applies when the data are approximately bell-shaped.

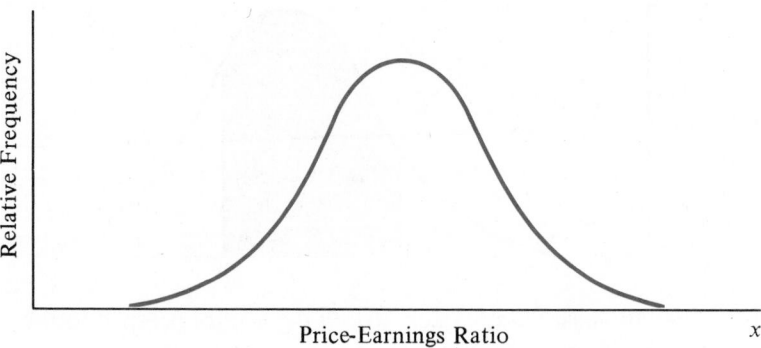

Figure 4.3   Relative Frequency Histogram for a Population

Let us construct a model for the probability distribution for a continuous random variable. Assume that the random variable x may take on any value on a real line, as in Figure 4.3. We then distribute 1 unit of probability along the line, much as a person might distribute a handful of sand, each measurement in the population corresponding to a single grain. The probability—grains of sand or measurements—will pile up in certain places, and the result will be the probability distribution shown in Figure 4.4. The depth or density of the probability, which varies with x, may be represented by a mathematical formula $f(x)$, called the *probability distribution*, or the *probability density function*, for the random variable x. The density function $f(x)$, represented graphically in Figure 4.4, provides a mathematical model for the population relative frequency histogram that exists in reality. The total area under the curve $f(x)$ is equal to 1. The area lying above a given interval equals the probability that x will fall in that interval. Thus, the probability that $a < x < b$ (a is less than x and x is less than b) is equal to the area under the density function between the two points a and b. This is the shaded area in Figure 4.4.

How do we choose the model—that is, the probability distribution $f(x)$—appropriate for a given physical situation? Many types of continuous curves are available for modeling, not all of which are mound-shaped as are those shown in Figures 4.3 and 4.4. Fortunately, we will find that many continuous random variables *have* mound-shaped frequency distributions, often very nearly bell-shaped. A probability model that provides a good approximation to such population distributions is the *normal probability distribution*, which we will study in detail in Chapter 6.

In actual practice, there will be a difference between the conceptual model $f(x)$ and the relative frequency histogram generated when the experiment is repeated an extremely large number of times. The model $f(x)$ that we adopt can be expected only to *approximate* the population relative frequency curve. This practice can lead to an invalid conclusion if an inappropriate model is chosen out of poor judgment or insufficient knowledge of the phenomenon under study. On the other hand, using a density function $f(x)$ to approximate the population relative frequency distribution for a continuous random variable is a strategy that, when properly applied, has been found

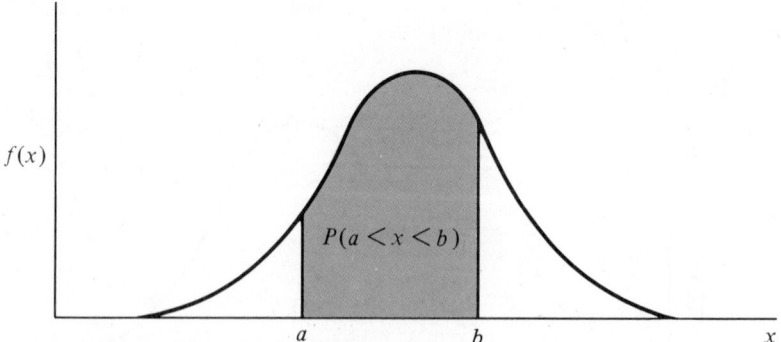

Figure 4.4   The Probability Distribution $f(x)$; $P(a < x < b)$ Is Equal to the Shaded Area Under the Curve

to be highly successful in scientific research. The equations, formulas, and various numerical expressions used in all the sciences are simply mathematical models that provide approximations to reality, the goodness of which is evaluated through experimental application. Hence, a model is evaluated in terms of the results of its application—which are decisions or predictions in business situations. Do the resulting inferences fit in with the body of accumulated evidence? Are the deductions that follow from these inferences verified experimentally? If so, the model has proved its worth.

4.5

Mathematical Expectation

The probability distributions described in Sections 4.3 and 4.4 provide models for the theoretical frequency distributions of discrete and continuous random variables. Hence, these distributions must possess a mean, variance, standard deviation, and other descriptive measures associated with the theoretical populations that they represent. Because both the mean and the variance are averages (Sections 2.7 and 2.8), we will confine our attention to the problem of calculating the average value of a random variable defined over a theoretical population. This average is called the *expected value* of the random variable.

The method for calculating the population mean or expected value of a random variable can be more easily understood by considering an example. Let x be the number of heads observed in the toss of two coins, as shown in Table 4.5. For convenience we also list $p(x)$ in the table.

Let us suppose that the experiment is repeated a large number of times, say $n = 4,000,000$ times. Intuitively, we would expect to observe approximately 1 million zeros, 2 million ones, and 1 million twos. Then the average value of x is

$$\frac{\text{sum of measurements}}{n} = \frac{(0)(1,000,000) + (1)(2,000,000) + (2)(1,000,000)}{4,000,000}$$

$$= \frac{(0)(1,000,000)}{4,000,000} + \frac{(1)(2,000,000)}{4,000,000} + \frac{(2)(1,000,000)}{4,000,000}$$

$$= (0)(1/4) + (1)(1/2) + (2)(1/4)$$

Table 4.5 Probability Distribution for x

x	$p(x)$
0	1/4
1	1/2
2	1/4

Note that the first term in this sum is equal to $(0)p(0)$, the second is equal to $(1)p(1)$, and the third is equal to $(2)p(2)$. The average value of x, then, is

$$\sum_{x=0}^{2} xp(x) = 1$$

This result is not an accident, and it provides some intuitive justification for the definition of the expected value of a discrete random variable x.

Definition

expected value
of x

Let x be a discrete random variable with probability distribution $p(x)$. Then $E(x)$, the **expected value of x**, is

$$E(x) = \sum_{x} xp(x)$$

where the elements are summed over all values of the random variable x.

Note that if $p(x)$ is an accurate description of the relative frequencies for a real population of data, then $E(x) = \mu$, the mean of the population. We will assume this result to be true and let $E(x)$ be synonymous with μ.

The method for calculating the expected value of x for a continuous random variable is rather similar from an intuitive point of view, but in practice, it involves the use of the calculus and is therefore beyond the scope of this text.

EXAMPLE 4.3   A discrete random variable x has the following probability distribution:

x	$p(x)$
10	.2
15	.3
20	.4
25	.1

Calculate the expected value of x, $E(x) = \mu$.

SOLUTION   Table 4.6 presents a systematic method for calculating expected values. It is similar to the approach used when calculating means and variances for grouped data in Section 2.11. The table contains three columns, the first for the values of x, the second for the probabilities $p(x)$. Column 3 contains the product $xp(x)$ for each value of x, and the sum of this column over all values of x is the expected value of x.
Hence,

$$\mu = E(x) = \sum_{x} xp(x) = 17.0$$

Table 4.6   Calculations for $E(x)$ in
            Example 4.3

x	$p(x)$	$xp(x)$
10	.2	2.0
15	.3	4.5
20	.4	8.0
25	.1	2.5
		$E(x) = 17.0$

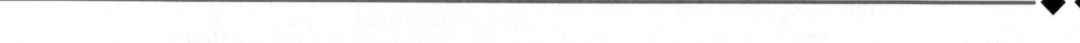

EXAMPLE 4.4   In competitive-bidding exercises contractors will often submit a bid on a project if their analysis of the project and of the opposing bidders suggests that their expected return for submitting a bid exceeds some limiting amount. Suppose a contractor is considering a bid on a project that would return $50,000. The cost of preparing a bid for the project is $5,000. The contractor feels that the probability that he will win the contract is .4. In addition, he will bid on all contracts with an expected return exceeding $12,000. Should he prepare a bid for the project?

SOLUTION   The contractor's net return x may take on one of two possible values. Either he will lose $5,000 (i.e., his return will be $-\$5,000$) or he will win ($50,000 - \$5,000) = \$45,000$, with probabilities .6 and .4, respectively. The probability distribution for his net gain x is shown in Table 4.7.

Table 4.7   Probability Distribution for x
            in Example 4.4

x	$p(x)$
$-\$ \ 5,000$	.6
$\$45,000$	.4

The expected net return is

$$E(x) = \sum_x xp(x) = (-\$5,000)(.6) + (\$45,000)(.4) = \$15,000$$

Since $E(x) = \$15,000$ exceeds $12,000, the contractor should prepare and submit a bid for the project.

EXAMPLE 4.5   An actuary, a statistician employed by an insurance company, determines the premiums the company should charge for its insurance. Consider the problem of determining the yearly premium for a $100,000 automobile liability insurance policy. The policy covers the type of events that, over a long period, have occurred at a yearly rate of three times per 5,000 drivers each year.

Let x be the yearly financial gain to the insurance company resulting from the sale of the policy and let C be the unknown yearly premium. We will calculate the value of C such that the expected gain $E(x)$ is 0. Then C is the premium required to break even. To this value the company would add administrative costs and a margin of profit.

SOLUTION   We assume that the expected gain $E(x)$ depends on C. Using the requirement that the expected gain must equal 0, we have

$$E(x) = \sum_x xp(x) = 0$$

We must solve this equation for C.

The first step in the solution is to determine the values that the gain x may take. Then we determine $p(x)$. If the event does not occur during the year, the insurance company will gain the premium, or $x = C$ dollars. If the event does occur, the gain will be negative—will be a loss—amounting to $x = -(100{,}000 - C)$ dollars. The probabilities associated with these two values of x are $4{,}997/5{,}000$ and $3/5{,}000$, respectively. The probability distribution for the gain is shown in Table 4.8.

Table 4.8   Probability Distribution for x in Example 4.5

x, Gain	$p(x)$
C	$4{,}997/5{,}000$
$-(100{,}000 - C)$	$3/5{,}000$

Setting the expected value of x equal to 0 and solving for C, we have

$$E(x) = \sum_x xp(x) = C\left(\frac{4{,}997}{5{,}000}\right) + [-(100{,}000 - C)]\left(\frac{3}{5{,}000}\right) = 0$$

or

$$\left(\frac{4{,}997}{5{,}000}\right)C + \left(\frac{3}{5{,}000}\right)C - 60 = 0 \qquad \text{so that} \qquad C = \$60$$

Thus, if the insurance company were to charge a yearly premium of \$60, the average gain calculated for a large number of similar policies would be 0. The actual premium the company would charge would be \$60 plus administrative costs and profit.
◆◆

The insurance premium problem can be extended to include any number of gains to the insurance company by solving for the premium C in the expression

$$E(x) = \sum_x xp(x) = 0$$

where x is the random variable representing the insurance company's gain. The difficulty in a practical problem of this type is to identify each gain and associate a meaningful probability with that gain. This task is the responsibility of the actuary.

Tips on Problem Solving

To find the expected value of a discrete random variable x, construct a table containing three columns, the first for x and the second for $p(x)$. Then multiply each x value by its corresponding probability and enter these values in the third column. The sum of this third column, the sum of $xp(x)$, will give you the expected value of x.

EXERCISES   Conceptual Level

4.16   Verbally describe the average or expected value of a random variable.

4.17   For a given hillside tract, a geologic consulting group assesses the probability of earth movement sufficient to cause structural damage to be .0001. If a contractor is considering the construction of structures valued at over $10 million, is it reasonable to speak of an average loss to the contractor based on her decision to proceed or not proceed with construction on this tract?

EXERCISES   Skill Level

4.18   A random variable x has the following probability distribution:

x	1.5	2.0	2.5	3.0	3.5
$p(x)$	.01	.49	.32	.15	.03

a.   Find the expected value of x.

b.   Graph $p(x)$ and locate μ on the graph.

c.   Does $\mu = E(x)$ fall near the center of the probability distribution?

4.19   A random variable x has the following probability distribution:

x	0	1	2	3	4	5
$p(x)$	.15	.35	.25	.10	.10	.05

a.   Find the expected value of x.

b.   Graph $p(x)$ and locate μ on the graph.

c.   Does $\mu = E(x)$ fall near the center of the probability distribution?

4.20   The accompanying table is the probability distribution of a random variable x. Find the expected value of x. Construct a graph of the probability distribution. Locate μ on the graph.

x	0	1	2	3
$p(x)$	.1	.2	.3	.4

EXERCISES Applications

4.21 Risky investment opportunities are often classified according to their expected profitability. Those with a positive expected profitability are subject to further analysis; those with negative expected profitability are rejected outright. As an example, consider an investment of $5,000 in an oil-drilling venture that is believed either to provide a return of $55,000 or to result in a loss of the original investment. If the probability of success of the oil-drilling venture is .1, find the expected profit for investing in the venture.

4.22 A manufacturing representative is considering the option of taking out an insurance policy to cover possible losses incurred by marketing a new product. If the product is a complete failure, the representative feels that a loss of $80,000 would be incurred; if it is only moderately success-ful, a loss of $25,000 would be incurred. Insurance actuaries have determined from market sur-veys and other available information that the probabilities that the product will be a failure or only moderately successful are .01 and .05, respectively. If the manufacturing representative is willing to ignore all other possible losses, what premium should the insurance company charge for the policy in order to break even?

4.23 The number N of residential homes that a fire company can serve depends on the distance r (in city blocks) that a fire engine can cover in a specified (fixed) period of time. If N is proportional to the area of a circle r blocks from the firehouse, then $N = C\pi r^2$, where C is a constant, $\pi = 3.1416\ldots$, and the random variable r is the number of blocks that a fire engine can move in the specified time interval. For a particular fire company, $C = 8$, and the probability distribu-tion for r is as shown in the accompanying table. [Note that $p(r) = 0$ for $r \leq 20$ and $r \geq 27$.] Find the expected value of N, the number of homes that the fire department can serve.

r	21	22	23	24	25	26
$p(r)$	.05	.20	.30	.25	.15	.05

4.6

The Variance of a Random Variable

The expected value of a random variable is a measure of central tendency for its probability distribution in the same way that $\bar{x}$ is a measure of central tendency for a relative frequency histogram. The term *expected value* possesses an intuitive meaning in the sense that we are more likely to draw a value of x near the center of the probability distribution.

However, just as $\bar{x}$ does not provide a complete description of a sample relative frequency histogram, the expected value μ does not provide a complete description of the behavior of a random variable. For a better description we need a measure of the spread of the probability distribution $p(x)$. To illustrate, suppose that you are managing a hospital and that you must decide on the number of shots of antitetanus vaccine that must be maintained in inventory. Furthermore, assume that you must

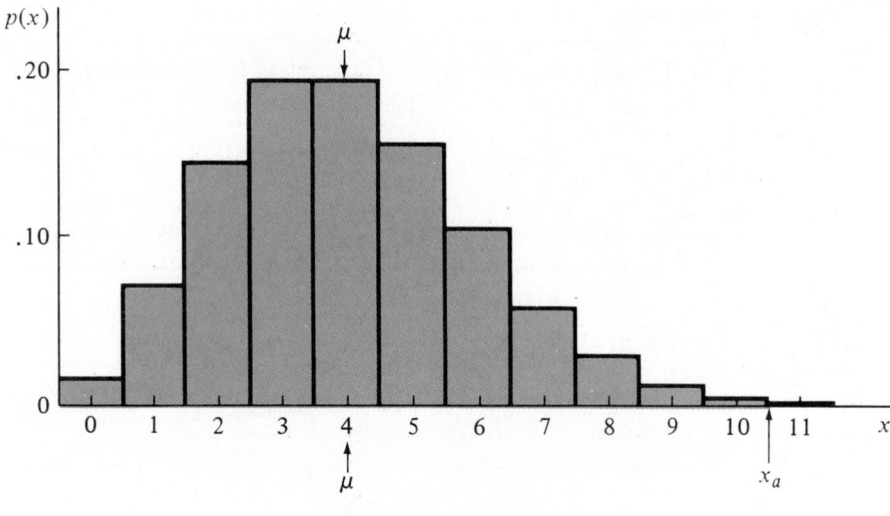

Figure 4.5   Probability Distribution for the Number x of Tetanus Shots Used in a Three-Day Period; $\mu = 4$

always have an adequate supply on hand to meet a three-day demand and that the probability distribution for x, the number of shots required for a three-day period, appears as shown in Figure 4.5.

You can see from Figure 4.5 that the expected demand μ (located on the figure) does not tell us the largest number of tetanus shots that might be required for a three-day period. Thus, we need to know the value of x, say x_a, in the right tail of $p(x)$ such that the probability of demand exceeding x_a is very small (see Figure 4.5).

Finding the number x_a of shots that should be maintained in inventory would be relatively easy if we knew $p(x)$, because we could move to the right along the x axis in Figure 4.5 until we found x_a such that $P(x \geq x_a)$ is small. If $p(x)$ is not known, we might be satisfied with a measure of spread for $p(x)$. This measure of variability would enable us to locate an approximate value for x_a.

Because a probability distribution is (in a sense) a theoretical relative frequency histogram for the random variable x, it is natural for us to describe the variability of x by finding its variance and its standard deviation. **Then we can interpret these descriptive measures by using Tchebysheff's Theorem and the Empirical Rule.** For the tetanus shot supply problem we would be able to find a value for x_a that would be in the upper tail of $p(x)$.

In Chapter 2 we defined the population variance to be the average of the squares of the deviations of the measurements from their mean. Because taking an expectation is equivalent to "averaging," we define the *variance* and the *standard deviation* as shown in the displays.

Definition

variance of x

> Let x be a discrete random variable with probability distribution $p(x)$ and expected value $E(x) = \mu$. The **variance of x** is
>
> $$\sigma^2 = E[(x - \mu)^2] = \sum_x (x - \mu)^2 p(x)$$
>
> where the elements are summed over all values of the random variable x.*

Definition

standard deviation
of a random
variable

> The **standard deviation σ of a random variable x** is equal to the square root of its variance.

EXAMPLE 4.6   Find the variance σ^2 for the population associated with Example 4.1, the coin-tossing problem. In Section 4.5 the expected value of x was shown to equal 1.

SOLUTION   The variance is equal to the expected value of $(x - \mu)^2$, or

$$\sigma^2 = E[(x - \mu)^2] = \sum_x (x - \mu)^2 p(x)$$
$$= (0 - 1)^2 p(0) + (1 - 1)^2 p(1) + (2 - 1)^2 p(2)$$
$$= (1)(1/4) + (0)(1/2) + (1)(1/4) = 1/2$$

Then $\sigma = \sqrt{1/2} = .707$. The values $\mu = 1$ and $\sigma = .707$ can be used to describe the probability distribution shown in Figure 4.1.

◆ ◆

EXAMPLE 4.7   A study of the mobility of corporate purchasing executives found that the distribution shown in Table 4.9 provides an adequate approximation to the probability distribution of x, the number of firms in which purchasing positions have been held by purchasing executives currently employed by industrial organizations. Find the mean and standard deviation of x.

SOLUTION   The mean is equal to $E(x)$, or

$$\mu = E(x) = \sum_x x p(x) = 1(.52) + 2(.22) + 3(.19) + 4(.04) + 5(.03) = 1.84$$

---

* It can be shown (proof omitted) that $\sigma^2 = E[(x - \mu)^2] = E(x^2) - \mu^2$, where $E(x^2) = \sum_x x^2 p(x)$.

Table 4.9   Probability Distribution for x in Example 4.7

x	$p(x)$
1	.52
2	.22
3	.19
4	.04
5	.03

Source: I. V. Fine and J. H. Westling, "Organizational Characteristics of Purchasing Personnel in Public and Private Hierarchies," *Journal of Purchasing*, August 1973. Reprinted by permission.

The variance is equal to the expected value of $(x - \mu)^2$, or

$$\sigma^2 = E[(x - \mu)^2] = \sum_x (x - \mu)^2 p(x)$$
$$= (1 - 1.84)^2(.52) + (2 - 1.84)^2(.22) + (3 - 1.84)^2(.19)$$
$$+ (4 - 1.84)^2(.04) + (5 - 1.84)^2(.03)$$
$$= (.7056)(.52) + (.0256)(.22) + (1.3456)(.19) + (4.6656)(.04)$$
$$+ (9.9856)(.03)$$
$$= 1.1144$$

The standard deviation of the distribution is

$$\sigma = \sqrt{\sigma^2} = \sqrt{1.1144} = 1.056$$

The mean $\mu = 1.84$ and the standard deviation $\sigma = 1.056$ can be used together with Tchebysheff's Theorem to describe the probability distribution for the number of firms in which purchasing positions have been held by purchasing executives currently employed by industrial organizations. Even though the distribution is not mound-shaped, you can see from Table 4.9 and Figure 4.6 that the probability that x will fall outside the interval $(\mu \pm 2\sigma)$, or $(-.27, 3.95)$, is very small. Since only $x = 4$ and $x = 5$ fall outside this interval, the probability that x will fall outside $(\mu \pm 2\sigma)$ is

$$P(x = 4) + P(x = 5) = .04 + .03 = .07$$

◆◆

EXAMPLE 4.8   In Chapter 5 we will discuss a particular type of discrete random variable called the *Poisson random variable*. You will find that the number x of people requiring a tetanus shot in a three-day period can be modeled by the Poisson probability distribution. **You will also learn that for a Poisson random variable the variance of x equals its mean.** Suppose that past experience shows that the average number of shots required for a

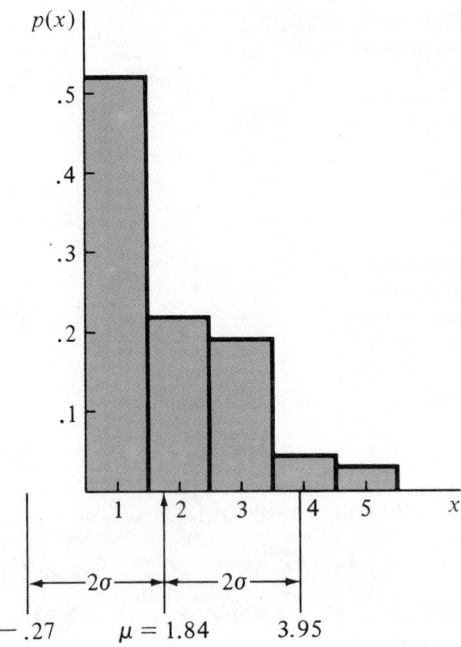

Figure 4.6   Probability Distribution for x in Example 4.7

three-day period is 4. Use this information to find an approximate value for x_a, the number of tetanus shots to stock for a three-day demand.

SOLUTION   The expected demand is given as 4. If the demand x follows a Poisson probability distribution (and we will assume that it does), we have

$$\sigma^2 = \mu = 4 \quad \text{and} \quad \sigma = \sqrt{4} = 2$$

From our knowledge of Tchebysheff's Theorem and the Empirical Rule, we would not expect x to fall more than three standard deviations from μ. In other words, we would not expect x to exceed

$$\mu + 3\sigma = 4 + 3(2) = 10$$

If we choose $x_a = 10$ as the number of shots to stock, we can be reasonably certain the demand x will not exceed this value.

Tips on Problem Solving

To find the variance of a random variable x, you must find the expected value of $(x - \mu)^2$. Start with a table containing four columns, the first for x, the second for $(x - \mu)^2$, the third for $p(x)$, and the fourth for the cross products, $(x - \mu)^2 p(x)$. First, calculate the value of $(x - \mu)^2$ for each value of x and enter these results in column 2. Then obtain the cross product $(x - \mu)^2 p(x)$ for each value of x and enter these results in column 4. The sum of column 4 will give the variance of x.

Alternatively, you can calculate the variance of x by using the formula

$$\sigma^2 = E(x^2) - \mu^2$$

which is less subject to rounding errors.

To find $E(x^2)$, follow the procedure described above, except enter the values of x^2, instead of $(x - \mu)^2$, in column 2 of the table. The sum of the elements in column 4 will give $E(x^2)$.

EXERCISES   Conceptual Level

4.24   Why might it be preferable to use the alternative formula, $\sigma^2 = E(x^2) - \mu^2$, in calculating the variance of x?

4.25   How can the mean μ and the variance σ^2 be used to describe the probability distribution of x?

EXERCISES   Skill Level

4.26   The random variable x given in Exercise 4.20 has the following probability distribution:

x	0	1	2	3
$p(x)$	.1	.2	.3	.4

   a.   Use the value of μ calculated in Exercise 4.20 to find the value of $(x - \mu)^2$ for each value of x.
   b.   Calculate σ^2, the variance of x.
   c.   Calculate σ.

4.27   Refer to Exercise 4.26. Use the alternative formula, $\sigma^2 = E(x^2) - \mu^2$, to calculate σ^2. Compare your results with those obtained in Exercise 4.26b.

4.28   The probability distribution for the random variable x described in Exercise 4.18 is reproduced below.

x	1.5	2.0	2.5	3.0	3.5
$p(x)$	.01	.49	.32	.15	.03

   a.   Which of the two formulas available for calculating σ^2 would be most appropriate in this situation?
   b.   Calculate σ^2 and σ.

c. Calculate the interval $(\mu \pm 2\sigma)$. Superimpose this interval on the graph of the probability distribution of x. What proportion of the measurements will fall within the interval $(\mu \pm 2\sigma)$? Does this result agree with Tchebysheff's Theorem?

4.29 The random variable x has the following probability distribution:

x	0	1	2	3	4
$p(x)$	.25	.30	.25	.15	.05

a. Find the expected value and variance of x.

b. Construct a graph of the probability distribution.

c. What is the probability that x exceeds $(\mu + 2\sigma)$? What is the probability that x exceeds $(\mu + 3\sigma)$?

EXERCISES   Applications

4.30 The extent of a firm's involvement in new-product research depends on the industries in which the firm competes. A survey was made of firms in high-technology fields to find the number of new products each firm introduced during 1988. From the survey the probability distribution shown in the table was determined for x, the number of new products introduced by each high-technology firm during 1988.

Number of New Products, x	3	4	5	6	7	8	9
$p(x)$	.08	.14	.22	.30	.14	.08	.04

a. Find the expected value and variance of x.

b. Give the proportion of high-technology firms for which x exceeds $(\mu + 2\sigma)$.

4.31 A sales representative for a firm that manufactures word-processing equipment can contact either one or two customers per day with probability 1/3 and 2/3, respectively. Each contact will result in no sale with probability .85, a $5,000 sale with probability .10, or a $10,000 sale with probability .05.

a. What is the expected value of the sales representative's daily sales?

b. What is the variance of her daily sales?

c. Suppose the manufacturing firm offers incentive bonuses to all sales representatives whose daily sales exceed $(\mu + 2\sigma)$. What is the minimal amount of daily sales volume the sales representative must achieve to qualify for a bonus?

4.32 Past experience has shown that the number of industrial accidents per month in a manufacturing plant has a Poisson probability distribution, with the expected value of x equal to 6. Use the information provided in Example 4.8 to give an upper limit to the number of accidents that we might expect in any month.

4.33 The manufacturer of a low-calorie dairy drink wishes to compare the taste appeal of a new formula (formula B) with that of the standard formula (formula A). Each of four judges is given three glasses in random order, two containing formula A and the other containing formula B.

Each judge is asked to state which glass he most enjoyed. Suppose that the two formulas are equally attractive. Let x be the number of judges stating a preference for the new formula.

a. Find the probability distribution for x.

b. What is the probability that at least three of the four judges state a preference for the new formula?

c. Find the expected value of x.

d. Find the variance of x.

<hr>

4.7

<hr>

Random Sampling

Since probability distributions are theoretical models for population relative frequency distributions, samples selected from populations can be viewed as observations on random variables. As noted in earlier chapters, the way a sample is selected from a population affects the quantity of information in the sample (the topic of Chapter 16) as well as the probabilities of observing particular samples. Since the probability of observed sample outcomes is the basis for inference making, a topic that is introduced in Chapters 5, 6, and 7, it is important that we define one of the most commonly employed sampling procedures at this point. It is called **simple random sampling**.

simple random sampling

In simple random sampling different samples of the same size in the population have an equal chance of being selected. To illustrate, suppose that we wish to select a sample of $n = 2$ from a population containing $N = 4$ elements (we are choosing a small value of N to simplify our discussion). If the four elements are identified by the symbols x_1, x_2, x_3, and x_4, then there are six different samples that could be selected from the population, as shown in Table 4.10. If the sample was selected in such a way that each of these six samples had an equal chance of selection (probability equal to 1/6), the sample would be called a simple random (or simply "random") sample.

It can be shown* that the number of ways of selecting $n = 2$ elements from a set of $N = 4$, denoted by the symbol C_2^4, is

$$C_2^4 = \frac{4!}{2!2!} = \frac{4 \cdot 3 \cdot 2 \cdot 1}{(2 \cdot 1)(2 \cdot 1)} = 6$$

Note: The symbol $n!$ (read "n factorial") is used to denote the product

$$n(n - 1)(n - 2) \cdots 3 \cdot 2 \cdot 1$$

Thus, $6! = 6 \cdot 5 \cdot 4 \cdot 3 \cdot 2 \cdot 1$. The quantity $0!$ is defined to be equal to 1.

In general, the number of ways of selecting n elements from a set of N is

$$C_n^N = \frac{N!}{n!(N - n)!}$$

<hr>

* A discussion of this result along with examples and applications is given in optional Section 3.9. An understanding of the derivation of the result is not essential to our discussion.

Table 4.10   Simple Random Sample

Sample	Observations in Sample
1	x_1, x_2
2	x_1, x_3
3	x_1, x_4
4	x_2, x_3
5	x_2, x_4
6	x_3, x_4

For example, if you wish to conduct an opinion poll of 5,000 people based on a sample of $n = 100$, there are C_{100}^{5000} different combinations of people who could be selected in the sample. If the sampling is conducted in such a way that each of these combinations has an equal probability of being selected, then the sample is called a simple random sample.

Definition

simple random
sample

> Let N and n represent the numbers of elements in the population and sample, respectively. If the sampling is conducted in such a way that each of the C_n^N samples has an equal probability of being selected, the sampling is said to be random and the resulting sample is said to be a **simple random sample**.

Perfect random sampling is difficult to achieve in practice. If the population is not too large, we might write each of the N numbers on a poker chip, mix the total, and select a sample of n chips. The numbers on the poker chips would specify the measurements to appear in the sample. Other techniques are available when the population is large. One of these, the use of a random number table, is discussed in Section 16.3.

In many situations the population is conceptual, as in an observation made during a laboratory experiment. Here, the population consists of the infinitely large number of measurements that would be generated if the experiment were repeated over and over again. If we wish a sample of $n = 10$ measurements from this population, we repeat the experiment 10 times and hope that the results represent, to a reasonable degree of approximation, a random sample.

Although the primary purpose of this discussion was to clarify the meaning of a random sample, we would like to mention that some sampling techniques are partly systematic and partly random. For instance, if we wish to determine the voting preference of the nation in a presidential election, we would not be likely to choose a random sample from the population of voters. Just because of pure chance, all the voters appearing in the sample might be drawn from a single city, say New York, which might not be at all representative of the population. We would prefer a random selection of voters from smaller political districts, perhaps states, allotting a specified number to each state. The information from the randomly selected subsamples drawn from the respective states would be combined to form a prediction concerning the entire population of voters in the country. This sampling procedure and others designed to decrease the cost of a specified amount of sample information are discussed in Chapter 16.

EXERCISES   Conceptual Level

4.34   Which of the following sampling techniques would produce a simple random sample?

   a.   All $n = 75$ members of a business law class are surveyed concerning their evaluation of a new judicial appointment.

   b.   People leaving a shopping mall who agree to be interviewed are queried concerning their brand preferences.

   c.   A telephone survey of $n = 100$ telephone customers having a fixed three-digit prefix is conducted by selecting the last four digits of a telephone number one at a time by drawing from 10 chips numbered $0, 1, 2, \ldots, 9$.

   d.   An energy questionnaire is included in every 100th utility bill sent out by a city in which utility bills are ordered by customer number.

EXERCISES   Skill Level

4.35   Evaluate these expressions.

   a. 6!   **b.** C_3^6 **c.** C_0^7 **d.** C_3^{10} **e.** C_1^{100}

4.36   How many different samples of $n = 2$ elements can be selected from $N = 5$? List them.

4.37   How many different samples of $n = 10$ could be selected from a population containing $N = 100$ elements?

EXERCISES   Applications

4.38   An auditor has been assigned the task of investigating the shipping invoices of a petroleum distribution facility to determine the extent of any recording errors that may exist. For the sake of time and efficiency the auditor will select a representative sample of $n = 10$ shipping invoices for review from among the $N = 120$ that the petroleum distribution facility retains on file. How many different combinations of invoices could be chosen by the auditor for review?

4.39   To estimate the costs of repairs and service for automatic washing machines sold by her store, the manager of an appliance store wishes to contact a sample of $n = 15$ from among the $N = 85$ individuals who have purchased an automatic washer from her store within the past 24 months. How many different samples could she select?

4.8

Summary

Random variables, representing numerical events defined over a sample space, may be classified as **discrete** or **continuous** random variables, depending on whether the number of sample points in the sample space is or is not countable. The theoretical population frequency distribution for a discrete random variable is called a **probability distribution** and may be derived by using the techniques of Chapter 3. The model for the frequency distribution of a continuous random variable is a mathematical function $f(x)$, called a **probability distribution** or **probability density function**. This function, usually a smooth curve, is defined over a line interval and is chosen so that the total area under the curve is equal to 1. The probabilities associated with a continuous random variable are given as areas under the probability distribution $f(x)$.

The **expected value of a random variable** is the average of that variable calculated for the theoretical population that is defined by its probability distribution. The standard deviation for a random variable can be interpreted in the same way as a sample standard deviation. Together with Tchebysheff's Theorem and the Empirical Rule, the standard deviation provides a measure of variability for a random variable.

As noted in earlier discussions, inferences about a population will be based on the probability of an observed sample, and this probability will depend on how the sample was selected from the population. The probability distribution for a random variable plays a role in inference making because it enables us to calculate the probability that a single observation, randomly selected from the population, will assume one or more specific values. The most common method for sampling more than one observation from a population is called simple random (or simply "random") sampling. A simple random sample is selected from a population in such a way that different samples of the same size in the population have an equal probability of selection. Simple random sampling will be used in Chapter 5 and in subsequent chapters in this text.

Supplementary Exercises

Applications

4.40   Identify the following as discrete or continuous random variables.

    a.   The number of defective light bulbs in a package containing four bulbs

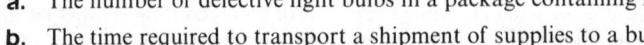
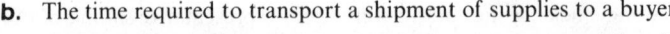

    b.   The time required to transport a shipment of supplies to a buyer

    c.   The weights of blue fin tuna sold each day to a commercial seafood cannery

    d.   The number of incoming telephone calls received during a day by an information operator for a large office building

    e.   The transaction price for single-family residences sold during a given week in Utica, New York

4.41   U.S. subcompact cars now account for 22% of all new-car sales in the United States [U.S. Department of Commerce, Bureau of the Census, *Statistical Abstract of the United States*,

107th ed. (Washington, D.C., 1987), p. 584]. Four new-car buyers are selected at random from a list supplied by a state automobile-licensing agency. Find the probability distribution for x, the number of new-car buyers who have purchased a U.S. subcompact car.

 4.42 An investor has decided to invest his money in three different stocks. He has his choice narrowed down to four common stocks and two preferred stocks. Unable to choose among these, he randomly chooses three stocks from the six. Find the probability distribution of x, the number of preferred stocks he chooses. Construct a graph of the probability distribution for x.

 4.43 Simulate the experiment described in Exercise 4.42 by marking six pieces of cardboard, or coins, so that four represent common stocks and two represent preferred stocks. Place them in a box, mix, draw three, and record x, the number of preferred stocks observed. Replace the three you have drawn and repeat the process until a total of $n = 100$ observations on x have been recorded. Construct a relative frequency histogram for this sample and compare it with the population probability distribution you constructed in Exercise 4.42. If the sampling process were repeated $n = 100,000$ times (instead of $n = 100$) and you constructed a relative frequency histogram for the data, how would your histogram compare with the population probability distribution of Exercise 4.42?

 4.44 Refer to Exercise 4.42. Find the expected value and variance of x. If you performed the simulation in Exercise 4.43, calculate $\bar{x}$ and s^2 for the 100 measurements. Compare these values with the expected value and variance of x.

 4.45 Refer to Exercise 4.43. Calculate σ and find the probability that x will lie within 2σ of μ. Does this result agree with Tchebysheff's Theorem? Do μ and σ characterize $p(x)$? What proportion of the 100 measurements lie within $2s$ of $\bar{x}$?

 4.46 To verify the accuracy of their financial accounts, companies utilize auditors on a regular basis to check accounting entries. Suppose that the company's employees make erroneous entries 5% of the time. Suppose also that an auditor randomly checks three entries.

    a. Find the probability distribution for x, the number of errors detected by the auditor.

    b. Construct a probability histogram for $p(x)$.

    c. Find the probability that the auditor will detect more than one error.

 4.47 The American economy is characterized by its mass purchasing power, to a great extent because of the rapid increase in family income we have enjoyed. In 1950 only 3% of American families had annual incomes exceeding $25,000, while records show that about 43% currently have incomes in excess of $25,000 [U.S. Department of Commerce, Bureau of the Census, *Statistical Abstract of the United States*, 107th ed. (Washington, D.C., 1987), p. 435]. Suppose that in a typical American city 43% of all families have incomes in excess of $25,000. In a consumer expenditure study in this city a sample of four families is randomly selected from a city directory in order to study the purchasing habits of the families. Find the probability distribution for x, the number of families in the sample whose current annual income exceeds $25,000. Construct a probability histogram for x.

 4.48 Competitive bids are required for the purchase of most services by governmental agencies. Suppose past experience has shown that a certain government contractor wins, on the average, three of every five contracts on which he submits a bid. Let x be the number of bid submissions until a contract is won by the contracting firm, and assume that successive bids are independent of one another.

    a. Find $p(1)$, $p(2)$, and $p(3)$.

    b. Give a formula for $p(x)$.

    c. Graph $p(x)$.

 4.49 Refer to Exercise 4.48. The probability distribution for x, the number of submissions until a contract is won, is known as a *geometric probability distribution*. It is known in the accompanying table, with calculations rounded to the nearest thousandth. Individual values of $p(x)$ for x larger than eight are omitted, but their total probability is .005.

x	$p(x)$	x	$p(x)$
1	.6	6	$(.4)^5(.6) = .006$
2	$(.4)(.6) = .24$	7	$(.4)^6(.6) = .002$
3	$(.4)^2(.6) = .096$	8	$(.4)^7(.6) = .001$
4	$(.4)^3(.6) = .038$	>8	.005
5	$(.4)^4(.6) = .015$		

a. It can be shown that for the distribution in the table $\mu = 1.67$ and $\sigma = \sqrt{10/9}$. Construct a graph of $p(x)$ by using these values. Locate μ on the graph.

b. Find the probability that x will be more than two standard deviations from its mean. Confirm that this result agrees with Tchebysheff's Theorem.

c. Find the probability that x will be more than three standard deviations from its mean. Confirm that this result agrees with Tchebysheff's Theorem.

d. Compare your answers to parts b and c with the approximate values given by the Empirical Rule. Note that despite the fact that $p(x)$ is highly skewed to the right, the Empirical Rule provides a satisfactory description of the variability of the random variable x.

MKTG   **4.50**   On the basis of past experience, a clothing manufacturer has developed the accompanying probability distribution for x, the length (in years) of the fashion period for women's dresses. Find the expected value and variance of x.

x	1	2	3	4
$p(x)$	1/12	3/12	4/12	4/12

MKTG   **4.51**   Refer to Exercise 4.50. Construct a probability histogram for x. Calculate σ and find the probability that x will lie within two standard deviations (2σ) of μ. Does this result agree with Tchebysheff's Theorem? Do μ and σ characterize $p(x)$?

MKTG   **4.52**   A sales agent is provided a \$50 commission in addition to her salary for each new energy-efficient convection oven she sells during a one-month introductory period. Based upon the response to the convection oven when introduced in other geographic areas, the distribution of daily sales x of the convection oven is defined as shown in the accompanying table.

x	0	1	2	3	4
$p(x)$	.80	.10	.05	.03	.02

a. Find the expected daily sales of convection ovens.

b. What is the sales agent's expected daily commission from convection oven sales?

4.53   A corporate personnel director wishes to select three of five candidates for three managerial trainee positions; all candidates should have the same chance of being selected. Three of the candidates are members of racial minority groups and two are not. Let x be the number of minority candidates appearing in the personnel director's selection, that is, $x = 1, 2$, or 3. Find the probability distribution for x. Construct a probability histogram for x. What is the probability that he selects at least two of the three minority group members for the managerial trainee positions?

4.54 Although nuclear power has the potential to solve many of our future energy needs, political considerations have stalemated the development of future nuclear power generating facilities. In a certain city that is considering public bonding for the construction of a nuclear power generating facility, five people have applied for two vacancies on the city's Power Advisory Commission. Although all candidates appear equally attractive to the city manager (who must fill the vacancies), three favor nuclear development and two are opposed. Let x be the number of nuclear opponents in the city manager's selection, that is, $x = 0, 1, 2$. If two candidates are selected at random from the five applicants, find the probability distribution for x. Construct a probability histogram for x. What is the probability that the city manager will select one nuclear supporter and one opponent?

4.55 A mail-order magazine subscription service receives orders for one-, two-, and three-year subscriptions, with probabilities 1/2, 1/4, and 1/4, respectively. For each subscription year it receives a $1 commission. What is the subscription service's expected commission for each order received?

4.56 A potential customer for a $100,000 fire insurance policy has a home in an area that, according to experience, may sustain a total loss in a given year with a probability of .001 and a 50% loss with a probability of .01. If you ignore all other partial losses, what premium should the insurance company charge for a yearly policy in order to break even?

4.57 Since farming is a very risky business owing to the uncertainties of the weather, many farmers now insure their crops each year against possible losses due to rain, hail, flood, wind, or freezing temperatures. Determine the annual premium for an insurance policy to cover a farmer's $600,000 wheat crop. Insurance actuaries suggest that chances are 1 in 50 that the crop will be destroyed by adverse weather conditions.

4.58 Although much of the world's oil supply remains undiscovered, the most accessible reserves were explored first, leaving only those in the harsh, remote areas as offering the greatest current opportunity for successful discovery. In Iceberg Alley off Canada's east coast, geologists report that successful, commercially profitable deposits are discovered in two out of three drilling efforts. Suppose Exxon, which holds several leases in that area, engages in five independent drilling efforts. Let x be the number of successful, commercially profitable deposits it discovers.

a. Find the probability distribution for $p(x)$.

b. Find the expected value and variance of x.

4.59 In a county containing many rural homes, 60% are thought to be insured against fire. Three rural homeowners are chosen at random from the entire population and a number x are found to be insured against a fire. Find the probability distribution for x. What is the probability that at least two of the three will be insured?

4.60 If you own common stock in a corporation, you can sometimes increase the return on your investment by selling an option. For a stated price the purchaser of the option gains the right to buy your stock at any time up to a specified expiration date. Suppose you purchased 200 shares of a stock at $25 per share and you sell an option for the option purchaser to buy the stock at any time within the next six months for $30 per share. If the stock reaches $30 per share within the next six months, your stock will be sold and you will gain $5 per share plus $2 per share (from the dividends and the sale of the option) less $109 commission to the broker for selling your stock. If you do not sell, you will gain $2 per share. If the probability of the stock reaching $30 per share within the next six months is .7, what is the expected return (in dollars) from your 200 shares of stock? What is the expected annual rate of return, in percentage, on your investment? (*Note:* We have ignored a $15 commission on the sale of the option.)

Supplementary Exercises

Interpretive

4.61 When investors or supervisors base decisions on variations in the records of funds or employees, they may attach too much importance to short-term performance. A slump or a single winning streak can be a true sign of deterioration or improvement, but this may not always be the case. Edward M. Purcell, a Harvard Nobel laureate, demonstrated that a prolonged business slump may be no more than a run of bad luck. He invented a baseball player with a lifetime batting average of .250 (this player averages one hit for every four times at bat), and he used computer simulation to track this player's career. This imaginary player suffered slumps as long as no hits in 15 times at bat (*Money*, February 1985, p. 12). Assume consecutive times at bat for a player with a lifetime batting average of .250 are independent.

a. What is the probability that this player's first hit occurs during his first time at bat?

b. What is the probability that this player's first hit occurs during his second time at bat?

c. Find the probability distribution for x, the number of times at bat until the player gets his first hit. Graph the probability distribution of x, for $x = 0, 1, 2, \ldots$.

d. What is the probability of suffering a batting slump of no hits in 15 or more times at bats? What implication does this result have for the interpretation of a business slump?

4.62 Risk management is an attempt to quantify the speculative and nonspeculative aspects of uncertain ventures. Usually, a subjective assessment of the probability of occurrence of each possible outcome in an uncertain venture is used as a measure of risk (E. G. Miller and H. W. Rubin, "The Changing Qualitative and Quantitative Approach to Risk Analysis," *Risk Management*, November 1979). Consider a situation faced by a manufacturer who has been offered three different facilities for use as a storage warehouse. Suppose the rental fees for the three are comparable, but uncertainties regarding future energy supplies and transportation to the region of each warehouse suggest that the distributions shown in Table 4.11 are descriptive of potential losses, and risks, associated with each selection. (Losses are projected additional costs and losses on a monthly basis.) Find the expected monthly loss for each warehouse. If the manufacturer wishes to minimize his expected monthly loss, which warehouse should he choose? Can you imagine a circumstance whereby the manufacturer might reject warehouse I apart from expected loss considerations?

Table 4.11   Loss Data for Exercise 4.62

Warehouse I		Warehouse II		Warehouse III	
Loss	Probability	Loss	Probability	Loss	Probability
$ 0	.90	$ 0	.25	$ 0	.80
500	.08	250	.50	100	.15
1500	.02	500	.25	250	.05

Case Study

Capital Budgeting Criteria

Estimated benefits from capital projects tend to be optimistic, and net present values of completed projects are often below those originally estimated. Over 80% of surveyed financial officers of *Fortune* 500 companies felt that revenue forecasts are typically overestimated. Therefore, an increasingly important task for the business strategist is to understand these biases and make decisions when basic estimates are known to be in error. Edward M. Miller ("The Competitive Market Assumption and Capital Budgeting Criteria," *Financial Management*, Winter 1987, pp. 22–28) proposes that economic theory concerning expected return from new investments has practical implications for capital budgeting and that the budgeting process should be viewed as a Bayesian decision problem.

For example, suppose that a project consists of finding a store location for which the present value of operating profits exceeds the cost of establishing the store, and that the cost of opening a new store is $90,000. Furthermore, suppose that only 4% of the proposed store sites will be good ones, 48% will be medium ones, and another 48% will be poor ones. The actual present value of the operating profits of a good project is $200,000, that of a medium project is $100,000, and that of a poor project is zero.

1. If 25% of all projects are overestimated, 50% correctly evaluated, and 25% underestimated, use Table 4.12 to find the expected value of the estimated present value of the operating profits; find the expected value of the actual present value of operating profits.

2. What is the difference in these two expected values? What are the implications for the decision maker in this situation?

3. If a project is estimated to have a present value of operating profits equal to $100,000, what is the probability that the project

Table 4.12   Present-Value Data

	Project		
Present Value	Good (.04)	Medium (.48)	Poor (.48)
Actual (.50)	$200,000	$100,000	0
Overestimate (.25)	300,000	200,000	$100,000
Underestimate (.25)	100,000	0	−100,000

is, in fact, a good one? What is the probability that the project is, in fact, a medium one? A poor one? Use these probabilities to find the expected present value of a project whose estimated present value is $100,000. On the average, will a project whose estimated present value is $100,000 cover the $90,000 costs of opening a new store?

References and Suggested Readings

Mendenhall, W., R. L. Scheaffer, and D. Wackerly. *Mathematical Statistics with Applications.* 3d ed. Boston: PWS-KENT Publishing Co., 1986.

Mosteller, F. R. E., K. Rourke, and G. B. Thomas Jr. *Probability with Statistical Applications.* 2d ed. Reading, Mass.: Addison-Wesley, 1970, Chap. 5.

Neter, J., W. Wasserman, and G. A. Whitmore. *Applied Statistics.* 3d ed. Boston: Allyn & Bacon, 1987.

Summers, G. W., W. S. Peters, and C. P. Armstrong. *Basic Statistics in Business and Economics.* 4th ed. Belmont, Calif.: Wadsworth, 1985.

TO
THE
READER

In Chapter 4 random variables were classified as being either continuous or discrete. In many business applications the variable of interest to a researcher is discrete, and its probability distribution can often be found by using the concepts of probability presented in Chapter 3.

The probability distributions for three discrete random variables that frequently arise in the areas of management and economics are the subject of Chapter 5. We examine situations in which a particular one of these random variables and its probability distribution can be used to adequately describe the outcomes associated with an experiment. The binomial, hypergeometric, and Poisson random variables presented in this chapter are frequently encountered when the random variable of interest results from recording the number of occurrences of a specified event in a fixed number of trials or a fixed unit of time or space. The binomial random variable is used to illustrate the role that the probability distribution plays in the process of making an inference about a population parameter using the information in a sample.

5

THREE USEFUL
DISCRETE PROBABILITY
DISTRIBUTIONS

5.1

Introduction

discrete random
variables

In Chapter 4 we found that random variables defined over a finite or countably infinite number of points are called **discrete random variables**. Examples of discrete random variables abound in business and economics, but three discrete probability distributions serve as *models* for a large number of these applications. These three distributions are **the binomial, the Poisson, and the hypergeometric probability distributions.** In this chapter we will study these distributions, discussing their development as logical models for discrete processes observed in different business settings.

Throughout the study of Chapter 5 we will refer to Chapter 4 and the definition of a probability distribution—a formula or model that assigns a probability to each possible numerical outcome of an experiment. Thus, the nature of the experiment itself and the numerical outcomes of the experiment must be considered before selecting the appropriate probability distribution that will serve as a model for the process.

5.2

The Binomial Experiment

One of the most elementary, useful, and interesting discrete random variables, the binomial random variable, is associated with the coin-tossing experiment described in Examples 3.5 and 4.1. In this experiment either one coin is tossed n times or n coins are each tossed once. One observation, "head" or "tail," is then recorded for each toss. In an abstract sense, numerous coin-tossing experiments of practical importance are conducted daily in the social sciences, physical sciences, and industry.

As an illustration, consider a sample survey conducted to predict voter preference in a political election. Interviewing a single voter bears a similarity, in many respects, to tossing a single coin, because the voter's response may be in favor of our candidate—a "head"—or it may be against (or indicate indecision)—a "tail." In most cases the fraction of voters favoring a particular candidate does not equal one-half, but even this similarity to the coin-tossing experiment is satisfied in national presidential elections. History demonstrates that the fraction of the total vote favoring the winning presidential candidate in most national elections is very near one-half.

Similar polls are conducted in the social sciences, in industry, and in education. The sociologist is interested in the fraction of rural homes that have been electrified; the soft-drink manufacturer wishes to know the fraction of consumers who prefer his brand; the teacher is interested in the fraction of students who pass the course. Each person sampled is analogous to the toss of an unbalanced coin for which the probability of a head is not one-half.

Firing a projectile at a target is similar to a coin-tossing experiment if the outcome "hit the target" and the outcome "miss the target" are regarded as a head and a tail, respectively. A single missile results in a successful or an unsuccessful launching. A new drug is effective or ineffective when administered to a single patient. A manufactured item selected from a production line is defective or nondefective. With each contact either a salesperson will consummate a sale or no sale will result. Although dissimilar in some respects, the experiments described above will often exhibit, to a reasonable degree of approximation, the characteristics of a *binomial experiment*.

Definition

binomial experiment

> A **binomial experiment** is an experiment that possesses the following properties:
>
> 1. The experiment consists of n identical trials.
> 2. Each trial results in one of two outcomes. For lack of a better nomenclature, we will call one outcome a success, S, and the other a failure, F.
> 3. The probability of success on a single trial is equal to p and remains the same from trial to trial. The probability of a failure is equal to $(1 - p) = q$.
> 4. The trials are independent.
> 5. The experimenter is interested in x, the number of successes observed during the n trials.

EXAMPLE 5.1   Suppose that there are approximately 1 million potential buyers for a manufacturer's product and that an unknown proportion p favor the product over all its competitors. A sample of 1,000 is selected in such a way that every one of the 1 million buyers has an equal chance of being selected, and each potential buyer is asked whether he or she prefers the manufacturer's product over all its competitors. (The ultimate objective of this market survey is to estimate the unknown proportion p, a problem that we will discuss in Chapter 8.) Is this a binomial experiment?

SOLUTION   To decide whether this is a binomial experiment, we must see whether the sampling satisfies the five characteristics described in the preceding definition.

1. The sampling consists of $n = 1,000$ identical trials. One trial represents the selection of a single person from the 1 million potential buyers.
2. Each trial results in one of two outcomes: a person prefers the product (call this a success) or does not (a failure).
3. The probability of a success is equal to the proportion of potential buyers. For example, if 400,000 of the 1 million potential buyers favor the product, then the probability of selecting a person favoring the product out of the 1 million potential buyers is $p = .4$. For all practical purposes, this probability will remain the same from trial to trial, even though persons selected in the earlier trials are not replaced as the sampling continues.
4. For all practical purposes, the probability of a success on any one trial is unaffected by the outcome on any of the others (it will remain very close to $p = .4$).
5. We are interested in the number x of people in the sample of 1,000 who favor the manufacturer's product.

Because the survey satisfies the five characteristics reasonably well, for all practical purposes, it can be viewed as a binomial experiment.

EXAMPLE 5.2   A purchaser who has received a boxcar containing 20 large computers wishes to sample 3 of the computers to see whether they are in working order before he unloads the shipment. The 3 nearest the door of the boxcar are removed for testing and, afterward, are declared either defective or nondefective. Unknown to the purchaser, 2 of the 20 computers are defective. Is this a binomial experiment?

SOLUTION   As in Example 5.1, we check the sampling procedure against the characteristics of a binomial experiment.

1. The experiment consists of $n = 3$ identical trials. Each trial represents the selection and testing of one computer from the total of 20.
2. Each trial results in one of two outcomes: Either a computer is defective (call this a success) or it is not (a failure).
3. Suppose that the computers were randomly loaded into the boxcar so that any one of the 20 computers could have been placed near the boxcar door. Then the unconditional probability of drawing a defective computer on a given trial is 2/20.
4. The condition of independence between trials *is not* satisfied because the probability of drawing a defective computer on the second and third trials is dependent on the outcome of the first trial. For example, if the first trial results in a defective computer, then there is only one defective left in the remaining 19 computers in the boxcar. Therefore, the conditional probability of success on

trial 2, given a success on trial 1, is 1/19. This probability differs from the unconditional probability of a success on the second trial (which is 2/20). Therefore, the trials are dependent and the sampling does not represent a binomial experiment.

Example 5.1 illustrates an important point. Very few real-life situations will completely satisfy the requirements for a binomial experiment, but this is of little consequence as long as the lack of agreement is moderate and does not affect the end result. For instance, the probability of drawing a buyer favoring a particular product in a marketing research survey remains approximately constant from trial to trial as long as the population of buyers is relatively large in comparison with the sample. If 50% of a population of 1,000 buyers prefer product A, then the probability of drawing an A on the first interview is 1/2. The probability of an A on the second draw is 499/999 or 500/999, depending on whether the first draw was favorable or unfavorable to A. Both are near 1/2 and would continue to be for the third, fourth, and nth trial as long as n is not too large. Hence, $P(A)$ remains approximately 1/2 from trial to trial, and, for all practical purposes, we could regard the trials as independent. Thus, when the sample is small in comparison with the number of elements in the population from which it is selected, the market survey will represent a binomial experiment. On the other hand, if the number of buyers in the population is 10, and 5 favor product A, then the probability of A on the first trial is 1/2; the probability of A on the second trial is 4/9 or 5/9, depending on whether A was or was not drawn on the first trial. For small populations the probability of A will vary appreciably from trial to trial, independence will not exist, and the resulting experiment will not be a binomial experiment.

EXERCISES Conceptual Level

5.1 Which of the following business problems can be modeled by the binomial distribution? For those that cannot be modeled by the binomial distribution, explain why.

a. Determination of the probability that a customer files a claim against a product warranty when past records show that 5% of all buyers file warranty claims

b. Determination of the probability that a manufacturer receives no defective assembly machines in a shipment of three machines from a firm that has three defective and seven nondefective assembly machines in its warehouse

c. Estimation of the probability that the gasoline mileage on a new car will exceed 25 miles per gallon if EPA reports suggest the average mileage rating for such a car is 28 miles per gallon

d. Computation of the probability that at least 20 people respond to the mailing of 100 advertising circulars when the usual response rate is 15%

e. Determination of the probability that no more than 1 of 10 items of machine output is defective when the items are selected over time and you know that the defective rate increases with excessive machine wear over time

5.2   Repeat the instructions for Exercise 5.1 for the following business problems.

   a.   Ten mill workers are chosen at random from a lumber manufacturing plant in which 70 mill workers are union members and 38 are not. Management wishes to determine the probability that at least 5 nonunion members are chosen in the sample.

   b.   A retailer currently has 35 almond, 27 bronze, and 24 white refrigerators in stock. Past records indicate that the store sells about 40 refrigerators per month. For planning purposes the manager would like to know the probability that demand for white refrigerators during the coming month will not exceed current supply in stock.

   c.   The shipping agent for a soft drink bottling company would like to determine the probability that no more than 1 bottle per case (24 bottles) is damaged in shipment. Past records indicate that the probability of a bottle being damaged during shipment is .03.

   d.   An insurance agency maintains a list of seven part-time secretaries whom they contact during periods of excessive work load. The manager of the agency randomly chooses three names from the list to contact for part-time duty. She would like to know the probability that at least two of the three secretaries she contacts will not be obligated elsewhere and will be able to assist her agency temporarily.

   e.   A large bin contains 10,000 standard No. 12 metal screws, 368 of which are defective and may give way under stress. An assembler randomly selects 100 metal screws from the bin and wishes to know the chances that no more than 2 are defective.

5.3   Why do we sometimes use the binomial experiment for computing probabilities when sampling from a finite set of objects? What are the conditions under which we are justified in using the binomial distribution in such a case?

5.3

The Binomial Probability Distribution

Having defined the binomial experiment and suggested several practical applications, we now turn to a derivation of the probability distribution for the random variable x, the number of successes observed in n trials. Rather than attempt a direct derivation, we will obtain $p(x)$ for experiments containing $n = 1, 2,$ and 3 trials and appeal to your intuition in presenting the general formula.

For $n = 1$ trial we have two sample points, E_1 representing a success S and E_2 representing a failure F, with probabilities p and $q = (1 - p)$, respectively. Since x is the number of successes for the one ($n = 1$) trial and since E_1 implies a success, we assign $x = 1$ to E_1. Similarly, since E_2 represents a failure for the single trial, we assign $x = 0$ to this sample point. The resulting probability distribution for x is given in Table 5.1.

The probability distribution for an experiment consisting of $n = 2$ trials is derived in a similar manner and is presented in Table 5.2. The four sample points associated with the experiment are presented in the first column; the notation SF in the second column denotes a success on the first trial and a failure on the second.

The probabilities of the sample points are easily calculated because each point is an intersection of two independent events, the outcomes of the first and second trials.

Table 5.1 $p(x)$ for a Binomial Experiment When $n = 1$

Sample Points	Outcomes*	$P(E_i)$	x		x	$p(x)$
E_1	S	p	1		0	q
E_2	F	q	0		1	p
					$\sum\limits_{x=0}^{1} p(x) = q + p = 1$	

*S represents a success on a single trial; F denotes a failure.

Table 5.2 $p(x)$ for a Binomial Experiment When $n = 2$

Sample Points	Outcomes	$P(E_i)$	x		x	$p(x)$
E_1	SS	p^2	2		0	q^2
E_2	SF	pq	1		1	$2pq$
E_3	FS	qp	1		2	p^2
E_4	FF	q^2	0			
					$\sum\limits_{x=0}^{2} p(x) = (q + p)^2 = (1)^2 = 1$	

Thus, $P(E_i)$ can be obtained by applying the Multiplicative Law of Probability:

$$P(E_1) = P(SS) = P(S)P(S) = p^2 \qquad P(E_3) = P(FS) = P(F)P(S) = qp$$
$$P(E_2) = P(SF) = P(S)P(F) = pq \qquad P(E_4) = P(FF) = P(F)P(F) = q^2$$

The value of x assigned to each sample point is given in the fourth column. Note that the numerical event $x = 0$ contains sample point E_4, the event $x = 1$ contains sample points E_2 and E_3, and the event $x = 2$ contains sample point E_1. The probability distribution $p(x)$, presented to the right in Table 5.2, reveals a most interesting consequence: The probabilities $p(x)$ are terms of the expansion of $(q + p)^2$.

Summing, we obtain

$$\sum\limits_{x=0}^{2} p(x) = q^2 + 2pq + p^2 = (q + p)^2 = 1$$

In general, the probability distribution for a binomial experiment consisting of n trials is obtained by expanding $(q + p)^n$. The proof of this statement is omitted, but we will present further evidence to justify this claim by deriving the probability distribution for a binomial experiment consisting of $n = 3$ trials. The computations for the derivation are presented in Table 5.3. Notice that $p(0)$, $p(1)$, $p(2)$, and $p(3)$ are terms of the expansion of $(q + p)^3$.

Since the probability associated with a particular value of x is the term involving p to the power x in the expansion of $(q + p)^n$, we use a result from algebra that involves

Table 5.3 $p(x)$ for a Binomial Experiment When $n = 3$

Sample Points	Outcomes	$P(E_i)$	x		x	$p(x)$
E_1	SSS	p^3	3		0	q^3
E_2	SSF	p^2q	2		1	$3pq^2$
E_3	SFS	p^2q	2		2	$3p^2q$
E_4	FSS	p^2q	2		3	p^3
E_5	SFF	pq^2	1			
E_6	FSF	pq^2	1			
E_7	FFS	pq^2	1			
E_8	FFF	q^3	0			

$$\sum_{x=0}^{3} p(x) = (q + p)^3 = 1$$

the expansion of a series so that we can write the probability distribution for the binomial experiment as given in the following display.

Binomial Probability Distribution

$$p(x) = C_x^n p^x q^{n-x} = \frac{n!}{x!(n-x)!} p^x q^{n-x}$$

where x may take values $0, 1, 2, 3, 4, \ldots, n$ and C_x^n is a symbol used to represent the expression

$$\frac{n!}{x!(n-x)!}$$

Recall from Section 3.9 that the factorial notation $n!$ is a short way of writing $n(n-1)(n-2)\cdots(3)(2)(1)$. Thus $3! = 3 \cdot 2 \cdot 1$, and $0!$ is defined to be equal to 1. The notation C_x^n is a short way of writing

$$\frac{n!}{x!(n-x)!}$$

This notation is useful because it occurs so often when working with the binomial probability distribution.

In the formula for $p(x)$ the quantity $p^x q^{n-x}$ represents the probability of observing a sample point with x successes and $(n-x)$ failures; C_x^n counts the number of such sample points. This pattern can be observed in Tables 5.1, 5.2, and 5.3. For example, the $C_2^3 = 3$ outcomes E_2, E_3, and E_4 result in $x = 2$ successes and $(n-x) = 1$ failure, each with probability p^2q. Hence, $p(3) = 3p^2q$.

Graphs of three binomial probability distributions are shown in Figure 5.1, the first for $n = 10, p = .1$, the second for $n = 10, p = .5$, and the third for $n = 10, p = .9$.

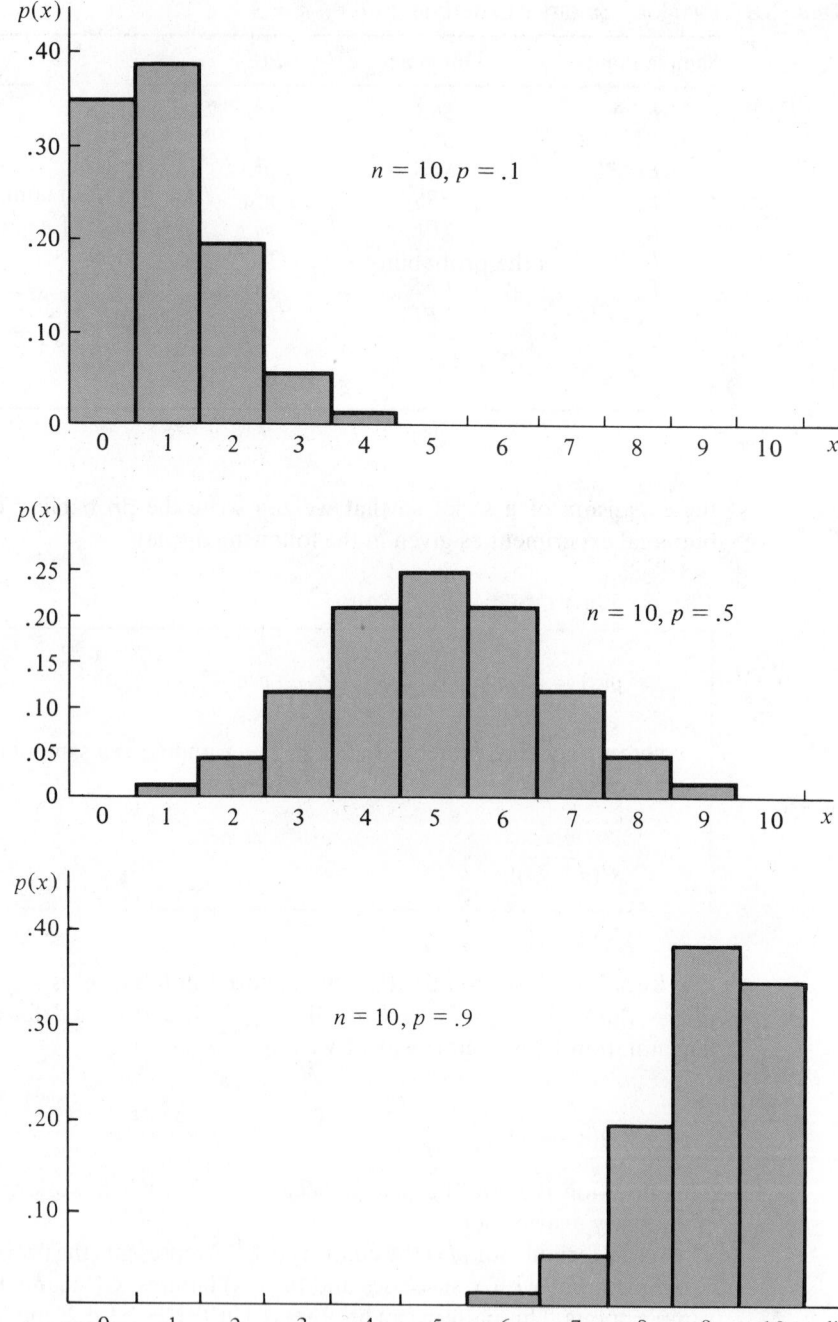

Figure 5.1   Examples of Binomial Probability Distributions

EXAMPLE 5.3   A study concerning the relative influence of husbands and wives in consumer purchasing, which appeared in a recent issue of *Time* magazine, reported that the husband exerts the primary influence in selecting the make of a new automobile in about 70% of all new-car purchases by families. Suppose four families have decided to buy a new car.

     a.   What is the probability that in *exactly* two of the four families the husband will exert the primary influence in choosing the make of a car?

     b.   What is the probability that the husband will exert the primary influence in choosing the make of a car in *at least* two of the four families?

     c.   What is the probability that the husband will select the make of car in all four families?

SOLUTION   Assuming that the family purchase decisions are independent and that p remains constant from one family to another, then $n = 4$ and $p = .7$. Let x denote the number of families in which the husband exerts the primary influence in selecting a new automobile. Then for $x = 0, 1, 2, 3, 4$, we have

$$p(x) = C_x^4(.7)^x(.3)^{4-x}$$

a. $p(2) = C_2^4(.7)^2(.3)^2 = \dfrac{4!}{2!2!}(.49)(.09) = .2646$

The probability is .2646 that in exactly two of the four families the husband will exert the primary influence in choosing the make of a car.

b. $P(\text{at least two}) = P(x \geq 2) = p(2) + p(3) + p(4)$
$$= 1 - p(0) - p(1)$$
$$= 1 - C_0^4(.7)^0(.3)^4 - C_1^4(.7)^1(.3)^3$$
$$= 1 - .0081 - .0756 = .9163$$

The probability is .9163 that the husband selects the make of car in at least two of the families.

c. $p(4) = C_4^4(.7)^4(.3)^0 = \dfrac{4!}{4!0!}(.7)^4(1) = .2401$

The probability is .2401 that the husband selects the make of car in all four families.

Note that these probabilities would be incorrect if the members of one family were in any way influenced by the purchase decision of one of the other three families. In that case the purchase decisions (trials) would be dependent upon one another and p would very likely be different from family to family.

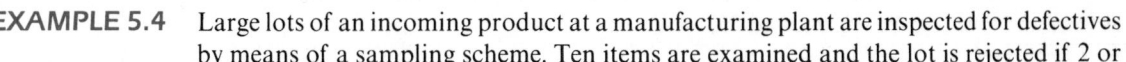

EXAMPLE 5.4   Large lots of an incoming product at a manufacturing plant are inspected for defectives by means of a sampling scheme. Ten items are examined and the lot is rejected if 2 or

more defectives are observed. If a lot contains exactly 5% defectives, what is the probability that the lot is accepted? Rejected? Assume independence between successive draws from the lot.

SOLUTION   Let x be the number of defectives observed. Then $n = 10$, and the probability of observing a defective on a single trial is $p = .05$. Hence,

$$p(x) = C_x^{10}(.05)^x(.95)^{10-x}$$

and

$$P(\text{accept}) = p(0) + p(1) = C_0^{10}(.05)^0(.95)^{10} + C_1^{10}(.05)^1(.95)^9$$
$$= .599 + .315 = .914$$
$$P(\text{reject}) = 1 - P(\text{accept}) = 1 - .914 = .086$$

◆ ◆

EXAMPLE 5.5   A taste-testing experiment was performed to compare a new sugar substitute with a sugar substitute already in widespread use. Each of 10 people who used both substitutes to sweeten their food and drinks was asked to state a preference for one of the two sugar substitutes with respect to its desirability as a sweetener as well as its lack of a bitter aftertaste. Eight of the 10 expressed a preference for the new sugar substitute. Suppose that the sugar substitutes do not differ with respect to the desired attributes, so that the probability that a person chooses the new sugar substitute is .5. Further assume that one person's preference is independent of any other person's preference. What is the probability of observing 8 or more preferences for the new sugar substitute when, in fact, the two substitutes are equally desirable?

SOLUTION   Assuming that the sugar substitutes are equally desirable, the probability of preferring the new substitute is $p = .5$. The probability distribution for x, the number of preferences for the new substitute, is

$$p(x) = C_x^{10}(.5)^x(.5)^{10-x} = C_x^{10}(.5)^{10}$$

Then we have

$$P(8 \text{ or more}) = p(8) + p(9) + p(10)$$
$$= C_8^{10}(.5)^{10} + C_9^{10}(.5)^{10} + C_{10}^{10}(.5)^{10}$$
$$= .0439 + .0098 + .0010 = .0547$$

◆ ◆

As illustrated in Examples 5.4 and 5.5, calculating binomial probabilities becomes a tedious task when n is large. To simplify our calculations, the sum of the binomial probabilities from $x = 0$ to $x = a$ is presented in Table 1 in the Appendix, for $n = 5$, 10, 15, 20, and 25.

Table 5.4   Portion of Table 1(a) for $n = 5$

							p							
a	0.01	0.05	0.10	0.20	0.30	0.40	0.50	0.60	0.70	0.80	0.90	0.95	0.99	a
0	—	—	—	—	—	—	—	—	—	—	—	—	—	0
1	—	—	—	—	—	—	—	—	—	—	—	—	—	1
2	—	—	—	—	—	—	—	—	—	—	—	—	—	2
3	—	—	—	—	—	—	—	.663	—	—	—	—	—	3
4	—	—	—	—	—	—	—	—	—	—	—	—	—	4

To illustrate the use of Table 1, suppose we wish to find the sum of the binomial probabilities from $x = 0$ to $x = 3$ for $n = 5$ trials and $p = .6$. That is, we wish to find

$$P(x \le 3) = \sum_{x=0}^{3} p(x) = p(0) + p(1) + p(2) + p(3)$$

where

$$p(x) = C_x^5 (.6)^x (.4)^{5-x}$$

Since the table values give

$$P(x \le a) = \sum_{x=0}^{a} p(x)$$

we seek the table value in the row corresponding to $a = 3$ and the column for $p = .6$. The table value, .663, is shown in Table 5.4 as it appears in Table 1(a). Therefore, the sum of the binomial probabilities from $x = 0$ to $x = a = 3$ (for $n = 5, p = .6$) is .663.

Table 1 can also be used to find individual binomial probabilities. For example, suppose we wish to find $p(3)$ when $n = 5$ and $p = .6$. Since $P(x = 3) = P(x \le 3) - P(x \le 2)$, we write

$$p(3) = \sum_{x=0}^{3} p(x) - \sum_{x=0}^{2} p(x) = .663 - .317 = .346$$

The values $\sum_{x=0}^{3} p(x)$ and $\sum_{x=0}^{2} p(x)$ are found directly from Table 1(a) with $p = .6$. In general, an individual binomial probability may be found by subtracting successive entries in the table for a given value of p.

EXAMPLE 5.6   Refer to Example 5.5. Use Table 1 in the Appendix to calculate the probability of 8 or more preferences for the new sugar substitute, given that the probability that a single person chooses the new substitute is $p = .5$.

SOLUTION   For this example $n = 10$ and $p = .5$. Consequently, we consult Table 1(b) in the Appendix. Since we know that the sum of the binomial probabilities from $x = 0$ to $x = 10$ is 1, we have

$$P(x \geq 8) = \sum_{x=8}^{10} p(x) = 1 - \sum_{x=0}^{7} p(x)$$

The quantity

$$\sum_{x=0}^{7} p(x)$$

can be found by moving across the top of Table 1(b) to the column for $p = .5$ and down that column to the row corresponding to $a = 7$. We read

$$\sum_{x=0}^{7} p(x) = .945$$

Then

$$P(x \geq 8) = \sum_{x=8}^{10} p(x) = 1 - \sum_{x=0}^{7} p(x) = 1 - .945 = .055$$

◆ ◆

Both individual and cumulative binomial probabilities are available through the MINITAB package. Individual binomial probabilities associated with each value of x for any combination of n and p can be found by using the probability density command PDF followed by a semicolon ; and then the subcommand BINOMIAL N P followed by a period. The variable N is the given sample size and P is the probability of success. Cumulative binomial probabilities can be found by using the cumulative distribution function command CDF followed by a semicolon ; and then the subcommand BINOMIAL N P followed by a period.

The MINITAB output for both the PDF and CDF commands when $n = 10$ and $p = .5$ is given in Table 5.5. The PDF command gives rise to the individual probabilities $P(x = K)$; the CDF command gives rise to the cumulative probabilities $P(x \leq K)$. (The letter K plays the role of the letter a in Table 1 of the Appendix.) Notice that in the MINITAB output $P(x \leq 7) = .9453$, so that $P(x \geq 8) = (1 - .9453) = .0547$, which, to three-decimal accuracy, agrees with our earlier results using Table 1.

The MINITAB output may not list the probabilities for all values of $x = 0, 1, 2, \ldots, n$ for various combinations of n and p, since the MINITAB package has an internal check that stops the calculations when $P(x = K) = 0$ [or equivalently, when $P(x \leq K) = 1$] to within a preassigned level of accuracy.

Examples 5.3, 5.4, and 5.5 illustrate the use of the binomial probability distribution in calculating the probability associated with values of x, the number of successes in n trials defined for the binomial experiment. The probability distribution, $p(x) = C_x^n p^x q^{n-x}$, provides a simple formula for calculating the probabilities of numerical events x applicable to a broad class of experiments that occur in everyday life. However, the experimenter must use some caution. Each physical application must be

Table 5.5   MINITAB Output of Binomial Probabilities When $n = 10$ and $p = .5$

```
MTB > PDF;                                    MTB > CDF;
SUBC> BINOMIAL 10 .5.                         SUBC> BINOMIAL 10 .5.

  BINOMIAL WITH N =  10  P = 0.500000           BINOMIAL WITH N =  10  P = 0.500000
    K        P(X  = K)                            K    P(X  LESS OR = K)
    0          0.0010                             0          0.0010
    1          0.0098                             1          0.0107
    2          0.0439                             2          0.0547
    3          0.1172                             3          0.1719
    4          0.2051                             4          0.3770
    5          0.2461                             5          0.6230
    6          0.2051                             6          0.8281
    7          0.1172                             7          0.9453
    8          0.0439                             8          0.9893
    9          0.0098                             9          0.9990
   10          0.0010                            10          1.0000
```

carefully checked against the defining characteristics of the binomial experiment (presented in Section 5.2) to determine whether the binomial experiment is a valid model for the given application.

EXERCISES   Skill Level

5.4 Let x be a binomial random variable with $n = 3$ and $p = 1/2$.

  a. Calculate the binomial probabilities $p(0)$, $p(1)$, $p(2)$, and $p(3)$.

  b. Construct a probability histogram for the random variable x.

5.5 Let x be a binomial random variable with $n = 4$ and $p = 1/2$.

  a. Calculate the binomial probabilities $p(0)$, $p(1)$, $p(2)$, $p(3)$, and $p(4)$.

  b. Construct a probability histogram for the random variable x.

5.6 Refer to the histograms in Exercises 5.4 and 5.5. Are these two distributions symmetric? (*Note:* It can be shown, in general, that when $p = 1/2$, the binomial distribution is symmetric.)

5.7 Evaluate the following binomial probabilities.

  a. $C_8^{10}(.2)^8(.8)^2$ **b.** $C_2^{11}(.4)^2(.6)^9$

  c. $C_3^5(.5)^3(.5)^2$ **d.** $C_0^9(.1)^0(.9)^9$

5.8 Let x be a binomial random variable with $p = .8$.

  a. For $n = 2$, find $P(x = 0)$.   **b.** For $n = 4$, find $P(x \leq 2)$.

5.9 Use Table 1 in the Appendix to evaluate the following binomial probabilities when $n = 15$ and $p = .3$.

  a. $P(x \geq 2)$ **b.** $P(x \leq 2)$ **c.** $P(x = 2)$

  d. $P(x < 2)$ **e.** $P(x > 2)$

5.10 Use Table 1 in the Appendix to evaluate the following binomial probabilities.

  a. $P(x \geq 5)$ when $n = 25$, $p = .05$

    b. $P(x = 6)$ when $n = 10, p = .7$

    c. $P(2 < x \le 8)$ when $n = 10, p = .2$

    d. $P(x < 3)$ when $n = 15, p = .2$

5.11 Let x be a binomial random variable with $n = 5$ and $p = .4$.

    a. Find $P(x \ge 3)$ using the binomial formula.

    b. Find $P(x \ge 3)$ using Table 1 in the Appendix.

    c. Compare the results of parts a and b.

5.12 Let x be a binomial random variable with $n = 20$ and $p = .3$.

Table 5.6 *MINITAB Output for Exercise 5.12*

```
MTB > PDF;
SUBC> BINOMIAL 20 .3.

      BINOMIAL WITH N   =   20   P = 0.300000
         K             P(X   = K)
         0               0.0008
         1               0.0068
         2               0.0278
         3               0.0716
         4               0.1304
         5               0.1789
         6               0.1916
         7               0.1643
         8               0.1144
         9               0.0654
        10               0.0308
        11               0.0120
        12               0.0039
        13               0.0010
        14               0.0002
        15               0.0000
```

    a. Calculate $P(x \le 4)$ using the binomial formula.

    b. Calculate $P(x \le 4)$ using Table 1 in the Appendix.

    c. Use the MINITAB output in Table 5.6 to calculate $P(x \le 4)$. Compare the results of parts a, b, and c.

EXERCISES Applications

  5.13 Some economists have proposed mandatory wage and price controls to combat inflation, but others claim such controls are ineffective because they deal with the effects, not the causes, of inflation. Suppose a recent national poll indicates that 40% of all American adults favor wage and price controls. If $n = 5$ adults are selected at random, find the probability that at least 3 favor wage and price controls. Find the probability that none of the 5 favor controls.

5.14   A survey (Exercise 4.12) conducted by Synergistics Research Corporation in Atlanta concerned the sources of financial advice most frequently used by consumers for investment information and information regarding the impact of the 1986 tax reform policies on investments. The survey found that, at that time, approximately 40% of the respondents did not consider themselves to be knowledgeable about what advantages were available under the new tax reform laws (*Journal of Accountancy*, January 1988, p. 24). Suppose that this survey is representative of all consumers and $n = 5$ consumers are randomly selected to be interviewed.

    a. Find the probability that all of those chosen did not consider themselves to be knowledgeable about the advantages of the new tax reform laws.

    b. Find the probability that one of the five consumers did not consider himself to be knowledgeable about the advantages of the new tax reform laws.

    c. Find the probability that two of the five consumers did not consider themselves to be knowledgeable about the advantages of the new tax reform laws.

5.15   Refer to Exercise 5.14. Let x be the number of consumers in the sample of $n = 5$ who did not consider themselves to be knowledgeable about the advantages of the new tax reform laws. Construct a probability histogram for x.

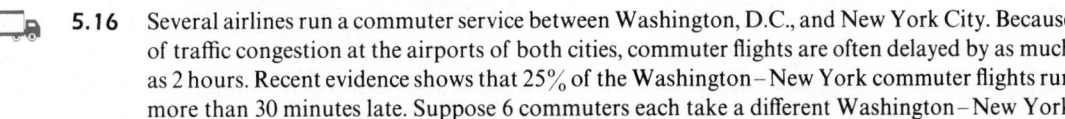

5.16   Several airlines run a commuter service between Washington, D.C., and New York City. Because of traffic congestion at the airports of both cities, commuter flights are often delayed by as much as 2 hours. Recent evidence shows that 25% of the Washington–New York commuter flights run more than 30 minutes late. Suppose 6 commuters each take a different Washington–New York commuter flight on 6 different days. (Assume that the flights are independent of each other.)

    a. Find the probability that all 6 arrive in New York within 30 minutes of their expected time of arrival.

    b. Find the probability that no more than 2 of the 6 commuters are more than 30 minutes late in arriving in New York.

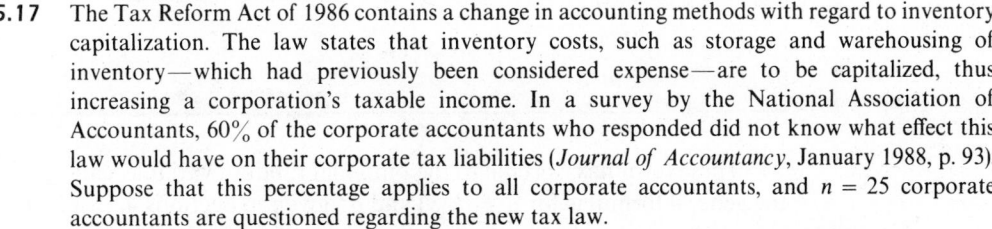

5.17   The Tax Reform Act of 1986 contains a change in accounting methods with regard to inventory capitalization. The law states that inventory costs, such as storage and warehousing of inventory—which had previously been considered expense—are to be capitalized, thus increasing a corporation's taxable income. In a survey by the National Association of Accountants, 60% of the corporate accountants who responded did not know what effect this law would have on their corporate tax liabilities (*Journal of Accountancy*, January 1988, p. 93). Suppose that this percentage applies to all corporate accountants, and $n = 25$ corporate accountants are questioned regarding the new tax law.

    a. Find the probability that at least 15 of the 25 do not know what effect the law will have on their corporate tax liabilities.

    b. Find the probability that no more than 10 do not know what effect the law will have.

    c. If only 10 of the 25 corporate accountants did not know what effect this law would have on their corporate liabilities, what would you conclude regarding the accuracy of the 60% figure given in the survey?

5.18   Maintenance records suggest that only 1 of every 100 electric typewriters of a certain brand requires major service and repair during the first year of use. An office manager has purchased 10 typewriters of this brand.

    a. Find the probability that none of the typewriters require major service and repair during the first year of use.

    b. Find the probability that 2 of the typewriters require major service and repair during the first year of use.

5.19 Refer to Exercise 5.18. If you are the office manager and find that 2 of the 10 typewriters require major service and repair during the first year of use, how would you feel about the manufacturer's claim that only 1 in 100 requires such repair? (We are asking here for an inference, a topic that will be discussed in Section 5.7.)

5.4

The Mean and Variance for the Binomial Random Variable

In Section 5.3 we saw that the calculation of $p(x)$ becomes very difficult for large values of n. Binomial tables in the Appendix simplified the task; however, these tables were available only for certain values of n and p. Nevertheless, even if the exact binomial probabilities are not calculated, we can still describe the binomial probability distribution by using its mean and standard deviation. This procedure will enable us to identify values of x that are highly improbable simply by using our knowledge of Tchebysheff's Theorem and the Empirical Rule. A more precise method for approximating binomial probabilities will be presented in Chapter 6, and this method will rely on knowledge of the mean and standard deviation of x, namely, μ and σ. Consequently, we need to know the expected value and variance of the binomial random variable x.

The formulas for the mean, the variance, and the standard deviation of the binomial random variable follow.

Mean, Variance, and Standard Deviation for a Binomial Random Variable

$$\mu = E(x) = np$$
$$\sigma^2 = npq$$
$$\sigma = \sqrt{npq}$$

These formulas, specific to the binomial random variable, can be derived by using the general formulas for $\mu = E(x)$ and σ^2 given in Sections 4.5 and 4.6. That is,

$$\mu = E(x) = \sum_x xp(x) \qquad \text{and} \qquad \sigma^2 = \sum_x (x - \mu)^2 p(x)$$

For the binomial random variable we substitute $p(x) = C_x^n p^x q^{n-x}$ in the formulas to obtain the results. Since the derivation requires some algebraic manipulation, we will omit it here and will simply use the results in describing a binomial probability distribution.

In Examples 4.1 and 4.6, dealing with the number of heads in the toss of two fair coins, we found by direct calculation that $\mu = 1$ and $\sigma^2 = 1/2$. Since this was in fact a binomial experiment with $n = 2$ and $p = 1/2$, these values can also be found by using the formulas specific to the binomial random variable:

$$\mu = np = 2(1/2) = 1 \qquad \text{and} \qquad \sigma^2 = npq = 2(1/2)(1/2) = 1/2$$

Notice that the two methods of calculation produce identical results.

EXAMPLE 5.7   A manufacturer believes that 30% of all consumers favor her product. To check her belief, she randomly samples 800 consumers and counts the number x favoring her product. If 30% of all consumers favor the manufacturer's product, within what limits would you expect x to fall?

SOLUTION   Calculating the probabilities for x would be difficult because n is so large. Consequently, we will describe the probability distribution by using μ and σ.
Since $n = 800$ and $p = .3$, we have

$$\mu = np = (800)(.3) = 240$$
$$\sigma = \sqrt{npq} = \sqrt{(800)(.3)(.7)} = \sqrt{168} = 12.96$$

Based on Tchebysheff's Theorem and the Empirical Rule, we would expect x to fall within the interval $(\mu \pm 2\sigma)$ with a high probability and almost certainly within the interval $(\mu \pm 3\sigma)$. The intervals are

$$(\mu \pm 2\sigma) = (240 \pm 25.92) \qquad \text{or} \qquad 214.08 \text{ to } 265.92$$
$$(\mu \pm 3\sigma) = (240 \pm 38.88) \qquad \text{or} \qquad 201.12 \text{ to } 278.88$$

(*Comment:* A histogram of the binomial probability distribution will be very mound-shaped for $n = 800$ and $p = .3$. Hence, we would expect the Empirical Rule to work very well. The justification for this statement will be given in Chapter 7.)   ◆ ◆

EXERCISES   Skill Level

5.20   Let x be a binomial random variable based on $n = 10$ trials. Calculate μ and σ^2 for the following probabilities.

a. $p = .1$ b. $p = .3$ c. $p = .5$

d. $p = .7$ e. $p = .9$ f. $p = .99$

5.21   Refer to Exercise 5.20.

a. How does the mean μ change with p when n remains fixed?

b. How does the variance σ^2 change with p when n remains fixed?

5.22   Let x be the number of successes in $n = 48$ trials for which $p = .25$. Calculate μ and σ^2, the mean and variance of x. What can be said about $P(6 \le x \le 18)$? (*Hint:* Use Tchebysheff's Theorem.)

5.23   A random variable x has a binomial distribution with $n = 3$ and $p = 1/2$. The probability distribution for x was found in Exercise 5.4.

a. Find the expected value and standard deviation of x, using the formulas $E(x) = np$ and $\sigma = \sqrt{npq}$.

b. Using the probability distribution in Exercise 5.4, find the fraction of the measurements in the population lying within one standard deviation of the mean. Within two standard deviations of the mean. Do your results agree with Tchebysheff's Theorem and the Empirical Rule? Of what practical use are these results?

5.24   Let x be a binomial random variable with $n = 3$ and $p = .1$.

    a.   Find the probability distribution for x, using the binomial formula.

    b.   Construct a probability histogram by using the results of part a.

    c.   Calculate μ and σ^2.

    d.   Using the probability distribution in part a, find the fraction of measurements lying within one standard deviation of the mean. Within two standard deviations of the mean. Do your results agree with Tchebysheff's Theorem and the Empirical Rule? Why or why not?

EXERCISES   Applications

5.25   One of the disadvantages of mail questionnaires is that their response rate is often low. They are, however, less expensive to administer than other types of surveys, such as telephone surveys or personal interviews. The manager of a publicly owned utility has sent a questionnaire to all rate payers requesting consideration of alternative future energy generation sources. From past experience the utility manager knows the return rate will be about 20%. If questionnaires are sent to 10,000 rate payers, find the expected value and variance of x, the number of completed questionnaires returned. Within what limits would x be expected to fall? (*Hint:* Use Tchebysheff's Theorem.)

5.26   The Energy Policy Center of the Environmental Protection Agency reports that 75% of the homes in New England are heated by oil-burning furnaces. If a certain New England community is known to have 2,500 homes, find the expected number of homes in the community that are heated by oil furnaces. If x is the number of homes in the community that are heated by oil, find the variance and standard deviation of x. Use Tchebysheff's Theorem to describe limits within which you could expect x to fall.

5.27   The annual report is one of the most important documents produced by publicly owned companies and a document that incurs considerable expense in its production. However, most stockholders read the report only cursorily, if at all. Suppose that 40% of stockholders spend 5 minutes or less reading their company's annual report and $n = 100$ stockholders of a publicly owned company are randomly selected from the firm's registry of stockholders.

    a.   Find the expected value of x, the number of stockholders who spend no more than 5 minutes reading their company's annual report.

    b.   Determine the standard deviation of x.

    c.   If the controller observed that of the 100 selected stockholders $x = 25$ spend no more than 5 minutes reading the annual report, does it appear that the proportion of all stockholders spending 5 minutes or less reading her company's annual report is really 40%? Explain.

5.28   A source of growing interest to economists and to government officials is a phenomenon called the "underground economy," which involves the economic activity of millions of people in the world who engage "in jobs that evade the relevant tax and labor laws—frequently with an official scowl but a tacit wink from their governments" (*Los Angeles Times*, February 16, 1988). In the *Los Angeles Times* article, Saskia Saseen-Koob, director of the urban planning program at Columbia University, estimates that there are 21,000 unlicensed cabs in Manhattan, twice the number of legal cabs. On a particular evening, 60 customers independently hail a cab in Manhattan. Let x be the number of unlicensed cabs that respond.

    a.   Find the expected value and variance of x.

b. What are the upper and lower limits for the number of unlicensed cabs hailed by the 60 customers?

c. If 45 of the 60 cabs are unlicensed, would we have reason to doubt the premise that there are twice as many unlicensed as licensed cabs in Manhattan?

5.5

The Poisson Probability Distribution (Optional)

Another discrete random variable that has numerous applications in business and economics is the *Poisson random variable*. Its probability distribution provides a good model for data that represent the number of occurrences of a specified event in a given unit of time or space. Here are some examples of experiments for which the random variable x can be modeled by the Poisson random variable:

1. The number of calls received by a switchboard during a given period of time
2. The number of claims against an insurance company during a given week
3. The number of arrivals at a checkout counter during a given minute
4. The number of machine breakdowns during a given day
5. The number of flaws on a 1-square-foot piece of material

In each example, x **represents the number of events occurring in a period of time or space during which an average of μ such events can be expected to occur.** The only assumptions needed when one uses the Poisson distribution to model experiments such as those described above are that the counts or events occur **randomly and independently** of one another. The formula for the Poisson probability distribution as well as its mean and variance are shown in the display.

The Poisson Probability Distribution

$$p(x) = \frac{\mu^x e^{-\mu}}{x!} \qquad x = 0, 1, 2, 3, \ldots$$

where

$\mu = E(x) = $ mean of random variable x

$\sigma^2 = \mu = $ variance of random variable x

$e = 2.71828 \ldots (e$ is the base of natural logarithms$)$

The value of $e^{-\mu}$ can be calculated by using an electronic calculator or by using Table 2 in the Appendix, which provides the values e^{-x} for values of x between 0 and 10 in increments of .05. Graphs of the Poisson probability distribution for $\mu = .5, 1,$ and 4 are shown in Figure 5.2.

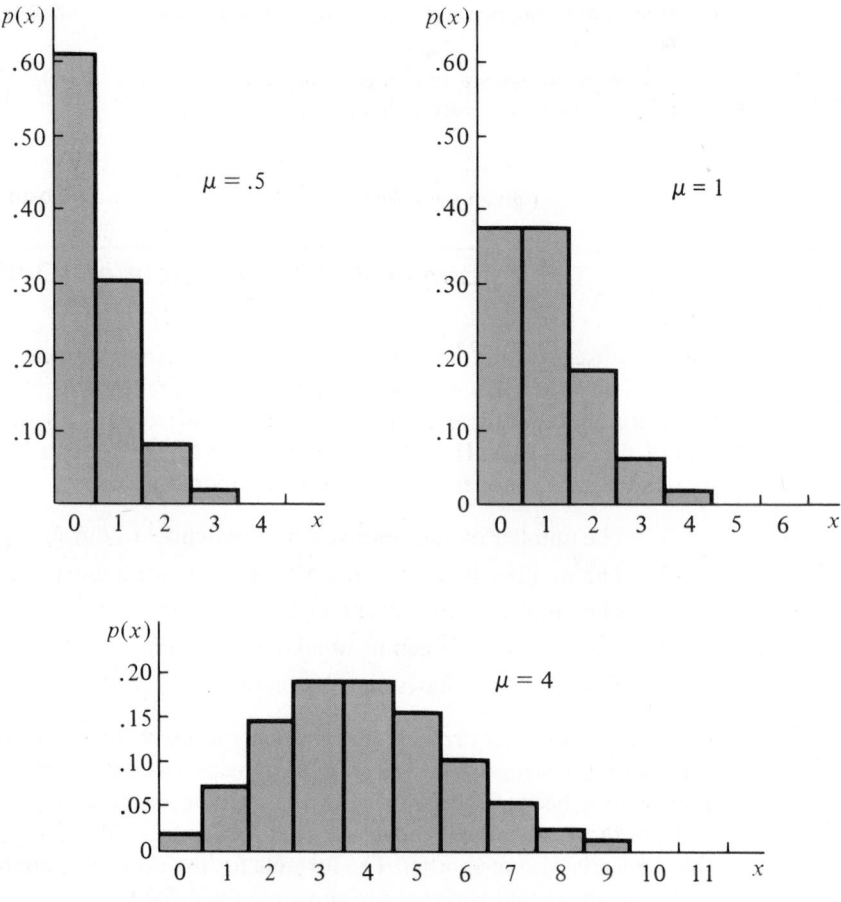

Figure 5.2   Poisson Probability Distributions for $\mu = .5, 1,$ and 4

EXAMPLE 5.8   Bank closures in the United States due to financial difficulties have occurred at the average rate of 7.2 closures per year from 1960 through 1981 [FDIC Annual Report (1983), p. 53]. Assume that the number of closures x in a given period possesses a Poisson probability distribution.

    a.   Find the probability of no bank closures during a given four-month period.
    b.   Find the probability of at least three bank closures during a given year.

SOLUTION   a.   If closures occur at the rate of 7.2 per year, then for four months (1/3 of a year) we could expect

$$\mu = 7.2(1/3) = 2.4$$

closures during any four-month period. Therefore, the probability of no closures in a given four-month period is

$$p(0) = \frac{2.4^0 e^{-2.4}}{0!} = \frac{e^{-2.4}}{1} = .091$$

(*Note:* $\mu^0 = 1$.)

b.   During a given year we could expect $\mu = 7.2$ bank closures. Since

$$P(x \geq 3) = \sum_{x=3}^{\infty} p(x) = 1 - p(0) - p(1) - p(2)$$

where

$$p(0) = \frac{7.2^0 e^{-7.2}}{0!} = .00075$$

$$p(1) = \frac{7.2^1 e^{-7.2}}{1!} = (7.2)(.00075) = .00538$$

$$p(2) = \frac{7.2^2 e^{-7.2}}{2!} = \frac{(51.84)(.00075)}{2} = .01935$$

then

$$P(x \geq 3) = 1 - .00075 - .00538 - .01935 = 1 - .02548 = .97452$$

◆ ◆

Recall from Section 5.3 that we were able to simplify the calculation of binomial probabilities by using Table 1 in the Appendix. However, binomial tables are seldom available for n greater than 100, and many applications of the binomial experiment with $n = 100$ or more arise in practical situations. Consequently, we need simple, easy-to-compute approximation procedures for calculating binomial probabilities. The Poisson probability distribution provides good approximations to binomial probabilities when n is large and $\mu = np$ is small, preferably with $np \leq 7$. An approximation procedure suitable for larger values of $\mu = np$ will be presented in Chapter 6.

As an illustration of the Poisson approximation procedure, consider the following application. Suppose that a life insurance company insures the lives of 5,000 men of age 42. If actuarial studies show the probability of any 42-year-old man dying in a given year to be .001, the exact probability that the company will have to pay $x = 4$ claims during a given year is given by the binomial distribution as

$$P(x = 4) = p(4) = \frac{5000!}{4!4996!}(.001)^4(.999)^{4996}$$

for which binomial tables are not available. To compute $P(x = 4)$ without the aid of a computer would be very time-consuming, but the Poisson distribution can be used to provide a good approximation to $P(x = 4)$. Computing $\mu = np = (5,000)(.001) = 5$ and substituting into the formula for the Poisson probability distribution, we have

$$p(4) \approx \frac{\mu^4 e^{-\mu}}{4!} = \frac{5^4 e^{-5}}{4!} = \frac{(625)(.0067)}{24} = .1745$$

EXAMPLE 5.9   Suppose that a large food-processing and canning plant has 20 automatic canning machines in operation at all times. If the probability that an individual canning machine breaks down during a given day is .05, find the probability that during a given day 2 canning machines fail. Use the binomial distribution to compute the exact probability, and then compute the Poisson approximation.

SOLUTION   This is a binomial experiment with $n = 20$ and $p = .05$, and the probability of interest is $P(x = 2)$. Using the binomial tables with $n = 20$—Table 1(d) in the Appendix—we have

$$P(x = 2) = p(2) = \sum_{x=0}^{2} p(x) - \sum_{x=0}^{1} p(x) = .925 - .736 = .189$$

For this binomial random variable, the expected number of machine breakdowns in a given day is $\mu = np = 20(.05) = 1.0$. Hence, we will use a Poisson distribution with $\mu = 1$ to approximate the probability that $x = 2$. From Table 2 in the Appendix, $e^{-\mu} = e^{-1} = .367879$. Then

$$P(x = 2) \approx p(2) = \frac{1^2 e^{-1}}{2!} = \frac{.367879}{2}$$

Correct to three decimal places, we have

$$P(x = 2) \approx p(2) = .184$$

You can see that the Poisson approximation, .184, is quite close to the exact value of the binomial probability, .189. The larger the value of n (for a fixed value of $\mu = np$), the better will be the Poisson approximation (see Feller 1968). In particular, we recommend that n be large and that $\mu = np$ be less than or equal to 7.

EXAMPLE 5.10   A manufacturer of power lawn mowers buys 1-horsepower, two-cycle engines in lots of 1,000 from a supplier. He then equips each of the mowers produced by his plant with one of the engines. Past history shows that the probability of any one engine purchased from the supplier proving unsatisfactory is .001. In a shipment of 1,000 engines, what is the probability that none are defective? One is defective? Two are? Three are? Four are?

SOLUTION   This is a binomial experiment with $n = 1,000$ and $p = .001$. The expected number of defectives in a shipment of $n = 1,000$ engines is $\mu = np = (1,000)(.001) = 1$. Since this is a binomial experiment with $np \leq 7$, the probability of x defective engines in the shipment may be approximated by

$$p(x) = \frac{\mu^x e^{-\mu}}{x!} = \frac{1^x e^{-1}}{x!} = \frac{e^{-1}}{x!}$$

(since $1^x = 1$ for any value of x). Therefore, we have

$$p(0) \approx \frac{e^{-1}}{0!} = \frac{.368}{1} = .368 \qquad p(3) \approx \frac{e^{-1}}{3!} = \frac{.368}{6} = .061$$

$$p(1) \approx \frac{e^{-1}}{1!} = \frac{.368}{1} = .368 \qquad p(4) \approx \frac{e^{-1}}{4!} = \frac{.368}{24} = .015$$

$$p(2) \approx \frac{e^{-1}}{2!} = \frac{.368}{2} = .184$$

◆ ◆

   The individual and cumulative probabilities for a Poisson distribution with mean μ can be found by using the PDF and the CDF MINITAB commands, followed by the subcommand POISSON μ. The binomial probabilities for $n = 1,000$ and $p = .001$, together with the Poisson probabilities for $\mu = 1$, are given in Table 5.7. Comparing the actual binomial probabilities with the corresponding probabilities found by using the Poisson approximation to binomial probabilities, we see that they are quite

Table 5.7   MINITAB Output of Binomial and Poisson Probabilities

```
MTB > PDF;                                    MTB > CDF;
SUBC> BINOMIAL 1000 .001.                     SUBC> BINOMIAL 1000 .001.

  BINOMIAL WITH N =1000  P = 0.001000           BINOMIAL WITH N =1000  P = 0.001000
      K         P(X  = K)                           K    P(X  LESS OR = K)
      0           0.3677                            0            0.3677
      1           0.3681                            1            0.7358
      2           0.1840                            2            0.9198
      3           0.0613                            3            0.9811
      4           0.0153                            4            0.9964
      5           0.0030                            5            0.9994
      6           0.0005                            6            0.9999
      7           0.0001                            7            1.0000
      8           0.0000

MTB > PDF;                                    MTB > CDF;
SUBC> POISSON 1.                              SUBC> POISSON 1.

  POISSON WITH MEAN =    1.000                  POISSON WITH MEAN =    1.000
      K         P(X  = K)                           K    P(X  LESS OR = K)
      0           0.3679                            0            0.3679
      1           0.3679                            1            0.7358
      2           0.1839                            2            0.9197
      3           0.0613                            3            0.9810
      4           0.0153                            4            0.9963
      5           0.0031                            5            0.9994
      6           0.0005                            6            0.9999
      7           0.0001                            7            1.0000
      8           0.0000
```

accurate in this case. Furthermore, we see that the POISSON command also terminates when an individual probability equals 0 (or when the cumulative probability equals 1) within a preassigned level of accuracy.

EXERCISES   Conceptual Level

5.29 Which of the following business problems can be modeled by the Poisson distribution? For those that cannot be modeled by the Poisson distribution, explain why.

    a. Estimation of the probability that there are more than five arrivals to a hospital emergency room during a given day when, on the average, arrivals occur randomly and independently at the rate of one every 6 hours

    b. Computation of the probability of selecting a batch of 10 nondefective lens filters from a crate of 10,000 lens filters of which 25 are defective

    c. Estimation of the probability that a United Way fund-raising agency receives more than two gifts of $10,000 or more in a single year when past experience has shown that, on the average, they receive one gift of this magnitude in a given year

    d. Estimation of the probability that a corporate relations officer receives more than 15 customer complaints during an 8-hour day if records show that customer complaints arrive at her office randomly and independently at the rate of 10 per day

    e. Calculation of the probability that 5 of 15 clients contacted by an insurance salesman will eventually purchase an insurance policy from the salesman when his records show that, on the average, 25% of his contacts result in sales

5.30 Under what conditions can the Poisson random variable be used to approximate the probabilities associated with a binomial random variable? What application does the Poisson distribution have other than to estimate certain binomial probabilities?

EXERCISES   Skill Level

5.31 Let x be a Poisson random variable with mean $\mu = 2$. Calculate the following probabilities.

    a. $P(x = 0)$ **b.** $P(x = 1)$ **c.** $P(x > 1)$ **d.** $P(x = 5)$

5.32 The MINITAB printout, using the command POISSON 2.5, in Table 5.8 shows the probability distribution for a Poisson random variable with $\mu = 2.5$. Find the following probabilities.

    a. $P(x \geq 5)$ **b.** $P(x < 6)$ **c.** $P(x = 2)$ **d.** $P(1 \leq x \leq 4)$

5.33 Let x be a binomial random variable with $n = 20$ and $p = .1$.

    a. Calculate $P(x \leq 2)$, using Table 1 in the Appendix to obtain the exact binomial probability.

    b. Use the Poisson approximation to calculate $P(x \leq 2)$.

    c. Compare the results of parts a and b. Is the approximation accurate?

5.34 To illustrate how well the Poisson probability distribution approximates the binomial probability distribution, calculate the Poisson approximate values for $p(0)$ and $p(1)$ for a binomial probability distribution with $n = 25$ and $p = .05$. Compare the answers with the exact values obtained from Table 1 in the Appendix.

Table 5.8   *MINITAB Output for Exercise 5.32*

```
MTB > PDF;
SUBC> POISSON 2.5.

    POISSON WITH MEAN =   2.500
       K             P(X  = K)
       0              0.0821
       1              0.2052
       2              0.2565
       3              0.2138
       4              0.1336
       5              0.0668
       6              0.0278
       7              0.0099
       8              0.0031
       9              0.0009
      10              0.0002
      11              0.0000
```

EXERCISES   *Applications*

5.35   The Labor-Management Reporting and Disclosure Act (LMRDA) of 1959 prescribes fiduciary responsibilities for union officials and makes embezzlement of union funds a federal offense. Since 1961 civil suits have been filed under this law randomly and independently of one another at the average rate of 2.9 suits per month.

   a. Find the probability of no suits being filed under the LMRDA during a given month.

   b. Find the probability of no more than 5 suits being filed under the law during a given two-month span. (*Hint:* See Example 5.8.)

5.36   The local manager of a rental car organization buys tires in lots of 500 to take advantage of volume price discounts from the supplier. From experience the manager knows that 1% of all new tires purchased from the supplier are defective and must be replaced within the first week of use. What is the probability that a shipment of 500 tires from the supplier will contain no defective tires? One defective tire? No more than 3 defective tires?

5.37   The increased number of small commuter planes in major airports has heightened concern over air safety. An eastern airport has recorded a monthly average of five near misses on landings and takeoffs in the past five years.

   a. Find the probability that during a given month there are no near misses on landings and takeoffs at the airport.

   b. Find the probability that during a given month there are five near misses.

   c. Find the probability that there are at least five near misses during a particular month.

5.38   The number x of people entering the intensive care unit at a particular hospital on any one day possesses a Poisson probability distribution with mean equal to 5 persons per day.

a. What is the probability that the number of people entering the intensive care unit on a particular day is equal to 2? Less than or equal to 2?

b. Is it likely that x will exceed 10? Explain.

5.39 A telephone switchboard for a medical office building can handle at most 5 incoming calls a minute. If past experience suggests that an average of 120 incoming calls per hour are received by the switchboard, find the probability that the switchboard is overloaded during any given minute.

5.40 Are audit rates the same in all 50 states? In a *Wall Street Journal* article ("Regional Risks: IRS's Audit Rate Shows That All States Aren't Equal," July 22, 1987), the Research Institute of America examined the audits conducted in 1986 that mainly concerned 1984 tax returns. They found that audit rates varied considerably from state to state, from a low rate of .47% in Rhode Island to a high rate of 2.61% in Nevada. Although the overall risk of an audit is quite low (1.1% in 1986), the IRS tends to audit more often people who function as partners and promoters of tax shelters, earn tip income, or are self-employed.

a. Suppose that 100 United States taxpayers are chosen at random. Use the Poisson approximation to the binomial to find the probability that at least two will be audited.

b. Now suppose that the 100 taxpayers are randomly selected from the state of Nevada. What is the approximate probability that at least 2 will be audited?

5.6

The Hypergeometric Probability Distribution (Optional)

Suppose you are selecting a sample of elements from a population and you record whether each element does or does not possess a certain characteristic. Consequently, you are dealing with the "success" or "failure" type of data encountered in the binomial experiment. The consumer preference survey of Example 5.1 and the sampling for defectives of Example 5.2 are practical illustrations of these sampling situations.

If the number of elements in the population is large in relation to the number in the sample (as in Example 5.1), the probability of selecting a success on a single trial is equal to the proportion p of successes in the population. Because the population is large in relation to the sample size, this probability will remain constant (for all practical purposes) from trial to trial, and the number x of successes in the sample will follow a binomial probability distribution. However, **if the number of elements in the population is small in relation to the sample size, the probability of a success for a given trial is dependent on the outcomes of preceding trials. Then the number x of successes follows what is known as a** *hypergeometric probability distribution.*

We define the following notation, which is necessary in order to present the formula for the hypergeometric probability distribution.

N = number of elements in population

k = number of elements in population that are successes (that is, number possessing one of two characteristics)

$N - k$ = number of elements in population that are not successes

$$n = \text{number of elements in sample, selected from } N \text{ elements in population}$$

$$x = \text{number of successes in sample}$$

The hypergeometric probability distribution for the random variable x is then as given in the display.

Hypergeometric Probability Distribution

$$p(x) = \frac{C_x^k C_{n-x}^{N-k}}{C_n^N}$$

where x can assume integer values $0, 1, 2, \ldots, n$ subject to the restrictions $x \leq k$ and $x \geq k + n - N$, and where

$$C_r^n = \frac{n!}{r!(n-r)!}$$

You may recall that several exercises in Chapter 3 portrayed situations for which the hypergeometric probability distribution would have been applicable. For those exercises N and n were kept small, and our intention was to use the exercises to develop your ability to solve probability problems. We can now solve similar but more complex probability problems by using the hypergeometric probability distribution.

EXAMPLE 5.11   An important problem encountered by personnel directors and others faced with the selection of the "best" in a finite set of elements is illustrated by the following situation. From a group of 20 PhD engineers, 10 are randomly selected for employment. What is the probability that the 10 selected include all the 5 best engineers in the group of 20?

SOLUTION   For this example $N = 20, n = 10, k = 5$, and $(N - k) = 15$. That is, there are only 5 in the set of 5 best engineers, and we seek the probability that $x = 5$, where x denotes the number of best engineers among the 10 selected. Then

$$p(x) = \frac{C_x^k C_{n-x}^{N-k}}{C_n^N}$$

$$p(5) = \frac{C_5^5 C_5^{15}}{C_{10}^{20}} = \frac{\left(\dfrac{5!}{5!0!}\right)\left(\dfrac{15!}{5!10!}\right)}{\dfrac{20!}{10!10!}} = \left(\frac{15!}{5!10!}\right)\left(\frac{10!10!}{20!}\right) = \frac{21}{1{,}292} = .0163$$

(Remember that $0! = 1$.)

◆ ◆

EXAMPLE 5.12 A particular industrial product is shipped in lots of 20. Testing to determine whether an item is defective is costly, and hence, the manufacturer samples production rather than using a 100% inspection plan. A sampling plan constructed to minimize the number of defectives shipped to customers calls for sampling 5 items from each lot and rejecting the lot if more than 1 defective is observed. (If rejected, each item in the lot is then tested.) If a lot contains 4 defectives, what is the probability that it will be accepted?

SOLUTION Let x be the number of defectives in the sample. Then $N = 20, k = 4, (N - k) = 16$, and $n = 5$. The lot will be rejected if $x = 2, 3,$ or 4. Then

$$P(\text{accept the lot}) = P(x \le 1) = p(0) + p(1) = \frac{C_0^4 C_5^{16}}{C_5^{20}} + \frac{C_1^4 C_4^{16}}{C_5^{20}}$$

$$= \frac{\left(\dfrac{4!}{0!4!}\right)\left(\dfrac{16!}{5!11!}\right)}{\dfrac{20!}{5!15!}} + \frac{\left(\dfrac{4!}{1!3!}\right)\left(\dfrac{16!}{4!12!}\right)}{\dfrac{20!}{5!15!}}$$

$$= \frac{91}{323} + \frac{455}{969} = .2817 + .4696 = .7513$$

◆ ◆

Notice that Example 5.12 is quite similar to Example 5.4. The only difference is that the number x of defectives possesses a hypergeometric probability distribution when the number of elements N in the population is small in relation to the sample size n.

Familiarity with discrete probability distributions and the properties of the experiments that generate them is extremely helpful. Rather than solve the same probability problem over and over again from first principles (as was done in Chapter 3), you need only recognize the type of random variable involved and then substitute into the formula for its probability distribution. The first three exercises in the Applications Exercises that follow have been selected from those given in Chapter 3.

EXERCISES Conceptual Level

5.41 Under what circumstances would one use the hypergeometric probability distribution rather than the binomial distribution in evaluating the probability of x successes in n trials?

5.42 How does the hypergeometric distribution differ from the Poisson distribution?

EXERCISES Skill Level

5.43 Evaluate the following probabilities.

  a. $\dfrac{C_1^3 C_1^2}{C_2^5}$ **b.** $\dfrac{C_2^4 C_1^3}{C_3^7}$ **c.** $\dfrac{C_4^5 C_0^3}{C_4^8}$

5.44 Let x be the number of successes observed in a sample of $n = 5$ items selected from $N = 10$. Suppose that of the $N = 10$ items, 6 are considered "successes."

    a. Find the probability of observing no successes.

    b. Find the probability of observing at least 2 successes.

    c. Find the probability of observing exactly 2 successes.

5.45 Let x be a hypergeometric random variable with $N = 15, n = 3$, and $k = 4$.

    a. Calculate $p(0)$, $p(1)$, $p(2)$, and $p(3)$.

    b. Construct the probability histogram for x.

    c. Use the formulas given in Sections 4.5 and 4.6 to calculate $\mu = E(x)$ and σ^2.

    d. What proportion of the population of measurements fall within the interval $(\mu \pm 2\sigma)$? Within the interval $(\mu \pm 3\sigma)$? Do these results agree with those given by Tchebysheff's Theorem?

EXERCISES   Applications

5.46 Many companies in the rapidly growing high-technology field are seeking to diversify their manufacturing base by locating branch facilities in small but progressive communities. In this way they can develop new sources of labor and improve their rate of retention of key management personnel. Suppose a preliminary analysis by an electronics firm has identified five communities that meet organizational objectives for site location. Of these five communities two are located in the state of Colorado and three are not. The president of the company plans to select two communities at random for in-depth discussions concerning the location of a manufacturing facility. We are interested in the geographic location of each chosen community.

    a. Let x be the number of communities from the state of Colorado chosen by the president. Explain why x possesses a hypergeometric probability distribution. Give the formula for $p(x)$.

    b. What is the probability that the president's choice will consist of at least one community from the state of Colorado? [*Hint:* $p(1) + p(2) = 1 - p(0)$.]

5.47 A large municipal bank chain has ordered six microcomputer systems to distribute among their six branches in a certain city. Unknown to the purchaser, three of the six systems are defective. Before installing the systems, an agent of the bank selects two of the six systems from the shipment, thoroughly tests them, and then classifies each as either defective or nondefective.

    a. Let x be the number of defective systems in the sample of $n = 2$. Explain why x possesses a hypergeometric probability distribution.

    b. What is the probability that both systems in the sample will be defective?

    c. Calculate the values of $p(x)$ for $x = 0, 1, 2$. Graph $p(x)$.

5.48 In recent years affirmative action commitments by industrial organizations have led to an increase in the number of women in executive positions. Suppose a company has vacancies for two positions of vice president. The vacancies are to be filled by randomly selecting two people from a list of five candidates. On the list are two women and three men, all of whom have worked for the company for a long time.

    a. Let x be the number of women selected to fill the two vice presidential positions. Explain why x possesses a hypergeometric probability distribution. Give the formula for $p(x)$.

    b. What is the probability that at least one of the women will be selected?

    c. What is the probability that neither woman will be selected?

5.49 During times when prevailing interest rates are high, bond prices with a low coupon value are depressed in price. The price of some bonds are more sensitive to interest rates than are others, offering difficulty to the investor seeking to capitalize on rate changes. At a time when prevailing rates are high, an investor selects at random 3 from among 50 publicly listed Aaa utility bonds. Unknown to the investor, 5 of the bonds under consideration will incur price gains exceeding 20% over the next year, 10 will result in a loss, and the remaining 35 will have a gain of less than 20% during the same period.

    a. Find the probability that none of the bonds chosen by the investor will offer a first-year price gain exceeding 20%.

    b. Find the probability that exactly 1 bond will offer a first-year price gain exceeding 20%.

    c. Find the probability that the investor avoids selecting a bond that will decrease in price the first year.

5.50 A particular antibiotic is shipped to drug stores in cases, each of which contains 24 bottles. Having doubts about the potency of the drug, the druggist decides to have 5 bottles of the drug tested. Suppose that actually 10 of the 24 bottles are understrength.

    a. Let x be the number of understrength bottles in the sample of 5. Explain why x is or is not a hypergeometric random variable.

    b. Find the probability that none of the bottles sent to the testing company are understrength.

    c. Find the probability that exactly 1 of the 5 is understrength.

5.7

The Role of the Probability Distribution in Making Inferences: A Test of a Researcher's Theory (Optional)

The taste-testing problem of Example 5.5 illustrates a *statistical test of an hypothesis*. The practical question to be answered concerns the relative desirability of the new sugar substitute. Do the data contained in the sample present sufficient evidence to indicate that the new sugar substitute is preferred over the substitute already in use?

    The reasoning employed in testing an hypothesis bears a striking resemblance to the procedure used in a court trial. In trying a person for theft, the court assumes the accused innocent until proved guilty. The prosecution collects and presents all available evidence in an attempt to contradict the "not guilty" hypothesis and hence to obtain a conviction. However, if the prosecution fails to disprove the "not guilty" hypothesis, this does not prove that the accused is "innocent" but merely that there is not sufficient evidence to conclude that the accused is "guilty."

    The statistical problem portrays the new sugar substitute as the accused. The *null hypothesis*   hypothesis to be tested, called the **null hypothesis**, is that the new sugar substitute is no better than that already in widespread use. The evidence in this case is contained in the sample drawn from the population of sugar substitute customers. The experimenter, *alternative*   playing the role of the prosecutor, believes that an **alternative hypothesis** is true—*hypothesis*   namely, that the new sugar substitute is really preferred. Hence, the experimenter attempts to use the evidence contained in the sample to reject the null hypothesis (no preference) and thereby to support the alternative hypothesis, the contention that the new sugar substitute is, in fact, preferred over the one in use. You will recognize this procedure as an essential feature of the scientific method, in which all proposed theories must be compared with reality.

Intuitively, we would select the number x of preferences for the new sugar substitute as a measure of the quantity of evidence in the sample. If x is very large, we would be inclined to reject the null hypothesis and conclude that the new sugar substitute is preferred. On the other hand, a small value of x would provide little evidence to support the rejection of the null hypothesis. As a matter of fact, if the null hypothesis were true and the sugar substitutes were equally desirable, the probability of preferring the new sugar substitute would be $p = .5$ and the average value of x would be

$$E(x) = np = 10(1/2) = 5$$

Most individuals, utilizing their own built-in decision makers, would have little difficulty arriving at a decision for the cases $x = 10$ or $x = 5, 4, 3$, or 1, which, on the surface, appear to provide substantial evidence to support rejection or acceptance, respectively. But what can be said concerning less obvious results, say $x = 7, 8$, or 9? Clearly, whether we employ a subjective or an objective decision-making procedure, we would choose the procedure that gives the smallest probability of making an incorrect decision.

As statisticians, we test the null hypothesis in an objective manner similar to our intuitive procedure. In arriving at a decision, we calculate a *test statistic* from information contained in the sample. In our example the number x of preferences for the new substitute would suffice as a test statistic. We then consider all possible values the test statistic may assume—for example, $x = 0, 1, 2, \ldots, 9, 10$. These values are divided into two groups, as shown in Figure 5.3—one called the **rejection region** and the other the **acceptance region**. An experiment is then conducted and the test statistic x is observed. If x takes a value in the rejection region, the null hypothesis is rejected. Otherwise, the null hypothesis is accepted. (*Caution:* As you will subsequently learn, you will reject or accept the null hypothesis only if the risks of a wrong decision are small for these two actions.)

For example, in our experiment we might choose $x = 8, 9$, or 10 as the rejection region and assign the remaining values of x to the acceptance region. Since we observed $x = 8$ preferences for the new substitute in the experiment, we reject the null hypothesis of no preference and conclude that the probability of preferring the new substitute is greater than $p = .5$.

What is the probability that we will reject the null hypothesis when, in fact, it is true? The probability of falsely rejecting the null hypothesis is the probability that x will equal $8, 9$, or 10 given that $p = .5$. This is the probability computed in Example 5.5

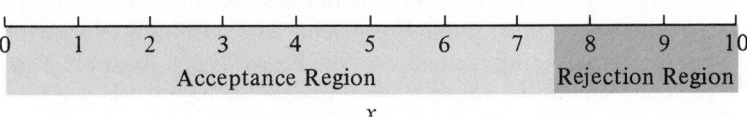

Figure 5.3 Possible Values for the Test Statistic x

and found to equal .055. Since we have decided to reject the null hypothesis and note that this probability is small, we are reasonably confident that we have made the correct decision.

Upon reflection, you will observe that the manufacturer of the new sugar substitute is faced with two possible types of error. On the one hand, he might reject the null hypothesis and falsely conclude that the new sugar substitute was preferred. Proceeding with a more thorough and expensive testing program or a pilot plant production of the sugar substitute would result in a financial loss. On the other hand, he might decide not to reject the null hypothesis and falsely conclude that the new substitute was no better than the one in use. This error would result in the loss of potential profits that could be derived through the sale of a better sugar substitute.

Definition

type I error

> Rejecting the null hypothesis when it is true is called a **type I error** for a statistical test. The probability of making a type I error is denoted by the symbol α (Greek letter alpha).

The probability α will increase or decrease as we increase or decrease the size of the rejection region. Then why not decrease the size of the rejection region and make α as small as possible? For example, why not choose $x = 10$ as the rejection region? Unfortunately, decreasing α increases the probability of not rejecting the null hypothesis when it is false and some alternative hypothesis is true. This second type of error is called the type II error for a statistical test and its probability is denoted by the symbol β.

Definition

type II error

> Accepting the null hypothesis when it is false is called a **type II error** for a statistical test. The probability of making a type II error when some specific alternative is true is denoted by the symbol β (Greek letter beta).

For a fixed sample size α and β are inversely related; as one increases, the other decreases. For a fixed value of α, increasing the sample size provides more information upon which to base the decision and hence reduces β. In an experimental situation the probabilities of the type I and type II errors for a test measure the risk of making an incorrect decision. The experimenter selects values for these probabilities, and the rejection region and sample size are chosen accordingly. Notice that both errors cannot be committed simultaneously. A type I error is possible only if the decision is to reject the null hypothesis; a type II error is possible only if the decision is to accept the null hypothesis.

EXAMPLE 5.13   Refer to the sugar substitute preference study and the statistical test based on the rejection region shown in Figure 5.3 (i.e., reject the null hypothesis if $x = 8, 9, 10$).

    a.   State the null hypothesis and the alternative hypothesis for the test.

    b.   Find α for the test.

    c.   Find β, the probability of accepting the null hypothesis when the probability of preference for the new sugar substitute is $p = .9$.

SOLUTION   a.   The null hypothesis is that $p = .5$ or, equivalently, that both substitutes are equally desirable. The alternative hypothesis is that there is a preference for the new substitute.

    b.   The probability of rejecting the null hypothesis when it is true ($p = .5$) is

$$\alpha = P(x = 8, 9, 10 \text{ given } p = .5) = \sum_{x=8}^{10} p(x)$$

where $p(x)$ is a binomial probability distribution with $p = .5$. Then

$$\alpha = \sum_{x=8}^{10} C_x^{10}(.5)^x(.5)^{10-x}$$

is found by using Table 1(b) in the Appendix; $\alpha = .055$. The probability distribution for $n = 10$ and $p = .5$ is shown in Figure 5.4; α is represented by the shaded portion of the probability distribution.

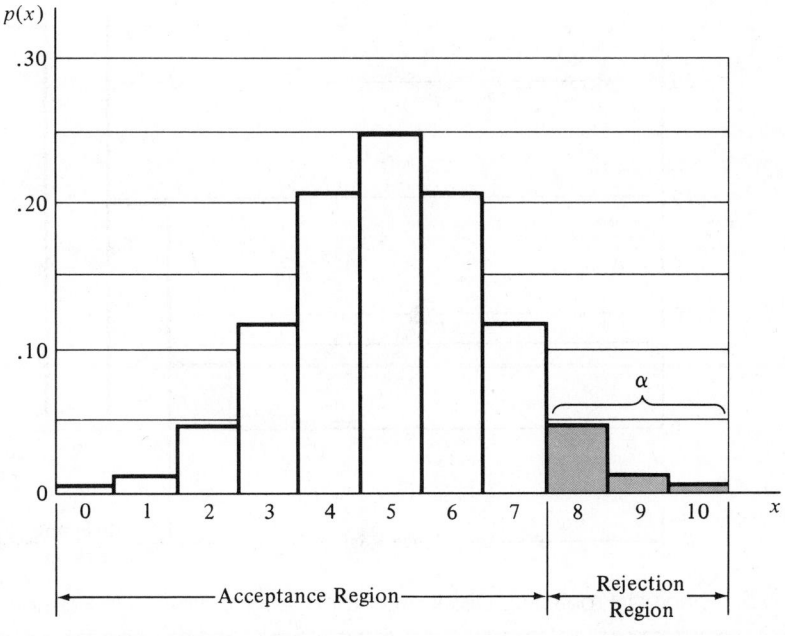

Figure 5.4   Binomial Probability Distribution for $n = 10$ and $p = .5$ in Example 5.13

c. $\beta = P(\text{accepting the null hypothesis when } p = .9)$

$$= P(x = 0, 1, 2, \ldots, 6, 7 \text{ given } p = .9) = \sum_{x=0}^{7} p(x)$$

where $p(x)$ is a binomial probability distribution with $p = .9$. Thus,

$$\beta = \sum_{x=0}^{7} C_x^{10}(.9)^x(.1)^{10-x} = .07$$

[This summation can be obtained directly from Table 1(b) in the Appendix.] The probability distribution for $n = 10$ and $p = .9$ is shown in Figure 5.5; β is represented by the portion of the probability distribution to the left of 7.5.

To summarize the implications of parts a and b, $\alpha = .055$ and $\beta = .07$ give measures of the risks of making the two (and only two) types of errors for this statistical test. The probability that the test statistic will, by chance, fall in the rejection region when the null hypothesis is true is only .055. That is, the probability of concluding that the new substitute is preferred, when in fact it is not, is only .055. But suppose that the new substitute is really preferred and that the probability of preference is .9. What is the probability of accepting the null hypothesis of no preference? We have shown that the risk of making this type II error is only $\beta = .07$.

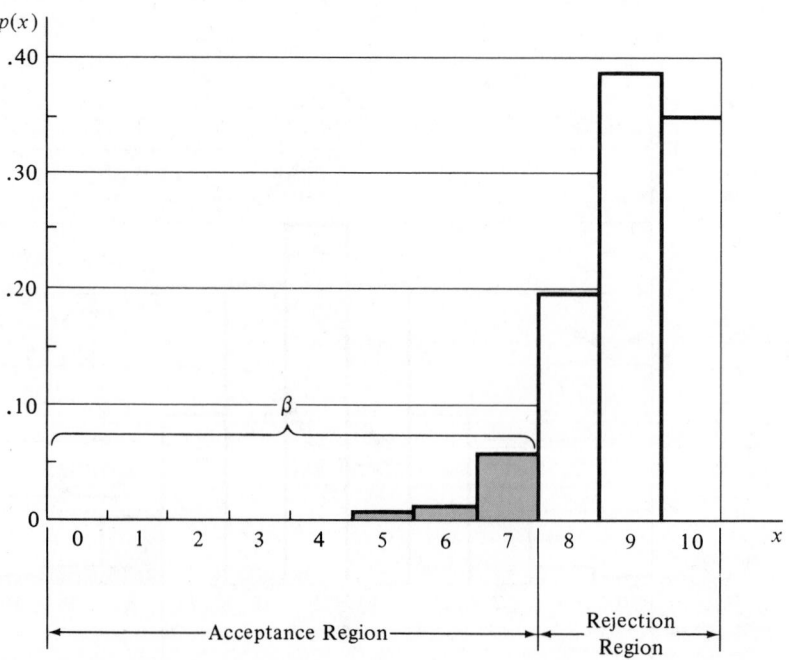

Figure 5.5   Binomial Probability Distribution for $n = 10$ and $p = .9$ in Example 5.13

> ## Tips on Problem Solving
>
> When you calculate α, you want to find the probability that x falls in the *rejection region* given the value of p as specified. When you calculate β, you want to find the probability that x falls in the *acceptance region* given some alternative value of p.

EXERCISES   Conceptual Level

5.51   Define the terms *null* and *alternative hypotheses*.

5.52   What is the relationship between the null hypothesis and the "rejection region"?

5.53   Define the two types of errors that can be made in the hypothesis-testing process.

5.54   If the sample size remains fixed, what happens to the probability of a type II error when the probability of a type I error is increased?

EXERCISES   Skill Level

5.55   A binomial experiment is conducted and the random variable x, the number of successes, is observed for $n = 15$ trials. The researcher would like to show that p, the unknown probability of success, is greater than .7.

   a.   State the null and alternative hypotheses to be tested.

   b.   What could the researcher use as a test statistic?

   c.   The researcher has chosen values of x greater than or equal to 14 as his rejection region. Calculate $\alpha = P(x \geq 14 \text{ when } p = .7)$.

   d.   If the observed value of the test statistic is $x = 11$, what conclusion would the researcher draw?

5.56   Refer to Exercise 5.55. The true value of the binomial proportion is $p = .8$. Calculate $\beta = P(x < 14 \text{ when } p = .8)$ by using Table 1 in the Appendix with $n = 15$.

5.57   State the null and alternative hypotheses appropriate for testing the binomial proportion p in the following situations.

   a.   The researcher wishes to show that p is greater than $1/3$.

   b.   The researcher wishes to show that p is different from $1/2$.

   c.   The researcher wishes to show that p is less than .25.

5.58   For a test of the null hypothesis $p = .5$ versus the alternative hypothesis $p \neq .5$, three possible rejection regions have been suggested when $n = 20$ and the test statistic is x, the number of successes in the $n = 20$ trials.

   I: $\{x \leq 2 \text{ or } x \geq 18\}$

   II: $\{x \leq 3 \text{ or } x \geq 17\}$

   III: $\{x \leq 4 \text{ or } x \geq 16\}$

a. Calculate α for each of the three possible rejection regions.

b. If the experimenter wants to fix α, the probability of falsely rejecting the null hypothesis, at $\alpha \approx .01$, which rejection region should he use?

c. The observed value of the test statistic is $x = 17$. What decision should be made when using the rejection region in part b?

EXERCISES   Applications

 5.59 A packaging experiment was conducted by placing two different package designs for a breakfast food side by side on a supermarket shelf. The objective of the experiment was to see whether buyers indicated a preference for one of the two package designs. In a given day six customers purchased a package from the supermarket, with one choosing package design 1 and five choosing design 2.

a. State the null hypothesis to be tested. (*Hint:* The null hypothesis should state equal preference for the two designs.)

b. Let x be the number of buyers who choose the second package design. What is the value of α for the test if the rejection region includes $x = 0$ and $x = 6$?

c. What is the value of β for the alternative $p = .8$ (that is, 80% of the buyers actually favor the second package design)?

d. In context of the problem, give a practical interpretation of the type I error and the type II error.

5.60 Brand preference studies are often conducted by providing a complimentary supply of two competing brands in similar unmarked packages to a selected group of consumers. After a trial period, each consumer states his or her brand preference. Under the hypothesis of no preference between brands, it is assumed that $p = 1/2$, where p is the probability a selected consumer favors brand A. Suppose a brand preference study of two brands is conducted among 15 consumers, and suppose further that there is actually no difference in the quality of the brands.

a. What is the probability that 10 or more consumers would state a preference for brand A?

b. What is the probability that 10 or more consumers would state a preference for either brand A or brand B?

5.61 Continuing Exercise 5.60, let p be the probability that a consumer will choose brand B in preference to A, and suppose that we wish to test the hypothesis that there is no observable difference between the brands—in other words, that $p = 1/2$. Let x, the number of times that B is preferred to A, be the test statistic.

a. Calculate the value of α for the test if the rejection region is chosen to include $x = 0, 1, 2, 3, 12, 13, 14,$ and 15.

b. If p is really equal to .8, what is the value of β for the test defined in a? (Note that β is the probability that $x = 4, 5, \ldots, 10, 11$ given that $p = .8$.)

5.62 Cable television franchises have become extremely attractive investments not only because of their current popularity but also because of their growth potential should cable systems be developed for emergency and commercial purposes. A cable television company has determined from past experience that offering cable services to a community is profitable if more than 20% of the community's households become initial subscribers. To examine the feasibility of entering a new community, a cable company will choose $n = 25$ households at random from the community with the decision to go ahead with the cable project if 5 or more indicate they would become subscribers.

a. What is the hypothesis to be tested in this case?

b. What is the probability of the type I error for this test?

c. If the true proportion of householders who would become initial subscribers is actually .1, what is the probability of the type II error for the test?

 5.63 In an effort to determine the effect of the 1986 Tax Reform Act on charitable giving by adults in upper-income households ($40,000 or more per year), a telephone survey of 500 households was conducted by the Gallup Organization for the Fund-Raising Counsel Trust for Philanthropy (*Journal of Accountancy*, January 1988, p. 93). Suppose that prior to the full survey a test sample of $n = 10$ households revealed that $x = 8$ of the households intended to contribute the same amount or more than they had in 1986. Based on the test sample, is there sufficient evidence to conclude that, in general, the proportion of households that intend to maintain or increase their charitable contributions exceeds 50%? Base your decision on a formal test of an hypothesis concerning p.

5.8

A General Comment

A discussion of the theory of tests of hypotheses may seem a bit premature at this point, but it provides an introduction to a line of reasoning that is sometimes difficult to grasp and that is best understood when it is allowed to incubate in your mind over a period of time. Thus, some of the exercises at the end of Chapter 5 involve the use of the binomial probability distribution and, at the same time, lead you to utilize the reasoning involved in statistical tests of hypotheses. We expand on these ideas through examples and exercises in Chapter 6, and we discuss in detail the topic of statistical tests of hypotheses in Chapter 8 and succeeding chapters.

5.9

Summary

Three useful discrete probability distributions were presented in this chapter—the binomial, the Poisson, and the hypergeometric. These probability distributions, discussed in Sections 5.1 through 5.6, enabled us to calculate the probabilities associated with several events that are of interest in business. More important, they provided the necessary mechanism to illustrate how statistical inferences are made. Thus, in Section 5.7 we used the binomial probability distribution to make inferences concerning the relative desirability of a sugar substitute. This example of a statistical inference resulted in a decision concerning the parameter p of a binomial population.

The binomial probability distribution allows us to calculate the probability of x successes in a series of n identical independent trials, where the probability of a success in a single trial is equal to p. The binomial experiment is an excellent model for many sampling situations, particularly market surveys that result in "yes" or "no" types of data.

The Poisson probability distribution is important because it can be used to approximate certain binomial probabilities when n is large and p is small. Consequently, it can greatly reduce the computations involved in calculating binomial probabilities. In addition, the Poisson probability distribution is important in its own right. It provides an excellent probabilistic model for the number of occurrences of rare events in time or space.

The hypergeometric probability distribution is also related to the binomial probability distribution. It gives the probability of drawing x elements of a particular type from a population where the number N of elements in the population is small in relation to the sample size n. The binomial probability distribution applies to the same situation except that it is appropriate only when N is large in relation to n.

Perhaps the most important aspect of Chapter 5 is the introduction to statistical inference—making an inference about a population based on information contained in a sample. For illustrative purposes we chose to make inferences about the binomial parameter p, the proportion of elements in a large population that possesses a specified characteristic. For the sugar substitute experiment we made a decision about whether the probability of preferring the new substitute was greater than the probability of preferring the old. This example shows how probability is used to make inferences (for this example the inference was a decision), and more important, it shows how to evaluate the "goodness" of the decision (the probability of making erroneous decisions). We will expand on these ideas in the following chapters.

Supplementary Exercises

Applications

5.64 Which of the following business problems can be modeled by the binomial distribution? For those that cannot be modeled by the binomial distribution, explain why.

    a. Determination of the probability that 5 of 10 assembly machines break down in a given day when the probability any 1 will break down in a day is .15

    b. Computation of the probability that at least one assembly machine will break down in a given day when the probability that any one will break down in a given day is .15

    c. Calculation of the probability of selecting 2 defective transistors in a sample of 5 transistors drawn from a bin containing 100 transistors, of which 10 are defective

    d. Determination of the probability of selecting two defective transistors in a sample of five transistors drawn from an assembly line in which each transistor produced is defective with a probability of .1

    e. Computation of the probability of selecting only 2 men in a group of 6 selected from a committee consisting of 10 men and 4 women

5.65 The proportion of residential households in Burlington, Vermont, that are heated by natural gas is approximately .2. A random sample of 25 residences is selected from the city of Burlington. Assume that the properties of a binomial experiment are satisfied.

    a. Find the probability that none of the households are heated by natural gas.

    b. Find the probability that no more than 7 of the 25 are heated by natural gas.

    c. Why might the binomial experiment not provide a good model for this sampling situation?

5.66 It is known that 10% of a brand of television tubes will burn out before their guarantee has expired. If 1,000 tubes are sold, find the expected value and variance of x, the number of original tubes that must be replaced. Within what limits would x be expected to fall? (*Hint:* Use Tchebysheff's Theorem.)

 5.67 Because of increased technical complexity in product lines, there has been a trend toward product line specialization and away from traditional general-line activities among independent wholesalers. *Industrial Distribution,* which conducts an annual census of industrial distributors, reports that currently only 10% of industrial distributors categorize themselves as "general-line" wholesalers. A manufacturer has randomly selected five wholesalers from a list of distributors to discuss with them the possibility of serving as the distribution agent for the firm's products.

    a. What is the probability that none of the wholesalers are general-line wholesalers?

    b. What is the probability that no more than one is a general-line wholesaler?

 5.68 Following federal deregulation of the airline industry, competition among carriers resulted in substantial decreases in airfares on most airlines as new carriers entered the market and established carriers extended service to new locations. Nonetheless, within this new environment flight delays and cancellations have caused complaints about service to soar to an all-time high; reports of midair near collisions have resulted in demands for increased safety; and mergers have given carriers sufficient market clout that fliers are losing their initial deregulation benefits. A Roper Organization poll of 1997 adults conducted in July 1987 revealed that approximately 50% of those polled felt that airlines should be required to get government permission to raise or lower fares (up from 35% in 1984), and 35% (up from 16% in 1984) felt that there was not enough government regulation of airlines (Laurie McGinley, "Bad Air Service Prompts Calls for Changes," *Wall Street Journal,* November 9, 1987, p. 29). Suppose that these proportions are true for the population at large. In a random sample of $n = 30$ adults, the number x in the sample who feel that government permission should be required before an airline raises or lowers fares, is recorded.

    a. What is the mean and standard deviation of x?

    b. Within what limits would you expect x to lie?

    c. What would you conclude about the value of p if $x = 28$ adults in the sample felt that airlines should get government permission to raise or lower fares?

5.69 Refer to Exercise 5.68. Suppose that among the $n = 30$ adults in the same sample, the number x in the sample who felt that there was not enough regulation of airlines was also recorded. Use the MINITAB CDF printout in Table 5.9 to answer the following questions.

    a. Find $P(x \leq 10)$ [which is the same as $P(x \leq \mu)!$].

    b. Find $P(8 \leq x \leq 12)$.

    c. Would you expect to find a value of x greater than or equal to 20 if, in fact, $p = .35$?

    d. Would the value of $p = .35$ be suspect if the observed value of x was as small or smaller than 4?

 5.70 A city commissioner claims that 80% of all people in the city favor garbage collection by contract to a private concern (rather than collection by city employees). To check the theory that the proportion of the people in the city favoring private collection is .8, you randomly sample 25 people and find that x, the number of people who support the commissioner's claim, is 22.

    a. What is the probability of observing at least 22 who support the commissioner's claim if, in fact, $p = .80$?

    b. What is the probability that x is exactly equal to 22?

Table 5.9   MINITAB Output for Exercise 5.69

```
MTB > CDF;
SUBC> BINOMIAL 30.35.

        BINOMIAL WITH N =   30   P = 0.350000
          K   P(X  LESS OR = K)
          1            0.0000
          2            0.0003
          3            0.0019
          4            0.0075
          5            0.0233
          6            0.0586
          7            0.1238
          8            0.2247
          9            0.3575
         10            0.5078
         11            0.6548
         12            0.7802
         13            0.8737
         14            0.9348
         15            0.9699
         16            0.9876
         17            0.9955
         18            0.9986
         19            0.9996
         20            0.9999
         21            1.0000
```

5.71   A large retailing chain that transacts much of its business by mail order has determined from past evidence that 90% of all orders are properly filled as specified by the customer. If 100 orders are received in a given week, find the expected value and variance of x, the number of orders that were properly filled. Within what limits would x be expected to fall?

5.72   Refer to Exercise 5.71. Suppose that the retailing chain wishes to explore in greater depth customer satisfaction with the firm's mail-order service. From a file of recent mail-order customers $n = 25$ customers are chosen at random for questioning.

a.   Find the probability that all 25 received their order as specified.

b.   Find the probability that no more than 2 orders were improperly filled by the mail-order department of the retailing chain.

5.73   Refer to Exercise 5.72. Suppose the firm wishes to examine the effect of a new policy designed to reduce the incidence of customer dissatisfaction with the firm's mail-order service. To measure the effectiveness of the new policy, the firm contacts $n = 20$ recent mail-order customers at random, with $x = 1$ stating that his order was not filled as specified.

a.   State the hypothesis to be tested.

b.   If we assume $\alpha = .10$ or less, do these data present sufficient evidence to indicate that the new policy is effective in reducing errors in filling orders from mail-order customers?

c. What are the implications to the retailing chain of committing a type I error? Of committing a type II error?

5.74 Imported cars comprise approximately 25% of all new-car sales in the United States [U.S. Department of Commerce, Bureau of the Census, *Statistical Abstract of the United States*, 107th ed. (Washington, D.C., 1988), p. 586]. Suppose we randomly select $n = 4$ people who have purchased a new car during the past week.

    a. Find the probability that all of them purchased an imported car.

    b. Find the probability that one of them purchased an imported car.

    c. Find the probability that none of them purchased an imported car.

5.75 Refer to Exercise 5.74, and let x be the number of buyers in the sample of $n = 4$ new-car buyers who purchase an imported car. Construct a probability histogram for $p(x)$.

5.76 A gasoline service station offers gasoline for sale at a 2-cents-per-gallon discount if the customer pays in cash and does not use a credit card. Past evidence has shown that 40% of all customers choose to pay in cash. During a given day 25 customers buy gasoline at the service station.

    a. Find the probability that at least 10 pay in cash.

    b. Find the probability that no more than 20 pay in cash.

    c. Find the probability that more than 10 but less than 15 pay in cash.

5.77 It is known that 90% of those who purchase a color television will not have claims against the guarantee during the duration of the guarantee. Suppose that each of 25 customers buys a color television set from a certain appliance dealer. What is the probability that at least 4 of these 25 customers will have claims against the guarantee? (Use Table 1 in the Appendix.)

5.78 Refer to Exercise 5.77. What is the expected value and standard deviation of x, the number of claims from 25 buyers? Within what limits would x be expected to fall?

5.79 Suppose that 1 of 10 undergraduate college textbooks is an outstanding financial success. A publisher has selected 10 new textbooks for publication.

    a. What is the probability that exactly 1 will be an outstanding financial success?

    b. What is the probability that at least 1 will be an outstanding financial success?

    c. What is the probability that at least 2 will be outstanding financial successes?

5.80 A manufacturing process that produces electronic components is known to have a 5% defective rate. Suppose a sample of $n = 25$ is selected from the manufacturing process.

    a. Find the probability that no more than 2 defectives are found.

    b. Find the probability that exactly 4 defectives are found.

    c. Find the probability that at least 3 defectives are found.

5.81 Over the past decade the median price for a new home has risen substantially. This increase in housing prices has led to a situation in which many American families seeking their first home are now priced out of homeownership. Suppose that only 20% of all American families seeking their first home are now able to afford a median-priced home, and that we randomly select $n = 5$ families who seek their first home. Let x be the number who are able to afford an $85,000 (median-priced) home [*Press Enterprise* (Riverside, Calif.), February 17, 1988].

    a. Find the probability that none of the 5 families are able to afford an $85,000 home.

    b. Find the probability that no more than 2 of the 5 families are able to afford an $85,000 home.

    c. Construct a probability histogram for $p(x)$.

5.82 Refer to Exercise 5.81. Suppose that in a certain city $n = 2,000$ families will seek their first home during the next 12 months. Assuming they are representative of all American families seeking their first home, find the expected value and variance of x, the number of families who will be able to afford a median-priced home. Within what limits would x be expected to fall? What do these data suggest regarding future housing trends in America?

5.83 A mail-order magazine subscription service considers any mail advertisement successful if at least 20% of those receiving an advertisement respond favorably by ordering a subscription. Let p be the probability that a single recipient of advertising will order a subscription. What is the smallest value for p in order to have a probability of 90% that at least 20% of 25 recipients (i.e., at least 5) of an advertisement respond favorably?

5.84 A retail variety store that advertises extensively by mail circulars expects a sale from 1 of every 20 mailings. Suppose 25 prospects are randomly selected from a citywide mailing.

   a. How many sales can the store expect to result from this sample of 25?

   b. What is the probability that no sales will result from mailings to this group of 25 prospects?

   c. What is the probability that at least 3 sales will result from mailings to the 25 prospects?

   d. Suppose that the 25 prospects had been selected from a single city neighborhood. Would this sample satisfy the properties of a binomial experiment?

5.85 In recent years the National Science Foundation (NSF) has offered considerable funding for research on detecting inferior products in the developmental phase and thus reducing the incidence of product failure in the marketplace. One area where considerable efforts have been expended in these studies is the electronics industry. Suppose past evidence suggests that the success rate of new products in the electronics industry is no greater than 20%. If $n = 25$ new products developed within the past year by the electronics industry are monitored in the marketplace and 8 are found to become commercial successes, do these data suggest the new-product success rate has improved for the electronics industry (perhaps owing to recent NSF studies)? State the null hypothesis to be tested and use x as a test statistic.

5.86 A certain machine is said to be in control if the proportion of defective items manufactured by the machine is not greater than 10%. In a check of whether the machine is in control, 10 finished items are randomly selected from its output. The null hypothesis that the machine is in control will be rejected if 3 or more defectives are found.

   a. What is the probability of the type I error for this test?

   b. If the machine is really out of control and the probability of a defective is .3, what is the probability of the type II error for the test?

5.87 A manufacturer of floor wax has developed two new brands, A and B, that he wishes to subject to a consumer evaluation to determine which of the two is superior. Both waxes A and B are applied to floor surfaces in each of 15 homes. If there is actually no difference in the quality of the brands, what is the probability that 10 or more consumers would state a preference for brand A?

5.88 Continuing Exercise 5.87, let p be the probability that a consumer will choose brand A in preference to B, and suppose that we wish to test the hypothesis that there is no observable difference between the brands—in other words, that $p = 1/2$. Let x, the number of times that A is preferred to B, be the test statistic.

   a. Calculate the value of α for the test if the rejection region is chosen to include $x = 0, 1, 14,$ and 15.

   b. If p is really equal to .8, what is the value of β for the test defined in part a? (Note that β is the probability that $x = 2, 3, \ldots, 12, 13$ given that $p = .8$.)

5.89 Continuing Exercise 5.87, suppose that the rejection region is enlarged to include $x = 0, 1, 2, 13, 14, 15$.

a. What is the value of α for the test? Should this probability be larger or smaller than the answer found in Exercise 5.88a?

b. If p is really equal to .8, what is the value of β for the test? Compare your answer here with your answer in part b of Exercise 5.88.

5.90 When evaluating competing capital items, a manufacturer must consider the cost and performance characteristics of each item. Suppose a manufacturer of electrical fuses must decide between two assembly machines A and B. Since the machines have identical costs, the manufacturer decides to base his choice on the performance characteristics of the machines. The number of defective electrical fuses produced by each of the two machines A and B is recorded daily for a period of ten days, with the results as given in the accompanying table. Assume that both assembly machines produce the same daily output. Compare the number of defectives produced by A and B each day, and let x be the number of days when B exceeds A. Do the data present sufficient evidence to indicate that the number of defectives per day from machine B exceeds the number from machine A on more than half of all days? State the null hypothesis to be tested, and use x as a test statistic. Let $\alpha = .05$.

Day	1	2	3	4	5	6	7	8	9	10
A	172	165	206	184	174	142	190	169	161	200
B	201	179	159	192	177	170	182	179	169	210

5.91 Which of the following business problems can be modeled by the Poisson distribution? For those that cannot be modeled by the Poisson distribution, explain why.

a. Determination of the probability that 2 of 10 city buses will break down during a given day when the probability that any 1 will break down is .01

b. Computation of the probability that an insurance company will not have to pay out on any fire damage claims during a year, given that the company has insured 1,000 firms against fire damage and the probability that any 1 of the firms incurs a fire during a given year is .002

c. Calculation of the probability that a telephone switchboard receives at least 5 incoming calls during a given hour, when incoming calls normally arrive randomly and independently of one another at an average rate of 1 every 15 minutes

MKTG **d.** Determination of the probability that a saleswoman consummates at least 25 sales in 100 contacts when the probability that she consummates a sale on any contact is .4

5.92 Highway engineers and patrol officers naturally tend to focus attention on roadways that have higher-than-average accident rates. In doing so, they consider policies such as reduced speed limits or the placement of caution or stop signals that may alleviate the problem. Assume that accidents occur randomly and independently of one another over a specified section of the highway at the average rate of two per week. If officials devote particular attention to the specified section of the highway during a given week, find the probability that they observe no accidents during that period.

5.93 Logging trucks have a special problem with tire failure because of the rough terrain they are often required to traverse. Suppose that a logging company with 100 trucks has reason to believe that the average number of trucks with at least one tire failure in a given day is 5.

a. Find the probability that during a given day none of the trucks have tire failure.

b. Find the probability that during a given day 5 have tire failure.

c. Find the probability that during a given day not more than 3 have tire failure.

5.94 Refer to Exercise 5.93. Suppose that a truck has a mechanical breakdown during any given day with a probability of .01.

    a. Find the probability that during a given day none of the trucks have a mechanical breakdown.

    b. Find the probability that during a given day at least 2 of the trucks have a mechanical breakdown.

    c. Assuming that tire failures and mechanical breakdowns are independent occurrences in the 100 logging trucks, find the probability that during a given day none of the 100 logging trucks has either a tire failure or a mechanical problem.

 5.95 Although labor disputes historically have focused almost exclusively on economic benefits, recent evidence suggests that workers are now equally concerned about matters related to the quality of their work environment. For example, in one recent incident a group of Weyerhaeuser employees in a Washington mill walked off the job because of expressed concerns about health hazards resulting from volcanic ash deposits. Wood products industry officials now claim that major mill closures over noneconomic issues occur at the rate of one every six months. Assume that this rate is accurate.

    a. Find the probability that no major mill closures will occur during a given year over noneconomic issues.

    b. Find the probability that noneconomic factors will be the cause for at least two major mill closures during a given year.

 5.96 In an attempt to minimize the chances of loan defaults, banks, credit unions, and other loan-granting institutions employ a set of rigid criteria to evaluate loan applications. Nevertheless, loan defaults still occur. A full-service commercial bank in Portland, Oregon, reports that defaults on personal loans of less than $2,500 have occurred randomly and independently of one another since January 1971 at the average rate of 1.5 defaults per month.

    a. Find the probability of no defaults on personal loans of less than $2,500 during a given month.

    b. Find the probability of no more than one loan default during a given two-month span.

 5.97 Because of the stalemate in the development of thermal (coal and nuclear) power generation facilities, many power company officials are concerned about the possibility of power shortages. The Department of Energy (DOE) suggested that we would average one major power outage a year in the highly industrialized eastern seaboard area. If the DOE projection is true, find the probability that in a given year no major power outages occur in the eastern seaboard region of the United States. What is the probability that none occur in two consecutive years?

 5.98 Since the 1979 accident at Three Mile Island, the Nuclear Regulatory Commission (NRC) has intensified its monitoring of existing nuclear power generating facilities. Suppose that during a given month the NRC randomly selects three nuclear power plants from among the seven nuclear facilities within the northwest power pool (Washington, Oregon, Idaho, and Montana) and subjects each to intensive investigation. Suppose further that, unknown to the NRC or the operating utilities, two of the seven facilities contain potentially dangerous mechanical flaws.

    a. Let x be the number of nuclear power plants in the sample of $n = 3$ chosen by the NRC for investigation that contain potentially dangerous mechanical flaws. Explain why x possesses a hypergeometric probability distribution.

    b. What is the probability that both of the plants containing potentially dangerous mechanical flaws will be chosen for investigation?

    c. What is the probability that at least one of the plants containing potentially dangerous mechanical flaws will be chosen?

5.99 An auditor for the Internal Revenue Service is provided with 50 tax returns for her review, 7 of which contain errors of misrepresentation of fact. As a matter of policy the auditor subjects 10% of all returns submitted for her review to an intensive audit, giving only a superficial review of arithmetic computations to the remaining tax returns. Let x be the number of returns subjected to intensive investigation that contain errors of misrepresentation of fact.

  a. Find the probability that at least 1 return containing misrepresentation of fact is subjected to an intensive audit.

  b. Find the probability that all 5 returns subjected to an intensive audit contain misrepresentations of fact.

  c. Find the probability that none of the returns with misrepresentations are subjected to intensive audit.

5.100 Critics of the Financial Accounting Standards Board (FASB) and the accounting profession demand a system of accounting that is as useful to government policymakers and consumer groups as it is to investors and creditors. To examine the importance of this issue among members of Congress, an FASB official randomly selects 5 from among the 20 members of his state's congressional delegation for a panel discussion of the proper role of the FASB. Unknown to the official, 12 members of the 20-person congressional delegation are strong critics of current FASB policy. Let x be the number of FASB critics chosen from the congressional delegation for the discussion panel.

  a. Explain why x possesses a hypergeometric probability distribution. Give the formula for $p(x)$.

  b. Find the probability that no more than 2 of the 5 members of Congress chosen for the discussion panel are FASB critics.

5.101 About 70% of all mutual funds are no-load funds; that is, they do not require a sales commission for a buy-sell transaction. This figure is up from 10% in 1970, largely owing to the current inflationary economy in which investors seek the flexibility of switching between stock mutual funds and high-yield money market funds ("Mutual Funds Resurge," *Business Week*, March 31, 1980). An investment advisor for a corporate pension fund has randomly chosen for consideration 5 mutual funds from 100 publicly listed funds, 70 of which are no-load funds.

  a. Find the probability that at least 2 of the funds chosen for consideration are no-load funds.

  b. Find the probability that all 5 are no-load funds.

  c. If no more than 1 of the mutual funds chosen for consideration is a no-load fund, would you question the investment advisor's claim that the 5 funds were chosen at random from the group of 100 publicly listed funds? Explain.

5.102 A shipment of 200 portable television sets is received by a retailer. To protect herself against a bad shipment, she will inspect 5 sets and accept the entire lot if she observes 0 or 1 defective. Suppose that there are actually 20 defective sets in the shipment.

  a. What is the probability that she accepts the entire shipment?

  b. Given that the retailer accepts the entire lot, what is the probability that she observed exactly 1 defective set?

5.103 A manufacturer of small electronic desk calculators knows from experience that 1% of all the calculators manufactured and sold by his firm are defective and will have to be replaced under the warranty. A large accounting firm purchases 500 calculators from the manufacturer for use by its employees.

  a. Find the probability that none of the calculators will have to be replaced.

  b. Find the probability that no more than 4 will have to be replaced.

c. Find the probability that at least 2 will have to be replaced.

d. What is the expected number of desk calculators purchased by the accounting firm that can be expected to fail and must be replaced under the warranty?

5.104 Following the October 1987 market plunge, approximately 90% of investors surveyed did not sell any stock or stock-oriented mutual funds, and of that number 34% said that they held their shares because they took a long-term approach to investments (*Wall Street Journal*, November 9, 1987, p. 10). In a test of whether the actual figure was less than 90%, a sample of $n = 25$ stockholders is interviewed concerning each one's stock transactions during this period.

a. Provide the null and alternative hypotheses consistent with the objective of this follow-up survey.

b. What is an appropriate rejection region if the probability of a type I error is to be no greater than .05?

c. Using the rejection region found in part b, evaluate the probability of a type II error if the actual proportion is, in fact, $p = .7$.

Supplementary Exercises

Interpretive

5.105 The manager of a large motor pool wished to compare the wearing qualities of two different types (type A and type B) of automobile tires. On each of 400 cars he replaced one rear tire with a new tire of type A and the other rear tire with a new tire of type B. When a given car had been driven 10,000 miles, he determined which of the two rear tires experienced the greater wear. Let x be the number of cars out of the 400 cars on which tire A showed the greater wear. Let p be the probability that on a given car the tire of type A will experience the greater wear. Using Tchebysheff's Theorem, find an upper bound to α for a test of the hypothesis that $p = 1/2$ if the rejection region includes the outcomes $x = 0, 1, 2, \ldots, 150, 250, 251, \ldots, 400$.

5.106 Early in the United States missile development program, government and industry defense officials were proclaiming that our missiles were highly reliable, and probabilities of successful firings in the neighborhood of .999 . . . were quoted. Such statements were made even though many missile firings resulted in failure (such as the Navy's Vanguard missile of the 1950s). If the reliability (probability of a successful launch) is even as high as .9, what is the probability of observing three or more failures in a total of four firings? One or more failures? If you observed two or more failures out of four, what would you think about the high claims of reliability of the missiles produced in the 1950s?

5.107 One cause of great concern within the American business community is the diminishing level of worker productivity observed in recent years. In a study of this issue Walton notes that work improvement efforts that have both productivity and quality of work life as goals are more likely to succeed than projects that emphasize one goal to the exclusion of the other (R. Walton, "Work Innovations in the United States," *Harvard Business Review*, July–August 1979). A survey is conducted to determine the response of the corporate sector to Walton's premise; 15 industrialists are contacted, with $x = 9$ indicating support for Walton's point of view. Do these data provide sufficient evidence to indicate that a majority of industrialists believe that both productivity and quality-of-work-life goals must be emphasized to favorably affect worker productivity?

Case Study

Western Energy Services

Section 17 of the Mineral Leasing Act of 1920, as amended, provides that public lands "not within any known geological structure of a producing oil or gas field" are awarded bimonthly on a lottery basis by the Bureau of Land Management (BLM). Under provisions of the law each individual or company may file only one application for each available parcel (accompanied by a nonrefundable $10 filing fee), thereby giving each applicant an equal chance of winning a lease.

Western Energy Services offers geological and managerial services to those interested in participating in the BLM Lottery program for a fee of $25 per application in addition to the $10 BLM filing fee (information obtained from "Western Energy Services Program Guide," Western Energy Services, Las Vegas, 1980). Thus, for a $14,000 investment Western Energy Services will place 400 applications for individuals in 400 BLM lotteries on parcels Western Energy has found to be promising in geological testing.

As of June 1980, Western Energy reports that it has submitted a total of 64,103 filings on leases for subscribing clients and has acquired 178 leases. Thus, the chance of a Western Energy client winning any given lottery is 1/360. In contrast, the success ratio for all applicants participating in BLM lotteries during 1979 was 1/552.

1. If an individual participates with Western Energy and files 400 separate lease applications, what is his probability of winning at least 1 lease? What is his expected number of lease acquisitions in 400 filings through Western Energy?

2. If the individual invests $14,000 privately (without the aid of an investment service) in 1,400 separate BLM lotteries, what is his probability of winning at least 1 lease?

3. Because of the cost and technicality of drilling for oil and gas, leases owned by individuals are almost always sold to a commercial organization. Western Energy reports that the average sales price (in 1980 dollars) of leases sold by its clients is $23,715. Find the expected monetary payoff to an individual who pays Western Energy $14,000 for 400 BLM lease filings.

4. Suppose the current market value of all leases sold by individuals to commercial organizations is $12,000. What is the expected monetary payoff to an individual who privately invests $14,000 in 1,400 separate BLM lease filings?

References and Suggested Readings

Feller, W. *An Introduction to Probability Theory and Its Applications*. Vol. 1, 3d ed. New York: Wiley, 1968, Chap. 6.

Hillier, F. S., and G. J. Liebermann. *Introduction to Operations Research*. 4th ed. San Francisco: Holden-Day, 1986.

National Bureau of Standards. *Tables of the Binomial Probability Distribution*. Washington, D.C.: Government Printing Office, 1949.

Saaty, T. L. *Elements of Queueing Theory: With Applications*. New York: Dover, 1983.

Winston, W. L. *Operations Research: Applications and Algorithms*. Boston: PWS-KENT Publishing Co., 1987.

TO
THE
READER

In the collection and recording of data resulting from an experiment, the random variable of interest may be either discrete or continuous. The general properties of discrete random variables were presented in Chapter 4. Discrete random variables having either a binomial, Poisson, or hypergeometric distribution were the subject of Chapter 5. However, not all random variables are discrete.

Data arising in the areas of business often result from observations on continuous random variables. The normal probability distribution can be used as a model to describe the behavior of a continuous random variable whose distribution is symmetrically mound-shaped with rapidly decreasing tails. In Chapter 6 we not only study the properties of the normal, uniform, and exponential distributions but also show how the normal distribution can be used to approximate binomial probabilities. In subsequent chapters we shall see that the normal probability distribution plays a fundamental role in both testing and estimation procedures.

6

THE NORMAL AND OTHER CONTINUOUS PROBABILITY DISTRIBUTIONS

6.1

Introduction

Continuous random variables, as described in Section 4.4, are associated with the sample spaces representing the infinitely many sample points associated with a line interval. Here are some examples of continuous random variables:

The heights or weights of a group of people

The length of life of a perishable item, such as a light bulb, a machine part, or a food product

The time it takes an individual to perform a task

The time between service calls on an office machine

The round-off error that results when a number is rounded to a given decimal accuracy

In general, any random variable whose values are measurements, as opposed to counts, is a continuous random variable.

In Section 4.4 we saw that the probabilistic model for the frequency distribution of a continuous random variable is represented by a curve, usually a smooth curve, called the *probability distribution* or the *probability density function*. Although these distributions may assume a variety of shapes, many random variables observed in nature possess a relative frequency distribution that is approximately bell-shaped or normal; others may possess a relatively uniform distribution between two points; and still others may possess distributions that begin at zero and are strongly skewed to the right. We will examine the normal, uniform, and exponential distributions and their utility as probability density models.

6.2

The Normal Distribution

Random variables such as heights, weights, distances, and completion times often have a bell-shaped frequency distribution that can be described by a normal probability distribution.

Normal Probability Density Function

$$f(x) = \frac{1}{\sigma\sqrt{2\pi}}\, e^{-(x-\mu)^2/2\sigma^2} \qquad -\infty \le x \le \infty$$

The symbols e and π are mathematical constants given approximately by 2.7183 and 3.1416, respectively; μ and σ ($\sigma > 0$) are parameters representing the population mean and standard deviation.

The normal density function has a total area under its curve equal to 1. The graph of a normal probability distribution with mean μ and standard deviation σ is given in Figure 6.1. From the form of the normal density function and from Figure 6.1, we see that the normal distribution is symmetric about its mean μ. Furthermore, the shape of the distribution is determined by σ, the population standard deviation. Large values of σ reduce the height of the curve and increase the spread; small values of σ increase the height and reduce the spread of the curve.

In practice, we seldom encounter variables that range from infinitely small negative values to infinitely large positive values. Nevertheless, many positive random variables such as heights, weights, and times generate a frequency histogram that is well approximated by a normal distribution. The approximation applies because almost all of the values of a normal random variable lie within three standard deviations of the mean, and in these cases, ($\mu \pm 3\sigma$) almost always encompasses positive values.

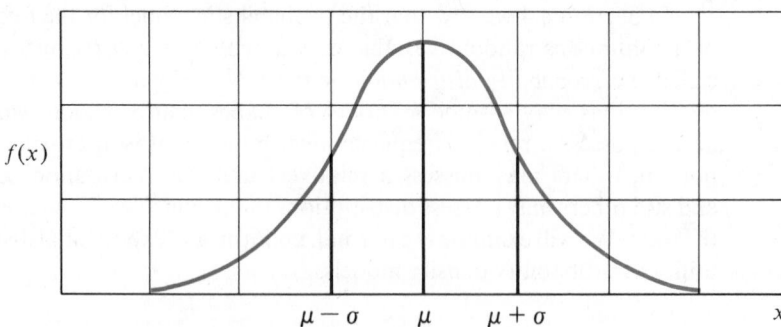

Figure 6.1   Normal Probability Density Function

6.3

Tabulated Areas of the Normal Probability Distribution

In Section 4.4 we explained that the probability that a continuous random variable assumes a value in the interval a to b is the area under the probability density function between the points a and b (see Figure 6.2). The probability model for a continuous random variable differs greatly from the model for a discrete random variable when we consider the probability that x equals some particular value, say a. Since **the area lying over any particular point, say $x = a$, is 0,** it follows from our probability model that the probability that $x = a$ is 0. Thus, the expression $P(x \leq a)$ is the same as $P(x < a)$ because $P(x = a) = 0$. Similarly, $P(x \geq a) = P(x > a)$. This statement is, of course, not true for a discrete random variable because $P(x = a)$ may not equal 0.

To find areas under the normal curve, we first note that the equation for the normal probability distribution (Section 6.2) is dependent on the numerical values of μ and σ and that by supplying various values for these parameters, we could generate an infinitely large number of bell-shaped normal distributions. A separate table of areas for each of these curves is obviously impractical; rather, we would like one table of areas applicable to all. The easiest way to use one table is to work with areas lying within a specified number of standard deviations of the mean, as was done in the case of the Empirical Rule. For instance, we know that approximately .68 of the area will lie within one standard deviation of the mean, .95 within two, and almost all within three. But what fraction of the total area will lie within .7 standard deviation, for instance? Questions of this type can be answered by using Table 3 in the Appendix.

Since the normal curve is symmetrical about the mean, half of the area under the curve lies to the left of the mean and half to the right (see Figure 6.3). Also, because of the symmetry, we can simplify the table of areas by listing the areas between the mean and a specified number z of standard deviations to the right of μ. An area to the left of the mean can be calculated by using the corresponding and equal area to the right of the mean. The distance from the mean to a given value of x is $(x - \mu)$. Expressing this

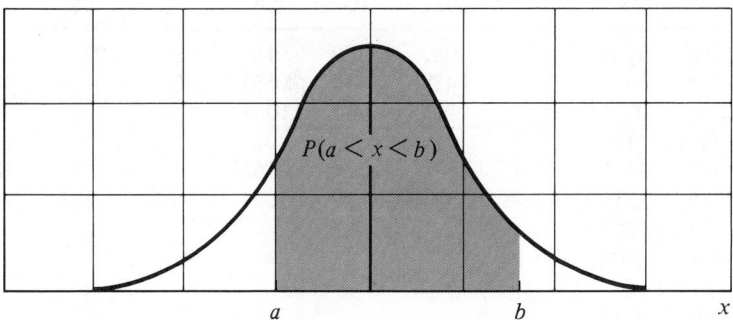

Figure 6.2   The Probability $P(a < x < b)$ for a Continuous Random Variable

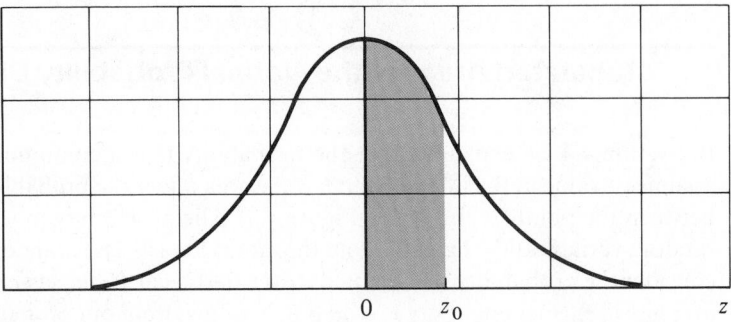

Figure 6.3   Standardized Normal Distribution

distance in units of the standard deviation σ, we obtain

$$z = \frac{x - \mu}{\sigma}$$

standardized normal distribution

Note that there is a one-to-one correspondence between the random variables z and x and, in particular, that $z = 0$ when $x = \mu$. The probability distribution for z is often called the **standardized normal distribution**, because its mean is 0 and its standard deviation is 1. It is shown in Figure 6.3. The area under the standard normal curve between the mean $z = 0$ and a specified value of z, say z_0, is the probability $P(0 \le z \le z_0)$. This area is recorded in Table 3 of the Appendix and is shown as the shaded area in Figure 6.3. An abbreviated version of Table 3 in the Appendix is shown in Table 6.1.

Note that z, correct to the nearest tenth, is recorded in the left-hand column. The second decimal place for z, corresponding to hundredths, is given across the top row. Thus, the area between the mean and $z = .7$ standard deviation to the right, read in the second column of the table opposite $z = .7$, is found to be .2580. Similarly, the area

Table 6.1   Abbreviated Version of Table 3 in the Appendix

z_0	.00	.01	.02	.03	.04	.05	.06	.07	.08	.09
0.0	.0000	.0040	.0080	.0120	.0160	.0199	.0239	.0279	.0319	.0359
0.1	.0398	.0438	.0478	.0517	.0557	.0596	.0636	.0675	.0714	.0753
0.2	.0793	.0832	.0871	.0910	.0948	.0987	.1026	.1064	.1103	.1141
0.3	.1179	.1217	.1255	.1293	.1331	.1368	.1406	.1443	.1480	.1517
0.4	.1554	.1591	.1628	.1664	.1700	.1736	.1772	.1808	.1844	.1879
0.5	.1915	.1950	.1985	.2019	.2054	.2088	.2123	**.2157**	.2190	.2224
0.6	.2257	:	:	:	:	:	:	:	:	:
0.7	**.2580**									
:	:									
1.0	**.3413**									
:	:									
2.0	**.4772**									

between the mean and $z = 1.0$ is .3413. The area lying within one standard deviation on either side of the mean would be two times .3413, or .6826. The area lying within two standard deviations of the mean, correct to four decimal places, is $2(.4772) = .9544$. These numbers agree with the approximate values, 68% and 95%, used in the Empirical Rule in Chapter 2.

To find the area between the mean and a point $z = .57$ standard deviation to the right of the mean, proceed down the left-hand column to the 0.5 row. Then move across the top row of the table to the .07 column. The intersection of this row-column combination gives the appropriate area, .2157.

Since the normal distribution is continuous, the area under the curve associated with a single point is equal to 0. Keep in mind that this result applies only to continuous random variables. Later in this chapter we will use the normal probability distribution to approximate the binomial probability distribution. The binomial random variable x is a discrete random variable. Hence, as you know, the probability that x takes some specific value, say $x = 10$, will not necessarily equal 0. Consequently, for discrete random variables $P(x \leq x_0)$ is not the same as $P(x < x_0)$.

Let us now consider some examples.

EXAMPLE 6.1   Find $P(0 \leq z \leq 1.63)$. This probability corresponds to the area between the mean ($z = 0$) and a point $z = 1.63$ standard deviations above the mean (see Figure 6.4).

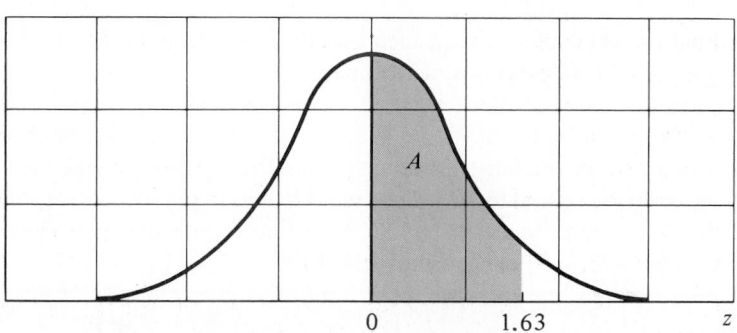

Figure 6.4   *Probability Required for Example 6.1*

SOLUTION   The area is shaded and indicated by the symbol A in Figure 6.4. Since Table 3 in the Appendix gives areas under the normal curve to the right of the mean, we need only find the table value corresponding to $z = 1.63$. Proceed down the left-hand column of the table to the row corresponding to $z = 1.6$ and across the top of the table to the column marked .03. The intersection of this row and column combination gives the area, $A = .4484$. Therefore, $P(0 \leq z \leq 1.63) = .4484$.

◆◆

EXAMPLE 6.2   Find $P(-.5 \leq z \leq 1.0)$. This probability corresponds to the area between $z = -.5$ and $z = 1.0$, as shown in Figure 6.5.

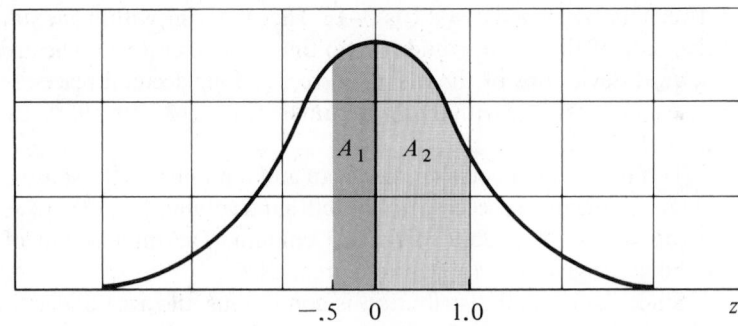

Figure 6.5   Area Under the Normal Curve in Example 6.2

SOLUTION   The area required is equal to the sum of A_1 and A_2 shown in Figure 6.5. From Table 3 in the Appendix we read $A_2 = .3413$. The area A_1 equals the area between $z = 0$ and $z = .5$, or $A_1 = .1915$. Thus, the total area is

$$A = A_1 + A_2 = .1915 + .3413 = .5328$$

That is, $P(-.5 \leq z \leq 1.0) = .5328$.

EXAMPLE 6.3   Find the value of z, say z_0, such that (to four decimal places) .95 of the area is within $\pm z_0$ standard deviations of the mean.

SOLUTION   Half of the area, $(1/2)(.95) = .475$, will lie to the left of the mean and half to the right, because the normal distribution is symmetrical. Thus, we seek the value z_0 corresponding to an area equal to .475. The area .475 falls in the row corresponding to $z = 1.9$ and the .06 column. Hence, $z_0 = 1.96$. Note that this result is very close to the approximate value, $z = 2$, used in the Empirical Rule.

EXAMPLE 6.4   Let x be a normally distributed random variable, with a mean of 10 and a standard deviation of 2. Find the probability that x lies between 11 and 13.6.

SOLUTION   As a first step, we must calculate the values of z corresponding to $x_1 = 11$ and $x_2 = 13.6$. Thus, we have

$$z_1 = \frac{x_1 - \mu}{\sigma} = \frac{11 - 10}{2} = .5 \qquad z_2 = \frac{x_2 - \mu}{\sigma} = \frac{13.6 - 10}{2} = 1.8$$

The desired probability is therefore $P(.5 \leq z \leq 1.8)$ and is the area lying between z_1 and z_2, as shown in Figure 6.6. The area between $z = 0$ and z_1 is $A_1 = .1915$, and the area between $z = 0$ and z_2 is $A_2 = .4641$; these areas are obtained from Table 3. The desired probability is equal to the difference between A_2 and A_1; that is,

$$P(.5 \leq z \leq 1.8) = A_2 - A_1 = .4641 - .1915 = .2726$$

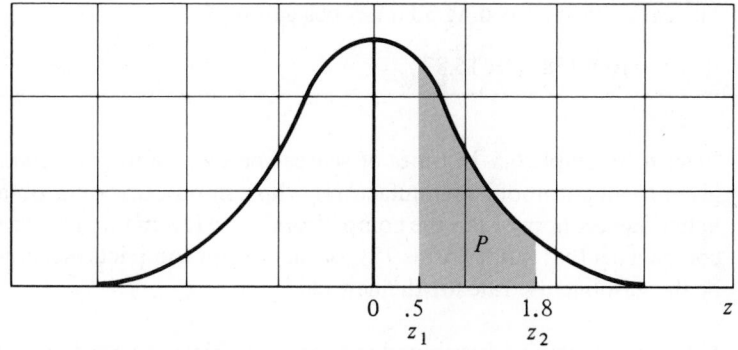

Figure 6.6   Area Under the Normal Curve in Example 6.4

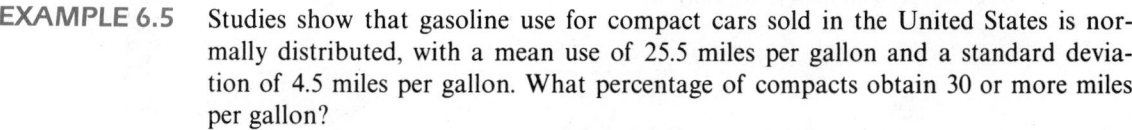

EXAMPLE 6.5   Studies show that gasoline use for compact cars sold in the United States is normally distributed, with a mean use of 25.5 miles per gallon and a standard deviation of 4.5 miles per gallon. What percentage of compacts obtain 30 or more miles per gallon?

SOLUTION   The proportion of compacts obtaining 30 or more miles per gallon is given by the shaded area in Figure 6.7.

We must find the z value corresponding to $x = 30$. Substituting into the formula for z, we obtain

$$z = \frac{x - \mu}{\sigma} = \frac{30 - 25.5}{4.5} = 1.0$$

The area A to the right of the mean, corresponding to $z = 1.0$, is .3413 (from Table 3). Then the proportion of compacts having a miles-per-gallon ratio equal to or greater than 30 is equal to the entire area to the right of the mean, .5, minus the area A:

$$P(x \geq 30) = .5 - P(0 \leq z \leq 1) = .5 - .3413 = .1587$$

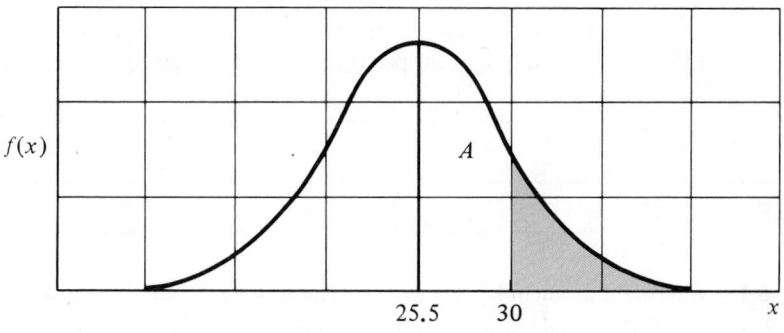

Figure 6.7   Area Under the Normal Curve for Example 6.5

The percentage exceeding 30 miles per gallon is

$$100(.1587) = 15.87\%$$

EXAMPLE 6.6   Refer to Example 6.5. In times of scarce energy resources a competitive advantage is given to an automobile manufacturer who can produce a car obtaining substantially better fuel economy than the competitors' cars. If a manufacturer wishes to develop a compact car that outperforms 95% of the current compacts in fuel economy, what must be the gasoline use rate for the new car?

SOLUTION   Let x be a normally distributed random variable, with a mean of 25.5 and a standard deviation of 4.5. We want to find the value x_0 such that

$$P(x \leq x_0) = .95$$

As a first step, we find

$$z_0 = \frac{x_0 - \mu}{\sigma} = \frac{x_0 - 25.5}{4.5}$$

and note that our required probability is the same as the area to the left of z_0 for the standardized normal distribution. Therefore,

$$P(z \leq z_0) = .95$$

The area to the left of the mean is .5. The area to the right of the mean between z_0 and the mean is $.95 - .5 = .45$. Thus, from Table 3 we find that z_0 is between 1.64 and 1.65. Notice that the area .45 is exactly halfway between the areas for $z = 1.64$ and $z = 1.65$. Thus, z_0 is exactly halfway between 1.64 and 1.65; that is, $z_0 = 1.645$.
Substituting $z_0 = 1.645$ into the equation for z_0, we have

$$1.645 = \frac{x_0 - 25.5}{4.5}$$

and solving for x_0, we obtain

$$x_0 = (1.645)(4.5) + 25.5 = 32.9$$

The manufacturer's new compact car must therefore obtain a fuel economy of 32.9 miles per gallon to outperform 95% of the compact cars currently available on the U.S. market.

EXERCISES   Conceptual Level

6.1   What is the distinction between a continuous random variable and a discrete random variable?

6.2   Why is the normal probability distribution useful for statistical analysis?

EXERCISES   Skill Level

6.3   Using Table 3 in the Appendix, calculate the area under the normal curve between these values.

 a. $z = 0$ and $z = 1.6$ **b.** $z = 0$ and $z = .39$

 c. $z = .86$ and $z = 1.75$ **d.** $z = -.34$ and $z = .84$

 e. $z = 0$ and $z = .74$ **f.** $z = -.62$ and $z = 1.25$

 g. $z = -.33$ and $z = -1.57$ **h.** $z = -1.47$ and $z = 2.52$

6.4   Use Table 3 to find z_0 for the following probabilities.

 a. $P(z > z_0) = .05$ **b.** $P(z > z_0) = .7054$

 c. $P(-z_0 < z < z_0) = .733$ **d.** $P(-z_0 < z < z_0) = .95$

 e. $P(z < z_0) = .01$ **f.** $P(z < z_0) = .1314$

 g. $P(-z_0 < z < z_0) = .90$

6.5   A normally distributed random variable x possesses a mean of 10 and a standard deviation of 5. Find these probabilities.

 a. That x falls between 10 and 12   **b.** That x exceeds 14.2

 c. That x falls between 5 and 12.5   **d.** That x falls between 4 and 14

 e. That x is less than 13   **f.** That x exceeds 19.5

6.6   Using Table 3 in the Appendix, calculate the following areas under the normal curve.

 a. Between $z = 0$ and $z = 1.4$ **b.** Between $z = 0$ and $z = 1.96$

 c. Between $z = 0$ and $z = -.60$ **d.** Between $z = 1.3$ and $z = 1.7$

 e. Between $z = -1.54$ and $z = 1.75$ **f.** Between $z = -2.0$ and $z = -.5$

 g. Between $z = 1.2$ and $z = 1.96$ **h.** Between $z = -1.96$ and $z = 1.96$

6.7   Find z_0 for the following probabilities.

 a. $P(z > z_0) = .10$ **b.** $P(-z_0 < z < z_0) = .10$

 c. $P(z < z_0) = .05$ **d.** $P(-z_0 < z < z_0) = .99$

6.8   Let x be a normal random variable with mean 310 and standard deviation 76. Find the following probabilities.

 a. That x exceeds 400   **b.** That x is between 300 and 400

 c. That x is greater than 450   **d.** That x is between 265 and 425

6.9   Refer to Exercise 6.8 and find the percentile corresponding to the following values of the random variable.

 a. $x = 280$ **b.** $x = 310$ **c.** $x = 350$ **d.** $x = 395$

EXERCISES   Applications

6.10   After the initiation of an energy conservation program, a public utility noted that the electrical use savings incurred by residential users averaged 10.4 kilowatt-hours (kWh) per month with a standard deviation of 7.8 kWh. Suppose the billing for one residential user has been selected at random from among those served by the utility. Find the following probabilities.

a. That the electrical use savings exceed 5 kWh

b. That the user incurred an electrical use saving

c. That the saving was between 5 and 15 kWh

d. That the residential user consumed at least 5 more kWh than used previously

 6.11 The length of life of a certain type of automatic washer is approximately normally distributed, with a mean of 3.1 years and a standard deviation of 1.2 years. If this type of washer is guaranteed for 1 year, what fraction of original sales will require replacement?

 6.12 Suppose you must establish regulations concerning the maximum number of people who can occupy an elevator. A study of elevator occupancies indicates that if eight people occupy the elevator, the probability distribution of the total weight of the eight people possesses a mean of 1,200 lb and a variance of 9,800 lb^2. What is the probability that the total weight of eight people exceeds 1,300 lb? What is the probability it exceeds 1,500 lb? (Assume that the probability distribution is approximately normal.)

 6.13 *Sales and Marketing Management* ("Survey of Selling Costs," February 16, 1987, Table III–8, p. 58) reported that the salaries for MBA graduates entering the field of marketing services (research and advertising) averaged approximately $38,000 with a standard deviation of $2,250. If the salaries of MBA graduates entering this field were normally distributed, find the proportion of MBA graduates entering the field of marketing services whose salaries exceed $40,900, the average salaries for those graduates entering the field of brand or product management.

6.14 In 1980, concerns began to surface concerning the possible association between caffeine in the human diet and certain health problems. A study in the *FDA Consumer* reports on the caffeine content of various foods and beverages, as well as the consumption rates for certain beverages containing caffeine. The findings concerning the caffeine content of brewed coffee and brewed tea are given in Table 6.2. Assume that the probability distributions of caffeine content for the four types of beverages are each approximately normal.

Table 6.2   Caffeine Data for Exercise 6.14

Item	Milligrams of Caffeine	
	Average	Standard Deviation
Brewed coffee (5-oz cup)		
Drip method	115	20
Percolator	80	21.7
Brewed tea (5-oz cup)		
Major U.S. brands	40	11.7
Imported brands	60	15

Source: Data adapted from Chris W. Lecos, "Caffeine Jitters: Some Safety Questions Remain," *FDA Consumer*, Vol. 21, No. 10 (December 1987 – January 1988).

a. What proportion of coffee brewed using the drip method will have more than 160 milligrams (mg) of caffeine per 5-oz cup?

b. What proportion of all percolator coffee exceeds the average caffeine content for coffee brewed using the drip method?

c. What is the probability that a 5-oz cup of U.S. brand brewed tea will have caffeine content exceeding the average caffeine content for a 5-oz cup of drip-brewed coffee? Of percolator-brewed coffee?

d. What conclusions would you draw concerning the difference in caffeine content between brewed coffee and brewed tea?

6.15 The average yield to maturity of industrial bonds issued during the quarter ending March 31, 1985, was 14.55, with a standard deviation of yield of .70. Suppose the bond yields were approximately normally distributed and that the bond yield for a certain firm during that quarter was 13.10. Give the percentile corresponding to a yield of 13.10. Recalling that the yield to maturity of an industrial bond is partly dependent on the issuing firm's bond rating, what can you infer about the financial state of the issuing firm during the first quarter of 1985?

6.16 Since the enactment of federal legislation on aerosal can propellants, manufacturers of hairsprays and deodorants have been introducing liquid sprays. A new machine used for filling cans of liquid hairspray can be set for any average fill. If the amount of fill is normally distributed around a setting with a standard deviation of .05 oz, what setting will cause 95% of the cans to contain 12.00 oz or less of liquid?

6.17 Ibbotson and Fall show that average rates of return for stocks listed in the over-the-counter (OTC) market exceed those of stocks for the more mature companies on the New York Stock Exchange (NYSE), a finding consistent with modern theories about the relationship between risk and return. Their data, collected for the period 1947–1978, is still used as a data base for current studies in the field of portfolio management. A portion of the findings of Ibbotson and Fall is shown in Table 6.3. Assume that the probability distributions of stock returns on the NYSE, of stocks in the OTC market, and of all publicly listed stocks are each approximately normal.

Table 6.3   Stock Return Data for Exercise 6.17

Market	Mean Annual Return	Standard Deviation
NYSE	11.56%	17.73%
OTC	14.79	21.79
Total (all listed stocks)	11.79	18.02

Source: R. Ibbotson and C. Fall, "The United States Market Wealth Portfolio," *Journal of Portfolio Management*, Fall 1979. Reprinted with permission.

a. What proportion of all publicly listed stocks offered annual returns in excess of 10% during this period?

b. What proportion of stocks listed on the NYSE outperformed the average for those listed on the OTC market?

c. Suppose one stock is chosen at random from the NYSE. Find the probability that the stock has incurred an annual return for the period 1947–1978 in excess of 20%. Find the probability that its annual return has been between 0% and 10%.

d. What proportion of stocks listed on the NYSE had negative annual returns over the period 1947–1978? What proportion of those listed on the OTC market had negative annual returns? What does this result suggest about the riskiness of stocks listed on the NYSE versus those listed on the OTC market?

6.4

The Normal Approximation to the Binomial Distribution

In Chapter 5 we considered several applications of the binomial probability distribution, all of which required that we calculate the probability that x, the number of successes in n trials, falls in a given region. Most examples involved small values of n because of the lengthy calculations involved in evaluating $p(x)$. However, when n was large and p was small with $np < 7$, the Poisson probability distribution with $\mu = np$ produced satisfactory approximations to binomial probabilities. When n is large and the conditions for using the Poisson approximation are not met, another approximation based on normal curve areas is available.

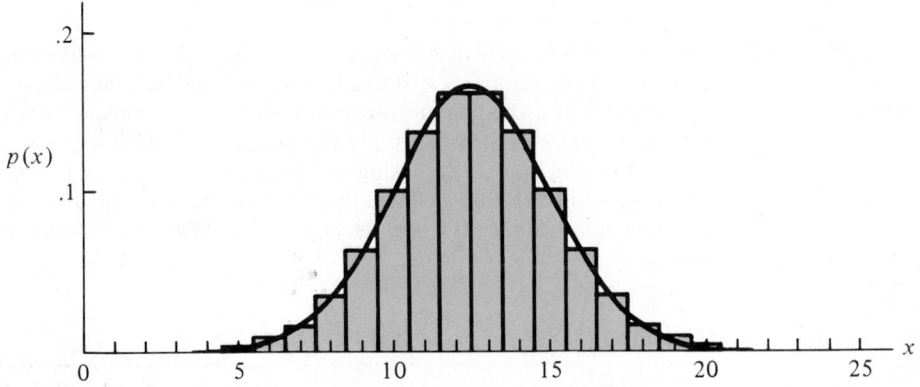

Figure 6.8   The Binomial Probability Distribution and the Approximating Normal Distribution for $n = 25$ and $p = .5$

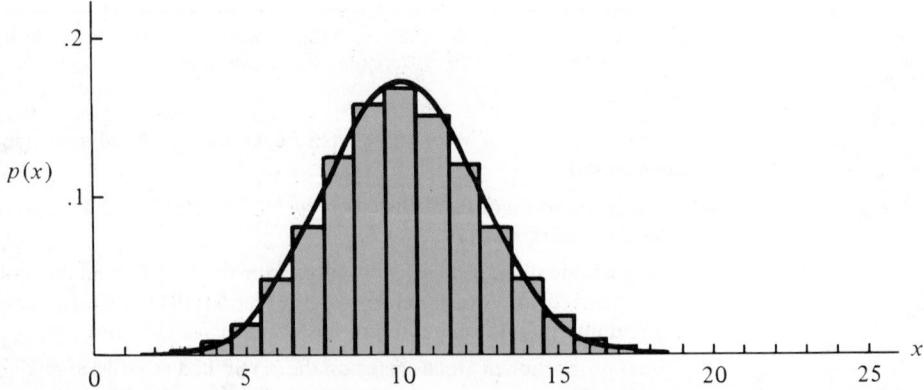

Figure 6.9   The Binomial Probability Distribution and the Approximating Normal Distribution for $n = 25$ and $p = .4$

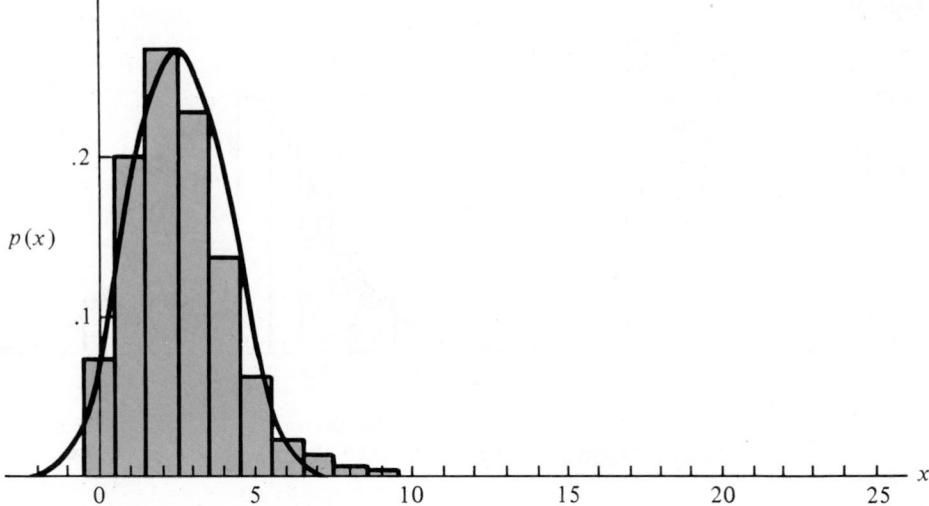

Figure 6.10   *The Binomial Probability Distribution and the Approximating Normal Distribution for*
n = 25 and p = .1

The binomial probability histogram is symmetrical and bell-shaped for $p = .5$ and is relatively so for values of p not close to 0 or 1 when n is large. In such cases the binomial probability histogram is well approximated by a normal curve with mean $\mu = np$ and variance $\sigma^2 = npq$. Figures 6.8, 6.9, and 6.10 show the binomial probability histograms for $n = 25$ and $p = .5$, $p = .4$, and $p = .1$, respectively, with the approximating normal curves superimposed. The correspondence between the areas for the binomial and normal distributions when $p = .5$ and $p = .4$ appears to be quite good. It is not so good for $p = .1$, for which the approximating symmetrical normal distribution poorly fits the nonsymmetrical binomial distribution. (However, when $n = 25$ and $p = .1$, $np = 2.5 < 7$, so that the Poisson approximation would be deemed appropriate in this particular case.)

Consider the binomial distribution for x when $n = 25$ and $p = .5$. For this distribution $\mu = np = 25(.5) = 12.5$ and $\sigma = \sqrt{npq} = \sqrt{6.25} = 2.5$. Therefore, we choose an approximating normal distribution with a mean $\mu = 12.5$ and a standard deviation $\sigma = 2.5$. The probability that $x = 8, 9,$ or 10 is equal to the area of the three rectangles lying over $x = 8, 9,$ and 10. This probability can be approximated by the area under the normal curve from $x = 7.5$ to $x = 10.5$, which is the shaded area in Figure 6.11. Therefore,

$$\sum_{x=8}^{10} p(x) \approx P(7.5 < x^* < 10.5)$$

where x^* is the approximating normal random variable. (The symbol $\approx$ means "approximately equal to.") Note that the area under the normal curve from 8 to 10 would not provide a good approximation to the probability that $x = 8, 9,$ or 10

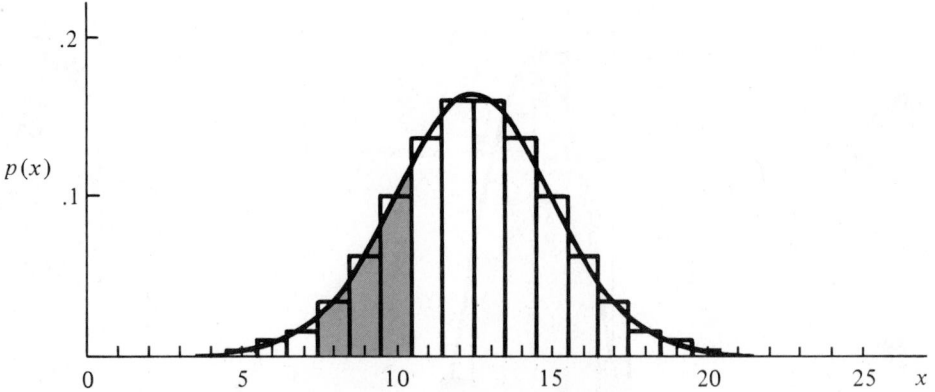

Figure 6.11   The Binomial Probability Distribution for $n = 25$ and $p = .5$ and the Approximating Normal Distribution with $\mu = 12.5$ and $\sigma = 2.5$

because it excludes half of the probability rectangles corresponding to $x = 8$ and $x = 10$. Thus, we must use the endpoints of the binomial probability rectangles, found by adding or subtracting .5 when calculating the approximating normal probabilities.

How can we tell whether the values of n and p are such that the normal approximation to binomial probabilities will be appropriate? From Section 6.2 (and from the Empirical Rule) we know that approximately 95% of the measurements associated with a normal curve lie within two standard deviations of the mean, and almost all lie within three. The binomial distribution would be nearly symmetrical if the distribution were able to spread out two standard deviations on both sides of the mean. Hence, to determine when the normal approximation is adequate, calculate $\mu = np$ and $\sigma = \sqrt{npq}$. If the interval $(\mu \pm 2\sigma)$ lies within the binomial bounds of 0 to n, the approximation is reasonably good. Note that this criterion is satisfied for the first two binomial distributions in Figures 6.8 and 6.9 but not for the third in Figure 6.10, for which $n = 25$ and $p = .1$.

EXAMPLE 6.7   To see how well the normal curve can be used to approximate binomial probabilities, refer to the binomial experiment illustrated in Figure 6.11, where $n = 25$ and $p = .5$. Calculate the probability that $x = 8, 9$, or 10, correct to three places, using Table 1(e) of binomial probabilities in the Appendix. Then calculate the corresponding normal approximation to this probability and compare the results.

SOLUTION   The exact probability P_1 can be calculated by using Table 1(e). Thus, we have

$$P_1 = \sum_{x=8}^{10} p(x) = \sum_{x=0}^{10} p(x) - \sum_{x=0}^{7} p(x) = .212 - .022 = .190$$

As noted earlier in this section, the normal approximation requires the area lying

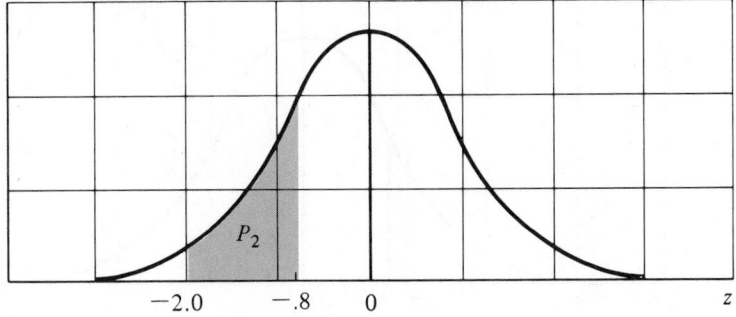

Figure 6.12   Area Under the Normal Curve for Example 6.7

between $x_1 = 7.5$ and $x_2 = 10.5$, where $\mu = 12.5$ and $\sigma = 2.5$. Thus, we have

$$\sum_{x=8}^{10} p(x) \approx P(z_1 \leq z \leq z_2) = P_2$$

where

$$z_1 = \frac{x_1 - \mu}{\sigma} = \frac{7.5 - 12.5}{2.5} = -2.0 \qquad z_2 = \frac{x_2 - \mu}{\sigma} = \frac{10.5 - 12.5}{2.5} = -.8$$

The probability P_2 is shown in Figure 6.12. The area between $z = 0$ and $z = 2.0$ is $A_1 = .4772$. Likewise, the area between $z = 0$ and $z = .8$ is $A_2 = .2881$. From Figure 6.12

$$P_2 = P(z_1 \leq z \leq z_2) = P(-2.0 \leq z \leq -.8) = A_1 - A_2$$
$$= .4772 - .2881 = .1891$$

The normal approximation is quite close to the binomial probability, .190, obtained from Table 1(e).

◆◆

EXAMPLE 6.8   According to some sources, 50% of all loans extended by consumer finance associations are taken for the purpose of consolidating existing bills. Find the probability that exactly 45 of 100 loans randomly selected from the files of a consumer loan agency were extended for the purpose of debt consolidation.

SOLUTION   The exact probability of observing $x = 45$ successes in $n = 100$ independent trials of a binomial experiment, with a probability of success of $p = .5$ at each trial, is found by evaluating

$$p(45) = C_{45}^{100}(.5)^{45}(.5)^{55}$$

Since this form is very tedious to evaluate, we will calculate the normal approximation to this probability. The mean and standard deviation of the binomial distribution are

$$\mu = np = 100(.5) = 50 \qquad \sigma = \sqrt{npq} = \sqrt{100(.5)(.5)} = 5$$

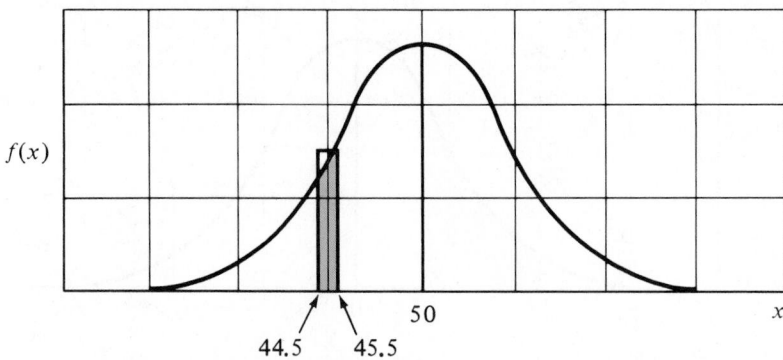

Figure 6.13   Area Under the Normal Curve for Example 6.8

The probability of exactly 45 successes is approximated by the area under the normal curve between 44.5 and 45.5 (see Figure 6.13). The z values corresponding to 44.5 and 45.5 are

$$z_1 = \frac{44.5 - 50}{5} = -1.1 \quad \text{and} \quad z_2 = \frac{45.5 - 50}{5} = -.9$$

The area between $z_1 = -1.1$ and $z_2 = -.9$, and hence the probability of exactly 45 loans for debt consolidation, is

$$p(45) \approx P(z_1 \leq z \leq z_2) = P(-1.1 \leq z \leq -.9) = .3643 - .3159 = .0484$$

The probability that exactly 45 of 100 loans were extended for the purpose of debt consolidation is approximately .0484. (The exact probability, computed using the binomial probability distribution, is .0484743.)

EXAMPLE 6.9   The reliability of an electrical fuse is the probability that a fuse, chosen at random from production, will function under the conditions for which it has been designed. A random sample of 1,000 fuses was tested and $x = 27$ defectives were observed. Calculate the probability of observing 27 or more defectives, assuming that fuse reliability is .98.

SOLUTION   The probability of observing a defective when a single fuse is tested is $p = .02$, given that fuse reliability is .98. Then

$$\mu = np = (1,000)(.02) = 20 \qquad \sigma = \sqrt{npq} = \sqrt{(1,000)(.02)(.98)} = 4.43$$

The probability of 27 or more defective fuses, given $n = 1,000$, is

$$P = P(x \geq 27) = p(27) + p(28) + p(29) + \cdots + p(999) + p(1,000)$$

The normal approximation to P is the area under the normal curve to the right of $x = 26.5$. (Note that we must use $x = 26.5$ rather than $x = 27$ in order to include

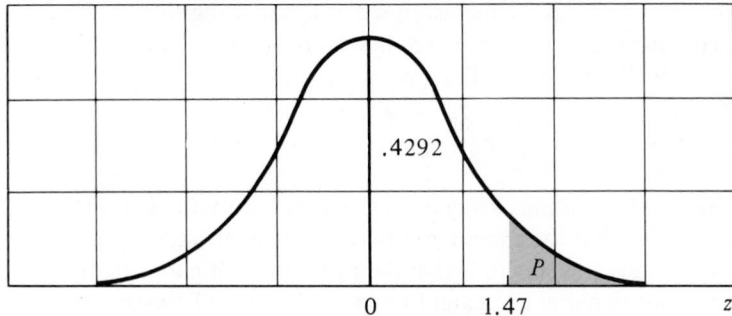

Figure 6.14   Normal Approximation to the Binomial for Example 6.9

the entire probability rectangle associated with $x = 27$.) The z value corresponding to $x = 26.5$ is

$$z = \frac{x - \mu}{\sigma} = \frac{26.5 - 20}{4.43} = \frac{6.5}{4.43} = 1.47$$

and the area between $z = 0$ and $z = 1.47$ is equal to .4292, as shown in Figure 6.14. Since the total area to the right of the mean is equal to .5, then

$$P = P(x \geq 27) \approx P(z \geq 1.47) = .5 - .4292 = .0708$$

Thus, the probability of observing 27 or more defectives is approximately .0708.

◆◆

EXAMPLE 6.10   A popular television advertisement claims that "taste experts" prefer the taste of a
(Optional)   certain brand of vegetable oil spread to butter. In a test of this claim 100 chefs were
given two pieces of bread to taste, one with the vegetable oil spread, the other with
butter. Sixty-eight of the chefs claimed that the vegetable oil spread had a richer, more
satisfying taste. If the difference in taste between the spread and butter is undetectable,
it is assumed that the probability is 1/2 that a chef will choose either piece as having
superior taste. On the basis of the results of this experiment, what conclusions would
you make about the truth of the advertisement?

SOLUTION   Translating the question into an hypothesis concerning the parameter of a binomial
population, we wish to test the null hypothesis that p, the probability of a chef choosing
the piece of bread with the vegetable oil spread as the piece with superior taste, is equal
to .5. Or equivalently, the content of the spread is such that it is undetectable in taste
from butter. The alternative hypothesis that the spread is preferred over butter would
imply rejection of the null hypothesis when x, the number of chefs preferring the
spread, is large.

Since the normal approximation to the binomial is adequate for this example, we
interpret a large and improbable value of x to be one that lies several standard
deviations away from the hypothesized mean, $\mu = np = (100)(.5) = 50$. Noting that

$$\sigma = \sqrt{npq} = \sqrt{(100)(.5)(.5)} = 5$$

we may arrive at a conclusion without bothering to locate a specific rejection region. The observed value of x, 68, lies more than 3σ away from the hypothesized mean $\mu = 50$. Specifically, x lies

$$z = \frac{x - \mu}{\sigma} = \frac{68 - 50}{5} = 3.6$$

standard deviations away from the hypothesized mean. This result is so improbable, assuming that the spread and butter are undetectable in taste, that we reject the null hypothesis and conclude that the probability of choosing the spread as having a taste superior to butter is greater than $p = .5$. (You will observe that the area above $z = 3.6$ is so small that it is not included in Table 3 of the Appendix.)

Rejecting the null hypothesis raises some questions in the minds of prospective wholesalers of margarine and butter. What proportion of consumers will prefer the taste of the vegetable oil spread, and is its taste sufficiently superior to that of butter to warrant a change in inventory policy? The former question leads to an estimation problem, a topic discussed in Chapter 8; the latter, involving an inventory decision, would utilize the results of a consumer preference experiment as well as a study of unit costs.

◆ ◆

EXAMPLE 6.11 (Optional)   The probability α of a type I error and the location of the rejection region for a statistical test of an hypothesis are usually specified before the data are collected. In the case of the previous example, a type I error is the error of assuming the vegetable oil spread is superior in taste to butter when, in fact, the taste difference between the two is undetectable (i.e., the probability of rejecting the null hypothesis $p = .5$ when it is true). For the taste-testing problem of Example 6.10, find the appropriate rejection region for the test of the null hypothesis $p = .5$ if we wish α, the probability of falsely concluding the vegetable oil spread has a superior taste to butter, to be approximately equal to .05 (see Figure 6.15.)

SOLUTION   We stated in Example 6.10 that x, the number indicating a preference for the vegetable oil spread, would be used as a test statistic and that the rejection region would be

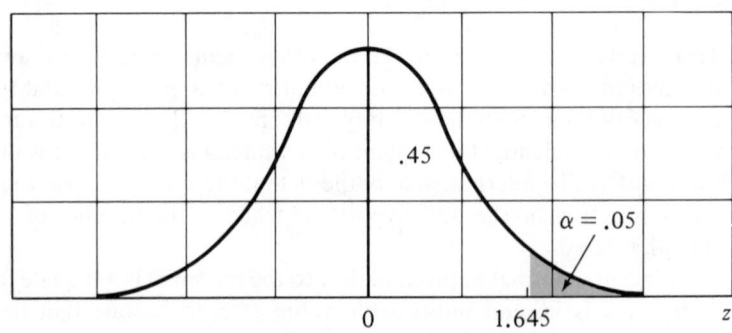

Figure 6.15   Location of the Rejection Region for Example 6.11

located in the upper tail of the probability distribution for x. Desiring α approximately equal to .05, we seek a value of x, say x_α, such that

$$P(x \geq x_\alpha) \approx .05$$

This value can be determined by first finding the corresponding z_α that gives the number of standard deviations between the mean $\mu = 50$ and x_α. Since the total area to the right of $z = 0$ is .5, the area between $z = 0$ and z_α equals .45 (see Figure 6.15). Checking Table 3, we find that $z = 1.64$ corresponds to an area equal to .4495 and $z = 1.65$ to an area of .4505. Thus, as we determined in Example 6.6,

$$z_\alpha = 1.645$$

Recall the relation between z and x,

$$z_\alpha = \frac{x_\alpha - \mu}{\sigma}$$

and substitute $z_\alpha = 1.645$ into the expression. Then

$$1.645 = \frac{x_\alpha - 50}{5}$$

Solving for x_α, we obtain

$$x_\alpha = 58.225$$

Since we cannot observe $x = 58.225$ chefs who indicate a preference for the vegetable oil spread, we must choose 58 or 59 as the point where the rejection region begins.

Suppose that we decide to reject when x is greater than or equal to 59. Then the actual probability α of the type I error for the test is

$$P(x \geq 59) = \alpha$$

which can be approximated by using the area under the normal curve above $x = 58.5$, a problem similar to that encountered in Example 6.9. The z value corresponding to $x = 58.5$ is

$$z = \frac{x - \mu}{\sigma} = \frac{58.5 - 50}{5} = 1.7$$

and the tabulated area between $z = 0$ and $z = 1.7$ is .4554. Then

$$\alpha = .5 - .4554 = .0446$$

While the method described above provides a more accurate value for α, there is very little practical difference between an α of .0446 and one of .05. When n is large, time and effort may be saved by using z as a test statistic rather than x. This method was employed in Example 6.10. We would then reject the null hypothesis when z is greater than or equal to 1.645.

◆ ◆

EXERCISES   Conceptual Level

6.18   Under what conditions can the normal probability distribution be used to approximate binomial probabilities?

6.19   If the normal distribution is not appropriate for approximating binomial probabilities, are there any other approximations that can be used?

6.20   Why is it appropriate to use a correction for continuity when approximating binomial probabilities with the normal distribution?

EXERCISES   Skill Level

6.21   Let x be a binomial random variable obtained from a binomial experiment with $n = 25$ and $p = .4$.

   a.   Use Table 1 in the Appendix to calculate $P(8 \leq x \leq 11)$.

   b.   Find μ and σ and use the normal approximation to the binomial distribution to approximate the probability $P(8 \leq x \leq 11)$. Note that this value is a good approximation to the exact value of $P(8 \leq x \leq 11)$.

6.22   Consider a binomial experiment with $n = 20$ and $p = .2$.

   a.   Calculate $P(x \geq 5)$ by using Table 1 in the Appendix.

   b.   Calculate $P(x \geq 5)$ by using the normal approximation to the binomial probability distribution. Compare the approximation with the exact value found in part a.

6.23   Consider a binomial experiment with $n = 25$ and $p = .3$.

   a.   Calculate $P(6 \leq x \leq 9)$ by using the binomial probabilities in Table 1 of the Appendix.

   b.   Calculate $P(6 \leq x \leq 9)$ by using the normal approximation to the binomial.

6.24   Consider a binomial experiment with $n = 25$ and $p = .2$.

   a.   Calculate $P(x \leq 4)$ by using Table 1 of the Appendix.

   b.   Calculate $P(x \leq 4)$ by using the normal approximation to the binomial.

EXERCISES   Applications

 6.25   Sensitivity training is designed to remove barriers of communication among employees in a firm and to create an atmosphere in which everyone feels free to speak openly about any issue, personal or otherwise, that affects the firm. A research note in an issue of the *Academy of Management Journal* reported that 50% of those managers who have participated in sensitivity training would personally recommend that other firms emphasize sensitivity training in their management development programs. Suppose a random sample of 100 managers who have participated in sensitivity training are selected.

   a.   What is the probability that 60 or more managers would not recommend sensitivity training to other firms?

   b.   What is the probability that more than 47 but less than 52 managers would recommend sensitivity training to other firms?

6.26 In a check of the truth of the claim reported in Exercise 6.25, suppose a sample of 100 managers who have participated in sensitivity training is selected, with $x = 38$ indicating they would recommend that other firms emphasize sensitivity training in their management development programs. Do these data provide sufficient evidence to reject the claim reported in the *Academy of Management Journal*?

 6.27 Actuarial data are used by the insurance industry to establish rates that effectively balance insurance costs with risk. Suppose that actuarial studies obtained from the trucking industry indicate that the probability that a truck driver will be involved in an injury accident during any given year is approximately 1/40. A random sample of $n = 400$ truck drivers from California indicates that 19 have been involved in injury accidents within the past year. Do these data suggest that truck drivers from California have a higher injury accident rate than those in the trucking industry in general? What does this result suggest regarding insurance pricing for truck drivers?

 6.28 Imported automobiles comprise approximately 25% of all new-car sales in the United States [U.S. Department of Commerce, Bureau of the Census, *Statistical Abstract of the United States*, 107th ed. (Washington, 1988), p. 586]. A state licensing office receives 100 requests for license plates for new cars during a given week.

   a. What is the probability that 25 or more of the license requests are for imported cars?

   b. What is the probability that no more than 10 of the requests are for imported cars?

   c. What is the probability that at least 10, but no more than 20, imported cars are represented within the group?

 6.29 Segmentation by age is becoming an important factor to be considered in planning marketing and advertising strategies. For example, today's teenagers have a discretionary income of $50 billion, and at least half of all college students have credit cards (Chester A. Swenson, "How to Sell to a Segmented Market," *Journal of Business Strategy*, January–February 1988, p. 19). In a test of the validity of this claim, 2,000 college students are surveyed. How few of the 2,000 students involved in the survey would have to have credit cards in order to reject this claim? Use $\alpha = .05$. [*Hint:* Find a number a such that $P(x \leq a) = .05$.]

6.5

The Uniform Distribution

The uniform distribution is a simple probability distribution useful in describing the behavior of a continuous random variable that can randomly assume any value between two points on a line, a and b $(a < b)$. The uniform probability function has a rectangular shape over the interval from a to b with height $1/(b - a)$, as shown in Figure 6.16. Unlike values for the normal random variable, values for the uniform distribution are not more concentrated near the mean and less concentrated in the tails. In fact, the uniform distribution provides a model for a continuous random variable whose values are evenly, or uniformly, distributed over an interval. For example, if buses arrive at a bus stop every 15 minutes and you arrive at the bus stop at a random time, the time that you will wait for the next bus could be described by a uniform distribution on the interval from 0 to 15. As another example, suppose that in a milling operation pieces of lumber less than 1 ft in length are considered scrap; then the

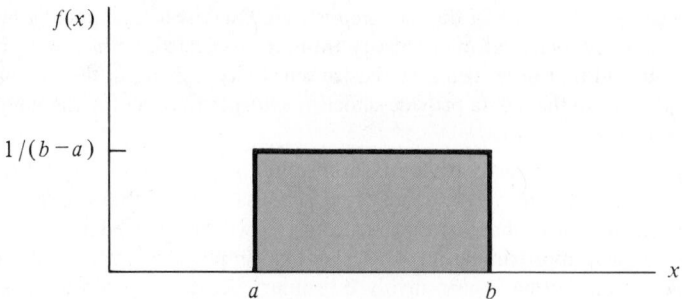

Figure 6.16   The Uniform Probability Distribution on the Interval $a \leq x \leq b$

distribution of the lengths of scrap lumber would have a uniform distribution on the interval from 0 to 1.

If x is a uniform random variable over the interval from a to b, the area under the probability density function is equal to 1; therefore, the height of the uniform probability density rectangle is $1/(b - a)$. Furthermore, the probability that the uniform random variable lies in the interval from c to d when the interval from c to d is contained within the interval from a to b is simply the area over the interval from c to d, given as $(d - c)/(b - a)$.*

Uniform Probability Density Function

$$f(x) = \frac{1}{b - a} \quad \text{for} \quad a \leq x \leq b$$

The mean and standard deviation are

$$\mu = \frac{a + b}{2} \qquad \sigma = \frac{b - a}{\sqrt{12}}$$

The mean of a uniform random variable is equal to the midpoint of the interval, and the standard deviation is directly proportional to the length of the interval.

EXAMPLE 6.12   Suppose that buses arrive at a bus stop every 15 minutes and that the waiting time for the next bus to arrive has a uniform probability distribution on the interval from 0 to 15 minutes.

---

* If you have had a calculus course, you may recall that the area under the curve $f(x) = 1/(b - a)$ between the points c and d is given by

$$\int_c^d \frac{1}{b - a} dx = \frac{d - c}{b - a}$$

a. Find the probability that x, a person's waiting time, will exceed 10 minutes.

b. Calculate the mean and standard deviation of x. Graph the probability density function of x, and indicate the location of the mean μ and the endpoints of the intervals $(\mu \pm \sigma)$ and $(\mu \pm 2\sigma)$.

c. Find the containment probabilities for these two intervals. How do they compare with those given by Tchebysheff's Theorem and the Empirical Rule?

SOLUTION

a. If a person's waiting time x is uniformly distributed over the interval from 0 to 15, then $P(x > 10)$ is found as the area above the interval from 10 to 15 and is given by

$$P(x > 10) = \frac{15 - 10}{15 - 0} = \frac{5}{15} = \frac{1}{3}$$

Hence, there is a 1/3 chance that a person will wait longer than 10 minutes (and a 2/3 chance of waiting less than 10 minutes).

b. To calculate the mean and standard deviation, we replace a and b by 0 and 15, respectively, in the given formulas to find

$$\mu = \frac{a + b}{2} = \frac{0 + 15}{2} = 7.5$$

and

$$\sigma = \frac{b - a}{\sqrt{12}} = \frac{15}{\sqrt{12}} = 4.33$$

Therefore, the interval $(\mu \pm \sigma)$, or (7.5 ± 4.3), has endpoints 3.2 and 11.8. The interval (7.5 ± 8.6) has endpoints -1.1 and 16.1. These interval endpoints and the mean are shown in Figure 6.17.

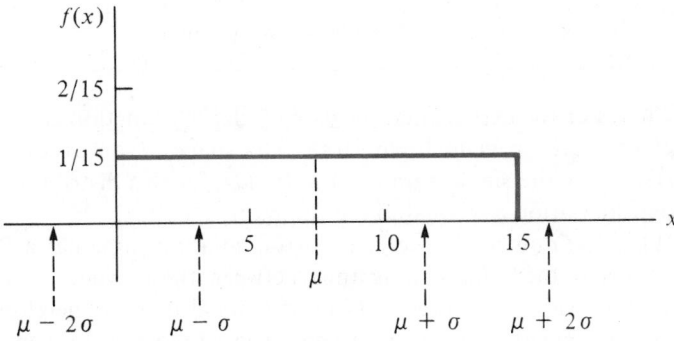

Figure 6.17   Endpoints and Mean for Example 6.12

c. The probability that the waiting time will be between 3.2 and 11.8 minutes is given as

$$P(3.2 < x < 11.8) = (\text{base})(\text{height}) = (11.8 - 3.2)\left(\frac{1}{15}\right)$$

$$= \frac{8.6}{15} = .5733$$

and

$$P(-1.1 < x < 16.1) = (15 - 0)\left(\frac{1}{15}\right) = 1$$

Tchebysheff's Theorem applies to any distribution, and hence, we have exceeded the minimum containment probabilities of 0 and .75. However, the Empirical Rule, which assumes that the distribution is mound-shaped, does not give a reasonable approximation to the actual containment probabilities.

◆◆

6.6

The Exponential Distribution

The exponential distribution is another continuous probability distribution that arises in business applications in which the random variable represents a waiting time. The time until the failure of a machine (or one of its components), the time between breakdowns of a mechanical piece of equipment, and the waiting time in a service line are all examples of random variables that often follow an exponential distribution.

The probability density function of an exponential random variable together with its mean and standard deviation are given in the display.

Exponential Probability Density Function

$$f(x) = \lambda e^{-\lambda x} \qquad x \geq 0; \quad \lambda > 0$$

The mean and standard deviation are

$$\mu = \frac{1}{\lambda} \qquad \sigma = \frac{1}{\lambda}$$

Curves of the exponential probability density function corresponding to several values of λ are given in Figure 6.18. The shape of an exponential distribution is determined by the single parameter λ. In fact, for this distribution the mean and the standard deviation are equal to the common value $1/\lambda$.

When the number of events occurring in a unit time has a Poisson distribution with mean λ, then the waiting time between these events follows an exponential distribution with an average waiting time equal to $1/\lambda$. Therefore, if the number of patients arriving at a hospital emergency room follows a Poisson distribution with an average of $\lambda = 5$ persons per hour, then the time between arrivals follows an

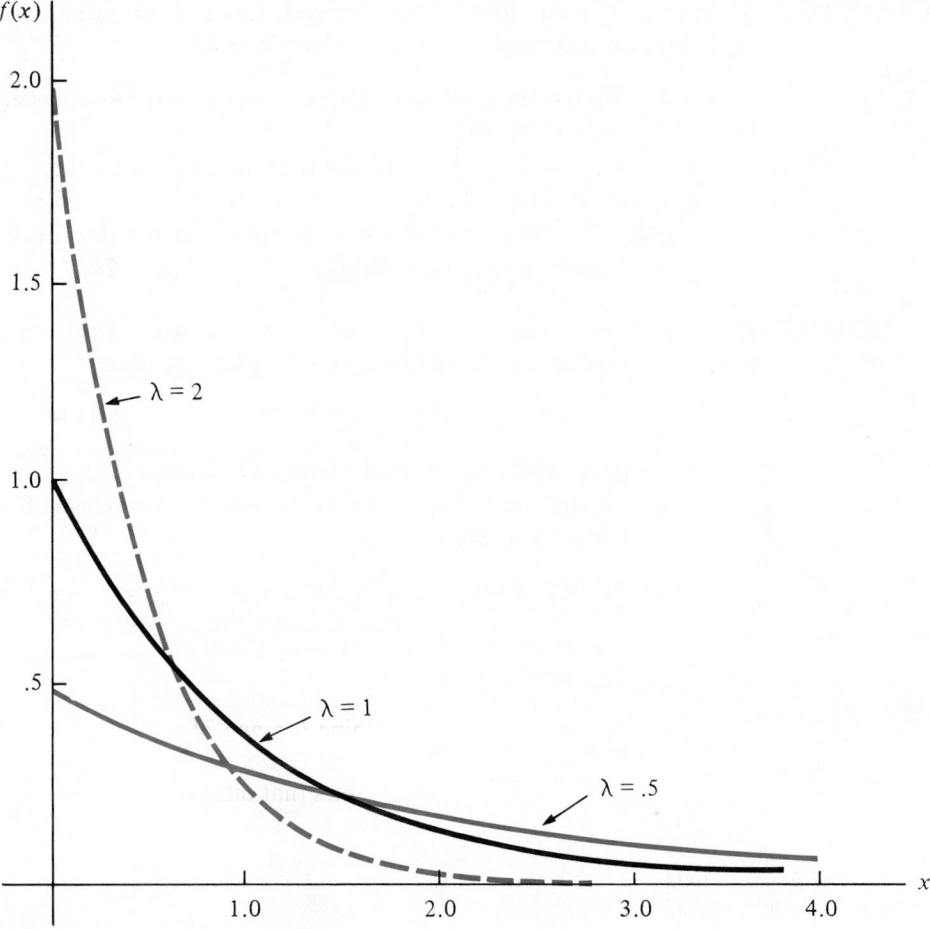

Figure 6.18   Exponential Probability Distributions for $\lambda = .5$, 1, and 2

exponential distribution with an average waiting time of $\mu = 12$ minutes (that is, $1/\lambda = 1/5 = .2$ hour).

In finding the probabilities associated with an exponential probability distribution, we can use the following relationship.

Finding Right-Tailed Probabilities for an Exponential Random Variable

$$P(x \geq a) = e^{-\lambda a} \qquad a \geq 0$$

After substituting the appropriate values of λ and a, the value of $e^{-\lambda a}$ can be found from Table 2 in the Appendix or by using a calculator that has an exponential function key as one of its features.

EXAMPLE 6.13   Suppose that the time in days between service calls on an office copying machine follows an exponential distribution with $\lambda = .02$.

   a.   What is the probability that the time until the machine again requires service exceeds 60 days?

   b.   What is the probability that the time until the machine again requires service is less than 20 days?

   c.   Find μ and σ. Find the probability that the time until the machine again requires service is between $(\mu - 2\sigma)$ and $(\mu + 2\sigma)$.

SOLUTION   a.   Let x represent the time in days between service calls; then x has an exponential distribution with $\lambda = .02$. Therefore,

$$P(x > 60) = e^{-(\lambda)(60)} = e^{-(.02)(60)} = e^{-1.2} = .301194$$

(Use Table 2 or your calculator to evaluate $e^{-1.2}$.)

   b.   Left-tail probabilities are easily found as complements of right-tail probabilities. Therefore,

$$P(x < 20) = 1 - P(x > 20) = 1 - e^{-\lambda(20)} = 1 - e^{-(.02)(20)}$$
$$= 1 - e^{-.4} = 1 - .670320 = .329680$$

   c.   The mean is given by

$$\mu = \frac{1}{\lambda} = \frac{1}{.02} = 50$$

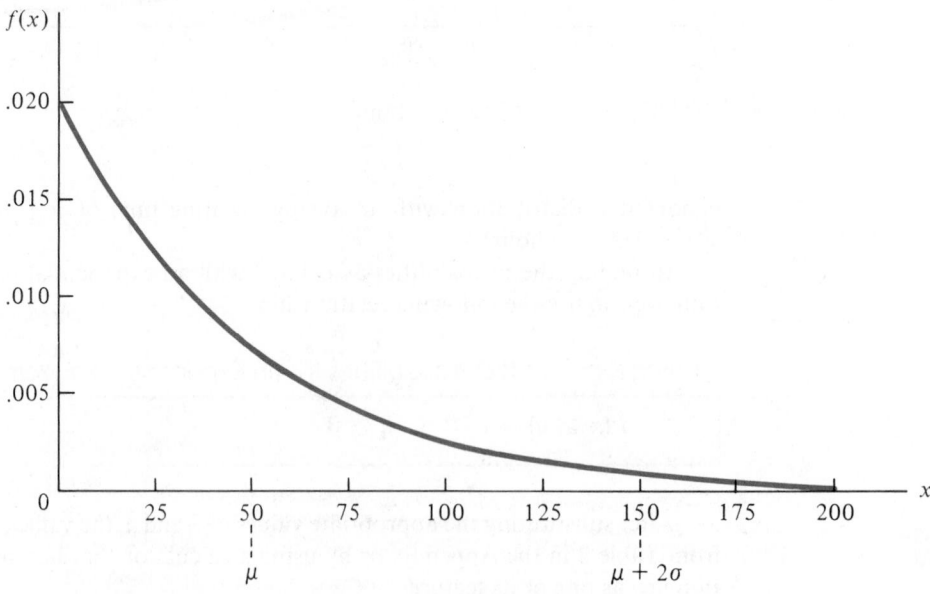

Figure 6.19   Exponential Distribution with $\lambda = .02$ for Example 6.13

For the exponential distribution $\mu = \sigma$; therefore, $\sigma = 50$. The interval $(\mu \pm 2\sigma)$ is (50 ± 100), or from -50 to 150. From Figure 6.19 we see that $P(-50 < x < 150)$ is equal to $P(0 < x < 150) = 1 - P(x > 150)$. Therefore,

$$P(\mu - 2\sigma < x < \mu + 2\sigma) = 1 - P(x > 150) = 1 - e^{-(.02)(150)}$$
$$= 1 - e^{-3} = 1 - .049787 = .950213$$

This value agrees with what you might expect from the Empirical Rule when the data are approximately normal, in spite of the fact that these data are strongly skewed to the right.

EXERCISES   Conceptual Level

6.30 Verbally describe the shape of the uniform and exponential probability distributions. For what types of variables are these distributions appropriate models?

6.31 How does the shape of the exponential probability distribution change as the value of λ becomes larger?

6.32 In a description of the distribution of the population of measurements that have a uniform distribution, can Tchebysheff's Theorem be used? The Empirical Rule? Would your answer change if the distribution were exponential?

EXERCISES   Skill Level

6.33 A random variable x has a uniform probability distribution with density function given as

$$f(x) = 1 \quad \text{for} \quad 0 \le x \le 1$$

a. Graph the probability distribution for x.

b. Find the probability that x falls between .5 and .75.

c. Calculate the mean and variance of x.

6.34 A random variable x has a uniform probability distribution with density function given as

$$f(x) = 1/2 \quad \text{for} \quad 2.5 \le x \le 4.5$$

a. Find the mean and standard deviation of x.

b. Graph the probability distribution of x. Calculate the intervals $(\mu \pm k\sigma)$ for $k = 1, 2,$ and 3, and superimpose these intervals on the graph.

c. Find the probability that x will fall within $(\mu \pm k\sigma)$ for $k = 1, 2, 3$. Do these results agree with those given by Tchebysheff's Theorem? The Empirical Rule?

6.35 A random variable x has an exponential probability distribution with density function given as

$$f(x) = (1/4)e^{-x/4} \quad \text{for} \quad x \ge 0$$

a. Find $P(x = 4)$.   **b.** Find $P(x \le 1)$.

c. Find $P(1.5 \le x \le 3)$.   **d.** Find $P(x \ge 1.5)$.

6.36 A random variable x has an exponential probability distribution with density function given as

$$f(x) = 25e^{-25x} \quad \text{for} \quad x \geq 0$$

a. Find the mean and standard deviation of x.

b. Graph the probability distribution of x. Calculate the intervals $(\mu \pm k\sigma)$ for $k = 1, 2,$ and 3, and superimpose these intervals on the graph.

c. Find the probability that x will fall within $(\mu \pm k\sigma)$ for $k = 1, 2, 3$. Do these results agree with those given by Tchebysheff's Theorem? The Empirical Rule?

EXERCISES Applications

6.37 The sales x of a gasoline distributor have a uniform probability distribution, as shown in Figure 6.20. Because of equipment limitations, daily sales will never be less than 10,000 gallons per day and never greater than 50,000 gallons per day. Use the information in the figure.

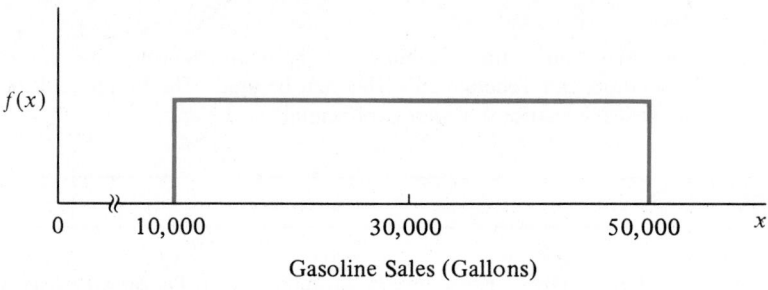

Figure 6.20   Sales Distribution for Exercise 6.37

a. Find the probability that the distributor sells at least 40,000 gallons per day.

b. Find the probability that the distributor sells between 20,000 and 30,000 gallons per day.

c. Find the probability that the distributor sells at most 35,000 gallons per day.

d. What is the mean and standard deviation of the sales (in gallons)?

e. What is the probability that the distributor's sales lie in the interval $(\mu \pm \sigma)$ gallons?

6.38 Large pieces of equipment, which periodically have worn parts replaced and are subject to periodic maintenance checks and service, will have a length of life that is exponentially distributed. The probability density function for such a piece of equipment with a mean life of 75,000 hours, given by

$$f(x) = \frac{1}{75,000} e^{-x/75,000} \quad x \geq 0$$

is shown in Figure 6.21.

a. What is the probability that one of these pieces of equipment will have an operating life exceeding 100,000 hours?

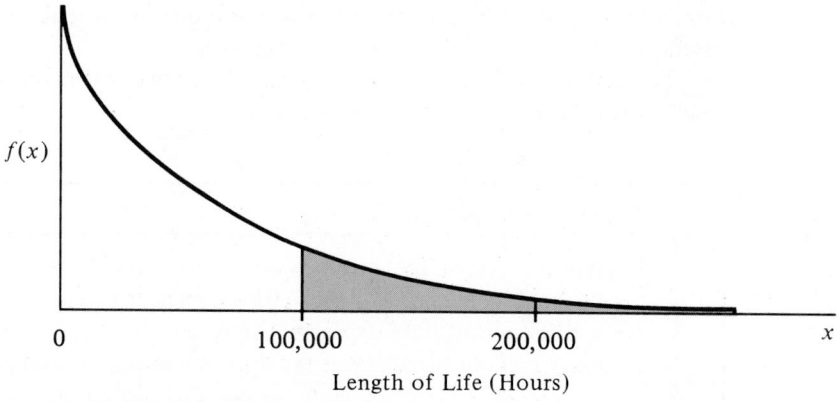

Figure 6.21   Lifetime Distribution for Exercise 6.38

b. What is the probability that one of these pieces of equipment has an operating life of 25,000 hours or less?

c. Between what two values would we find the useful operating lifetimes for 95% of such pieces of equipment if we exclude the largest 2.5% and the smallest 2.5%?

6.39 Sales of United States–made microchips in Japan has been hampered by long delivery times, according to a Japanese survey conducted by the Ministry of International Trade and Industry involving 63 major Japanese semiconductor users (*Japan Economic Journal*, February 13, 1988, p. 14). U.S. chips arrive from 6 to 12 months after an order is placed, compared with 45 to 90 days for Japanese chips. Suppose the delivery times for United States–made microchips has an exponential distribution with an average of 6 months.

a. What is the probability that delivery of a U.S. shipment takes 12 months or longer?

b. What is the probability that delivery of a U.S. shipment takes less than 6 months?

c. Use parts a and b to find the probability that delivery time for a U.S. order is between 6 months and 1 year.

6.7

Summary

Many continuous random variables observed in nature possess a probability distribution that is bell-shaped and may be approximated by the normal probability distribution discussed in Section 6.2. The normal probability distribution is actually the basis for the Empirical Rule (Chapter 2), which can be applied when a distribution of measurements is symmetrical and bell-shaped.

When n is large, the probabilities associated with the binomial probability distribution can be approximated with reasonable accuracy by using corresponding areas under the normal probability distribution. Since the opportunities for the application of the binomial probability distribution in the field of business are numerous, the normal approximation to the binomial probability distribution is an essential topic in our study of business statistics.

The uniform and exponential distributions are two other continuous distributions that have applications in business statistics.

Tips on Problem Solving

1. Always sketch the appropriate density function and locate the probability areas pertinent to the exercise. If you are approximating a binomial probability distribution, sketch in the binomial probability rectangles as well as the approximating normal curve.

2. Read each exercise carefully to see whether the data come from a binomial experiment or whether they possess a normal, uniform, or exponential distribution. If you are approximating a binomial probability distribution, do not forget to make a half-unit correction so that you will include the half rectangles at the ends of the interval. If the distribution is not binomial, *do not* make the half-unit correction. If you make a sketch (as suggested in step 1), you will see why the half-unit correction is or is not needed.

Supplementary Exercises

Applications

6.40 An auditor has found that the credit records of a large mail-order house are approximately normally distributed and show an average account billing error of $0 and a standard deviation of $1. (Billing errors may be positive or negative according to whether the purchaser was overcharged or undercharged.) Suppose one credit account is randomly selected from the files of the mail-order house.

   a. Find the probability that it contains a billing error between $0 and $1.50.

   b. Find the probability that it contains a billing error between −$2.00 and $0.

   c. Find the probability that it contains a billing error of at least $1.75.

   d. Find the probability that it contains a billing error that is not an overcharge.

   e. Find the probability that it contains a billing error between −$1.50 and $1.25.

   f. Find the probability that it contains a billing error between −$2.00 and −$1.00.

6.41 The loan departments of a large system of state banks found that home loans they issued over the past year were approximately normally distributed, with a mean of $83,000 and a standard deviation of $8,500. Suppose loan conditions at the banks remain essentially the same next year.

   a. What proportion of new loans would you expect to be $75,000 or less?

   b. What proportion of new loans would you expect to be greater than $90,000?

c. What proportion of new loans would you expect to be between $70,000 and $85,000?

d. What proportion of new loans would you expect to be between $75,000 and $80,000?

e. What proportion of new loans would you expect to be between $81,000 and $85,000?

f. What proportion of new loans would you expect to be less than $70,000?

6.42 The majority of entrepreneurs are satisfied with their accounting firms, according to a new study conducted by the Technology Executive Roundtable ("Accounting Services Evaluated by Technology Sector," *Journal of Accountancy*, January 1988, p. 23). Companies in the survey paid more than 2 1/2 times as much for an audit performed by one of the nation's larger firms than by any other type of accounting firm. They also waited longer to receive an audited statement—an average of 74 days, compared with 57 days for a regional firm. Assume that the waiting times to receive an audited statement from one of the larger firms are exponentially distributed with a mean waiting time of 74 days.

a. What is the probability that a company waits longer than 90 days?

b. What is the probability that a company waits no longer than 30 days before receiving its audited statement?

c. If two companies request an audit from one of the larger firms, what is the probability that they both wait longer than 90 days for the audited statement?

6.43 Refer to Exercise 6.42. A particular company used a regional firm with an average waiting time of 57 days to prepare an audit statement.

a. What is the probability that the company waits longer than 90 days? Less than 30 days?

b. How do the probabilities in part a compare with those for the same periods found for the larger firms whose average waiting time was 74 days?

6.44 Pay scales of federal government employees are determined according to each employee's government service (GS) rating. GS-10 experienced accountants with the Government Accounting Office (GAO) were reported to earn salaries that are approximately normally distributed, with a mean of $28,440 and a standard deviation of $1,236.

a. What proportion of GS-10 accountants with the GAO earn more than $30,000?

b. What is the fraction who earn less than $26,000 in annual salary?

6.45 The average length of time required to complete the civil service examination for prospective U.S. Treasury Department employees is found to equal 70 minutes, with a standard deviation of 12 minutes. When should the examination be terminated if the examination supervisor wishes to allow sufficient time for 90% of the applicants to complete the test? (Assume that the time required to complete the examination is normally distributed.)

6.46 All prospective employees of the U.S. Treasury Department are required to pass a special civil service examination before they are hired. Over a period of time the examination scores have been normally distributed, with a mean of $\mu = 85$ and a standard deviation of $\sigma = 4$. Suppose that a large number of applicants for jobs with the Treasury Department sit for the test.

a. What proportion can be expected to obtain an examination score of 90 or greater?

b. An examination score of less than 80 is considered as failing. What is the probability that an individual will fail the examination?

c. Of the large number sitting for the examination, what proportion would you expect to pass?

6.47 The owner of a fast-food restaurant has found the daily demand for ground beef at her restaurant to be normally distributed, with a mean of 240 lb and a standard deviation of 23 lb. Since the fast-food business is very competitive, the manager would like to ensure that sufficient ground beef is available in her restaurant each day so that the probability is no greater than 1% that the day's supply is exhausted. How many pounds of ground beef should the manager have available for use each day?

6.48   An artificial intelligence system (AI) has been developed by Fuji Xerox that will enable users to repair workstation problems by themselves (*Japan Economic Journal*, February 13, 1988, p. 14). The system asks questions about the apparent symptoms exhibited at the workstation and provides solutions by "electronic mail." Fuji claims that this system will allow users to repair up to 45% of system-related problems. On a given day $n = 25$ users at independent workstations attempt to repair a system-related problem by means of the AI system. Suppose, in fact, that 45% of such problems can be repaired by using AI.

Table 6.4   MINITAB Output for Exercise 6.48

BINOMIAL WITH N =	25 P = 0.450000
K	P(X LESS OR = K)
1	0.0000
2	0.0001
3	0.0005
4	0.0023
5	0.0086
6	0.0258
7	0.0639
8	0.1340
9	0.2424
10	0.3843
11	0.5426
12	0.6937
13	0.8173
14	0.9040
15	0.9560
16	0.9826
17	0.9942
18	0.9984
19	0.9996
20	0.9999
21	1.0000

a. Use the MINITAB printout in Table 6.4 to evaluate the probability that at most 10 problems are solved by using the AI system's approach to workstation repairs.

b. Use the normal approximation to binomial probabilities to evaluate the required probability in part a. (Use the correction for continuity in your calculations.)

c. What is the probability that at most 10 workstation problems remain unsolved by using the AI system? How does this value compare with the value found by using the normal approximation?

6.49   Most graduate schools of business require applicants to submit a score on the GMAT (Graduate Management Admissions Test), which is administered by the Educational Testing Service. Since 1965 GMAT scores have averaged 480, with a standard deviation of 100. Past evidence shows the distribution of GMAT scores to be normal.

a. What fraction of scores would you expect to find in the interval from 400 to 580?

b. What fraction of scores would you expect to find in the interval from 300 to 600?

c. A certain prestigious graduate business school automatically accepts all applicants whose GMAT scores exceed 650. Approximately what fraction of all those taking the GMAT would qualify for admission under this criterion?

 6.50 A rather large and well-known company has been giving serious consideration to adopting flextime, a system in which employees can work any hours during the week as long as they work 40 hours from Monday through Friday. The results of a preliminary survey show that 30% of the employees are not in favor of adopting flextime procedures. In a check of these results a sample of $n = 20$ employees is selected. Use the normal approximation to the binomial to calculate the following probabilities.

a. That at least 2 but no more than 4 of the 20 employees oppose flextime

b. That no more than 4 of the employees are opposed to flextime

 6.51 Although the standard mileage rate for business use of a private automobile is 22¢ per mile for income tax purposes, this rate is still not high enough to compensate for the skyrocketing costs of automobile ownership. Assume that 40% of those who use their personal car in the conduct of business take the standard mileage rate in figuring their deductions, and 60% separately record all costs attributable to the use of their car in business affairs. Fifty insurance agents who use their own car in the conduct of their business are contacted.

a. Find the probability that no more than 20 use the standard mileage rate for tax purposes.

b. Find the probability that at least half of them use the standard mileage rate.

c. Find the probability that at least 18, but no more than 22, use the standard mileage rate.

 6.52 A salesperson for a pollution abatement equipment manufacturer has found that, on the average, the probability of a sale of a certain piece of equipment on any given contact is equal to .7. If the salesperson contacts 50 customers, what is the probability that at least 10 will buy? (Assume that x, the number of sales, follows a binomial probability distribution.)

 6.53 Voters in a certain city were sampled concerning their voting preference in a primary election. Suppose that candidate A could win if he could poll 40% of the vote. If 920 of a sample of 2,500 voters favored A, does this contradict the hypothesis that A will win?

 6.54 A manufacturer has developed a new line of Plexiglas storm doors and windows to better protect homes against heat loss during the winter. In an evaluation of market acceptability of the product, advertising brochures describing the products and their costs were mailed to 1,000 New England homeowners. Sixty-three homeowners responded to the brochure, indicating a positive interest in the products. Does this sample present sufficient evidence to indicate that the new product line meets the firm's criteria for success—namely, that more than 5% of the homeowners show a positive interest in the product line?

 6.55 Market segmentation is concerned with the recognition of significant differences in buyer characteristics. Generally, the market is partitioned into distinct groups for the purpose of designing products and marketing programs to satisfy the distinctive consumer characteristics of each group. Since the market is often segmented according to the sex of the consumer, a newly designed portable radio was styled by an electronics manufacturer on the assumption that 50% of all purchasers are female. If the manufacturer's assumption is correct and if a random sample of 400 purchasers is selected, what is the probability that the number of female purchasers in the sample will be greater than 175?

 6.56 During periods of economic recession, experience shows that 10% of the people with loans outstanding at a certain commercial bank will become delinquent in their loan repayment schedule. To examine this issue and the characteristics of borrowers from her bank, the bank president randomly selects $n = 100$ account files at a time in which the economic climate of the community is very uncertain.

a. Find the probability that 10 or more of the borrowers are delinquent in their loan repayment schedule.

b. Find the probability that no more than 5 are delinquent.

6.57 Airlines and hotels often grant reservations in excess of capacity to minimize losses due to no-shows. Reservation agents must use discretion, however, because the loss potential may be great when they cannot accommodate an arriving customer with a valid reservation. (Consumer advocate Ralph Nader won a substantial sum in a class action suit after he was bumped from a flight although he had a valid reservation.) An airline reservation office has found that 5% of the passengers reserving a place on a particular flight will not show up for the flight. If the reservation office accepts 160 reservations for the flight and there are only 155 available seats on the plane, what is the probability that a seat will be available for every passenger who arrives with a valid reservation?

6.58 Refer to Exercise 6.57. If the airline wishes to grant sufficient reservations so that the probability is .99 that a seat is available for each arriving passenger with a valid reservation, how many reservations should it grant if the airplane capacity is 200? If capacity is 100?

6.59 Most manufacturers of machinery or appliances offer an optional warranty to protect buyers against repair costs for product breakdowns during the early stages of product life. In doing so, the manufacturer faces one difficult issue: to price the warranty so that it covers expected maintenance costs without raising the product price to the point of dissuading buyers. A manufacturer of nonintelligent computer terminals has found that, in continuous operation, the time before required repair of the terminals is normally distributed with a mean of 324 days and a standard deviation of 47 days. How long should the warranty time be for the terminals if the manufacturer is willing to cover the repair costs of at most 10% of the terminals under warranty?

6.60 Generally speaking, preventive maintenance is less expensive to perform than maintenance after failure or breakdown, because preventive maintenance can be scheduled at noncritical, or nonpeak, time intervals. A manufacturing plant utilizes 3,000 electric light bulbs that have a length of life that is normally distributed with a mean of 500 hours and a standard deviation of 50 hours. To minimize the number of bulbs that burn out during operating hours, the maintenance department replaces all the bulbs after a given period of operation. How often should the bulbs be replaced if the plant wishes no more than 1% of the bulbs to burn out between replacement periods?

6.61 Thirty percent of all calls coming into a telephone exchange are long-distance calls. If 200 calls come into the exchange, what is the probability that at least 50 will be long-distance calls?

6.62 A machine operation produces bearings whose diameters are normally distributed, with a mean of .498 and a standard deviation of .002. If specifications require that the bearing diameter be .500 inch $\pm$ .004 inch, what fraction of the production will be unacceptable?

6.63 Refer to Exercise 6.62. Three bearings are independently and randomly selected from the production process. What is the probability that all three conform to specifications?

6.64 In the study by Ibbotson and Fall (see Exercise 6.17), annual returns on investments in real estate for the period 1947–1978 averaged 8.19% with a standard deviation of annual returns of 3.53%. Assume the distribution of annual returns for real estate to be normally distributed.

a. What proportion of investments in real estate during this period outperformed the average annual rate of inflation, 4.5%?

b. What proportion provided a positive return on investment?

c. What proportion of investments in real estate outperformed the average annual return for common market stocks, 11.79%?

d. Using information provided here and in Exercise 6.17, compare the risk for investing in common stocks listed on the NYSE, common stocks listed on the OTC market, and real estate for the period 1947–1978.

6.65 Many state legislatures have been closely monitoring the effects of the Oregon bottle bill, an enactment that, in an attempt to reduce litter, prohibits the use of nonreturnable containers for beer or soft drinks. A prominent official in another state claims that at least 50% of the residents of her state favor such legislation. In a check of her claim 64 residents are randomly selected from the state, with $x = 24$ indicating they favor the enactment of a bottle bill in their state. Do these data present sufficient evidence to reject the official's claim?

6.66 When a doctor writes a prescription, the patient often takes it to the nearest drugstore to have it filled. He learns what the pills cost only when they are handed to him, because pharmacies are prohibited from advertising prices for prescription drugs. A study of pharmacies in the San Francisco area found the prices charged for 100 tablets of ampicillin to be approximately normally distributed, with a mean price of $8.50 and a standard deviation of $2.00. Find the probability that a particular San Francisco pharmacy charges between $10.00 and $12.00 for 100 tablets of ampicillin.

6.67 Credit cards are a multibillion-dollar business in the United States. In 1984 the top four card issuers (Visa, MasterCard, American Express, and Sears) had over $160 billion in billings (*Fortune*, February 4, 1985).

a. Individual banks that issue Visa and MasterCard are concerned about the cost of maintaining small accounts. A particular bank wishes to analyze these costs for accounts under $200. If this bank had a large number of Visa accounts with an average billing for 1984 of $785 and a standard deviation of $363, what proportion of these accounts are $200 or less in annual billing? Assume that the annual billings for 1984 are approximately normally distributed.

b. Banks also wish to keep track of large accounts. As a policy, the bank described in part a wishes to pay special attention to the top 15% of the accounts. How large must an account be to qualify as part of the top 15%?

6.68 The survival of many of the programs presented by the major television networks depends on the weekly Nielsen ratings. Last week's Nielsen ratings showed that 20% of all television viewers watch a particular program. In a random sample of $n = 1,000$ viewers, $x = 184$ viewers watch the program. Do these data present sufficient evidence to contradict last week's Nielsen ratings?

6.69 Suppose the owner of a delicatessen knows from past experience that the daily demand for fresh pastrami is uniformly distributed over the range from 20 to 40 lb. That is, all demand levels, from 20 to 40 lb per day, are equally likely to occur.

a. Find the probability that on a given day the demand for pastrami by the customers of the delicatessen does not exceed 30 lb.

b. Find the probability that the demand is at least 25 lb.

c. Find the probability that the demand is between 22 and 30 lb.

6.70 Refer to Exercise 6.69. The owner realizes that if demand falls below her inventory, she must sell leftover pastrami at a loss in the delicatessen's prepared sandwich section. How many pounds of pastrami should the owner prepare (or purchase from a supplier) each morning if she wishes the chances to be no more than 25% that she is left with excess inventory at the day's end? (Note: This problem illustrates an elementary form of the demand-inventory model that we will explore in Chapter 19. This model has many valuable applications to retail inventory management.)

6.71   Emissions of lead into the atmosphere declined sharply in 1986 because of the reduction of lead in gasoline, according to a report released by the Environmental Protection Agency [*Press Enterprise* (Riverside, Calif.), February 18, 1988]. Lead emissions, which peaked at 169,000 tons in 1976, have steadily decreased to the 9,500 tons reported in 1986. In 1985 the EPA began reducing the permissible lead concentration in gasoline from 1.1 grams (g) per gallon to .1 g per gallon over a period of two years. Assume that the lead concentration in gasoline is normally distributed with a standard deviation of .01 g per gallon. If the permissible concentration of .1 g per gallon is interpreted as the 99th percentile of the distribution, what average lead concentration is required of gasoline refiners?

6.72   Efforts to reduce energy use through conservation have allowed many utility companies to meet energy demands despite continued commercial and residential expansion. A public utility that serves a growing, midsized community in the southern United States has noticed that monthly demand over the past eight years has averaged 250 kilowatts (kW), with a standard deviation of 60 kW. Assume that monthly energy use in the community can be described by a normal distribution.

a.   Find the probability that monthly demand exceeds 250 kW.

b.   Find the probability that monthly demand is less than 190 kW.

6.73   Refer to Exercise 6.72 and suppose that the maximum capacity of the public utility is a supply of 370 kW of power during any given month. What is the probability that demand will exceed supply during a given month—that is, the probability that the community will face a power shortage?

6.74   The monthly rental rates for residential apartments in a certain city are normally distributed with an average rent of $346 and a standard deviation of $49.

a.   What proportion of apartments rent for $400 or more per month?

b.   Rent subsidies are available in the city for senior citizens and for low-income properties for which the monthly rent exceeds $375. What percentage of apartments fall into this category?

6.75   In the past, marketing and sales promotions were aimed at the "average" or "typical" consumer. Today, the opposite approach is employed, whereby marketing strategists emphasize targeting emerging market segments to ensure future market shares (Chester A. Swenson, "How to Sell to a Segmented Market," *Journal of Business Strategy*, January–February 1988). Women, who comprise the largest consumer segment of the market, have entered the labor force in dramatic numbers in recent years. Even among the nation's 33 million women with children under the age of 18, 63% work outside the home. A sample of $n = 100$ women is randomly selected from among those women with children under 18.

a.   What is the mean and standard deviation of the number of women in the sample working outside the home?

b.   What is the probability that the sample will contain at least 55, but not more than 75, women who work outside the home?

c.   What is the probability that the sample contains no more than 50 women who work outside the home?

6.76   Refer to Exercise 6.75. Surveys are a simple and fairly inexpensive way to keep informed of changes in characteristics of the various consumer segments and their marketing preferences. In a recent survey $n = 493$ persons were women with children under the age of 18, of whom $x = 351$ were employed outside the home. Do the results of this survey indicate that there has been a significant increase in the 63% working outside the home for the consumer segment consisting of women with children under the age of 18?

Supplementary Exercises

Interpretive

6.77 A common theme of advertising is to state that a certain product is superior to its competitors in some superficial way (e.g., shinier floors, brighter teeth, richer flavor). A similar theme offers a product as a panacea for certain real problems without reference to a competitor (e.g., using brand X reduces cavities). Experimental evidence is needed but seldom offered to check the veracity of such advertised claims. Consider, for example, a new serum that a commercial laboratory claims is effective in preventing the common cold. Suppose 40 people were injected with the serum and observed for a period of one year. Twenty-eight survived the winter without a cold. From prior information it is known that the probability of surviving the winter without a cold is .5 when the serum is not used. On the basis of the results of this experiment, what conclusions would you make regarding the effectiveness of the serum?

6.78 Acceptance of the fact that the world's energy resources are not limitless has caused many to wonder whether economic growth and increased productivity are doomed. However, business executives generally view government as insensitive to this issue. In fact, S. Mehra and A. Kefalas ("Energy Crisis—From the Manufacturing Executives' Viewpoint," *Proceedings of the American Institute for Decision Sciences*, November 1979) note that at least 70% of our nation's manufacturing executives consider the government's current energy policies to be unrealistic and unworkable. Realizing that the Mehra and Kefalas study was conducted in 1979, a researcher surveyed $n = 50$ manufacturing executives for evidence to see whether attitudes toward government economic policy might have changed in recent years. Of those surveyed, 17 offered support for current government energy policy and 33 found current policy unworkable. Do these data suggest that attitudes of manufacturing executives toward government energy policy have changed since 1979?

6.79 College bookstores face a delicate inventory management problem when ordering from a publisher. If the order is too small, they incur costly emergency procurement charges; if they overorder, they may be left with useless inventory if the text is never used again. Records show that attendance in Accounting 1 at a particular university possesses a probability distribution that is approximately normal, with a mean of 150 students a semester and a standard deviation of 20 students. How many Accounting 1 textbooks should be ordered by the university bookstore? (*Hint:* The bookstore cannot be absolutely sure that it will have sufficient books, but it can order enough so that the chance of exceeding the ordered amount is very small.)

6.80 Advertising agencies compete vigorously with one another to sponsor network television programs that attract a significant portion of the television-viewing audience. The producer of a certain television program has stated that 20% of all television viewers watch his program. An advertising agency interested in sponsoring the program would like to check the accuracy of the producer's claim. In a random sample of 1,600 viewers contacted during a showing of the program, $x = 284$ were watching the program. Do these data present sufficient evidence to contradict the producer's claim?

References and Suggested Readings

Mendenhall, W. *A Course in Business Statistics, 2nd ed.* Boston: PWS-KENT Publishing Co., 1988.

Neter, J., W. Wasserman, and G. A. Whitmore. *Applied Statistics, 3rd ed.* Boston: Allyn & Bacon, 1987.

Summers, G. W., W. S. Peters, and C. P. Armstrong. *Basic Statistics in Business and Economics, 4th ed.* Belmont, Calif.: Wadsworth, 1985.

TO
THE
READER

The primary objective of most statistical investigations is to make inferences about a population from information contained in a sample from that population. Because inferences are usually made in terms of one or more population parameters such as the population mean μ or population variance σ^2, the pertinent information in the sample is summarized by the corresponding sample statistics such as the sample mean $\bar{x}$ or sample variance s^2.

However, since the value of a statistic depends upon the observed values in the sample, a statistic is actually a random variable that may be either discrete or continuous. The probability distribution of a sample statistic is called its *sampling distribution*, since it describes the distribution of the values of the statistic when samples of a fixed size are repeatedly drawn from the population. In general, the sampling distribution of a statistic is used to assess the reliability of a parameter estimate based on the observed value of a statistic. In Chapter 7 we shall see that under fairly general conditions many statistics possess sampling distributions that are approximately normal.

7

SAMPLING AND SAMPLING DISTRIBUTIONS

7.1

Sampling and Statistics

In Chapter 4 we became familiar with random variables and their probability distributions, which serve as theoretical models for population relative frequency distributions. The binomial, Poisson, and hypergeometric probability distributions were introduced in Chapter 5 as possible models for discrete random variables; the normal, uniform, and exponential probability distributions presented in Chapter 6 are widely applicable models for many continuous random variables. Knowing the probability distribution allowed us to calculate various descriptive parameters, such as the population mean or variance.

In situations that arise in practice, however, we may be able to decide which type of probability distribution would apply, but the values of the parameters that specify the distribution exactly are unavailable. For example, we may feel reasonably certain that a binomial probability distribution would adequately model the responses to a preference test involving two products, without knowing which preference probabilities apply. Similarly, we may be willing to assume that the useful life of an appliance is normally distributed, without knowing its mean and variance. In such situations we rely on the information contained in a sample to determine which parameter values would be most appropriate in describing the sampled population. For instance, the sample proportion of preferences for product A should reflect the actual value of p in the population at large. Information about the population mean and variance of the useful life of an appliance would be contained in the sample mean and variance. Therefore, the sample information, which is summarized in the values of statistics computed from the sample measurements, would be used to make inferences about the sampled population in terms of its parameters.

Because the sample measurements vary from sample to sample, however, the value of a statistic will vary from sample to sample. Therefore, to assess the reliability of an inference based on a statistic computed from sample values, we must be able to determine or approximate the *sampling distribution* of that statistic, which is its probability distribution in repeated sampling.

Sampling Distribution of a Statistic

> The probability distribution of a statistic that results when random samples of size n are repeatedly drawn from a given population is called the **sampling distribution** of the statistic.

The sampling distribution of a statistic depends upon the population sampled and may be derived mathematically or approximated empirically. When we approximate the sampling distribution of a statistic empirically, a large number of samples of size n are drawn from the population of interest, the value of the statistic is calculated for each sample, and the results are tabulated in the form of a *relative frequency histogram*. When the number of samples is large, the relative frequency histogram should closely approximate the actual sampling distribution.

As we shall see, the probability distributions discussed in Chapters 5 and 6 arise in this chapter as sampling distributions of statistics commonly used as estimators of population parameters. Knowledge of the sampling distributions of statistics allows us to choose the best among several competing statistics available for estimating a population parameter. Furthermore, the sampling distribution of a statistic allows us to determine limits within which we would expect the value of an estimator to lie with a specified probability. To clarify the concept of a sampling distribution, in the next section we derive the sampling distribution for three different statistics when randomly sampling $n = 3$ elements from a finite population of $N = 5$ elements.

7.2

Sampling Distributions

In Chapter 2 we discussed the sample mean, the sample median, the sample standard deviation, and other numerical descriptive measures computed from the sample. Not only can these *statistics* be used to describe the sample, but they can also be used to make inferences about the corresponding population parameters in the form of estimates or tests of hypotheses. However, making these inferences requires that we know the sampling distribution of the statistic—which is to say we must know how the statistic behaves in repeated sampling. When used as an estimator, does a statistic consistently under- or overestimate its corresponding population parameter? Is one statistic more variable, and therefore possibly less useful, than another? We shall try to address these questions by finding the sampling distributions for several statistics within the simplest possible framework.

In Chapter 4 we defined a *simple random sample* of size n to be a sample chosen in such a way that each of the possible samples of size n has the same probability of

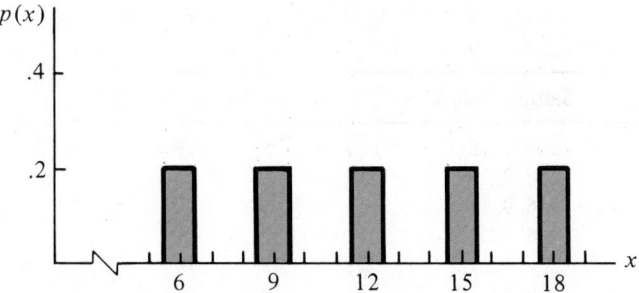

Figure 7.1   Probability Histogram for the $N = 5$ Population Values

being selected. In the selection of a sample of n observations from a population consisting of N elements, there are C_n^N possible samples. A simple random sample would be chosen by assigning a selection probability of $1/C_n^N$ to each sample.

Let us begin with a population of $N = 5$ elements whose values are 6, 9, 12, 15, and 18. Since the five population values are distinct, the population probability distribution assigns an equal probability of $1/5$ to each value of x in the population, so that

$$p(x) = .2 \quad \text{for} \quad x = 6, 9, 12, 15, 18$$

The probability distribution displayed as a probability histogram is given in Figure 7.1. Since these five measurements constitute a population with probabilities equal to $1/N$, the population mean is

$$\mu = \frac{\sum\limits_{i=1}^{N} x_i}{N} = \frac{60}{5} = 12$$

In calculating the population variance, we first find the sum of squared deviations about the mean:

$$\sum_{i=1}^{N} (x_i - \mu)^2 = \sum_{i=1}^{N} x_i^2 - \frac{\left(\sum\limits_{i=1}^{N} x_i\right)^2}{N} = 810 - \frac{(60)^2}{5} = 810 - 720 = 90$$

The population variance is then found to be

$$\sigma^2 = \sum_{i=1}^{N} \frac{(x_i - \mu)^2}{N} = \frac{90}{5} = 18$$

By inspection, we see that the population median is $M = 12$ and that $M = \mu$.

In the selection of a random sample of size $n = 3$ from this population of $N = 5$ elements, the number of possible samples is

$$C_3^5 = \frac{5!}{3!2!} = 10$$

Each of the ten possible samples is given in Table 7.1. For each sample we have calculated the sample mean, the sample median, and the sample variance. Since each

Table 7.1   The Values of $\bar{x}$, m, and s^2 Using Simple Random
            Sampling When $n = 3$, $N = 5$

Sample	Sample Values	$\bar{x}$	m	s^2
1	6, 9, 12	9	9	9
2	6, 9, 15	10	9	21
3	6, 9, 18	11	9	39
4	6, 12, 15	11	12	21
5	6, 12, 18	12	12	36
6	6, 15, 18	13	15	39
7	9, 12, 15	12	12	9
8	9, 12, 18	13	12	21
9	9, 15, 18	14	15	21
10	12, 15, 18	15	15	9

Table 7.2   The Sampling Distribution of $\bar{x}$ and m Obtained
            from Table 7.1: (a) Sample Mean; (b) Sample Median

(a) $\bar{x}$	$p(\bar{x})$		(b) m	$p(m)$
9	.1		9	.3
10	.1		12	.4
11	.2		15	.3
12	.2			
13	.2			
14	.1			
15	.1			

sample in Table 7.1 has an equal probability of being selected under simple random sampling, the sample value of a statistic associated with each distinct sample is assigned probability 1/10. Thus, when simple random sampling is used to select a sample of size $n = 3$, the value of $\bar{x} = 9$ will be observed only if sample 1 is selected, which will occur with probability .1. Similarly, the value $\bar{x} = 12$ will be observed if either sample 5 or sample 7 is observed; therefore, the probability of observing $\bar{x} = 12$ is .1 + .1 = .2.

Proceeding in this manner, we find the sampling distribution of the sample mean $\bar{x}$, which is given in Table 7.2. The sampling distribution of the sample median m, found in the same way, is also given in Table 7.2. Notice that although there are $N = 5$ values in the population, there are eight possible values of $\bar{x}$, only three (9, 12, and 15) of which are population values, whereas the sample median has three possible values, all of which are values in the population.

Suppose that we were interested in choosing between the sample mean and the sample median as estimators of the parameter $M = \mu = 12$. Which of the two would we choose? The probability histograms for the sample mean and median based on Table 7.2 are given in Figure 7.2. One simple way of comparing $\bar{x}$ and m as possible

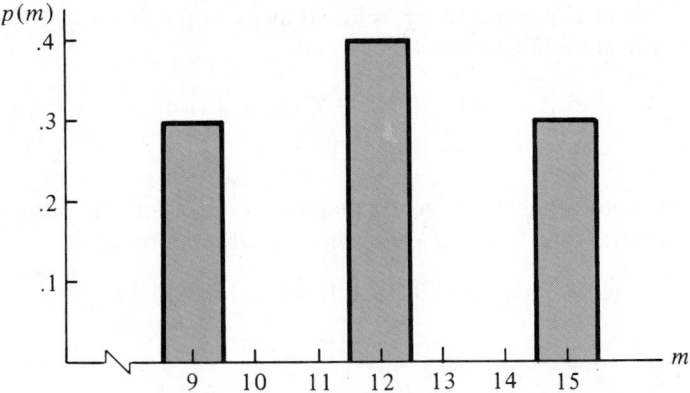

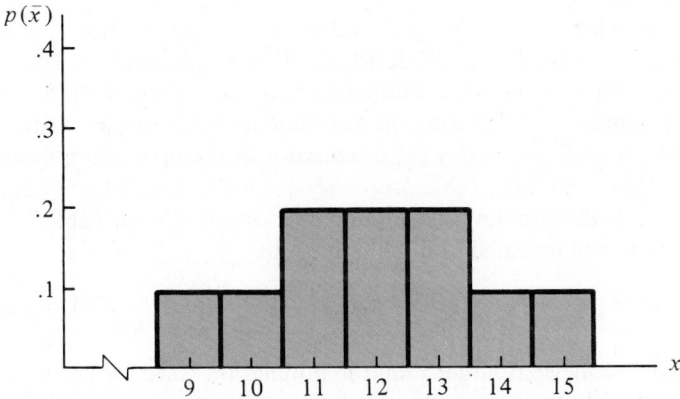

Figure 7.2   Probability Histograms for the Sampling Distributions of the Sample Mean $\bar{x}$ and the Sample Median m

estimators of $\mu = 12$ is to notice that in using the median, we would be in error by $(9 - 12) = -3$ with probability .3 or by $(15 - 12) = 3$ also with probability .3. Therefore, we would have an error of *at least* 3 with probability .6, which is quite high. In the use of $\bar{x}$ as an estimator, the probability of an error of 3 or larger is only .2. On these grounds we might prefer to use $\bar{x}$ rather than m to estimate μ.

Since sampling distributions are valid probability distributions in their own right, random variables such as $\bar{x}$ and m also have means and variances that can be calculated from the sampling distribution, using the expectation formulas given in Chapter 4. For the sampling distribution of $\bar{x}$ given in Table 7.2 we find that the mean of $\bar{x}$, denoted by $\mu_{\bar{x}}$, is

$$\mu_{\bar{x}} = \sum_{\bar{x}} \bar{x}p(\bar{x}) = 9(.1) + 10(.1) + \cdots + 15(.1) = 12$$

which is exactly equal to μ, the mean of the population that we sampled. This is no accident and is true in general when simple random sampling has been employed. The variance of $\bar{x}$, denoted by $\sigma_{\bar{x}}^2$, is found by using the shortcut formula for a population variance given in Chapter 4 and is

$$\sigma_{\bar{x}}^2 = \sum_{\bar{x}} \bar{x}^2 p(\bar{x}) - (\mu_{\bar{x}})^2 = 9^2(.1) + 10^2(.1) + \cdots + 15^2(.1) - (12)^2$$
$$= 147 - 144 = 3$$

Suppose that we wish to compare $\bar{x}$ and m on the basis of their means and variances. From Table 7.2 the mean of m, denoted by μ_m, is

$$\mu_m = \sum_m mp(m) = 9(.3) + 12(.4) + 15(.3) = 12$$

and its variance σ_m^2 is

$$\sigma_m^2 = \sum_m m^2 p(m) - \mu_m^2 = 9^2(.3) + 12^2(.4) + 15^2(.3) - (12)^2 = 5.4$$

Therefore, we see that $\mu_{\bar{x}}$ is exactly 12 and $\mu_m = 12$, so that on the average, both estimators would be equally desirable. However, since $\sigma_{\bar{x}}^2 = 3$ whereas $\sigma_m^2 = 5.4$, using the median would produce estimates almost twice as variable as estimates based on the sample mean $\bar{x}$. Comparing the sampling distributions for $\bar{x}$ and m, we see that $\bar{x}$ has more desirable properties as an estimator of the population mean than does m.

Does s^2 have desirable properties as an estimator of σ^2? Let us turn to Table 7.3, which lists the sampling distribution of s^2 obtained from Table 7.1. What is the average value of s^2 in repeated sampling? We find

$$E(s^2) = \sum_{s^2} s^2 p(s^2) = 9(.3) + 21(.4) + \cdots + 39(.2) = 22.5$$

which is somewhat larger than the population variance given as $\sigma^2 = 18$. However, when sampling from a finite population of size N,

$$E(s^2) = \left(\frac{N}{N-1}\right)\sigma^2$$

so that for our problem

$$E(s^2) = \left(\frac{5}{4}\right)(18) = 22.5$$

Table 7.3   The Sampling Distribution of s^2 Obtained from Table 7.1

s^2	$p(s^2)$
9	.3
21	.4
36	.1
39	.2

Notice that when the population size is large or infinite, the value of $N/(N-1)$ is 1 for all practical purposes, and $E(s^2) = \sigma^2$. More about the sampling distribution of this important statistic and its use as an estimator of σ^2 will be explained in Chapter 9.

When N is not too large, sampling distributions can be derived directly as we have done. Generally, the sampling distribution can be derived mathematically. The sampling distribution can also be approximated by holding the sample size n fixed, drawing repeated samples, and calculating the value of the statistic for each sample. The resulting relative frequency distribution would approximate the sampling distribution of the statistic and characterize its behavior. In the next section we will introduce a very important theorem that will allow us to approximate the sampling distributions of statistics that are sums or means of the sample values when the sample size is large.

EXERCISES Conceptual Level

7.1 Why is the sampling distribution of a statistic important to inferential statistics?

7.2 What is the difference between a sampling distribution and a population distribution?

EXERCISES Skill Level

7.3 We wish to randomly select a sample of size $n = 3$ from a population consisting of $N = 10$ elements.

    a. Sampling is said to be *without replacement* if we sample in such a way that, once selected for inclusion in the sample, an element cannot be selected again for the same sample. How many different samples of size $n = 3$ can be selected from this population if we sample without replacement?

    b. What is the probability of selecting any one sample?

7.4 A population consists of the following four elements: 4, 7, 9, 12.

    a. How many different samples of size $n = 2$ can be selected from this population if we sample without replacement?

    b. List the possible samples of size $n = 2$.

    c. Compute the sample mean for each of the samples given in part b.

    d. Find the sampling distribution of $\bar{x}$. Use a probability histogram to graph the sampling distribution of $\bar{x}$.

    e. If all four population values are equally likely, calculate the value of the population mean μ. Do any of the samples listed in part b produce a value of $\bar{x}$ exactly equal to μ? What is the probability that a randomly selected sample will produce a value of $\bar{x}$ exactly equal to μ?

    f. Verify that $E(\bar{x})$, the average value of $\bar{x}$, is equal to the value of the population mean μ.

7.5 A population consists of the following five elements: 1, 4, 7, 10, 16.

    a. How many different samples of size three are possible when sampling without replacement?

    b. List all of the possible different samples.

    c. Compute the sample mean for each of the samples given in part b.

    d. Find the sampling distribution of the sample mean $\bar{x}$. Use a probability histogram to graph the sampling distribution of $\bar{x}$.

    e. If all five population values are equally likely, compute the value of the population mean μ. What is the probability that the sample value of $\bar{x}$ will exactly equal μ?

    f. Verify that $E(\bar{x}) = \mu$.

7.6   Refer to Exercise 7.5.

    a. Determine the sample median for each of the possible samples of size $n = 3$.

    b. Find the sampling distribution of the median and graph the resulting distribution.

    c. Determine the value of the population median. Are there any samples that produce a sample median whose value is exactly equal to the population median? What is the probability that the sample median will equal the value of the population median?

7.7   Refer to Exercise 7.5.

    a. Compute the sample variance s^2 for each of the possible samples of size $n = 3$.

    b. Find the sampling distribution of s^2 and graph the results.

    c. Compute the value of the population variance σ^2. What is the probability that the sample variance s^2 will exactly equal the population variance σ^2?

7.3

The Central Limit Theorem

Central Limit Theorem   The **Central Limit Theorem** states that under rather general conditions sums and means of samples of random measurements drawn from a population tend to possess, approximately, a bell-shaped sampling distribution. The significance of this statement is perhaps best illustrated by an example.

    Consider a population of die throws generated by tossing a die infinitely many times, with a resulting probability distribution as given by Figure 7.3. Draw a sample of $n = 5$ measurements from the population by tossing a die five times and recording each of the five observations, as indicated in Table 7.4. Note that the numbers observed in

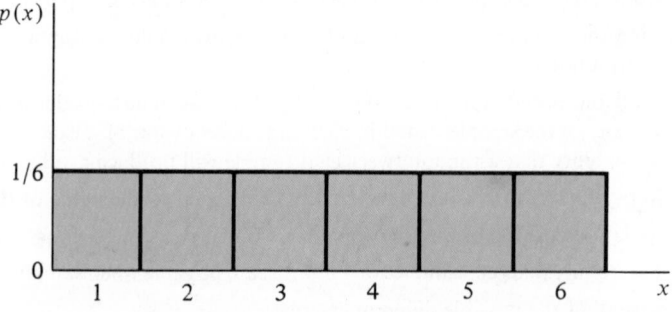

Figure 7.3   Probability Distribution for x, the Number Appearing on a Single Toss of a Die

Table 7.4   Sampling from the Population of Die Throws

Sample Number	x, Sample Measurements	$\sum x_i$	$\bar{x}$	Sample Number	x, Sample Measurements	$\sum x_i$	$\bar{x}$
1	3, 5, 1, 3, 2	14	2.8	51	2, 3, 5, 3, 2	15	3.0
2	3, 1, 1, 4, 6	15	3.0	52	1, 1, 1, 2, 4	9	1.8
3	1, 3, 1, 6, 1	12	2.4	53	2, 6, 3, 4, 5	20	4:0
4	4, 5, 3, 3, 2	17	3.4	54	1, 2, 2, 1, 1	7	1.4
5	3, 1, 3, 5, 2	14	2.8	55	2, 4, 4, 6, 2	18	3.6
6	2, 4, 4, 2, 4	16	3.2	56	3, 2, 5, 4, 5	19	3.8
7	4, 2, 5, 5, 3	19	3.8	57	2, 4, 2, 4, 5	17	3.4
8	3, 5, 5, 5, 5	23	4.6	58	5, 5, 4, 3, 2	19	3.8
9	6, 5, 5, 1, 6	23	4.6	59	5, 4, 4, 6, 3	22	4.4
10	5, 1, 6, 1, 6	19	3.8	60	3, 2, 5, 3, 1	14	2.8
11	1, 1, 1, 5, 3	11	2.2	61	2, 1, 4, 1, 3	11	2.2
12	3, 4, 2, 4, 4	17	3.4	62	4, 1, 1, 5, 2	13	2.6
13	2, 6, 1, 5, 4	18	3.6	63	2, 3, 1, 2, 3	11	2.2
14	6, 3, 4, 2, 5	20	4.0	64	2, 3, 3, 2, 6	16	3.2
15	2, 6, 2, 1, 5	16	3.2	65	4, 3, 5, 2, 6	20	4.0
16	1, 5, 1, 2, 5	14	2.8	66	3, 1, 3, 3, 4	14	2.8
17	3, 5, 1, 1, 2	12	2.4	67	4, 6, 1, 3, 6	20	4.0
18	3, 2, 4, 3, 5	17	3.4	68	2, 4, 6, 6, 3	21	4.2
19	5, 1, 6, 3, 1	16	3.2	69	4, 1, 6, 5, 5	21	4.2
20	1, 6, 4, 4, 1	16	3.2	70	6, 6, 6, 4, 5	27	5.4
21	6, 4, 2, 3, 5	20	4.0	71	2, 2, 5, 6, 3	18	3.6
22	1, 3, 5, 4, 1	14	2.8	72	6, 6, 6, 1, 6	25	5.0
23	2, 6, 5, 2, 6	21	4.2	73	4, 4, 4, 3, 1	16	3.2
24	3, 5, 1, 3, 5	17	3.4	74	4, 4, 5, 4, 2	19	3.8
25	5, 2, 4, 4, 3	18	3.6	75	4, 5, 4, 1, 4	18	3.6
26	6, 1, 1, 1, 6	15	3.0	76	5, 3, 2, 3, 4	17	3.4
27	1, 4, 1, 2, 6	14	2.8	77	1, 3, 3, 1, 5	13	2.6
28	3, 1, 2, 1, 5	12	2.4	78	4, 1, 5, 5, 3	18	3.6
29	1, 5, 5, 4, 5	20	4.0	79	4, 5, 6, 5, 4	24	4.8
30	4, 5, 3, 5, 2	19	3.8	80	1, 5, 3, 4, 2	15	3.0
31	4, 1, 6, 1, 1	13	2.6	81	4, 3, 4, 6, 3	20	4.0
32	3, 6, 4, 1, 2	16	3.2	82	5, 4, 2, 1, 6	18	3.6
33	3, 5, 5, 2, 2	17	3.4	83	1, 3, 2, 2, 5	13	2.6
34	1, 1, 5, 6, 3	16	3.2	84	5, 4, 1, 4, 6	20	4.0
35	2, 6, 1, 6, 2	17	3.4	85	2, 4, 2, 5, 5	18	3.6
36	2, 4, 3, 1, 3	13	2.6	86	1, 6, 3, 1, 6	17	3.4
37	1, 5, 1, 5, 2	14	2.8	87	2, 2, 4, 3, 2	13	2.6
38	6, 6, 5, 3, 3	23	4.6	88	4, 4, 5, 4, 4	21	4.2
39	3, 3, 5, 2, 1	14	2.8	89	2, 5, 4, 3, 4	18	3.6
40	2, 6, 6, 6, 5	25	5.0	90	5, 1, 6, 4, 3	19	3.8
41	5, 5, 2, 3, 4	19	3.8	91	5, 2, 5, 6, 3	21	4.2
42	6, 4, 1, 6, 2	19	3.8	92	6, 4, 1, 2, 1	14	2.8
43	2, 5, 3, 1, 4	15	3.0	93	6, 3, 1, 5, 2	17	3.4
44	4, 2, 3, 2, 1	12	2.4	94	1, 3, 6, 4, 2	16	3.2
45	4, 4, 5, 4, 4	21	4.2	95	6, 1, 4, 2, 2	15	3.0
46	5, 4, 5, 5, 4	23	4.6	96	1, 1, 2, 3, 1	8	1.6
47	6, 6, 6, 2, 1	21	4.2	97	6, 2, 5, 1, 6	20	4.0
48	2, 1, 5, 5, 4	17	3.4	98	3, 1, 1, 4, 1	10	2.0
49	6, 4, 3, 1, 5	19	3.8	99	5, 2, 1, 6, 1	15	3.0
50	4, 4, 4, 4, 4	20	4.0	100	2, 4, 3, 4, 6	19	3.8

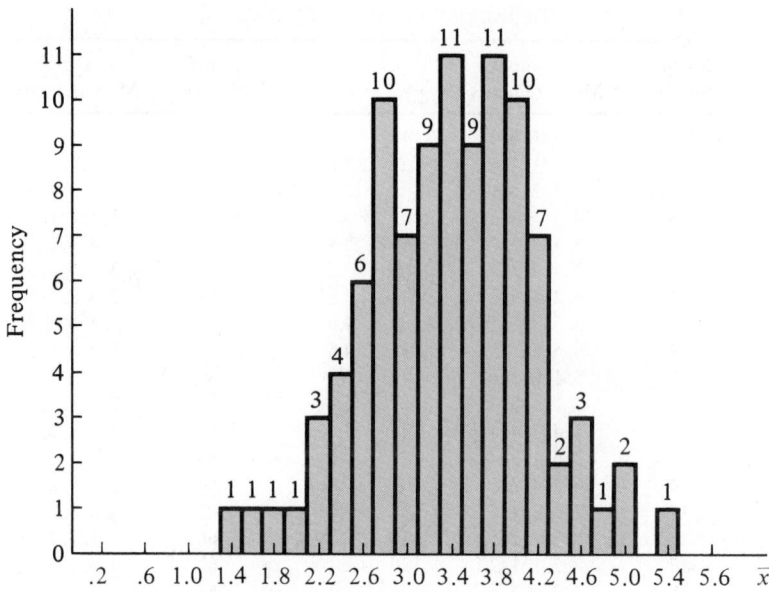

Figure 7.4   Histogram of Sample Means for the Die-Tossing Experiment

the first sample are $x = 3, 5, 1, 3, 2$. Calculate the sum of the five measurements as well as the sample mean $\bar{x}$.

For experimental purposes, repeat the sampling procedure 100 times, or preferably an even larger number of times. The results for 100 samples are given in Table 7.4 along with the corresponding values of the sum $\sum_{i=1}^{5} x_i$ and the sample mean $\bar{x}$. Construct a frequency histogram for $\bar{x} \left(\text{or for } \sum_{i=1}^{5} x_i \right)$ for the 100 samples and observe the resulting distribution, shown in Figure 7.4. You will observe an interesting result: Although the values of x in the population ($x = 1, 2, 3, 4, 5, 6$) are equiprobable and hence possess a probability distribution that is perfectly horizontal, the distribution of the sample **bell-shaped** means (or sums) chosen from the population forms a **bell-shaped distribution**. We will **distribution** add one additional comment without proof. If we should repeat the study outlined here by using larger samples of size $n = 10$, we would find that the distribution of the sample means tends to become more nearly bell-shaped.

The relative frequency distribution (Figure 7.4) provides only a rough approximation to the sampling distribution of the sample mean. An accurate evaluation of the form of the sampling distribution would require an infinitely large number of samples or, at the very least, far more than the 100 samples contained in our experiment. Nevertheless, the relative frequency distribution constructed from the 100 samples provides a good indication of the form of the sampling distribution of the mean of samples of $n = 5$ observations for the die-tossing experiment and it illustrates the basic idea involved in the Central Limit Theorem.

Central Limit Theorem

If random samples of n observations are drawn from a nonnormal population with finite mean μ and standard deviation σ, then when n is large, the sampling distribution of the sample mean $\bar{x}$ is approximately normally distributed, with mean and standard deviation

$$\mu_{\bar{x}} = \mu \qquad \text{and} \qquad \sigma_{\bar{x}} = \frac{\sigma}{\sqrt{n}}$$

The approximation will become more and more accurate as n becomes large.

The Central Limit Theorem can be restated to apply to the **sum of the sample measurements**

$$\sum_{i=1}^{n} x_i$$

which, as n becomes large, would also tend to possess a normal distribution, in repeated sampling, with mean $n\mu$ and standard deviation $\sigma\sqrt{n}$.

It can be shown (proof omitted) that the mean and standard deviation of the sampling distribution of $\bar{x}$ are always related to the mean and standard deviation of the sampled population as well as to the sample size n. The two distributions have the same mean μ, and the standard deviation of the sampling distribution of $\bar{x}$ is equal to the population standard deviation σ divided by $\sqrt{n}$. (It can be shown that this relationship is true regardless of the sample size n.) Consequently, the spread of the distribution of sample means is considerably less than the spread of the population distribution.

The significance of the Central Limit Theorem is twofold. First, it explains the rather common occurrence of normally distributed random variables in nature. We might imagine the height of a person as being composed of a number of effects, each random, and associated with such things as the height of the mother, the height of the father, the activity of a particular gland, the environment, and diet. If each of these effects tends to add to the others to yield the measurement of height, then height is the sum of a number of random variables, and according to the Central Limit Theorem, the distribution of heights is approximately normal.

The second and most important contribution of the Central Limit Theorem is in statistical inference. Many estimators that are used to make inferences about population parameters are sums or averages of the sample measurements. When sums or averages are used and the sample size n is sufficiently large, according to the Central Limit Theorem, we expect the estimator to possess (approximately) a normal probability distribution in repeated sampling. We can then use the normal distribution discussed in Chapter 6 to describe the behavior of the inference maker. This aspect of the Central Limit Theorem will be utilized in later chapters dealing with statistical inference.

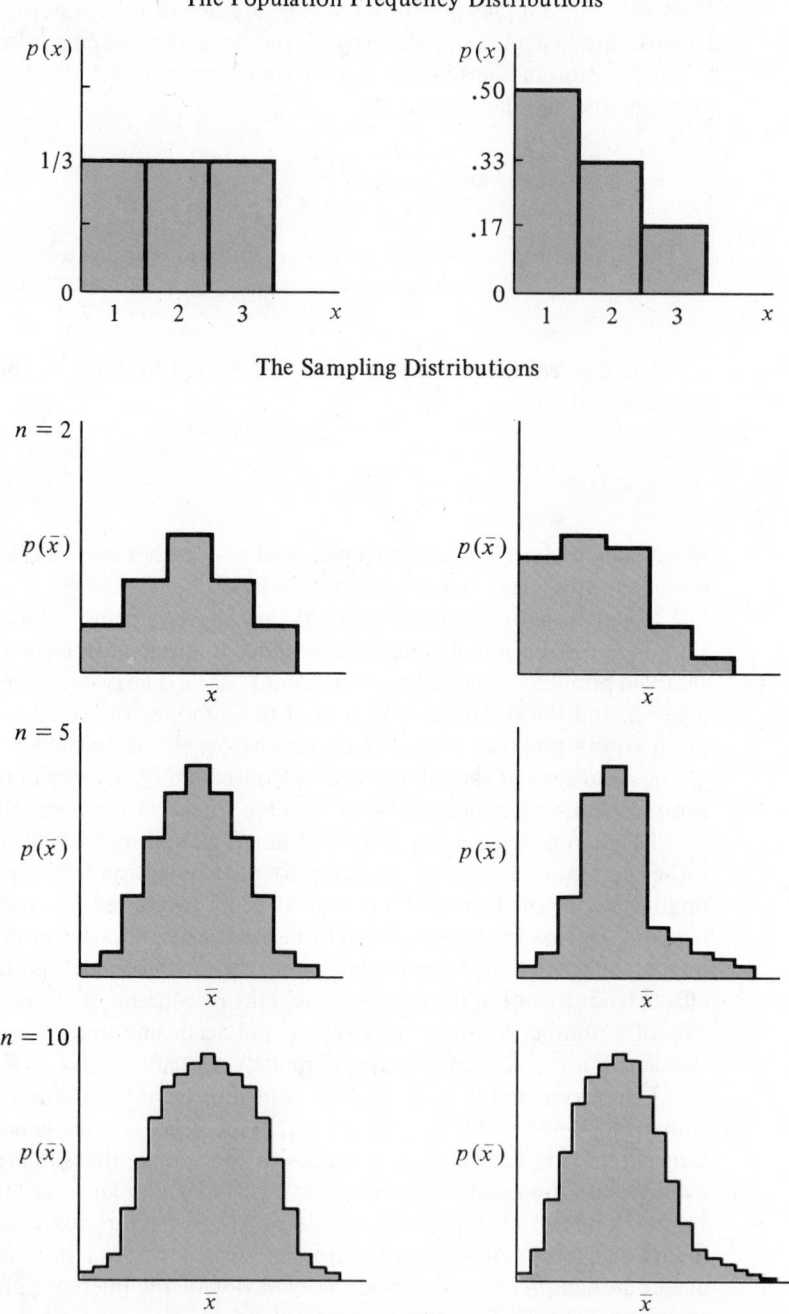

Figure 7.5   Frequency Distribution for $\bar{x}$ for Two Different Probability Distributions; $n = 2, 5, 10, 25$

$n = 25$

$p(\bar{x})$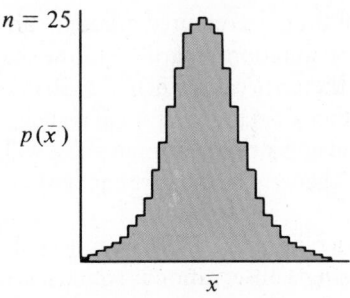

$\bar{x}$

$p(\bar{x})$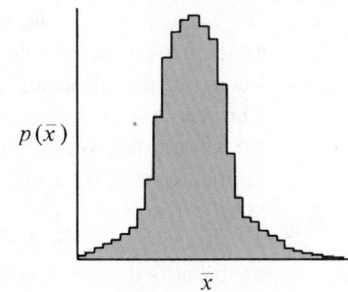

$\bar{x}$

Figure 7.5 (Continued)

One disturbing feature of the Central Limit Theorem, and of most approximation procedures, is that we must have some idea of how large the sample size n must be for the approximation to be valid. Unfortunately, there is no clear-cut solution to this problem, since the appropriate value for n will depend on the population probability distribution as well as how the approximation will be used. Although the preceding comment sidesteps the difficulty and suggests that we must rely solely on experience, we may take comfort in the results of the die-tossing experiment discussed earlier in this section. In repeated sampling the distribution of $\bar{x}$, based on a sample of only $n = 5$ measurements, was approximately bell-shaped. This approximation would be even better for larger values of n.

In Figure 7.5 the frequency distributions for $\bar{x}$ are shown for two different probability distributions, the uniform distribution on the left and a nonsymmetrical distribution on the right, for sample sizes ranging (down the page) from $n = 2$ to $n = 25$. The sample means for each frequency histogram were obtained by drawing a large number of samples from the population, each of fixed sample size, and then computing $\bar{x}$ for each. Figure 7.5 illustrates two significant characteristics of the Central Limit Theorem. The first is the somewhat remarkable fact that the Central Limit Theorem applies regardless of the shape of the underlying population frequency distribution. Clearly, neither of the population frequency distributions depicted in the first row of Figure 7.5 is bell-shaped. But in both case 1 and case 2 the sampling distributions of the sample mean $\bar{x}$ are effectively approximated by the normal distribution when the sample size is sufficiently large. Note how the sampling distribution of $\bar{x}$ approaches a bell shape as the sample size increases from $n = 2$ to $n = 25$ (moving from top to bottom of Figure 7.5).

A second point to note is that the sampling distribution of $\bar{x}$ for case 1 is closely approximated by a smooth bell-shaped curve for samples as small as $n = 10$, but a much larger sample size is required to gain an effective normal approximation to the sampling distribution of $\bar{x}$ for case 2. This result is explained by the fact that when the probability distribution of x is symmetrical about its mean μ, the Central Limit Theorem will apply very well to small sample sizes, often as small as $n = 10$. However, when the population frequency distribution is skewed (nonsymmetrical), larger sample sizes are required to yield an effective approximation to the distribution of $\bar{x}$ by the normal probability distribution.

Some authors offer as a rule of thumb a required minimal sample size of $n = 30$ to guarantee an effective normal approximation, regardless of the shape of the population frequency distribution. However, there are cases when $n = 30$ may be much too small: The binomial experiment when either p or $(1 - p)$ is small is such a case. Consequently, we will not rely on such a rule. The appropriate sample size n will be given for specific applications of the Central Limit Theorem as they are encountered in this chapter and later in the text.

In using the sample mean $\bar{x}$ as a random variable, we must differentiate between the probability density function for a single observation x, which has mean μ and standard deviation σ, and the probability density function for $\bar{x}$, which has mean μ but standard deviation $\sigma/\sqrt{n}$.

EXAMPLE 7.1   Most manufacturing and production operations employ quality control procedures to monitor changes in the quality of production items with respect to some preset standard or operating criterion. Suppose that the past records of a firm producing $4' \times 8'$ sheets of hardwood paneling indicate that the average number of imperfections per panel is $\mu = 5$ with a standard deviation of $\sigma = 2.5$ when the process is operating properly. The manufacturing process is deemed to be in control if the mean $\bar{x}$ of a sample of n panels is within three standard deviations ($\sigma_{\bar{x}} = \sigma/\sqrt{n}$) of the mean μ.

If a random sample of $n = 30$ panels selected for thorough scrutiny contained an average of $\bar{x} = 6$ imperfections per panel, what is the probability of observing a value of $\bar{x} \geq 6$ if the manufacturing process is operating properly? Would the process be deemed in control?

SOLUTION   The Central Limit Theorem can be used to approximate $P(\bar{x} \geq 6)$ if the sample size is sufficiently large. The number of imperfections per panel, which is the variable being measured, might be expected to follow a Poisson distribution. Variables having Poisson distributions tend to exhibit skewness when the mean is small. However, in examining case 2 in Figure 7.5, we see that the distribution of $\bar{x}$ in sampling a skewed distribution appears approximately normal for samples of size $n = 25$. Therefore, the approximation based on the Central Limit Theorem should be adequate in this case, since the sample size is $n = 30$. Hence, the distribution of the sample mean $\bar{x}$ is approximately normal with mean $\mu = 5$ and standard deviation

$$\sigma_{\bar{x}} = \frac{\sigma}{\sqrt{n}} = \frac{2.5}{\sqrt{30}} = .456$$

The required probability is the shaded area under the normal curve in Figure 7.6. To find the area under the curve between $\mu = 5$ and $\bar{x} = 6$, we calculate

$$z = \frac{\bar{x} - \mu}{\sigma/\sqrt{n}} = \frac{6 - 5}{.456} = 2.19$$

From Table 3 in the Appendix the area between $z = 0$ and $z = 2.19$ is $A = .4857$. Then

$$P(\bar{x} > 6.0) = .5 - A = .5 - .4857 = .0143$$

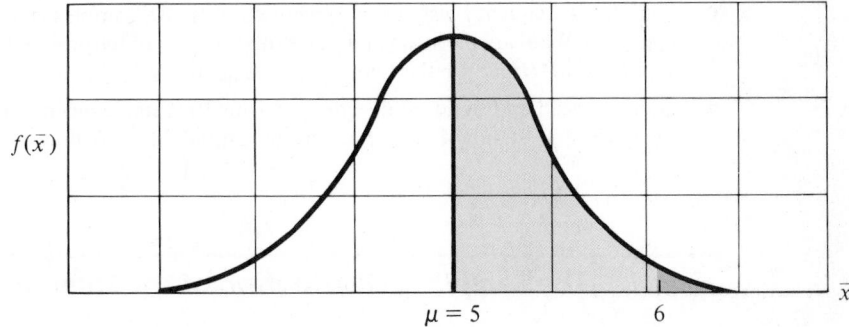

Figure 7.6   Sampling Distribution of $\bar{x}$, the Average Number of Imperfections for $n = 30$
Hardwood Panels in Example 7.1

Although the probability of observing a sample mean of $\bar{x} = 6$ or greater is quite small, the process is not deemed to be out of control, since $\bar{x}$ lies less than three standard deviations from the mean. However, the process should be monitored closely, because we may have observed an unusual sample or, on the other hand, the process may soon need to be stopped and adjusted.

EXERCISES   Conceptual Level

7.8   What is the role of the Central Limit Theorem in statistics?

7.9   Under what conditions would the results obtained using the Central Limit Theorem be considered reasonably accurate?

7.10   What important information does the Central Limit Theorem provide concerning the sampling distribution of the sample mean $\bar{x}$ when the sampled population is not normal?

EXERCISES   Skill Level

7.11   A sample of size $n = 100$ is selected from a nonnormal population whose mean is $\mu = 50$ and whose standard deviation is $\sigma = 10$.

   a.   What is the mean and standard deviation of the sampling distribution of $\bar{x}$?

   b.   According to the Central Limit Theorem, what is the approximate distribution of the sample mean $\bar{x}$?

   c.   Use the Central Limit Theorem to approximate $P(\bar{x} > 52)$; $P(47.5 < \bar{x} < 52.5)$.

7.12   Let x be the number of dots appearing on the upper face when a single six-sided die is tossed.

   a.   Verify that the mean and the standard deviation of x are $\mu = 3.5$ and $\sigma = 1.71$, respectively.

   b.   Repeat the sampling experiment of Table 7.4 by tossing a die $n = 5$ times and recording the 5 observations. Repeat this procedure 100 times so that you have 100 samples, each consisting of $n = 5$ observations. Compute the sample mean for the $n = 5$ observations in each of the 100 samples.

c. Construct a frequency histogram to represent your 100 sample means and compare it with Figure 7.4. What are the mean and standard deviation of the probability distribution for the sample mean? (*Hint:* See the Central Limit Theorem.)

d. Find the mean and standard deviation of your 100 sample means and compare them with the mean and standard deviation of the probability distribution for $\bar{x}$ determined in part c.

7.4

The Sampling Distribution of the Sample Mean

Decision-making problems encountered in business often involve inferences about the value of a population mean. For example, we may be concerned with the average daily production of workers on an assembly line, with the average strength of a new type of steel, with the average number of accidents per month, or with the average demand for a new product.

Many estimators are available for estimating the population mean, including the median, the trimmed mean, and the midrange (the average of the largest and smallest observation in the set), as well as the sample mean $\bar{x}$. Each generates a sampling distribution in repeated sampling, and depending upon the population and the problem involved, each possesses certain advantages and disadvantages. Some statistics are easier to calculate than others; some statistics produce estimates that are consistently too large or too small; others produce estimates that are highly variable in repeated sampling. The sampling distributions for $\bar{x}$ and m with $n = 3$ for the population of $N = 5$ elements given in Section 7.2 showed that when we use criteria such as these, the sample mean seemed to perform better than the median as an estimator of the population mean μ. In many situations the sample mean $\bar{x}$ has desirable properties as an estimator that are not shared by other competing estimators; therefore, it is more widely used.

When repeated samples of size n are randomly selected from a *finite* population with N elements whose mean is μ and whose variance is σ^2, the sampling distribution of the sample mean has the following properties.

Properties of the Sample Mean $\bar{x}$

1. The mean or the expected value of $\bar{x}$ is equal to the population mean μ.
2. The standard deviation of $\bar{x}$ is

$$\sigma_{\bar{x}} = \frac{\sigma}{\sqrt{n}} \sqrt{\frac{N - n}{N - 1}}$$

where σ^2 is the population variance. When N is large relative to the sample size n, $\sqrt{(N - n)/(N - 1)}$ is approximately equal to 1. Then

$$\sigma_{\bar{x}} = \frac{\sigma}{\sqrt{n}}$$

Not only does $\bar{x}$ have a mean equal to the mean of the population sampled, but its standard deviation is directly proportional to the population standard deviation σ and inversely proportional to the square root of the sample size. Although we give no proof of these results, we suggest that they are intuitively reasonable. Certainly, the more variable the population data as measured by σ, the more variable will be $\bar{x}$. On the other hand, more information is available as n becomes large. Therefore, the values of $\bar{x}$ should be less variable, and $\sigma_{\bar{x}}$ should decrease.

EXAMPLE 7.2   Find the mean and variance of the sample mean $\bar{x}$ for the sampling distribution given in Table 7.2. Compare these values with those derived by using the relationships just presented.

SOLUTION   In Section 7.2 we found the mean and variance of $\bar{x}$ to be $\mu_{\bar{x}} = 12$ and $\sigma_{\bar{x}}^2 = 3$. The value of $\mu_{\bar{x}} = 12$ is exactly equal to the population mean $\mu = 12$. Using the formula just presented with $N = 5$, $n = 3$, and $\sigma^2 = 18$, we obtain

$$\sigma_{\bar{x}}^2 = \frac{\sigma^2}{n}\left(\frac{N-n}{N-1}\right) = \frac{18}{3}\left(\frac{5-3}{5-1}\right) = 3$$

which agrees with our result obtained through direct calculation.

◆◆

finite population
correction factor
The quantity $(N-n)/(N-1)$ is referred to as a **finite population correction factor** and is sometimes ignored in calculations when its value exceeds .95 or, equivalently, when $N \geq 20n$.

Can anything more be said about the distribution of the sample mean with respect to its form? The answer to this question depends upon the form of the population distribution and upon the sample size. When we sample a population with mean μ and standard deviation σ (which are assumed to be finite numbers), the Central Limit Theorem becomes applicable when the sample size is large. Samples of size 30 or more from continuous populations are usually considered large unless the population exhibits extreme skewness, in which case a much larger sample of, say, 100 or more might be required. In discrete populations such as the binomial, sample sizes larger than 30 are usually required unless $p \approx 1/2$.

The Sampling Distribution of the Sample Mean for Large Samples

> When the sample size is large, the sample mean $\bar{x}$ is approximately normally distributed according to the Central Limit Theorem, with mean μ and standard deviation $\sigma/\sqrt{n}$.

Since we have taken the standard deviation of $\bar{x}$ to be $\sigma/\sqrt{n}$, we have tacitly assumed that the size of the population is large or infinite and that $(N-n)/(N-1)$ is approximately 1.

The standard deviation of a statistic used as an estimator of a population parameter is often called the standard error of the estimator, since it refers to the precision of the estimator. Therefore, the standard deviation of $\bar{x}$, given by $\sigma/\sqrt{n}$, is referred to as the **standard error of the mean**.

standard error of the mean

EXAMPLE 7.3   Acid rain, produced by the chemical reaction of water vapor and chemical pollutants in the atmosphere, is a serious environmental problem in areas with highly developed industrial capacities. Acid rain is responsible for annihilating aquatic life in many of the lakes and streams in the Northeast. Acidity is measured in terms of pH, which varies from 0 to 14. A pH value of 7.0 is neutral; values less than 7.0 indicate increasing acidity. Suppose that past records indicate that the pH values of rainfall in the Northeast have a mean of 4.8 and a standard deviation of .5. What is the probability that the average pH value of 36 randomly selected future rainstorms in the area will have a pH value greater than 5.0?

SOLUTION   The Central Limit Theorem would ensure the approximate normality of the distribution of $\bar{x}$ when the sample size is as large as 36. Since the mean and standard deviation of rainfalls in this area are $\mu = 4.8$ and $\sigma = .5$, $\bar{x}$ is approximately normally distributed with mean 4.8 and standard deviation

$$\sigma_{\bar{x}} = \frac{\sigma}{\sqrt{n}} = \frac{.5}{\sqrt{36}} = .083$$

The probability that $\bar{x}$ is greater than 5.0 is approximately equal to the dark shaded area in Figure 7.7. This area is equal to the difference $.5 - A$, where A is the area between $\mu = 4.8$ and $\bar{x} = 5.0$. Expressing the distance between 4.8 and 5.0 in terms of standard deviations, we have

$$z_1 = \frac{\bar{x} - \mu}{\sigma/\sqrt{n}} = \frac{5.0 - 4.8}{.083} = 2.41$$

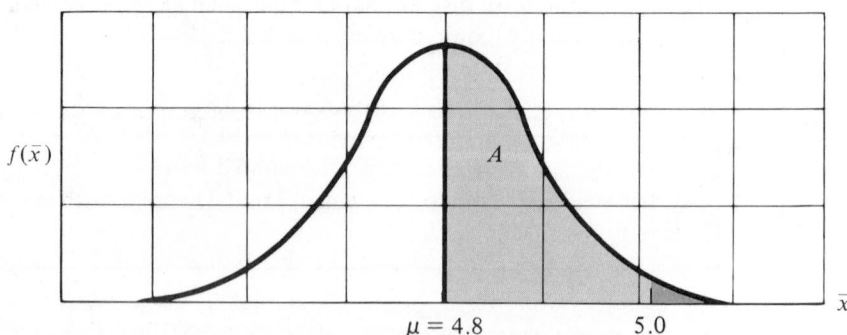

Figure 7.7   Sampling Distribution of $\bar{x}$, the Mean of $n = 36$ pH Determinations on Rainfall in Example 7.3

Notice that $\sigma_{\bar{x}}$ and not σ was used in calculating the value of z because we are finding the area under the sampling distribution for $\bar{x}$ and not the area under the probability distribution for x. From Table 3 of the Appendix the area over the interval from $z = 0$ to $z = 2.41$ is $A = .4920$. The probability that $\bar{x}$ will exceed 5.0 is

$$P(\bar{x} > 5.0) = .5 - A = .5 - .4920 = .0080$$

or approximately .01. Therefore, the probability of an increase in the mean pH value of rainfall in this region, indicating a decrease in the average acidity of rainfall, is very small, if past conditions remain unchanged.

In the special situation in which the sampled population has a normal distribution, even more can be said about the sampling distribution of the sample mean.

The Distribution of $\bar{x}$ When Sampling a Normal Population

> When sampling from a normal population with mean μ and standard deviation σ, the sample mean $\bar{x}$ is normally distributed with mean μ and standard deviation $\sigma/\sqrt{n}$ for *all* values of n.

When the sample size is large, the probabilities found by using the statistic $z = (\bar{x} - \mu)/\sqrt{\sigma^2/n}$ will be appropriate whether or not the sampled population is normal. **However, when the sample size is small, we must be relatively certain that the sample has been selected from a normal population in order to use our results with any degree of confidence.**

EXAMPLE 7.4   To avoid difficulties with the Federal Trade Commission or state and local consumer protection agencies, a beverage bottler must make reasonably certain that 12 oz bottles actually contain 12 oz of beverage. To infer whether a bottling machine is working satisfactorily, one bottler randomly samples 10 bottles per hour and measures the amount of beverage in each bottle. The mean $\bar{x}$ of the 10 measurements is used to decide whether to readjust the amount of beverage delivered per bottle by the filling machine. If records show that the amount of fill per bottle is normally distributed with a standard deviation of .2 oz and if the bottling machine is set to produce a mean fill per bottle of 12.1 oz, what is the probability that the sample mean $\bar{x}$ of the 10 test bottles is less than 12 oz?

SOLUTION   The mean of the sampling distribution of $\bar{x}$ is identical to the mean of the population of bottle fills—namely, $\mu = 12.1$ oz—and the standard deviation (or standard error) of $\bar{x}$ is

$$\sigma_{\bar{x}} = \frac{\sigma}{\sqrt{n}} = \frac{.2}{\sqrt{10}} = .063$$

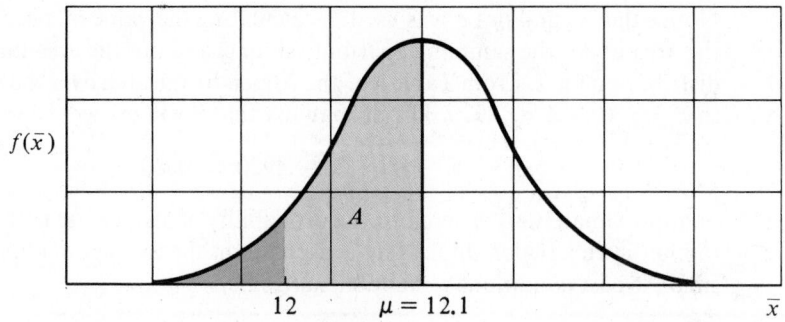

Figure 7.8   Probability Distribution of $\bar{x}$, the Mean of the $n = 10$ Bottle Fills in Example 7.4

(*Note:* σ is the standard deviation of the population of bottle fills and n is the number of bottles in the sample.) Since the amount of fill is normally distributed, $\bar{x}$ is also normally distributed. Then the probability distribution of $\bar{x}$ will appear as shown in Figure 7.8.

The probability that $\bar{x}$ will be less than 12 oz will equal $(.5 - A)$, where A is the area between 12 and the mean $\mu = 12.1$. Expressing this distance in standard deviations, we have

$$z = \frac{\bar{x} - \mu}{\sigma_{\bar{x}}} = \frac{12 - 12.1}{.063} = -1.59$$

Then the area A over the interval $12 < \bar{x} < 12.1$, found in Table 3 of the Appendix, is .4441, and the probability that $\bar{x}$ will be less than 12 oz is

$$P(\bar{x} < 12) = .5 - A = .5 - .4441 = .0559 \approx .056$$

Thus, if the machine is set to deliver an average fill of 12.1 oz, the mean fill $\bar{x}$ of a sample of 10 bottles will be less than 12 oz with probability equal to .056. When this danger signal occurs ($\bar{x}$ is less than 12), the bottler takes a larger sample to recheck the setting of the filling machine.

◆ ◆

EXERCISES   Conceptual Level

7.13   What can be said about the distribution of $\bar{x}$ when sampling a population with mean μ and standard deviation σ? When sampling a normal population with mean μ and standard deviation σ?

7.14   What is the relationship between the quantity of information as measured by the sample size n and the variability of the sampling distribution of $\bar{x}$?

7.15   When sampling a finite population, the standard deviation of the sample mean, $\sigma/\sqrt{n}$, is multiplied by $\sqrt{(N - n)/(N - 1)}$ to correct for the finite population size. When can this correction be ignored?

7.16   Does using the finite population correction factor increase or decrease the standard deviation of $\bar{x}$? Why should this be the case?

EXERCISES   *Skill Level*

7.17   A sample of size $n = 200$ observations is randomly selected from a population consisting of 12 million observations with a mean $\mu = 6$ and variance $\sigma^2 = 81$.

   a.   What is the probability that the sample value of $\bar{x}$ is greater than or equal to 7?

   b.   If the size of the population were 1200 rather than 12 million, what is the probability that the sample mean $\bar{x}$ is greater than or equal to 7?

   c.   If the population size is 12 million and a sample of size $n = 600$ is randomly selected from this population, what is the probability that $\bar{x}$ is greater than or equal to 7?

7.18   A sample of size $n = 10$ is randomly selected from a normal population with mean $\mu = 52$ and variance $\sigma^2 = 22.5$. The sample value $\bar{x}$ is calculated.

   a.   Find $P(\bar{x} > 55)$.   **b.**   Find $P(50 < \bar{x} < 60)$.   **c.**   Find $P(\bar{x} \leq 49)$.

EXERCISES   *Applications*

 7.19   A production process for steel rods used to reinforce concrete is known to produce rods whose lengths have a variance of 64 cm. The production machinery has been set to produce rods with a mean of $\mu = 6$ meters (600 centimeters, cm) in length. These rods are tied into bundles of 40 for shipment to construction sites.

   a.   What is the probability that the average length of a randomly selected bundle is less than 598 cm?

   b.   What is the probability that the average length of a randomly selected bundle is less than 598 cm or more than 601 cm?

   c.   Under what assumptions are the probabilities found in parts a and b valid?

   d.   If the bundle size is 25 rather than 40, what additional assumption is necessary in order to use the sampling distribution of $\bar{x}$ that was appropriate in parts a and b?

   e.   Suppose that the bundle size is 25. What is the probability of selecting a bundle that has an average length of less than 598 cm?

 7.20   Refer to Exercise 7.19. Suppose that only $N = 500$ of the 6-m steel rods are produced by this process.

   a.   What is the probability of randomly selecting a bundle of 40 rods with a mean of less than 598 cm?

   b.   What is the probability of selecting a bundle with an average length of less than 598 cm or more than 601 cm?

c. Compare the answers obtained in parts a and b with the answers obtained in parts a and b of Exercise 7.19. How does the finite population size ($N = 500$) affect the probabilities?

7.21 In Exercise 6.13 we examined the distribution of salaries for MBA graduates entering the field of marketing services (research and advertising). The distribution of salaries was assumed to be normal, with mean $38,000 and standard deviation $2,250 (adapted from *Sales and Marketing Management*, "Survey of Selling Costs," February 16, 1987. Table III–8, p. 58). Suppose that 10 students graduate from a small midwestern university with MBA degrees and enter the field of marketing services, and that the starting salaries of these 10 students are independent.

a. What is the probability that the average salary of these 10 students exceeds $40,000?

b. What might you conclude if the average salary of these 10 students were $34,280?

7.22 Refer to Exercise 2.89 in which a study reports on the caffeine content of various foods and beverages. The means and standard deviations of the caffeine content for several beverages are shown in Table 7.5. An experimenter in the research and development department of a major coffee manufacturer, in an attempt to verify or disprove the results of this survey, takes a random sample of $n = 45$ of the 5-oz cups of her company's brand of drip-brewed coffee and measures the caffeine content of each cup.

Table 7.5   Caffeine Data for Exercise 7.22

	Milligrams of Caffeine	
Item	Average	Standard Deviation
Brewed coffee (5-oz cup)		
Drip method	115	20
Percolator	80	21.7
Brewed tea (5-oz cup)		
Major U.S. brands	40	11.7
Imported brands	60	15

Source: Data from Chris W. Lecos, "Caffeine Jitters: Some Safety Questions Remain," *FDA Consumer*, Vol. 21, No. 10 (December 1987 – January 1988), pp. 24–26.

a. What is the probability that the sample average will exceed 121?

b. What is the probability that the sample mean will lie within 5 milligrams (mg) of the average quoted in the table?

c. If the sample mean is $\bar{x} = 125$, what would you conclude?

7.5

The Sampling Distribution of the Sample Proportion

Many sampling problems are concerned with estimating the proportion of individuals in the population that possess a particular attribute. Lot acceptance sampling for defectives (Example 5.4) and marketing research surveys involving consumer preferences are two situations in which the primary objective is to estimate a population proportion. As we have seen, these and similar situations can be modeled by a binomial probability distribution if the sampling procedure is conducted in a manner that satisfies the requirements of the binomial experiment.

In a binomial experiment with n trials, the number of successes x, or, equivalently, the sample proportion of successes, x/n, contains all the relevant sample information concerning the population proportion p. It is no surprise that the sample proportion, which we will denote by

$$\hat{p} = \frac{x}{n}$$

is a statistic whose sampling distribution plays an important role in making inferences about the population parameter p. [The "hat" or "carat" ($\hat{}$) over a parameter is used to denote the estimator of that parameter.] Since each distinct value of x results in a distinct value of $\hat{p} = x/n$, the probabilities associated with $\hat{p}$ are equal to the probabilities associated with the corresponding values of x. Although evaluating binomial probabilities directly is always possible, it is impractical to do so for large values of n.

In Chapter 6 we used the normal distribution to approximate binomial probabilities when the interval $(np \pm 2\sqrt{npq})$ was contained within the binomial limits of 0 and n. The Central Limit Theorem now provides the basis for this approximation within the following framework. In a binomial experiment with n trials, each trial is itself a binomial experiment with a sample size of 1. If we denote a success as a "1" and a failure as a "0," then *for each trial* the mean is $(1)(p) = p$ and the variance is $(1)(p)(q) = pq$. In this context the number of successes x in n trials is simply a sum of n observations that are either 1s or 0s and the sample proportion $\hat{p}$ is the mean of these n observations. Therefore, according to the Central Limit Theorem, when n is large, x is approximately normally distributed with mean np and variance npq. The large-sample approximation to the distribution of the sample proportion, based on the Central Limit Theorem, is given in the following display.

The Sampling Distribution of $\hat{p}$ for Large Samples

> When the sample size n is large, the sampling distribution of $\hat{p}$ is approximately normal with mean p and standard deviation $\sqrt{pq/n}$.

In the use of this approximation to the sampling distribution of $\hat{p}$, the interval $(p \pm 2\sqrt{pq/n})$ should be contained within the interval from 0 to 1. When we compare this result with the large-sample distribution of $\bar{x}$, we see that the mean μ and variance σ^2 are replaced by the mean p and variance pq of the binomial population.

EXAMPLE 7.5   A decline in the demand for a specific brand of a product may result from either a decline in demand for the product in general or a reduction in the percentage of the purchasers selecting this specific brand over its competitors. Suppose that to remain competitive in the marketplace, a company producing brand A must maintain at least an equal share of the market with its three major competitors. If 100 consumers are randomly selected and interviewed regarding their brand preferences, what is the probability of observing a sample proportion preferring brand A as small as .15 or less if, in fact, one-fourth, or 25%, of the consumers prefer brand A?

SOLUTION   Since the sample size is large, the distribution of $\hat{p}$ will be approximately normal with mean $p = .25$ and standard deviation (standard error) given as

$$\sigma_{\hat{p}} = \sqrt{\frac{pq}{n}} = \sqrt{\frac{(.25)(.75)}{100}} = .0433$$

Since the interval $(.25 \pm .09)$ is within the interval from 0 to 1, the normal approximation should be adequate. To find $P(\hat{p} < .15)$, we calculate

$$z_1 = \frac{.15 - .25}{\sqrt{\frac{(.25)(.75)}{100}}} = -2.31$$

The area A between $z = 0$ and $z_1 = -2.31$ is .4896; the required probability is $(.5 - A) = (.5 - .4896) = .0104$, or approximately .01. Hence, it is very unlikely that the sample preference proportion would be as small as .15 if brand A actually enjoys 25% of the market.

◆ ◆

In employing the normal distribution to approximate the binomial probabilities associated with x, we used a correction of $\pm.5$ to improve the approximation. The equivalent correction here would be $\pm(1/2n)$. For example, for $\hat{p}$ the value of z with the correction would be

$$z_1 = \frac{(.15 + .005) - .25}{\sqrt{\frac{(.25)(.75)}{100}}} = -2.19$$

with $A = .4857$ and $(.5 - A) = .0143$. To two-decimal accuracy, this value agrees with our earlier result. When n is large, the effect of using the correction is generally negligible. You should solve problems in this and the remaining chapters *without* the correction factor unless specifically instructed to use it.

EXERCISES   Conceptual Level

7.23   Under what conditions can we assume that the sampling distribution for the sample proportion $\hat{p}$ is approximately normal?

7.24   List the values of pq for $p = .1, .2, .3, \ldots, .9$. For what value of p is pq largest? For a given sample size, when is the variance of $\hat{p}$ a maximum?

EXERCISES   Skill Level

7.25   In a sample of $n = 100$ observations the number of successes was found to be $x = 38$.

    a.   If the population proportion is $p = .45$, find $P(\hat{p} \leq .38)$.

    b.   If $p = q = .5$, find $P(\hat{p} \leq .38)$.

7.26   If a sample of $n = 200$ observations is randomly selected from a binomial population with $p = .7$, find $P(|\hat{p} - p| > .05)$.

EXERCISES   Applications

 7.27   A marketer of a nondairy whitener for coffee felt that her product was so good that most coffee drinkers could not tell the difference between her product and real cream in a cup of coffee. Before making this claim in advertising, the product manager for the coffee whitener felt that some tests should be conducted to find out whether this claim could be substantiated if it were challenged. The test involved having coffee drinkers who regularly use whitener in their coffee taste a cup of coffee with the nondairy whitener and a cup with real cream, then indicate which one they thought had real cream. If the claim is really correct, then each coffee drinker is equally likely to pick the whitener as the cream ($p = .5$) because he or she really cannot tell the difference.

    a.   If the claim is correct, what is the probability that out of a randomly selected sample of size 60, 25 or fewer identify the cream correctly?

    b.   If the sample size were 120 instead of 60, what would be the probability of 50 or fewer subjects selecting cream?

    c.   Under what assumptions are the probabilities found in parts a and b valid?

    d.   The probabilities in parts a and b are different even though the population proportion (p) and the sample proportion ($\hat{p}$) values are unchanged. What impact did a change in the sample size have on the variance of the sampling distribution?

7.28   Tests have shown that consumers generally tend to agree on the strength and sweetness of various fragrances but that they are far less likely to agree regarding preference for fragrances (Howard R. Moskovitz, "Segmenting Consumers: Uncovering Distinct Preferences for Fragrance Sensory Characteristics," *Aerosol Age*, January 1985). Thirty-five consumers were randomly selected and tested regarding their preference for a given fragrance over another.

    a.   Suppose that there was no difference in preference regarding the two fragrances. Then the proportion of consumers in the population who prefer a given fragrance is $p = .5$. What is the probability that less than 15 of the 35 consumers would prefer a given fragrance?

    b.   What is the probability that less than 25 of the 35 consumers sampled would prefer a given fragrance?

    c.   Under what assumption(s) are the probabilities obtained in parts a and b valid?

    d.   If the sample size is small and the normal distribution cannot be used to approximate the sampling distribution of $\hat{p}$, what is the appropriate form of the sampling distribution of $\hat{p}$?

7.29   A report by McGraw-Hill Research's Laboratory of Advertising Performance indicated that approximately 10% of decision makers receive product information directly from salespeople (*Sales and Marketing Management*, February 1988, p. 24). If this percentage is, in fact, representative of the population at large, what is the probability that in a sample of $n = 200$ persons the sample proportion $\hat{p}$ of those receiving product information directly from salespeople would exceed 13%?

7.30   Exercise 6.75 refers to a report that of the 33 million women in the United States with children under the age of 18, 63% work outside the home (Chester A. Swenson, "How to Sell to a Segmented Market," *Journal of Business Strategy*, January–February 1988, p. 18). In a test of this claim a random sample of $n = 100$ women with children under 18 was selected and $x = 84$ were found to work outside the home.

a.   If the reported proportion is correct and $p = .63$, describe the sampling distribution of $\hat{p}$.

b.   If the reported proportion is incorrect and, in fact, $p = .75$, describe the sampling distribution of $\hat{p}$.

c.   Find the probability of observing a sample value of $\hat{p}$ as large as or larger than the observed sample value, $\hat{p} = .84$, for the two cases in parts a and b.

d.   On the basis of part c, which is the more likely decision—that $p = .63$ or that $p = .75$?

7.6

The Sampling Distribution of the Difference Between Two Sample Means

A comparison of two populations quite often focuses on the difference between the population means. We may be interested in the difference between the average production rates for each of two production lines or the difference in the average saving in heating costs for households with and without passive-heating devices (such as solar collectors). Intuitively, the difference between two sample means would provide the maximum information about the actual difference between two population means, and this is in fact the case.

Suppose now that the first population of interest has a mean equal to μ_1 and standard deviation equal to σ_1, and the second population has a mean μ_2 and standard deviation σ_2. If two random and independent samples of size n_1 and n_2 measurements are repeatedly drawn from the two populations and the value of $(\bar{x}_1 - \bar{x}_2)$ is calculated for these samples, the resulting relative frequency distribution of $(\bar{x}_1 - \bar{x}_2)$ will approximate the sampling distribution of $(\bar{x}_1 - \bar{x}_2)$. What properties will this sampling distribution possess? When sample sizes n_1 and n_2 are large, both $\bar{x}_1$ and $\bar{x}_2$ will have approximately normal sampling distributions with means μ_1 and μ_2 and standard deviations $\sigma_1/\sqrt{n_1}$ and $\sigma_1/\sqrt{n_2}$. **If both populations have normal distributions, then both $\bar{x}_1$ and $\bar{x}_2$ will be normally distributed regardless of the sample sizes.** It can be shown that the sum or difference of two independent normally distributed statistics is also normally distributed with a mean equal to the sum or difference of their means and a variance equal to the *sum* of their variances. Since, when n_1 and n_2 are large, $\bar{x}_1$ and $\bar{x}_2$ are either normally distributed or approximately so, the difference between the sample means $(\bar{x}_1 - \bar{x}_2)$ is also normally distributed, or approximately so, with mean $(\mu_1 - \mu_2)$ and variance $(\sigma_1^2/n_1 + \sigma_2^2/n_2)$.

The Sampling Distribution of $\bar{x}_1 - \bar{x}_2$ for Large Samples

When repeated samples of size n_1 and n_2 are randomly and independently drawn from populations with means μ_1 and μ_2 and variances σ_1^2 and σ_2^2 respectively, the sampling distribution of $(\bar{x}_1 - \bar{x}_2)$ is approximately normal with mean

$$\mu_{(\bar{x}_1 - \bar{x}_2)} = \mu_1 - \mu_2$$

and standard deviation

$$\sigma_{(\bar{x}_1 - \bar{x}_2)} = \sqrt{\frac{\sigma_1^2}{n_1} + \frac{\sigma_2^2}{n_2}}$$

when n_1 and n_2 are large.

EXAMPLE 7.6   Patterns of increasing electricity use indicate that more power plants will have to be built to meet peak-demand power use. Critics have suggested that electric rates, like telephone rates, should reflect the time of use as well as the amount of use. Lower electric rates could be provided to households during the early morning and evening hours and higher rates during peak-demand hours. The hope is that residential power use associated with household appliances such as washers, dryers, and vacuums would be reduced during peak-demand hours, thereby delaying the need for construction of new power plants. Suppose that of 200 randomly selected residences, comparable with respect to total square footage and type of construction, $n_1 = 100$ were assigned fixed electric rates and the remaining $n_2 = 100$ were assigned variable electric rates. Suppose also that we can assume that the average electric bill for fixed electric rates is $20 higher than the average electric bill with variable rates and that the standard deviations for both populations are equal to $15.

During the third month after beginning the experimental program, the 100 residences with fixed electric rates had an average electric bill of $104.35, while the 100 residences with variable electric rates had an average bill of $78.92, so that $(\bar{x}_1 - \bar{x}_2) =$ ($104.35 − $78.92) = $25.43. What is the probability of observing a sample difference as large or larger than $(\bar{x}_1 - \bar{x}_2) = \$25.43$?

SOLUTION   If the average bill is $20 higher for a fixed electric rate than for the variable rate, then $(\mu_1 - \mu_2) = \$20$. With samples as large as $n_1 = n_2 = 100$, the sampling distribution of $(\bar{x}_1 - \bar{x}_2)$ is approximately normal with mean $(\mu_1 - \mu_2) = 20$ and standard deviation equal to

$$\sigma_{(\bar{x}_1 - \bar{x}_2)} = \sqrt{\frac{\sigma_1^2}{n_1} + \frac{\sigma_2^2}{n_2}}$$

$$= \sqrt{\frac{15^2}{100} + \frac{15^2}{100}}$$

$$= 2.12$$

Therefore, $P(\bar{x}_1 - \bar{x}_2 > 25.43) = P(z > z_1)$, where

$$z_1 = \frac{(\bar{x}_1 - \bar{x}_2) - (\mu_1 - \mu_2)}{\sqrt{\dfrac{\sigma_1^2}{n_1} + \dfrac{\sigma_2^2}{n_2}}}$$

$$= \frac{25.43 - 20}{\sqrt{\dfrac{15^2}{100} + \dfrac{15^2}{100}}} = 2.56$$

From Table 3 the area between 0 and $z_1 = 2.56$ is .4948, and the required probability is $(.5 - .4948) = .0052$. The probability of observing our sample value or something more extreme indicates that it is highly unlikely that the average saving from using variable rates is actually as small as $20.

◆ ◆

EXERCISES Conceptual Level

7.31 In a comparison of two population means we use the difference between the two sample means, $\bar{x}_1 - \bar{x}_2$, to make inferences about the difference in the two population means, $\mu_1 - \mu_2$. What are the mean and standard deviation of the statistic $(\bar{x}_1 - \bar{x}_2)$?

7.32 Under what conditions is the sampling distribution of $(\bar{x}_1 - \bar{x}_2)$ normal? Approximately normal?

EXERCISES Skill Level

7.33 Independent random samples of size $n_1 = n_2 = 30$ are drawn from populations having identical population means, $\mu_1 = \mu_2$, and with identical population standard deviations, $\sigma_1 = \sigma_2 = 5$. The sample means are $\bar{x}_1 = 14$ and $\bar{x}_2 = 16$. Evaluate $P(|\bar{x}_1 - \bar{x}_2| > 2)$.

7.34 Refer to Exercise 7.33. Evaluate $P(\bar{x}_1 - \bar{x}_2 < 0)$ if, in fact, $\mu_1 = 15$ and $\mu_2 = 14$.

EXERCISES Applications

7.35 Packaging can be one of the most important attributes of a product. A producer of honey, who has a problem with a container coming open in shipment, is considering a differently designed plastic container. The producer is interested in the impact resistance of the new containers, which have a screw cap, as compared with the present containers, which have snap-on caps. Random samples of 40 new containers and 40 present containers were selected. Each container was filled with honey and subjected to measured impact until it failed, usually as a result of the cap popping off. The average impact resistance was 30 for the screw cap and 28.5 for the snap-on cap. For purposes of comparison the producer is willing to assume that the variance of the impact resistance for each type of container is 36.

a. If there is really no difference in the impact resistance for the two containers, what is the probability that the screw cap would outperform the snap-on cap by as much as or more than it did in this test?

b. The manufacturer of the new container claims that his screw cap container is more impact-resistant than the snap-on cap by 1 point on the measured impact resistance scale. If this claim is correct, what is the probability of the screw cap outperforming the snap-on cap by as much or more than it did in this test?

c. Under what assumptions are the probabilities obtained in parts a and b valid?

d. The honey producer was pleased to find a demonstrably better container, but he was not sure that the test results were strong enough to justify a total changeover in his production process. What would you suggest to strengthen the test and give the honey producer more confidence in the results?

7.36 A study comparing the images of products from six different countries was conducted in New Zealand (T., Barker, *Australian Marketing Research*, February 1985). On a scale in which lower scores indicated a greater willingness to buy, four groups of 197 respondents independently rated products from Australia, Great Britain, Japan, Taiwan, West Germany, and the United States. Assume that the variances for the scores from each of the countries was equal to .77.

a. On the willingness-to-buy scale, products from West Germany received a score of 3.36 and those from the United States received 3.54. If there is really no difference in the willingness of consumers to purchase products from these two countries, what is the probability of West Germany scoring as low as or lower than the United States?

b. The score for Japan was 3.74. If there is really no difference in the willingness of consumers to buy from these two countries, what is the probability of Japan scoring as high as or higher than the United States?

7.37 The results of a consumer survey concerning annual income and expenditures of urban consumer units delineated gasoline and motor oil costs as different from all other transportation costs for four regions of the United States [U.S. Department of Commerce, Bureau of the Census, *Statistical Abstract of the United States*, 107th ed. (Washington, D.C., 1988), p. 428]. Gasoline expenditures in western states, when compared with expenditures in northeastern states, would reflect not only possible differences in costs per gallon but also differences in distances traveled. A sample survey involved $n_1 = 100$ consumer units in western states and $n_2 = 120$ consumer units in northeastern states, for which expenditures for gasoline and motor oil are recorded as part of the survey. Use the information in the accompanying table to answer the following questions.

Western	Northeastern
$\mu_1 = \$1085$	$\mu_2 = \$872$
$\sigma_1 = \quad 130$	$\sigma_2 = \quad 100$
$n_1 = \quad 100$	$n_2 = \quad 120$

a. Find the probability that the difference in the sample means, $\bar{x}_1 - \bar{x}_2$, exceeds $200.

b. What is the probability that the sample difference $(\bar{x}_1 - \bar{x}_2)$ is less than $-\$200$ or greater than \$200?

7.7

The Sampling Distribution of the Difference Between Two Sample Proportions

The sampling distribution of the difference between two sample proportions $\hat{p}_1 - \hat{p}_2$, based on two independent random samples from binomial populations with parameters p_1 and p_2, can be obtained by using the fact that a sample proportion is approximately normally distributed when the sample size is large (Section 7.5) and that the sum or difference of two normal random variables is also normally distributed (Section 7.6). Using this information, we have the following result.

The Sampling Distribution of $\hat{p}_1 - \hat{p}_2$ for Large Samples

> When repeated samples of size n_1 and n_2 are randomly and independently drawn from two binomial populations with parameters p_1 and p_2, the sampling distribution of $(\hat{p}_1 - \hat{p}_2)$ is approximately normal, with mean $(p_1 - p_2)$ and standard deviation
>
> $$\sigma_{(\hat{p}_1 - \hat{p}_2)} = \sqrt{\frac{p_1 q_1}{n_1} + \frac{p_2 q_2}{n_2}}$$
>
> when n_1 and n_2 are large.

When we use a normal distribution to approximate binomial probabilities, the interval $(p_1 - p_2) \pm 2\sigma_{(\hat{p}_1 - \hat{p}_2)}$ should be contained within the range of $(\hat{p}_1 - \hat{p}_2)$, which varies from -1 to 1 and not from 0 to 1, as in the case of a single proportion.

EXAMPLE 7.7   A bond proposal for school construction is to be submitted to the voters during the next municipal election. A major portion of the money derived from this bond issue will be used to build new schools in a rapidly developing section of the city, and the remainder will·be used for renovating and updating school buildings in the rest of the city. The local newspaper reported that 75% of the residents in the developing section and 60% of the residents in other parts of the city favor passage of the proposed bond issue. Random samples of $n_1 = 50$ residents in the developing section of the city and $n_2 = 100$ residents in other parts of the city are selected, and the residents in the sample are asked whether or not they favor the bond proposal. What is the probability that the difference in magnitude between the sample proportions favoring the bond proposal does not exceed 10%?

SOLUTION   If we assume that $p_1 = .75$ and $p_2 = .60$, the sampling distribution of $(\hat{p}_1 - \hat{p}_2)$ is approximately normal, with mean $(p_1 - p_2) = (.75 - .60) = .15$ and standard deviation (or standard error)

$$\sigma_{(\hat{p}_1 - \hat{p}_2)} = \sqrt{\frac{p_1 q_1}{n_1} + \frac{p_2 q_2}{n_2}} = \sqrt{\frac{(.75)(.25)}{50} + \frac{(.60)(.40)}{100}} = .0784$$

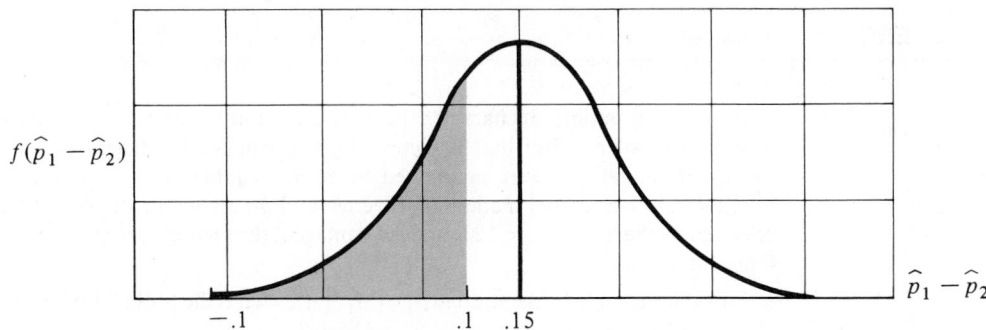

Figure 7.9   The Sampling Distribution of $(\hat{p}_1 - \hat{p}_2)$ Based on Sample Sizes of $n_1 = 50$ and $n_2 = 100$ for Example 7.7

We wish to find $P(-.1 < \hat{p}_1 - \hat{p}_2 < .1)$, which corresponds to the shaded area in Figure 7.9. With the normal approximation this probability corresponds to $P(z_1 < z < z_2)$, where

$$z_1 = \frac{(-.1) - .15}{.0784} = -3.19 \qquad z_2 = \frac{.1 - .15}{.0784} = -.64$$

The area between 0 and $z_1 = -3.19$ is .5 for all practical purposes, and the area between 0 and $z_2 = -.64$ is .2389. Therefore, the required probability is equal to the difference in these areas, $(.5 - .2389) = .2611$.

◆ ◆

EXERCISES   Conceptual Level

7.38   What are the mean and variance of $(\hat{p}_1 - \hat{p}_2)$?

7.39   Under what conditions is the sampling distribution of the difference between $\hat{p}_1$ and $\hat{p}_2$ approximately normally distributed? Is the distribution of $(\hat{p}_1 - \hat{p}_2)$ ever exactly normal?

EXERCISES   Skill Level

7.40   Independent random samples of size $n_1 = n_2 = 50$ are drawn from each of two binomial populations for which $p_1 = p_2 = .8$. The sample proportions are calculated as $\hat{p}_1 = .66$ and $\hat{p}_2 = .58$.

    a.   What is the probability of obtaining a difference as large as or larger in magnitude than the observed difference of .08?

    b.   What is the probability of obtaining a difference less than or equal to the observed difference?

7.41   Refer to Exercise 7.40. Suppose we had selected samples of size $n_1 = n_2 = 100$ from the two binomial populations and had obtained the same sample results.

    a.   Find the probability described in Exercise 7.40a.

    b.   How does $P(|\hat{p}_1 - \hat{p}_2| > .08)$ change with increasing sample sizes?

EXERCISES   Applications

   7.42   There are many important characteristics that a firm must consider in selecting a transportation strategy. Consider a firm that is comparing two equal-cost carriers on the basis of freight damage. Both carriers were contracted to deliver shipments to a randomly selected set of locations. The first carrier made 180 deliveries, and the second made 160. Of the 180 shipments delivered by the first carrier, 12% arrived damaged; the second carrier delivered only 6% of his shipments damaged.

a. If the firm is able to assume that $p_1 = p_2 = .1$, what is the probability that the second carrier outperforms the first carrier by 6% or more?

b. Suppose that the firm had contracted for only 40 deliveries with each of the carriers. If the firm is still willing to assume that $p_1 = p_2 = .1$, and if the sample proportions remain unchanged, calculate the probability given in part a.

c. What effect does a reduction in the sample size have on the variance of the statistic $(\hat{p}_1 - \hat{p}_2)$? What effect does this reduction have on the sampling distribution of $(\hat{p}_1 - \hat{p}_2)$?

   7.43   According to the *Wall Street Journal*, an Institute of Social Research survey found an erosion of consumer confidence following the stock market crash in October 1987 (John Bussey, "Survey Finds Sharp Drop in Confidence of Consumers After Stock Market Crash," *Wall Street Journal*, November 9, 1987, p. 12). In October 1987, 61% of the $n_1 = 347$ families surveyed prior to the crash thought that it was a good time to buy a vehicle; but following "Black Monday," only 54% of the $n_2 = 153$ families surveyed thought so. If the actual percentage of people willing to purchase a vehicle remained constant at 60% throughout October 1987, and if the "before" and "after" samples represent two independent samples from the same population, what is the probability that we would observe a difference greater than $(\hat{p}_1 - \hat{p}_2) = (.61 - .54) = .07$?

7.8

Sampling and Sample Surveys

The sampling distributions given in Sections 7.1 through 7.7 are based upon simple random sampling. Other types of sampling procedures are often used in practice, especially in the area of sample surveys. In addition to simple random sampling, stratified, systematic, and cluster sampling also involve random selection of the elements in the sample and thus provide a probabilistic basis for developing the sampling distributions of the estimators of population parameters, which are usually means, totals, or proportions.

stratified random sample   **Stratified random sampling** involves selecting a simple random sample from each of a given number of subpopulations called *strata*. Cluster sampling can be used when the available sampling unit is a collection of population elements called a *cluster*. For

cluster sample   example, a household represents a cluster of individuals living together. A **cluster sample** is a simple random sample of clusters from among the possible clusters. When a cluster is selected for inclusion in the sample, a census of every element in the cluster

systematic random sample   is taken. **Systematic random sampling** involves the random selection of one of the first k elements in an ordered population and then the systematic selection of every kth element thereafter.

Not all sampling plans involve random selection. A *convenience sample* is one that can be obtained easily and simply without random selection. Advertising for subjects who will be paid a fee for participation in an experiment produces a convenience sample. *Judgment sampling* allows the sampler to make a judgment about which individuals will or will not be included in the sample. *Quota sampling*, in which the makeup of the sample must reflect the makeup of the population on some preselected characteristics, very often has a nonrandom component in the selection process.

Those sampling plans that involve random selection of individuals produce a probabilistic base for determining the sampling distribution of estimators. When sample sizes are large, the Central Limit Theorem assures us that estimators that are sums, means, or proportions will have an approximately normal sampling distribution, and hence, valid inferences about population parameters can be made by using sample data. Chapter 16, dealing with survey sampling, contains the pertinent formulas and further details for working with these random sampling plans. When the sample is a nonrandom sample, such as a convenience or judgment sample, then only descriptive measures for the sample can be calculated; **no valid inferences about population parameters can be made from nonrandom samples.**

Tips on Problem Solving

1. Begin by reading each exercise carefully to determine the nature of the sampled population or populations. Identify and record the values of population parameters given in the problem, as well as the values of any statistics computed from sample measurements.

2. In using a normal approximation to a sampling distribution, check to see whether the sample size or sizes are large enough to ensure that the approximation will be adequate for the problem. Depict the problem pictorially by drawing a normal curve and locating the areas under the curve that will be used in solving the problem.

3. The accuracy of your answer will depend on the number of significant digits used in the calculations. For example, in calculating a value of z, use full accuracy for the mean and standard deviation and round your result only after the calculation is completed. This step is especially important in working with $\hat{p}$ and $(\hat{p}_1 - \hat{p}_2)$.

7.9

Summary

Inferences concerning population parameters are made in terms of the sample information obtained through random sampling. The sample information about a population parameter is summarized in terms of statistics computed from the sample values. To effectively use these statistics in making inferences, we need to have information about their sampling distributions—which is to say we need to be able to describe their behavior when repeated samples of a fixed size are drawn from a population of interest.

The *sampling distribution* of a statistic may be found mathematically or approximated empirically. For example, statisticians are able to prove that if the sampled population is normal, the sampling distribution of $\bar{x}$ is also normal. When the sampling distribution cannot be found mathematically, or mathematical results need to be verified, computers can be used to simulate the sampling and to construct a relative frequency histogram as an approximation to the actual sampling distribution. (The MINITAB package has several random data generation programs that can be used for this purpose.) Alternatively, when a statistic is a sum or mean of sample values, according to the Central Limit Theorem, its sampling distribution can be approximated by a normal distribution when the sample size is sufficiently large.

We have examined the sampling distributions for $\bar{x}$, $\hat{p}$, $(\bar{x}_1 - \bar{x}_2)$, and $(\hat{p}_1 - \hat{p}_2)$ because these statistics and their behavior in repeated sampling will serve as the basis for making inferences about the corresponding population parameters. Chapter 8 concerns large-sample statistical inferences and deals specifically with the concepts involved in using these four statistics as estimators. The obvious question at this point might be, "What happens when the sample size is small?" This question will be answered in Chapter 9, which explores the area of small-sample inference.

Supplementary Exercises

Applications

7.44 A population of $N = 10$ is to be sampled by using a sample size of $n = 3$. How many different samples can be selected if the sampling is done so that each element of the population can be selected only once in a particular sample?

7.45 The following population of $N = 4$ values is to be sampled by using a sample size of $n = 2$: 3, 7, 6, 9.

    a. If sampling is done without replacement, how many different samples could be obtained?

    b. Determine the value of $\bar{x}$ for the samples.

    c. Find the sampling distribution of $\bar{x}$ and graph the results.

    d. Determine the mean of the population, μ.

    e. Compare the population mean to the sample means. What is the probability of selecting a sample that provides an exact estimate of the population mean?

7.46 Define a population consisting of the following five elements: 12, 7, 5, 15, 11.

    a. How many different samples of size $n = 2$ can be selected from this population if we sample without replacement?

    b. How many different samples of size $n = 4$ can be selected if we sample without replacement?

    c. Calculate the value of the population mean μ.

    d. What is the value of $\bar{x}$ for each of the samples of size $n = 2$?

e. What is the value of $\bar{x}$ for each of the samples of size $n = 4$?

f. Which sampling distribution has the least variability (as measured by the smallest standard deviation)?

7.47 Refer to Exercise 7.46. Consider a sample of size $n = 4$ chosen without replacement from this population. For each of the samples listed in Exercise 7.46e, calculate the value of the sample variance s^2. Use these values to find the sampling distribution of s^2, and compare the expected value of s^2 with the population variance σ^2.

7.48 If we sample without replacement from a population of 50 and select a sample of size 10, how many different samples could we select?

7.49 A recent Department of Commerce report indicates that personal income in Connecticut averages \$19,600, with a standard deviation of \$2,450.

a. A sample of 64 people is randomly chosen from Connecticut. Find the probability that the mean income for the sample exceeds \$20,000.

b. A second sample of 64 people is randomly chosen from Connecticut. Find the probability that both sample means exceed \$20,000.

7.50 A random sample is drawn from a very large population with mean equal to 54 and variance equal to 128.

a. If the size of the sample is $n = 40$, what is the probability of observing a sample mean less than or equal to 52?

b. If the sample size is increased to $n = 100$, what is the probability of observing a sample mean less than or equal to 46?

c. No assumptions have been made about the shape of the population distribution described above, and yet we can calculate probabilities by using the sampling distribution of $\bar{x}$. What important theoretical result has been used?

7.51 Credit cards are a multibillion dollar business. As of 1984, MasterCard had issued 60 million cards in the United States with \$49.7 million in billings for that year (*Fortune*, February 4, 1984). The average billing amount was \$828.34. Suppose that the standard deviation of the 1984 billings is \$408, and that a random sample of 60 accounts is chosen for analysis. What is the probability that the mean billing amount for the sample accounts exceeds \$900?

7.52 Refer to Exercise 7.51. During the same time period Visa had issued 77.2 million cards in the United States, with an average billing of \$784.97 per account (*Fortune*, February 4, 1984). Suppose that the standard deviation of the 1984 Visa billings is \$363. If a random sample of 50 Visa accounts is drawn, what is the probability that the mean billing amount for the sample is less than or equal to \$800?

7.53 Refer to Exercises 7.51 and 7.52. Suppose that two independent random samples of size $n_1 = n_2 = 40$ were drawn from the MasterCard and Visa account records. What is the probability that the average billing for MasterCard exceeds the average billing for Visa by \$50 or more?

7.54 Calculate the probability of the event described in Exercise 7.53 if the sample sizes are 400 rather than 40.

7.55 Fragrance is often an important part of a product's appeal to a customer. A study designed to compare preferences for various fragrances used a quantitative measure of preference in an attempt to locate differences between fragrances. This study concluded that a fragrance called fragrance *D*, which received a preference rating of 42, was preferable to fragrance *A*, which received a preference rating of 40 (*Aerosol Age*, January 1985). Suppose that these results were based on independent random samples of 30 customers each and that the variances of the preference ratings for fragrances *A* and *D* were 4.7 and 4.3, respectively. If there really is no difference in preference for the two fragrances, what is the probability of the samples showing fragrance D exceeding fragrance A by 2 or more preference points?

7.56 Productivity has become an important concern for all business firms. In an effort to improve productivity, a firm has instituted a combination of increased investment and improved training of employees and has negotiated with a union for a trial relaxation of some work rules in one facility. After one year the firm compares employee productivity in the facility with the relaxed work rules to productivity at the other facilities within the firm. The firm has 4,320 employees covered by the union contract, of which 851 work under the relaxed work rules. A random sample of 30 employees working under the relaxed rules and 50 employees working under the traditional rules was selected for the comparison.

    a. Assume that there is an equal probability of an employee increasing or decreasing in productivity over the one-year period. What is the probability that the sample from the relaxed work rules group shows 60% or more of the employees having increased productivity?

    b. What is the probability that the sample from the traditional work rules group shows 60% or more of the employees having increased productivity?

7.57 Refer to Exercise 7.56 and assume that there is no difference in the proportion of employees showing increased productivity for the relaxed and traditional work rules groups. If we continue to assume that there is an equal probability of an employee's increasing or decreasing in productivity over the one-year period, what is the probability that the selected samples indicate that the proportion of increased productivity for the relaxed work rules group exceeds that for the traditional group by 10% or more?

7.58 The firm described in Exercise 7.56 was able to devise a quantitative measure of productivity that could be applied to each worker in the study. From previous testing, the firm is reasonably sure that the standard deviation of this measure is 24.

    a. If the mean productivity level for the traditional work rules group as measured by this scale is 150, what is the probability of selecting a sample from the traditional work rules group that has a mean of 160 or more?

    b. If, in fact, the relaxed work rules do not affect productivity, so that the workers enjoying the relaxed work rules can be considered no different from any other workers in the firm, what is the probability of selecting a sample from the relaxed work rules group that has a mean of 160 or more?

7.59 Suppose that the true proportion p in favor of wage and price controls is the same for labor as it is for management. Independent random samples are selected, one consisting of 800 laborers and the other of 800 managers. Find the probability that the sample proportion of laborers favoring controls exceeds that of the managers by more than .03 if the true proportion who favor controls is $p = .5$. What is this probability if $p = .1$?

7.60 A manufacturer claims that at least 25% of the public prefers his product to others on the market. A random sample of 30 individuals is drawn, and 6 express a preference for the manufacturer's product. If the manufacturer's claim is correct, and if in fact the actual preference proportion is equal to the minimum claimed ($p = .25$), what is the probability of obtaining a sample proportion as small as or smaller than the one obtained?

7.61 A supplier of industrial products claims that at least 95% of the products it supplies meet product specifications. A random sample of 600 products was tested, and 46 failed to meet the required specifications. If the supplier's claim is correct, and if in fact the proportion failing to meet specifications is the maximum possible ($p = .05$), what is the probability of obtaining a sample proportion as large as or larger than the one obtained?

7.62 The average salary for MBA graduates entering the fields of brand or product management and marketing services are shown in the accompanying table. Samples of $n_1 = 50$ and $n_2 = 50$ MBA graduates entering these respective fields are taken.

Brand or Product Management	Marketing Services
$\mu_1 = \$40,900$	$\mu_2 = \$38,000$
$\sigma_1 = \$\ 5,850$	$\sigma_2 = \$\ 2,250$

Source: Data from *Sales and Marketing Management*, "Survey of Selling Costs," February 16, 1987, p. 58.

a. What is the probability that $(\bar{x}_1 - \bar{x}_2)$ exceeds \$5,600?

b. What is the probability that $(\bar{x}_1 - \bar{x}_2)$ is less than 0?

c. If the samples resulted in a difference $(\bar{x}_1 - \bar{x}_2)$ of less than 0 or greater than \$5,600, would you agree that there is statistical evidence tending to refute the underlying difference, $(\mu_1 - \mu_2) = \$2,900$?

Case Study

Reality Versus Theory: Sampling Money Market Funds

Money market funds are an important part of personal investment portfolios. The inclusion of one or more money market funds in a portfolio is equivalent to sampling from a large population of available money market funds; in most cases, however, the sampling technique is purposive and does not meet the criteria of random sampling. How would a sample of money market funds behave relative to standard measures of performance? The quotations presented in Table 7.6, collected by the National Association of Securities Dealers, Inc., represent the average of the annualized yields and the dollar-weighted portfolio maturities over the seven previous days for 424 money market funds. Consecutive entries in the columns should not be taken to be in random order, since many firms manage more than one money market fund and the funds are listed alphabetically.

A summary of each of these two variables for the 424 money market funds with available information was found by using the MINITAB commands DESCRIBE and HISTOGRAM. The output is presented in Table 7.7.

1. From the information given, how would you describe the distribution of average maturities? Of average yield? Are there any distinct characteristics that you should include in your description?

2. Use random number tables (or a random device) to select as many as 20 samples of size $n = 5$ observations for each of the variables. Find the sample means and sample variances for each of these samples. Now, find the average and the variance of the sample means, and compare your values with the theoretical values predicted by statistical sampling theory.

3. Repeat the instructions of part 2, using samples of size $n = 10$. Discuss the effect of increasing the sample size on the sampling distribution of the sample mean.

4. The theoretical average of the sample variances is $N/(N - 1)$ times the value of the population variance σ^2. Discuss your results.

Table 7.6   Seven-Day Money Market Data for 424 Funds

FUNDS	AVG MAT.	AVG YLD.	FUNDS	AVG MAT.	AVG YLD.	FUNDS	AVG MAT.	AVG YLD.	FUNDS	AVG MAT.	AVG YLD.
AARP Money	40	5.72	Cardinal TEMT	15	3.86	DailyTaxFree c	20	4.22	FidDly Tax Ex	42	4.21
ActvAsst GovSc	60	6.17	Carillon Csh	23	6.05	DBL GvtSc Prtf	60	6.17	FidDly US Tr	5	5.95
ActvAsst Money	58	6.59	CarngieGov Sec	14	5.79	DBL MM Portf	62	6.31	FidUS Govt Res	55	6.21
ActvAsst TxFr	27	4.28	Carnegie Tax Fr	48	4.34	DBL Tax Free	44	4.28	FidMM USTrea	33	6.14
AlexBCash Gvt	44	6.13	CashAssett Tr	26	6.41	DWS Liqd Asset	65	6.52	FidMM Domstc	23	6.37
AlexBCash Prm	35	6.40	Cash Equiv MM	32	6.23	DWS TF Daily	26	4.11	FidMM Govmnt	31	6.29
Alliance Capital	66	6.26	CashEq GovSec	13	6.18	DWS USGvtMM	58	6.14	FidMass TaxFr	55	4.15
AllianceGvt Res	69	6.04	CshMgt TrAm	18	6.21	DelaCashRsrv f	63	6.31	FidNYTxF MM	49	4.11
Alliance TaxEx	40	4.17	Cash Rsv Mgt a	(z)	(z)	DelaTaxFr MF	32	3.92	FidelPA TFMM	56	4.36
AllianceTE NY	34	3.48	CBA Money Fd	45	6.17	DelaTreas Rsrv	43	5.43	FidTaxExmpt c	39	4.44
AMA PrimePrt	93	6.22	Centenl GovtTr	21	5.68	Dry Cal TE	74	4.19	FnclDlyinc Shr	25	6.01
AMA TreasPort	88	5.66	CentennlMM Tr	30	5.94	Dreyfsinst Govt	76	6.07	FnclPlanFed Sc	(z)	(z)
AmCap Resrv a	47	6.34	Centennial Tax	63	4.52	Dreyfsinst MM	45	6.47	Fincl Resv	34	6.35
Amer Natl MM	17	6.18	Churchill Cash	24	6.25	DreyfsLiq Asst	52	6.33	FnclTxF Money	42	4.18
AMEV Money	45	6.08	Cigna Cash Fd	50	6.45	DryfusGvt Sers	66	5.88	FstAm Money	32	6.35
AT Ohio Tax Fr	40	4.11	CignaMM Fd b	61	6.37	DryfMnyMk Ser	41	6.30	FirstinstTax Ex	85	3.90
AutomCash Mgt	30	6.45	Cigna TxEx	59	4.22	DreyfsNY TE	43	4.00	FstinvCshMgt f	20	5.76
AutomGvt MTr	37	6.10	CimcoMM Trst	26	5.88	DryTaxEx MM	32	4.09	FstLakeshr Gov	33	5.67
Axe Hghtn MM	21	5.91	CMA GovtSec a	52	6.66.	DryUS GrtMM	99	6.21	FirstLkshr MM	47	6.16
Babson Prime	36	5.91	CMA MnyFd a	56	6.37	EatonVan Cash	30	6.18	FirstLkshr TE	37	4.26
BenhamCal TF	7	4.02	CMA Tax	47	4.26	EatonV TF Res	28	4.14	FstTrMM GenP	(z)	(z)
BenhamNatl TF	23	4.34	Col Daily Inc af	29	5.98	EGT MMTrust f	28	5.42	FstTrMMFd Gv	(z)	(z)
BirrWilson MFd	25	5.57	ColGovtMM Tr	53	5.53	Emblem Prime	28	6.22	FstTrust TaxFr	(z)	(z)
Boston Co Cash	49	6.21	Comp Cash M a	24	5.79	Emblem US Gv	47	5.86	First Variable	39	6.22
BostonCo Gvt	42	5.48	ConnDaily TF	62	3.79	Empire TxFr	83	4.11	Flex Fund MM	12	6.52
BostonCo Mass	65	4.25	CoreFund CshR	34	6.35	Evergrn MM Tr	49	6.78	Ft Wash	43	6.34
Bull&Bear DRs	59	6.46	Cortland GnMM	42	5.98	FBL MnMkFd b	26	5.98	Founders MMk	20	5.93
CA TFMF	70	4.07	Cortland TxFr c	26	4.04	FedMaster Trst	47	6.56	Frnk CATEMF	35	4.66
CalvrtSocInv af	37	6.07	Cortland USGv	51	5.83	FedShtUS Gov	24	6.52	Frankin FedMF	1	5.45
CalvertTF Rsrv	58	4.77	Counsellor CR	46	6.62	FedrtTaxFree c	34	4.30	FrnklFT Money	25	6.85
CAM Fund	7	5.49	Counsellor NY	20	3.92	FFB Cash	58	6.48	FrnklnMnyFd a	26	6.11
CapCash MgtTr	19	5.97	CountryCp MM	23	5.55	FFB Govt	46	6.17	Frnk NYTEMF	32	3.91
Cap Preservin	57	5.66	Current Interest	36	5.65	FFB TxFrMM	44	4.05	Frnkln TaxEx c	31	4.48
Cap Preservin 2	3	5.74	Currentin TxFr	39	4.05	FFB USTreas	35	5.87	Fund For TxFr	61	3.96
Capital T MM	8	6.03	Currentint USG	36	5.28	FidelCal TaxFr	52	4.02	FundGov Invst	6	5.76
Cap T ins MM	6	5.65	Daily cash Fd1	29	6.14	FidelCash Resv	44	6.24	FundSource MT	20	6.16
Capitl T TxFr	2	1.70	DailyDollar Rs	40	6.26	FidelDlyIncm b	42	6.12	FS Wash MT	36	6.29
CardGovt SecTr	31	5.84	Daily Incm Fd	38	6.21	FidDly MM Prt	27	6.24	Galaxy Gov	33	6.41

Notes: Eligible funds not included declined to provide information; a = yield includes capital gains or losses; b = fixed charges vary yield based upon account size; c = investments primarily in federally tax exempt securities; f = as of previous day; z = no quote.

Table 7.6 (Continued)

FUNDS	AVG MAT.	AVG YLD.	FUNDS	AVG MAT.	AVG YLD.	FUNDS	AVG MAT.	AVG YLD.	FUNDS	AVG MAT.	AVG YLD.
Galaxy MM	30	6.41	Lazard TxFree	57	4.44	NY TFMF	75	3.74	ScuddCashinv f	44	6.33
GenlCal TEMM	51	4.22	LeggMas TE	24	4.11	ThMcK NtlGovt	42	5.91	ScuddCaTFM	18	4.19
GenlGovt Secur	50	5.93	LehmnM CshRs	39	6.33	ThMcK Ntl MM	29	6.24	ScuddNYTFM	34	3.61
GenlMoney Mkt	55	6.18	LehmnM GvSec	14	5.73	ThMcK Ntl TxE	50	4.13	ScuddrGvt Mny	60	5.49
GenlNY TEMM	25	3.61	LehmM TF Rsv	58	3.97	Nationwide MM	35	6.13	ScuddrTaxFr cf	46	4.06
GenlTxEx MM	32	4.09	LexGvtScMM a	4	5.55	NeubrgBer Gov	66	5.51	Secur Cash Fd	19	5.88
GovtinvstTrst a	23	5.91	LexMoneyMkt a	37	6.02	Neuberger TxF	42	4.49	Select Mny Mkt	24	5.61
GovtSecur Cash	40	5.92	LexTaxFrDly c	81	4.31	Newton Money	46	6.33	Selig CalTE	5	3.49
Gradison Cash	39	5.95	LFRoth EarnLq	3	5.85	NYTFT	41	4.37	SelgmnCM Prm	29	5.90
GradisonUS Gv	23	5.54	LF Roth Exmpt	4	3.37	NLR Cash Port	38	6.22	SeligmnGov Prt	2	5.46
Granit MM	20	6.02	LibrtyCash Mgt	(z)	(z)	NLR Govt Prt	40	5.98	Sentinel Cash	39	6.24
GuardCsh MM	24	5.68	LibertyUS Govt	30	5.88	Nuveen Cal TF	27	4.10	Shrsn Cal Dly	30	4.16
Guard CashFd	30	6.16	LiquidCapitl Tr	24	5.95	Nuveen TE MM	49	4.52	ShrsnDailyDiv f	40	6.13
Harbor MM	107	5.68	LiquidCAshTR f	1	6.72	Nuveen TF Res	41	4.20	Shrsn Daily TxF	46	4.34
Heritg Cash Tr	46	6.10	LiquidGrn TxFr	67	4.59	NuveenMas TF	74	4.42	ShrsnFMA Cash	35	6.30
HiMrkCal TF	25	3.97	LiquidGreen f	40	6.28	Oppn Mny Mrkt	26	6.17	ShrsnFMA Govt	61	6.46
HiMrk Div	37	6.31	LordAbbet Cash	28	5.94	Oxford CshMgt	2	6.31	ShrsnFMA Mun	49	4.36
HiMrkUS Gv	33	6.15	LuthBrMon Mkt	44	5.99	PacHrzGov MM	50	6.12	ShrsnGovt Agen	65	6.38
HiMrkUS Treas	32	6.15	MacKay Sh MM	33	6.05	PacHrzMM Prt	33	6.31	Shrsn NY Dly	30	3.87
Hilliard Gov't	16	5.38	MAP GovtFund	55	6.36	PacHrz TaxEx	49	3.99	ShortTerm Govt	83	5.73
HomeCsh Resv	34	6.47	Mariner Cash	46	6.39	PW Cash	36	6.33	ShortTm Asset	22	6.15
HomeGvt Resv	26	6.29	Mariner Govt	34	6.31	ParkAve NYTE	25	3.55	ShortTerm MM	38	6.11
HomeFed TF	53	4.31	Mariner NYTF	83	3.99	Parkway Cash	50	6.15	ShortTermYld a	(z)	(z)
Horizon Pr	35	6.73	Mariner TaxFree	43	4.35	Phoenix MMkt	36	6.21	Sigma Moneyfd	10	5.58
Horizon TE	32	4.63	MarinerUS Trs	25	6.10	Pilgrim GyMM	1	5.49	SthFrmBur Csh	42	5.99
Horizn Tre	48	6.41	MassCashMgt a	34	6.23	Pioneer MM Tr	16	6.08	Standby Reserv	43	6.29
HYCT	54	6.60	MasaCshM Trst	30	5.72	Piper MM Fund	39	5.94	StandbyTE Res	46	4.20
Hutton AMA Fd	(z)	(z)	MassMtl Liquid	22	5.89	PrimeCash Fd	32	6.22	SteinRoeCsh Rs	44	6.36
Hutton Govt Fd	(z)	(z)	Mass TE MM	59	3.96	Princor CashM	17	5.68	SteinRoe Govt	43	5.83
IDS Cash Mngt	50	6.35	McDonald MM	64	6.37	PruBacheC Gvt	57	6.23	SteinRoe Tax	23	4.23
IDS Stratgy Fd	56	5.30	McDonald TxE	56	4.18	PruBacheC Mn	46	6.35	StrongMM Fnd	108	6.79
IDS TxFtMny c	69	4.12	Merit MM	24	6.13	PruBacheC Tax	56	4.28	Strong TF MM	79	4.52
IMG TE Liquid	8	4.20	MerrLGovtFd a	49	6.05	PrudB GvtSec	56	6.30	Summit Cash	48	6.36
IMG Liqd Asset	16	5.52	MerrLinstFd af	49	6.50	PruBache MMt	44	6.32	TaxExmpMM c	28	4.24
Instit LqA Gov	49	6.52	Merlinst TxEx	49	4.34	PrudB NYMM	83	4.17	TF CshRsv Gen	28	4.16
InstLqdGPr Aff	49	6.56	MerrLRdyAst a	56	6.33	PrudB TaxFr/c	74	4.56	Tx Ex CA MM	21	4.00
InstLqATrs af	44	6.37	MerrLRetRsv a	58	6.34	Pru InstLiq M	45	6.91	TaxFrinst Tr	32	4.13
Instit MM Pt	23	6.36	Merrl USA Gr	52	6.61	Pru InstLiq TE	5	4.36	TaxFree Mny c	38	4.11
InstTEAsst afc	41	4.39	MetLf State St	18	5.89	PutnamDly Div	44	6.06	Tempinvest Fd	48	6.77
Intgrtd Cash	24	5.78	Met NYTF	32	3.65	Putnam CA TE	45	4.34	Temple Money	10	5.75
IntegrtMM Sec	30	5.91	MFS LifeMM	32	5.35	Putnam NY TE	87	4.19	Thoro Pr Ob	40	6.48
Intgrid TxFr	6	3.38	MidwstGrp TF	65	4.63	Putnam TaxEx	52	4.51	Thoro US Gv	67	6.21
Inv Csh Resv	28	5.60	Midwst Incm Tr	56	5.52	QuestFor Value	28	6.22	Trnsom CashRs	42	6.38
JohnHanc Cash	32	5.92	MnyMgtP Govt	50	5.53	Renaisnc GvFd	17	6.30	Trinity LiqdAst	32	6.38
JonesDly Pssp f	46	6.10	MnyMgtP Pr	39	6.17	Renaissnc MM	10	6.41	Trust Fds Fedl	36	6.29
Kempr Gvt MM	12	6.46	MnyMgtPl TxF	53	4.32	ReservConn TE	48	3.85	Trust Fds MM	48	6.85
Kemper MnyM	31	6.60	MoneyMrkt Fd	59	8.24	Reserv GOVT	1	5.86	Trust CashRes f	47	6.27
Kempr MM TE	28	4.51	MoneyMktMgt f	47	6.17	Reserv PRIMR	13	6.08	Trustfd East	31	4.20
Keystone LiqTr	31	5.37	MnyMkt Tr MM	32	5.98	Reserv INTRST	46	4.30	Trustfd West	20	4.13
KidderP CalTE	33	4.15	Money Mkt Trst	42	6.62	Reserv NY TxE	40	3.77	TrFdTreas Prtf	28	6.29
KidderP GovtM	40	6.19	Monitor Gov	31	6.13	PW RMA Mony	36	6.07	TrFdUS Agency	34	6.54
KidderP PrmAc	43	6.35	Monitor MM	40	6.51	PW RMA TaxF	26	4.12	TrFd Commercl	45	6.79
KidderP TaxEx	44	4.33	Monitor TF	27	4.39	PW RMA USGv	59	5.78	TrFd Obligation	45	6.76
Lndmrk Cash	39	6.20	MorganKeegn f	46	5.95	RNC LiquidAsst	35	6.33	TrShtTr FedFd	31	6.54
LndmrkTF Rsv	18	4.11	Muni CA Inv	26	4.30	RodneySq MMP	35	6.30	TrShtTFed TFd	33	6.41
Lndmrk NYTF	36	3.70	MuniCashRsv c	(z)	(z)	RodneySq TE	22	4.32	TrShTer US Gvt	33	6.20
Lazardins Cash	18	6.70	Munifundinv c	29	4.48	RodneySq USG	21	6.14	TrstUSTrea Obl	39	6.21
Lazardinst Gov	16	6.68	Muni NY Inv	33	4.23	RowePrPrRes f	41	6.42	TuckerA CashM	31	6.12
Lazardinst TF	48	4.66	MtlOmCash Res	37	6.01	RowePTxEx cf	82	4.46	TuckerA GvtSc	37	5.74
Lazardinst Trs	2	6.30	MtlOmMny Mkt	49	6.11	RowePUSTrea f	39	5.80	TuckerA TxEx	51	4.06
LeggMasn Csh f	44	6.22	Natl Cash Resv	32	6.26	Safeco MnyMk f	37	6.27	20thCent Cash	28	5.88
Lazard CashM	24	6.29	NECashMT MM	55	6.41	Safeco TF MM	30	4.23	UMB Federal	38	6.21
Lazardinst Pr	21	6.68	NE TxEx MM	69	4.25	StClair Prime	37	6.26	UMB Prime	36	6.37
Lazard Govt	15	6.24	NE Cash Mgmt	81	5.71	StClair •TxFr	32	4.34	UMB TxF	52	4.35

(continues)

Table 7.6 (Continued)

FUNDS	AVG MAT.	AVG YLD.	FUNDS	AVG MAT.	AVG YLD.	FUNDS	AVG MAT.	AVG YLD.	FUNDS	AVG MAT.	AVG YLD.
UnitedCshMgt a	46	6.08	ValueLine Cash	31	6.29	Vang MM Cal	63	4.52	Vista TF M	43	4.13
USAA MutlMM	31	6.39	ValLineMM Prt	69	3.69	VanKampn MM	54	6.12	WayneHum MM	25	5.91
USAA TxExMM	73	4.61	VanEck US MM	44	5.46	VanKampn TF	85	4.65	WebstrCash Rs	42	6.35
UST MSTR Gvt	31	6.28	VangdMMFed f	37	6.43	VantageCsh Prt	37	6.20	Working Assets	39	5.84
USTMSTR Mny	17	6.39	VangrdMM Pr f	39	6.66	VantageGvt Prt	39	5.88	WPG Sht Term	69	6.29
UST MSTR TxE	48	4.79	Vangrd Insure f	21	6.10	Viking MM Fnd	38	5.92			
US Treas SecFd	58	6.13	VangMunMM cf	62	4.53	Vista USGv	36	6.04			

Source: National Association of Securities Dealers, Inc., data from *The Press Enterprise* (Riverside, Calif.) February 25, 1988. Reprinted with permission.

Table 7.7 MINITAB Output for the Case Study on Money Market Funds

```
MTB > DESCRIBE C1 C2

                N      MEAN    MEDIAN    TRMEAN     STDEV    SEMEAN
AVG.MAT       424    39.243    38.000    38.686    18.331     0.890
AVG.YLD       424    5.5292    5.9450    5.5654    0.9756    0.0474

               MIN       MAX        Q1        Q3
AVG.MAT      1.000   108.000    28.000    49.000
AVG.YLD     1.7000    8.2400    4.3975    6.2875

MTB > HISTOGRAM C1

HISTOGRAM OF AVG.MAT    N = 424
EACH * REPRESENTS 2 OBS.

MIDPOINT     COUNT
       0        12   ******
      10        19   *********
      20        51   *************************
      30        96   ************************************************
      40        99   *************************************************
      50        69   ********************************
      60        41   *********************
      70        19   *********
      80        10   *****
      90         5   ***
     100         1   *
     110         2   *

MTB > HISTOGRAM C2

HISTOGRAM OF AVG.YLD    N = 424
EACH * REPRESENTS 5 OBS.

MIDPOINT     COUNT
     1.5         1   *
     2.0         0
     2.5         0
     3.0         0
     3.5        11   ***
     4.0        65   *************
     4.5        51   ***********
     5.0         3   *
     5.5        42   *********
     6.0       136   ***************************
     6.5       106   *********************
     7.0         8   **
     7.5         0
     8.0         1   *
```

References and Suggested Readings

Bain, L. J., and M. Englehart. *Introduction to Probability and Mathematical Statistics.* Boston: PWS-KENT Publishing Co.,

Hogg, R. V., and A. T. Craig. *Introduction to Mathematical Statistics, 4th ed.* New York: Macmillan, 1978.

Mendenhall, W. *A Course in Business Statistics, 2nd ed.* Boston: PWS-KENT Publishing Co., 1988.

Mendenhall, W., R. L. Scheaffer, and D. D. Wackerly. *Mathematical Statistics with Applications. 3rd ed.* Boston: PWS-KENT Publishing Co., 1986.

TO
THE
READER

In the preceding chapters we have focused attention on the probability distributions of random variables, and in Chapter 7 we paid particular attention to the probability distributions of several statistics that are prime candidates for use as estimators of unknown population parameters.

When dealing with large sample sizes, the Central Limit Theorem assures us that the sampling distributions of many statistics used as estimators will be approximately normal. Chapter 8 addresses the problem of estimating population parameters and testing hypotheses about the values of population parameters when the sampling distributions can be assumed to be normal or approximately so. The techniques involved in the estimation and testing procedures presented in this chapter reflect the basic concepts underlying inferential statistics; they provide the user with a conceptual background upon which further inferential procedures can be based. You will find in subsequent chapters that the normal probability distribution is but one of several sampling distributions that arise in inferential statistics.

8

LARGE-SAMPLE STATISTICAL INFERENCE

8.1

A Brief Summary

The preceding seven chapters set the stage for the objective of this text, developing an understanding of statistical inference and the role it plays in business decision making. In Chapter 1 we stated that statisticians are concerned with making inferences about populations of measurements based on information contained in samples. We showed you how you phrase an inference—that is, how you describe a set of measurements—in Chapter 2. We discussed probability, the mechanism for making inferences, in Chapter 3, and we followed that with three chapters about probability distributions—a general presentation in Chapter 4, useful discrete probability distributions in Chapter 5, and the uniform, exponential, and normal distributions in Chapter 6. In Chapter 7 we presented the sampling distributions for four commonly used statistics, $\bar{x}$, $(\bar{x}_1 - \bar{x}_2)$, $\hat{p}$, and $(\hat{p}_1 - \hat{p}_2)$.

To get you thinking about statistical inference, in Chapter 5 we introduced you to the useful application of the binomial probability distribution to the test of an hypothesis concerning the desirability of a new sugar substitute. We touched lightly on this topic again in the examples and exercises of Chapter 6 and introduced the concept of a sampling distribution. Now we are ready to utilize the foundation we have laid—to study the basic concepts involved in statistical inference.

Perhaps the most important contribution to our preparation for a study of statistical inference is the Central Limit Theorem of Chapter 7. When we sample from a nonnormal population, this theorem justifies the approximate normality of the sampling distribution of sample means for large samples and also justifies the normal approximation to the binomial probability distribution in Chapter 6. But more important, it is used in Chapter 7 to justify the approximate normality of the sampling distributions of estimators and decision makers encountered in this chapter. These statistics will be used to make inferences about population parameters, and their sampling distributions provide a means of assessing the reliability of inferences.

8.2

Inference: The Objective of Statistics

Inference, specifically decision making and prediction, is centuries old and plays an important role in our individual lives. Each of us is faced with daily personal decisions and situations that require predictions about the future. The government is concerned with predicting the flow of goods to and from a foreign market. The broker seeks knowledge concerning the behavior of the stock market. The metallurgist wishes to use the results of an experiment to infer whether a new type of steel is more resistant to temperature changes than another. The consumer wishes to know whether detergent *A* is more effective than detergent *B*. Hopefully, these inferences are based on relevant bits of available information, which we call observations or data.

In many practical situations the relevant information is abundant, seemingly inconsistent, and, in many respects, overwhelming. As a result, our carefully considered decision or prediction is often little better than an outright guess. You need only refer to the "Market Views" section of the *Wall Street Journal* to observe the diversity of expert opinion concerning future stock market behavior. Similarly, a visual analysis of data by scientists and engineers will often yield conflicting opinions regarding conclusions to be drawn from an experiment. Although many individuals tend to feel that their own built-in inference-making equipment is quite good, experience suggests that most people are incapable of utilizing large amounts of data, mentally weighing each bit of relevant information, and arriving at a good inference. (You may test your individual inference-making equipment by using the exercises in this chapter and the next. Scan the data and make an inference before using the appropriate statistical procedure. Compare the results.) Certainly, a study of inference-making systems is desirable, and this is the objective of the mathematical statistician.

Although we have purposely touched upon some of the notions involved in statistical inference in preceding chapters, organizing our knowledge is beneficial at this point as we attempt an elementary presentation of some of the basic ideas involved in statistical inference.

> The **objective of statistics** is to make inferences about a population based on information contained in a sample.

parameters   Since populations are characterized by numerical descriptive measures called **parameters**, statistical inference is concerned with making inferences about population parameters. Typical population parameters are the mean, the standard deviation, the area under the probability distribution that is above or below some value of the random variable, or the area between two values of the variable. Indeed, all the practical problems mentioned in the first paragraph of this section can be restated in the framework of a population with a specified parameter of interest.

make decisions

estimate

Methods for making inferences about parameters fall into one of two categories. We may **make decisions** concerning the value of the parameter, as exemplified by the lot acceptance sampling (Example 5.4) and test of an hypothesis described in Chapter 5. Or we may **estimate** or predict the value of the parameter. Although some statisticians view estimation as a decision-making problem, we will retain the two categories and, in particular, concentrate separately on estimation and on tests of hypotheses.

measure of goodness

A statistical problem, which involves planning, analysis, and making inferences, would be incomplete without reference to a **measure of the goodness** of inferential procedures. We may define numerous objective methods for making inferences, in addition to our own individual procedures based on intuition. Therefore, a measure of goodness must be defined so that one procedure may be compared with another. More than that, we wish to state the goodness of a particular inference in a given physical situation. Thus, to predict that the price of a stock will be $80 next Monday would be insufficient and would stimulate few of us to take action to buy or sell. We would also wish to know whether the estimate is correct to within plus or minus $1, $2, or $10.

Elements of Statistical Inference

> Statistical inference in a practical situation contains two elements: (1) the inference and (2) a measure of its goodness.

Which method of inference should be used; that is, should the parameter be estimated or should we test an hypothesis concerning its value? The answer to this question is dictated by the practical question posed and is often determined by personal preference. Some people like to test theories concerning parameters; others prefer to express their inference as an estimate. Inasmuch as both estimation and tests of hypotheses are frequently used in scientific literature, we will include both methods in our discussion.

The remainder of Chapter 8 is divided into two parts. Part I, consisting of Section 8.3 through 8.12, is concerned with estimation. The sampling distributions presented in Chapter 7 form the foundation for the large-sample estimation procedures in the first part of the chapter. Sections 8.13 through 8.17 comprise Part II of Chapter 8, in which the basic concepts involved in a statistical test of an hypothesis are developed and presented, using sampling distributions that are normal or approximately normal because of the Central Limit Theorem.

PART I
LARGE-SAMPLE ESTIMATION

8.3

Types of Estimators

Estimation procedures may be divided into two types, *point estimation* and *interval estimation*. Suppose that, as a merchandising executive, you are concerned with the shrinkage rate (disappearance of merchandise due to internal and external theft) of items in your store. To estimate the rate, you sample n items from inventory and record the ratio

$$\text{shrinkage rate} = \frac{(\text{book inventory value}) - (\text{value on hand})}{\text{book inventory value}}$$

$$= \frac{\text{loss}}{\text{book inventory value}}$$

From these sample data you could form two types of estimates of the mean shrinkage rate μ for all items in the store. For example, if the sample mean is .20, you use this single number as an estimate of μ. Thus, you estimate the mean loss to be 20% of inventory value. Or you might estimate the mean shrinkage loss to lie in the interval from .15 to .25; or, equivalently, you estimate that the mean loss is between 15% and 25% of inventory value. The first type of estimate is called a *point estimate* because a single number, representing the estimate, may be associated with a point on a line. The second type, involving two points and defining an interval on a line, is called an *interval estimate*. We will consider each of these methods of estimation in turn.

A point estimation procedure utilizes information in a sample to arrive at a single number or point that estimates the population parameter of interest. The actual estimation is accomplished by an **estimator**. An estimator is a rule that tells us how to calculate the estimate by using information in the sample; it is generally expressed as a formula. For example, the sample mean

estimator

$$\bar{x} = \frac{\sum_{i=1}^{n} x_i}{n}$$

is an estimator of the population mean μ, and its formula explains exactly how the actual numerical value of the estimate may be obtained once the sample values $x_1, x_2, \ldots, x_n$ are known. On the other hand, an interval estimator uses the data in the sample to calculate two points that are intended to enclose the value of the population parameter estimated. Both point and interval estimators are statistics that are calculated from sample data.

Definition

point estimator

point estimate

> A **point estimator** of a population parameter is a rule that tells you how to calculate a single number based on sample data. The resulting number is called a **point estimate** of the parameter.

Definition

interval estimator

> An **interval estimator** of a population parameter is a rule that tells you how to calculate two numbers based on sample data.

Definition

interval estimate or
confidence interval

upper confidence
limit

lower confidence
limit

> When an interval estimator is employed to estimate a population parameter, the pair of numbers obtained from the estimator is called an **interval estimate** or **confidence interval** for the parameter. The larger number, which locates the upper end of the interval, is called the **upper confidence limit** and is denoted by UCL. Similarly, the number that locates the lower extreme of the interval is called the **lower confidence limit** and is denoted by LCL.

Both point estimates and interval estimates are employed in surveys of consumer preference, but point estimates seem to be the most common. When estimation arises in connection with industrial experimentation, confidence intervals seem to be more commonly used.

8.4

Evaluating the Goodness of a Point Estimator

Returning to the estimation of the shrinkage rate, suppose your sample mean is .20 and you report this value to the store manager. Would your manager be satisfied with this estimate or would he or she ask you, "Plus or minus what?" In other words, how much faith can the manager place in your estimate? Is it an accurate estimate of μ? How far away from μ might you expect this estimate to be?

To answer this question, let us consider an analogy that may help explain the reasoning employed in evaluating the goodness of a point estimator. Point estimation is similar, in many respects, to firing a revolver at a target. The estimator, which generates estimates, is analogous to the revolver; a particular estimate is analogous to the bullet; and the parameter of interest is analogous to the bull's-eye. Drawing a sample from the population and estimating the value of the parameter is equivalent to firing a single shot at the target.

Suppose that a man fires a single shot at a target and the shot pierces the bull's-eye. Do we conclude that he is an excellent shot? The answer is no, because not one of us would consent to hold the target while a second shot is fired. On the other hand, if one million shots in succession hit the bull's-eye, we might acquire sufficient confidence in the marksman to hold the target for the next shot, if the compensation were adequate. The point we wish to make is that we cannot evaluate the goodness of an estimation procedure on the basis of a single estimate. Rather, we must observe the results when the estimation procedure is used over and over again, many, many times—we then observe how closely the shots are distributed about the bull's-eye. In fact, since the estimates are numbers, we would evaluate the goodness of the estimator by constructing a frequency distribution of the estimates obtained in repeated sampling and noting how closely the distribution centers about the parameter of interest. This relative frequency distribution would be the sampling distribution of the estimator.

As an illustration, consider the die-tossing experiment used in Chapter 7 where we generated 100 samples of $n = 5$ measurements each and calculated the mean for each sample. Since we know the mean value μ of the number showing on a die toss ($\mu = 3.5$), we can use the results of the die-tossing experiment to see how well the mean of a sample of $n = 5$ measurements estimates μ.

The frequency histogram of the 100 sample means (shown in Figure 8.1) is an approximation to the sampling distribution of the mean $\bar{x}$ of a sample of 5 observations on 5 die tosses. Notice how the estimates group about the population

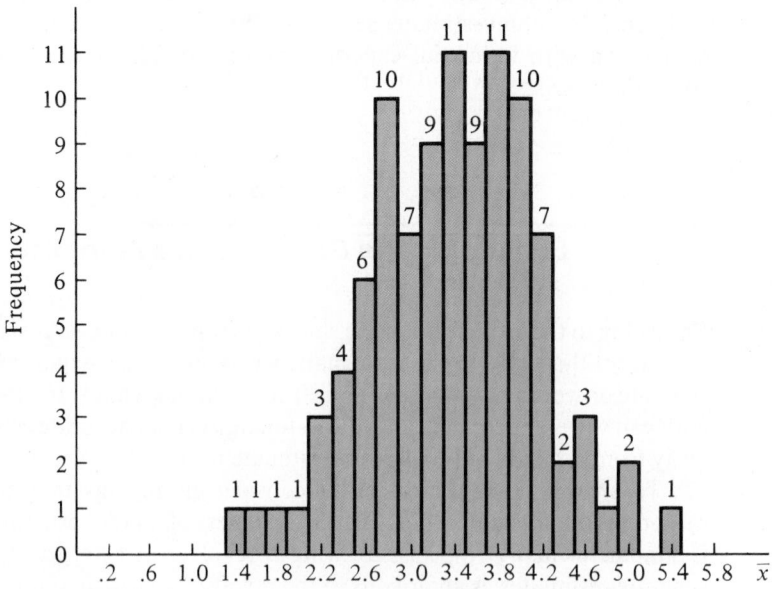

Figure 8.1   Histogram of Sample Means for the Die-Tossing Experiments in Section 7.3

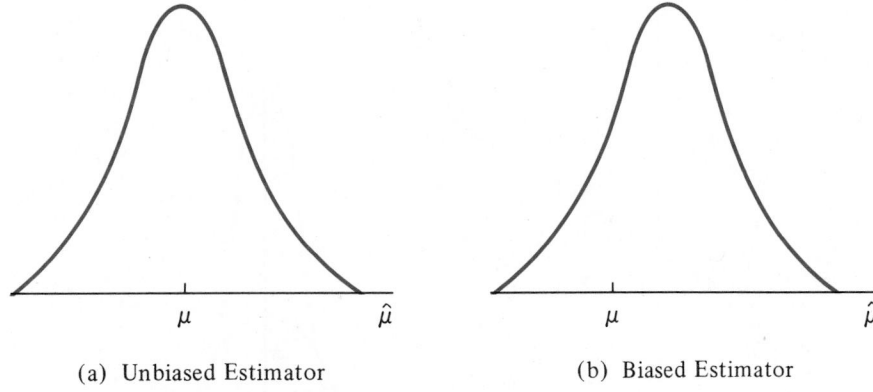

(a) Unbiased Estimator   (b) Biased Estimator

Figure 8.2   *Distributions for Unbiased and Biased Estimators*

mean $\mu = 3.5$, and that they range from 1.4 to 5.4. Surely this distribution of estimates tells us something about how good a new estimate of μ would be if we were to draw one more sample of $n = 5$ measurements and compute the sample mean $\bar{x}$.

We would know the characteristics of every point estimator if for each estimator we knew its sampling distribution. Finding the sampling distributions for various estimators is the work of research statisticians. Fortunately, the sampling distributions of most commonly used point estimators are known.

Since the properties of a point estimator are revealed by its sampling distribution, we might ask what properties are most desirable. Essentially there are two, and they can be seen by viewing Figure 8.2. First, we want the sampling distribution of estimates to center on the parameter of interest. For example, if we are estimating μ, we would like the sampling distribution of the estimator to center on μ, as shown in Figure 8.2a. Such an estimator is said to be *unbiased*. The sampling distribution for a *biased* estimator is shown in Figure 8.2b. [Recall that a hat (^) over a parameter is a symbol used to denote an estimator of that parameter.]

Definition

unbiased estimator

biased estimator

> An estimator of a population parameter is said to be **unbiased** if the mean of its sampling distribution is equal to the parameter. Otherwise, the estimator is said to be **biased**.

The second desirable property of a point estimator is that the standard deviation of its sampling distribution be small. Thus, we would like the spread of the distribution of estimates (see Figure 8.3) to be as small as possible. For most estimators the standard deviation of the sampling distribution is controllable. That is, we can make the standard deviation (which measures the spread of the distribution) as small as we wish by increasing the sample size.

For example, in Chapter 7 we learned from the Central Limit Theorem that the sampling distribution of the sample mean $\bar{x}$ was approximately normally distributed,

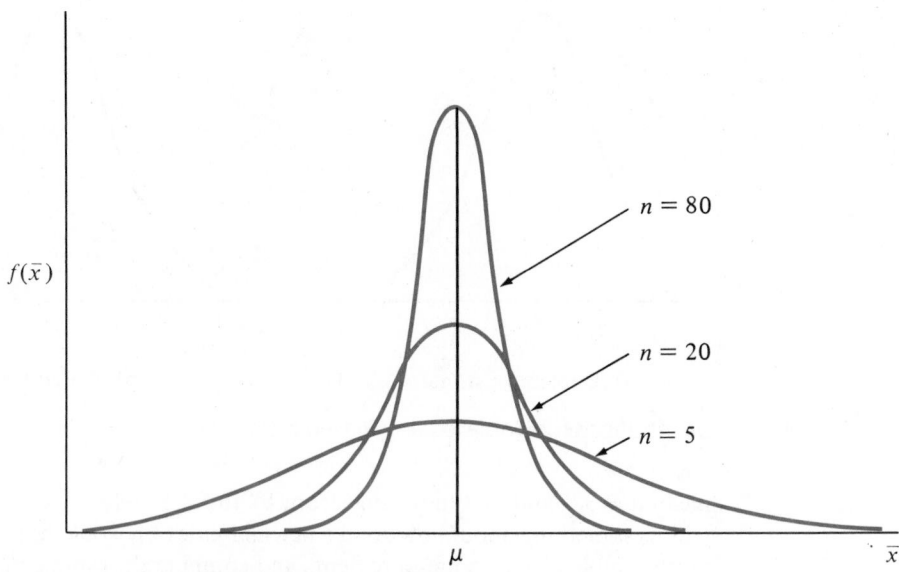

Figure 8.3   Sampling Distributions for $\bar{x}$; $n = 5, 20,$ and 80

with mean μ and standard deviation $\sigma/\sqrt{n}$ (μ and σ are the mean and standard deviation of the population from which the sample was selected and n is the sample size). Consequently, we can make the error of $\bar{x}$, denoted by the symbol $\sigma_{\bar{x}}$, as small as we choose by increasing the value of n. In Figure 8.3 we show three sampling distributions for $\bar{x}$, corresponding to $n = 5, n = 20,$ and $n = 80$, sketched on the same graph. The figure shows how the spread of the sampling distributions decreases as n increases. An estimate based on $n = 80$ measurements will, with a high probability, lie much closer to μ than one based on a sample of $n = 5$ measurements, since the sampling distribution of $\bar{x}$ for $n = 80$ is much less variable then the sampling distribution for $n = 5$.

In a real-life sampling situation you may know that the sampling distribution of an estimator centers about the parameter that you are attempting to estimate, but you may not know the value of the parameter. All that you have is the estimate computed from the n measurements contained in the sample. How far will your particular estimate be from the estimated parameter? Since the parameter usually lies in the center of the sampling distribution (it is usually the mean of the distribution), the distance between the estimate and the parameter, called the *error of estimation*, is less than or equal to the distance between the center and the tails of the distribution.

Definition

error of estimation

> The distance between an estimate and the estimated parameter is called the **error of estimation**.

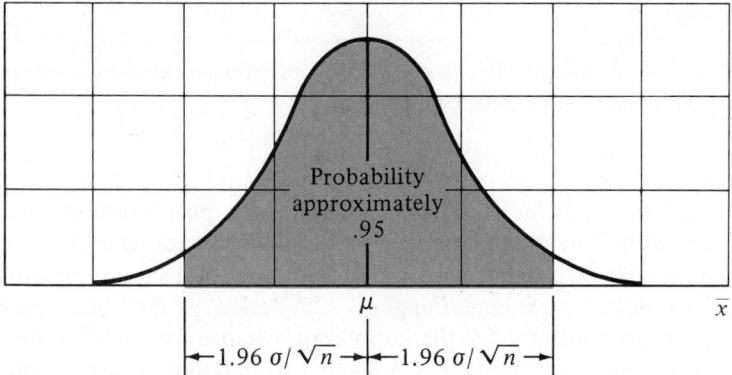

Figure 8.4   The Sampling Distribution of $\bar{x}$

In fact, the error of estimation will be less than two standard deviations of the sampling distribution, with a probability equal to at least .75 (Tchebysheff's Theorem) and most likely very close to .95 (the Empirical Rule). For example, the sampling distribution of the sample mean $\bar{x}$ is approximately normal, with mean μ and standard deviation $\sigma/\sqrt{n}$ (from the Central Limit Theorem), and appears as shown in Figure 8.4. You can see that the probability that an estimate falls within $1.96\sigma/\sqrt{n}$, either above or below μ, is approximately equal to .95. Equivalently, the probability that the error of estimation is less than $1.96\sigma/\sqrt{n}$ is approximately equal to .95.

In summary, the properties of a point estimator are characterized by its sampling distribution. We prefer estimators that are unbiased and have small standard errors. If we are dissatisfied with the size of the standard error of an estimator, we can reduce it by increasing the sample size n. The error of estimation—the distance between the estimate and the estimated parameter—should be less than two standard deviations of the sampling distribution, with probability at least equal to .75 and very likely near .95. If the sampling distribution is approximately normal (as will be the case for all estimators discussed in this chapter), we can be more precise and state that the error of estimation will be less than 1.96 standard deviations of the estimator, with probability approximately equal to .95.

8.5

Evaluating the Goodness of an Interval Estimator

Constructing an interval estimate is like attempting to throw a lariat around a fence post. In this case the parameter that you wish to estimate corresponds to the post and the interval corresponds to the loop formed by the cowboy's lariat. Each time you draw a sample, you construct a confidence interval for a parameter and you hope to "rope it," that is, include it in the interval. But you will not be successful for every sample.

Definition

confidence coefficient

The probability that a confidence interval will enclose the estimated parameter is called the **confidence coefficient**.

The confidence coefficient measures the proportion of samples that produce a confidence interval containing the population parameter. For example, suppose that we wish to estimate the mean profit per week of a small company. If we were to draw 10 samples, each containing $n = 20$ weekly profit observations, and construct a confidence interval for the population mean μ for each sample, the intervals might appear as shown in Figure 8.5. The horizontal line segments represent the 10 intervals and the vertical line represents the location of the true mean weekly profit. Note that all but one of the intervals enclose μ for these particular samples.

A good confidence interval is one that is as narrow as possible and has a large confidence coefficient, near 1. The narrower the interval, the more exactly we have located the estimated parameter. The larger the confidence coefficient, the more confidence we have that a particular interval encloses the estimated parameter. Remember that the confidence coefficient gives the probability that the interval estimator will produce confidence limits that enclose the estimated parameter. It gives you a measure of the confidence you can place in the confidence limits constructed from the data contained in a sample. In that sense the width of an interval and its associated confidence coefficient measure the goodness of the confidence interval.

What is the effect of larger samples on the width of a confidence interval? Larger samples provide more information to use in forming the interval estimate. Therefore,

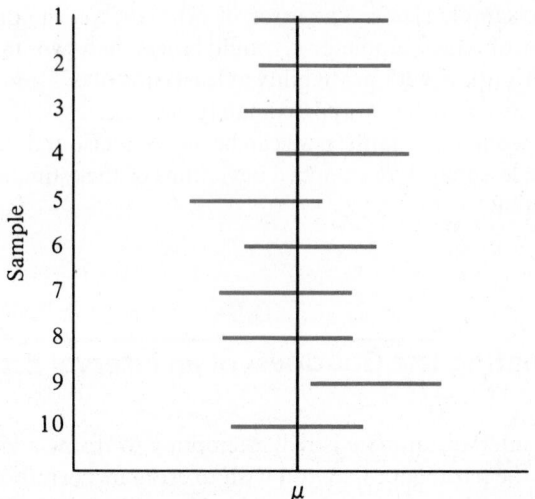

Figure 8.5   Ten Confidence Intervals for Mean Weekly Profit
(Each Based on a Sample of $n = 20$ Observations)

for a given confidence coefficient, the larger the sample, the narrower will be the resulting confidence interval.

8.6

Point Estimation of a Population Mean

Many business problems are concerned with estimating μ, the mean of a population. For example, we may be interested in the average monthly balance per credit account or the average cost per square foot for residential construction. The estimation of a population mean μ is a practical application of statistical inference and serves as an excellent illustration of the principles of point estimation discussed in Section 8.4.

When we sample a large or infinite population, the sampling distribution of the sample mean has a mean μ and a standard deviation $\sigma/\sqrt{n}$ (Section 7.4). **If the sampled population is normal, then regardless of the sample size, $\bar{x}$ always has a normal distribution.**

Sampling Distribution of $\bar{x}$

> According to the Central Limit Theorem, when we sample a *nonnormal* population, the **sampling distribution of $\bar{x}$ is approximately normal**, with mean μ and standard deviation $\sigma/\sqrt{n}$ when the sample size is large.

In any case, the sample mean $\bar{x}$ is always an *unbiased estimator* of the population mean μ. When the sample size is large, the sampling distribution of $\bar{x}$ is approximately normal, with a standard error that decreases as the sample size increases. The large-sample distribution of the sample mean is given in Figure 8.6. Hence, as an estimator of μ, $\bar{x}$ has the desirable properties of unbiasedness and minimum variance.

bound on the error of estimation   The point estimate of a population mean μ is the calculated value of the sample mean. Then, as explained in Section 8.4, a **bound on the error of estimation** is found as two—or, more exactly, as 1.96—standard deviations of the estimator $\bar{x}$.

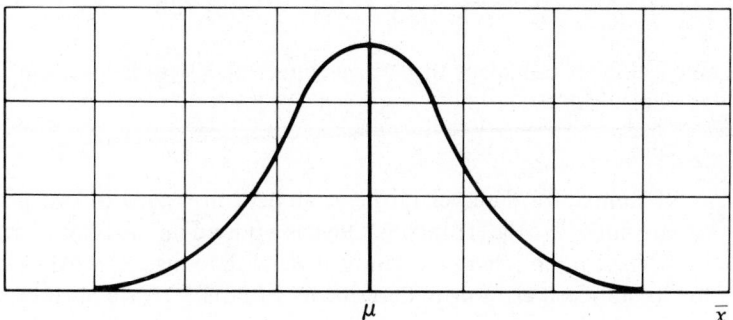

Figure 8.6   Distribution of $\bar{x}$ for Large n

Point Estimator of a Population Mean μ for Large Samples

Estimator: $\bar{x}$.

Bound on error: $1.96\sigma_{\bar{x}} = 1.96 \times$ (standard error)

$$= 1.96 \frac{\sigma}{\sqrt{n}}$$

The probability that the error of estimation is less than the bound on error is approximately .95.

Note: If σ is unknown and n is 30 or more, you can use the sample standard deviation s to approximate the value of σ.

Most often, we do not know the value of the population standard deviation σ that appears in the formula for the bound on error. Thus, we must approximate it, substituting the sample standard deviation s for σ. In applications of the Central Limit Theorem this approximation is adequate if the sample size is 30 or larger.

EXAMPLE 8.1   Suppose that we wish to estimate the average daily yield of a chemical manufactured in a chemical plant. The daily yield, recorded for $n = 50$ days, produced a mean and a standard deviation of

$$\bar{x} = 871 \text{ tons} \qquad \text{and} \qquad s = 21 \text{ tons}$$

Estimate the average daily yield μ.

SOLUTION   The estimate of the daily yield is $\bar{x} = 871$ tons. The bound on the error of estimation is

$$1.96\sigma_{\bar{x}} = 1.96 \frac{\sigma}{\sqrt{n}} = 1.96 \frac{\sigma}{\sqrt{50}}$$

Although σ is unknown, we can approximate its value by using s, the estimator of σ. Thus, the bound on the error of estimation is approximately

$$1.96 \frac{s}{\sqrt{n}} = 1.96 \frac{21}{\sqrt{50}} = 1.96(2.97) = 5.82$$

We feel fairly confident that our estimate of 871 tons is within 5.82 tons of the true average yield.

◆ ◆

Example 8.1 deserves further comment in regard to two points. You should be careful not to use the erroneous 1.96σ as a bound on the error of estimation rather than the correct value $1.96\sigma_{\bar{x}}$. Certainly, if we wish to discuss the distribution of $\bar{x}$, we must use its standard error $\sigma_{\bar{x}}$ to describe its variability. Care must be taken not to confuse the descriptive measures of one distribution with another.

The second point concerns the use of s to approximate σ. This approximation is reasonably good when n is large—say 30 or greater. If the sample size is small, two techniques are available. Sometimes, experience or data obtained from previous experiments will provide a good estimate of σ. You can then substitute this value for σ in the formula for $\sigma_{\bar{x}}$. **When an approximate value for σ is not available and when the sample has been selected from a population that has an approximately normal** (or at least mound-shaped) **relative frequency distribution, we may resort to a small-sample procedure to be described in Chapter 9.** The choice of $n = 30$ as the division between "large" and "small" samples is arbitrary. The reasoning for its selection will be discussed in Chapter 9.

Tips on Problem Solving

1. Be careful to use the standard error of $\bar{x}$, which is $\sigma_{\bar{x}} = \sigma/\sqrt{n}$, to evaluate the variability of $\bar{x}$. *Do not* use the standard deviation σ of the random variable x.

2. Since you will rarely know σ, you will need a good approximation of its value to substitute into the formula for $\sigma_{\bar{x}}$. The sample standard deviation s can be used for this purpose for large sample sizes. As a rule of thumb, we suggest $n \geq 30$.

EXERCISES Conceptual Level

8.1 What is the basic objective of inferential statistics?

8.2 Define the term *parameter* as it applies to inferential statistics.

8.3 Explain the difference between parameters and statistics.

8.4 List the two elements of a statistical inference. What role does the sampling distribution play in statistical inference?

8.5 Explain what is meant by an unbiased estimator. Why is unbiasedness a good characteristic for an estimator?

EXERCISES Skill Level

8.6 A random sample of size $n = 45$ is drawn from a population with mean μ and variance σ^2. Calculate the bound on error for estimating μ for the following values of σ.

   a. $\sigma = 10$ **b.** $\sigma = 20$ **c.** $\sigma = 50$ **d.** $\sigma = 100$

8.7 Refer to Exercise 8.6. For a fixed sample size n, what is the effect of an increasing value of σ on the bound on the error of estimation?

8.8 A random sample of size n is drawn from a population with mean μ and standard deviation $\sigma = 25$. Calculate the bound on the error for estimating μ for the following values of n.

   a. $n = 50$ **b.** $n = 100$ **c.** $n = 200$ **d.** $n = 500$

8.9 Refer to Exercise 8.8. For a fixed value of σ, what is the effect of increasing the value of n on the bound on the error of estimation?

8.10 A random sample of size n is drawn from a population with mean μ and variance σ^2. Estimate μ and place a bound on the error of estimation, using the following sample information:

a. $n = 45, \bar{x} = 10, s = 5.5$ **b.** $n = 89, \bar{x} = 1.5, s = .25$

c. $n = 100, \bar{x} = 102.7, s = 10.8$

EXERCISES   Applications

 8.11 After reviewing a manufacturer's claim of increased gasoline efficiency for its midsized line of passenger cars, an independent testing agency randomly selected a sample of $n = 64$ of the new cars and tested them to determine the gasoline mileage for each. The results showed an average rating of 34.5 miles per gallon (mpg) with a standard deviation of $s = .8$. Estimate μ, the average mpg rating for the manufacturer's midsized line, and place a bound on the error of estimation.

 8.12 The Environmental Protection Agency has noted that for the past several years a large East Coast community has been plagued with severe summer water shortages. To monitor water use, city officials randomly selected and monitored $n = 100$ residential water meters to estimate the average daily water consumption per household over a certain dry spell. The sample mean and standard deviation were found to be 117.5 gallons and 16.8 gallons, respectively. Estimate μ, the average daily household consumption for the community, and place a bound on the error of estimation.

 8.13 Several researchers have found that evaluating the internal control procedures of an organization by studying its accounting records is impractical because businesses are usually too large to allow for an examination of all, or even a quarter, of the year's transactions. An audit was conducted by randomly selecting $n = 50$ bills of lading from among those transacted during the past year by a trucking firm. The mean and standard deviation of the bills were found to be $2,160 and $575, respectively. Estimate μ, the average bill of lading transacted by the firm during the past year. Place a bound on the error of estimation.

 8.14 In a survey of 94 major corporations the average annual earnings of the top executive was $105,400. Assuming a standard deviation of $10,000 to apply to the earnings of the executives, estimate the average earnings for top executives in all major corporations, and give a bound on your error of estimation. What assumption must you make in order that this estimate be unbiased and that the bound on the error of estimation be valid?

 8.15 With increased concern over maintaining adequate cash flow, a builders' supply company wishes to estimate the average amount owed to it by its credit customers. For this purpose 49 of the company's invoices are randomly sampled and are found to possess an average credit balance of $3,200 with a standard deviation of $350. Estimate the average credit balance for all credit customers of the supply company, and place a bound on the error of estimation.

8.7

Interval Estimation of a Population Mean

confidence interval The interval estimate, or **confidence interval**, for a population mean may be easily obtained from the results presented in Section 8.6. The sample mean $\bar{x}$ might lie either above or below the population mean, although we would not expect it to deviate more

lower confidence limit

upper confidence limit

large-sample confidence interval

than approximately $2\sigma_{\bar{x}}$ from μ. Hence, if we choose $(\bar{x} - 2\sigma_{\bar{x}})$ as the lower point of the interval, called the **lower confidence limit**, or LCL, and $(\bar{x} + 2\sigma_{\bar{x}})$ as the upper point, or **upper confidence limit**, UCL, the interval most probably will enclose the true population mean μ. In fact, if n is large and the distribution of $\bar{x}$ is approximately normal, we would expect approximately 95% of the intervals obtained in repeated sampling to enclose the population mean μ.

The confidence interval just described is called a **large-sample confidence interval** (or confidence limits) because n must be large enough for the Central Limit Theorem to be effective and hence for the distribution of $\bar{x}$ to be approximately normal. Since σ is usually unknown, the sample standard deviation s must be used to estimate σ. **As a rule of thumb, this confidence interval is appropriate when $n = 30$ or more.**

The confidence coefficient .95 corresponds to $\pm 2\sigma_{\bar{x}}$, or, more exactly, $1.96\sigma_{\bar{x}}$. Recalling that .90 of the measurements in a normal distribution fall within $z = 1.645$ standard deviations of the mean (Table 3 in the Appendix), we could construct 90% confidence intervals by using

$$\text{LCL} = \bar{x} - 1.645\sigma_{\bar{x}} = \bar{x} - 1.645\frac{\sigma}{\sqrt{n}}$$

$$\text{UCL} = \bar{x} + 1.645\sigma_{\bar{x}} = \bar{x} + 1.645\frac{\sigma}{\sqrt{n}}$$

In general, we may construct confidence intervals corresponding to any desired confidence coefficient—say $(1 - \alpha)$—by using the formula given in the next display.

A $(1 - \alpha)$100% Large-Sample Confidence Interval for μ

$$\bar{x} \pm z_{\alpha/2} \times (\text{standard error of } \bar{x})$$

given by

$$\bar{x} \pm z_{\alpha/2}\frac{\sigma}{\sqrt{n}}$$

The normal curve value, $z_{\alpha/2}$, that appears in the formula for the confidence interval is located as shown in Figure 8.7. For example, if you want a confidence coefficient $(1 - \alpha)$ equal to .95, then the tail-end area is $\alpha = .05$, and half of α (.025) is placed in each tail of the distribution. Then $z_{.025}$ is the table z value corresponding to an area of .475 to the right of the mean, or

$$z_{.025} = 1.96$$

Confidence limits corresponding to some of the commonly used confidence coefficients are shown in Table 8.1.

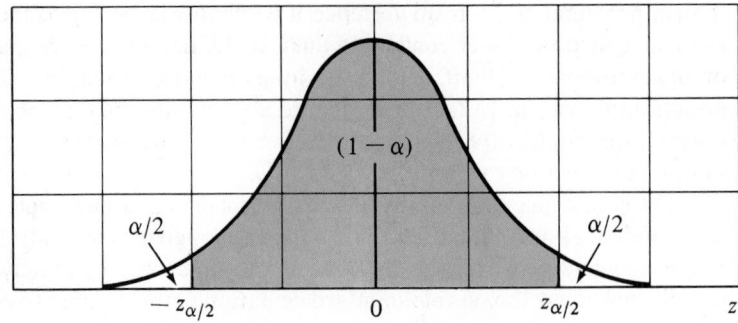

Figure 8.7   Location of $z_{\alpha/2}$.

Table 8.1   Confidence Limits for μ

Confidence Coefficient, $(1 - \alpha)$	α	$z_{\alpha/2}$	LCL	UCL
.90	.10	1.645	$\bar{x} - 1.645 \dfrac{\sigma}{\sqrt{n}}$	$\bar{x} + 1.645 \dfrac{\sigma}{\sqrt{n}}$
.95	.05	1.96	$\bar{x} - 1.96 \dfrac{\sigma}{\sqrt{n}}$	$\bar{x} + 1.96 \dfrac{\sigma}{\sqrt{n}}$
.99	.01	2.58	$\bar{x} - 2.58 \dfrac{\sigma}{\sqrt{n}}$	$\bar{x} + 2.58 \dfrac{\sigma}{\sqrt{n}}$

EXAMPLE 8.2   Find a 90% confidence interval for the population mean of Example 8.1. Recall that $\bar{x} = 871$ tons and $s = 21$ tons.

SOLUTION   The 90% confidence limits are

$$\bar{x} \pm 1.645 \frac{\sigma}{\sqrt{n}}$$

Using s to estimate σ, we obtain

$$871 \pm (1.645) \frac{21}{\sqrt{50}} \qquad \text{or} \qquad 871 \pm 4.89$$

Therefore, we estimate that the average daily yield μ lies in the interval from 866.11 to 875.89 tons. The confidence coefficient .90 implies that in repeated sampling 90% of the confidence intervals similarly formed would enclose μ.

◆◆

EXAMPLE 8.3   A sample of $n = 100$ employees from a company was selected and the annual salary for each was recorded. The mean and standard deviation of their salaries were found to be

$$\bar{x} = \$17{,}750 \quad\text{and}\quad s = \$900$$

Construct the 95% confidence interval for the population average salary μ.

SOLUTION   The 95% confidence limits are

$$\bar{x} \pm 1.96 \frac{\sigma}{\sqrt{n}}$$

Using s to estimate σ, we obtain

$$\$17{,}750 \pm (1.96)\frac{\$900}{\sqrt{100}} \quad\text{or}\quad \$17{,}750 \pm \$176.40$$

In repeated sampling, intervals constructed in such a manner will enclose μ 95% of the time. Hence, we are fairly confident that the average salary for employees of the company is contained within the interval from \$17,573.60 to \$17,926.40.

The confidence interval of Example 8.3 is approximate because we substituted s as an approximation for σ. That is, instead of the confidence coefficient being .95, the value specified in the example, the true value of the coefficient may be .92, .94, or .97. But this discrepancy is of little concern from a practical point of view; as far as our "confidence" is concerned, there is little difference among these confidence coefficients. Most interval estimators employed in statistics yield approximate confidence intervals, because the assumptions upon which they are based are not satisfied exactly. Having made this point, we will not continue to refer to confidence intervals as "approximate." It is of little practical concern as long as the actual confidence coefficient is near the value specified.

Note in Table 8.1 that for a fixed sample size the width of the confidence interval increases as the confidence coefficient increases, a result that is in agreement with our intuition. Certainly, if we wish to be more confident that the interval will enclose μ, we would increase the width of the interval. Since we prefer narrow confidence intervals and large confidence coefficients, we must reach a compromise in choosing the confidence coefficient.

The choice of the confidence coefficient to be used in a given situation is made by the experimenter and depends on the degree of confidence the experimenter wishes to place in the estimate. Most confidence intervals are constructed by using one of the three confidence coefficients shown in Table 8.1. The most popular seem to be 95% confidence intervals. Use of 99% confidence intervals is less common because of the wider interval width that results. Of course, you can always decrease the width by increasing the sample size n.

Note the fine distinction between point estimators and interval estimators. Note also that in placing bounds on the error of a point estimate, for all practical purposes we are constructing an interval estimate when a population mean is being estimated. Although this close relationship exists for most of the parameters estimated in this text, the two methods of estimation are not equivalent. For instance, it is not obvious that the best point estimator falls in the middle of the best interval estimator—in many cases it does not. Furthermore, it is not necessarily true that the best interval estimator is even a function of the best point estimator. Although these problems are of a theoretical nature, they are important and worth mentioning. From a practical point of view, the two methods are closely related, and **the choice between the point and the interval estimator in an actual problem depends on the preference of the experimenter.**

EXERCISES Conceptual Level

8.16 What part does the Central Limit Theorem play in large-sample estimation?

8.17 What rule of thumb is used to decide whether or not it is appropriate to use large-sample techniques for statistical estimation?

8.18 Explain the relationship between the width of a confidence interval and the level of significance when constructing a confidence interval.

8.19 Assuming that α remains constant, explain the relationship between the width of a confidence interval and the sample size n.

EXERCISES Skill Level

8.20 A random sample of size $n = 67$ is drawn from a population with mean μ and standard deviation σ. The mean and standard deviation of the sample are $\bar{x} = 408.9$ and $s = 31.9$.

   a. Find a 90% confidence interval for the mean μ. Interpret this interval.

   b. Find a 95% confidence interval for the mean μ. Interpret this interval.

   c. Find a 99% confidence interval for the mean μ. Interpret this interval.

   d. Compare the widths of the three confidence intervals. Which interval is the widest? The narrowest?

8.21 A random sample of size n is drawn from a population with mean μ and standard deviation σ. The mean and standard deviation of the sample are $\bar{x} = 45$ and $s = 5.8$. Calculate a 95% confidence interval for μ by using the sample sizes indicated in parts a–c.

   a. $n = 30$ **b.** $n = 60$ **c.** $n = 90$

   d. Compare the widths of the three intervals. Which interval is the widest? The narrowest?

8.22 A random sample from a population with mean μ and standard deviation σ produced the following sample information:

$$n = 110 \qquad \bar{x} = 699 \qquad s = 20.4$$

   a. Find a 95% confidence interval for the mean μ. Interpret this interval.

   b. Find a 99% confidence interval for the mean μ. Interpret this interval.

EXERCISES   Applications

 8.23 A study designed to compare Japanese and American patterns of corporate finance and ownership structure used information drawn from 344 Japanese companies and 452 U.S. companies in 27 different industries. The researchers observed that the Japanese use of bank debt as a low-cost source of capital allows Japanese companies to charge lower prices, bear higher costs in other areas, and hence become more competitive than U.S. companies. Leverage, the company's level of debt financing used to raise capital, is measured by a debt-equity ratio that has been adjusted for such factors as growth, profitability, risk, size, and industry classification. For the sample of $n = 344$ Japanese companies, one type of debt-equity ratio (book value equity and gross debt) had a mean of 2.703 with a standard deviation of 3.822 (W. C. Kester, "Capital and Ownership Structure: A Comparison of United States and Japanese Manufacturing Corporations," *Financial Management*, Spring 1986, pp. 5–15). Find 90% and 95% confidence intervals for the mean debt-equity ratio for all Japanese companies.

 8.24 An increase in the rate of consumer savings is frequently tied to a lack of confidence in the economy and is said to be an indicator of a recessional tendency in the economy. A random sampling of $n = 200$ savings accounts in a local community showed a mean increase in savings account values of 7.2% over the past 12 months with a standard deviation of 5.6%. Estimate the mean percentage increase in savings account values over the past 12 months for depositors in the community with a 95% confidence interval.

 8.25 Forecasting sales revenue in a given geographical region is a difficult task and often requires an estimate of consumer expenditures for a particular product class. A market research study was undertaken by a major oil company to determine the amount spent on gasoline and heating oil during a particular year by the residential households of a certain city. The city directory was used to select a random sample of $n = 64$ households from the city. The sample mean and standard deviation of their expenditures on gasoline and heating oil during the year were $1,136 and $178, respectively. Find a 90% confidence interval for the average annual expenditure on gasoline and heating oil by the residential households of the city.

 8.26 Ocean Routes, a ship-routing company based in Palo Alto, California, uses sophisticated weather-forecasting techniques, including satellite pictures, to recommend routes to captains of ships on transatlantic and transpacific crossings. The company report indicates that for 6,215 Atlantic crossings the mean time saved by ships using the Ocean Routes advisory service was 4.6 hours, with a standard deviation of .88 hour; for 19,777 Pacific crossings the mean time saved was 7.6 hours with a standard deviation of 2.07 hours (data provided courtesy of Ocean Routes, Inc., Palo Alto, Calif.).

  a. Construct a 95% confidence interval for the mean time saved for all transatlantic crossings.

  b. Construct a 95% confidence interval for the mean time saved for all transpacific crossings.

  c. If a ship's daily operating cost is $14,750, find a 95% confidence interval for the expected amount of money saved on a Pacific crossing when consulting Ocean Routes.

8.27 An audit of the inventory of a retailer was conducted by randomly selecting the purchase invoices from $n = 100$ unsold units in stock. The average purchase price per unit was found to be $17.50, with a standard deviation of $6.75.

  a. Find a 95% confidence interval for the mean purchase price of all units of inventory held by the retailer.

  b. Audit results are interpreted in terms of the stated precision and the materiality of the derived estimate. The precision of the estimate is the level of confidence in the confidence interval; materiality is synonymous with accuracy—the absolute difference between μ and its estimate $\bar{x}$. With the sample size held constant, how can the estimate derived in part a be made more precise? With n fixed, how can greater materiality be achieved?

 8.28 Most studies indicate that for long periods of time mean rates of return on common stocks have exceeded mean rates on bonds and other fixed-dollar assets. Since stocks are more risky than fixed-dollar assets, we are thus offered clear evidence of the assumed relationship between return and market risk. The data in the accompanying table provide the mean and standard deviation of annual returns for randomly chosen securities within each of four asset classes for the past nine years.

Asset Class	Number of Chosen Securities	Mean Annual Return	Standard Deviation of Annual Return (Risk)
Common stocks	50	10.57	19.05
Corporate bonds	35	4.38	5.52
U.S. government securities	30	3.47	3.87
Municipal bonds	35	2.51	8.76

a. Find a 95% confidence interval for the mean annual return for each asset class for the past nine years.

b. Which asset class offered the greatest chance of a negative return during the past nine years? Which offered the greatest opportunity for gain?

8.8

Estimation from Large Samples

Estimation of a population mean, presented in Sections 8.6 and 8.7, sets the stage for the other estimation problems to be discussed in this chapter. Like the sample mean, every estimator presented in this chapter possesses a sampling distribution that is approximately normal, owing all or in part to the Central Limit Theorem. All are unbiased estimators, and the standard deviations of their sampling distributions are known. Consequently, the point estimators and confidence intervals will be constructed in exactly the same way as for the population mean μ.

We will encounter other estimators in later chapters that will possess approximately normal sampling distributions, but here we will concentrate on four situations that occur frequently in practice. In particular, we consider inferences concerning population means (Sections 8.6 and 8.7), a comparison of two means, a binomial parameter p, and a comparison of two binomial parameters.

Business surveys frequently have as their objective the estimation of a population mean μ or a binomial parameter p. For example, estimating the mean shrinkage of items of merchandise in a department store is an example of estimating a population mean μ. The random selection of a sample of people to determine the proportion who favor a particular household product has as its objective the estimation of a binomial parameter p.

Other business surveys may seek a comparison of these parameters for two or more populations. For example, suppose you plan to purchase one of two automobiles and wish to choose the one that will give the more economical fuel consumption (the greater miles per gallon, mpg). Since you know that the Environmental Protection Agency (EPA) ratings are far from realistic, and that the mpg ratings vary considerably from one automobile to another, you plan to base your decision on a comparison of the measurements for two randomly selected samples of automobiles, n_1 of the first type and n_2 of the second. Your decision would be based on the estimated difference in the mean mpg ratings for the two automobiles. Similarly, you might wish to estimate the difference in the proportions of defectives for two large lots of industrial products based on random samples selected from each.

The four situations described above identify the four estimation problems covered in this chapter. We give estimates for these situations:

1. A single population mean μ
2. The difference between two population means
3. A single population proportion p
4. The difference between two population proportions

Since we have already stated that the sampling distributions for these estimators possess the same properties, we can give the general form of the point and interval estimators for all four estimation problems. Although the formulas for the point estimators differ, the bound on the error for all is as given in the display.

Bound on Error of Estimation for a Large-Sample Point Estimator

> bound on error = 1.96 standard deviations of the sampling distribution of the point estimator
>
> = 1.96 × (standard error of the point estimator)

Similarly, the large-sample confidence intervals for each parameter are as given here.

A $(1 - \alpha)100\%$ Large-Sample Confidence Interval

> point estimate $\pm\ z_{\alpha/2}\ \times$ (standard error of point estimator)
>
> where $z_{\alpha/2}$ is obtained from Table 8.1.

The specific formulas for estimating the difference between two means, for estimating a single binomial proportion, and for estimating the difference between two proportions will be presented in Sections 8.9, 8.10, and 8.11.

8.9

Estimating the Difference Between Two Means

A problem of equal importance to the estimation of a population mean is the estimation of the difference between two population means. For instance, we might wish to estimate the difference in mean lengths of time required to assemble an industrial product using two different methods of assembly. Assemblers could be randomly divided into two groups, the first using assembly method 1 and the second using method 2. We could then make inferences concerning the difference in mean time to assemble the device for the two assembly methods. Or we might wish to compare the average yield in a chemical plant using raw materials furnished by two suppliers, A and B. Samples of daily yield, one for each of the two raw materials, could be recorded and used to make inferences concerning the difference in mean yield.

For each of these examples there are two populations, the first with mean and variance μ_1 and σ_1^2 and the second with mean and variance μ_2 and σ_2^2. A random sample of n_1 measurements is drawn from population 1 and n_2 from population 2, where the samples are assumed to have been drawn independently of one another. Finally, the estimates of the population parameters are calculated from the sample data using the estimators $\bar{x}_1$, s_1^2, $\bar{x}_2$, and s_2^2.

The point estimator of the difference $(\mu_1 - \mu_2)$ between the population means is $(\bar{x}_1 - \bar{x}_2)$. In Chapter 7 we stated without proof that the sampling distribution of the difference between two normally distributed sample means is normally distributed with a mean equal to $(\mu_1 - \mu_2)$, the difference of the means for the two populations sampled, with a standard deviation given by

$$\sigma_{(\bar{x}_1 - \bar{x}_2)} = \sqrt{\frac{\sigma_1^2}{n_1} + \frac{\sigma_2^2}{n_2}}$$

where σ_1^2 and σ_2^2 are the two population variances and n_1 and n_2 are the corresponding sample sizes. The sampling distribution of $(\bar{x}_1 - \bar{x}_2)$ is shown in Figure 8.8.

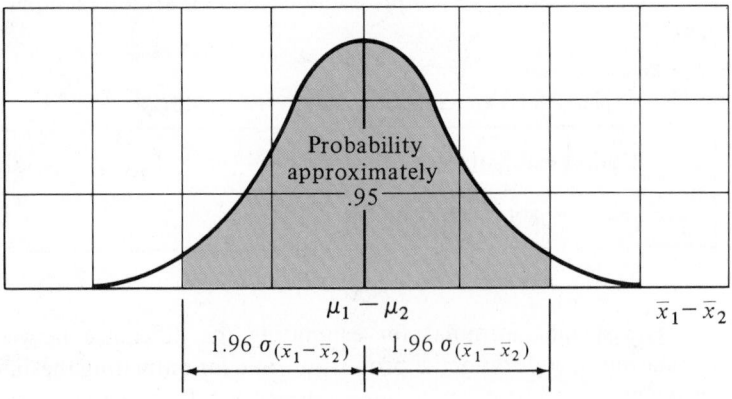

Figure 8.8   *The Distribution of $(\bar{x}_1 - \bar{x}_2)$ for Large Samples*

To summarize, the point estimator of the difference between two population means is as given in the next display.

Point Estimation of $(\mu_1 - \mu_2)$

Estimator: $(\bar{x}_1 - \bar{x}_2)$.

Bound on error: $1.96\sigma_{(\bar{x}_1 - \bar{x}_2)} = 1.96\sqrt{\dfrac{\sigma_1^2}{n_1} + \dfrac{\sigma_2^2}{n_2}}$.

Note: If σ_1^2 and σ_2^2 are unknown, but both n_1 and n_2 are 30 or more, you can use the sample variances s_1^2 and s_2^2 to estimate σ_1^2 and σ_2^2.

We illustrate this point estimation procedure with an example.

EXAMPLE 8.4   A comparison of the wearing quality of two types of automobile tires was obtained by road-testing samples of $n_1 = n_2 = 100$ tires for each type. The number of miles until wear-out was recorded, where wear-out was defined as a specific amount of tire wear. The test results are given in Table 8.2. Estimate $(\mu_1 - \mu_2)$, the difference in mean miles to wear-out, and place bounds on the error of estimation.

Table 8.2   Test Results for Example 8.4

Tire 1	Tire 2
$\bar{x}_1 = 26{,}400$ miles	$\bar{x}_2 = 25{,}100$ miles
$s_1^2 = 1{,}440{,}000$	$s_2^2 = 1{,}960{,}000$

SOLUTION   The point estimate of $(\mu_1 - \mu_2)$ is

$$(\bar{x}_1 - \bar{x}_2) = 26{,}400 - 25{,}100 = 1{,}300 \text{ miles}$$

The standard error of $(\bar{x}_1 - \bar{x}_2)$ is

$$\sigma_{(\bar{x}_1 - \bar{x}_2)} = \sqrt{\frac{\sigma_1^2}{n_1} + \frac{\sigma_2^2}{n_2}} \approx \sqrt{\frac{s_1^2}{n_1} + \frac{s_2^2}{n_2}}$$

$$= \sqrt{\frac{1{,}440{,}000}{100} + \frac{1{,}960{,}000}{100}} = \sqrt{34{,}000} = 184.4 \text{ miles}$$

where s_1^2 and s_2^2 are used to estimate σ_1^2 and σ_2^2, respectively. We would expect the error of estimation to be less than $(1.96)(184.4) = 361.4$ miles. If the point estimate of the difference in mean miles to wear-out is 1,300 miles and the error of estimation is less than 361.4 miles (with a high probability), it seems fairly conclusive that there is a substantial difference in the mean miles to wear-out for the two types of tires. In fact, tire 1 is apparently superior to tire 2 in wearing quality when subjected to the road test.

◆ ◆

Following the procedure of Section 8.7, we can construct a confidence interval for $(\mu_1 - \mu_2)$ with confidence coefficient $(1 - \alpha)$.

A $(1 - \alpha)100\%$ Confidence Interval for $(\mu_1 - \mu_2)$

$$(\bar{x}_1 - \bar{x}_2) \pm z_{\alpha/2}\sqrt{\frac{\sigma_1^2}{n_1} + \frac{\sigma_2^2}{n_2}}$$

As a rule of thumb we will require both n_1 and n_2 to be equal to 30 or more in order that s_1^2 and s_2^2 provide good estimates of their respective population variances.

EXAMPLE 8.5   Place a confidence interval on the difference in the mean miles to wear-out for the problem described in Example 8.4. Use a confidence coefficient of .99.

SOLUTION   The confidence interval is

$$(\bar{x}_1 - \bar{x}_2) \pm 2.58\sqrt{\frac{\sigma_1^2}{n_1} + \frac{\sigma_2^2}{n_2}}$$

Using the results of Example 8.4, we find that the confidence interval is

$$1,300 \pm (2.58)(184.4)$$

Therefore, LCL = 824.2, UCL = 1,775.8, and the difference in the mean miles to wear-out is estimated to lie between these two points. Note that the confidence interval is wider than the $\pm 1.96\sigma_{(\bar{x}_1 - \bar{x}_2)}$ bound used in Example 8.4, because we have chosen a larger confidence coefficient.

EXERCISES   Conceptual Level

8.29   Under what conditions is it reasonable to assume that the sampling distribution of the difference between two means is normally distributed?

8.30   If σ_1 and σ_2 are unknown, what values are used as estimates of σ_1 and σ_2 in calculating the standard error of $(\mu_1 - \mu_2)$?

8.31   If the $(1 - \alpha)100\%$ confidence interval for the difference $(\mu_1 - \mu_2)$ contains the value 0, what inference can be made about the two values μ_1 and μ_2?

EXERCISES   Skill Level

8.32   Independent random samples of size n_1 and n_2 are drawn from two populations with means μ_1 and μ_2 and standard deviations σ_1 and σ_2, respectively. The following sample information is available:

Sample 1	Sample 2
$n_1 = 50$	$n_2 = 50$
$\bar{x}_1 = 91.1$	$\bar{x}_2 = 92.3$
$s_1 = 5.4$	$s_2 = 7.6$

Estimate the difference ($\mu_1 - \mu_2$), and place a bound on the error of estimation.

8.33   Refer to Exercise 8.32. Find a 99% confidence interval for the difference ($\mu_1 - \mu_2$). Interpret this interval.

8.34   Independent random samples of size n_1 and n_2 are drawn from two populations with means μ_1 and μ_2 and standard deviations σ_1 and σ_2, respectively. The following sample information is available:

Sample 1	Sample 2
$n_1 = 100$	$n_2 = 150$
$\bar{x}_1 = 40$	$\bar{x}_2 = 32$
$s_1 = 13$	$s_2 = 16$

a.   Calculate a 95% confidence interval for the difference ($\mu_1 - \mu_2$).

b.   Calculate a 99% confidence interval for the difference ($\mu_1 - \mu_2$).

EXERCISES   Applications

8.35   The *San Francisco Chronicle* (June 23, 1980) reported the results of an extensive study presented to the Society of Epidemiological Research, which showed that men and women experiencing similar business world life-styles have considerably different life spans. The study examined the careers of 2,000 men and 2,000 women and found that the average age at death for men is 69.2 years, and the life span of the women involved in the study averaged 77.3 years. Sample standard deviations were 6.9 years and 5.3 years, respectively, for the men and women involved in the study. Estimate the mean difference in life span between men and women for those involved in similar business world life-styles, and place bounds on the error of estimation.

MKTG   **8.36**   Consumer activism, legislation, and heightened consumer expectations have placed more responsibility on manufacturers for the performance of their goods. Consequently, a manufacturer of two distinct products is interested in estimating the difference in mean monthly complaints received concerning the two products. Over a period of four years (48 months) the number of complaints concerning each product produced the results shown in the accompanying table. Find a 90% confidence interval for the difference in the mean monthly complaints received concerning the two products. (Assume that the numbers of complaints for the two products can be regarded as independent random samples.)

Product 1	Product 2
$\bar{x}_1 = 17.2$	$\bar{x}_2 = 25.1$
$s_1 = 4.6$	$s_2 = 5.3$

8.37 Traditional sales training programs have directed considerable attention to the sales agent's role in closure. A recent article reports on an investigation of a new training program that minimizes attention to the pressure applied by the sales agent in closing a sale ("Why Closing Is Bad for Sales," *Marketing*, April 16, 1980). The article reports that 40 retail sales agents who were subject to traditional sales training had an average transaction time of 12.58 minutes with a standard deviation of 2.31, while 51 sales agents who had received the new low-pressure sales training program averaged 8.67 minutes per transaction, with a standard deviation of 2.09 minutes. Find a 95% confidence interval for the difference in mean transaction times for retail sales agents who have received the two types of training programs.

8.38 Refer to Exercise 8.23. For the sample of $n = 452$ U.S. companies, the mean debt-equity ratio based on book value equity and gross debt was 0.745, with a standard deviation of 1.332 (W. C. Kester, "Capital and Ownership Structure: A Comparison of United States and Japanese Manufacturing Corporations," *Financial Management*, Spring 1986, pp. 5–15). Find a 95% confidence interval for the difference in mean debt-equity ratios for Japanese and U.S. companies. Interpret this interval.

8.39 Since the income from municipal bonds is generally free from federal, state, and local income tax obligations, municipal rates have trailed those of the corporate sector. Refer to Exercise 8.28. Find a 90% confidence interval for the difference in mean annual returns between corporate bonds and municipal bonds for the past nine years.

8.10

Estimating the Parameter of a Binomial Population

Many surveys have as their objective the estimation of the proportion of people or objects in a large group that possess a particular attribute. Such a survey is a practical example of the binomial experiment discussed in Chapter 5. Estimating the proportion of sales that can be expected in a large number of customer contacts is a practical problem requiring the estimation of a binomial parameter p.

The best point estimator of the binomial parameter p is also the estimator that would be chosen intuitively. That is, the estimator of p, denoted by the symbol $\hat{p}$, is the total number x of successes divided by the total number n of trials. That is,

$$\hat{p} = \frac{x}{n}$$

where x is the number of successes in n trials. [Recall that a "hat" ($\hat{\ }$) over a parameter is the symbol used to denote the estimator of the parameter.] By "best" we mean that $\hat{p}$ is unbiased and possesses a smaller variance than other possible estimators.

As noted in Chapter 7, the estimator $\hat{p}$ possesses a sampling distribution that is approximately normally distributed because of the Central Limit Theorem. It is an unbiased estimator of the population proportion p, with a standard error of

$$\sigma_{\hat{p}} = \sqrt{\frac{pq}{n}}$$

where p is the unknown population proportion, $q = (1 - p)$, and n is the sample size. Then from Section 8.8 the point estimation procedure for p is as given in the display.

Point Estimator for p

> Estimator: $\hat{p} = \dfrac{x}{n}$.
>
> Bound on error: $1.96\sigma_{\hat{p}} = 1.96\sqrt{\dfrac{pq}{n}}$.
>
> Estimated bound on error: $1.96\sqrt{\dfrac{\hat{p}\hat{q}}{n}}$.

The corresponding large-sample confidence interval with confidence coefficient $(1 - \alpha)$ is shown in the next display.

A $(1 - \alpha)100\%$ Confidence Interval for p

> $$\hat{p} \pm z_{\alpha/2}\sqrt{\dfrac{\hat{p}\hat{q}}{n}}$$

The sample size will be considered large when we can assume that the distribution of $\hat{p}$ is approximately normal. These conditions were discussed in Section 7.5.

The only difficulty encountered in our procedure will be in calculating $\sigma_{\hat{p}}$, which involves the unknown value of p (and $q = 1 - p$). You will note that we have substituted $\hat{p}$ for the parameter p in the standard deviation $\sqrt{pq/n}$. When n is large, little error will be introduced by this substitution. As a matter of fact, the standard deviation changes only slightly as p changes. This feature can be observed in Table 8.3, where $\sqrt{pq}$ is recorded for several values of p. Note that $\sqrt{pq}$ changes very little as p changes, especially when p is near .5.

Table 8.3 Some Calculated Values of $\sqrt{pq}$

p	$\sqrt{pq}$
.5	.50
.4	.49
.3	.46
.2	.40
.1	.30

EXAMPLE 8.6   A random sample of $n = 100$ voters in a community produced $x = 59$ voters in favor of candidate A. Estimate the fraction of the voting population favoring A, and place a bound on the error of estimation.

SOLUTION   The point estimate of p is

$$\hat{p} = \frac{x}{n} = \frac{59}{100} = .59$$

and the bound on the error of estimation is estimated to be

$$1.96\sigma_{\hat{p}} = 1.96\sqrt{\frac{\hat{p}\hat{q}}{n}} = 1.96\sqrt{\frac{(.59)(.41)}{100}} = .096$$

A 95% confidence interval for p is

$$\hat{p} \pm 1.96\sqrt{\frac{\hat{p}\hat{q}}{n}} \quad \text{or} \quad .59 \pm (1.96)(.049)$$

Thus, we estimate that the interval from .494 to .686 includes p, with confidence coefficient .95.

EXERCISES   Conceptual Level

8.40   Under what conditions can the sampling distribution of the sample proportion be approximated by a normal distribution?

8.41   What part does the Central Limit Theorem play in the use of the normal distribution to approximate the sampling distribution of a sample proportion?

8.42   If p is unknown, what value is used to estimate p in order to calculate the standard error of $\hat{p}$ for a large-sample estimation problem?

EXERCISES   Skill Level

8.43   A random sample of size $n = 100$ is drawn from a binomial population with parameter p, the proportion of successes in the population. The sample produces $x = 65$ successes. Estimate p and place a bound on the error of estimation.

8.44   Refer to Exercise 8.43.

   a.   Calculate a 95% confidence interval for p.

   b.   Calculate a 98% confidence interval for p.

8.45   A random sample of size $n = 400$ is drawn from a binomial population with parameter p, the proportion of successes in the population. The sample produces $x = 180$ successes.

   a.   Calculate a 90% confidence interval for p.

   b.   Calculate a 95% confidence interval for p.

   c.   Calculate a 99% confidence interval for p.

EXERCISES   Applications

 8.46   With the decreasing number of deductions allowed for individuals under the new tax laws, fund-raisers for philanthropic organizations are concerned about the effect the new tax law will have on the generosity of Americans with regard to charitable gifts. A survey of 500 households having adults in the upper-income category ($40,000 or more per year) revealed that 42% contributed more to charity in 1986 than they did in 1985 (*Journal of Accountancy*, January 1988, p. 96). Find a 99% confidence interval for p, the proportion of all upper-income Americans who contributed more to charity in 1986 than they did in 1985. What assumptions must you make in order for this confidence interval to be valid?

 8.47   The National Labor Relations Board requires that a certain percentage of workers favor union affiliation before workers vote on which union they wish to have representing their concerns. Sixty out of 200 workers polled in a nonunion shop favored union affiliation. Estimate the proportion of workers favoring union affiliation, and place a bound on the error of estimation.

 8.48   According to a survey based on over 3,000 responses, small-business owners are definitely not in favor of increased government spending, even for the national defense. In particular, 2,012 of 2,990 small-business owners said that increased defense spending would have a negative rather than a positive impact on the country (*Journal of Accountancy*, January 1988, p. 93). If this sample represents a random sampling of all small-business owners, find a 95% confidence interval for p, the proportion of all small-business owners who feel that increased defense spending will have a negative impact on the country.

 8.49   Although many television commercials appear to be more entertaining than informative, their ultimate test of effectiveness is positive consumer response and increased sales. To examine the effectiveness of a proposed television commercial for a new brand of laundry detergent, an advertising firm selects 100 homemakers and separately shows the commercial to each on closed-circuit television. Of those involved in the study, 63 indicate willingness to purchase the product after viewing the commercial. Using the sample data, construct a 95% confidence interval for the proportion of all homemakers in the TV-viewing audience who will purchase the new laundry detergent as a result of the TV commercial. What assumptions must be made in order for the confidence interval to be valid?

 8.50   Generally, feasibility studies include some measure of demand to ascertain the potential profitability of the given product or service. In studying the feasibility of expanding public television programming, an investigator found that 76 of 180 randomly chosen households with television sets watch at least 2 hours of public television programming per week. Find a 90% confidence interval for p, the proportion of households in the population that watch at least 2 hours of public television programming per week.

8.51   In late 1975 many U.S. newspapers contained the following article.

NEARLY 30% CAN'T PINPOINT IMPORTANCE OF 1776

With the nation about to celebrate its 200th anniversary, nearly 3 Americans in 10 are unable to say what important event occurred in the year 1776. In a random sample of 550 Americans, only 396 respondents were able to correctly identify 1776 as the year our country declared its independence from Great Britain by the signing of the Declaration of Independence.

Source: Oregonian, November 30, 1975.

Using these findings, estimate the proportion of all Americans in 1975 who could identify the importance of 1776, and place a bound on the error of estimation.

8.11

Estimating the Difference Between
Two Binomial Parameters

The fourth and final estimation problem considered in this chapter is the estimation of the difference between the parameters of two binomial populations. Assume that the two populations, 1 and 2, possess parameters p_1 and p_2, respectively. Independent random samples consisting of n_1 and n_2 trials are drawn from the population and the estimates $\hat{p}_1$ and $\hat{p}_2$ are calculated.

From Chapter 7, Section 7.7, the difference in sample proportions $(\hat{p}_1 - \hat{p}_2)$, like the difference between two sample means, is approximately normally distributed when both n_1 and n_2 are large. Furthermore, we state (without proof) that the estimator $(\hat{p}_1 - \hat{p}_2)$ is an unbiased estimator of the difference $(p_1 - p_2)$ in the population proportions, and that it possesses a standard error of

$$\sigma_{(\hat{p}_1 - \hat{p}_2)} = \sqrt{\frac{p_1 q_1}{n_1} + \frac{p_2 q_2}{n_2}}$$

where p_1 and p_2 are the two population proportions, $q_1 = (1 - p_1)$, $q_2 = (1 - p_2)$, and n_1 and n_2 are the sizes of the samples selected from the two populations.

Using the procedure of Section 8.8, we define the point estimator for $(p_1 - p_2)$ as given in the following display.

Point Estimator of $(p_1 - p_2)$

Estimator: $(\hat{p}_1 - \hat{p}_2)$.

Bound on error: $1.96\sigma_{(\hat{p}_1 - \hat{p}_2)} = 1.96\sqrt{\dfrac{p_1 q_1}{n_1} + \dfrac{p_2 q_2}{n_2}}$.

Note: The estimates $\hat{p}_1$ and $\hat{p}_2$ must be substituted for p_1 and p_2 to estimate the bound on the error of estimation.

The $(1 - \alpha)100\%$ confidence interval, appropriate when n_1 and n_2 are large, is shown in the next display.

A $(1 - \alpha)100\%$ Confidence Interval for $(p_1 - p_2)$

$$(\hat{p}_1 - \hat{p}_2) \pm z_{\alpha/2}\sqrt{\frac{\hat{p}_1 \hat{q}_1}{n_1} + \frac{\hat{p}_2 \hat{q}_2}{n_2}}$$

EXAMPLE 8.7   A manufacturer of fly sprays wishes to compare two new formulations, 1 and 2. Two rooms of equal size, each containing 1,000 flies, are employed in the experiment, one treated with fly spray 1 and the other treated with an equal amount of fly spray 2.

Totals of 825 and 760 flies succumb to sprays 1 and 2, respectively. Estimate the difference in the rate of kill for the two sprays when used in the test environment.

SOLUTION   The point estimate of $(p_1 - p_2)$ is

$$(\hat{p}_1 - \hat{p}_2) = .825 - .760 = .065$$

The bound on the error of estimation is estimated to be

$$1.96 \sqrt{\frac{\hat{p}_1 \hat{q}_1}{n_1} + \frac{\hat{p}_2 \hat{q}_2}{n_2}} = 1.96 \sqrt{\frac{(.825)(.175)}{1,000} + \frac{(.76)(.24)}{1,000}} = .035$$

The corresponding confidence interval, using confidence coefficient .95, is

$$(\hat{p}_1 - \hat{p}_2) \pm 1.96 \sqrt{\frac{\hat{p}_1 \hat{q}_1}{n_1} + \frac{\hat{p}_2 \hat{q}_2}{n_2}}$$

The resulting confidence interval is

$$.065 \pm .035 \qquad \text{or} \qquad .030 \text{ to } .100$$

Hence, we estimate that the difference $(p_1 - p_2)$ between the rates of kill falls in the interval from .030 to .100. That is, we estimate that p_1 exceeds p_2 by as little as .030 or as much as .100. We are fairly confident of this estimate, because we know that if our sampling procedure were repeated over and over again, each time generating an interval estimate, approximately 95% of the estimates would enclose the quantity $(p_1 - p_2)$.

◆◆

EXERCISES Conceptual Level

8.52 When is it appropriate to assume that the sampling distribution of the difference between two sample proportions is approximately normally distributed?

8.53 When one samples from two binomial populations, what is assumed about the sampling procedures used?

EXERCISES Skill Level

8.54 Independent random samples of size $n_1 = n_2 = 1,000$ were drawn from two binomial populations with parameters p_1 and p_2, respectively. The samples produced $x_1 = 280$ and $x_2 = 250$ successes. Estimate the difference $(p_1 - p_2)$, and place a bound on the error of estimation.

8.55 Refer to Exercise 8.54.
  a. Find a 90% confidence interval for the difference $(p_1 - p_2)$. Interpret this interval.
  b. Find a 99% confidence interval for the difference $(p_1 - p_2)$. Interpret this interval.

8.56   Independent random samples of size $n_1 = 450$ and $n_2 = 500$ were drawn from two binomial populations with parameters p_1 and p_2, respectively. The samples produced $x_1 = 355$ and $x_2 = 325$ successes. Estimate the difference $(p_1 - p_2)$.

   a. Use a 90% confidence interval.   **b.** Use a 95% confidence interval.

   c. Use a 98% confidence interval.

EXERCISES   Applications

8.57   Participative decision making, sometimes called *codetermination*, is a managerial strategy involving worker participation in management decisions and is intended as a means of improving both the performance and participation of individuals in organizations. This strategy is relatively new in America, but many European firms have considerable experience with codetermination. Two groups of employees that differed substantially in the opportunity for employee participation in decision making were asked whether or not they were satisfied with their present jobs. Seventy-seven of 110 employees from a group where employee participation was encouraged indicated they were satisfied with their jobs, whereas 52 of 125 employees from a group where employee participation was discouraged indicated they were satisfied with their jobs.

   a. Estimate the difference in the fraction of employees satisfied with their jobs, and place a bound on the error of estimation.

   b. Find a 90% confidence interval for the difference in the proportion of employees who are satisfied with their jobs.

8.58   To demonstrate its effectiveness as a medium for advertising consumer goods, the *New York Post* conducted a survey among 118 *Post* subscribers and 253 subscribers to the *Daily News*. The survey showed that 13.6% of the *Post* subscribers shop at least once a week in one of the major retailing centers of New York City versus 11.4% for subscribers of the *Daily News* (*New York Post*, June 26, 1980). Estimate the difference in the proportion of subscribers who regularly patronize a New York City retail center, and place a bound on the error of estimation.

8.59   In the competitive area of retail consumer goods, advertising serves to clarify product distinctiveness and increase market penetration. Before adopting a new advertising campaign, a large-volume vintner conducted a product preference survey among 1,000 regular buyers of wine in the supermarket chain that serves as his primary channel of distribution. From that survey he found that 33% of those contacted were regular purchasers of his wine. Six months after the institution of a revised advertising campaign, 1,200 buyers were surveyed, with 44% indicating preference for the vintner's product. Find a 95% confidence interval for the gain in market penetration (percentage adopting the vintner's product) as a result of the revised advertising campaign.

8.60   In October 1986, 81% of families had favorable attitudes about buying a home. A year later, during the first part of October 1987, among the $n_1 = 347$ households interviewed, only 72% had favorable attitudes about buying a house. This figure fell further to 65% of the $n_2 = 153$ households interviewed after the October 1987 market crash (John Bussey, "Survey Finds Sharp Drop in Confidence of Consumers After Stock Market Crash," *Wall Street Journal*, November 9, 1987, p. 12). If the "before" and "after" samples have been randomly and independently drawn, construct a 95% confidence interval for the difference in the "before" and "after" proportions interested in buying a house. Does this interval suggest that the proportion interested in buying a house has dropped since the October crash? Explain.

8.12

Choosing the Sample Size

The design of an experiment is essentially a plan for purchasing a quantity of information. This information, like any other commodity, may be acquired at varying prices, depending on the manner in which the data are obtained. Some measurements contain a large amount of information about the parameter of interest; others may contain little or none. Since the sole product of research is information, we should try to purchase it at minimum cost.

The *sampling procedure*—or *experimental design*, as it is usually called—affects the quantity of information per measurement. This procedure, along with the sample size *n*, controls the total amount of relevant information in a sample. With few exceptions, we will be concerned with the simplest sampling situation—random sampling from a relatively large population—and will devote our attention to the selection of the sample size *n*.

The researcher makes little progress in planning an experiment before encountering the problem of selecting the sample size. Indeed, perhaps one of the most frequent questions asked of the statistician is, How many measurements should be included in the sample? Unfortunately, the statistician cannot answer this question without knowing how much information the experimenter wishes to buy. Certainly, the total amount of information in the sample will affect the measure of goodness of the method of inference and must be specified by the experimenter. Referring specifically to estimation, we would like to know how accurate the experimenter wishes his estimate to be. This accuracy may be stated by specifying a bound on the error of estimation.

For instance, suppose that we wish to estimate the average daily yield μ of a chemical (see Example 8.1), and we wish the error of estimation to be less than 4 tons with a probability of .95. Since approximately 95% of the sample means will lie within $1.96\sigma_{\bar{x}}$ of μ in repeated sampling, we are asking that $1.96\sigma_{\bar{x}}$ equal 4 tons (see Figure 8.9).

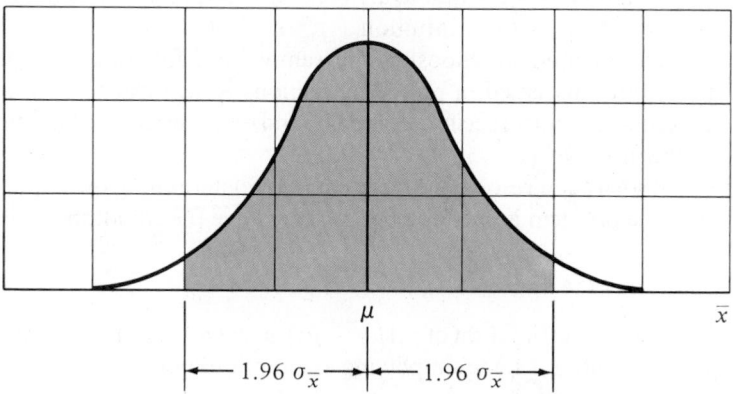

Figure 8.9   Approximate Distribution of $\bar{x}$ for Large Samples

Then

$$1.96\sigma_{\bar{x}} = 4 \quad \text{or} \quad 1.96\frac{\sigma}{\sqrt{n}} = 4$$

Solving for n, we obtain

$$\sqrt{n} = \frac{1.96\sigma}{4} \quad \text{or} \quad n = .24\sigma^2$$

This formula gives the minimum sample size such that the error of estimation will be less than $1.96\sigma_{\bar{x}}$ with probability approximately equal to .95.

You will quickly note that we cannot obtain a numerical value for n unless the population standard deviation σ is known. And certainly, this is exactly what we should expect, because the variability of the sample mean $\bar{x}$ depends on the variability of the population from which the sample was drawn. Lacking an exact value for σ, we would use the best approximation available, such as an estimate s obtained from a previous sample or knowledge of the range in which the measurements fall. Since the range is approximately equal to 4σ (the Empirical Rule), one-fourth of the range provides an approximate value for σ. For our example we would use the results of Example 8.1, which provided a reasonably accurate estimate of σ equal to $s = 21$. Then

$$n = .24\sigma^2 = (.24)(21)^2 \approx 106$$

Using a sample size $n = 106$, we would be reasonably certain (with probability approximately equal to .95) that our estimate lies within $1.96\sigma_{\bar{x}} = 4$ tons of the true average daily yield.

Actually, we should expect the error of estimation to be much less than 4 tons. According to the Empirical Rule, the probability is approximately equal to .68 that the error of estimation is less than $\sigma_{\bar{x}} = 2$ tons. You will note that the probabilities .95 and .68 used in these statements are inexact, since s was substituted for σ. Although this method of choosing the sample size is only approximate for a specified desired accuracy of estimation, it is the best available and is certainly better than selecting the sample size on the basis of our intuition.

The method of choosing the sample size for all the large-sample estimation procedures discussed in preceding sections is identical to that described above. The experimenter must specify a desired bound on the error of estimation and an associated confidence level $(1 - \alpha)$.

Rather than resolve the basic equation determining the required sample size every time the problem is encountered, we can solve the equation

$$z_{\alpha/2}\sigma_{\hat{\theta}} = B$$

in general for estimation of $\mu, (\mu_1 - \mu_2), p$, or $(p_1 - p_2)$. A summary of these solutions with and without some simplifying assumptions (for example, $n_1 = n_2 = n$) is given in the following display.

Sample Size Formulas

Let B represent the bound on the error of estimation and $(1 - \alpha)$ the confidence coefficient.

1. Estimation of μ:

$$n = \frac{z_{\alpha/2}^2 \sigma^2}{B^2}$$

2. Estimation of $(\mu_1 - \mu_2)$ when $n_1 = n_2 = n$:

$$n = \frac{z_{\alpha/2}^2 (\sigma_1^2 + \sigma_2^2)}{B^2}$$

3. Estimation of p:

a. $n = \dfrac{z_{\alpha/2}^2 \hat{p}\hat{q}}{B^2}$

if a prior estimate $\hat{p}$ of p is available.

b. $n = \dfrac{z_{\alpha/2}^2 (.25)}{B^2}$

if no prior estimate of p is available.

4. Estimation of $(p_1 - p_2)$ if no prior estimates are available and assuming maximum variation with $n_1 = n_2 = n$:

$$n = \frac{z_{\alpha/2}^2 (2)(.25)}{B^2}$$

EXAMPLE 8.8   One of the more baffling problems of organizations that depend on consumer information to direct organizational policy is the failure of selected consumers to respond to a survey questionnaire. N. L. Chervany and J. S. Heinlen ("The Structure of a Student Project Course," *Decision Sciences*, January 1975) reported their experience with pilot surveys to estimate the response rate for different versions of a questionnaire. Suppose an organizational executive would like to estimate the response rate for a newly designed survey questionnaire by observing the response rate in a pilot survey. How many consumers should be selected for the pilot survey if the executive wishes to estimate p, the response rate, with a probability of .90 that the error of estimation is less than .06?

SOLUTION   Since the confidence coefficient is $(1 - \alpha) = .90$, α must equal .10, and $\alpha/2 = .05$. The z value corresponding to an area of .05 in the upper tail of the z distribution is

$z_{\alpha/2} = 1.645$. We then require $z_{\alpha/2}\sigma_{\hat{p}} = B$. Substituting into this equation, we have

$$1.645\sigma_{\hat{p}} = .06 \qquad \text{or} \qquad 1.645\sqrt{\frac{pq}{n}} = .06$$

Since the variability of $\hat{p}$ is dependent on p, which is unknown, we can use the value of $p = .5$, for which the value of pq is a *maximum*. Then

$$1.645\sqrt{\frac{(.5)(.5)}{n}} = .06$$

Solving for n, we obtain

$$\sqrt{n} = \frac{1.645}{.06}\sqrt{(.5)(.5)} = 13.71 \qquad n = 187.9$$

Hence, the pilot survey should include 188 consumers. If we knew p to differ from .5, then using that value would result in a somewhat smaller sample size required to achieve the same bound on the error of estimation, equal to .06.

◆◆

EXAMPLE 8.9 An experimenter wishes to compare the effectiveness of two methods of training industrial employees to perform a certain assembly operation. Selected employees are to be divided into two equal groups, the first receiving training method 1 and the second, training method 2. Each will perform the assembly operation, and the length of assembly time will be recorded. The experimenter expects that the measurements for both groups will have a range of approximately 8 minutes. If the estimate of the difference in mean time to assemble is desired correct to within 1 minute, with probability approximately equal to .95, how many workers must be included in each training group?

SOLUTION Equating $1.96\sigma_{(\bar{x}_1 - \bar{x}_2)}$ to $B = 1$ minute, we obtain

$$1.96\sqrt{\frac{\sigma_1^2}{n_1} + \frac{\sigma_2^2}{n_2}} = 1$$

In order that this equation have a unique solution, we let $n_1 = n_2 = n$ and obtain the equation

$$1.96\sqrt{\frac{\sigma_1^2}{n} + \frac{\sigma_2^2}{n}} = 1$$

As noted above, the variability of each method of assembly is approximately the same, and hence, $\sigma_1^2 = \sigma_2^2 = \sigma^2$. Since the range, equal to 8 minutes, is approximately equal to 4σ, then we have

$$4\sigma \approx 8 \qquad \text{or} \qquad \sigma \approx 2$$

Substituting this value for σ_1 and σ_2 in the equation above, we obtain

$$1.96\sqrt{\frac{(2)^2}{n} + \frac{(2)^2}{n}} = 1 \qquad \sqrt{n} = 1.96\sqrt{8}$$

or $n = 31$. Thus, each group should contain $n = 31$ members.
Using formula 2 given in the display, the same result can be obtained as

$$n = \frac{(1.96)^2(4 + 4)}{1} = 31$$

EXERCISES   Conceptual Level

8.61   Why is the choice of a sample size important? Why not simply choose large samples in all cases?

8.62   What decisions must be made about an estimation process before the appropriate sample size can be determined?

8.63   Suppose you are interested in determining an appropriate sample size to estimate the binomial proportion p, but no information exists regarding p. Explain how you would estimate the variance pq to use in the sample size determination formula. (*Hint:* See Table 8.3.)

EXERCISES   Skill Level

8.64   Find the sample size necessary to estimate μ to within a bound B on the error of estimation with confidence coefficient $(1 - \alpha)$. Assume that $\sigma = 4$.

  a. $B = .5, (1 - \alpha) = .95$ **b.** $B = 1, (1 - \alpha) = .95$ **c.** $B = 1, (1 - \alpha) = .99$

8.65   Find the sample size necessary to estimate the binomial parameter p to within $B = .1$ with confidence coefficient .95.

  a.   Assume that there is no prior knowledge of the approximate value of p.

  b.   Assume that it is known that $.2 < p < .4$.

8.66   Find the sample size necessary to estimate $(\mu_1 - \mu_2)$ to within $B = 200$ with confidence coefficient .99. Assume that $n_1 = n_2 = n$. Although σ_1 and σ_2 are unknown, the range of both populations is approximately 1500.

8.67   Find the sample size necessary to estimate the difference $(p_1 - p_2)$ to within $B = .05$ with 95% confidence. Assume that $n_1 = n_2 = n$ and that no prior estimates of p_1 and p_2 are available.

EXERCISES   Applications

8.68   In 1988 great interest in the state of California was focused on the controversial slow-growth initiative. The initiative, opposed by land developers among others, is favored by many Californians as a proposal that will stem rapid growth, reduce smog and traffic congestion, and have a positive effect on the quality of life and the economy of California. A poll involving a random sample of 600 Orange County, California, adults provided an estimate of p, the proportion of registered voters favoring the slow-growth initiative ("Support of Slow-Growth Plan Soars," *Los Angeles Times*, February 7, 1988, p. 28). The poll claimed to have a margin of error of 4% either way.

a. Verify that the margin of error claimed by this polling organization is correct.

b. If the pollsters wished to decrease the margin of error to 3% either way, how many registered voters should be sampled?

8.69 Are subcompacts noticeably less expensive to operate than compact cars? To explore this issue, a private testing agency wishes to estimate the difference in the average operating cost per mile between subcompacts and compacts, accurate to within 1 cent per mile, with a probability of approximately .95. The agency knows from past experience that the standard deviation of total operating costs per mile (including depreciation, maintenance, gas and oil, insurance, and taxes) is about 3 cents. How many cars should the testing agency include in each group? (Assume the same number of cars will be used in each group.)

8.70 The maintenance of charge accounts may become too costly if the average account purchase falls below a certain level. A department store manager would like to estimate the average amount of account purchasers per month by its customers with charge accounts to within $2.50, with a probability of approximately .95. How many accounts should be selected from the store's records if the standard deviation of monthly account balances is known to be $5.50?

8.71 Refer to Exercise 8.70. If no knowledge exists regarding the standard deviation (or variance) of credit account purchases, how can the store manager estimate σ to allow for the determination of an appropriate number of accounts for review?

8.72 To bid competitively for the lumbering rights on a certain tract of land, a wood products firm needs to know the mean diameter of the trees on the tract to within 2 in., with a probability of .90. How large a sample of trees should be selected from the tract to estimate the mean diameter if it is known that the diameters of usable timber on the tract range from 10 to 38 in.?

8.73 A typical assumption of accountants involved in commercial auditing practice is that 5 out of every 100 corporate accounts contain some type of procedural or recording error. In the audit of the records of a manufacturing company, how many accounts should be reviewed to estimate p, the proportion of accounts in error, to within .04?

PART II
LARGE-SAMPLE HYPOTHESIS TESTING

8.13

A Statistical Test of an Hypothesis

The basic reasoning employed in a statistical test of an hypothesis was outlined in Section 5.7 in connection with the test of the desirability of a new sugar substitute. In this section we will attempt to emphasize the basic points involved. We refer you to Section 5.7 for an intuitive presentation of the subject.

Parts of a Statistical Test

> The objective of a statistical test is to test an hypothesis concerning the values of one or more population parameters. A statistical test involves four elements:
>
> 1. Null hypothesis
> 2. Alternative hypothesis
> 3. Test statistic
> 4. Rejection region

The specification of these four elements defines a particular test; changing one or more of the parts creates a new test.

alternative (or research) hypothesis
null hypothesis

The **alternative** (or **research**) **hypothesis** is the hypothesis that the researcher wishes to support. The **null hypothesis** is a contradiction of the alternative hypothesis; that is, if the null hypothesis is false, the research hypothesis must be true. For reasons you will subsequently see, it is easier to show support for the research hypothesis by presenting evidence (sample data) that indicates that the null hypothesis is false. Thus, we are building a case in support of the research hypothesis by using a method that is analogous to proof by contradiction.

Even though we wish to gain evidence in support of the alternative hypothesis (denoted by the symbol H_a), the null hypothesis (indicated by the symbol H_0) is the hypothesis to be tested. Thus, H_0 will specify hypothesized values for one or more population parameters. For example, we might wish to test the null hypothesis that a population mean is equal to 50, hoping to show, in fact, that the mean exceeds 50. Or we might wish to test the null hypothesis that two population means, say μ_1 and μ_2, are equal, hoping to show, perhaps, that μ_1 is larger than μ_2.

test statistic

The decision to reject or accept the null hypothesis is based on information contained in a sample drawn from the population of interest. The sample values are used to compute a single number, corresponding to a point on a line, which operates as a decision maker. This decision maker is called the **test statistic**. The entire set of values that the test statistic may assume is divided into two sets, or regions, one corresponding to the **rejection region** and the other to the **acceptance region**. If the test statistic computed from a particular sample assumes a value in the rejection region, the null hypothesis is rejected and the alternative hypothesis H_a (the research hypothesis) is accepted. If the test statistic falls in the acceptance region, either the null hypothesis is accepted or the test is judged to be inconclusive. The circumstances leading to this latter decision are explained subsequently.

rejection region
acceptance region

The decision procedure described above is subject to two types of errors, which are prevalent in a two-choice decision problem.

Definition

type I error

> A **type I error** for a statistical test is the error made by rejecting the null hypothesis when it is true. The probability of making a type I error is denoted by the symbol α.

type II error

A **type II error** for a statistical test is the error made by accepting (not rejecting) the null hypothesis when it is false and some alternative hypothesis is true. The probability of making a type II error is denoted by the symbol β.

The two possibilities for the null hypothesis—that is, true or false—along with the two decisions the experimenter can make, are indicated in the two-way table, Table 8.4. The occurrences of the type I and type II errors are indicated in the appropriate cells.

The goodness of a statistical test of an hypothesis is measured by the probabilities of making a type I or a type II error, denoted by the symbols α and β, respectively. These probabilities, calculated for the elementary statistical tests presented in the exercises for Chapter 5, illustrate the basic relationship among α, β, and the sample size n. Since α is the probability that the test statistic falls in the rejection region, assuming H_0 to be true, **an increase in the size of the rejection region increases** α and, at the same time, decreases β for a fixed sample size. Reducing the size of the rejection region decreases α and increases β. If the sample size n is increased, more information is available upon which to base the decision, and for fixed α, β will decrease.

The probability β of making a type II error varies depending upon the true value of the population parameter. For instance, suppose that we wish to test the null hypothesis that the binomial parameter p is equal to $p_0 = .4$. (We will use a subscript 0 to indicate the parameter value specified in the null hypothesis H_0.) Furthermore, suppose that H_0 is false and that p is really equal to an alternative value, say p_a. Which will be more easily detected, a $p_a = .4001$ or a $p_a = 1.0$? Certainly, if p is really equal to 1.0, every single trial will result in a success and the sample results will produce strong evidence to support a rejection of $H_0 : p_0 = .4$. On the other hand, $p_a = .4001$ lies so close to $p_0 = .4$ that it would be extremely difficult to detect p_a without a very large sample. In other words, the probability β of accepting H_0 will vary depending on the difference between the true value of p and the hypothesized value p_0. Ideally, the further p_a lies from p_0, the higher the probability is of rejecting H_0. This probability is measured by $(1 - \beta)$, which is called the *power* of the test.

Definition

power of a
statistical test

The **power of a statistical test**, given as

$$1 - \beta = P(\text{reject } H_0 \text{ when } H_0 \text{ is false})$$

measures the ability of the test to perform as required.

Table 8.4   Decision Table

Decision	Null Hypothesis	
	True	False
Reject H_0	Type I error	Correct decision
Accept H_0	Correct decision	Type II error

power curve   A graph of $(1 - \beta)$, the probability of rejecting H_0 when, in fact, H_0 is false, as a function of the true value of the parameter of interest is called the **power curve** for the statistical test. For constant values of n and α the power of a test should increase as the distance between the true and hypothesized values of the parameter increases. An increase in the sample size n will increase the power, $1 - \beta$, for all alternative values of the parameter being tested. Thus, we can create a power curve corresponding to each sample size.

The experimenter should be able to specify values of α and β, measuring the risks of the respective errors he is willing to tolerate, as well as some deviation from the hypothesized value of the parameter he considers of practical importance and wishes to detect. The rejection region for the test will be located in accordance with the specified value of α; the sample size will be chosen large enough to achieve an acceptable value of β for the specified deviation the experimenter wishes to detect. This choice could be made by consulting the power curves, corresponding to various sample sizes, for the chosen test.

In practice, β is often unknown, either because it was never computed before the test was conducted or because it may be extremely difficult to compute for the test. Then rather than accept the null hypothesis when the test statistic falls in the acceptance region, you should withhold judgment. That is, you should not accept the null hypothesis unless you know the risk (measured by β) of making an incorrect decision. Notice that you will never be faced with this "no conclusion" situation when the test statistic falls in the rejection region. Then you can reject the null hypothesis (and accept the alternative hypothesis) because you always know the value of α, the probability of rejecting the null hypothesis when it is true. The fact that β is often unknown explains why we attempt to support the alternative hypothesis by rejecting the null hypothesis. When we reach this decision, the probability α that such a decision is incorrect is known.

8.14

A Large-Sample Statistical Test

Large-sample tests of hypotheses concerning the population parameters μ, p, $(\mu_1 - \mu_2)$, and $(p_1 - p_2)$ are each based on a normally distributed test statistic and for that reason may be regarded as one and the same test procedure. We will present the reasoning involved in the test in a very general manner, referring to the parameter of interest as θ. Thus, we could imagine θ as representing μ, $(\mu_1 - \mu_2)$, p, or $(p_1 - p_2)$. The specific test for each parameter will be illustrated by examples.

Suppose that we wish to test an hypothesis concerning a parameter θ and that an unbiased point estimator $\hat{\theta}$ is available and known to be normally distributed, with standard deviation $\sigma_{\hat{\theta}}$. If the null hypothesis

$$H_0: \theta = \theta_0$$

is true, then the sampling distribution for $\hat{\theta}$ is normally distributed, with a mean equal to θ_0, as shown in Figure 8.10.

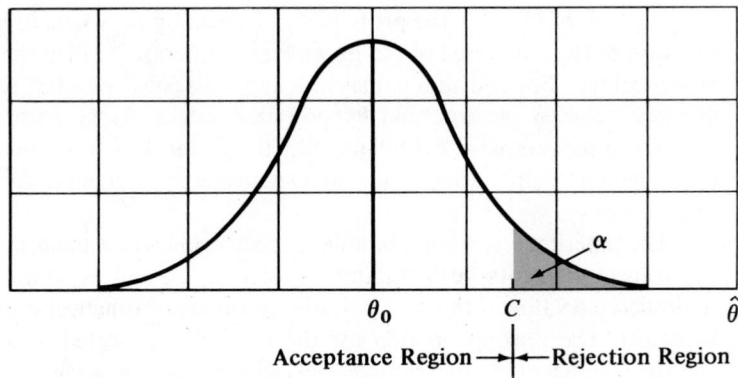

Figure 8.10   Distribution of $\hat{\theta}$ When H_0 Is True

Suppose that the objective of our experiment is to gain evidence to show that θ is greater than θ_0. The alternative hypothesis is $H_a: \theta > \theta_0$, and we would reject the null hypothesis H_0 when $\hat{\theta}$ is too large. "Too large," of course, means too many standard deviations, $\sigma_{\hat{\theta}}$, away from θ_0. The rejection region for the test is shown in Figure 8.10. The value of $\hat{\theta}$ that separates the rejection and acceptance regions, denoted by the

critical value   symbol C, is called the **critical value** of the test statistic. The probability of rejecting the null hypothesis when the null hypothesis is true is equal to the area under the normal curve lying above the rejection region. This area α is shaded in Figure 8.10. Thus, if we desire $\alpha = .05$, we would reject the null hypothesis when $\hat{\theta}$ is more than $1.645\ \sigma_{\hat{\theta}}$ to the right of θ_0.

Definition

one-tailed
statistical test   A **one-tailed statistical test** is one that locates the rejection region in only one tail of the sampling distribution of the test statistic. To detect $\theta > \theta_0$, place the rejection region in the upper tail of the distribution of $\hat{\theta}$. To detect $\theta < \theta_0$, place the rejection region in the lower tail of the distribution of $\hat{\theta}$.

If we wish to detect departures either greater than or less than θ_0, the alternative hypothesis is

$$H_a: \theta \neq \theta_0$$

that is,

$$\theta > \theta_0 \quad \text{or} \quad \theta < \theta_0$$

In this case α, the probability of a type I error, is equally divided between the two tails of the normal distribution, resulting in a *two-tailed statistical test*.

Definition

A **two-tailed statistical test** is one that locates the rejection region in both tails of the sampling distribution of the test statistic. Two-tailed tests are used to detect either $\theta > \theta_0$ or $\theta < \theta_0$.

All the point estimators discussed in the preceding section satisfy the requirements of the test described above when the sample size n is large. That is, the sample size must be large enough so that the point estimator is approximately normally distributed, according to the Central Limit Theorem, and permits a reasonably good estimate of its standard deviation. We may therefore test hypotheses concerning $\mu, p, (\mu_1 - \mu_2)$, and $(p_1 - p_2)$.

The mechanics of testing are simplified by using

$$z = \frac{\hat{\theta} - \theta_0}{\sigma_{\hat{\theta}}}$$

as a test statistic, as discussed in Example 6.11. Remember that z is simply the deviation of a normally distributed random variable $\hat{\theta}$ from θ_0 expressed in units of $\sigma_{\hat{\theta}}$. Thus, for a two-tailed test with $\alpha = .05$, we would reject H_0 when $z > 1.96$ or $z < -1.96$, since $P(z < -1.96$ or $z > 1.96) = .05$ when H_0 is true.

The next display outlines the large-sample statistical tests for the population parameters $\mu, p, (\mu_1 - \mu_2)$, and $(p_1 - p_2)$.

Large-Sample Statistical Test

Note: In these tests θ may equal $\mu, p, (\mu_1 - \mu_2)$, or $(p_1 - p_2)$.

Null hypothesis $H_0: \theta = \theta_0$.
Alternative hypothesis: For a one-tailed test

$$H_a: \theta > \theta_0 \qquad \text{or} \qquad H_a: \theta < \theta_0$$

For a two-tailed test

$$H_a: \theta \neq \theta_0$$

Test statistic: $z = \dfrac{\hat{\theta} - \theta_0}{\sigma_{\hat{\theta}}}$.

Rejection region: For a one-tailed test with $H_a: \theta > \theta_0$, reject H_0 if $z > z_\alpha$; with $H_a: \theta < \theta_0$, reject H_0 if $z < -z_\alpha$. For a two-tailed test, reject H_0 if $z > z_{\alpha/2}$ or if $z < -z_{\alpha/2}$.

The calculation of β for the one-tailed statistical test described previously can be facilitated by considering Figure 8.11. When H_0 is false and $\theta = \theta_a$, the test statistic

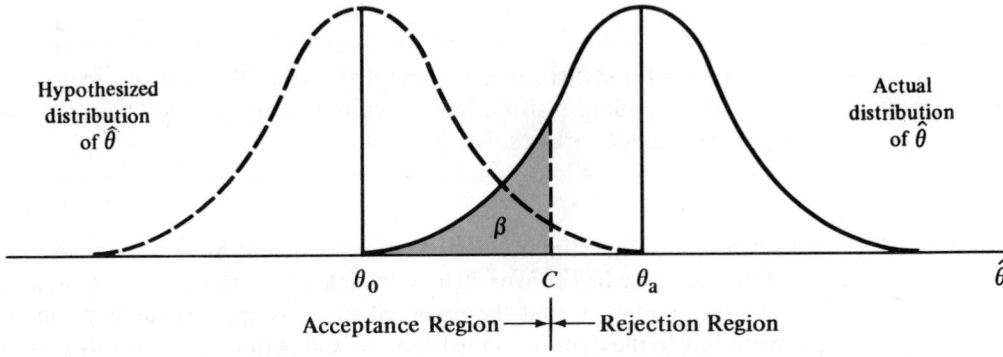

Figure 8.11   Distribution of $\hat{\theta}$ When H_0 Is False and $\theta = \theta_a$

will be normally distributed about a mean θ_a rather than θ_0. The distribution of $\hat{\theta}$, assuming $\theta = \theta_a$, is shown by the solid normal curve. The hypothesized distribution of $\hat{\theta}$ (that is, the distribution of $\hat{\theta}$ under the null hypothesis), shown by a dashed normal curve, is used to locate the rejection region and the critical value of $\hat{\theta}$, C. Since β is the probability of accepting H_0, given $\theta = \theta_a$, β is equal to the area under the solid curve located above the acceptance region. This area, which is shaded, could be calculated by using the methods described in Chapter 6.

The following example demonstrates the close relationship between the statistical test and the large-sample confidence intervals discussed in the preceding sections.

EXAMPLE 8.10   Refer to Example 8.1. Test the hypothesis that the average daily yield of the chemical is $\mu = 880$ tons per day against the alternative that μ is either greater or less than 880 tons per day. The sample (Example 8.1), based on $n = 50$ measurements, yielded $\bar{x} = 871$ and $s = 21$ tons.

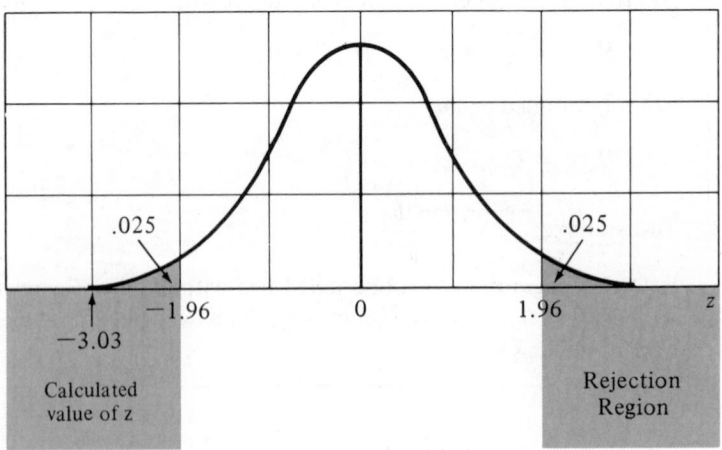

Figure 8.12   Rejection Region for Example 8.10

SOLUTION   The null and alternative hypotheses are

$$H_0 : \mu = 880 \qquad \text{and} \qquad H_a : \mu \neq 880$$

The point estimate for μ is $\bar{x}$. Therefore, the test statistic is

$$z = \frac{\bar{x} - \mu_0}{\sigma_{\bar{x}}} = \frac{\bar{x} - \mu_0}{\sigma / \sqrt{n}}$$

Using s to approximate σ, we obtain

$$z = \frac{871 - 880}{21/\sqrt{50}} = -3.03$$

For $\alpha = .05$ the rejection region consists of values of $z > 1.96$ and values of $z < -1.96$ (see Figure 8.12). Since the calculated value -3.03 of z falls in the rejection region, we reject the hypothesis that $\mu = 880$ tons and conclude that it is less. The probability of rejecting H_0 when H_0 is true is $\alpha = .05$. Hence, we are reasonably confident that our decision is correct.

◆ ◆

The statistical test based on a normally distributed test statistic, with a given α, and the $(1 - \alpha)100\%$ confidence interval of Section 8.7 are related. The interval $[\bar{x} \pm (1.96\sigma/\sqrt{n})]$, or approximately (871 ± 5.82) for Example 8.10, is constructed so that in repeated sampling $(1 - \alpha)$ of the intervals will enclose μ. Noting that $\mu = 880$ does not fall in the interval, we would be inclined to reject $\mu = 880$ as a likely value and conclude that the mean daily yield was, indeed, less.

There is another similarity between this test and the confidence interval of Section 8.7. The test is "approximate" because we substituted s, an approximate value, for σ. That is, the probability α of a type I error selected for the test is not .05. It is close to .05, close enough for all practical purposes, but it is not exact. This will be true for most statistical tests because one or more assumptions will not be satisfied exactly.

EXAMPLE 8.11   Refer to Example 8.10. Calculate the power of the test, $1 - \beta$, when μ is actually equal to 870 tons.

SOLUTION   The acceptance region for the test of Example 8.10 is located in the interval $(\mu_0 \pm 1.96\sigma_{\bar{x}})$. Substituting numerical values, we obtain

$$880 \pm 1.96 \frac{21}{\sqrt{50}} \qquad \text{or} \qquad 874.18 \text{ to } 885.82$$

The probability of accepting H_0, given $\mu = 870$, is equal to the area under the frequency distribution for the test statistic $\bar{x}$ in the interval from 874.18 to 885.82. Since $\bar{x}$ is normally distributed with a mean of 870 and $\sigma_{\bar{x}} \approx 21/\sqrt{50} = 2.97$, β is equal to the area under the normal curve located between 874.18 and 885.82 (see Figure 8.13).

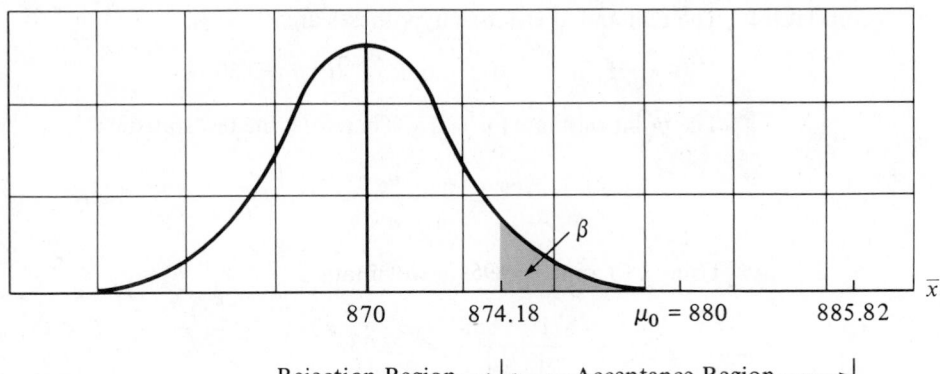

Figure 8.13   Calculating β in Example 8.11

Calculating the z values corresponding to 874.18 and 885.82, we obtain

$$z_1 = \frac{\bar{x} - \mu}{\sigma/\sqrt{n}} \approx \frac{874.18 - 870}{21/\sqrt{50}} = 1.41$$

$$z_2 = \frac{\bar{x} - \mu}{\sigma/\sqrt{n}} \approx \frac{885.82 - 870}{21/\sqrt{50}} = 5.33$$

Then

$$\beta = P(\text{accept } H_0 \text{ when } \mu = 870) = P(874.18 < \bar{x} < 885.82 \text{ when } \mu = 870)$$
$$= P(1.41 < z < 5.33)$$

You can see from Figure 8.13 that the area under the normal curve above $\bar{x} = 885.82$ (or $z = 5.33$) is negligible. Therefore,

$$\beta = P(z > 1.41)$$

From Table 3 in the Appendix we find that the area between $z = 0$ and $z = 1.41$ is .4207 and

$$\beta = .5 - .4207 = .0793$$

Hence, the power of the test is

$$1 - \beta = 1 - .0793 = .9207$$

The probability of correctly rejecting H_0, given that μ is really equal to 870, is .9207, or approximately 92 chances in 100.

Values of $(1 - \beta)$ can be calculated for various values of μ_a different from $\mu_0 = 880$ to measure the power of the test. For example, if $\mu_a = 885$,

$$\beta = P(874.18 < \bar{x} < 885.82 \text{ when } \mu = 885)$$
$$= P(-3.64 < z < .28) = .5 + .1103 = .6103$$

Table 8.5   Values of $(1 - \beta)$ for Various
Values of μ_a, Example 8.11

μ_a	$1 - \beta$	μ_a	$1 - \beta$
865	.9990	883	.1726
870	.9207	885	.3897
872	.7673	888	.7673
875	.3897	890	.9207
877	.1726	895	.9990
880	.0500		

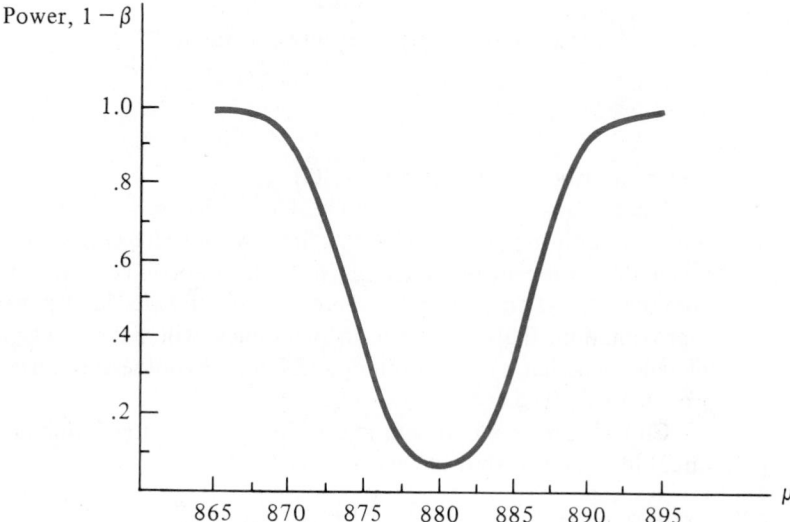

Figure 8.14   Power Curve for Example 8.11

and the power is $(1 - \beta) = .3897$. Table 8.5 shows the power of the test for various values of μ_a, and a power curve is graphed in Figure 8.14. Note that the power of the test increases as the distance between μ_a and μ_0 increases. The result is a U-shaped curve for this two-tailed test.

EXAMPLE 8.12   Approximately 1 in 10 smokers favors cigarette brand A. After a promotional campaign in a given sales region, 200 cigarette smokers were interviewed to determine the effectiveness of the campaign. The result of this sample survey showed that a total of 26 people expressed a preference for brand A. Do these data present sufficient evidence to indicate an increase in the acceptance of brand A in the region? (Note that for all practical purposes this problem is identical to the vegetable oil problem in Example 6.10.)

SOLUTION   We assume that the sample satisfies the requirements of a binomial experiment. The question posed may be answered by testing the hypothesis

$$H_0:p = .10 \text{ (the program is ineffective)}$$

against the alternative

$$H_a:p > .10 \text{ (the program is effective)}$$

A one-tailed statistical test will be utilized, because we are primarily concerned with detecting a value of p greater than .10.

The point estimator of p is $\hat{p} = x/n$, and the test statistic is

$$z = \frac{\hat{p} - p_0}{\sigma_{\hat{p}}} = \frac{\hat{p} - p_0}{\sqrt{p_0 q_0 / n}}$$

(*Note:* This test statistic is algebraically equivalent to

$$z = \frac{x - np_0}{\sqrt{np_0 q_0}}$$

the test statistic used in Example 6.10.)

Once again, we require a value of p so that $\sigma_{\hat{p}} = \sqrt{pq/n}$, which appears in the denominator of z, may be calculated. Since we have hypothesized that $p = p_0$, it seems reasonable to use p_0 in the standard deviation of $\hat{p}$. Note that this approach differs from the estimation procedure where, lacking knowledge of p, we chose $\hat{p}$ as the best approximation. This apparent inconsistency will have a negligible effect on the inference, whether it is the result of a test of an hypothesis or an estimation procedure, provided n is large.

Choosing $\alpha = .05$, we will reject H_0 when $z > 1.645$. Substituting the numerical values into the test statistic, we obtain

$$z = \frac{\hat{p} - p_0}{\sqrt{\dfrac{p_0 q_0}{n}}} = \frac{.13 - .10}{\sqrt{\dfrac{(.10)(.90)}{200}}} = 1.41$$

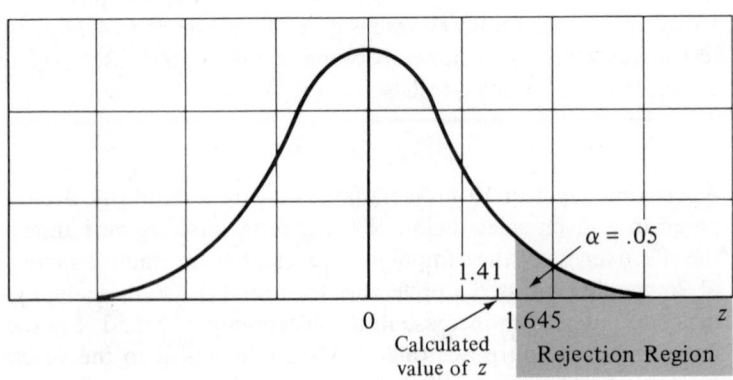

Figure 8.15   Rejection Region for Example 8.12

The calculated value $z = 1.41$ does not fall in the rejection region (see Figure 8.15), and hence, we do not reject H_0, the hypothesis that the proportion of smokers of cigarette brand A is .10. There is not sufficient evidence to indicate that the advertising campaign has been effective and that the percentage of smokers has increased as a result of the advertising effort.

Do we accept H_0? No, not until we have stated some alternative value of p that is larger than $p_0 = .10$ and is considered to be of practical significance. The probability β of a type II error should be calculated for this alternative. If β is sufficiently small, we would accept H_0 and would do so with the risk of an erroneous decision fully known.

◆ ◆

The calculation of β or the power of the test is not too difficult for the statistical test procedures outlined in this section but may be extremely difficult, if not beyond the capability of the beginner, in other test situations. A much simpler procedure is **to not reject H_0** rather than to accept it; then **estimate** p by using a confidence interval. The interval will give a range of possible values for p.

EXAMPLE 8.13   With hopes of reducing its massive 350-million-gallons-a-year-gasoline bill, the U.S. Postal Service has ordered a number of electric mail trucks on an experimental basis. Officials estimate that savings will result if they use electric trucks on flat routes and use a special off-peak-hours rate for recharging. In an experiment to compare operating costs, $n_1 = 100$ conventional gasoline mail trucks and $n_2 = 100$ electric mail trucks were placed in service for a period of time. The cost per mile for operating the gasoline trucks averaged $\bar{x}_1 = 6.70$ cents per mile with variance $s_1^2 = .36$; for the electric trucks the mean and variance of the cost per mile were $\bar{x}_2 = 6.54$ and $s_2^2 = .40$, respectively. Do these data present sufficient evidence to indicate a difference in the mean operating cost between conventional gasoline and electric-powered mail trucks?

SOLUTION   We wish to test the null hypothesis that the difference $(\mu_1 - \mu_2)$ between two population means equals some specified value—say D_0. (For our example we would hypothesize that $D_0 = 0$; that is, there is no difference in mean operating costs for the two types of trucks.)

Recall that $(\bar{x}_1 - \bar{x}_2)$ is an unbiased point estimator of $(\mu_1 - \mu_2)$, which will be approximately normally distributed in repeated sampling when n_1 and n_2 are large. Furthermore, the standard deviation of $(\bar{x}_1 - \bar{x}_2)$ is

$$\sigma_{(\bar{x}_1 - \bar{x}_2)} = \sqrt{\frac{\sigma_1^2}{n_1} + \frac{\sigma_2^2}{n_2}}$$

Then

$$z = \frac{(\bar{x}_1 - \bar{x}_2) - D_0}{\sqrt{\dfrac{\sigma_1^2}{n_1} + \dfrac{\sigma_2^2}{n_2}}}$$

will serve as a test statistic when σ_1^2 and σ_2^2 are known or when s_1^2 and s_2^2 provide good approximations for σ_1^2 and σ_2^2 (that is, when n_1 and n_2 are larger than 30).

For our example we have

$$H_0: \mu_1 - \mu_2 = D_0 = 0 \qquad \text{and} \qquad H_a: \mu_1 - \mu_2 \neq 0$$

Substituting the numerical values into the formula for the test statistic, we obtain

$$z = \frac{6.70 - 6.54}{\sqrt{\dfrac{.36}{100} + \dfrac{.40}{100}}} = 1.84$$

Using a two-tailed test with $\alpha = .05$, we will reject H_0 when $z > 1.96$ or $z < -1.96$. Since z does not fall in the rejection region, we do not reject the null hypothesis.

Note, however, that if we choose $\alpha = .10$, the rejection region will be $z > 1.645$ or $z < -1.645$, and the null hypothesis will be rejected.

The decision to reject or accept depends on the risk that we are willing to tolerate. If we choose $\alpha = .05$, the null hypothesis is not rejected; but we could not accept H_0 (that is, $\mu_1 = \mu_2$) without investigating the probability of a type II error.

On the other hand, if α were chosen equal to .10, the null hypothesis would be rejected. With no other information given, we would be inclined to reject the null hypothesis that there is no difference in the mean operating cost for conventional gasoline and electric-powered mail trucks. The chance of rejecting H_0, assuming H_0 true, is only $\alpha = .10$; and hence, we would be inclined to think that we had made a reasonably good decision.

◆◆

EXAMPLE 8.14 Hospital administrators are often responsible for gathering and computing certain medical statistics that are vital to doctors and hospital policymakers. The records of a particular hospital show that 52 men in a sample of 1,000 men versus 23 women in a sample of 1,000 women were admitted because of heart disease. Do these data present sufficient evidence to indicate a higher rate of heart disease among men admitted to the hospital?

SOLUTION We will assume that the number of patients admitted for heart disease follows approximately a binomial probability distribution for both men and women, with parameters p_1 and p_2, respectively. We wish to test the hypothesis that a difference exists between p_1 and p_2—say $(p_1 - p_2) = D_0$. (For our example we wish to test the hypothesis that $D_0 = 0$.) Recall that for large samples the point estimator of $(p_1 - p_2)$, namely $(\hat{p}_1 - \hat{p}_2)$, is approximately normally distributed in repeated sampling, with a mean of $(p_1 - p_2)$ and a standard deviation of

$$\sigma_{(\hat{p}_1 - \hat{p}_2)} = \sqrt{\frac{p_1 q_1}{n_1} + \frac{p_2 q_2}{n_2}}$$

Then

$$z = \frac{(\hat{p}_1 - \hat{p}_2) - (p_1 - p_2)}{\sigma_{(\hat{p}_1 - \hat{p}_2)}}$$

possesses a standardized normal distribution in repeated sampling. Hence, z can be employed as a test statistic to test

$$H_0 : p_1 - p_2 = D_0$$

when suitable approximations are used for p_1 and p_2, which appear in $\sigma_{(\hat{p}_1 - \hat{p}_2)}$. Approximations are available for two cases, as discussed below.

Case 1: If we hypothesize that p_1 equals p_2, that is,

$$H_0 : p_1 = p_2 \quad \text{or} \quad p_1 - p_2 = 0$$

then $p_1 = p_2 = p$, and the best estimate of p is obtained by pooling the data from both samples. Thus, if x_1 and x_2 are the numbers of successes obtained from the two samples, then consistent with H_0, the best estimate of p is

$$\hat{p} = \frac{x_1 + x_2}{n_1 + n_2}$$

The test statistic is

$$z = \frac{(\hat{p}_1 - \hat{p}_2) - 0}{\sqrt{\dfrac{\hat{p}\hat{q}}{n_1} + \dfrac{\hat{p}\hat{q}}{n_2}}} = \frac{\hat{p}_1 - \hat{p}_2}{\sqrt{\hat{p}\hat{q}\left(\dfrac{1}{n_1} + \dfrac{1}{n_2}\right)}}$$

Case 2: On the other hand, if we hypothesize that D_0 is not equal to zero, that is,

$$H_0 : p_1 - p_2 = D_0$$

where $D_0 \neq 0$, then the best estimates of p_1 and p_2 are $\hat{p}_1$ and $\hat{p}_2$, respectively. The test statistic is

$$z = \frac{(\hat{p}_1 - \hat{p}_2) - D_0}{\sqrt{\dfrac{\hat{p}_1 \hat{q}_1}{n_1} + \dfrac{\hat{p}_2 \hat{q}_2}{n_2}}}$$

For most practical problems involving the comparison of two binomial populations, the experimenter will wish to test the null hypothesis that $(p_1 - p_2) = D_0 = 0$. For our example we test

$$H_0 : p_1 - p_2 = 0$$

against the alternative

$$H_a : p_1 - p_2 > 0$$

Note that a one-tailed statistical test will be employed, because if a difference exists, we wish to detect $p_1 > p_2$. Therefore, we will reject the null hypothesis in favor of the alternative hypothesis if $(\hat{p}_1 - \hat{p}_2)$—or, equivalently, if the calculated value of z—is large. In fact, if we choose $\alpha = .05$, we will reject H_0 when $z > 1.645$.

The pooled estimate of p required for $\sigma_{(\hat{p}_1 - \hat{p}_2)}$ is

$$\hat{p} = \frac{x_1 + x_2}{n_1 + n_2} = \frac{52 + 23}{1,000 + 1,000} = .0375$$

The test statistic is

$$z = \frac{\hat{p}_1 - \hat{p}_2}{\sqrt{\hat{p}\hat{q}\left(\dfrac{1}{n_1} + \dfrac{1}{n_2}\right)}} = \frac{.052 - .023}{\sqrt{(.0375)(.9625)\left(\dfrac{1}{1,000} + \dfrac{1}{1,000}\right)}} = 3.41$$

Since the computed value of z falls in the rejection region, we reject the hypothesis that $p_1 = p_2$ and conclude that the data present sufficient evidence to indicate that the percentage of men entering the hospital because of heart disease is higher than that of women. Note that this conclusion does not imply that the incidence of heart disease is higher in men. Perhaps fewer women enter the hospital when afflicted with the disease.

◆◆

EXERCISES Conceptual Level

8.74 Clearly identify and define the four parts of a statistical problem.

8.75 Define and discuss the following concepts with regard to their role in hypothesis testing.

a. Type I and type II errors

b. The probabilities α and β

c. The level of significance

d. The power of the test

e. The test statistic

f. The acceptance region

g. The rejection region

8.76 Explain the role of the Central Limit Theorem in the statistical test of an hypothesis.

8.77 Distinguish between a one-tailed and a two-tailed test of an hypothesis.

8.78 If the sample size remains unchanged, what relationship exists between the probabilities of the two types of error?

8.79 What information is provided by the power curve?

8.80 If we were to compare a one-tailed test and a two-tailed test for a particular population parameter, what difference would we find in the form of the null and alternative hypotheses?

8.81 How large a sample is required in order to apply large-sample, statistical-testing procedures? What is the value of having a large sample?

EXERCISES Skill Level

8.82 In each of the following hypothesis-testing situations, suggest whether a one-tailed test or a two-tailed test of an hypothesis is appropriate. In each case, justify your answer, and then write the null and alternative hypotheses.

a. A power company executive conducts a study to determine whether the implementation of an energy conservation program has been effective in reducing average residential consumption.

b. A marketing research study is conducted to determine whether the proportion of buyers who favor a particular product is greater than the proportion who favor the product of a competing manufacturer.

c. A study is conducted to determine whether unionized employees in the electronics industry receive benefits different from those who do not belong to a union.

d. A study is proposed to examine the efficient-market hypothesis and determine whether average returns on stocks chosen according to specific performance criteria differ from the average return of the stock market as a whole.

e. An examination is made of the effect of codetermination on productivity and whether the chances are greater than 50–50 that codetermination will increase industrial productivity.

8.83   An experimenter is interested in testing

$$H_0 : \mu = 650 \qquad \text{versus} \qquad H_a : \mu < 650$$

A random sample of size $n = 50$ produced $\bar{x} = 643$ and $s = 14.5$.

a. Is this a one- or a two-tailed test?

b. Locate the rejection region for the test if $\alpha = .05$.

c. Calculate the value of the test statistic, using the sample information.

d. Do the data support rejection of the null hypothesis? Why or why not? Verbally describe your conclusions in this test of hypothesis.

8.84   Refer to Exercise 8.83. Construct a power curve for the test.

a. Calculate β for the alternative values of μ of interest to the experimenter, $\mu = 641, 643, 645,$ 647, 649.

b. Use the results of part a to calculate $(1 - \beta)$.

c. Graph the power curve. Interpret your results.

d. If the experimenter wishes to reject H_0 when $\mu = 647$ with probability .90 or greater, is the test powerful enough? If not, what can be done?

8.85   Refer to Exercise 8.83. The experimenter has decided to test

$$H_0 : \mu = 650 \qquad \text{versus} \qquad H_a : \mu \neq 650$$

a. Is this a one- or a two-tailed test?

b. Locate the rejection region for the test if $\alpha = .05$.

c. Use the calculated value of the test statistic found in Exercise 8.83 to draw a conclusion for this test of hypothesis.

8.86   Independent random samples drawn from each of two populations produced the following sample information:

Sample 1	Sample 2
$n_1 = 160$	$n_2 = 180$
$\bar{x}_1 = 98$	$\bar{x}_2 = 96$
$s_1 = 16.6$	$s_2 = 17.2$

The experimenter wishes to determine whether there is a difference between the two population means.

a. Is this a one- or a two-tailed test?

b. Locate the rejection region for the test if $\alpha = .10$.

c. Calculate the value of the test statistic, using the sample information.

d. Do the data support rejection of the null hypothesis? Why or why not? Verbally describe your conclusions in this test of hypothesis.

8.87   A random sample of size $n = 700$ drawn from a binomial population produced $x = 282$ successes. Test the hypothesis $H_0: p = .4$ versus $H_a: p > .4$, using a large-sample statistical test of hypothesis. Use $\alpha = .01$.

8.88   Independent random samples drawn from two binomial populations produced the following information:

Sample 1	Sample 2
$n_1 = 200$	$n_2 = 200$
$x_1 = 145$	$x_2 = 135$

Test to determine whether there is a difference between the two population proportions with $\alpha = .05$.

a. State the null and alternative hypotheses.

b. Locate the rejection region for the test.

c. Calculate the observed value of the test statistic, using the sample information.

d. Do the data provide sufficient evidence to indicate a difference in the population proportions p_1 and p_2?

EXERCISES   Applications

8.89   Food and Drug Administration (FDA) scientists are conducting studies to assess the possible association between caffeine and various health problems (Exercise 2.89). Although many results have been inconclusive with regard to implicating caffeine in a causal role, the consumption of caffeine *in moderation* seems to be important [C. W. Lecos, "Caffeine Jitters: Some Safety Questions Remain," *FDA Consumer*, Vol. 21, No. 10 (December 1987–January 1988), pp. 24–26]. A follow-up test is conducted concerning two brands of cola, A and B, using independent random samples of $n_1 = n_2 = 36$ 6-oz servings. The results of the testing are shown in the accompanying table. Do the data provide sufficient evidence to conclude that there is a difference between the mean caffeine content for the two brands of cola? Use $\alpha = .05$.

Statistic	Caffeine Content (Milligrams)	
	Brand A	Brand B
$\bar{x}$	18	20
s	1.25	1.35

8.90   Refer to Exercise 8.89. In its summary article the FDA reported that the average caffeine content of regular cola was approximately 19 mg per 6-oz serving. In a test of this claim, after the initial investigation, a second random sample of $n = 40$ 6-oz servings of brand B cola reveals a mean caffeine content of 20.5 mg with a standard deviation of 1.40 mg. Do the data provide sufficient evidence to indicate that regular cola produced by brand B has a mean caffeine content in excess of 19 mg per 6-oz serving? Use $\alpha = .01$.

8.91   The National Aeronautics and Space Administration (NASA) has been working with utilities throughout the nation to find sites for large wind machines for generating electric power. Wind speeds must average at least 15 miles per hour (mph) for a site to be acceptable. Thirty-six wind speed recordings were taken at random intervals on a site under consideration for a wind machine; the wind speeds averaged 14.2 mph, with a standard deviation of 3 mph. Do these data indicate that the site fails to meet NASA requirements for acceptability as a site for the location of an electric power–generating wind machine? Use $\alpha = .10$.

8.92   Continuing Exercise 8.91, discuss the meaning of type I and type II errors in the site selection problem. If you were a NASA official, would you recommend the selection of an extremely small (.01) value for α, a moderate value (.05), or a reasonably large value (.10) when testing for the acceptability of a site on the basis of its average wind speed? Explain.

8.93   It is often said that those who invest in government bonds do so out of a sense of patriotism, since return rates on U.S. government securities are not competitive with those of the private sector. Refer to Exercise 8.28 and examine this long-accepted premise; that is, do these data suggest that the mean annual return on corporate bonds surpassed that of U.S. government securities for the past nine years? Use $\alpha = .05$.

8.94   A northern California health insurance company sponsors the following advertisement, which appears on all Bay Area Rapid Transit (BART) trains: "One out of every seven people on this train is headed for the hospital this year." To test the validity of the claim, the insurance company surveys a random sample of 70 BART passengers as they depart a BART train, and 8 indicate that they had spent at least one day in a hospital within the past year. Do these data present sufficient evidence to reject the insurance company's claim as too high? Test by using $\alpha = .01$.

8.95   In a study to assess various effects of using a female model in automobile advertising, each of 100 male subjects was shown photographs of two automobiles matched for price, color, and size but of different makes. One automobile was shown with a female model and one without a model to 50 of the subjects (group A), and both automobiles were shown without a model to the other 50 subjects (group B). In group A the automobile shown with the model was judged as more expensive by 37 subjects, while in group B the same automobile was judged as the more expensive by 23 subjects. Do these results indicate that using a female model influences the perceived expensiveness of an automobile? Use a one-tailed test with $\alpha = .05$.

8.96   An industrial device called an ionizing air gun was recalled in 1988 when it was suspected of leaking particles of radioactive polonium into the atmosphere. The Nuclear Regulatory Commission requires that the amount of radiation present at any industrial site not exceed .005 microcurie [*The Press Enterprise* (Riverside, Calif.), February 19, 1988]. In a check for excessive radiation a random sample was taken of 118 plants using the ionizing air gun, and the average radiation was reported to be .0056 with a standard deviation of .0012. Does this sample provide evidence to indicate that plants using the ionizing air gun are exceeding the allowable limit? Use $\alpha = .05$.

8.97   A study of corporate mergers, sponsored by the International Institute for Management and Administration (IIMA), shows, surprisingly, that when companies merge, they are as likely to experience a decline or no perceptible change in sales volume and profits as they are to experience an increase ("Conglomerate Takeovers," *Washington Post*, April 19, 1980). The IIMA study involved 765 mergers in seven countries and found that in 225 cases profitability declined after the merger, in 192 cases it remained the same, and in 348 cases it increased. Do these results contradict the traditional view that mergers are undertaken for the sole purpose of enhancing profitability? That is, test the hypothesis $H_0:p = 1/2$ versus $H_a:p < 1/2$, where p is the probability that profitability increases after a merger. Test by using $\alpha = .05$.

8.15

Choosing the Null and Alternative Hypotheses

The reasoning employed in a statistical test of an hypothesis runs counter to our everyday method of thinking. It is similar to the mathematical method of proof by contradiction. The hypothesis that scientists wish to prove true is called the *research hypothesis* and is taken to be the alternative hypothesis. To this end, they test the negation of the research hypothesis, called the *null hypothesis*. They hope that the data will support the rejection of the null hypothesis, because this implies support for the alternative or research hypothesis, which was the research objective.

Why employ this reverse type of thinking, gaining support for a theory by showing that there is little evidence to support its negation? Why not test the alternative or research hypothesis directly? The answer lies in the problem of evaluating the probabilities of incorrect decisions.

If the research hypothesis is true, testing the null hypothesis, which is the negation of the research hypothesis, should lead to its rejection. Then the probability of making the incorrect decision of rejecting the null hypothesis is readily available. It is α, a probability that was specified in setting up the rejection region. Thus, if we reject the null hypothesis, which is what we hope will occur, we immediately know the probability of making an incorrect decision. This probability gives us a measure of confidence in our conclusion.

Suppose that we had taken the opposite tack and had used the research hypothesis as the null hypothesis. If the research hypothesis is true, the test statistic will most probably fall in the acceptance region instead of the rejection region. Now to find the probability of an incorrect decision, we must evaluate β, the probability of accepting the null hypothesis when it is false. Although evaluating β is not an insurmountable task for some tests, for others it is very difficult for meaningful values of the parameter(s) under test.

Administrators, analysts, and others in decision-making capacities are often faced with deciding for or against a certain action or deciding which strategy among several available strategies should be followed. Such decisions are made only after the risks involved with each possible action are studied and evaluated. A statistical test of an hypothesis is one way of approaching a decision problem when there are two available courses of action. In the use of the formal structure of a test of an hypothesis in these circumstances, there is no research hypothesis per se. However, the probability α of falsely rejecting the null hypothesis in favor of the alternate hypothesis can always be fixed in advance and the rejection region selected accordingly. In a decision problem H_0 and H_a should be specified so that the most serious error in the decision process is measured by α.

For example, if faced with a decision to implement a new production process or to continue using the current process, we would first consider the financial impact of each decision. The costs of implementing the new process together with the possible increase or decrease in actual revenues must be weighed against the possible loss of revenue that could be incurred if the new process is not substantially better than the process currently in use. In this situation we would most likely be interested in guarding against the decision to implement the new production process when in fact it is no better than the current one. In order that the probability α of falsely rejecting H_0 reflect the probability of deciding to implement the new process, when in fact it is no better than the current one, we would select the null hypothesis to be "there is no difference in the two processes" and the alternative hypothesis as "the new process is better than the current process." The next step would be to interpret these hypotheses in terms of population parameters such as the population means μ_1 and μ_2, or perhaps the population proportion of defectives p_1 and p_2. When the sample data has been analyzed and the test performed, rejecting H_0 supports the decision to implement the new process, and the probability that this decision is in error is given by α, which is known exactly. Not rejecting H_0 supports the decision to remain with the current process, and the probability that this decision is in error is given by β, which must be evaluated for parameter values of interest to the decision maker.

In summary, statisticians use the route of proof by contradiction in implementing a test of an hypothesis. If there is a research hypothesis, the null hypothesis is taken as its negation. In a decision-making context the null hypothesis is chosen so that the type I error of falsely rejecting H_0 reflects the error associated with the more serious incorrect decision. When the null and alternative hypotheses are chosen in this manner and the test leads to rejecting the null hypothesis, the statistician knows α and has a measure of the confidence he or she can place in this conclusion.

8.16

Another Way to Report the Results
of Statistical Tests: *p*-Values

significance level · · · The probability α of making a type I error is often called the **significance level** of the statistical test, a term that originated in the following way. The probability of the observed value of the test statistic, or some value even more contradictory to the null hypothesis, measures, in a sense, the weight of evidence favoring rejection. Some experimenters report test results as being significant (we would reject) at the 5% significance level but not at the 1% level. This statement means that we would reject H_0 if α were .05 but not if α were .01.

p-value · · · The **p-value** represents the probability of observing a sample outcome more contradictory to H_0 than the observed sample result if, in fact, H_0 is true. The smaller the value of this probability, the heavier is the weight of the sample evidence for rejecting H_0. For example, a statistical test with a level of significance of .01 has more evidence for the rejection of H_0 than another statistical test with a level of significance of .20. The smallest value of α for which test results become statistically significant is often called the *p-value* for the test.

Definition

p-value

> The **p-value** for a test of an hypothesis is the probability of obtaining a value of the test statistic as extreme or more extreme than the actual sample value when H_0 is true.

The p-value is also called the observed significance level of the test, since it represents the smallest value of α for which we could reject H_0 using the observed sample results. Some statistical computer programs compute *p*-values correct to four or five decimal places. But if researchers use statistical tables to determine a *p*-value, they usually are able only to approximate its value, because most statistical tables give the critical values of test statistics only for large differential values of α (for example, .01, .025, .05, .10, etc.). Consequently, the *p*-value reported by most experimenters is the *smallest* tabulated value of α for which the test remains statistically significant. For example, if a test result is statistically significant for $\alpha = .10$ but not for $\alpha = .05$, then the *p*-value for the test would be given as .10. (More correctly, of course, the *p*-value would be reported as being between .05 and .10.)

Many scientific journals require researchers to report the *p*-values associated with statistical tests because these values provide a reader with *more information* than simply stating that a null hypothesis is or is not to be rejected for some value of α chosen by the experimenter. In a sense, it allows the reader of published research to evaluate the extent to which the data disagree with the null hypothesis. In particular, it enables each reader to choose his or her own personal value for α and then decide whether or not the data lead to rejection of the null hypothesis.

The procedure for finding the *p*-value for a test is illustrated in the following examples.

EXAMPLE 8.15   Find the *p*-value for the statistical test of Example 8.10. Interpret your results.

SOLUTION   Example 8.10 presents a test of the null hypothesis $H_0: \mu = 880$ against the alternative hypothesis $H_a: \mu \neq 880$. The value of the test statistic, computed from the sample data, was $z = -3.03$. Therefore, the *p*-value for this two-tailed test is the probability that $z \leq -3.03$ or $z \geq 3.03$ (the upper- and lower-tail areas; see Figure 8.16); that is, the *p*-value will be double the value we find in the table.

From Table 3 in the Appendix we find that the tabulated area under the normal curve between $z = 0$ and $z = 3.03$ is .4988 and the area to the right of $z = 3.03$ is $.5 - .4988 = .0012$. Then since this was a two-tailed test, the value of α corresponding to a rejection region of $z > 3.03$ or $z < -3.03$ is $2(.0012) = .0024$. Consequently, we would report the *p*-value for the test as *p*-value $= .0024$.

This *p*-value is interpreted as follows: The probability that we would observe a value of z as large as 3.03 or as small as -3.03, given that H_0 is true (i.e., that $\mu = 880$), is only *p*-value $= .0024$.

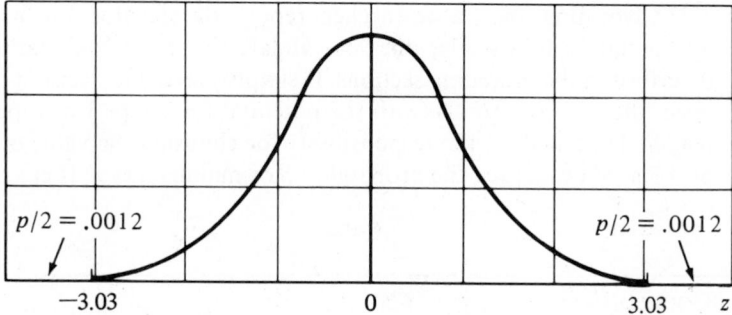

Figure 8.16   Locating the p-Value for the Test of Example 8.10

EXAMPLE 8.16   Find the p-value for the statistical test of Example 8.12.

SOLUTION   Example 8.12 presented a one-tailed test of the null hypothesis $H_0 : p = .10$ against the alternative hypothesis $H_a : p > .10$, and the observed value of the test statistic was $z = 1.41$. Therefore, the p-value for the test is the probability of observing a value of the z statistic larger than 1.41. This value is the area under the normal curve to the right of $z = 1.41$ (the shaded area in Figure 8.17).

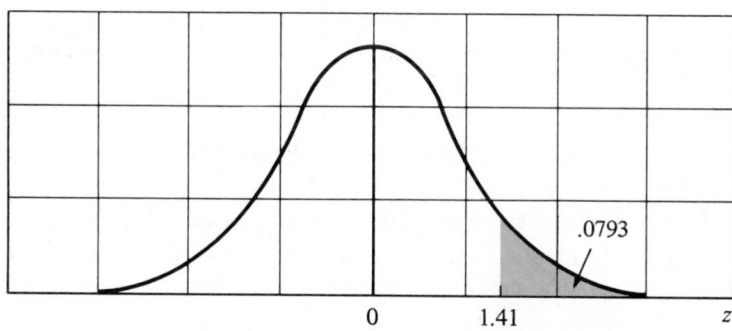

Figure 8.17   Finding the p-Value for the Test of Example 8.12

From Table 3 in the Appendix we find that the area under the normal curve between $z = 0$ and $z = 1.41$ is .4207. Therefore, the area under the normal curve to the right of $z = 1.41$, the p-value for the test, is $p\text{-value} = .5 - .4207 = .0793$.

How do we judge the importance of a reported p-value? Many researchers evaluate p-values along the following lines: If $p\text{-value} > .05$, the results are said to be *not significant*; if $.01 < p\text{-value} < .05$, the results are said to be *significant*; if $.001 < p\text{-value} < .01$, the results are said to be *highly significant*; if $p\text{-value} < .001$, the results are said to be *very highly significant*. Reported p-values between .05 and .10, although not significant, are often noted as tending toward significance.

Advocating that a researcher report the p-value for a test and leave its interpretation to a reader does not violate the traditional statistical test procedure described in the preceding sections. It simply leaves the decision of whether or not to reject the null hypothesis (with the potential for a type I or type II error) up to the reader. Thus, it shifts the responsibility for choosing the value of α, and possibly the problem of evaluating the probability β of making a type II error, to the reader.

EXERCISES Conceptual Level

8.98 What is meant by the term *significance level* in regard to hypothesis testing?

8.99 What is the distinction, if any, between the significance level and the p-value for a statistical test?

8.100 Assume you were to determine the p-value for an observed value of a test statistic and that you also were to select an α level for the test. What would be the size of the p-value relative to the α level if you rejected the null hypothesis? What would the relationship be if you could not reject the null hypothesis?

EXERCISES Skill Level

8.101 Find the p-values associated with the test statistics given below, resulting from a one-tailed test of hypothesis.

  a. $z = -2.31$ **b.** $z = 3.08$ **c.** $z = 1.62$ **d.** $z = -4.12$

8.102 Find the p-values associated with the test statistics given below, resulting from a two-tailed test of hypothesis.

  a. $z = .87$ **b.** $z = -2.06$ **c.** $z = -3.8$ **d.** $z = 1.43$

8.103 Calculate the observed level of significance for the test of hypothesis given in Exercise 8.83. Use this p-value to determine whether $\mu < 650$. Does this result agree with the conclusion in Exercise 8.83?

8.104 A random sample of size $n = 300$ from a binomial population produced $x = 156$ successes for testing to determine whether p differs from $1/2$.

  a. State the null and alternative hypotheses.

  b. Calculate the observed value of the test statistic, based on the sample information.

  c. Calculate the observed level of significance for this test.

  d. Use the p-value found in part c to draw a conclusion concerning the test of hypothesis.

8.105 A random sample from a population with mean μ and standard deviation σ produced the following sample information:

$$n = 110 \qquad \bar{x} = 699 \qquad s = 20.4$$

Find the p-value for a test of the null hypothesis $H_0 : \mu = 695$ against the alternative hypothesis $H_a : \mu \neq 695$. What conclusions would you draw?

EXERCISES   Applications

8.106   Refer to Exercises 8.89 and 8.90. Find the *p*-values for the two statistical tests used in these exercises.

8.107   Refer to Exercise 8.93. Find the *p*-value for the statistical test examining the difference in annual returns for government and corporate bonds.

8.108   To protect franchisees, several states have passed laws that prohibit a franchisor from terminating or failing to renew a franchise agreement without good cause. These laws have had a profound effect on the retail gasoline industry, in which 84% of all service stations are franchise dealerships. Nevin, Hunt, and Ruekert suggest that 30% of all franchise relationships in the retail gasoline industry require the franchisee to adhere to specified hours of operation (J. Nevin, S. Hunt, and R. Ruekert, "The Impact of Fair Practice Laws on a Franchise Channel of Distribution," *MSU Business Topics*, Summer 1980). In a state without a fair-practices act for franchise agreements, a sample of 50 franchise gasoline dealerships showed that 20 involved a requirement in the franchise agreement specifying hours of operation. Find the *p*-value for this test, and determine whether sufficient evidence exists to assume that the rate of control of operating hours by franchisors in the state without a fair-practices act exceeds the national rate for the retail gasoline industry.

8.109   The manufacturer of a pollution control device claims that, on the average, when the device is attached to an automobile exhaust system, carbon emissions will not exceed 10 parts per million (ppm). In an experiment involving the application of the pollution control device to 40 automobiles, carbon emissions were found to average 11 ppm with a standard deviation of 3.6 ppm. Find the *p*-value for the statistical test. Do these data provide sufficient evidence to refute the manufacturer's claim?

8.17

Some Comments on the Theory of Tests of Hypotheses

As outlined in Section 8.13, the theory of a statistical test of an hypothesis is indeed a very clear-cut procedure, enabling the experimenter to either reject or accept the null hypothesis with measured risks α and β. Unfortunately, as we noted, the theoretical framework does not suffice for all practical situations.

The crux of the theory requires that we be able to specify a meaningful alternative hypothesis that permits the calculation of β, the probability of a type II error, for all alternative values of the parameter(s). This calculation can indeed be done for many statistical tests, including the test discussed in Section 8.13, although the calculation of β for various alternatives and sample sizes may, in some cases, be a formidable task. On the other hand, in some test situations it is extremely difficult to clearly specify alternatives to H_0 that have practical significance. This difficulty may arise when we wish to test an hypothesis concerning the values of a set of parameters, a situation that we will encounter in Chapter 17 in analyzing enumerative data.

The obstacle that we mention does not invalidate the use of statistical tests. Rather, it urges caution in drawing conclusions when insufficient evidence is available to reject the null hypothesis. The difficulty of specifying meaningful alternatives to the null hypothesis, together with the difficulty encountered in the calculation and tabulation of β for other than the simplest statistical tests, justifies skirting this issue in an introductory text. Hence, we can adopt one of two procedures. We can present the p-value associated with a statistical test and leave the interpretation to the reader. Or we can agree to adopt the procedure described in Section 8.14 when tabulated values of β or $1 - \beta$ (the power curve) are unavailable for the test. **When the test statistic falls in the acceptance region, we will "not reject" rather than "accept" the null hypothesis.** Further conclusions may be made by calculating an interval estimate for the parameter or by consulting one of several published statistical handbooks for tabulated values of β. We will not be too surprised to learn that these tabulations are inaccessible, if not completely unavailable, for some of the more complicated statistical tests.

Finally, we might comment on the choice between a one- or a two-tailed test for a given situation. We emphasize that this choice is dictated by the practical aspects of the problem and will depend on the alternative value of the parameter—say θ—the experimenter is trying to detect. If we were to sustain a large financial loss if θ were greater than θ_0, but not if it were less, we would concentrate our attention on the detection of a value of θ greater than θ_0. Hence, we would reject in the upper tail of the distribution for the test statistics previously discussed. On the other hand, if we are equally interested in detecting a value of θ that is either less than or greater than θ_0, we would employ a two-tailed test.

8.18

Summary

The material presented in this chapter has been directed toward two objectives. First, we discussed the various methods of inference along with procedures for evaluating their goodness. Second, we presented a number of estimation procedures and statistical tests of hypotheses that, owing to the Central Limit Theorem, make use of the results of Chapter 7. The resulting techniques possess practical value and, at the same time, illustrate the principles involved in statistical inference. Inferences concerning the parameter(s) of a population may be made by estimating or by testing hypotheses about their value.

In Part I of this chapter we presented procedures for constructing either point or interval estimators for population parameters. The measure of goodness, or confidence, that we can place in an estimator is given by the bound on the error of estimation. Similarly, the confidence coefficient and the width of the interval measure the goodness of an interval estimator.

A statistical test of an hypothesis or theory concerning one or more population parameters that results in its acceptance or rejection was the major topic of Part II. In practice, we may be forced to view this decision in terms of rejection or nonrejection. The probabilities of making the two possible incorrect decisions, resulting in type I and type II errors, measure the goodness of the decision procedure. Although a test of an hypothesis may be best suited for some physical situations, estimation is often the eventual goal of many experimental investigations and hence would be desirable if we were permitted an option in our choice of a method of inference.

All the confidence intervals and statistical tests described in this chapter are based on the Central Limit Theorem and hence apply to large samples. When n is large, each of the respective estimators and test statistics possesses, for all practical purposes, a normal distribution in repeated sampling. This result, along with the properties of the normal distribution studied in Chapter 6, permits the construction of the confidence intervals and the calculation of α and β for the statistical tests.

Tips on Problem Solving

In solving the exercises in this chapter, you will be required to answer a practical question of interest to a businessperson. To find the answer to the question, you will need to make an inference about one or more population parameters. Consequently, the first step in solving a problem is deciding on the objective of the exercise. What parameters do you wish to make an inference about? Answering the following two questions will help you to reach a decision.

1.   What *type of data* is involved? The answer will help you decide the type of parameters about which you will wish to make inferences, binomial proportions (p's) or population means (μ's). Check to see whether the data are of the yes–no (two-possibility) variety. If they are, the data are probably binomial and you will be interested in proportions. If not, the data probably represent measurements on one or more quantitative random variables and you will be interested in means. As an aid, look for key words such as *proportions, fractions*, and so on, which indicate binomial data. Binomial data often (but not exclusively) evolve from a "sample survey."

2.   Do I wish to make an inference about a *single parameter*, p or μ, or about the *difference between two parameters*, $(p_1 - p_2)$ or $(\mu_1 - \mu_2)$? This is an easy question to answer. Check on the number of samples involved. One sample implies an inference about a single parameter; two samples imply a comparison of two parameters. The answers to questions 1 and 2 identify the parameter.

After identifying the parameter(s) involved in the exercise, you must identify the exercise objective. It will be one of these three:

1.   Choosing the sample size required to estimate a parameter with a specified bound on the error of estimation

2.   Estimating a parameter (or difference between two parameters)

3.   Making a decision about one or more parameters (a test of an hypothesis)

The objective will be very clear if it is 1 because the question will ask for or direct you to find the "sample size." Objective 2 will be clear because the exercise will specifically direct you to estimate a parameter (or the difference between two parameters). If you are required to make a decision (other than a choice of sample size), the objective is most likely a test of an hypothesis.

To summarize these tips, your thought process should follow the decision tree shown in Figure 8.18.

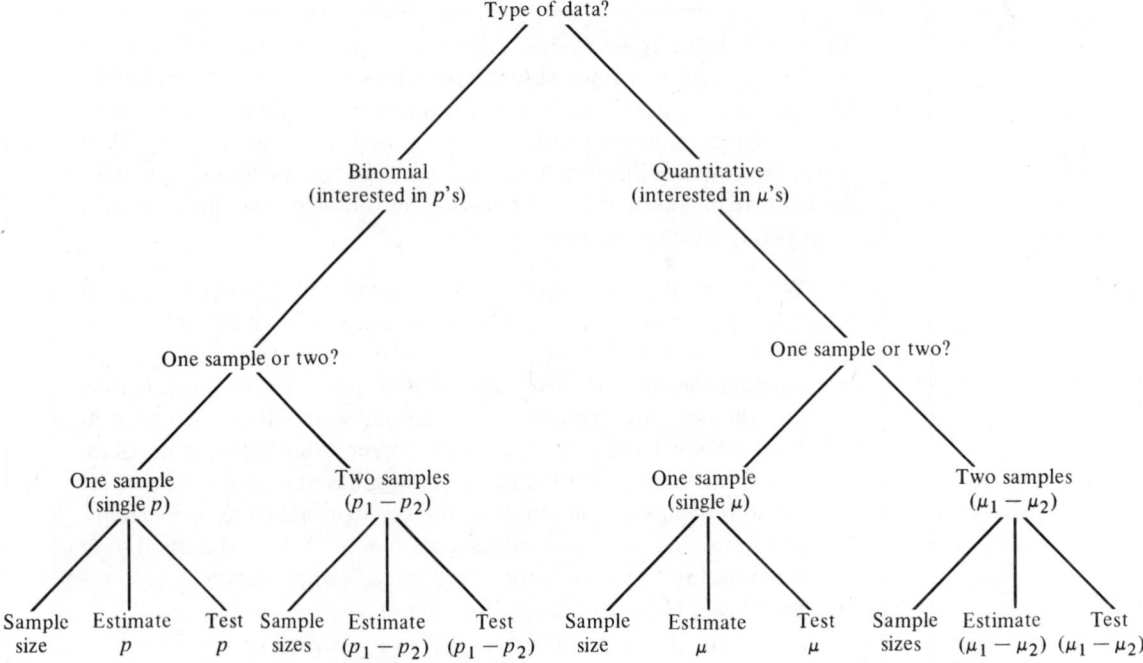

Figure 8.18   Decision Tree for Problem Solving

Supplementary Exercises

Applications

PART I   Estimation

8.110   The mean and standard deviation of the after-tax profits of 100 manufacturers for a certain period were found to be 4.3 cents per dollar revenue and 1.5 cents per dollar revenue, respectively. Estimate the mean after-tax profits for this period for the population of manufacturers from which this sample was drawn. Place a bound on the error of estimation.

8.111 If the population mean after-tax profits for all manufacturers during a certain period is really 4.7 cents per dollar revenue, with $\sigma = 1.3$ cents, what is the probability that the mean after-tax profits of a random sample of $n = 100$ manufacturers would exceed 5 cents?

 8.112 The mean and standard deviation for the life of a random sample of 100 light bulbs were calculated to be 1,960 hours and 142 hours, respectively. Estimate the mean life of the population of light bulbs from which the sample was drawn, and place bounds on the error of estimation.

 8.113 If the population mean in Exercise 8.112 were really 1,965 hours, with $\sigma = 150$ hours, what is the probability that the mean of a random sample of $n = 100$ measurements would exceed 2,000 hours?

 8.114 Using a confidence coefficient of .90, place a confidence interval on the mean life of the light bulbs of Exercise 8.112.

 8.115 In industrial selling, negotiation costs represent a significant percentage of a firm's sales because the seller must formulate a sales strategy before negotiating with a business or governmental client. Several researchers interviewed 500 firms involved in industrial selling and found their average negotiation cost per transaction to be $11,336, with a standard deviation of $1,951. Estimate the average negotiation cost per transaction for all industrial sellers, and place a bound on the error of estimation.

 8.116 In a summary of a landmark study of energy utilization and resources, Stobaugh and Yergin report that conservation and solar power are our best answers to the problem of becoming independent of foreign suppliers (R. Stobaugh and D. Yergin, "The Energy Outlook: Combining the Options," *Harvard Business Review*, January–February 1980). Because the impact of conservation and changing life-styles is so uncertain, the authors report that much uncertainty exists regarding expected energy consumption in the United States by the year 2000. Suppose that $n = 30$ experts report that the total energy consumption in the United States in the year 2000 will average 64.5 million barrels of oil equivalent per day with a standard deviation of 9.8 million barrels. Using this information, estimate the average daily energy consumption in the United States in the year 2000, and place a bound on the error of estimation.

8.117 Use a 90% confidence interval to estimate the average daily energy consumption in the United States in the year 2000, using the mean and standard deviation of estimates provided in Exercise 8.116.

 8.118 Investors who purchase income-producing stocks are generally interested only in the annual yield produced by this class of common stocks. To assist investors of this type, a leading stock brokerage firm randomly sampled 50 income-producing common stocks. The average annual yield and standard deviation of the yields of the 50 securities over the past five years were observed to be $\bar{x} = 9.71\%$ and $s = 2.4\%$. Estimate the true average annual yield μ for this class of common stocks, using a 90% confidence interval.

 8.119 Meaningful union-management wage negotiations require a precise estimate of current union member wages. A random sample of 60 unionized elevator operators in New York City was selected by a labor arbitrator. The mean and standard deviation of the weekly wage rates for the unionized elevator operators were found to be $174.45 and $12.40, respectively. Find a 95% confidence interval for the average weekly wage for all unionized New York City elevator operators.

 8.120 Manufacturers of photographic equipment have introduced many new easy-to-use cameras, film types, and flash equipment in recent years. A new type of flashbulb was tested to estimate the probability p that the new bulb would produce the required light output at the appropriate time. A sample of 1,000 bulbs was tested and 920 were observed to function according to specifications. Estimate p, and place bounds on the error of estimation.

8.121 To draw more tourists and conventioneers into the stores, many department stores are now accepting Visa and MasterCard cards. In a random sample of 50 department stores in the Los Angeles area, 45 stores are now accepting both Visa and MasterCard cards for retail purchases. Find a 98% confidence interval for the proportion of all Los Angeles area department stores that accept Visa and MasterCard cards for retail transactions.

 8.122 During a period of high interest rates bank profitability suffers owing to the reduction in loan size and loan demand. For instance, when rates reached 20% in 1980, loan demand came to a virtual standstill in most commercial banks. To examine the effect of a reduction in interest rates on loan size, a large commercial bank examined 49 loan requests for consumer credit and found the average loan to be $2,650 with a standard deviation of $837. Find a 90% confidence interval for the expected size of all consumer loans at the new reduced rate of interest. Find a 99% confidence interval for the expected size of all loans.

8.123 Almost all graduate business schools require a score on the Graduate Management Admission Test (GMAT) from each applicant for admission. The GMAT, like the Graduate Record Examination, measures a student's verbal ability and quantitative ability on separate 800-point scales. Two graduate schools decided to compare the quantitative scores for students admitted to their graduate programs. Random samples of the GMAT scores of 75 students were selected for each school. The averages and standard deviations were as follows:

School 1 $\bar{x}_1 = 525$ $s_1 = 52$
School 2 $\bar{x}_2 = 564$ $s_2 = 70$

Find a 90% confidence interval for the difference in GMAT scores between the two schools.

 8.124 A recent Gallup poll reported that 82% of 1,200 residents of greater São Paulo, Brazil, polled in a recent survey consider air pollution in their city to be "very serious." Find a 95% confidence interval for the proportion of the population of São Paulo who consider air pollution in their city to be very serious.

 8.125 The marketing research department of a soap and detergent manufacturing company conducted a survey of housewives to determine the fraction who prefer the company's brand, detergent A. Sixty of 87 housewives prefer detergent A. If the 87 housewives represent a random sample from the population of all potential purchasers, estimate the fraction of total housewives favoring detergent A. Use a 90% confidence interval.

 8.126 In examining the credit accounts of a department store, an auditor selected a random sample of 64 accounts and found the average account error to be $-$37 with a standard deviation of $15. Construct a confidence interval for the true average account error, using a confidence coefficient of .95.

 8.127 A manufacturing firm recently instituted an occupational safety program designed to reduce time lost due to on-the-job accidents. In the 48 months since the institution of the program, employee time lost due to on-the-job accidents has averaged 91 hours per month, with a standard deviation of 14 hours per month. In the 50 months prior to the safety program, time lost due to accidents averaged 108 hours per month, with a standard deviation of 12 hours. Estimate the difference in mean employee time lost per month due to on-the-job accidents before and after the occupational safety program. Place bounds on the error of estimation.

 8.128 Refer to Exercise 8.127. Estimate the difference in the mean employee time lost per month by using a confidence coefficient of .90.

 8.129 In a poll taken among the stockholders of a company, 300 of 500 men favored the decision to adopt a new product line, whereas 64 of 100 women favored the new product. Estimate the difference in the fractions favoring the decision, and place a bound on the error of estimation.

8.130 Manufacturers of golf balls are concerned with a scientific concept called the coefficient of restitution, defined as the ratio of the relative velocity of the ball and club after impact to the relative velocity before impact. A manufacturer has developed a new solid-core golf ball it wishes to test and compare with the firm's standard brand. Fifty of the new golf balls and 50 of the standard brand are subjected to a test. The results are given in the accompanying table. Estimate the difference in the mean coefficient of restitution for the new solid-core ball and the firm's standard brand of golf ball. Place bounds on the error of estimation.

Solid-Core Ball	Standard Brand
$\bar{x}_1 = .69$	$\bar{x}_2 = .55$
$s_1 = .18$	$s_2 = .22$

 8.131 A dry goods retailer is interested in buying a small computer to facilitate inventory and accounting requirements. Since both of two brands she is considering have equal operational capabilities, the retailer has requested information concerning computer reliability. Manufacturing representatives from each of the computer firms provide the following reliability data:

Computer A: In 64 independent tests time until first failure averaged 129 hours with a standard deviation of 8 hours.

Computer B: In 108 independent tests time until first failure averaged 124 hours with a standard deviation of 17 hours.

Find a 95% confidence interval for the difference in expected time until failure between the two brands of computers.

 8.132 One method for solving the electric power shortage employs the construction of floating nuclear power plants located a few miles offshore in the ocean. Because there is concern about the possibility of a ship colliding with the floating (but anchored) plant, an estimate of the density of ship traffic in the area is needed. The number of ships passing within 10 miles of the proposed power plant location per day, recorded for $n = 60$ days during July and August, possessed a sample mean and variance of $\bar{x} = 7.2$ and $s^2 = 8.8$.

a. Find a 95% confidence interval for the mean number of ships passing within 10 miles of the proposed power plant location during a 24-hour time period.

b. The density of ship traffic was expected to decrease during the winter months. A sample of $n = 90$ daily recordings of ship sightings for December, January, and February gave a mean and variance of $\bar{x} = 4.7$ and $s^2 = 4.9$. Find a 90% confidence interval for the difference in mean density of ship traffic between the summer and winter months.

c. What is the population associated with your estimate in part b? What could be wrong with the sampling procedure in parts a and b?

 8.133 Faulty products are costly to a manufacturer in terms of replacement costs and loss of product image among the consuming public. A manufacturer of portable tape recorders believes that no more than 10% of the products produced by the firm are faulty. If the manufacturer wishes to estimate the actual fraction of faulty recorders to within .03, how large a sample of finished recorders should be selected?

 8.134 Ideally, inventory control avoids both excessive carrying costs and stock-out costs. A large trade association wishes to estimate the difference in the proportion of retail firms that utilize formal inventory control techniques and the proportion of wholesalers that utilize such techniques. In a random sample of 91 retail firms the association found that 62 employ formal inventory control techniques; 37 of 65 firms primarily involved in wholesale trade were found to use such procedures. Estimate the difference in the proportions using formal inventory control procedures, and place bounds on the error of estimation.

 8.135 A General Motors personnel executive wishes to investigate employee attitudes toward the workplace and the extent to which they have changed in the past 25 years. She surveys a group of GM employees who have been with the firm for at least 25 years. How many employees should she select if she wishes to estimate the proportion of employees who are more dissatisfied with GM now than they were 25 years ago, to within 3%? Assume that no information exists regarding p.

 8.136 How many voters must be included in a sample collected to estimate the fraction of the popular vote favorable to a presidential candidate in a national election if the estimate is desired correct to within .005? Assume that the true fraction will lie somewhere in the neighborhood of .5.

 8.137 Expert opinion is quite divided among corporate executives about the most important training topic for purchasing managers. However, traditional theory has held that the most important training topic for purchasing managers is the development of negotiation skills. In light of this opinion, a researcher would like to see whether the traditional theory still holds true. From past evidence he knows that the proportion who believe the development of negotiation skills to be the most important training topic is about .6. How large a sample of corporate executives should the researcher select if he wishes to estimate the proportion who hold to the traditional theory, correct to within 5%?

 8.138 The manager of the chamber of commerce of a convention city wishes to estimate the average amount spent by each convention attendee correct to within $25.00, with a probability of .95. The manager knows only that the range of expenditures will vary, approximately, from $250.00 to $700.00. If a random sample of convention attendees is to be selected and asked to keep financial expenditure data, how many must be included in the sample?

 8.139 You wish to estimate the difference in the average time to assemble an electronic component for two factory workers to within 2 minutes, with a probability of .95. The standard deviation of the assembly times is approximately equal to 6 minutes for each worker. How many component assembly times must be observed for each worker? (Assume that the number of component assembly times observed for each worker will be the same.)

PART II   Hypothesis Testing

 8.140 A manufacturer claims that at least 95% of the equipment that it supplied to a factory conformed to specifications. An examination of 700 pieces of equipment reveals that 53 are faulty. Do these results provide sufficient evidence to reject the manufacturer's claim? Use $\alpha = .05$.

    a. State the null hypothesis to be tested.

    b. State the alternative hypothesis.

    c. Conduct a statistical test of the null hypothesis and state your conclusions.

 8.141 In the past a chemical plant has produced an average of 1,100 lb of chemical per day. A random sample of 260 operating days from the past year shows that $\bar{x} = 1,040$ and $s = 360$. The plant manager wishes to test whether the average daily production $\bar{x}$ has dropped significantly over the past year.

    a. Give the appropriate null and alternative hypotheses.

    b. If z is used as a test statistic, determine the rejection region corresponding to a level of significance of $\alpha = .05$.

    c. Do the data provide sufficient evidence to indicate a drop in average daily production?

8.142 A manufacturer of automatic washers provides a particular model in one of three colors, *A*, *B*, or *C*. Of the first 1,000 washers sold, 400 of the washers were of color *A*. Find the *p*-value for the statistical test that examines whether sufficient evidence exists to assume that more than 1/3 of all customers prefer color *A*. Interpret your results.

8.143 A manufacturer claims that at least 20% of the public prefers her product. A sample of 100 persons is taken to check her claim. With $\alpha = .05$, how small would the sample percentage need to be before the claim could be rightfully refuted? (*Note:* This problem requires a one-tailed test of an hypothesis.)

8.144 Weekly ratings published by the A.C. Nielsen Company have made television producers and advertisers very sensitive to claims regarding the proportion of the total viewing audience reached by a particular program. A television station claims that its six o'clock evening news reaches 50% of the viewing audience in the area. A firm considering the purchase of advertising time during the six o'clock time slot wishes to test the validity of the station's claim. How large a sample should the firm select if it wants the bound on the error of estimation to be 5%?

8.145 Refer to Exercise 8.144. A sample of 100 television viewers is selected from the potential viewing audience of the television station and 38 indicate that they watch the station's six o'clock evening news. Is this sufficient evidence to refute the claim of the television station that its six o'clock evening news reaches 50% of the viewing audience? (Test at the 1% level of significance.)

8.146 Refer to Exercise 8.145. Find the *p*-value for the statistical test. Interpret your results.

8.147 A test of the breaking strengths of two different types of cables was conducted, using samples of $n_1 = n_2 = 100$ pieces of each type of cable. The data are shown in the accompanying table. Do the data provide sufficient evidence to indicate a difference in the mean breaking strengths of the two cables? Use $\alpha = .10$.

Cable 1	Cable 2
$\bar{x}_1 = 1{,}457$	$\bar{x}_2 = 1{,}405$
$s_1 = 40$	$s_2 = 30$

8.148 Refer to Exercise 8.127. Find the *p*-value for the statistical test that examines whether the data indicate that the firm's occupational safety program has been effective in reducing on-the-job accidents. Interpret your results.

8.149 The yields to maturity on 30 recently issued Treasury notes and 36 recently issued Treasury bonds were recorded by an investor; see the accompanying table. Find the *p*-value for the statistical test that examines whether these data present sufficient evidence to indicate a difference in the average yield to maturity between U.S. Treasury notes and bonds. Interpret your results.

Treasury Notes	Treasury Bonds
$\bar{x}_1 = 6.91\%$	$\bar{x}_2 = 7.12\%$
$s_1 = 4.04\%$	$s_2 = 3.71\%$

8.150 Refer to Exercise 8.58. Find the *p*-value for the statistical test that examines whether *New York Post* subscribers are more likely to be regular patrons of one of the New York City retailing centers than are subscribers of the *Daily News*. Interpret your results.

8.151 A union official believes that the fraction p_1 of laborers in favor of wage and price controls is greater than the fraction p_2 of managers in favor of wage and price controls. The official acquired independent random samples of 300 laborers and 300 managers and found 69 laborers and 51 managers favoring controls. Does this evidence provide statistical support at the .05 level of significance for the union official's belief?

8.152 Refer to Exercise 8.91. Use the procedure described in Example 8.11 to calculate the power $(1 - \beta)$ for several alternative values of μ. For example, you might calculate $(1 - \beta)$ for $\mu = 13$, 13.5, and 14.5. Use these computed values to construct a power curve for the statistical test.

8.153 All corporate directors need timely and accurate information about the company each serves. However, the amount of information needed by an executive may depend on the industry within which the firm operates. Suppose that a random sample of 36 corporate directors of manufacturing firms shows that they receive an average of 6 reports during the course of a month from top management, with a standard deviation of 1.8, whereas a random sample of 30 corporate directors from the transportation industry shows that they receive, on the average, 3.1 reports per month, with a standard deviation of 1.4. Do these data suggest that the mean number of reports received by corporate directors from these two industries differs? Use $\alpha = .05$.

8.154 Presently, 20% of potential customers buy a certain brand of soap—say brand A. To increase sales, the company conducts an extensive advertising campaign. At the end of the campaign a sample of 300 potential customers is interviewed to determine whether the campaign was successful.

a. State H_0 and H_a in terms of p, the probability that a customer prefers brand A.

b. The company decides to conclude that the advertising campaign was a success if at least 70 of the 300 customers interviewed prefer brand A. Find α. (Use the normal approximation to the binomial distribution to evaluate the desired probability α.)

8.155 Laptop computers are the hottest new item in the personal computer (PC) business. These lightweight portable computers are no longer a novelty but are becoming standard equipment, especially for field sales forces (T. C. Taylor, "Make Way for the Salesman's New Friend," *Sales and Marketing Management*, February 1988, pp. 53–56). In fact, a researcher in an issue of *Sales and Marketing Management* claims that one out of seven marketers uses laptops in the sales and marketing area. In a test of this claim a random sample of 45 marketers was chosen and 13 of the marketers used laptops. Find the p-value for a test to determine whether sufficient evidence exists to dispute the researcher's claim. What conclusion would you draw?

8.156 Refer to Exercise 8.154. Calculate the power of the test, $1 - \beta$, for several alternative values of p that might be of interest to the experimenter. In particular, calculate the power of the test if $p = .21, .23, .25, .27,$ and $.30$. Use these computed values to construct a power curve for the test.

Supplementary Exercises

Interpretive

8.157 When evaluating a loan applicant, a financial officer is faced with the problem of granting loans to those who are good risks and denying loans to those who appear to be poor risks. In a sense, one could say that the financial officer is testing the statistical hypothesis

H_0: the applicant is a good risk

against the alternative

> H_a: the applicant is a poor risk

for each loan applicant. A type I error is then committed by the loan officer if she rejects an applicant who is actually a good risk; a type II error is committed if she grants a loan to an applicant who is a poor risk. Discuss the selection of a significance level α in the following instance.

a. Lending money is tight; interest rates are high and loan applicants are numerous.

b. Lending money is plentiful; interest rates are moderate and there is competition for loan applicants.

8.158 A small soft drink bottle carries a claim on its label that the bottle contains 6 1/2 fluid ounces of soda. To examine the validity of this claim, a consumer group randomly selects a sample of 50 bottles of the soft drink and finds an average content of 6.4 fluid ounces with a standard deviation of .14 fluid ounce. Considering the possible legal implications of claiming incorrectly that the manufacturer is engaging in deceptive advertising, select an appropriate α and then test to see whether the data present sufficient evidence to support a claim by the consumer group that the bottles are underfilled.

8.159 A magazine subscription service in a city containing 25,000 households is interested in estimating the average number of magazine subscriptions held by the residents of each household. Suppose that we let x equal the number of magazine subscriptions held by the residents of a sample household. In a random sample of 401 households selected from the city,

$$\sum_{i=1}^{401} x_i = 785 \qquad \text{and} \qquad \sum_{i=1}^{401} x_i^2 = 2{,}015$$

a. Find a point estimate for the average number of magazine subscriptions per resident household in the city. Does the estimate supply the reader with any measure of confidence in its closeness to μ, the unknown mean number of magazine subscriptions per resident household in the entire city? Provide another estimate that bounds μ with a certain degree of confidence, and discuss the meaning of this estimate.

b. If a separate sample of 401 households was collected, would the estimate for μ based on the new sample be the same as the estimate computed in part a?

c. If the two sets of sample observations are combined, what will be the effect on the width of the resulting confidence interval estimate (for a fixed confidence level)?

d. Use the sample results to find an estimate for the *total* number of magazine subscriptions held by all the residents of the city.

e. The director of the magazine subscription service has issued the following statement: "We shall undertake an extensive sales campaign in the city unless there is sufficient evidence to indicate that the average number of subscriptions per household is 1.8 or greater. We are further willing to assume no more than a 5% chance of failing to undertake the sales campaign when we should have." From the sample information, would you say that the subscription service should undertake a sales campaign in the city?

8.160 A company marketing representative wishes to determine the acceptability of a new product in a particular community. If the company could feel confident that about 50% of the community's residents would buy its product, the marketing representative would suggest that it be marketed in that community. A random sample of $n = 64$ is selected from the target community, and 24 of those sampled state that they would buy the product. What conclusions can you draw concerning the marketability of the product in the target community?

Case Study

West Coast Container Corporation

The marketing concept is generally defined as a means of organizing the plans and actions of a firm to satisfy the consumer at a profit. Some authors suggest that adoption of the marketing concept results in increased reliance on consumer responses—as opposed to research and development—for new-product ideas. Furthermore, Lawton and Parasuraman note that "since the marketing concept emphasizes a consumer orientation, adoption of the concept implies greater reliance on (costly) marketing research" (L. Lawton and A. Parasuraman, "The Impact of the Marketing Concept on New Product Development," *Journal of Marketing*, Winter 1980).

For example, consider the case of the West Coast Container Corporation (WCCC), a manufacturer of paperboard containers, with primary service to the food service industry. In response to requests from customers WCCC has prepared a new container to protect bulk shipments of taco shells in transit to various food service operations around the country.

The new container is more costly than competitive containers for transporting taco shells, and the additional cost must be offset by a reduction in product breakage. Thus, sales personnel representing WCCC products must be able to provide convincing evidence from independent use tests to encourage adoption of the new container.

To obtain such evidence, the marketing director of WCCC contacted a private research testing firm with a request to provide an accurate estimate of the breakage rate when using the new 500-shell-capacity container. His request was for an estimate of the average number of broken taco shells per container, accurate to within one taco shell, with a probability of .99.

A preliminary report from the testing service stated that an estimate that satisfied the marketing director's requirements would cost $100,000. The testing procedure was fairly complex, involving the shipment of taco shells through the distribution channels customarily used by the industry and then a report of the number of broken shells per container at the point of destination. Using results of preliminary studies, the research firm estimated that testing costs would average about $200 per container.

In its preliminary studies the testing firm examined the breakage in 20 containers; the results are listed in the accompanying table.

Container	Broken Shells	Container	Broken Shells
1	29	11	12
2	25	12	38
3	18	13	33
4	28	14	24
5	23	15	38
6	14	16	35
7	27	17	20
8	23	18	11
9	40	19	15
10	20	20	27

1. Using these preliminary data, would you say that the cost estimate provided by the testing service appears to be reasonable?

2. Suppose the marketing director of WCCC establishes a limit of $20,000 for testing and research for the new containers. How might he compromise his requirements of confidence and accuracy (size of the allowable bound) to accommodate this budget?

References and Suggested Readings

Berenson, M. L., and D. M. Lavine. *Basic Business Statistics: Concepts and Applications*, 3d ed. Englewood Cliffs, N.J.: Prentice-Hall, 1986.

Harnett, D. L., and J. L. Murphy. *Introductory Statistical Analysis*, 2d ed. Reading, Mass.: Addison-Wesley, 1980.

Hoel, P. G., and R. J. Jessen. *Basic Statistics for Business and Economics*, 3d ed. New York: Wiley, 1982.

Kenkel, J. L. *Introductory Statistics for Management and Economics*, 2d ed. Boston: PWS-KENT Publishers, 1984.

Kohler, H. *Statistics for Business and Economics*, 2d ed. Glenview, Ill.: Scott Foresman, 1988.

Lapin, L. *Statistics for Modern Business Decisions*, 3d ed. New York: Harcourt Brace Jovanovich, 1987.

Summers, G. W., W. S. Peters, and G. P. Armstrong. *Basic Statistics in Business and Economics*, 4th ed. Belmont, Calif.: Wadsworth, 1985.

TO
THE
READER

The inferential procedures in Chapter 8 relied heavily on the Central Limit Theorem, which assures us that the sampling distributions of statistics that are sample means or differences between sample means are approximately normal when the sample sizes are large. However, selecting large sample sizes may not always be feasible because of factors such as time, cost, or the availability of experimental materials.

When the results of an investigation or experiment are based on small sample sizes, the sampling distribution of a single mean or the difference between two means is known if the sampled populations themselves have a normal distribution. In fact, the sampling distributions of s^2 as well as the ratio s_1^2/s_2^2 are also known in this case. The sampling distributions known as Student's t distribution, the chi-square distribution, and the F distribution arise when the sampled populations have normal distributions. The role of these distributions in small-sample testing and estimation procedures for means and variances is the subject of Chapter 9. You will find that the inferential concepts and procedures remain the same in these situations and that it is only the sampling distribution that has changed.

9

INFERENCES FROM
SMALL SAMPLES

9.1

Introduction

Large-sample methods for making inferences about population means and the difference between two means were discussed, with examples, in Chapter 8. Frequently, however, cost and available time limit the size of the sample that can be acquired. In this case the large-sample procedures of Chapter 8 are inappropriate, and other tests and estimation procedures must be used. In this chapter we will study several small-sample inferential procedures that are closely related to the large-sample methods presented in Chapter 8. Specifically, we will consider methods for estimating and testing hypotheses about population means, the difference between two means, a population variance, and a comparison of two population variances. (Small-sample tests about the binomial parameter p are discussed in Chapter 5.)

9.2

Student's *t* Distribution

We introduce our topic by considering a problem. A very costly experiment has been conducted to evaluate a new process for producing synthetic diamonds. Six diamonds have been generated by the new process, with recorded weights of .46, .61, .52, .48, .57, and .54 carat.

A study of the process costs indicates that the average weight of the diamonds must be greater than .5 carat if the process is to be operated at a profitable level. Do the six diamond weight measurements present sufficient evidence to indicate that the average weight of the diamonds produced by the process is in excess of .5 carat? That is, we wish to test the null hypothesis that $\mu = .5$ against the alternative hypothesis that $\mu > .5$.

According to the Central Limit Theorem, the test statistic

$$z = \frac{\bar{x} - \mu}{\sigma/\sqrt{n}}$$

possesses, approximately, a normal distribution in repeated sampling when n is large. For $\alpha = .05$ we could employ a one-tailed statistical test and reject the null hypothesis if $z > 1.645$. This procedure, of course, assumes that σ is known or that a good estimate s is available and is based on a reasonably large sample (we have suggested $n \geq 30$). Unfortunately, the latter requirement is not satisfied for the $n = 6$ diamond weight measurements. How, then, may we test the hypothesis that $\mu = .5$ against the alternative that $\mu > .5$ when we have a small sample?

The problem we pose is not new; it received serious attention from statisticians and experimenters at the turn of the century. If a sample standard deviation s is substituted for σ in z, does the resulting test statistic possess, approximately, a standardized normal distribution in repeated sampling? More specifically, is the rejection region $z > 1.645$ appropriate; that is, do approximately 5% of the values of the test statistic, computed in repeated sampling, exceed 1.645 when H_0 is true?

The answer to these questions, not unlike many of the problems encountered in the sciences, may be resolved by experimentation. In other words, we could draw a small sample, say $n = 6$ measurements, and compute the value of the test statistic. Then we would repeat this process many, many times and construct a frequency distribution for the computed values of the test statistic. The general shape of the distribution and the location of the rejection region would then be evident.

The distribution of the test statistic

$$t = \frac{\bar{x} - \mu}{s/\sqrt{n}}$$

for samples drawn from a normally distributed population was discovered by W. S. Gosset and published (in 1908) under the pen name "Student." He referred to the **Student's *t* statistic** quantity under study as t and it has ever since been known as **Student's *t***. We omit the complicated mathematical expression for the sampling distribution for t but we describe some of its characteristics.

In repeated sampling the distribution of the test statistic

$$t = \frac{\bar{x} - \mu}{s/\sqrt{n}}$$

is, like z, mound-shaped and perfectly symmetrical about $t = 0$. Unlike z, it is much more variable, tailing rapidly out to the right and left, a phenomenon that may be readily explained. The variability of z in repeated sampling is due solely to $\bar{x}$; the other quantities appearing in z (n and σ) are nonrandom. On the other hand, the variability of t is contributed by two random quantities, $\bar{x}$ and s, which can be shown to be independent of one another. When $\bar{x}$ is very large, s may be very small, and vice versa. As a result, t is more variable than z in repeated sampling (see Figure 9.1). Finally, as we

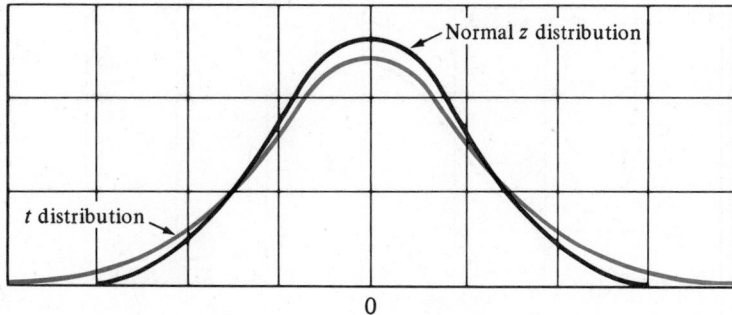

Figure 9.1   Standard Normal z and the t Distribution Based on n = 6 Measurements (5 d.f.)

might surmise, the variability of t decreases as n increases because the estimate s of σ is based on more and more information. When n is infinitely large, the t and z distributions are identical. Thus, Gosset discovered that the distribution of t depended on the sample size n.

degrees of freedom   The divisor $(n - 1)$ of the sum of squares of deviations that appears in the formula for s^2 is called the number of **degrees of freedom** (d.f.) associated with s^2. The origin of the term *degrees of freedom* is linked to the statistical theory underlying the probability distribution of s^2 and refers to the number of independent squared deviations available for estimating σ^2. We will not pursue this point further except to say that the test statistic t based on a sample of n measurements possesses $(n - 1)$ degrees of freedom.

The critical values of t that separate the rejection and acceptance regions for a statistical test are presented in Table 4 of the Appendix. A portion of Table 4 showing the format and giving some of the entries is reproduced in Table 9.1. The tabulated value t_α is the value of t such that an area α lies to its right, as shown in Figure 9.2. The

Table 9.1   Format of the Student's t Table, Table 4 in the Appendix

d.f.	$t_{.100}$	$t_{.050}$	$t_{.025}$	$t_{.010}$	$t_{.005}$	d.f.
1	3.078	6.314	12.706	31.821	63.657	1
2	1.886	2.920	4.303	6.965	9.925	2
3	1.638	2.353	3.182	4.541	5.841	3
4	1.533	2.132	2.776	3.747	4.604	4
5	1.476	**2.015**	2.571	3.365	4.032	5
6	1.440	1.943	2.447	3.143	3.707	6
7	1.415	1.895	2.365	2.998	3.499	7
8	1.397	1.860	2.306	2.896	3.355	8
$\vdots$	$\vdots$	$\vdots$	$\vdots$	$\vdots$	$\vdots$	$\vdots$
27	1.314	1.703	2.052	2.473	2.771	27
28	1.313	1.701	2.048	2.467	2.763	28
29	1.311	1.699	2.045	2.462	2.756	29
inf.	1.282	1.645	1.960	2.326	2.576	inf.

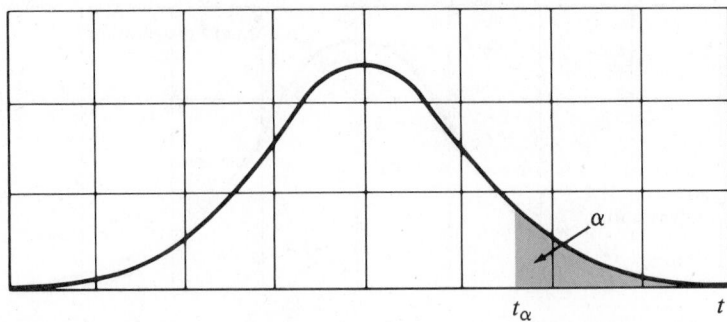

Figure 9.2   Tabulated Values of Student's t

degrees of freedom (d.f.) associated with s^2 are shown in the first and last columns of the table, and the t_α's corresponding to various values of α appear in the top row. Thus, if we wish to find the value of t such that 5% of the area lies to its right, we would use the column marked $t_{.05}$. The critical value of t for our example, found in the $t_{.05}$ column opposite the row corresponding to d.f. $= n - 1 = 6 - 1 = 5$, is $t_{.05} = 2.015$. This entry is shown in color in Table 9.1. Thus, we would reject $H_0 : \mu = .5$ when $t > 2.015$. Since the distribution of t is symmetrical about $t = 0$, a left-tailed critical value of t is merely the negative of the corresponding right-tailed critical value. For example, with 5 degrees of freedom, the area to the left of $t = -2.015$ is equal to .05.

Note that the critical value of t is always larger than the corresponding critical value of z for a specified α. For example, when $\alpha = .05$, the critical value of t for $n = 2$ (d.f. $= 1$) is $t = 6.314$, which is very large when compared with the corresponding $z = 1.645$. Proceeding down the $t_{.05}$ column, we note that the critical value of t decreases, reflecting the effect of a larger sample size on the estimation of σ. Finally, when n is infinitely large, the critical value of t equals 1.645.

The reason for choosing $n = 30$ (an arbitrary choice) as the dividing line between large and small samples becomes apparent when you examine Table 9.1. For $n = 30$ the critical value of $t_{.05} = 1.699$ is numerically quite close to $z_{.05} = 1.645$. For a two-tailed test based on $n = 30$ measurements and $\alpha = .05$, we would place .025 in each tail of the t distribution and reject $H_0 : \mu = \mu_0$ when $t > 2.045$ or $t < -2.045$. The value of $t_{.025} = 2.045$ is very close to the value $z_{.025} = 1.96$ employed in the z test.

Note that Student's t and the corresponding tabulated critical values are based on the assumption that the sampled population possesses a normal probability distribution. This, indeed, is a very restrictive assumption because in many sampling situations the properties of the population will be completely unknown and may be nonnormal. If this restriction were to seriously affect the distribution of the t statistic, the application of the t test would be very limited. Fortunately, this point is of little consequence, since it can be shown that the distribution of the t statistic possesses nearly the same shape as the theoretical t distribution for populations that are nonnormal but possess a mound-shaped probability distribution. This property of the t statistic and the common occurrence of mound-shaped distributions of data in nature enhance the value of the Student's t for use in statistical inference.

We should also remember that $\bar{x}$ and s^2 must be *independent* (in a probabilistic sense) in order that the quantity

$$\frac{\bar{x} - \mu}{s/\sqrt{n}}$$

possess a t distribution in repeated sampling. As mentioned previously, this requirement will automatically be satisfied when the sample has been randomly drawn from a normal population.

Having discussed the origin of the Student's t and the tabulated critical values (Table 4 of the Appendix), we now return to the problem of making an inference about the mean diamond weight based on our sample of $n = 6$ measurements. Prior to considering the solution, you may wish to test your built-in inference-making equipment by glancing at the six measurements and arriving at a conclusion about the significance of the data.

9.3

Small-Sample Inferences About a Population Mean

The statistical test of an hypothesis about a population mean may be stated as follows:

Small-Sample Test for a Population Mean

Null hypothesis $H_0: \mu = \mu_0$.

Alternative hypothesis H_a: Specified by the experimenter, depending on the alternate value of μ he or she wishes to detect.

Test statistic: $t = \dfrac{\bar{x} - \mu_0}{s/\sqrt{n}}$.

Rejection region: For a one-tailed test when H_a is $\mu > \mu_0$, reject H_0 if $t > t_\alpha$; when H_a is $\mu < \mu_0$, reject H_0 if $t < -t_\alpha$. For a two-tailed test, reject H_0 if $t > t_{\alpha/2}$ or $t < -t_{\alpha/2}$.

The mean and standard deviation for the six diamond weights are .53 and .0559, respectively, and the elements of the test as defined above are as follows:

Null hypothesis $H_0: \mu = .5$.

Alternative hypothesis $H_a: \mu > .5$.

Test statistic: $t = \dfrac{\bar{x} - \mu_0}{s/\sqrt{n}} = \dfrac{.53 - .5}{.0559/\sqrt{6}} = 1.31.$

Rejection region: With $\alpha = .05$ and d.f. $= 5$, we will reject H_0 if $t > 2.015$ (see Figure 9.2).

Since the calculated value of the test statistic does not fall in the rejection region, we do not reject H_0. Thus, the data do not present sufficient evidence to indicate that the mean diamond weight exceeds .5 carat.

The calculation of the probability β of a type II error for the t test is difficult and is beyond the scope of this text. Therefore, we will avoid this problem and obtain an interval estimate for μ, as noted in Section 8.17.

The large-sample confidence interval for μ given in Chapter 8 is

$$\bar{x} \pm z_{\alpha/2} \frac{\sigma}{\sqrt{n}}$$

where $z_{\alpha/2} = 1.96$ for a confidence coefficient of .95. This result assumes that σ is known and simply involves a measurement of $1.96\sigma_{\bar{x}}$ (or approximately $2\sigma_{\bar{x}}$) on either side of $\bar{x}$, in conformity with the Empirical Rule. When σ is unknown and must be estimated by a small-sample standard deviation s, the large-sample confidence interval based on z will not enclose μ 95% of the time in repeated sampling. To account for the added variability introduced through s, the estimate of σ, the appropriate confidence interval for μ is obtained by replacing the critical value $z_{\alpha/2}$ with the critical value $t_{\alpha/2}$ based on $(n-1)$ degrees of freedom. The small-sample confidence interval for μ is given in the next display.

Small-Sample $(1 - \alpha)$ 100% Confidence Interval for μ

$$\bar{x} \pm t_{\alpha/2} \frac{s}{\sqrt{n}}$$

standard error of the mean

where $s/\sqrt{n}$ is the estimated standard error of $\bar{x}$ but is often referred to as the **standard error of the mean**.

For our example a 95% confidence interval for μ is

$$\bar{x} \pm t_{\alpha/2} \frac{s}{\sqrt{n}} \qquad \text{or} \qquad .53 \pm 2.571 \frac{.0559}{\sqrt{6}} \qquad \text{or} \qquad .53 \pm .059$$

Therefore, the interval estimate for μ is .471 to .589, with confidence coefficient of .95. If the experimenter wishes to detect a small increase in mean diamond weight in excess of .5 carat, the width of the interval must be reduced by obtaining more diamond weight measurements. Additional measurements will decrease both $1/\sqrt{n}$ and $t_{\alpha/2}$ and thereby decrease the width of the interval. From the standpoint of a statistical test of an hypothesis, an increase in n increases the available information upon which to base a decision and decreases the probability of making a type II error.

EXAMPLE 9.1   A manufacturer of gunpowder has developed a new powder that is designed to produce a muzzle velocity of 3,000 feet per second. Eight shells are loaded with the charge and the muzzle velocities measured. The resulting velocities are shown in Table 9.2. Do the data present sufficient evidence to indicate that the average velocity differs from 3,000 feet per second?

Table 9.2   Data for Example 9.1

Muzzle Velocity (Feet per Second)	
3,005	2,995
2,925	3,005
2,935	2,935
2,965	2,905

SOLUTION   Testing the null hypothesis that $\mu = 3{,}000$ feet per second against the alternative that μ is either greater than or less than 3,000 feet per second results in a two-tailed statistical test. Thus,

$$H_0:\mu = 3{,}000 \quad \text{and} \quad H_a:\mu \neq 3{,}000$$

Using $\alpha = .05$ and placing .025 in each tail of the t distribution, we find that the critical value of t for $n = 8$ measurements [or $(n - 1) = 7$ d.f.] is $t_{.025} = 2.365$. Hence, we will reject H_0 if $t > 2.365$ or $t < -2.365$. (Recall that the t distribution is symmetrical about $t = 0$.)

The sample mean and standard deviation for the recorded data are

$$\bar{x} = 2{,}958.75 \quad \text{and} \quad s = 39.26$$

Then

$$t = \frac{\bar{x} - \mu_0}{s/\sqrt{n}} = \frac{2{,}958.75 - 3{,}000}{39.26/\sqrt{8}} = -2.97$$

Since the observed value of t falls in the rejection region, we reject H_0 and conclude that the average velocity is less than 3,000 feet per second. Furthermore, we are reasonably confident that we have made the correct decision. Using our procedure, we should erroneously reject H_0 only $\alpha = .05$ of the time in repeated applications of the statistical test.

A 95% confidence interval provides additional information about μ. Calculating

$$\bar{x} \pm t_{\alpha/2} \frac{s}{\sqrt{n}}$$

we obtain

$$2{,}958.75 \pm (2.365)\frac{39.26}{\sqrt{8}} \quad \text{or} \quad 2{,}958.75 \pm 32.83$$

Thus, we estimate that the average muzzle velocity lies in the interval from 2,925.92 to 2,991.58 feet per second. A more accurate estimate can be obtained by increasing the sample size.

EXAMPLE 9.2   If you planned to report the results of the statistical test in Example 9.1, what p-value would you report?

SOLUTION   The p-value for this test is the probability of observing a value of the t statistic as contradictory to the null hypothesis as the one observed for this set of data, namely, $t = -2.97$. Since this is a two-tailed test, the p-value is the probability that either $t \leq -2.97$ or $t \geq 2.97$.

Unlike the table of areas under the normal curve (Table 3 of the Appendix), the table fot t (Table 4 of the Appendix) does not give the areas corresponding to various values of t. Rather, it gives the values of t corresponding to upper-tail areas equal to .10, .05, .025, .010, and .005. Consequently, we can only approximate the upper-tail area that corresponds to the probability that $t > 2.97$. Since the t statistic for this test is based on 7 degrees of freedom, we refer to the d.f. $= 7$ row of Table 4 and find that 2.97 falls between $t_{.025} = 2.365$ and $t_{.010} = 2.998$. The right-tail area corresponding to the probability that $t > 2.97$ lies between .025 and .010. Since this value represents only half of the required area, the actual p-value lies between $2(.025) = .05$ and $2(.010) = .020$. Using the tabulated values of t, we could reject H_0 with $\alpha = .05$ but not with $\alpha = .02$. Therefore, the p-value for this test would be reported as p-value $= .05$. Since most researchers would be using a t table similar or identical to Table 4, the reader of a published research report would realize that for a reported p-value of .05, the exact p-value for the test was probably less than .05 and was between .05 and .02. In fact, some researchers report these results as $.02 < p$-value $< .05$, indicating significance at the .05 level but not at the .02 level.

◆◆

The MINITAB package contains commands that can be used to implement a small-sample test of a population mean or to produce a small-sample confidence interval for a population mean. The command TTEST requires the user to provide the value of the population mean to be tested, together with the column number in which the data are stored. The subcommand ALTERNATIVE, followed by a 1, -1, or 0, will implement a right-tailed, left-tailed, or two-tailed test procedure. If this subcommand is not used, a two-tailed test is implemented by default.

The results of using TTEST to implement the test in Example 9.1 are given in Table 9.3. In addition to the observed value of $t = -2.97$, the output gives the sample mean $\bar{x} = 2,958.7$, the sample standard deviation $s = 39.3$, and the standard

Table 9.3   MINITAB Printout for the Data in Table 9.2

```
MTB > TTEST 3000 C1

TEST OF MU = 3000.0 VS MU N.E. 3000.0

            N      MEAN    STDEV   SE MEAN         T     P VALUE
C1          8    2958.7     39.3      13.9     -2.97       0.021

MTB > TINTERVAL C1

            N      MEAN    STDEV   SE MEAN    95.0 PERCENT C.I.
C1          8    2958.7     39.3      13.9   ( 2925.9,   2991.6)
```

error of the mean (SE MEAN), $s/\sqrt{n} = 13.9$. When compared with our results in Example 9.1, the only noticeable difference is in the reported decimal accuracy of the results. In addition, the exact p-value for the test appears in the printout and is given as P VALUE $= .021$. In using Table 4 to approximate this value, we found that the p-value was between .05 and .02, consistent with the MINITAB result.

The MINITAB command TINTERVAL is available for constructing a confidence interval for a population mean. The command requires that the user supply the confidence coefficient and the column location of the data. A 95% confidence interval for μ using the data in Example 9.1 also appears in Table 9.3. Apart from the reported decimal accuracy, the results agree with those found in Example 9.2.

EXERCISES   Conceptual Level

9.1   When does the statistic $(\bar{x} - \mu)/(\sigma/\sqrt{n})$ have a normal distribution? When does the statistic $(\bar{x} - \mu)/(s/\sqrt{n})$ have a Student's t distribution? An approximate normal distribution?

9.2   Is the t distribution more or less variable than the z distribution? Can you explain this behavior in terms of the statistic $(\bar{x} - \mu)/(s/\sqrt{n})$?

9.3   How would you define the term *degrees of freedom*? How does the t distribution change with increasing degrees of freedom?

EXERCISES   Skill Level

9.4   Use Table 4 in the Appendix to find the following values of t_α, the value such that $P(t > t_\alpha) = \alpha$.

	α	Degrees of Freedom
a.	.05	10
b.	.05	20
c.	.025	10
d.	.01	28
e.	.10	13

9.5   A random sample of $n = 10$ observations from a normal population produced $\bar{x} = 10.33$ and $s = 3.25$.

a.   Find a 95% confidence interval for the population mean.

b.   Test the hypothesis $H_0: \mu = 10$ against $H_a: \mu \neq 10$, using $\alpha = .05$.

c.   What is the observed level of significance for the test in part c?

9.6   The following $n = 12$ observations resulted when sampling from a normal distribution.

2.4	2.5	1.2	1.9
2.1	2.6	2.5	2.7
1.5	2.2	1.3	1.7

a. Find the mean and standard deviation of these data.

b. Find a 99% confidence interval for the population mean μ.

c. Test $H_0 : \mu = 3$ versus $H_a : \mu \neq 3$ with $\alpha = .01$.

d. Do the results in part b support your conclusion in part c?

EXERCISES   Applications

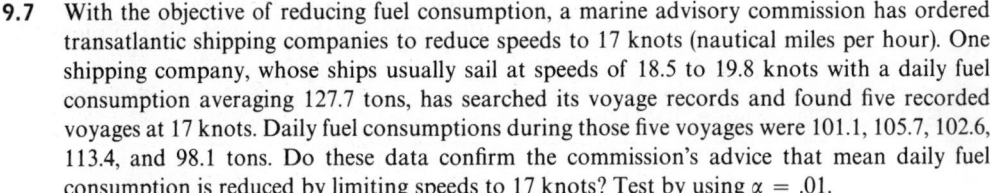

9.7 With the objective of reducing fuel consumption, a marine advisory commission has ordered transatlantic shipping companies to reduce speeds to 17 knots (nautical miles per hour). One shipping company, whose ships usually sail at speeds of 18.5 to 19.8 knots with a daily fuel consumption averaging 127.7 tons, has searched its voyage records and found five recorded voyages at 17 knots. Daily fuel consumptions during those five voyages were 101.1, 105.7, 102.6, 113.4, and 98.1 tons. Do these data confirm the commission's advice that mean daily fuel consumption is reduced by limiting speeds to 17 knots? Test by using $\alpha = .01$.

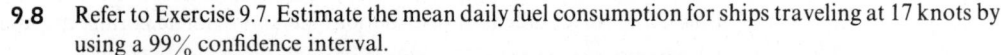

9.8 Refer to Exercise 9.7. Estimate the mean daily fuel consumption for ships traveling at 17 knots by using a 99% confidence interval.

9.9 Bausch and Lomb Corporation has recently developed a new type of extended-life contact lens, made of silicone, which it claims has a useful life exceeding that of other types of contact lenses (*Economist*, June 21, 1980). During the research and development period six people were asked to use the new silicone-based contact lenses. The mean and standard deviation of the useful life of the six pairs of lenses were 4.6 and .49 years, respectively. Construct a 90% confidence interval for the mean life of the new silicone-based contact lenses developed by Bausch and Lomb.

9.10 A building contractor has built a large number of houses of about the same size and value. The contractor claims that the average value of these houses (or similar houses that he might have built) does not exceed $95,000. A real estate appraiser randomly selected five of the new homes built by the contractor and assessed their values as $94,500, $97,000, $96,000, $95,000, and $95,500. Do the five appraisals contradict the building contractor's claim regarding the mean value of his houses? Test at the $\alpha = .05$ level of significance.

9.11 Refer to Exercise 9.10. Estimate the p-value associated with the statistical test regarding the contractor's claim.

9.12 Fast-food service companies try to devise wage plans that provide incentive and produce salaries for their managers that are competitive with corresponding positions in competing companies. A random sampling of 12 unit managers for one company shows that they earn an average salary of $26,750, with a standard deviation of $3,100. Do these data suggest that the mean salary earned by the company's unit managers differs from $28,500, the mean annual salary paid by a competitive firm to its managers? Test the null hypothesis that $\mu = \$28,500$ against the alternative that $\mu \neq \$28,500$ at the 5% level of significance.

9.13 Industrial wastes and sewage dumped into our rivers and streams absorb oxygen and thereby reduce the amount of dissolved oxygen available for fish and other forms of aquatic life. One state agency requires a minimum of 5 parts per million (ppm) of dissolved oxygen in order that the oxygen content be sufficient to support aquatic life. Six water specimens taken from a river at a specific location during the low-water season (July) gave readings of 4.9, 5.1, 4.9, 5.0, 5.0, and 4.7 ppm of dissolved oxygen. Estimate the p-value associated with the statistical test that examines whether the data provide sufficient evidence to indicate that the dissolved oxygen content is less than 5 ppm. What are your conclusions?

9.14 With the skyrocketing prices of gold and silver, companies that use precious metals in their manufacturing processes have become increasingly concerned with possible employee pilferage. One recent article reports on the case of a Dutch manufacturer of silver-plated tableware who found that boxes of industrial silver, labeled as containing 4.5 kilograms (kg) of silver each, did not check out in review as expected. Suppose 9 boxes of industrial silver were weighed by the manufacturer, revealing a mean weight of 4.21 kg with a standard deviation of .39 kg. Do these data suggest that the mean silver content per box is less than 4.5 kg (the advertised content) and, hence, that the manufacturer's suspicions are confirmed? Test by using $\alpha = .05$.

 9.15   Refer to Exercise 9.14. If the Dutch manufacturer is concerned with the possible legal implications of false accusation, how would you advise him in the conduct of his statistical analyses? That is, would you advise him to base his inference on a level of significance of $\alpha = .10$, $\alpha = .05$, or $\alpha = .01$ before concluding that pilferage is responsible for the underweight contents in the boxes of industrial silver?

 9.16   Gasohol (trademark of Nebraska Agricultural Products) is a blend of ethyl alcohol (ethanol) and unleaded gasoline; it has been used in recent years in an effort to stretch available gasoline supplies. Development of gasohol has been most prominent in the midwestern Corn Belt because corn appears to be the most productive source of ethanol ("Corn States Uncork Gasohol Gusher," *Machine Design*, March 10, 1980). A research group has conducted ethanol content studies among nine different samples of a certain variety of corn and finds the following yields, in pounds of ethanol per bushel: 15.9, 17.1, 16.1, 16.3, 15.8, 15.3, 16.5, 16.2, 17.3. Find a 95% confidence interval for the average ethanol content per bushel from this particular variety of corn.

9.4

Small-Sample Inferences About the
Difference Between Two Means

The physical setting for the problem we consider here is identical to that discussed in Section 8.9. Independent random samples of n_1 and n_2 measurements, respectively, are drawn from two populations, which possess means and variances μ_1, σ_1^2 and μ_2, σ_2^2. Our objective is to make inferences concerning the difference $(\mu_1 - \mu_2)$ between the two population means.

The following small-sample methods for testing hypotheses and placing a confidence interval on the difference between two means are, like the case for a single mean, founded on assumptions regarding the probability distributions of the sampled populations. Specifically, **we will assume that both populations possess a normal probability distribution and also that the population variances σ_1^2 and σ_2^2 are equal.** In other words, we assume that the variability of the measurements in the two populations is the same and can be measured by a common variance, which we will designate as σ^2; that is, $\sigma_1^2 = \sigma_2^2 = \sigma^2$. Although the assumption of equal population variances may be surprising, it is quite reasonable for many sampling situations.

The point estimator of $(\mu_1 - \mu_2)$ is $(\bar{x}_1 - \bar{x}_2)$, the difference between the sample means. In Section 8.9 we found that in repeated sampling $(\bar{x}_1 - \bar{x}_2)$ was an unbiased estimator of $(\mu_1 - \mu_2)$, with a standard deviation

$$\sigma_{(\bar{x}_1 - \bar{x}_2)} = \sqrt{\frac{\sigma_1^2}{n_1} + \frac{\sigma_2^2}{n_2}}$$

This result was used in placing bounds on the error of estimation, in constructing a large-sample confidence interval, and in testing an hypothesis concerning the difference between two population means. The large sample test of the hypothesis $H_0 : \mu_1 - \mu_2 = D_0$, where D_0 is the hypothesized difference between the means, was based on the z statistic given by

$$z = \frac{(\bar{x}_1 - \bar{x}_2) - D_0}{\sqrt{\dfrac{\sigma_1^2}{n_1} + \dfrac{\sigma_2^2}{n_2}}}$$

When we sample two normal populations with equal variances, and $\sigma_1^2 = \sigma_2^2 = \sigma^2$, then the z statistic can be simplified as follows:

$$z = \frac{(\bar{x}_1 - \bar{x}_2) - D_0}{\sqrt{\dfrac{\sigma^2}{n_1} + \dfrac{\sigma^2}{n_2}}} = \frac{(\bar{x}_1 - \bar{x}_2) - D_0}{\sigma \sqrt{\dfrac{1}{n_1} + \dfrac{1}{n_2}}}$$

For small-sample tests of the hypothesis $H_0 : \mu_1 - \mu_2 = D_0$, it seems reasonable to use the test statistic given next.

Small-Sample Test Statistic t for the Difference Between Two Means

$$t = \frac{(\bar{x}_1 - \bar{x}_2) - D_0}{s \sqrt{\dfrac{1}{n_1} + \dfrac{1}{n_2}}}$$

That is, we use a sample standard deviation s as an estimator of σ. This test statistic possesses a Student's t distribution in repeated sampling when the stated assumptions are satisfied, a fact that can be proved mathematically or verified by experimental sampling from two normal populations.

The estimate s used in the t statistic could be either s_1 or s_2, the standard deviation for each of the two samples. However, the use of either one alone would be wasteful, since both sample standard deviations are independent estimators of σ. The information from both sample variances can be combined by using a weighted average in which the weights reflect the relative amount of information in s_1^2 and s_2^2. Since s_1^2 is based on $(n_1 - 1)$ degrees of freedom and s_2^2 is based on $(n_2 - 1)$ degrees of freedom, a **pooled estimator of σ^2** using the degrees of freedom as weights is

pooled estimator of σ^2

$$s^2 = \frac{(n_1 - 1)s_1^2 + (n_2 - 1)s_2^2}{(n_1 - 1) + (n_2 - 1)}$$

with $(n_1 - 1) + (n_2 - 1) = n_1 + n_2 - 2$ degrees of freedom.

The pooled estimator s^2 can also be expressed as a sum of the squared deviations in each sample. If x_{1i} and x_{2i} represent the ith observations in samples 1 and 2, then

$$s_1^2 = \frac{\sum\limits_{i=1}^{n_1} (x_{1i} - \bar{x}_1)^2}{(n_1 - 1)} \quad \text{and} \quad s_2^2 = \frac{\sum\limits_{i=1}^{n_2} (x_{2i} - \bar{x}_2)^2}{(n_2 - 1)}$$

Using these expressions, we can write the pooled estimator of σ^2 in either of the following two ways.

Pooled Estimator of σ^2

$$s^2 = \frac{\displaystyle\sum_{i=1}^{n_1}(x_{1i} - \bar{x}_1)^2 + \sum_{i=1}^{n_2}(x_{2i} - \bar{x}_2)^2}{(n_1 - 1) + (n_2 - 1)}$$

$$s^2 = \frac{(n_1 - 1)s_1^2 + (n_2 - 1)s_2^2}{(n_1 - 1) + (n_2 - 1)}$$

The first form for calculating s^2 shows that the numerator is the pooled sum of squared deviations from each sample, and the denominator is the pooled sum of the degrees of freedom from each sample. Regardless of the form used for calculation, it can be proved that s^2 is an unbiased estimator of σ^2 and hence is an unbiased estimator of the common population variance. If the pooled estimator s^2 is used to estimate σ^2 and the samples are randomly and independently drawn from normal populations with a common variance, then the statistic

$$t = \frac{(\bar{x}_1 - \bar{x}_2) - (\mu_1 - \mu_2)}{s\sqrt{\dfrac{1}{n_1} + \dfrac{1}{n_2}}}$$

will have a Student's t distribution with $n_1 + n_2 - 2$ degrees of freedom.

The small-sample test for the difference between two means is given in the next display.

Small-Sample Test for the Difference Between Two Means Based on Independent Random Samples

Null hypothesis $H_0: \mu_1 - \mu_2 = D_0$.

Alternative hypothesis H_a: One- or two-tailed hypothesis determined by the experimenter.

Test statistic:

$$t = \frac{(\bar{x}_1 - \bar{x}_2) - D_0}{s\sqrt{\dfrac{1}{n_1} + \dfrac{1}{n_2}}}$$

with $(n_1 + n_2 - 2)$ degrees of freedom.

Rejection region: For a one-tailed test when H_a is $\mu_1 - \mu_2 > D_0$, reject H_0 if $t > t_\alpha$; when H_a is $\mu_1 - \mu_2 < D_0$, reject H_0 if $t < -t_\alpha$. For a two-tailed test, reject H_0 when $t > t_{\alpha/2}$ or $t < -t_{\alpha/2}$.

Here, s is the square root of s^2, the pooled estimator of σ^2.

Critical values of t can be obtained from Table 4 of the Appendix. For example, if $n_1 = 10$ and $n_2 = 12$, we use the tabled values of t corresponding to $n_1 + n_2 - 2 = 20$ degrees of freedom.

EXAMPLE 9.3   Manufacturing organizations incur considerable cost in the training of new employees. Not only is there a direct cost involved in the training program, but there is also an indirect cost to the firm since employees in training do not contribute directly to the firm's manufacturing process. Hence, such organizations seek training programs that can bring new employees to maximum efficiency in the shortest possible time.

An assembly operation in a manufacturing plant requires approximately a one-month training period for a new employee to reach maximum efficiency. A new method of training was suggested and a test conducted to compare the new method with the standard procedure. Two groups of nine new employees were trained for a period of three weeks, one group using the new method and the other following the standard training procedure. The length of time (in minutes) required for each employee to assemble the device was recorded at the end of the three-week period. These measurements appear in Table 9.4. Do the data present sufficient evidence to indicate that the mean time to assemble at the end of the three-week training period is less for the new training procedure?

Table 9.4   Length of Time (Minutes) to Assemble Device, for Example 9.3

Standard Procedure		New Procedure	
32	44	35	40
37	35	31	27
35	31	29	32
28	34	25	31
41		34	

SOLUTION   Let μ_1 and μ_2 be the mean time to assemble for the standard and the new assembly procedures, respectively. Then since we seek evidence to support the theory that $\mu_1 > \mu_2$, we will test the null hypothesis $H_0: \mu_1 = \mu_2$ (i.e., $\mu_1 - \mu_2 = 0$) against the alternative hypothesis $H_a: \mu_1 > \mu_2$ (i.e., $\mu_1 - \mu_2 > 0$). To conduct this test, assume that the population distributions of measurements are approximately normal, and that the variability for the two populations of measurements is approximately the same.

The sample means and sums of squared deviations are

$$\bar{x}_1 = 35.22 \quad \text{and} \quad \sum_{i=1}^{9} (x_{1i} - \bar{x}_1)^2 = 195.56$$

$$\bar{x}_2 = 31.56 \quad \text{and} \quad \sum_{i=1}^{9} (x_{2i} - \bar{x}_2)^2 = 160.22$$

Then the pooled estimate of the common variance is

$$s^2 = \frac{\sum\limits_{i=1}^{9}(x_{1i} - \bar{x}_1)^2 + \sum\limits_{i=1}^{9}(x_{2i} - \bar{x}_2)^2}{n_1 + n_2 - 2} = \frac{195.56 + 160.22}{9 + 9 - 2} = 22.24$$

and the standard deviation is $s = 4.72$.

The alternative hypothesis $H_a : \mu_1 > \mu_2$, or, equivalently, $\mu_1 - \mu_2 > 0$, implies that we should use a one-tailed statistical test and that the rejection region for the test is located in the upper tail of the t distribution. Referring to Table 4 in the Appendix, we find that the critical value of t for $\alpha = .05$ and $(n_1 + n_2 - 2) = 16$ degrees of freedom is 1.746. Therefore, we will reject the null hypothesis when the calculated value of t is greater than 1.746.

The calculated value of the test statistic is

$$t = \frac{(\bar{x}_1 - \bar{x}_2)}{s\sqrt{\dfrac{1}{n_1} + \dfrac{1}{n_2}}} = \frac{35.22 - 31.56}{4.72\sqrt{\dfrac{1}{9} + \dfrac{1}{9}}} = 1.64$$

Comparing this value with the critical value, $t_{.05} = 1.746$, we note that the calculated value does not fall in the rejection region. Therefore, we must conclude that there is insufficient evidence to indicate that the new method of training is superior, at the .05 level of significance.

◆◆

EXAMPLE 9.4   Find the p-value that would be reported for the statistical test of Example 9.3.

SOLUTION   The observed value of t for this one-tailed test was $t = 1.64$. Therefore, the p-value for the test would be the probability that $t > 1.64$. Since we cannot obtain this probability from Table 4 of the Appendix, we would report the p-value for the test as the smallest tabulated value for α that leads to the rejection of H_0. Consulting the row in Table 4 corresponding to 16 degrees of freedom, we see that the observed value, $t = 1.64$, lies between $t_{.10} = 1.337$ and $t_{.05} = 1.746$. Therefore, the probability that $t > 1.64$ is between .05 and .10, and the p-value for this test would be reported as $.05 < p\text{-value} < .10$.

◆◆

The small-sample confidence interval for $(\mu_1 - \mu_2)$ is based on the same assumptions as the statistical test procedure. This confidence interval, with confidence coefficient $(1 - \alpha)$, is given by the formula in the display.

Small-Sample $(1 - \alpha)100\%$ Confidence Interval for $(\mu_1 - \mu_2)$ Based on Independent Random Samples

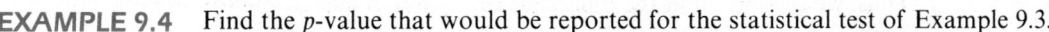

$$(\bar{x}_1 - \bar{x}_2) \pm t_{\alpha/2}s\sqrt{\frac{1}{n_1} + \frac{1}{n_2}}$$

The value of $t_{\alpha/2}$ is found in Table 4, using $(n_1 + n_2 - 2)$ degrees of freedom. Note the similarity in the procedures for constructing the confidence intervals for a single mean (Section 9.3) and for the difference between two means. In both cases the interval is constructed by using the appropriate point estimator and then adding and subtracting an amount equal to $t_{\alpha/2}$ times the estimated standard error of the point estimator.

EXAMPLE 9.5   Refer to Example 9.3. Since there is expense as well as risk associated with abandoning a working procedure, management would like to obtain more information concerning the difference in mean assembly times for the two training procedures. In particular, they would like to estimate the difference $(\mu_1 - \mu_2)$ in mean times to assemble with 95% confidence. Find an interval estimate for $(\mu_1 - \mu_2)$ by using a confidence coefficient of .95.

SOLUTION   Substituting into the formula

$$(\bar{x}_1 - \bar{x}_2) \pm t_{\alpha/2}s\sqrt{\frac{1}{n_1} + \frac{1}{n_2}}$$

we find that the interval estimate (or 95% confidence interval) is

$$(35.22 - 31.56) \pm (2.120)(4.72)\sqrt{\frac{1}{9} + \frac{1}{9}} \qquad \text{or} \qquad 3.66 \pm 4.71$$

Thus, we estimate that the difference $(\mu_1 - \mu_2)$ in mean time to assemble falls in the interval from -1.05 to 8.37. Note that the interval width is considerable; it seems advisable to increase the size of the samples and reestimate $(\mu_1 - \mu_2)$ by using this additional information.

To implement a two-sample t procedure with a pooled estimate of variance using the MINITAB package, enter the main command TWOSAMPLE, the columns identifying the two sets of sample values, and a semicolon. The subcommand POOLED, followed by a period, completes the instructions. The calculated value of t, its p-value, and its degrees of freedom are given as output together with a 95% confidence interval estimate of $(\mu_1 - \mu_2)$. Using the subcommand ALTERNATIVE, the user may also specify whether the test is to be left-tailed (-1), right-tailed (1), or two-tailed (0). The test is performed as two-tailed unless another alternative is specified.

The MINITAB output for the TWOSAMPLE command using the data in Table 9.3 appears in Table 9.5. Notice that the entry SE MEAN (standard error of the mean), given for each column, is calculated as $s/\sqrt{n}$. For example, the standard error of the mean for C1 is $4.94/\sqrt{9} = 1.646$, or 1.6. The remaining entries are self-explanatory and can be compared with the results of Examples 9.3, 9.4, and 9.5.

Before concluding our discussion, we should comment on the two assumptions upon which our inferential procedures are based. Moderate departures from the assumption that the populations possess normal probability distributions do not seriously affect the distribution of the test statistic and the confidence coefficient for the

Table 9.5   MINITAB Output for Two Independent Samples, Using the Data in Table 9.4

```
MTB > PRINT C1 C2
  ROW     C1      C2

   1      32      35
   2      37      31
   3      35      29
   4      28      25
   5      41      34
   6      44      40
   7      35      27
   8      31      32
   9      34      31

MTB > TWOSAMPLE C1 C2;
SUBC> POOLED;
SUBC> ALTERNATIVE 1.

TWOSAMPLE T FOR C1 VS C2
     N        MEAN       STDEV    SE MEAN
C1   9       35.22        4.94        1.6
C2   9       31.56        4.48        1.5

95 PCT CI FOR MU C1 - MU C2: (-1.0, 8.4)
TTEST MU C1 = MU C2 (VS GT): T=1.65 P=0.059 DF=16.0
```

corresponding confidence interval. **On the other hand, the population variances should be nearly equal to ensure that the procedures given above are valid.**

If there is reason to believe that the population variances are far from being equal, two changes must be made in the testing and estimation procedures. Since the pooled estimator s^2 is no longer appropriate when the population variances are not equal, the sample variances s_1^2 and s_2^2 are used as estimators of σ_1^2 and σ_2^2. The resulting test statistic is

$$\frac{(\bar{x}_1 - \bar{x}_2) - D_0}{\sqrt{\dfrac{s_1^2}{n_1} + \dfrac{s_2^2}{n_2}}}$$

When the sample sizes are *small*, critical values for this statistic are found in Table 4 of the Appendix, using degrees of freedom approximated by the formula

$$\text{d.f.} \approx \frac{\left(\dfrac{s_1^2}{n_1} + \dfrac{s_2^2}{n_2}\right)^2}{\dfrac{\left(\dfrac{s_1^2}{n_1}\right)^2}{(n_1 - 1)} + \dfrac{\left(\dfrac{s_2^2}{n_2}\right)^2}{(n_2 - 1)}}$$

Obviously, this result must be rounded to the nearest integer. In the MINITAB package the TWOSAMPLE t command without the subcommand POOLED implements this procedure.

In Section 9.7 we will present a procedure for testing an hypothesis concerning the equality of two population variances that can be used to determine whether or not the underlying population variances are equal.

EXERCISES   Conceptual Level

9.17 Under what assumptions can a Student's t statistic be used in making inferences about the difference between two population means?

9.18 How does the two-sample Student's t statistic differ from the large-sample z statistic used in Chapter 8?

9.19 When the population variances are both equal to σ^2, what are the advantages, if any, of using a two-sample pooled estimator of σ^2 rather than either sample variance, s_1^2 or s_2^2 alone?

EXERCISES   Skill Level

9.20 Find the p-values associated with the test statistics given below, resulting from a two-tailed test of an hypothesis with a specified number of degrees of freedom.

	Observed t	Degrees of Freedom
a.	2.318	18
b.	1.861	13
c.	2.784	27
d.	3.218	8

9.21 Observations from two independent random samples are given in the accompanying table.

Sample 1	Sample 2
1	4
3	7
2	6
3	8
5	6

a. Summarize these data by finding the mean, variance, and standard deviation for each of the samples.

b. Find a 90% confidence interval for $(\mu_1 - \mu_2)$.

c. Test the hypothesis $H_0: \mu_1 = \mu_2$ versus $H_a: \mu_1 < \mu_2$ at the $\alpha = .05$ level of significance.

d. Calculate the approximate observed significance level or p-value for the test in part c.

9.22 Two independent random samples of size $n = 8$ were selected from each of two normal populations. The data are as follows:

Sample 1		Sample 2	
20.1	22.6	14.6	18.5
18.4	20.0	13.7	17.7
20.9	18.1	21.0	14.6
18.6	18.7	16.9	16.1

The MINITAB printout describing sample 1 (C1) and sample 2 (C2) is given in Table 9.6.

a. What is the pooled estimate of σ^2?

b. Find a 95% confidence interval estimate for $(\mu_1 - \mu_2)$.

c. Do the data present sufficient evidence to indicate a difference in the population means? Test by using $\alpha = .05$.

d. Do the results of part b substantiate the results of part c?

Table 9.6   MINITAB Output for the Data of Exercise 9.22

```
MTB > PRINT C1 C2
  ROW       C1       C2

    1      20.1     14.6
    2      18.4     13.7
    3      20.9     21.0
    4      18.6     16.9
    5      22.6     18.5
    6      20.0     17.7
    7      18.1     14.6
    8      18.7     16.1

MTB > DESCRIBE C1 C2

              N      MEAN    MEDIAN    TRMEAN     STDEV    SEMEAN
  C1          8    19.675    19.350    19.675     1.538     0.544
  C2          8    16.638    16.500    16.638     2.418     0.855

            MIN       MAX        Q1        Q3
  C1     18.100    22.600    18.450    20.700
  C2     13.700    21.000    14.600    18.300

MTB >
```

EXERCISES Applications

9.23 In an attempt to increase sales to French export markets, British apple growers initiated a "Le Crunch" advertising campaign (*Marketing*, April 23, 1980). To examine the effectiveness of this campaign, a representative of the British apple growers conducted a survey among 11 French apple importers and found that since the Le Crunch campaign began, these importers purchased an average of 9.12 tons of apples annually, with a standard deviation of 1.06 tons. From records predating the Le Crunch campaign, the representative found that 14 French importers purchased an average of 7.39 tons annually, with a standard deviation of 1.28 tons. Do these data present sufficient evidence to indicate that exports to France have increased since the initiation of the Le Crunch advertising campaign? Test at a 5% level of significance.

9.24 Five vice presidents and four market research analysts of a large industry were asked to estimate what they consider to be the optimal market share for their company. Their responses are given in the accompanying table. Do these data suggest that corporate vice presidents and market research analysts tend to disagree when estimating their firm's optimal market share? Test at a 5% level of significance.

Vice Presidents	22.5	25.0	30.0	27.5	20.0
Market Analysts	21.0	17.5	17.0	20.0	

9.25 Refer to Exercise 9.24. Estimate the *p*-value associated with the statistical test concerning agreement between corporate vice presidents and market research analysts in estimating their firm's optimal market share.

9.26 Flextime and other alternatives to the standard five-day, 9-to-5 workweek are not new, but a growing number of firms are moving toward rescheduling work hours in an attempt to improve worker morale and productivity. To examine the effects of an experimental flextime program involving 50% of the assembly workers of a California-based electronics manufacturer, a manager recorded the weekly production rates of 10 flextime employees and 13 employees working a standard 40-hour workweek; all employees in the survey performed an identical task. The means and standard deviations of the observed data are given in Table 9.7. Estimate the *p*-value for the statistical test that examines whether the data present sufficient evidence to indicate that the productivity rate of the flextime employees exceeds that of the employees involved in a standard work schedule. What are your conclusions?

Table 9.7   Production Rate Data for Exercise 9.26

Statistic	Weekly Production Rates	
	Flextime Employees	Standard-Time Employees
n	10	13
$\bar{x}$	59.3	53.8
s	4.7	4.5

9.27 The product life cycle (PLC) historically has been a basis for a variety of marketing tactics and strategies; however, the PLC has gained renewed acceptance in strategic marketing planning in general and corporate product portfolio analysis in particular. PLC theory suggests that the price of durable goods is affected by forces that tend to drive real prices down. D. J. Curry and P. C. Reisz ["Prices and Price/Quality Relationships: A Longitudinal Analysis," *Journal of Marketing*, Vol. 52 (January 1988), pp. 36–51] collected information on comparative test studies reported from January 1961 through December 1980 to produce a data base containing information on 938 comparative product test studies of 14,000 individual brands or items. In the short data summary in Table 9.8, the average and standard deviation of product prices are given in terms of 1980 dollars.

a. Do the data concerning FM roof antennas support the hypothesis put forward by Curry and Reisz concerning the PLC, namely, that the average price of brands for a fixed product declines over time when expressed in constant dollars? Use $\alpha = .05$.

b. What is the *p*-value of the test in part a?

Table 9.8   Product Data for Exercise 9.27

Product	Year	Number of Brands	Average Price	Standard Deviation
FM antennas	1961	16	$74.59	$10.40
	1973	17	52.62	10.50
Blenders	1965	22	89.37	14.80
	1973	20	62.26	11.10

9.28 Refer to Exercise 9.27. Do the data concerning the price of blenders for the years 1965 and 1973 substantiate the hypothesis that the mean price of brands of blenders is declining over time? What is the *p*-value of this test?

9.29 Worker productivity is highly dependent on a number of factors, such as wage rates, task complexity, and the work environment. But often the design of the job (i.e., the ordered sequencing of worker movements and material inputs) is the most crucial factor influencing worker productivity. Two work designs are under consideration for adoption in a plant. A time and motion study shows that 12 workers using design *A* have a mean assembly time of 324 seconds with a standard deviation of 18 seconds and that 15 workers using design *B* have a mean assembly time of 335 seconds with a standard deviation of 14 seconds. Do these data present sufficient evidence to indicate a difference in the rate of worker productivity for the two job designs? Test by using $\alpha = .01$.

9.30 Refer to Exercise 9.29. The manufacturing plant currently uses work design *B* in its assembly operations. Design *A* is an experimental work design and one that should be considered as a possible replacement for design *B* if shown to be effective in reducing assembly time. Does this additional information change the statistical test you would use? If so, how?

9.31 Producers of home heating systems are increasingly aware of consumer demands for energy efficiency in new heating systems. How much combustion efficiency should a homeowner expect from an oil furnace? A home heating contractor who sells two makes of oil heaters, *A* and *B*, decided to compare their mean efficiencies on a percentage scale of 0 to 100. An analysis was made of the efficiencies for 8 heaters of type *A* and 6 of type *B*. The efficiency ratings, in percentages, for the 14 heaters are shown in the table.

Type *A*	72	78	73	69	75	74	69	75
Type *B*	78	76	81	74	82	75		

a. Do the data provide sufficient evidence to indicate a difference in mean efficiencies for the two makes of home heaters?

b. Find a 90% confidence interval for $(\mu_A - \mu_B)$, and interpret the result.

9.32 Although union and nonunion wages tend to rise at the same rate in the long run, union wages usually advance faster during recessions and early in periods of recovery, and nonunion wages tend to advance more rapidly later in the business cycle when labor markets are tight. To examine this issue, an economist records the average hourly wages (including employee benefits)

of employees with two years' experience for 11 randomly chosen consumer products manufacturing firms, 6 of which have nonunion shops and 5 of which have union shops. The data are as follows:

Nonunion Shops	$8.26	$8.17	$8.45	$9.09	$8.85	$8.31
Union Shops	$7.92	$8.39	$8.64	$8.04	$8.24	

Do these data suggest that union and nonunion wages differ for employees with two years' experience in the consumer products manufacturing industry? Use the MINITAB printout in Table 9.9 and test by using a level of significance of $\alpha = .05$.

Table 9.9   MINITAB Output for the Data of Exercise 9.32

```
MTB > TWOSAMPLE C1 C2;
SUBC> POOLED.

TWOSAMPLE T FOR C1 VS C2
       N       MEAN      STDEV     SE MEAN
C1     6       8.522     0.367       0.15
C2     5       8.246     0.285       0.13

95 PCT CI FOR MU C1 - MU C2: (-0.18, 0.73)
TTEST MU C1 = MU C2 (VS NE): T=1.37 P=0.20 DF=9.0
```

9.33 If you planned to report the results of the statistical test in Exercise 9.32, what *p*-value would you report?

9.34 The development of synthetic materials such as nylon, polyester, and latex and their introduction into the marketplace stirred much debate about the comparative wearing quality and strength of human-made materials versus natural materials. A manufacturer of a new synthetic fiber claims that her product has greater tensile strength than natural fibers. Samples of 10 synthetic fibers and 10 natural fibers were randomly selected and tested for breaking strength. The means and variances for the two samples are given in the accompanying table. Do these data support the manufacturer's claim? Test by using $\alpha = .10$.

Natural Fiber	Synthetic Fiber
$\bar{x}_1 = 272$ lb	$\bar{x}_2 = 335$ lb
$s_1^2 = 1{,}636$ lb^2	$s_2^2 = 1{,}892$ lb^2

9.35 Refer to Exercise 9.34. Find a 95% confidence interval for the difference in mean breaking strength between the natural fiber and the synthetic fiber.

9.36 Continuing Exercises 9.34 and 9.35, suppose you wish to estimate the difference in mean breaking strength correct to within 10 lb, with a probability of approximately .95.

    a. Assuming that the sample sizes for natural and synthetic fibers are equal, approximately how large a sample would be required for each fiber?

    b. To conduct the experiment using the sample sizes in part a will require additional expenditures of time and money. Can the sample sizes be reduced from those specified in part a and still allow us to achieve the 10 lb bound on the error of estimation?

9.5

A Paired-Difference Test

A manufacturer wishes to compare the wearing qualities of two different types of automobile tires, A and B. For the comparison a tire of type A and one of type B are randomly assigned and mounted on the rear wheels of each of five automobiles. The automobiles are then operated for a specified number of miles and the amount of wear is recorded for each tire. These measurements appear in Table 9.10. Do the data present sufficient evidence to indicate a difference in average wear for the two tire types?

Analyzing the data, we note that the difference between the two sample means is $(\bar{x}_1 - \bar{x}_2) = .48$, a rather small quantity considering the variability of the data and the small number of measurements involved. At first glance it would seem that there is little evidence to indicate a difference between the population means, a conjecture that we may check by the method outlined in Section 9.4.

The pooled estimate of the common variance σ^2 is

$$s^2 = \frac{\sum_{i=1}^{n_1} (x_{1i} - \bar{x}_1)^2 + \sum_{i=1}^{n_2} (x_{2i} - \bar{x}_2)^2}{n_1 + n_2 - 2} = \frac{6.932 + 7.052}{5 + 5 - 2} = 1.748$$

and thus

$$s = 1.32$$

The calculated value of t used to test the hypothesis that $\mu_1 = \mu_2$ is

$$t = \frac{\bar{x}_1 - \bar{x}_2}{s\sqrt{\dfrac{1}{n_1} + \dfrac{1}{n_2}}} = \frac{10.24 - 9.76}{1.32\sqrt{\dfrac{1}{5} + \dfrac{1}{5}}} = .58$$

a value that is not nearly large enough to reject the hypothesis that $\mu_1 = \mu_2$.

The corresponding 95% confidence interval is

$$(\bar{x}_1 - \bar{x}_2) \pm t_{\alpha/2} s\sqrt{\frac{1}{n_1} + \frac{1}{n_2}} \qquad (10.24 - 9.76) \pm (2.306)(1.32)\sqrt{\frac{1}{5} + \frac{1}{5}}$$

Table 9.10   Tire Wear for Two Types of Tires

Automobile	Tire Type	
	A	B
1	10.6	10.2
2	9.8	9.4
3	12.3	11.8
4	9.7	9.1
5	8.8	8.3
	$\bar{x}_1 = 10.24$	$\bar{x}_2 = 9.76$

Table 9.11   Differences in Tire Wear, Using the Data
of Table 9.10

Automobile	A	B	$d = A - B$
1	10.6	10.2	.4
2	9.8	9.4	.4
3	12.3	11.8	.5
4	9.7	9.1	.6
5	8.8	8.3	.5
			$\bar{d} = .48$

or -1.45 to 2.41. This interval is quite wide, considering the small difference between the sample means.

A second glance at the data reveals a marked inconsistency with the conclusion. We note that the wear measurement for type A is larger than the corresponding value for type B for each of the five automobiles. These differences, recorded as $d = A - B$, are shown in Table 9.11.

Suppose that we were to use x, the number of times that A is larger than B, as a test statistic. Then the probability that A is larger than B for a given automobile, assuming no difference between the wearing quality of the tires, is $p = 1/2$, and so x would be a binomial random variable.

If we choose $x = 0$ and $x = 5$ as the rejection region for a two-tailed test, then $\alpha = p(0) + p(5) = 2(1/2)^5 = 1/16$. We would then reject $H_0: \mu_1 = \mu_2$ with a probability of a type I error equal to $\alpha = 1/16$. Certainly, this result is evidence to indicate that a difference exists in the mean wear of the two tire types.

You will note that we have employed two different statistical tests to test the same hypothesis. Is it not peculiar that the t test, which utilizes more information (the actual sample measurements) than the binomial test, fails to supply sufficient evidence for rejection of the hypothesis $\mu_1 = \mu_2$?

There is an explanation for this inconsistency. The t test described in Section 9.4 is not the proper statistical test to be used for our example. The statistical test procedure of Section 9.4 requires that the two samples be *independent and random*. Certainly, the independence requirement was violated by the manner in which the experiment was conducted. The (pair of) measurements, an A and a B tire, for a particular automobile are definitely related. A glance at the data shows that the readings have approximately the same magnitude for a particular automobile but vary markedly from one automobile to another. This, of course, is exactly what we might expect. Tire wear, in a large part, is determined by driver habits, the balance of the wheels, and the road surface. Since each automobile has a different driver, we would expect a large amount of variability in the data from one automobile to another.

The familiarity we have gained with interval estimation has shown us that the width of the large-sample and small-sample confidence intervals depends on the magnitude of the standard deviation of the point estimator of the parameter. The smaller its value, the better is the estimate and the more likely it is that the test statistic will provide evidence to reject the null hypothesis if it is, in fact, false. Knowledge of this phenomenon was utilized in *designing* the tire wear experiment. The experimenter realized that the wear measurements would vary greatly from auto to auto and that this variability could not be separated from the data if the tires were assigned to the ten wheels in a random manner. (A random assignment of the tires would have implied that the data be analyzed according to the procedure of Section 9.4.) Instead, a comparison of the wear between the tire types A and B made on each automobile resulted in the five difference measurements. This design eliminates the effect of the car-to-car variability and yields more information on the mean difference in the wearing quality for the two tire types.

A proper analysis of the data would utilize the five difference measurements to test the hypothesis that the average difference is equal to zero, a statement that is equivalent to the null hypothesis $\mu_1 = \mu_2$.

You may verify that the average and standard deviation of the five difference measurements are

$$\bar{d} = .48 \quad \text{and} \quad s_d = .0837$$

Then

$$H_0 : \mu_d = 0$$

and

$$t = \frac{\bar{d} - 0}{s_d/\sqrt{n}} = \frac{.48}{.0837/\sqrt{5}} = 12.8$$

The critical value of t for a two-tailed statistical test with $\alpha = .05$ and 4 degrees of freedom is 2.776. Certainly, the observed value of $t = 12.8$ is extremely large and highly significant. In fact, with 4 degrees of freedom the observed value of $t = 12.8$ exceeds $t_{.005} = 4.604$, so that the p-value for this test would be less than $2(.005) = .01$. Hence, we would conclude that the average amount of wear for tire type A is greater than that for type B.

A 95% confidence interval for the difference between the mean wear is

$$\bar{d} \pm t_{\alpha/2} \frac{s_d}{\sqrt{n}} \quad \text{or} \quad .48 \pm (2.776) \frac{.0837}{\sqrt{5}} \quad \text{or} \quad .48 \pm .10$$

or from .38 to .58.

When the units used to compare two or more procedures exhibit marked variability before any experimental procedures are implemented, the effect of this variability can be minimized by comparing the procedures *within* groups of relatively **blocks** homogeneous units called **blocks**. In this way the effects of the procedures are not masked by the initial variability among the units in the experiment. An experiment **randomized** conducted in this manner is called a **randomized block design**. In an experiment **block design** involving daily sales, blocks may represent days of the week; in an experiment involving product marketing, blocks may represent geographic areas. (Randomized block designs are discussed in more detail in Section 10.7 of Chapter 10.)

The statistical design of the tire experiment is a simple example of a randomized **paired-difference test** block design, and the resulting statistical test is often called a **paired-difference test**. You will note that the **pairing occurred when the experiment was planned and not after the data were collected.** Comparisons of tire wear were made within relatively homogeneous blocks (automobiles), with the tire types randomly assigned to the two automobile wheels.

The small-sample test of an hypothesis for a paired-difference experiment and a $(1 - \alpha)100\%$ confidence interval for $(\mu_1 - \mu_2)$ are given in the next displays.

Small-Sample Statistical Test for the Difference Between Two Means Based on a Paired-Difference Experiment

Null hypothesis $H_0 : \mu_1 - \mu_2 = \mu_d = D_0$.
Test statistic:

$$t = \frac{\bar{d} - D_0}{s_d/\sqrt{n}}$$

where n is equal to the number of paired differences and s_d, given by

$$s_d = \sqrt{\frac{\sum_{i=1}^{n}(d_i - \bar{d})^2}{n - 1}}$$

is the standard deviation of the sample of n paired differences.

The alternative hypothesis H_a and α are specified by the experimenter and are used to locate the critical value of t for the rejection region.

Small-Sample $(1 - \alpha)100\%$ Confidence Interval for $(\mu_1 - \mu_2)$ Based on a Paired-Difference Experiment

$$\bar{d} \pm t_{\alpha/2} \frac{s_d}{\sqrt{n}}$$

where n is equal to the number of paired differences and s_d is as defined in the preceding display.

The amount of information gained by blocking the tire experiment may be measured by comparing the calculated confidence interval for the unpaired (and incorrect) analysis with the interval obtained for the paired-difference analysis. The confidence interval for $(\mu_1 - \mu_2)$ that might have been calculated, had the tires been randomly assigned to the ten wheels (unpaired), is unknown but probably would have been of the same magnitude as the interval -1.45 to 2.41, which was calculated by analyzing the observed data in an unpaired manner. Pairing the tire types on the automobiles (blocking) and the resulting analysis of the differences produced the interval estimate .38 to .58. Note the difference in the width of the intervals, which indicates the very sizable increase in information obtained by blocking in this experiment.

Although blocking proved to be very beneficial in the tire experiment, it may not always be. We observe that the degrees of freedom available for estimating σ^2 are less for the paired than for the corresponding unpaired experiment. If there were actually no differences among the blocks, the reduction in the degrees of freedom would produce a moderate increase in the $t_{\alpha/2}$ employed in the confidence interval and hence would increase the width of the interval. This, of course, did not occur in the tire experiment because the large reduction in the standard error of $\bar{d}$ more than compensated for the loss in degrees of freedom.

Except for notation, the analysis of a paired-difference experiment is the same as that for a single sample presented in Section 9.3. This similarity enables us to use the MINITAB commands TTEST and TINTERVAL to analyze the differences in a paired-difference experiment. Table 9.12 displays the MINITAB output for the commands PRINT, TTEST, and TINTERVAL, using the paired-difference data in Tables 9.10 and 9.11. The differences that appear in column 3 are easily generated by using the MINITAB command LET C3 = C1 − C2.

Before concluding, we want to reemphasize a point. Once you have used a paired design for an experiment, you no longer have the option of using the unpaired analysis of Section 9.4. The assumptions upon which that test is based have been violated. Your only alternative is to use the·correct method of analysis, the paired-difference test (and associated confidence interval) of this section.

Table 9.12   MINITAB Output for Paired Samples, Using the Data in Table 9.10

```
MTB > PRINT C1 C2 C3
  ROW      C1       C2       C3

    1     10.6     10.2     0.4
    2      9.8      9.4     0.4
    3     12.3     11.8     0.5
    4      9.7      9.1     0.6
    5      8.8      8.3     0.5

MTB > TTEST 0 C3

TEST OF MU = 0.0000 VS MU N.E. 0.0000

             N      MEAN     STDEV    SE MEAN        T    P VALUE
C3           5    0.4800    0.0837     0.0374    12.83     0.0002

MTB > TINTERVAL C3

             N      MEAN     STDEV    SE MEAN    95.0 PERCENT C.I.
C3           5    0.4800    0.0837     0.0374    ( 0.3761,   0.5839)
```

EXERCISES   Conceptual Level

9.37   When is it appropriate to use a paired-difference analysis rather than the analysis presented in Section 9.4?

9.38   What is the result of effective pairing in a paired-difference experiment?

9.39   Is it correct to use the paired and the unpaired analysis on the same data?

9.40   What is meant by the term *block* in the context of a paired-difference experiment?

EXERCISES   Skill Level

9.41   The data in the accompanying table are from a paired-difference experiment with $n = 7$ pairs of observations.

	Treatment	
Pair	1	2
1	6.1	4.8
2	9.2	7.4
3	4.1	4.2
4	10.2	8.9
5	9.6	8.6
6	7.6	6.4
7	8.7	7.1

a.   Calculate the differences d_i for each pair of observations.

b.   Calculate $\bar{d}$ and s_d^2 for the $n = 7$ differences.

c.   State the null and alternative hypotheses used to determine whether there is a difference in the two treatment means, μ_1 and μ_2.

d.   Calculate the observed value of the test statistic and the approximate p-value for the test.

e.   Based on the results of part d, what is your conclusion about the value of $(\mu_1 - \mu_2)$?

9.42   The data in the accompanying table are from a paired-difference experiment with $n = 6$ pairs of observations.

	Treatment	
Pair	1	2
1	7	8
2	10	8
3	12	11
4	2	3
5	15	12
6	8	9

a. Do the data provide sufficient evidence to indicate a difference between the two treatment means? Test using $\alpha = .05$.

b. Find a 95% confidence interval for the difference between the two treatment means. Do the results substantiate your conclusion in part a?

EXERCISES Applications

9.43 Concerned with the practice by Japanese manufacturers of dumping merchandise in the retail markets of the United States, American manufacturers lobbied for retaining meaningful import quotas on certain Japanese goods. In a study of the effect on color television inventories of lifting a particular import quota, a survey was conducted among eight retail outlets. For each outlet a recording was made of the inventory of imported color TVs one month before and one month after removal of the import quota. The data are given in Table 9.13. Do these data present sufficient evidence to indicate that the average inventory of imported TVs increased among retail outlets after removal of the import quota? Test using $\alpha = .05$.

Table 9.13   Import Quota Data for Exercise 9.43

| | Retail Outlet | | | | | | | |
Status	1	2	3	4	5	6	7	8
Quota in effect	161	192	219	91	160	132	57	87
No quota	212	200	221	87	158	143	86	91

9.44 If you planned to report the results of the statistical test in Exercise 9.43, what p-value would you report?

9.45 With increasing concern about the ability of several Third World countries to feed their growing populations, a recent study reported on the economic effects of urbanization in a South American country. For this study nine locations were chosen within the country, and the typical size (in acres) of an agricultural unit was recorded for each for the years 1970 and 1980. The results are shown in Table 9.14. Find a 95% confidence interval for the average reduction in size of the typical agricultural unit in this South American country from 1970 to 1980.

Table 9.14   Agricultural Size Data for Exercise 9.45

| | Location | | | | | | | | |
Year	1	2	3	4	5	6	7	8	9
1970	44.7	103.6	89.0	73.7	34.5	22.8	30.5	22.6	46.3
1980	33.7	44.1	49.3	44.1	26.9	19.3	21.3	18.3	41.1

9.46 Numerous methods are available for evaluating inventories. Two of the more common methods are LIFO (last-in, first-out) and FIFO (first-in, first-out), with inherent advantages and disadvantages for each. However, in times of inflation LIFO tends to reduce taxes and improve cash flow. A multiproduct firm is contemplating changing from FIFO to LIFO as a method for evaluating inventories. Five different finished-goods inventories are evaluated at the end of the year by using both FIFO and LIFO. The results are given in Table 9.15. Do these data suggest that LIFO is effective in reducing the inventory value of the firm's finished-goods inventories? Test with an $\alpha = .05$ level of significance, and use the MINITAB printout in Table 9.16.

Table 9.15   Inventory Data for Exercise 9.46

| Product | Inventory Value (× 1,000) | |
	FIFO	LIFO
1	121	117
2	217	198
3	92	105
4	98	86
5	52	49

Table 9.16   MINITAB Output for the Data of Exercise 9.46

```
MTB > SET C1
DATA> 121 217 92 98 52
DATA> END
MTB > SET C2
DATA> 117 198 105 86 49
DATA> END
MTB > LET C3 = C1-C2
MTB > TTEST 0 C3;
SUBC> ALTERNATIVE = 1.

TEST OF MU = 0.000 VS MU G.T. 0.000

              N      MEAN     STDEV    SE MEAN       T    P VALUE
C3            5     5.000    11.979      5.357    0.93       0.20

MTB >
```

9.47 Why would you use paired observations to estimate the difference between two population means in preference to estimation based on independent random samples selected from the two populations? For example, in Exercise 9.46, what advantage is gained by applying both LIFO and FIFO inventory valuation techniques to the same 5 products rather than using 10 different products, 5 for valuation by LIFO and 5 for valuation by FIFO? Is a paired experiment always preferable? Explain.

9.48 Although the Occupational Safety and Health Act (OSHA) is not very popular with management because of the cost of implementing its requirements, some sources claim that it has been effective in reducing industrial accidents. The data in Table 9.17 were collected on lost-time accidents (the figures given are mean man-hours lost per month over a period of one year), both before and after OSHA came into effect. Data were recorded for six industrial plants. Do these data provide sufficient evidence to indicate that OSHA has been effective in reducing lost-time accidents? Test at the $\alpha = .10$ level of significance.

Table 9.17 Lost-Time Data for Exercise 9.48

| | Plant Number | | | | | |
Period	1	2	3	4	5	6
Before OSHA	38	64	42	70	58	30
After OSHA	31	58	43	65	52	29

9.49 Consider the test of hypothesis in Exercise 9.48 from the point of view of (1) management, which wishes to be absolutely certain of the positive effects of OSHA to justify its cost of implementation, and (2) government, which wishes to provide workers with all available health and personal injury safeguards at any reasonable cost. Recommend the selection of a significance level α for the test of hypothesis concerning the effectiveness of OSHA from these points of view:

a. Management   **b.** Government

9.6

Inferences About a Population Variance

We have seen in the preceding sections that an estimate of the population variance σ^2 is fundamental to procedures for making inferences about population means. Moreover, there are many practical situations where σ^2 is the primary objective of an experimental investigation; thus, σ^2 assumes a position of far greater importance than that of the population mean.

Scientific measuring instruments must provide unbiased readings with a very small error of measurement. An aircraft altimeter that measures the correct altitude *on the average* would be of little value if the standard deviation of the error of measurement were 5,000 ft. Indeed, bias in a measuring instrument can often be corrected for, but the precision of the instrument, measured by the standard deviation of the error of measurement, is usually a function of the design of the instrument itself and cannot be controlled.

Machined parts in a manufacturing process must be produced with minimum variability to reduce out-of-size, and hence defective, products. In general, manufacturers wish to maintain a minimum variance in the measurements of the quality characteristics of an industrial product to achieve process control and therefore minimize the percentage of poor-quality product.

The sample variance

$$s^2 = \frac{\sum_{i=1}^{n}(x_i - \bar{x})^2}{n - 1}$$

is an unbiased estimator of the population variance σ^2. The distribution of sample variances generated by repeated sampling will have a probability distribution that begins at $s^2 = 0$ (since s^2 cannot be negative), with a mean equal to σ^2. Unlike the distribution of $\bar{x}$, the distribution of s^2 is nonsymmetrical, the exact form being dependent on the probability distribution of the population from which the sample measurements were drawn.

For the methodology that follows **we will assume that the sample is drawn from a normal population and that s^2 is based on a random sample of n measurements.** Using the terminology of Section 9.2, we would say that s^2 possesses $(n - 1)$ degrees of freedom.

The next step is to consider the distribution of s^2 in repeated sampling from a specified normal distribution—one with a specific mean and variance—and to tabulate the critical values of s^2 for some of the commonly used tail areas. We will find that the distribution of s^2 is independent of the population mean μ but possesses a different distribution for each sample size and each value of σ^2. This task is quite laborious, but fortunately, it may be simplified by **standardizing**, as we did with normal random variables. The quantity we use for standardizing is shown in the display.

standardizing

A Chi-square Random Variable

$$\chi^2 = \frac{(n - 1)s^2}{\sigma^2}$$

chi-square
variable

The quantity χ^2, called a **chi-square variable** by statisticians (χ is the Greek letter chi), admirably suits our purposes. Its distribution in repeated sampling is called, as we might suspect, a chi-square sampling distribution. The equation of the sampling distribution for the chi-square variable is well known to statisticians, who have tabulated critical values corresponding to various tail areas of the distribution. These values are presented in Table 5 of the Appendix. The format of Table 5 in the Appendix is shown in Table 9.18.

The shape of the chi-square distribution, like that of the t distribution, varies with the sample size or, equivalently, with the degrees of freedom associated with s^2. Thus,

Table 9.18   Format of the Chi-square Table, Table 5 in the Appendix

d.f.	$\chi^2_{0.995}$	$\cdots$	$\chi^2_{0.950}$	$\chi^2_{0.900}$	$\chi^2_{0.100}$	$\chi^2_{0.050}$	$\cdots$	$\chi^2_{0.005}$	d.f.
1	0.0000393		0.0039321	0.0157908	2.70554	3.84146		7.87944	1
2	0.0100251		0.102587	0.210720	4.60517	5.99147		10.5966	2
3	0.0717212		0.351846	0.584375	6.25139	7.81473		12.8381	3
4	0.206990		0.710721	1.063623	7.77944	9.48773		14.8602	4
5	0.411740		1.145476	1.61031	9.23635	11.0705		16.7496	5
6	0.675727		1.63539	2.20413	10.6446	12.5916		18.5476	6
$\vdots$	$\vdots$		$\vdots$	$\vdots$	$\vdots$	$\vdots$		$\vdots$	$\vdots$
15	4.60094		7.26094	8.54675	22.3072	24.9958		32.8013	15
16	5.14224		7.96164	9.31223	23.5418	26.2962		34.2672	16
17	5.69724		8.67176	10.0852	24.7690	27.5871		35.7185	17
18	6.26481		9.39046	10.8649	25.9894	28.8693		37.1564	18
19	6.84398		10.1170	11.6509	27.2036	30.1435		38.5822	19

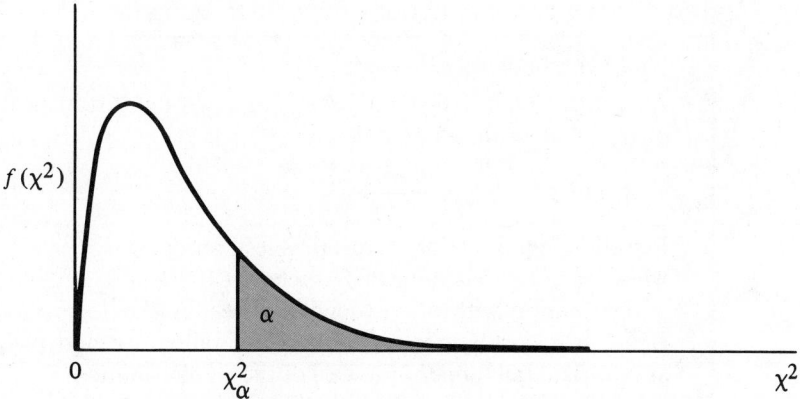

Figure 9.3   A Chi-square Distribution

Table 5 is constructed in exactly the same manner as the t table, with the degrees of freedom shown in the first and last columns. The symbol χ_α^2 indicates that the tabulated χ^2 value is such that an area α lies to its right (see Figure 9.3). Stated in probabilistic terms,

$$P(\chi^2 > \chi_\alpha^2) = \alpha$$

Thus, 99% of the area under the χ^2 distribution lies to the right of $\chi_{.99}^2$. We note that the extreme values of χ^2 must be tabulated for both the lower and upper tails of the distribution because it is nonsymmetrical.

You can check your ability to use the table by verifying the following statements. The probability that χ^2, based on $n = 16$ measurements (d.f. $= 15$), will exceed 24.9958 is .05. For a sample of $n = 6$ measurements (d.f. $= 5$), 95% of the area under the χ^2 distribution lies to the right of $\chi^2 = 1.145476$. These entries are shown in color in Table 9.18.

The statistical test of a null hypothesis concerning a population variance,

$$H_0:\sigma^2 = \sigma_0^2$$

employs the test statistic

$$\chi^2 = \frac{(n-1)s^2}{\sigma_0^2}$$

Notice that when H_0 is true, s^2/σ_0^2 should be near 1, so that χ^2 should be close to $(n-1)$, the degrees of freedom. If σ^2 is really greater than the hypothesized value σ_0^2, the test statistic will tend to be larger than $(n-1)$ and will probably fall toward the upper tail of the distribution. If $\sigma^2 < \sigma_0^2$, the test statistic will tend to be smaller than $(n-1)$ and will probably fall toward the lower tail of the chi-square distribution. As in other testing situations, we may use either a one- or a two-tailed statistical test, depending upon the alternative hypothesis under test.

Test of an Hypothesis About a Population Variance

Null hypothesis $H_0: \sigma^2 = \sigma_0^2$.
Alternative hypothesis $H_a: \sigma^2 \neq \sigma_0^2$, for a two-tailed statistical test; $H_a: \sigma^2 > \sigma_0^2$ or $H_a: \sigma^2 < \sigma_0^2$ for a one-tailed test.

Test statistic: $\chi^2 = \dfrac{(n-1)s^2}{\sigma_0^2}$.

Rejection region: For a two-tailed test, reject H_0 if $\chi^2 < \chi_{1-\alpha/2}^2$ or $\chi^2 > \chi_{\alpha/2}^2$, where $\chi_{1-\alpha/2}^2$ is that value of χ^2 such that $100(1 - \alpha/2)\%$ of the area under the χ^2 distribution, with $(n-1)$ degrees of freedom, lies to its right. For alternatives $H_a: \sigma^2 > \sigma_0^2$ or $H_a: \sigma^2 < \sigma_0^2$, use a one-tailed test and place all of α in the appropriate tail (upper or lower) of the χ^2 distribution.

We illustrate a test of an hypothesis about a population variance with an example.

EXAMPLE 9.6 A cement manufacturer claims that concrete prepared from its product possesses a relatively stable compressive strength and that the strength, measured in kilograms per square centimeter (kg/cm^2), lies within a range of 40 kg/cm^2. A sample of $n = 10$ measurements produced a mean and variance of

$$\bar{x} = 312 \quad \text{and} \quad s^2 = 195$$

Do these data present sufficient evidence to reject the manufacturer's claim?

SOLUTION As stated, the manufacturer claims that the range of the strength measurements equals 40 kg/cm^2. We will suppose that the manufacturer meant that the measurements lie within this range 95% of the time and therefore that the range equals approximately 4σ and that $\sigma = 10$. Then we wish to test the null hypothesis

$$H_0: \sigma^2 = (10)^2 = 100$$

against the alternative

$$H_a: \sigma^2 > 100$$

The alternative hypothesis requires a one-tailed statistical test, with the entire rejection region located in the upper tail of the χ^2 distribution. The critical value of χ^2 for $\alpha = .05$ and $n = 10$ is $\chi^2 = 16.9190$, which implies that we will reject H_0 if the test statistic exceeds this value.

Calculating the value of the test statistic, we obtain

$$\chi^2 = \frac{(n-1)s^2}{\sigma_0^2} = \frac{1{,}755}{100} = 17.55$$

Since the value of the test statistic falls in the rejection region, we conclude that the null hypothesis is false and that the range of concrete strength measurements actually exceeds the manufacturer's claim.

Notice that the observed value of χ^2 with 9 degrees of freedom lies between $\chi^2_{.05} = 16.9190$ and $\chi^2_{.025} = 19.0228$. Therefore, the p-value for this test is less than .05 but greater than .025, so that

$$.025 < p\text{-value} < .05$$

◆◆

A confidence interval for σ^2 with a $(1 - \alpha)$ confidence coefficient is given in the display.

A $(1 - \alpha)100\%$ Confidence Interval for σ^2

$$\frac{(n - 1)s^2}{\chi^2_U} < \sigma^2 < \frac{(n - 1)s^2}{\chi^2_L}$$

where χ^2_L and χ^2_U are the lower and upper χ^2 values that locate one-half of α in each tail of the chi-square distribution.

For example, a 90% confidence interval for σ^2 in Example 9.6 with $(n - 1) = 9$ degrees of freedom would use

$$\chi^2_L = \chi^2_{.95} = 3.32511 \qquad \text{and} \qquad \chi^2_U = \chi^2_{.05} = 16.9190$$

Then the interval estimate for σ^2 would be

$$\frac{(9)(195)}{16.9190} < \sigma^2 < \frac{(9)(195)}{3.32511} \qquad \text{or} \qquad 103.73 < \sigma^2 < 527.80$$

EXAMPLE 9.7   An experimenter is convinced that his measuring equipment possesses a variability measured by a standard deviation $\sigma = 2$. During an experiment he recorded the measurements 4.1, 5.2, and 10.2. Do these data contradict his assumption? Test the hypothesis $H_0: \sigma = 2$, or $\sigma^2 = 4$, and place a 90% confidence interval on σ^2.

SOLUTION   The calculated sample variance is $s^2 = 10.57$. Since we wish to detect $\sigma^2 > 4$ as well as $\sigma^2 < 4$, we should employ a two-tailed test. Using $\alpha = .10$ and placing .05 in each tail, we will reject H_0 when $\chi^2 > 5.99147$ or $\chi^2 < .102587$.
   The calculated value of the test statistic is

$$\chi^2 = \frac{(n - 1)s^2}{\sigma_0^2} = \frac{(2)(10.57)}{4} = 5.29$$

Since the test statistic does not fall in the rejection region, the data do not provide sufficient evidence to reject the null hypothesis $H_0: \sigma^2 = 4$.
   The corresponding 90% confidence interval is

$$\frac{(n - 1)s^2}{\chi^2_U} < \sigma^2 < \frac{(n - 1)s^2}{\chi^2_L}$$

The values of χ_L^2 and χ_U^2 are

$$\chi_L^2 = \chi_{.95}^2 = .102587 \quad \text{and} \quad \chi_U^2 = \chi_{.05}^2 = 5.99147$$

Substituting these values into the formula for the interval estimate, we obtain

$$\frac{2(10.57)}{5.99147} < \sigma^2 < \frac{2(10.57)}{.102587} \quad \text{or} \quad 3.53 < \sigma^2 < 206.1$$

Thus, we estimate that the population variance falls in the interval from 3.53 to 206.1. This very wide confidence interval indicates how little information about the population variance is obtained in a sample of only three measurements. Consequently, it is not surprising that there was insufficient evidence to reject the null hypothesis $\sigma^2 = 4$. To obtain more information on σ^2, the experiment needs to increase the sample size.

EXERCISES   Conceptual Level

9.50 Under what conditions does the statistic

$$\chi^2 = \frac{(n-1)s^2}{\sigma^2}$$

have a chi-square distribution in repeated sampling?

9.51 What is meant by the term *standardizing*? How do we standardize a normal random variable? How do we standardize s^2? What is gained by standardizing the value of s^2?

9.52 Since $E(s^2) = \sigma^2$ (that is, the average value of s^2 is σ^2), what is the expected value of $\chi^2 = (n-1)s^2/\sigma^2$? How do the degrees of freedom characterize a chi-square distribution?

EXERCISES   Skill Level

9.53 Find the critical value of χ^2 having an area of α to its right based on the given degrees of freedom.

	α	Degrees of Freedom
a.	.05	23
b.	.01	8
c.	.95	30
d.	.99	13
e.	.025	29
f.	.975	20

9.54 A random sample of $n = 12$ observations from a normal population produced a sample mean and standard deviation of $\bar{x} = 13.1$ and $s = 2.7$. Use this information to construct a 95% confidence interval estimate of σ^2.

9.55 Refer to Exercise 9.54. Consider a test of the null hypothesis $H_0: \mu = 5$ against the alternative hypothesis $H_a: \sigma^2 > 5$ at the $\alpha = .05$ level of significance.

   a. What is the critical value of the test statistic necessary for rejection of the null hypothesis?

   b. Does the sample provide sufficient evidence to reject H_0?

   c. What is the approximate p-value of this test?

EXERCISES   Applications

9.56 The uniformity of light emitted by cathode ray tubes (CRTs) used in television sets, computer terminals, and certain electronic equipment is measured in terms of resolution, with high resolution implying low light spot variation ("Bright View of Data," *Financial Times*, June 26, 1980). The manufacturer of CRTs for use in computer terminals claims that light spot variation does not exceed $\sigma^2 = .24$ square millimeter (mm²). A test is carried out to measure the resolution of a particular CRT by recording the light spot intensity at a random spot on the screens of each of 12 different CRTs; the recorded light intensity variation is .28 mm². Do these data refute the manufacturer's claim?

9.57 A manufacturer of steel bars used as reinforcement in concrete for building construction claims that its product provides a yield strength that ranges from 5,000 to 9,000 lb. Each bar in a sample of 8 steel bars was subjected to stress tests, providing the following maximum yield strengths:

6,420   7,465   8,240   5,780   6,800   4,875   8,050   7,170

Do these data present sufficient evidence to reject the manufacturer's claim regarding yield strength of steel bars manufactured by its firm? Test by using a 10% level of significance.

9.58 Refer to Exercise 9.57.

   a. Find a 90% confidence interval for the variance of yield strength of the steel bars.

   b. Using the rule that provides an approximate relationship between the range and the standard deviation, find an approximate 90% confidence interval for the range of yield strength of the steel bars.

9.59 Precision instruments such as those designed to measure volume, temperature, pressure, meat fat content, or content mixture must be designed to provide a measure that not only is correct on the average but also possesses very little variation around the true value. As an example, consider the volume meter on a gasoline station fuel pump. The station operator insists that the meter not underregister the dispensed volume, and the buyer demands that it not register in excess of the true dispensed amount. A manufacturer of volume meters claims that its product registers accurately to within .1 gallon (gal) of the actual amount of gasoline dispensed through the meter. The manufacturer's meter is installed in a gasoline pump and 5 different samples of exactly 10 gal of gasoline are dispensed through the meter and the pump. The recorded volumes on the meter for the 5 samples are as follows: 10.05, 10.00, 9.90, 9.95, 10.15. Do these data support or refute the meter manufacturer's claim? Conduct the test at a 5% level of significance. (*Hint:* If the manufacturer claims its product is accurate to within .1 gal of the actual amount, the manufacturer is specifying that the recorded amount will vary by no more than .1 gal *in either direction* from the actual amount. Therefore, the manufacturer is specifying that the *range* of readings for a particular amount dispensed will be no more than .2 gal.)

9.7

Comparing Two Population Variances

The need for statistical methods to compare two population variances is readily apparent from the discussions in Sections 9.4 and 9.6. We may frequently wish to compare the precision of one measuring device with that of another, the stability of one manufacturing process with that of another, or the variability in the grading procedure of one college professor with that of another. We may wish to determine whether two population variances can be assumed to be equal prior to using the pooled two-sample t procedure.

Intuitively, we might compare two population variances σ_1^2 and σ_2^2 by using the ratio of the sample variances s_1^2/s_2^2. If s_1^2/s_2^2 is nearly equal to 1, we would find little evidence to indicate that σ_1^2 and σ_2^2 are unequal. On the other hand, a very large or small value for s_1^2/s_2^2 would provide evidence of a difference in the population variances.

How large or small must s_1^2/s_2^2 be to provide sufficient evidence for rejecting the null hypothesis $H_0: \sigma_1^2 = \sigma_2^2$? The answer to this question may be obtained by studying the distribution of s_1^2/s_2^2 in repeated sampling.

When independent random samples are drawn from two normal populations with equal variances—that is, $\sigma_1^2 = \sigma_2^2$—then s_1^2/s_2^2 possesses a sampling distribution that is known to statisticians as an **F distribution**. We need not concern ourselves with the equation for the sampling distribution for F except to state that, as we might surmise, it is reasonably complex. For our purposes it will suffice to accept the fact that the distribution is well known and that critical values have been tabulated. These values appear in Table 6 of the Appendix.

The shape of the F distribution is nonsymmetrical and depends on the number of degrees of freedom associated with s_1^2 and s_2^2. We represent these quantities as v_1 and v_2, respectively. (An F distribution with $v_1 = 10$ numerator degrees of freedom and $v_2 = 10$ denominator degrees of freedom is shown in Figure 9.4.) This fact complicates

F distribution

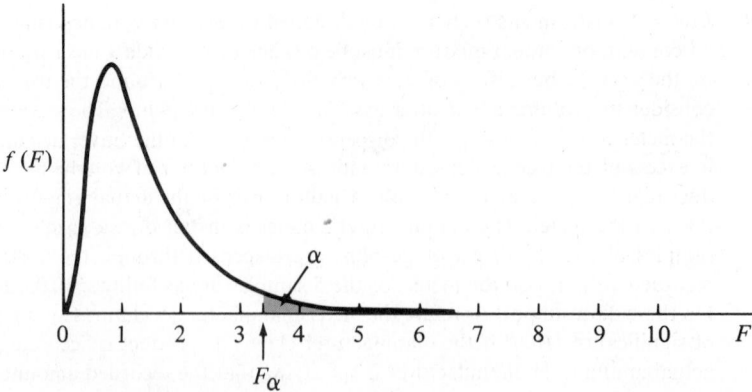

Figure 9.4   An *F* Distribution with $v_1 = 10$ and $v_2 = 10$

the tabulation of critical values of the F distribution and necessitates the construction of a table to accommodate differing values of v_1, v_2, and α.

In Table 6 of the Appendix critical values of F for right-tailed areas corresponding to $\alpha = .10, .05, .025, .010,$ and $.005$ are tabulated for various combinations of v_1 numerator degrees of freedom and v_2 denominator degrees of freedom. A portion of Table 6 is reproduced in Table 9.19. The numerator degrees of freedom v_1 are listed across the top margin and the denominator degrees of freedom v_2 are listed along both side margins. The values of α are listed in the second column from the left as well as from the right. For a fixed combination of v_1 and v_2 the appropriate critical values of F are found in the line indexed by the value of α required.

From Table 9.19, when $v_1 = 5$ and $v_2 = 7$, the critical value of $F_{.10}$ is equal to 2.88 and $F_{.05} = 3.97$. For $v_1 = 8$ and $v_2 = 12$, $F_{.025} = 3.51$ and $F_{.01} = 4.50$. These entries are shown in color in Table 9.19.

The statistical test of the null hypothesis

$$H_0 : \sigma_1^2 = \sigma_2^2$$

utilizes the test statistic

$$F = \frac{s_1^2}{s_2^2}$$

When the alternative hypothesis implies the one-tailed test

$$H_a : \sigma_1^2 > \sigma_2^2$$

the rejection region is in the right tail and we may use Table 6 directly. If the alternative hypothesis is given by

$$H_a : \sigma_1^2 < \sigma_2^2$$

and we define the test statistic as

$$F = \frac{s_2^2}{s_1^2}$$

then the rejection region is again in the right tail and Table 6 can again be used directly.

When the alternative hypothesis requires a two-tailed test, given by

$$H_a : \sigma_1^2 \neq \sigma_2^2$$

the rejection region will be divided between the lower and upper tails of the F distribution. However, tables of the critical values for the lower tail are conspicuously missing. The reason for their absence is easily explained.

We are at liberty to identify either of the two populations as population 1. If the population with the larger sample variance is associated with population 2, then $s_2^2 > s_1^2$; and in using $F = s_1^2 / s_2^2$, we will be concerned with rejection in the lower tail of the F distribution. Since the identification of the population is arbitrary, we may

Table 9.19   Format of the *F* Table, Table 6 in the Appendix

v_2	α	v_1 1	2	3	4	5	6	7	8	9	...	40	60	120	∞	v_2
1	.100	39.86	49.50	53.59	55.83	57.24	58.20	58.91	59.44	59.86	...	62.53	62.79	63.06	63.33	1
	.050	161.4	199.5	215.7	224.6	230.2	234.0	236.8	238.9	240.5	...	251.1	252.2	253.3	254.3	
	.025	647.8	799.5	864.2	899.6	921.8	937.1	948.2	956.7	963.3	...	1006	1010	1014	1018	
	.010	4052	4999.5	5403	5625	5764	5859	5928	5982	6022	...	6287	6313	6339	6366	
	.005	16211	20000	21615	22500	23056	23437	23715	23925	24091	...	25148	25253	25359	25465	
2	.100	8.53	9.00	9.16	9.24	9.29	9.33	9.35	9.37	9.38	...	9.47	9.47	9.48	9.49	2
	.050	18.51	19.00	19.16	19.25	19.30	19.33	19.35	19.37	19.38	...	19.47	19.48	19.49	19.50	
	.025	38.51	39.00	39.17	39.25	39.30	39.33	39.36	39.37	39.39	...	39.47	39.48	39.49	39.50	
	.010	98.50	99.00	99.17	99.25	99.30	99.33	99.36	99.37	99.39	...	99.47	99.48	99.49	99.50	
	.005	198.5	199.0	199.2	199.2	199.3	199.3	199.4	199.4	199.4	...	199.5	199.5	199.5	199.5	
3	.100	5.54	5.46	5.39	5.34	5.31	5.28	5.27	5.25	5.24	...	5.16	5.15	5.14	5.13	3
	.050	10.13	9.55	9.28	9.12	9.01	8.94	8.89	8.85	8.81	...	8.59	8.57	8.55	8.53	
	.025	17.44	16.04	15.44	15.10	14.88	14.73	14.62	14.54	14.47	...	14.04	13.99	13.95	13.90	
	.010	34.12	30.82	29.46	28.71	28.24	27.91	27.67	27.49	27.35	...	26.41	26.32	26.22	26.13	
	.005	55.55	49.80	47.47	46.19	45.39	44.84	44.43	44.13	43.88	...	42.31	42.15	41.99	41.83	
4	.100	4.54	4.32	4.19	4.11	4.05	4.01	3.98	3.95	3.94	...	3.80	3.79	3.78	3.76	4
	.050	7.71	6.94	6.59	6.39	6.26	6.16	6.09	6.04	6.00	...	5.72	5.69	5.66	5.63	
	.025	12.22	10.65	9.98	9.60	9.36	9.20	9.07	8.98	8.90	...	8.41	8.36	8.31	8.26	
	.010	21.20	18.00	16.69	15.98	15.52	15.21	14.98	14.80	14.66	...	13.75	13.65	13.56	13.46	
	.005	31.33	26.28	24.26	23.15	22.46	21.97	21.62	21.35	21.14	...	19.75	19.61	19.47	19.32	
5	.100	4.06	3.78	3.62	3.52	3.45	3.40	3.37	3.34	3.32	...	3.16	3.14	3.12	3.10	5
	.050	6.61	5.79	5.41	5.19	5.05	4.95	4.88	4.82	4.77	...	4.46	4.43	4.40	4.36	
	.025	10.01	8.43	7.76	7.39	7.15	6.98	6.85	6.76	6.68	...	6.18	6.12	6.07	6.02	
	.010	16.26	13.27	12.06	11.39	10.97	10.67	10.46	10.29	10.16	...	9.29	9.20	9.11	9.02	
	.005	22.78	18.31	16.53	15.56	14.94	14.51	14.20	13.96	13.77	...	12.53	12.40	12.27	12.14	

Critical values of the F distribution (portion of table). Row bands give the denominator degrees of freedom ν_2; within each band the rows give the upper‑tail area α. Columns give the numerator degrees of freedom ν_1.

ν_2	α	1	2	3	4	5	6	7	8	9	⋯	40	60	120	∞
6	.100	3.78	3.46	3.29	3.18	3.11	3.05	3.01	2.98	2.96	⋯	2.78	2.76	2.74	2.72
	.050	5.99	5.14	4.76	4.53	4.39	4.28	4.21	4.15	4.10	⋯	3.77	3.74	3.70	3.67
	.025	8.81	7.26	6.60	6.23	5.99	5.82	5.70	5.60	5.52	⋯	5.01	4.96	4.90	4.85
	.010	13.75	10.92	9.78	9.15	8.75	8.47	8.26	8.10	7.98	⋯	7.14	7.06	6.97	6.88
	.005	18.63	14.54	12.92	12.03	11.46	11.07	10.79	10.57	10.39	⋯	9.24	9.12	9.00	8.88
7	.100	3.59	3.26	3.07	2.96	2.88	2.83	2.78	2.75	2.72	⋯	2.54	2.51	2.49	2.47
	.050	5.59	4.74	4.35	4.12	3.97	3.87	3.79	3.73	3.68	⋯	3.34	3.30	3.27	3.23
	.025	8.07	6.54	5.89	5.52	5.29	5.12	4.99	4.90	4.82	⋯	4.31	4.25	4.20	4.14
	.010	12.25	9.55	8.45	7.85	7.46	7.19	6.99	6.84	6.72	⋯	5.91	5.82	5.74	5.65
	.005	16.24	12.40	10.88	10.05	9.52	9.16	8.89	8.68	8.51	⋯	7.42	7.31	7.19	7.08
⋯															
12	.100	3.18	2.81	2.61	2.48	2.39	2.33	2.28	2.24	2.21	⋯	1.99	1.96	1.93	1.90
	.050	4.75	3.89	3.49	3.26	3.11	3.00	2.91	2.85	2.80	⋯	2.43	2.38	2.34	2.30
	.025	6.55	5.10	4.47	4.12	3.89	3.73	3.61	3.51	3.44	⋯	2.91	2.85	2.79	2.72
	.010	9.33	6.93	5.95	5.41	5.06	4.82	4.64	4.50	4.39	⋯	3.62	3.54	3.45	3.36
	.005	11.75	8.51	7.23	6.52	6.07	5.76	5.52	5.35	5.20	⋯	4.23	4.12	4.01	3.90
13	.100	3.14	2.76	2.56	2.43	2.35	2.28	2.23	2.20	2.16	⋯	1.93	1.90	1.88	1.85
	.050	4.67	3.81	3.41	3.18	3.03	2.92	2.83	2.77	2.71	⋯	2.34	2.30	2.25	2.21
	.025	6.41	4.97	4.35	4.00	3.77	3.60	3.48	3.39	3.31	⋯	2.78	2.72	2.66	2.60
	.010	9.07	6.70	5.74	5.21	4.86	4.62	4.44	4.30	4.19	⋯	3.43	3.34	3.25	3.17
	.005	11.37	8.19	6.93	6.23	5.79	5.48	5.25	5.08	4.94	⋯	3.97	3.87	3.76	3.65
14	.100	3.10	2.73	2.52	2.39	2.31	2.24	2.19	2.15	2.12	⋯	1.89	1.86	1.83	1.80
	.050	4.60	3.74	3.34	3.11	2.96	2.85	2.76	2.70	2.65	⋯	2.27	2.22	2.18	2.13
	.025	6.30	4.86	4.24	3.89	3.66	3.50	3.38	3.29	3.21	⋯	2.67	2.61	2.55	2.49
	.010	8.86	6.51	5.56	5.04	4.69	4.46	4.28	4.14	4.03	⋯	3.27	3.18	3.09	3.00
	.005	11.06	7.92	6.68	6.00	5.56	5.26	5.03	4.86	4.72	⋯	3.76	3.66	3.55	3.44

avoid this difficulty by designating the population with the larger sample variance as population 1. In other words, always place the larger sample variance in the numerator of

$$F = \frac{s_1^2}{s_2^2}$$

and designate that population as 1. This procedure ensures that H_0 will be rejected only if the sample value is greater than a right-tailed critical value of F. However, this right-tailed portion of the rejection region represents only $\alpha/2$. Therefore, in following this procedure for a two-tailed test with a level of significance equal to α, we would reject H_0 if $F > F_{\alpha/2}$.

Test of an Hypothesis About the Equality of Two Population Variances

Null hypothesis $H_0: \sigma_1^2 = \sigma_2^2$.

Alternative hypothesis $H_a: \sigma_1^2 \neq \sigma_2^2$ for a two-tailed statistical test.

Test statistic: $F = s_1^2/s_2^2$, where s_1^2 is the larger sample variance.

Rejection region: For a two-tailed test, reject H_0 if $F > F_{\alpha/2}$, where $F_{\alpha/2}$ is based on $(n_1 - 1)$ and $(n_2 - 1)$ degrees of freedom.

We illustrate these ideas with some examples.

EXAMPLE 9.8   The risk of alternative investments is generally evaluated by the variance of returns associated with each investment [W. F. Sharpe, *Portfolio Theory and Capital Markets* (New York: McGraw-Hill, 1970)]. The distribution of returns for two alternative investments A and B is shown in Figure 9.5. The expected rate of return on each investment is 17.8%, but on the basis of returns over the past 10 years for investment A and 8 years for investment B, the variances of returns for the two investments are 3.21 and 7.14, respectively. Do these variances present sufficient evidence to indicate that the risks of investments A and B are unequal? (That is, is there a difference in the population variances?)

SOLUTION   Assume that the populations possess probability distributions that are reasonably mound-shaped and hence will satisfy, for all practical purposes, the assumption that the populations are normal.

We wish to test the null hypothesis

$$H_0: \sigma_A^2 = \sigma_B^2$$

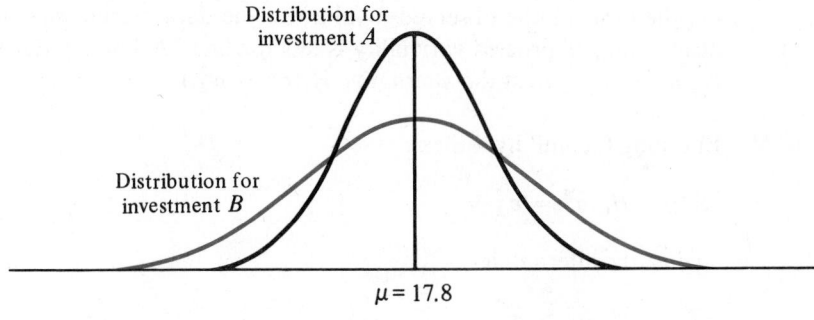

Figure 9.5   *Distribution of Rates of Return for Investments A and B*

against the alternative

$$H_a:\sigma_A^2 \neq \sigma_B^2$$

at an $\alpha = .05$ significance level. The right-tailed critical value of F with $v_1 = 8 - 1 = 7$ and $v_2 = 10 - 1 = 9$ degrees of freedom is $F_{.025} = 4.20$.

The calculated value of the test statistic is

$$F = \frac{s_B^2}{s_A^2} = \frac{7.14}{3.21} = 2.22$$

Since the test statistic does not fall in the rejection region, we do not reject $H_0:\sigma_A^2 = \sigma_B^2$. Thus, there is insufficient evidence to conclude that the population variances differ.

◆ ◆

EXAMPLE 9.9   Consistency in the taste of beer is an important quality in retaining customer loyalties. The variability in the taste of a given beer can be affected by the length of brewing, variations in ingredients, and differences in the equipment used in the brewing process. A brewery with two production lines, 1 and 2, has made a slight adjustment to line 2, hoping to reduce the variability as well as the average taste index. Random selections of $n_1 = 25$ and $n_2 = 25$ eight-ounce glasses of beer were selected from the two production lines and were measured by using an instrument designed to index the beer taste. The two samples produced means and variances as follows:

$$\bar{x}_1 = 3.2 \qquad \bar{x}_2 = 3.0$$
$$s_1^2 = 1.04 \qquad s_2^2 = .51$$

On the basis of the observed p-value, do the data present sufficient evidence to indicate that the process variability is less for line 2? (That is, test the null hypothesis $H_0:\sigma_1^2 = \sigma_2^2$ against the alternative $H_a:\sigma_1^2 > \sigma_2^2$.)

SOLUTION   In testing the null hypothesis

$$H_0:\sigma_1^2 = \sigma_2^2$$

against the alternative

$$H_a:\sigma_1^2 > \sigma_2^2$$

we observe that the calculated value of the test statistic is

$$F = \frac{s_1^2}{s_2^2} = \frac{1.04}{.51} = 2.04$$

For $v_1 = v_2 = 24$ degrees of freedom, the observed value of $F = 2.04$ lies between $F_{.05} = 1.98$ and $F_{.025} = 2.27$. Therefore, the p-value for this test lies between .025 and .05. If we are willing to use a significance level of $\alpha = .05$, then H_0 is rejected, and we conclude that the variability of line 2 is less than that for line 1.

◆◆

The test for the equality of two population variances does not assume that the means are equal, but we might notice that the process averages in both Example 9.8 and Example 9.9 are nearly equal. What the test does examine is the **comparative uniformity of individual values within each population**. For instance, in Example 9.8 if we had rejected our hypothesis, we would have concluded that the closing prices of one stock are significantly more volatile than the closing prices of the other. Thus, we would conclude that even though the average prices of the two stocks are nearly equal, one stock involves more risk than the other, as indicated by its larger variance. Since the hypothesis of equal variances was not rejected for that example, there is not sufficient evidence to indicate a difference in risks for the two stocks. Although a difference in risks may exist, it was not evident based on samples of sizes $n_A = 10$ and $n_B = 8$.

EXERCISES   Conceptual Level

9.60   What assumptions must be met in order that the sampling distribution for the ratio of two sample variances have an F distribution in repeated sampling?

9.61   Briefly describe the characteristics of the F distribution.

9.62   Why do we follow the convention of placing the larger of the two variance estimators in the numerator of F?

9.63   If the larger sample variance is used as the numerator of an F statistic in testing $H_0:\sigma_1^2 = \sigma_2^2$ versus $H_a:\sigma_1^2 \neq \sigma_2^2$, how is the critical value of F found?

EXERCISES Skill Level

9.64 Find the tabulated value of F based on v_1 and v_2 degrees of freedom having a right-tailed area of α.

	v_1	v_2	α
a.	10	20	.05
b.	7	25	.01
c.	24	12	.10
d.	8	6	.025
e.	15	19	.05
f.	4	15	.005

9.65 Two independent random samples selected from normal populations produced the accompanying data summary.

Sample 1	Sample 2
$\bar{x}_1 = 22.1$	$\bar{x}_2 = 18.2$
$s_1 = 4.8$	$s_2 = 3.5$
$n_1 = 16$	$n_2 = 12$

a. What statistic will you use for testing $H_0: \sigma_1^2 = \sigma_2^2$ versus $H_a: \sigma_1^2 \neq \sigma_2^2$?

b. What is the critical value of the test statistic if $\alpha = .05$?

c. Do these data contain sufficient evidence to conclude that the two population variances are different?

d. What is the observed significance level (p-value) of the test?

9.66 Repeat the instructions for Exercise 9.65, given the following data summary.

Sample 1	Sample 2
$\bar{x}_1 = 89.7$	$\bar{x}_2 = 103.2$
$s_1 = 5.2$	$s_2 = 7.8$
$n_1 = 9$	$n_2 = 11$

EXERCISES Applications

9.67 A manufacturer of electronic equipment is considering the selection of one of two potential suppliers to provide silicone chips, which are essential components in a piece of equipment manufactured by the company. Both proposed chips are claimed to possess an identical life span, but information concerning the variability of the life spans of the chips is unavailable. In a test conducted to identify the chip with the least variability, 10 chips are chosen from each potential supplier and subjected to continual use until failure. The results are shown in the accompanying table. Do these data present sufficient evidence to indicate a difference in life span variability between the two kinds of silicone chips? Test by using a 10% level of significance.

Statistic	Supplier A	Supplier B
n	10	10
$\bar{x}$	1,684.3	1,675.9
s	173.4	95.7

9.68   Refer to Exercise 9.67. Find an approximate 95% confidence interval for the life span range of silicone chips provided by supplier B.

9.69   The stability of measurements of the characteristics of a manufactured product is important in maintaining product quality. In fact, it is sometimes better to have a small variation in the measured value of some important characteristic of a product and have the process mean slightly off target than to suffer wide variation with a mean value that perfectly fits requirements. The latter situation may produce a higher percentage of defective products than the former. A manufacturer of light bulbs suspects that one of the production lines is producing bulbs with a higher variation in length of life. To test this theory, she compares the lengths of life of $n = 50$ bulbs randomly sampled from the suspect line and $n = 50$ from a line that seems to be in control. The sample means and variances for the two samples are given in the accompanying table. Do these data present sufficient evidence to indicate that bulbs produced by the suspect line possess a larger variance in length of life than those produced by the line that is assumed to be in control? Use $\alpha = .05$.

Suspect Line	Line in Control
$\bar{x}_1 = 1,520$	$\bar{x}_2 = 1,476$
$s_1^2 = 92,000$	$s_2^2 = 37,000$

9.70   In Exercise 8.28 we referred to a study that measured the mean and standard deviation of annual returns for randomly chosen securities within each of four asset classes for the past nine years. The data shown in Exercise 8.28 are reproduced in Table 9.20.

Table 9.20

Asset Class	Number of Chosen Securities	Mean Annual Return	Standard Deviation of Annual Return (Risk)
Common stocks	50	10.57	19.05
Corporate bonds	35	4.38	5.52
U.S. government securities	30	3.47	3.87
Municipal bonds	35	2.51	8.76

a. Do these data indicate a difference in risk (as measured by the standard deviation of annual returns) between municipal bonds and corporate bonds as investments over the past nine years? Test by using $\alpha = .10$.

b. Do the data indicate a difference in risk between common stocks and corporate bonds? Between corporate bonds and U.S. government securities?

c. Does the study tend to support the belief of financial scholars that the expected return of an investment consists of a rate of interest plus a risk premium?

9.8

Summary

The t, χ^2, and F statistics employed in the small-sample statistical methods discussed in the preceding sections are based on the assumption that the sampled **populations possess a normal probability distribution**. This requirement will be satisfied for many types of experimental measurements. However, when this requirement is not satisfied, actual significance levels and confidence coefficients may differ strongly from the nominal levels of α and $(1 - \alpha)$. **When the sampled populations are nonnormal, the nonparametric techniques given in Chapter 18 can be used as alternative tests in place of the t, χ^2, and F tests given in this chapter.**

You will observe the close relationship connecting the Student's t and the z statistic and therefore the similarity of the methods for testing hypotheses and the construction of confidence intervals. The χ^2 and F statistics employed in making inferences about population variances do not, of course, follow this pattern, but the reasoning employed in the construction of the statistical tests and confidence intervals is identical for all the methods we have presented.

Tips on Problem Solving

To help you decide whether the techniques of this chapter are appropriate for the solution of a problem, ask yourself the following questions:

1. Does the problem imply that an inference should be made about a population mean or the difference between two means? Are the samples small—say, $n < 30$? If the answers to both questions are "yes," you may be able to use one of the methods of Sections 9.3, 9.4, or 9.5. If the sample sizes are large—say, $n \geq 30$—you can use the methods of Chapter 8. In practice, you would also need to verify that the assumptions underlying each procedure are satisfied. Are the population distributions nearly normal, and have the sampling procedures conformed to those prescribed for the statistical method?

2. In a comparison of population means, were the observations from the two populations selected in a paired manner? If they were, you must use the paired-difference analysis of Section 9.5. If the samples were selected independently and in a random manner, use the methods of Section 9.4.

3. Is data variation the primary objective of the problem? If it is, you may be required to make an inference about a population variance σ^2 (Section 9.6) or to compare two population variances σ_1^2 and σ_2^2 (Section 9.7).

    Note: The Tips on Problem Solving following Section 8.18 will also be helpful in solving the exercises in this chapter.

Supplementary Exercises

Applications

9.71 Owing to the variability of trade-in allowance, the profit per new car sold by an automobile dealer varies from car to car. The profit per sale, tabulated for the past week, was (in hundreds of dollars)

2.1   3.0   1.2   6.2   4.5   5.1

Find a 90% confidence interval for the average profit per sale.

9.72 Continuing Exercise 9.71, suppose that the profit goal of the dealership is to average at least $480 on the sale of each new car. On the basis of the available data, is there evidence to indicate that the dealership has not met its profit goal? Test at the 5% level of significance.

9.73 A chemical process has produced, on the average, 800 tons of chemical per day. The daily yields (in tons) for the past week are

785   805   790   793   802

Do these data indicate that the average yield is less than 800 tons and hence that something is wrong with the process? Test at the 5% level of significance.

9.74 Find a 90% confidence interval for the mean yield in Exercise 9.73.

9.75 Refer to Exercises 9.73 and 9.74. How large should the sample be in order that the width of the confidence interval be reduced to approximately 5 tons?

9.76 A manufacturer can tolerate .05 milligram per liter (.05 mg/l) of impurities in a raw material needed for manufacturing its product. Because the laboratory test for the impurities is subject to experimental error, the manufacturer tests each batch 10 times. Assume that the mean value of the experimental error is zero and hence that the mean value of the 10 test readings is an unbiased estimate of the true amount of the impurities in the batch. For a particular batch of the raw material the mean of the 10 test readings is .058 mg/l, with a standard deviation of .012 mg/l. Estimate the *p*-value for the statistical test that examines whether sufficient evidence exists to indicate that the amount of impurities in the batch exceeds .05 mg/l. What are your conclusions?

9.77 Since the deregulation of commercial aviation by the Federal Aeronautics Administration, many airlines have developed discount fares with the objective of increasing passenger use. One airline, which historically had observed an average of 198 passengers on its San Francisco–New York flight, offered a reduced-fare program to increase its share of the air traffic market between these two cities. On its first 16 flights after initiation of the program, the airline observed an average of $\bar{x} = 213$ passengers per flight between San Francisco and New York with a standard deviation of $s = 8$. Do these data present sufficient evidence to indicate that the reduced-fare program has been effective in increasing the airline's share of passenger traffic between San Francisco and New York? Test by using $\alpha = .05$.

9.78 A West Coast firm claims that its gasoline additive will result in fuel savings of more than 15%. A city police agency conducted an experiment by using the additive in eight of its cars for a one-week period. The recorded gasoline savings, as measured by the percentage decrease in fuel use per mile after using the additive, are

15.2   14.1   13.7   15.2   18.6   15.0   14.5   13.8

Do these data support the developer's claim that the additive will reduce fuel use by more than 15%? Use $\alpha = .05$.

9.79 If you planned to report the results of the statistical test in Exercise 9.78, what p-value would you report?

9.80 Since 1963, when the surgeon general first reported the results of tests linking smoking to several types of physical ailments, cigarette manufacturers have sought to decrease the amount of harmful impurities in their products. One cigarette manufacturer claims that the mean amount of nicotine in cigarettes sold under its brand does not exceed 15 mg of nicotine. A sample of 16 cigarettes from this manufacturer yields a mean and standard deviation of 16.4 and 2 mg of nicotine, respectively. Do these data provide sufficient evidence to refute the manufacturer's claim? Use $\alpha = .10$.

9.81 In their study of average price of brands for a fixed product line (Exercise 9.27), D. J. Curry and P. C. Reisz ["Prices and Price/Quality Relationships: A Longitudinal Analysis," *Journal of Marketing*, Vol. 52 (January 1988), pp. 36–51], examined one aspect of the product life cycle theory, which, in general terms, predicts that because of the impact of buyer learning during the PLC, the price variance of a product line declines in magnitude over time. Do the data given in Table 9.21 provide sufficient evidence to contradict the PLC theory as it relates to the variances of the prices of 10-speed bicycles? Use $\alpha = .05$. What is the observed significance level of the test?

Table 9.21   Product Data for Exercise 9.81

Product	Year	Number of Brands	Average Price	Standard Deviation
10-speed bicycles	1974	16	$209.51	$38.70
	1980	30	206.23	65.50

9.82 In 1986 the United States 3.5% gain in manufacturing labor productivity was matched only by the United Kingdom, among 11 other industrial countries studied by A. Neef and J. Thomas in their comparative study of labor productivity and costs through 1986. The data in Table 9.22

reflect the annual percentage changes in manufacturing productivity reported by Neef and Thomas for the period 1960–1986.

Table 9.22   Manufacturing Productivity Data for Exercise 9.82

Year	United States	Canada	United Kingdom	Year	United States	Canada	United Kingdom
1975	2.5	2.9	2.5	1981	2.2	1.4	2.4
1976	4.6	4.5	2.9	1982	2.2	2.9	5.3
1977	3.0	3.2	1.9	1983	5.8	2.9	6.1
1978	1.5	1.2	.5	1984	5.5	2.3	4.4
1979	−.1	−.3	−.2	1985	5.1	2.5	3.8
1980	0	.4	1.3	1986	3.7	−.2	3.6

Source: United States figures taken from U.S. Department of Labor, Bureau of Labor Statistics, 1988. Other figures reflect the averages reported by A. Neef and J. Thomas, "Trends in Manufacturing Productivity and Labor Costs in the U.S. and Abroad," *Monthly Labor Review*, Vol. 110, No. 12 (December 1987), pp. 25–30.

a.   If the reported figures are taken to be sample values over the years reported, test whether a significant difference in average percentage change in manufacturing labor productivity exists between the United States and Canada.

b.   What is the observed significance level of the test performed in part a?

9.83   Refer to Exercise 9.82 in which the percentage change in manufacturing labor productivity was reported for the United States, Canada, and the United Kingdom during the period 1975–1986. If the reported figures are taken to be sample values during this time period, test for a significant difference in average percentage change in manufacturing labor productivity between the United States and the United Kingdom. What is the observed significance level of this test? How would you interpret your results?

9.84   A manufacturer of television sets claims that her product possesses an average defect-free life of three years. Three households in a community have purchased the sets and all three sets are observed to fail before three years, with failure times of 2.5, 1.9, and 2.9 years. Do these data present sufficient evidence to contradict the manufacturer's claim? Test at the $\alpha = .05$ level of significance.

9.85   Refer to Exercise 9.84. Approximately how many observations would be required to estimate the mean life of the television sets correct to within .2 year with a probability of .90?

9.86   A cannery prints "weight 16 ounces" on its label. The quality control supervisor selects 9 cans at random and weighs them. He finds $\bar{x} = 15.7$ and $s = .5$. Do the data present sufficient evidence to indicate that the mean weight of the cans is less than the amount claimed on the label? Use $\alpha = .05$.

9.87   A manufacturing plant has two assembly machines that perform identical operations on different assembly lines. As a result of their constant use, the machines break down quite frequently. The time between 10 consecutive breakdowns was recorded for each machine. Assume that the time between breakdowns for each machine is normally distributed, with a common variance σ^2. The sample means and variances of recorded times between breakdowns are given in Table 9.23. Do these data present sufficient evidence to indicate a difference in the population mean machine breakdown times? Test at the $\alpha = .10$ level of significance.

Table 9.23   Breakdown Data for Exercise 9.87

Machine 1	Machine 2
$\bar{x}_1 = 60.4$ minutes	$\bar{x}_2 = 65.3$ minutes
$s_1^2 = 31.40$ minutes (squared)	$s_2^2 = 44.82$ minutes (squared)

9.88 Proven natural resource reserves are defined to be the amount of deposits known to exist that can profitably be extracted within an existing technology and price structure. Thus, the amount of "proven reserves" may vary according to the assumptions made by reporting geologists regarding technology and price. In 1987 and 1988 six experts provided estimates of the level of proven oil reserves in the United States. Their estimates are provided in Table 9.24. Find the p-value for the statistical test that examines whether sufficient evidence exists to indicate that proven oil reserves in the United States increased from 1987 to 1988. What are your conclusions?

Table 9.24   Oil Reserves Data for Exercise 9.88

Expert	Estimated Reserves (Billions of Barrels) 1987	1988
1	31.7	34.1
2	30.6	32.7
3	32.0	30.9
4	31.6	31.8
5	33.4	33.9
6	33.1	34.2

9.89 When applied to a random sample of 16 people, a comparison of the time before recognition of the product for two different-colored advertising layouts of a common product produced the results (in seconds) shown in the accompanying table.

Layout 1	1	3	2	1	2	1	3	2
Layout 2	4	2	3	3	1	2	3	3

a. Do the data present sufficient evidence to indicate a difference in mean recognition time for the two layouts? Test at the $\alpha = .05$ level of significance.

b. Construct a 90% confidence interval for $(\mu_1 - \mu_2)$.

c. Give a practical interpretation for the confidence interval in part b.

9.90 Refer to Exercise 9.89. Suppose that the product recognition experiment had been conducted by using people as blocks and by making a comparison of recognition time within each person; that is, each of the 8 persons would be subjected to both layouts in a random order. The data for this experiment (in seconds) are given in the accompanying table.

Person	1	2	3	4	5	6	7	8
Layout 1	3	1	1	2	1	2	3	1
Layout 2	4	2	3	1	2	3	3	3

a. Do the data present sufficient evidence to indicate a difference in mean recognition time for the two layouts? Test at the $\alpha = .05$ level of significance.

b. Construct a 95% confidence interval for $(\mu_1 - \mu_2)$.

c. Give a practical interpretation for the confidence interval obtained in part b.

9.91 Analyze the data in Exercise 9.90 as though the experiment had been conducted in an unpaired manner. Calculate a 95% confidence interval for $(\mu_1 - \mu_2)$ and compare it with the interval obtained in part b of Exercise 9.90. Does it appear that blocking increased the amount of information available in the experiment?

9.92 What is the real value of a company's stock? Although the market value may be highly inflated by investor speculation, book value represents the amount that would be distributed on a per-share basis according to the value of the firm's assets. An investment committee for a union pension fund wishes to invest a large sum of money by purchasing the shares of one of two companies, A or B. The market values of the stocks are equal for both companies, but the book values, as determined by six independent financial analysts, are as given in Table 9.25. Do these data present sufficient evidence to assume that a difference exists in book value between the two companies? Test by using $\alpha = .01$.

Table 9.25   Stock Book Value Data for Exercise 9.92

Company	Analyst					
	1	2	3	4	5	6
A	$65.80	$60.90	$71.20	$72.00	$69.70	$65.00
B	78.10	71.30	75.40	69.20	78.40	81.20

9.93 If you planned to report the results of the statistical test in Exercise 9.92, what p-value would you report?

9.94 Since shelf space is a limited resource of a retail store, product selection, shelf space allocation, and shelf space placement decisions must be made according to a careful analysis of profitability and inventory turnover. The manager of a chain of variety stores wishes to see whether shelf location affects the sales of a certain product. She believes that placing the product at eye level will result in greater sales than will placing the product on a lower shelf. The data shown in Table 9.26 represent the number of sales of the product in four different stores. Sales were observed over two weeks, with product placement at eye level one week and on a lower shelf the other week. Test, at the 5% level of significance, to determine whether placement of the product at eye level significantly increases sales.

Table 9.26   Sales Data for Exercise 9.94

Store	Number Sold	
	Lower Shelf	Eye-Level Shelf
1	27	33
2	22	23
3	32	38
4	32	33

9.95 The cost of moving between major United States cities increased in 1987, but at a slower rate than the previous year. The data in Table 9.27 are based on an average move of 8,000 lb, excluding container, packing, or unpacking charges. Estimate the average cost of moving 8,000 lb, excluding container, packing, or unpacking charges, between two major cities, using a 95% confidence interval. Interpret your results.

Table 9.27   Cost Data for Exercise 9.95

Cities	Cost
New York–Miami	$4,046
New York–Los Angeles	6,757
Los Angeles–Chicago	5,721
Chicago–New York	3,284
Houston–New York	4,641
Houston–Los Angeles	4,453

Source: "1988 Survey of Selling Costs," *Sales and Marketing Management,* January 1988. Reprinted by permission.

9.96 An investor wishes to determine whether growth stocks associated with utility companies tend to outperform industrial high-yield stocks. From a list of high-yield stocks recommended by several leading brokerage houses, the investor randomly selects 6 industrial stocks and 5 utilities. Using Standard and Poor's value line estimate of yield (dividend as a percentage of price) for each chosen stock, the investor obtained the data shown in the table. Do these data suggest that growth stocks of utilities outperform industrial growth stocks? Test at the 5% level of significance.

Industrials	7.8	10.3	7.9	8.7	9.2	8.9
Utilities	9.2	9.1	11.1	8.8	9.6	

9.97 Refer to Exercise 9.96.

a. Estimate the difference in mean yield between industrial growth stocks and utility growth stocks, using a 95% confidence interval.

b. Find a 95% confidence interval for the average yield of utility growth stocks.

9.98 Many manufacturers of large household appliances have recently changed from riveting procedures to spot welding in an attempt to cut costs. However, shear strength and strength uniformity must be maintained with spot welding. A manufacturer of an arc welder claims his product can produce spot welds on household appliances that range in shear strength from 400 to 500 lb. A sample of $n = 25$ spot welds produced by the manufacturer's arc welder was subjected to a shear strength test. The mean and standard deviation of the recorded shear strengths were $\bar{x} = 438$ lb and $s = 29$ lb. Do these data present sufficient evidence to reject the arc welder manufacturer's claim? Test by using $\alpha = .05$.

9.99 Refer to Exercise 9.98.

a. Estimate the mean shear strength of spot welds produced by the manufacturer's arc welder, using a 90% confidence interval.

b. Find a 90% confidence interval for the variance of the shear strengths of spot welds produced by the arc welder.

9.100   A manufacturer of a machine that packages soap powder claims that his machine can fill cartons at a given weight with a range of no more than 2/5 oz. The mean and variance of a sample of eight 3-lb boxes are found to be 3.1 lb and .018 oz (squared), respectively. Do these results tend to refute the manufacturer's claim? Test by using a 5% level of significance. (*Hint:* If the manufacturer's claim is that the range is less than or equal to .4 oz, then, since the range is approximately 4σ, an hypothesis consistent with the manufacturer's claim is $\sigma^2 = .01$ against the alternative $\sigma^2 > .01$.)

9.101   Continuing Exercise 9.100, find a 90% confidence interval for σ^2, the variance of fill using the manufacturer's fill machine. Use these results to find a 90% confidence interval for the *range* of fill using the manufacturer's fill machine.

9.102   A dairy is in the market for a new bottle-filling machine and is considering models A and B manufactured by company A and company B, respectively. If ruggedness, cost, and convenience are comparable in the two models, the deciding factor is the variability of fills (the model producing fills with the smaller variance being preferred). Wishing to demonstrate that the variability of fills is less for model A than for model B, a salesman for a company A acquired a sample of 30 fills from a machine of model A and a sample of 10 fills from a machine of model B. The sample variances were $s_A^2 = .027$ and $s_B^2 = .065$. Do these sample variances provide statistical support at the .05 level of significance for the salesman's belief?

9.103   A chemical manufacturer claims that the purity of her product never varies more than 2%. Five batches were tested and gave purity readings of 98.2%, 97.1%, 98.9%, 97.7%, and 97.9%. Do these data provide sufficient evidence to contradict the manufacturer's claim? Test by using a 5% level of significance.

9.104   The closing prices of two common stocks were recorded for a period of 15 days. The means and variances are

$$\bar{x}_1 = 37.58 \qquad \bar{x}_2 = 38.24$$
$$s_1^2 = 1.54 \qquad s_2^2 = 2.96$$

Do these data present sufficient evidence to indicate a difference in variability (and hence risk) of the two stocks for the populations associated with the two samples? Use $\alpha = .10$.

9.105   In the metal-fabricating industry productivity and subsequent profitability are highly dependent on the quality and uniformity of needed raw materials. Suppose there are two principal sources of raw materials under consideration for use by a metal fabricator in a certain process. Both sources appear to have similar quality characteristics, but the manufacturer is not certain about their respective uniformity of impurities content. Ten 100-lb samples of each source are selected and the amount, in pounds, of impurities is measured for each sample. The results are given in the accompanying table. Do these data suggest that a difference exists in the uniformity of impurities content in the two raw materials? Test with a 10% level of significance.

Raw Material A	Raw Material B
$\bar{x}_1 = 41.3$	$\bar{x}_2 = 39.6$
$s_1^2 = 18.75$	$s_2^2 = 7.85$

9.106   Continuing Exercise 9.105, test to see whether a difference exists in the mean impurities content in the 100-lb samples of the two raw materials. Use $\alpha = .05$.

9.107   The frequency with which the Internal Revenue Service (IRS) conducts intensive audits of individual tax returns is dependent on several factors, one of which is the region in which the individual lives (Rose Gutfeld, "Regional Risks: IRS's Audit Rate Shows That All States Aren't Equal," *Wall Street Journal*, July 22, 1987). In 1986 the IRS audited only .47% of individual returns in Rhode Island; in contrast, it examined 2.61% of returns in Nevada. The IRS computers flag returns that trigger statistical alarms for each income level. In recent years the IRS has publicized campaigns to give close scrutiny to "partners and promoters of tax shelters, people who earn tip income, and the self-employed, among others." Hence, the IRS is more likely to audit the return of an individual who lives in an area where the tax paid by individuals is highly variable. A tax analyst for the IRS seeks to compare the variability between federal tax collections from individuals in Idaho and Nevada. A sample of 121 individual returns from Idaho revealed a standard deviation of tax paid of $s_1 = \$46.22$; 61 returns from Nevada provided $s_2 = \$77.84$. Do these data indicate that a difference exists between the variability of federal taxes collected from individuals in Idaho and Nevada? Test by using $\alpha = .02$.

 9.108 Strong fluctuations in the price of a gallon of gasoline have occurred during the last several years in the United States and other industrial nations of the world. Table 9.28 gives the average price of a gallon of gasoline (in United States dollars) during 1984 and 1985 for six industrial nations.

    a. Do these data provide sufficient evidence to indicate that there was a significant change in the average price of gasoline during this two-year period? What is the observed significance level of your test?

    b. Estimate the average change in the price per gallon of gasoline from 1984 to 1985 with a 95% confidence interval.

Table 9.28   Gasoline Price Data for Exercise 9.108

City, Country	1984	1985
United States	$1.37	$1.34
Paris, France	2.27	2.18
Rome, Italy	2.77	2.52
Bonn, GFR	1.85	1.66
Mexico City, Mexico	1.20	1.36
Tokyo, Japan	2.42	2.37

Source: Data from U.S. Department of Commerce, Bureau of the Census, *Statistical Abstract of the United States*, 107th ed. (Washington, D.C., 1987), p. 592.

Supplementary Exercises

Interpretive

 9.109 The Food and Drug Administration requires precise measurements of meat fat content for labeling and pricing procedures. A meat-packing company is considering the use of two different methods to determine the percentage of fat content in samples of meat. Both methods were used to evaluate the fat content in eight different meat samples. The results are given in Table 9.29. Interpret the results of this experiment by using the MINITAB output in Table 9.30. The printout was produced by implementing the command TTEST (of $H_0: \mu_d = 0$ versus $H_a: \mu_d \neq 0$) 0 (for the differences corresponding to method 1 − method 2, stored in) C3.

Table 9.29   Fat Content Data for Exercise 9.109

Percentage Fat Content, Using	Meat Sample							
	1	2	3	4	5	6	7	8
Method 1	23.1	27.1	25.0	27.6	22.2	27.1	23.2	24.7
Method 2	22.7	27.4	24.9	27.2	22.5	27.4	23.6	24.4

Table 9.30   MINITAB Output for the Data of Exercise 9.109

```
MTB > SET C1
DATA> 23.1 27.1 25 27.6 22.2 27.1 23.2 24.7
DATA> END
MTB > SET C2
DATA> 22.7 27.4 24.9 27.2 22.5 27.4 23.6 24.4
DATA> END
MTB > LET C3 = C1-C2
MTB > TTEST 0 C3

TEST OF MU = 0.000 VS MU N.E.  0.000

            N       MEAN    STDEV    SE MEAN         T   P VALUE
C3          8     -0.012    0.348      0.123     -0.10      0.92

MTB >
```

Can the two methods be deemed equivalent? If so, are there any other factors that you might use in selecting one method over the other?

9.110   The office of the Lane County assessor is considering the use of a computer valuation model for obtaining the assessed valuations for each of the residential dwellings of Lane County. The assessor is interested in seeing how assessed valuations obtained from the model compare with assessments made by an "expert" valuer. To note the comparison, the assessor randomly selects 10 residential dwellings from Lane County, computes their assessed valuation using the computer model, and obtains an assessed valuation for each from an expert valuer. The results are given in Table 9.31. What is the advantage, if any, of using the two different methods of assessing valuation for each of 10 dwellings, compared with an alternative procedure that involves selecting 10 dwellings whose valuations are obtained by using the computer valuation model and another 10 different dwellings whose valuations are obtained from an expert valuer? Using the experimental results given in Table 9.31, can you conclude that these two methods of obtaining assessed valuations produce different valuations, on the average? If so, what is the estimated average gain (or loss) in assessed valuation in Lane County if the computer model for valuation is used instead of an expert valuer? What is the estimated gain (or loss) in property tax revenue if property taxes are assessed at the annual rate of 3% of the assessed valuation of the residential dwelling? What is the bound on the error of your estimate?

Table 9.31   Valuation Data for Exercise 9.110

	Assessed Valuations	
Dwelling	Computer Model	Expert Valuer
1	$71,000	$70,000
2	87,500	86,000
3	92,000	90,000
4	78,000	78,500
5	80,000	81,000
6	86,500	85,000
7	94,500	94,000
8	73,000	74,500
9	96,000	94,000
10	80,000	79,500

Case Study

Custom Precast

Consumer information systems are indispensable elements of consumer policy for the protection and education of consumers about product capabilities. Over the past decade the emphasis of consumer information programs has been on consumer protection—that is, "public trustee-ship." Now the aim of such programs is to "foster a self-reliant, self-actualizing consumer who can make the most of decisions and play an equal role with sellers in the market place" (H. Thorelli and J. Engledow, "Information Seekers and Information Systems: A Policy Perspective," *Journal of Marketing*, Spring 1980). The key here is better information. This case study explores the way one consumer, Custom Precast, dealt with the consumer information it was given.

Custom Precast manufacturers precast concrete products principally for the construction industry. Its product line includes a variety of precast beams for bridges and buildings. In the past the products of Custom Precast have met acceptable quality standards but have not proved to be superior in strength to those of the competition. In an attempt to gain a competitive advantage, Custom Precast has long sought a way of improving the strength of its precast concrete products. The marketing director for a chemical company claims that an additive available from his firm will substantially improve the strength of Custom Precast's large beams. To support his claim, the marketing director noted the results of several breakage tests conducted by his firm.

If his information is correct, it would prove advantageous to Custom Precast to sign an exclusive licensing agreement with the chemical company for its additive.

To test the claim, Custom Precast obtained sufficient additive from the chemical company to produce 12 test beams. The firm realized that results would be more conclusive with a larger sample size, but costs prohibited experimentation with a larger number of beams. Custom Precast then prepared 12 separate batches of concrete, each large enough to pour two beams. Each batch was then split in half, and the additive was mixed into one of the halves. The beams were then poured, allowed to set, and cured. The process was repeated 12 times, producing 12 pairs of beams. The results of the breakage tests are given in the following table.

	Breaking Strength of Concrete Beams (Pounds)	
Batch	Without Additive	With Additive
1	4,550	4,600
2	4,950	4,900
3	6,250	6,650
4	5,700	5,950
5	5,350	5,700
6	5,300	5,400
7	5,150	5,400
8	5,800	5,850
9	4,900	4,850
10	6,050	6,450
11	5,550	5,850
12	5,750	5,600

Comment on the efficiency of the experimental design used by Custom Precast by comparing the two sample variances, s^2 (without pairing) and s_d^2 (with pairing). On the basis of these experimental results, what course of action should Custom Precast take?

References and Suggested Readings

Johnson, R. J. *Elementary Statistics*, 5th ed. Boston: PWS-KENT Publishing Company, 1988.

Koopmans, L. H. *An Introduction to Contemporary Statistics*, 2d ed. Boston: PWS-KENT Publishing Company, 1987.

Neter, J., W. Wasserman, and G. A. Whitmore. *Applied Statistics, 3rd ed.* Boston: Allyn & Bacon, 1987.

Ryan, B. F., B. L. Joiner, and T. A. Ryan. *Minitab Handbook*, 2d ed. Boston: PWS-KENT Publishing Company, 1985.

Summers, G. W., W. S. Peters, and C. P. Armstrong. *Basic Statistics in Business and Economics, 4th ed.* Belmont, Calif.: Wadsworth, 1985.

TO
THE
READER

In Chapters 8 and 9 we addressed the problem of using sample information to make inferences about populations in terms of population descriptive measures, called *parameters*. Testing and estimation procedures for a single population mean using the z or t sampling distributions were easily extended to the situations involving two population means.

This chapter generalizes the methods given in Chapters 8 and 9 to allow for the simultaneous comparison of more than two population means. A procedure called an *analysis of variance* is extremely useful for comparing two or more means under various sampling designs, specifically for those designs that are analogous to the independent sampling and paired-difference designs of Chapter 9. You will find that an analysis of variance is a general technique that has applications in other situations as well.

10

THE ANALYSIS
OF VARIANCE

10.1

Introduction

In Chapters 8 and 9 we presented techniques for estimating and testing the value of a single population mean, or the difference between two population means. In many investigations we may wish to estimate or compare **several population means simultaneously**. In general, the populations to be compared correspond to the values of one or more experimental variables that may or may not affect the response under investigation. For example, we may wish to determine what effect, if any, three different promotional campaigns exert on the sales of a new product. In this situation the experimental variable under investigation is "promotional campaigns." As a second example, suppose that a manufacturing firm, interested in determining an optimal strategy in minimizing the maintenance costs of its production line machines, conducts a study of the maintenance costs of three different makes of machines. The study utilizes machines of each make that have been in continuous operation for three months, six months, and one year. In this investigation the experimental variables are "make of machine" and "time of continuous operation."

These examples indicate that independent experimental variables can be categorized as being one of two types. **Quantitative variables** are those whose values can be located on a line interval. For example, time of continuous operation, advertising expenditures, and interest rates are quantitative variables. **Qualitative variables** are those whose values result in assignment to a category. Promotional campaigns, make of machine, and geographic areas are examples of qualitative variables.

In this chapter we present a technique that is useful in comparing two or more means. In the context of the previous discussion this technique allows us to study the effect of a single experimental variable (which may be either quantitative or qualitative) on a response of interest. We also extend this procedure to include the analysis of a

quantitative variables

qualitative variables

designed experiment involving two independent experimental variables. A generalization to experiments involving more than two experimental variables can be found in Chapter 12 and in the references at the end of this chapter.

10.2

The Analysis of Variance

The analysis of variance is a general method of analysis that can be used to analyze data from designed experiments in which the independent experimental variables have been controlled (that is, fixed at preassigned values) or from experiments in which the independent variables cannot be or have not been controlled. Once the experiment has been designed, the quantities needed to test hypotheses concerning underlying population parameters can be found by using relatively simple formulas involving the response measurements. Experiments in which the independent variables are recorded but not controlled can nevertheless be analyzed by means of an analysis of variance; however, failure to control the independent variables results in calculational formulas that are best implemented on a computer.

In an analysis of variance the variation in the response measurements is partitioned into components that reflect the effects of one or more independent variables. The reasoning behind this procedure is that the total variation in the data can be attributed to random error and the variability among measurements under constant conditions, as well as variability due to the lack of uniformity in the values of the independent variables. For example, the variation in the amount of impurities in steel would reflect the natural variability in the material processed, as well as the reaction temperature maintained during the refining process. **The objective in an analysis of variance is to isolate and assess sources of variation associated with independent experimental variables and to determine how these variables interact and affect the response.**

The variability of a set of measurements is proportional to the sum of squares of deviations

$$\sum_{i=1}^{n} (x_i - \bar{x})^2$$

used to calculate the sample variance. The analysis of variance partitions the sum of squares of deviations, called the **total sum of squares**, into parts associated with one or more variables in the experiment, plus a remainder that is associated with random error. In a **well-designed experiment** one can partition the total sum of squares into components associated with each independent variable in the experiment. This partitioning is shown diagrammatically in Figure 10.1 for three experimental variables that act independently (as opposed to jointly) in affecting the response measurement.

In the cases we consider, when the experimental variables are unrelated to or have no effect on the response, each part of the total sum of squares divided by an appropriate constant provides an independent and unbiased estimator of σ^2. When a variable is highly related to the response, its part of the total sum of squares will be

total sum of squares

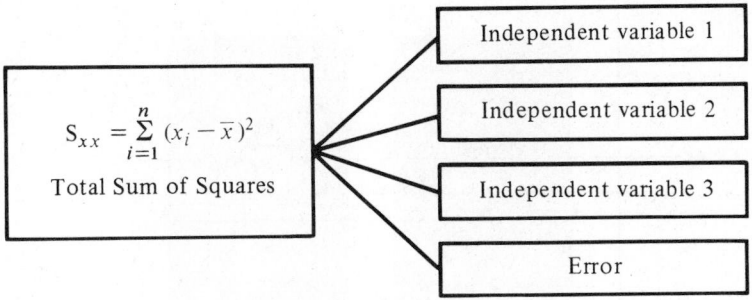

Figure 10.1   Partitioning of the Total Sum of Squares of Deviations

highly inflated. This condition can be detected by comparing the estimate of σ^2 for a particular independent variable with that obtained from the part associated with random error using an F test (see Section 9.7). If the observed value of F is significantly large, the hypothesis of "no effect for the independent variable" is rejected, indicating that the independent variable is in some way related to the response.

To illustrate the logic behind an analysis of variance, we consider the comparison of two population means based on two independent (unpaired) samples. In Chapter 9 we based our analysis on the difference in the observed sample means using Student's t statistic. We begin by showing graphically that an analysis of variance utilizes this same information recast in the framework of variability of observations within samples and variability of observations between samples.

Suppose that we have selected random samples of five observations each from two populations, 1 and 2, and that the x values are plotted as shown in Figure 10.2. Note that the $n_1 = 5$ observations from population 1 lie to the left; the $n_2 = 5$ observations

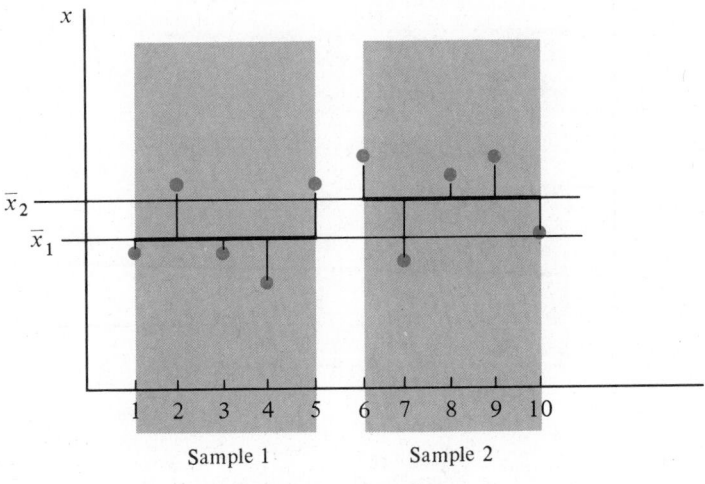

Figure 10.2   Graphical Portrayal of the Deviations of the x Values About Their Means

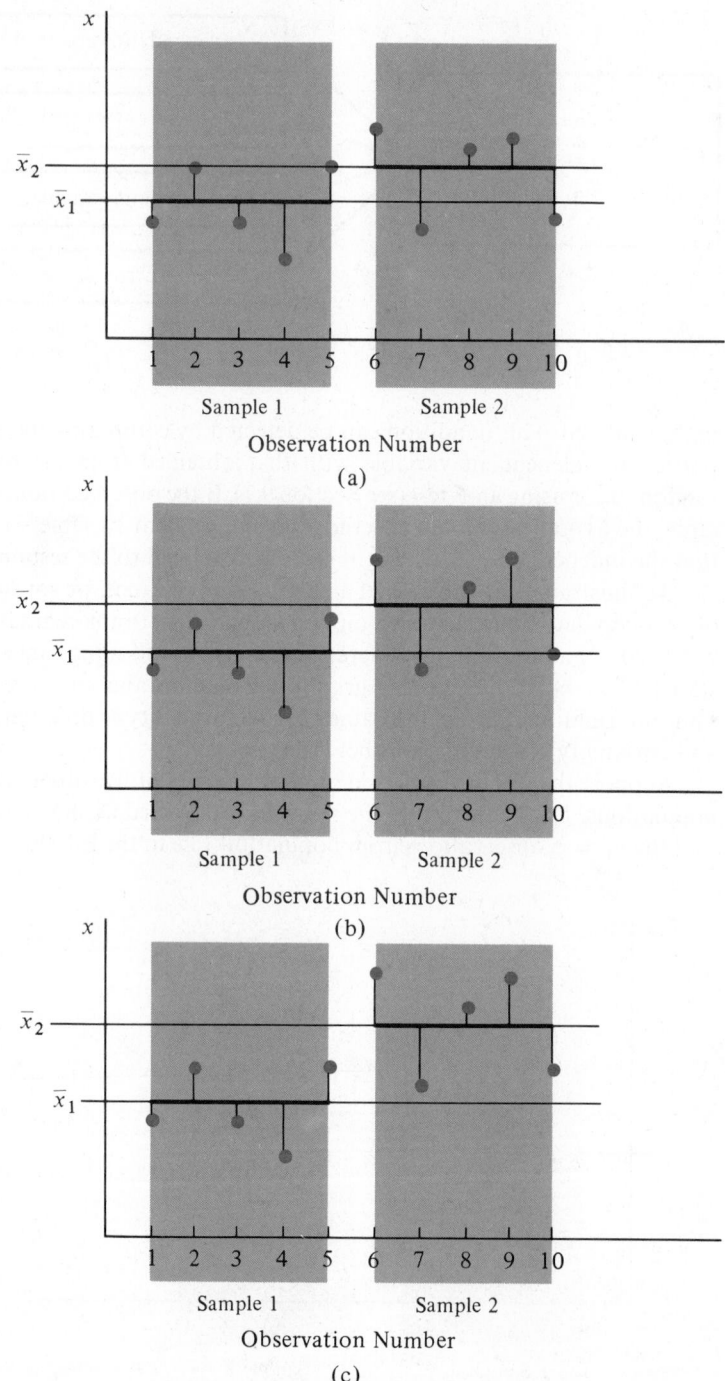

Figure 10.3   Three Fictitious Sets of Measurements, $n_1 = n_2 = 5$ (the Relative
Positions of Points Within Each Set Held Constant)

from population 2 lie to the right. The sample means $\bar{x}_1$ and $\bar{x}_2$ are shown as horizontal lines in the figure, and the deviations of the x values about their respective means are the vertical line segments. Now examine Figure 10.2. Do you think the data provide sufficient evidence to indicate a difference between the two population means? Before we explain, let us look at another figure.

The same two sets of five points are plotted in Figure 10.3 except that the distance between the two sets (as measured by the distance between $\bar{x}_1$ and $\bar{x}_2$) is greater in Figure 10.3b than in Figure 10.3a and is even greater in Figure 10.3c. Therefore, the distance between $\bar{x}_1$ and $\bar{x}_2$ increases as you move from Figure 10.3a to Figure 10.3b and then to Figure 10.3c, but the variation within each set is held constant.

Now view the three plots of Figure 10.3—a, b, and c—and decide which situation, a, b, or c, provides the greatest evidence to indicate a difference between μ_1 and μ_2. We think you will choose Figure 10.3c because that plot shows the greatest difference between sample means in comparison with the variation of the points about their respective sample means. This latter variation was held constant for all three plots.

Note that the population means actually may differ for Figure 10.3a, but this fact would not be apparent, because the variation of the points about their respective sample means is too large in comparison with the difference between $\bar{x}_1$ and $\bar{x}_2$. Figure 10.4 shows the same difference between sample means as for Figure 10.3a, but the variation within samples has been reduced. Now it appears that a difference does exist between μ_1 and μ_2.

Now let us leave our intuitive discussion and consider the two-sample comparison of means for sample sizes n_1 and n_2. In particular, we are interested in determining how the total sum of squares of deviations can be partitioned into portions corresponding to the difference between the means and another to the variation within the two samples.

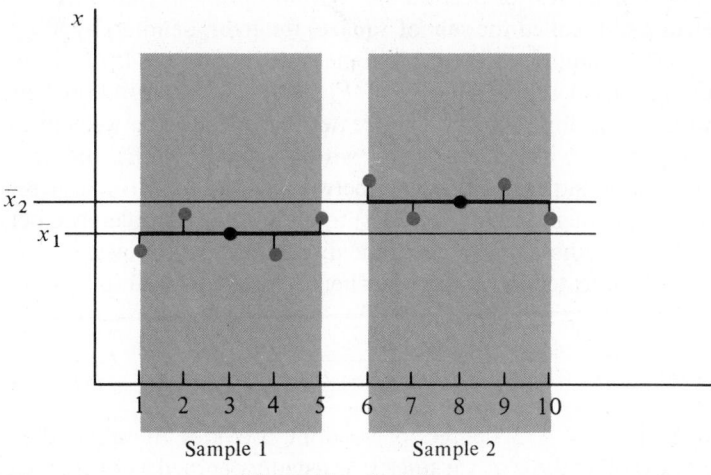

Figure 10.4   Data Showing the Same Difference Between Means as Shown in Figure 10.3a but with Less Within-Sample Variation

The total sum of squares of deviations of all $(n_1 + n_2)$ x values about the general mean is

$$\text{Total SS} = \sum_{i=1}^{2} \sum_{j=1}^{n_i} (x_{ij} - \bar{x})^2$$

where x_{ij} is the jth observation in the ith sample, and $\bar{x}$ is the average of all $(n_1 + n_2)$ observations contained in the two samples. Algebraically, we can show that

$$\text{Total SS} = \underbrace{\sum_{i=1}^{2} \sum_{j=1}^{n_i} (x_{ij} - \bar{x})^2 = \sum_{i=1}^{2} n_i (\bar{x}_i - \bar{x})^2}_{\text{SST}} + \underbrace{\sum_{i=1}^{2} \sum_{j=1}^{n_i} (x_{ij} - \bar{x}_i)^2}_{\text{SSE}}$$

where $\bar{x}_i$ is the average of the observations in the ith sample, $i = 1, 2$.

The first quantity to the right of the equal sign represents the variation of the sample means about the grand mean $\bar{x}$. Since the samples represent groups that have been "treated" at different settings of one or more independent variables, this quantity is called the **sum of squares for treatments** and denoted by the symbol SST. It can be shown that for two samples

sum of squares for treatments

$$\text{SST} = \frac{n_1 n_2}{n_1 + n_2} (\bar{x}_1 - \bar{x}_2)^2$$

Hence, SST, which measures the variation between the sample means, increases as the difference between $\bar{x}_1$ and $\bar{x}_2$ increases.

In partitioning the Total SS the second quantity to the right of the equal sign represents the variation of the individual measurements about their respective means. This quantity is the pooled sum of squares of deviations from both samples used in the two-sample t test of Section 9.4. Within-sample variation is associated with random error and is called the **sum of squares for error**, denoted by SSE.

sum of squares for error

The quantities SST and SSE measure the two kinds of variation that we viewed in the graphical representation of Figure 10.3, the variation between means and the variation within samples. The greater the variation between means (the larger SST) in comparison with the variation within samples (SSE), the greater is the weight of evidence to indicate a difference between μ_1 and μ_2. How large is large? When will SST be large enough (relative to SSE) to indicate a real difference between μ_1 and μ_2? We will answer these questions in the discussion that follows.

As indicated in Chapter 9, when there are two samples and $\sigma_1^2 = \sigma_2^2$,

$$s^2 = \text{MSE} = \frac{\text{SSE}}{n_1 + n_2 - 2}$$

with $(n_1 + n_2 - 2)$ degrees of freedom, provides an unbiased estimator of σ^2. In the context of analysis of variance s^2 is usually denoted as MSE, meaning **mean square for error**. Also, when the null hypothesis is true (that is, $\mu_1 = \mu_2$), SST divided by the appropriate number of degrees of freedom yields a second unbiased estimator of σ^2, which we denote as MST, **mean square for treatments**. For this example the number of degrees of freedom for MST is equal to 1.

mean square for error

mean square for treatments

When the null hypothesis is true (that is, $\mu_1 = \mu_2$), MSE (the mean square for error) and MST (the mean square for treatments) estimate the same quantity and should be "roughly" of the same magnitude. When the null hypothesis is false and $\mu_1 \neq \mu_2$, MST will almost always be larger than MSE.

The preceding discussion, along with a review of the variance ratio given in Section 9.7, suggests the use of

$$\frac{\text{MST}}{\text{MSE}}$$

as a test statistic to test the hypothesis $\mu_1 = \mu_2$ against the alternative $\mu_1 \neq \mu_2$. Indeed, when both populations are normally distributed, it can be shown that MST and MSE are independent in a probabilistic sense and that they can be used in a test statistic, as shown in the display.

Test Statistic for the Null Hypothesis $H_0: \mu_1 = \mu_2$

$$F = \frac{\text{MST}}{\text{MSE}}$$

The test statistic F follows the F probability distribution of Section 9.7. Disagreement with the null hypothesis is indicated by a large value of F, and hence the rejection region for a given α is

$$F \geq F_\alpha$$

Thus, the analysis of variance test results in a one-tailed F test. The degrees of freedom for F will be those associated with MST and MSE, which we denote as v_1 and v_2, respectively. Although we have not indicated how we determine v_1 and v_2, in general, for the two-sample experiment described earlier, $v_1 = 1$ and $v_2 = (n_1 + n_2 - 2)$.

EXAMPLE 10.1   The coded values for the hours of life of two brands of light bulbs are given in Table 10.1 for samples of six bulbs drawn randomly from each of the two brands. The true values, in hundreds of hours, were coded by multiplying each by 1/100 to give the coded values shown in Table 10.1. Do the data present sufficient evidence to indicate a difference in mean lifetime for the two brands of light bulbs?

SOLUTION   Although the Student's t could be used as the test statistic for this example, we will use the analysis of variance F test, since it is more general and can be used to compare more than two means.

The two desired sums of squares of deviations are

$$\text{SST} = \sum_{i=1}^{2} n_i(\bar{x}_i - \bar{x})^2$$
$$= 6(7.2833 - 7.6583)^2 + 6(8.0333 - 7.6583)^2$$
$$= .84375 + .84375 = 1.6875$$

$$\text{SSE} = \sum_{i=1}^{2}\sum_{j=1}^{6}(x_{ij} - \bar{x}_i)^2 = \sum_{j=1}^{6}(x_{1j} - \bar{x}_1)^2 + \sum_{j=1}^{6}(x_{2j} - \bar{x}_2)^2 = 5.8617$$

Table 10.1   Hours of Life for Two Brands of
Light Bulbs for Example 10.1

Bulb	Brand		
	A	B	
1	6.1	9.1	
2	7.1	8.2	
3	7.8	8.6	
4	6.9	6.9	
5	7.6	7.5	
6	8.2	7.9	
Total	43.7	48.2	91.9
Mean	7.2833	8.0333	7.6583

(You can verify that SSE is the pooled sum of squares of the deviations for the two samples discussed in Section 9.4. Also, note that Total SS = SST + SSE.) The mean squares for treatment and error are

$$\text{MST} = \frac{\text{SST}}{1} = 1.6875$$

$$\text{MSE} = \frac{\text{SSE}}{n_1 + n_2 - 2} = \frac{5.8617}{10} = .5862$$

To test the null hypothesis $\mu_1 = \mu_2$, we compute the test statistic

$$F = \frac{\text{MST}}{\text{MSE}} = \frac{1.6875}{.5862} = 2.88$$

The critical value of the F statistic for $\alpha = .05$ is 4.96 (see Table 6 in the Appendix). Although the mean square for treatments is almost three times as large as the mean square for error, it is not large enough to reject the null hypothesis. Consequently, there is not sufficient evidence to indicate a difference between μ_1 and μ_2.

The purpose of the preceding example was to illustrate the computations involved in a simple analysis of variance. The F test for comparing two means is equivalent to a Student's t test, because an F statistic with one degree of freedom in the numerator is equal to t^2. Had the t test been used for Example 10.1, we would have found $t = -1.6967$, which satisfies the relationship $t^2 = (-1.6967)^2 = 2.88 = F$. This relationship also holds for the critical values. You can verify that the square of $t_{.025} = 2.228$ (used for the two-tailed test with $\alpha = .05$ and $v = 10$ degrees of freedom) is equal to $F_{.05} = 4.96$. Since each value of F corresponds to two values of t, one positive and one negative, the F test with one degree of freedom in the numerator always corresponds to a two-tailed t test.

Of what value is the Total SS? The answer is that it provides an easy way to compute SSE. Since the Total SS partitions into SST and SSE, that is,

$$\text{Total SS} = \text{SST} + \text{SSE}$$

then we have

$$\text{SSE} = \text{Total SS} - \text{SST}$$

Both the Total SS and SST are easy to compute. Hence, we can easily find SSE by substituting into the expression above. For Example 10.1 we have

$$\text{Total SS} = \sum_{i=1}^{2}\sum_{j=1}^{6}(x_{ij} - \bar{x})^2 = \sum_{i=1}^{2}\sum_{j=1}^{6}x_{ij}^2 - \frac{\left(\sum_{i=1}^{2}\sum_{j=1}^{6}x_{ij}\right)^2}{12}$$

$$= (\text{sum of squares of all } x \text{ values}) - \frac{(\text{total of all } x \text{ values})^2}{12}$$

$$= 711.35 - \frac{(91.9)^2}{12} = 7.5492$$

Then

$$\text{SSE} = \text{Total SS} - \text{SST} = 7.5492 - 1.6875 = 5.8617$$

This is exactly the same value obtained by the tedious computation and pooling of the sums of squares of deviations from the individual samples.

EXERCISES   Conceptual Level

10.1   Analysis of variance is a technique that can be used to test for differences in means. Explain the logic of this process that allows us to test means by analyzing variances.

10.2   Discuss the effect on the test statistic F (described in Section 10.2) if all the data are coded by the addition of the constant a. Discuss the effect if all data are coded by multiplying each observation by the constant b.

EXERCISES   Skill Level

10.3   Two samples were randomly and independently selected from normal populations with common variance σ^2. The resulting data is shown in the accompanying table.

Sample 1		Sample 2	
4.5	5.3	7.6	10.6
5.6	4.2	3.4	8.3
5.4	9.7	8.4	12.4
4.7	2.9	12.1	9.8
4.4	3.3	8.4	5.2

a. Find the mean and variance for each sample.

b. Use Student's t test (Section 9.4) to test for a significant difference between the population means. Use $\alpha = .05$.

c. Use the F test described in Section 10.2 to test for a significant difference between the population means, using $\alpha = .05$.

d. Verify that $t_{.025}^2 = F_{.05}$ and that the observed values of the test statistics in parts b and c satisfy the relationship $t^2 = F$.

EXERCISES   Applications

10.4   To the investor, risk and the rate of return are the elements of concern when selecting an investment. An investor is interested in developing a portfolio of either bank issues or industrial bonds. She selects a sample of $n_1 = 7$ recent bank issues and $n_2 = 5$ recent industrial bond issues and records the yield to maturity on each. The results are shown in the accompanying table. Do these data present sufficient evidence to indicate a difference in average yield to maturity between bank and industrial bond issues? Use $\alpha = .05$.

Bank	9.14	8.85	9.52	10.16	8.90	9.65	9.85
Industrial	9.69	8.94	8.85	9.45	9.15		

a. Use the Student's t test from Section 9.4.

b. Use the procedure outlined in Section 10.2. (Note that both methods lead to the same conclusion and that the computed values of F and t for the two methods are related. That is, $F = t^2$; a fact that holds true only for the comparison of two population means.)

10.5   Many recent studies of industrial relations suggest that an improvement in the quality of the work environment can reduce absenteeism and improve worker productivity. In an examination of this issue a large manufacturing company with nine separate manufacturing facilities was divided into two groups. In five of the facilities several physical improvements were made in the workrooms; in the other four no improvements were undertaken. The accompanying table gives the absentee rate per employee for the manufacturing facilities within each group. Use both the t test from Section 9.4 and the procedure outlined in Section 10.2 to determine whether there is evidence to indicate that the absentee rate in improved workrooms differs from that in facilities without such improvements. Test at the $\alpha = .05$ level of significance.

Facilities with Workroom Improvements	4.3	5.1	4.7	3.5	5.4
Facilities Without Workroom Improvements	7.1	4.9	6.3	6.8	

10.3

A Comparison of More Than Two Means

An analysis of variance to detect a difference in a set of more than two population means is a simple generalization of the analysis of variance of Section 10.2. The

completely
randomized
experimental design

random selection of independent samples from t populations is known as a **completely randomized experimental design**.

Assume that independent random samples have been drawn from t normal populations with means $\mu_1, \mu_2, \ldots, \mu_t$, respectively, and with a common variance σ^2. Thus, all populations are assumed to possess equal variances. To be completely general, we allow the sample sizes to be unequal and we let n_i, $i = 1, 2, \ldots, t$, be the number in the sample drawn from the ith population. The total number of observations in the experiment is $n = n_1 + n_2 + \cdots + n_t$.

Let x_{ij} denote the measured response on the jth experimental unit in the ith sample and let T_i and $\bar{T}_i$ represent the total and the mean, respectively, for the observations in the ith sample. (The modification in the symbols for sample totals and averages will simplify the computing formulas for the sums of squares.) Then, as in the analysis of variance involving two means,

$$\text{Total SS} = \text{SST} + \text{SSE}$$

where

$$\text{Total SS} = \sum_{i=1}^{t} \sum_{j=1}^{n_i} (x_{ij} - \bar{x})^2 = \sum_{i=1}^{t} \sum_{j=1}^{n_i} x_{ij}^2 - \text{CM}$$
$$= (\text{sum of squares of all } x \text{ values}) - \text{CM}$$

$$\text{CM} = \frac{(\text{total of all observations})^2}{n} = \frac{\left(\sum_{i=1}^{t} \sum_{j=1}^{n_i} x_{ij} \right)^2}{n} = n\bar{x}^2$$

(the term CM denotes "correction for the mean"),

$$\text{SST} = \sum_{i=1}^{t} n_i (\bar{T}_i - \bar{x})^2 = \sum_{i=1}^{t} \frac{T_i^2}{n_i} - \text{CM}$$
$$= \left\{ \begin{array}{l} \text{sum of squares of treatment totals, with each square divided} \\ \text{by the number of observations in that particular total} \end{array} \right\} - \text{CM}$$

$$\text{SSE} = \text{Total SS} - \text{SST}$$

Although the easy way to compute SSE is by subtraction, as shown above, it is interesting to note that SSE is the pooled sum of squares for all t samples and is equal to

$$\text{SSE} = \sum_{i=1}^{t} \sum_{j=1}^{n_i} (x_{ij} - \bar{T}_i)^2$$

The unbiased estimator of σ^2, based on $(n_1 + n_2 + \cdots + n_t - t)$ degrees of freedom, is

$$s^2 = \text{MSE} = \frac{\text{SSE}}{n_1 + n_2 + \cdots + n_t - t}$$

The mean square for treatments possesses $(t - 1)$ degrees of freedom—that is, one less than the number of means, and is given by

$$\text{MST} = \frac{\text{SST}}{t - 1}$$

To test the null hypothesis

$$H_0 : \mu_1 = \mu_2 = \cdots = \mu_t$$

against the alternative that at least one of the equalities does not hold, MST is compared with MSE by using the F statistic, based on $v_1 = t - 1$ and

$$v_2 = \sum_{i=1}^{t} n_i - t = n - t$$

degrees of freedom. The null hypothesis will be rejected if

$$F = \frac{\text{MST}}{\text{MSE}} > F_\alpha$$

where F_α is the critical value of F, based on $(t - 1)$ and $(n - t)$ degrees of freedom, for a probability α of a type I error.

Intuitively, **the greater the difference between the observed treatment means** $\bar{T}_1, \bar{T}_2, \ldots, \bar{T}_t$, **the greater is the evidence to indicate a difference between their corresponding population means.** We can see from the formula for SST that SST $= 0$ when all the observed treatment means are identical because then $\bar{T}_1 = \bar{T}_2 = \cdots = \bar{T}_t = \bar{x}$, and the deviations appearing in SST, $(\bar{T}_i - \bar{x})$, $i = 1, 2, \ldots, t$, equal zero. As the treatment means get farther apart, the deviations $(\bar{T}_i - \bar{x})$ increase in absolute value and SST increases in magnitude. Consequently, **the larger the value of SST, the greater is the weight of evidence favoring a rejection of the null hypothesis.** This same line of reasoning applies to the F tests employed in the analysis of variance for all designed experiments.

The test is summarized in the display.

F Test for Comparing t Population Means

Null hypothesis $H_0 : \mu_1 = \mu_2 = \cdots = \mu_t$.

Alternative hypothesis H_a: One or more pairs of population means differ.

Test statistic:

$$F = \frac{\text{MST}}{\text{MSE}}$$

where F is based on $v_1 = (t - 1)$ and $v_2 = (n - t)$ degrees of freedom.

Rejection region: Reject H_0 if $F > F_\alpha$, where F_α lies in the upper tail of the F distribution (with $v_1 = t - 1$ and $v_2 = n - t$) and satisfies the expression

$$P(F > F_\alpha) = \alpha$$

The assumptions underlying the analysis of variance F tests deserve particular attention. **The samples are assumed to have been randomly selected from the t**

populations in an independent manner. The populations are assumed to be normally distributed, with equal variances σ^2 and means $\mu_1, \mu_2, \ldots, \mu_t$. Moderate departures from these assumptions will not seriously affect the properties of the test. This statement is particularly true of the normality assumption.

EXAMPLE 10.2   Four groups of salespeople for a magazine sales agency were subjected to different sales training programs. Because there were some dropouts during the training programs, the number of trainees varied from group to group. At the end of the training programs each salesperson was randomly assigned a sales area from a group of sales areas that were judged to have equivalent sales potentials. The number of sales made by each person in each of the four groups of salespeople during the first week after completing the training program is listed in Table 10.2. Do the data present sufficient evidence to indicate a difference in the mean achievement for the four training programs?

Table 10.2   Number of Sales Made by Each Person in Each Training Group

	Training Group			
	1	2	3	4
	65	75	59	94
	87	69	78	89
	73	83	67	80
	79	81	62	88
	81	72	83	
	69	79	76	
		90		
T_i	454	549	425	351
$\bar{T}_i$	75.67	78.43	70.83	87.75

SOLUTION   We must compute the following quantities:

$$CM = \frac{\left(\sum_{i=1}^{4} \sum_{j=1}^{n_i} x_{ij}\right)^2}{n} = \frac{(\text{total of all observations})^2}{n}$$

$$= \frac{(1{,}779)^2}{23} = 137{,}601.8$$

$$\text{Total SS} = \sum_{i=1}^{4} \sum_{j=1}^{n_i} x_{ij}^2 - CM = (\text{sum of squares of all } x \text{ values}) - CM$$

$$= (65)^2 + (87)^2 + (73)^2 + \cdots + (88)^2 - CM$$

$$= 139{,}511 - 137{,}601.8 = 1{,}909.2$$

$$SST = \sum_{i=1}^{4} \frac{T_i^2}{n_i} - CM$$

$$= \left\{ \begin{array}{l} \text{sum of squares of treatment totals, with each square} \\ \text{divided by the number of observations in that} \\ \text{particular total} \end{array} \right\} - CM$$

$$= \frac{(454)^2}{6} + \frac{(549)^2}{7} + \frac{(425)^2}{6} + \frac{(351)^2}{4} - CM$$

$$= 138{,}314.4 - 137{,}601.8 = 712.6$$

$$SSE = \text{Total SS} - SST = 1{,}196.6$$

The mean squares for treatment and error are

$$MST = \frac{SST}{t-1} = \frac{712.6}{3} = 237.5$$

$$MSE = \frac{SSE}{n_1 + n_2 + \cdots + n_t - t} = \frac{SSE}{n-t} = \frac{1{,}196.6}{19} = 63.0$$

The test statistic for testing the hypothesis $\mu_1 = \mu_2 = \mu_3 = \mu_4$ is

$$F = \frac{MST}{MSE} = \frac{237.5}{63.0} = 3.77$$

where

$$v_1 = (t-1) = 3 \quad \text{and} \quad v_2 = \sum_{i=1}^{t} n_i - 4 = 19$$

The critical value of F for $\alpha = .05$ is $F_{.05} = 3.13$. Since the computed value of F, 3.77, exceeds $F_{.05} = 3.13$, we reject the null hypothesis and conclude that the evidence is sufficient to indicate a difference in mean achievement for the four training programs. Since the observed value of $F = 3.77$ lies between $F_{.05} = 3.13$ and $F_{.025} = 3.90$, the observed significance level of the test lies between .025 and .05, so that

$$.025 < p\text{-value} < .05$$

You may feel that the above conclusion could have been made on the basis of visual observation of the treatment means. However, it is not difficult to construct a set of data that will lead the "visual" decision maker to erroneous results.

10.4

An Analysis of Variance Table for a Completely Randomized Design

The calculations of the analysis of variance are usually displayed in an analysis of variance (ANOVA or AOV) table. The table for the design of Section 10.3 involving t treatment means is shown in Table 10.3. Column 1 shows the sources of variation

Table 10.3   ANOVA Table for a Comparison of Means, Completely Randomized Design

Source	d.f.	SS	MS	F
Treatments	$t-1$	SST	$\text{MST} = \text{SST}/(t-1)$	MST/MSE
Error	$n-t$	SSE	$\text{MSE} = \text{SSE}/(n-t)$	
Total	$n-1$	Total SS		

corresponding to each sum of squares of deviations; column 2 gives the respective degrees of freedom; columns 3 and 4 give the corresponding sums of squares and mean squares, respectively. A calculated value of F, comparing MST and MSE, is usually shown in column 5. Note that the degrees of freedom and sums of squares add to their respective totals.

The ANOVA table for Example 10.2, shown in Table 10.4, gives a compact presentation of the appropriate computed quantities for the analysis of variance.

Table 10.4   ANOVA Table for Example 10.2

Source	d.f.	SS	MS	F
Treatments	3	712.6	237.5	3.77
Error	19	1,196.6	63.0	
Total	22	1,909.2		

10.5

Estimation for the Completely Randomized Design

Confidence intervals for a single treatment mean and the difference between a pair of treatment means are identical to those given in Chapter 9. The confidence intervals for the mean of treatment i or the difference between treatments i and j are given in the display.

Completely Randomized Design: $(1 - \alpha)100\%$ Confidence Intervals

A single treatment mean:

$$\bar{T}_i \pm t_{\alpha/2} \frac{s}{\sqrt{n_i}}$$

The difference between two treatment means:

$$(\bar{T}_i - \bar{T}_j) \pm t_{\alpha/2} \sqrt{s^2 \left(\frac{1}{n_i} + \frac{1}{n_j} \right)}$$

where

$$s = \sqrt{s^2} = \sqrt{\mathrm{MSE}} = \sqrt{\frac{\mathrm{SSE}}{n-t}}$$

$$n = n_1 + n_2 + \cdots + n_t$$

and $t_{\alpha/2}$ is based on $(n-t)$ degrees of freedom.

Notice that confidence intervals are, in general, found as

(point estimator) $\pm\ t_{\alpha/2}$(standard error of estimator)

The confidence intervals in the display are appropriate for single treatment means or a comparison of a pair of means **selected prior to observing the data**. The stated confidence coefficients are based on random sampling. If you were to look at the data and then compare the largest and smallest sample means, the assumption of randomness would be disturbed. Certainly, the difference between the largest and smallest sample means is expected to be larger than for a pair selected at random.

EXAMPLE 10.3   Find a 95% confidence interval for the mean number of sales for those trained in training program 1 of Example 10.2.

SOLUTION   The 95% confidence interval for the mean number of sales is

$$\bar{T}_1 \pm t_{.025}\frac{s}{\sqrt{n_i}} \qquad \text{or} \qquad 75.67 \pm 2.093\frac{7.94}{\sqrt{6}} \qquad \text{or} \qquad 75.67 \pm 6.78$$

where $t_{.025}$ is based on $(n-t) = 19$ degrees of freedom, and $s = \sqrt{\mathrm{MSE}} = \sqrt{63.0} = 7.94$. Then we estimate the mean number of sales to be contained in the interval from 68.89 to 82.45.

◆ ◆

EXAMPLE 10.4   Find a 95% confidence interval for the difference in mean sales for training programs 1 and 4 of Example 10.2.

SOLUTION   The 95% confidence interval for $(\mu_1 - \mu_4)$ is

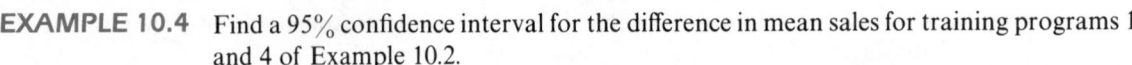

$$(75.67 - 87.75) \pm (2.093)\sqrt{63\left(\frac{1}{6} + \frac{1}{4}\right)} \qquad \text{or} \qquad -12.08 \pm 10.72$$

Therefore, we estimate that the interval from -22.80 to -1.36 encloses the difference in mean sales for training programs 1 and 4. Because all points in the interval are negative, we infer that μ_4 is larger than μ_1. Also, note that the variability in the number of sales among salespersons is rather large. Consequently, the sample sizes should be increased if we want to reduce the width of the confidence interval.

◆ ◆

> ## Tips on Problem Solving
>
> The following suggestions apply to all the analyses of variance in this chapter.
>
> 1. When calculating sums of squares, be certain to carry at least six significant figures before performing subtractions.
> 2. Remember that sums of squares can never be negative. If you obtain a negative sum of squares, you have made a mistake in arithmetic.
> 3. Always check your analysis of variance table to make certain that the degrees of freedom sum to the total degrees of freedom, $n - 1$, and that the sums of squares sum to the Total SS.

10.6

Two Computer Printouts for a Completely Randomized Design

Computer packages for analysis of variance are readily available through your computer center (or personal microcomputer). The purpose of this section is to familiarize you with a typical printout of the analysis of variance for a completely randomized design in case you would like to perform your computations on a computer. We will use both the MINITAB and the SAS® (Statistical Analysis System)* packages, although other packages are available and would produce similar output.

So that you will more readily identify the elements on the printout, we have used the data from Example 10.2. A completely randomized design corresponds to a one-way classification of the observations by treatment groups. The MINITAB command ONEWAY for an analysis of variance produced the output that appears in Table 10.5. The values to be analyzed were stored in column 1, and the corresponding treatment group designations (1, 2, 3, or 4) were stored in column 2. The analysis of variance table, which appears as the first section of the printout, is identical to that in Table 10.4, which was found by direct calculations. Notice that the source of variation due to treatments is identified by row C2, which contains the treatment designations. The second section of the printout provides the sample size, the sample mean, and the standard deviation for each treatment group (identified as level 1, 2, 3, or 4). The pooled standard deviation is equal to $s = \sqrt{\text{MSE}} = 7.936$. The program also provides a graphical display locating the sample mean as well as the 95% lower and upper confidence limits for each treatment mean using the pooled standard deviation.

The printout from an equivalent SAS program for implementing a one-way analysis of variance for these same data is given in Table 10.6. You will note that the

---

* SAS is the registered trademark of SAS Institute, Inc., Cary, NC, USA.

Table 10.5   MINITAB Printout for Example 10.2

```
MTB > ONEWAY C1 C2

ANALYSIS OF VARIANCE ON C1
SOURCE      DF          SS          MS          F
C2           3        712.6       237.5       3.77
ERROR       19       1196.6        63.0
TOTAL       22       1909.2
```

```
                               INDIVIDUAL 95 PCT CI'S FOR MEAN
                               BASED ON POOLED STDEV
LEVEL      N        MEAN      STDEV    ------+---------+---------+---------+
  1        6      75.667      8.165          (------*-----)
  2        7      78.429      7.115            (-----*------)
  3        6      70.833      9.579    (------*------)
  4        4      87.750      5.795                  (--------*-------)
                                       ------+---------+---------+---------+

POOLED STDEV =     7.936                   70        80        90       100
```

Table 10.6   SAS Printout for Example 10.2

```
                        ANALYSIS OF VARIANCE PROCEDURE

DEPENDENT VARIABLE X
                     SUM OF          MEAN
SOURCE      DF      SQUARES         SQUARE    F VALUE   PR>F     R-SQUARE      C.V.
MODEL        3    712.58643892   237.52881297   3.77   0.0280   0.373235    10.2602
ERROR       19   1196.63095238    62.98057644
CORRECTED                                              ROOT MSE             X MEAN
TOTAL       22   1909.21739130                         7.93603027         77.34782608

SOURCE      DF    ANOVA SS     F VALUE     PR>F
TRTMENTS     3  712.58643892    3.77      0.0280
```

analysis of variance is broken down into two parts. The upper part of the table partitions that total sum of squares into two sources, MODEL and ERROR. The numbers appearing in the ERROR row of Table 10.6 are identical (except for rounding) to the numbers appearing in the "Error" row of the analysis of variance given in Table 10.4. The MODEL row of the printout in Table 10.6 corresponds to all effects other than error. In this case there is only one other source of variation, namely, treatments. Consequently, the row identified as MODEL gives the values corresponding to treatments, as well as the calculated value of F. These values, apart from rounding, agree with those in the "Treatments" row of Table 10.4.

The lower part of the table breaks down the MODEL source of variation into its components. In this particular case, there is only one source, TRTMENTS. Consequently, this line repeats the value of SST (712.58 . . .), gives the computed F value (3.77) for a test of the null hypothesis "no difference between treatment means," and gives the probability of observing a value of F as large as or larger than 3.77, given that the null hypothesis is true. This probability, 0.0280, is the significance level (p-value) for the test. With p-value $= 0.0280$, we could reject H_0 with $\alpha = .05$ but not with $\alpha = .01$.

The value of $s = \sqrt{\text{MSE}} = 7.936\ldots$, given at the right in the top part of the printout, can be used to construct confidence intervals. The last two columns of the SAS printout are not relevant to the analysis of variance that we undertake in this chapter.

EXERCISES   Conceptual Level

10.6   Explain what is meant by a completely randomized experimental design.

10.7   What assumptions underlie the F test used in an analysis of variance?

10.8   Why is a large value of SST indicative of underlying differences among the population means?

EXERCISES   Skill Level

10.9   The data in the accompanying table resulted from an experiment run in a completely randomized design in which each of four treatments was replicated five times.

Sample 1	Sample 2	Sample 3	Sample 4
6.9	8.3	8.0	5.8
5.4	6.8	10.5	3.8
5.8	7.8	8.1	6.1
4.6	9.2	6.9	5.6
4.0	6.5	9.3	6.2

a. Perform an analysis of variance for these data.

b. Test for significant differences among the population means. Use $\alpha = .05$.

c. What is the approximate p-value for the test in part b? Does this p-value substantiate the results obtained in part b?

d. Find the mean for each sample. Estimate the difference ($\mu_1 - \mu_2$), using a 95% confidence interval.

10.10   An experiment run in a completely randomized design resulted in $n_1 = 4$, $n_2 = 4$, $n_3 = 5$, and $n_4 = 3$ observations from each of four treatment populations, respectively.

Source	d.f.	SS	MS	F
Treatments			56.4	
Error		102.7		
Total				

    a. Complete the accompanying analysis of variance table.

    b. Calculate the F statistic for testing for significant differences among the population means. Use the approximate observed significance level of the test to determine whether the population means are significantly different.

10.11 The MINITAB printout in Table 10.7 resulted from an analysis of data from a completely randomized design involving five treatments, each with four replicates. When the data are stored by columns, and each column represents a distinct treatment, the MINITAB command AOVONEWAY, followed by the column numbers containing the data, can be used to implement an analysis of variance for a completely randomized design. Independent of the manner in which the data were stored, the output for the commands AOVONEWAY and ONEWAY are identical.

    a. Find the p-value associated with the reported F in testing for differences among treatment means. What is your conclusion?

    b. Verify that the value of the pooled standard deviation is given by $s = \sqrt{MSE}$.

    c. Use a 95% confidence interval to estimate μ_3.

    d. Use the printout to construct a 99% confidence interval estimate of $(\mu_3 - \mu_4)$. Does it appear that these two population means are different?

Table 10.7   MINITAB Printout for Exercise 10.11

```
MTB > PRINT C1-C5
 ROW      C1       C2       C3       C4       C5

   1     31.9     23.5     23.4     28.3     22.5
   2     24.8     30.8     39.3     21.9     12.8
   3     33.0     25.3     41.1     30.3     26.1
   4     30.1     32.8     40.0     30.4     15.0

MTB > AOVONEWAY C1-C5

ANALYSIS OF VARIANCE
SOURCE        DF          SS         MS         F
FACTOR         4        584.7      146.2       4.62
ERROR         15        474.8       31.7
TOTAL         19       1059.4
```

```
                                     INDIVIDUAL 95 PCT CI'S FOR MEAN
                                     BASED ON POOLED STDEV
LEVEL      N       MEAN      STDEV    ----+---------+---------+---------+---------+--
C1         4     29.950      3.635                       (------*------)
C2         4     28.100      4.411                   (------*------)
C3         4     35.950      8.399                           (-------*------)
C4         4     27.725      4.002                 (-------*------)
C5         4     19.100      6.247     (-------*------)
                                     ----+---------+---------+---------+---------+--
POOLED STDEV =    5.626             16.0      24.0      32.0      40.0
MTB >
```

EXERCISES Applications

10.12 What types of advertising best hold the attention of children? An article by Horst H. Stipp ("Children as Consumers," *American Demographics*, February 1988) investigates the behavior of children as consumers in American society and studies their reactions to products and advertising. Although most children ages 6 to 11 spend their own money on candy and gum, more than a fourth frequently buy toys, soft drinks, and presents with their own funds. An advertising executive wishes to know whether current advertisements for these products are successful in capturing the attention of young viewers. She observes 15 children: 5 children during the showing of an advertisement featuring toys, 5 during the showing of an advertisement featuring food and gum, and 5 during the showing of an advertisement of soft drinks. All advertisements are exactly 60 seconds in length. Recorded in Table 10.8 are the times of attention to the advertisements for the 15 children.

a. From a visual inspection of the data, does it seem that the food and gum advertisement appears best able to capture the attention of young viewers?

b. Perform an analysis of variance for this experiment.

c. Do these data provide sufficient evidence to indicate a difference in mean time of attention by the children for the three classes of advertisements?

Table 10.8   Advertisement Data for Exercise 10.12

Advertisement	Time of Attention				
Toys	45	40	30	25	45
Food, gum	50	25	55	45	50
Soft drinks	30	45	40	50	35

10.13 Refer to Exercise 10.12. Let μ_A and μ_B denote, respectively, the mean attention time of the children to advertisements directed toward toys (A) and food and gum (B).

a. Find a 95% confidence interval for μ_A.

b. Find a 95% confidence interval for μ_B.

c. Find a 95% confidence interval for $(\mu_A - \mu_B)$.

10.14 In Exercise 10.5 we referred to studies indicating a relationship between the quality of work life and productivity. In a follow-up to these studies 3 groups of workers involved in a common assembly operation in a major electronics firm were chosen for investigation. One group was chosen at random as a control group (no improvement in the work environment), and the daily production rate of its 10 employees was recorded after a four-week period of investigation. Nine employees from the second group and 9 from the third were separately subjected to two different work improvement programs, again recording the daily production rates after four weeks. The mean gains in daily production for the three groups are as follows:

Control: −.68   Program A: 2.64   Program B: 9.38

A partially completed analysis of variance table for the data is shown in Table 10.9.

a. Fill in the missing numbers in the analysis of variance table.

b. Do these data provide sufficient evidence to indicate a difference in the population means for the three groups? Explain the implications of the test results.

    c. Find a 95% confidence interval for the difference in mean gain in daily productivity for workers in the control group and those in work improvement program *B*.

    d. Find a 95% confidence interval for the mean gain in daily productivity for workers in work improvement program *B*.

Table 10.9 ANOVA for Exercise 10.14

Source	d.f.	SS	MS
Improvement programs		257.24	
Error			
Total		1609.32	

10.15 An experiment was conducted to compare the price of a loaf of bread (a particular brand) in four city locations. Eight stores were randomly sampled in locations 1, 2, and 3, but owing to an omission, only 7 were selected from location 4. A completely randomized design was employed. Conduct an analysis of variance for the data presented in Table 10.10. Use the MINITAB printout given in Table 10.11.

Table 10.10 Price Data for Exercise 10.15

Location	Bread Price (Cents)							
1	139	143	145	141	144	138	140	141
2	138	141	144	143	137	140	143	140
3	134	139	135	138	139	136	140	135
4	149	150	148	150	146	151	149	

Table 10.11 MINITAB Printout for the Data of Exercise 10.15

```
ANALYSIS OF VARIANCE
SOURCE      DF        SS        MS          F
FACTOR       3     557.46    185.82      36.52
ERROR       27     137.38      5.09
TOTAL       30     694.84



                                    INDIVIDUAL 95 PCT CI'S FOR MEAN
                                    BASED ON POOLED STDEV
LEVEL       N       MEAN      STDEV   ----------+---------+---------+------
C1          8     141.38      2.45                 (---*--)
C2          8     140.75      2.49                (---*--)
C3          8     137.00      2.27        (--*--)
C4          7     149.00      1.63                           (--*--)
                                    ----------+---------+---------+------
POOLED STDEV =     2.26                     140.0     145.0     150.0
MTB >
```

a. Do these data provide sufficient evidence to indicate a difference in the mean price of the bread in stores located in the four areas of the city?

b. Suppose that prior to seeing the data, we wished to compare the mean prices between locations 1 and 4. Estimate the difference in mean prices, using a 95% confidence interval.

 10.16 Although equity instruments range from precious art to real estate to common stocks, the internal rate of return offers a common measure for comparison of the investment value of competing equity instruments. Table 10.12 lists the annual internal rates of return for several different investment portfolios managed by three separate investment firms. An SAS computer printout of the data is shown in Table 10.13. Use the information in the printout to answer the following questions.

a. Do these data present sufficient evidence to indicate a difference in the mean annual internal rate of return earned on portfolios managed by the three investment firms?

b. Use a 95% confidence interval to estimate the difference in mean annual internal rates of return earned on portfolios between firm A and firm C; between firm A and firm B; between firm B and firm C.

c. If you were an investor, which firm would you select if your investment objective is to maximize the mean annual internal rate of return?

Table 10.12   Rate of Return Data for Exercise 10.16

Firm A	Firm B	Firm C
16.9	10.0	15.2
15.0	13.1	12.5
16.2	12.3	13.0
15.8	10.2	17.4
17.1	8.9	11.7

Table 10.13   SAS Printout for the Data of Exercise 10.16

ANALYSIS OF VARIANCE PROCEDURE

DEPENDENT VARIABLE X

SOURCE	DF	SUM OF SQUARES	MEAN SQUARE	F VALUE	PR>F	R-SQUARE	C.V.
MODEL	2	70.785333	35.392667	11.63	0.0016	0.659589	12.7482
ERROR	12	36.532000	3.044333				
CORRECTED TOTAL	14	107.317333					

SOURCE	DF	ANOVA SS	F VALUE	PR>F
TRTMENTS	2	70.785333	11.63	0.0016

 10.17 Workers' unemployment insurance programs have been hailed as providing benefits that carry a worker through a spell of unemployment resulting from involuntary layoffs; at the same time, they have been criticized as being subsidized times of leisure between periods of employment. To determine whether a program of monetary rewards might shorten the period of unemployment and hence the amount of unemployment benefits paid, S. A. Woodbury and R. G. Spiegelman provided $500 bonuses to unemployed workers (or their employers) who found employment within 11 weeks of filing for unemployment benefits. In their study Woodbury and Spiegelman

["Bonuses to Workers and Employers to Reduce Unemployment: Randomized Trials in Illinois," *American Economic Review*, Vol. 77, No. 4 (September 1987), p. 513] examined the effect of awards to the employee versus their employer. They found that the average number of weeks of insured unemployment was lower for the group in which the employee received the bonus, when compared with unemployment time for a control group for which no monetary award was given to employee or employer, but was not statistically different from unemployment time for the group in which the employer was paid a bonus. The data in Table 10.14 reflect the results of the large-scale study of over 12,000 individuals conducted by Woodbury and Spiegelman. The table entries are the number of weeks of unemployment following the initial application for unemployment benefits for five individuals in each of the experimental groups.

a. Interpret the results of the MINITAB printout in Table 10.15 (C1 = control; C2 = employee bonus; C3 = employer bonus).

b. Use the data to construct a 95% confidence interval for the difference in mean duration of unemployment between the employee versus employer bonus plans.

Table 10.14   Unemployment Data for Exercise 10.17

Control	Employee Bonus	Employer Bonus
12	6	28
25	20	23
18	14	16
23	18	13
26	17	14

Table 10.15   MINITAB Printout for the Data of Exercise 10.17

```
MTB > AOVONEWAY C1-C3

ANALYSIS OF VARIANCE
SOURCE      DF        SS       MS        F
FACTOR       2      86.8     43.4     1.24
ERROR       12     421.6     35.1
TOTAL       14     508.4

                                 INDIVIDUAL 95 PCT CI'S FOR MEAN
                                 BASED ON POOLED STDEV
LEVEL       N      MEAN    STDEV   --+---------+---------+---------+----
C1          5    20.800    5.805                (----------*----------)
C2          5    15.000    5.477   (----------*----------)
C3          5    18.800    6.458         (----------*----------)
                                 --+---------+---------+---------+----
POOLED STDEV =     5.927    10.0      15.0      20.0      25.0
MTB >
```

10.7

The Randomized Block Design

The completely randomized design discussed in Sections 10.3 and 10.4 is the
appropriate design to use when the experimental material or experimental units are
relatively homogeneous. When the experimental material is not homogeneous, we may
be able to find groups of homogeneous units, called **blocks**, within which the means
associated with the treatments under investigation may be compared. With such a
design the comparisons among the treatments are made within homogeneous blocks of
material; therefore, the effects of the blocks as well as the block-to-block variability are
removed from the error variation.

blocks

Definition

**randomized block
design**

> A **randomized block design** consists of b blocks, each containing t experimental
> units. The t treatments are randomly assigned to the units in each block, and each
> treatment appears once in every block.

For example, in assessing the effects of three package designs on the amount of
product sales, we might decide to use a completely randomized design and select 12
supermarkets, 4 of which would be assigned to display and promote one of the three
package designs. One possible design is given in Table 10.16. However, the difference in
sales could be due to more than just differences among the packaging designs.
Supermarkets with a large volume of business would be expected to have a large overall
sales of the product, but supermarkets with a small volume of business would be
expected to have a smaller volume of product sales. Hence, an alternative design that
uses the blocking principle is the randomized block design in which only 4
supermarkets are used but all three packaging designs are displayed and available for
sale in each of the 4 stores. In this design for which $b = 4$ supermarkets (blocks) and
$t = 3$ package designs (treatments), the store-to-store variability has been eliminated
from uncontrolled error through the choice of design. The resulting randomized block
design would appear symbolically as shown in Figure 10.5.

The word *randomized* in the name of the design implies that the treatments are
randomly assigned within the block. For our experiment the position within the block
pertains to the position in the sequence when assigning a particular package design to a
given supermarket over time. For instance, the experiment may be conducted over a
period of three weeks, with each store being randomly assigned a different package

Table 10.16   Completely Randomized
Experimental Design

Treatments	Supermarkets
1	4, 9, 1, 8
2	11, 3, 5, 7
3	2, 6, 12, 10

Supermarkets (Blocks)

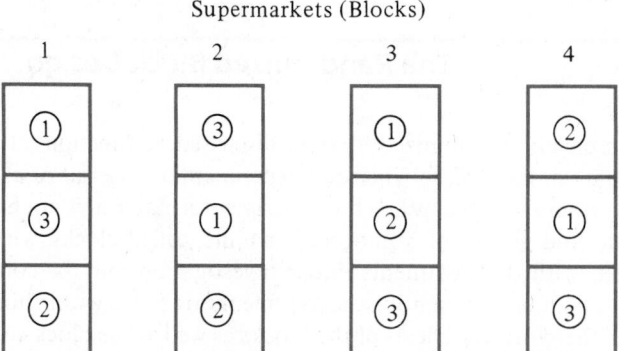

Figure 10.5   Randomized Block Design for Supermarket Experiment

design to sell each week. The purpose of the randomization (that is, the position in the block) is to eliminate bias due to time. That is, in case the demand for the product in any one week is higher or lower than in another week, it gives every package an equal chance of being displayed during that week.

Blocks may represent time, location, or experimental material. If three treatments are to be compared and there is a suspected trend in the mean response over time, a substantial part of the time variation may be removed by blocking. All three treatments would be randomly applied to experimental units in one small block of time. This procedure would be repeated in succeeding blocks of time until the required amount of data is collected.

As we have seen, a comparison of the sale of competitive or differently designed products in supermarkets should be made within supermarkets, thus using the supermarkets as blocks and removing store-to-store variability. An experiment designed to test and compare subject response to a set of stimuli uses the subjects as blocks to remove subject-to-subject variability. That is, each person is subjected to the complete set of stimuli, with the stimuli spaced in time (to reduce the possibility of a residual effect from the previous stimulus) and assigned in random order. Experiments in medicine often utilize children within a single family as blocks, applying all the treatments, one each, to children within a family. Because of heredity, children within a family are more homogeneous than those in different families. This type of blocking removes the family-to-family variation, just as the stimulus-response experiment is designed to remove the subject-to-subject variation and the market preference experiment is designed to remove the store-to-store variation.

10.8

The Analysis of Variance for a Randomized Block Design

The randomized block design implies the presence of two independent variables, "blocks" and "treatments." Consequently, the total sum of squares of deviations of the response measurements about their mean can be partitioned into three parts: the sums of squares for blocks, treatments, and error.

We denote the total and average of all observations in block i as B_i and $\bar{B}_i$, respectively. Similarly, we let T_j and $\bar{T}_j$ denote the total and mean of all observations receiving treatment j. Then for a randomized block design involving b blocks and t treatments, we have

$$\text{Total SS} = \text{SSB} + \text{SST} + \text{SSE}$$

with

$$\text{Total SS} = \sum_{i=1}^{b} \sum_{j=1}^{t} (x_{ij} - \bar{x})^2 = \sum_{i=1}^{b} \sum_{j=1}^{t} x_{ij}^2 - \text{CM}$$
$$= (\text{sum of squares of all } x \text{ values}) - \text{CM}$$

$$\text{SSB} = t \sum_{i=1}^{b} (\bar{B}_i - \bar{x})^2 = \frac{\sum_{i=1}^{b} B_i^2}{t} - \text{CM}$$
$$= \frac{\text{sum of squares of all block totals}}{\text{number of observations in a single total}} - \text{CM}$$

$$\text{SST} = b \sum_{j=1}^{t} (\bar{T}_j - \bar{x})^2 = \frac{\sum_{j=1}^{t} T_j^2}{b} - \text{CM}$$
$$= \frac{\text{sum of squares of all treatment totals}}{\text{number of observations in a single total}} - \text{CM}$$

$$\text{SSE} = \text{Total SS} - \text{SSB} - \text{SST}$$

where

$$\bar{x} = (\text{average of all } n = bt \text{ observations}) = \frac{\sum_{i=1}^{b} \sum_{j=1}^{t} x_{ij}}{n}$$

$$\text{CM} = \frac{(\text{total of all observations})^2}{n} = \frac{\left(\sum_{i=1}^{b} \sum_{j=1}^{t} x_{ij}\right)^2}{n}$$

The analysis of variance for the randomized block design is presented in Table 10.17. The degrees of freedom associated with each sum of squares are shown in the second column. Mean squares are calculated by dividing the sums of squares by their respective degrees of freedom. Note that the degrees of freedom for blocks, treatments, and error always sum to $(n - 1) = (bt - 1)$.

To test the null hypothesis "there is no difference in treatment means," we use the following test.

Table 10.17   ANOVA Table for a Randomized Block Design

Source	d.f.	SS	MS
Blocks	$b - 1$	SSB	$\text{MSB} = \text{SSB}/(b-1)$
Treatments	$t - 1$	SST	$\text{MST} = \text{SST}/(t-1)$
Error	$(b-1)(t-1)$	SSE	$\text{MSE} = \text{SSE}/(b-1)(t-1)$
Total	$bt - 1$	Total SS	

F Test for Comparing t Treatments, Using a Randomized Block Design

Null hypothesis H_0:   The population treatment means are equal.
Alternative hypothesis H_a: One or more pairs of population means differ.
Test statistic:

$$F = \frac{\text{MST}}{\text{MSE}}$$

where F is based on $v_1 = (t - 1)$ and $v_2 = (b - 1)(t - 1)$ degrees of freedom.
Rejection region: Reject H_0 if $F > F_\alpha$.

Blocking not only may reduce the experimental error but also may provide an opportunity to see whether evidence exists to indicate a difference in the mean response for blocks, if this is of interest to the experimenter. Under the null hypothesis that there is no difference in mean response for blocks, MSB provides an unbiased estimator of σ^2, based on $(b - 1)$ degrees of freedom. Where a real difference exists in the block means, MSB will probably be inflated in comparison with MSE, and then

$$F = \frac{\text{MSB}}{\text{MSE}}$$

provides a test statistic. As in the test for treatments, the rejection region is

$$F > F_\alpha$$

based on $v_1 = (b - 1)$ and $v_2 = (b - 1)(t - 1)$ degrees of freedom.

EXAMPLE 10.5   A consumer preference study involving three different package designs (treatments) was laid out in a randomized block design among four supermarkets (blocks). The data shown in Figure 10.6 represent the number of units sold for each package design within each supermarket during each of three given weeks. Do the data present sufficient evidence to indicate a difference in the mean sales for each package design? Do they present sufficient evidence to indicate a difference in mean sales for the supermarkets?

SOLUTION   The treatment and block totals are as follows:

$$T_1 = 39 \qquad T_2 = 105 \qquad T_3 = 68$$
$$B_1 = 74 \qquad B_2 = 62 \qquad B_3 = 32 \qquad B_4 = 44$$

The sums of squares for the analysis of variance are shown individually in the following equations and jointly in the analysis of variance table (Table 10.18).

$$\text{CM} = \frac{(\text{total})^2}{n} = \frac{(212)^2}{12} = 3{,}745.33$$

Supermarkets

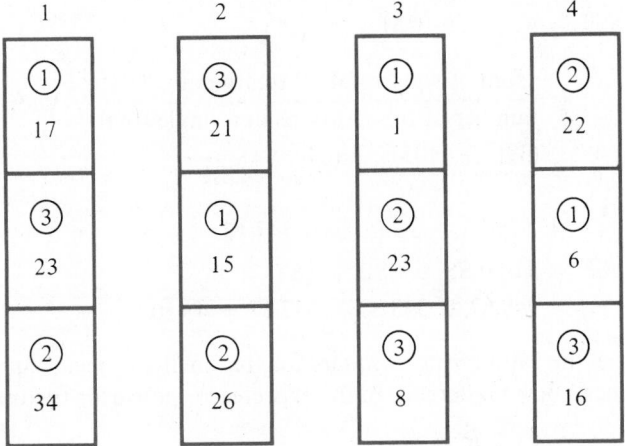

Figure 10.6   Number of Units Sold for Each Package Design in Each Supermarket for Example 10.5

Table 10.18   ANOVA Table for Example 10.5

Source	d.f.	SS	MS	F
Blocks	3	348.00	116.00	15.30
Treatments	2	547.17	273.58	36.09
Error	6	45.50	7.58	
Total	11	940.67		

$$\text{Total SS} = \sum_{i=1}^{4} \sum_{j=1}^{3} (x_{ij} - \bar{x})^2 = \sum_{i=1}^{4} \sum_{j=1}^{3} x_{ij}^2 - CM$$
$$= (\text{sum of squares of all } x \text{ values}) - CM$$
$$= (17)^2 + (23)^2 + \cdots + (16)^2 - CM$$
$$= 4{,}686 - 3{,}745.33 = 940.67$$

$$SSB = \frac{\sum_{i=1}^{4} B_i^2}{3} - CM$$
$$= \frac{\text{sum of squares of all block totals}}{\text{number of observations in a single total}} - CM$$
$$= \frac{(74)^2 + (62)^2 + (32)^2 + (44)^2}{3} - CM$$
$$= 4{,}093.33 - 3{,}745.33$$
$$= 348.00$$

$$SST = \frac{\sum_{j=1}^{3} T_j^2}{4} - CM$$

$$= \frac{\text{sum of squares of all treatment totals}}{\text{number of observations in a single total}} - CM$$

$$= \frac{(39)^2 + (105)^2 + (68)^2}{4} - CM$$

$$= 4,292.5 - 3,745.33 = 547.17$$

$$SSE = \text{Total SS} - \text{SSB} - \text{SST}$$

$$= 940.67 - 348.00 - 547.17 = 45.50$$

We use the ratio of mean square for treatments to mean square for error to test an hypothesis of no difference in the expected response for treatments. Thus,

$$F = \frac{MST}{MSE} = \frac{273.58}{7.58} = 36.09$$

The critical value of the F statistic ($\alpha = .05$) for $v_1 = 2$ and $v_2 = 6$ degrees of freedom is $F_{.05} = 5.14$. Since the computed value of F exceeds the critical value, there is sufficient evidence to reject the null hypothesis and conclude that a real difference does exist in the expected sales for the three package designs. In fact, the computed value of F exceeds $F_{.005} = 14.54$ so that the p-value for this test is less than .005. Therefore, the null hypothesis could be rejected for a value of α as small as .005.

A similar test can be conducted for the null hypothesis that no differences exist in the mean sales for supermarkets. Rejection of this hypothesis would imply that store-to-store variability does exist and that blocking is desirable. The computed value of F based on $v_1 = 3$ and $v_2 = 6$ degrees of freedom is

$$F = \frac{MSB}{MSE} = \frac{116.00}{7.58} = 15.30$$

Since this value of F exceeds the tabulated critical value $F_{.005} = 12.92$, the p-value of this test is less than .005. Hence, we reject the null hypothesis and conclude that a real difference exists in the expected sales in four supermarkets; that is, the data present sufficient evidence to support our decision to use supermarkets as blocks.

10.9

Estimation for the Randomized Block Design

The confidence interval for the difference between a pair of means is similar to that for the completely randomized design given in Section 10.5. It is given in the following display.

Randomized Block Design: A $(1 - \alpha)100\%$ Confidence Interval for the Difference Between Two Treatment Means

$$(\bar{T}_i - \bar{T}_j) \pm t_{\alpha/2} \sqrt{\frac{2s^2}{b}}$$

where $n_i = n_j = b$, the number of observations contained in a treatment mean, t is the number of treatments, and $t_{\alpha/2}$ is based on d.f. $= (b - 1)(t - 1)$, the error degrees of freedom from the corresponding analysis of variance.

The difference between the confidence intervals for the completely randomized and the randomized block designs is that s^2, appearing in the expression in the display, will probably be smaller for the randomized block design.

Similarly, we can construct a $(1 - \alpha)\,100\%$ **confidence interval for the difference between a pair of block means.** Each block contains t observations corresponding to the t treatments. Therefore, the confidence interval is

$$(\bar{B}_i - \bar{B}_j) \pm t_{\alpha/2} \sqrt{\frac{2s^2}{t}}$$

EXAMPLE 10.6 For the consumer preference study described in Example 10.5, construct a 90% confidence interval for the difference between the expected sales from package designs 1 and 2—that is, for the difference between treatments 1 and 2.

SOLUTION From Example 10.5 we have

$$\bar{T}_1 = \frac{T_1}{b} = \frac{39}{4} = 9.75 \qquad \bar{T}_2 = \frac{T_2}{b} = \frac{105}{4} = 26.25$$

$$s^2 = \frac{\text{SSE}}{(b - 1)(t - 1)} = \frac{45.50}{6} = 7.5833$$

The confidence interval for the difference in mean response for a pair of treatments is

$$(\bar{T}_i - \bar{T}_j) \pm t_{\alpha/2} \sqrt{\frac{2s^2}{b}}$$

where, for our example, $t_{.050}$ is based on $(b - 1)(t - 1) = (4 - 1)(3 - 1) = 6$ degrees of freedom. Substituting the appropriate quantities into this formula, we have

$$(26.25 - 9.75) \pm (1.943) \sqrt{\frac{2(7.5833)}{4}} \qquad \text{or} \qquad 16.50 \pm 3.78$$

Thus, we estimate that the difference between mean sales for the two package designs lies in the interval from 12.72 to 20.28. A narrower confidence interval can be acquired by increasing b, the number of supermarkets included in the experiment.

◆◆

Tips on Problem Solving

Be careful of this point: Unless the blocks have been randomly selected from a population of blocks, you cannot obtain a confidence interval for a single treatment mean. This limitation occurs because the sample treatment mean is biased by the positive and negative effects that the blocks have on the response.

10.10

Two Computer Printouts for a Randomized Block Design

A randomized block design corresponds to a two-way classification of the observations by treatments and blocks. The TWOWAY analysis of variance command in the MINITAB package was used to analyze the data in Example 10.5. For this program each observation must be identified according to the block in which it is located and the treatment applied. The values to be analyzed were stored in column 1, the corresponding treatment designation (1, 2, or 3) was stored in column 2, and the corresponding block designation (1, 2, 3, or 4) was stored in column 3. For ease in reading, columns 2 and 3 were labeled as TRTS and BLOCKS, respectively. A listing of the data and the resulting MINITAB output is given in Table 10.19.

Notice that the computed values of the F statistics are not given and must be computed by the user. Since the mean squares associated with blocks and treatments are identical to those given earlier in Table 10.18, computed values of the F statistics for testing significant differences among treatments and among blocks (supermarkets) will be those given in that table.

The SAS computer printout for the randomized block data of Example 10.5 is shown in Table 10.20. The upper part of the table gives a breakdown of the Total SS in two parts, one corresponding to MODEL and the second to ERROR. The numbers appearing in the row corresponding to ERROR are identical (except for rounding error) to the numbers in the "Error" row of the ANOVA table, Table 10.18. The row corresponding to MODEL gives the totals of d.f. and SS corresponding to blocks and treatments. A breakdown of these totals is given in the lower part of the table. There you see that the degrees of freedom and sums of squares for blocks and treatments correspond to the quantities given in the ANOVA table, Table 10.18.

Notice that MSB and MST are not given in Table 10.20. However, they can easily be calculated from the sums of squares if they are desired. The lower part of the table gives the F values and significance levels for tests of the null hypotheses "no difference between block means" ($F = 15.30$) and "no difference between treatment means" ($F = 36.08$). The value of s, $s = 2.75 \ldots$, given in the sixth column, can be used to

Table 10.19   MINITAB TWOWAY Printout
for Example 10.5

```
MTB>NAME C2    'TRTS' C3    'BLOCKS'
MTB>PRINT C1-C3

   ROW   C1    TRTS    BLOCKS

    1    17     1        1
    2    34     2        1
    3    23     3        1
    4    15     1        2
    5    26     2        2
    6    21     3        2
    7     1     1        3
    8    23     2        3
    9     8     3        3
   10     6     1        4
   11    22     2        4
   12    16     3        4

MTB>TWOWAY C1 C2 C3

ANALYSIS OF VARIANCE ON C1

   SOURCE        DF       SS         MS
   TRTS           2    547.17     273.58
   BLOCKS         3    348.00     116.00
   ERROR          6     45.50       7.58
   TOTAL         11    940.67


MTB>
```

Table 10.20   SAS Printout for Example 10.5

```
                    ANALYSIS OF VARIANCE PROCEDURE

DEPENDENT VARIABLE X
                       SUM OF        MEAN
SOURCE          DF     SQUARES       SQUARE     F VALUE   PR>F      R-SQUARE     C.V.
MODEL            5   895.166667    179.033333    23.61    0.0007    0.951630   15.5875
ERROR            6    45.500000      7.583333
CORRECTED TOTAL 11   940.666667                           ROOT MSE            X MEAN
                                                          2.7537853          17.666667


SOURCE          DF    ANOVA SS    F VALUE    PR>F
BLOCKS           3   348.000000     15.30    0.0032
TRTMENTS         2   547.166667     36.08    0.0005
```

calculate confidence intervals for the difference between any pair of treatment or block means. The slight discrepancies between the numbers in the ANOVA table, Table 10.18, and those in the computer printout of Table 10.20 are due to rounding errors.

EXERCISES   Conceptual Level

10.18   State the assumptions underlying the following experimental design models.
  a.   A completely randomized design
  b.   A randomized block design

10.19   Suppose you must choose an experimental design.
  a.   Discuss the advantages of blocking.
  b.   What happens to these advantages as you increase the size of the blocks (the number of experimental units per block)?

EXERCISES   Skill Level

10.20   The data in Table 10.21 resulted from an experiment run in a randomized block design in which each of four treatments were randomly applied to experimental units in each of six blocks.
  a.   Perform an analysis of variance for these data.
  b.   Test for a significant difference among treatment means. Use $\alpha = .05$.
  c.   Do the data present sufficient evidence to indicate that blocking was effective? Test by using $\alpha = .05$.
  d.   Estimate the difference between the population means for treatments 1 and 3, using a 99% confidence interval. Interpret this interval.

Table 10.21   Data for Exercise 10.20

	Treatments			
Block	1	2	3	4
1	9.6	11.3	8.3	11.6
2	9.2	9.0	14.1	15.0
3	9.7	10.3	12.8	15.5
4	11.5	18.3	18.8	15.4
5	12.4	14.8	14.2	12.3
6	13.2	10.4	15.3	15.4

10.21   The ANOVA table for an experiment run in a randomized block design is partially reproduced in Table 10.22.
  a.   How many treatments were involved in the experiment? How many blocks were used?
  b.   Complete the ANOVA table.

c. Do the data present sufficient evidence to indicate a difference in the treatment means? Use $\alpha = .05$.

d. Find the approximate p-value in testing for differences among the block means. Use this value to determine whether the block means are significantly different.

Table 10.22   ANOVA for Exercise 10.21

Source	d.f.	SS	MS	F
Blocks			10.81	
Treatments	2	12.52		
Error	6			
Total		62.54		

10.22 An analysis of the data given in Table 10.23 from a randomized block design involving four treatments and five blocks produced the MINITAB printout presented in Table 10.24.

a. Calculate the F statistic used in testing for significant differences among the treatment means. Find the p-value associated with the observed value of F. What is your conclusion?

b. Calculate the F statistic used in testing for significant differences among the block means. Find the p-value associated with the observed value of F. What is your conclusion about the effectiveness of blocking?

Table 10.23   Data for Exercise 10.22

	Treatment			
Block	1	2	3	4
1	28.7	53.0	57.8	54.8
2	38.8	63.7	49.1	54.4
3	48.0	54.8	48.9	64.8
4	53.0	59.8	53.9	69.8
5	52.4	52.7	63.2	61.9

Table 10.24   MINITAB Printout for the Data of Exercise 10.23

```
MTB > TWOWAY C1 C2 C3

ANALYSIS OF VARIANCE   C1

SOURCE          DF          SS        MS
TRTS             3       777.3     259.1
BLOCKS           4       297.6      74.4
ERROR           12       544.8      45.4
TOTAL           19      1619.7

MTB >
```

c. Calculate the value of s, the sample standard deviation used as an estimate of σ.

d. Construct a 95% confidence interval for the difference between the means for treatments 2 and 3.

e. Construct a 95% confidence interval for the difference between the means for blocks 1 and 2.

EXERCISES   Applications

10.23 Reinmuth and Barnes studied the bidding behavior of three drilling contractors who opposed one another on a number of invitations to bid. Their studies suggest that bidding contractors must not be too cautious, since conservative bidders, whose bid prices range on the high side, win few contract awards and thus take a chance of draining their revenues. The bid data provided by Reinmuth and Barnes are the bids recorded on five randomly selected invitations to bid. Each recorded bid represents the bid price for the hourly operating costs when using a 4,000-ft-capacity drilling rig with a four-man crew. The data are shown in Table 10.25.

a. Do these data provide sufficient evidence to indicate a difference in the mean bid prices submitted by the three contractors?

b. Is there evidence to suggest a difference in mean bid price over the bid trials?

c. Prior to examining the data, the researchers decided to compare the mean bids submitted by contractors A and C. Find a 90% confidence interval for this difference.

Table 10.25   *Bid Data for Exercise 10.23*

	Bidding Trial				
Bidding Contractor	1	2	3	4	5
A	$45.00	$46.00	$43.75	$44.50	$46.50
B	42.50	45.50	43.50	40.90	47.55
C	39.75	40.00	40.20	43.75	47.50

Source: Data from J. E. Reinmuth and J. D. Barnes, "A Strategic Competitive Bidding Approach to Pricing Decisions for Petroleum Industry Drilling Contractors," *Journal of Marketing Research*, August 1975. Reprinted with permission.

10.24 A study was undertaken to determine the relative typing speeds that could be obtained when using four different brands of typewriters. Each typewriter was assigned to each of eight secretaries, the order of assignment conducted in a random manner. The typing speed, in words per minute, for 10 minutes of typing was recorded for each secretary-typewriter combination. The data obtained are given in Table 10.26.

a. Identify the design used for this experiment, and justify your diagnosis.

b. Perform an analysis of variance on the data.

c. Do the data provide sufficient evidence to indicate that the mean typing speed for the secretaries varies with the brand of typewriter used? Test by using $\alpha = .05$.

d. Why was the order of assignment of typewriters to each secretary conducted in a random manner? In general, what is the advantage of randomly assigning the treatments to the blocks?

Table 10.26   Typing Speed Data for Exercise 10.24

Typewriter Brands	Secretary							
	1	2	3	4	5	6	7	8
A	79	80	77	75	82	77	78	76
B	74	79	73	70	76	78	72	74
C	82	86	80	79	81	80	80	84
D	79	81	77	78	82	77	77	78

10.25   Refer to Exercise 10.24. Let μ_C and μ_D, respectively, denote the mean typing speeds when a secretary uses typewriter C and typewriter D. Find a 99% confidence interval for $(\mu_C - \mu_D)$. Interpret this interval.

10.26   An experiment was conducted to compare the effect of four different chemicals, $A, B, C,$ and D, in producing water resistance in textiles. A strip of material, randomly selected from a bolt, was cut into four pieces, and the prices were randomly assigned to receive one of the four chemicals, $A, B, C,$ or D. This process was replicated three times, thus producing a randomized block design. The design, with moisture resistance measurements, is as shown in Figure 10.7 (low readings indicate low moisture penetration). An SAS computer printout of the analysis of variance for the data is presented in Table 10.27. Use the information in the printout to answer the following questions.

Blocks (Bolt Samples)

1	2	3
C 9.9	D 13.4	B 12.7
A 10.1	B 12.9	D 12.9
B 11.4	A 12.2	C 11.4
D 12.1	C 12.3	A 11.9

Figure 10.7   Moisture Resistance Data for Exercise 10.26

Table 10.27   SAS Printout for the Data of Exercise 10.26

```
                        ANALYSIS OF VARIANCE PROCEDURE

DEPENDENT VARIABLE X
                        SUM OF          MEAN
SOURCE           DF     SQUARES         SQUARE      F VALUE   PR>F      R-SQUARE    C.V.
MODEL            5      12.37166667     2.47433333    27.75   0.0004    0.958549    2.5023
ERROR            6      0.53500000      0.08916667
CORRECTED TOTAL  11     12.90666667                           ROOT MSE              X MEAN
                                                              0.29860788            11.93333333


SOURCE           DF     ANOVA SS     F VALUE   PR>F
BLOCKS           2      7.17166667    40.21    0.0003
TRTMENTS         3      5.20000000    19.44    0.0017
```

a. Do the data provide sufficient evidence to indicate a difference in the mean moisture penetration for fabric treated with the four chemicals?

b. Do the data provide evidence to indicate that blocking increased the amount of information in the experiment?

c. Find a 95% confidence interval for the difference in mean moisture penetration for fabric treated by chemicals A and D. Interpret the interval.

 10.27 Three federally chartered commercial banks with branch offices in several locations jointly conducted an experiment to examine the six-month retention rate of demand-deposit accounts, both with and without premium incentives. Each banking chain offered three different types of incentive premiums, while withholding incentive premiums in some branch banks. Table 10.28 gives the six-month retention rates for the first 100 new demand-deposit accounts obtained by each bank under each of four different incentive premiums.

Table 10.28   Retention Rate Data for Exercise 10.27

	Incentive Premium			
Bank	None	Cash	Small Appliance	Gasoline Credit
A	92	86	76	75
B	86	84	72	80
C	89	82	74	77

a. Do these data provide sufficient evidence to indicate that the retention rate varies according to the type of incentive premium offered by a bank?

b. Use a 95% confidence interval to estimate the difference in mean retention rate of new demand-deposit accounts between banks offering no incentive premium and those offering gasoline credit as a premium.

c. Is there evidence to suggest that the mean retention rate of new demand-deposit accounts differs among banks?

 10.28 The average annual overhead per loan, excluding advertising, for three consumer finance companies is given in Table 10.29. An SAS printout of the analysis of variance for the randomized block data is presented in Table 10.30.

a. Do these data provide sufficient evidence to indicate a difference in mean overhead cost per loan for the three finance companies?

b. Use a 90% confidence interval to estimate the difference in mean overhead cost per loan between consumer finance companies A and C.

Table 10.29   Annual Overhead Data for Exercise 10.28

	Year									
Company	1980	1981	1982	1983	1984	1985	1986	1987	1988	1989
A	9.68	9.80	12.11	12.13	12.46	14.31	15.37	16.64	19.50	18.92
B	14.26	14.10	13.95	14.57	14.04	14.76	15.20	15.62	15.01	15.67
C	9.73	10.12	11.30	12.05	12.59	13.00	12.83	13.94	13.53	16.05

Table 10.30   SAS Printout for the Data of Exercise 10.28

```
                        ANALYSIS OF VARIANCE PROCEDURE

DEPENDENT VARIABLE X
                         SUM OF          MEAN
SOURCE            DF     SQUARES         SQUARE     F VALUE   PR>F      R-SQUARE    C.V.
MODEL             11     125.73053333    11.43004848   4.69   0.0019    0.74116     11.3387
ERROR             18     43.90961333     2.43942296
CORRECTED TOTAL   29     169.64014667                          ROOT MSE            X MEAN
                                                               1.56186522          13.77466667


SOURCE            DF     ANOVA SS      F VALUE   PR>F
BLOCKS             9     99.93194667     4.55    0.0031
TRTMENTS           2     25.79858667     5.29    0.0156
```

10.11

Some Cautionary Comments on Blocking

There are two major steps in designing an experiment and you need to be careful to distinguish between the two. The first step is deciding what treatments you wish to include in an experiment and how many observations to select per treatment. The treatments often may be the levels of a single qualitative or quantitative variable. For example, you might wish to compare the mean sales per store for three different brands of coffee. Then the three brands of coffee represent three levels of the qualitative variable "brands" and are the three treatments employed in the experiment. Suppose you wish to see whether the mean sales per brand varies with the type of product display in a store. If you had two types of product displays, say A_1 and A_2, and three brands, B_1, B_2, and B_3, then the six treatments would be the combinations of product displays and brands: A_1B_1, A_1B_2, A_1B_3, A_2B_1, A_2B_2, A_2B_3. The best strategies to employ in selecting treatments for an experiment and deciding on the number of observations per treatment are contained in a course on the design of experiments. You will find more information on this topic in the texts listed in the references.

The second step in designing an experiment is deciding how to apply the treatments to the experimental units. For the six combinations of displays and brands this step in designing an experiment would consist of deciding how these display-brand combinations are actually to be used in one or more stores.

If the stores to be included in the experiment are relatively homogeneous with respect to type, size, sales volume, and perhaps some other factors as well, then a completely randomized design would be used, and a store would represent an experimental unit. A completely randomized design would consist of randomly assigning one display-brand combination per store so that in the total experiment each display-brand combination would appear the required number of times. For example, if each treatment combination were to appear four times, then 24 stores would be used in the experiment.

On the other hand, if the stores are not homogeneous with respect to type, size, or sales, then a randomized block design should be used with a store representing a block, and all six of the display-brand combinations would be run within a store. For this design the experimental unit might be a time or an area within a store, not the store itself. For the randomized block design the number of blocks would equal the number of stores in the experiment.

Blocking is not always beneficial. It produces a gain in information if the between-block variation is larger than the within-block variation. In this case blocking removes this larger source of variation from SSE, and s^2 assumes a smaller value (as does the population variance σ^2) because of the design. At the same time, however, some information is lost because blocking reduces the degrees of freedom with SSE and s^2. For example, if the completely randomized design used 24 stores, and **each of the six display-brand combinations were run in 4 stores**, then the degrees of freedom for error would be $(n - t) = (24 - 6) = 18$. If the randomized block design were used and **all six treatment combinations were run within each of** $b = 4$ **stores**, then the degrees of freedom for error would be $(b - 1)(t - 1) = (3)(5) = 15$. As you can see, each treatment combination was replicated four times using either design. However, the randomized block has $(b - 1) = (4 - 1) = 3$ fewer degrees of freedom for error than does the completely randomized design.

Consequently, if blocking is to be beneficial, the gain in information due to the elimination of block variation must outweigh the loss due to a reduction in the number of degrees of freedom associated with SSE. Unless blocking leaves you with only a small number of degrees of freedom for SSE, the loss in degrees of freedom causes only a small reduction in information in the experiment. Consequently, if you have a reason to suspect that there is block-to-block variation, it will usually be beneficial to block.

10.12

Two-Way Classifications: The Factorial Experiment

The randomized block design is an example of an experiment involving two independent variables that results in a two-way classification, whereby an observation is classified as belonging to a specific block and receiving a specific treatment. The block classification corresponds to an independent variable introduced into the experiment in an attempt to reduce the error variation, whereas the treatment classification corresponds to an independent experimental variable under investigation.

In many situations an experimenter may wish to investigate the effects of one or more independent treatment variables, called **factors**, on a response of interest. The settings of values of a factor in an experiment are called **levels**. For example, the tensile strength of carbon steel is assumed to depend upon the amount of carbon added and the firing temperature. In this case an experiment to determine an optimum combination would involve the two quantitative experimental factors: amount of carbon and firing temperature. If three different amounts of carbon were to be compared at four temperature settings, the experiment would involve three levels of carbon and four levels of temperature. Or consider the situation in which a manufacturer wishes to compare the purity of five chemicals, each available from three suppliers, before making any contractual commitments with a supplier. In this case the investigation would involve two qualitative experimental factors: chemicals at five levels and suppliers at three levels.

Definition

<table>
<tr><td>factorial experiment</td><td>An experiment that utilizes every combination of factor levels as treatments is called a **factorial experiment**.</td></tr>
</table>

A factorial experiment involving factor A at a levels and factor B at b levels is called an $a \times b$ *factorial experiment* with $t = ab$ *treatments*. Notice that a randomized block design is a two-way classification, but it does not constitute a factorial experiment because only one factor is a treatment factor. The blocking factor is, in effect, a factor whose effect we wish to control and remove from experimental error.

A factorial experiment represents one way of combining several experimental factors simultaneously to arrive at the t treatments to be investigated in an experiment. The factorial treatment combinations could be used in a completely randomized design or in a randomized block design. However, we will restrict attention to factorial experiments involving two factors run in a completely randomized design.

Although we could run a series of experiments in which all factors but one are held constant and investigate the effects of the factors separately, a factorial experiment allows the investigator to study the effect of each factor as well as their joint effects, called **interactions**. Interaction occurs when the effect of one factor varies as the levels of other factors are changed. The interaction of two factors can be explained within the following context.

interactions

Suppose that an investigator, interested in determining how the amount of carbonation and the amount of sweetener affect the desirability of a soft drink, ran a 2×2 factorial experiment involving two levels of carbonation and two levels of sweetener. If additional carbonation enhances the taste of the drink, we would observe the response pattern shown in Figure 10.8. If additional sweetener tends to detract from the taste, we would observe the response pattern shown in Figure 10.9. If the factors do not interact, we would expect to observe the same pattern for increased sweetener at

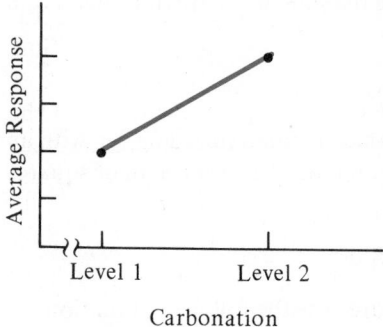

Figure 10.8   Average Response for Soft Drink at Two Levels of Carbonation

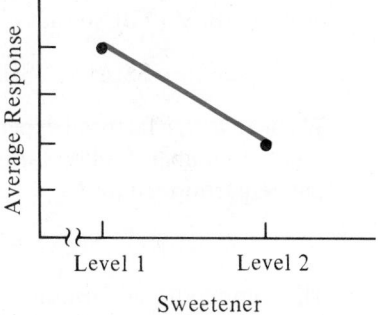

Figure 10.9   Average Response for Soft Drink at Two Levels of Sweetener

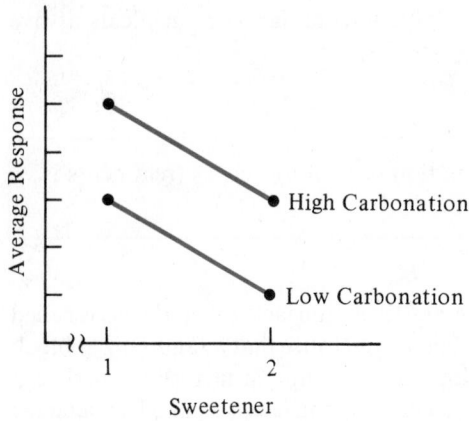

Figure 10.10   Average Responses If Factors
Do Not Interact

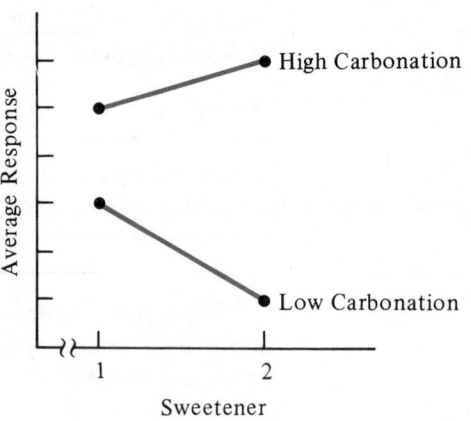

Figure 10.11   Average Responses If Factors
Do Interact

both levels of carbonation, as shown in Figure 10.10. If the factors do interact in such a way that increased carbonation enhances the desirability of a drink with increased sweetener, we would observe a pattern similar to the one in Figure 10.11. If the two factors do not interact, a factorial experiment provides information about each factor individually. However, if two factors do interact, a factorial experiment is designed to detect the interaction and shed light on its nature.

10.13

The Analysis of Variance and Estimation for an $a \times b$ Factorial Experiment

In a factorial experiment involving factor A at a levels and factor B at b levels, the sum of squares for treatments can be further partitioned into the sum of squares for factor A, SS(A); the sum of squares for factor B, SS(B); and the sum of squares for interaction of A and B, SS(AB), so that

$$SST = SS(A) + SS(B) + SS(AB)$$

When an $a \times b$ factorial experiment is run in a completely randomized design with an equal of number of observations per treatment combination, the total sum of squares can be partitioned into

$$\text{Total SS} = SS(A) + SS(B) + SS(AB) + SSE$$

The computational formulas for these sums of squares use the following notation:

A_i denotes the sum of all observations recorded at the ith level of factor A, $i = 1, 2, \ldots, a$.

Table 10.31   ANOVA Table for an $a \times b$ Factorial Experiment

Source	d.f.	SS	MS
Factor A	$(a - 1)$	SS(A)	MS(A) = SS(A)/$(a - 1)$
Factor B	$(b - 1)$	SS(B)	MS(B) = SS(B)/$(b - 1)$
Interaction AB	$(a - 1)(b - 1)$	SS(AB)	MS(AB) = SS(AB)/$(a - 1)(b - 1)$
Error	$(n - ab)$	SSE	MSE = SSE/$(n - ab)$
Total	$(n - 1)$	Total SS	

B_j denotes the sum of all observations recorded at the jth level of factor B, $j = 1, 2, \ldots, b$.

$(AB)_{ij}$ denotes the sum of the observations recorded at the ith level of factor A and the jth level of factor B, $i = 1, 2, \ldots, a; j = 1, 2, \ldots, b$.

Notice that $(AB)_{ij}$ is simply the total for the treatment corresponding to the combination of factor A at level i and factor B at level j. Finally, let r be the number of times each factorial treatment combination appears in the experiment, so that each $(AB)_{ij}$ total contains r observations. Consequently, each A_i total will contain rb observations, each B_j total will contain ra observations, and there will be a total of $n = rab$ observations in the experiment.

When each factorial combination appears an equal* number of times in the experiment, the computational formulas for the appropriate sums of squares are

$$\text{Total SS} = \sum(\text{each observation})^2 - \text{CM}$$

$$\text{SS(A)} = \frac{\sum\limits_{i=1}^{a} A_i^2}{rb} - \text{CM} \qquad \text{SS(B)} = \frac{\sum\limits_{j=1}^{b} B_j^2}{ra} - \text{CM}$$

$$\text{SS(AB)} = \frac{\sum\limits_{i=1}^{a} \sum\limits_{j=1}^{b} (AB)_{ij}^2}{r} - \text{CM} - \text{SS(A)} - \text{SS(B)}$$

$$\text{SSE} = \text{Total SS} - \text{SS(A)} - \text{SS(B)} - \text{SS(AB)}$$

where

$$\text{CM} = \frac{(\text{total of all observations})^2}{rab}$$

In calculating the sum of squares corresponding to a source of variation, the sum of squared totals is always divided by the number of observations in each total.

The analysis of variance for an $a \times b$ factorial experiment is given in Table 10.31. The line labeled "Interaction AB" under sources of variation corresponds to the interaction of A and B. Three hypotheses can be tested within the analysis of variance.

---

* The case of unequal replication can be found in the texts listed in the references at the end of this chapter.

To test the null hypothesis of "no interaction," we will use the test statistic

$$F = \frac{MS(AB)}{MSE}$$

which has an F distribution with $v_1 = (a - 1)(b - 1)$ and $v_2 = (n - ab)$ degrees of freedom when the null hypothesis is true. The null hypothesis of "no interaction" is rejected at the α level of significance if $F > F_\alpha$, since large values of F indicate that MS(AB) is larger than would be expected if no interaction were present.

In testing the null hypothesis of "no differences among the levels of B," we use

$$F = \frac{MS(B)}{MSE}$$

with $v_1 = (b - 1)$ and $v_2 = (n - ab)$ degrees of freedom, rejecting H_0 if $F > F_\alpha$.

Testing the null hypothesis of "no differences among the levels of A" is based on the statistic

$$F = \frac{MS(A)}{MSE}$$

with $v_1 = (a - 1)$ and $v_2 = (n - ab)$ degrees of freedom.

When no interaction is detected, we can conclude that factors A and B are acting independently. Therefore, testing for differences among the levels of A and among the levels of B provides independent information about each factor. However, when a significant interaction is present, it is the joint effect of the factors that is important, not the effect of either factor taken alone. A plot of the factorial treatment means is an effective way of determining the nature of the interaction.

EXAMPLE 10.7   Federal regulations require that certain materials, such as those used for children's pajamas, be treated with a flame retardant. An evaluation of a flame retardant applied to three different materials was conducted at two different laboratories. Each laboratory tested three samples from each of the treated materials. Part of the data collected was the length of the charred portion of each sample. These data appear in Table 10.32. In addition to determining whether the effect of the retardant varies from

Table 10.32   Data for Example 10.7

Laboratory	Materials 1	2	3
1	4.1	3.1	3.5
	3.9	2.8	3.2
	4.3	3.3	3.6
2	2.7	1.9	2.7
	3.1	2.2	2.3
	2.6	2.3	2.5

material to material, we also wish to determine whether the laboratories are consistent in their test results. Provide an analysis of variance for these data, testing for significant differences due to materials, laboratories, and their interaction.

SOLUTION   The six treatments constitute a 2×3 factorial experiment with laboratories (factor A) at $a = 2$ levels, and materials (factor B) at $b = 3$ levels. Each factor combination occurs $r = 3$ times in the experiment, and there are $n = rab = 3(2)(3) = 18$ observations in all.

The totals required for calculating appropriate sums of squares are given in Table 10.33. The calculations follow.

$$CM = \frac{(54.1)^2}{18} = 162.6006$$

$$Total\ SS = (2.7^2 + 3.1^2 + \cdots + 3.6^2) - CM$$
$$= 170.53 - 162.6006 = 7.9294$$

$$SS(A) = \frac{(31.8^2 + 22.3^2)}{9} - CM$$
$$= 167.6144 - 162.6006 = 5.0139$$

$$SS(B) = \frac{(20.7^2 + 15.6^2 + 17.8^2)}{6} - CM$$
$$= 164.7817 - 162.6006 = 2.1811$$

$$SS(AB) = \frac{(12.3^2 + 9.2^2 + \cdots + 7.5^2)}{3} - CM - SS(A) - SS(B)$$
$$= 169.93 - 162.6006 - 5.0139 - 2.1811 = .1344$$

$$SSE = Total\ SS - SS(A) - SS(B) - SS(AB)$$
$$= 7.9294 - 5.0139 - 2.1811 - .1344 = .6000$$

The resulting ANOVA is given in Table 10.34.

In testing the null hypothesis of "no interaction," we use the statistic

$$F = \frac{MS(AB)}{MSE} = \frac{.0672}{.0500} = 1.34$$

which is not significant when compared with $F_{.05} = 3.89$, the 5% critical value of F with $v_1 = 2$ and $v_2 = 12$ degrees of freedom.

Table 10.33   Totals for Calculating Sums of Squares for Example 10.7

Laboratory (A)	Material (B) 1	2	3	Total (A_i)
1	12.3	9.2	10.3	31.8
2	8.4	6.4	7.5	22.3
Total (B_j)	20.7	15.6	17.8	54.1

Table 10.34   ANOVA for Example 10.7

Source	d.f.	SS	MS	F
Laboratories (A)	1	5.0139	5.0139	100.28
Materials (B)	2	2.1811	1.0906	21.81
Interaction (AB)	2	.1344	.0672	1.34
Error	12	.6000	.0500	
Total	17	7.9294		

Testing the hypothesis of "no differences among materials" uses the statistic

$$F = \frac{MS(B)}{MSE} = \frac{1.0906}{.0500} = 21.81$$

which, with $v_1 = 2$ and $v_2 = 12$ degrees of freedom, exceeds $F_{.005} = 8.51$. Since the p-value associated with this test is less than .005, F is significant at the $\alpha = .005$ level.

A test of the null hypothesis "no difference between laboratories" is based on the statistic

$$F = \frac{MS(A)}{MSE} = \frac{5.0139}{.0500} = 100.28$$

with $v_1 = 1$ and $v_2 = 12$ degrees of freedom. Since the calculated value of $F = 100.28$ exceeds $F_{.005} = 11.75$, we conclude that the p-value for this test is much less than .005 and that our results are highly significant.

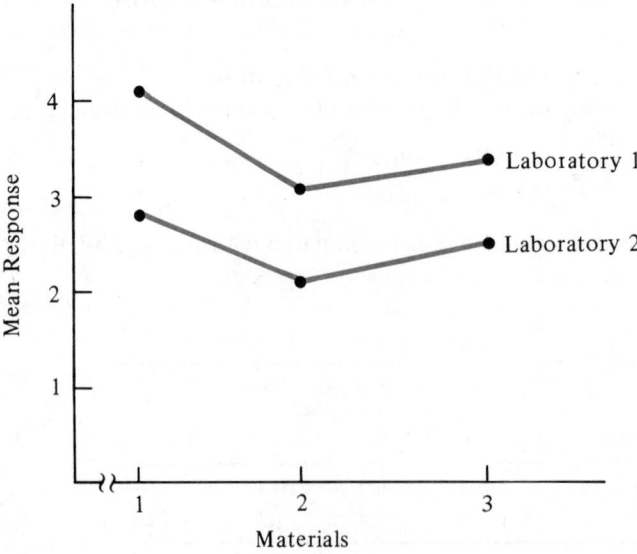

Figure 10.12   Plot of the Treatment Means for Example 10.7

Since the interaction of the factors "laboratories" and "materials" was not significant, we can focus our attention on the difference between laboratories or the differences among the materials tested. A plot of the factorial treatment means given in Figure 10.12 provides some insight into the experimental results. The pattern of average charring for the three materials is consistent for both laboratories, resulting in a nonsignificant interaction. Notice that the average readings for laboratory 1 are consistently higher than those for laboratory 2, a conclusion substantiated by the highly significant F test of the null hypothesis "no difference between laboratories." Similar statements can be made about the different materials tested.

A $(1 - \alpha)100\%$ confidence interval for the difference between two treatment means within a factorial experiment is found in the usual way. Let $\bar{x}_{ij}$ denote the mean response for factor A at level i and factor B at level j, where

$$\bar{x}_{ij} = \frac{(AB)_{ij}}{r}$$

The confidence interval is as given in the display.

A $(1 - \alpha)100\%$ Confidence Interval for the Difference Between Two Treatment Means in a Factorial Experiment

$$(\bar{x}_{ij} - \bar{x}_{kl}) \pm t_{\alpha/2}\sqrt{\frac{2s^2}{r}}$$

where (i, j) and (k, l) correspond to two distinct factor combinations, $s^2 = \text{MSE}$ with $(n - ab)$ degrees of freedom, and r is the number of observations in each mean.

EXAMPLE 10.8   Using the information in Example 10.7, construct a 95% confidence interval for the difference between the average charred length for materials 1 and 2 tested at laboratory 1.

SOLUTION   From Example 10.7, we have

$$\bar{x}_{11} = \frac{(AB)_{11}}{r} = \frac{12.3}{3} = 4.10$$

$$\bar{x}_{12} = \frac{(AB)_{12}}{r} = \frac{9.2}{3} = 3.07$$

$$s^2 = \text{MSE} = .0500$$

The degrees of freedom are $(n - ab) = 18 - (2)(3) = 12$; therefore, $t_{.025} = 2.179$. The confidence interval for the difference in mean length of charring for these two

treatments is

$$(\bar{x}_{11} - \bar{x}_{12}) \pm t_{.025}\sqrt{\frac{2s^2}{r}} \qquad (4.10 - 3.07) \pm 2.179\sqrt{\frac{2(.05)}{3}} \qquad 1.03 \pm .40$$

Therefore, we estimate that the difference in average charred length for materials 1 and 2 tested at laboratory 1 lies in the interval from .63 to 1.43.

When no significant interaction is present, it may be of interest to estimate the difference in means corresponding to a single factor. These means would be based on either ra or rb observations; hence, r in the formula for the estimated standard deviation of the difference between two means is replaced by the number of observations in each of the means.

EXAMPLE 10.9   Using the information in Example 10.7, construct a 95% confidence interval for the difference in average charred length of materials tested for the two laboratories.

SOLUTION   The mean for laboratory 1 is $\bar{x}_{A_1} = A_1/rb = 31.8/9 = 3.53$; the mean for laboratory 2 is $\bar{x}_{A_2} = A_2/rb = 22.3/9 = 2.48$. With the results from Example 10.8, the confidence interval is

$$(\bar{x}_{A_1} - \bar{x}_{A_2}) \pm t_{.025}\sqrt{\frac{2s^2}{rb}} \qquad (3.53 - 2.48) \pm 2.179\sqrt{\frac{2(.05)}{(3)(3)}} \qquad 1.05 \pm .23$$

or from .82 to 1.28. Notice that this interval does not enclose the value zero, which is consistent with the conclusion that there is a highly significant difference between laboratories.

10.14

Two Computer Printouts for a Factorial Experiment

An $a \times b$ factorial experiment results in a two-way classification and can be analyzed by using the MINITAB TWOWAY analysis of variance command provided that each factorial treatment combination is replicated an equal number of times in the experiment. The data from Example 10.7 and the resulting MINITAB output are given in Table 10.35. The observations are listed in column 1, the laboratory designation (1 or 2) is listed in column 2, and the material designation (1, 2, or 3) is listed in column 3. For clear identification of these columns in the analysis of variance table, columns 2 and 3, corresponding to laboratories and materials, have been labeled as LABS and MATRLS. In the analysis of variance portion of Table 10.35 the sums of squares and mean square entries are identical to those in Table 10.34, which were found by direct calculation. However, the values of appropriate F statistics are not

Table 10.35   MINITAB TWOWAY Printout for Example 10.7

```
MTB>NAME C2='LABS' C3='MATRLS'
MTB>PRINT C1 C2 C3
ROW     C1      LABS      MATRLS

 1      4.1       1          1
 2      3.9       1          1
 3      4.3       1          1
 4      2.7       2          1
 5      3.1       2          1
 6      2.6       2          1
 7      3.1       1          2
 8      2.8       1          2
 9      3.3       1          2
10      1.9       2          2
11      2.2       2          2
12      2.3       2          2
13      3.5       1          3
14      3.2       1          3
15      3.6       1          3
16      2.7       2          3
17      2.3       2          3
18      2.5       2          3

MTB>TWOWAY C1 C2 C3

ANALYSIS OF VARIANCE ON C1

SOURCE          DF        SS         MS
LABS             1      5.0139     5.0139
MATRLS           2      2.1811     1.0906
INTERACTION      2      0.1344     0.0672
ERROR           12      0.6000     0.0500
TOTAL           17      7.9294

MTB>
```

given and must be calculated by the user. For the mean squares in Table 10.35 the resulting F statistics would be identical to those given in Table 10.34.

The SAS computer printout for the factorial experiment in Example 10.7 is shown in Table 10.36. The results of the analysis of variance procedure are always given in two parts. In the upper part of the table the total sum of squares is partitioned into two sources, one corresponding to the MODEL and the second to ERROR. The line labeled MODEL contains the degrees of freedom and the sums of squares corresponding to factor A (laboratories), factor B (materials), and their interaction. In the lower portion of the table the degrees of freedom and the sum of squares for MODEL are

Table 10.36   SAS Printout for Example 10.7

ANALYSIS OF VARIANCE PROCEDURE

DEPENDENT VARIABLE: LENGTHS

SOURCE	DF	SUM OF SQUARES	MEAN SQUARE	F VALUE	PR>F	R-SQUARE	C.V.
MODEL	5	7.32944444	1.46588889	29.32	0.0001	0.924333	7.4398
ERROR	12	0.60000000	0.05000000		ROOT MSE		LENGTHS MEAN
CORRECTED TOTAL	17	7.92944444			0.22360680		3.00555556

SOURCE	DF	ANOVA SS	F VALUE	PR>F
LAB	1	5.01388889	100.28	0.0001
MATRL	2	2.18111111	21.81	0.0001
LAB*MATRL	2	0.13444444	1.34	0.2973

furthered partitioned into components associated with laboratories (LAB), materials (MATRL), and their interaction (LAB * MATRL).

Although the mean squares are not given in this lower portion of the printout, columns 4 and 5 of this portion of the printout show the values of the F statistics used in testing the null hypotheses of "no difference between laboratories" ($F = 100.28$); "no difference among materials" ($F = 21.81$); and "no interaction" ($F = 1.34$) and their respective p-values. In the test for significant interaction the observed value of $F = $ MS(AB)/MSE with $v_1 = 2$ and $v_2 = 12$ degrees of freedom was 1.34. The p-value for this test, given as $P(F > 1.34) = .2973$, is larger than .05; therefore, consistent with our earlier results, the null hypothesis of "no interaction" cannot be rejected at the $\alpha = .05$ level of significance. The remaining two p-values indicate that the differences between laboratories and among materials are very highly significant.

The quantity labeled ROOT MSE is the estimate of σ given as $s = \sqrt{MSE} = .22360680$. It is useful in constructing confidence intervals for the factorial treatment means. Notice that within rounding errors the analysis of variance results given in Table 10.34 agree with those in the SAS printout.

EXERCISES   Conceptual Level

10.29   In what ways are a two-factor factorial experiment and an experiment run in a randomized block design similar? In what ways are they different?

10.30   What is meant by the interaction of two factors A and B? Are tests of main effects important in an analysis in which a significant interaction of factors A and B is present?

EXERCISES   Skill Level

10.31   A factorial experiment involving two factors, A and B, each at two levels, was replicated five times. The resulting data are given in Table 10.37.

a. Perform an analysis of variance for this data.

b. Test for a significant interaction between factors A and B, using $\alpha = .05$.

c. Test for significant main effects, using $\alpha = .05$.

d. Calculate the four treatment means. Plot the mean responses for the two levels of factor A as a function of factor B. (See Figure 10.12.) Does the plot reinforce the results found in parts b and c?

e. Construct a 95% confidence interval for the difference in factor A means.

Table 10.37 Data for Exercise 10.31

Factor B	Factor A	
	1	2
1	13	14
	15	17
	14	17
	16	20
	16	16
2	14	17
	16	16
	15	22
	18	20
	16	19

10.32 A factorial experiment involving two factors, A and B, each at four levels, was replicated twice. The ANOVA table for the analysis of the resulting data is partially reproduced in Table 10.38.

a. Complete the ANOVA table.

b. Do the data provide sufficient evidence to indicate that there is an interaction between factors A and B? Test by using $\alpha = .05$.

c. Refer to part b. If there is no evidence of a significant interaction, test to determine whether significant differences exist among the levels of factor A. Perform a similar test for differences among the levels of factor B. Use $\alpha = .05$.

Table 10.38 ANOVA for Exercise 10.32

Source	d.f.	SS	MS	F
A		60.2		
B		12.3		
AB			4.10	
Error				
Total		237.8		

10.33 A factorial experiment involving factor A at two levels and factor B at three levels with four replications of each treatment combination resulted in the analysis of variance table presented in Table 10.39.

Table 10.39   MINITAB Printout for Exercise 10.33

```
MTB > TWOWAY C1-C3

ANALYSIS OF VARIANCE   C1

SOURCE            DF           SS          MS
FACTOR A           1         66.7        66.7
FACTOR B           2       1278.6       639.3
INTERACTION        2       1372.6       686.3
ERROR             18       1140.0        63.3
TOTAL             23       3857.8

MTB >
```

a. Using the analysis of variance printout, test for a significant interaction between factors A and B. Use $\alpha = .01$.

b. Test for significant main effects, using $\alpha = .01$. Is it reasonable to conclude that factor A has no effect on the response?

c. The means for the six treatment combinations are shown in Table 10.40. Plot the treatment means as in Figure 10.12. Does the plot reflect an interaction between factors A and B? Does your conclusion agree with the test results in parts a and b?

d. What is the half width of a 95% confidence interval estimate for the difference between any two treatment means?

Table 10.40   Means for the Data of Exercise 10.33

| | Factor B | | |
Factor A	1	2	3
1	62.75	78.75	91.75
2	80.50	61.75	81.00

EXERCISES   Applications

 10.34 A study was undertaken to investigate the importance of two training techniques and three word-processing software packages for new typists. Two student typists were randomly assigned to each combination of training technique and software package. Table 10.41 presents the scores on a standard typing test at the end of the training session for each of the 12 new typists.

a. Perform an analysis of variance for this experiment, and test for significant differences in mean typing scores due to training techniques, software packages, and their interaction.

b. What are the approximate p-values corresponding to the observed values of F found in testing for differences due to training techniques, software packages, and their interaction?

c. What can you conclude about the effect of these factors and their interaction on mean typing scores for new typists?

Table 10.41   Score Data for Exercise 10.34

Software Package	Training Techniques	
	A	B
1	38	28
	42	25
2	33	13
	37	16
3	31	4
	29	7

10.35 A factorial experiment was used to examine the effect of three advertising levels and two types of advertising appeal. Two subjects were randomly assigned to each of the six factor combinations of advertising level and type of appeal. An advertising recall score for each of the 12 subjects in the study are presented in Table 10.42. Perform an analysis of variance on these data, and determine whether advertising level, type of advertising appeal, or their interaction have any significant effect on recall scores.

Table 10.42   Recall Score Data for Exercise 10.35

Type of Appeal	Advertising Level		
	1	2	3
A	1.1	2.5	7.8
	.8	2.8	9.4
B	1.4	2.1	2.1
	1.6	1.8	2.8

10.36 In studying the effect of the numbers of package features and product price on sales levels, a researcher designed a factorial experiment with 3 levels of package features (very few, some, many) and 2 levels of price (high, low) in which 3 local test markets were randomly assigned to each of the 6 factor combinations. The sales levels (in thousands of dollars) for each of the 18 test markets are given in Table 10.43.

Table 10.43   Sales Data for Exercise 10.36

Price	Number of Packaging Features		
	Very Few	Some	Many
High	22.8	20.9	23.1
	21.4	22.9	21.9
	23.5	21.6	22.5
Low	15.3	17.7	16.6
	18.4	16.8	18.1
	16.2	15.9	15.7

a. Do these data indicate that the number of packaging features, price, or their interaction have any affect on sales levels?

b. What are the p-values for the observed F ratios in each case?

10.37 After considering the results of the experiment reported in Exercise 10.36, the researcher felt that package features were more important than the results indicated, and their impact might be more evident if price were not aggregated into only 2 levels. A second experiment was conducted in which the same number of package levels was used, but price was broken into 3 levels. Again, 3 local test markets were randomly assigned to each of the 9 factor combinations. The resulting sales levels (in thousands of dollars) for the 27 test markets are presented in Table 10.44.

a. Do these data indicate that the researcher was correct?

b. What additional information seems to be gained by using 3 levels of price?

c. What are the approximate p-values for the observed F ratios?

Table 10.44   Sales Data for Exercise 10.37

	Number of Packaging Features		
Price	Very Few	Some	Many
High	23.2	21.3	21.6
	21.7	20.8	22.8
	22.6	21.7	22.4
Medium	16.3	16.7	16.6
	15.7	16.6	17.9
	18.1	17.3	16.3
Low	12.1	16.8	21.7
	11.6	17.8	22.2
	11.8	17.2	21.6

10.15

Assumptions of Analysis of Variance

The assumptions about the probability distribution of the random response x that result in a valid analysis of variance are given in the following display.

Assumptions

1. For any treatment or block combination or combination of factor levels, x is normally distributed with variance equal to σ^2.

2. The observations have been selected independently so that any pair of x values, x_i and x_j, are independent in a probabilistic sense for all i and $j(i \neq j)$.

In a practical situation you can never be certain that the assumptions are satisfied, but you will often have a fairly good idea whether the assumptions are reasonable for your data. To illustrate, the inferential methods of Chapters 8, 9, and 10 are not seriously affected by moderate departures from the assumptions of normality, but you would want the probability distribution of x to be at least mound-shaped. So if x is a discrete random variable that can assume only three values, say $x = 10, 11, 12$, then it is *unreasonable* to assume that the probability distribution of x is approximately normal.

The assumption of a constant variance for x for the various experimental conditions should be approximately satisfied, although violation of this assumption is not too serious if the sample sizes for the various experimental conditions are equal. However, suppose the response is binomial—say the proportion p of people who favor a particular type of investment. We know that the variance of a proportion (Section 7.5) is

$$\sigma_{\hat{p}}^2 = \frac{pq}{n} \qquad \text{where} \qquad q = 1 - p$$

and therefore that the variance is dependent on the expected (or mean) value of $\hat{p}$, namely, p. Then the variance of $\hat{p}$ will change from one experimental setting to another, and the assumptions of the analysis of variance have been violated.

A similar situation occurs when the response measurements are Poisson data (say the number of industrial accidents per month in a manufacturing plant). If the response possesses a Poisson probability distribution, the variance of the response equals the mean (Section 5.5). Consequently, Poisson response data also violate the analysis of variance assumptions.

Many kinds of data are not measurable and hence are unsuitable for an analysis of variance. For example, many responses cannot be measured but can be ranked. Product preference studies yield data of this type. You know you like product A better than B and B better than C but you have difficulty assigning an exact value to the strength of your preferences.

What do you do when the assumptions of an analysis of variance are not satisfied? For example, suppose the variances of the responses for various experimental conditions are not equivalent. This situation can sometimes be remedied by transforming the response measurements. That is, instead of using the original response measurements, we might use their square roots, logarithms, or some other function of the response x. Transformations that tend to stabilize the variance of the response have been found to make the probability distributions of the transformed responses more nearly normal. See the texts listed in the references for discussions of these topics.

When nothing can be done to satisfy (even approximately) the assumptions of the analysis of variance, or if the data are rankings, you should use nonparametric testing and estimation procedures. These procedures, which rely on the comparative magnitudes of measurements (often ranks), are almost as powerful in detecting treatment differences as the tests presented in this chapter. When the assumptions are not satisfied, the nonparametric procedures may be more powerful. Nonparametric tests—what they are and how to use them—are presented in Chapter 18.

10.16

Summary

The completely randomized and the randomized block designs are procedures that specify the manner in which the treatments are assigned to the units to be used in an experiment. The **completely randomized design** is used to investigate the effect of one independent variable corresponding to treatments when the experimental units are relatively homogeneous. The **randomized block design**, which involves two independent variables corresponding to treatments and blocks, is used when the experimental units are not homogeneous but can be grouped into blocks such that the variability within blocks is less than the variability between blocks. The randomized block design is used to control the effect of the block variable and remove its effect from experimental error.

The **analysis of variance** is a technique that partitions the total sum of squares of deviations of the observations about their mean into portions associated with the independent variables in the experiment and a portion associated with error. The mean squares corresponding to the independent variables are tested against the mean square for error by using the F statistic. A significantly large value of F indicates that the mean square under test is larger than would be expected if the corresponding independent variable had no effect on the response of interest.

An experiment in which every combination of factor levels is included as treatments is called a **factorial experiment**. In a factorial experiment the sum of squares corresponding to the independent variable "treatments" can be further partitioned into components associated with each factor and their interactions. Each component can then be tested for significance.

This chapter represents a brief introduction to the design of experiments and the widely used technique for analyzing designed experiments called the analysis of variance. Designs are available for experiments involving several design variables as well as several treatment factors. **Design variables** corresponds to factors whose effect we wish to control and hence remove from experimental error; **treatment factors** are the experimental variables whose effects we wish to investigate. A properly designed experiment can be analyzed by using an analysis of variance. Experiments in which the levels of one or more variables are measured (as opposed to being set at preselected values) can be analyzed by using another general approach called *regression analysis*, which is the subject of Chapters 11 and 12.

Supplementary Exercises

Applications

10.38   A study was undertaken to compare the productivity of the operators of four identical assembly machines. Production records were examined for three randomly selected days, where the days of record were not necessarily the same for any two assembly machine operators. The data are given in Table 10.45. Assuming that the requirements for a completely randomized design are met, analyze the data. State whether there is statistical support at the $\alpha = .05$ level of significance for the conclusion that the four assembly machine operators differ in average daily productivity.

Table 10.45   Productivity Data for Exercise 10.38

Operator 1	Operator 2	Operator 3	Operator 4
230	220	215	225
220	210	215	215
225	220	220	225

10.39 Refer to Exercise 10.38. Let μ_1 and μ_2 denote the mean production rates of operators 1 and 2, respectively.

a. Find a 90% confidence interval for μ_1. Interpret this interval.

b. Find a 95% confidence interval for $(\mu_1 - \mu_2)$. Interpret this interval.

10.40 The data-processing capabilities of Japanese and U.S.-made supercomputers and mainframes are measured in terms of megaFLOPS (million floating-point operations per second). In a study conducted by Nikkei–McGraw Hill, Inc., 13 models were compared, using a loop program developed by the Lawrence Livermore Laboratory ("Japan Tops Supercomputer Study," *Japan Economic Journal*, February 13, 1988). The vector process speed of each model was measured for 14 test loops, and the averages and approximated standard deviations are given in Table 10.46 for the 5 fastest models.

Table 10.46   Computer Speed Data for Exercise 10.40

Model	Average	Standard Deviation
Hitachi S820/80	417.9	39.7
NEC SX2	285.4	27.2
Fujitsu VP400	163.9	21.6
Hitachi S810/20	130.6	16.1
Fujitsu VP200	121.2	14.9

a. What type of experimental design has been used in this situation?

b. Use the values of s_i given in the table to calculate s^2, the pooled estimate of σ^2.

c. Calculate SSE $= \left(\sum n_i - k\right)s^2$.

d. Calculate SST by first finding T_i, $\sum T_i$ = grand total, and CM and then using the appropriate formula for SST.

e. Construct an ANOVA table for this experiment.

f. Find the *p*-value for the test concerning the equality of treatment means. Do these data provide sufficient evidence to indicate a difference in mean vector process speed for the 5 models?

g. Does it appear that any of the assumptions underlying the analysis of variance have been violated?

10.41 A trucking company wished to compare three makes of trucks before ordering an entire fleet of one of the makes. Purchase prices for each of the makes were about the same and thus were ignored in the comparison. Five trucks of each make were run 5,000 miles each, and the average variable cost of operation per mile was noted for each truck. However, because of tire failure, accidents, and driver illness, two make B and two make C trucks did not complete the 5,000-mile test. For those that did finish, the results are as given in Table 10.47.

Table 10.47   Cost Data for Exercise 10.41

Make A	Make B	Make C
27.3	25.4	27.9
28.3	27.4	29.5
27.6	27.1	28.7
26.8		
28.0		

a. Perform an analysis of variance for this experiment.

b. Do these data provide sufficient evidence to indicate a difference in the average variable cost per mile of operation for the three makes of trucks? Use $\alpha = .05$.

c. Is there an advantage in having the same number of measurements within each treatment in a completely randomized design? Explain.

 10.42 Refer to Exercise 10.41. Let μ_A and μ_B denote, respectively, the mean variable cost per mile of operating a truck of make A and make B.

a. Find a 95% confidence interval for μ_A.

b. Find a 95% confidence interval for μ_B.

c. Find a 95% confidence interval for $(\mu_A - \mu_B)$.

d. Is it correct to assume that the confidence interval computed in part c can be obtained as the difference between the confidence intervals found in parts a and b? Explain.

 10.43 Precision assembly operations, such as those required in the assembly of electronic circuitry in the manufacture of television components and computers, require a high degree of specialized training. The objective of all such training programs is to prepare assemblers to perform their tasks properly in as short a time as possible. A manufacturer of computer machinery has proposed three different training programs for its employees involved in circuitry assembly operations. Fifteen assembly employees were divided equally among the three training programs. After training, the average time for the proper assembly of a circuit board was recorded for each employee (three employees had resigned from the firm during the course of the training program). The data are shown in Table 10.48. Use the information in the printout of Table 10.49 to answer the following questions.

Table 10.48   Data for Exercise 10.43

Training Program	Average Assembly Time (Minutes)				
A	59	64	57	62	
B	52	58	54		
C	58	65	71	63	64

Table 10.49   SAS Printout for the Data of Exercise 10.43

```
                      ANALYSIS OF VARIANCE PROCEDURE

DEPENDENT VARIABLE TIME
                      SUM OF           MEAN
SOURCE        DF      SQUARES          SQUARE       F VALUE   PR>F      R-SQUARE      C.V.
MODEL         2       170.45000000     85.22500000    5.70    0.0251   0.559005      6.3802
ERROR         9       134.46666667     14.94074074
CORRECTED                                                     ROOT MSE            TIME MEAN
  TOTAL       11      304.91666667                            3.86532544          60.583333


SOURCE        DF      ANOVA SS         F VALUE    PR>F
TRTMENTS      2       170.45000000       5.70     0.0251
```

a. Do the data provide sufficient evidence to indicate a difference in mean assembly time for people trained by the three programs?

b. Find a 90% confidence interval for the difference in mean assembly time between persons trained by programs A and B.

c. Find a 90% confidence interval for the mean assembly time for persons trained in program A.

d. Do you think the data will satisfy (approximately) the assumption that they have been selected from normal populations? Explain.

MKTG **10.44** A pricing experiment was conducted at the retail store level by a manufacturer of packaged breakfast foods. Currently the company manufactures three products, A, B, and C. Product C was considered adequately profitable, but price increases of 4¢ per package were sought for products A and B. To explore the effects of increasing the price on one or both of the products, management chose four retail stores where the sales for A and B were observed to be relatively constant and approximately equal for each product for a period of time. Each store was then assigned to sell product A at one of the price levels (same or 4¢ increase) and product B at one of the price levels (same or 4¢ increase). The number of units sold were recorded for each store for five randomly selected weeks after the pricing experiment was initiated. The results are shown in Table 10.50.

Table 10.50   Sales Data for Exercise 10.44

| | Product B | |
Product A	Same	4¢ Increase
Same	54 62 46 57 68	57 59 51 56 58
4¢ Increase	53 59 50 54 55	51 50 45 45 52

a. Do these data provide sufficient evidence to indicate that there is an interaction between products A and B and their prices? Use $\alpha = .05$.

b. If there is no interaction present, does the increase in price affect sales for either of the two products? Use $\alpha = .05$.

10.45 Refer to Exercise 10.44. Using a 95% confidence interval, estimate the average number of units of products A and B sold per week for the following conditions:

a. The prices of A and B are both increased by 4¢.

b. The price of A is left unchanged while the price of product B is increased by 4¢ per package.

10.46 A cab company conducted a study of 3 brands of tires before determining which brand to order for all its cabs. The study involved selecting 4 different tires from each brand and randomly assigning them to the left front wheel of 12 different cabs. The wear was recorded after 10,000 miles of use. The wear noted at the end of the test period is given in Table 10.51 in terms of the millimeters of tread wear. Assume that the requirements for a completely randomized design are met, and analyze the data. State whether there is statistical support at the $\alpha = .05$ level for the conclusion that the 3 brands of tires differ in resistance to wear.

Table 10.51   Tread Wear Data for Exercise 10.46

Brand A	Brand B	Brand C
462	250	319
421	336	425
470	322	460
411	268	380

10.47 A study was conducted to compare automobile gasoline mileage for five different brands of gasoline, A, B, C, D, and E. Six automobiles, all of the same make and model, were employed in the experiment, and each gasoline brand was tested in each automobile. Using each brand within the same automobile has the effect of eliminating (blocking out) automobile-to-automobile variability. The results of the experiment, in miles per gallon, are shown in Table 10.52. An SAS computer printout for the analysis of variance of the gasoline mileage recordings is presented in Table 10.53. Use the information in the printout to answer the following questions.

a. Do these data provide sufficient evidence to indicate a difference in the mean mileage per gallon for the five brands of gasoline?

b. Is there evidence of a difference in mean mileage per gallon for the six automobiles?

c. Prior to examining the data, the researcher decided to compare the mean mileage per gallon for gasoline brands A and B. Find a 90% confidence interval for this difference.

Table 10.52   Mileage Data for Exercise 10.47

Gasoline	Automobile					
	1	2	3	4	5	6
A	35.3	31.0	32.7	36.8	37.2	33.1
B	30.7	32.2	31.4	31.7	35.0	32.7
C	38.2	33.4	33.6	37.1	37.3	38.2
D	34.9	36.1	35.2	38.3	40.2	36.0
E	32.4	28.9	29.2	30.7	33.9	32.1

Table 10.53   SAS Printout for the Data of Exercise 10.47

ANALYSIS OF VARIANCE PROCEDURE

DEPENDENT VARIABLE X

SOURCE	DF	SUM OF SQUARES	MEAN SQUARE	F VALUE	PR>F	R-SQUARE	C.V.
MODEL	9	210.81166667	23.42351852	12.22	0.0001	0.846152	4.0499
ERROR	20	38.33000000	1.91650000				
CORRECTED					ROOT MSE		X MEAN
TOTAL	29	249.14166667			1.38437712		34.18333333

SOURCE	DF	ANOVA SS	F VALUE	PR>F
BLOCKS	5	68.14166667	7.11	0.0006
TRTMENTS	4	142.67000000	18.61	0.0001

10.48 How do large multiproduct firms manage the diversity of their operations? Some experts have suggested that multidivisional (M form) structure, which evolved as a response to managing growth and diversity, positively affects corporate performance. In his study of M form structure and performance R. E. Hoskisson ["Multidivisional Structure and Performance: The Contingency of Diversification Strategy," *Academy of Management Journal*, Vol. 30, No. 4 (1987), pp. 625–644] examined the return on assets (ROA) 10 years prior to and 10 years following the implementation of M form structure for firms that he categorized as implementing one of three diversification strategies: vertical integration, related diversification, and unrelated diversification. Hoskisson found that, after correcting and/or controlling for several variables that could affect the response, unrelated diversified firms increased their rate of return (measured as the ROA) and decreased their risk (measured as variability in the ROA) after implementing M form structure, but vertically integrated and related diversified firms did not fare as well. A similar study concerning diversification strategies involved 10 firms in each of the three diversification categories whose rates of return were examined for 10 years prior to the change to M form structure and another 10 firms in each category for the 10 years following change to M form structure. The data are given in Table 10.54. The analysis of variance is presented in Table 10.55.

Table 10.54   Data for Exercise 10.48

		Diversification Strategy		
Period	Statistic	Vertical Integration	Related Diversification	Unrelated Diversification
Before	Mean	6.82	7.75	4.10
	Standard deviation	2.54	2.81	2.58
After	Mean	5.81	7.13	5.21
	Standard deviation	2.17	2.56	2.12

a. Use the information given to determine whether there are significant differences in the average ROA for the integration strategies, for the periods, or their interaction. Explain your results.

b. Provide a 95% confidence interval estimate of the difference between average ROA before versus after implementing M form structure for unrelated diversified firms.

c. Find a 95% confidence interval estimate for the difference in average ROA for related diversified firms and unrelated diversified firms.

Table 10.55   Analysis of Variance for Exercise 10.48

Source	d.f.	SS	MS
Strategies (S)	2	78.5163	39.7286
Periods (P)	1	.4507	.4507
$S \times P$	2	12.7320	6.3660
Error	54	330.8472	6.1268

10.49 A portion of a questionnaire was constructed to enable judges to evaluate four proposed site locations. The judges, selected from the executive group and upper and middle management of the company, were asked to respond concerning their perceptions of accessibility of each site to primary markets, transportation facilities, state corporation regulations, and living desirability of the area relative to each proposed site. Their responses were then collated and coded on a 0-to-20 scale. The data obtained are shown in Table 10.56.

a. The primary objective of this experiment was to compare site locations. State the type of design employed for this experiment, and justify your diagnosis.

b. Perform an analysis of variance on the data.

c. Do the data provide sufficient evidence to indicate that the mean coded scores vary from site to site? Test by using $\alpha = .05$.

d. Suppose that the data did provide sufficient evidence to indicate differences among the mean coded questionnaire scores associated with the four sites. Would this evidence imply that the questionnaire was able to detect a difference in preferences for the building sites?

Table 10.56   Site Evaluation Data for Exercise 10.49

Site	Judge 1	2	3	4	5	6	7	8
1	9	10	7	5	12	7	8	6
2	4	9	3	0	6	8	2	4
3	12	16	10	9	11	10	10	14
4	9	11	7	8	12	7	7	8

10.50 The Environmental Protection Agency and certain state agencies have established rigid regulations on the output of polluting effluents from manufacturing plants. A wood products firm has four branch plants located in a certain western state. One plant, plant A, has recently passed an EPA investigation of its water effluents, but management is uncertain about how the other three plants will fare in similar investigations. In a study of this issue five samples of liquid waste were selected from each plant and the polluting effluents measured in each sample. The results of the experiment are shown in Table 10.57. An SAS computer printout of the analysis of variance for the data is shown in Table 10.58. Use the information in the printout to answer the following questions.

a. Do the data provide sufficient evidence to indicate a difference in the mean amount of effluents discharged by the four plants?

b. If the maximum allowable mean discharge of effluents is 1.5 lb/gal, do the data provide sufficient evidence to indicate that the limit is exceeded at plant B?

c. Estimate the difference in the mean discharge of effluents between plants A and D, using a 95% confidence interval.

Table 10.57   Pollution Data for Exercise 10.50

Plant	Polluting Effluents (Pounds of Waste per Gallon)				
A	1.65	1.72	1.50	1.37	1.60
B	1.70	1.85	1.46	2.05	1.80
C	1.40	1.75	1.38	1.65	1.55
D	2.10	1.95	1.65	1.88	2.00

Table 10.58   SAS Printout for the Data of Exercise 10.50

```
                          ANALYSIS OF VARIANCE PROCEDURE

DEPENDENT VARIABLE X
                   SUM OF          MEAN
SOURCE        DF   SQUARES         SQUARE        F VALUE   PR>F      R-SQUARE    C.V.
MODEL         3    0.46489500      0.15496500    5.20      0.0107    0.493679    10.1515
ERROR         16   0.47680000      0.02980000
CORRECTED                                                  ROOT MSE              X MEAN
  TOTAL       19   0.94169500                              0.17262677            1.70050000

SOURCE        DF   ANOVA SS        F VALUE   PR>F
TRTMENTS      3    0.46489500      5.20      0.0107
```

 10.51 Legislation affecting credit allows banks to become dual issuers of credit cards, offering banks the opportunity to serve a broader variety of their customers' needs. To develop proper policy regarding the distribution of credit cards, however, bankers must first understand the rate of credit use by those who prefer different types of cards. In response to this issue a Toledo banker has commissioned a study to compare the use of credit by those who hold only a Visa card, those who hold only a MasterCard, and those who hold both. The data in Table 10.59 represent the credit use over a six-month period by the 19 subjects involved in the experiment.

Table 10.59   Credit Card Data for Exercise 10.51

Visa	MasterCard	Both Cards
$ 984.20	$1,253.05	$1,315.50
763.52	1,169.18	1,096.82
1,014.25	1,005.50	1,264.75
860.56	984.37	1,482.61
927.15	1,358.40	1,183.75
1,124.09	1,262.12	
911.13		
1,176.68		

a. Use the computer printout in Table 10.60 to perform an analysis of variance for these data.

b. Do the data provide sufficient evidence to indicate a difference in the average credit use by the three different groups of cardholders? Use $\alpha = .05$.

c. Estimate the average credit difference (in a six-month period) between holders of both Visa and MasterCard and those who hold only Visa. That is, find a 95% confidence interval for $(\mu_{both} - \mu_{Visa})$.

Table 10.60   MINITAB Printout for the Data of Exercise 10.51

```
MTB > AOVONEWAY C1-C3

ANALYSIS OF VARIANCE
SOURCE       DF          SS         MS            F
FACTOR        2      305286     152643         7.50
ERROR        16      325626      20352
TOTAL        18      630912

                                    INDIVIDUAL 95 PCT CI'S FOR MEAN
                                    BASED ON POOLED STDEV
LEVEL        N        MEAN      STDEV    ---+---------+---------+---------+---
C1           8       970.2      135.5    (------*------)
C2           6      1172.1      149.9               (-------*-------)
C3           5      1268.7      145.5                       (--------*--------)
                                         ---+---------+---------+---------+---
POOLED STDEV =      142.7                900       1050      1200      1350

MTB >
```

 10.52 Paper machines distribute a thin mixture of wood fibers and water to a wide wire mesh belt that is traveling at a very high speed. Thus, the distribution of fibers, thickness, porosity, and so on, may vary along the belt and produce variations in the strength of the final paper product. A paper company designed an experiment to compare the strength of four coatings intended to improve the appearance of packaging paper. Because the uncoated paper strength could vary down a roll, the experiment was conducted as a randomized block experiment. The strength measurements for the four coatings A, B, C, and D were as shown in Table 10.61.

Table 10.61   Strength Data for Exercise 10.52

Position Down the Roll	Coating			
	A	B	C	D
1	10.4	12.4	13.1	11.8
2	10.9	12.4	13.4	11.8
3	10.5	12.3	12.9	11.4
4	10.7	12.0	13.3	11.4

a. Do these data present sufficient evidence to indicate a difference in mean strength for paper treated with the four paper coatings?

b. Do these data present sufficient evidence to indicate a difference in mean strength for locations down the roll?

c. Prior to seeing the data, the experimenter had decided to compare the mean strength between paper coated with coatings A and C. Estimate the difference, using a 95% confidence interval.

MKTG **10.53** A completely randomized design was employed to compare the effect of five different advertising layouts on product recognition time. Twenty-seven people were employed in the experiment. Regardless of the results of the analysis of variance, the experimenter wishes to compare layouts A and D. The results of the experiment were as shown in Table 10.62 (time, in seconds).

a. Conduct an analysis of variance, and test for a difference in mean recognition time resulting from the five advertising layouts.

b. Compare layouts A and D to see whether there is a difference in mean recognition time.

Table 10.62   Comparison Data for Exercise 10.53

	Layout A	Layout B	Layout C	Layout D	Layout E
	.8	.7	1.2	1.0	.6
	.6	.8	1.0	.9	.4
	.6	.5	.9	.9	.4
	.5	.5	1.2	1.1	.7
		.6	1.3	.7	.3
		.9	.8		
		.7			
Total	2.5	4.7	6.4	4.6	2.4
Mean	.625	.671	1.067	.920	.480

MKTG **10.54** The experiment in Exercise 10.53 might have been more effectively conducted by using a randomized block design with subjects as blocks, since we would expect mean recognition time to vary from one person to another. Four people were used in a new experiment, and each person was shown each of the five advertising layouts in a random order. The results were as shown in Table 10.63 (time, in seconds). Conduct an analysis of variance, and test for differences in treatments (advertising layouts). Test at the $\alpha = .05$ level.

Table 10.63   Comparison Data for Exercise 10.54

Subject	Layout				
	A	B	C	D	E
1	7	.8	1.0	1.0	.5
2	.6	.6	1.1	1.0	.6
3	.9	1.0	1.2	1.1	.6
4	.6	.8	.9	1.0	.4

Supplementary Exercises

Interpretive

10.55   A study has been initiated to investigate the cleaning ability of 3 laundry detergents. Four different brands of automatic washing machines are to be used in the experiment, with each of 3 laundry detergents tested in each of the 4 washers. Thus, 12 combinations exist within the experiment. Twelve stacks of laundry, containing an equal amount of soil, are to be laundered. At the completion of each wash load, the laundry is to be tested by a meter for "whiteness" and the results are to be recorded.

   a.   Is this a randomized block design? Explain.

   b.   Suppose that 2 stacks of soiled laundry are to be subjected to each of the 3 detergents in each of the 4 washing machines. What type of experimental design is this?

10.56   The effectiveness of advertising is measured from a variety of perspectives. Such issues as evaluating relative effectiveness of various advertising layouts and determining strengths and weaknesses of different media are often as important as the immediate impact of advertising on sales. A marketing executive has undertaken a study to examine the comparative effect of three different media sources (treatments) in four different sales areas (blocks) and has obtained the results shown in Table 10.64. State the conclusions you would draw from the analysis of variance table. Use $\alpha = .05$ for all tests.

Table 10.64   Analysis of Variance for Exercise 10.56

Source	d.f.	SS
Blocks	3	24.15
Treatments	2	76.38
Error	6	30.66
Total	11	131.19

10.57   A zoning commission has been formed to estimate the average appraisal value of houses in a residential suburb of a city. The commission is considering using one of three different appraisal models. In a test for consistency among the three appraisal models each model is separately used to generate an appraisal value for each of five different residential dwellings. The results are given in Table 10.65. Without any specific directives, perform an analysis of the data. What are your conclusions? What recommendations would you make to the zoning commission about the relative merits of the three appraisal models? (*Hint:* Code the data.)

Table 10.65   Appraisal Value Data for Exercise 10.57

Appraisal Model	Dwelling 1	2	3	4	5
A	$91,000	$107,000	$98,000	$107,000	$100,000
B	92,500	110,000	97,500	109,000	103,000
C	89,000	105,000	97,500	106,000	101,000

10.58   Measuring effective advertising is of more than academic interest, since positive brand identification is usually translated into increased sales in the marketplace. Advertising effectiveness, measured in terms of learning and memory, uses either *recall*, in which an individual describes a stimulus, or *recognition*, in which an individual simply identifies a stimulus as having been seen or heard previously. Recall measures have been popular with the broadcast media, even though researchers have shown that recognition scores are more sensitive and more discriminating than unaided recall scores [S. N. Singh, M. L. Rothschild, and G. A. Churchill, Jr., "Recognition Versus Recall as Measures of Television Commercial Forgetting," *Journal of Marketing Research*, Vol. 25 (February 1988), pp. 72–80]. Suppose that an experiment concerning the effectiveness of television advertising during network newscasts is designed to evaluate the effect of message length (10 seconds versus 20 seconds) and repetition number (once versus twice). The five subjects assigned to each of the four (2 × 2) experimental conditions were shown a 30-minute tape. Two tapes had commercials 10 seconds long; the other two had commercials 20 seconds long. On two tapes commercials were repeated once; on the other, commercials were repeated twice. Each subject was given an unaided recall test of product, brand, and advertising claim. Data representing subject recall scores for the product are given in Table 10.66. Analyze these data by using an analysis of variance. Discuss the effects of the two experimental factors and their possible interaction. Use confidence interval estimation to estimate the difference in means where you deem appropriate. Summarize the results of this experiment with respect to how these factors affect commercial forgetting, and discuss the resulting implications for product advertising.

Table 10.66   Recall Score Data for Exercise 10.58

Number of Repetitions	Message Length	
	10 Seconds	20 Seconds
1	.45	.52
	.38	.38
	.20	.42
	.31	.22
	.26	.31
2	.35	.83
	.52	.71
	.57	.58
	.47	.65
	.59	.78

Case Study

Exxon's Response to the Regulators

As a result of several major studies conducted by the Environmental Protection Agency, the federal government has issued regulations mandating a gradual phaseout of tetraethyl lead (TEL) for gasoline sold in the United States. Separate legislation developed by the state of California provided even stiffer limits for the TEL content of gasoline.

For years refiners had used TEL as an addition to gasoline as a cheap and convenient way to improve the octane rating of the gasoline, thus reducing the potential for harm to motor vehicle engines. In the absence of TEL a refinery must reprocess some of the low-octane components of gasoline to increase the octane ratings. Reprocessing can be accomplished either by breaking apart the hydrocarbon chains, through processes known in the trade as "cat cracking" or "hydrocracking," or by rearranging the bonding in the chains, through processes called re-forming or alkylation. All four processes are very costly and efficient but provide more variability in results than the simple addition of TEL to improve the octane rating of a blend.

Faced with the dual impact of federal regulations on TEL and the more stringent California limits, officials at Exxon's Benicia, California, refinery initiated a crash program to determine the most efficient means of addressing the new requirements (L. Golovin, "Product Blending: A Simulation Case Study in Double-Time," *TIMS Interfaces*, November 1979). Experiments were undertaken using each of the four known methods of increasing octane ratings without exceeding mandated limits on the use of TEL. In their study gasoline from Exxon's Benicia refinery was reprocessed in such a way that costs were equalized by using each experimental procedure. The data in the following table show the octane ratings resulting from the application of the four reprocessing procedures to gasoline derived from each of eight storage tanks.

Storage Tank	Octane Rating When Reprocessing By			
	Cat Cracking	Hydrocracking	Re-forming	Alkylation
1	89.4	88.6	95.5	89.6
2	88.3	91.3	94.0	90.2
3	87.2	88.2	86.7	90.0
4	89.8	89.0	87.5	88.9
5	90.1	88.9	90.7	90.2
6	87.7	88.8	94.8	92.5
7	84.6	86.0	87.3	87.1
8	88.3	89.1	91.5	92.0

1. Do these data suggest that a difference exists in the ability of the four reprocessing procedures to increase octane ratings?
2. Does a difference exist between the octane efficiency of processes that break apart the hydrocarbon chains (cat cracking and

hydrocracking) and processes that rearrange the bonding in chains (re-forming and alkylation)?

3. If you were advising Exxon, which reprocessing procedure would you recommend for increasing octane ratings? (*Hint:* Consider the variability of octane rating when using each reprocessing procedure.)

References and Suggested Readings

Dixon, W. J., and M. B. Barnes (editors). *BMDP-83; Biomedical Computer Programs, P-Series.* Berkeley, Calif.: University of Calif. Press, 1983.

Dunn, O. J., and V. A. Clark. *Applied Statistics: Analysis of Variance and Regression, 2nd ed.* New York: John Wiley and Sons, 1987.

Mendenhall, W. *A Course in Business Statistics, 2nd ed.* Boston: PWS-KENT Publishing Co., 1988.

Mendenhall, W. *An Introduction to Linear Models and the Design and Analysis of Experiments.* Belmont, Calif.: Wadsworth, 1968.

Miller, R. B. *Minitab Handbook for Business and Economics.* Boston: PWS-KENT Publishing Co., 1988.

Ott, L. *An Introduction to Statistical Methods and Data Analysis, 2nd ed.* Boston: PWS-KENT Publishing Co., 1984.

Ryan, B. F., B. L. Joiner, and T. A. Ryan. *Minitab Handbook, 2nd ed.* Boston: PWS-KENT Publishing Co., 1985.

SAS User's Guide: Basics, 1982 Edition. SAS Institute, Inc., Box 800, Cary, North Carolina, 27511.

TO
THE
READER

Previous chapters were concerned with estimating the value of a population parameter or testing an hypothesis about its value. In many business applications, however, additional information may be available through other variables that are related to the variable of interest. For example, the costs of shipping goods to the market would vary according to the size (weight or volume) of the shipment, distance, and the mode of shipment. Hence, in using the techniques of previous chapters, we could select a random sample of freight invoices and estimate an average shipping cost. Alternatively, we could use the auxiliary information concerning the size, distance, and mode of shipment in estimating the average shipping cost for a fixed size, distance, and mode of transportation.

In this chapter we will begin by considering situations in which the observation of interest, y, is linearly related to one auxiliary variable, x. In addition to assessing the strength of the linear relationship, we will also show how to use this linear relationship for prediction and estimation.

11

LINEAR REGRESSION
AND CORRELATION

11.1

Introduction

forecasting
predicting

An estimation problem of particular importance in almost every field of study is the problem of **forecasting**, or **predicting**, the value of a process variable from known, related variables. Practical examples of prediction problems are numerous in business, industry, and the sciences. The stockbroker wishes to predict stock market behavior as a function of a number of observable key indices. The sales manager of a chain of retail stores wishes to predict the monthly sales volume of each store from the number of credit customers and the amount spent on advertising. The manager of a manufacturing plant would like to relate the yield of a chemical to a number of process variables. He will then use the prediction equation to find the settings for the controllable process variables that will provide the maximum yield of the chemical. The personnel director of a corporation, like the admissions director of a university, wishes to test and measure individual characteristics so that she may hire the person best suited for a job. The political scientist may wish to relate success in a political campaign to the characteristics of a candidate, the opposition, and various campaign issues and promotional techniques. Certainly, all these prediction problems are identical in many respects.

In a sense, the statistical approach to each problem is a formalization of the procedure we might follow intuitively. Suppose that a security analyst is attempting to predict the price of a firm's securities on the securities market from the Dow-Jones industrial average, the prime interest rate, and the sales volume for the firm over the past month. He could expect the security price to rise as the Dow-Jones average and the sales volume rise. However, a rise in the prime interest rate would tend to drive money out of the securities market into more risk-free investments and hence would be accompanied by a decrease in security prices. Each of the variables above seems to individually influence the security price in some way. But when observed together, these variables may have an interactive effect that influences the security price differently from the way either variable does alone. For instance, how would a firm's security price be affected if a rise in sales volume were accompanied by an increase in the prime interest rate? The true relationship in this case could probably best be seen by introducing another variable into the model—say the amount of new debt undertaken by the firm in the past month—that would be related to both the prime interest rate and the sales volume. Carrying this line of thought to the ultimate and idealistic **function** extreme, we would expect the price of a firm's securities to be a mathematical **function** of the preceding variables plus any others that may influence price and be easily measurable. Ideally, we would like to possess a mathematical equation that relates the price of a particular firm's securities to all relevant variables so that the equation could be used for prediction.

independent Observe that the problem we have defined is of a very general nature. We are **predictor** interested in a random variable y that is related to a number of **independent predictor** **variables** **variables** $x_1, x_2, x_3, \ldots, x_k$. The variable y for our example is the price of a particular firm's securities at a certain time, and the independent predictor variables might be

$$x_1 = \text{Dow-Jones industrial average}$$
$$x_2 = \text{prime interest rate}$$
$$x_3 = \text{last month's retail sales volume}$$

and so on. The ultimate objective is to measure $x_1, x_2, x_3, \ldots, x_k$ for a particular firm, substitute these values into the prediction equation, and thereby predict the price of the firm's securities. To accomplish this objective, we must first locate the related variables $x_1, x_2, x_3, \ldots, x_k$ and obtain a measure of the strength of their relationship to y. Then we must construct a good prediction equation that expresses y as a function of the selected independent predictor variables.

In this chapter we will restrict our attention to the simple problem of predicting y as a linear function of a single variable x. The solution for the multivariable problem—for example, predicting the price of a firm's securities as a function of more than one predictor variable—will be the subject of Chapter 12.

11.2

A Simple Linear Probabilistic Model

For purposes of illustration we introduce our topic by considering the problem of predicting the gross monthly sales volume y for a corporation that is not subject to substantial seasonal variation in its sales volume. As the predictor variable x, we will use the amount spent by the company on advertising during the month of interest. As noted in Section 11.1, we wish to determine whether advertising is actually worthwhile—that is, whether advertising is actually related to the firm's sales volume—and in addition, we wish to obtain an equation that will be useful for predicting the monthly sales y as a function of advertising expenditures x. The

Table 11.1   Advertising Expenditures and Sales Volumes for a
Corporation During 10 Randomly Selected Months

Month	Advertising Expenditure, $x(\times \$10{,}000)$	Sales Volume, $y(\times \$10{,}000)$
1	1.2	101
2	.8	92
3	1.0	110
4	1.3	120
5	.7	90
6	.8	82
7	1.0	93
8	.6	75
9	.9	91
10	1.1	105

evidence, presented in Table 11.1, represents a sample of advertising expenditures and the associated sales volumes for the company during 10 randomly selected months. We will assume that the advertising expenditures and sales volumes for these 10 months constitute a random sample of measurements for all past and current months' operations for the company.

      Our initial approach to the analysis of the data of Table 11.1 is to plot the data as points on a graph, representing a month's sales volume as y and the corresponding advertising expenditure as x. The graph, shown in Figure 11.1, is called a **scatter diagram**. You will probably observe that y appears to increase as x increases. (Could this arrangement of the points occur by chance even if x and y were unrelated?)

scatter
diagram

      One method of obtaining a prediction equation relating y to x is to place a ruler on the graph and move it about until it seems to pass through the points, thus providing what we might regard as the "best fit" to the data. Indeed, if we draw a line through the

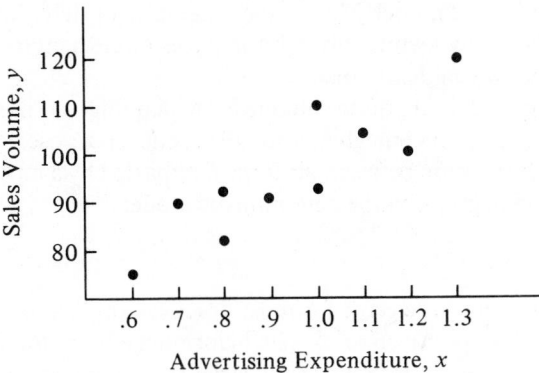

Figure 11.1   Plot of the Data of Table 11.1

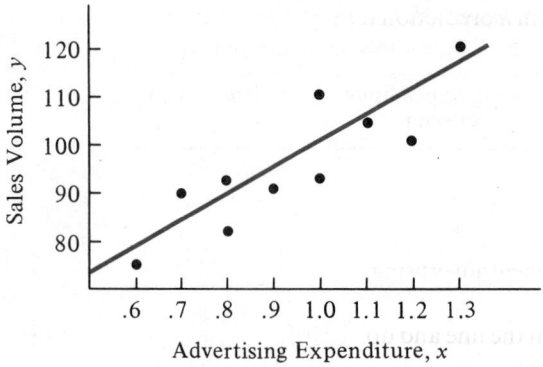

Figure 11.2   Fitting a Line by Eye

points, it would appear that our prediction problem had been solved (see Figure 11.2). Certainly, we can now use the graph to predict the company's monthly sales volume y as a function of the amount x budgeted for advertising during that month.

Let us review several facts concerning the graphing of mathematical functions. First, the mathematical **equation of a straight line** is

equation of a straight line

$$y = \beta_0 + \beta_1 x$$

y intercept
slope

where β_0 is the **y intercept** and β_1 is the **slope** of the line. Second, the line that we may graph for each linear equation is unique. Each equation corresponds only to one line, and vice versa. Thus, when we draw a line through the points, we automatically choose a mathematical model for the response y:

$$y = \beta_0 + \beta_1 x$$

where β_0 and β_1 have unique numerical values.

deterministic mathematical model

The linear model $y = \beta_0 + \beta_1 x$ is said to be a **deterministic mathematical model**, because when a value of x is substituted into the equation, the value of y is determined and no allowance is made for error. Fitting a straight line through a set of points by eye produces a deterministic model. Many other examples of deterministic mathematical models may be found by leafing through the pages of elementary chemistry, physics, economics, or engineering textbooks.

Deterministic models are quite suitable for explaining and predicting phenomena when the error of prediction is negligible for all practical purposes. Thus, Newton's law, which expresses the relation between the force F imparted by a moving body with mass m and acceleration a, given by the deterministic model

$$F = ma$$

predicts force with very little error for most practical applications. "Very little" is, of course, a relative concept. An error of .1 in. in forming a beam for a bridge is extremely small, but the same error is impossibly large in the manufacture of parts for a wristwatch. Thus, in many physical situations the error of prediction cannot be ignored. Indeed, consistent with our stated philosophy, we would hesitate to place much confidence in a prediction unaccompanied by a measure of its goodness. For this reason, a visual choice of a line to relate the advertising expenditures to the sales volume would have limited value.

Now let us return to our example. What is wrong with using a deterministic model to relate sales volume to advertising expenditures? The answer is that although deterministic models permit us to predict y for various values of x (by substituting values of x into the prediction equation), they do not provide us with a way to evaluate the error of prediction. For example, if we use the line in Figure 11.2 to predict sales volume y when advertising expenditure is $x = 1.2$, we obtain a predicted value $y = 111$. But plus or minus what? You can see that the plotted points deviate substantially from the line and do so in a seemingly random pattern. What is the bound on the error of prediction? A businessperson would need this information to be able to decide whether it would be profitable to make a particular advertising expenditure. Consequently, we need a method for fitting a model to data, a method that will enable us to place bounds on the errors of estimating β_0 and β_1 and bounds on the error of predicting y for given values of x. More important, we want a method that can be used to fit models when more than one predictor variable x is involved.

The probabilistic model we use to relate sales volume y and advertising expenditures x is a simple modification of the deterministic model. Rather than saying that y and x are related by the deterministic model

$$y = \beta_0 + \beta_1 x$$

mean value of y for a given value of x we say that the **mean** (or expected) **value of y for a given value of x**, denoted by the symbol $E(y\,|\,x)$, has a graph that is a straight line. That is, we let

$$E(y\,|\,x) = \beta_0 + \beta_1 x$$

For any given values of x, y values will vary in a random manner about the mean $E(y\,|\,x)$. For example, if the company spends \$10,000 $(x = 1.0)$ per month on advertising, the gross monthly sales y could assume some value from a population of possible values. But the *mean* gross monthly sales is precisely determined by substituting $x = 1.0$ into the equation

$$E(y\,|\,x) = \beta_0 + \beta_1 x$$

probabilistic model   So to summarize, we write the **probabilistic model** for any particular observed value of y as

$$y = (\text{mean value of } y \text{ for given value of } x) + (\text{random error})$$

$$= \overbrace{\beta_0 + \beta_1 x}^{E(y\,|\,x)} + \epsilon$$

random error   where ϵ is a **random error**, the difference between an observed value of y and the mean value of y for a given value of x. Thus, we assume that for any given value of x the observed value of y varies in a random manner and possesses probability distribution with a mean value $E(y\,|\,x)$. For assistance in conveying this idea, the probability

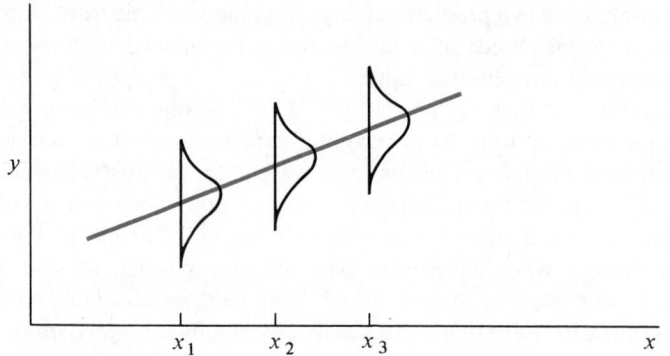

Figure 11.3   Linear Probabilistic Model

distributions of y about the line of means $E(y\,|\,x)$ are shown in Figure 11.3 for three hypothetical values of x, x_1, x_2, and x_3, although, as noted, we imagine a distribution of y values for every value of x.

What properties will we assume for the probability distribution of y for a given value of x? Assumptions that seem to fit many practical situations, and upon which the methodology of Sections 11.4, 11.5, 11.6, and 11.7 is based, are given in the display.

Assumptions for the Probabilistic Model

For any given value of x, y possesses a normal distribution, with a mean value given by the equation

$$E(y\,|\,x) = \beta_0 + \beta_1 x$$

and with a variance of σ^2. Furthermore, any one value of y is independent of every other value.

These assumption permit us to construct tests of hypotheses and confidence intervals for β_0, β_1, and $E(y\,|\,x)$. In addition, we are able to give a prediction interval for y when we use the fitted model (the line that we find to fit a particular set of data) to predict some new value of y for a particular set of x. The practical problems we can solve by using these procedures will become apparent when we have worked some examples.

In the next section we will consider the problem of finding the "best-fitting" line for a given set of data. This line will give us a prediction equation that is also called a *regression line*.

EXERCISES   Conceptual Level

11.1   What is the difference between a deterministic and a probabilistic model?

11.2   Explain the purpose of the random error term ϵ in a probabilistic model.

EXERCISES   Skill Level

11.3   The equation of a straight line is given as $y = 5 - 4x$.
  a. Give the y intercept and the slope of the line.
  b. Graph the line corresponding to the equation.

11.4   The equation of a straight line is given as $2y = 3x + 1$.
  a. Give the y intercept and the slope of the line.
  b. Graph the line corresponding to the equation.

11.5   The equation of a straight line is given as $y = 6 + 3x$.
  a. Give the y intercept and the slope of the line.
  b. Graph the line corresponding to the equation.

11.3

The Method of Least Squares

The statistical procedure for finding the "best-fitting" straight line for a set of points is, in many respects, a formalization of the procedure employed when we fit a line by eye. For instance, when we visually fit a line to a set of data, we move the ruler until we think

deviations that we have minimized the **deviations** of the points from the prospective line. If we denote the predicted value of y obtained from the fitted line as $\hat{y}$, then the prediction equation is

$$\hat{y} = \hat{\beta}_0 + \hat{\beta}_1 x$$

where $\hat{\beta}_0$ and $\hat{\beta}_1$ represent estimates of the true β_0 and β_1. This line for the data of Table 11.1 is shown in Figure 11.4. The vertical lines drawn from the prediction line to each point represent the deviations of the points from the predicted value of y.

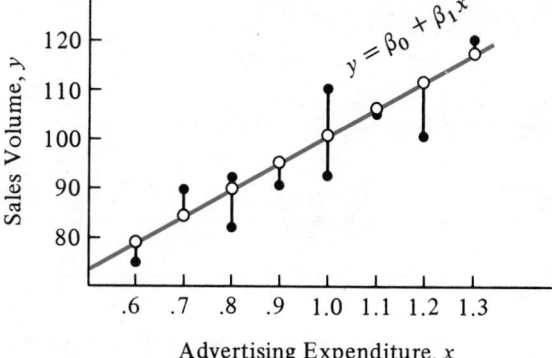

Figure 11.4   *Linear Probabilistic Equation*

Thus, the deviation of the ith point is

$$y_i - \hat{y}_i$$

where

$$\hat{y}_i = \hat{\beta}_0 + \hat{\beta}_1 x_i$$

Having decided that in some manner or other we will attempt to minimize the deviations of the points in choosing the best-fitting line, we must now define what we mean by *best*. That is, we wish to define a criterion for "best fit" that seems intuitively reasonable, is objective, and under certain conditions gives the best prediction of y for a given value of x.

principle of least squares

We will employ a criterion of goodness that is known as the **principle of least squares**, which may be stated as follows: **Choose as the best-fitting line that line that minimizes the sum of squares of the deviations of the observed values of y from those predicted.** * Expressed mathematically, we wish to minimize the **sum of squared errors**

sum of squared errors

given by

$$SSE = \sum_{i=1}^{n} (y_i - \hat{y}_i)^2$$

Since the predicted value of y_i corresponding to $x = x_i$ is $\hat{\beta}_0 + \hat{\beta}_1 x_i$, we can substitute this quantity for $\hat{y}_i$ in SSE and obtain the following expression.

Sum of Squared Errors

$$SSE = \sum_{i=1}^{n} [y_i - (\hat{\beta}_0 + \hat{\beta}_1 x_i)]^2$$

The method for finding the numerical values of $\hat{\beta}_0$ and $\hat{\beta}_1$ that minimize SSE utilizes differential calculus and hence is beyond the scope of this text. It can be shown that the values of $\hat{\beta}_0$ and $\hat{\beta}_1$ that minimize SSE are given by the following formulas.

Least Squares Estimators of β_0 and β_1

$$\hat{\beta}_1 = \frac{S_{xy}}{S_{xx}} \quad \text{and} \quad \hat{\beta}_0 = \bar{y} - \hat{\beta}_1 \bar{x}$$

where

$$S_{xx} = \sum_{i=1}^{n} (x_i - \bar{x})^2 = \sum_{i=1}^{n} x_i^2 - \frac{\left(\sum_{i=1}^{n} x_i\right)^2}{n}$$

$$S_{xy} = \sum_{i=1}^{n} (x_i - \bar{x})(y_i - \bar{y}) = \sum_{i=1}^{n} x_i y_i - \frac{\left(\sum_{i=1}^{n} x_i\right)\left(\sum_{i=1}^{n} y_i\right)}{n}$$

---

* It can be shown (proof omitted) that the sum of deviations of the points about the least squares line will always equal 0; that is,

$$\sum_{i=1}^{n} (y_i - \hat{y}_i) = 0$$

Note that S_{xx} (sum of squares for x) is computed by using the familiar shortcut formula for calculating sums of squares of deviations that was used in calculating s^2 in Chapter 2. The sum S_{xy} is computed by using a very similar formula. Once $\hat{\beta}_0$ and $\hat{\beta}_1$ have been computed, we substitute their values into the equation of a line to obtain the least squares prediction equation, or **regression line**,

regression line

$$\hat{y} = \hat{\beta}_0 + \hat{\beta}_1 x$$

The use of these formulas for finding $\hat{\beta}_0, \hat{\beta}_1$, and the least squares line is illustrated by an example.

EXAMPLE 11.1   Obtain the least squares prediction line for the data of Table 11.1.

SOLUTION   The calculation of $\hat{\beta}_0$ and $\hat{\beta}_1$ for the data of Table 11.1 is simplified by the use of Table 11.2.

Substituting the appropriate sums from Table 11.2 into the least squares equations, we obtain the calculations shown next.

$$S_{xx} = \sum_{i=1}^{n} x_i^2 - \frac{\left(\sum_{i=1}^{n} x_i\right)^2}{n} = 9.28 - \frac{(9.4)^2}{10} = .444$$

$$S_{xy} = \sum_{i=1}^{n} x_i y_i - \frac{\left(\sum_{i=1}^{n} x_i\right)\left(\sum_{i=1}^{n} y_i\right)}{n} = 924.8 - \frac{(9.4)(959)}{10} = 23.34$$

$$\bar{y} = \frac{\sum_{i=1}^{n} y_i}{n} = \frac{959}{10} = 95.9 \qquad \bar{x} = \frac{\sum_{i=1}^{n} x_i}{n} = \frac{9.4}{10} = .94$$

Hence,

$$\hat{\beta}_1 = \frac{S_{xy}}{S_{xx}} = \frac{23.34}{.444} = 52.5676 \approx 52.57$$

$$\hat{\beta}_0 = \bar{y} - \hat{\beta}_1 \bar{x} = 95.9 - (52.5676)(.94) \approx 46.49$$

Table 11.2   Calculations for the Data of Table 11.1

y_i	x_i	x_i^2	$x_i y_i$	y_i^2
101	1.2	1.44	121.2	10,201
92	.8	.64	73.6	8,464
110	1.0	1.00	110.0	12,100
120	1.3	1.69	156.0	14,400
90	.7	.49	63.0	8,100
82	.8	.64	65.6	6,724
93	1.0	1.00	93.0	8,649
75	.6	.36	45.0	5,625
91	.9	.81	81.9	8,281
105	1.1	1.21	115.5	11,025
Sum 959	9.4	9.28	924.8	93,569

Then according to the principle of least squares, the best-fitting straight line (or the regression line) relating the advertising expenditures to the sales volume is

$$\hat{y} = \hat{\beta}_0 + \hat{\beta}_1 x \qquad \text{or} \qquad \hat{y} = 46.49 + 52.57x$$

The graph of this prediction equation was shown in Figure 11.4.

We may now predict y for a given value of x by referring to Figure 11.4 or by substituting into the prediction equation. For example, if the corporation has budgeted $10,000 for advertising in a month, its predicted sales volume is

$$\hat{y} = \hat{\beta}_0 + \hat{\beta}_1 x = 46.49 + (52.57)(1.0) = 99.06$$

or expressing the sales volume in dollars, $990,600.

Keep in mind that the best-fitting straight line, the prediction equation

$$\hat{y} = \hat{\beta}_0 + \hat{\beta}_1 x$$

also estimates the line of means

$$E(y \mid x) = \beta_0 + \beta_1 x$$

That is, in Example 11.1 the best estimate of the *mean* gross monthly sales for a $10,000 advertising expenditure ($x = 1.0$) is also equal to $990,600. Thus, the prediction equation can be used either to predict some future value of y or to estimate the mean value of y for a given advertising expenditure x.

The next step in our procedure is to place a bound on our error of estimation. We will consider this and related problems in succeeding sections.

Tips on Problem Solving

1. Be careful of rounding errors. Carry at least six significant figures in computing sums of squares of deviations or the sums of squares of cross products of deviations.
2. Always plot the data points and graph your least squares line. If the line does not provide a reasonable fit to the data points, you may have committed an error in your calculations.

EXERCISES   Conceptual Level

11.6   What property does the least squares line have when compared with any other line that might be used to describe the data?

11.7   What assumptions are necessary to use the prediction equation $\hat{y} = \hat{\beta}_0 + \hat{\beta}_1 x$ to predict the value of a dependent variable y from a predictor variable x?

EXERCISES Skill Level

11.8 The accompanying table displays the coordinates for $n = 5$ pairs of observations (x, y).

x	-2	-1	0	1	2
y	0	0	1	1	3

a. Find the least squares line for the data.

b. As a check on the calculations in part a, plot the five points and graph the least squares line. Does the line appear to provide a good fit to the data points?

11.9 The accompanying table displays the coordinates for $n = 5$ pairs of observations (x, y).

x	-3	-1	1	1	2
y	6	4	3	1	1

a. Find the least squares line for the data.

b. As a check on the calculations in part a, plot the five points and graph the least squares line. Does the line appear to provide a good fit to the data points?

11.10 The accompanying table displays the coordinates for $n = 7$ pairs of observations (x, y).

x	7	8	2	3	5	3	7
y	2	0	5	6	4	9	2

a. Find the least squares line for the data.

b. As a check on the calculations in part a, plot the seven points and graph the least squares line. Does the line appear to provide a good fit to the data points?

EXERCISES Applications

11.11 The accompanying data represent the profit y per sales dollar earned by a construction company on nine separate projects and the number x of years of experience of the construction superintendent assigned to each project.

x	4	4	2	6	2	2	4	6	6
y	2.0	3.5	8.5	4.5	7.0	7.0	2.0	6.5	8.0

a. Find the least squares line for the data.

b. As a check on the calculations in part a, plot the nine points and graph the least squares line. Does the line appear to provide a good fit to the data points?

11.12 The flexible budget is an expression of management's expectations concerning revenues and costs for some future period and serves to communicate top management goals to the various managers of the organization. The management of a manufacturing company is interested in establishing a flexible budget for purposes of estimating overhead costs over a certain range of production. Historical cost and production data are given in the table.

Production ($\times \$10,000$)	3	4	5	6	7	8	9
Overhead Costs ($\times \$1,000$)	12	10.5	13	12	13	13.3	16.5

a. Find the least squares line to allow for the estimation of overhead costs from production (that is, the least squares line relating cost to production).

b. As a check on your calculations, plot the seven points and graph the least squares line. Does the line appear to provide a good fit to the data points?

 11.13   As the national budget continues to grow, one spending category that has come under increased scrutiny is military spending. Is the amount spent on the military consuming a disproportionate amount of the total available budget? Table 11.3 shows the military expenditure per capita and the GNP (to which the total available budget amount is related) for the years 1976 to 1982.

a. Plot the seven data points.

b. Find the least squares line for the data.

c. Graph the least squares line, and see how well it fits the data.

d. Using the least squares equation, estimate the per capita military spending in a year when the GNP is 3,000.

Table 11.3   Spending Data for Exercise 11.13

Year	Military Expenditure per Captita (Millions of Dollars)	GNP (1982) (Billions of Dollars)
1976	423	2,827
1977	465	2,959
1978	499	3,115
1979	553	3,192
1980	631	3,187
1981	739	3,249
1982	846	3,166

Source: Data from Value Line's Investment Survey, December 25, 1987; Bobby E. Apostolakis, "The Buy-American Practices of the U.S. Defense Department and Their Repercussions," *Journal of Economic Studies*, Vol. 14, No. 3 (1987).

11.4

Calculating s^2, an Estimator of σ^2

Recall that the probabilistic model for y in Section 11.2 assumes that y is related to x by the equation

$$y = \beta_0 + \beta_1 x + \epsilon$$

For a given value of x the y values are independent and possess a normal probability distribution with mean $E(y \mid x)$ and variance σ^2. The variance σ^2 measures the random variation of the y values about the line $E(y \mid x) = \beta_0 + \beta_1 x$. The greater the variation, the larger will be the value of σ^2. Thus, each observed value of y is subject to a random error ϵ that enters into the computations of $\hat{\beta}_0$ and $\hat{\beta}_1$ and introduces error into these estimates. Furthermore, if we use the least squares line

$$\hat{y} = \hat{\beta}_0 + \hat{\beta}_1 x$$

to predict some future value of y, the random errors affect the error of prediction. Consequently, the variability of the random errors, measured by σ^2, plays an important role when estimating or predicting by using the least squares line.

The first step toward acquiring a bound on a prediction error requires that we estimate σ^2, the variance of y for a given value of x. For this purpose it seems reasonable to use SSE, the sum of squares of deviations (sum of squares for error) about the predicted line. Indeed, it can be shown that the formula given in the display provides an estimator for σ^2 that is unbiased, based on $(n - 2)$ degrees of freedom.

An Estimator for σ^2

$$\hat{\sigma}^2 = s^2 = \frac{\text{SSE}}{n - 2}$$

The sum of squares of deviations SSE may be calculated directly by using the prediction equation to calculate $\hat{y}$ for each point, then calculating the deviations $(y_i - \hat{y}_i)$, and finally calculating

$$\text{SSE} = \sum_{i=1}^{n} (y_i - \hat{y}_i)^2$$

This procedure tends to be tedious and is rather poor from a computational point of view because the numerous subtractions tend to introduce computational rounding errors. An easier and computationally better procedure is to use the formula given in the display.

Formula for Calculating SSE

$$\text{SSE} = S_{yy} - \hat{\beta}_1 S_{xy} = S_{yy} - \frac{(S_{xy})^2}{S_{xx}}$$

where

$$S_{yy} = \sum_{i=1}^{n} (y_i - \bar{y})^2 = \sum_{i=1}^{n} y_i^2 - \frac{\left(\sum_{i=1}^{n} y_i\right)^2}{n}$$

$$S_{xy} = \sum_{i=1}^{n} x_i y_i - \frac{\left(\sum_{i=1}^{n} x_i\right)\left(\sum_{i=1}^{n} y_i\right)}{n}$$

$$S_{xx} = \sum_{i=1}^{n} x_i^2 - \frac{\left(\sum_{i=1}^{n} x_i\right)^2}{n}$$

Observe that S_{xy} and S_{xx} were used in the calculation of $\hat{\beta}_1$ and hence have already been computed. Furthermore, note that if you must round a number, you should retain six or more significant figures in the calculations to avoid serious rounding errors in the final answer.

EXAMPLE 11.2   Calculate an estimate of σ^2 for the data of Table 11.1.

SOLUTION   First, we calculate

$$S_{yy} = \sum_{i=1}^{n} (y_i - \bar{y})^2 = \sum_{i=1}^{n} y_i^2 - \frac{\left(\sum_{i=1}^{n} y_i\right)^2}{n} = 93,569 - \frac{(959)^2}{10} = 1,600.9$$

Substituting this value of S_{yy} and the values of $\hat{\beta}_1$ and S_{xy} calculated in Example 11.1 into the formula for SSE, we obtain

$$\text{SSE} = S_{yy} - \hat{\beta}_1 S_{xy} = 1,600.9 - (52.5676)(23.34) = 373.97$$

Then

$$s^2 = \frac{\text{SSE}}{n-2} = \frac{373.97}{8} = 46.75$$

How can you interpret these values of SSE and s^2? Refer to Figure 11.4 and note the deviations of the $n = 10$ points from the least squares line (shown as the vertical line segments between the points and the line). The sum $\text{SSE} = 373.97$ is equal to the sum of squares of the numerical values of these deviations. This quantity is then used to calculate $s^2 = 46.75$ and $s = \sqrt{46.75} = 6.84$, estimates of σ^2 and σ.

The practical interpretation that can be given to s ultimately rests on the meaning of σ. Since σ measures the spread of the y values about the line of means $E(y \mid x) = \beta_0 + \beta_1 x$ (see Figure 11.3), we would expect (from the Empirical Rule) approximately 95% of the y values to fall within 2σ of that line. Since we do not know σ, $2s$ provides an approximate value for the half width of this interval. Now return to Figure 11.4 and note the location of the data points about the least squares line. Since we used the $n = 10$ data points to fit the least squares line, you would not be too surprised to find that most of the points fall within $2s = 2(6.84) = 13.68$ of the line. If you check Figure 11.4, you will see that all 10 points fall within $2s$ of the least squares line. (You will find that, in general, most of the data points used to fit the least squares line will fall within $2s$ of the line. This estimate provides you with a rough check for your calculated value of s.)

But s will play a much more important role in this chapter than the application described. As mentioned at the beginning of this section, the less the variability of the y values about the line of means (i.e., the smaller the value of σ), the closer the least squares line will be to the line of means. Consequently, s will play an important role in evaluating the goodness of all of the inferential methods described in this chapter.

Tips on Problem Solving

1. To reduce rounding error, always carry at least six significant figures when calculating S_{yy} and SSE. You can round when you obtain the answer for SSE if you desire.

2. As a check on your calculated value of s, remember that s measures the spread of the points about the least squares line. Therefore, you would expect (by the Empirical Rule) most of the points to fall within $2s$ of the least squares line. For example, if the points appear to fall in a band roughly equal to 4 units in width on the scale of the y variable and if your calculated value of s is 10, your value of s is too large. You have made an error. For example, perhaps you forgot to divide SSE by $(n - 2)$.

EXERCISES   Conceptual Level

11.14   Explain what variability is being measured by s^2 in a regression analysis.

11.15   For what configurations of sample points will s^2 be zero?

11.16   For what parameter is s^2 an unbiased estimator? Explain how this parameter enters into the description of the probabilistic model.

EXERCISES   Skill Level

11.17   Compute SSE and s^2 for the data in Exercise 11.8.

11.18   Compute SSE and s^2 for the data in Exercise 11.9.

11.19   Compute SSE and s^2 for the data in Exercise 11.10.

11.20   Many modern calculators contain a least squares regression routine that allows the user to perform a two-dimensional analysis for sample sizes of 99 or less. If you possess such a calculator, use it to compute SSE and s^2 for Exercises 11.8, 11.9, and 11.10. Compare the results with those obtained by hand calculation in Exercises 11.17, 11.18, and 11.19.

EXERCISES   Applications

11.21   Compute SSE and s^2 for the data in Exercise 11.11.

11.22   Compute SSE and s^2 for the data in Exercise 11.12.

11.23   Compute SSE and s^2 for the data in Exercise 11.13.

11.5

Inferences Concerning the Slope β_1 of a Line

In studying the relationship between y and x, our first concern is to determine whether or not y and x are linearly related. That is, do the data present sufficient evidence to indicate that x contributes information for the prediction of y over the region of observation? Or is it quite probable that when y and x are completely unrelated, the points would fall on the graph in a manner similar to that observed in Figure 11.1?

The practical question we pose concerns the value of β_1, which is the average change in y for a one-unit change in x. Stating that y does not increase (or decrease) linearly as x increases is equivalent to saying that $\beta_1 = 0$. (If $\beta_1 = 0$, we always predict the same value of y regardless of the value of x.) We should first test the hypothesis that $\beta_1 = 0$ against the alternative that $\beta_1 \neq 0$. As you might suspect, the estimator $\hat{\beta}_1$ is extremely useful in constructing a test statistic for this hypothesis. Therefore, we wish to examine the distribution of the estimates $\hat{\beta}_1$ that would be obtained when samples, each containing n points, are repeatedly drawn from the population of interest.

From our earlier assumptions concerning the probability distribution of y for a given value of x, it can be shown that both $\hat{\beta}_0$ and $\hat{\beta}_1$ are normally distributed in repeated sampling and that the expected value and variance of $\hat{\beta}_1$ are

$$E(\hat{\beta}_1) = \beta_1 \qquad \text{and} \qquad \sigma_{\hat{\beta}_1}^2 = \frac{\sigma^2}{S_{xx}}$$

Thus, $\hat{\beta}_1$ is an unbiased estimator of β_1; we know its standard error, and hence, we can construct a z statistic in the manner described in Section 8.14. Then

$$z = \frac{\hat{\beta}_1 - \beta_1}{\sigma_{\hat{\beta}_1}} = \frac{\hat{\beta}_1 - \beta_1}{\sigma/\sqrt{S_{xx}}}$$

possesses a standardized normal distribution in repeated sampling. Since the actual value of σ^2 is unknown, we should obtain the estimated standard error of $\hat{\beta}_1$, which is

$$s_{\hat{\beta}_1} = \frac{s}{\sqrt{S_{xx}}}$$

By substituting s for σ in z, we obtain, as in Chapter 9, the test statistic

$$t = \frac{\hat{\beta}_1 - \beta_1}{s_{\hat{\beta}_1}} = \frac{\hat{\beta}_1 - \beta_1}{s/\sqrt{S_{xx}}}$$

which can be shown to follow a Student's t distribution in repeated sampling, with $(n - 2)$ degrees of freedom. Note that the number of degrees of freedom associated with s^2 determines the number of degrees of freedom associated with t.

We observe that the test of an hypothesis that β_1 equals some particular numerical value, say $\beta_1 = 0$, involves the t test encountered in Chapter 9. Let β_{10} be the hypothesized value of β_1. Then the test is as given in the display.

Test of an Hypothesis Concerning the Slope of a Line

Null hypothesis $H_0: \beta_1 = \beta_{10}$.

Alternative hypothesis: Specified by the experimenter, depending on the values of β_1 he or she wishes to detect.

Test statistic: $t = \dfrac{\hat{\beta}_1 - \beta_{10}}{s/\sqrt{S_{xx}}}$.

Rejection region: See the critical values of t, Table 4 of the Appendix, for $(n - 2)$ degrees of freedom.

EXAMPLE 11.3   Use the data of Table 11.2 to determine whether there is evidence to indicate that β_1 differs from 0 by using a linear relationship between advertising expenditure x and monthly sales volume y.

SOLUTION   We wish to test the null hypothesis

$$H_0: \beta_1 = 0$$

against the alternative hypothesis

$$H_a: \beta_1 \neq 0$$

for the sales volume and advertising expenditure data in Table 11.2. The test statistic is

$$t = \frac{\hat{\beta}_1 - 0}{s/\sqrt{S_{xx}}}$$

and if we choose $\alpha = .05$, we will reject H_0 when $t > 2.306$ or $t < -2.306$. The critical value of t is obtained from the t table, using $(n - 2) = 8$ degrees of freedom. Substituting values determined in Examples 11.1 and 11.2 into the test statistic, we obtain

$$t = \frac{\hat{\beta}_1}{s/\sqrt{S_{xx}}} = \frac{52.57}{6.84/\sqrt{.444}} = 5.12$$

Observing that the test statistic exceeds the critical value of t, $t_{.025} = 2.306$, we reject the null hypothesis $\beta_1 = 0$ and conclude that there is evidence to indicate that advertising expenditures provide information for the prediction of gross monthly sales volume. In fact, with 8 degrees of freedom, $t = 5.12$ exceeds $t_{.005} = 3.355$. Therefore, the p-value for this test is less than $2(.005) = .01$, indicating that our results are highly significant.

◆ ◆

Once we have decided that β_1 differs from 0, we are interested in examining this relationship in detail. If x increases by 1 unit, what is the estimated change in y, and how much confidence can be placed in the estimate? In other words, we require an estimate of the slope β_1. You probably will not be surprised to observe a continuity in the procedures of Chapter 9 and 11. That is, the $(1 - \alpha)100\%$ confidence interval for β_1 can be shown to be $\hat{\beta}_1 \pm t_{\alpha/2} \times$ (standard error of $\hat{\beta}_1$), as given in the display.

A $(1 - \alpha)100\%$ Confidence Interval for β_1

$$\hat{\beta}_1 \pm t_{\alpha/2} \frac{s}{\sqrt{S_{xx}}}$$

EXAMPLE 11.4   Find a 95% confidence interval for β_1 by using the data of Table 11.1.

SOLUTION   The 95% confidence interval for β_1, using values calculated previously, is

$$\hat{\beta}_1 \pm t_{.025} \frac{s}{\sqrt{S_{xx}}}$$

Substituting, we obtain

$$52.57 \pm 2.306 \frac{6.84}{\sqrt{.444}} \qquad \text{or} \qquad 52.57 \pm 23.67$$

Intervals constructed by using this procedure will enclose the true value of β_1 95% of the time. Hence, we are fairly certain that the increase in monthly sales volume for a 1-unit ($10,000) increase in advertising expenditure will fall in the interval from 28.90 to 76.24, or in the original units for y, from $289,000 to $762,400. ◆ ◆

Several points concerning the interpretation of our results deserve particular attention. As we have noted, β_1 is the slope of the assumed line over the region of observation and indicates the linear change in $E(y\,|\,x)$ for a 1-unit change in x. Even if we do not reject the null hypothesis that the slope of the line β_1 equals zero, it does not necessarily mean that x and y are unrelated. In the first place, we must be concerned with the probability of committing a type II error—that is, of accepting the null hypothesis that the slope equals zero when this hypothesis is false. Second, it is possible that x and y might be perfectly related in a curvilinear, but not linear, manner. For example, Figure 11.5 depicts a curvilinear relationship between y and x over the domain of $x: a \leq x \leq f$. We note that a straight line would provide a good predictor of y if fitted over a small interval in the x domain, say, $b \leq x \leq c$. The resulting line is line 1. On the other hand, if we attempt to fit a line over the region $c \leq x \leq d$, then β_1 equals zero and the best fit to the data is the horizontal line 2. This result would occur even though all the points fell perfectly on the curve and y and x possessed a functional relation as defined in Section 11.2. We must take care in drawing conclusions if we do not find evidence to indicate that β_1 differs from zero. Perhaps we have chosen the wrong type of probabilistic model for the physical situation.

Note that the comments contain a second implication. **If the data provide values of**

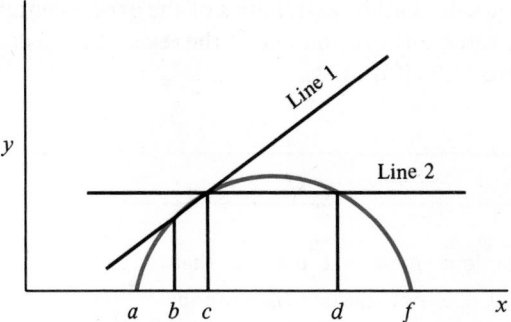

Figure 11.5   Curvilinear Relation

x in an interval $b \le x \le c$, then the calculated prediction equation is appropriate only over this region. Extrapolation in predicting y for values of x outside the region $b \le x \le c$ for the situation indicated in Figure 11.5 would result in a serious prediction error.

If the data present sufficient evidence to indicate that β_1 differs from zero, we do not conclude that the true relationship between y and x is linear. Undoubtedly, y is a function of a number of variables, which demonstrate their existence to a greater or lesser degree in terms of the random error ϵ that appears in the model. This random error, of course, is why we have been obliged to use a probabilistic model in the first place. Large errors of prediction imply curvatures in the true relation between y and x, the presence of other important variables that do not appear in the model, or both, as most often is the case. All we can say is that we have evidence to indicate that y changes as x changes and that we may obtain a better prediction of y by using x and the linear predictor than by simply using $\bar{y}$ and ignoring x. Note that this statement **does not imply a causal relationship between x and y**. Some third variable may have caused the change in both x and y, producing the relationship that we have observed.

The standard deviation of the estimator of β_1,

$$\sigma_{\hat{\beta}_1} = \frac{\sigma}{\sqrt{S_{xx}}} = \frac{\sigma}{\sqrt{\sum_{i=1}^{n}(x_i - \bar{x})^2}}$$

sheds information on the way to select the x values—that is, on the way to design an experiment to obtain the best estimate of the slope β_1 and of the mean value of y for a given value of x, $E(y \mid x)$, which will be discussed in Section 11.6. To illustrate, note that

$$\sum_{i=1}^{n}(x_i - \bar{x})^2$$

appears in the formula for $\sigma_{\hat{\beta}_1}$. This quantity measures the spread (variation) of the x values. The greater the spread of the x values, the larger will be $\sum_{i=1}^{n}(x_i - \bar{x})^2$. Now examine the formula for $\sigma_{\hat{\beta}_1}$. This observation provides the clue to constructing a good design of an experiment to find the best-fitting straight line to fit a set of data. Locate the majority of the x values at the extremities of the experimental region, half at each end. Locate a few x values near the middle of the region to detect curvature (similar to that shown in Figure 11.5), if it exists.

EXERCISES   Conceptual Level

11.24   What value of the slope β_1 would indicate that y and x are not linearly related?

11.25   In a test of $H_0: \beta_1 = 0$, does not rejecting H_0 mean that x and y are not related? Under what conditions might x and y be related, even though the null hypothesis is not rejected?

EXERCISES   Skill Level

11.26   Refer to Exercise 11.8. Test to see whether the data present sufficient evidence to indicate that x and y are linearly related. Use $\alpha = .05$.

11.27   Refer to Exercise 11.26. Find a 95% confidence interval for the slope of the regression line.

11.28   Refer to Exercise 11.9. Test to see whether the data present sufficient evidence to indicate that x and y are linearly related. Use $\alpha = .10$.

11.29   Refer to Exercise 11.28. Find a 99% confidence interval for β_1. Interpret this interval.

11.30   Refer to Exercise 11.10. Test to see whether the data present sufficient evidence to indicate that x and y are linearly related. Use $\alpha = .01$.

EXERCISES   Applications

11.31   For each of the following exercises, test to see whether the data present sufficient evidence to indicate that y and x are linearly related. That is, in each case test the hypothesis that $\beta_1 = 0$; use $\alpha = .05$.

  a.   Exercise 11.11   **b.**   Exercise 11.12   **c.**   Exercise 11.13

11.32   Frequently, the results of a statistical analysis are improperly utilized or are ignored because of an inability to interpret the results of such an analysis in the context of the problem under investigation. Refer to Exercise 11.31 and explain, in the context of Exercise 11.11, the decision you have reached as a result of the test of an hypothesis performed in Exercise 11.31. Repeat these instructions by explaining the results of Exercise 11.31 in the context of the problem described in Exercise 11.13.

 11.33   J. L. Treynor has developed a capital asset pricing model in which he uses a "characteristic line" device to evaluate competing investment funds. Treynor's characteristic line, the regression of the fund's rate of return to the average market rate of return, contains information about the fund's inherent risk. If the slope coefficient of the line is significantly different from zero, the fund is said to be sensitive to fluctuations in the securities market and thus is a risky investment. Funds with a slope coefficient near zero are more stable investments and hence less risky. The rates of return for a growth fund, the Penn Square Mutual Fund, and the average market rate of return (averaged over five groups of stocks monitored by *Value Line Investment Surveys*) for the period 1977 through 1986 were as given in Table 11.4.

a. Find the characteristic line for the Penn Square Mutual Fund (that is, the least squares line relating the return on Penn Square to the market rate of return).

b. Plot the points, and graph the least squares line as a check on your calculations.

c. Describe the risk characteristics of the Penn Square Mutual Fund (that is, test the hypothesis that $\beta_1 = 0$; use $\alpha = .05$).

Table 11.4   Rate of Return Data for Exercise 11.33

					Year					
Item	1977	1978	1979	1980	1981	1982	1983	1984	1985	1986
Penn Square Mutual Fund	12.8	26.6	.5	29.3	20.7	1.1	25.6	21.6	4.1	−6.3
Average market return	10.2	27.0	−4.7	27.5	25.0	.9	23.4	28.0	9.6	5.8

Source: Data from Wiesenberger Financial Services, *Investment Companies* (1987); *Value Line's Investment Survey*, January 22, 1988.

11.34 Find a 95% confidence interval for the slope of the characteristic line for the Penn Square Mutual Fund of Exercise 11.33. In the context of the use of a characteristic line in a capital asset pricing model, explain the confidence interval you have derived.

11.35 Find a 90% confidence interval for the slope of the line relating overhead costs to production in Exercise 11.12. Interpret your results so that they might be understood by a production manager who is unfamiliar with modern statistics.

11.36 A marketing research experiment was conducted to study the relationship between the length of time necessary for a buyer to reach a decision and the number of alternative package designs of a product presented. Brand names were eliminated from the packages to reduce the effects of brand preferences. The buyers made their selections by using the manufacturer's product descriptions on the packages as the only buying guide. The length of time necessary to reach a decision was recorded for 15 participants in the marketing research study; see the accompanying table.

Length of Decision Time (Seconds)	5, 8, 8, 7, 9	7, 9, 8, 9, 10	10, 11, 10, 12, 9
Number of Alternatives	2	3	4

a. Find the least squares line appropriate for these data.

b. Plot the points, and graph the line as a check on your calculations.

c. Calculate s^2.

d. Do the data present sufficient evidence to indicate that the length of decision time is linearly related to the number of alternative package designs? (Test at the $\alpha = .05$ level of significance.)

11.6

Estimating $E(y|x)$, the Expected Value of y for a Given Value of x

In Chapters 8 and 9 we studied methods for estimating a population mean μ and encountered numerous practical applications of these methods in the examples and exercises. Now let us consider a generalization of this problem.

Estimating the mean value of y for a given value of x [that is, estimating $E(y|x)$] can be a very important practical problem. A corporate safety director might wish to estimate the mean number of accidents (of a particular type) given the number of hours of safety education each employee receives. Or a company personnel director might wish to estimate the mean number of years a new employee will stay with the company given the score on a test designed to test the employee's job compatibility. If a corporation's profit y is linearly related to advertising expenditures x, the marketing director may wish to estimate the mean profit for a given expenditure x. For example, if the corporation invests \$10,000 in advertising, what can it expect the mean sales volume to be? Finding a confidence interval for $E(y|x)$ will be the topic of this section.

Let us assume that x and y are linearly related according to the probabilistic model defined in Section 11.2 and therefore that $E(y|x) = \beta_0 + \beta_1 x$ represents the expected value of y for a given value of x. The fitted line

$$\hat{y} = \hat{\beta}_0 + \hat{\beta}_1 x$$

attempts to estimate the line of means $E(y|x)$ (that is, to estimate β_0 and β_1). Thus, $\hat{y}$ can be used to estimate the expected value of y as well as to predict some value of y that might be observed in the future. It seems quite reasonable to assume that the errors of estimation and prediction differ for these two cases. Consequently, the two estimation procedures differ. In this section we consider the estimation of the expected value of y for a given value of x.

Observe the two lines in Figure 11.6. The first line represents the line of means

$$E(y|x) = \beta_0 + \beta_1 x$$

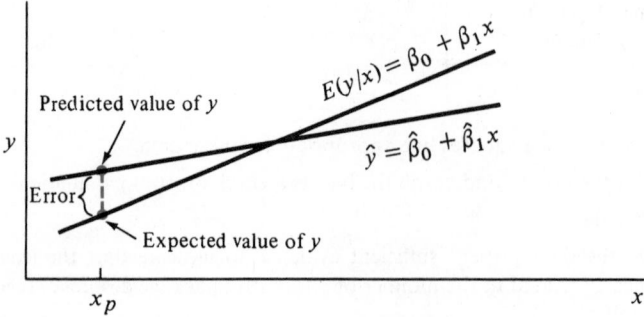

Figure 11.6   Expected and Predicted Values of y

and the second is the fitted prediction equation

$$\hat{y} = \hat{\beta}_0 + \hat{\beta}_1 x$$

We observe from the figure that the error in estimating the expected value of y when $x = x_p$ is the deviation between the two lines above the point x_p. Also, this error increases as we move to the endpoints of the interval over which x has been measured. It can be shown that the predicted value,

$$\hat{y} = \hat{\beta}_0 + \hat{\beta}_1 x$$

is an unbiased estimator of $E(y|x)$—that is, $E(\hat{y}) = \beta_0 + \beta_1 x$—and that $\hat{y}$ is normally distributed, with standard deviation

$$\sigma_{\hat{y}} = \sqrt{\sigma^2 \left[\frac{1}{n} + \frac{(x_p - \bar{x})^2}{S_{xx}} \right]}$$

The corresponding estimated standard error of $\hat{y}$, denoted by $s_{\hat{y}}$, uses s^2 in place of σ^2 in the expression above.

The results outlined above may be used to test an hypothesis about the average or expected value of y for a given value of x, say x_p.* (We may also, of course, test an hypothesis concerning the y intercept β_0, which is the special case where $x_p = 0$.) The null hypothesis is

$$H_0: E(y|x = x_p) = E_0$$

where E_0 is the hypothesized numerical value of $E(y)$ when $x = x_p$. Once again, it can be shown that when H_0 is true, the quantity

$$t = \frac{\hat{y} - E_0}{s_{\hat{y}}} = \frac{\hat{y} - E_0}{\sqrt{s^2 \left[\frac{1}{n} + \frac{(x_p - \bar{x})^2}{S_{xx}} \right]}}$$

follows a Student's t distribution in repeated sampling with $(n - 2)$ degrees of freedom. The statistical test is conducted in exactly the same manner as the other t test discussed previously.

A Test Concerning the Expected Value of y

> Null hypothesis $H_0: E(y|x = x_p) = E_0$.
>
> Alternative hypothesis: Specified by the experimenter, depending on the values of $E(y|x)$ that he or she wishes to detect.
>
> Test statistic: $t = \dfrac{\hat{y} - E_0}{\sqrt{s^2 \left[\dfrac{1}{n} + \dfrac{(x_p - \bar{x})^2}{S_{xx}} \right]}}$.
>
> Rejection region: See the critical values of t, Table 4 in the Appendix, for $(n - 2)$ degrees of freedom.

---

* See cautionary comments in Section 11.5. Predicting y for x_p outside the range of x is called **extrapolation**. It is best if x_p lies within the range of the observed values of x.

The corresponding confidence interval, with confidence coefficient $(1 - \alpha)$, for the expected value of y given $x = x_p$ is given in the next display.

A Confidence Interval for $E(y|x)$

$$\hat{y} \pm t_{\alpha/2} \sqrt{s^2 \left[\frac{1}{n} + \frac{(x_p - \bar{x})^2}{S_{xx}} \right]}$$

EXAMPLE 11.5   Using the data of Table 11.1, find a 95% confidence interval for the expected monthly sales volume for an advertising expenditure of $x = 1.0$ ($10,000).

SOLUTION   To estimate the mean monthly sales volume for an advertising expenditure $x_p = 1.0$, we use

$$\hat{y} = \hat{\beta}_0 + \hat{\beta}_1 x_p$$

to calculate $\hat{y}$, the estimate of $E(y|x = 1.0)$. Then using values calculated in previous examples, we find

$$\hat{y} = 46.49 + (52.57)(1.0) = 99.06 \qquad \text{or} \qquad \$990{,}600$$

The formula for the 95% confidence interval is

$$\hat{y} \pm t_{.025} \sqrt{s^2 \left[\frac{1}{n} + \frac{(x_p - \bar{x})^2}{S_{xx}} \right]}.$$

Substituting the appropriate quantities into this expression, we find that the 95% confidence interval for the expected (mean) monthly sales volume, given an advertising expenditure of 1.0, is

$$99.06 \pm (2.306) \sqrt{46.75 \left[\frac{1}{10} + \frac{(1.0 - .94)^2}{.444} \right]}$$

Performing the indicated calculations, we have

$$99.06 \pm 5.18 \qquad \text{or} \qquad 93.88 \text{ to } 104.24$$

Recall that each unit of sales volume represents $10,000. Therefore, we estimate that the mean monthly sales volume for the population of months for which the corporation invests $10,000 in advertising falls in the interval from $938,800 to $1,042,400.

◆ ◆

EXERCISES   Conceptual Level

11.37   Extrapolation occurs when we predict a value of y for a value of x that is outside the experimental range of x. Why should extrapolation be avoided?

11.38   How does the width of the confidence interval for $E(y|x_p)$ change as the distance between x_p and $\bar{x}$ increases?

EXERCISES   Skill Level

11.39   Refer to Exercise 11.8. Estimate the expected value of y when $x = 0$. Use a 95% confidence interval.

11.40   Refer to Exercise 11.9. Estimate the expected value of y when $x = -1$. Use a 90% confidence interval.

11.41   Refer to Exercise 11.10. Estimate the expected value of y when $x = 4.5$. Use a 99% confidence interval.

EXERCISES   Applications

11.42   Refer to Exercise 11.11. Estimate the expected profit per sales dollar earned by a construction company on projects for which the construction superintendent possesses five years of experience. Use a 90% confidence interval.

11.43   Refer to Exercise 11.12. Estimate the mean overhead cost associated with the production of 55,000 units ($x = 5.5$). Use a 95% confidence interval.

11.44   Refer to Exercise 11.13. Using a 95% confidence interval, estimate the expected per capita military spending during years when the GNP is $3,000 in the United States.

11.45   As you will learn in Chapters 13 and 14, a time series analysis involves the investigation of a process over time, in which time is assumed to be an independent variable. Suppose the accompanying data represent the investment in new-product development by a manufacturing company over a 10-year period.

Year, t	1	2	3	4	5	6	7	8	9	10
Investment, y ($\times$ \$10,000)	1.7	2.3	3.1	2.9	3.2	3.4	3.1	3.8	3.6	4.0

a.   Find the least squares lines for estimating investment in new-product development by the company as a function of time.

b.   Explain the assumptions necessary to use the least squares line as a prediction equation for the purpose of, say, predicting investment in new-product development in year $t = 14$.

c.   Estimate the company's expected investment in new-product development in year $t = 11$, using a 95% confidence level.

11.46   The rising price of petroleum products has led to continually increasing costs to the manufacturer for shipping goods to the market. These costs have led manufacturers to seek cheaper, but often slower, means of goods shipment, such as substituting rail freight for air freight services. In a study of shipping costs incurred by his firm, a company controller has randomly selected $n = 9$ air freight invoices from current shippers in order to estimate the relationship between shipping costs and distance for a given volume of goods. The results of his sample are given in the table.

Distance ($\times$ 100 Miles)	6	13	27	15	9	11	21	14	12
Invoice Charges	\$49	\$93	\$159	\$115	\$66	\$90	\$139	\$98	\$88

a. Find the least squares line for estimating invoice charges (y) from distance (x) when using current air freight shippers.

b. Logically, if the distance traveled is 0 miles, the invoice charge should be $0, since no service has been rendered. Explain why the least squares line does not go through the origin. Should it?

c. Estimate the mean invoice charge for air freight shipped a distance of 1,700 miles ($x = 17$), using a 90% confidence interval.

11.7

Predicting a Particular Value of *y* for a Given Value of *x*

Suppose that the prediction equation obtained for the 10 measurements in Table 11.1 were used to predict the corporation's sales volume for a month selected at random. Although the expected value of y for a particular value of x is of interest for our example (Table 11.1), we are primarily interested in *using* the prediction equation $\hat{y} = \hat{\beta}_0 + \hat{\beta}_1 x$, based on our observed data, to predict the sales volume for a month during which the corporation is or has been in operation. If the corporation's advertising expenditures during the month of interest are x_p, we intuitively see that the error of prediction (the deviation between $\hat{y}$ and the actual sales volume y that will occur during that month) is composed of two elements. The quantity $(y - \hat{y})$ equals the deviation between $\hat{y}$ and the expected value of y, described in Section 11.6 (and shown in Figure 11.6), plus a random error ϵ that represents the difference between the actual value of y and its expected value (see Figure 11.7). Thus, the variability in the error for predicting a single value of y exceeds the variability for estimating the expected value of y.

It can be shown that the variance of the error $(y - \hat{y})$ of predicting a particular value of y when $x = x_p$ is

$$\sigma^2_{(y-\hat{y})} = \sigma^2_y + \sigma^2_{\hat{y}} = \sigma^2 + \sigma^2 \left[\frac{1}{n} + \frac{(x_p - \bar{x})^2}{S_{xx}} \right]$$

$$= \sigma^2 \left[1 + \frac{1}{n} + \frac{(x_p - \bar{x})^2}{S_{xx}} \right]$$

When n is very large, the second and third terms in the brackets become small and the variance of the prediction error approaches σ^2. These results may be used to construct the prediction interval for y, given $x = x_p$. The confidence coefficient for the prediction interval is $(1 - \alpha)$.

A $(1 - \alpha)$100% Prediction Interval for *y*

$$\hat{y} \pm t_{\alpha/2} s_{(y-\hat{y})}$$

where

$$s_{(y-\hat{y})} = \sqrt{s^2 \left[1 + \frac{1}{n} + \frac{(x_p - \bar{x})^2}{S_{xx}} \right]}$$

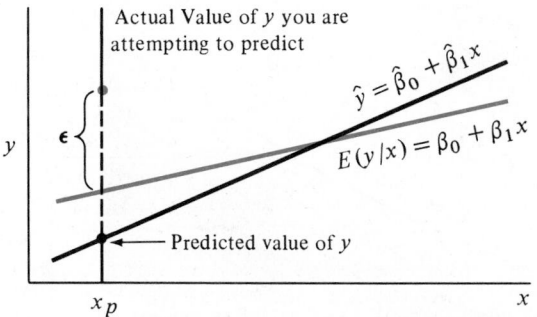

Figure 11.7   Error in Predicting a Particular Value of y

EXAMPLE 11.6   Find a 95% prediction interval for the next month's sales for the corporation if the advertising expenditure is $10,000, assuming that other economic conditions remain approximately the same as during the months included in Table 11.1.

SOLUTION   If in a particular month the advertising expenditures were $10,000, then $x_p = 1.0$, and we would predict that the sales volume would be

$$99.06 \pm (2.306)\sqrt{46.75\left[1 + \frac{1}{10} + \frac{(1.0 - .94)^2}{.444}\right]} \qquad \text{or} \qquad 99.06 \pm 16.60$$

or 82.46 to 115.66. Keep in mind that each unit of sales volume represents $10,000. Then the 95% prediction interval for the next month's sales volume is $990,600 $\pm$ $166,000, or

$824,600 to $1,156,600

Note that in a practical situation we would probably have data on the sales volume and advertising expenditures from more than the $n = 10$ months indicated in Table 11.1. More data would reduce somewhat the width of the prediction interval by decreasing the quantity under the square root sign in the expression above.

Again, note the distinction between the confidence interval for $E(y\,|\,x)$ discussed in Section 11.6 and the prediction interval presented in this section. The quantity $E(y\,|\,x)$ is a mean, a parameter of a population of y values, and y is a random variable that varies in a random manner about $E(y\,|\,x)$. The mean value of y when $x = 1.0$ is vastly different from some value of y chosen at random from the set of all y values for which $x = 1.0$. To make this distinction when making inferences, we always estimate the value of a parameter and predict the value of a random variable. As noted in our earlier discussion and as shown in Figures 11.7 and 11.8, the error of predicting y is different from the error of estimating $E(y\,|\,x)$. This point is evident in the difference in widths of the prediction and confidence intervals.

A graph of the confidence interval for $E(y\,|\,x)$ and the prediction interval for a particular value of y for the data of Table 11.1 is shown in Figure 11.8. The plot of the confidence interval is shown by solid lines; the prediction interval is identified by

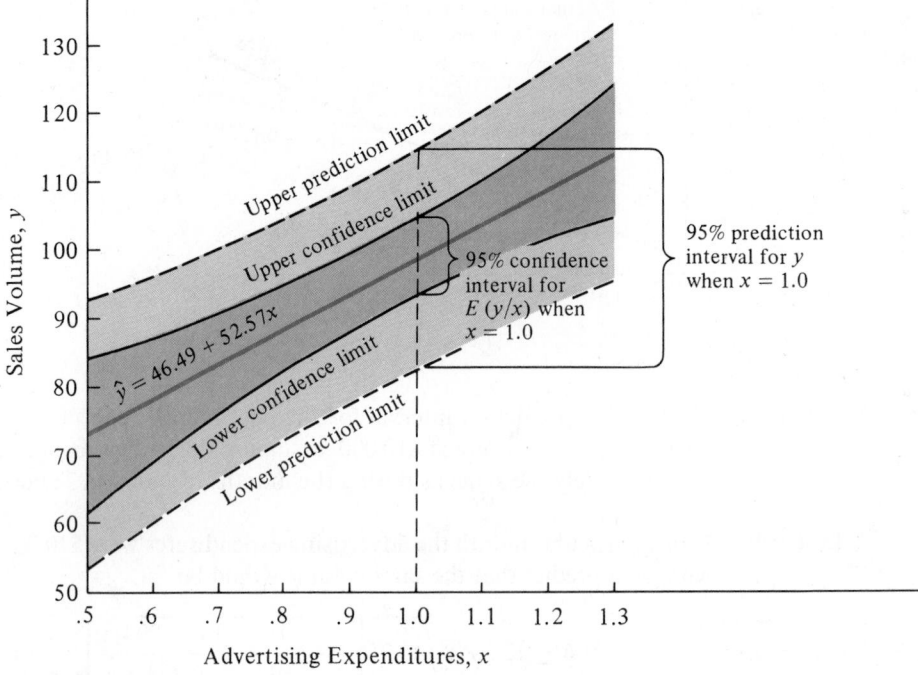

Figure 11.8   Confidence Intervals for $E(y|x)$ and Prediction Intervals for y from Data of Table 11.1

dashed lines. Note how the widths of the intervals increase as you move to the right or left of $\bar{x} = .94$. In particular, note the confidence interval and prediction interval for $x = 1.0$, which were calculated in Examples 11.5 and 11.6.

EXERCISES   Conceptual Level

11.47   Why does the variability in predicting a particular value of y when $x = x_p$ exceed the variability in estimating the expected value of y when $x = x_p$?

11.48   Is there a difference between the width of the confidence interval for $E(y|x = x_p)$ and the width of the prediction interval for a particular value of y when $x = x_p$? If so, how do they differ?

EXERCISES   Skill Level

11.49   Refer to Exercise 11.8. Find a 95% prediction interval for a particular value of y when $x = 0$. How does the width of this interval compare with the width of the interval computed in Exercise 11.39?

11.50   Refer to Exercise 11.9. Find a 90% prediction interval for a particular value of y when $x = -1$.

11.51   Refer to Exercise 11.10. Find a 95% prediction interval for a particular value of y when $x = 4$.

EXERCISES Applications

11.52 A life insurance company has introduced a term life insurance policy for its current whole life policyholders to compensate for the loss in whole life coverage due to the inflationary economy. In an examination of the relationship between annual income and current coverage, the information presented in Table 11.5 was extracted from the files of $n = 14$ policyholders.

a. Find the least squares line that best describes the relationship between annual income and whole life insurance coverage.

b. Find a 95% confidence interval for the mean annual income for policyholders with $50,000 in whole life insurance coverage.

c. Suppose a certain individual possesses a $50,000 whole life insurance policy with the company. Find a 95% prediction interval for that individual's annual income.

d. Explain the difference between the intervals computed in part b and part c. Is a prediction interval always wider than an interval estimate for the expected values of y at a given value of x? Explain.

Table 11.5   Income and Coverage Data for Exercise 11.52

Annual Income, y (× $1,000)	Whole Life Coverage, x (× $1,000)	Annual Income, y (× $1,000)	Whole Life Coverage, x (× $1,000)
27.5	20	44.3	50
38.4	30	56.2	100
46.2	60	29.3	36
22.1	20	18.4	24
19.3	10	27.6	30
29.4	30	54.8	80
16.8	20	33.3	40

11.53 In a follow-up study the life insurance company mentioned in Exercise 11.52 examined the relationship between life insurance coverage in the United States and the amount of disposable income (as opposed to total income). The data in Table 11.6 show the average amount of life insurance and the average disposable income (in thousands of dollars) for the years 1970 to 1985.

a. Find the least squares line for predicting average life insurance coverage using average amount of disposable income.

b. Find a 99% confidence interval for the average life insurance coverage over all years in which the average disposable income is $25,000.

c. Find a 99% prediction interval for the average life insurance coverage in a particular year in which the average disposable income is $25,000. Explain the difference between this interval and the interval calculated in part b.

d. Is it appropriate to use the data for the period 1970 to 1985 to predict average life insurance coverage in 1992 by using average disposable income? Why or why not?

11.54 Refer to Exercise 11.12. Suppose that the production during a certain period is 55,000 units. Find a 95% prediction interval for y, the overhead cost incurred during the period. Compare this prediction interval with the confidence interval for $E(y \mid x = 5.5)$ computed in Exercise 11.43.

Table 11.6   Insurance and Income Data for Exercise 11.53

Year	Life Insurance	Disposable Income	Year	Life Insurance	Disposable Income
1970	20.7	10.3	1978	35.1	17.9
1971	21.7	10.8	1979	38.5	19.6
1972	22.9	11.4	1980	41.5	21.4
1973	24.4	12.6	1981	45.7	22.7
1974	26.5	13.3	1982	49.3	23.9
1975	28.1	14.4	1983	54.2	25.5
1976	30.1	15.3	1984	58.7	27.5
1977	32.4	16.5	1985	63.4	29.6

Source: Data from U.S. Department of Commerce, Bureau of the Census, *Statistical Abstract of the United States*, 107th ed. (Washington, D.C., 1988), p. 586.

11.55 Refer to Exercise 11.13. Suppose that the GNP in the United States is $3,000 next year. Find a 95% prediction interval for next year's per capita military expenditures.

11.56 Refer to Exercise 11.36. Suppose a particular buyer is shown three alternative package designs. Find a 90% prediction interval for y, the length of time for the buyer to reach a decision on choice of design.

11.57 An experiment was conducted in a supermarket to observe the relationship between the amount of display space allotted to a brand of coffee (brand A) and its weekly sales. The amount of space allotted to brand A was varied over 3-, 6-, and 9-square-foot displays in a random manner over 12 weeks; the space allotted to competing brands was maintained at a constant 3 square feet for each. The data are given in the table.

Weekly Sales, y (Dollars)	526	421	581	630	412	560	434	443	590	570	346	672
Space Allotted, x (Square Feet)	6	3	6	9	3	9	6	3	9	6	3	9

a. Find the least squares line appropriate for these data.

b. Plot the points and graph the least squares line as a check on your calculations.

c. Calculate s^2.

d. Find a 95% confidence interval for the mean weekly sales if the space allotted is 6 square feet.

e. Suppose you intend to allot 6 square feet next week. Find a 95% prediction interval for the weekly sales. Explain the difference between this interval and the interval obtained in part d.

11.8

A Coefficient of Correlation

Sometimes, we wish to obtain an indicator of the strength of the linear relationship between two variables y and x that is independent of their respective scales of

linear correlation   measurement. We call this indicator a measure of the **linear correlation** between y and x.

A measure of linear correlation commonly used in statistics is called the *Pearson product-moment coefficient of correlation* between y and x. This quantity, denoted by the symbol r, is computed as shown in the display.

Pearson Product-Moment Coefficient of Correlation

$$r = \frac{S_{xy}}{\sqrt{S_{xx}S_{yy}}}$$

EXAMPLE 11.7   Calculate the coefficient of correlation for the advertising expenditure and sales volume data of Table 11.1.

SOLUTION   The coefficient of correlation for the advertising expenditure and sales volume data of Table 11.1 may be obtained by using the formula for r and the quantities

$$S_{xy} = 23.34 \qquad S_{xx} = .444 \qquad S_{yy} = 1{,}600.9$$

which were computed previously. Then

$$r = \frac{S_{xy}}{\sqrt{S_{xx}S_{yy}}} = \frac{23.34}{\sqrt{.444(1{,}600.9)}} = .88$$

◆◆

A study of the coefficient of correlation r yields rather interesting results and explains the reason for its selection as a measure of linear correlation. We note that the denominators used in calculating r and $\hat{\beta}_1$ will always be positive since they both involve sums of squares of numbers. We also note that the numerator used in calculating r is identical to the numerator of the formula for the slope $\hat{\beta}_1$. Therefore, the coefficient of correlation r will assume exactly the same sign as $\hat{\beta}_1$ and will equal zero when $\hat{\beta}_1 = 0$. **Thus, $r = 0$ implies no linear correlation between y and x. A positive value for r implies that the line slopes upward to the right; a negative value indicates that it slopes downward to the right.**

Figure 11.9 shows six typical scatter diagrams and their associated correlation coefficients. Note that **$r = 0$ implies no linear correlation**, not simply "no correlation." A pronounced curvilinear pattern may exist, as in Figure 11.9b, but its linear correlation coefficient may equal 0. In general, we can say that r measures the linear association of the two variables y and x. When $r = 1$ or -1, all the points fall on a straight line; when $r = 0$, they are scattered and give no evidence of a *linear* relationship. Any other value of r suggests the degree to which the points tend to be linearly related.

The interpretation of nonzero values of r may be obtained by comparing the errors of prediction for the prediction equation

$$\hat{y} = \hat{\beta}_0 + \hat{\beta}_1 x$$

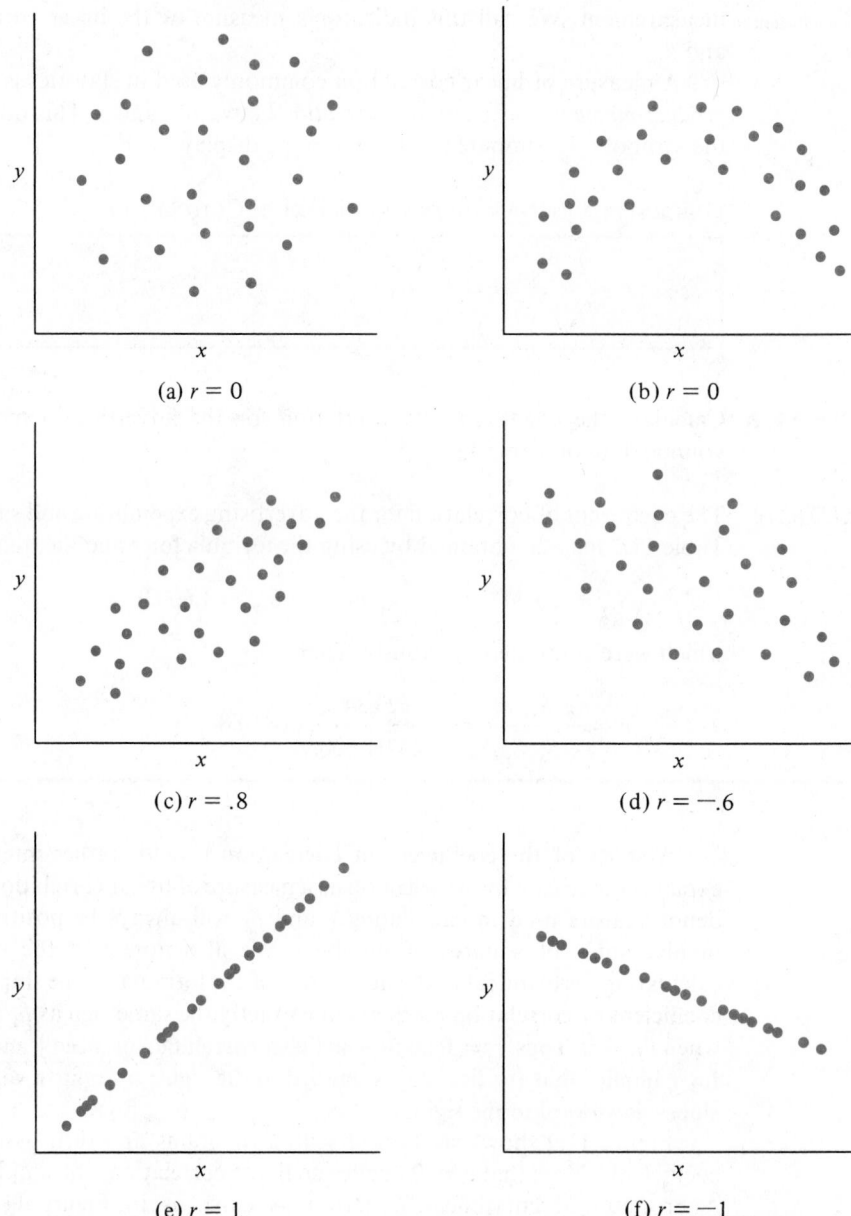

(a) $r = 0$

(b) $r = 0$

(c) $r = .8$

(d) $r = -.6$

(e) $r = 1$

(f) $r = -1$

Figure 11.9   Some Typical Scatter Diagrams and Their Associated Correlation Coefficients

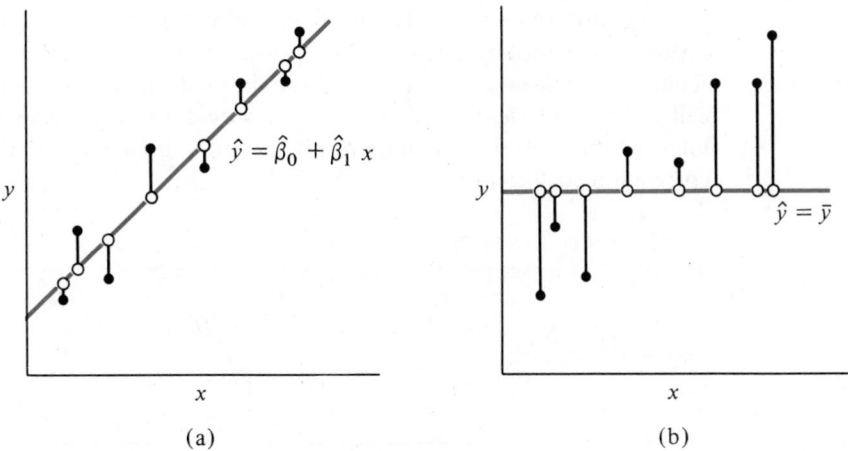

Figure 11.10   Two Models Fitted to the Same Data

with the predictor of y, $\bar{y}$, that would be employed if x were ignored. Figures 11.10a and 11.10b show the lines $\hat{y} = \hat{\beta}_0 + \hat{\beta}_1 x$ and $\hat{y} = \bar{y}$ fit to the same set of data. Certainly, if x is of any value in predicting y, then SSE, the sum of squares of deviations of y about the linear model, should be less than the sum of squares of deviations about the predictor $\bar{y}$, which is

$$S_{yy} = \sum_{i=1}^{n} (y_i - \bar{y})^2$$

Indeed, we see that SSE can *never* be larger than

$$S_{yy} = \sum_{i=1}^{n} (y_i - \bar{y})^2$$

because

$$\text{SSE} = S_{yy} - \hat{\beta}_1 S_{xy} = S_{yy} - \left(\frac{S_{xy}}{S_{xx}}\right) S_{xy} = S_{yy} - \frac{(S_{xy})^2}{S_{xx}}$$

Therefore, SSE is equal to S_{yy} minus a non-negative quantity. Consequently, SSE must always be less than or equal to S_{yy}.

Furthermore, with the aid of a bit of algebraic manipulation, we can show that

$$r^2 = 1 - \frac{\text{SSE}}{S_{yy}} = \frac{S_{yy} - \text{SSE}}{S_{yy}}$$

In other words, r^2 lies in the interval

$$0 \le r^2 \le 1$$

and r will equal $+1$ or -1 only when all the points fall exactly on the fitted line—that is, when SSE equals zero.

Actually, we see that r^2 is **equal to the ratio of the reduction in the sum of squares of deviations obtained by using the linear model to the total sum of squares of deviations about the sample mean $\bar{y}$, which would be the predictor of y if x were ignored. Thus r^2,** called the *coefficient of determination*, would seem to give a more meaningful interpretation of the strength of the relation between y and x than would the correlation coefficient r.

Coefficient of Determination

$$r^2 = \frac{S_{yy} - \text{SSE}}{S_{yy}} = \frac{\sum_{i=1}^{n}(y_i - \bar{y})^2 - \text{SSE}}{\sum_{i=1}^{n}(y_i - \bar{y})^2}$$

You will observe that the sample correlation coefficient r is an estimator of a population correlation coefficient ρ (Greek letter rho), which would be obtained if the coefficient of correlation were calculated by using all the points in the population. A formal discussion of a test of an hypothesis concerning the value of ρ, as well as appropriate estimation procedures, is omitted here. Ordinarily, we would be interested in testing the null hypothesis that $\rho = 0$. In fact, a common test statistic for testing the null hypothesis $\rho = 0$ is the Student's t statistic:

$$t = \frac{r\sqrt{n-2}}{\sqrt{1-r^2}}$$

where t is based on $(n-2)$ degrees of freedom. Using ordinary algebra, we can show that

$$t = \frac{r\sqrt{n-2}}{\sqrt{1-r^2}} = \frac{\hat{\beta}_1}{s/\sqrt{S_{xx}}}$$

and hence that the t test of the null hypothesis $\rho = 0$ is equivalent to the t test of Section 11.5, the test of the null hypothesis $\beta_1 = 0$. Although this test is sometimes of interest, most investigations will have as their objective the estimation of the mean of y or the prediction of a particular value of y for a given value of x.

Although r gives a rather nice measure of the goodness of fit of the least squares line to the fitted data, its use in making inferences concerning ρ appears to be of dubious practical value in many situations. It seems unlikely that a phenomenon y observed in the physical sciences and especially in economics would be a function of a single variable. Thus, the correlation coefficient between the monthly sales volume of a firm and any one variable probably would be quite small and of questionable value. A larger reduction in SSE could possibly be obtained by constructing a predictor of y based on a set of variables $x_1, x_2, \ldots, x_k$.

One further reminder is worthwhile concerning the interpretation of r. It is not uncommon for researchers in some fields to speak proudly of sample correlation coefficients r in the neighborhood of .5 (and, in some cases, as low as .1) as being indicative of a "relation" between y and x. Certainly, even if these values were accurate estimates of ρ, only a very weak relation would be indicated. A value $r = .5$ implies that the use of x in predicting y reduces the sum of squares of deviations about the prediction line by only $r^2 = .25$, or 25%. A correlation coefficient of $r = .1$ implies only an $r^2 = .01$, or a 1% reduction in the total sum of squares of deviations that could be explained by x.

If the linear coefficient of correlation between y and each of two variables x_1 and x_2 were calculated to be .4 and .5, respectively, it does not follow that a predictor using both variables would account for a $(.4)^2 + (.5)^2 = .41$, or a 41%, reduction in the sum of squares of deviations. Actually, x_1 and x_2 might be highly correlated and therefore contribute virtually the same information for the prediction of y.

Finally, we remind you that r is a measure of *linear correlation* and that x and y could be perfectly related by a *curvilinear* function even when the observed value of r is equal to zero.

EXERCISES   Conceptual Level

11.58   What type of relationship between x and y is measured by the coefficient of correlation?

11.59   Does the scale of measurement used for x and y have any effect on the value of the correlation coefficient?

11.60   What values can the correlation coefficient r assume? What is implied when the value of r is very close to zero?

11.61   What value does r assume if all the sample points fall on the same straight line with positive slope? With negative slope?

11.62   What additional information is provided by the coefficient of determination?

EXERCISES   Skill Level

11.63   A linear regression analysis for $n = 15$ pairs of observations produced a sample value of $r = .94$.

    a.   Calculate r^2 and use this value to describe the importance of x as a predictor of y.

    b.   Do the data present sufficient evidence to indicate a significant correlation between x and y? Use $\alpha = .05$. (*Hint:* Test the hypothesis $H_0 : \rho = 0$).

11.64   Refer to Exercise 11.8.

    a.   Calculate the sample correlation coefficient r.

    b.   Calculate r^2. By what percentage is the sum of squares of deviations of y about the mean reduced by using $\hat{y}$ rather than $\bar{y}$ as a predictor of y?

    c.   Test the hypothesis $H_0 : \rho = 0$ versus $H_a : \rho \neq 0$, using the test statistic $t = r\sqrt{n-2}/\sqrt{1-r^2}$. Do the data present sufficient evidence to indicate a significant correlation between x and y? Find the observed level of significance for the test.

    d.   Compare the results of part c with the results of the test performed in Exercise 11.26. Are the results identical?

11.65   Calculate the sample correlation coefficient r for the data in Exercise 11.9. Do these data suggest that the underlying population coefficient of correlation ρ is significantly different from zero? Use $\alpha = .05$.

11.66   Refer to Exercise 11.65. By what percentage is the sum of squared deviations of y reduced by using the auxiliary information provided by x rather than using $\bar{y}$ as a predictor of y?

EXERCISES   Applications

11.67   The Consumer Price Index (CPI) is an important economic indicator designed to characterize the living costs of all families in the United States. As the CPI increases, the cost of living also rises. Table 11.7 shows the CPI for the years 1979 through 1986, along with the number of cases of soft drinks consumed during the period.

Table 11.7   Data for Exercise 11.67

Year	CPI	Cases Consumed (Millions)
1979	11.3	4980
1980	13.5	5180
1981	10.4	5345
1982	6.2	5510
1983	3.2	5780
1984	4.3	6130
1985	3.6	6500
1986	1.9	6770

Source: Data from *Value Line's Investment Survey*, December 25, 1987; *Standard & Poor's Industry Surveys, Food and Beverages*, October 8, 1987.

a.   Compute the correlation r between CPI and soft drink consumption for the period 1979–1986.

b.   Do the data present sufficient evidence to indicate that there is a significant correlation between CPI and soft drink consumption? Test the hypothesis that $\rho = 0$, using $\alpha = .05$.

c.   In the context of the problem, explain the findings you derived in part b. Is there another variable that might be highly correlated with soft drink consumption?

11.68   Calculate the coefficient of correlation r between the annual rates of return for the Penn Mutual Fund and the average market rate of return given in Exercise 11.33. If you have a calculator that possesses a linear correlation routine, use your calculator to compute r from these data, and compare your results with those derived by using the formula for r given in Section 11.8.

11.69   Refer to Exercise 11.57. Compute the coefficient of determination r^2 for the data describing the relationship between the amount of display space allocated by a supermarket to a particular brand of coffee and its weekly sales. Discuss the meaning of this term in the context of the marketing problem presented in Exercise 11.57.

11.70   Refer to Exercise 11.12. By what percentage is the sum of squares of deviations in overhead costs reduced by using the prediction equation $\hat{y}$ rather than $\bar{y}$ as a predictor of y?

11.71 An independent variable that shows a strong negative relationship with y is as useful as one that exhibits a positive relationship. The important feature is the absolute magnitude of the correlation between y and x, not the direction of the relationship. Consider, for instance, interest rates and housing starts. Interest rates (x) provide an excellent leading indicator for predicting housing starts (y). As interest rates decline, housing starts increase, and vice versa. The data given in Table 11.8 represent the prevailing interest rates on new home mortgages and the recorded housing starts over a 10-year span.

  a. Find the least squares line to allow for the estimation of housing starts from interest rates. Plot the data points, and graph the least squares line as a check on your calculations.

  b. Calculate the correlation coefficient r for these data. Is the correlation between housing starts and interest rates significantly different from zero? Use $\alpha = .05$.

  c. By what percentage is the sum of squares of deviations of housing starts reduced by using interest rates as a predictor rather than using the average annual housing starts $\bar{y}$ as a predictor of y for these data?

  d. If economic indicators suggest that the prevailing interest rate on new home mortgages will be 10.5% next year, predict the number of housing starts during the year, using a 95% prediction interval.

Table 11.8   Data for Exercise 11.71

Year	Mortgage Rate	Number of Housing Starts (Millions)
1977	9.02	1.96
1978	9.56	2.00
1979	10.78	1.72
1980	12.66	1.30
1981	14.70	1.10
1982	15.14	1.06
1983	12.57	1.71
1984	12.38	1.77
1985	11.55	1.74
1986	10.17	1.82

Source: *Value Line's Investment Survey*, December 25, 1987.
Reprinted by permission.

11.9

The Additivity of Sums of Squares: An Analysis of Variance

An important property of a regression analysis is that it partitions the total sum of squares of deviations

$$S_{yy} = \sum_{i=1}^{n} (y_i - \bar{y})^2$$

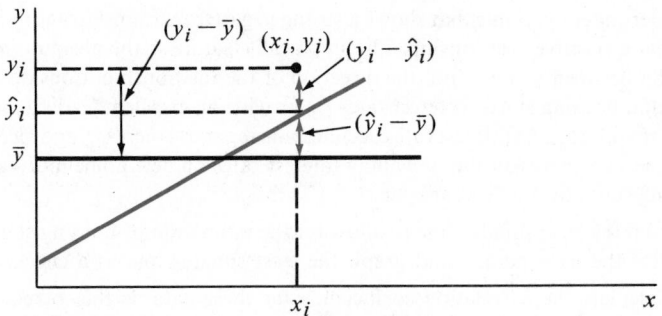

Figure 11.11 Partitioning of $(y_i - \bar{y})$ into $(y_i - \hat{y}_i)$ and $(\hat{y}_i - \bar{y})$

into two parts. One part is attributable to the sum of squares of deviations of the y values about the fitted line SSE. The other part represents the reduction in the sum of squares of deviations that results from information contributed by the auxiliary variable x.

To understand the partitioning of sums of squares, note that a regression analysis tends to partition the deviation of each measurement from its mean, $(y_i - \bar{y})$, into two parts. Thus,

$$(y_i - \bar{y}) = (y_i - \hat{y}_i) + (\hat{y}_i - \bar{y})$$

The partitioning of $(y_i - \bar{y})$ is shown in Figure 11.11.

Taking the sum of squared deviations over all observations for each expression within the partitioning of $(y_i - \bar{y})$, we can show that

$$S_{yy} = \sum_{i=1}^{n} (y_i - \bar{y})^2 = \sum_{i=1}^{n} (\hat{y}_i - \bar{y})^2 + \sum_{i=1}^{n} (y_i - \hat{y}_i)^2$$

Thus, the total sum of squares of y, denoted by the symbol S_{yy} (called Total SS), can be partitioned into two components.

Partitioning the Total Sum of Squares of y

> Total SS = SSR + SSE
>
> where
>
> $$SSR = \sum_{i=1}^{n} (\hat{y}_i - \bar{y})^2 = \text{sum of squares due to regression}$$
>
> which is the amount of total variation explained by the auxiliary variable x, and
>
> $$SSE = \sum_{i=1}^{n} (y_i - \hat{y}_i)^2 = \text{sum of squares for error}$$
>
> which is the amount of total variation unexplained by the auxiliary variable x.

The partitioning of the sums of squares is important for several reasons. It gives an important additive relationship for the sums of squares,

$$\text{Total SS} = \text{SSR} + \text{SSE}$$

that can be used to reduce computational effort, since we can compute two quantities and obtain the third by subtraction. Second, it helps explain the contribution of the auxiliary variable in providing information for the prediction of y. In fact, r^2, the coefficient of determination, is given as

$$r^2 = \frac{S_{yy} - \text{SSE}}{S_{yy}} = \frac{\text{Total SS} - \text{SSE}}{\text{Total SS}} = \frac{\text{SSR}}{\text{Total SS}} = \frac{\sum_{i=1}^{n} (\hat{y}_i - \bar{y})^2}{\sum_{i=1}^{n} (y_i - \bar{y})^2}$$

Finally, using the additivity of the sums of squares, a regression analysis can be presented as an analysis of variance. The quantities required for an ANOVA table have already been computed, either directly or indirectly. For example, in calculating SSE, we used the relationship

$$\text{SSE} = S_{yy} - \hat{\beta}_1 S_{xy}$$

or, equivalently, the relationship

$$S_{yy} = \hat{\beta}_1 S_{xy} + \text{SSE}$$

with

$$\text{SSR} = \hat{\beta}_1 S_{xy} = \frac{(S_{xy})^2}{S_{xx}}$$

The analysis of variance for a simple linear regression analysis is given in Table 11.9. A test of the null hypothesis of no linear relationship between y and x, given as $H_0: \beta_1 = 0$ versus $H_a: \beta_1 \neq 0$, is based on the statistic

$$F = \frac{\text{MSR}}{\text{MSE}}$$

which has an F distribution with $v_1 = 1$ and $v_2 = (n - 2)$ degrees of freedom when H_0 is true. If H_a is true and $\beta_1 \neq 0$, then MSR will tend to be larger than MSE. Therefore,

Table 11.9   ANOVA for Simple Linear Regression

Source	d.f.	SS	MS
Regression	1	SSR	$\text{MSR} = \text{SSR}/1$
Error	$(n - 2)$	SSE	$\text{MSE} = \text{SSE}/(n - 2)$
Total	$(n - 1)$	Total SS	

the null hypothesis is rejected when F exceeds a right-tailed critical value of the F distribution based on $v_1 = 1$ and $v_2 = (n - 2)$ degrees of freedom. The calculated value of F will be equal to the square of the value of the t statistic in Section 11.5 used for testing $H_0: \beta_1 = 0$. It is not difficult to see that

$$t^2 = \left(\frac{\hat{\beta}_1}{s/\sqrt{S_{xx}}} \right)^2 = \frac{SSR}{MSE} = F$$

since $s = \sqrt{MSE}$ and SSR with one degree of freedom is equal to MSR.

EXAMPLE 11.8   With the information given in Examples 11.1 and 11.2, use the analysis of variance procedure to test for a significant linear relationship between sales and advertising expenditures.

SOLUTION   In Examples 11.1 and 11.2 we found the following quantities:

$$S_{xx} = .444 \qquad S_{xy} = 23.34 \qquad S_{yy} = \text{Total SS} = 1600.9$$

Therefore,

$$SSR = \frac{(S_{xy})^2}{S_{xx}} = \frac{(23.34)^2}{.444} = 1226.93$$

and

$$SSE = \text{Total SS} - SSR = 1600.90 - 1226.93 = 373.97$$

The resulting analysis of variance is displayed in Table 11.10.

In testing $H_0: \beta_1 = 0$ versus $H_a: \beta_1 \ne 0$, we use

$$F = \frac{MSR}{MSE} = \frac{1226.93}{46.75} = 26.25$$

With $v_1 = 1$ and $v_2 = 8$ degrees of freedom, the observed value of $F = 26.25$ exceeds the critical value of $F_{.005} = 14.96$. The p-value for this test is less than .005, and hence, we can reject the null hypothesis with a value of α as small as .005. Notice that within rounding errors the value of $F = 26.25$ is equal to the square of the value $t = 5.12$ found in testing $H_0: \beta_1 = 0$ versus $H_a: \beta_1 \ne 0$ in Section 11.5. Once again, we see that

Table 11.10   ANOVA Table for the Data in Example 11.1

Source	d.f.	SS	MS	F
Regression	1	1,226.93	1,226.93	26.25
Error	8	373.97	46.75	
Total	9	1,600.90		

the square of a t statistic with v degrees of freedom has the same distribution as an F statistic with one numerator and v denominator degrees of freedom. You should remember this relationship, since some packaged computer programs report values of t statistics, but others report the equivalent values of F statistics.

11.10

Computer Printouts for a Regression Analysis

Although a simple linear regression problem can be analyzed as we have done in Sections 11.3 through 11.5, these same calculations can be implemented on a computer, using one of several packaged computer programs. Table 11.11 displays the output that resulted when the data in Table 11.1 were analyzed by using the REGRESS command in the MINITAB package. The values of y and x were stored in columns 1 and 2 of a data array, so that the regression equation found at the top of the printout is given as

$$C1 = 46.5 + 52.6\ C2$$

Table 11.11   MINITAB Output for the Sales Data of Table 11.1

```
MTB>REGRESS C1 1 C2

THE REGRESSION EQUATION IS
C1=46.5 + 52.6 C2

                              ST. DEV.      T-RATIO=
COLUMN      COEFFICIENT     OF COEF.      COEF/S.D.
              46.486          9.885          4.70
   C2         52.57          10.26          5.12

S = 6.837

R-SQUARED=76.6 PERCENT
R-SQUARED=73.7 PERCENT, ADJUSTED FOR D.F.

ANALYSIS OF VARIANCE

   DUE TO            DF            SS         MS=SS/DF
REGRESSION           1          1226.9        1226.9
RESIDUAL             8           374.0          46.7
TOTAL                9          1600.9

DURBIN-WATSON STATISTIC = 1.16

MTB>
```

The estimated intercept $\hat{\beta}_0$ and slope $\hat{\beta}_1$ are found in the column labeled COEFFICIENT. The estimated standard deviations of $\hat{\beta}_0$ and $\hat{\beta}_1$ are found in the column labeled ST. DEV. OF COEF.; values of the t statistics used in testing the null hypothesis "coefficient equals zero" are found in the column labeled T-RATIO = COEF/S.D. For example, in a test of the null hypothesis $H_0: \beta_1 = 0$ the value of the t statistic is 5.12, which from Example 11.3 has a p-value less than .01.

In addition to providing the value $S = 6.837$, the sample estimate of σ, the printout also provides the value of the coefficient of determination expressed as a percentage (R-SQUARED = 76.6 PERCENT).

The ANOVA table, found in the lower portion of the printout, provides the degrees of freedom, the sums of squares and mean squares corresponding to the sources of variation DUE TO REGRESSION and error (RESIDUAL). The value of the F statistic used in testing for significant regression is not given but is easily calculated as

$$F = \frac{\text{MSR}}{\text{MSE}} = \frac{1226.9}{46.7} = 26.27$$

This value differs slightly from the value of 26.25 found in Example 11.8 because of the rounded values of MSR and MSE.

The output for the general linear models procedure within the SAS package is given in Table 11.12. The analysis of variance portion of the printout is similar to that produced by the SAS analysis of variance procedure used in Chapter 10.

The source of variation due to the model is identical to that due to regression in that the model contains one independent variable, advertising expenditures (ADV). The value of the F statistic used in testing for significant regression is 26.25. The p-value for this test, given as PR > F, is .0009, indicating that we could reject $H_0: \beta_1 = 0$ with α as small as .0009 and that the value of $F = 26.25$ is very highly significant. The coefficient of determination, which assesses the practical significance of the regression, is given as R-SQUARE = .766398. The estimate of σ is reported under ROOT MSE as $s = 6.83715011$.

Table 11.12 SAS Output for the Sales Data of Table 11.1

```
                        GENERAL LINEAR MODELS PROCEDURE

DEPENDENT VARIABLE: SALES
                    SUM OF
SOURCE      DF      SQUARES        MEAN SQUARE    F VALUE   PR>F      R-SQUARE    C.V.
MODEL       1       1226.92702703  1226.92702703  26.25    0.0009    0.766398    7.1295
ERROR       8       373.97297297   46.74662162
CORRECTED                                                  ROOT MSE              SALES MEAN
  TOTAL     9       1600.90000000                          6.83715011            95.90000000


SOURCE      DF      TYPE I SS      F VALUE   PR>F      DF      TYPE III SS    F VALUE   PR>F
ADV         1       1226.92702703  26.25     0.0009    1       1226.92702703  26.25     0.0009


                            T FOR H0:     PR>|T|    STD ERROR OF
PARAMETER       ESTIMATE    PARAMETER=0             ESTIMATE
INTERCEPT       46.48648649    4.70      0.0015      9.88456628
ADV             52.56756757    5.12      0.0009      10.26085688
```

The lower portion of the printout gives the estimates of the intercept $\hat{\beta}_0$ and slope $\hat{\beta}_1$, the values of the t statistics used in testing the null hypothesis H0: PARAMETER $= 0$, and the two-tailed p-values (PR $> |T|$) for these tests. The column labeled STD ERROR OF ESTIMATE gives the values of $s_{\hat{\beta}_0}$ and $s_{\hat{\beta}_1}$, the estimated standard deviations of $\hat{\beta}_0$ and $\hat{\beta}_1$.

Although both computer printouts contain the same basic information, the SAS printout reports results with more decimal accuracy than MINITAB and contains the p-values associated with calculated test statistics. Having p-values available on the printout allows us to assess the significance of the testing results without resorting to tables.

11.11

Assumptions

The assumptions for a regression analysis are given in the display.

Assumptions for a Regression Analysis

1. The response y can be represented by the probabilistic model

   $$y = \beta_0 + \beta_1 x + \epsilon$$

2. x is measured without error.
3. ϵ is a random variable such that, for a given value of x,

   $$E(\epsilon) = 0 \qquad \text{and} \qquad \sigma_\epsilon^2 = \sigma^2$$

   and all pairs, ϵ_i, ϵ_j, are independent in a probabilistic sense.
4. ϵ possesses a normal probability distribution.

At first glance you might fail to understand the significance of the first assumption. Models, deterministic or otherwise, are, as the name implies, only models for real relationships that occur in nature. Consequently, model misspecification is always a possibility. Even if you have obtained a good fit to the data, a large error in prediction is possible if you use the model to predict y for some value of x outside the range of values used to fit the least squares equation. Of course, this problem will always occur if x is time and you attempt to forecast y at some point in the future. This problem occurs with any model for future time predictions; consequently, you make the forecast but keep the model limitations in mind.

The assumption that the variance of the ϵ's is constant and equal to σ^2 will not be true for all types of data. Furthermore, if x is time, it is possible that y values measured over adjacent time periods will tend to be dependent (an overly large value of y in 1988 might signal a large value of y in 1989). Substantial departure from either of these assumptions will affect the confidence coefficients of interval estimates and significance levels of tests of hypotheses.

If the normality assumption (4) is not satisfied, confidence coefficients and significance levels will not be what we expect them to be. However, modest departures from normality will not seriously disturb these values.

11.12

Summary

Although it was not stressed, you will observe that predicting the value of a random variable y was considered for the most elementary situation in Chapters 8 and 9. Thus if we possessed no information concerning variables related to y, the sole information available for predicting y would be provided by its probability distribution. If we were to select one value as representative of the population, we would most likely choose μ, or some other measure of central tendency. The estimation of the mean was considered in Chapters 8 and 9.

Chapter 11 was concerned with the problem of predicting y when auxiliary information is available on other variables, say $x_1, x_2, x_3, \ldots, x_k$, which are related to y and hence assist in its prediction. We have concentrated primarily on the problem of predicting y as a linear function of a single variable x, which provides the simplest extension of the prediction problem beyond that considered in Chapters 8 and 9. The more interesting case, where y is a linear function of a set of independent variables, is the subject of Chapter 12.

Supplementary Exercises

11.72 If you try to rent an apartment or buy a house, you will find that real estate representatives establish apartment rents and house prices on the basis of the square footage of the heated floor space. The data in Table 11.13 give the square footages and sales prices of $n = 12$ houses randomly selected from those sold in a small city.

a. Find the least squares line for these data. Use the MINITAB printout given in Table 11.14.

b. Do these data present sufficient evidence to indicate that y and x are linearly related? (Use $\alpha = .05$.)

Table 11.13   Data for Exercise 11.72

Square Feet, x	Price, y	Square Feet, x	Price, y
1,460	$68,700	1,977	$85,400
2,108	89,300	1,610	77,000
1,743	81,400	1,530	72,400
1,499	71,100	1,759	78,200
1,864	82,400	1,821	84,300
2,391	94,900	2,216	91,700

Table 11.14   MINITAB Output for the Data of Exercise 11.72

```
MTB > REGRESS C1 1 C2

THE REGRESSION EQUATION IS
C1 = 31206 + 27.4 C2

PREDICTOR          COEF          STDEV        T-RATIO
CONSTANT           31206          3389           9.21
C2                 27.406         1.828         14.99

S = 1793          R-SQ = 95.7%        R-SQ(aDJ) = 95.3%

ANALYSIS OF VARIANCE

SOURCE         DF            SS              MS
REGRESSION      1       722001728       722001728
ERROR          10        32138280         3213828
TOTAL          11       754140032

MTB > CORRELATION C1 C2

CORRELATION OF C1 AND C2 = 0.978
```

11.73   Continuing Exercise 11.72, find a 90% confidence interval for the expected selling price of houses with 1,800 square feet of heated floor space.

11.74   The owner of a particular home in the city from which the data of Exercise 11.72 were collected has decided to list her home for sale. If her home contains 1,800 square feet of heated floor space, find a 90% prediction interval for its selling price. Explain the difference between the estimates derived in Exercise 11.73 and this exercise.

11.75   Since long-run forecasts of temperature are better indicators of fuel use than are direct forecasts of heating oil sales and demand for other heating fuels, most distributors of fuels use the relationship between temperature and fuel sales to determine their proper levels of inventory. A regional distributor of heating oils has recorded the monthly sales volume and the average daily high temperature for 9 randomly selected months. The results are given in the table.

Sales Volume, y ($\times 100$ Gallons)	26.2	17.4	7.8	12.3	35.9	42.1	26.4	19.0	10.1
Average Daily High Temperature, x (°F)	46.5	54.6	65.2	62.3	41.9	38.6	43.7	52.0	59.8

a.   Find the least squares line appropriate for these data.

b.   Plot the points, and graph the least squares line as a check on your calculations.

c.   Calculate s^2.

d.   Do these data present sufficient evidence to indicate that fuel oil sales are linearly related to temperature? Test by using $\alpha = .05$.

11.76   Refer to Exercise 11.75. Partition the total sum of squares into its two components, regression and the sum of squares for error. What is the percentage of variation in monthly sales volume that is explained by the average daily high temperature?

 11.77 An experiment was conducted by a pharmaceutical manufacturer to observe the effect of an increase in temperature on the potency of an antibiotic. Three 1-oz portions of the antibiotic were stored for equal lengths of time at each of the following temperatures: 30°, 50°, 70°, and 90°F. The potency readings observed at the temperature of the experimental period are given in the table.

Potency Readings	38, 43, 29	32, 26, 33	19, 27, 23	14, 19, 21
Temperature	30	50	70	90

   a. Find the least squares equation appropriate for these data.

   b. Plot the points and graph the line as a check on your calculations.

   c. Calculate s^2.

 11.78 Refer to Exercise 11.77. Estimate the change in potency for a 1-unit change in temperature, using a 90% confidence interval.

 11.79 Refer to Exercise 11.77. Estimate the mean potency corresponding to a temperature of 50°F, using a 90% confidence interval.

 11.80 Refer to Exercise 11.77. Suppose a batch of the antibiotic was stored at 50°F for the same length of time as the experimental period. Predict the potency of the batch at the end of the storage period, using a 90% prediction interval.

 11.81 Calculate the coefficient of correlation for the data in Exercise 11.77. How much reduction in SSE is obtained by using the least squares predictor rather than using $\bar{y}$ in predicting y for the data given in Exercise 11.77?

 11.82 A comparison of the undergraduate grade point averages (GPAs) of 12 corporate employees with their scores on a managerial trainee examination produced the results shown in the table.

Exam Score, y	76	89	83	79	91	95	82	69	66	75	80	88
GPA, x	2.2	2.4	3.1	2.5	3.5	3.6	2.5	2.0	2.2	2.6	2.7	3.3

   a. Find the least squares prediction equation appropriate for the data.

   b. Graph the points and the least squares line as a check on your calculations.

   c. Calculate s^2.

   d. Do the data present sufficient evidence to indicate that x (undergraduate GPA) is useful in predicting y (managerial trainee exam score)? Test by using $\alpha = .05$.

 11.83 Calculate the coefficient of correlation for the data in Exercise 11.82. By what percentage is the sum of squares of deviations reduced by using the least squares predictor $\hat{y} = \hat{\beta}_0 + \hat{\beta}_1 x$ rather than $\bar{y}$ as a predictor of y for the data in Exercise 11.82?

 11.84 Refer to Exercise 11.82. Obtain a 90% confidence interval for the expected exam score for all managerial trainees whose undergraduate GPA was 2.6.

 11.85 Use the least squares equation derived in Exercise 11.82 to predict the exam score for a particular managerial trainee whose undergraduate GPA was 2.6. Obtain a 90% prediction interval for this individual's true exam score.

11.86   What is the difference in the inferential objectives of Exercises 11.84 and 11.85?

11.87   The Public Health Cigarette Smoking Act of 1969 established a federal program to inform the public of the health hazards of smoking and to regulate advertising of tobacco products. A major feature of this bill was a ban on broadcast cigarette advertising, effective January 2, 1971. The intent of this bill was to improve public health by reducing the consumption of cigarettes.

   a.   Find the least squares line relating cigarette consumption to the size of the adult population. Do the data suggest the presence of a linear relationship between cigarette consumption and the size of the adult population? Test the appropriate hypothesis, using $\alpha = .05$. (See Table 11.15.)

   b.   Suppose that the adult population in the United States next year will number about 245 million. Using a 95% prediction interval, estimate the number of packs of cigarettes that will be sold in the United States during the period.

   c.   For the purpose of evaluating the rate of cigarette consumption from 1979 through 1986, compute an index of the per capita packs per year. Analyze these data to determine whether there is a decrease, an increase, or no change in the rate of consumption over this time period.

Table 11.15   Cigarette Consumption Data for Exercise 11.87

Year	Adult Population (Millions)	Cigarette Packs Consumed (Millions)
1979	160.950	31,071.4
1980	164.062	31,590.1
1981	166.854	32,036.0
1982	169.567	31,819.2
1983	172.019	30,120.5
1984	174.218	30,148.4
1985	176.269	29,701.3
1986	179.228	29,339.6

Source: Data adapted from U.S. Department of Commerce, Bureau of the Census, *Statistical Abstract of the United States*, 107th ed. (Washington, D.C., 1988); *Standard and Poor's Industry Surveys, Food and Beverages*, October 8, 1987.

Supplementary Exercises

Interpretive

11.88   The principal reason most people choose to invest part of their capital is to hedge against inflation and protect, or enhance, the buying power of their savings. But how do various forms of investment fare against inflation? To study this issue, Ibbotson and Fall separately regressed the annual rates of return from 17 different capital investment markets on the inflation rate in the United States for the years 1947–1978. The results are shown in Table 11.16.

a. Which capital investment markets have proven to be most sensitive to the rate of inflation? That is, which markets show a return that is linearly related to inflation? (Use $\alpha = .05$.)

b. Which capital investment markets are least sensitive (not linearly related) to inflation?

c. Of those capital investment markets identified within part a, which have fared best against inflation? Which have fared the worst against inflation?

d. Which single capital market investment is most likely to fluctuate in direct proportion to the rise and fall of inflation?

e. Suppose the level of inflation over the next few years is expected to be high, say about 10%. Compute the expected annual rate of return for each capital investment market (except AMEX) with this assumed level of inflation.

f. Estimate the expected annual rate of return for each capital investment market (except AMEX) if the level of inflation is low, say 4%.

g. For this study, under what conditions are inflation-sensitive investments best? When would it be best to invest in investment markets that seem reasonably insensitive to the inflation rate?

Table 11.16   Investment Data for Exercise 11.88

Dependent Variable, y	Independent Variable, x	$\hat{\beta}_0$	$\hat{\beta}_1$	t	r^2	s
NYSE	Inflation*	20.60	−2.44	−2.69	.194	16.18
AMEX (1963–1978 only)	Inflation	32.09	−4.20	−2.13	.246	23.08
OTC	Inflation	25.21	−2.82	−2.49	.171	20.17
Preferreds	Inflation	6.88	−.97	−1.95	.113	8.81
LT corporate bonds	Inflation	3.77	−.36	−.96	.030	6.72
International corporate bonds	Inflation	4.50	−.13	−.43	.006	5.55
Commercial paper	Inflation	2.56	.47	4.48	.401	1.87
Banks; savings & loans	Inflation	3.90	.05	.23	.002	3.98
Total corporate	Inflation	16.38	−1.97	−2.81	.208	12.52
Farms	Inflation	6.43	1.47	5.29	.482	4.97
Housing	Inflation	3.82	.84	7.81	.671	1.92
T bills	Inflation	1.95	.43	4.67	.421	1.63
T notes	Inflation	3.00	.19	.91	.027	3.73
T bonds	Inflation	3.36	−.21	−.61	.012	6.23
ST municipals	Inflation	1.50	.26	4.13	.362	1.11
LT municipals	Inflation	3.99	−.54	−1.18	.044	8.15
Total market	Inflation	7.86	−.24	−.91	.027	4.67

Source: R. Ibbotson and C. Fall, "The United States Market Wealth Portfolio," *Journal of Portfolio Management*, Fall 1979. Reprinted with permission.
Note: The column of t values gives test statistics for the null hypothesis H_0: $\beta_1 = 0$; $s = \sqrt{SSE/(n-2)}$.
* The average annual rate of inflation during this period was 5.6.

 11.89 In Exercise 11.33 we discussed Treynor's characteristic line device for evaluating investment funds. The fund studied in Exercise 11.33 was a growth fund, established with the objective of capital value appreciation. Another common class of funds is balanced funds, which, through greater diversification than growth funds, usually are less risky and place more emphasis on dividend accrual. One of the larger balanced funds is the George Putnam Fund of Boston. Listed in Table 11.17 are its annual returns and the average market rate of return for the years 1977 through 1986. Use a linear regression analysis to find the characteristic line for the Putnam Investors' Fund. Describe its risk characteristics.

Table 11.17   Rate of Return Data for Exercise 11.89

					Year					
	1977	1978	1979	1980	1981	1982	1983	1984	1985	1986
Putnam Investors' Fund	15.7	29.2	.2	10.6	33.6	−8.8	44.3	20.4	13.7	−3.2
Average Market Return	10.2	27.0	−4.7	27.5	25.0	.9	23.4	28.0	9.6	5.8

Source: Data from Wiesenberger Financial Services, *Investment Companies* (1987); *Value Line's Investment Survey*, January 22, 1988.

11.90   Colleges and universities depend upon state and federal governments as well as the private sector as outside sources of research funding. How does the amount of research and development (R&D) funding for colleges and universities vary as the amount of federal funding for R&D changes from year to year? So that the effect of inflation is eliminated, the amount of total R&D monies from federal sources and the amount of R&D monies received by colleges and universities are recorded in constant 1982 dollars in Table 11.18. Use the MINITAB printout provided in Table 11.19 to discuss the estimated linear relationship between y, the amount of R&D funds awarded to colleges and universities, and x, the amount of R&D funds provided by the federal government. Interpret the results of the regression subcommand PREDICT 42000.

Table 11.18   R&D Funds Data for Exercise 11.90

R&D Funds Provided by Federal Government	R&D Funds Awarded to Colleges and Universities
28,191	2,077
35,636	5,629
30,986	5,927
34,546	7,151
35,685	7,316
36,505	7,288
39,090	7,464
41,813	7,818
45,481	8,463
47,722	9,140

Source: Data from U.S. Department of Commerce, Bureau of the Census, *Statistical Abstract of the United States*, 107th ed. (Washington, D.C., 1988), p. 564.

Table 11.19   MINITAB Output for the Data of Exercise 11.90

```
MTB > REGRESS C1 1 C2;
SUBC> PREDICT 42000.

THE REGRESSION EQUATION IS
C1 = - 3494 + 0.275 C2

PREDICTOR          COEF          STDEV       T-RATIO
CONSTANT          -3494           2264         -1.54
C2              0.27474        0.05956          4.61

S = 1090        R-SQ = 72.7%      R-SQ(ADJ) = 69.3%

ANALYSIS OF VARIANCE

SOURCE          DF           SS              MS
REGRESSION       1       25275714        25275714
ERROR            8        9502922         1187865
TOTAL            9       34778636

UNUSUAL OBSERVATIONS
OBS.        C2            C1      FIT STDEV.FIT   RESIDUAL    ST.RESID
  1      28191          2077     4252       656      -2175      -2.50R

R DENOTES AN OBS. WITH A LARGE ST. RESID.

    FIT   STDEV.FIT         95% C.I.             95% P.I.
   8046         434   (  7044,   9047)  (  5339,   10752)

MTB >
```

Case Study

Product Line Decisions for Automobile Dealerships

In 1980 there were over thirty thousand automobile dealerships within the United States, half of which were single-line dealerships and half of which were multiple-line dealerships. A study by Marx discussed the advantages and disadvantages of each type of dealership (T. Marx, "The Economics of Single- and Multiple-Line Retail Automobile Dealerships," *Journal of Marketing*, Spring 1980). In particular, Marx notes that the basic economic advantage of a single-line dealership is that it minimizes capital investment in sales and service facilities, staff training, and inventories. Thus, if demand were adequate, Marx argues that the dealer would stock only a single model of a chosen automotive line. The greatest disadvantage of a single-line dealership (as expressed by Marx) is the risk of public unacceptability of the single line. This risk, then, becomes the primary motivating factor for the multiple-line dealership, which, as a consequence, suffers the economic disadvantages listed as advantages for the single-line dealership. The choice by a dealer is therefore one of risk reduction versus economics.

Marx suggests that one of the most effective means of determining the product lines for a multiple-line dealership is to examine the correlations of sales between various product lines for the dealer's region. Listed in Table 11.20 are the correlations of total U.S. domestic car sales for the period 1970–1987.

1. If a dealer seeks to minimize sales fluctuation risk by establishing a multiple-line dealership, should he choose product lines with strongly negative, strongly positive, or negligible (near zero) correlations? Explain.

2. From Table 11.20, which pairings would you recommend for a multiple-line dealer who wishes to minimize sales fluctuation risk? Which pairings would you recommend for a multiple-line dealer who wishes to minimize sales fluctuation risk *and* capital investment in sales and service facilities, staff training, and inventories?

3. Many multiple-line dealerships seem to be expanding their product line to include foreign or foreign-American-made compacts or sports cars. How would your answer to question 2 change if such pairings were considered?

Table 11.20   Correlation Data for the Case Study

	American Motors	Plymouth	Dodge	Chrysler	Ford	Mercury	Lincoln	Buick	Cadillac	Chevy	Olds
Plymouth	.788										
Dodge	.611	.911									
Chrysler	−.393	−.355	−.002								
Ford	.708	.897	.937	−.095							
Mercury	−.330	−.218	.168	.704	.159						
Lincoln	−.350	−.275	.104	.775	.109	.909					
Buick	−.566	−.498	−.220	.580	−.358	.651	.526				
Cadillac	−.233	−.268	.072	.702	.088	.888	.876	.641			
Chevy	.602	.674	.701	−.192	.857	.257	.158	−.217	.254		
Olds	−.619	−.593	−.249	.731	−.345	.795	.771	.888	.814	−.172	
Pontiac	−.148	.129	.476	.521	.443	.841	.778	.488	.724	.526	.627

Source: Data from *Automotive News*, 1987 Market Data Book issue, April 29, 1987.

References and Suggested Readings

Dixon, W. J., and M. B. Barnes (editors). *BMDP-83; Biomedical Computer Programs, P-Series.* Berkeley, Calif.: University of Calif. Press, 1983.

Draper, N., and H. Smith. *Applied Regression, 2nd. ed.* New York: Wiley, 1981.

Dunn, O. J., and V. A. Clark. *Applied Statistics: Analysis of Variance and Regression, 2nd ed.* New York: Wiley, 1987.

Kleinbaum, D. G., L. L. Kupper, and K. E. Muller. *Applied Regression Analysis and Other Multivariable Methods, 2nd. ed.* Boston: PWS-KENT Publishing Co., 1988.

Mendenhall, W. *An Introduction to Linear Models and the Design and Analysis of Experiments.* Belmont, Calif.: Wadsworth, 1968.

Mendenhall, W., and J. T. McClave. *A Second Course in Business Statistics: Regression Analysis.* San Francisco: Dellen, 1981.

Miller, R. B. *Minitab Handbook for Business and Economics.* Boston: PWS-KENT Publishing Co., 1988.

Neter, J., and W. Wasserman. *Applied Linear Regression Analysis.* Homewood, Ill.: Irwin, 1983.

Ryan, B. F., B. L. Joiner, and T. A. Ryan. *Minitab Handbook, 2nd ed.* Boston: PWS-KENT Publishing Co., 1985.

Younger, M. S. *A Handbook for Linear Regression, 2nd. ed.* Boston: PWS-KENT Publishing Co., 1985.

TO
THE
READER

The preceding chapter presented a model for predicting, or estimating, the mean value of y on the basis of some known related variable x. In this instance sample observations on y and the x variable are used to estimate the linear relationship between y and x. Such analyses are especially useful for examining the sensitivity of y to adjustments in the independent (or control) variable x.

In many instances several independent variables may contribute to the prediction, or estimation, of y. For example, in the estimation of the value of a residential dwelling, the dwelling's size, site characteristics, age, and location offer useful information. Multiple linear regression provides an effective procedure for developing a multi-variable prediction equation. Chapter 12 describes multiple linear regression, avoiding the cumbersome mathematics associated with the derivation of such models; attention is directed to the use and interpretation of common multiple linear regression computer programs. Many of these programs, if not all, should be available within your college or university computing center.

12

MULTIPLE LINEAR REGRESSION

12.1

Introduction

multiple linear regression

Multiple linear regression is an extension of the methodology of Chapter 11 to more than one independent variable. That is, instead of using only a single independent variable to explain the variation in y, multiple linear regression allows for the simultaneous use of several independent (or predictor) variables. By using more than one independent variable, we should do a better job of explaining the variation in y and hence provide more accurate predictions.

A common application of multiple linear regression is residential property value assessment by municipal assessors or private appraisers. In this instance the objective is to estimate (or predict) the value of a residence y, based on certain descriptive information. The assessor may begin by recording the size of the residence, x_1, as measured by the number of square feet of living space in the residence. But size alone is an imprecise determinant of value. Two residences of equal size may have different characteristics with respect to total number of rooms (x_2), the number of bedrooms (x_3), number of bathrooms (x_4), and age since initial construction (x_5). All are important determinants of value and, when considered collectively with size in a multiple regression model, will almost certainly improve our ability to estimate market value accurately in comparison with a model that only uses size.

As a further illustration, think of your special area of business (or your anticipated special area) and then think of some criterion variable y that measures success in the performance of that specialty. For example, if you are majoring in marketing, you might think of sales volume as a measure of success. A person operating a small business would probably use profit, and the director of security for a large department store might measure performance by the value of merchandise lost by theft.

Now suppose that you possessed a multivariable prediction equation that gave accurate predictions of values of y for given values of the x's. Think of the benefits to be derived from this tool. You would be able to predict values of the criterion variable for various values of the x's, and just by noting when x variables enter into the equation, you would likely develop a better understanding of how to control the criterion variable y and make it take values advantageous to you.

Finding a multivariable prediction equation is the subject of this chapter. The application of this method—the statistical tests and estimation procedures—to a set

multiple regression
analysis

of data is often called **multiple regression analysis**.

12.2

Multiple Linear Regression: Estimation and Prediction

A prediction equation based on a number of variables $x_1, x_2, \ldots, x_k$ can be obtained by the method of least squares in exactly the same manner as that employed for the simple linear model. For example, we might wish to fit the model

$$y = \beta_0 + \beta_1 x_1 + \beta_2 x_2 + \beta_3 x_3 + \epsilon$$

where y is the price of a firm's securities at the end of some month, x_1 is earnings per share during the past fiscal year, x_2 is the gross sales volume for the firm during the preceding month, x_3 is the firm's profit margin during the past month, and ϵ is a random error. (Note that we could add other variables, as well as the squares, cubes, and cross products of x_1, x_2, and x_3.)

We would need a random sample of the recorded values for y, x_1, x_2, and x_3 for n randomly selected months during which the firm was in operation. The set of measurements y, x_1, x_2, x_3 for each of the n months could be regarded as the coordinates of a point in four-dimensional space. Then ideally, we would like to possess a multidimensional "ruler" (in our case, a plane) that we could visually move about among the n points until the deviations of the observed values of y from the predicted values would in some sense be a minimum. Although we cannot graph points in four dimensions, you can readily see that this device is provided by the method of least squares, which, mathematically, performs this task for us.

The sum of squares of deviations of the observed value of y from the fitted model is

$$\text{SSE} = \sum_{i=1}^{n} (y_i - \hat{y}_i)^2 = \sum_{i=1}^{n} [y_i - (\hat{\beta}_0 + \hat{\beta}_1 x_{1i} + \hat{\beta}_2 x_{2i} + \hat{\beta}_3 x_{3i})]^2$$

where $\hat{y} = \hat{\beta}_0 + \hat{\beta}_1 x_1 + \hat{\beta}_2 x_2 + \hat{\beta}_3 x_3$ is the fitted model and $\hat{\beta}_0, \hat{\beta}_1, \hat{\beta}_2$, and $\hat{\beta}_3$ are estimates of the model parameters. We then use calculus to find the estimates $\hat{\beta}_0, \hat{\beta}_1, \hat{\beta}_2$, and $\hat{\beta}_3$ that make SSE a minimum. For practical purposes, it is simple and efficient to employ a computer to estimate the unknown regression parameters $\beta_0, \beta_1, \beta_2$, and β_3. Almost every computing facility has access to at least one packaged regression analysis program, such as MINITAB, which requires only that the user execute the proper commands to activate the program and then to submit the problem data. Remember, however, that different computing systems using the same program may provide dissimilar answers because of differences in the use of computer subroutines and machine compilers. Therefore, for some cases in which we have used a particular computer program in the solution to an example in this text, the results you obtain using the same program may be slightly different.

The assumptions associated with the multiple linear regression model are a logical extension of those for the simple linear regression model of Chapter 11. For the multiple linear regression model

$$y = \beta_0 + \beta_1 x_1 + \beta_2 x_2 + \cdots + \beta_k x_k + \epsilon$$

with k separate independent variables, the assumptions are provided in the display.

Assumptions of the Multiple Linear Regression Model

1. The random error term ϵ has an expected value of zero and a constant variance. That is,

$$E(\epsilon) = 0 \qquad \text{and} \qquad \sigma_\epsilon^2 = \sigma^2$$

for each recorded value of the dependent variable y.

2. The error components are uncorrelated with one another.
3. The regression coefficients $\beta_0, \beta_1, \beta_2, \ldots, \beta_k$ are parameters (and hence constant).
4. The independent variables $x_1, x_2, \ldots, x_k$ are known constants.
5. The inferential procedures given in this text require that the random errors ϵ be normally distributed.

The "best fit" equation based on the sample data is the one that minimizes

$$\text{SSE} = \sum_{i=1}^{n} [y_i - (\hat{\beta}_0 + \hat{\beta}_1 x_{1i} + \hat{\beta}_2 x_{2i} + \cdots + \hat{\beta}_k x_{ki})]^2$$

12.3

Explaining the Computer Output for the Multiple Linear Regression Model

Because of the complexity involved in computing by hand a "best fit" multiple linear regression equation for a set of data, we will use computer programs for that task for the examples and exercises within this chapter. Six multiple linear regression computer programs tend to be used much more frequently than others. These programs, their suppliers, and available options are listed in Table 12.1. The available options and output from each program. will be explained in association with worked examples throughout this chapter.

Table 12.1   Available Options for Some Common Multiple Regression Programs

Program and Supplier	Type of Output	t or F Tests	Confidence Intervals for Mean of y	Prediction Intervals for Particular Values of y	R or R^2	Residual Plots
BMDP1R*: BMDP Statistical Software, Inc.	Standard	t	No	No	Both	Yes
BMDP2R*: BMDP Statistical Software, Inc.	Stepwise	t	No	No	Both	Yes
BMDO2R*: BMDP Statistical Software, Inc.	Stepwise	F	No	No	R	Yes
SPSS®-Regression[†]: SPSS, Inc., Chicago, Ill.	Standard or stepwise	F	No	No	Both	Yes
SAS: SAS Institute, Cary, N.C.	Backward, forward, or stepwise	F	Yes	Yes	R^2	Yes
MINITAB: MINITAB, Inc.	Standard or stepwise	t	Yes	Yes	R^2	Yes

*W. J. Dixon et al. (ed.), *BMDP Statistical Software Manual*, 1985 reprint (Berkeley, Calif.: University of California Press, 1985).
[†]SPSS is the trademark of SPSS, Inc., for its proprietary software product.

EXAMPLE 12.1   Consider a study designed to examine the role of television viewing in the lives of a selected group of people over 65 years of age. The purpose of the study was to provide guidelines for developing television programming that would adequately meet the special needs of this audience. A sample of $n = 25$ senior citizens was selected and from each senior citizen the following data were obtained: y = the average number of hours per day an interviewee spends watching television; x_1 = the marital status of the interviewee ($x_1 = 1$ if the interviewee is living with his or her spouse, $x_1 = 0$ if not); x_2 = the age of the interviewee; and x_3 = the number of years of education of the interviewee.* The data are listed in Table 12.2.

The objective of this study is to relate y, the average daily hours an interviewee spends watching television, to the independent variables x_1, x_2, and x_3. For purposes of simplicity we select the prediction model

$$y = \beta_0 + \beta_1 x_1 + \beta_2 x_2 + \beta_3 x_3 + \epsilon$$

Use the MINITAB regression program to find the least squares prediction equation for the data of Table 12.2, and explain the program output.

---

* Variable x_1 is an example of a *dummy variable*, a frequently employed independent variable designed to include the effect of a *qualitative* factor in a regression model. Dummy variables are discussed in detail in Section 12.6 of this chapter.

Table 12.2 Daily Hours Spent Watching Television, Marital Status, Age, and Years of Education of 25 Randomly Selected Senior Citizens; Example 12.1

Individual	Hours, y	Marital Status, x_1	Age, x_2	Education, x_3
1	.5	1	73	14
2	.5	1	66	16
3	.7	0	65	15
4	.8	0	65	16
5	.8	1	68	9
6	.9	1	69	10
7	1.1	1	82	12
8	1.6	1	83	12
9	1.6	1	81	12
10	2.0	0	72	10
11	2.5	1	69	8
12	2.8	0	71	16
13	2.8	0	71	12
14	3.0	0	80	9
15	3.0	0	73	6
16	3.0	0	75	6
17	3.2	0	76	10
18	3.2	0	78	6
19	3.3	1	79	6
20	3.3	0	79	4
21	3.4	1	78	6
22	3.5	0	76	9
23	3.6	0	65	12
24	3.7	0	72	12
25	3.7	0	80	6

SOLUTION See Table 12.3 for the MINITAB output. The paragraphs that follow explain the output.

Finding the least squares prediction equation: Table 12.3 reproduces the computer output obtained by applying a MINITAB regression program to the data of Table 12.2. In this example we are concerned only with the estimates of β_0, β_1, β_2, and β_3, which are illustrated in color in Table 12.3. The other portions of the output are explained in Section 12.4.

The estimates of β_0, β_1, β_2, and β_3 that define the least squares prediction equation for the senior citizens data are given in the column with the heading "COEF" (which stands for "coefficient"). The column entitled "PREDICTOR" displays the names the programmer assigned to the variables. Thus,

$$\hat{\beta}_0 = 1.495 \qquad \hat{\beta}_1 = -1.1757 \qquad \hat{\beta}_2 = .03876 \qquad \hat{\beta}_3 = -.15228$$

Table 12.3   MINITAB Output for the Data of Table 12.2

```
THE REGRESSION EQUATION IS
C10 = 1.50 - 1.18 X1 + 0.0388 X2 - 0.152 X3


PREDICTOR           COEF        STDEV        T-RATIO
CONSTANT           1.495        2.637          0.57
X1                -1.1757       0.3156        -3.73
X2                 0.03876      0.03193        1.21
X3                -0.15228      0.05011       -3.04


S = 0.7536      R-SQ = 62.6%      R-SQ(ADJ) = 57.2%

ANALYSIS OF VARIANCE

SOURCE          DF            SS             MS
REGRESSION       3        19.9327         6.6442
ERROR           21        11.9273         0.5680
TOTAL           24        31.8600


SOURCE          DF         SEQ SS
X1               1         8.6400
X2               1         6.0468
X3               1         5.2459


UNUSUAL OBSERVATIONS
OBS.        X1         C10       FIT   STDEV.FIT   RESIDUAL   ST.RESID
 23        0.00       3.600     2.187    0.298       1.413      2.04R

R DENOTES AN OBS. WITH A LARGE ST. RESID.
```

and it follows that the prediction equation is

$$\hat{y} = 1.495 - 1.1757x_1 + .03876x_2 - .15228x_3$$

For this particular model β_1, β_2, and β_3 represent the change in the mean value of y, $E(y)$, for a one-unit change in x_1, x_2, and x_3, respectively, when the other two variables remain constant. For example, $\hat{\beta}_2 = .03876$ is the estimated mean change in television viewer time for a one-year increase in x_2 (age of the interviewee). For those interviewees of the same age and educational level the coefficient $\hat{\beta}_1$ of the dummy variable x_1 represents the difference in mean viewing time between interviewees who are living with spouses and those who are not. The estimate of β_1 is -1.1757 hours. Thus, we estimate that unmarried interviewees spend, on the average, approximately 1.18 more hours per day watching television than the interviewees who live with their spouses.

Estimation of σ^2, the variance of ϵ: The variance σ^2 of the y values for any given set of $x_1, x_2, \ldots, x_k$ is estimated by s^2, which is equal to SSE divided by the appropriate number of degrees of freedom. Recall that for simple linear regression, which estimates two parameters (β_0 and β_1), SSE was divided by ($n - 2$). In the general case, ($k + 1$) parameters ($\beta_0, \beta_1, \beta_2, \ldots, \beta_k$) are estimated, and s^2 equals SSE divided by $[n - (k + 1)]$. In other words, one degree of freedom is lost for each additional parameter estimated. The value for s, the square root of s^2, is given in the MINITAB printout right after the parameter estimates. From Table 12.3, then, $s = .7536$ for the television viewer data. The formula for s^2 is given in the next display.

Estimate of Variance for Multiple Regression

$$s^2 = \frac{\text{SSE}}{n - (\text{number of } \beta \text{ parameters in the model})}$$
$$= \frac{\text{SSE}}{n - (k + 1)}$$

For our example,

$$s^2 = \frac{11.9273}{21} = .5680 \qquad \text{or} \qquad s = .7536$$

Some computer outputs print SSE, some print s^2, and some print s. Once you have one of the quantities, you can find any other. As in simple linear regression, s or s^2 appears in all the formulas for both confidence intervals and prediction intervals.

Estimation of the β parameters: The procedure for constructing confidence intervals for the β parameters is identical to the procedure employed in Section 11.5 for the simple linear model except that the computing formulas for $s_{\hat{\beta}_1}$, $s_{\hat{\beta}_2}$, and $s_{\hat{\beta}_3}$ are much more complex than the corresponding quantity calculated in Section 11.5. These values are given in the column headed STDEV. Thus, the formula for a $(1 - \alpha)100\%$ confidence interval for a regression coefficient, say β_i, is given in the display.

$(1 - \alpha)100\%$ Confidence Interval for β_i

$$\hat{\beta}_i \pm t_{\alpha/2} s_{\hat{\beta}_i}$$

The degrees of freedom for t are the same as those associated with s, or $[n - (k + 1)]$. To find the 95% confidence interval for β_1, refer to Table 12.3 and note that $s_{\hat{\beta}_1} = .3156$; and from Table 4 of the Appendix $t_{.025}$ with 21 degrees of freedom is 2.080. Thus, a 95% confidence interval for β_1 is

$$\hat{\beta}_1 \pm t_{\alpha/2} s_{\hat{\beta}_1} \qquad \text{or} \qquad -1.1757 \pm (2.080)(.3156)$$

or from -1.83215 to $-.51925$. Since $x_1 = 1$ if interviewees live with their spouses and 0 if not, we estimate that the "loners" watch television, on the average, between .52 and 1.83 more hours per day than those living with their spouses.

We can find confidence intervals for β_2 and β_3 in a similar manner.

Tests of significance for individual β parameters: A test of an hypothesis that a particular parameter, say β_i, equals zero can be conducted by using a t statistic:

$$t = \frac{\hat{\beta}_i - 0}{s_{\hat{\beta}_i}} = \frac{\hat{\beta}_i}{s_{\hat{\beta}_i}}$$

The procedure is identical to the procedure employed in testing an hypothesis about the slope β_1 in the simple linear model (Section 11.5) except that the formula for computing $s_{\hat{\beta}_1}$ in a multiple regression analysis is much more complex.

The test can also be conducted by using the F statistic, since the square of a t statistic (with v degrees of freedom) is equal to an F statistic with 1 degree of freedom in the numerator and v degrees of freedom in the denominator. That is,

$$t_v^2 = F_v^1$$

Some multiple regression computer programs, like MINITAB, use a t statistic to test the null hypothesis

$$H_0 : \beta_i = 0$$

against the two-sided alternative hypothesis

$$H_a : \beta_i \neq 0$$

Others use the F statistic (see Table 12.1). The next display summarizes the conclusions for the tests of significance.

Conclusions Resulting from Tests of Hypotheses for the β Parameters

Reject $H_0 : \beta_i = 0$:   Predictor variable x_i is a *significant* predictor of the response variable y in the presence of the other predictor variables.

Do not reject $H_0 : \beta_i = 0$:   Predictor variable x_i is not a *significant predictor of y in the presence of the other predictor variables.*

To test $H_0 : \beta_1 = 0$ against $H_a : \beta_1 \neq 0$, we find

$$t = \frac{\hat{\beta}_1}{s_{\hat{\beta}_1}} = \frac{-1.1757}{.3156} = -3.73$$

Then because the alternative hypothesis implies a two-tailed test, we will reject the null hypothesis if $t > t_{.025}$ or $t < -t_{.025}$, where $t_{.025}$ is based on (n − number of β parameters in the model) $= 21$ degrees of freedom. From Table 4 of the Appendix $t_{.025} = 2.080$.

Since the test statistic $t = -3.73$ is less than the critical value $-t_{.025} = -2.080$, we reject the null hypothesis that $\beta_1 = 0$. Thus, marital status, when considered in the presence of age (x_2) and education (x_3), is an important factor in predicting television-viewing time among senior citizens and that "loners" spend more time watching

Table 12.4   *Results of Analysis of t Values for β_2 and β_3*

Null Hypothesis	t	Inference	Conclusion
$H_0: \beta_2 = 0$	1.21	Do not reject H_0	In the presence of x_1 and x_3, x_2 is a poor predictor of y
$H_0: \beta_3 = 0$	-3.04	Reject H_0	In the presence of x_1 and x_2, x_3 is a good predictor of y

television (on the average) than interviewees of the same age and educational level who live with their spouses. An analysis of the T-RATIO values for β_2 and β_3 offers the results summarized in Table 12.4. Thus, when considered together as predictors of time spent watching television by senior citizens, marital status and education are important factors but age is not.

EXERCISES   Conceptual Level

12.1   What advantage(s) may be gained by using more than one independent variable in a regression model?

12.2   What are the assumptions upon which the multiple linear regression model is based?

EXERCISES   Skill Level

12.3   A multiple regression analysis based on $n = 17$ data points was employed to predict a dependent variable y using three independent variables x_1, x_2, and x_3. The following least squares estimates were calculated:

$$\hat{\beta}_0 = 16.38 \quad \hat{\beta}_2 = -2.85 \quad s_{\hat{\beta}_0} = 3.179 \quad s_{\hat{\beta}_2} = .492$$
$$\hat{\beta}_1 = 1.42 \quad \hat{\beta}_3 = .46 \quad s_{\hat{\beta}_1} = .268 \quad s_{\hat{\beta}_3} = .401$$

a.   Write the least squares equation relating y to x_1, x_2, and x_3.

b.   What is the predicted value of y if $x_1 = 5$, $x_2 = 3.5$, and $x_3 = -.2$?

c.   Do the data provide sufficient evidence to indicate that, in the presence of x_1 and x_2, x_3 is useful in predicting y? Test by using $\alpha = .05$.

d.   Find the approximate observed significance level for the test in part c.

e.   Find a 95% confidence interval for the change in y for a 1-unit change in x_2, for fixed values of x_1 and x_3.

12.4   Use the MINITAB printout in Table 12.5 to answer questions about the regression analysis in which y was a function of two predictor variables, x_1 and x_2.

a.   What is the least squares equation relating y, x_1, and x_2?

b.   What is the value of t used for testing $H_0: \beta_0 = 0$? For $H_0: \beta_1 = 0$? For $H_0: \beta_2 = 0$? Are the values of these test statistics significant at $\alpha = .05$?

Table 12.5   MINITAB Output for Exercise 12.4

```
THE REGRESSION EQUATION IS
Y = 19.6 + 0.471 X1 + 0.089 X2

PREDICTOR           COEF            STDEV        T-RATIO
CONSTANT          19.576           5.250          3.73
X1                 0.4713          0.1554         3.03
X2                 0.0893          0.4404         0.20

S = 3.333           R-SQ = 91.5%      R-SQ(ADJ) = 89.0%

ANALYSIS OF VARIANCE

SOURCE          DF              SS             MS
REGRESSION       2           834.63         417.32
ERROR            7            77.77          11.11
TOTAL            9           912.40

SOURCE          DF          SEQ SS
X1               1          834.18
X2               1            0.46
```

c. Given the value of $t = .20$ in testing $H_0: \beta_2 = 0$, can we conclude that the variable x_2 is of no value in predicting y? If not, what is the proper interpretation of this test?

d. Find a 95% confidence interval estimate for β_1.

EXERCISES Applications

12.5 The marketing director for a major softwood timber products company has developed a model to predict monthly lumber orders from her firm's domestic markets. Data were recorded for the past 24 months for each of the following variables:

y = lumber orders (in thousands of board feet)
x_1 = building permits issued in market region
x_2 = interest rate on conventional first mortgages
x_3 = federal defense spending (in billions of dollars)

A BMDP2R regression analysis program was then used to provide the results shown in Table 12.6. Write the prediction equation for the linear model relating lumber orders to building permits, interest rates, and federal defense spending.

12.6 Refer to Exercise 12.5, and find a 95% confidence interval for the expected increase in lumber orders for each unit increase in building permits issued in the market region in the presence of x_2 and x_3.

12.7 Repeat Exercise 12.6 for an increase of 100 building permits issued in the market region.

Table 12.6   BMDP2R Output for Exercise 12.5

```
MULTIPLE R              .7092
R SQUARE                .5030
STD. ERROR OF EST.     2.3181

ANALYSIS OF VARIANCE

                  DF     SUM OF SQUARES     MEAN SQUARE     F RATIO
    REGRESSION     3         108.740          36.247         6.745
    RESIDUAL      20         107.474           5.374

                VARIABLES IN EQUATION

   VARIABLE      COEFFICIENT    STD. ERROR    F TO REMOVE
  (CONSTANT      36.54597)
  BLDG. PMTS.     0.80106        0.42556        3.5433
  INT. RATE      -5.30878        2.53936        4.3706
  DEF. SPDG.      2.28814        1.18411        3.7340
```

12.8   Using Table 12.6, test the null hypothesis that the expected increase (or decrease) in lumber orders for a \$1 billion increase in federal defense spending is zero in the presence of x_1 and x_2. That is, test $H_0 : \beta_3 = 0$ against the alternative hypothesis $H_a : \beta_3 \neq 0$. Test at the $\alpha = .05$ level of significance, using the t test. (Since the t test statistic is not listed within this program, you must compute it from the available output.)

12.9   Repeat Exercise 12.8, using the F test. (In this case, the value of the F statistic is listed in the column headed by F TO REMOVE). Confirm that $t^2 = F$.

12.10   On the basis of the regression printout listed in Table 12.6, which of the three variables x_1, x_2, and x_3 are significant predictors of y (lumber orders) in the presence of the other variables? Which are insignificant? How might you explain why some variables are insignificant predictors and others are good (significant) predictors when they are all considered together in a multiple linear regression model?

12.11   A land developer was interested in creating a model to use for estimating the selling price of beach lots on the Oregon coast. To do so, he recorded the following items for each of 20 beach lots recently sold:

y = sale price of beach lot (in \$1,000 units)
x_1 = area of lot (in hundreds of square feet)
x_2 = elevation of lot
x_3 = slope of lot

The land developer then employed a regression analysis computer program and obtained the output in Table 12.7.

a.   Give the prediction equation for the linear model relating selling price to the area, elevation, and slope of a beach lot.

b.   Which of the predictor variables contributes information for the prediction of y? Determine the answer by using the appropriate statistical test. Use $\alpha = .05$.

Table 12.7   BMDP2R Output for Exercise 12.11

```
MULTIPLE R              .8854
R SQUARE               .7838
STD. ERROR OF EST.     .6075

ANALYSIS OF VARIANCE

                    DF    SUM OF SQUARES    MEAN SQUARE    F RATIO
     REGRESSION      3        21.409           7.136       19.345
     RESIDUAL       16         5.903            .369

INDIVIDUAL ANALYSIS OF VARIABLES

  VARIABLE      COEFFICIENT    STD. ERROR    F VALUE
 (CONSTANT        -2.491)
 AREA               .099          .058        2.935
 ELEVATION          .029          .006       23.327
 SLOPE              .086          .031        7.705
```

12.12   Suppose that before you collected the data for the analysis of Exercise 12.11, you had a theory that sloping beach lots were preferred over those with lesser slope. Do the data provide sufficient evidence to indicate that sales price increases as the slope increases in the presence of the other variables?

12.13   Refer to Exercise 12.11. Find a 90% confidence interval for the regression parameter relating area to selling price in the presence of elevation and slope. For given values of elevation and slope, provide an interpretation of this confidence interval.

12.4

Measuring the Goodness of Fit of a Model

We noted in Section 11.9 that an important property of a regression analysis is that the total sum of squares of deviations of the y values about their mean can be partitioned into two quantities,

$$\text{SSE} = \sum_{i=1}^{n}(y_i - \hat{y}_i)^2 = \text{sum of squares for error}$$

$$\text{SSR} = \sum_{i=1}^{n}(\hat{y}_i - \bar{y})^2 = \text{sum of squares due to regression}$$

That is,

$$\text{Total SS} = \sum_{i=1}^{n}(y_i - \bar{y})^2 = \text{SSR} + \text{SSE}$$

SSE, the sum of squares of deviations of the y values about their predicted values (those values calculated from the prediction equation), divided by the appropriate number of degrees of freedom is equal to s^2, an estimate of σ^2.

In addition, we showed that r^2, the coefficient of determination, is equal to

$$r^2 = \frac{\text{Total SS} - \text{SSE}}{\text{Total SS}} = \frac{\text{SSR}}{\text{Total SS}}$$

You will recall that r^2 measures the proportion of the Total SS that can be explained by the single predictor variable x. Consequently, r^2, which assumes values in the interval $0 \leq r^2 \leq 1$, measures the goodness of fit of the simple linear regression model.

In a multiple regression analysis the Total SS,

$$\sum_{i=1}^{n}(y_i - \bar{y})^2$$

is partitioned in exactly the same way. Thus,

$$\text{Total SS} = \text{SSR} + \text{SSE}$$

and SSR and SSE are defined in exactly the same way as they were for a simple linear regression analysis. The only difference here is that y is a function of more than one predictor variable.

Suppose you fit the multiple regression model

$$y = \beta_0 + \beta_1 x_1 + \beta_2 x_2 + \cdots + \beta_k x_k + \epsilon$$

to a set of data. Then the quantity

$$R^2 = \frac{\text{Total SS} - \text{SSE}}{\text{Total SS}} = \frac{\text{SSR}}{\text{Total SS}}$$

gives the proportion of the Total SS that is explained by the predictor variables $x_1, x_2, \ldots, x_k$. The remainder is explained by the omission of important information-contributing variables from the model, an incorrect formulation of the model, and experimental error. Just like r^2, the simple linear coefficient of determination, the

multiple coefficient of determination R^2 takes values in the interval

multiple coefficient of determination

$$0 \leq R^2 \leq 1$$

A small value of R^2 means that $x_1, x_2, \ldots, x_k$ contribute very little information for the prediction of y; a value of R^2 near 1 means that $x_1, x_2, \ldots, x_k$ provide almost all the information necessary for the prediction of y. Thus, just as r^2 provides a measure of the fit of the simple linear model, R^2 provides a measure of the fit of a more complex regression model.

One must be careful when interpreting the coefficient of determination R^2 in a regression study since one can very easily inflate R^2 by simply adding more and more independent variables to the model. In fact, it can be shown mathematically that adding *any* additional variables *must* increase R^2, regardless of whether these variables are good predictors of the y variable. Thus, simply trying to find a regression model where R^2 is as close to 1 as possible is not a good practice, since this goal can be always be achieved by adding a few more independent variables, even when these variables do not have statistically significant t ratios.

adjusted R^2 To remedy this problem, many researchers also use R_a^2, the **adjusted R^2** value, when evaluating a regression printout. The adjusted R^2 is calculated by

$$R_a^2 = 1 - \left(\frac{n-1}{n-k-1}\right)\left(\frac{\text{SSE}}{\text{Total SS}}\right) = 1 - (n-1)\left(\frac{s^2}{\text{Total SS}}\right)$$

Unlike R^2, R_a^2 may either increase or decrease as more variables are added to the model. In general, if the additional variables are good predictors of y, then they should cause s^2 to decrease, and in that case R_a^2 will increase. Otherwise, if new variables don't decrease s^2, then the R_a^2 value will reflect this by decreasing. Thus, R_a^2 penalizes you when poor predictor variables are added and rewards you when good ones are added.

To illustrate, let us return to the computer output (Table 12.3) for Example 12.1, to television viewer data analysis. Midway through Table 12.3, the "R-SQ" label stands for R^2, the multiple coefficient of determination. This value, $R^2 = 62.6\%$, indicates that only 62.6% of the total variation of the y values about their mean can be explained by the variables used in the model. The remainder, 34.4%, is unexplained variation. Similarly, the "R-SQ(ADJ)" value of 57.2% refers to the adjusted R^2 value. If the values of R^2 and R_a^2 were very different, then we would have reason to suspect that at least one of the x_i variables was contributing very little to the prediction of y. The relatively poor fit of this model could be due to the fact that x_1, x_2, and x_3 are not entered properly into the model (perhaps we should include terms involving x_2^2, x_3^2, $x_1 x_2$, $x_1 x_3$, $x_2 x_3$, etc.). Or perhaps y, the average daily viewing time of interviewees, is a function of many other variables besides x_1, x_2, and x_3. For example, you might wish to include a variable x_4 that measures an interviewee's propensity to read and a dummy variable x_5 that takes a value 1 if the interviewee is employed and a value 0 if not employed. You may think of many other variables that might affect the length of television viewer time. Actually, the fact that R^2 is as low as .626 is probably due to both reasons. Perhaps x_1, x_2, and x_3 are not entered into the model in the best way (some comments on model formulation are given in Section 12.7), and perhaps the model does not include an adequate number of predictor variables related to y.

12.5

Testing the Utility of the Regression Model

The multiple regression analysis can be presented as an analysis of variance by using the additivity of the sums of squares described in Section 12.4, by which

Total SS = SSR + SSE

The ANOVA table for a multiple regression analysis is shown in Table 12.8.

The mean squares associated with the two sources of variation are calculated as the corresponding sum of squares divided by the appropriate degrees of freedom. In particular, the degrees of freedom for error were given in Section 12.3 as $[n - (k + 1)]$, and the degrees of freedom for Total SS are $(n - 1)$. By subtraction, the degrees of

Table 12.8   ANOVA for Multiple Regression Analysis

Source	d.f.	SS	MS
Regression	k	SSR	MSR $=$ SSR$/k$
Error	$n - (k + 1)$	SSE	MSE $=$ SSE$/[n - (k + 1)]$
Total	$n - 1$	Total SS	

freedom for regression must be

$$(n - 1) - [n - (k + 1)] = k$$

which is one less than the number of β parameters in the model.

For the multiple regression model

$$y = \beta_0 + \beta_1 x_1 + \beta_2 x_2 + \cdots + \beta_k x_k + \epsilon$$

we use MSR and MSE to test the null hypothesis that $x_1, x_2, \ldots, x_k$ contribute no information for the prediction of y. This hypothesis is equivalent to hypothesizing that

$$\beta_1 = \beta_2 = \cdots = \beta_k = 0$$

If the data provide sufficient information to reject this hypothesis, then at least one of the predictor variables $x_1, x_2, \ldots, x_k$ contributes significant information for the prediction of y.

When the null hypothesis is true, and $x_1, x_2, \ldots, x_k$ contribute no information for the prediction of y, each of the quantities MSR and MSE provides an independent (in a probabilistic sense) estimate of σ^2, the variance of y for given values of $x_1, x_2, \ldots, x_k$, and the test statistic $F =$ MSR/MSE possesses an F distribution with $v_1 = k$ and $v_2 = [n - (k + 1)]$ degrees of freedom. When the null hypothesis is false, and the model is useful in predicting y, MSR will tend to be larger than expected and hence F will be large. Thus, we will reject $H_0: \beta_1 = \beta_2 = \cdots = \beta_k = 0$ for values of F that exceed F_α, an upper-tailed value in the F distribution with $v_1 = k$ and $v_2 = [n - (k + 1)]$ degrees of freedom (see Figure 12.1).

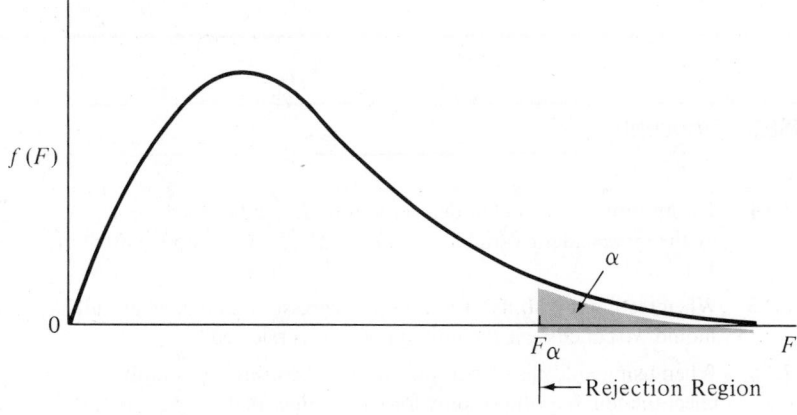

Figure 12.1   Rejection Region for the F Test of $H_0: \beta_1 = \beta_2 = \cdots = \beta_k = 0$

EXAMPLE 12.2   Test the adequacy of the regression model for predicting daily television viewing (Example 12.1). Use the computer output of Table 12.3 to conduct the test.

SOLUTION   Refer to the portion of the computer output for the television viewer data of Table 12.3 headed by the title "ANALYSIS OF VARIANCE." For convenience, that portion of Table 12.3 is given here as Table 12.9.

Table 12.9   Analysis of Variance Portion of MINITAB Output for the Data of Table 12.2

```
ANALYSIS OF VARIANCE

SOURCE          DF            SS          MS
REGRESSION       3       19.9327      6.6442
ERROR           21       11.9273      0.5680
TOTAL           24       31.8600
```

The value of the F statistic for testing the hypothesis that at least one of the predictor variables contributes significant information for the prediction of television viewing time, y, is

$$F = \frac{MSR}{MSE} = \frac{MSR}{s^2} = \frac{6.6442}{.5680} = 11.698$$

(Some regression programs print this statistic and label it as the F ratio.)

We compare this calculated value of F with the critical value of $F_{.05}$ based on $v_1 = 3$ and $v_2 = 21$ degrees of freedom, and we reject H_0 if $F > F_{.05}$. Checking Table 6 of the Appendix, we find $F_{.05} = 3.07$. Since the value of F calculated from the data $F = 11.698$ exceeds $F_{.05} = 3.07$, we reject the null hypothesis

$$H_0 : \beta_1 = \beta_2 = \beta_3 = 0$$

In fact, $F = 11.698$ exceeds $F_{.005} = 5.73$ and is significant at p-value $< .005$. It appears that at least one of the predictor variables contributes information for the prediction of y. ◆ ◆

EXERCISES   Conceptual Level

12.14   The multiple coefficient of determination (R^2) indicates the portion of Total SS that is explained by the independent variables in the model. To what do we attribute the remaining portion of Total SS?

12.15   When testing the utility of a multiple regression model, what null hypothesis do we use? What should we conclude if the null hypothesis is rejected?

12.16   When using analysis of variance to test the utility of a multiple regression model, why do we reject the null hypothesis only for large values of F?

EXERCISES   Skill Level

12.17   The MINITAB computer printout from Exercise 12.4 is reproduced in Table 12.10.

   a.   What proportion of the variation in y is explained by the predictor variables x_1 and x_2? How different is this value from R^2 adjusted?

   b.   Use the values of MSR and MSE to test for a significant regression of y on x_1 and x_2. What is the p-value associated with the observed value of the test statistic?

   c.   Show that the F statistic in part b can also be written as a function of R^2, given by

$$F = \left(\frac{n - k - 1}{k}\right)\left(\frac{R^2}{1 - R^2}\right)$$

   How would you describe the relationship between F and R^2?

Table 12.10   MINITAB Output for Exercise 12.17

```
THE REGRESSION EQUATION IS
Y = 19.6 + 0.471 X1 + 0.089 X2

PREDICTOR        COEF           STDEV        T-RATIO
CONSTANT       19.576          5.250          3.73
X1              0.4713         0.1554         3.03
X2              0.0893         0.4404         0.20

S = 3.333        R-SQ = 91.5%       R-SQ(ADJ) = 89.0%

ANALYSIS OF VARIANCE

SOURCE        DF            SS              MS
REGRESSION     2          834.63          417.32
ERROR          7           77.77           11.11
TOTAL          9          912.40

SOURCE        DF          SEQ SS
X1             1          834.18
X2             1            0.46
```

12.18   A multiple regression analysis, based on $n = 25$ data points, attempts to predict y using the model

$$y = \beta_0 + \beta_1 x_1 + \beta_2 x_2 + \beta_3 x_3 + \beta_4 x_4 + \epsilon$$

and produces

   SSE $= 1586.2$ SSR $= 6311.7$ Total SS $= 7897.9$

   a.   Complete the ANOVA table for the regression analysis.

b. Do the data indicate that the independent variables x_1, x_2, x_3, and x_4 jointly contribute significant information for the prediction of y? Test by using $\alpha = .01$.

c. Calculate the multiple coefficient of correlation R^2, and interpret its value.

d. Calculate R_a^2 and compare the results with the value of R^2 found in part c. What implications does this result have in terms of the fit of the model?

EXERCISES Applications

12.19 A marketing representative for a company that sells soybeans as a meat supplement in the United States is interested in constructing a model to predict the sales of soybeans in different market areas. Sales data (in thousands of dollars) were obtained from 25 different market areas and were related to each area's measurement of the following:

x_1 = coefficient of cross-elasticity between soybeans and beef

x_2 = per capita income (in thousands of dollars)

x_3 = average utilities consumption index

x_4 = unit price of 1-lb of packaged soybeans

x_5 = proportion of total expenses devoted to advertising

x_6 = 1 if area is major beef-producing area, 0 if it is not

The marketing representative then employed a **BMDP2R** regression analysis computer program, which listed the results given in Table 12.11.

Table 12.11   BMDP2R Output for Exercise 12.19

```
MULTIPLE R              .9825
R SQUARE                .9654
STD. ERROR OF EST.    1.7098

ANALYSIS OF VARIANCE
```

	DF	SUM OF SQUARES	MEAN SQUARE	F RATIO
REGRESSION	6	1468.034	244.672	
RESIDUAL	18	52.614	2.923	

```
INDIVIDUAL ANALYSIS OF VARIABLES
```

VARIABLE	COEFFICIENT	STD. ERROR	F VALUE
(CONSTANT	-51.034)		
CROSS-ELAS.	57.600	13.813	17.356
PER CAP. INC.	2.956	2.548	1.347
UTIL. IDX.	-.934	1.129	.682
PRICE	-46.542	31.661	2.160
ADV. EXP.	46.355	9.499	23.843
BEEF AREA	-1.805	.775	5.431

a. Calculate the missing F RATIO, and test the hypothesis associated with this missing value, using $\alpha = .05$.

b. What proportion of the variation of soybean sales is explained by the six predictor variables in the model?

c. Give the regression equation for predicting the sales of soybeans in a particular market area.

d. In the presence of the other variables in the model, which is the most significant predictor of sales? Which is the least significant?

e. Suppose it is very difficult to obtain a precise measure of the per capita income of a market area. Does the prediction model lose much information if we eliminate per capita income as a predictor of sales? If we do so, how would the revised model be established?

f. Find a 90% confidence interval for the difference in predicted sales between areas that are major beef-producing regions and those that are not (i.e., for β_6).

 12.20 Table 12.12 lists the sales price y as a substitute for value and seven assumed related predictor variables $x_1, x_2, \ldots, x_7$ for each of 50 single-family residences for the purpose of developing a residential value appraisal model. The linear model $y = \beta_0 + \beta_1 x_1 + \beta_2 x_2 + \cdots + \beta_7 x_7 + \epsilon$ was fitted to the data set by using a MINITAB regression analysis program. The results of the analysis are given in Table 12.13. Explain the computer output for this analysis by discussing the same topics that were addressed in the television-viewing example.

Table 12.12   Home Measurement Data for Exercise 12.20

Residence, i	Sales Price, $y(\times \$1,000)$	Square Feet, $x_1(\times 100)$	Bedrooms, x_2	Bathrooms, x_3	Total Rooms, x_4	Age, x_5	Attached Garage, x_6	View, x_7
1	50.6	8.0	2	1	5	5	0	0
2	51.5	9.5	2	1	5	8	0	0
3	53.3	9.1	3	1	6	2	0	0
4	65.9	9.5	3	1	6	6	0	0
5	67.4	12.0	3	2	7	5	0	0
6	68.9	10.0	3	1	6	11	0	0
7	71.6	11.8	3	2	7	8	0	0
8	73.1	10.0	2	1	7	15	1	0
9	74.0	13.8	3	2	7	10	0	0
10	74.3	12.5	3	2	7	11	0	0
11	75.2	15.0	3	2	7	12	0	0
12	75.2	12.0	3	2	7	8	0	0
13	76.7	16.0	3	2	7	9	1	1
14	77.9	16.5	3	2	7	15	0	0
15	78.5	16.0	3	2	7	11	1	0
16	79.7	16.8	2	2	7	12	0	0
17	80.9	15.0	3	1	7	8	1	0
18	80.9	17.8	3	2	8	13	1	0
19	82.4	17.9	3	2	7	18	1	0
20	83.0	19.0	2	2	7	22	0	0
21	84.5	17.6	3	1	6	17	0	0
22	86.0	18.5	3	2	8	11	1	0

(*continues*)

Table 12.12   (Continued)

Residence, i	Sales Price, $y(\times \$1,000)$	Square Feet, $x_1(\times 100)$	Bedrooms, x_2	Bathrooms, x_3	Total Rooms, x_4	Age, x_5	Attached Garage, x_6	View, x_7
23	86.3	18.0	3	2	7	5	0	0
24	87.5	17.0	2	3	8	2	1	0
25	88.4	18.7	3	1	6	6	0	0
26	88.4	20.0	3	2	7	16	0	0
27	88.7	20.0	3	2	7	12	0	0
28	89.6	21.0	3	2	7	10	1	0
29	90.5	20.5	2	2	7	11	1	0
30	94.7	19.9	3	1	7	13	1	1
31	95.0	21.5	2	2	7	8	0	0
32	95.3	20.5	3	1	7	9	1	0
33	99.8	22.0	3	2	7	10	0	0
34	100.7	22.0	3	2	7	6	1	1
35	100.7	21.8	2	1	6	15	1	0
36	103.4	22.5	3	2	7	11	1	0
37	104.0	24.0	3	2	7	17	0	0
38	106.1	23.5	3	2	8	12	0	0
39	107.0	25.0	3	2	7	11	1	0
40	110.3	25.6	3	2	7	15	1	0
41	116.0	25.0	4	2	8	12	1	0
42	121.4	25.0	2	2	8	8	0	1
43	125.9	26.8	3	2	7	6	1	0
44	131.3	22.1	3	2	8	18	1	0
45	132.5	27.5	3	2	8	12	1	0
46	134.0	25.0	4	2	8	10	1	0
47	135.2	24.0	3	2	8	13	1	1
48	137.0	31.0	4	3	9	25	1	0
49	149.0	21.0	4	2	9	18	1	0
50	185.0	40.0	5	3	12	22	1	0

 12.21 Refer to Exercise 12.20. Use the model derived in that exercise to establish assessed valuations for each of five residences. The data are given in Table 12.14.

 12.22 We are all aware of the effect of inflation on property values; that is, they tend to appreciate at least as rapidly as the local rate of inflation. Inflation requires local assessment officers to periodically update the assessed valuations of properties in their jurisdiction. There are three approaches that might be adopted by the assessor to update property valuations:

Each listed valuation could be increased by the percentage increase in inflation since the previous assessment.

New sales data could be obtained and pooled with past available sales data to develop a new regression assessment model.

A new regression assessment model could be developed based only on current sales data, ignoring all past sales data.

Which approach would you suggest? Explain.

Table 12.13   MINITAB Output for the Data of Exercise 12.20

```
THE REGRESSION EQUATION IS
PRICE = - 21.8 + 2.93 X1 + 2.40 X2 - 8.75 X3 + 9.45 X4 - 0.000 X5
          + 1.77 X6 + 2.55 X7


PREDICTOR          COEF        STDEV      T-RATIO
CONSTANT          -21.83       10.14       -2.15
X1                2.9326      0.3335        8.79
X2                 2.401       2.902        0.83
X3                -8.747       3.820       -2.29
X4                 9.449       2.568        3.68
X5               -0.0001      0.3169       -0.00
X6                 1.772       3.040        0.58
X7                 2.555       4.544        0.56


S = 9.092       R-SQ = 89.9%      R-SQ(ADJ) = 88.2%

ANALYSIS OF VARIANCE

SOURCE         DF           SS            MS
REGRESSION      7       30885.0        4412.1
ERROR          42        3472.1          82.7
TOTAL          49       34357.0


SOURCE         DF        SEQ SS
X1              1       28650.4
X2              1         788.1
X3              1           8.1
X4              1        1374.7
X5              1           2.0
X6              1          35.5
X7              1          26.1

UNUSUAL OBSERVATIONS
OBS.       X1      PRICE       FIT STDEV.FIT    RESIDUAL    ST.RESID
 44      22.1     131.30    110.05      2.58       21.25        2.44R
 47      24.0     135.20    118.18      4.18       17.02        2.11R
 49      21.0     149.00    118.67      3.83       30.33        3.68R
 50      40.0     185.00    196.39      6.69      -11.39       -1.85 X

R DENOTES AN OBS. WITH A LARGE ST. RESID.
X DENOTES AN OBS. WHOSE X VALUE GIVES IT LARGE INFLUENCE.
```

Table 12.14   Home Measurement Data for Exercise 12.21

Residence, i	Square Feet, x_1	Bedrooms, x_2	Bathrooms, x_3	Total Rooms, x_4	Age, x_5	Attached Garage, x_6	View, x_7
1	22.4	4	2	7	18	1	1
2	15.3	3	2	7	6	0	0
3	17.2	4	1	7	4	1	0
4	31.7	5	3	9	24	0	0
5	20.0	4	2	8	11	1	1

12.23   Is there a consistent relationship between capital budgeting practices and earnings performance? If so, this evidence would support the practice of many firms that use capital budgeting techniques in their investment programs. To study this issue, S. H. Kim and N. K. Kwak ("Capital Budgeting Practices and Their Impact on Earnings Performance," *Proceedings of the American Institute for Decision Sciences*, November 1976) regressed y, the estimated value of the average earnings per share, on the following variables:

x_1 = degree of sophistication of capital budgeting system (0–100)
x_2 = size of firm (average annual total assets)
x_3 = capital intensity (depreciation/total assets)
x_4 = risk (standard deviation of annual earnings per share)
x_5 = capitalization (debt/total assets)
x_6 = price-earnings ratio

Data were taken from each of $n = 114$ machinery firms with revenue in excess of $50 million in 1974. Data were used for the period from 1969 through 1974. The results of their analysis are given in Table 12.15.

a. Does the Kim-Kwak study support the notion that there is a significant relationship between capital budgeting practices and earnings performance for the class of firms studied in their analysis? Explain.

b. Which of the capital budgeting variables used in the analysis contribute information for the prediction of earnings performance? Determine the answer by using appropriate tests of hypotheses. Use $\alpha = .05$.

c. Explain and interpret the $R^2 = .776$ value listed for this analysis.

Table 12.15   Computer Output for Exercise 12.23

$R^2 = .776$
$F = 61.991$

VARIABLE	CONSTANT	X1	X2	X3	X4	X5	X6
COEFFICIENT	−1.613	.040	.001	.090	−.072	.018	.010
T VALUE		10.282	.464	2.232	−.559	4.143	.806

12.6

Selecting Quantitative Independent
Variables and Dummy Variables

The independent variables that contribute information for the prediction of y can be of two types, quantitative and qualitative. As you will subsequently see, the way you enter an independent variable into a prediction equation depends on its type.

Definition

quantitative
independent variable

qualitative variables

> A **quantitative independent variable** is a variable that can take values corresponding to the points on the real line. Independent variables that are not quantitative are said to be **qualitative variables**.

Interest rate, employment rate, numbers of employees, and number of machines are four examples of quantitative independent variables. In contrast, suppose you own four similar plants manufacturing the same product and that the response of interest is the plant profit per unit of time. Four different plant supervisors, call them A, B, C, and D, manage the four plants, one assigned to each plant. Certainly, "plant supervisor" is an independent variable that may affect plant profit. Consequently, "plant supervisor" is a qualitative independent variable that we enter into a prediction equation to model the plant profit y.

Definition

level

> The intensity setting of an independent variable is called a **level**.

For quantitative independent variables the levels correspond to the values that these independent variables may assume and therefore correspond to points on a line. For example, if you think that interest rates may affect your response, and the response y is recorded for three interest rates, 6%, 8%, and 9.2%, then you have observed the independent variable "interest rate" at three levels, 6%, 8%, and 9.2%.

The levels for qualitative independent variables are not quantifiable and therefore do not correspond to points on a line. They can only be defined by describing them. For the manufacturing profit example given above, the independent variable "plant supervisor" is observed at four levels, each level corresponding to one of the supervisors A, B, C, or D.

A good way to portray a model for a response y that is a function of a single quantitative predictor variable is to graph $E(y)$ (or $\hat{y}$) as a function of x, as we did in Chapter 11. The resulting curve (or line) is called a **response curve**.

response curve

Just as the equation expressing $E(y)$ (or $\hat{y}$) as a function of a single quantitative predictor variable traces a curve on a sheet of paper, the corresponding equation response surface

involving two (or more) quantitative predictor variables graphs a **response surface** in a three- (or higher) dimensional space.

Qualitative independent variables are entered into a model by using *dummy variables*, the number of dummy variables always being one less than the number of levels (categories) associated with the independent variable.

Definition

dummy variable

> A **dummy variable** is an independent variable used to include or exclude the effect of a qualitative factor.

For example, if the sales y of a retailing firm depend on the variable "location," and if there are three locations, A, B, and C, the first few terms of the model are

$$y = \beta_0 + \beta_1 x_1 + \beta_2 x_2 + \left\{\begin{matrix} \text{terms associated with} \\ \text{other predictor variables} \end{matrix}\right\} + \epsilon$$

where

$x_1 = 1$ if response is at location B $x_1 = 0$ if not
$x_2 = 1$ if response is at location C $x_2 = 0$ if not

Then the codings for the three locations are as shown in Table 12.16. Thus, when a response measurement is taken at location A, we let $x_1 = 0$ and $x_2 = 0$. A dummy variable was used in Example 12.1 to account for marital status in the prediction of television-viewing time, and dummy variables also appear in other worked examples in this chapter.

The effect of dummy variables is to produce several regression models simultaneously. For example, in the case of the senior citizen television-viewing example, we used the prediction equation

$$\hat{y} = \hat{\beta}_0 + \hat{\beta}_1 x_1 + \hat{\beta}_2 x_2 + \hat{\beta}_3 x_3$$

to characterize the data. In this case x_1 is a dummy variable representing marital status such that

$x_1 = 1$ if the subject is married   and $x_1 = 0$ if the subject is not married

Table 12.16   Codings for the Three Locations

Location	x_1	x_2
A	0	0
B	1	0
C	0	1

Estimates of the two separate regression models representing this relationship are then

$$Married\ (x_1 = 1): \quad \hat{y} = (\hat{\beta}_0 + \hat{\beta}_1) + \hat{\beta}_2 x_2 + \hat{\beta}_3 x_3$$
$$Unmarried\ (x_1 = 0): \quad \hat{y} = \hat{\beta}_0 + \hat{\beta}_2 x_2 + \hat{\beta}_3 x_3$$

Substituting the estimated regression coefficients for Example 12.1 as listed in Table 12.3, we find these fitted models are

$$\hat{y} = (1.4950 - 1.1757) + .03876x_2 - .15228x_3$$

or

$$\hat{y} = .3193 + .03876x_2 - .15228x_3 \qquad \text{if } married$$

and

$$\hat{y} = 1.4950 + .03876x_2 - .15228x_3 \qquad \text{if } not\ married$$

As you can see, the dummy variable provides an *additive effect*, or an adjustment, to the intercept term β_0, which is intended to account for the qualitative factor, marital status.

EXAMPLE 12.3 The conservation director for a public utility wishes to develop a multiple linear regression model to predict the monthly use (kilowatt-hours) y by a residential user on the basis of (a) the number of major electrical appliances in the household, (b) the number of electrical outlets, (c) the number of residents in the household, (d) whether the residence passes local insulation code requirements, and (e) the type of dwelling (part of a multiple residential unit, detached single-family unit with two levels, or detached single-family unit with one level). Write a multiple linear regression model to represent this relationship.

SOLUTION Independent variables represented by (a), (b), and (c) are *quantitative* independent variables, because their values are uniquely measured quantitative values. Predictive factors (d) and (e) are *qualitative* variables and must be included in the model as *dummy variables*. To develop an appropriate model to represent this relationship, we denote

y = monthly use (kilowatt-hours) of residential power subscriber to utility

x_1 = number of major electrical appliances in use in residence

x_2 = number of electrical outlets in residence

x_3 = number of people living in residence

$$x_4 = \begin{cases} 1 \text{ if residence passes local insulation code requirements} \\ 0 \text{ if it doesn't} \end{cases}$$

$$x_5 = \begin{cases} 1 \text{ if residence is part of multiple-unit dwelling} \\ 0 \text{ if it isn't} \end{cases}$$

$$x_6 = \begin{cases} 1 \text{ if residence is detached single-family unit with two levels} \\ 0 \text{ if it isn't} \end{cases}$$

An appropriate multiple linear regression model to represent this relationship can then be written as

$$y = \beta_0 + \beta_1 x_1 + \beta_2 x_2 + \beta_3 x_3 + \beta_4 x_4 + \beta_5 x_5 + \beta_6 x_6 + \epsilon$$

where $y, x_1, x_2, \ldots, x_6$ are defined as above. To estimate the regression parameters, the conservation director must gather data sets $(y, x_1, x_2, \ldots, x_6)$ for a large number of residential users and apply these data to an appropriate multiple regression analysis computer program.

Note that the model *did not* include a separate variable to accommodate the effect of single-family unit with one level since this effect is automatically included when $x_5 = 0$ and $x_6 = 0$.

EXERCISES   Conceptual Level

12.24   What is the distinction between a quantitative and a qualitative variable? What difference is there in the way these variables are entered into a multiple regression model?

12.25   Discuss the following terms as they relate to the multivariable predictor model.

   a.   Quantitative independent variable

   b.   Qualitative independent variable

   c.   A dummy variable

   d.   A response surface

EXERCISES   Skill Level

12.26   How would you incorporate a qualitative variable with two levels into a multiple regression model? A qualitative variable with three levels? With four levels?

12.27   Write a multiple linear regression model relating y to two quantitative variables and one qualitative variable at three levels.

EXERCISES   Applications

12.28   Multiple linear regression is frequently used to develop demand prediction models as an aid in product and manufacturing planning. Suppose you wish to predict the monthly sales (y) of plywood by a diversified forest products company. Indicate whether each prospective independent variable is quantitative or qualitative.

   a.   The price per 4×8 panel charged to retail distributors

   b.   The prevailing mortgage interest rate

   c.   The U.S. region where the retail sales outlet under consideration is located

   d.   The number of housing starts observed during the previous month

   e.   The type of wood veneer used to construct the plywood (e.g., pine, fir, birch)

   f.   The thickness of the plywood under consideration (e.g., 1/4 in., 1/2 in., 3/4 in.)

12.29   The sales manager for a national fire and casualty insurance company would like to develop a regression model to predict y, an agent's annual sales volume, from the following information:

the agent's age, the agent's experience in years, and the sales region to which the agent has been assigned (Northwest, Southwest, Northeast, Southeast).

a. Write the regression model to represent this relationship.

b. Define the independent variables used in the model.

c. Explain how the information included in the various independent variables to be used will be incorporated in the analysis.

 12.30 Refer to Exercise 12.29. Suppose the sales manager wishes to expand his predictive model by incorporating information to include the agent's education, categorizing each agent as a high school graduate, a college attendee but not a college graduate, or a college graduate. Write the regression model to represent this additional information, and define the independent variables used in the model.

12.7

Formulating a Multiple Linear Regression Model

---

first-order linear model

second-order linear model

response plane

The two most common multiple linear regression models used to characterize the relationship between a response y and a set of independent variables are called **first-order** and **second-order linear models**. A first-order model, given by the equation in the display, graphs as a **response plane**.

A First-Order Linear Model

$$y = \beta_0 + \beta_1 x_1 + \beta_2 x_2 + \cdots + \beta_k x_k + \epsilon$$

where $x_1, x_2, \ldots, x_k$ are quantitative predictor variables and ϵ is a random error.

A fitted first-order response surface depicting the relationship between the price y of a stock and two quantitative predictor variables

$$x_1 = \text{annual dividends of stock}$$
$$x_2 = \text{earnings per share of stock}$$

is shown in Figure 12.2.

Second-order linear models in k quantitative predictor variables $x_1, x_2, \ldots, x_k$ include all the terms contained in a first-order model plus all two-way cross product terms $x_1 x_2, x_1 x_3, \ldots, x_2 x_3, \ldots$ and all terms involving the squares, $x_1^2, x_2^2, \ldots, x_k^2$. Thus, a second-order model in two predictor variables is given by the equation in the following display.

A Second-Order Linear Model in Two Predictor Variables

$$y = \beta_0 + \beta_1 x_1 + \beta_2 x_2 + \beta_3 x_1 x_2 + \beta_4 x_1^2 + \beta_5 x_2^2 + \epsilon$$

where x_1 and x_2 are quantitative predictor variables and ϵ is a random error.

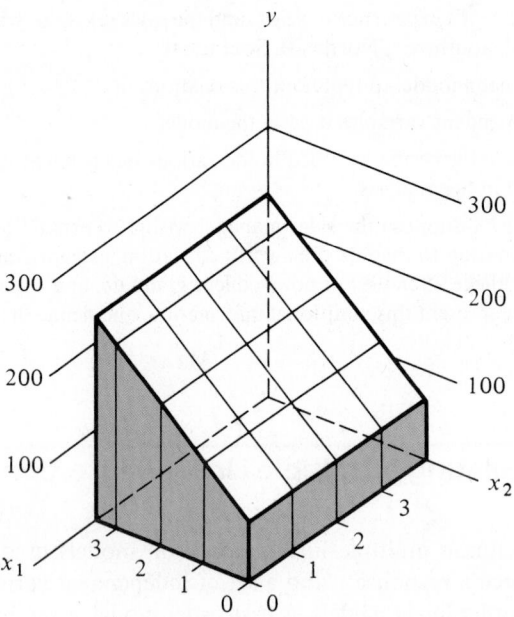

Figure 12.2   The Response Surface for a First-Order Linear Model

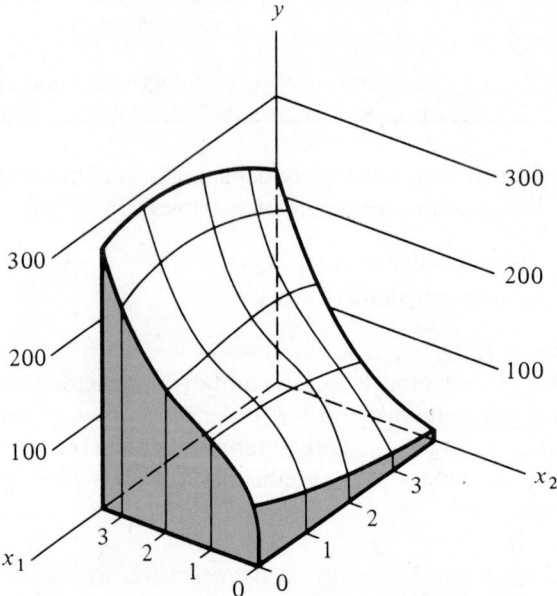

Figure 12.3   The Response Surface for a Second-Order Linear Model

The response surface for a second-order model can be curved (induced primarily by the terms involving $x_1^2, x_2^2, \ldots, x_k^2$) and it also can be warped (or twisted). Warping of the surface is caused by the terms (called *interaction* terms) containing the cross products $x_1 x_2, x_1 x_3, \ldots$. A second-order response surface, again reflecting the relationship between stock price y and the two predictor variables x_1, the annual dividends, and x_2, the earnings per share, is shown in Figure 12.3.

The formulation of the probabilistic model is perhaps the most important part of a regression analysis. Why? Because even if you have all the information contributing to the predictor variables in the model, you may obtain a very poor fit to a set of data if you have not properly formulated the model.

For example, suppose you wish to fit a linear model to the data points shown in Figure 12.4a and you think one predictor variable x contributes most of the information for the prediction of y. If you fit the first-order model

$$y = \beta_0 + \beta_1 x + \epsilon$$

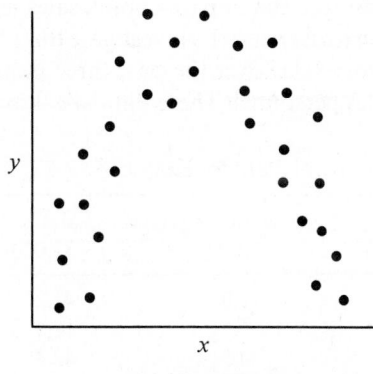

(a) The Data

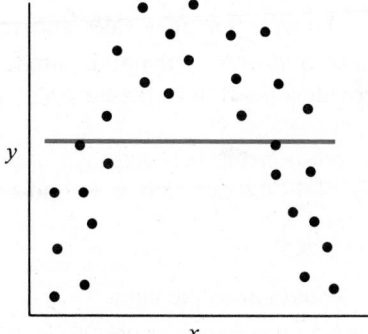

(b) A First-Order Model Fit to the Data   (c) A Second-Order Model Fit to the Data

Figure 12.4   *A Comparison of First-Order and Second-Order Models Fit to the Same Set of Data*

to the data, you can see in Figure 12.4b that you obtain a very poor fit. In contrast, a second-order model

$$y = \beta_0 + \beta_1 x + \beta_2 x^2 + \epsilon$$

provides a good fit to the data (Figure 12.4c).

Similarly, if you fit a first-order linear model to a set of data associated with a curved or warped response surface, you will obtain a poor fit. The only way you can improve the fit is to use a second- (or higher) order linear model that will adjust to the curvatures and warpages of the actual response surface.

Learning to formulate the appropriate linear model for a given application takes experience. You will gain some insight into model building by studying the worked examples in this chapter. More information on this important topic can be found in the texts listed in the references.

EXAMPLE 12.4 The sales manager for a pharmaceutical company is concerned about the complacency that seems to exist among the more experienced sales personnel employed by the firm. She has noticed that, up to a point, sales tend to level off with experience and in some cases even to decrease. To investigate this phenomenon, the sales manager has recorded the territory sales over the past three months and the work experience for 10 of the firm's sales personnel. These data are shown in Table 12.17.

Table 12.17   Sales Personnel Data for Example 12.4

Sales, y ($\times$ \$1,000)	Experience, x (Years)	Sales, y ($\times$ \$1,000)	Experience, x (Years)
36.7	2.0	41.2	4.5
22.9	1.5	18.5	1.0
30.5	4.5	43.4	3.0
9.2	.8	25.5	2.3
38.4	3.5	28.4	5.5

a. Plot the relationship between territory sales and years of work experience.

b. Fit y to x by using a second-order polynomial model. Does the second-order model appear to provide a good fit to these data?

SOLUTION   a. A plot depicting the relationship between sales y and experience x is shown in Figure 12.5. This plot suggests that a second-order model of the form

$$y = \beta_0 + \beta_1 x + \beta_2 x^2 + \epsilon$$

should provide an adequate fit to the data.

b. The BMDP1R regression program was used to fit the second-order model to the sales and work experience data. The results are shown in Table 12.18. The most important quantities appearing in the output are explained in the paragraphs that follow.

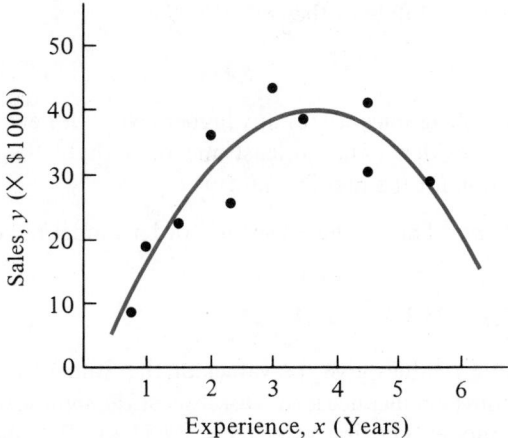

Figure 12.5   A Plot of the Sales and Work Experience Data of Example 12.4 with the Fitted Second-Order Model Superimposed

R^2 (*coefficient of determination*): The proportion of variation in sales explained by the second-order model involving the explanatory variable "work experience" is

$$R^2 = .8029$$

Thus, approximately 80% of the total variation in sales is explained by this model.

F ratio: The F ratio is 14.259 and the associated P(TAIL) probability is .0034. This value is the probability of observing a value of F as large as 14.259 or larger; consequently, the P(TAIL) probability is the significance level of the observed F value.

Table 12.18   BMDP1R Output for the Data of Example 12.4

```
REGRESSION TITLE..............................  BMD REGRESSION #2
DEPENDENT VARIABLE............................  1Y
TOLERANCE.....................................  .0100

ALL DATA CONSIDERED AS A SINGLE GROUP

MULTIPLE R              .8961    STD. ERROR OF EST.    5.4526
MULTIPLE R-SQUARE      .8029

ANALYSIS OF VARIANCE
```

	SUM OF SQUARES	DF	MEAN SQUARE	F RATIO	P(TAIL)
REGRESSION	847.880	2	423.940	14.259	.00340
RESIDUAL	208.120	7	29.731		

VARIABLE		COEFFICIENT	STD. ERROR	STD. REG. COEFF.	T	P(2TAIL)
INTERCEPT		-6.173				
X1	1	25.332	5.439	3.770	4.657	.002
X2	2	-3.499	.872	-3.249	-4.014	.005

Then we would reject the null hypothesis

$$H_0: \beta_1 = \beta_2 = 0$$

at the $\alpha = .0034$ level of significance (or any higher level—for example, $\alpha = .05$). This result provides strong evidence that at least one, or both, of the terms $\beta_1 x$ and $\beta_2 x^2$ contributes information for the prediction of y.

Variable coefficients: The estimated second-order model relating sales y to work experience x is

$$\hat{y} = -6.173 + 25.332x - 3.499x^2$$

t value: The listed t values give the values of the Student's t and the associated P(2TAIL) probabilities (significance levels) for tests of the significance of the individual β parameters in the model. The first values, $t = 4.657$ and P(2TAIL) $= .002$, indicate that we would reject the hypothesis

$$H_0: \beta_1 = 0$$

at a level of significance of $\alpha = .002$. The second t value has an analogous interpretation for β_2.

As noted previously, the most important portion of the computer output is the analysis of variance F test. This test tells us that the model used in this analysis contributes information for the prediction of y. At the same time, a value of R^2 as low as .8029 indicates there is much room for improving the model. As shown by the plot of the data, it is doubtful that the addition to the model of a few higher-order terms in x will greatly increase R^2 (and improve the fit). Apparently, we should add to the model some of the many other variables that are related to sales.

EXAMPLE 12.5 A county assessor wishes to develop a model to relate the market value of single-family residences in a community to the size (in square feet) of each residence and the number of bedrooms contained in the residence. The resulting prediction equation will then be used for assessing the values of single-family residences in the county to establish the amount each homeowner owes in property taxes. The market sales price y, the square feet x_1 of floor space, and the number x_2 of bedrooms were recorded for 20 single-family residences recently sold at fair market value. The data are given in Table 12.19.

An SAS regression program was used to fit the second-order model

$$y = \beta_0 + \beta_1 x_1 + \beta_2 x_2 + \beta_3 x_1^2 + \beta_4 x_2^2 + \beta_5 x_1 x_2 + \epsilon$$

to the data set. The printout obtained from the analysis is shown in Table 12.20. Give an analysis of the printout, and construct a contour diagram for the second-order model fitted to the data by the computer.

SOLUTION R-SQUARE (*coefficient of determination*): Since the coefficient of determination R^2 is .8561, we can say that 85.61% of the variability in sales price for the 20 residences

Table 12.19   Assessor's Data for Example 12.5

$y\ (\times\ \$1,000)$	$x_1\ (\times\ 100)$	x_2	$y\ (\times\ \$1,000)$	$x_1\ (\times\ 100)$	x_2
41.8	8.40	2	65.2	14.75	3
43.4	8.65	2	69.4	17.50	4
45.0	10.25	3	71.6	19.20	4
48.0	9.60	2	77.8	18.70	4
52.2	10.50	2	80.0	18.50	3
54.6	11.25	3	84.2	19.00	4
55.0	14.70	3	89.2	17.50	3
58.4	12.80	2	95.8	19.50	4
60.8	13.85	3	98.6	19.00	4
61.0	15.00	4	104.8	19.45	4

is explained by a second-order model involving x_1 = square footage and x_2 = the number of bedrooms in the residence.

F value: The F VALUE = 16.665 indicates that we reject the null hypothesis $H_0: \beta_1 = \beta_2 = \cdots = \beta_5 = 0$ in the second-order model

$$y = \beta_0 + \beta_1 x_1 + \beta_2 x_2 + \beta_3 x_1^2 + \beta_4 x_2^2 + \beta_5 x_1 x_2 + \epsilon$$

since PROB > F, the observed significance level, is .0001. Thus, at least one of the parameters is significantly different from zero.

Table 12.20   SAS Output for the Data of Example 12.5

```
DEP VARIABLE: Y       SALES PRICE

                   SUM OF        MEAN
SOURCE      DF     SQUARES       SQUARE      F VALUE    PROB>F

MODEL        5     6035.350      1207.070    16.665     0.0001
ERROR       14     1014.058      72.432736
C TOTAL     19     7049.408

       ROOT MSE     8.510742     R-SQUARE    0.8561
       DEP MEAN     67.840000    ADJ R-SQ    0.8048
       C.V.         12.54532

                   PARAMETER     STANDARD    T FOR H0:                VARIABLE
VARIABLE    DF     ESTIMATE      ERROR       PARAMETER=0   PROB>|T|   LABEL

INTERCEPT    1     52.645976     42.464669      1.240      0.2354     INTERCEPT
X1           1     -3.750419      6.150974     -0.610      0.5518     FLOOR SPACE
X2           1      3.510346     30.918417      0.114      0.9112     BEDROOMS
X3           1      0.305505      0.422104      0.724      0.4811     X1 SQ
X4           1     -1.305688     10.188010     -0.128      0.8998     X2 SQ
X5           1      0.026101      3.198031      0.008      0.9936     X1 X2
```

Parameter estimates: The fitted second-order model relating sales price to square footage and number of bedrooms for the 20 residences is

$$\hat{y} = 52.646 - 3.750x_1 + 3.510x_2 + .306x_1^2 - 1.306x_2^2 + .026x_1x_2$$

T FOR HO: PARAMETER = 0: Each t value tests the hypothesis

$$H_0: \beta_j = 0$$

for its associated regression parameter β_j. Recall, however, that this is not simply a test of significance of the variable x_i as a predictor of y. Since the two-tailed p-values given in the printout all exceed .05, we would not reject H_0 for any of the five parameters of the model, a clear contradiction of the result suggested by the analysis of variance F test. What is implied is that none of the terms x_1, x_2, x_1^2, x_2^2, and x_1x_2 is significant as a predictor of y *in the presence of the other four terms* (probably because of a duplication of information content). However, when considered individually, any or all of the five terms may be valuable predictors of market sales price y. A contour diagram for the fitted second-order model

$$\hat{y} = 52.646 - 3.750x_1 + 3.510x_2 + .306x_1^2 - 1.306x_2^2 + .026x_1x_2$$

relating sales price to square footage and the number of bedrooms is shown in Figure 12.6.

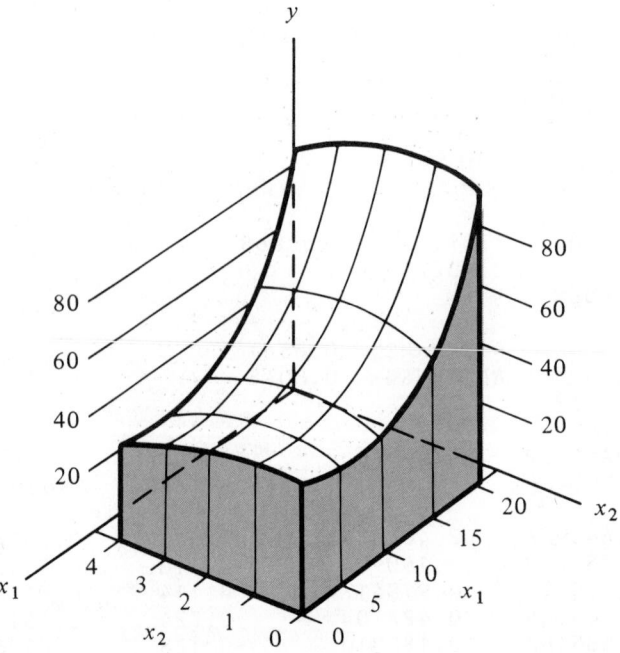

Figure 12.6   Contour Diagram for the Second-Order Model Fitted to the Data of Example 12.6

EXERCISES Conceptual Level

12.31 What is the distinction between a first-order and a second-order regression model? When would it be appropriate to use each of these models?

12.32 What is meant by model building in regression analysis?

EXERCISES Skill Level

12.33 Suppose that the mean response for given values of x_1 and x_2 is given by the equation

$$E(y) = 4 - 2x_1 + x_2 + 4x_1^2 + 2x_2^2$$

Graph the contour curves of $E(y)$ as a function of x_1 (in the interval $0 \le x_1 \le 5$) for values of x_2 equal to 0, 1, and 2. Note that except for vertical shifts the contour curves are portions of identical parabolas.

12.34 Continuing Exercise 12.33, add a cross product term to the model. Note that the shapes of the contour curves depend on the value assigned to x_2. For example, let

$$E(y) = 4 - 2x_1 + x_2 - 3x_1x_2 + 4x_1^2 + 2x_2^2$$

Graph $E(y)$ for $x_2 = 0$, 1, and 2.

EXERCISES Applications

 12.35 Corporate marketing strategists continually search for new products, since even the most popular products tend to have a limited profitable life span. The marketing director for a manufacturer of household appliances recognizes that the profits realized from appliances produced by his company are quadratic with time. That is, profits initially increase during the introduction of a new product but then taper off in later years owing to product obsolescence and competition. Suppose the manufacturing company produces household appliances that are categorized as either large appliances or small appliances. An appropriate profit estimation model for each appliance item is then

$$y = \beta_0 + \beta_1 x_1 + \beta_2 x_2 + \beta_3 x_1 x_2 + \beta_4 x_1^2 + \epsilon$$

where

y = profits
x_1 = time since inital market introduction of appliance
$x_2 = \begin{cases} 1 \text{ if appliance is "large" appliance} \\ 0 \text{ if it is "small" appliance} \end{cases}$

a. What hypothesis would the marketing director test if he wishes to determine whether the relationship between profits and time since introduction is really quadratic and not simply linear?

b. In the context of this exercise, what is the meaning if the marketing director *falls to reject* the hypothesis that $\beta_4 = 0$?

   c. In the context of this exercise, what is the meaning if the marketing director *falls to reject* the hypothesis that $\beta_2 = 0$?

   d. In the context of this exercise, what is the meaning if the marketing director *rejects* the hypothesis that $\beta_3 = 0$?

 12.36 The relationship between interest rates and the residential housing industry has been observed for some time. High interest rates cause the monthly payment on a home mortgage to be noticeably increased. For example, a 1% increase in the interest-rate causes a \$20.83 increase in the monthly payment on a 30-year, \$25,000 mortgage. Interest rates, in turn, are affected by economic factors such as the total money supply and the yield on U.S. government securities. Typically, interest rates rise when the money supply is tightened and when Treasury bill yields rise.

     Suppose you have been assigned the task of investigating the relationship between new residential housing starts y and the interest rates x_1 on covential 30-year mortgages, the total money supply x_2, and the yield x_3 on U.S. government securities for each of the past 36 months. Write a second-order model to represent this relationship. Would you expect the interaction terms to be important factors in this analysis? Explain.

12.8

Estimating $E(y)$, Predicting y, and Explaining Relationships Among the Variables

Until now, we have directed our attention to building and constructing an appropriate multiple linear regression model. Once established, a best-fit prediction equation resulting from a multiple linear regression analysis can be used for any or all of the purposes listed in the following display.

Uses of a Multiple Linear Regression Analysis

1. **Estimation.** The prediction equation can be used to *estimate* $E(y)$, the mean value of y, for given values of the predictor variables.
2. **Prediction.** The prediction equation can be used to *predict* some future value of y for given values for $x_1, x_2, \ldots, x_k$.
3. **Explanation.** If a prediction equation provides a good fit to a set of data (R^2 is large) and the number of predictor variables is not too large, then the equation may help you *understand* and *explain* the process you are investigating.

     Estimates of the mean value of y for given values of $x_1, x_2, \ldots, x_k$ or predictions of specific values of y(to be observed in the future) for given values of $x_1, x_2, \ldots, x_k$ can be obtained by substituting numerical values of $x_1, x_2, \ldots, x_k$ into the prediction equation.

     For example, suppose that a mail-order firm wishes to relate the Christmas holiday sales y to two predictor variables (we will keep the number of predictor variables small for purposes of illustration), the number of mailings x_1 and the number of months x_2

that the mailing preceded Christmas. Furthermore, suppose the company varied x_1 and x_2 over 20 somewhat similar sales regions and measured the amount of sales y for each. Then the firm fitted a prediction equation to the data it collected and obtained the following result:

$$\hat{y} = 1.3 + 1.7x_1 - .1x_1^2 + 1.0x_1x_2 - .3x_1^2x_2$$

where

x_1, expressed in units of 100,000, is measured over the interval $.5 \leq x_1 \leq 2.5$

x_2, expressed in months, is measured over the interval $1 \leq x_2 \leq 3$

y is measured in units of $100,000

The best estimate of the mean sales $E(y)$ for a given combination of values of x_1 and x_2—say, $x_1 = 1$ (100,000 mailings) and $x_2 = 2$ (mailings sent two months before Christmas)—is obtained by substituting $x = 1$ and $x_2 = 2$ into the prediction equation. Then

$$\begin{aligned} \hat{y} &= 1.3 + 1.7x_1 - .1x_1^2 + 1.0x_1x_2 - .3x_1^2x_2 \\ &= 1.3 + 1.7(1) - .1(1)^2 + 1.0(1)(2) - .3(1)^2(2) = 4.3 \end{aligned}$$

or $\hat{y} = \$430,000$.

As noted in Sections 11.6 and 11.7, not only is $\hat{y}$ the best estimate of the mean value of y for given values of x_1 and x_2, but it also gives the best prediction of some value of y to be observed in the future. That is, if we were to select a region, send out 100,000 mailings ($x_1 = 1$) two months ($x_2 = 2$) prior to Christmas, the predicted holiday sales for the region would be $430,000.

We can construct a confidence interval for $E(y)$ and a prediction interval for y using a procedure similar to that employed for the simple linear model in Sections 11.6 and 11.7. However, the formulas for these intervals are much too complex to present in a text at this level. Fortunately, the computed intervals are included as an option in some computer program packages (see Table 12.1). If you do not have such a computer program available, you will need to become familiar with the actual formulas, which, because of their complexity, are always expressed in matrix notation. This notation and the formulas for the confidence interval for $E(y)$ and the prediction interval for some future value of y are presented and explained in *An Introduction to Linear Models and the Design and Analysis of Experiments*, by W. Mendenhall; the matrix approach to regression analysis is described in *A Handbook for Linear Regression*, by M. S. Younger (see the references).

EXAMPLE 12.6   Although we would expect demand to decline as price increases when competing products are available at a lower price, that is not always the case. In fact, the relationship between demand and price can often be approximated by a second-order model; slight price increases result in decreased demand, and significant price increases result in perception of increased quality and demand. In an attempt to study the relationship between price and demand and to develop an appropriate pricing policy, a distributor of a common brand of whiskey, which ordinarily sells for $5.00 per 750-milliliter (ml) bottle, conducted a pricing experiment in 15 different market areas over a 12-month period, using five different price levels. The results of the experiment are shown in Table 12.21.

Table 12.21   Pricing Data for Example 12.6

No. of Cases Sold per Month per 10,000 Population, y	Price per 750 ml x
23, 20, 21	$5.00
19, 21, 18	5.50
18, 17, 20	6.00
21, 19, 20	6.50
25, 24, 22	7.00

a. Fit the second-order model $y = \beta_0 + \beta_1 x + \beta_2 x^2 + \epsilon$ to these data.

b. Predict y, the number of cases likely to be sold per month per 10,000 population in a sales area in which the price per bottle is $5.00. Predict y when $x = \$6.00$; when $x = \$7.00$.

c. Construct a 95% prediction interval for y when $x = \$5.00$; when $x = \$6.00$; when $x = \$7.00$.

d. Construct a 95% confidence interval for $E(y)$, the average number of cases sold per month per 10,000 population, when $x = \$5.00$; when $x = \$6.00$; when $x = \$7.00$.

SOLUTION   A common regression program, with an option providing the computations of prediction intervals for y and confidence intervals for $E(y)$, provided the results in Table 12.22 follow.

a. The prediction equation for estimating y, the predicted number of cases that will be sold per month per 10,000 population in a sales area in which the price per unit is given in terms of x, is

$$\hat{y} = 156.1333 - 46.9333x + 4.0000x^2$$

b. At $x = \$5.00$, we have

$$\hat{y} = 156.1333 - 46.9333(5.0) + 4.0000(25.0) = 21.4668$$

or more than 21 cases. Similarly, at other unit prices we find the following as estimates of y:

$$\hat{y} = 18.5335 \qquad \text{when} \qquad x = \$6.00$$
$$\hat{y} = 23.6002 \qquad \text{when} \qquad x = \$7.00$$

The model for estimating a particular outcome y or the mean response $E(y)$ is the same. For example, the estimated mean number of cases sold per month per 10,000 population when the price per bottle is $5.00 is $\hat{E}(y) = 21.4668$. Similarly, when $x = \$6.00$ and $x = \$7.00$, the mean estimates are, respectively, $\hat{E}(y) = 18.5335$ and $\hat{E}(y) = 23.6002$.

c.,d. The 95% prediction intervals for y and the 95% confidence intervals for $E(y)$ can be read directly from the computer printout. They are given in Table 12.23.

Table 12.22   Computer Output for the Data of Example 12.6

```
MULTIPLE R              .8393
R SQUARE               .7045
STD. ERROR OF EST.    1.3292
```

ANALYSIS OF VARIANCE

	DF	SUM OF SQUARES	MEAN SQUARE	F RATIO
REGRESSION	2	50.533	25.266	14.301
RESIDUAL	12	21.200	1.767	

INDIVIDUAL ANALYSIS OF VARIABLES

VARIABLE	COEFFICIENT	STD. ERROR	F VALUE
(CONSTANT	156.1333)		
PRICE (X)	-46.9333	9.8564	22.6739
PRICE SQ. (X2)	4.0000	.8204	23.7722

OBSERVATION	OBS. X VALUE	OBS. Y VALUE	PREDICTED Y VALUE	LOWER 95% CL FOR E(Y)	UPPER 95% CL FOR E(Y)
1	5.0000	23.0000	21.4667	19.8931	23.0402
2	5.0000	20.0000	21.4667	19.8931	23.0402
3	5.0000	21.0000	21.4667	19.8931	23.0402
4	5.5000	19.0000	19.0000	17.9810	20.0190
5	5.5000	21.0000	19.0000	17.9810	20.0190
6	5.5000	18.0000	19.0000	17.9810	20.0190
7	6.0000	18.0000	18.5333	17.3681	19.6986
8	6.0000	17.0000	18.5333	17.3681	19.6986
9	6.0000	20.0000	18.5333	17.3681	19.6986
10	6.5000	21.0000	20.0667	19.0477	21.0857
11	6.5000	19.0000	20.0667	19.0477	21.0857
12	6.5000	20.0000	20.0667	19.0477	21.0857
13	7.0000	25.0000	23.6000	22.0264	25.1736
14	7.0000	24.0000	23.6000	22.0264	25.1736
15	7.0000	22.0000	23.6000	22.0264	25.1736

OBSERVATION	OBS. X VALUE	OBS. Y VALUE	PREDICTED Y VALUE	LOWER 95% CL FOR Y	UPPER 95% CL FOR Y
1	5.0000	23.0000	21.4667	18.4230	24.5104
2	5.0000	20.0000	21.4667	18.4230	24.5104
3	5.0000	21.0000	21.4667	18.4230	24.5104
4	5.5000	19.0000	19.0000	16.2024	21.7976
5	5.5000	21.0000	19.0000	16.2024	21.7976
6	5.5000	18.0000	19.0000	16.2024	21.7976
7	6.0000	18.0000	18.5333	15.6792	21.3874
8	6.0000	17.0000	18.5333	15.6792	21.3874
9	6.0000	20.0000	18.5333	15.6792	21.3874
10	6.5000	21.0000	20.0667	17.2691	22.8643
11	6.5000	19.0000	20.0667	17.2691	22.8643
12	6.5000	20.0000	20.0667	17.2691	22.8643
13	7.0000	25.0000	23.6000	20.5563	26.6437
14	7.0000	24.0000	23.6000	20.5563	26.6437
15	7.0000	22.0000	23.6000	20.5563	26.6437

Table 12.23   Prediction and Confidence Intervals for Example 12.6

Price per Bottle	95% Prediction Interval for y	95% Confidence Interval for $E(y)$
$5.00	18.4230 to 24.5104	19.8931 to 23.0402
6.00	15.6792 to 21.3874	17.3681 to 19.6986
7.00	20.5563 to 26.6437	22.0264 to 25.1736

Notice that in each case the 95% prediction interval for y is wider than the 95% confidence interval for $E(y)$. As we noted in Sections 11.6 and 11.7, this reflects the fact that the variance of the error of predicting a particular value y exceeds the variance of the error of estimating the mean value $E(y)$. In addition, since each of these variances depends on the particular values selected for the independent variables in computing y, the three prediction intervals and the three confidence intervals vary in width.

◆ ◆

As we have seen, the prediction equation is of value for estimating $E(y)$ and for predicting some future value of y. But it may also help us understand the process under study. For example, consider the prediction equation for the sales of the mail-order firm. An easy way to view this relationship is to graph the holiday sales y as a function of the number of mailings x_1 for various dates of mailings—say, $x_2 = 1$, 2, or 3 months prior to Christmas. For example, substituting $x_2 = 1$ month into the prediction equation, we obtain

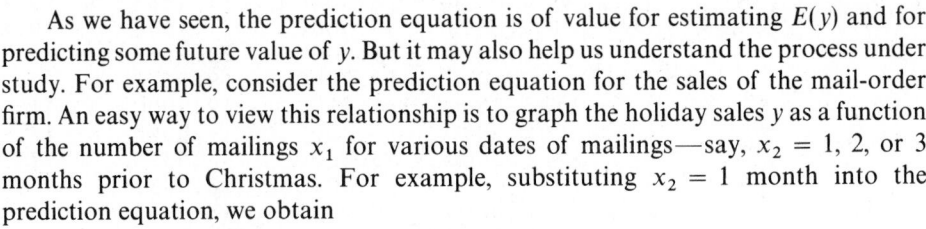

$$\hat{y} = 1.3 + 1.7x_1 - .1x_1^2 + 1.0x_1x_2 - .3x_1^2x_2$$
$$= 1.3 + 1.7x_1 - .1x_1^2 + 1.0x_1(1) - .3x_1^2(1)$$
$$= 1.3 + 2.7x_1 - .4x_1^2$$

This equation gives the predicted sales if the mailings are made $x_2 = 1$ month before Christmas. Similarly, for $x_2 = 2$ months and $x_2 = 3$ months, we obtain the following two prediction equations:

$$\textit{For } x_2 = 2: \quad \hat{y} = 1.3 + 3.7x_1 - .7x_1^2$$
$$\textit{For } x_2 = 3: \quad \hat{y} = 1.3 + 4.7x_1 - 1.0x_1^2$$

These three equations give the predicted holiday sales as a function of the number of mailings x_1. Graphs of these three sales curves are shown in Figure 12.7.

Note how the shapes of the sales curves change for the three values of x_2. This result tells us that the relationship between predicted holiday sales y and the number of mailings x_1 is dependent on the time of mailing x_2. When this situation occurs, we say
interaction that x_1 and x_2 **interact**. Or putting it another way, the effect of x_1 on the predicted value of y is dependent on the value of x_2 (or vice versa). This example illustrates how graphs of $\hat{y}$ help us understand the relationship between the predicted value of y and a set of predictor variables.

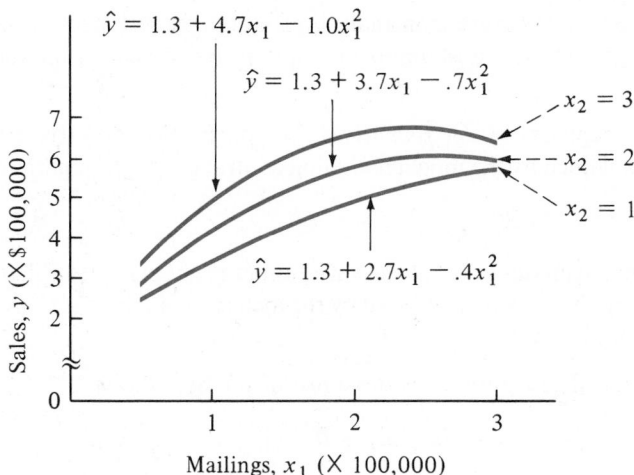

Figure 12.7   A Graph of Holiday Sales y as a Function of the Number of Mailings x_1

EXAMPLE 12.7   A study was undertaken to examine the profit y per sales dollar earned by a construction company and its relationship to the size x_1 of the construction contract and the number x_2 of years of experience of the construction superintendent. An additional purpose of the study was to investigate the interaction effect of the size of the contract and the experience of the construction superintendent on profit. Data were obtained from a sample of $n = 18$ construction projects undertaken by the construction company over the past two years. These data are shown in Table 12.24. Fit the model

$$y = \beta_0 + \beta_1 x_1 + \beta_2 x_2 + \beta_3 x_1^2 + \beta_4 x_1 x_2 + \epsilon$$

to the construction contract data. Carefully interpret your analysis. Graph the profit y per sales dollar as a function of contract size x_1 for the three levels for experience x_2. What does this graph suggest about the effect of the interaction between contract size and superintendent experience?

Table 12.24   Construction Contract Data for Example 12.7

Profit, y	Contract Size, x_1 ($\times \$1000$)	Superintendent Experience, x_2 (Years)	Profit, y	Contract Size, x_1 ($\times \$1000$)	Superintendent Experience, x_2 (Years)
2.0	5.1	4	5.0	4.3	6
3.5	3.5	4	6.0	2.9	2
8.5	2.4	2	7.5	1.1	2
4.5	4.0	6	4.0	2.6	4
7.0	1.7	2	4.0	4.0	6
7.0	2.0	2	1.0	5.3	4
2.0	5.0	4	5.0	4.9	6
5.0	3.2	2	6.5	5.0	6
8.0	5.2	6	1.5	3.9	4

SOLUTION   Using a BMDP2R regression analysis program, we obtained the results in Table 12.25. An analysis of the most important quantities appearing in the computer output follows.

R^2 (*coefficient of determination*): The proportion of variation in profit per sales dollar explained by the model involving contract size and superintendent experience is

$$R^2 = .8628$$

Thus, approximately 14% of the variation in profit per sales dollar is left unexplained, an amount that might be reduced by the inclusion of additional predictor variables not used in our analysis.

F ratio: The F ratio provides a test of the hypothesis

$$H_0 : \beta_1 = \beta_2 = \beta_3 = \beta_4 = 0$$

The calculated F ratio from the data, $F = 20.432$, exceeds the critical value of F based on $v_1 = 4$ and $v_2 = 13$ degrees of freedom, which is $F_{.05} = 3.18$ and in fact is significant at the *p*-value $< .005$ level of significance. We reject the null hypothesis and conclude that our chosen model contributes information for the prediction of y.

Variable coefficients: The prediction equation relating profit per sales dollar to the size x_1 of the contract and the experience x_2 of the construction superintendent is

$$\hat{y} = 19.3048 - 1.4874x_1 - 6.3707x_2 - .7521x_1^2 + 1.7169x_1x_2$$

F value: Each computed F value provides a separate test of the hypothesis

$$H_0 : \beta_j = 0 \qquad j = 1, 2, 3, 4$$

Table 12.25   BMDP2R Output for the Data of Example 12.7

```
MULTIPLE R              .9289
R SQUARE               .8628
STD. ERROR OF EST.     .9708
```

ANALYSIS OF VARIANCE

	DF	SUM OF SQUARES	MEAN SQUARE	F RATIO
REGRESSION	4	77.026	19.256	20.432
RESIDUAL	13	12.252	.942	

INDIVIDUAL ANALYSIS OF VARIABLES

VARIABLE	COEFFICIENT	STD. ERROR	F VALUE
(CONSTANT	19.3048)		
CONT. SIZE (X1)	-1.4874	1.1779	1.5944
EXPER. (X2)	-6.3707	1.0424	37.3506
SIZE SQ. (X1 SQ.)	-.7521	.2253	11.1460
SIZE X EXP. (X1X2)	1.7169	.2538	45.7593

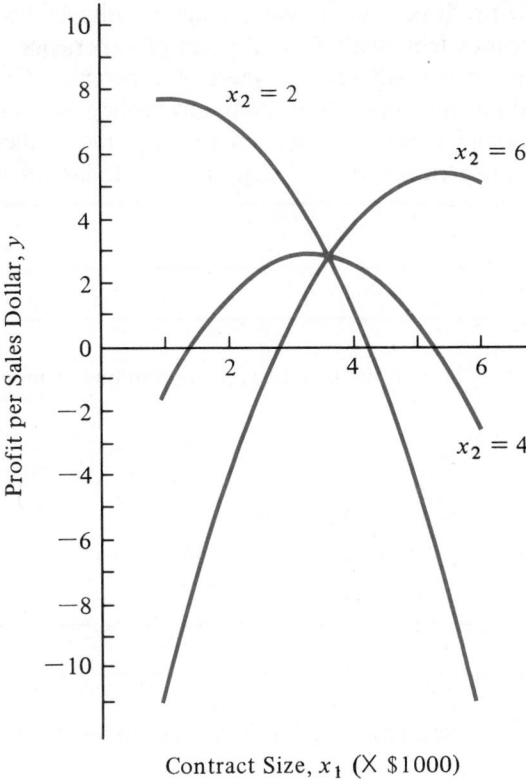

Contract Size, x_1 (X $1000)

Figure 12.8   A Graph of the Profit y per Sales Dollar as a Function of the Size of the Contract x_1

for its associated regression parameter β_j. The tabulated $F_{.05}$ with $v_1 = 1$ and $v_2 = 13$ degrees of freedom is $F_{.05} = 4.67$. Thus, the parameters $\beta_2, \beta_3,$ and β_4 are assumed to be significantly different from 0, but β_1 is not. Recall that this result does *not* imply that the term x_1 contributes no information for the prediction of y. It could mean that the contribution of x_1 is small when in the presence of the other predictors x_1^2, x_2, and $x_1 x_2$.

The F value associated with the contract size and construction superintendent term $(x_1 x_2)$ is $F = 45.7593$, which greatly exceeds the critical value $F_{.05} = 4.67$. There appears to be a strong interaction between these two factors. This interaction is especially evident when we view Figure 12.8, which provides a graph of the separate functions relating profits y to the size x_1 of the contract for the three different levels of superintendent experience x_2.

The three separate functions appearing in Figure 12.8 are as follows:

$$\text{For } x_2 = 2: \quad \hat{y} = 6.5634 + 1.9464x_1 - .7521x_1^2$$
$$\text{For } x_2 = 4: \quad \hat{y} = -6.1780 + 5.3802x_1 - .7521x_1^2$$
$$\text{For } x_2 = 6: \quad \hat{y} = -18.9194 + 8.8140x_1 - .7521x_1^2$$

Notice that the profit per sales dollar declines rapidly as the size of the contract increases for superintendents with $x_2 = 2$ years of experience. The opposite effect occurs for superintendents with $x_2 = 6$ years of experience. On the basis of these sample contract data, an appropriate company policy is to assign construction superintendents with a few years of experience only to the smaller jobs and to assign those with the most experience to construction jobs with larger-sized contracts.

EXERCISES   Conceptual Level

12.37   What are the differences in each of the following applications of multiple linear regression analysis?

    a.   Estimating $E(y)$

    b.   Predicting y

    c.   Explaining relationships among the variables

12.38   What is meant when we say that two predictor variables interact?

EXERCISES   Skill Level

12.39   The least squares equation relating the dependent variable y to two independent variables x_1 and x_2 is given as

$$\hat{y} = 3.47 + .2x_1 - 1.64x_2 + .63x_1x_2$$

    a.   Write the least squares equation as a function of x_1 when $x_2 = 1, 2,$ and 3.

    b.   Graph the three equations found in part a on the same sheet of graph paper.

    c.   What type of curve is represented by the three equations? Does there appear to be an interaction between x_1 and x_2?

12.40   The least squares equation relating the dependent variable y to two independent variables x_1 and x_2 is given as

$$\hat{y} = -2 + 3x_1 - x_2 + 3x_1^2 + 2x_2^2 - x_1x_2$$

    a.   Write the least squares equation as a function of x_1 when $x_2 = 1, 2,$ and 3.

    b.   Graph the three equations found in part a on the same sheet of graph paper.

    c.   What type of curves is represented by the three equations? Does there appear to be an interaction between x_1 and x_2?

12.41   A dependent variable y is recorded along with two independent variables x_1 and x_2 in Table 12.26.

    a.   Plot the dependent variable y as a function of x_1 when $x_2 = 0, 2,$ and 4. What type of relationship appears to exist between y and x_1?

    b.   Plot y as a function of x_2 when $x_1 = 0, 1,$ and 2. What type of relationship appears to exist between y and x_2?

Table 12.26   Data for Exercise 12.41

Observation	y	x_1	x_2	Observation	y	x_1	x_2
1	50	0	0	9	15	1	3
2	40	0	1	10	10	1	4
3	30	0	2	11	40	2	0
4	25	0	3	12	30	2	1
5	30	0	4	13	10	2	2
6	35	1	0	14	15	2	3
7	30	1	1	15	23	2	4
8	20	1	2				

c. A MINITAB regression program was used to fit the second-order model

$$y = \beta_0 + \beta_1 x_1 + \beta_2 x_2 + \beta_3 x_1^2 + \beta_4 x_2^2 + \beta_5 x_1 x_2 + \epsilon$$

to the data set. The printout is shown in Table 12.27. Does the second-order model contribute information for the prediction of y? Test by using $\alpha = .05$.

d. Interpret the value of R^2 given in the printout.

e. Use the regression equation to predict y when $x_1 = 0$ and $x_2 = 3$. Compare this value with the value given in the printout. Find the 95% prediction interval for y.

Table 12.27   MINITAB Output for the Data of Exercise 12.41

```
MTB > REGRESS C1 5 C2-C6;
SUBC> PREDICT 0 3 0 9 0.

THE REGRESSION EQUATION IS
Y = 51.7 - 20.9 X1 - 15.6 X2 + 7.30 X1SQ + 2.40 X2SQ + 0.30 X1X2

PREDICTOR          COEF          STDEV          T-RATIO
CONSTANT          51.676         3.761           13.74
X1               -20.900         5.809           -3.60
X2               -15.552         3.232           -4.81
X1SQ               7.300         2.598            2.81
X2SQ              2.4048         0.7319           3.29
X1X2               0.300         1.061            0.28

S = 4.743        R-SQ = 89.4%        R-SQ(ADJ) = 83.4%

ANALYSIS OF VARIANCE

SOURCE         DF              SS              MS
REGRESSION      5          1699.25          339.85
ERROR           9           202.49           22.50
TOTAL          14          1901.73
```

(*continues*)

Table 12.27   (Continued)

SOURCE	DF	SEQ SS
X1	1	324.90
X2	1	952.03
X1SQ	1	177.63
X2SQ	1	242.88
X1X2	1	1.80

UNUSUAL OBSERVATIONS

OBS.	X1	Y	FIT	STDEV.FIT	RESIDUAL	ST.RESID
13	2.00	10.00	18.79	2.58	-8.79	-2.21R

R DENOTES AN OBS. WITH A LARGE ST. RESID.

FIT	STDEV.FIT	95% C.I.	95% P.I.
26.66	2.63	(20.71, 32.61)	(14.39, 38.93)

EXERCISES   Applications

12.42 The administrator for an organization that conducts management seminar programs is interested in examining the relationship between seminar enrollments (y), the number of mailings (x_1), and the lead time of mailings (x_2) of seminar announcements. Data were obtained from a sample of $n = 25$ management seminars offered by the organization and are listed in Table 12.28. To allow for an analysis of all possible considerations, the administrator fit a second-order model

$$y = \beta_0 + \beta_1 x_1 + \beta_2 x_2 + \beta_3 x_1^2 + \beta_4 x_2^2 + \beta_5 x_1 x_2 + \epsilon$$

to the data set by using an SAS regression computer program. The results in Table 12.29 were obtained.

Table 12.28   Seminar Program Data for Exercise 12.42

Enrollment, y	No. of Mailings, x_1 ($\times 1,000$)	Lead Time, x_2 (Weeks)	Enrollment, y	No. of Mailings, x_1 ($\times 1,000$)	Lead Time, x_2 (Weeks)
27	6.5	3	35	7.5	12
29	6.5	2	27	4.9	9
41	13.0	15	19	3.7	6
36	8.1	13	36	9.1	12
22	4.0	6	43	23.0	13
40	11.5	13	40	23.5	10
52	18.0	17	38	9.0	9
39	10.0	12	40	7.0	12
27	7.1	4	42	12.5	16
28	6.5	10	21	5.0	6
24	7.0	5	29	6.8	12
29	7.3	11	35	7.2	14
33	7.5	12			

Table 12.29   SAS Output for the Data of Exercise 12.42

DEPENDENT VARIABLE: ENROLLMENT

SOURCE	DF	SUM OF SQUARES	MEAN SQUARE	F VALUE
MODEL	5	1391.723645	278.344724	31.26
ERROR	19	169.179451	8.904182	
CORRECTED TOTAL	24	1560.903099		

R-SQUARE	0.8916

VARIABLE	DF	PARAMETER ESTIMATE	STANDARD ERROR	T RATIO	PROB>|T|
INTERCEPT	1	8.267885	5.518542	1.4982	0.1505
NO. MAILINGS (X1)	1	3.362369	0.901247	3.7308	0.0014
LEAD TIME (X2)	1	-0.103668	0.771424	-0.1344	0.8945
X1 SQ	1	-0.084138	0.032465	-2.5917	0.0179
X2 SQ	1	0.052671	0.063488	0.8296	0.4171
(X1)(X2)	1	0.019078	0.064719	-0.2948	0.7714

a. What percentage of the variation in enrollments is explained by the second-order model?

b. Write the equation for the second-order model that relates enrollments to the number of mailings and mailing lead time.

c. Separately test for the significance of each of the five terms used in the second-order model. That is, separately test $H_0 : \beta_j = 0$ for $j = 1, 2, 3, 4,$ and 5.

d. From the results of part c, what simplified model would you recommend for the prediction of seminar enrollments?

12.43   Refer to Example 12.7 and the study concerning the profits of a construction company.

a. Estimate the gain in profits if the contract size is increased from $3,000 to $4,000 for a construction superintendent with two years of experience.

b. Repeat the instructions of part a for a superintendent with four years of experience; with six years of experience.

c. Explain why you cannot estimate the gain in profits for a $1,000 increase in the contract size, from $3,000 to $4,000, without specifying the work experience of the construction superintendent?

MKTG   **12.44**   Suppose we wish to estimate $E(y)$, the weekly sales of a retail consumer product, as a function of its retail price (in thousands of dollars), the number of competing products offered in the same marketplace, and location. In this case location refers to the retail outlet (store) in which the product is sold and involves *three* possible store sites.

a. Write a *first-order* linear regression model to relate expected sales $E(y)$ to retail price, the number of competing products, and location.

b. Now rewrite the model to estimate $E(y)$ assuming that retail price is dependent upon location (that is, an interaction exists between price and location).

c. How can we determine whether a significant interaction between retail price and location does, in fact, exist?

d. For the model developed in b, explain how we could use the model to estimate the change in expected sales for an increase of $1,000 in sales price for a given location and a fixed number of competing products.

12.9

Testing Portions of a Model

This section is concerned with a test of an hypothesis that one or more β parameters in the model equals zero. For example, one of the tests given in the computer printout for the television viewer data of Section 12.3 used the F statistic to test the hypothesis that $\beta_1 = \beta_2 = \beta_3 = 0$. The F statistic was also used to test an hypothesis that an individual β parameter equals zero. In this section we will explain in greater detail the reasoning behind the test and describe several situations in which the test can be applied.

Suppose you have a model

$$y = \beta_0 + \beta_1 x_1 + \beta_2 x_2 + \cdots + \beta_k x_k + \epsilon$$

or, equivalently,

$$E(y) = \beta_0 + \beta_1 x_1 + \beta_2 x_2 + \cdots + \beta_k x_k$$

and you want to know whether certain variables contribute information for the prediction of y. Or putting it another way, you are asking whether those terms should be included in the model. If a set of x's in the model contributes no information for the prediction of y, then the β parameters associated with those x's should equal zero. Consequently, testing to determine whether a set of terms should be included in the model is a test of an hypothesis that a set of β parameters equals zero.

Suppose we have two models for $E(y)$, one that we will call the "complete model" (call this model 2) and another that we will call the "reduced model" (call this model 1). The reduced model includes only a portion of the terms in the complete model. Thus, the complete model contains all the terms in the reduced model plus some additional terms. We wish to see whether the additional terms contribute information for the prediction of y. That is equivalent to testing the hypothesis that the β parameters associated with these additional terms equal zero.

We represent the reduced and complete models as follows:

$$\text{Model 1 } (reduced\ model): \quad E(y) = \beta_0 + \beta_1 x_1 + \beta_2 x_2 + \cdots + \beta_g x_g$$
$$\text{Model 2 } (complete\ model): \quad E(y) = \beta_0 + \beta_1 x_1 + \beta_2 x_2 + \cdots + \beta_g x_g$$
$$+ \beta_{g+1} x_{g+1} + \cdots + \beta_k x_k$$

Note that the complete model contains, in addition to all the terms in the reduced model, the terms $\beta_{g+1} x_{g+1}, \beta_{g+2} x_{g+2}, \ldots, \beta_k x_k$.

The test we describe here is intuitive. We use the method of least squares to fit the reduced model and calculate the sum of squares of error SSE_1 (the sum of squares of the deviations between the y values and the fitted model). Then we fit the complete model and find the sum of squares for error SSE_2. Then we compare the two sums of squares for error, SSE_1 and SSE_2. If the variables $x_{g+1}, x_{g+2}, \ldots, x_k$ contribute information for the prediction of y, then SSE_2 should be significantly smaller than SSE_1. That is, retaining these variables in the model should reduce the sum of squares of the errors of prediction. Consequently, the larger the difference ($SSE_1 - SSE_2$), the greater is the weight of evidence to indicate that the terms should be included in the model. Or the greater is the weight of evidence indicating that at least one of the parameters $\beta_{g+1}, \beta_{g+2}, \ldots, \beta_k$ differs from zero.

It can be shown that whenever we add terms to a model, we reduce the sum of squares for error. The question is whether the reduction in the sums of squares for error is due merely to chance or whether it is due to information contributed by $x_{g+1}, x_{g+2}, \ldots, x_k$.

To test the hypothesis that these x variables contribute no information for the prediction of y (i.e., $\beta_{g+1} = \beta_{g+2} = \cdots = \beta_k = 0$), we use the test statistic

$$F = \frac{(SSE_1 - SSE_2)/(k - g)}{SSE_2/(n - k - 1)} = \frac{(\text{drop in SSE})/(k - g)}{s^2_{\text{complete model}}}$$

When the assumptions of the earlier sections of this chapter are satisfied—that is, the y values are independent and normally distributed, with mean $E(y)$ and variance σ^2— the F statistic has an F distribution, with $v_1 = (k - g)$ and $v_2 = (n - k - 1)$ degrees of freedom. Note that $v_1 = (k - g)$ is equal to the difference in the number of parameters in the complete and reduced models. Also, $v_2 = n - (k + 1)$ is equal to the number n of data points less the number of parameters in the complete model.

As we stated earlier, the larger the drop in SSE (the numerator of the F statistic), the greater is the weight of evidence favoring rejection of the null hypothesis and acceptance of the alternative hypothesis that at least one or more of the parameters $\beta_{g+1}, \beta_{g+2}, \ldots, \beta_k$ differ from zero. Consequently, we will reject the null hypothesis

$$H_0: \beta_{g+1} = \beta_{g+2} = \cdots = \beta_k = 0$$

when F is too large. That is, we use a one-tailed test and reject H_0 when F exceeds some critical value F_α, as shown in Figure 12.9.

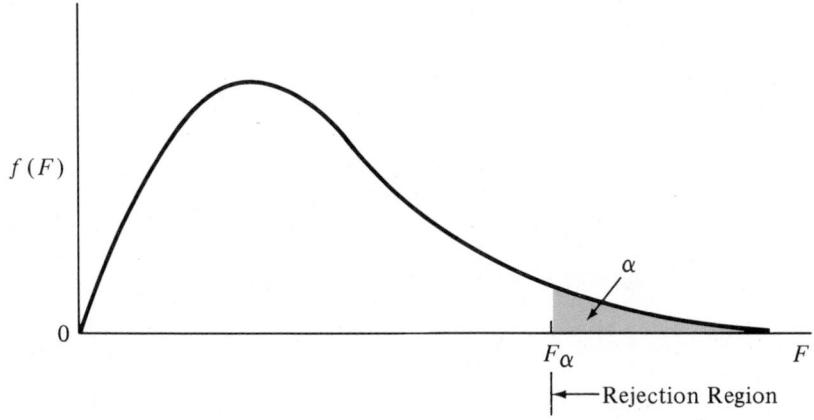

Figure 12.9   Rejection Region for the F Test for $H_0: \beta_{g+1} = \beta_{g+2} \ldots = \beta_k = 0$

EXAMPLE 12.8   Refer to Example 12.7. The second-order model

$$y = \beta_0 + \beta_1 x_1 + \beta_2 x_2 + \beta_3 x_1^2 + \beta_4 x_1 x_2 + \epsilon$$

was fit to the data relating the profit y per sales dollar to the size x_1 of the construction contract and the years x_2 of experience of the construction superintendent for $n = 18$ different projects undertaken by a construction firm. Test the hypothesis

$$H_0: \beta_3 = \beta_4 = 0$$

that is, that the second-order model does not provide an improvement over the first-order model for the prediction of profit y per sales dollar.

SOLUTION   A BMDP2R regression analysis program was used to fit the first-order (reduced) model

$$y = \beta_0 + \beta_1 x_1 + \beta_2 x_2 + \epsilon$$

to the data set of Example 12.7. The results are given in Table 12.30. The computer printout for the *reduced* model shows that

$$SSE_1 = 59.957$$

From the computer printout for the *full* model, shown in Table 12.25, we find

$$SSE_2 = 12.252$$

Since $n = 18$, $k = 4$, and $g = 2$, our test statistic is

$$F = \frac{(SSE_1 - SSE_2)/(k - g)}{SSE_2/(n - k - 1)} = \frac{(59.957 - 12.252)/(4 - 2)}{12.252/13}$$

$$= \frac{23.8525}{.9425} = 25.31$$

Table 12.30   BMDP2R Output for Example 12.8

MULTIPLE R		.5731
R SQUARE		.3284
STD. ERROR OF EST.		1.9993

ANALYSIS OF VARIANCE

	DF	SUM OF SQUARES	MEAN SQUARE	F RATIO
REGRESSION	2	29.321	14.660	3.668
RESIDUAL	15	59.957	3.997	

INDIVIDUAL ANALYSIS OF VARIABLES

VARIABLE	COEFFICIENT	STD. ERROR	F VALUE
(CONSTANT	8.0136)		
CONT. SIZE	-1.3548	.5530	6.0012
EXPER.	.4626	.4346	1.1333

This calculated value of F greatly exceeds the critical value $F_{.05} = 3.81$ (with $v_1 = 2$ and $v_2 = 13$ degrees of freedom). So we reject the null hypothesis and conclude that the second-order model does in fact provide an improvement over the first-order model for the prediction of profit y per sales dollar as a function of the size x_1 of the construction contract and the experience x_2 of the construction superintendent.

◆ ◆

EXERCISES   Conceptual Level

12.45   If the null hypothesis that all $\beta_i = 0$ is rejected, what course of action should then be followed?

12.46   As additional predictor variables are added to a model, what happens to the sum of squares for error? What happens to R^2?

12.47   The Lane County, Oregon, assessor uses a regression model to relate the assessed value (y) of a single-family residence to square footage (x_1), the number of rooms (x_2), the age of the dwelling (x_3), and whether or not the residence has a view (x_4), for single-family residences in the cities of Eugene and Springfield. Suppose the assessor is interested in determining whether or not city differences exist. That is, he wishes to determine whether one assessment model will suffice for both cities or whether a separate assessment model is required for each city. Explain how the assessor's question can be resolved.

12.48   Suppose we wish to test an hypothesis that certain variables in a regression model are insignificant in their collective ability to predict the dependent variable y in the presence of the other variables. Why do misleading conclusions sometimes result if we test this hypothesis solely on the basis of an examination of the F value (or t values) associated with each individual parameter?

EXERCISES   Skill Level

12.49   **a.**   Refer to Exercise 12.41. Use the MINITAB printout given in that exercise (Table 12.27) to determine whether there is a significant interaction between x_1 and x_2. Test by using $\alpha = .05$.

Table 12.31   MINITAB Output for Exercise 12.49

```
MTB > REGRESS C1 4 C2-C5

THE REGRESSION EQUATION IS
Y = 51.1 - 20.3 X1 - 15.3 X2 + 7.30 X1SQ + 2.40 X2SQ

PREDICTOR          COEF          STDEV          T-RATIO
CONSTANT         51.076          2.959           17.26
X1              -20.300          5.153           -3.94
X2              -15.252          2.909           -5.24
X1SQ              7.300          2.476            2.95
X2SQ             2.4048          0.6974           3.45

S = 4.520         R-SQ = 89.3%       R-SQ(ADJ) = 85.0%
```

(continues)

Table 12.31   (Continued)

```
ANALYSIS OF VARIANCE
```

SOURCE	DF	SS	MS
REGRESSION	4	1697.45	424.36
ERROR	10	204.29	20.43
TOTAL	14	1901.73	

SOURCE	DF	SEQ SS
X1	1	324.90
X2	1	952.03
X1SQ	1	177.63
X2SQ	1	242.88

```
UNUSUAL OBSERVATIONS
```

OBS.	X1	Y	FIT	STDEV.FIT	RESIDUAL	ST.RESID
13	2.00	10.00	18.79	2.46	−8.79	−2.32R

```
R DENOTES AN OBS. WITH A LARGE ST. RESID.
```

b. The MINITAB printout shown in Table 12.31 is the regression analysis of the data using the reduced model

$$y = \beta_0 + \beta_1 x_1 + \beta_2 x_2 + \beta_3 x_1^2 + \beta_4 x_2^2 + \epsilon$$

Test for a significant interaction term in the model using the drop-in-sum-of-squares technique. Do your conclusions agree with those found in part a? Does $t^2 = F$, to within rounding error?

12.50 Refer to Exercise 12.41. The MINITAB printout shown in Table 12.32 is the regression analysis of the data using the reduced model

$$y = \beta_0 + \beta_1 x_1 + \beta_2 x_2 + \epsilon$$

Test to determine whether the second-order terms in the model are significant. Use $\alpha = .05$. Comment on the results of your test in light of the results of Exercises 12.41 and 12.49.

Table 12.32   MINITAB Output for Exercise 12.50

```
MTB > REGRESS C1 2 C2 C3

THE REGRESSION EQUATION IS
Y = 43.8 - 5.70 X1 - 5.63 X2
```

PREDICTOR	COEF	STDEV	T-RATIO
CONSTANT	43.833	3.952	11.09
X1	−5.700	2.282	−2.50
X2	−5.633	1.317	−4.28

```
S = 7.216      R-SQ = 67.1%      R-SQ(ADJ) = 61.7%
```

(continues)

Table 12.32   *(Continued)*

ANALYSIS OF VARIANCE

SOURCE	DF	SS	MS
REGRESSION	2	1276.93	638.47
ERROR	12	624.80	52.07
TOTAL	14	1901.73	

SOURCE	DF	SEQ SS
X1	1	324.90
X2	1	952.03

UNUSUAL OBSERVATIONS

OBS.	X1	Y	FIT	STDEV.FIT	RESIDUAL	ST.RESID
15	2.00	23.00	9.90	3.95	13.10	2.17R

R DENOTES AN OBS. WITH A LARGE ST. RESID.

EXERCISES   Applications

12.51   The general sales manager of a firm that sells hotel and restaurant supplies is interested in the predictive relationship of advertising and the size of the sales force on the monthly sales volume. To study this issue, the manager recorded the sales volume y, the amount x_1 spent in direct-mail advertising, and the number x_2 of sales representatives in each of 12 randomly selected sales territories for the past year. The data are shown in Table 12.33. The model

$$y = \beta_0 + \beta_1 x_1 + \beta_2 x_2 + \beta_3 x_1 x_2 + \epsilon$$

was then fit to the data by using the BMDP1R program, providing the results in Table 12.34.

a.   Provide a complete analysis of the printout.

b.   Use the methods of Chapter 11 to fit the simple first-order model

$$y = \beta_0 + \beta_1 x_1 + \epsilon$$

to the data for y and x_1. Compute SSE for this first-order model.

c.   Use the computer analysis and the results from part b to determine whether sales are adequately described by the amount spent on advertising. That is, for the model

$$y = \beta_0 + \beta_1 x_1 + \beta_2 x_2 + \beta_3 x_1 x_2 + \epsilon$$

test the hypothesis $H_0: \beta_2 = \beta_3 = 0$.

Table 12.33   Sales Data for Exercise 12.51

y (× \$10,000)	x_1 (× \$1,000)	x_2	y (× \$10,000)	x_1 (× \$1,000)	x_2
25.9	9.75	3	30.3	12.20	3
27.1	9.20	2	31.0	13.86	4
27.9	10.00	3	31.7	13.50	3
29.0	11.04	3	33.0	14.20	4
29.8	10.80	2	34.1	16.30	4
30.2	12.00	3	36.2	17.50	4

Table 12.34   BMDP1R Output for the Data of Exercise 12.51

```
REGRESSION TITLE...............................  BMD REGRESSION #3
DEPENDENT VARIABLE.............................  1Y
TOLERANCE......................................  .0100

ALL DATA CONSIDERED AS A SINGLE GROUP

MULTIPLE R             .9809    STD. ERROR OF EST.    .6734
MULTIPLE R-SQUARE      .9621

ANALYSIS OF VARIANCE

                    SUM OF            MEAN
                    SQUARES    DF    SQUARE    F RATIO    P(TAIL)
      REGRESSION    92.110     3    30.703     67.717     .0001
      RESIDUAL       3.627     8      .453

                                             STD. REG.
    VARIABLE      COEFFICIENT    STD. ERROR    COEFF.·        T        P(2TAIL)
  INTERCEPT        14.9600
  X1        1       1.5321         .5910       1.3582     2.5925       .0345
  X2        2       -.4323        1.7964       -.1052      .2407       .8870
  X1X2      3       -.0553         .1554       -.3165      .3557       .7146
```

12.10

Stepwise Regression

Throughout this chapter we have used a variety of common multiple linear regression programs. These programs are generally inexpensive and easy to use; they provide an abundant amount of useful descriptive information as output. Explanation provided in association with several worked examples within this chapter should help you understand how to interpret the output you will obtain when using a commercially prepared multiple linear regression program.

In general, we wish to find the best linear regression model of the form

$$y = \beta_0 + \beta_1 x_1 + \beta_2 x_2 + \cdots + \beta_k x_k + \epsilon$$

to describe the relationship between y and the independent variables, $x_1, x_2, \ldots, x_k$. Recall that some independent variables can, in fact, involve transformations of other independent variables—for example, in the case of a second-order model with quadratic and interaction terms included as separate variables. Up to this point our process of model building has involved (a) the selection of a set of variables (both quantitative and qualitative) that we believe to be appropriate predictors of y and (b) the inclusion of certain variables to represent second-order relationships we believe to exist within the data. We have then used one of several regression programs to assist us in finding the best general linear regression model that expresses y as a function of all quantitative and qualitative variables, quadratic terms, and interaction terms we believe are important for our analysis.

Test of hypotheses involving individual β parameters (Section 12.3) or tests of portions of the model (Section 12.9) have allowed us to **screen out** or remove redundant variables **after** formulation of the complete model. An alternative is to screen out redundant variables **during the model-building process**, thus providing a model in which all variables, individually and collectively, provide a meaningful contribution toward the explanation of the response variable, y. This goal can be accomplished by using a

stepwise regression computer program with a **stepwise regression** analysis option.

Stepwise regression employs a systematic screening procedure designed to target redundant predictor variables *before* they are entered into the model. Table 12.1 lists the common regression analysis programs that offer a stepwise option. As is apparent from its name, stepwise regression provides a multistep screening procedure in which the independent variables are individually admitted to the model according to their marginal ability to explain the response variable y.

At the **first step** of analysis a simple linear model is formed by regressing y against the independent variable **with which it is most highly correlated**. This step provides a model of the form

$$y = \beta_0 + \beta_1 x_1 + \epsilon$$

where the absolute correlation between y and x_1 is greater than that between y and any of the other $(k - 1)$ independent variables under consideration in the analysis. The computer then examines the explanatory ability of each remaining independent variable **in the presence of x_1**, and a new model is formed by adding the variable that possesses the **greatest** marginal explanatory ability.

At the **second step** in the stepwise procedure an analysis is provided for the two-variable model of the form

$$y = \beta_0 + \beta_1 x_1 + \beta_2 x_2 + \epsilon$$

where x_1 is the independent variable chosen at the first step, and x_2 is the variable found to possess the greatest marginal explanatory ability in the presence of x_1. Note that the estimated β parameters at the second step are in general different from those obtained in the first step since the influence of x_2 will affect all three estimates, not just $\hat{\beta}_2$. (For an explanation, see Draper and Smith in the references.)

We then continue the stepwise procedure as before by examining the marginal explanatory ability of each of the remaining $(k - 2)$ independent variables. A new model is computed by adding a third variable (let's call it x_3) that is found to possess the greatest marginal explanatory ability in the presence of x_1 and x_2.

In addition to selecting a new variable to enter into the model, at **the third step and all succeeding steps** we must reexamine all variables previously entered. If the F value or t value associated with any independent variable *within* the model is found to be insignificant, the associated variable is removed from the analysis. **Thus, the stepwise regression procedure is both a selection and an elimination algorithm.** This selection and elimination screening procedure is continued until the point where all remaining variables not entered into the model are judged to be insignificant in their ability to contribute to the process under investigation.

Does the application of a stepwise regression analysis procedure provide an "ideal" prediction model? Not necessarily, since stepwise regression *cannot* identify new variables for inclusion other than those we contribute for review and analysis. It is possible, for instance, that quadratic and interaction terms can improve our model or that we have overlooked important quantitative or qualitative variables. Therefore, we may wish to improve our model further by adding such variables to the short list of variables resulting from an earlier application of stepwise regression. A second application of stepwise regression could then provide for a screening of these new variables in conjunction with those resulting from an initial application of stepwise regression.

EXAMPLE 12.9   In Exercise 12.20 we provided a MINITAB computer analysis (Table 12.13) to fit a linear regression model relating sales price y to seven independent variables for 50 single-family residences. The seven independent variables used in this analysis as descriptors of sales price (value) were as follows:

$$x_1 = \text{square feet of floor space } (\times 100)$$
$$x_2 = \text{number of bedrooms}$$
$$x_3 = \text{number of bathrooms}$$
$$x_4 = \text{total number of rooms}$$
$$x_5 = \text{age (years since initial construction)}$$
$$x_6 = \text{attached garage (1 if yes, 0 if no)}$$
$$x_7 = \text{view (1 if yes, 0 if no)}$$

The data for Exercise 12.20 are listed with that exercise in Table 12.12. Use a **stepwise regression** procedure to identify which among the seven independent variables of Exercise 12.20 should be included in a model for valuing residential properties. Which variables are redundant and therefore contribute little marginal improvement to the model when considered in the presence of the others?

SOLUTION   Table 12.35 lists the stepwise computer output for the data of Table 12.12 from a BMDP2R stepwise regression program. We will provide a brief interpretation of the three steps involved in this analysis.

Step 0: Before any predictor variables are entered. BMDP2R summarizes the information about the y variable. With no variables in the model the residual sum of squares (SSE) is the same as the total sum of squares, so MSE must be equal to the sample variance of the y data. The square root of MSE is shown in the STD. ERROR OF EST. (standard error of the estimate) line of the printout. The F values for potential variables are given in the VARIABLES NOT IN EQUATION columns of the printout. The variable with the largest significant F value will be the one added in the next step.

Table 12.35 BMDP2R Output for Example 12.9

STEP NO. 0

STD. ERROR OF EST. 26.4795

ANALYSIS OF VARIANCE

	SUM OF SQUARES	DF	MEAN SQUARE
RESIDUAL	34357.039	49	701.1641

VARIABLES IN EQUATION FOR Y

VARIABLE	COEFFICIENT	STD. ERROR OF COEFF	STD REG COEFF	TOLERANCE	F TO REMOVE	LEVEL
(Y-INTERCEPT	94.30400)					

VARIABLES NOT IN EQUATION

VARIABLE		PARTIAL CORR.	TOLERANCE	F TO ENTER	LEVEL
X1	2	.91318	1.00000	240.99	1
X2	3	.57766	1.00000	24.04	1
X3	4	.51686	1.00000	17.50	1
X4	5	.80707	1.00000	89.68	1
X5	6	.48971	1.00000	15.14	1
X6	7	.50732	1.00000	16.64	1
X7	8	.14542	1.00000	1.04	1

STEP NO. 1

VARIABLE ENTERED 2 X1

MULTIPLE R   .9132
MULTIPLE R-SQUARE   .8339
ADJUSTED R-SQUARE   .8304

STD. ERROR OF EST. 10.9036

ANALYSIS OF VARIANCE

	SUM OF SQUARES	DF	MEAN SQUARE	F RATIO
REGRESSION	28650.391	1	28650.39	240.99
RESIDUAL	5706.6486	48	118.8885	

VARIABLES IN EQUATION FOR Y

VARIABLE		COEFFICIENT	STD. ERROR OF COEFF	STD REG COEFF	TOLERANCE	F TO REMOVE	LEVEL
(Y-INTERCEPT		20.04721)					
X1	2	3.88698	.2504	.913	1.00000	240.99	1

VARIABLES NOT IN EQUATION

VARIABLE		PARTIAL CORR.	TOLERANCE	F TO ENTER	LEVEL
X2	3	.37163	.76206	7.53	1
X3	4	-.00307	.67837	0.00	1
X4	5	.51039	.47071	16.56	1
X5	6	.11806	.75956	0.66	1
X6	7	.21101	.77669	2.19	1
X7	8	.08141	.98482	0.31	1

Table 12.35 (Continued)

STEP NO. 2

VARIABLE ENTERED 5 X4

MULTIPLE R   .9366
MULTIPLE R-SQUARE .8772
ADJUSTED R-SQUARE .8719

STD. ERROR OF EST. 9.4757

ANALYSIS OF VARIANCE

	SUM OF SQUARES	DF	MEAN SQUARE	F RATIO
REGRESSION	30136.971	2	15068.49	167.82
RESIDUAL	4220.0686	47	89.78869	

VARIABLES IN EQUATION FOR Y

VARIABLE	COEFFICIENT	STD. ERROR OF COEFF	STD REG COEFF	TOLERANCE	F TO REMOVE	LEVEL
(Y-INTERCEPT	-16.23317)					
X1 2	2.94809	.3172	.693	.47071	86.40	1
X4 5	7.55165	1.8559	.303	.47071	16.56	1

VARIABLES NOT IN EQUATION

VARIABLE	PARTIAL CORR.	TOLERANCE	F TO ENTER	LEVEL
X2 3	-.17938	.60100	1.53	1
X3 4	-.39041	.47661	8.27	1
X5 6	-.02282	.73109	0.02	1
X6 7	.16186	.76069	1.24	1
X7 8	.11210	.98398	0.59	1

STEP NO. 3

VARIABLE ENTERED 4 X3

MULTIPLE R   .9465
MULTIPLE R-SQUARE .8959
ADJUSTED R-SQUARE .8891

STD. ERROR OF EST. 8.8180

ANALYSIS OF VARIANCE

	SUM OF SQUARES	DF	MEAN SQUARE	F RATIO
REGRESSION	30780.196	3	10260.07	131.95
RESIDUAL	3576.8435	46	77.75747	

VARIABLES IN EQUATION FOR Y

VARIABLE	COEFFICIENT	STD. ERROR OF COEFF	STD REG COEFF	TOLERANCE	F TO REMOVE	LEVEL
(Y-INTERCEPT	-22.62741)					
X1 2	3.02471	.2963	.711	.46690	104.17	1
X3 4	-10.04316	3.4919	-.198	.47661	8.27	1
X4 5	10.78355	2.0605	.433	.33071	27.39	1

VARIABLES NOT IN EQUATION

VARIABLE	PARTIAL CORR.	TOLERANCE	F TO ENTER	LEVEL
X2 3	-.10728	.57393	0.52	1
X5 6	-.02137	.72248	0.02	1
X6 7	-.09493	.73176	0.41	1
X7 8	-.07902	.97379	0.28	1

Step 1: Variable x_1 (square footage) has been included in the analysis as the single best predictor of y, sales price (or value). The model representing this relationship is

$$\hat{y} = 20.0472 + 3.8870x_1$$

For this simple linear regression model square footage explains about 83.4% of the variation in value.

Step 2: Variable x_4 (total number of rooms) has been included as the variable offering the greatest marginal improvement after square footage. The model relating y to x_1 and x_4 is then

$$\hat{y} = -16.2372 + 2.9481x_1 + 7.5516x_4$$

This model explains 87.7% of the variation in value. Thus, the inclusion of total number of rooms with square footage offers a marginal gain of 4.3% in addition to variation in value explained by the model using x_1 alone.

Step 3: Variable x_3 (number of bathrooms) enters as the next best predictor of y (value) in the presence of x_1 (square footage) and x_4 (total number of rooms). The least squares model representing this relationship is

$$\hat{y} = -22.6274 + 3.0247x_1 - 10.0432x_3 + 10.7836x_4$$

and we see that the model explains an additional 1.9% of the variation of y over that explained by the two-variable model in step 2.

The stepwise algorithm stops at step 3 since none of the remaining predictor variables have significant F values (see the F TO ENTER column of the printout), and the three predictor variables already chosen still have significant F values (see the F TO REMOVE column). Thus, the stepwise method has identified x_1, x_3, and x_4 as the best predictors of sales price.

Perhaps the county assessor described in Exercise 12.20 is not fully satisfied with the ability of the model at step 3 to provide residential value appraisals. Attention should then be directed to incorporating second-order terms (quadratic or interaction terms) involving x_1, x_3, and x_4 or variables not yet considered in the analysis (e.g., neighborhood quality). The second-order terms and new variables should then be considered together with x_1, x_3, and x_4 for determining whether we can identify a model to further improve our ability to explain the variability of y.

12.11

Residual Analysis

residual analysis

So far, we have described the means by which multiple regression computer programs can assist in the selection of the best variables from among a specified set to define a multiple linear regression relationship. **Residual analysis, a capability of many multiple regression computer programs, examines the degree to which a specified model satisfies**

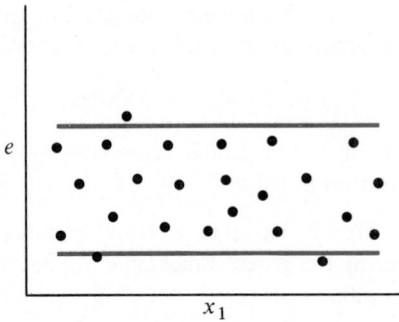

(a) Residuals Suggest Model Is Properly Specified

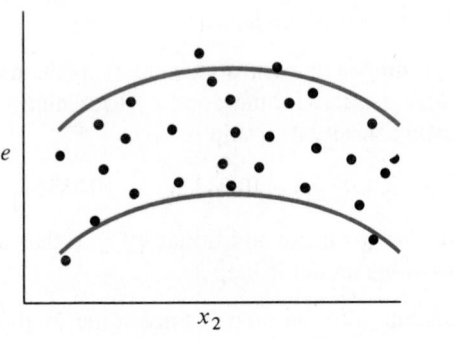

(b) Residuals Suggest the Quadratic Term x_2^2
       Should be Added

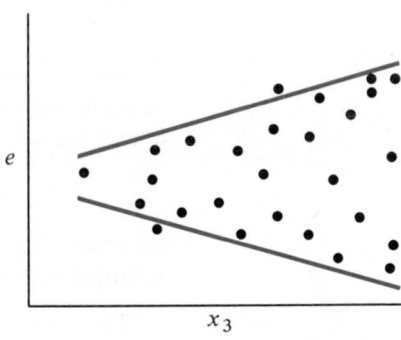

(c) Residuals Suggest the Variance
       Increases with x_3

Figure 12.10   *Patterns That May Result from the Plot of Multiple Regression Residuals Against the Independent Variables*

the assumptions of the multiple linear regression model. (See Section 12.2 for a review of these assumptions.) Expressed differently, residual analysis is a diagnostic technique that helps us determine whether a particular prediction model is appropriate for characterizing a given data set.

The set of model residuals is the set of n errors of prediction, $e_i = y_i - \hat{y}_i$. The basic premise underlying residual analysis is quite simple. If we have adequately specified our regression model, then this model should explain the systematic patterns that exist among the variables. Thus, the residuals, or errors of prediction, should appear as random error and not exhibit any systematic trends. If they do not appear as random error, systematic patterns or trends in the residuals suggest the inclusion of additional variables that may improve the fit of the prediction model to the data set.

Computer programs that offer a residual analysis option generally provide a plot of the residuals against each independent variable. Three general patterns usually appear in such plots. Figure 12.10a is an illustration of the type of pattern to be expected when no adjustment is needed. Figure 12.10b illustrates a case in which the residual plot suggests the inclusion of a second-order term x_2^2 into the analysis as an additional independent variable. Figure 12.10c depicts a case in which the variance of y increases proportionally with x_3. Usually, the substitution of the variable $\log(x_3)$ in place of x_3 will accommodate this problem; the reader is referred to Neter and Wasserman (see the references) for a more general discussion of required variable transformations when this effect occurs. Note that residual plots are often inconclusive, especially when working with a small sample size. Thus, as in all activities involving statistical analysis, caution should be used when making judgments.

12.12

Problems in Using Multiple Linear Regression Analysis

Several problems can result when multiple linear regression analysis is used to fit a prediction model to a set of data. Some were discussed in the previous section with respect to the analysis of residuals. In particular, residual analysis can assist with problems concerning the specification of our model. Other problems commonly encountered when using multiple linear regression analysis are discussed in the remainder of this section.

Multicollinearity

If you fit a multiple regression model

$$y = \beta_0 + \beta_1 x_1 + \beta_2 x_2 + \cdots + \beta_k x_k + \epsilon$$

multicollinearity

you must be very careful in interpreting the results of t tests on the β parameters. One reason for this caution is that the estimators may be highly correlated. We call this phenomenon **multicollinearity**. That is, some of the information contributed by two or more of the independent variables for predicting y may be different, but some information may be identical.

To illustrate, suppose you wish to construct a model to assess (predict) the market value of a single-family residence as a function of the two independent variables x_1 and x_2 and that

x_1 = square feet of living space in residence
x_2 = number of bedrooms in residence

In general, you would expect residences with more bedrooms and a greater amount of living space to cost more. That is, both of these variables contribute information for the assessment of market value but some of the information (although not all) is the same; in other words, the independent variables "square footage" and "number of bedrooms" are correlated. Larger residences tend to have more bedrooms.

When two or more of the independent variables are correlated, you cannot determine their respective individual contributions to the reduction in SSE, the sum of squares of deviations between the observed and the predicted values of y. How much information a particular independent variable contributes to the prediction of y depends on the other independent variables included in the model. If two variables contribute overlapping information—for instance, in the previous illustration square footage (x_1) and number of bedrooms (x_2) may contribute overlapping information regarding market value— we may reject the hypothesis that $\beta_1 = 0$ (indicating the significance of x_1 as a prediction variable) while failing to reject the hypothesis that $\beta_2 = 0$. This may occur even though we are certain that x_2 is causally related to y–that is, that residences with more bedrooms have a greater market value. In fact, multicollinearity may even cause the algebraic sign of one or more regression parameter estimates to be contrary to logic. Thus, when multicollinearity exists, the explanation of the causal effects of individual variables on y should be undertaken with great caution.

Multicollinearity also tends to confuse the interpretation of confidence interval estimates for the β parameters, since it is impractical to describe the effect of a particular independent variable without also describing the effects of other variables related to it. However, if one is interested primarily in the use of a multiple variable regression model for purposes of prediction of y from $x_1, x_2, \ldots, x_k$, rather than in an interpretation of the individual effects $x_1, x_2, \ldots, x_k$, multicollinearity should present no problem. In the presence of multicollinearity it is the complete model that is important, not the individual β parameters.

Prediction Outside the Experimental Region

As we noted in Section 11.5, a prediction equation should be used only when the value of the independent variable is within the range of independent variables in the sample data set. Why? Because we cannot be certain that the relationship between the response variable y and the independent variable x is the same for values of x outside the sample range.

Consider, for example, an agricultural experiment that measures the relationship between the output (in bushels) per acre of feed corn, y, and the application (in pounds) of a nitrogen-based fertilizer, x. Within an acceptable range of application, increased use of fertilizer should increase the output per acre of corn. If data within this range are used to establish the prediction model, a researcher may derive misleading results by using the prediction model to estimate output for applications that *exceed* the acceptable range, since excess amounts of fertilizer may overstimulate or "burn" the corn crop.

In a similar manner, we can infer from Example 12.5 that the prediction model relating market value y to square feet of floor space x_1 and number of bedrooms x_2 is valid only for values of x_1 and x_2 within the following ranges:

$$8.40 \le x_1 \le 19.50 \quad \text{and} \quad 2 \le x_2 \le 4$$

So the county assessor of Example 12.5 should not use the derived prediction model to value a residence with 5,000 square feet of floor space $(x_1 = 50.00)$ and seven bedrooms

($x_2 = 7$). The prediction model would provide \$596,200 as an estimate of the market value for this residence. However, such a large residence may prove to be a "white elephant" in its community and not have a market value any greater than that of homes of more moderate size. Thus, the value derived from the prediction equation may be highly inflated.

Overfitting the Model

Within both Chapters 11 and 12 we have used the coefficient of determination (r^2 in Chapter 11, R^2 in Chapter 12) as our primary measure of the success of the regression analysis. Our objective has been to find a prediction equation that minimizes SSE, the *unexplained* sum of squares associated with the prediction equation. The coefficient of determination provides a convenient measure of this effect, since

$$R^2 = 1 - \frac{\text{SSE}}{\text{Total SS}}$$

approaches 1.0 as SSE diminishes.

To see where a problem may arise, let's consider the case of the single-variable model of Chapter 11. Refer to Figure 11.9 and note that $r^2 = 1$ for the single-independent-variable model when all the (x, y) points fall in a straight line. If the underlying population of (x, y) points is as appears in Figure 12.11, we would still derive a prediction equation with $r^2 = 1$ if, by chance, our chosen sample points are all collinear. In fact, we could "guarantee" a sample set with $r^2 = 1$ if we limit our sample data set to two data points, because only two points are needed to define a straight-line equation. Similarly, we need only three points to define the plane that perfectly expresses the relationship among three variables, y, x_1, and x_2.

In multiple linear regression analysis our objective is to derive a prediction model that accurately portrays the relationship between the response variable y and the independent variables $x_1, x_2, \ldots, x_k$ for the entirety of all such data sets in the underlying population. To avoid misleading evidence regarding the fit of the model,

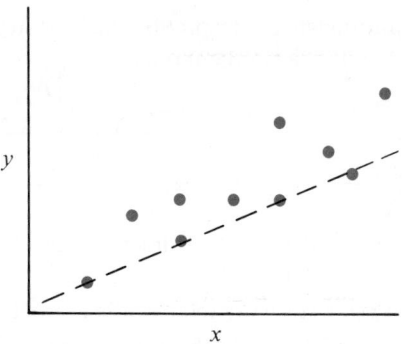

Figure 12.11   *The Scatter Diagram of (x, y) Points for a Population*

we must ensure that our data set is sufficiently large to effectively characterize the population under study. When n is small in comparison with $(k + 1)$, the number of parameters in the multiple linear regression model, R^2 tends to provide an inflated estimate of our success in explaining the variability in the response variable, y. This problem, called **overfitting**, can be avoided by making certain that our sample size n is large in relationship to k. **A common rule of thumb to avoid overfitting is to require that $n \geq 4k$, where n is our sample size and k is the number of independent variables in the model.**

overfitting

Serial Correlation

time series data

An additional problem sometimes results when the multiple regression model is sequenced over time and involves **time series data**, as in the use of the model for applications to sales forecasting, demand analysis, and econometric studies. When one or more key variables have been omitted from the multiple regression model for time series data, the residuals are often dependent upon one another—said to be **autocorrelated**, or **serially correlated**. For example, suppose that the number of new housing starts for each of the past 60 months is regressed against the prime lending rate. The omission of population size as a separate independent variable may lead to serially correlated residuals if population size is correlated with the prime lending rate over this 60-month span.

autocorrelation
serial correlation

Serial correlation affects the precision but not the accuracy of the estimation of the β parameters in a multiple regression model. The estimates are unbiased, but their true variances are underestimated when serial correlation is present. As a result, the SSE may seriously underestimate the true unexplained variation, causing the t values (or F values) to be larger than they should be. This situation could lead to the conclusion that certain β parameters are statistically significant when in fact they are not. Thus, the effect of serial correlation is opposite to that of multicollinearity.

The most common test for the presence of serial correlation is the Durbin-Watson test, which is described in Neter and Wasserman (see the references). An easier test to apply, and one that overcomes the limitations of the Durbin-Watson test, is a runs test applied to the residuals (deviations). This test will be discussed in Section 18.7 of this text.

To summarize, exercise caution when interpreting the t tests (or F tests) concerning the individual β parameters that appear in the model.

EXERCISES Conceptual Level

12.52 What is the principal advantage and the principal disadvantage of the stepwise multiple regression procedure?

12.53 How does the stepwise procedure decide which variables to include in or exclude from the model?

12.54 What is meant by the term *residual*? What value is there in doing residual analysis?

12.55   What is multicollinearity, and how can it affect multiple regression analysis?

12.56   Explain the effect of multicollinearity on each of the following.

    a.   The use of a multivariable regression model for the prediction of a response from a set of predictor variables

    b.   The use of a multivariable regression model to explain the relationship of a set of predictor variables on response

12.57   What potential problems do we face if we use a regression model to predict y for values of the independent variable(s) outside the observed range of the independent variable(s)?

12.58   What does "overfitting" a regression model mean? What should be done to avoid overfitting?

EXERCISES   Applications

12.59   Refer to the computer output of Exercise 12.5 (Table 12.6) and answer the following questions.

    a.   In the presence of the other independent variables, which predictor has the greatest effect on lumber orders? What is the basis for your decision?

    b.   Do the data in the computer output support the theory that rising interest rates have a negative effect on lumber orders? Test the null hypothesis $H_0 : \beta_2 = 0$ against the one-sided alternative $H_a : \beta_2 < 0$. Test by using $\alpha = .05$.

    c.   Does evidence exist to indicate that multicollinearity exists among the predictor variables? In the context of Exercise 12.5, explain the effect of multicollinearity among the predictor variables.

12.60   Refer to Example 12.5 and answer the following questions.

    a.   How do you explain the apparent contradiction that the t values are small, finding each individual variable to be insignificant, while the F value indicates that at least one of the predictor variables is significant?

    b.   Under what conditions can this model be depended on to provide reasonably accurate property valuations?

    c.   Suppose a homeowner is contemplating the addition of a 150-square-foot bedroom to her home. Can the model be used to estimate the gain in the home's market value for this addition?

    d.   Using the model to value a house with 2,500 square feet of floor space and 5 bedrooms ($x_1 = 25$; $x_2 = 5$), we find $\hat{y} = \$137,998$; but a house with 2,500 square feet and only 2 bedrooms is valued by the model at \$152,929. Explain how the model can predict greater value for a house with 2 bedrooms than one with 5 bedrooms when both have the same amount of floor space.

12.61   The illustration within Section 12.7 proposed the use of a multivariable regression model to assess the market value of a single-family residence (y) from the square feet of living space (x_1) and number of bedrooms (x_2). If multicollinearity exists between x_1 and x_2, would it be possible to fail to reject (accept) *both* hypotheses $\beta_1 = 0$ and $\beta_2 = 0$ even though the model provides a good fit to the data base? Explain.

12.62   After considerable research into traits of leadership, management theorists have concluded that leadership arises more from situational factors than inherent traits. One exception to this rule has been provided by evidence that persons scoring high in Machiavellian traits tend to emerge as leaders in small groups. To examine this issue, two researchers conducted a controlled study

Table 12.36   Leadership Data for Exercise 12.62

Variable	Coefficient	F Ratio
Machiavellianism	1.4109	9.8169
Task orientation	2.2030	8.6648
Interaction	−17.9755	3.7158
$R^2 = .4677$		

Source: R. Jones and C. White, "Machiavellianism and Task Orientation as Predictors of Group Effectiveness," in *Proceedings of the American Institute for Decision Sciences, Western Regional Conference*, ed. S. Wood and K. Coney (San Diego, 1978). Reprinted by permission.

involving 115 corporate decision makers, regressing corporate earnings (as a measure of corporate effectiveness) against measurements of Machiavellianism and task orientation for each subject. The results are shown in Table 12.36.

a. Separately test for the significance of Machiavellianism, task orientation, and their interaction as predictors of effectiveness.

b. How useful is this model in explaining corporate effectiveness?

c. How do you account for the fact that R^2 is small even though the F ratios separately associated with the three predictor variables are each reasonably large?

12.13

Summary

A multiple regression analysis is an extension of the simple regression analysis of Chapter 11 to the case in which the response variable y is related to a number of predictor variables $x_1, x_2, \ldots, x_k$. All the tests and estimation and prediction procedures of Chapter 11 are applicable to the general linear model of this chapter. Even the multiple coefficient of correlation R and the multiple coefficient of determination R^2 have meanings similar to those for the simple coefficient of correlation r and coefficient of determination r^2 encountered in Chapter 11.

The major difference between simple and multiple regression analyses is the utility of the latter. Very few response variables in business are adequately modeled by the simple linear probabilistic model

$$y = \beta_0 + \beta_1 x + \epsilon$$

of Chapter 11. In contrast, the multiple regression model of this chapter, when properly constructed, provides a good model for many business response variables. We must be careful, however, not to "overfit" the least squares model. Typically, such problems are negligible if $n \geq 4k$, where k is the number of predictor variables used in the model. The resulting prediction equation often provides a good estimator of the mean response for given values of the predictor variables or provides a good predictor of values of y that might be observed in the future.

Supplementary Exercises

Applications

MKTG

12.63 The marketing director for a manufacturer of small home kitchen appliances has developed a multivariable model to examine the relationship between sales and four related predictor variables in the sales regions served by the firm. Sales data (in thousands of dollars) for each of 68 different sales regions were regressed against the following variables:

$$x_1 = \text{promotional budget (in hundreds of dollars)}$$
$$x_2 = \text{number of retail outlets}$$
$$x_3 = \text{per capita income (in thousands of dollars)}$$
$$x_4 = \text{average unit retail price}$$

A BMDP2R regression analysis computer program was then employed to provide the results given in Table 12.37.

a. What proportion of the variation in sales is explained by the four predictor variables in the model?

b. Give the regression equation for predicting the sales in a particular region knowing the promotional budget, number of retail outlets, per capita income, and the average unit retail price.

c. In the presence of the other variables in the model, which is the most significant predictor of sales? Which is the least significant?

d. Estimate the gain in sales by adding an additional retail outlet in a particular sales region.

e. How might multicollinearity affect the answer to part d?

Table 12.37   BMDP2R Output for Exercise 12.63

```
MULTIPLE R                  .8256
R SQUARE                    .6816
STD. ERROR OF EST.       23.9722

ANALYSIS OF VARIANCE

                    DF     SUM OF SQUARES    MEAN SQUARE    F RATIO
     REGRESSION      4        90867.625       22716.906     39.530
     RESIDUAL       63        36204.168         574.669

INDIVIDUAL ANALYSIS OF VARIABLES

   VARIABLE      COEFFICIENT     STD. ERROR    F VALUE
  (CONSTANT       -1.25272)
  PRM. BGT.         .42721         .48139       .7876
  RET. STRS.       1.89678         .41451      20.9398
  INC.             2.23333         .63499      17.4269
  PRICE            1.11673         .39231       8.1029
```

12.64 A labor union representative is interested in creating a multivariable prediction model to predict the hourly wage of union clerical workers by using the employee's age, her or his years of work experience, and the number of years the employee has been a union member. Data are available on 75 male and female clerical workers employed by five different firms in two different states. Discuss the type of independent variables you would recommend to use in the study, and describe the nature of the levels that will be measured on each.

 12.65 Does product competition always result in lower prices to the consumer? Some marketing specialists have suggested that the relationship between price and the number of competing products is of second order—that is, that prices are high in the presence of either few or many competitors and lowest when there is a moderate number of competitors. In a study of this issue a record was made of the average price for a six-pack of beer and the number of competing brands sold in a certain chain of supermarkets in 12 different cities. The data are shown in Table 12.38. Fit the second-order model $y = \beta_0 + \beta_1 x + \beta_2 x^2 + \epsilon$ to these data by using an available regression analysis program. Carefully interpret your results. Do these data suggest that the number x of competing brands can be used to predict the price y for the product involved in this study?

Table 12.38 Price and Competition Data for Exercise 12.65

Price, y	Number of Competing Brands, x	Price, y	Number of Competing Brands, x
$1.25	4	$1.56	10
1.40	7	1.34	5
1.60	2	1.45	8
1.52	9	1.41	2
1.40	3	1.27	7
1.35	6	1.55	8

 12.66 A study by A. Allaway, J. B. Mason, and G. Brown ["An Optimal Decision Support Model for Department-Level Promotion Mix Planning," *Journal of Retailing*, Vol. 63, No. 2 (Fall 1987)] proposes an optimal control model for predicting sales as a function of advertising and other variables. The objective is to achieve specific storewide performance levels for sales during a given period of time by using the least costly mixture of promotional techniques. The model was developed and tested in a large western wear specialty outlet, using weekly sales records by department for three years, a total of 156 weeks of information. Prior to the development of the optimal control model, sales responses were analyzed separately for each of 13 departments by using ordinary least squares regression. The results of this prior analysis for the Levi's jeans department are shown in Table 12.39.

a. Identify the seven variables in the upper portion of the table and the three variables in the lower portion of the table as either qualitative or quantitative, defining dummy variables if necessary.

b. Are all of the variables used in this regression analysis present in the table? Why or why not?

c. What are the values given in the column labeled "Coefficient (b)"? What is the significance of the superscript on each of these values?

d. Does the data indicate that the model contributes significant information for the prediction of y? Test the appropriate hypothesis, using an approximate critical value for F.

e. What proportion of the total variation in sales is explained by the independent predictor variables used in the model?

Table 12.39   Sales Response Model for Levi's Jeans; Exercise 12.66

Explanatory Variable (x)	Coefficient (b)
Previous week's sales dollars of Levi's jeans	.55*
Constant, or y intercept term	148.00[†]
Current-week column-inches of Levi's jeans advertising	9.34*
Number of newspaper column-inches featuring men's boots at a 25% discount	12.63*
Number of newspaper column-inches featuring men's clothing at a 35% discount	7.74[‡]
Column-inches of men's clothing advertising at no price discount	10.71[‡]
Number of billboards on display during the week	75.18*
Number of radio advertising seconds aired during the week	.41[‡]
No other variables were significant at the .10 level	

In addition, specific nonpromotion variables that assisted in fitting the
model to the data included:

Christmas season	4250.00*
Week of school start-up	2907.00*
Annual Boy's Ranch special sale	1869.00[†]

Overall R square: .9006
Overall F statistic: 87

Source: A. Allaway, J. Barry Mason, and Gene Brown, "An Optimal Decision
Support Model for Department-Level Promotion Mix Planning," *Journal of
Retailing*, Vol. 63, No. 2 (Fall 1987). Reprinted by permission.
* $p \leq .01$.   [†] $p \leq .05$.   [‡] $p \leq .10$.

12.67 The production manager of a plant that manufactures a chemical fertilizer recorded the marginal cost of production at various levels of output for 12 randomly selected months. The results
are shown in Table 12.40. A third-order model

$$y = \beta_0 + \beta_1 x + \beta_2 x^2 + \beta_3 x^3 + \epsilon$$

Table 12.40   Cost Data for Exercise 12.67

Cost, y (Per 100 lb)	Output, x ($\times$ 1,000 lb)	Cost, y (Per 100 lb)	Output, x ($\times$ 1,000 lb)
30.0	1.9	40.5	5.5
29.6	4.5	23.4	3.9
15.1	3.5	15.0	2.4
41.2	7.1	34.9	6.0
22.6	2.6	20.5	1.1
21.0	1.9	34.6	4.5

Table 12.41   BMDP1R Output for the Data of Exercise 12.67

```
REGRESSION TITLE............... BMD REGRESSION #3
DEPENDENT VARIABLE............. 1Y
TOLERANCE...................... .0100


ALL DATA CONSIDERED AS A SINGLE GROUP


MULTIPLE R              .7840     STD. ERROR OF EST.    6.2573
MULTIPLE R-SQUARE       .6146


ANALYSIS OF VARIANCE
```

	SUM OF SQUARES	DF	MEAN SQUARE	F RATIO	P(TAIL)
REGRESSION	562.002	2	281.001	7.177	.01369
RESIDUAL	352.383	9	39.154		

VARIABLE		COEFFICIENT	STD. ERROR	STD. REG. COEFF.	T	P(2TAIL)
INTERCEPT		21.495				
X1	1	.019	2.466	.004	.008	.994
X2	2			REDUNDANT VARIABLE		
X3	3	.065	.044	.780	1.482	.172

was selected to describe the marginal cost y as a function of ouput x. A BMDP1R analysis is presented in Table 12.41. Does the third-order model appear to provide a good fit to these data? (In your answer, address the issues discussed in the preceding sections.)

 12.68   The sales manager for a firm that sells packaged soybeans through a nationwide supermarket chain is interested in examining the relationship of the retail price charged for the product and advertising on retail sales. To study this issue, the sales manager recorded the sales y (in thousands of units), the average unit price x_1 per package, and the percentage x_2 of total expenses devoted to advertising for the past year in each of $n = 25$ sales regions. An SPSS regression program was used to fit the second-order model

$$y = \beta_0 + \beta_1 x_1 + \beta_2 x_2 + \beta_3 x_1^2 + \beta_4 x_2^2 + \beta_5 x_1 x_2 + \epsilon$$

to the data. The results are given in Table 12.42. Give a complete analysis of the printout, and construct a contour diagram for the second-order model fitted to the sample data.

12.69   An SPSS regression program was used to fit the first-order model

$$y = \beta_0 + \beta_1 x_1 + \beta_2 x_2 + \epsilon$$

to the data of Exercise 12.68. The results are given in Table 12.43. Test to see whether the second-order model fitted in Exercise 12.68 provides an improvement over the first-order model. That is, for the second-order model of Exercise 12.68, test the hypothesis

$$H_0: \beta_3 = \beta_4 = \beta_5 = 0$$

Table 12.42   SPSS Output for Exercise 12.68

VARIABLE	MEAN	STANDARD DEVIATION
1(Y)	30.6519	7.9602
2(X1)	.3604	.0357
3(X2)	6.3600	1.7049
4(X1)SQ.	.1311	.0251
5(X2)SQ.	43.2400	22.5153
6(X1)(X2)	2.2840	.6285

		ANAL. OF				
		VARIANCE	DF	SS	MS	F
MULTIPLE R	.8396	REGRESSION	5	1072.065	214.413	9.080
R SQUARE	.7049	RESIDUAL	19	448.678	23.615	
STD. ERROR	4.8598					

VARIABLES IN THE EQUATION

VARIABLE	B	BETA	STD. ERROR B	F
(CONSTANT)	134.0779			
X1	-628.4045	-2.8183	667.1304	.8873
X2	2.4161	.5175	7.5804	.1016
X1SQ.	841.6853	2.6540	907.5220	.8602
X2SQ.	.2286	.6466	.3071	.5541
(X1)(X2)	-5.4993	-.4342	21.9262	.0629

Note: In the SPSS regression program the column headed by B lists the least squares regression coefficients; the column headed by BETA lists the standardized regression coefficients. For example, the standardized coefficient associated with the *j*th independent variable is

$$\hat{\beta}_j \left(\frac{s_{x_j}}{s_y} \right)$$

where s_{x_j} and s_y are respectively, the standard deviations of x_j and y.

Table 12.43   SPSS Output for Exercise 12.69

		ANAL. OF				
		VARIANCE	DF	SS	MS	F
MULTIPLE R.	.8281	REGRESSION	2	1042.917	521.458	24.009
R SQUARE	.6857	RESIDUAL	22	477.826	21.719	
STD. ERROR	4.6604					

VARIABLES IN EQUATION

VARIABLE	B	BETA	STD. ERROR B	F
(CONSTANT)	35.6170			
X1	-72.8205	-.3271	26.8613	7.3494
X2	3.3458	.7166	.5635	35.2592

12.70   Refer to Exercises 12.68 and 12.69.

    a.   Use each model to separately estimate the expected annual sales of the firm's product in a sales region where the average unit price is $.40 and 7% of total expenses are devoted to advertising. How do you account for differences between your two estimates?

    b.   If your computer package will calculate confidence intervals for $E(y)$, find 95% confidence intervals for the estimates of mean response in part a.

12.71   Write a multiple linear regression model relating household energy use y and the following variables:

    i) Living area (in square feet)
    ii) Age of house
    iii) Whether or not house has wall and attic insulation
    iv) Type of fuel for heating–electric, gas, oil, solar, coal, or wood

    Which variables might exhibit quadratic effects? Which variables might interact?

Table 12.44   Investment Data for Exercise 12.72

Rate of Return, y	Investment, x ($\times \$100,000$)	Rate of Return, y	Investment, x ($\times \$100,000$)
4.0	1.1	5.6	2.5
9.2	3.5	8.0	3.1
6.0	4.7	6.1	5.9
1.8	1.0	3.2	2.1
7.7	5.2	7.8	4.6

Table 12.45   MINITAB Output for the Data of Exercise 12.72

```
THE REGRESSION EQUATION IS
C10 = - 1.73 + 4.51 X1 - 0.537 X2

PREDICTOR          COEF             STDEV           T-RATIO
CONSTANT          -1.730            1.998            -0.87
X1                 4.505            1.365             3.30
X2                -0.5373           0.1997           -2.69

S = 1.410          R-SQ = 72.1%      R-SQ(ADJ) = 64.2%

ANALYSIS OF VARIANCE

SOURCE          DF          SS              MS
REGRESSION       2        36.063          18.031
ERROR            7        13.921           1.989
TOTAL            9        49.984
```

12.72 New-product research provides the lifeblood for most manufacturing organizations. But how much should the firm spend to develop and promote new products? The answer depends on the rate of return expected from new ventures and the amount invested in the development and promotion of the new product. The marketing research director for a food products company had recorded the rate of return and the amount invested in the development and promotion of 10 new snack food products introduced by his firm over the past 10 years. The data are shown in Table 12.44. A second-order model of the form $y = \beta_0 + \beta_1 x + \beta_2 x^2 + \epsilon$ was fit to the data, using the MINITAB regression program. Carefully interpret the output provided in Table 12.45.

12.73 Sales force management requires the proper adjustment of sales territories to equalize sales potential among territories as well as the establishment of sales force goals and compensation plans. To examine this issue, the sales manager for a company that sells office machines and supplies has recorded the territory sales y for the past month, the number x_1 of accounts, and the number x_2 of years of work experience for a random sample of $n = 25$ of her firm's sales personnel. The data are shown in Table 12.46. She assumed that a second-order model for $k = 2$ independent variables appropriately describes the relationship between sales potential and the independent variables "number of accounts" and "sales potential." An SAS regression program was used to fit the second-order model to the data set. The results are given in Table 12.47. Carefully interpret the results of this analysis.

Table 12.46   Sales Data for Exercise 12.73

y (× \$1,000)	x_1 (× 100)	x_2 (Years)	y (× \$1,000)	x_1 (× 100)	x_2 (Years)
36.70	15	1.7	23.37	10	1.5
34.74	14	1.7	45.87	19	2.3
22.95	12	1.5	27.29	12	1.6
46.76	18	2.6	32.89	14	1.8
61.26	24	4.7	28.01	13	1.6
21.35	9	1.3	32.64	13	1.9
50.32	22	2.5	34.54	15	1.7
33.67	14	1.9	17.41	7	1.3
65.19	25	4.3	20.36	9	1.4
48.76	21	2.4	15.78	6	1.2
24.68	11	1.7	41.68	16	2.0
25.33	11	1.6	28.00	11	1.6
24.08	12	1.4			

12.74 An economist for an interstate bank-holding company is interested in estimating the monthly number of housing starts in the region served by his bank from (a) the prevailing mortgage interest rate, (b) the prevailing unemployment rate, (c) the state of the national economy (recession, stable, robust), and (d) the season of the year in which the month under consideration occurs (spring, summer, fall, winter).

a. Write a regression model to relate y, monthly number of housing starts, to the independent variables.

b. Explain how all the information contained in the independent variables can be incorporated into the regression analysis.

c. Explain the steps the economist must follow to actually compute a model to predict housing starts from the suggested independent variables.

Table 12.47   SAS Output for the Data of Exercise 12.73

DEP VARIABLE: Y   SALES

SOURCE	DF	SUM OF SQUARES	MEAN SQUARE	F VALUE	PROB>F
MODEL	5	4064.105	812.821	216.213	0.0001
ERROR	19	71.427606	3.759348		
C TOTAL	24	4135.532			

ROOT MSE	1.938904	R-SQUARE	0.9827	
DEP MEAN	33.741600	ADJ R-SQ	0.9782	
C.V.	5.74633			

VARIABLE	DF	PARAMETER ESTIMATE	STANDARD ERROR	T FOR H0: PARAMETER=0	PROB>|T|	VARIABLE LABEL
INTERCEPT	1	-10.768792	6.330026	-1.701	0.1052	INTERCEPT
X1	1	1.418484	0.651275	2.178	0.0422	ACCOUNTS
X2	1	15.598396	7.294636	2.138	0.0457	EXPERIENCE
X3	1	-0.028221	0.064340	-0.439	0.6659	X1 SQ
X4	1	-3.386117	2.420499	-1.399	0.1779	X2 SQ
X5	1	0.495918	0.854659	0.580	0.5686	X1 X2

Case Study

The Wood Stove Industry

Wood stoves have become increasingly popular for home heating since the advent of higher energy prices. In many urban areas where wood stoves have become popular, they have contributed to local air pollution, thus drawing considerable attention from governmental agencies charged with monitoring air quality. Proposals have been offered to legislative bodies that wood stoves, like automobiles, require pollution control devices or standards.

Some members of the wood stove industry have become concerned that imposed standards might require difficult and expensive modifications that would make their products less competitive in the market for home heating equipment. The claim is made that a substantial reduction in the air pollution produced by a wood stove can be made by a slight modification of most stoves and by informing wood stove users of better methods for operating their stoves.

As a result of concerns over the impact of such regulation, one wood stove manufacturer has undertaken research to determine what product or operational modifications would minimize the air pollution produced by its brand of wood stoves. The research involved the measurement of particulate matter, as observed by a photoelectric cell, shown by the gases venting out of a wood stove chimney. This method, used in conjunction with a measure of air flown out of the stack, provided a relative measure of particulate matter in terms of the percentage of light blocked by (or passing through) the venting gases.

Since the manufacturer's product could be used with two sizes of flue pipe, the experiment involved varying the air intake setting (1/4, 1/2, 3/4, and fully open) and the flue size, and measuring the temperature and the relative level of particulate matter exiting the stack. The experiment was conducted by first building a fire in the stove and then closing the stove and adjusting the air intake valve to the appropriate level. After half an hour (to allow the fire to stabilize at the set air flow level), measurements of temperature and particulate matter were recorded, as shown in the following table.

Observation Number	Relative Particulate Matter Concentration	Air Intake Setting	Flue Size	Flue Temperature
1	44	1/4	S	106
2	28	1/2	S	248
3	26	3/4	S	385
4	31	Open	S	534
5	42	1/4	L	124
6	26	1/2	L	211
7	29	3/4	L	374
8	34	Open	L	487
9	42	1/4	S	131
10	28	1/2	S	230
11	27	3/4	S	353
12	36	Open	S	515
13	40	1/4	L	144
14	26	1/2	L	255
15	27	3/4	L	286
16	33	Open	L	517
17	42	1/4	S	117
18	27	1/2	S	235
19	25	3/4	S	378
20	34	Open	S	510
21	39	1/4	L	139
22	30	1/2	L	248
23	29	3/4	L	302
24	34	Open	L	521

Analyze these data by the development of an appropriate multivariable regression model relating relative particulate matter concentration to the air intake setting, flue size, and flue temperature. Interpret your results.

References and Suggested Readings

Dixon, W. J., and M. B. Barnes, eds. *BMDP-83: Biomedical Computer Programs, P-Series.* Berkeley, Calif.: University of California Press, 1983.

Draper, N., and H. Smith. *Applied Regression.* 2d. ed. New York: Wiley, 1981.

Dunn, O. J., and V. A. Clark. *Applied Statistics: Analysis of Variance and Regression.* 2d ed. New York: Wiley, 1987.

Kleinbaum, D. G., L. L. Kupper, and K. E. Muller. *Applied Regression Analysis and Other Multivariable Methods.* 2d ed. Boston: PWS-KENT Publishing Co., 1988.

Mendenhall, W. *An Introduction to Linear Models and the Design and Analysis of Experiments.* Belmont, Calif.: Wadsworth, 1968.

Mendenhall, W., and J. T. McClave. *A Second Course in Business Statistics: Regression Analysis.* San Francisco: Dellen, 1981.

Miller, R. B. *Minitab Handbook for Business and Economics.* Boston: PWS-KENT Publishing Co., 1988.

Neter, J. and W. Wasserman. *Applied Linear Regression Analysis.* Homewood, Ill.: Irwin, 1983.

Ryan, B. F., B. L. Joiner, and T. A. Ryan. *MINITAB Handbook, 2nd ed.* Boston: PWS-KENT Publishing Co., 1985.

SAS User's Guide: Basics, 1982 Edition. SAS Institute, Inc., Box 800, Cary, N. C., 27511.

Younger, M. S. *A Handbook for Linear Regression, 2nd. ed.* Boston: PWS-KENT Publishing Co., 1985.

TO
THE
READER

The statistical techniques presented in earlier chapters have typically been illustrated by applications to business problems involving cross-sectional data gathered at a single point in time. Frequently, business problems involve time as an important variable. For example, one may wish to examine interest rate fluctuations over time and detect patterns (if any) that may exist in the time series defined by the sequence of observed results.

Chapter 13 examines the effect of time as a variable in business decision-making situations. First, we focus on the components of a time series and how certain components suggest the presence or absence of certain implied characteristics (e.g., a linear growth trend or a seasonal component) of the process that is generating the time series. Smoothing methods are offered to assist in the discovery of the hidden components. Second, we discuss the use of index numbers for making meaningful comparisons over time. We show how to derive and interpret price indexes, devoting particular attention to the interpretation of some of the more common published indexes.

13

ELEMENTS OF TIME
SERIES ANALYSIS

13.1

Introduction

We are all frequently confronted with decision-making situations in which time is an important variable. This situation usually occurs in problems in which we are attempting to estimate the expected value of a random process or to predict a new value at a future point in time after having observed the historical pattern of outcomes resulting from the process. For example, an investor is interested in predicting security prices, a store manager is interested in the effect of time on demand for products, and the marketing manager is interested in the pattern of sales over time. In a sense, everyone who plans for the future by attempting to budget time and resources is concerned with processes that are random over time. If we believe that interest rates will drop in the near future, we may be wise to rent now and buy a home later. The skier plans vacations during the winter because the seasonal weather pattern calls for the greatest snowfall during the winter.

Time may also be a hindrance in the decision-making process. We have seen in previous chapters that many unmeasured and uncontrolled variables may cause a response to vary over time and thereby inflate the experimental error. The undesirable effect of time can be reduced by using blocking designs (such as the randomized block design of Chapter 10) and by making experimental treatment comparisons within relatively homogeneous blocks of time.

time series   Any sequence of measurements taken on a response that is variable over time is called a time series. The time series is usually represented by the mathematical equation listing the values of the response as a function of time or, equivalently, as a figure on a graph whose vertical coordinate gives the value of the random response plotted against time on the horizontal axis. In Figure 13.1 we show a time series that plots U.S. Treasury bill rates as a function of time between the years 1935 and 1988. The same information can be shown in tabular form, but the pattern of change over time would be much more

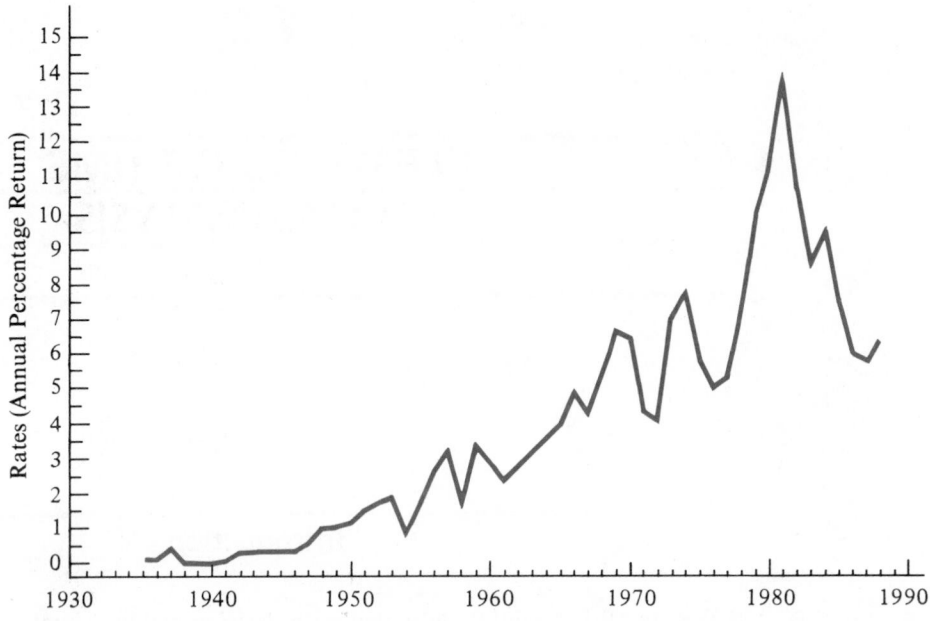

Figure 13.1   U.S. Treasury Bill Rates from 1935 to 1988

series pattern   obscure. It is the **pattern** generated by the time series and not necessarily the individual values that offers the planning device. An individual who is planning a long-term investment portfolio will want to buy Treasury bills only at times when the rates are projected to be high. Projecting the Treasury bill rates into the future implies extending the trends, cycles, and other elements of the pattern from the time series of Figure 13.1, as well as anticipating future government economic policy.

   Time series analysis is not limited solely to business problems concerned with economic data. Time series methods have been successfully applied in digital-signal-processing areas such as those found in telephone, radar, and sonar transmissions. Physical quantities, such as chemical concentrations, have been studied with the aid of the Box-Jenkins time series methodology (see Section 14.9). Quality control data, which mostly consists of time series data, can be studied with time series methods. For example, the exponential-smoothing method of Section 14.7 has recently been applied by Hunter to quality control data (see the references).

   The analysis of time series (that is, the utilization of sample data for the purposes of estimation, prediction, and decision making) can be complicated and difficult. For example, time series data often defy the usual assumption of independence required in regression analysis. The measurements $y_1, y_2, \ldots, y_n$ appearing in a time series are usually correlated, with the correlation decreasing as the time interval between a pair of measurements increases. In addition, the variation in the series may increase with time, which adds more difficulties to the analysis. The mathematical background required for handling these problems can quickly exceed the grasp of many nonmathematically trained forecasters. For this reason, many published time series analyses and forecasts are based on relatively primitive, subjective techniques.

The preceding comments are intended to introduce the subject of time series analysis and also to explain the absence of a single method of analysis. We will begin by first exploring the components of a time series, which together determine its pattern. In Chapter 14 we will examine some of the available techniques for extending the pattern of the time series and forecasting future values. Because of the importance of forecasting in modern business decision making, our emphasis in the study of time series methods will be placed on the analytical methods of Chapter 14 rather than on the descriptive methods of this chapter.

13.2

Components of Time Series

Statisticians often think of a time series as the addition of four meaningful *component series*:

1. Long-term trend
2. Cyclical effect
3. Seasonal effect
4. Random variation

long-term trends   **Long-term trends** are often present in time series because of a steady increase in population, gross national product, the effect of competition, or other factors that fail to produce sudden changes in response but produce a steady and gradual change over time. A time series with an upward long-term trend is similar to the increase in the operating revenues of U.S. oil pipeline companies from 1955 to 1985, as shown in Figure 13.2.

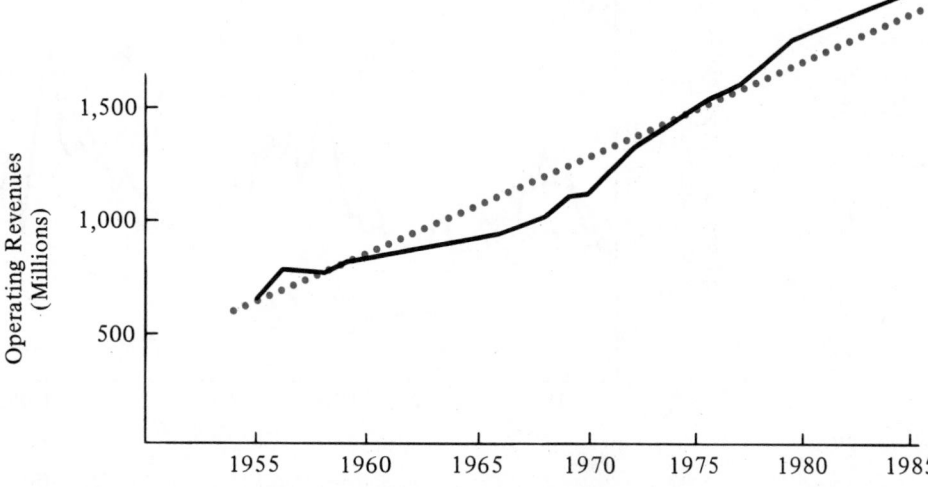

Figure 13.2   Operating Revenues of U.S. Oil Pipeline Companies from 1955 to 1985; Long-Term Trend Indicated as Dotted Line

cyclical effects   **Cyclical effects** in a time series are apparent when the response rises and falls in a gentle, wavelike manner about a long-term trend curve, like the average yields of new Aa corporate bonds offered by American business firms from 1969 to 1984 as illustrated in Figure 13.3. The average yield appears to follow a gradually increasing trend over time and tends to fluctuate with general conditions of the economy. Note that each complete cycle within the time series representing corporate bond yields covers more than one year (in this case, five years), a fact that we will later see distinguishes cyclic effects from seasonal effects. Generally, cyclic effects in a time series are caused by changes in the demand for a product (for instance, a decrease in sales of large cars when gasoline prices increase), by business cycles (for instance, business inventories decline during times of economic recession and rise during times of expansion), and, in particular, by the inability of supply to meet exactly the requirements of customer demand. In many other instances, especially those involving an analysis of financial or monetary factors, cyclic effects are caused by governmental economic or political policies.

seasonal effects   **Seasonal effects** in a time series are those rises and falls that always occur at a particular time of the year. For example, auto sales tend to decrease during August and September because of the changeover to new models, retail sales rise during the months of November and December, the sales of beer and soft drinks fluctuate in direct proportion to the temperature, and sales of skiing equipment are the greatest during the fall and winter months. **The essential difference between seasonal and cyclical effects is that seasonal movements are predictable, occurring at regular intervals of time from year to year; cyclical movements tend to be irregular and occur over a period of many years.** A time series that exhibits seasonal movement is illustrated by the department store sales shown in Figure 13.4.

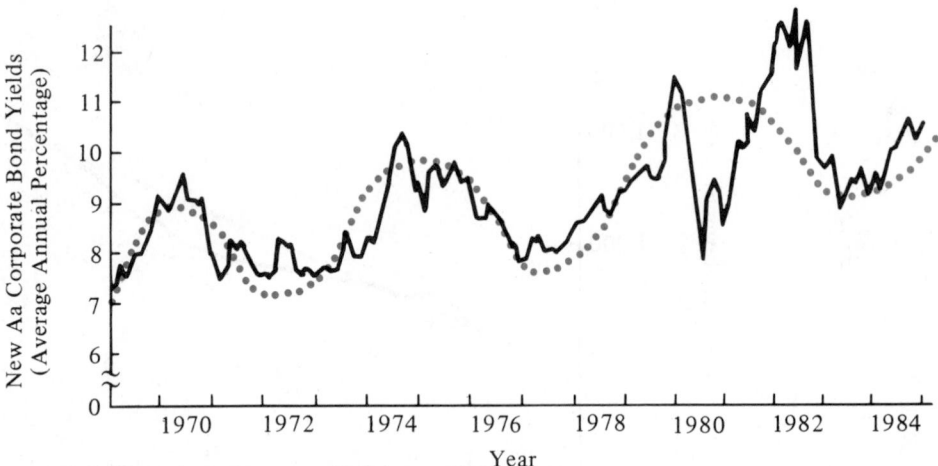

Figure 13.3   New Aa Corporate Bond Yields from 1969 to 1984; Cyclical Effect Illustrated by Dotted Line

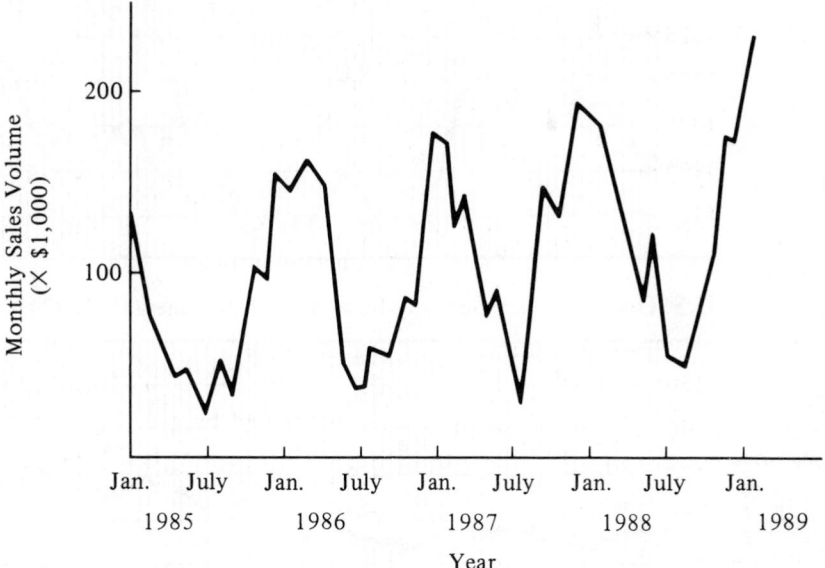

Figure 13.4   Monthly Retail Sales Volume of an Urban Department Store from 1985 to 1989

random variation   The fourth component of a time series is **random variation**. This component represents the random upward and downward movement of the series after accounting for the long-term trend, the cyclic effect, and the seasonal effect. Random variation, which might appear as the short-term (weekly) fluctuation in the Dow-Jones industrial average and the U.S. Dollar Index (as published daily in the *Wall Street Journal* and shown in Figure 13.5), is the unexplained shifting and bobbing of the series over a short period. Political events, unpredictable weather, and an amalgamation of many human actions tend to cause random and unexpected changes in a time series.

All time series contain random variation. In addition, a time series may contain none, one, two, or all of the three components long-term trend, cyclic effect, and seasonal effect. The objective of a time series analysis is to identify the components that do exist in order to identify their causes and to forecast future values of the time series.

With most time series processes distinguishing between the components is not easy. Often, seasonal and cyclic effects—or the three components, long-term trend and cyclic and seasonal effects—have become so integrated that they are inseparable. On the other hand, if the components appear to be distinguishable, separating them is not difficult. For instance, the monthly sales of the urban department store shown in Figure 13.4 illustrate a seasonal effect and a long-term trend with superimposed random variation. The long-term trend and seasonal effect, when identified, can be subtracted from the response values. The remainder is attributable to random variation. An illustration of how this separation might be accomplished is shown in Figure 13.6.

Stocks Dow Jones Industrial Average

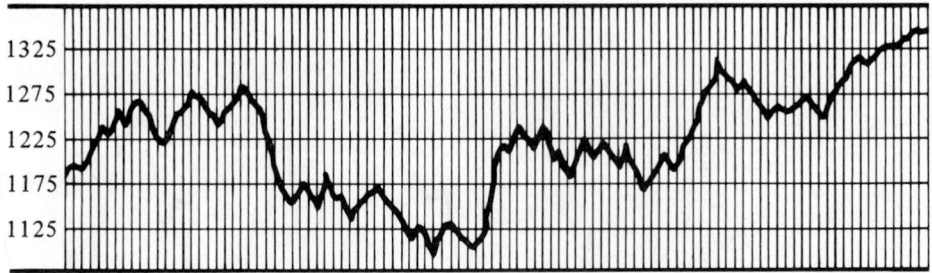

U.S. Dollar Morgan Guaranty Index vs. 15 Currencies (1980–82=100)

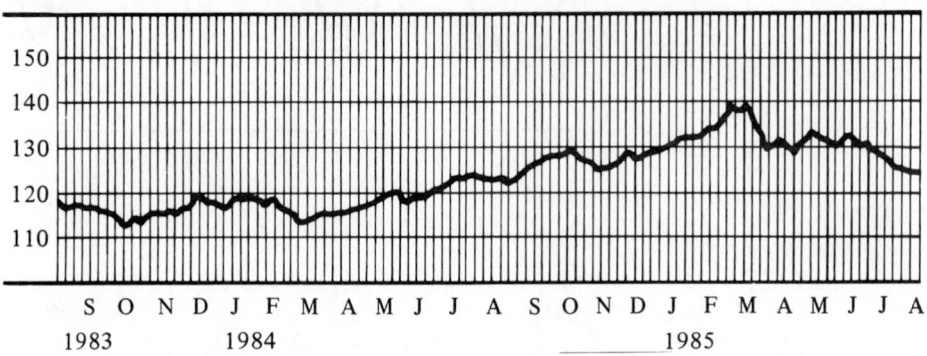

Figure 13.5   Weekly Changes in the Dow-Jones Industrial Average and the U.S. Dollar Index from September 1983 to August 1985.
Source: Wall Street Journal, August 9, 1985. Reprinted by permission of the *Wall Street Journal*, © Dow Jones & Company, Inc. 1985. All Rights Reserved.

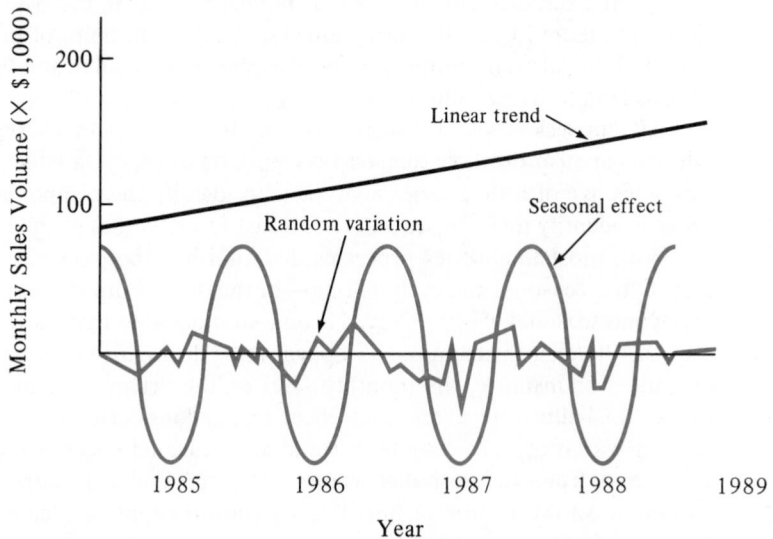

Figure 13.6   Monthly Retail Sales Volume of an Urban Department Store from 1985 to 1989

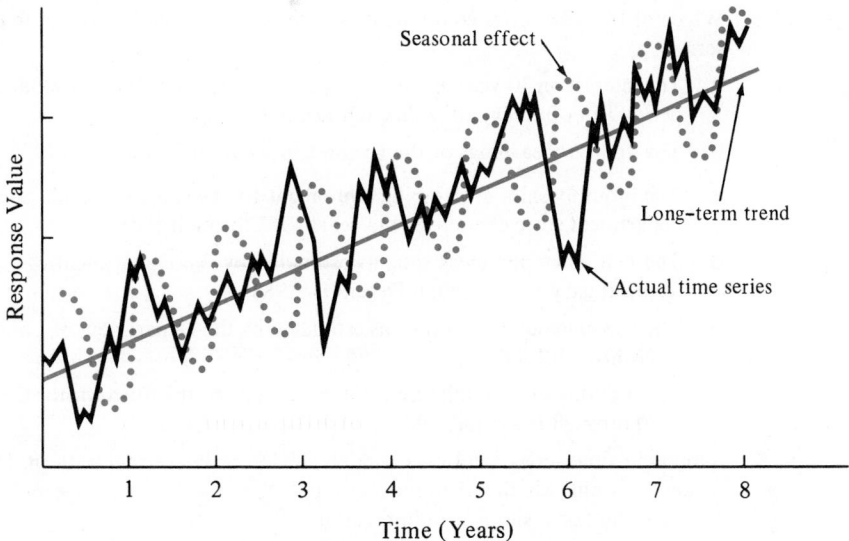

Figure 13.7   *Time Series Whose Seasonal Effect Has Been Hidden by the Presence of Random Variation with Excessively Large Variance*

signal of the time series

The communications engineer refers to the long-term trend and the cyclic and seasonal effects as the **signal of the time series**. We will use this terminology for purposes of discussion. Since the signal is the part of the time series that is deterministic, for purposes of prediction we must be able to separate the signal from random variation, called **noise** by the communications engineer. Determination of future values would simply require the statistician to add the extended patterns of the known components of the signal to the projected random variation.

noise

Since the random variation is at best probabilistic, accurate estimation of future values can be expected only when the magnitude of the random variation, as measured by its variance, is small. Otherwise, the oscillations (fluctuations) of the random variation over time may overwhelm the effect of the signal components or even cancel them out entirely. Such a process is illustrated in Figure 13.7, where the seasonal effect and the long-term trend from Figure 13.6 are combined with a random variation with a large variance. The resultant time series illustrates only the long-term trend, the seasonal effect having been "hidden" by the random variation.

Even if the proper signal is discovered and projected, the predicted values may be inaccurate if the magnitude of the random variation is great. In that case the best we can do is to give a probability interval for the predicted value, where the probability interval is based on the particular probability distribution of the random variation.

EXERCISES   Conceptual Level

13.1   List and define the components of a time series.

13.2   Why is it easier to forecast future values of a time series containing a seasonal effect than of one that possesses a cyclic effect?

13.3   Which of the time series components would you expect to be present in each of the following series?

   a.   The interest on 30-year first mortgages offered by First Interstate Bank over the 120-month period from January 1979 through December 1988

   b.   The quarterly earnings of the Exxon Corporation for the years 1979 through 1988

   c.   The monthly sales of camping equipment by the sporting goods division of a large retail department store chain over the years 1965 through 1988

   d.   The U.S. unemployment rate, as estimated by the Department of Labor, for each *month* from January 1970 through December 1988

   e.   The U.S. unemployment rate, as estimated by the Department of Labor, for each *year* from 1970 through 1988

   f.   The monthly sales of full-sized station wagons by the Ford Motor Company from January 1970 through December 1988

13.4   Suppose a time series exhibits a sudden shift from its normal pattern. How can the decision maker determine whether the shift is an indication of a lasting change in the time series (a new or revised signal) or is simply random variation?

EXERCISES   Skill Level

13.5   The data given in Table 13.1 represent a time series recorded monthly for a period of two years.

   a.   Plot the data against time.

   b.   Which time series components appear to exist?

Table 13.1   Time Series Data for Exercise 13.5

Month	Year 1	Year 2	Month	Year 1	Year 2
January	15.8	16.4	July	16.1	16.6
February	15.7	16.2	August	15.9	16.5
March	15.3	15.9	September	15.9	16.6
April	15.5	16.1	October	16.4	17.0
May	16.0	16.2	November	16.6	17.4
June	16.2	16.7	December	17.1	18.0

EXERCISES   Applications

13.6   For each of the following settings, explain how the decision maker might determine whether a shift observed in a time series under investigation is the result of a new signal or random variation.

   a.   A building contractor has observed that conventional 30-year residential mortgage rates have risen to an all-time high, surpassing the 16.5% rate achieved in 1980. If he believes that history will repeat itself and that low rates will soon return, he will not lay off his construction workers.

b. After the overthrow of the government of a middle eastern country, world market gold prices suddenly skyrocket. An investor is wondering whether this trend will be short-lived, indicating that he should sell his gold holdings, or whether he should retain his equity in gold in anticipation of additional price increases.

c. The owner of a marina has noticed a dramatic decline in the sales of pleasure boats since 1980. If she believes this trend is permanent, she will diversify into lower-priced, and less profitable, outdoor and camping equipment.

d. Because of an exceedingly cold winter and the resultant loss of many beef cattle, there has been an increase in beef prices, which has caused a meat processor to consider mixing ground beef with an inexpensive soybean flour to reduce the price of the product to consumers. To obtain sufficient soybean flour, the meat processor must sign a three-year contract with a supplier.

13.7 The time series in Figure 13.8 shows the GNP per capita in 1972 dollars in the United States from 1910 through 1985. Which time series components appear to exist in this time series?

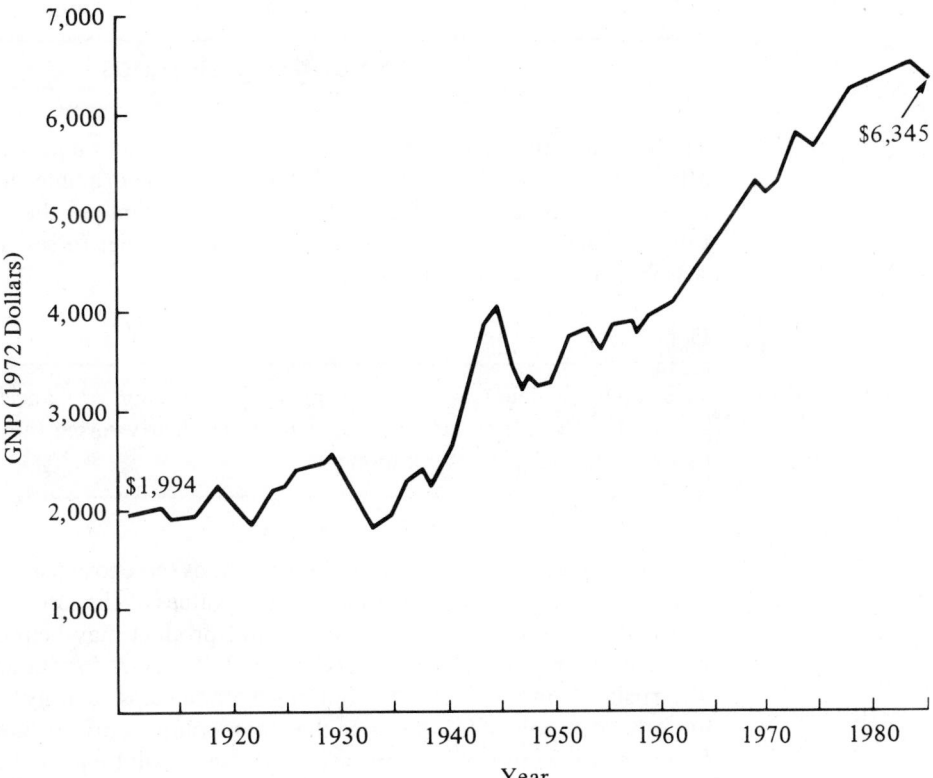

Figure 13.8   GNP Time Series for Exercise 13.7

13.8 The data given in Table 13.2 represent the average prices of steers (in dollars per 100 lb) in Kansas City recorded monthly for the period January 1981 through February 1986.

Table 13.2   Steer Price Data for Exercise 13.8

Year	Jan.	Feb.	Mar.	Apr.	May	June	July	Aug.	Sept.	Oct.	Nov.	Dec.	Average
1981	68.56	68.41	65.47	66.28	63.10	63.51	61.51	64.15	64.58	62.52	61.77	58.96	64.26
1982	59.22	62.37	63.96	64.72	66.07	63.70	64.17	66.42	63.55	62.21	61.24	59.17	62.79
1983	63.70	66.34	66.71	65.90	63.88	60.41	58.21	59.58	55.81	56.97	58.12	61.00	61.39
1984	64.39	65.97	66.30	64.15	60.82	59.28	62.17	61.34	62.01	62.74	63.96	64.26	63.11
1985	66.00	67.02	66.66	66.06	64.25	59.11	57.43	57.81	56.27	59.12	60.05	62.04	62.08
1986	61.34	61.68											

Source: *1986 CRB Commodity Year Book*, Commodity Research Bureau, Jersey City, N.J., p. 25. Reprinted by permission.

a. Plot the average price of steers against time for the period January 1984 through February 1986.

b. Which time series components appear to exist in the price pattern?

13.3

Smoothing Methods

Traditionally, time series methods have rested heavily on *smoothing techniques* that attempt to filter out the effect of the random variation in a time series. Most smoothing methods are based on some simple averaging technique that tends to reduce the random fluctuations in a series, making it much easier to see the underlying trend, seasonal, and cyclical components of the series.

Definition

smoothing techniques

> **Smoothing techniques** are methods for reducing or canceling the random variation in a time series in order to more clearly reveal the underlying trend, seasonal, and cyclical components.

Like regression analysis, smoothing methods serve to assist in both the *explanation* of a time series and the *forecasting* of future values of the series. Discovering a strong seasonal component in the sales of a retail product may help explain the cause of fluctuations in demand and, hence, suggest strategies for staffing, production, and advertising. If, in addition, the smoothing method helps quantify the seasonal effect and any other components of the series, then these components can be extrapolated into the future. A reassembly of the forecasts of the seasonal, trend, and cyclical components can then be used to generate a forecast of the series itself. In this chapter we will concentrate on the use of smoothing methods to decompose a time series into its fundamental components and save for Chapter 14 the discussion of forecasting with smoothing techniques.

centered moving average

The simplest smoothing technique is to calculate a **centered moving average** of the series, using a fixed number M of time periods. For example, we might average the

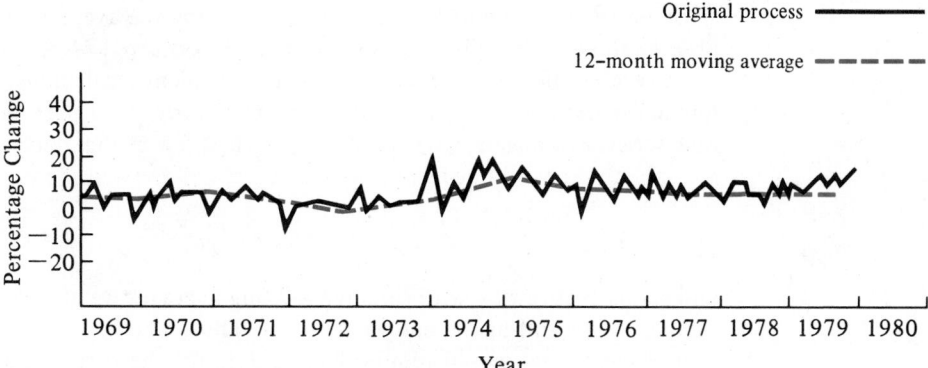

Figure 13.9   *Percentage Change in the Department of Labor's Medical Care Index from 1969 to 1980, with the 12-Month Moving Average Curve Superimposed*

monthly sales of a company over four successive time periods (so $M = 4$). Since the average we calculate is supposed to represent the series over four time periods, we plot it at the midpoint of the particular interval. In this sense, the average is "centered" in the middle of the time period it represents. The smoothing process is continued by advancing *one* period and calculating another centered average. Thus, if data from January through April were used in the first average, then the second average would use the data from February through May. Similarly, the third average would use the data from March through June, and so forth. In this way, the averages "move" through the entire time series a month at a time. In Figure 13.9 we show a 12-month centered moving average (i.e., $M = 12$) for the annual percentage change in the Department of Labor's medical care index. Note that the irregular fluctuations observed in the original time series have been smoothed out, revealing a fairly flat long-term trend line. Thus, although the cost of medical care has certainly increased over time, the rate of increase has been fairly constant.

Equation for a Centered Moving Average

At time point t the **centered moving average** $\bar{y}_t$ of the time series values over M periods is defined to be

$$\bar{y}_t = \frac{y_{t-(M-1)/2} + y_{t+1-(M-1)/2} + \cdots + y_{t+(M-1)/2}}{M}$$

where y_t is the value of the time series at time t, y_{t-1} is the value of the series at time $(t - 1)$, and so forth.

The centered moving average formula can be simplified by rewriting it in the form shown here (called a *recursive form*):

$$\bar{y}_t = \bar{y}_{t-1} + \frac{\text{next observation} - \text{most remote observation}}{M}$$

since all we are doing at each step is recomputing the average by adding in the next observation and dropping the observation that occurred M periods in the past.

Centered moving averages are most conveniently calculated when M is an odd number since then the center point will coincide with one of the time periods used in the time series. For instance, when $M = 3$, the first few centered averages for a series y_1, $y_2, \ldots, y_n, \ldots$ would be as follows:

$$\bar{y}_2 = \frac{y_1 + y_2 + y_3}{3} \qquad \bar{y}_3 = \frac{y_2 + y_3 + y_4}{3} \qquad \bar{y}_4 = \frac{y_3 + y_4 + y_5}{3}$$

and so on. Notice that with centered moving averages we always lose a few points at the beginning and at the end of the data. Thus, when $M = 3$, we cannot find a $\bar{y}_1$ average since that would require having data for the period just prior to $t = 1$. As a general rule, for M odd we lose exactly $(M - 1)/2$ points at the beginning and $(M - 1)/2$ at the end of the time series data.

With economic time series, values of M such as 4 or 12 are very frequently used since they correspond to quarterly and monthly data. As mentioned above, this situation creates a slight inconvenience since the centered averages for $M = 4$ or 12 will be plotted between time periods (e.g., at $t = 2.5, 3.5, \ldots$) rather than at the original time periods. So that this problem is avoided, most moving average procedures used in practice simply smooth the data one more time, using a small even value like $M = 2$. For example, a set of $M = 12$-month centered averages is computed and then a set of $M = 2$-month centered averages is computed from the 12-month averages. Thus, we smooth the smoothed values. The result is a final smoothed series centered at the same periods as the original data. With monthly data, for instance, the 12-month centered averages would be centered at times $t = 6.5, 7.5, \ldots$. The 2-month centered averages of the 12-month averages would then be centered at times $t = 7, 8, 9, \ldots$. Having the averages centered at points coinciding with the original data greatly facilitates the estimation of the trend, seasonal, and cyclical components.

exponential smoothing   Another smoothing scheme, called **exponential smoothing**, is sometimes used to smooth a time series. Besides its basic property of producing a smoother series, exponential smoothing is primarily used to produce forecasts of the series. The forecasting applications of exponential smoothing will be discussed in Chapter 14. The exponentially smoothed response value at time period t is denoted by S_t. The smoothing scheme begins by assigning $S_1 = y_1$ at the first period. For the second time period,

$$S_2 = \alpha y_2 + (1 - \alpha)S_1$$

and so on.

Basic Equation of Exponential Smoothing

For any time period t the smoothed value S_t is found by computing

$$S_t = \alpha y_t + (1 - \alpha)S_{t-1} \qquad 0 \le \alpha \le 1$$

This equation is called the **basic equation of exponential smoothing**, and the **smoothing constant**   constant α is called the **smoothing constant**.

Note to the Reader

> The symbol α is used exclusively throughout Chapters 13 and 14 to represent the smoothing constant, despite the fact the α was defined as the probability of making a type I error, or the significance level of a statistical test, in Chapter 8. Since smoothing is discussed only in Chapters 13 and 14, this double usage of the symbol α should not present a problem. We have chosen not to use a symbol different from α in either case since both uses of α are traditional in statistical literature.

The centered moving average smoothing scheme forms averages over M time periods, but S_t computes an average from all past values $y_t, y_{t-1}, \ldots, y_1$, where y_t is the value at the time period t, y_{t-1} is the value at time period $t-1$, and y_1 is the value from the first time period in which data are available. This averaging process can be seen if we expand the basic equation by first substituting

$$S_{t-1} = \alpha y_{t-1} + (1 - \alpha)S_{t-2}$$

into the equation for S_t to obtain

$$S_t = \alpha y_t + (1 - \alpha)\alpha y_{t-1} + (1 - \alpha)^2 S_{t-2}$$

By substituting for S_{t-2}, then for S_{t-3}, and so forth, until we substitute y_1 for S_1, we can show (details not given here) that the expanded equation can be written as

$$S_t = \alpha \sum_{i=0}^{t-2}(1 - \alpha)^i y_{t-i} + (1 - \alpha)^{t-1} y_1$$

Even though remote responses are not dropped in an exponential-smoothing scheme as they are in a moving average, their contribution to the smoothed value S_t becomes less at each successive time point. The speed at which remote responses are dampened (smoothed) out is determined by the selection of the smoothing constant α. For values of α near 1 remote responses are dampened out quickly; for α near 0 they are dampened out slowly.

The most important problem when applying exponential smoothing is to find the "best" smoothing constant α for a particular set of data. Unfortunately, there does not exist a simple formula for finding such a value of α. Instead, practitioners usually employ a computer to "search" for some good values of α by simply looking at the smoothed series generated by various values of α in the range from 0 to 1. In general, the more "noisy" or volatile a time series is, the smaller the value of α should be. Otherwise, a large α will give too much weight to the most recent "noisy" measurement y_t. Similarly, for more stable series large values of α would be used.

The following example illustrates the use of three different smoothing models on a rather volatile time series.

EXAMPLE 13.1   The week's end closing prices for the securities of the Color-Vision Company, a manufacturer of color television sets, have been recorded over a period of 30 consecutive weeks. Find the 5-week centered moving average time series and the exponentially smoothed time series, using smoothing constants $\alpha = .1$ and $\alpha = .5$.

SOLUTION   The original process values y_t (week's end closing prices for each of the 30 weeks), the 5-week centered moving averages $\bar{y}_t$, and the exponentially smoothed time series S_t for $\alpha = .1$ and $\alpha = .5$ are listed in Table 13.3. The moving averages were found by computing

$$\bar{y}_t = \frac{y_{t-2} + y_{t-1} + y_t + y_{t+1} + y_{t+2}}{5}$$

Table 13.3   Original and Smoothed Week's End Closing Prices for the Securities of the Color-Vision Company Over 30 Consecutive Weeks; Example 13.1

t	y_t	$\bar{y}_t$	$S_t\ (\alpha = .1)$	$S_t\ (\alpha = .5)$
1	71		71.0	71.0
2	70		70.9	70.5
3	69	68.4	70.7	69.8
4	68	67.2	70.4	68.9
5	64	67.6	69.8	66.5
6	65	69.4	69.3	65.8
7	72	70.8	69.6	68.9
8	78	73.0	70.4	73.5
9	75	75.0	70.9	74.3
10	75	74.6	71.3	74.7
11	75	74.0	71.7	74.9
12	70	74.0	71.5	72.5
13	75	73.8	71.9	73.8
14	75	74.4	72.2	74.4
15	74	77.6	72.4	74.2
16	78	79.0	73.0	76.1
17	86	79.0	74.3	81.1
18	82	78.8	75.1	81.6
19	75	77.6	75.1	78.3
20	73	75.2	74.9	75.7
21	72	73.0	74.6	73.4
22	73	73.4	74.4	73.5
23	72	75.4	74.2	72.8
24	77	77.2	74.5	74.9
25	83	78.8	75.4	79.0
26	81	81.4	76.0	80.0
27	81	83.0	76.5	80.5
28	85	83.2	77.4	82.8
29	85		78.2	83.9
30	84		78.8	84.0

Note: All smoothed values have been rounded to the nearest tenth of a unit.

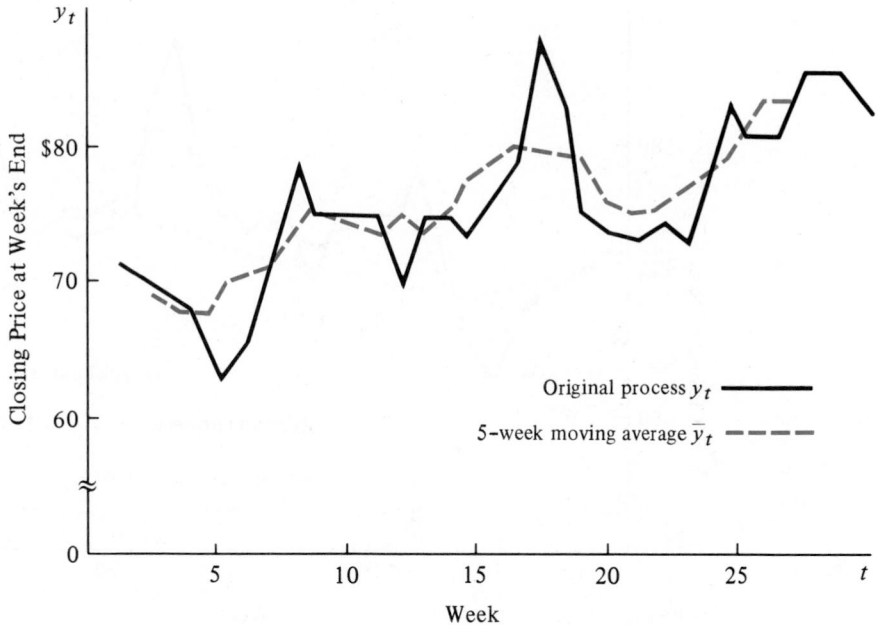

Figure 13.10   Week's End Closing Security Prices for the Color-Vision Company Over 30 Weeks, with the 5-Week Centered Moving Average Curve Superimposed

for each of the time periods $t = 3, 4, \ldots, 28$. For instance, the seventh moving average value $\bar{y}_7$ was found by computing

$$\bar{y}_7 = \frac{y_5 + y_6 + y_7 + y_8 + y_9}{5} = \frac{64 + 65 + 72 + 78 + 75}{5} = 70.8$$

The original time series and the moving average time series are shown together in Figure 13.10.

The exponentially smoothed time series employing a smoothing constant $\alpha = .1$ was computed by first setting

$$S_1 = 71.0$$

and then computing

$$S_2 = (.1)(70) + (1 - .1)(71.0) = 70.9$$
$$S_3 = (.1)(69) + (1 - .1)(70.9) = 70.7$$

and so forth.

Similarly, each of the values for the exponentially smoothed time series with $\alpha = .5$ are found by first setting

$$S_1 = 71.0$$

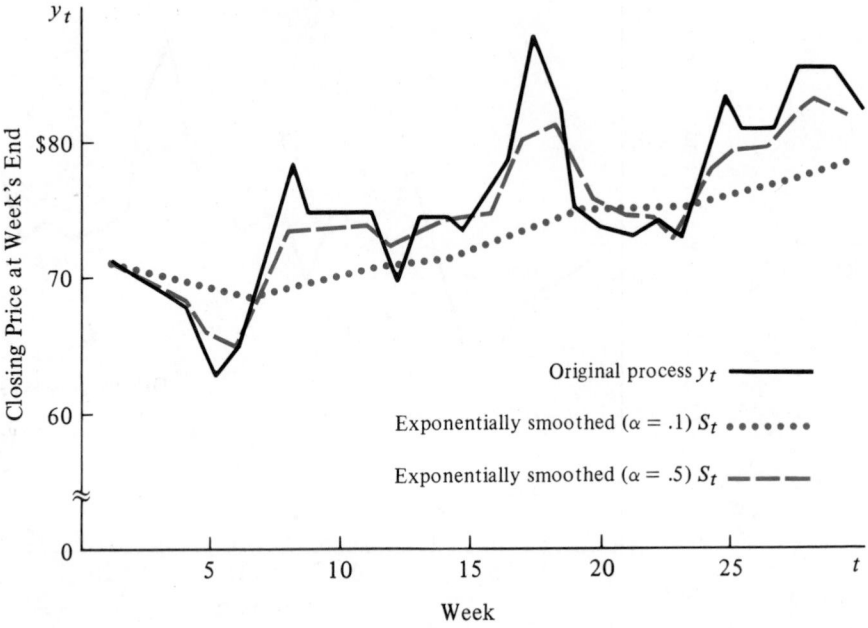

Figure 13.11   Week's End Closing Security Prices for the Color-Vision Company Over 30 Weeks, with the Two Exponentially Smoothed Processes Superimposed

and then computing

$$S_t = (.5)y_t + (1 - .5)S_{t-1}$$

for each of the time periods $t = 2, 3, 4, \ldots, 30$. The original time series is plotted together with both of the exponentially smoothed time series in Figure 13.11.

Observe that in both Figures 13.10 and 13.11 the smoothed time series appear more stable than the original series. However, the 5-week centered moving average series and the exponentially smoothed time series with $\alpha = .5$ appear to be much less stable than the exponentially smoothed series with $\alpha = .1$. The latter series, although undershooting the original series much of the time, appears to suggest the presence of a linear trend with a cyclic effect. Hence, the true components of the original series (if a linear trend and a cyclic effect are the true components) most readily become apparent when the original series is smoothed by an exponential-smoothing scheme employing a small smoothing constant. This statement is not a generalization; however, the small smoothing constant happens to yield the best results for the data used here.

◆ ◆

Perhaps the major advantage of smoothing techniques is typified by the old saying that a picture is worth a thousand words. Moving averages and exponentially smoothed time series sometimes make trends, cycles, and seasonal effects more visible to the eye and consequently lead to a simple and useful description of the time series process for the businessperson or economist. When presented simply as a description of the time series (as is often the case), smoothing techniques often ignore the basic objective of statistics inference and leave this difficult task to the reader. Thus, we still seek techniques for estimation and prediction that are accompanied by measures of goodness. We will investigate such methods in Chapter 14.

EXERCISES Conceptual Level

13.9 Describe the centered moving average techniques for smoothing a time series.

13.10 When you compute centered moving averages, should you select a value for M that is odd or even? Why?

13.11 What is the basic difference between the centered moving average and the exponential-smoothing technique?

13.12 What advantage does exponential smoothing have over the centered moving average technique?

13.13 What advantage does the centered moving average technique have?

EXERCISES Skill Level

13.14 Refer to Exercise 13.5.

a. Smooth the time series by computing a three-month centered moving average. Plot the smoothed series and the original series on the same sheet of graph paper.

b. Repeat the process in part a for a four-month centered moving average, using the two-stage centering procedure.

c. Compare the results of parts a and b. Which centered moving average appears most useful in stabilizing the series?

13.15 Refer to Exercise 13.5.

a. Smooth the monthly series by using exponential smoothing with $\alpha = .1$.

b. Smooth the monthly series by using exponential smoothing with $\alpha = .3$.

c. Smooth the monthly series by using exponential smoothing with $\alpha = .5$.

d. Plot the original time series on a sheet of graph paper. Superimpose the smoothed series for the three values of α. What conclusions can you draw?

EXERCISES Applications

13.16 Refer to Exercise 13.8. Smooth the steer price time series for January 1984 through February 1986 by computing a three-month centered moving average. Plot the smoothed series and the original series. Now repeat this process for a five-month centered moving average. What conclusions do you draw concerning the inherent components within the time series? Which centered moving average model was most useful in detecting these inherent components?

13.17   The data in Table 13.4 represent the gross monthly sales volume (in thousands of dollars) of a pharmaceutical company from January 1985 to January 1988.

    a.   Plot the sales values against time, and construct the time series.

    b.   Which time series components appear to exist in the sales pattern?

    c.   Smooth the monthly sales values by computing a three-month centered moving average series. Plot the smoothed series and the original series together on the same sheet of graph paper. What conclusions do you draw?

Table 13.4   Sales Volume Data for Exercise 13.17

	Year			
Month	1985	1986	1987	1988
January	18.0	23.3	24.7	28.3
February	18.5	22.6	24.4	27.5
March	19.2	23.1	26.0	28.8
April	19.0	20.9	23.2	22.7
May	17.8	20.2	22.8	19.6
June	19.5	22.5	24.3	20.3
July	20.0	24.1	27.4	20.7
August	20.7	25.0	28.6	21.4
September	19.1	25.2	28.8	22.6
October	19.6	23.8	25.1	28.3
November	20.8	25.7	29.3	27.5
December	21.0	26.3	31.4	28.1

13.18   Refer to Exercise 13.17.

    a.   Smooth the monthly sales volumes by computing an exponentially smoothed series, employing the smoothing constant $\alpha = .1$.

    b.   Smooth the monthly sales volumes by computing an exponentially smoothed series, employing the smoothing constant $\alpha = .25$.

    c.   On a sheet of graph paper, superimpose the two smoothed series on the original series. What conclusions do you draw?

13.19   Because interest from municipal bonds is exempt from federal, state, and local income tax obligations, their yields have traditionally been less than those for investment devices for which earnings are not tax-exempt. Nevertheless, municipals continue to offer an attractive alternative for those who seek tax shelters. Table 13.5 lists the average yields of high-grade municipal bonds (S&P) issued over the eight-year period 1980–1987.

    a.   Smooth the monthly municipal bond yield data by computing a five-month centered moving average series.

    b.   Smooth the monthly municipal bond yield data by computing an exponentially smoothed series, employing the smoothing constant $\alpha = .1$.

    c.   Smooth the monthly municipal bond yield data by computing an exponentially smoothed series, employing the smoothing constant $\alpha = .4$.

    d.   On a sheet of graph paper, superimpose the three smoothed series on the original series. What conclusions do you draw?

Table 13.5 Bond Yield Data for Exercise 13.19

	Year							
Month	1980	1981	1982	1983	1984	1985	1986	1987
Jan.	7.21	9.65	13.16	9.45	9.61	9.55	8.06	6.63
Feb.	8.04	10.03	12.81	9.48	9.63	9.66	7.44	6.66
Mar.	9.09	10.12	12.72	9.16	9.92	9.79	7.07	6.71
Apr.	8.40	10.55	12.45	8.96	9.98	9.48	7.32	7.62
May	7.37	10.73	11.99	9.03	10.55	9.08	7.67	8.10
June	7.60	10.56	12.42	9.51	10.71	8.78	7.98	7.89
July	8.08	11.03	12.11	9.46	10.50	8.90	7.62	7.83
Aug.	8.62	12.13	11.12	9.72	10.03	9.18	7.31	7.90
Sept.	8.95	12.86	10.61	9.57	10.17	9.37	7.14	8.36
Oct.	9.11	12.67	9.59	9.64	10.34	9.24	7.12	8.84
Nov.	9.55	11.71	9.97	9.79	10.27	8.64	6.86	8.09
Dec.	10.09	12.77	9.91	9.90	10.04	8.51	6.93	8.07

Source: Data from U.S. Council of Economic Advisors, *Economic Indicators*, January 1981–1988.

13.4

Adjustment of Seasonal Data

Suppose we are interested in examining the trend and cyclical movements in a time series of economic data. This task is difficult if the time series exhibits a pronounced seasonal component, since seasonal fluctuations often tend to overwhelm the other components of the series. If, however, the seasonal component can first be removed from the series, then the interpretation of trends and cycles can be greatly simplified.

Definition

deseasonalized time series

> A **deseasonalized time series** is one in which the seasonal component has been estimated and then removed from the series.

In most seasonal time series the seasonal effect repeats itself every 4 or 12 periods, depending on whether the data was measured *quarterly* or *monthly*. For example, monthly sales of soft drinks, beer, and golf equipment tend to be high during the warm months and low during the cool months, a pattern that repeats itself yearly (i.e., every 12 periods). The most common method for removing the seasonal component from seasonal data is to use centered moving averages to smooth out the seasonal variation and the noise. The number of terms M in the moving average is always taken to be equal to the number of time periods in one complete season. That is, for monthly data $M = 12$, and for quarterly data $M = 4$. Thus, each average will contain all 12 months (or all 4 quarters), so these *yearly* averages should not exhibit any seasonal fluctuations.

ratio-to-moving-
average method
The method used in this section for deseasonalizing a time series is called the **ratio-to-moving-average method** since it is based on calculating ratios of the original series to a moving average series. This method was invented in 1922 by the National Bureau of Economic Research for the purpose of deseasonalizing economic data so that changes in the business cycle could be monitored. Many refinements have been added over the years, and computer programs have been written to handle the calculations. One of these refinements, the X-11 Variant of the Census Method II Seasonal Adjustment Program, is currently used by most businesses and governmental agencies to deseasonalize time series data.

The ratio-to-moving-average method is carried out by using the following two-stage moving average procedure:

1. Let M equal the number of time periods in one complete seasonal period.
2. Compute the M-period centered moving averages of the series.
3. Compute the 2-period centered moving averages of the series resulting from step 2.
4. Using the series from step 3, calculate a set of seasonal indexes (which measure the seasonal effect in each of the M periods).
5. Finally, deseasonalize the original series by dividing each y_t by the appropriate seasonal index.

Even though, as mentioned before, a few observations will be lost at the beginning and end of the data when we use moving averages, this loss will not create a problem when we calculate the seasonal indexes as long as there is a sufficient amount of data. As a rule of thumb, we should have at least three complete seasons represented in the data before applying the ratio-to-moving-average method.

As an illustration of the ratio-to-moving-average technique, consider the time series of monthly sales of a regional brewing company for the years 1986 to 1988, as shown in the second column of Table 13.6. By examining Figure 13.12, we can see that this time series follows a 12-month seasonal pattern. Thus, we let $M = 12$.

The first moving average value falls between June and July of 1986 since $M = 12$ is an even number. We find this first moving average value to be

$$\bar{y}_{\text{June-July}} = 19.6 + 18.6 + 23.2 + 24.5 + 27.7 + 30.0 + 28.7 + 33.8$$
$$\frac{+ 25.1 + 22.1 + 21.8 + 20.9}{12}$$

$$= \frac{296}{12} = 24.667$$

Similarly, we find

$$\bar{y}_{\text{July-Aug}} = \frac{296}{12} + \frac{23.3 - 19.6}{12} = \frac{299.7}{12} = 24.975$$

Notice that the July–August value is computed by using the shortcut procedure introduced in Section 13.3. The shortcut procedure cannot be used to compute the first moving average but can be used to compute every moving average after the first.

Table 13.6   Actual Sales and Specific Seasonal Indexes for the Sales of a Brewing Company, 1986–1988 (Thousands of Barrels)

Time, t	Actual Sales, y_t	12-Month Moving Average	2-Month Moving Average	Specific Seasonal Index, s_t
1986				
(1) Jan.	19.6			
(2) Feb.	18.6			
(3) Mar.	23.2			
(4) Apr.	24.5			
(5) May	27.7			
(6) June	30.0	24.667		
(7) July	28.7	24.975	24.821	1.156
(8) Aug.	33.8	25.100	25.038	1.350
(9) Sept.	25.1	25.508	25.304	.992
(10) Oct.	22.1	25.683	25.596	.863
(11) Nov.	21.8	25.758	25.721	.848
(12) Dec.	20.9		25.896	.807
1987				
(13) Jan.	23.3	26.033	26.267	.887
(14) Feb.	20.1	26.500	26.300	.764
(15) Mar.	28.1	26.100	26.154	1.074
(16) Apr.	26.6	26.208	26.333	1.010
(17) May	28.6	26.458	26.479	1.080
(18) June	33.3	26.500	26.475	1.258
(19) July	34.3	26.450	26.504	1.294
(20) Aug.	29.0	26.558	26.671	1.087
(21) Sept.	26.4	26.783	26.796	.985
(22) Oct.	25.1	26.808	26.833	.935
(23) Nov.	22.3	26.858	26.858	.830
(24) Dec.	20.3	26.858	26.692	.761
1988				
(25) Jan.	24.6	26.520	26.692	.922
(26) Feb.	22.8	26.858	26.983	.845
(27) Mar.	28.4	27.108	27.046	1.050
(28) Apr.	27.2	26.983	27.092	1.004
(29) May	28.6	27.200	27.196	1.052
(30) June	29.3	27.192	27.242	1.076
(31) July	38.3	27.292		
(32) Aug.	32.0			
(33) Sept.	24.9			
(34) Oct.	27.7			
(35) Nov.	22.2			
(36) Dec.	21.5			

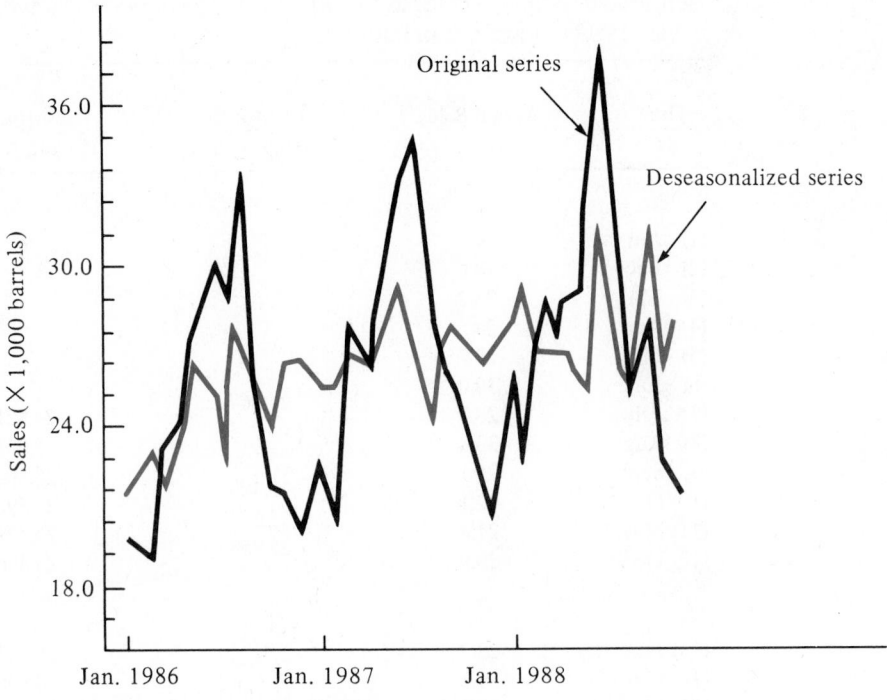

Figure 13.12   Sales Volume for a Brewing Company for the Years 1986–1988, with the Deseasonalized Series Superimposed

We wish to have a moving average uniquely associated with each observation in the time series (except the first six and last six observations, which are lost by employing the moving average). Thus, we **center** the moving average values by computing

$$\bar{y}_{\text{July}} = \frac{24.667 + 24.975}{2} = 24.821$$

Taking the process one step further, we find

$$\bar{y}_{\text{Aug–Sept}} = \frac{299.7}{12} + \frac{20.1 - 18.6}{12} = \frac{301.2}{12} = 25.1$$

$$\bar{y}_{\text{Aug}} = \frac{\bar{y}_{\text{July–Aug}} + \bar{y}_{\text{Aug–Sept}}}{2} = \frac{24.975 + 25.1}{2} = 25.038$$

The remaining 22 moving averages have been computed and are listed in Table 13.6.

The next step is to find the seasonal indexes for each month. This step can be done by first calculating the *specific seasonal indexes* and then applying a normalizing transformation.

Definition

specific seasonal
index

> The **specific seasonal index** s_t for period t is found by dividing the original value of the series y_t by the corresponding two-stage centered moving average at period t.

In Table 13.6 the first specific seasonal index we can calculate is for July 1986, where we find

$$s_7 = \frac{28.7}{24.821} = 1.156$$

The remaining indexes are calculated similarly and are recorded in the last column of Table 13.6. Essentially, the two-stage centered moving average represents an estimate of the trend and cycle at period t since the averaging procedure should have eliminated most of the noise and seasonality. Thus, the specific seasonal index measures how much y_t falls above or below the trend in period t. Indexes greater than 1 mean that the seasonal component is pulling the series above the long-term trend; indexes smaller than 1 occur when the seasonal effect causes the series to drop below the trend.

Table 13.7 shows the specific seasonals of Table 13.6 arranged by month. To calculate a *single* index for each month, we average the specific seasonal indexes for each individual month and then normalize them so that they sum to M.

Definition

seasonal index

> The **seasonal index** $\bar{s}_i$ for a particular month (quarter) is found by averaging all the specific seasonal indexes associated with that month (quarter) and then normalizing the resulting M indexes so that they sum to M.

In our example the (unnormalized) seasonal index for January is

$$\bar{s}_1 = \frac{.887 + .922}{2} = .905$$

Table 13.7   Seasonal Indexes for the Data of Table 13.6

	Jan.	Feb.	Mar.	Apr.	May	June	July	Aug.	Sept.	Oct.	Nov.	Dec.
	—	—	—	—	—	—	1.156	1.350	.992	.863	.848	.807
	.887	.764	1.074	1.010	1.080	1.258	1.294	1.087	.985	.935	.830	.761
	.922	.845	1.050	1.004	1.052	1.076	—	—	—	—	—	—
Seasonal index	.905	.805	1.062	1.007	1.066	1.167	1.225	1.219	.989	.899	.839	.784

February's index is

$$\bar{s}_2 = \frac{.764 + .845}{2} = .805$$

and so forth. These averages are shown at the bottom of Table 13.7.

The final step is to normalize the indexes by forcing them to sum to either 12 (for monthly data) or 4 (for quarterly data). Normalizing is done by dividing each unnormalized index by the sum of all the indexes and multiplying the resulting numbers by 12 (or 4, for quarterly data). For our example the sum of the unnormalized indexes is

$$.905 + .805 + 1.062 + \cdots + .784 = 11.967$$

The normalized seasonal indexes are then

$$\bar{s}_1 = \left(\frac{12}{11.967}\right)(.905) = .907 \qquad \bar{s}_7 = \left(\frac{12}{11.967}\right)(1.225) = 1.228$$

$$\bar{s}_2 = \left(\frac{12}{11.967}\right)(.805) = .807 \qquad \bar{s}_8 = \left(\frac{12}{11.967}\right)(1.219) = 1.222$$

$$\bar{s}_3 = \left(\frac{12}{11.967}\right)(1.062) = 1.065 \qquad \bar{s}_9 = \left(\frac{12}{11.967}\right)(.989) = .992$$

$$\bar{s}_4 = \left(\frac{12}{11.967}\right)(1.007) = 1.010 \qquad \bar{s}_{10} = \left(\frac{12}{11.967}\right)(.899) = .901$$

$$\bar{s}_5 = \left(\frac{12}{11.967}\right)(1.066) = 1.069 \qquad \bar{s}_{11} = \left(\frac{12}{11.967}\right)(.839) = .841$$

$$\bar{s}_6 = \left(\frac{12}{11.967}\right)(1.167) = 1.170 \qquad \bar{s}_{12} = \left(\frac{12}{11.967}\right)(.784) = .786$$

To deseasonalize the time series in Table 13.6, we divide each y_t by the appropriate normalized seasonal index. Table 13.8 shows the deseasonalized series in the last column. Notice that the few points lost at the beginning and end of the data by the averaging process only affect our estimates of the seasonal indexes, and that there are *no* points lost in the deseasonalized series.

In Figure 13.12 we have plotted the original time series of sales of the brewing company for the years 1986 through 1988 and have superimposed the deseasonalized series. Since the seasonal fluctuations have been smoothed out, the seasonally adjusted data exhibits only a linear growth trend. No short-term trends or business cycles appear for the period 1986 through 1988. In the final analysis, the brewing company's sales manager can assume that sales of the firm's products are increasing relatively over time but fluctuate seasonally over the year, exhibiting greater-than-average sales during the warm months and lesser-than-average sales during the cooler months. The only other factors affecting the firm's time series are of minimal importance and very short-lived, allowing for their categorization as random variation.

Table 13.8   Data from Table 13.6 Deseasonalized

Month, t	Actual Sales, y_t	Normalized Seasonal Index	Deseasonalized Series	Month, t	Actual Sales, y_t	Normalized Seasonal Index	Deseasonalized Series
1	19.6	.907	21.6	19	34.3	1.228	27.9
2	18.6	.807	23.0	20	29.0	1.222	23.7
3	23.2	1.065	21.8	21	26.4	.992	26.6
4	24.5	1.010	24.3	22	25.1	.901	27.9
5	27.7	1.069	25.9	23	22.3	.841	26.5
6	30.0	1.170	25.6	24	20.3	.786	25.8
7	28.7	1.228	23.4	25	24.6	.907	27.1
8	33.8	1.222	27.7	26	22.8	.807	28.3
9	25.1	.992	25.3	27	28.4	1.065	26.7
10	22.1	.901	24.5	28	27.2	1.010	26.9
11	21.8	.841	25.9	29	28.6	1.069	26.8
12	20.9	.786	26.6	30	29.3	1.170	25.0
13	23.3	.907	25.7	31	38.3	1.228	31.2
14	20.1	.807	24.9	32	32.0	1.222	26.2
15	28.1	1.065	26.4	33	24.9	.992	25.1
16	26.6	1.010	26.3	34	27.7	.901	30.7
17	28.6	1.069	26.8	35	22.2	.841	26.4
18	33.3	1.170	28.5	36	21.5	.786	27.4

EXERCISES   Conceptual Level

13.20   Why is it desirable to remove the seasonal component from a time series?

13.21   Describe the ratio-to-moving-average method for deseasonalizing a time series. Why is this method preferred to simply using the two-stage centered moving averages as the deseasonalized series?

EXERCISES   Skill Level

13.22   Refer to Exercise 13.5.

a. Apply a 12-month, two-stage centered moving average to the data.

b. Calculate the specific seasonal indexes for the 12 available data points.

c. Use the results of part b to calculate the seasonal indexes.

d. Normalize the seasonal indexes, and obtain the deseasonalized time series.

e. Plot the original and the deseasonalized series on the same graph. Are any inherent components apparent?

EXERCISES   Applications

13.23   A company that operates four hotels in a large midwestern city has recorded the number of occupied rooms in the four hotels it operates for each day for a five-year period. From these data average daily occupancies were obtained for each month of the five-year period; these data are listed in Table 13.9. Remove the seasonal component from these data by using the ratio-to-moving-average method with $M = 12$. Plot the original series and the deseasonalized series together on the same piece of graph paper. After removing the seasonal component from the series, describe the inherent characteristics of the room occupancy time series.

Table 13.9   Hotel Occupancy Data for Exercise 13.23

| | Year | | | | |
Month	1	2	3	4	5
January	691	723	758	748	811
February	649	655	709	731	732
March	656	658	715	748	745
April	735	761	788	827	844
May	748	768	794	788	833
June	837	885	893	937	935
July	995	1,067	1,046	1,076	1,110
August	1,040	1,038	1,075	1,125	1,124
September	809	812	812	840	868
October	793	790	822	864	860
November	692	692	714	717	762
December	763	782	802	813	877

Source: Adapted from B. Bowerman and R. O'Connell, *Forecasting and Time Series* (Boston: PWS-Kent Publishing Company, 1979), p. 424.

13.24   Refer to Exercise 13.23. Smooth the room occupancy time series by computing an exponentially smoothed series, using the smoothing constant $\alpha = .4$. Now repeat this process by using the smoothing constant $\alpha = .1$. On the same piece of graph paper, plot the two exponentially smoothed series together with the original room occupancy time series. Did either of the exponentially smoothed series appear to eliminate the seasonal component from the room occupancy time series?

13.25   Texas Chemical Products manufactures an agricultural chemical that is applied to farmlands after crops have been harvested. Since the chemical tends to deteriorate in storage, Texas Chemical cannot stockpile quantities in advance of the winter season demand for the product. Sales of the product over four consecutive years are shown in Table 13.10 (the recorded sales values are listed in thousands of pounds). Remove the seasonal component from these data by using the ratio-to-moving-average method with $M = 12$. Plot the original series and the deseasonalized series together on the same piece of graph paper. After you have removed the seasonal component from the series, what inherent components does the time series appear to possess?

13.26   The data in Table 13.11 represent the quarterly common share earnings (in dollars) for Hilton Hotel stockholders for the years 1979 through 1987. Remove the seasonal component from the

Table 13.10   Sales Data for Exercise 13.25

	Year			
Month	1	2	3	4
January	123.327	133.708	143.747	145.151
February	129.585	146.156	159.360	146.323
March	157.480	174.000	168.129	164.262
April	155.027	162.574	152.642	157.848
May	161.040	176.280	178.682	181.914
June	169.076	154.033	164.432	169.352
July	142.196	165.715	160.469	165.623
August	156.731	167.835	169.940	174.069
September	169.057	165.715	160.469	165.623
October	185.070	223.205	208.081	215.474
November	208.645	238.217	220.516	212.594
December	238.468	251.588	243.519	258.063

Table 13.11   Common Share Earnings Data for Exercise 13.26

	Year								
Quarter	1979	1980	1981	1982	1983	1984	1985	1986	1987
1	0.75	1.00	1.02	0.82	0.54	0.71	0.94	0.70	0.96
2	1.21	1.04	1.12	0.87	1.05	0.92	1.11	1.22	1.40
3	0.90	0.97	0.99	0.73	1.67	0.93	0.82	0.95	0.89
4	0.90	0.99	1.10	0.70	0.94	1.79	1.16	1.05	1.15*

Source: Data from *Stock Market Encyclopedia*, Standard & Poor's Corporation.
*Estimated.

Hilton Hotel's earnings data by using the ratio-to-moving-average method with $M = 4$. Plot the original series and the deseasonalized series together on the same piece of graph paper, and describe the results.

13.5

Index Numbers

Because of the variability in the buying power of the dollar over time, we must deflate some values and inflate others to make meaningful comparisons. For example, to compare the relative cost of a four-year college education today with its cost in 1940, we must first determine the buying power of the dollar today as compared with the buying power of a 1940 dollar. *Index numbers* are computed for such purposes and are used every day by businesspeople and economists to make meaningful comparisons over time. Their use is not limited strictly to monetary comparisons; but for business problems, application of index numbers to other than monetary processes is uncommon.

Definition

index number

> An **index number** is a ratio or an average of ratios expressed as a percentage. Two or more time periods are involved, one of which is the base time period. The value at the base time period serves as the standard point of comparison; the values at the other time periods are used to show the percentage change in value from the standard value of the base period.

The concept of an index number is best illustrated by an example.

EXAMPLE 13.2   Suppose that we wish to compare the average hourly wages for a journeyman electrician in 1950, 1955, 1960, 1965, 1970, 1975, 1980, and 1985, using 1950 as the base year. The average hourly wages and their computed index numbers for each of the eight years are listed in Table 13.12.

Table 13.12   Average Hourly Wage Data for Example 13.2

Year	Average Hourly Wages	Wage Index (1950 Base)
1950	$ 2.00	100
1955	2.85	142.5
1960	3.90	195
1965	5.25	262.5
1970	6.00	300
1975	8.85	442.5
1980	11.70	585
1985	13.90	695

The wage index for each year is computed by evaluating the ratio

$$I_k = \frac{(\text{average hourly wages in year } k)(100)}{\text{average hourly wages in 1950}}$$

Each wage index is a percentage that indicates the percentage of 1950 wages that were earned in the year of interest. For example, the wage index for 1985 indicates that the hourly wage in 1985 was 695% the 1950 hourly wage. Since the value of a 1950 dollar was worth $6.95 in 1985 (as measured by the change in the Consumer Price Index over that period), one can assume that wages of journeyman electricians have kept pace with inflation since 1950.

◆◆

The preceding example is an index time series for the years 1950, 1955, 1960, 1965, 1970, 1975, 1980, and 1985. An *index time series* is simply a transformation of the original time series to one giving each year's value as a percentage of the value for the base year.

Definition

index time series

> An **index time series** is a list of index numbers for two or more periods of time, where each index number employs the same base year.

A commonly used index for comparing two sets of prices from a wide variety of items is called a *simple aggregate index*.

Definition

simple aggregate index

> A **simple aggregate index** is the ratio of an aggregate (sum) of commodity prices for a given year k to an aggregate of the prices of the same commodities in some base year, expressed as a percentage.

This index is computed by evaluating the formula given in the next display.

Simple Aggregate Index

$$I_k = \frac{\sum_{i=1}^{n} p_{ki}}{\sum_{i=1}^{n} p_{0i}} (100)$$

where p_{ki} is the price in year k of item i, and p_{0i} is the base-year price of item i, $i = 1, 2, \ldots, n$.

EXAMPLE 13.3   The average consumer prices (in cents per pound) for certain staple food items in 1958 and 1988 are given in Table 13.13. Find the aggregate price index for 1988.

Table 13.13   Consumer Price Data for Example 13.3

Item	1958	1988
Sugar	10	30
Wheat flour	11	23
Butter	71	189
Sirloin steak	91	359
Ground beef	39	139
Frying chicken	51	109

SOLUTION The aggregate price index for 1988 is

$$I_{1988} = \frac{30 + 23 + 189 + 359 + 139 + 109}{10 + 11 + 71 + 91 + 39 + 51}(100) = \frac{849}{273}(100) = 311$$

which implies that the prices of these six items in 1988 are 211% higher than they were in 1958.

The greatest weakness of the simple aggregate index is that changes in the measuring units may affect the value of the index. Suppose that in Example 13.3 we had considered the prices of 10-lb bags of sugar and wheat flour instead of the prices of these items per pound. If the 10-lb prices were $1.00 and $3.00 for the sugar and $1.10 and $2.30 for the flour, then our price index would be

$$I_{1988} = \frac{1326}{462}(100) = 287$$

a decrease of 32 percentage points from our former index value. By changing the scale of measurement of some of the units being measured within the simple aggregate index, the statistician could derive almost any index value at his or her discretion. This lack of objectivity of the simple aggregate index tends to lessen its usefulness for making meaningful comparisons.

An index that gives a more uniform measure of comparison is called a *weighted aggregate index*.

Definition

weighted aggregate index

> A **weighted aggregate index** is the ratio of an aggregate of weighted commodity prices for a given year k to an aggregate of the weighted prices of the same commodities in some base year, expressed as a percentage.

In a weighted aggregate index the prices do not necessarily contribute equally to the value of the index. Each price is weighted (multiplied) by the quantity of the item produced or the number of units purchased or consumed. Thus, each item is included according to its importance in the aggregate of prices of the items being described by the index. The index is found by computing the formula given in the display.

Weighted Aggregate Index

$$I_k = \frac{\sum_{i=1}^{n} p_{ki} q_{ki}}{\sum_{i=1}^{n} p_{0i} q_{0i}}(100)$$

where the q_{ki}'s and the q_{0i}'s are the quantities associated with the n prices for the reference year and the base year, respectively.

Laspeyres index The U.S. Department of Labor uses a special form of a weighted aggregate index for several of its published indexes. This index is called the **Laspeyres index** and is found by computing

$$L = \frac{\sum\limits_{i=1}^{n} p_{ki}q_{0i}}{\sum\limits_{i=1}^{n} p_{0i}q_{0i}} (100)$$

The rationale for using base-year quantities as the weights for the reference year prices is that the base-year quantities do not change from year to year. Thus, we can make more meaningful comparisons of the change in prices and buying power over time, since we are considering only the change in price per given number of units without changing the number of units. By using the base-year quantities, though, we tend to give too much weight to the commodities whose prices have increased, since an increase in price will often be accompanied by a decrease in the quantity consumed or purchased. However, if essential or staple items are being considered, the effect of using only the base-year weights on the value of the index should be small. Since it is often difficult and expensive to obtain the quantities for each time period, it may be a worthwhile trade-off to give up some accuracy of the computed index value by employing the Laspeyres index.

EXAMPLE 13.4 The average consumer prices (in cents per pound) for certain staple food items from Example 13.3 are listed in Table 13.14. Also, we are given the average amount of each item recommended as necessary to sustain a family of four in 1958. Find the Laspeyres index to measure the amount of change of the 1988 prices on these items from 1958 to 1988.

Table 13.14   Consumer Price Data for Example 13.4

Item	1958 Price, p_0	1988 Price, p_k	1958 Quantity (Pounds), q_0
Sugar	10	30	25
Wheat flour	11	23	60
Butter	71	189	50
Sirloin steak	91	359	25
Ground beef	39	139	120
Frying chicken	51	109	40

SOLUTION The Laspeyres index is

$$L = \frac{30(25) + 23(60) + 189(50) + 359(25) + 139(120) + 109(40)}{10(25) + 11(60) + 71(50) + 91(25) + 39(120) + 51(40)} (100)$$

$$= \frac{41,595}{13,455} (100) = 309$$

Thus, the cost for these staple food items in 1988 is 209% higher than the cost for these items in 1958, when considering the total annual food expenditures.

◆◆

Paasche index
Fisher's ideal index

Other less frequently used weighted aggregate price indexes are the **Paasche index** and **Fisher's ideal index**. The Paasche index uses the reference-year quantities rather than the base-year quantities as weights for the weighted index. Otherwise, the computational procedure is the same for the Laspeyres and the Paasche indexes. Previously, we mentioned that the Laspeyres index tends to "overweigh" commodities whose prices have increased. Analogously, the Paasche index tends to "underweigh" commodities whose prices have increased. Hence, we might suspect that the price index should be somewhere between these two indexes. This is the logic behind the use of Fisher's ideal index.

The Fisher index is computed from the Laspeyres and the Paasche indexes as follows:

$$\text{Fisher's index} = \sqrt{(\text{Laspeyres index})(\text{Paasche index})}$$

$$= 100 \sqrt{\frac{\sum p_{ki} q_{0i}}{\sum p_{0i} q_{0i}} \frac{\sum p_{ki} q_{ki}}{\sum p_{0i} q_{ki}}}$$

Although the Fisher index might seem to measure the price index more accurately than the Laspeyres or the Paasche indexes, it is seldom used in practice. The Fisher index, since it is a function of the Paasche index, requires a new set of quantities at each time period. These quantities are often difficult and expensive to obtain. Also, the Fisher index does not give a uniform index for purposes of comparison in an index time series. That is, like the Paasche index, the Fisher index does not hold the quantity measure constant, as does the Laspeyres index. Thus, the Laspeyres index, or some form of the Laspeyres index, is used at the practical exclusion of other types of indexes when the business statistician chooses to use a weighted aggregate index.

Consumer Price
Index (CPI)

Two important price indexes computed regularly by the Bureau of Labor Statistics are the *Consumer Price Index* and the *Wholesale Price Index*. The **Consumer Price Index (CPI)** is a weighted aggregate index that is computed and published monthly. Historically, the CPI was constructed to measure the cost of living for city wage earner and clerical worker families. More recently, the CPI has been broadened to include urban families other than those of wage earners and clerical workers and a sampling of rural families. Monthly, the costs of about 300 different items, from food products to clothing to rent to luxuries to fees paid to doctors and dentists, are sampled for each family involved in the survey. The base year for the CPI was 1967; however, beginning with the release of data for January 1988, the CPI base period was changed from 1967 to 1982–1984.

The CPI is used as a factor to cancel out the effect of inflation or deflation. For instance, automatic escalator clauses in union contracts and in the wage and salary contracts of federal employees, as well as the benefits received by Social Security recipients, are tied to the CPI. Even though it has limitations, since it does not fairly characterize the living costs of all families living in the United States, the CPI has become our most widely quoted and frequently used economic index.

Wholesale Price
Index (WPI)

The **Wholesale Price Index (WPI)** is computed from information obtained by sampling the producers' selling prices of about 2,000 different items in the primary markets. Agricultural commodities, raw materials, fabricated products, and many manufactured items are included. The WPI is a weighted aggregate index computed and published monthly by the Bureau of Labor Statistics. Its base period is 1967–1969.

Many industrial contracts, especially those with the Department of Defense, allow for an adjustment of the contract price and payments according to the change in the WPI. This adjustment allows the contractor and the buyer to negotiate and plan in terms of some constant-dollar amounts. The primary difficulty with the WPI is that it is not really an indicator of wholesale prices at all but represents the *change* in producers' selling prices. In many instances this makes no difference; but the name of the index, the Wholesale Price Index, may in itself be misleading and may cause the businessperson to form unwarranted conclusions.

Security market indexes, such as the Dow-Jones industrial average, are computed differently from the wage and price indexes previously described. They might better be described as averages of a group of individual time series rather than as indicators of change of value or price.

The Dow-Jones industrial average purports to compute the average of the daily closing prices of the securities of 30 predetermined industrial firms. The computations become involved, though, when financial transactions such as mergers and stock splits take place. For the sake of discussion, suppose that the Dow-Jones average consists of the securities of only two firms, one whose stock is valued at $20 per share, the other valued at $30. The Dow-Jones average is

$$(1/2)(\$20 + \$30) = \$25$$

During the next day, suppose that a stock split occurs for the firm whose securities were valued at $20 per share. Now each security from the first firm is worth only $10, and the Dow-Jones average, ignoring the stock split, is

$$(1/2)(\$10 + \$30) = \$20$$

To consider the effect of the split, the Dow-Jones average finds the divisor d such that

$$\frac{1}{d}(\$10 + \$30) = \$25$$

where $25 was the average before the split. Here, $d = 1.6$. If, during the day of the stock split, the first security increased in price to $12 per share while the other remained at $30, the new Dow-Jones average would be

$$\frac{1}{1.6}(\$12 + \$30) = \$26.25$$

Each time stock dividends are declared, a merger occurs, or a stock split takes place, the Dow-Jones industrial average becomes less meaningful because of the continual adjustment of the divisor term d. Although the divisor started out as 30 (since the Dow-Jones average is computed from the security prices of 30 firms), it is now less than 1 and will continue to decrease as more stock splits, mergers, and other financial happenings occur. Actually, the Dow-Jones average is not an average at all. Its usefulness as a measure of market value has diminished and will continue to diminish with time.

EXERCISES   Conceptual Level

13.27   Why is it often useful to express realizations of a time series as index numbers, rather than in their original form?

13.28   Differentiate among each of the following types of index.

   a.   Index time series

   b.   Simple aggregate index

   c.   Weighted aggregate index

13.29   The Bureau of Labor Statistics publishes the values of the Consumer Price Index and the Wholesale Price Index. What do these indexes measure, and of what importance are they to business and economics?

13.30   For each of the following decision-making situations, explain how an individual might construct and use index numbers as an aid in arriving at a conclusion.

   a.   A public utility company wishes to appeal to the public utility commissioner that its rates be increased by the level of inflation in the economy of its area of service since rates were last set. The current rates were established in 1982.

   b.   The contractor of a construction project to build an elementary school would like to renegotiate the contract price to accommodate increased production costs resulting from an inflationary economy.

   c.   The owner of a small hardware store would like to know how the growth in sales of her store the past five years compares with the growth in sales of a large competing hardware chain.

   d.   The manager of a local chamber of commerce is interested in comparing the aggregate level of business activity of a city at present with the aggregate level of activity in 1980.

   e.   The directors of a Dallas-based manufacturing company are exploring the possibility of establishing a branch facility in Colorado Springs. A key element in their decision is the cost of living in Colorado Springs in comparison with that in Dallas.

EXERCISES   Skill Level

13.31   Listed in Table 13.15 are the prices and quantities consumed for five commodities in three different years.

Table 13.15   Price and Quantity Data for Exercise 13.31

Commodity	Price			Quantity		
	Year 1	Year 2	Year 3	Year 1	Year 2	Year 3
1	5.60	5.90	5.80	152	186	192
2	89.00	92.00	87.00	26	25	24
3	1.80	1.85	1.92	400	500	300
4	4.40	4.50	4.50	100	150	150
5	189.20	201.10	199.95	10	15	14

a. Using year 1 as the base year, compute the simple price index for each of the five commodities for years 2 and 3.

b. Using year 1 as the base year, compute the simple aggregate price index for years 2 and 3.

13.32   Refer to Exercise 13.31.

a. Using year 1 as the base year and the quantities consumed as weights, compute the Laspeyres index for the prices of these commodities in years 2 and 3.

b. Using year 1 as the base year, compute the Paasche index for the prices of these five commodities in years 2 and 3.

c. Compute Fisher's ideal index for years 2 and 3. Compare it with the indexes calculated in parts a and b.

EXERCISES   Applications

13.33   The cost indexes for the five primary categories of consumer expenditures are shown in Table 13.16 for the years 1970, 1975, 1980, and 1985 (1967 = 100).

a. Compute the percentage gain in the prices of items within each of the five main categories of consumer expenditures from 1970 to 1975, 1975 to 1980, and 1980 to 1985.

b. Suppose the average family of four spent the following amounts in each of the five primary categories of consumer expenditures in 1970: housing, $3,300; food, $2,380; apparel and upkeep, $940; transportation, $1,150; and health and recreation, $1,470. Estimate their expenditures in each of these five categories for the years 1980 and 1985.

Table 13.16   Cost Index Data for Exercise 13.33

Year	Housing	Food	Apparel and Upkeep	Transportation	Health and Recreation
1970	140.1	132.8	135.9	135.5	147.4
1975	172.2	180.7	145.2	157.6	157.5
1980	274.3	243.8	171.0	233.5	253.9
1985	341.9	306.6	205.2	316.1	301.4

Source: Data from U.S. Department of Labor, Bureau of Labor Statistics, *CPI Detailed Report*, selected issues 1970–1985.

13.34   For the period 1974–1985 Table 13.17 lists the annual salaries offered by the University of Oregon to new assistant professors of accounting and the CPI. Using 1974 as the base year,

Table 13.17   Salary Data for Exercise 13.34

Year	Annual Salary	CPI (1967 = 100)	Year	Annual Salary	CPI (1967 = 100)
1974	$17,000	139.1	1980	23,500	233.3
1975	18,000	156.1	1981	25,500	260.4
1976	18,500	166.7	1982	28,000	276.0
1977	19,000	175.3	1983	31,000	287.1
1978	19,500	187.7	1984	35,500	295.7
1979	21,500	204.7	1985	36,500	304.6

compare the actual salaries offered to new assistant professors of accounting with the real (deflated) salaries.

 13.35   Listed in Table 13.18 are the prices in seven different years of five common family market basket items.

    a.   Using 1955 as the base year, compute the simple price index for each food product for each of the remaining six years.

    b.   Using 1955 as the base year, compute the simple aggregate index of food prices for each of the six years.

Table 13.18   Price Data for Exercise 13.35

	Year						
Item	1955	1960	1965	1970	1975	1980	1985
Milk (quart)	22.4	25.1	26.0	33.8	39.6	57.3	65.9
Wheat flour (pound)	10.4	11.2	12.3	14.1	20.3	26.0	23.3
Hamburger	39.0	51.3	54.2	78.0	88.1	173.8	139.2
Coffee (pound)	93.0	74.6	70.7	76.5	122.2	342.7	247.8
Potatoes (pound)	4.7	6.7	11.1	27.3	20.5	25.4	24.4

Source: Data from U.S. Department of Labor, Bureau of Labor Statistics, *CPI Detailed Report*, selected issues 1955–1985.

 13.36   Refer to Exercise 13.35. In 1955 the average family of four bought the following:

        350 quarts of milk

        60 lb of wheat flour

        120 lb of hamburger

        40 lb of coffee

        140 lb of potatoes

Compute the Laspeyres index for the prices of this market basket for each of the six years under study, using 1955 as the base year.

 13.37   Does much variance exist in the cost of living in the major metropolitan areas of the United States? The *CPI Detailed Report* offers some insight into this issue by including within each monthly report a listing of cost indexes for several items in five major metropolitan areas. Listed in Table 13.19 are the 1987 CPI values (1967 = 100) for each cost group in each of five metropolitan areas and the measures of relative importance (weights) of each cost group.

    a.   Compute the overall CPI for each metropolitan area for 1987.

    b.   With 1967 as a base, by what percentage has the increase in the cost of living in New York exceeded that of Detroit? By what percentage has it exceeded that of Philadelphia?

    c.   With 1967 as a base, by what percentage has the increase in fuel and utilities and transportation costs in New York exceeded that of Los Angeles?

    d.   Do these data provide sufficient evidence to determine which metropolitan area among the five offers the lowest cost of living and which offers the greatest cost of living? Explain.

Table 13.19   Cost Index Data for Exercise 13.37

		Metropolitan Area				
Group	Importance	Chicago	Detroit	L.A.	N.Y.	Philadelphia
Food, beverage	17.6	313.1	301.6	320.2	349.8	329.2
Housing	27.8	466.0	428.9	455.8	407.5	411.1
Fuel, utilities	7.7	318.1	395.7	344.9	363.8	369.3
Household operations	7.0	247.8	216.2	242.1	263.6	256.5
Apparel, upkeep	6.3	185.0	184.5	196.0	196.2	199.5
Transportation	17.5	321.0	318.2	342.7	356.9	351.0
Medical care	5.8	475.3	502.7	502.8	487.7	511.1
Entertainment	4.4	307.6	250.2	241.8	308.0	265.5
Other goods and services	5.9	368.9	350.1	364.4	394.6	409.2

Source: Data from U.S. Department of Labor, Bureau of Labor Statistics, *CPI Detailed Report*, May 1988.

13.6

Summary

A **time series** is a sequence of measurements taken on a response that varies over time. It is usually represented in graphical form, with its response values listed on the vertical axis plotted against time on the horizontal axis. We may think of a time series as consisting of components such as a long-term trend, a cyclical effect, or a seasonal effect. Any or none of these components may be present in a time series. In addition, all time series possess random variation, which tends to obscure the nonrandom components of the series. **Smoothing methods** may sometimes reveal the underlying components in a time series by canceling out the effects of random variation.

The values in a time series are often misleading, especially those that occur in monetary units. Index numbers are used to deflate some monetary values and inflate others so that meaningful monetary comparisons can be made over time. **Index numbers** measure the change in price or value of a single or an aggregate of commodities from some base time period to a reference time period. Weighted indexes, which multiply the commodity prices by a predetermined factor, are the most commonly used type of index number. Most price indexes computed regularly by the Bureau of Labor Statistics are weighted indexes.

Care must be taken in interpreting an index number or a series. Many commonly used indexes, such as the Consumer Price Index, the Wholesale Price Index, and the Dow-Jones industrial average, do not really indicate what their titles might imply. We must examine the exact intent of an index, the identity of the commodities that are included within the index, and the weights associated with each commodity included within the index before we can draw meaningful conclusions.

Supplementary Exercises

Applications

13.38   Table 13.20 shows the number of miles flown by air carriers on international routes for each month over five consecutive years. To simplify recording, the data have been coded in millions of passenger miles.

a.   Smooth the monthly international airline traffic data by computing a five-month centered moving average series.

b.   Smooth the monthly international airline traffic data by computing an exponentially smoothed series, employing the smoothing constant $\alpha = .1$.

c.   Smooth the monthly international airline traffic data by computing an exponentially smoothed series, employing the smoothing constant $\alpha = .4$.

d.   On a sheet of graph paper, superimpose the three smoothed series on the original series. What components appear to exist in the international airline traffic time series data?

Table 13.20   Airline Mileage Data for Exercise 13.38

Month	Year				
	1	2	3	4	5
January	284.0	315.1	340.1	359.9	416.9
February	276.9	300.9	317.9	342.0	391.1
March	317.0	356.0	362.1	405.8	419.0
April	312.9	347.9	347.9	395.8	460.8
May	317.9	354.9	362.8	419.8	472.0
June	373.9	421.9	434.8	472.0	534.8
July	412.8	466.8	489.8	547.8	622.0
August	405.0	466.8	505.2	558.9	606.0
September	354.9	403.8	403.8	463.1	507.7
October	306.1	346.8	358.8	407.0	460.8
November	270.9	304.9	310.1	362.1	389.9
December	306.1	335.9	336.9	405.0	431.8

13.39   Refer to Exercise 13.38. Use the ratio-to-moving-average method with $M = 12$ for the international airline traffic data. Does the ratio-to-moving-average model appear to more effectively remove the seasonal pattern from the airline traffic data than either of the exponential-smoothing models computed in Exercise 13.38? Would a ratio-to-moving-average model with M other than 12 have eliminated the seasonal component as well? Explain.

13.40   Because of the seasonal nature of styles and the impact of the Christmas season on retail sales, the sales of department stores typically follow a distinctly seasonal pattern. Table 13.21 lists the monthly credit sales (in thousands of dollars) for a department store over five consecutive years. Remove the seasonal component from these data by using the ratio-to-moving-average method. Plot the original series and the deseasonalized series together on the same piece of graph paper. After you remove the seasonal component, what inherent components do these data appear to possess?

Table 13.21   Sales Data for Exercise 13.40

Month	Year				
	1	2	3	4	5
January	12.3327	13.3708	14.3747	14.5151	15.1144
February	12.9585	14.6156	15.9360	14.6323	15.3555
March	15.7480	17.4000	16.8129	16.4262	18.6998
April	15.5027	16.2574	15.2642	15.7848	17.1827
May	16.1040	17.6280	17.8682	18.1914	18.2053
June	16.9076	16.4033	16.4432	16.9352	18.6406
July	14.2196	14.8934	15.1211	16.3979	17.2713
August	15.6731	16.7835	16.9940	17.4069	17.1966
September	16.9057	16.5715	16.0469	16.5623	17.4080
October	18.5070	22.3205	20.8081	21.5474	21.2315
November	20.8645	23.8217	22.0516	21.2594	21.5733
December	23.8468	25.1588	24.3519	25.8063	27.5733

Source: J. E. Reinmuth and D. R. Wittink, "Recursive Models for Forecasting Seasonal Processes," *Journal of Financial and Quantitative Analysis*, September 1974. Reprinted by permission.

13.41   Refer to the credit sales data listed in Exercise 13.40.

   a. Smooth the monthly sales volumes over the five-year period by computing an exponentially smoothed series, employing the smoothing constant $\alpha = .1$.

   b. Smooth the monthly sales volumes over the five-year period by computing an exponentially smoothed series, employing the smoothing constant $\alpha = .4$.

   c. On a sheet of graph paper, superimpose the two smoothed series on the original series. What conclusions do you draw?

13.42   Refer to Exercises 13.40 and 13.41. Does the 12-month ratio-to-moving-average model used in Exercise 13.40 appear to more effectively eliminate the seasonal pattern from the credit sales data than either of the exponential-smoothing models from Exercise 13.41? Would a ratio-to-moving-average model with an M other than 12 have eliminated the seasonal component as well? Explain.

13.43   The December 1987 Consumer Price Index for five major metropolitan areas are listed below (1967 = 100).

Chicago	329.5
Detroit	323.4
Los Angeles	342.0
New York	339.0
Philadelphia	345.6

Do these data suggest that 1987 living costs in Philadelphia were higher than in the other four metropolitan areas? Explain.

13.44   The costs (in dollars per hundred weight) for rice in two different regions of the United States are listed in Table 13.22 for six consecutive years, along with the total production (in 1,000 hundred weight) for the two regions.

Table 13.22   Rice Cost and Production Data for Exercise 13.44

	Production		Wholesale Price	
Crop Year	Southern States	California	Southern States	California
1979–1980	97,905	34,042	20.05	22.05
1980–1981	109,764	36,386	25.30	25.55
1981–1982	141,818	40,924	19.40	21.15
1982–1983	117,740	35,848	16.80	18.70
1983–1984	76,631	23,089	17.35	19.88
1984–1985*	106,750	32,060	16.23	18.69

Source: Adapted from *1986 CRB Commodity Year Book*, Commodity Research Bureau, Jersey City, N.J.
*Preliminary.

   a. Using the 1979–1980 crop year as the base year, compute the simple price index for each region for each of the five remaining crop years.

   b. Using the 1979–1980 crop year as the base year, compute the simple aggregate index of rice prices for each of the five remaining crop years.

13.45 Refer to Exercise 13.44.

   a. Using the production amounts as weights and 1979–1980 as the base year, compute the Laspeyres index for the rice prices over the remaining five crop years.

   b. Compute the Paasche index for the rice prices in the 1984–1985 crop year.

13.46 Refer to Exercise 13.45. Compute Fisher's ideal index for the rice prices in the 1984–1985 crop year, using the 1979–1980 crop year as the base year. Discuss the significance of this index.

13.47 An executive of a power company is interested in examining the relative costs of heating fuels and the impact of a statewide energy conservation program over a 15-year period. In her study the executive contacted a sample of householders who heat by electricity, by oil, or by natural gas and obtained the average quantity each householder used per month. (Quantities were obtained from utility company records.) The results are given in Table 13.23.

Table 13.23   Heating Fuel Data for Exercise 13.47

	Unit Cost				Average Monthly Use			
Fuel	1970	1975	1980	1985	1970	1975	1980	1985
Electricity (kilowatt-hours)	$.025	$.046	$.070	$.104	1,672	1,595	1,607	1,521
Oil (gallons)	.31	.50	.81	.96	263	260	238	204
Gas (million cubic feet)	9.26	10.20	12.19	13.44	7.1	6.7	6.5	5.8

   a. Find the Laspeyres index for the 1975, 1980, and 1985 average monthly heating costs, using 1970 as the base year.

   b. Find the Paasche index for the 1975, 1980, and 1985 average monthly heating costs, using 1970 as the base year.

c. If the executive believes the conservation program has been effective in permanently reducing the use of heating fuels, should she use the Laspeyres index or the Paasche index? Explain.

13.48 Suppose that the security price movements of the common stock of four companies, American Biscuit Corporation, National Motors, Southwestern Oil, and Texas Chemical Products, are considered to be representative of the behavior of all industrial securities traded on the New York Stock Exchange. The closing prices of the securities of the four companies for four consecutive market days in October are given in Table 13.24. Using the procedure discussed in Section 13.5, compute a Dow-Jones type of market average for each of the four listed market days, using the security prices of these four representative companies.

Table 13.24   Price Data for Exercise 13.48

Day	American Biscuit	National Motors	Southwestern Oil	Texas Chemical
October 10	27	30	62	14
October 11	26	15*	60	13
October 12	28	17	30*	15
October 13	29	18	32	14

*National Motors declared a 2-for-1 stock split after the close of the market on October 10, while Southwestern Oil declared a 2-for-1 stock split on October 11.

Supplementary Exercises

Interpretive

13.49 Which of the time series components would you expect to be present in each of the following series?

    a. The number of new residential housing starts in the United States in *monthly* intervals from January 1970 to January 1988

    b. The number of new residential housing starts in the United States in *annual* intervals from 1970 to 1988

    c. The annual sales of term life insurance policies in the United States from 1950 to 1988

    d. The monthly farm employment in Canada from January 1955 to January 1988

    e. The gross monthly sales volume of a brewing company from June 1960 to June 1988

    f. The quarterly level of industrial production in the state of Michigan from January 1955 to January 1988.

13.50 Criticize the following uses of index numbers.

    a. A union official notes that the Consumer Price Index has risen from 166.7 in 1976 to 303.3 in 1985. He concludes that living costs have risen by 136.6% during that time.

    b. During a period in which the CPI increased by 40%, the average wages for workers in a particular industry have increased by that same amount. An industry official thus claims that the workers' standard of living has not deteriorated over that period.

c. Noting that the CPI index for fuel oil has doubled from 1976 to 1980, a manufacturer has decided to locate a branch facility in Phoenix, where little fuel oil will be required for heating.

d. A government contractor notices that the Wholesale Price Index has increased by 18% over the past two years. Thus, she asks the government for an 18% increase in the contractual payoff.

e. A lumber wholesaler is interested in the real increase in sales he has incurred over the past five years. To measure the increase, he computes a Laspeyres index, considering all types of lumber he handles, using the current and five-year-old board foot selling prices and the quantities of each type of lumber he sold five years ago.

References and Suggested Readings

Bowerman, B. L., and R. T. O'Connell. *Time Series Forecasting: Unified Concepts and Computer Implementation.* 2d ed. Boston: PWS-KENT Publishing Co., 1987.

Doody, F. S. *Introduction to the Use of Economic Indicators.* New York: Random House, 1965.

Hunter, J. S. "The Exponentially Weighted Moving Average," *Journal of Quality Technology*, Vol. 18, No. 4 (1986), pp. 203–210.

Spurr, W. A., and C. P. Bonini, *Statistical Analysis for Business Decisions.* Rev. ed. Homewood, Ill.: Irwin, 1973.

U.S. Department of Commerce. *Business Statistics*, biennial supplement to the *Survey of Current Business.* Washington, D.C.: Government Printing Office, 1967 et seq.

TO
THE
READER

In Chapter 13 smoothing techniques were introduced as a method for identifying the components of a time series. In this chapter we use smoothing methods similar to those in Chapter 13 to generate forecasts of a time series. We also discuss econometric methods, which use regression analysis to model the relationships between a time series and several predictor variables.

Some fundamental forecasting concepts, such as the autocorrelation function, the method of differencing, and measures of forecast accuracy, are also introduced. Measures of accuracy are used to evaluate the various forecasting models used throughout the chapter. Differencing and autocorrelation appear in many forecasting applications, especially the Box-Jenkins methodology discussed later in the chapter.

14

FORECASTING MODELS

Introduction

We all exist in an environment governed by time. Business organizations, public organizations, and individuals thus have the common goal of allocating available time among competing resources in some optimal manner. This goal is accomplished by making forecasts of future activities and taking the proper actions as suggested by these forecasts.

In business and public administration the organization is concerned with both short-term and long-term forecasts. The short-term forecast usually looks no more than one year into the future and involves forecasting sales, price changes, and customer demand, which, in turn, reflect the need for seasonal employment, short-term capital expenditures, and inventory management procedures. The long-term forecast usually looks from 2 to 10 years into the future and is used as a planning model for product line and capital investment decisions, as indicated by changing demand patterns.

Naturally, the further a forecast is projected into the future, the more speculative it becomes. But since the future is *always* uncertain, we cannot expect complete accuracy for any forecast. The time series underlying the process to be forecast is bound to be influenced by many causal factors—some forcing the time series up while conflicting factors act to force the series down. Nevertheless, businesspeople must make forecasts of future business activity in order to budget their time and resources efficiently. They cannot hope to account for every possible factor that may cause the response of interest to rise or fall over time. All that can be expected is that the benefits gained by forecasting offset the opportunity cost for not forecasting. Note that such benefits are not limited to real monetary savings but may imply a sharpening of the businessperson's thinking to consider the interplay of the events that affect the movement of the time series.

We recall from Chapter 13 that smoothing methods sometimes reveal trends and seasonal and cyclic effects in the time series. Forecasting by extending these patterns is a very speculative procedure. We must first assume that the past is a mirror of the future—that past trends and cycles will continue into the future. This is seldom the case. In the end, mathematical forecasting procedures and judgment must work hand in hand. Thus, we not only must smooth the data and try to extend the signal components into the future but must also predict the impact of unknown factors such as political events, research, changing buyer behavior, and new-product development. These subjective evaluations must, in turn, be used to condition the forecast obtained from the mathematical forecasting model. So long as uncertainty is involved with future business and economic activity, forecasting must be recognized as an art that becomes more perfect as the forecaster gains experience and the ability to adapt procedures to meet the changing environment of the firm.

14.2

Choosing an Appropriate Forecasting Model: The Organization of This Chapter

Choosing a forecasting model involves the selection of an estimation procedure. A forecast is, after all, an estimate of a future outcome of a random process. In Chapter 8 we learned that estimation procedures can be associated with a measure of the goodness of our estimate. Unfortunately, this technique is not possible with most time series. (See Section 13.1.) The decision maker must therefore rely heavily on experience when forecasting and on his or her ability to judge the reasonableness of a forecast in light of all surrounding circumstances related to the time series under investigation.

econometric models
times series models
qualitative
forecasting models

Forecasting models are generally classified as **econometric models, time series models**, or **qualitative forecasting models**. The first two are projection techniques that initially involve the fitting of a theoretical model to a sample data set. An assumption underlying both econometric models and time series models is that sample observations from a random process provide reliable evidence of future activity. That is, econometric models and time series models provide a characterization of past and current activity of a random process and then project forward assuming that past and current activity is a reliable mirror of future activity. The difference between these two classes of models is that econometric models use auxiliary variables as predictors and time series models do not. Time series models are "pattern fitters," relying on an extension of inherent components. Qualitative models are designed for forecasting the sales of a newly introduced product and for other cases where a relevant historical sample data base is not available. We will devote most of our attention to econometric and time series models and give only a few brief passing remarks to qualitative forecasting models.

There is no such thing as a single best forecasting model to use in all instances. A forecasting model that may be appropriate for estimating future levels of sales for an established product may be totally inappropriate for forecasting the sales of a new product not yet introduced to the marketplace. Thus, one of the primary tasks associated with forecasting is matching an appropriate forecasting model to the time series to be forecast. The forecaster becomes more proficient at this task through experience gained from the study of time series behavior and from trial and error in the use of various forecasting procedures.

In addition to the task of matching an appropriate forecasting model to the time series under study, the choice involves trade-offs among cost, timeliness, and reliability. For example, some forecasting models are simple and reasonably inexpensive to apply, especially those for which data and appropriate computer programs are available. Alternative forecasting models may improve forecast accuracy but require expensive data collection procedures or the acquisition of a costly computer program. In such instances the forecaster must decide whether expected improvements in forecast accuracy warrant additional expenditure. Another consideration in the selection of a forecast model is the time horizon of the forecast period. Some models are more accurate for short-term time horizons (six months or less), and others are more reliable for long-term horizons (one year or more). In general, the forecaster should choose the forecasting model that makes best use of available data.

Our purpose throughout this chapter will be to illustrate the use of various forecasting models. As a result, there may be instances where an example or an exercise shows that a particular model provides a poor fit to a sample data set. We believe that if you are aware of the characteristics of the various models and know how to use them in practice, then you will be able to properly tailor an appropriate model to an underlying time series to obtain an adequate level of forecast accuracy.

We will first introduce the concept of forecast accuracy and provide four common measures of forecast accuracy. These measures will then be used throughout the chapter as comparative indicators of forecast accuracy for the application of forecast models to real data sets.

Econometric models are discussed in Section 14.5. This discussion shows that econometric models are deterministic in nature and, in regression form, attempt to express the future value of a time series from the known relationship between the time series and other factors.

time series forecasting models

The next four sections will examine the use of smoothing techniques for forecasting. With this class of models, called **time series forecasting models**, forecasts are developed by projecting forward the inherent components revealed by the application of smoothing. We will present four common forecasting models based on smoothing techniques. Other time series models are left for you to explore. (See Farnum and Stanton in the references.)

qualitative forecasting models

Section 14.10 introduces several common **qualitative forecasting models** that are used in instances when a sample data set is either unavailable or unreliable. Finally, we will discuss the improvement of forecast accuracy through the combination of forecasts (Section 14.11).

Exercises are offered throughout the chapter to assist you in understanding the application and use of the various forecast models we discuss. In many cases the exercises require considerable computational effort. Therefore, it may be advisable to use a computer to assist you in some of the exercises.

14.3

Four Common Measures of Forecast Accuracy

When choosing between competing forecasting models, or when evaluating an existing model, we need to use measures that summarize the overall accuracy provided by the model(s). Generally speaking, the closer the forecasts $\hat{y}_t$ are to the actual values y_t of the series, the more accurate the forecasting model is. Thus, the quality of a model can be evaluated by examining the series of forecast errors $(y_t - \hat{y}_t)$. All of the measures that follow are based on some simple function of the forecast errors.

The most commonly used measures of forecast accuracy are the **mean absolute deviation (MAD)**, the **mean square error (MSE)**, the **root mean square error (RMSE)**, and the **mean absolute percentage error (MAPE)**. Formulas for the computation of these measures are given in the displays.

Computation of the MAD

$$\text{MAD} = \frac{1}{n} \sum_{t=1}^{n} |y_t - \hat{y}_t|$$

Computation of the MSE

$$\text{MSE} = \frac{1}{n} \sum_{t=1}^{n} (y_t - \hat{y}_t)^2$$

Computation of the RMSE

$$\text{RMSE} = \sqrt{\frac{1}{n} \sum_{t=1}^{n} (y_t - \hat{y}_t)^2}$$

Computation of the MAPE

$$\text{MAPE} = \frac{1}{n} \sum_{t=1}^{n} \left| \frac{y_t - \hat{y}_t}{y_t} \right| (100\%)$$

The basic difference between the MAD and the MSE (or RMSE) is that the MSE (and RMSE) penalize extreme errors more heavily than does the MAD. Thus, the MAD is an appropriate measure of forecast accuracy when the costs of forecast errors increase linearly with the size of the error. The MSE (and RMSE) is best if costs for large errors are prohibitively expensive.

Since the MAPE is measured as a percentage and is therefore "unitless," it is particularly useful for comparing the performance of a model on many different time series. One drawback to using the MAPE arises when a series has any extremely small terms, since the division by those terms will tend to seriously inflate the MAPE. For this reason, it is also not wise to use the MAPE as a model selection tool.

When we use the MAD, MSE, or RMSE to select a good forecasting model, the time series data is usually split into two parts. The first part of the data is used to estimate the parameters of the particular model. Then with these estimates the model is used to forecast the *remaining* data points. The MAD, MSE, and RMSE are calculated from the forecast errors for this second part of the series.

Thus, if 72 consecutive monthly observations are available from a time series we wish to forecast, we could use the data for the first 48 months to fit the model. With this derived model we could use the remaining data to forecast the monthly outcomes for the succeeding two years. The MAD or MSE computed from the forecasts and the outcomes over the final 24 months would then be used as the measure of model forecast accuracy.

14.4

Naive Models

naive models

There are some forecasting models that are very intuitive and easy to apply to any time series. These models are usually examined prior to beginning the search for a more sophisticated forecasting model. Because of their simplicity, the models we discuss in this section are often referred to as naive models. **Naive models** provide a baseline against which we can compare the forecasts generated by the more sophisticated models. At the very least, for example, we would want the forecasting accuracy provided by a more complicated model to be substantially better than the accuracy provided by a naive model. We would have little justification for using a sophisticated, time-consuming analysis if it did not yield better forecast accuracy than a simple, inexpensive model.

In this section we present two of the most commonly used naive models, the no-change and the percent-change models.

Definition

no-change
forecasting model

> The **no-change forecasting model** simply uses y_t as the forecast for y_{t+1}. That is, $\hat{y}_{t+1} = y_t$. The model assumes that the current value of the series is a good estimate of what the next value will be.

Definition

percent-change
model

> The **percent-change model** forecasts that y_{t+1} will increase (or decrease) by some fixed percentage of y_t. That is, $\hat{y}_{t+1} = (1 + k)y_t$, where k represents the percentage change expressed in decimal form.

Many businesses use these simple models, sometimes without realizing that they are doing so. For example, if a company believes next year's sales ought to be about the same as this year's, then the company is implicitly using the no-change forecasting model. If the company believes that sales ought to increase by about 10% per year, then it is using a percent-change model with $k = .10$.

The no-change model can be used as a basis of comparison for those time series that do not exhibit any long-term growth or decline. The percent-change model, on the other hand, can be used for evaluating models that grow or decline in an exponential fashion. With these distinctions in mind, you may find it instructive to calculate various accuracy measures such as the MAD and MSE for these naive models when applied to the time series data in this chapter.

EXERCISES   Conceptual Level

14.1   Describe the importance of a good forecasting system in the context of a large manufacturing company. Discuss the considerations in choosing an appropriate model to forecast the sales of large sawmill saws by a manufacturer of sawmill equipment.

14.2   State and defend your choice of the MAD or the MSE as the measure of forecast accuracy in each of the following instances.

   a.   Forecast of the demand for a specialty bakery item to be sold by a bakery

   b.   Forecast of the sales for a new luxury-class, intermediate-sized automobile to be distributed by a European automobile manufacturer

   c.   Forecast of the future energy requirements by the public and private users in the state of Wisconsin

   d.   Forecast of the sales of soft drinks sold by a regional bottler over the next several months

   e.   Forecast of the demand for passenger space on an airline's Los Angeles–San Francisco flight for each month of the next two years

14.3   Businesses depend on the ability to raise money from the financial markets so that they can expand activities and continue to prosper. However, the continual rise in interest rates from 1977 through 1980 caused a dilemma for businesses seeking to raise capital—do they borrow now in hopes that rates have peaked, or do they wait and seek equity capital (with resultant loss of ownership) if rates continue to climb? Many who assumed rates would continue to climb, basing their assumption on the continual rise from 1977 through 1980, found themselves committed to excessively high interest rates. Figure 14.1 shows the time series representing the prime rate in the United States from January 1977 to August 1980. Why were the trend projection forecasting procedures employed by certain business economists in 1977–1980 especially risky? In general, discuss possible dangers of forecasting by extrapolating beyond the interval of time in which the sample data were collected.

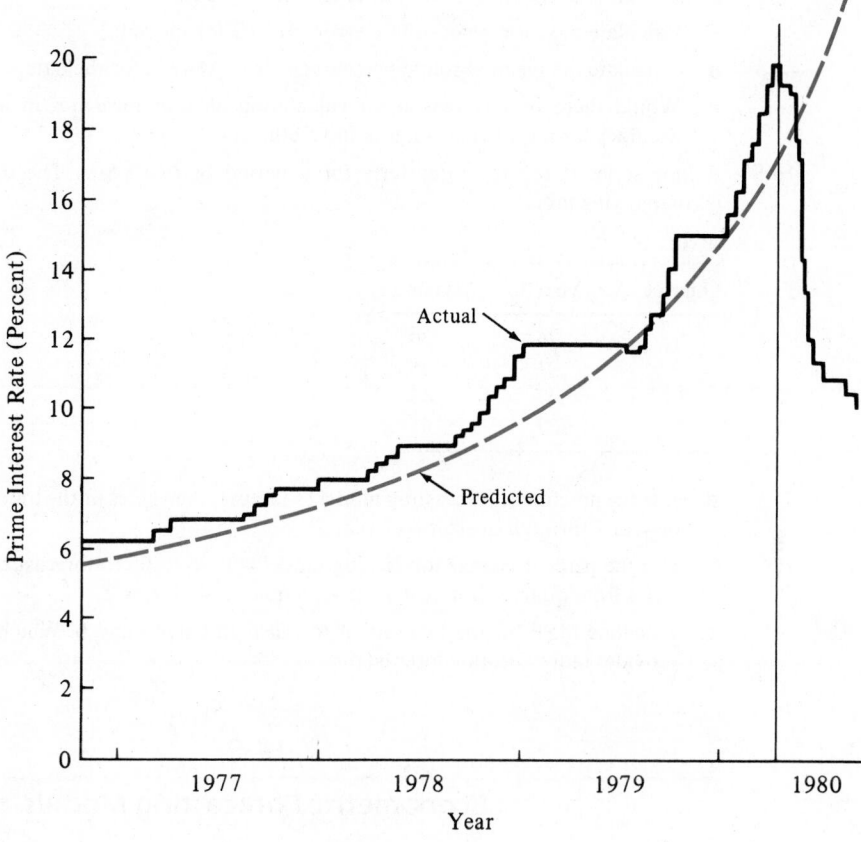

Figure 14.1   Prime Rate Time Series for Exercise 14.3

EXERCISES   Skill Level

14.4   The data given in the accompanying table represent the forecasts for a time series at time $t = 1, 2, 3, 4,$ and 5, along with the actual observed values.

t	y_t	$\hat{y}_t$
1	32.8	33.1
2	35.6	32.4
3	33.6	30.6
4	29.3	29.9
5	31.5	33.7

a. Calculate the mean absolute deviation, MAD, for the data.

b. Calculate the mean square error, MSE, for the data.

c. Calculate the root mean square error, RMSE, for the data.

d. Calculate the mean absolute percentage error, MAPE, for the data.

e. Would there be any reason for calculating all four measures in assessing the forecast accuracy for a particular forecasting method?

14.5 A time series is recorded quarterly for a period of two years. The data are given in the accompanying table.

Quarter	Year 1	Year 2
1	.26	.30
2	.31	.35
3	.47	.45
4	.37	.41

a. Use the no-change forecasting model to forecast the values of the time series from quarter 2 of year 1 through quarter 4 of year 2.

b. Use the percent-change forecasting model with $k = .05$ to forecast the values of the time series from quarter 2 of year 1 through quarter 4 of year 2.

c. Calculate MSE for the two sets of forecasts in parts a and b. Which of the naive models provides more accurate forecasts?

14.5

Econometric Forecasting Models

An econometric model is a system of one or more equations that describe the relationship among several economic and time series variables. Econometric models are probabilistic models and capitalize on the probabilistic relationship that exists between a dependent variable representing the time series and any of a number of independent variables.

The primary feature that distinguishes econometric forecasting models from time series models is their use of economic and demographic variables that are thought to be *causally* related to y. **Econometric models attempt to describe the relationship among such variables by use of one or more regression equations, but time series models ignore these causal variables and rely on a projection of the time series components inherent in y.**

In building an econometric forecasting model, we usually begin with a large number of variables that might be closely related to the response. We then combine these variables to form models that are fitted to the sample data by using the method of least squares (see Chapters 11 and 12). However, a model that fits past data very well may be insensitive to the uncertainties associated with future events and may lead to inaccurate forecasts. Since forecasting is concerned with *future* events, we should select a forecasting model that demonstrates the best ability to forecast the future, not fit the past.

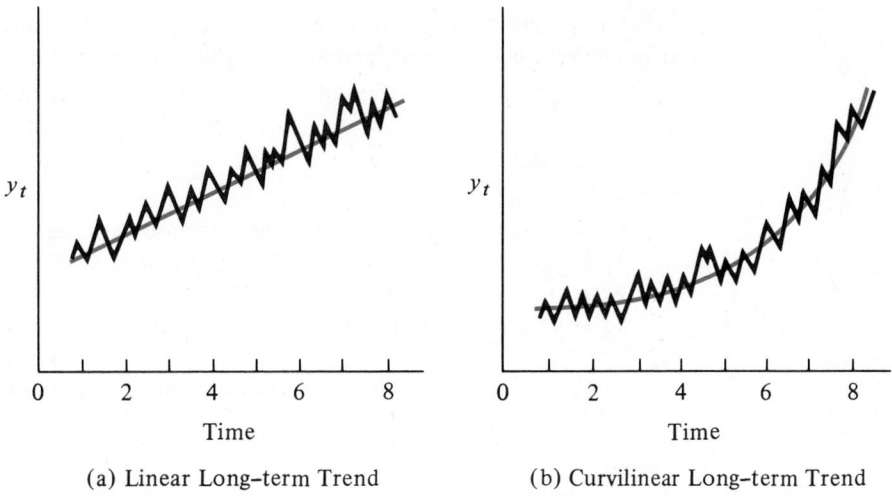

(a) Linear Long–term Trend   (b) Curvilinear Long–term Trend

Figure 14.2   *Times Series with Linear and Curvilinear Long-Term Trends*

The linear regression model of Chapters 11 and 12 is an econometric model that sometimes provides a suitable probabilistic model for establishing the long-term trend for a time series. For example, a long-term upward or downward trend similar to that shown in Figure 14.2a might be isolated by using the procedures of Chapter 11 to fit the straight line.

$$y = \beta_0 + \beta_1 x + \epsilon$$

where the independent variable x represents time. A curvilinear long-term trend, as shown in Figure 14.2b, could be modeled by using a second-order function such as

$$y = \beta_0 + \beta_1 x + \beta_2 x^2 + \epsilon$$

The corresponding prediction equation, $\hat{y} = \hat{\beta}_0 + \hat{\beta}_1 x + \hat{\beta}_2 x^2$, could be determined by using the method of least squares, as described in Chapter 12.

The assumption of independence of the random error ϵ associated with successive measurements will not usually be satisfied. Consequently, the probabilistic statements associated with the estimation of $E(y)$ or the prediction of y will be incorrect. We would suspect that they would be conservative and that knowledge of the actual pattern of correlation would permit more accurate estimation and prediction. If the response is an average over a period of time, the correlation of adjacent response measurements will be reduced and will quite possibly satisfy adequately the assumption of independence implied in the least squares inferential procedures.

Other linear models (linear in the unknown terms, or "weights," $\beta_0, \beta_1, \beta_2$, etc.) can be constructed and fitted to data generated from economic time series by using the method of least squares. For example, the yearly production y of steel is a function of its price, the price of competitive structural materials, the production of competitive products during the preceding year, the amount of steel purchased during the immediately preceding years (to measure current inventory), and other variables. A linear model relating these independent variables to steel production might be

$$y = \beta_0 + \beta_1 x_1 + \beta_2 x_2 + \beta_3 x_3 + \cdots + \beta_k x_k + \epsilon$$

where

$x_1 = $ time

$x_2 = $ price of steel

$x_3 = x_2^2$ (allowing curvature in the response curve as a function of price)

$x_4 = $ production of aluminum during previous year

$x_5 = $ price of aluminum

$x_6 = $ steel production during previous year

$\vdots$

$x_k = x_2 x_5$ (an interaction effect between steel and aluminum prices)

We could even include the variable $x_7 = (\text{time})^2$ to capture a possible curvilinear long-term trend.

dummy variables   If we were analyzing, for example, seasonal monthly data, we could create seasonal **dummy variables** to check the effect that that month might have on the dependent variable. (See Section 12.6.) For example, a January dummy variable would have a value of 1 for every January and 0 for all other months. The total number of dummy variables to be included in a model is *always* one less than the number of seasonal categories, because all effects are measured in comparison with the base category.

In other words, the model builder has unlimited room for ingenuity in constructing the linear model. It should be readily apparent that the major benefit associated with this type of modeling, if properly structured, is in providing the analyst with the ability to understand the impact of each variable on the dependent variable. The major drawback, however, is that forecasts for the explanatory variables are necessary inputs for the forecast of the dependent variable. The forecasts of the explanatory variables may have to be purchased from major forecasting organizations such as DRI, Chase Econometrics, or Mapcast (G.E.), or they may be acquired internally from company budgets or sales analyses.

transfer function   A different single-equation method is **transfer function analysis**, the time series
analysis   analog of regression analysis. In addition to possessing all the features of regression analysis, this method allows for the data to indicate whether lagged dependent or
lagged variable   explanatory variables are needed in the structure of the equation. A **lagged variable** is an observation from a variable recorded from an earlier time period, such as y_{t-1} or $x_{1,t-1}$. Furthermore, the method minimizes the biases due to correlated errors and omitted variables because one may model the residuals in these "regressionlike" equations as autoregressive moving average processes. Transfer function analysis is a sophisticated single-equation technique that requires trained analysts using computers to formulate the models.

Finally, we should add that all single-equation models are susceptible to misspecification biases because they all require the analyst to postulate a priori which variables explain the dependent variable. The possibility of the dependent variable
feedback among   influencing any of the explanatory variables, technically known as **feedback among**
variables   **variables**, is not allowed in single-equation models. Multiple-equation systems explore these types of relationships. (Refer to Pindyck and Rubinfeld in the references for a discussion of multiple-equation econometric models.)

EXERCISES   Conceptual Level

14.6   What is the distinction between econometric models and time series models used for forecasting?

14.7   Econometric model building involves the use of the method of least squares, which assumes independence of the error terms (ϵ) associated with successive measurements. What are the implications if this assumption is not met?

14.8   Describe briefly the major benefit associated with econometric modeling.

14.9   What is meant by the following terms?

   a.   Transfer function analysis

   b.   Misspecification biases

   c.   Feedback among variables

EXERCISES   Skill Level

14.10   The data from Exercise 13.5, representing a time series recorded monthly for two years, is reproduced in Table 14.1.

   a.   Use the simple linear econometric model, $y = \beta_0 + \beta_1 t + \epsilon$, to fit the data.

   b.   Calculate MAD as a measure of fit for the linear model in part a. Does the model provide accurate forecasts?

Table 14.1   Times Series Data for Exercise 14.10

Month	Year 1	Year 2	Month	Year 1	Year 2
Jan.	15.8	16.4	July	16.1	16.6
Feb.	15.7	16.2	Aug.	15.9	16.5
Mar.	15.3	15.9	Sept.	15.9	16.6
Apr.	15.5	16.1	Oct.	16.4	17.0
May	16.0	16.2	Nov.	16.6	17.4
June	16.2	16.7	Dec.	17.1	18.0

14.11   Refer to Exercise 14.10. The plot of the data against time suggests that a quadratic model might be a more realistic forecasting tool.

   a.   A MINITAB computer program was used to fit the quadratic econometric model, $y = \beta_0 + \beta_1 t + \beta_2 t^2 + \epsilon$, to the data. Use the printout in Table 14.2 to identify the forecast equation.

   b.   Calculate MAD as a measure of fit for the quadratic model in part a. Does the model fit better than the linear model in Exercise 14.10?

   c.   Might any of the assumptions necessary for the use of multiple regression analysis been violated? If so, which ones?

14.12   Refer to Exercise 14.10. A more sophisticated model sometimes used to fit data that has seasonal or cyclic effects is the model

$$y = \beta_0 + \beta_1 t + \beta_2 \sin\left(\frac{2\pi t}{6}\right) + \beta_3 \cos\left(\frac{2\pi t}{6}\right) + \epsilon$$

Table 14.2   MINITAB Output for Exercise 14.11

```
MTB > REGRESS C1 2 C2 C3

THE REGRESSION EQUATION IS
Y = 15.7 + 0.0123 T + 0.00223 TSQ

PREDICTOR         COEF           STDEV        T-RATIO
CONSTANT       15.7294          0.2558         61.48
T               0.01226         0.04715         0.26
TSQ             0.002228        0.001831        1.22

S = 0.3835      R-SQ = 64.2%      R-SQ(ADJ) = 60.8%

ANALYSIS OF VARIANCE

SOURCE          DF            SS             MS
REGRESSION       2         5.5285         2.7642
ERROR           21         3.0878         0.1470
TOTAL           23         8.6163

SOURCE          DF         SEQ SS
T                1         5.3108
TSQ              1         0.2177

OBS.        T           Y         FIT   STDEV.FIT   RESIDUAL    ST.RESID
  1        1.0      15.8000    15.7439     0.2165     0.0561       0.18
  2        2.0      15.7000    15.7628     0.1823    -0.0628      -0.19
  3        3.0      15.3000    15.7862     0.1539    -0.4862      -1.38
  4        4.0      15.5000    15.8141     0.1317    -0.3141      -0.87
  5        5.0      16.0000    15.8464     0.1164     0.1536       0.42
  6        6.0      16.2000    15.8832     0.1079     0.3168       0.86
  7        7.0      16.1000    15.9244     0.1051     0.1756       0.48
  8        8.0      15.9000    15.9701     0.1062    -0.0701      -0.19
  9        9.0      15.9000    16.0202     0.1094    -0.1202      -0.33
 10       10.0      16.4000    16.0748     0.1129     0.3252       0.89
 11       11.0      16.6000    16.1338     0.1158     0.4662       1.28
 12       12.0      17.1000    16.1973     0.1174     0.9027       2.47R
 13       13.0      16.4000    16.2653     0.1174     0.1347       0.37
 14       14.0      16.2000    16.3377     0.1158    -0.1377      -0.38
 15       15.0      15.9000    16.4146     0.1129    -0.5146      -1.40
 16       16.0      16.1000    16.4959     0.1094    -0.3959      -1.08
 17       17.0      16.2000    16.5817     0.1062    -0.3817      -1.04
 18       18.0      16.7000    16.6719     0.1051     0.0281       0.08
 19       19.0      16.6000    16.7666     0.1079    -0.1666      -0.45
 20       20.0      16.5000    16.8657     0.1164    -0.3657      -1.00
 21       21.0      16.6000    16.9693     0.1317    -0.3693      -1.03
 22       22.0      17.0000    17.0774     0.1539    -0.0774      -0.22
 23       23.0      17.4000    17.1899     0.1823     0.2101       0.62
 24       24.0      18.0000    17.3069     0.2165     0.6931       2.19R
```

    a. Use the MINITAB printout in Table 14.3 to identify the econometric forecast equation.

    b. Calculate MAD as a measure of fit for the model in part a. Does the model fit better than the linear or quadratic models derived in Exercises 14.10 and 14.11?

EXERCISES   Applications

14.13 Suggest some independent variables that might be used to assist in the prediction of future values of each of the following time series.

    a. The annual sales of ski clothing and equipment

    b. The domestic sales of U.S. manufactured automobiles

    c. The monthly number of applicants for unemployment assistance from a state welfare agency

    d. The monthly price per 1,000 board feet of construction-grade dimensioned lumber

    e. The monthly demand for rooms at a large resort hotel

14.14 The *capital asset pricing theory* states that the expected return on a security consists of the rate of interest (or dividend) plus a risk premium that is proportional to the stock's sensitivity to market movement. Using the capital asset pricing theory and other factors you consider relevant, create an econometric model to forecast one-step-ahead market prices for a publicly listed utility stock. How would your econometric model change if you wished to forecast one-step-ahead market prices for the publicly listed stock of an electronics manufacturer?

14.15 To create better awareness of the potential of fire damage and to better protect its customers, a large fire insurance company has offered home smoke detection sensors to policyholders for a price of $20. The number of smoke sensors sold to policyholders over the first eight months of the offer is shown in the accompanying table. The sales manager for the insurance company claims that the simple linear econometric model $\hat{y}_t = 200 + 66.67t$, where t is the month as numbered since the beginning of the sales of the smoke sensors, describes the sales of smoke sensors by the insurance company over time.

Month, t	1	2	3	4	5	6	7	8
Number Sold, y_t	410	290	420	310	410	580	600	990

    a. On a piece of graph paper, plot the time series and superimpose the linear model $\hat{y}_t = 200 + 66.67t$. Does this model appear to provide an adequate fit to the time series?

    b. Use the procedures of Chapter 12 to fit the second-order model $\hat{y}_t = \hat{\beta}_0 + \hat{\beta}_1 t + \hat{\beta}_2 t^2$ to the time series for the smoke sensor sales. Compare the fit of the second-order model with that of the simple linear model $\hat{y}_t = 200 + 66.67t$ to the time series, using the MSE as the measure of fit.

    c. Use both the simple linear model $\hat{y}_t = 200 + 66.67t$ and the second-order model you have derived to forecast the sales for the smoke sensors during months $t = 9$ and $t = 10$.

    d. What other variables might be important for forecasting sales of smoke detectors?

14.16 One way of determining whether a model provides an adequate fit to an underlying data set is to examine the residuals $(y_t - \hat{y}_t)$. If a plot of the residuals against time appears to be random (i.e., scattered, or not patterned), the model is said to provide a good fit to the underlying time series. Refer to Exercise 14.15 and plot the residuals against time for both the linear model $\hat{y}_t = 200 + 66.67t$ and the second-order model derived as part of the exercise. What do these plots suggest?

Table 14.3   MINITAB Output for Exercise 14.12

```
MTB > REGRESS C1 3 C2 C4 C5

THE REGRESSION EQUATION IS
Y = 15.5 + 0.0635 T - 0.0344 SIN + 0.370 COS

PREDICTOR        COEF        STDEV       T-RATIO
CONSTANT       15.5441        0.1239      125.49
T             0.063476      0.008712        7.29
SIN          -0.03439        0.08484       -0.41
COS           0.36986        0.08394        4.41

S = 0.2892      R-SQ = 80.6%      R-SQ(ADJ) = 77.7%

ANALYSIS OF VARIANCE

SOURCE        DF          SS            MS
REGRESSION     3       6.9434        2.3145
ERROR         20       1.6728        0.0836
TOTAL         23       8.6163

SOURCE        DF      SEQ SS
T              1       5.3108
SIN            1       0.0088
COS            1       1.6239

OBS.        T          Y          FIT   STDEV.FIT    RESIDUAL    ST.RESID
  1        1.0     15.8000     15.7627     0.1372      0.0373       0.15
  2        2.0     15.7000     15.4563     0.1262      0.2437       0.94
  3        3.0     15.3000     15.3646     0.1262     -0.0646      -0.25
  4        4.0     15.5000     15.6428     0.1315     -0.1428      -0.55
  5        5.0     16.0000     16.0761     0.1315     -0.0761      -0.30
  6        6.0     16.2000     16.2948     0.1213     -0.0948      -0.36
  7        7.0     16.1000     16.1435     0.1095     -0.0435      -0.16
  8        8.0     15.9000     15.8371     0.1045      0.0629       0.23
  9        9.0     15.9000     15.7455     0.1045      0.1545       0.57
 10       10.0     16.4000     16.0237     0.1067      0.3763       1.40
 11       11.0     16.6000     16.4570     0.1067      0.1430       0.53
 12       12.0     17.1000     16.6756     0.1031      0.4244       1.57
 13       13.0     16.4000     16.5244     0.1031     -0.1244      -0.46
 14       14.0     16.2000     16.2180     0.1067     -0.0180      -0.07
 15       15.0     15.9000     16.1263     0.1067     -0.2263      -0.84
 16       16.0     16.1000     16.4045     0.1045     -0.3045      -1.13
 17       17.0     16.2000     16.8379     0.1045     -0.6379      -2.37R
 18       18.0     16.7000     17.0565     0.1095     -0.3565      -1.33
 19       19.0     16.6000     16.9052     0.1213     -0.3052      -1.16
 20       20.0     16.5000     16.5989     0.1315     -0.0989      -0.38
 21       21.0     16.6000     16.5072     0.1315      0.0928       0.36
 22       22.0     17.0000     16.7854     0.1262      0.2146       0.82
 23       23.0     17.4000     17.2187     0.1262      0.1813       0.70
 24       24.0     18.0000     17.4373     0.1372      0.5627       2.21R
```

14.17 The following regression model was constructed to describe monthly sales of a nationally advertised breakfast cereal:

$$\text{sales}_t = \beta_0 + \beta_1 t + \beta_2 \text{Prom}A_t + \beta_3 \text{Prom}A_{t-1} + \beta_4 \text{Prom}A_{t-2}$$
$$+ C_1 \text{Prom}B_t + C_2 \text{Prom}B_{t-1} + C_3 \text{Prom}B_{t-2}$$
$$+ d_1 \text{Feb.} + d_2 \text{Mar.} + \cdots + d_{11} \text{Dec.} + \epsilon_t$$

where

sales_t = monetary volume of monthly sales of product

$\text{Prom}A_{t-i}$ = special consumer-oriented promotion, measured in cases, at time $t - i, i = \{0, 1, 2\}$

$\text{Prom}B_{t-i}$ = special dealer-oriented promotion, measured in dollars, at $t - i$, $i = \{0, 1, 2\}$

t = month numbered consecutively: Jan. of first year = 1; Feb. of first year = 2; ...; Dec. of first year = 12; Jan. of second year = 13; and so on

Feb. = February dummy variable: 1 for Feb.; 0 elsewhere

Mar. = March dummy variable: 1 for Mar.; 0 elsewhere

$\vdots$

Dec. = December dummy variable: 1 for Dec.; 0 elsewhere

The two lagged periods in the promotion variables were incorporated into the model to accommodate borrowing effects—for example, consumers stocking up during a particular month because of lower prices but continuing to consume the product on a constant basis. Estimation of the above regression based on 96 periods (Jan. 1981–Dec. 1988) yielded the results given in Table 14.4. Data for the 12 months of 1988 are presented in Table 14.5.

Table 14.4   Estimation Data for Exercise 14.17

Parameter	Parameter Estimate	Standard Error of Estimate
$\beta_0 =$	61,006	34,577
$\beta_1 =$	161.0	73.6
$\beta_2 =$	2.6	.9
$\beta_3 =$	.7	.8
$\beta_4 =$	$-.3$	.1
$C_1 =$	2.1	.5
$C_2 =$	$-.4$	.2
$C_3 =$	$-.1$	.3
$d_1 =$	40,691	12,576
$d_2 =$	63,296	30,971
$d_3 =$	12,641	4,222
$d_4 =$	85,375	37,080
$d_5 =$	32,782	15,076
$d_6 =$	31,644	12,839
$d_7 =$	9,720	3,780
$d_8 =$	42,697	15,983
$d_9 =$	52,002	14,222
$d_{10} =$	28,006	13,698
$d_{11} =$	16,241	7,289

Table 14.5   Sales Data for Exercise 14.17

Sales	PromA	PromB	t
Jan. 506,132	83,006	94,600	85
Feb.	82,241	106,720	86
Mar. 419,026	43,298	120,890	87
Apr.	71,600	28,600	88
May	104,600	142,896	89
June	98,667	48,115	90
July	164,296	49,984	91
Aug.	42,006	8,000	92
Sept.	62,888	89,794	93
Oct.	45,978	103,698	94
Nov.	15,060	48,222	95
Dec. 267,983	66,686	21,096	96

a. Interpret the coefficients associated with each variable.

b. What would have been your sales forecast for Dec. 1988?

c. Assuming that the budget for 1989 calls for 77,000 equivalent cases of PromA and $100,000 in PromB allowances in January, calculate the sales forecast for Jan. 1989.

14.6

Moving Average Forecasting Models

In Sections 13.3 and 13.4 we examined the use of smoothing techniques for "averaging out" the effects of random variation and seasonality in a time series. The methods used in those sections were based on **centered moving averages** and **exponential smoothing**. In this section and in the next two sections we discuss how these smoothing techniques can also be modified for use as forecasting methods.

When a time series pattern does not exhibit trends or seasonal characteristics, calculation of a moving average may be useful in canceling out the random variation so that short-term forecasts can be generated. This method consists of simply averaging the available observations over the most recent M periods and using this average as the forecast of the next observation. This procedure is similar to the moving average method used in Section 13.3, except now we do not center the averages; we simply use them as forecasts. For example, if a regional distribution manager were interested in forecasting quarterly shipments of office supplies to a specific geographical area, he or she could average the observations for the next M quarters and use that average as a forecast for the next quarter. Since we are only considering time series that do not have trends or seasonal movements, the moving average forecast also serves as a forecast for any number of periods ahead.

Moving Average Forecasting Model

The **moving average forecasting model** uses the average of the most recent M observations as the forecast of the next observation in the time series:

$$\hat{y}_{t+1} = \frac{y_t + y_{t-1} + y_{t-2} + \cdots + y_{t-M+1}}{M}$$

Since we apply this model only to series with no trend or seasonality, this forecast is also a forecast for any number of observations ahead.

We often use the abbreviation MA(2) when referring to a moving average model with $M = 2$, MA(3) for the case $M = 3$, and so forth. The choice of the best M to use in the averages is similar to the choice of the appropriate smoothing constant to use in exponential smoothing in that there is no explicit formula to follow. In practice, we usually pick an M that generates forecasts that minimize some measure of forecast accuracy, such as the MAD or the MSE.

EXAMPLE 14.1   The data in Table 14.6 represent the monthly sales revenue data over a two-year period for a retail hardware store.

   a.   Compute the one-step-ahead moving average forecasts, beginning with January of year 2. Let $M = 5$ (five-period moving average).
   b.   Compute the two-steps-ahead moving average forecasts, beginning with February of year 2. Let $M = 5$.
   c.   Using the MAD, compare the one-step-ahead forecasts and the two-steps-ahead forecasts with the actual sales for year 2.

Table 14.6   Sales Revenue Data for Example 14.1

Month (t)	Sales Revenue ($\times$ \$1,000)	
	Year 1	Year 2
January	19.3	21.8
February	20.6	22.5
March	18.4	21.6
April	17.6	19.9
May	21.5	23.7
June	27.8	24.1
July	26.2	28.6
August	27.1	30.0
September	23.9	25.7
October	24.0	26.1
November	22.8	25.3
December	24.0	28.8

SOLUTION If t is the index of time referencing the months ($t = 1, 2, \ldots, 24$), forecasts are expressed symbolically as

$$\hat{y}_{t+1} = \frac{y_t + y_{t-1} + y_{t-2} + y_{t-3} + y_{t-4}}{5}$$

for $t = 12, 13, \ldots, 23$. Thus, for January of year 2 ($t = 13$), we have

$$\hat{y}_{13} = \frac{y_{12} + y_{11} + y_{10} + y_9 + y_8}{5}$$

$$= \frac{24.0 + 22.8 + 24.0 + 23.9 + 27.1}{5} = 24.36$$

Similarly, the one-step-ahead forecast for February of year 2 ($t = 14$) is

$$\hat{y}_{14} = \frac{y_{13} + y_{12} + y_{11} + y_{10} + y_9}{5}$$

$$= \frac{21.8 + 24.0 + 22.8 + 24.0 + 23.9}{5} = 23.30$$

Similar to the computation of a smoothed series of moving averages in Section 13.3, we could find

$$\hat{y}_{t+1} = \hat{y}_t + \frac{\text{next observation} - \text{most remote observation}}{M}$$

or $$\hat{y}_{14} = \hat{y}_{13} + \frac{21.8 - 27.1}{5} = 23.30$$

The one-month-ahead forecasts for March through December are listed in Table 14.7.

Table 14.7   One-Step- and Two-Steps-Ahead Forecasts for Example 14.1 Using a Moving Average Forecast Model with $M = 5$

		Forecasts		
t	Month	One Month Ahead	Two Months Ahead	Actual Sales
13	January	24.36	24.80	21.8
14	February	23.30	24.36	22.5
15	March	23.02	23.30	21.6
16	April	22.54	23.02	19.9
17	May	21.96	22.54	23.7
18	June	21.90	21.96	24.1
19	July	22.36	21.90	28.6
20	August	23.58	22.36	30.0
21	September	25.26	23.58	25.7
22	October	26.42	25.26	26.1
23	November	26.90	26.42	25.3
24	December	27.14	26.90	28.8

In defining a moving average forecasting model, we stated that the two-steps-ahead moving average forecasts are exactly the same as the one-step-ahead forecasts except that they are moved ahead two periods instead of one. That is, the two-steps-ahead forecast for February of year 2 (period 14) would be 24.36, which is the five-term moving average that was calculated two periods prior at period 12. Similarly, the two-steps-ahead forecast for period 13 is 24.80, the five-term moving average calculated at period 11; and so forth. The two-steps-ahead forecasts are shown in Table 14.7. The original series and the one-step- and two-steps-ahead forecasts are shown together in Figure 14.3.

The MAD for the one-step-ahead moving average forecasts for year 2 is calculated by

$$\text{MAD} = \frac{|21.8 - 24.36| + |22.5 - 23.30| + |21.6 - 23.02| + \cdots + |28.8 - 27.14|}{12}$$

$$= 2.3367$$

For the two-steps-ahead forecasts it is

$$\text{MAD} = \frac{|21.8 - 24.80| + |22.5 - 24.36| + |21.6 - 23.30| + \cdots + |28.8 - 26.90|}{12}$$

$$= 2.775$$

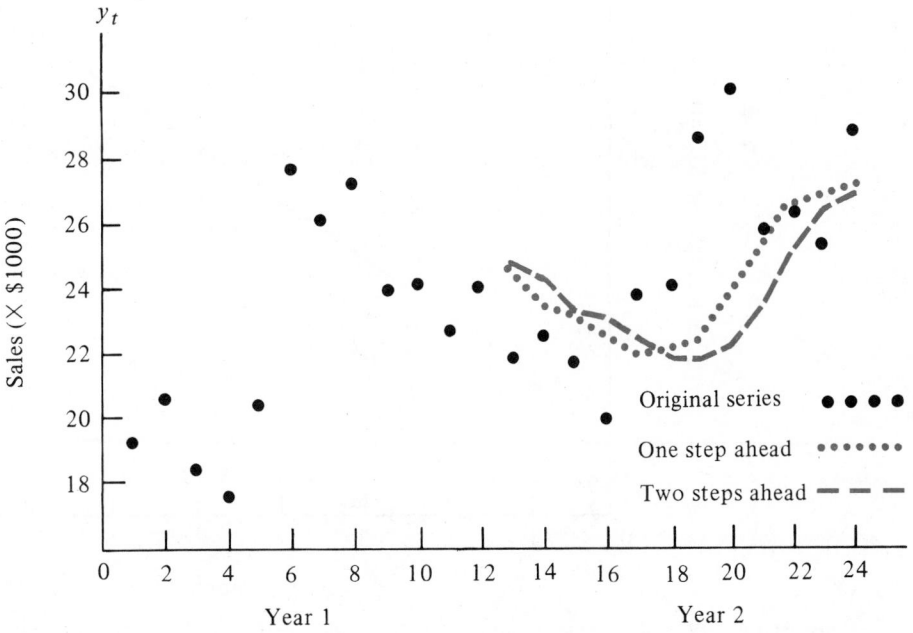

Figure 14.3   Actual Sales, One-Step- and Two-Steps-Ahead Forecasts for Example 14.1

Thus, the one-step-ahead forecasts appear to be slightly better since they give rise to a smaller MAD than do the two-steps-ahead forecasts. This result makes sense since we generally expect forecasting performance to deteriorate as we try to project further and further into the future.

upward or
downward trend

If a time series contains a pronounced **upward or downward trend**, a moving average forecast model will likely provide misleading results. For example, consider the sales of a new exterior paint remover by a paint supply manufacturer, and assume that the sales for the first eight months of operation are as shown in the following table.

Month	1	2	3	4	5	6	7	8
Sales (× $1,000)	16	22	29	35	42	48	55	60

A plot of these data in Figure 14.4 reveals a distinct linear growth trend for sales. Projection of the best straight line through these data suggests a sales forecast for month 9 of about $66,000. However, calculation of a three-period, simple moving average (Table 14.8) yields a sales forecast of $54,330 for month 9. Further analysis would show that increasing M in the moving average computation would provide little improvement to the accuracy of our sales forecast for month 9.

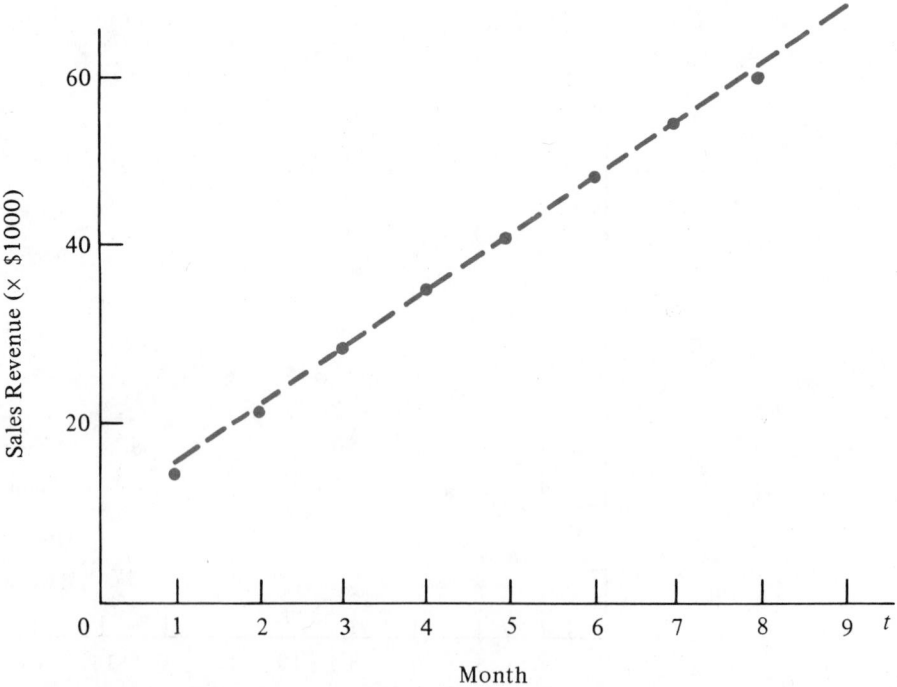

Figure 14.4   Plot of Paint Remover Sales Data, Illustrating a Linear Growth Trend

Table 14.8   Three-Period Moving Average Sales
Forecast for the Paint Remover Data

t	y_t	MA(3) Forecast
1	16	—
2	22	—
3	29	—
4	35	22.33
5	42	28.67
6	48	35.33
7	55	41.67
8	60	48.33
9	—	54.33

The low forecast value arises because we are ignoring the trend component in the data and are trying to "average" it instead of modeling it explicitly. For data with pronounced trends we must first remove them and then apply an appropriate method to the detrended (filtered) data. Then after generating forecasts for the detrended series, we must reincorporate the trend back into the data.

One simple method used to filter out a trend component in a time series is to apply **first-order difference transformation** a **first-order difference transformation**.

Definition

> **first differences** The time series formed by taking differences of successive terms in the y_t series is called the series of **first differences** of y_t. It is denoted by ∇y_t and calculated by
>
> $$\nabla y_t = y_t - y_{t-1}$$

The del symbol (∇), used to denote a difference operation, is actually an inverted Greek letter delta.

A differencing transformation can involve second-order differences, where differences are computed from the first difference series, for removing nonlinear trends from a series. We will reserve this issue for later discussion.

To illustrate the application of a first-order difference transformation, Table 14.9 lists the first differences computed for the paint remover sales data. Note that the series of first differences ∇y_t has completely filtered out the upward trend inherent in the original series. Notice also that the differencing transformation exacts a small price, namely, the loss of one observation. That is, the ∇y_t series always has one less observation than does the y_t series.

Calculation of a three-period moving average forecast for the detrended data for time period 9 yields $\hat{\nabla} y_9 = 6.0$. Consequently, the forecast for month 9, $\hat{y}_9$, can be

Table 14.9   Three-Period Moving Average Forecast of Differenced
Series for the Paint Remover Sales Data

Month, t	Sales ($\times \$1,000$)	∇y_t	MA(3) Forecast of ∇y_t
1	16	—	—
2	22	6	—
3	29	7	—
4	35	6	—
5	42	7	6.33
6	48	6	6.67
7	55	7	6.33
8	60	5	6.67
9	—	—	6.00

derived from the basic definition of differences, namely,

$$\hat{\nabla} y_9 = 6.0 = (\hat{y}_9 - y_8)$$

or

$$\hat{y}_9 = 6.0 + y_8 = 66$$

Thus, the sales forecast for month 9 obtained by applying a three-period, moving average forecast to the series of first differences is $66,000.

Although a first-order differencing transformation will remove a pronounced linear trend from a time series, second-order differencing is required for removal of a quadratic trend.

Definition

second
differences

> The series of **second differences** of y_t is found by taking first differences of the series of first differences. That is,
>
> $$\nabla^2 y_t = \nabla(\nabla y_t) = \nabla(y_t - y_{t-1}) = (y_t - y_{t-1}) - (y_{t-1} - y_{t-2})$$
> $$= y_t - 2y_{t-1} + y_{t-2}$$

After removing the quadratic trend from the y_t series by using second differences, we can proceed to forecast the $\nabla^2 y_t$ series and, from that, the y_t series itself. Of the various methods for forecasting a series with no trend or seasonality, we will again use the moving average method. The procedure can be summarized as follows:

1. Apply a moving average model to find a forecast $\hat{\nabla}^2 y_{t+1}$ of the next value $\nabla^2 y_{t+1}$ of the series of second differences.
2. Using the defining formula for second differences, write the forecast from step 1 as $\hat{\nabla}^2 y_{t+1} = \hat{y}_{t+1} - 2y_t + y_{t-1}$.
3. Finally, solve the equation in step 2 for $\hat{y}_{t+1}$.

The next example illustrates this method of using second differences in conjunction with a moving average forecasting model.

EXAMPLE 14.2 In many successful new-product introductions, sales tend to rise rapidly before competition and product obsolescence cause sales growth to moderate. Consider the sales of a medical services computer software program during the first eight months after introduction of the program to the marketplace. These sales data are listed in Table 14.10 and illustrated as Figure 14.5a. Apply a second-order difference transformation to the software sales data. Then use a moving average forecasting model with $M = 3$ [MA(3)] to derive a sales estimate for month 9.

Table 14.10   Original Series, First and Second Differences and MA(3) Forecasts for the Computer Software Sales Data

t (Month)	y_t ($\times$ \$1,000)	∇y_t	$\nabla^2 y_t$	MA(3) for $\nabla^2 y_t$
1	1	—	—	—
2	7	6	—	—
3	17	10	4	—
4	30	13	3	—
5	47	17	4	—
6	68	21	4	3.67
7	94	26	5	3.67
8	124	30	4	4.33
9	—	—	—	4.33

SOLUTION Applying a second-order difference transformation to the software sales time series first requires the application of a first-order difference transformation. The first differences are computed as $\nabla y_t = (y_t - y_{t-1})$, taking differences of successive observations of the time series. The second differences $\nabla^2 y_t = (y_t - y_{t-1}) - (y_{t-1} - y_{t-2})$ are computed by taking differences of successive first differences. These first and second differences are listed in Table 14.10.

Since the series of second differences $\nabla^2 y_t$ has no trend, we can apply a moving average forecast with $M = 3$ to derive a second-difference forecast for month $t = 9$. Computationally,

$$\hat{\nabla}^2 y_{t+1} = \frac{\nabla^2 y_t + \nabla^2 y_{t-1} + \nabla^2 y_{t-2}}{3}$$

with the one-month-ahead second-difference forecasts listed in Table 14.10 for months 6, 7, 8, and 9. Next, we find

$$\hat{\nabla}^2 y_9 = \hat{y}_9 - 2y_8 + y_7$$

or

$$4.33 = \hat{y}_9 - 2(124) + 94$$

Solving, we find $\hat{y}_9 = 158.33$, or \$158,330.

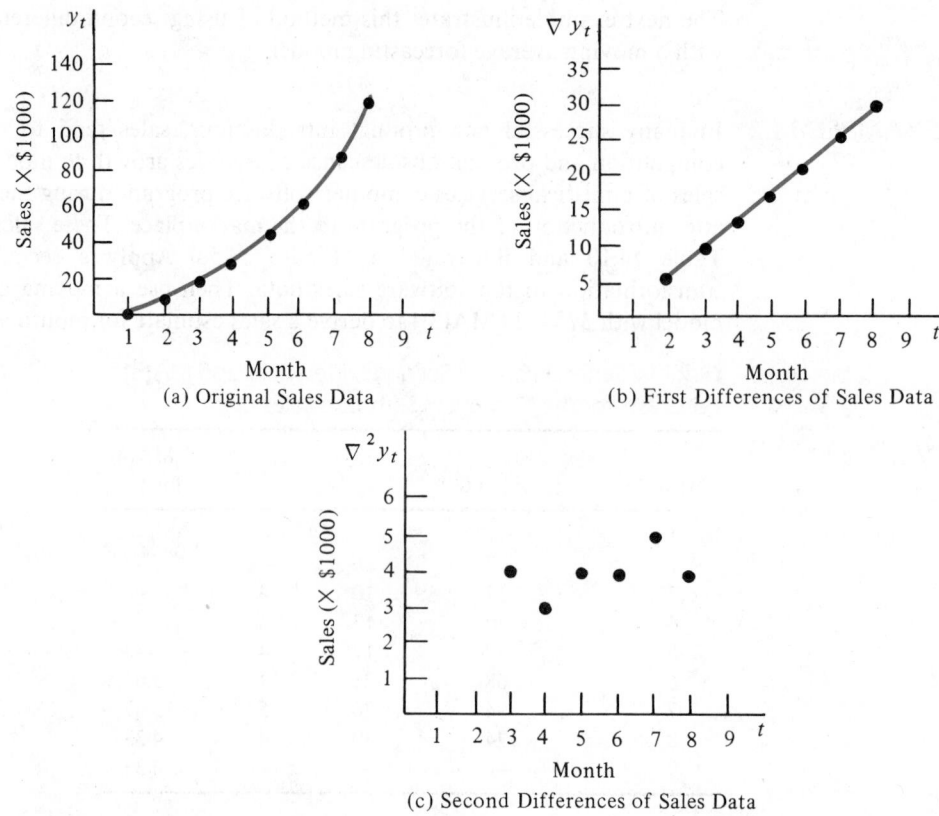

Figure 14.5   Plot of Original Sales Data, First Differences, and Second Differences for the Computer Software Example

The method described in this section can be extended to include higher-order differencing (for handling other types of trends) and seasonal differencing (to cancel the seasonal movement in a series). We will not pursue these topics further in this chapter. Instead, we note that Section 14.8 provides an alternative approach to forecasting time series that have a seasonal component.

EXERCISES   Conceptual Level

14.18   When is it appropriate to use a moving average forecasting model?

14.19   What difficulties might arise in using a moving average forecasting model with a time series that contains a pronounced trend? How can this difficulty be dealt with?

14.20   When would it be appropriate to use second-difference transformations?

EXERCISES   Skill Level

14.21   A time series recorded for 10 consecutive time periods is shown in the accompanying table.

t	1	2	3	4	5	6	7	8	9	10
y_t	9	10	12	11	15	15	17	19	21	20

    a.   Compute a one-step-ahead forecast for time $t = 5$ through $t = 10$, using a moving average model with $M = 3$.

    b.   Compute a two-steps-ahead forecast for time $t = 5$ through $t = 10$, using the moving average model with $M = 3$.

    c.   Calculate MAD for the two sets of forecasts in parts a and b. Which forecasting method is more accurate? Explain.

14.22   Refer to Exercise 14.21. Since the series contains a pronounced upward trend, a first-order difference transformation may be useful in filtering out the trend.

    a.   Calculate the first differences ∇y_t.

    b.   Calculate the three-period moving average MA(3) of the first differences.

    c.   Forecast the value of the time series at time $t = 5$ through $t = 10$.

    d.   Calculate MAD for the forecasts in part c. Compare this result with the MAD calculated for the forecasts from Exercises 14.21a and 14.21b. Which forecasting method is the most accurate?

14.23   The following data represent the demand over a five-month period for home computers as recorded by the manager of a retail computer store.

Month	Jan.	Feb.	Mar.	Apr.	May
Demand	80	102	58	80	91

    a.   Use the available data to compute a moving average forecast of June demand. Let $M = 3$.

    b.   If the actual demand for June was 100 microcomputers, find the one-step-ahead moving average forecast for July demand. Again, let $M = 3$.

    c.   Using the data through May, find the two-steps-ahead moving average forecast for July demand. Let $M = 3$.

EXERCISES   Applications

14.24   The treasurer of Logic Technologies, a manufacturer of computer hardware, has been asked by the board of directors to forecast the month-ending market trading price of Logic Technologies stock in her monthly report to the board. The month's ending price of Logic Technology stock over 28 consecutive months is listed in Table 14.11.

    a.   Use a five-month moving average forecast model to forecast the one-step-ahead prices for Logic Technologies stock for months 20 through 28.

Table 14.11   Stock Price Data for Exercise 14.24

Month	Price	Month	Price	Month	Price	Month	Price
1	63	8	66	15	65	22	65
2	62	9	68	16	64	23	68
3	64	10	62	17	59	24	68
4	63	11	64	18	59	25	64
5	61	12	61	19	60	26	66
6	65	13	64	20	61	27	67
7	66	14	64	21	66	28	65

b. Repeat the instructions in part a, using a ten-month moving average forecast model to generate one-step-ahead forecasts.

c. Using MSE as a measure of forecast accuracy, compare the accuracy of the five-month moving average model with that of the ten-month moving average model for generating one-step-ahead forecasts for the security prices of Logic Technologies.

14.25   Refer to Exercise 13.17.

a. Compute the first differences for the sales values listed in Exercise 13.17. Then derive the one-step-ahead forecasts for the period January 1987 through December 1988, using a three-month moving average model.

b. Compute both the first differences and the second differences for the sales data of Exercise 13.17. Derive one-step-ahead forecasts for the period January 1987 through December 1988, using a three-month moving average model.

c. Using MSE as a measure of forecast accuracy, compare the accuracy of the forecasts developed using first differences with those developed using second differences.

14.26   The data below represent the gross monthly sales volume (in thousands of dollars) for a firm that manufactures energy-saving thermostats and rheostats.

Month	Jan.	Feb.	Mar.	Apr.	May	June	July	Aug.	Sept.	Oct.	Nov.	Dec.
Sales	51.2	50.7	52.6	54.9	55.5	56.2	57.1	56.9	58.3	58.6	59.4	60.0

a. Based on an analysis of these data, would you recommend applying first differencing, second differencing, or neither before using a moving average forecasting model to forecast future sales? Explain.

b. Use the data for January through August as assumed past data, then generate one-step-ahead forecasts for September through December, using a five-month moving average model but no differencing.

c. Repeat the instructions of part b, but first apply a first-order differencing transformation to the data.

d. Repeat the instructions of part b, but first apply a second-order differencing transformation to the data.

e. Use MSE to compare the forecast accuracy using the original data, the first-differenced data, and the second-differenced data.

14.7

An Exponential-Smoothing
Forecasting Model

In Section 13.3 we examined the use of smoothing techniques for "averaging out" the effects of random variation. We mentioned that the smoothed series is a series of averages, each computed from available past data. In the case of the centered moving average series, only the data from the past M periods are used for each value; with the exponentially smoothed series all available past data are used for the computation of each smoothed value.

Just as with moving averages, exponential-smoothing methods can also be used for forecasting. We will present two time series forecasting models based on exponential smoothing.

multiple exponential- smoothing forecasting model

The first model we introduce is the **multiple exponential-smoothing forecasting model** developed by R. G. Brown. Suppose that we have the observations $y_1, y_2, \ldots, y_t$ from a time series. Our objective is to compute a forecast of the process value y_{t+T}, which is T time points ahead of our available data. (This period is sometimes referred to as a *forecast lead time of T*.)

Brown's method gives a convenient way of expressing the forecast y_{t+T} in terms of the exponentially smoothed statistics discussed in Section 13.3. If the time series appears *constant over time*, such as perhaps the average annual rainfall in Portland, Oregon, over the past 25 years, then we use the simple exponentially smoothed value S_t to forecast y_{t+T}.

First-Order Exponential-Smoothing Forecasting Model

$$\hat{y}_{t+T} = S_t = \alpha y_t + (1 - \alpha)S_{t-1}$$

If the time series is not flat but, instead, has a *linear trend*, then first-order exponential smoothing will not work well. For time series with linear trends the exponential-smoothing technique can be altered to model the trend component. This modification is called *second-order exponential smoothing*.

Second-Order Exponential-Smoothing Forecasting Model

$$\hat{y}_{t+T} = \left(2 + \frac{\alpha T}{1 - \alpha}\right)S_t - \left(1 + \frac{\alpha T}{1 - \alpha}\right)S_t(2)$$

$$\text{where} \quad S_t(2) = \alpha S_t + (1 - \alpha)S_{t-1}(2)$$

double-smoothed statistic

The statistic $S_t(2)$ is called the **double-smoothed statistic** and is a smoothing of the smoothed values. That is, the series of $S_t(2)$ values is a smoothing of the series of S_t

values, where the observations, the y_t values, in the S_t series are the counterparts of the S_t values in the $S_t(2)$ series. The statistic $S_t(2)$ gives an indication of the trend of the averages S_t over time. Hence, its inclusion in the model will account for a linear trend of y_t with time.

If the time series is neither constant nor linear over time, it is best to use a triple exponential-smoothing forecasting model. Higher-order exponential-smoothing forecasting models exist, but the difficulties in computing the forecasting equation for models of an order higher than the triple exponential-smoothing models are considerable. (See the text by R. G. Brown listed in the references.) Unless a time series is extremely volatile, triple exponential-smoothing forecasting models work quite well.

The suggested forecasting equation when the **time series is neither constant nor linear with time** is given in the display.

Third-Order Exponential-Smoothing Forecasting Model

$$
\begin{aligned}
\hat{y}_{t+T} = {} & [6(1 - \alpha)^2 + (6 - 5\alpha)\alpha T + \alpha^2 T^2] \frac{S_t}{2(1 - \alpha)^2} \\
& - [6(1 - \alpha)^2 + 2(5 - 4\alpha)\alpha T + 2\alpha^2 T^2] \frac{S_t(2)}{2(1 - \alpha)^2} \\
& + [2(1 - \alpha)^2 + (4 - 3\alpha)\alpha T + \alpha^2 T^2] \frac{S_t(3)}{2(1 - \alpha)^2}
\end{aligned}
$$

triple-smoothed statistic   The **triple-smoothed statistic** $S_t(3)$ is, in a sense, describing the average rate of change of the average rates of change. It is found by computing

$$
S_t(3) = \alpha S_t(2) + (1 - \alpha)S_{t-1}(3)
$$

The forecasting equations listed earlier for the constant model, the linear model, and the second-order model were presented without an explanation of their derivation. The derivations, in each case, are very involved and are eliminated from our discussion. Those interested in the derivation of these forecasting equations are referred to Chapter 9 in the text by Brown (1963).

The purpose of using the smoothed statistics S_t, $S_t(2)$, and $S_t(3)$ in Brown's method is to develop estimates for the coefficients of a model that adequately describes the relationship of the value y_t with time. This procedure is performed recursively by continually updating the coefficients in the forecasting model as more data become available. That is, a different forecasting equation is acquired at every point in time, each based on all the available past and present response values. The recursions are initiated by assuming some value for $S_t(2)$ and $S_t(3)$ at the first time period $t = 1$. A wise choice for these initial values is the first observation y_1. The values selected in the beginning are not critical, though, since their contribution to the forecasting equation will decrease as more data are incorporated into the model.

The selection of the smoothing constant α is arbitrary, as explained in the discussion of smoothing methods in Section 13.3. The selection rule suggested for data smoothing is also recommended for the use of smoothing methods for forecasting. That is, if the process is volatile, a small value of α is selected; if the process is stable, a large α will probably give the most accurate forecasts. Some authors have suggested that the smoothing constant α change with time to reflect any new trends in the process. The benefits of such a plan, though, would probably be outweighed by the additional computational difficulties.

You might ask why Brown's method should be used when a least squares model can be fitted to the data to represent the same polynomial relationship between y_t and time that is being assumed for Brown's method. The advantage of Brown's method is that it is a recursive method that develops a new forecasting model each time an additional observation is observed. We could also compute a new least squares model with each new observation, but the computational difficulties then are substantial. Least squares models are usually of the type described in Sections 14.2 and 14.5, where one forecasting model is constructed from which forecasts are made, without updating the model each time new observational information is obtained. Thus, the multiple-smoothing forecasting method of Brown is more efficient because it incorporates more information into the forecasting model.

Care should be taken when we are attempting to forecast further than one time point ahead ($T > 1$). Unforeseeable events may cause the time series to react differently than it has reacted before. Thus, regardless of how much past data are incorporated into the model, we may not be able to forecast future values accurately. If you are interested in the details of Brown's method of forecasting, we again refer you to Brown (1963).

EXAMPLE 14.3   In Figure 13.10 of Chapter 13, the week's ending closing prices for the Color-Vision Company were plotted against time. Use Brown's method of multiple smoothing to forecast the week's ending security prices of the Color-Vision Company. Assume that the relationship of the security prices to time is neither constant nor linear. Let $T = 1$, meaning we will develop one-week-ahead forecasts.

SOLUTION   Since we are assuming that our process is neither constant nor linear with time, we will employ the triple-smoothing method to forecast the prices one time period ahead of available data. Since the process is rather volatile, as illustrated in Figure 13.9, we select $\alpha = .2$. Since $T = 1$, our forecasts are computed from the equation

$$\hat{y}_{t+1} = [6(1 - .2)^2 + \{6 - 5(.2)\}(.2) + (.2)^2] \frac{S_t}{2(1 - .2)^2}$$

$$- [6(1 - .2)^2 + 2\{5 - 4(.2)\}(.2) + 2(.2)^2] \frac{S_t(2)}{2(1 - .2)^2}$$

$$+ [2(1 - .2)^2 + \{4 - 3(.2)\}(.2) + (.2)^2] \frac{S_t(3)}{2(1 - .2)^2}$$

$$= 3.8125S_t - 4.375S_t(2) + 1.5625S_t(3)$$

At $t = 1$ we have $S_1 = 71$, $S_1(2) = 71$, and $S_1(3) = 71$, since $y_1 = 71$. At each succeeding time period $t = 2, 3, \ldots, 30$, the smoothed statistics are found by computing

$$S_t = (.2)y_t + (1 - .2)S_{t-1}$$
$$S_t(2) = (.2)S_t + (1 - .2)S_{t-1}(2)$$
$$S_t(3) = (.2)S_t(2) + (1 - .2)S_{t-1}(3)$$

The smoothed statistics are then reentered into the forecasting equation to obtain the forecast for the next time period.

The forecasts obtained by Brown's method and the true values are listed in Table 14.12 and illustrated in Figure 14.6. They have been obtained with the aid of a computer program. There is a noticeable one-period lag effect to the multiple

Table 14.12   Actual and Forecasted Week's End Security Prices for the Color-Vision Company; Brown's Triple-Smoothing Forecasting Method Used to Compute the Forecasts

Time, t	Actual Price, y_t	Forecast, $\hat{y}_t$	Forecast Error, $e_t = y_t - \hat{y}_t$
1	71		
2	70	71.00	−1.00
3	69	70.40	−1.40
4	68	69.44	−1.44
5	64	68.28	−4.28
6	65	65.22	−.22
7	72	64.06	7.94
8	78	67.70	10.30
9	75	73.64	1.36
10	75	75.45	−.45
11	75	76.42	−1.42
12	70	76.84	−6.84
13	75	73.93	1.07
14	75	75.01	−.01
15	74	75.60	−1.60
16	78	75.26	2.74
17	86	77.36	8.64
18	82	88.35	−6.35
19	75	84.42	−9.42
20	73	80.59	−7.59
21	72	76.82	−4.82
22	73	73.83	−.83
23	72	72.61	−.61
24	77	71.35	5.65
25	83	73.68	9.32
26	81	78.81	2.19
27	81	80.73	.27
28	85	81.79	3.21
29	85	84.70	.30
30	84	86.29	−2.29

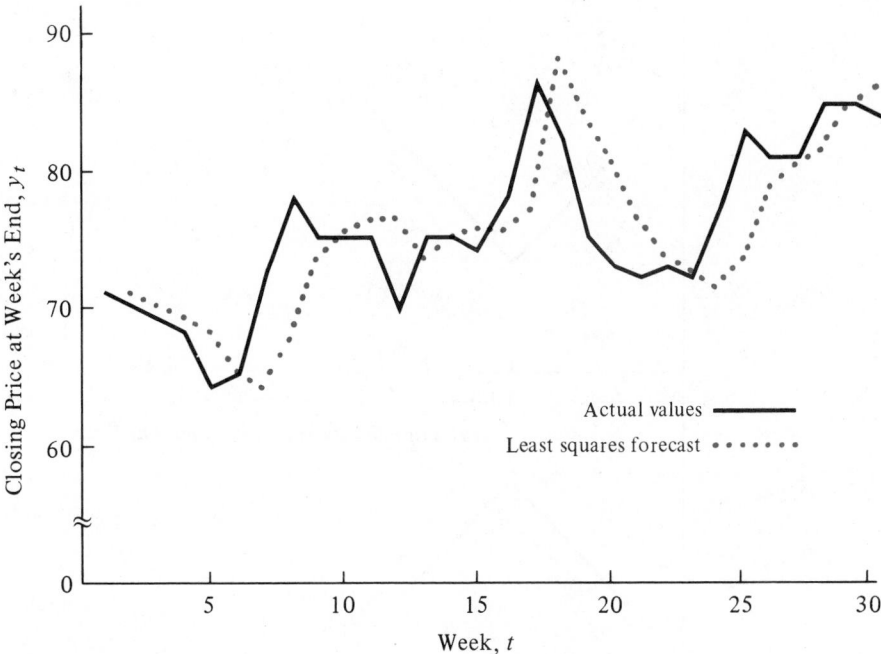

Figure 14.6   Exponential-Smoothing Forecasts for the Week's End Security Prices of the Color-Vision Company

exponential-smoothing forecasts. The forecasts are, however, responsive to changes in the direction of the security price time series. In fact, in only 5 of the 30 time periods does the forecast error percentage exceed 10% of the actual price y_t.

◆ ◆

Although Brown's method can be employed with the aid of an electronic calculator, it is certainly more convenient to use a computer program specifically prepared for the first-, second-, and third-order exponential-smoothing models. One such program is provided in the Study Guide for this textbook. You may find this program helpful in solving certain exercises in this chapter that call for the application of Brown's method.

turning points   The multiple exponential-smoothing model is often used as a tracking model to detect **turning points** in a time series. Identification of turning points is very useful when studying such time series as price movements, consumer buying habits, economic indicators like the floating value of currencies, or security price movements over time. The multiple exponential-smoothing model is generally interpreted as having noted a time series that has bottomed out when the true values *cut over* the smoothing forecasts, as illustrated in Figure 14.7a. A process that has peaked out is noted when the true values *cut under* the smoothing model forecasts. (See Figure 14.7b.)

In Figure 14.6 the multiple exponential-smoothing model indicates that the time series of Color-Vision prices has bottomed out at periods 6, 13, 16, and 24. When prices

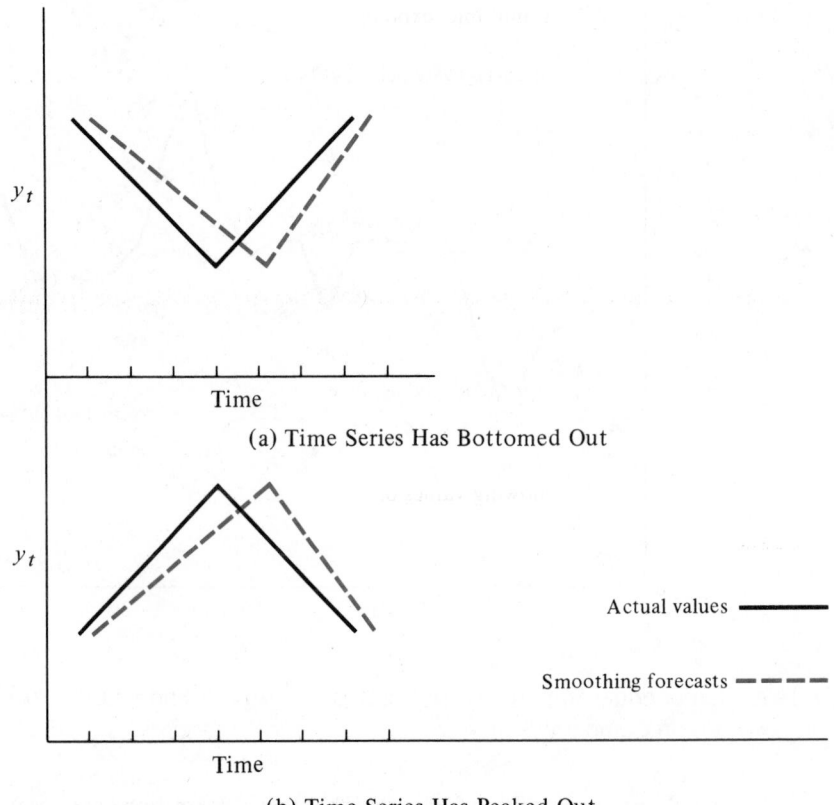

(a) Time Series Has Bottomed Out

(b) Time Series Has Peaked Out

Figure 14.7   Detection of Turning Points in a Time Series by the Multiple Exponential-Smoothing Model

have reached a high point and should be on the way down, we see that the series has peaked out at periods 10, 14, 18, and 29. If an investor with a typical buy-low-and-sell-high objective had purchased Color-Vision securities each time a low point was indicated by the model and had sold the securities each time the prices peaked out, he would have made considerable profit over the period indicated. In fact, his average gain per share would have been $25 over the 24-week period $t = 5$ to $t = 30$.

EXERCISES   Conceptual Level

14.27   What rule of thumb should be applied to selecting the level of α for exponential-smoothing forecasting models?

14.28   How should one decide if it is appropriate to use a first-order, second-order, or third-order exponential-smoothing model for forecasting?

14.29   What is a "turning point"? How can exponential-smoothing models be used to identify turning points?

14.30 Explain how the multiple exponential-smoothing forecasting model can be adjusted to accommodate sudden shifts in a time series. For example, consider the case of a business owner in the late-1979-to-early-1980 period who may have been using the multiple exponential-smoothing model to forecast interest rates. As we can see from Figure 14.1 (Exercise 14.3), interest rates skyrocketed during this period and then fell off just as rapidly. How could the business owner have adjusted the multiple exponential-smoothing model to accommodate the sudden shifts in interest rates during this period?

EXERCISES Skill Level

14.31 Evaluate the first-order exponential-smoothing forecasting model for forecasting T time points ahead of the data for the following values of α.

a. $\alpha = .1$ **b.** $\alpha = .5$ **c.** $\alpha = .8$

14.32 Evaluate the second-order exponential-smoothing model for forecasting $T = 2$ points ahead of the data for the following values of α.

a. $\alpha = .2$ **b.** $\alpha = .4$ **c.** $\alpha = .7$

14.33 Evaluate the third-order exponential-smoothing model for forecasting $T = 1$ point ahead of the data for the following values of α.

a. $\alpha = .1$ **b.** $\alpha = .3$ **c.** $\alpha = .6$

EXERCISES Applications

14.34 An opinion by the Accounting Principles Board allows companies with considerable investment in securities to overcome the problem of earnings and income fluctuation by smoothing income and spreading market fluctuation over a wider period of time. With the use of smoothing methods, realized and unrealized gains and losses are combined so that each period's income includes a pro rata portion of the gains or losses arising in prior periods. The data in Table 14.13

Table 14.13   Sales Volume Data for Exercise 14.34

	Year			
Month	1985	1986	1987	1988
January	18.0	23.3	24.7	28.3
February	18.5	22.6	24.4	27.5
March	19.2	23.1	26.0	28.8
April	19.0	20.9	23.2	22.7
May	17.8	20.2	22.8	19.6
June	19.5	22.5	24.3	20.3
July	20.0	24.1	27.4	20.7
August	20.7	25.0	28.6	21.4
September	19.1	25.2	28.8	22.6
October	19.6	23.8	25.1	28.3
November	20.8	25.7	29.3	27.5
December	21.0	26.3	31.4	28.1

represent the gross monthly sales volume (in thousands of dollars) of a pharmaceutical company from January 1985 to December 1988.

a. Use the data for 1985 and 1986 as assumed past data, then use the triple exponential-smoothing model with $\alpha = .1$ to generate one-month-ahead forecasts for each of the months of 1987 and 1988. Compute MSE as a measure of forecast accuracy.

b. Repeat part a, but generate four-months-ahead forecasts for each of the months of 1987 and 1988. Compute MSE as a measure of forecast accuracy.

c. Compare the accuracy when using the triple exponential-smoothing model to generate one- and four-months-ahead forecasts for the data.

 14.35 Refer to Exercise 13.19. Use a multiple exponential-smoothing forecasting computer program (see the Study Guide) to forecast the yields on municipal bonds one month ahead of available data. Use $\alpha = .2$.

a. From the values for MSE, which model appears to provide the greatest forecast accuracy?

b. On a piece of graph paper, plot the single-, double-, and triple-smoothed forecasts with the actual data. Do the forecast models appear sensitive to the changes in the underlying time series?

14.36 As was mentioned in the text, the multiple exponential-smoothing forecasting model is sometimes used to detect turning points in a time series. In production management the turning point means that when a peak in demand is observed, we should curtail production and reduce inventories; when a bottom is observed, inventories should be replenished. Table 14.14 shows the monthly demand for an air purification unit produced by a small manufacturing firm over a three-year period. Use the triple exponential-smoothing forecasting model, with $\alpha = .2$ and $T = 1$, to find the months during which production should have either been increased or curtailed. You may wish to use a multiple exponential-smoothing computer program for this exercise.

Table 14.14   Demand Data for Exercise 14.36

Month	Year 1	Year 2	Year 3
January	57	63	58
February	59	62	64
March	62	66	66
April	60	69	63
May	63	64	65
June	66	62	68
July	63	70	66
August	67	71	72
September	70	59	69
October	68	59	62
November	64	61	65
December	66	61	65

14.8

The Exponentially Weighted Moving Average Forecasting Model

One time series forecasting method has proved to be especially effective for generating forecasts of a process with a pronounced *seasonal component*. This method, called the **exponentially weighted moving average (EWMA) forecasting model**, operates by separately estimating, at each point in time, the smoothed average, the average trend gain, and the seasonal factor, and then combining these three components to compute a forecast.

exponentially
weighted moving
average (EWMA)
forecasting model

As we noted in Section 13.2, the seasonal effect is independent of a long-term trend and cyclic effects. Furthermore, since the seasonal effect is recurrent and periodic, it is predictable. Seasonality is a very common effect in business forecasting problems, especially as a component of the sales pattern of a firm. In most retailing enterprises sales tend to be high around December; breweries and soft drink companies tend to have a sales pattern that follows the temperature—high sales when the weather is warm and low sales during the cool weather; ski manufacturers have a sales pattern that peaks every winter. We could create an endless list of examples of time series in business that exhibit a seasonal pattern. Thus, the exponentially weighted moving average model is an extremely valuable aid to the business forecaster.

At each point in time the separate components of the time series that are estimated by the EWMA model are as follows:

smoothed average

1. The **smoothed average** at time t:

$$S_t = (\alpha)\frac{y_t}{F_{t-L}} + (1 - \alpha)(S_{t-1} + R_{t-1})$$

trend gain

2. The updated **trend gain** at time t:

$$R_t = (\beta)(S_t - S_{t-1}) + (1 - \beta)R_{t-1}$$

seasonal factor

3. The updated **seasonal factor** at time t:

$$F_t = (\gamma)\frac{y_t}{S_t} + (1 - \gamma)F_{t-L}$$

smoothing constants

In these equations α, β, and γ (Greek letter gamma) are **smoothing constants**, arbitrarily selected by the statistician and satisfying the properties

$$0 \leq \alpha \leq 1 \qquad 0 \leq \beta \leq 1 \qquad 0 \leq \gamma \leq 1$$

The index L in the seasonal factor F_{t-L} is the *seasonal period*—the number of time points required for a repeat of the seasonal effect present in the time series—y_t is the true value at time t, and the statistics S_{t-1} and R_{t-1} are the smoothed average and trend gain, respectively, which are estimated at time period $(t - 1)$. These component equations are combined to give the forecast of the response value T time periods ahead of the most recent data (time period t).

EWMA Forecasting Equation

$$\hat{y}_{t+T} = [S_t + (T)R_t]F_{t-L+T}$$

In each of the component equations the model computes an exponentially weighted moving average—from which the name of the model arises. Each component equation estimates a factor that will be used in the forecasting equation by weighting an estimate of that factor based on the most recent response value with the previous value of the factor. We can see how this procedure is performed by examining Table 14.15. Educational psychologists refer to such updating schemes as those illustrated by the three component equations of our model as **learning curves**, since we are learning more about each component by incorporating new information into the estimation equations while still utilizing the available past information.

initial values **Initial values** must be determined for the statistics S and R so that we can compute S_t and R_t at time $t = 1$. Furthermore, initial values are needed for the seasonal factor F_t at **each time point** (each month, each quarter, etc.) **of the entire seasonal period**. The initial values are usually improved by using two or more complete seasonal periods of the data as a "warm-up" period before any forecasts are computed. For instance, if we are interested in analyzing and then forecasting the monthly sales of a retail department store, initial values for S and R are selected and 12 seasonal factors (one for each month) are selected. These values are used to initiate the recursion for each equation. After two or more complete seasonal periods the erroneous effects of the initial estimates for S, R, and the 12 values of F should be sufficiently dampened out. The rate at which the initial values are dampened out depends on the selection of smoothing constants α, β, and γ. **As a general rule, α and β are small, usually about .1, and the seasonal factor smoothing constant γ is best set at about .4.** The statistician should not arbitrarily select those values, though, but should try many combinations of values for α, β, and γ until a combination is found that generates sufficiently accurate forecasts. Note, however, that the number of combinations of possible values for α, β, and γ is exceedingly large.

Although the smoothing constants are usually not changed as the EWMA model is continuously applied over time, they can be adjusted to accommodate sudden shifts in the time series. For instance, if a temporary shift in the level of a time series is expected because of a sales promotion campaign, a labor strike, the weather, or some other atypical situation, the constants α and β can be temporarily increased to some very large value—say .8—and then returned to their original values when the atypical

Table 14.15 Using the EWMA Forecasting Equation

Component (Factor)	Estimate Based on Most Recent Response Value	Estimate Based on Remote Value
Smoothed average	y_t/F_{t-L}	$S_{t-1} + R_{t-1}$
Trend gain	$S_t - S_{t-1}$	R_{t-1}
Seasonal factor	y_t/S_t	F_{t-L}

situation has subsided. If some effect has dampened the seasonal effect or increased its amplitude, the smoothing constant γ should likewise be temporarily increased. In short, the forecaster should use ingenuity and experience to continuously adapt the model to its environment to guarantee that the resulting forecasts are as accurate as possible.

EXAMPLE 14.4   The monthly sales volumes for a regional brewing company are listed in Table 14.16 for the period from January 1, 1984, to January 1, 1989. Use the sales data from 1984 and 1985 to generate estimates for the smoothed average, the trend factor, and the monthly seasonal factors. Beginning on January 1, 1986, use the exponentially weighted moving average method to forecast the sales of the brewery one month ahead of available data.

Table 14.16   Monthly Sales (Thousands of Barrels) Data for Example 14.4

| | Sales Volume | | | | |
Month	1984	1985	1986	1987	1988
January	18.7	18.3	19.6	23.3	24.6
February	15.6	17.6	18.6	20.1	22.8
March	18.3	24.1	23.2	28.1	28.4
April	19.6	21.8	24.5	26.6	27.2
May	21.4	23.3	27.7	28.6	28.6
June	28.9	28.7	30.0	33.3	29.3
July	24.5	30.0	28.7	34.3	38.3
August	24.5	29.1	33.8	29.0	32.0
September	21.9	23.5	25.1	26.4	24.9
October	20.1	21.6	22.1	25.1	27.7
November	17.9	21.6	21.8	22.3	22.2
December	17.9	19.8	20.9	20.3	21.5

SOLUTION   We will use the previously suggested smoothing constants, $\alpha = .1$, $\beta = .1$, and $\gamma = .4$, in the three recursive component equations. Initial estimates for the factors S and R and the 12 seasonal factors may be obtained as follows:

a.   Let S_0, the initial value for the smoothed average, be represented by the first actual value y_1. Thus, $S_0 = 18.7$.

b.   On a sheet of graph paper, plot the available data for the time series. (In our problem we are assuming that the monthly data from 1984 and 1985 are available as sample data.) With the aid of a ruler, draw a line through a plot of the available time series data, which, with eyeball accuracy, appears to depict the linear trend of the data. Such a line is illustrated for the brewing company sales data in Figure 14.8.*

---

* As an alternative to this approximation procedure, you may choose to find the simple linear regression model that relates sales volume to time over the 24-month period from January 1, 1984, to January 1, 1986.

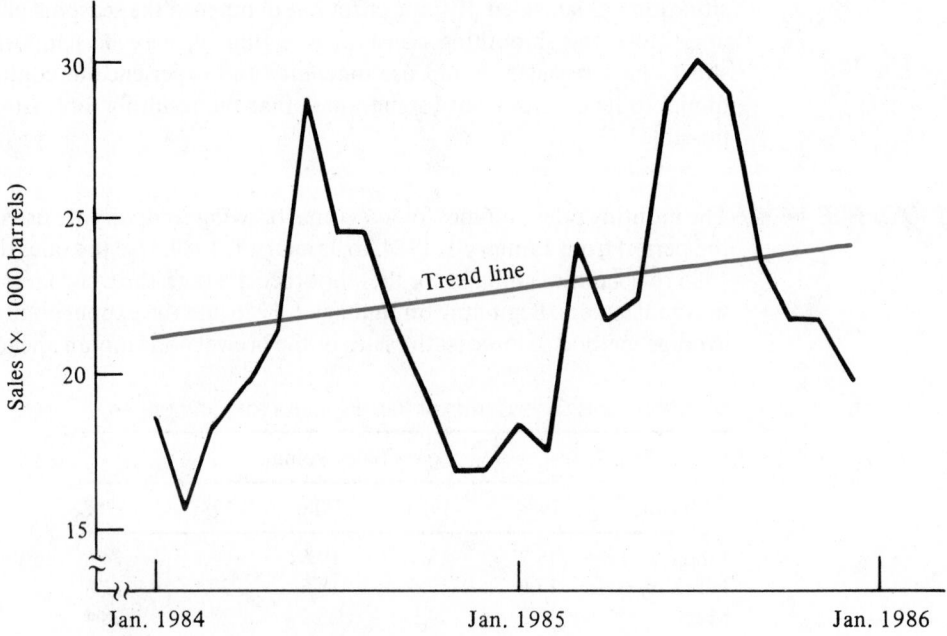

Figure 14.8   Sales Volume of a Regional Brewing Company for the Years 1984 and 1985, Illustrating an Arbitrarily Drawn Trend Line

c. Let R_0, the initial value for the trend gain, be represented by the slope of the arbitrarily drawn trend line, that is, the average gain in trend (sales) per unit of time. The value of the trend line at $t = 1$ is 21; the value at $t = 24$ is 24. Thus,

$$R_0 = \frac{24 - 21}{24 - 1} = \frac{3}{23} = .13$$

d. The initial monthly seasonal factors are found by evaluating the ratio

$$\frac{\text{actual sales volume for the month}}{\text{value of the trend for the month}}$$

for each month of the first entire seasonal pattern. For the brewing company the first three seasonal indexes are

$$F_{\text{January}} = \frac{18.7}{21.0} = .89 \qquad F_{\text{February}} = \frac{15.6}{21.13} = .74 \qquad F_{\text{March}} = \frac{18.3}{21.26} = .86$$

The other monthly indexes are calculated in a similar fashion.

Using the initial values for the smoothed average, the trend factor, and the 12 seasonal factors, we initiate the recursive component equations. At the first time point

$t = 1$, we compute these statistics:

$$S_1 = \alpha\left(\frac{y_1}{F_{\text{Jan}}}\right) + (1 - \alpha)(S_0 + R_0) = .1\left(\frac{18.7}{.89}\right) + .9(18.7 + .13) = 19.05$$

$$R_1 = \beta(S_1 - S_0) + (1 - \beta)R_0 = .1(19.05 - 18.7) + .9(.13) = .15$$

$$F_1 = \gamma\left(\frac{y_1}{S_1}\right) + (1 - \gamma)F_{\text{Jan}} = .4\left(\frac{18.7}{19.05}\right) + .6(.89) = .93$$

At each succeeding month of the smoothing period (the first 24 months of data), we compute these statistics:

$$S_t = .1\left(\frac{y_t}{F_{t-12}}\right) + .9(S_{t-1} + R_{t-1}) \qquad R_t = .1(S_t - S_{t-1}) + .9(R_{t-1})$$

$$F_t = .4\left(\frac{y_t}{S_t}\right) + .6(F_{t-12})$$

Forecasting is begun at the 24th time period by computing the forecast $\hat{y}_{25} = (S_{24} + R_{24})F_{13}$ of the sales during the 25th time period. Thereafter, forecasts are made one month ahead of available data by the forecasting equation

$$\hat{y}_{t+1} = (S_t + R_t)F_{t-11}$$

for the time periods $t = 25, 26, \ldots, 60$. The one-month-ahead forecasts and the actual monthly sales values for the brewing company for the years 1986, 1987, and 1988 are listed in Table 14.17 and are plotted in Figure 14.9. These forecasts have been obtained with the aid of a computer program.

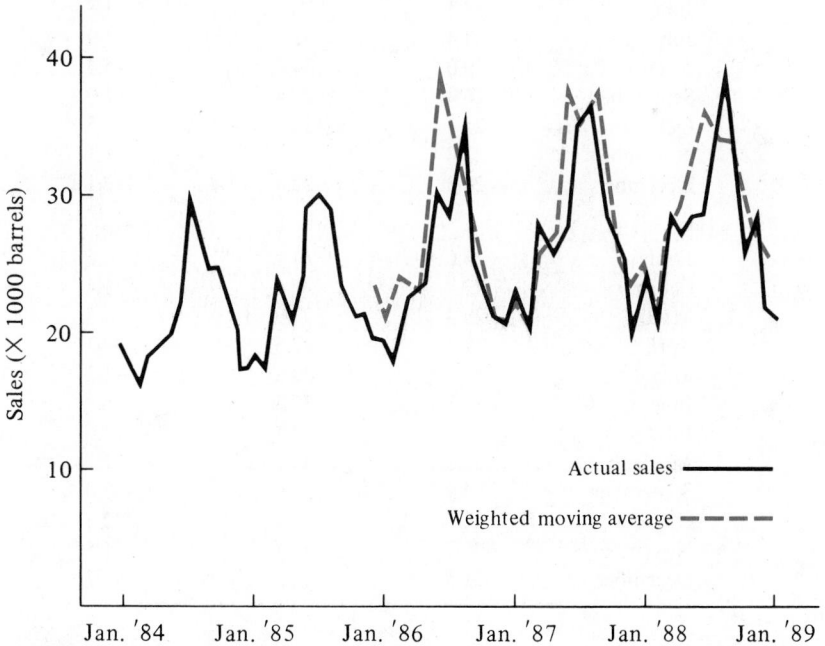

Figure 14.9   Weighted Moving Average Forecasts for the Monthly Sales of the Brewing Company

Table 14.17   Actual and One-Month-Ahead Forecasts of Monthly Sales of the Brewing Company, Using Exponentially Weighted Moving Average (Listed Values in Thousands of Barrels)

Time, t	Actual Sales, y_t	Forecast, $\hat{y}_t$	Forecast Error, $e_t = y_t - \hat{y}_t$
1986			
January	19.6	22.7	-3.1
February	18.6	19.7	-1.1
March	23.2	24.2	-1.0
April	24.5	24.0	.5
May	27.7	26.0	1.7
June	30.0	34.2	-4.2
July	28.7	30.9	-2.2
August	33.8	30.1	3.7
September	25.1	26.2	-1.1
October	22.1	23.9	-1.8
November	21.8	22.0	$-.2$
December	20.9	21.2	$-.3$
1987			
January	23.3	22.6	.7
February	20.1	20.5	$-.4$
March	28.1	25.4	2.7
April	26.6	26.2	.4
May	28.6	28.7	$-.1$
June	33.3	35.0	-1.7
July	34.4	32.5	1.9
August	29.0	34.4	-5.4
September	26.4	27.4	-1.0
October	25.1	24.6	.5
November	22.3	23.4	-1.1
December	20.3	22.4	-2.1
1988			
January	24.6	24.1	.5
February	22.8	21.4	1.4
March	28.4	27.9	.5
April	27.2	27.5	$-.3$
May	28.6	29.8	-1.2
June	29.3	35.5	-6.2
July	38.3	33.7	4.6
August	32.0	33.1	-1.1
September	24.9	27.9	-3.0
October	27.7	25.3	2.4
November	22.2	23.7	-1.5
December	21.5	22.2	$-.7$

From Figure 14.9 and the forecasting equation, we can note the dependence of the EWMA forecasting model on the seasonal factors. Notice, for example, the erratic up-and-down behavior of the true sales pattern during the summer of 1986. This same behavior is exhibited by the sales forecasts for the summer months of 1987, because the seasonal indexes used in the forecast equations for the summer months of 1987 were those computed in 1986 based on actual sales through that year.

The brewing company example emphasizes an earlier remark. The EWMA forecasting model is an appropriate forecasting model only if the seasonal pattern is pronounced, regular, and predictable. Otherwise, because of the inclusion of the seasonal factor as a multiplicative term in the forecasting equation, an erratic behavior in the seasonal pattern will cause a similarly erratic and possibly erroneous response in the pattern of the forecasts during the following season.

Finally, we should note that the EWMA forecasting model, like the multiple exponential-smoothing forecast models, is greatly facilitated with the aid of an appropriate computer program. A general EWMA program, written in Fortran IV language, is included in the Study Guide for this textbook. You may find it useful in solving certain exercises in this chapter or in other forecasting applications.

EXERCISES   Conceptual Level

14.37   Discuss the advantages of the EWMA forecasting model over the other models presented in this chapter.

14.38   Discuss the selection of initial values for the three recursive equations involving S, R, and the L values of F. Why is a warm-up period needed for the values of S, R, and F before any forecasts are generated?

14.39   What rules of thumb should be applied to the selection of values for α, β, and γ?

EXERCISES   Applications

14.40   Refer to Exercise 13.8 in which the average price of steers in Kansas City was recorded for the period January 1981 through February 1986. Use the first three years as available historical data to develop the initiating values. Then, forecast steer prices starting with January 1984, one month ahead of available data. Use $\alpha = .1$, $\beta = .1$, and $\gamma = .4$.

14.41   Refer to Exercise 13.23. Using an EWMA computer program, consider the hotel occupancy data from the first three years as historical data to develop initiating values for the model. Then beginning with January of year 4, forecast the monthly occupancy rate one month ahead of available data for year 4 and year 5. Use $\alpha = .1$, $\beta = .1$, and $\gamma = .4$.

14.42   The data in Table 14.18 represent the number of long-distance calls (in thousands of calls) placed by the subscribers of a small regional telephone service. Consider the data from the first four years as available historical data for the purpose of developing a forecast model. Then, use an EWMA computer program appropriately modified for quarterly data to develop initiating values for the model and to forecast the quarterly number of long-distance calls for years 5 and 6, one quarter ahead of available data and four quarters ahead of available data. (Two separate forecasts are required.) Use $\alpha = .1$, $\beta = .1$, and $\gamma = .4$.

Table 14.18   Long-Distance Calls Data for Exercise 14.42

Quarter	Year 1	Year 2	Year 3	Year 4	Year 5	Year 6
1	7.640	9.321	12.004	10.712	14.162	12.905
2	5.296	6.173	9.507	6.817	9.995	8.702
3	6.380	6.911	7.894	7.563	8.429	8.018
4	12.604	13.057	14.663	14.129	17.534	16.187

14.9

The Box-Jenkins Forecasting Procedure

The Box-Jenkins forecasting methodology consists of two main parts: a large class of time series models (called ARIMA models) and a set of specific procedures for choosing one of these models to use for forecasting. The class of **autoregressive integrated moving average (ARIMA) models** contains all time series of the form

autoregressive integrated moving average (ARIMA) models

$$y_t = \phi_1 y_{t-1} + \phi_2 y_{t-2} + \cdots + \phi_p y_{t-p} + \epsilon_t - \theta_1 \epsilon_{t-1} - \theta_2 \epsilon_{t-2} - \cdots - \theta_q \epsilon_{t-q}$$

$$\underbrace{\qquad\qquad}_{\substack{\text{autoregressive} \\ \text{part of the model}}} \qquad\qquad \underbrace{\qquad\qquad}_{\substack{\text{moving average} \\ \text{part of the model}}}$$

where $\epsilon_t, \epsilon_{t-1}, \epsilon_{t-2}, \ldots, \epsilon_{t-q}$ are the current and past q error terms of the series. The constants $\phi_0, \phi_1, \phi_2, \ldots, \phi_p$ are called the autoregressive parameters of the model, and $\theta_1, \theta_2, \ldots, \theta_q$ are the moving average parameters. Note that, here, the phrase "moving average" is not used in the same manner as in the rest of this chapter. In Box-Jenkins analysis "moving average" refers to the weighted sum of the last q errors terms in an ARIMA model.

In order to choose the best ARIMA model for a given set of data, we must first remove any trend in the data by performing an appropriate amount of differencing. The order of the differencing required is denoted by d. Thus, if only first-order differencing is needed, then $d = 1$; if second differences are required, then $d = 2$; and so forth. After that, the appropriate number p of autoregressive terms and/or the number q of moving average terms must be selected. The resulting model is then referred to as an ARIMA (p, d, q) model. ARIMA models can be purely autoregressive $(q = 0)$, purely moving average $(p = 0)$, or a mixture of the two $(q \neq 0$ and $p \neq 0)$.

One of the primary tools used in determining the parameters p and q in an ARIMA model is the **autocorrelation function (acf)** of the time series. The acf summarizes the various correlations that a time series has with *itself*. The components of the acf are calculated by the formula

autocorrelation function (acf)

$$r_k = \frac{\sum_{t=k+1}^{n} (y_t - \bar{y})(y_{t-k} - \bar{y})}{\sum_{t=1}^{n} (y_t - \bar{y})^2} \qquad \text{for} \qquad k = 1, 2, 3, \ldots$$

Table 14.19   MINITAB Output of the Autocorrelation Function for the 24 Observations in Example 14.1

```
ACF OF C1

           -1.0 -0.8 -0.6 -0.4 -0.2  0.0  0.2  0.4  0.6  0.8  1.0
            +----+----+----+----+----+----+----+----+----+----+
    1   0.614                              XXXXXXXXXXXXXXXX
    2   0.344                              XXXXXXXXXX
    3   0.029                              XX
    4  -0.124                          XXXX
    5  -0.211                        XXXXXX
    6  -0.228                       XXXXXXX
    7  -0.211                        XXXXXX
    8  -0.187                        XXXXXX
    9  -0.093                          XXX
   10   0.014                              X
   11   0.155                              XXXX
   12   0.318                              XXXXXXXXX
   13   0.268                              XXXXXXXX
   14   0.185                              XXXXX
```

where r_k, the autocorrelation coefficient of lag k, measures the correlation between values of the y_t series that are k periods apart. The acf simply lists all such autocorrelations $r_1, r_2, r_3, \ldots$ together.

The acf can be easily calculated by using the ACF command in MINITAB. For example, suppose we examine the two years of sales data given in Example 14.1. After entering these 24 observations into a MINITAB column called C1, we then type ACF C1 to obtain the acf of this data. Table 14.19 shows the MINITAB acf of this data.

To interpret the acf, we need a method for testing which of the r_k's are statistically different from zero. Since the exact probability distributions of the r_k's are rarely known, we must satisfy ourselves in practice with *approximate* tests. One such test is given in the display.

A Test Concerning Autocorrelation Coefficients

Null hypothesis $H_0: \rho_k = 0$.

Alternative hypothesis $H_a: \rho_k \neq 0$, where ρ_k is the *population* autocorrelation coefficient of lag k.

Test statistic: r_k, the sample autocorrelation coefficient of lag k.

Rejection region: At an approximate significance level of $\alpha \approx .05$, reject H_0 if $|r_k| > 2/\sqrt{n}$.

Applying this test to the 24 observations of Example 14.1, only those r_k's whose absolute values exceed $2/\sqrt{24} = .408$ would be considered significantly different from zero at $\alpha \approx .05$. From Table 14.19 only the first-order autocorrelation coefficient, r_1, satisfies this requirement. We conclude that there seems to be a positive correlation between values of the y_t series that are one period (since $k = 1$) apart.

There is much more to the analysis of the acf in the Box-Jenkins technique than we have presented so far. A thorough treatment of this topic would take an entire chapter in itself. Since this treatment is beyond the scope of an introductory text, we must necessarily restrict our discussion to a more general overview of the Box-Jenkins method and some of its associated tools.

The Box-Jenkins procedure is usually depicted as having four basic steps:

1. *Model identification:* Since the Box-Jenkins procedure encompasses a broad range of different models, we must identify the particular (p, d, q) combinations that seem to provide an adequate fit to the underlying time series. This identification is done by matching sample autocorrelations (internal correlations between members of a time series separated by a constant interval of time) computed from the data against theoretical autocorrelation functions for various (p, d, q) autoregressive moving average models.

2. *Estimation of the model parameters:* For the particular (p, d, q) combination identified in stage 1, the method of least squares is used to fit the tentatively entertained model to the underlying time series. Coefficients are thus obtained for the autoregressive or moving average components of the model.

3. *Diagnostic checking:* A diagnostic check is made of the adequacy of fit of the estimated model by analyzing the residuals it generates. If these residuals do not appear to exhibit any distinct pattern but appear random over time, the fitted model is assumed to provide an adequate fit to the underlying time series. If at this stage the fit is found to be less than adequate, we return to the first stage and entertain a new class of models.

4. *Forecasting:* The acceptable forecasting model is used to generate forecasts of future values.

As an example of how a Box-Jenkins analysis is performed, we will again consider the data from Example 14.1. To begin the analysis, we examine the graph of this data in Figure 14.3 for possible trends that would have to be eliminated by a differencing transformation. Since the graph of this series appears fairly flat, no differencing appears to be needed, and we set $d = 0$. Next, patterns in the autocorrelation function and the partial autocorrelation function (a tool we do not discuss in this section) are used to identify appropriate values for p and q. For this data such an analysis shows that $p = 1$ and $q = 0$ might provide a good model. To explore this model, in the next step we use a computer program to estimate the model parameters. With MINITAB this estimation can be done by means of the ARIMA command. Specifically, we would type ARIMA 1 0 0 C1, where the three numbers following the ARIMA command refer to the values $p = 1, d = 0$, and $q = 0$ of our proposed model. Table 14.20 shows part of a MINITAB printout resulting from applying the ARIMA command to the data of Example 14.1.

From the printout the estimated model is $y_t = 6.1369 + .7427y_{t-1} + \epsilon_t$. To generate a forecast of the next observation, y_{25}, in the time series, we substitute $t = 25$

Table 14.20   MINITAB Output for the Estimation of an ARIMA $(1, 0, 0)$ Model to the Data of Example 14.1

```
MTB > ARIMA 1 0 0 C1

ESTIMATES AT EACH ITERATION
ITERATION          SSE        PARAMETERS
    0           230.932      0.100    21.514
    1           194.251      0.250    17.948
    2           167.245      0.400    14.378
    3           149.899      0.550    10.800
    4           142.339      0.693     7.375
    5           141.794      0.726     6.557
    6           141.728      0.737     6.287
    7           141.719      0.741     6.188
    8           141.718      0.742     6.150
    9           141.718      0.743     6.137
RELATIVE CHANGE IN EACH ESTIMATE LESS THAN   0.0010

FINAL ESTIMATES OF PARAMETERS
TYPE          ESTIMATE      ST. DEV.    T-RATIO
AR    1        0.7427        0.1596        4.65
CONSTANT       6.1369        0.5124       11.98
MEAN          23.851         1.991

NO. OF OBS.:   24
RESIDUALS:     SS = 136.595   (BACKFORECASTS EXCLUDED)
               MS =   6.209   DF = 22
```

into the estimated model:

$$y_{25} = 6.1369 + .7427y_{24} + \epsilon_{25}$$

Since we are forecasting as of period $t = 24$, the error term ϵ_{25} cannot be known yet, so it is set equal to its expected value of 0 in order to generate the forecast. Thus, the forecast for period $t = 25$ would be $\hat{y}_{25} = 6.1369 + .7427(28.8) + 0 = 27.53$, which agrees fairly well with the moving average forecast of this data in Example 14.1.

Although the Box-Jenkins procedures have been found to be very efficient for handling time series with complex inherent patterns, recent studies have indicated some shortcomings in this new methodology. Computer programs designed to facilitate the Box-Jenkins procedure are often very expensive to apply. In addition, Geurts and Ibrahim (1975) and Newbold and Granger (1974) (see the references) have noted that the **Box-Jenkins procedures are seldom capable of outperforming the EWMA model when applied to a time series with a distinct seasonal component.** An additional difficulty arises in the model identification stage: Matching of the sample autocorrelations with theoretical autocorrelation functions is done subjectively and thus relies heavily on the experience of the forecaster. Further developments in the theory of the Box-Jenkins procedure may perhaps overcome this latter shortcoming.

14.43   From the information provided in this section, under what conditions should you use a Box-Jenkins forecasting procedure?

14.44   If you were assigned the task of generating forecasts for future values of a process with a distinct seasonal component, such as the sales of skiing equipment, soft drinks, beer, or recreational equipment, would you use a Box-Jenkins forecasting procedure or the EWMA forecasting model? Explain.

14.45   In an ARIMA(p, d, q) model, what do the values of p, d, and q represent?

14.46   The autocorrelation function for the data in Exercise 13.5 found by using the MINITAB command ACF is presented in Table 14.21.

   a.   What is the approximate rejection region for testing $H_0 : \rho_k = 0$ versus $H_a : \rho_k \neq 0$?

   b.   Which autocorrelations are significantly different from zero?

Table 14.21   MINITAB Output for Exercise 14.46

```
MTB > ACF C31
ACF OF C31

          -1.0 -0.8 -0.6 -0.4 -0.2  0.0  0.2  0.4  0.6  0.8  1.0
          +----+----+----+----+----+----+----+----+----+----+
   1    0.665                              XXXXXXXXXXXXXXXXXXX
   2    0.369                              XXXXXXXXXX
   3    0.119                              XXXX
   4    0.097                              XXX
   5    0.143                              XXXXX
   6    0.146                              XXXXX
   7   -0.025                            XX
   8   -0.136                          XXXX
   9   -0.180                        XXXXXX
  10   -0.025                            XX
  11    0.115                              XXXX
  12    0.237                              XXXXXXX
  13    0.071                              XXX
  14   -0.069                          XXX
```

14.47   The MINITAB printout in Table 14.22 gives the fitted Box-Jenkins ARIMA(1, 1, 0) model for the data in Exercise 13.5. The subcommand FORECAST 22 2 asks for the forecasted values of y_{23} and y_{24}, using period 22 as the starting value for the forecasts. Notice that the fit has been accomplished by using first differences, so the model fitted is

$$y_t = \phi_0 + \phi_1 \nabla y_{t-1} + \epsilon_t$$

a. What is the prediction equation in terms of the first differences? Use this equation to find the prediction equation for y_t.

b. What is the predicted value of y_{23} using only the data values up to and including y_{22}? Verify this answer by using the equation found in part a.

c. Show that the predicted value for time period 24 is found by using the predictor

$$\hat{y}_{24} = \hat{\phi}_0 + \hat{\phi}_1(\hat{\nabla} y_{23}) + \hat{y}_{23}$$

with $\hat{\nabla} y_{23} = \hat{y}_{23} - y_{22}$.

Table 14.22   MINITAB Output for Exercise 14.47

```
MTB > ARIMA 1 1 0 C31;
SUBC> CONSTANT;
SUBC> FORECAST 22 2.

ESTIMATES AT EACH ITERATION
ITERATION          SSE      PARAMETERS
     0          2.62663    0.100    0.176
     1          2.42661    0.180    0.092
     2          2.42281    0.195    0.080
     3          2.42278    0.198    0.079
     4          2.42278    0.198    0.079
     5          2.42278    0.198    0.079
RELATIVE CHANGE IN EACH ESTIMATE LESS THAN  0.0010

FINAL ESTIMATES OF PARAMETERS
TYPE        ESTIMATE     ST. DEV.    T-RATIO
AR    1      0.1980       0.2267       0.87
CONSTANT    0.07931      0.07106      1.12

DIFFERENCING: 1 REGULAR DIFFERENCE
NO. OF OBS.:   ORIGINAL SERIES 24, AFTER DIFFERENCING 23
RESIDUALS:     SS = 2.42129  (BACKFORECASTS EXCLUDED)
               MS = 0.11530  DF = 21

MODIFIED BOX-PIERCE CHISQUARE STATISTIC
LAG                  12            24            36            48
CHISQUARE    26.6(DF=11)    * (DF= *)    * (DF= *)    * (DF= *)

FORECASTS FROM PERIOD 22
                         95 PERCENT LIMITS
PERIOD      FORECAST       LOWER       UPPER       ACTUAL
   23       17.1585      16.4928     17.8242     17.4000
   24       17.2692      16.2304     18.3080     18.0000
```

EXERCISES   Applications

 14.48   The prices of steers (in dollars per pound) from January 1981 through February 1986 are given in Exercise 13.8. In the analysis in Table 14.23 the first 60 data points were used to fit an ARIMA(2, 0, 0) model. Use the computer printout to answer the following questions.

Table 14.23   MINITAB Output for Exercise 14.48

```
                MTB > ARIMA 2 0 0 C45;
                SUBC> CONSTANT;
                SUBC> FORECAST 2.

FINAL ESTIMATES OF PARAMETERS
TYPE         ESTIMATE      ST. DEV     T-RATIO
AR    1        1.0647       0.1231        8.65
AR    2       -0.3635       0.1233       -2.95
CONSTANT      18.7677       0.2426       77.36
MEAN          62.8242       0.8121

NO. OF OBS.:   60
RESIDUALS:     SS = 200.969  (BACKFORECASTS EXCLUDED)
               MS =   3.526   DF = 57

MODIFIED BOX-PIERCE CHISQUARE STATISTIC
LAG                  12              24              36              48
CHISQUARE   18.0(DF=10)   35.9(DF=22)   50.2(DF=34)   62.9(DF=46)

FORECASTS FROM PERIOD 60
                              95 PERCENT LIMITS
PERIOD      FORECAST        LOWER         UPPER        ACTUAL
   61       62.9976        59.3165       66.6786
   62       63.2938        57.9169       68.6707
```

a.   Give the fitted model.

b.   What is the value of MSE? What are the degrees of freedom associated with MSE?

c.   Use the t ratios on the printout to determine whether ϕ_0, ϕ_1, and ϕ_2 are significantly different from zero.

d.   Forecast the value of y_t at times $t = 61$ and $t = 62$. Comment on the accuracy of these forecasts when compared with the actual values of $y_{61} = 61.34$ and $y_{62} = 61.68$.

14.10

Qualitative Forecasting Models

When a relevant data base is not available, as in the case of a new product being considered for market introduction, the methods and models we have introduced up to this point cannot be used. The forecaster must then rely upon other evidence, usually in

qualitative
forecasting models

the form of subjective or judgmental information, as the basis for a market forecast. Forecasting models that do not rely on a historical data base are called **qualitative forecasting models.*** We briefly introduce four of the more common qualitative models here and discuss their use in the context of sales forecasting for a new-product introduction.

panel consensus
method

The **panel consensus method** assumes that the organization or firm contains experts who have special knowledge or experience that enables them to effectively evaluate the uncertain effects of the future. It further assumes that these experts will recognize one another's special areas of expertise and, by supplementing each other's knowledge, arrive at a consensus about the appropriate sales forecast for the firm's product.

One obvious difficulty with the panel consensus method is that certain social factors may make a consensus agreement impossible. Perhaps certain members of the panel are simply not willing to compromise. Also, a hierarchical bias may exist within the group, causing lower-level experts to be reluctant in their criticism of superiors, even when they believe their opinions to be of more value than those offered by their superiors.

Delphi method

The **Delphi method** attempts to eliminate the bandwagon effect of majority opinion by employing a sequence of questionnaires that are designed to filter out the factors that can best be of assistance in focusing attention on the resultant forecast. The method employs the services of a panel of experts, either in-house experts or experts hired as consultants from the outside. The responses of the experts to a first questionnaire are used to generate a second questionnaire. These same experts or new experts may be used to respond to the second questionnaire. Their responses are then used to generate questions for a third questionnaire, and so forth, until the experts have sufficient information to sharpen their focus on the expected level of sales.

Expert information is passed on from one expert or set of experts to another by using the Delphi method. Successive experts may modify previous expert opinion, but since their evaluation is made separately from the expression of opinion of earlier experts, snowballing and hierarchical influencing are not likely to occur.

historical analogy

The **historical analogy** is a qualitative model for forecasting the sales of a new product that assumes we can use the sales history of some product previously introduced to gauge the success of our current product. A natural assumption underlying this approach is that the earlier product must have had an economic and market environment during its introductory stage that is similar to the environment for the current product. If this assumption is not valid, we have no justification for the analogy.

market research
methods

A sample survey of opinions from those best able to assess the potential sales of a product—the potential buyers—seems to be the most logical approach for obtaining an accurate new-product sales forecast. Typical **market research methods** identify the population of prospective buyers of the product, select a representative sample of size n from this population, then from the sample find the probability p that any one individual would buy the product if presented with the opportunity. The sales forecast is then Np, where N is the number of prospective buyers in the population.

Estimation of the population proportion p in a market research survey is an

---

* See the article by Geurts and Reinmuth (1980) listed in the references.

application of survey sampling. An appropriate survey design should be selected according to the characteristics of the population being sampled (see Chapter 16).

EXERCISES   Conceptual Level

14.49   Discuss the relative advantages and disadvantages of the panel consensus method and the Delphi method of forecasting.

14.50   When is it appropriate to use historical analogy as a qualitative model for forecasting? Marketing research methods?

EXERCISES   Applications

 14.51   For years airplane manufacturing companies have sought to develop an economical and efficient airplane for transporting large numbers of people between short-distance city pairs, such as San Francisco–Los Angeles, Dallas-Houston, Washington–New York, and London-Paris. Recent developments suggest that helicopters may prove to be economical and efficient for short-distance city pair transportation. Suppose you have been hired by a manufacturer of military and commercial helicopters to forecast the two-year sales of a new 300-passenger, 500-mile-range helicopter that the company has just developed for the commercial market. Explain exactly how you might conduct this forecasting exercise when using each of the following four forecasting methods.

    a. The panel consensus method   **b.** The delphi method

    c. An historical analogy   **d.** A market research survey

 14.52   In some cases an historical data base may exist for a product, but because of a change in the market environment of the product, the data base becomes worthless. For instance, the demand for large motor homes was unalterably changed after the energy crisis of 1973 and the dramatic increase in fuel costs in 1979. Suppose you were acting in an advisory capacity to a company in an attempt to forecast future demand for its motor homes. How would you recommend that the firm approach this problem? Would you suggest it use past sales and modify them to accommodate new trends in buying behavior? Or would you recommend it discard past sales data and use a qualitative forecasting procedure? If the latter, which method(s) would you recommend the firm explore?

14.11

Combining Forecasts

When developing a forecasting model, most analysts examine several different forecasting techniques, finally choosing the one generating the most satisfactory measure of forecast accuracy. The principal objective of this single-model approach to forecasting is to identify the forecasting model among the competing models that best fits the underlying time series and that generates the most accurate projections of future process values.

Recent evidence from several researchers suggests that the forecast accuracy of any single forecast model can almost always be improved by combining forecasts generated by two or more competing forecast models. This combination approach is based on the premise that a discarded forecast model almost always contains some useful information independent of that provided by the chosen forecast model. As the quantity of information supplied by an estimator is inversely related to its variance, the multiple-model approach should return forecasts with smaller error variances than the forecasts of either component model.

The simplest, and probably most efficient, manner of combining forecasts is to let the separate forecast values assume the role of independent variables in a multiple regression model. For example, if we wish to use a combination approach to provide one-month-ahead forecasts for petroleum tax revenues, we could let

y_t = actual petroleum tax revenues for month t

x_{1t} = one-month-ahead EWMA forecast for month t

x_{2t} = one-month-ahead Box-Jenkins forecast for month t

We could then determine a multiple regression model that relates y_t, the actual revenues, to the separate one-month-ahead forecasts x_{1t} and x_{2t}:

$$y_t = \beta_0 + \beta_1 x_{1t} + \beta_2 x_{2t} + \epsilon_t$$

To compute the *combined* forecast for the next period $(t + 1)$, we first compute $x_{1,t+1}$ and $x_{2,t+1}$ and find

$$\hat{y}_{t+1} = \hat{\beta}_0 + \hat{\beta}_1 x_{1,t+1} + \hat{\beta}_2 x_{2,t+2}$$

This approach need not be limited to the combination of two forecasts; any number can be involved in the combinatorial procedure. In fact, the forecast in a regression model is *assured* of providing further reduction in the forecast error variance. As in any regression analysis, additional independent variables (in this case additional competing model forecasts) help explain a portion of the time series not explained by other independent variables. The result is an increase in R^2 as a measure of "fit" of the combined model to the underlying data set and, hence, a reduction in error variance.

The one difficulty in applying this procedure is that the regression equation must be computed anew at each point in time. This problem can be facilitated by using a recursive regression approach [see Reinmuth and Geurts (1979)].

EXAMPLE 14.5   Table 14.24 shows the monthly Hawaii tourist traffic (in thousands of tourists) for 1971. Monthly data for the period 1952 through 1971 were available for this study, with 1971 held back for purposes of testing and comparison. The EWMA and Box-Jenkins (1, 1, 1) forecasts were thus initiated, with tourist traffic beginning in January 1952; the listed EWMA and Box-Jenkins forecasts represent one-month-ahead forecasts. Both this example and the next use MSE as a measure of forecast accuracy.

As shown in Table 14.24, the regression model combining the EWMA and Box-Jenkins forecasts provides a 59.2% improvement in forecast accuracy (reduction in

Table 14.24   Comparison of Forecasts for Hawaii Tourist Traffic Data (×1,000); Example 14.5

Month, t	Actual, y_t	EWMA Forecast, $x_{1,t}$	Box-Jenkins, $x_{2,t}$	Combined (Regression), y_t
January	18.024	15.395	17.332	16.079
February	18.806	17.109	20.069	19.150
March	21.707	19.162	25.976	23.259
April	13.463	13.424	13.512	13.643
May	14.930	15.844	13.961	14.908
June	12.287	13.618	11.046	12.499
July	16.248	15.583	16.664	16.124
August	11.466	13.944	9.287	11.738
September	12.672	15.979	10.291	13.029
October	13.630	16.394	9.769	12.790
November	14.422	14.479	14.262	14.449
December	14.441	13.660	15.043	14.473
MSE	—	29.825	31.018	12.169

MSE) over that provided by the EWMA model alone, and a 60.8% improvement over the Box-Jenkins model. This enhanced precision in forecasting may, in fact, be translated into cost savings resulting from improved planning for the Hawaii tourist industry.

EXAMPLE 14.6   The gross monthly sales from the retail establishments of Salt Lake City (SLC) were recorded for 1973 and 1974 and used to develop an econometric forecasting model and a number of different exponential-smoothing and Box-Jenkins forecasting models. These models were used to generate one-month-ahead forecasts for the SLC retail sales for the first six months of 1975. The five single models generating the greatest measure of forecast accuracy are shown in Table 14.25.

As noted earlier, the regression approach for the combination of individual forecast models lends itself well to the combination of any number of models. Table 14.26 lists the MSE as a measure of forecast accuracy, for five different combinations of

Table 14.25   Comparison of Five Forecast Models: Salt Lake City Retail Sales

Forecast Model	MSE
Second-order exponential smoothing ($\alpha = .1$)	36.837
Third-order exponential smoothing ($\alpha = .1$)	80.040
Box-Jenkins (1, 1, 1)	70.958
Box-Jenkins (1, 1, 0)	69.769
Econometric	26.312

Table 14.26   Comparison of Five Combined Forecast
Models: Salt Lake City Retail Sales

Model	MSE
Second-order exponential smoothing ($\alpha = .1$) Third-order exponential smoothing ($\alpha = .1$) Box-Jenkins (1, 1, 1) Box-Jenkins (1, 1, 0)	4.838
Second-order exponential smoothing ($\alpha = .1$) Box-Jenkins (1, 1, 0)	10.610
Box-Jenkins (1, 1, 1) Box-Jenkins (1, 1, 0) Econometric	7.979
Second-order exponential smoothing ($\alpha = .1$) Third-order exponential smoothing ($\alpha = .1$) Econometric	3.904
Second-order exponential smoothing ($\alpha = .1$) Box-Jenkins (1, 1, 0) Econometric	2.971

the best individual models listed in Table 14.25. Note that the best combination provides an 88% improvement in forecast accuracy over the best individual model (MSE = 26.312 vs. MSE = 2.971). Also note that all models involving the combination of three or more individual forecasts outperform the one involving the combination of only two forecasts. Thus, one might conclude that the inclusion of additional auxiliary information into an existing model cannot decrease its measure of forecast accuracy. But that is not surprising because it is a natural result of an increase in the number of independent variables in a general linear regression model. One must, however, be cautioned against the usual problems of interpretation in a regression analysis that occur when the number of independent variables is increased while the sample size is held constant.

◆◆

14.12

Summary

Forecasting is a necessary task for the businessperson who wishes to budget time and resources. Forecasts provide a plan, however tentative, for the businessperson to follow in order to achieve objectives and still remain competitive.

Forecasting methods are many and varied. Frequently, forecasting has been quite subjective, without relying on rigorous mathematical forecasting models. The use of mathematical forecasting models has been suggested in this chapter for the following reasons:

1. Mathematical models can be designed to track the specific components (long-term trend, seasonal effects, etc.) in the time series under study.
2. Mathematical models can be adapted to conform to the intuitive knowledge of the businessperson about the time series.

The last point is especially important; it implies that the ingenuity of the business statistician is an essential factor in the selection of an appropriate forecasting model.

The ultimate criterion for the value of a forecasting model is how well it forecasts the future. Inherent in this criterion is the assumption that past behavior of the time series mirrors its future behavior. Dormant variables and other factors may cause a time series to react differently in the future than it has reacted before. Thus, a model that fits the sample data well might not be a good model for forecasting the future behavior of the time series.

Statistics provides only a starting point in the analysis of a time series and the adoption of a forecasting procedure. Statistical and mathematical models cannot provide a complete solution to forecasting problems so long as uncertainty exists in the problem. Personal judgment and subjective evaluations concerning the forecast environment must be used to condition the forecast obtained from a mathematical forecasting model.

Supplementary Exercises

Applications

 14.53 For the past four years an ice cream manufacturer has been having problems forecasting sales of gallon units of vanilla ice cream. Table 14.27 lists the quarterly sales figures for 1985 through 1988. Use the EWMA forecasting model to generate one-quarter-ahead forecasts for 1988, assuming the data for 1985–1987 are known in advance. Let $\alpha = .2$, $\beta = .2$, and $\gamma = .4$.

Table 14.27   Sales Data for Exercise 14.53

Quarter	1985	1986	1987	1988
1	45,000	50,000	54,000	59,000
2	5,000	6,000	7,000	7,000
3	9,000	10,000	11,000	13,000
4	37,000	40,000	44,000	46,000

 14.54 Refer to Exercise 13.26. Use a second-order multiple exponential-smoothing model to forecast the quarterly common share earnings one quarter ahead of available data. (Let $\alpha = .2$.) Is this model a good choice for forecasting the quarterly common share earnings for Hilton Hotel stockholders? Explain. Can you suggest a better forecasting model?

 14.55 The following data represent the sales, in hundreds of cases, of a new brand of wine cooler during the first 10 months after introduction of the product to the consumer marketplace.

Month	1	2	3	4	5	6	7	8	9	10
Sales	21	27	30	45	56	61	88	109	126	195

a. Use the first 8 months as assumed past data and find the one-step-ahead forecasts for months 9 and 10, using a three-month moving average forecast model. *Do not* apply a differencing transformation to the data.

b. Repeat part a, but first apply a second-difference transformation to the sales data.

c. Compare the forecasts generated in a and b, using MSE as a measure of forecast accuracy.

 14.56 Internal auditing is used to examine and evaluate a business's activities. One of its chief roles is to identify evidence of fraud and theft (see J. Nocera and J. Lovelace, "Strengthening Fraud Detection," *Internal Auditor*, April 1980). Suspecting shoplifting and theft as the cause of inventory shrinkage within a large urban department store, the store manager has commissioned an internal auditor to assess inventory shrinkage for each month of the past three years. The data in Table 14.28 were obtained. The manager of the department store would like to establish a model to forecast the level of inventory shrinkage each month so that she might undertake proper advance precautions to avoid losses.

a. Assume the data from year 1 and year 2 are available past data, and generate one-step-ahead forecasts for year 3, using a five-month moving average forecast model. Use the original data.

b. Repeat part a, but generate three-steps-ahead forecasts for year 3, using a five-month moving average forecast model.

c. Compare the accuracy of the one-step-ahead forecasts with that of the three-steps-ahead forecasts for year 3 by computing MSE for each set of forecasts.

Table 14.28   Inventory Shrinkage Data for Exercise 14.56

Month	Losses (× $1,000)		
	Year 1	Year 2	Year 3
January	6.4	5.9	6.5
February	3.8	4.1	4.0
March	2.9	3.0	3.6
April	3.1	2.9	3.2
May	3.0	3.3	2.8
June	2.4	2.7	2.9
July	2.5	2.8	3.0
August	2.8	3.1	3.4
September	3.9	4.3	4.3
October	4.0	5.1	5.2
November	5.8	6.2	6.4
December	8.2	8.5	8.3

14.57 The data in Table 14.29 represent the monthly revenues (in thousands of dollars) of a resort restaurant. Assume that the pattern of revenue is neither constant nor linear. Use Brown's multiple exponential-smoothing forecasting method to forecast the monthly revenue for 1988

Table 14.29   Revenue Data for Exercise 14.57

Month	1986	1987	1988
January	12.0	13.3	15.3
February	13.2	14.8	16.4
March	14.5	17.0	18.9
April	15.6	18.1	18.4
May	20.0	21.8	21.2
June	20.5	21.1	23.8
July	21.9	24.0	25.4
August	19.2	21.2	24.5
September	16.1	20.7	23.4
October	16.7	17.4	20.4
November	14.8	16.8	18.0
December	14.2	14.7	16.6

one month ahead of available data. Plot the forecast revenue against the true monthly revenue. Use the smoothing constant $\alpha = .1$.

 14.58   Refer to Exercise 14.57. Use the exponentially weighted moving average forecasting method to forecast the monthly revenue one month ahead of available data for the year 1988. Use the sales data from 1986 and 1987 to recursively generate values for the smoothed statistics S, R, and the 12 values of F. Plot the forecast revenues for 1988 against the true revenues. Use the smoothing constants $\alpha = .1$, $\beta = .1$, and $\gamma = .4$.

 14.59   Repeat Exercise 14.58 by using the exponentially weighted moving average forecasting method to forecast the monthly revenue for 1988 three months ahead of available data. Compare the accuracy of the one-step-ahead forecasts generated in Exercise 14.58 with the accuracy of the three-steps-ahead forecasts of this exercise by computing MSE for each model. What does this result suggest about the relationship between forecast accuracy and the length of the forecast lead time?

 14.60   The accompanying table gives the annual gross revenue for a manufacturer of a water purification unit for the years 1981 through 1988.

Year	1981	1982	1983	1984	1985	1986	1987	1988
Revenue (× $1,000)	21.1	19.4	23.0	26.9	29.1	40.5	64.3	101.2

a.   Apply a second-difference transformation to the data.

b.   Estimate the gross revenue for the company for 1989 by using an MA(3) model.

 14.61   Refer to Exercise 14.60 in which annual gross revenues are given for the years 1981 through 1988. Use the MINITAB computer output in Table 14.30 for an ARIMA(1, 2, 0) model fitted to these data to answer the following questions.

a.   What is the estimated prediction equation?

b.   Comment on the model fit, using the observed t ratios.

c.   Use the fitted model to predict the gross revenues for 1989. Compare this estimate with that found in Exercise 14.60 using an MA(3) model.

 14.62   Refer to Exercise 13.40. Use the exponentially weighted moving average forecasting model to forecast the monthly sales for the department store one month ahead of available data for year 3,

Table 14.30   MINITAB Output for Exercise 14.61

```
MTB > ARIMA 1 2 0 C5;
SUBC> CONSTANT;
SUBC> FORECAST 1.

FINAL ESTIMATES OF PARAMETERS
TYPE          ESTIMATE      ST. DEV     T-RATIO
AR    1        0.6237        0.4898       1.27
CONSTANT       2.792         2.592        1.08

DIFFERENCING: 2 REGULAR DIFFERENCES
NO. OF OBS.:  ORIGINAL SERIES 8, AFTER DIFFERENCING 6
RESIDUALS:    SS = 134.563  (BACKFORECASTS EXCLUDED)
              MS =  33.641  DF = 4

MODIFIED BOX-PIERCE CHISQUARE STATISTIC
LAG                  12               24              36               48
CHISQUARE     * (DF= *)       * (DF= *)       * (DF= *)       * (DF= *)

FORECASTS FROM PERIOD 8
                              95 PERCENT LIMITS
PERIOD        FORECAST        LOWER           UPPER           ACTUAL
   9          149.062         137.691         160.432

MTB >
```

year 4, and year 5. Use the sales data from year 1 and year 2 to recursively generate values for the smoothed statistics S, R, and the 12 values of F. Plot the forecast sales for year 3, year 4, and year 5 against the true sales volumes. Use the smoothing constants $\alpha = .1$, $\beta = .1$, and $\gamma = .4$. (You may wish to compare your results with those derived by Reinmuth and Wittink; see the reference given in Exercise 13.40.)

14.63   Refer to Exercises 14.57 and 14.58, which provide one-step-ahead forecasts of the revenues of a restaurant. Use the 1988 one-step-ahead forecasts to develop a combined forecast model (see Section 14.11), where x_1 is the forecast derived by using Brown's method (Exercise 14.57) and x_2 is the EWMA forecast (Exercise 14.58). Regress the actual 1988 revenues on x_1 and x_2. Compare the accuracy of the forecasts generated by the combined model with the accuracy of the forecasts generated by the models of Exercises 14.57 and 14.58 for the 12 months of 1988 by computing MSE for each model.

14.64   Energy consumption in the United States has more than doubled from 1952 to 1978. The total energy consumption (in quadrillion British thermal units), which includes energy from coal, natural gas, petroleum, hydroelectric power, nuclear electric power and geothermal energy, is given in Table 14.31 for the years 1952–1986. Two different Box-Jenkins models were fitted to these data, both of which use the first-order differences. See Table 14.32.

a.   What exactly are the two fitted models? How do they differ? How are they alike?

b.   Which model appears to provide the best fit to these data?

c.   Comment on the significance or nonsignificance of the estimated parameters in these models.

Table 14.31   Energy Consumption Data for Exercise 14.64

Year	Energy Consumption	Year	Energy Consumption	Year	Energy Consumption
1952	35.30	1964	50.50	1976	74.36
1953	36.27	1965	52.68	1977	76.29
1954	35.27	1966	55.66	1978	78.09
1955	38.82	1967	57.57	1979	78.90
1956	40.38	1968	61.00	1980	75.96
1957	40.48	1969	64.19	1981	73.99
1958	40.35	1970	66.43	1982	70.84
1959	42.14	1971	67.89	1983	70.50
1960	43.80	1972	71.26	1984	74.06
1961	44.46	1973	74.28	1985	73.96
1962	46.53	1974	72.54	1986	73.93
1963	48.32	1975	70.55		

Source: U.S. Department of Energy, Energy Information Administration, *Annual Energy Review 1986* (Washington, D.C., 1988).

Table 14.32   MINITAB Output for the Data of Exercise 14.64

```
MTB > ARIMA 1 1 0 C1;
SUBC> CONSTANT.

FINAL ESTIMATES OF PARAMETERS
TYPE        ESTIMATE      ST. DEV.    T-RATIO
AR   1       0.3442       0.1670       2.06
CONSTANT     0.7320       0.3160       2.32

DIFFERENCING: 1 REGULAR DIFFERENCE
NO. OF OBS.:  ORIGINAL SERIES 35, AFTER DIFFERENCING 34
RESIDUALS:    SS = 108.566  (BACKFORECASTS EXCLUDED)
              MS =    3.393   DF = 32

MODIFIED BOX-PIERCE CHISQUARE STATISTIC
LAG               12              24               36             48
CHISQUARE    7.0(DF=11)    14.0(DF=23)      * (DF= *)      * (DF= *)

MTB > ARIMA 1 1 1 C1;
SUBC> CONSTANT.

FINAL ESTIMATES OF PARAMETERS
TYPE        ESTIMATE      ST.DEV.     T-RATIO
AR   1      -0.3545       0.2564      -1.38
MA   1      -0.8195       0.1609      -5.09
CONSTANT     1.5866       0.5513       2.88

DIFFERENCING: 1 REGULAR DIFFERENCE
NO. OF OBS.:  ORIGINAL SERIES 35, AFTER DIFFERENCING 34
RESIDUALS:    SS = 97.5095  (BACKFORECASTS EXCLUDED)
              MS =  3.1455   DF = 31

MODIFIED BOX-PIERCE CHISQUARE STATISTIC
LAG               12              24               36             48
CHISQUARE    6.1(DF=10)    14.1(DF=22)      * (DF= *)      * (DF= *)

MTB >
```

Supplementary Exercises

Interpretive

14.65 Discuss the importance of the following factors as they relate to our discussion of forecasting concepts and methods.

    a. Seasonal factor   **b.** Curvilinear long-term trend

    c. Smoothing constant   **d.** Box-Jenkins procedures

    e. Autocorrelation function   **f.** Qualitative forecasting methods

14.66 Distinguish among the MAD, MSE, RMSE, and MAPE as measures of forecast accuracy. Under what conditions is each appropriate to use as a measure of forecast accuracy when choosing a forecast model?

 14.67 A special section in the May 1, 1978, issue of *U.S. News & World Report* contained an article entitled "How to Be Your Own Forecaster," which offered advice on how the ordinary citizen can predict where the U.S. economy is leading. The advice of *USN&WR* was to follow several key indicators: gross national product, industrial production, retail sales, the Consumer Price Index, the Wholesale Price Index, the level of employment, and housing starts. Discuss the advice offered by the magazine. How would you modify the advice it has provided its readers?

14.68 Explain the difference in the conditions underlying a forecasting problem that would suggest the use of a forecasting method from each of the following classes of forecasting methods: econometric methods, time series methods, and qualitative forecasting methods.

14.69 For each of the following forecasting problems, suggest the forecasting method you would recommend, defend the selection of your chosen method, and suggest the relevant data you would use in the forecast.

    a. The unemployment rate in the United States for each of the next 12 months

    b. The total number of housing starts in the United States in the next calendar year

    c. The demand for electric power for each of the next 20 years by the resident and commercial users in Topeka, Kansas

    d. The sales of skis and skiing equipment by a large sporting goods chain during each of the next 24 months

    e. The sales over the next six months of a new diet drink being offered for the first time by the Coca-Cola Company

    f. The sales of term life insurance in the state of Georgia during the next calendar year

 14.70 What part does judgment play in business forecasting? In particular, explain how the judgments of different people within a business firm (sales force, production manager, executives) can be used in the process of business forecasting.

 14.71 Since 1972, enrollment in schools of business has increased at a much more rapid rate than the overall growth in college enrollment in the United States. To properly allocate resources to accommodate the students and to adjust for other effects of enrollment, college and university administrators must be able to predict the anticipated number of students who will enroll in business courses as well as the number who will enroll in other areas. Design an econometric forecasting model to predict the enrollment in all undergraduate business courses at your college or university during each of the next two terms (or semesters). In particular, suggest the predictor variables you believe are related to business enrollment and indicate where you could obtain measurements for these variables and for business enrollment for each of the past several terms.

14.72 An econometric forecasting model is basically a multiple regression model. As we discussed in Chapter 12, the multiple regression model can be used to explain the relationship among variables in addition to its capacity as a prediction model. R. Sinche has developed an econometric model to predict capital formation for nonfinancial corporations in the United States ("Stimulating Capital Stock Growth," *Business Economics*, September 1979, National Association of Business Economists). His econometric model, based on data gathered over 40 quarters, is

$$\text{percentage change in K stock} = 14.16 + .40 \text{ ROR} + .12 \text{ UCAP} - 22.27 \text{ REL}$$
$$(3.64) \qquad (2.93) \qquad (4.51)$$

where

      ROR = rate of return on stockholders' equity for nonfinancial corporations
      UCAP = Federal Reserve Board index of capacity utilization in manufacturing industries
      REL = relative price of capital goods

The values in parentheses are the *t* values measuring the significance of the associated regression parameter. According to Sinche's model, what have been the primary contributing factors in the disappointing rate of capital stock growth in recent years? Discuss Sinche's econometric model.

14.73 In Section 14.8 we mentioned that the smoothing constants in the EWMA model should be appropriately adjusted to accommodate sudden shifts in a time series due to atypical situations such as labor strikes, price cut promotions, and the weather. Adjustments to accommodate atypical situations should not, however, be limited to the EWMA model but should be applied to all forecasting models. Explain how relevant subjective information can, in a general sense, be incorporated into a total forecasting model to its total environment.

14.74 When we develop a forecasting model, why should we test the model using "holdback" data? That is, why should we see how well the model forecasts during a period outside the data set used to derive the model, and why is it insufficient to rely on a model that simply fits the sample data set well?

 14.75 A manufacturer of industrial equipment has developed an efficient but expensive energy-monitoring device for dramatically reducing energy consumption in large industrial or-

Table 14.33   Sample Response Data for Exercise 14.75

Descriptive Word	Number Responding	Purchase Intent Probability
Certain	0	.99
Almost sure	1	.9
Very probable	1	.8
Probable	0	.7
Good possibility	4	.6
Fairly good possibility	0	.5
Fair possibility	7	.4
Some possibility	2	.3
Slight possibility	2	.2
Very slight possibility	2	.1
No chance	1	.01

ganizations. To estimate demand for the product, the manufacturer has offered an in-house performance demonstration for each of 20 potential customers who will serve as a test market sample. After the performance demonstrations each industrialist involved in the test market sample was asked to indicate his or her likelihood of purchase by selecting a descriptive word from a list of descriptive words. The responses and the associated purchase intent probabilities are offered in Table 14.33. If the manufacturer of the energy-monitoring device believes that there are currently $N = 2,060$ industrial firms in the distribution area of this firm that are potential customers for the product, estimate the total demand from this market for the product.

Case Study

Middletown Ice and Cold Storage

Middletown Ice and Cold Storage is in a business that involves substantial risks, such as the risk associated with the high fixed costs of production, storage, and delivery. In addition, the sales of ice in Middletown's marketing area fluctuate dramatically throughout the year. Most of Middletown's customers are either seasonal businesses or businesses (such as supermarkets and restaurants) that have ice-making equipment of their own. This latter group buy from Middletown only when their need for ice outpaces their ice-making capacity. Middletown Ice faces a situation like many other businesses with high fixed costs: As sales fall below the break-even point, losses accumulate quickly. But as sales rise, profits also accumulate quickly.

On several occasions in recent years Middletown Ice has been unable to meet peak seasonal demand, thus generating ill will among some customers and creating an opportunity cost of lost sales. Therefore, Middletown's general manager assigned more cold storage space to ice storage during the winter of 1985–1986, thereby creating a larger inventory for the following year and effectively increasing Middletown's monthly production capacity to 440,000 lb. This policy has been continued since 1986 and has allowed the company to avoid shortages during the peak demand periods of 1986, 1987, and 1988.

Now, however, there is an increased demand for the cold storage space in which the stockpiled inventory is held. Thus, there is now an opportunity cost associated with stockpiling ice. As a result, Middletown's general manager wishes to develop an improved forecast of sales for the coming year to avoid excessive inventory. To do so, he collected the data (monthly sales, in thousands of pounds of ice) shown in Table 14.34.

1. From the available sales data, develop a forecast of monthly ice sales for 1989 by using the EWMA forecast model.

2. From the forecasts you have derived, you will find that during the peak months the monthly sales will exceed production capacity (440,000 lb). Hence, during the months previous to these, Middletown Ice would have had to produce more ice than necessary in order to have sufficient ice to meet demand in the peak months. Using this premise, determine the required level of production for each month of 1989.

3. From the required levels of production determined in part 2, determine the amount of ice Middletown Ice should maintain in storage each month during 1989. Will any extra storage be necessary during the year if Middletown Ice expects to meet demand?

Table 14.34   Sales Data for the Case Study

Month	1986	1987	1988
January	39	41	38
February	24	28	27
March	26	33	35
April	87	99	97
May	307	339	301
June	489	447	487
July	509	540	571
August	501	536	564
September	508	550	581
October	441	465	485
November	119	158	179
December	38	31	35

References and Suggested Readings

General-Use Forecasting Articles and Books

Armstrong, J. S. *Long-Range Forecasting: From Crystal Ball to Computer.* 2d ed. New York: Wiley, 1985.

Bowerman, B. L. and R. T. O'Connell. *Time Series Forecasting: Unified Concepts and Computer Implementation.* 2d ed. Boston: PWS-KENT Publishing Co., 1987.

Butler, W. F., R. A. Kavesh, and R. B. Platt. *Methods and Techniques of Business Forecasting.* Englewood Cliffs, N. J.: Prentice-Hall, 1974.

Chambers, J., S. Mullick, and D. Smith. "How to Choose the Right Forecasting Model." *Harvard Business Review*, Vol. 49 (July–August 1971), pp. 45–74.

Farnum, N. R., and L. W. Stanton. *Quantitative Forecasting Methods.* Boston: PWS-KENT Publishing Co., 1989.

Makridakis, S., and S. C. Wheelwright. *The Handbook of Forecasting: A Manager's Guide.* New York: Wiley, 1982.

Makridakis, S., and S. C. Wheelwright. "Forecasting: Framework and Overview." *TIMS Studies in Management Sciences,* Vol. 12 (1979), pp. 1–16.

Forecasting Using Time-Series Techniques

Box, G. E. P., and G. M. Jenkins. *Time Series Analysis, Forecasting and Control.* 2d ed. San Francisco: Holden-Day, 1976.

Brown, R. G. *Smoothing, Forecasting, and Prediction of Discrete Time Series.* Englewood Cliffs, N. J.: Prentice-Hall, 1963.

Geurts, M. D., and I. B. Ibrahim. "Comparing the Box-Jenkins Approach with the Exponential Smoothed Forecast Model." *Journal of Marketing Research,* Vol. 12 (1975), pp. 182–188.

Johnson, L. A., and D. C. Montgomery. "Forecasting with Exponential Smoothing and Related Methods." *TIMS Studies in Management Sciences,* Vol. 12 (1979), pp. 18–31.

Nelson, C. R. *Applied Time Series Analysis for Managerial Forecasting.* San Francisco: Holden-Day, 1973.

Newbold, P., and C. W. J. Granger. "Experience with Forecasting Invariant Time Series and the Combination of Forecasts." *Journal of the Royal Statistical Society* (A), Vol. 137, part 2 (1974), pp. 131–164.

Econometric Forecasting Techniques

McNees, S. K. "Lessons from the Track Record of Macroeconomic Forecasts in the 1970s." *TIMS Studies in Management Sciences,* Vol. 12 (1979), pp. 227–239.

Pindyck, R. S., and D. L. Rubinfeld. *Econometric Models and Economic Forecasts,* 2d ed. New York: McGraw-Hill, 1980.

Qualitative Forecasting Techniques

Geurts, M. D., and J. E., Reinmuth. "New Product Sales Forecasting Without Past Sales Data." *European Journal of Operational Research,* Vol. 4 (1980), pp. 84–94.

Kahneman, D., and A. Tversky. "Intuitive Predictions: Biases and Corrective Procedures." *TIMS Studies in Management Sciences,* Vol. 12 (1979), pp. 313–329.

Spetzler, C. S., and C. A. S. Stäel von Holstein. "Probability Encoding in Decision Analysis." *Management Science,* Vol. 22 (1975), pp. 340–358.

Other Forecasting References

Beckenstein, A. R. "Forecasting Considerations in a Rapidly Changing Economy." *TIMS Studies in Management Sciences,* Vol. 12 (1979), pp. 247–264.

Reinmuth, J. E., and M. D. Geurts. "A Bayesian Approach to Forecasting the Effects of Atypical Situations." *Journal of Marketing Research,* Vol. 9 (August 1972), pp. 292–297.

Reinmuth, J. E., and M. D. Geurts. "A Multideterministic Approach to Forecasting." *TIMS Studies in Management Sciences,* Vol. 12 (1979), pp. 203–212.

TO
THE
READER

Quality control is the name given to the collection of procedures and statistical techniques used to help attain and maintain the production of high-quality goods and services. The statistical methods used in quality control fall into two broad categories: techniques concerned with the quality of design and techniques used to ensure that the production process puts out products that consistently conform to the design requirements.

In this chapter we do not present statistical methods used in the product design phase, since these techniques are more appropriately covered in a second course in statistics. Instead, we introduce statistical methods used for monitoring production processes. Most of these techniques depend only on basic data analysis tools, such as those described in Chapter 2, and upon the concept of sampling distributions, discussed in Chapter 7.

The chapter begins by introducing some of the necessary terminology and notation used and then proceeds to a short discussion of goals of modern quality control. Next, simple problem identification tools such as histograms, Pareto charts, and fishbone diagrams are presented, followed by an introduction to the construction and use of control charts. Process capability indexes, which are relatively recent additions to the field and which are used in most modern quality control programs, are treated next. Finally, acceptance sampling plans are discussed.

15

QUALITY CONTROL

15.1

Introduction

The field of quality control is concerned with the design of products and with techniques for attaining consistently high quality in their production. Until recently, quality control was applied almost exclusively in manufacturing environments, where the critical characteristics of a product were monitored or inspected for conformance to some quality standards, and the defective items were thereby separated from the good ones. The focus of modern quality control, however, has expanded far beyond this use.

The scope of quality control has been widened in two ways. First, modern quality techniques are now applied in all areas of business activity, not just those concerned with manufacturing. Quality control is used to monitor computer transactions, accounting records, customer service records, and other service-related functions. In other words, whether one is making a good or providing a service, just about every business activity can be viewed as providing a "product" to the consumer, and the use of quality control helps achieve the reliability and consistency that causes consumers of all goods and services to recognize high-quality products. Second, quality control is now actively used in the design of new products. Achieving a good design is considered to be just as important as the actual manufacture of the product, since a badly designed product, no matter how well made, will still be viewed as being of poor quality in the eyes of the consumer.

Quality control had its origins in the United States and Great Britain in the 1920s but did not gain prominence on a large scale until World War II. With the war came production goals that taxed the capabilities of industries. Quality control techniques were needed in order to avoid drops in quality that could occur with the increased production requirements. Many of the modern concepts of quality control evolved during this wartime period.

After the war many countries curtailed their use of quality control. One notable exception was Japan, which, with the guidance of the American statistician W. Edwards Deming, embarked on a large-scale program of training in the use of statistical quality control methods in their manufacturing plants. The success of this endeavor, as measured by Japan's current reputation for building inexpensive products of very high quality, has caused a resurgence of interest in statistical quality control in the United States and other countries.

15.2

Modern Terminology and Concepts

As a result of the modern view that quality control techniques are applicable to all business activities, some new terminology has evolved. This terminology helps standardize how we talk about quality control techniques, much as using the terminology *success* and *failure* standardized the presentation of the binomial distribution in Chapter 5.

process

The concept of a process is fundamental to quality control terminology. Most simply, every step in a manufacturing line or in a service-related business is referred to as a **process**, with its own specific inputs and outputs, as depicted in Figure 15.1. For example, building an automobile requires thousands of individual steps or processes. Welding two door parts together, preparing the car surface for painting, and then applying the paint would constitute three such processes, each having specific inputs and outputs. The two door parts would be the inputs to the welding process, whose output would be a welded unit. In turn, the welded units would be inputs to the surface preparation process, and so on. Producing a good or service can then be pictured, as in Figure 15.2, as a large series of interrelated processes. Statistical methods can then be used to monitor the quality of the output of any process.

Definition

statistical process control (SPC)

> The use of statistical quality control techniques is called **statistical process control (SPC)**.

Some authors refer to these techniques as statistical quality control (SQC) methods, but this terminology is being used less and less since it can mistakenly give the impression that the statistical techniques are limited to the quality control (QC) department in an organization.

detection approach

There are two basic approaches to monitoring the output quality of a given process. One is called the **detection approach** and is shown in Figure 15.3. A company using the detection approach waits until the products are made, sorts out the good ones from the bad, and then tries to use the flaws found in the bad items as indicators of how

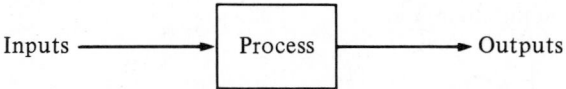

Figure 15.1   Single Process in Quality Control

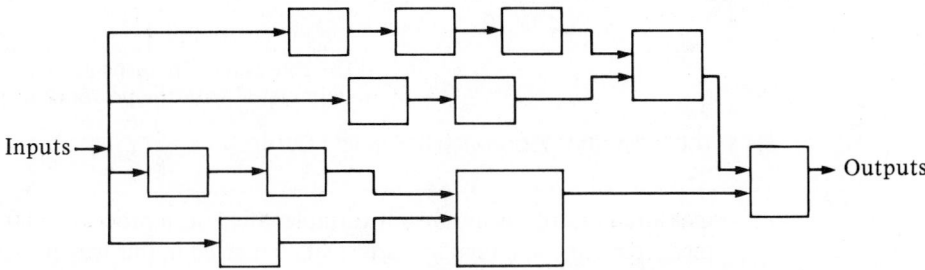

Figure 15.2   Typical Breakdown of a Good or Service into a Series of Interrelated Subprocesses

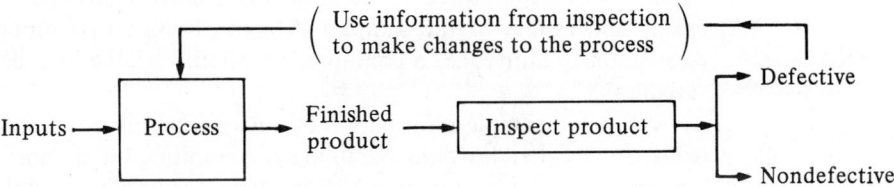

Figure 15.3   The Detection Approach to Quality Control

or where to fix the process. The problem with this approach is that the information obtained from examining the defective items is usually too old to be of much value in fixing or adjusting the process. For example, key factors affecting the process, such as the workers involved, the work shift, materials used, and equipment used, can easily change from day to day, which makes it difficult to determine exactly which factor was responsible for the defective products.

prevention approach   The second approach to monitoring a process is called the **prevention approach** and is diagrammed in Figure 15.4. With this approach a company monitors some of the key process variables *while* the process is running and *before* the final product is made. In this way potential problems can be spotted quickly and the process can be adjusted immediately, before defective products are made.

Until recently, many companies operated primarily in the detection mode. For example, many manufacturers use warranty data or data on customer returns to monitor quality. Such data is of some value in identifying major design flaws but is of little value in fixing the problems on the manufacturing line.

The prevention mode is by far the preferrable approach and serves as the basis for most quality control programs today. This approach demands that efforts be made to

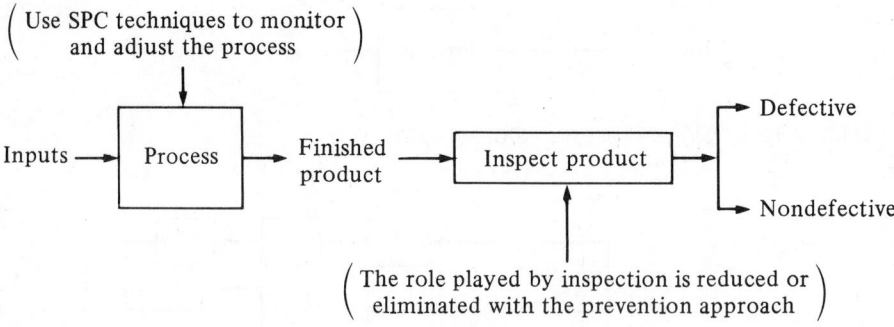

Figure 15.4   The Prevention Approach to Quality Control

measure some of the important variables affecting a process and then monitoring these measurements on a regular basis. Data collected in this way is usually monitored by a

control chart   control chart. The **control chart** is based on sample information, measurable or qualitative, extracted from the process at different points in time. For example, if the product can be measured by its length, weight, potency, cost, and so forth, then its quality can be monitored by constructing control charts that show the successive means and ranges of the samples. A control chart that monitors a continuously

control chart   measurable quantity (i.e., a continuous random variable) is called a **control chart for**
for variables   **variables**.

When the sample information is not continuous but discrete, then the data is referred to as attribute data. Sampling and counting the number of defective items in the sample is the most common type of attribute sampling. Control charts designed to

control chart   monitor such data are called **control charts for attributes**, the most common of which
for attributes   are the p chart, c chart, and u chart. Before reading the section on control charts for attributes, you may wish to review the discussion of the binomial and Poisson distributions in Chapter 5.

Control charts are designed to separate the usual, uncontrollable variation in a process from the controllable variation. Such factors as machine wear, machine adjustment, and operator skill may contribute to the controllable variation and may be detected by the control chart. Ideally, if the controllable variation is detected and corrective action is taken, the total variation affecting the process can be reduced to

statistical control   only random variation. In this event the process is said to be in a state of **statistical control**.

on-line quality   Control charting is an example of what is called an **on-line quality control**
control technique   **technique**, since the charting is done at the same time the product is being made. There

off-line techniques   are many off-line SPC techniques also in use. **Off-line techniques** consist mainly of experimental design methods such as the factorial designs introduced in Chapter 10 or the so-called Taguchi methods recently developed in Japan (see the references). For the most part, off-line techniques are used to *optimize* a process, resulting in settings or levels of key process variables that yield the highest possible quality in the output of a process. Discussing experimental design techniques is beyond the scope of this chapter, and the interested reader is referred to texts on industrial experimentation and experimental design (see the references).

15.3

Simple Graphical Techniques

Recording data is a necessary and frequent activity in companies. Some data, such as records of accounts receivable, accounts payable, payroll, and inventory, are fundamental to the normal conduct of business. Other data is collected primarily to monitor activity within the company. Such internal data might reside on log sheets that monitor daily production levels and task completion times or on exception reports that highlight defective rates and other problem areas.

Creating a visual display of data serves a number of purposes in quality control. For one thing, graphing data can produce a clear summary of a situation so that workers can quickly pinpoint problems without having to sort through pages of records. Where language and other differences exist, graphs also allow a common medium for communicating ideas. Some graphs are useful for sorting out which process variables are the critical ones; other graphs help to focus problem-solving efforts. Lastly, graphs tend to be impartial, allowing people to make decisions more on the basis of factual evidence and less on the basis of their opinions or hunches.

Of the many graphical techniques available today, four are widely used in quality control: histograms (and, to some extent, stem-and-leaf displays), Pareto charts, fishbone diagrams (also called cause-and-effect diagrams), and scatter plots. These simple graphs are usually employed in the early stages of problem solving, before control charting or more complicated procedures are used.

histograms · Histograms have already been discussed in Chapter 2. In quality control, **histograms** are used to graph data in order to display the central tendency and variation in a particular process. Most often, the variable being studied has a specified range within which it is required to lie. The width of a car door, for example, cannot be too small or the door won't close snugly enough to keep out rain and noise; on the other hand, the door can't be too wide or it won't close at all. Design engineers usually state specification limits · the **specification limits** within which a variable should lie if it is to meet functional and quality standards.

Definition

upper specification limit (USL)
lower specification limit (LSL)

> The largest allowable value of a variable is called the **upper specification limit (USL)** and the lowest allowable value is the **lower specification limit (LSL)**.

Histograms allow one to easily determine the extent to which a process is capable of staying within its specification limits.

EXAMPLE 15.1   Printed-circuit (PC) boards are used in almost all electronic equipment such as computers, TVs, radios, and appliances. PC boards are the (usually green-colored) cards upon which various electronic components are mounted, each board performing some specific function within the appliance. One of the steps in manufacturing PC

Table 15.1   Copper-plating Thickness Data (in Inches) for Example 15.1

Board	Thickness	Board	Thickness	Board	Thickness
1	.00125	18	.00170	35	.00345
2	.00110	19	.00255	36	.00335
3	.00170	20	.00155	37	.00195
4	.00235	21	.00295	38	.00205
5	.00100	22	.00240	39	.00295
6	.00110	23	.00255	40	.00320
7	.00025	24	.00225	41	.00415
8	.00135	25	.00205	42	.00350
9	.00205	26	.00155	43	.00395
10	.00300	27	.00230	44	.00400
11	.00115	28	.00315	45	.00320
12	.00275	29	.00280	46	.00345
13	.00255	30	.00185	47	.00395
14	.00210	31	.00300	48	.00475
15	.00210	32	.00270	49	.00360
16	.00260	33	.00310	50	.00350
17	.00245	34	.00290		

boards involves putting the boards in a copper solution and using electrolysis to build up the thin layers of copper that eventually form the circuits on the boards.

Lab measurements on 50 PC boards recently put through the plating process are shown in Table 15.1. Create a histogram using this data. The manufacturing specifications require that acceptable PC boards have copper thicknesses between .001 and .003 in. Does the copper-plating process appear to be capable of meeting these specifications?

SOLUTION   Table 15.2 shows a MINITAB histogram of the data in Table 15.1. The upper and lower specification limits have also been included. A cursory examination of the table indicates that many of the PC boards have plating thicknesses exceeding the specifications. Upon closer examination, since the class boundaries of the histogram are

Table 15.2   MINITAB Histogram of the Data in Table 15.1

```
MTB> HIST 'COPPER';
MTB> START 0;
MTB> INCREMENT .0008.

HISTOGRAM OF COPPER     N = 50

MIDPOINT     COUNT
0.000000       1    *
0.000800       4    ****                {.001 = LSL (LOWER SPECIFICATION LIMIT)
0.001600       8    ********
0.002400      16    ****************
0.003200      15    ***************     {.003 = USL (UPPER SPECIFICATION LIMIT)
0.004000       5    *****
0.004800       1    *
```

located at .0004, .0012, .0020, .0028, .0036, ... , we see that there are 26 boards (about 52%) that do not meet specifications, which clearly indicates that there is a problem with the plating process.

◆ ◆

Another technique for identifying process problems is to use a Pareto chart.

Definition

Pareto chart

> A **Pareto chart** is a bar chart where each bar is associated with a particular area of concern and the bars are drawn, from left to right, in order of decreasing height.

Thus, in a Pareto chart the most frequently occurring concerns are toward the left side of the chart; the less significant problems are toward the right.

Pareto charts are so named because they usually show a Pareto effect, first noticed by the Italian economist Vilfredo Pareto (1848–1923) in his studies of the distribution of incomes. The Pareto effect refers to the fact that, in most cases, only a small group of bars in the chart tend to account for a majority of the problems. The Pareto effect also occurs in inventory theory, where the *80–20 rule* states that about 80% of a company's sales can be attributed to only 20% of the inventoried items. In quality control Pareto charts help sort out the few, serious problems from the many minor ones.

EXAMPLE 15.2   The manufacturer of PC boards, discussed in Example 15.1, also inspects finished boards before shipping them to customers. During this inspection some defective boards are found, and the causes of the nonconformities are recorded. For the month of January, Table 15.3 shows the types and numbers of boards rejected at final inspection. Create a Pareto chart for this data. Which defect categories account for over 80% of the rejects?

Table 15.3   PC Rejection Data for Example 15.2

Type of Defect	Number of Rejected Boards
Poor electroless coverage	35
Lamenation problems	10
Low copper plating	112
Plating separation	8
Etching problems	5
Miscellaneous	12

SOLUTION   The Pareto chart for this data appears in Figure 15.5, where the categories of defects have simply been sorted according to decreasing frequency of occurrence. The only exception is the "Miscellaneous" category, which is listed last. The reason for this exception is that the miscellaneous category usually consists of a small number of rejects that are not worth the effort of putting into separate categories. So since this

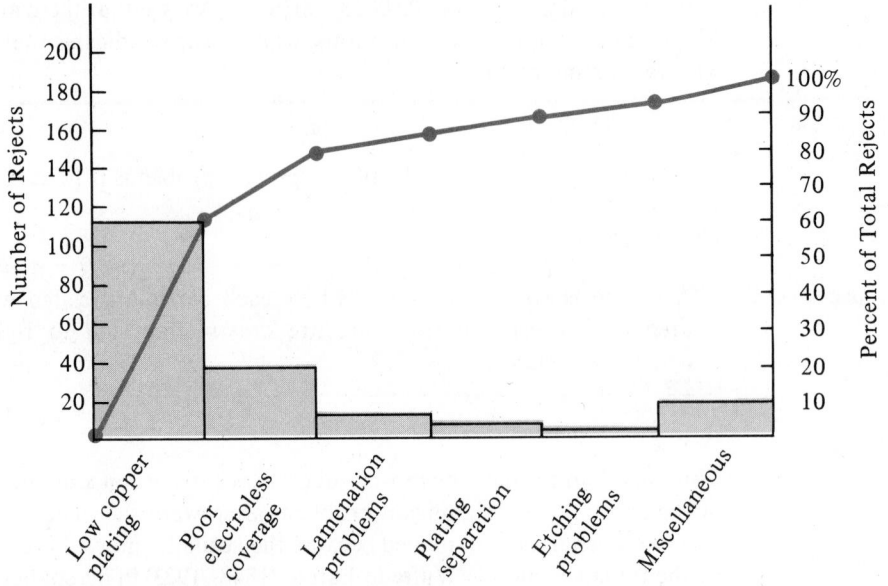

Figure 15.5   Pareto Chart of the Data in Table 15.3

category actually represents a group of infrequently occurring smaller categories, it properly belongs to the right side of the chart.

Pareto charts often include a broken line depicting the cumulative frequency over the various bars in the chart. This line has been included in Figure 15.5, where we can see that the first two bars (low copper plating and poor electroless coverage) account for about 81% of the rejects.

◆ ◆

The vertical scale on a Pareto chart represents either frequency or money. Sometimes, managers are interested in seeing a Pareto chart that is based on frequencies of rejects converted into one where the vertical scale is in dollars. This conversion may completely change the ordering of the categories in the chart, since the categories contributing most to the dollar volume may turn out to be quite different from the categories contributing to the total numbers of rejects.

Once a Pareto chart has pointed to the more serious problem areas, something has to be done to fix these problems. Since it isn't always obvious how to do this, another graphical procedure, the fishbone diagram (also called the *cause-and-effect diagram*) can be used at this stage.

Definition

fishbone diagram

> The **fishbone diagram** is a graphical way of displaying the possible reasons, or causes, of a particular problem.

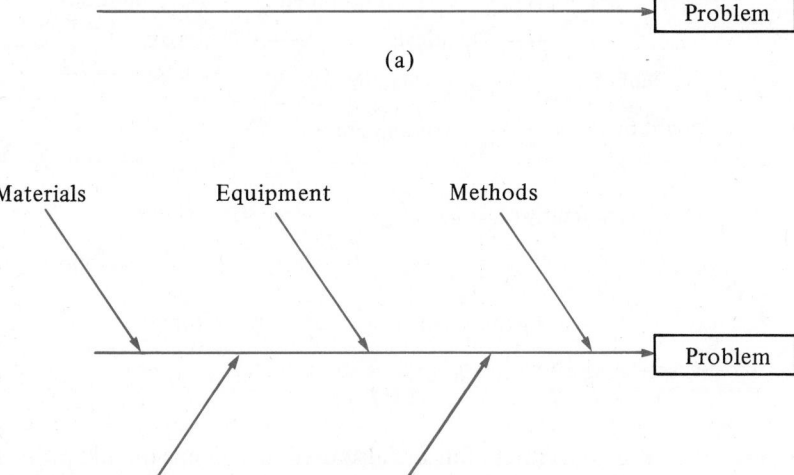

Figure 15.6   Construction of a Cause-and-Effect (Fishbone) Diagram

To construct a fishbone chart, one must first choose a problem area to attack. This problem area then becomes the *head* of the diagram, as shown in Figure 15.6a. Next, some of the possible causes are drawn on the diagram, as in Figure 15.6b. A convenient technique is to start the diagram with the five basic "causes" shown in Figure 15.6b, since they are common to many, if not most, problems. They also provide a good starting point for discussions that eventually lead to more and more detailed causes, which are then added to the chart. Additional levels of causes are represented as additional branches on the diagram. The fundamental idea is that things on the smaller branches cause the things on the next level of branches, so each cause can eventually be traced back through the chain to the primary problem being studied. The herringbone appearance of a completed cause-and-effect diagram is what gives rise to its alternative name of fishbone diagram.

EXAMPLE 15.3   In Examples 15.1 and 15.2 wide variations in copper-plating thickness was shown to be a major problem area in the manufacture of PC boards. A group of production personnel was formed to look into possible reasons for this problem. Starting with the five basic reasons (machines, materials, workforce, methods, and environment), they created the cause-and-effect diagram in Figure 15.7. After the diagram was finished, the group decided to begin studying the questions of proper anode placement and how the amperage estimate was obtained. Other branches on the diagram would also eventually be studied, but only a couple at a time because of time constraints.

◆ ◆

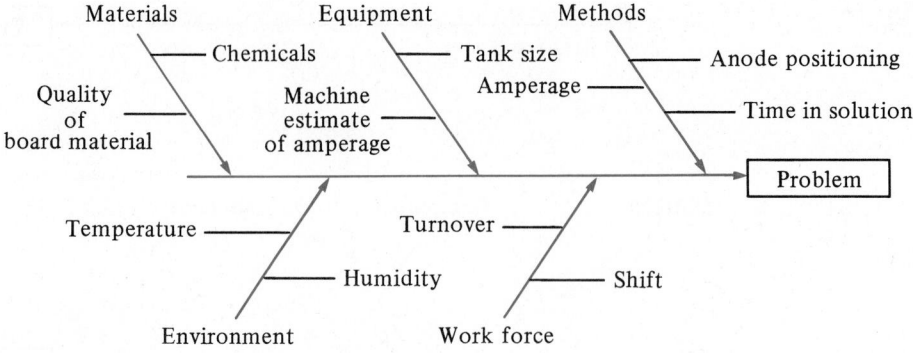

Figure 15.7   Cause-and-Effect Diagram for Example 15.3

scatter diagrams   Finally, scatter plots are also used to examine quality problems. **Scatter diagrams**, as described in Chapter 11, are graphs of a response variable y versus an explanatory variable x. If, for example, some quantitative variable x can be found that is thought to affect copper thickness y, then a simple scatter plot should be sufficient to reveal this fact. A regression analysis may also be performed on the data in the scatter plot if a more detailed description of the relationship between x and y is needed.

EXERCISES   Conceptual Level

15.1   Explain the concept of a process, together with its inputs and outputs. Give a specific example of a process, and explain its inputs and outputs.

15.2   Discuss the differences and similarities between the detection and prevention approaches to monitoring a process. Which approach is preferable?

15.3   Distinguish between controllable and uncontrollable variation in a process. Which type of variation can be detected by control charts?

15.4   Discuss the following graphical techniques as they relate to statistical process control. Give a specific situation in which each might be useful.

   a. Histograms   **b.** Pareto charts

   c. Fishbone diagrams   **d.** Scatter plots

15.5   Discuss the advantages of using graphical techniques in statistical process control.

15.4

Control Charts

Control charts are used to separate the controllable from the uncontrollable variation in a process. Every repetitive process, such as manufacturing cars or typing computerized mailing labels, involves some inherent degree of uncontrollable or natural variation. It is caused by the innumerable small events that occur when performing any task, events that eventually have a small but measurable effect on the outcome of that task. Walter A. Shewhart, who invented the concept of control charts in the 1920s, referred to such variation as **chance variation** [see Shewhart (1926, 1931) in the references]. The other type of variation in a process is due to what Shewhart termed **assignable causes**. Assignable causes might include, for example, changes in the types of raw material used, differences in the workers used, the slow wearing down of the machinery used, or changes in environmental factors such as temperature or humidity. Assignable causes, as opposed to chance causes, can often be discovered, and steps can be taken to correct the problems. When all the assignable causes have been found and eliminated, a process is then said to be in statistical control, or, more simply, in control.

Control charts are constructed by plotting a sample statistic for many successive samples collected from a process. For example, suppose that samples of five parts are taken hourly from the output of a manufacturing process and that some important quality characteristic, such as the part's length, is measured for each item in the sample. From each sample of five parts statistics such as the mean ($\bar{x}$), median, range (R), standard deviation (s), or proportion of defectives ($\hat{p}$) can be calculated. Plotting any one of these statistics versus the sample number gives rise to a control chart for that particular statistic, as shown in Figure 15.8. The various types of control charts are classified by the sample statistic charted and whether variables or attribute data are used (see Table 15.4).

Every control chart has a centerline and control limits (see Figure 15.8). The **centerline**, which is just the grand average of all the sample statistics plotted, estimates the mean of the sampling distribution of the statistic being charted. The **upper and lower control limits** are also determined by the sampling distribution and are positioned three standard deviations above and below the centerline. Points on the control chart that fall outside the control limits are said to be *out of control*; those within the limits are *in control*. In this way a control chart separates chance variation from assignable causes. If a process is running smoothly, then the natural (chance) variation in the

(margin notes) chance variation

assignable causes

centerline
upper and lower
control limits

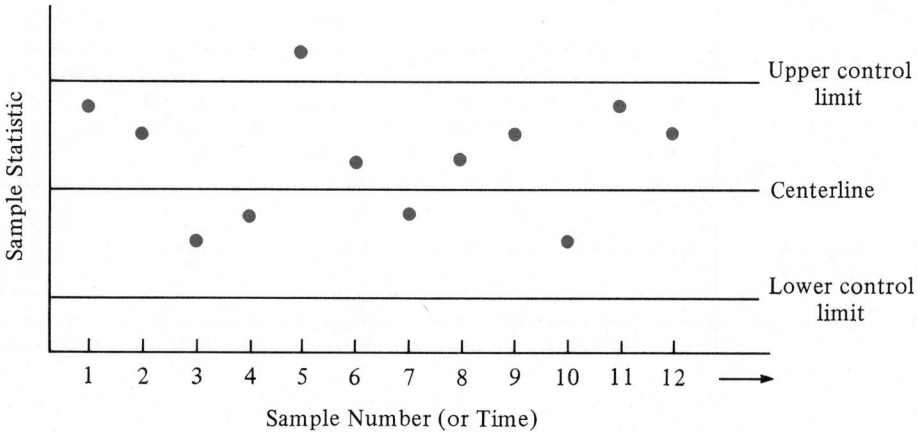

Figure 15.8   A Typical Control Chart

Table 15.4   Classification of Control Charts

Control Charts for Variables	Control Charts for Attributes
$\bar{x}$ (mean)	p (proportion defective)
R (range)	np (number defective)
s (standard deviation)	c (number of defects)
Median	u (number of defects per unit)

Note: A *defective item* is one that does not meet all the specifications placed on it. A *defect* refers to the failure of a particular characteristic of the item to meet its specifications. Thus, a defective item may contain many defects. The *p* and *np* charts are concerned with monitoring defective items; the *c* and *u* charts monitor the number of defects on such items.

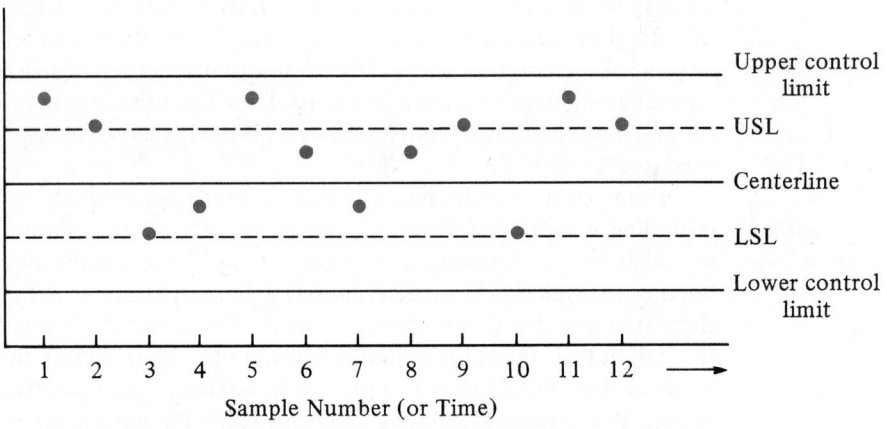

(a) Process in Control but Not Within Specification

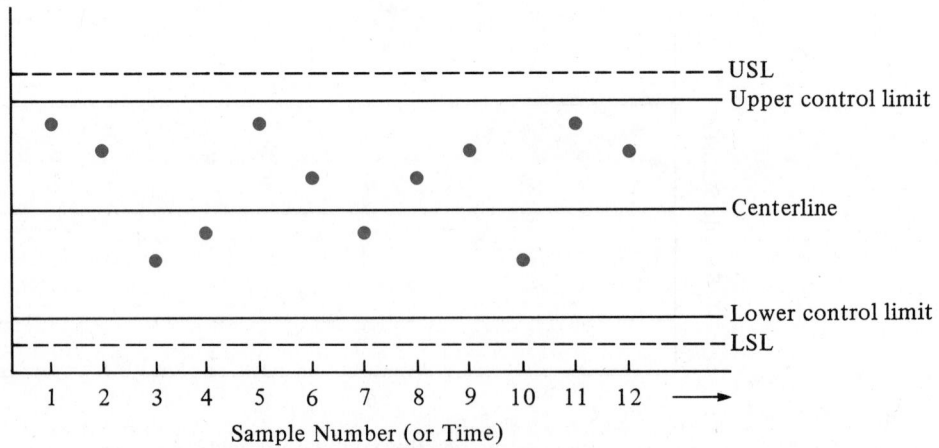

(b) Process in Control and Within Specification

Figure 15.9   Distinction Between Control Limits and Specification Limits

process is represented by the region between the control (three-standard-deviation) limits; and points in the out-of-control region are thought to be due to assignable causes. Knowing which sample or samples are out of control then allows workers to investigate—and eliminate—the reasons (causes) for these problems.

Control limits and specification limits are very different things. Control limits show what a process is actually doing: specification limits describe what we would *like* it to do. For example, a process may very possibly be in statistical control and yet fail to meet the specifications. The ideal situation, of course, is to have the process be simultaneously both in control and within specification. Figure 15.9 illustrates the distinction between these two types of limits.

EXAMPLE 15.4 Suppose in Example 15.1 that the design engineers, after reconsidering the copper-plating requirements, decide to relax the specification limits on plating thickness by establishing new limits of USL = .0005 and LSL = .0050. Using the data from Table 15.1 and the associated histogram in Table 15.2, reassess whether the copper-plating process is capable of meeting these new specifications. Next, plot the data in order of production (i.e., by board number). Does the plating process appear to be in control?

SOLUTION From Table 15.2 we can see that all but one data point lie between .0005 and .0050 in., so the plating process appears to be capable of meeting the new specification limits. However, this does not mean that the process is necessarily in statistical control. In fact, when the data is plotted in order of production (Figure 15.10), a strong trend is evident,

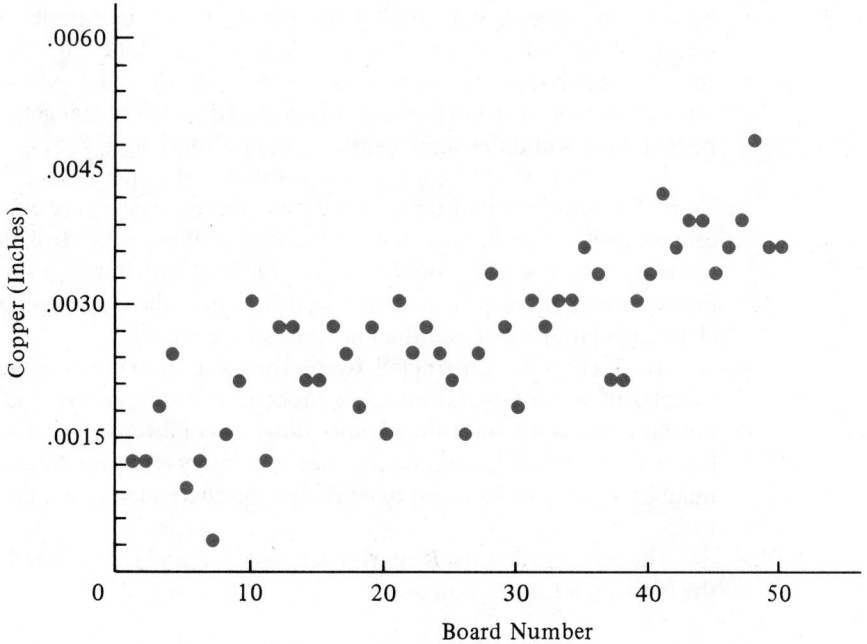

Figure 15.10   Plot of the Data in Table 15.1 in Order of Production

indicating that the process will soon start producing PC boards whose copper thicknesses exceed the upper specification limit. Even without showing control limits, the chart indicates that this process is not in control. Looking at the histogram alone would not have shown this potential problem.

◆◆

15.5

$\bar{x}$ and R Charts

The $\bar{x}$ and R charts are the most commonly used control charts for variables data. Indeed, many who have had only brief exposure to charting methods sometimes think that these are the only charts used in SPC.

$\bar{x}$ chart   The $\bar{x}$ **chart** monitors the means of small samples taken from a process. When the means are plotted, the chart gives a clear picture of how they tend to vary around their long-run mean. For example, if length measurements are made each hour on five sampled parts from a production line, we expect the means of these samples to exhibit some natural variation around their average. If one of the $\bar{x}$'s exceeds the three-standard-deviation control limits, however, we conclude that something has gone wrong. We should immediately find the cause before the process goes so far out of control that it begins to produce defective parts.

From the same samples that are used to make the $\bar{x}$ chart we can also calculate the
R chart   sample ranges and plot them on an **R chart**. The sample range is just the difference between the largest and smallest measurements in the sample, and thus, it gives a picture of the variability within the process. The smaller the sample ranges are, the smaller the part-to-part variation is, which means that each part will tend to look and function like the next. On the other hand, the larger the ranges get, the more likely it will be that parts will differ significantly from one another.

The $\bar{x}$ chart should *not* be used without first constructing the corresponding R chart. The reason is that the control limits on the $\bar{x}$ chart are calculated by using the centerline, $\bar{R}$, of the R chart. Thus, if the R chart is not in control, then the average $\bar{R}$ of the sample ranges will probably not be reliable, which will in turn cause the control limits on the $\bar{x}$ chart to be incorrect. In this section, then, we begin by introducing the R chart calculations first and then proceed to the $\bar{x}$ chart.

An R chart is constructed by plotting the individual ranges R_i for successive samples of size n taken from some process. The sample size n is usually kept small, typically below 10 since the sample range becomes very volatile and unreliable for larger sample sizes. A common practice in industry is to use values of $n = 3, 4,$ or 5. The number of samples k needed to construct the chart should, at minimum, be around 20 to 25.

The centerline for the R chart is denoted by $\bar{R}$ and is calculated by simply averaging the individual sample ranges:

$$\bar{R} = \frac{1}{k} \sum_{i=1}^{k} R_i$$

The average $\bar{R}$ serves to estimate μ_R, the mean of the sampling distribution of the ranges. The logic behind the R chart (or any control chart) is that if the process is in control, the sample ranges should vary about their expected value, μ_R, and that almost all of the R_i's should fall within three standard deviations $(3\sigma_R)$ of μ_R. Ideally, then, the control limits would be placed at

$$\text{UCL} = \mu_R + 3\sigma_R \quad \text{and} \quad \text{LCL} = \mu_R - 3\sigma_R$$

In practice, the control limits must be estimated from the available data by using the simple formulas

$$\text{UCL} = D_4\bar{R} \quad \text{and} \quad \text{LCL} = D_3\bar{R}$$

The values of D_4 and D_3 are given in Table 12 of the Appendix and are based on the assumption that we are sampling from a normally distributed population.

The R Chart

Centerline: $\bar{R} = \dfrac{1}{k}\displaystyle\sum_{i=1}^{k} R_i.$

Upper control limit: $\text{UCL} = D_4\bar{R}.$

Lower control limit: $\text{LCL} = D_3\bar{R}.$

Note: The values of D_4 and D_3 depend on the sample size n and are given in Table 12 of the Appendix.

EXAMPLE 15.5 A manufacturing process that makes precision ball bearings is sampled every hour in order to monitor the variation in bearing diameters being produced. The engineering design requires that the bearings have a diameter of .500 ± .010 in. That is, the specification limits are USL = .510 and LSL = .490. Table 15.5 shows diameter measurements from 25 hourly samples of five bearings. So that the data are easier for the production personnel to record, the measurements in the table are recorded as the difference (in thousandths of an inch) between the actual measured diameter and the target value of .500 in. Use this data to construct an R chart. Does the variation in the bearing diameters appear to be in control?

SOLUTION The sample means and ranges for each of the 25 samples are shown in the last two columns of Table 15.5. The average of all 25 sample ranges is 6.56, which appears at the bottom of the table.

The centerline of the R chart is then $\bar{R} = 6.56$ in. Since each sample is of size $n = 5$, the control limits are calculated by using $D_4 = 2.115$ and $D_3 = 0$ (found in Table 12 of the Appendix). (*Note:* For samples of six or less the lower control limit is negative, so D_3 is taken to be zero in those cases.) The upper and lower control limits are then

$$\text{UCL} = D_4\bar{R} = (2.115)(6.56) = 13.874$$

and $\quad \text{LCL} = D_3\bar{R} = (0)(6.56) = 0$

Figure 15.11 shows the R chart for this data. The 25 sample ranges are plotted in the order the samples were taken, and the centerline and control limits have also been

Table 15.5   Diameter Data for Example 15.5

Sample Number (Hour)	Sample Measurements					$\bar{x}$	R
	1	2	3	4	5		
1	5	3	−1	0	3	2.0	6
2	−1	1	2	1	−3	.0	5
3	−7	−4	1	−2	−3	−3.0	8
4	−2	1	3	−1	2	.6	5
5	2	−1	−1	0	−2	−.4	4
6	3	5	1	−1	2	2.0	6
7	1	−4	−2	0	−3	−1.6	5
8	−2	1	2	−1	−3	−.6	5
9	1	−4	−6	−1	2	−1.6	8
10	0	2	−1	3	1	1.0	4
11	2	−1	4	0	−2	.6	6
12	7	0	−3	2	1	1.4	10
13	3	3	−2	1	2	1.4	5
14	−1	−4	2	6	1	.8	10
15	2	0	−5	−2	1	−.8	7
16	−1	4	−2	0	−3	−.4	7
17	−3	−1	−1	−2	0	−1.4	3
18	−2	2	3	−1	6	1.6	8
19	1	1	−2	−3	−2	−1.0	4
20	6	−2	2	4	1	2.2	8
21	3	1	−4	−2	0	−.4	7
22	−2	5	6	−1	−2	1.2	8
23	−8	−5	0	−1	1	−2.6	9
24	0	−4	2	3	−1	.0	7
25	5	−2	−4	0	2	.2	9
						$\bar{\bar{x}} = .048$	$\bar{R} = 6.56$

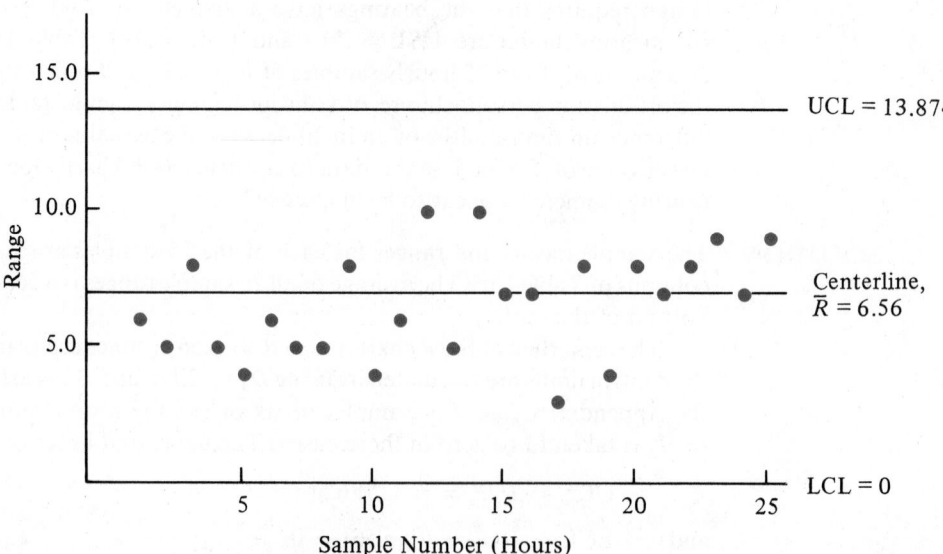

Figure 15.11   *R* Chart of the Data in Table 15.5

drawn. Since none of the points on the chart fall outside these control limits, we conclude that the chart (and therefore the process variation) appears to be in control.
$\blacklozenge\blacklozenge$

Process variation can equally well be monitored by plotting the sample standard deviation s instead of the range. In fact, some practitioners have recommended using the s chart instead of the R chart because the standard deviation makes more efficient use of the data than does the range. If production personnel have access to a computer that can perform the necessary calculation of s, then this recommendation can be followed. In practice, though, there are two reasons why most people still prefer to use the R chart. First, although workers usually have little trouble understanding and calculating the sample range, they often find the calculation of s time-consuming and difficult. Second, with the small samples normally used with data for variables, the sample range and standard deviation differ very little in how efficiently they use the information in the data, so the added value in using the s chart is very small compared with the additional computational work it requires.

We turn next to the $\bar{x}$ chart. The centerline $\bar{\bar{x}}$ is calculated by averaging all the sample means:

$$\bar{\bar{x}} = \frac{1}{k} \sum_{i=1}^{k} \bar{x}_i$$

where each of the k samples is of size n. Note that the two bars above x are intended to indicate that the centerline is the result of averaging a collection of sample averages.

From Chapter 7 we also know that the standard deviation of the sample means is given by

$$\sigma_{\bar{x}} = \frac{\sigma}{\sqrt{n}}$$

so the three-sigma (three-standard-deviation) control limits should be at $(\mu \pm 3\sigma_{\bar{x}})$. One way to estimate σ (and, ultimately, $\sigma_{\bar{x}}$) is simply to use the sample standard deviation calculated from the *combined* set of data in the k samples. If a calculator or computer is available, then the estimate s obtained in this manner will result in approximate control limits of

$$\text{UCL} = \bar{\bar{x}} + 3\frac{s}{\sqrt{n}} \qquad \text{and} \qquad \text{LCL} = \bar{\bar{x}} - 3\frac{s}{\sqrt{n}}$$

In practice, though, it is usually easier for production workers to use a different method for estimating σ and, thereby, the control limits. This method is based on the estimate

$$\hat{\sigma} = \frac{\bar{R}}{d_2}$$

where $\bar{R}$ is the centerline from the corresponding R chart and d_2 is a constant that depends on the sample size n. Values of d_2 can be found in Table 12 of the Appendix.

To illustrate, recall that in Example 15.5 the centerline was found to be $\bar{R} = 6.56$. In that example n was 5, so the appropriate d_2 factor (from Table 12) is $d_2 = 2.326$ and σ is then estimated by

$$\hat{\sigma} = \frac{\bar{R}}{d_2} = \frac{6.56}{2.326} = 2.820$$

By comparison, the sample standard deviation of all $kn = 125$ measurements turns out to be $s = 2.819$.

Using $\bar{R}/d_2$ to estimate σ, we obtain approximate three-sigma limits of

$$3\hat{\sigma}_{\bar{x}} = 3\frac{\hat{\sigma}}{\sqrt{n}} = 3\frac{\bar{R}}{d_2\sqrt{n}} = A_2\bar{R}$$

where the values of

$$A_2 = \frac{3}{d_2\sqrt{n}}$$

are also found in Table 12 of the Appendix. Thus, the control limits for an $\bar{x}$ chart can quickly be calculated as ($\bar{\bar{x}} \pm A_2\bar{R}$).

The $\bar{x}$ Chart

Centerline: $\bar{\bar{x}} = \dfrac{1}{k}\displaystyle\sum_{i=1}^{k} \bar{x}i.$

Upper control limit: $\text{UCL} = \bar{\bar{x}} + A_2\bar{R}.$
Lower control limit: $\text{LCL} = \bar{\bar{x}} - A_2\bar{R}.$

Note: The value of A_2 depends on the sample size n and is given in Table 12 of the Appendix.

EXAMPLE 15.6   Construct an $\bar{x}$ chart for the data in Table 15.5. Do the points on the chart indicate that the process is in control?

SOLUTION   The sample means, which are also included in Table 15.5, have a grand mean of $\bar{\bar{x}} = .048$. Since the R chart appeared to be in control in Example 15.5, the value of $\bar{R} = 6.56$ can reliably be used to calculate the control limits of the $\bar{x}$ chart. From Table 12 the value of A_2 is .577 (since the sample size is $n = 5$). Therefore,

$$\text{UCL} = \bar{\bar{x}} + A_2\bar{R} = .048 + (.577)(6.56) = 3.83$$

and

$$\text{LCL} = \bar{\bar{x}} - A_2\bar{R} = .048 - (.577)(6.56) = -3.74$$

Recall that the data in Table 15.5 were measured as the *differences* between the actual diameters and the target of .500, which explains why negative measurements and, therefore, a negative LCL occur.

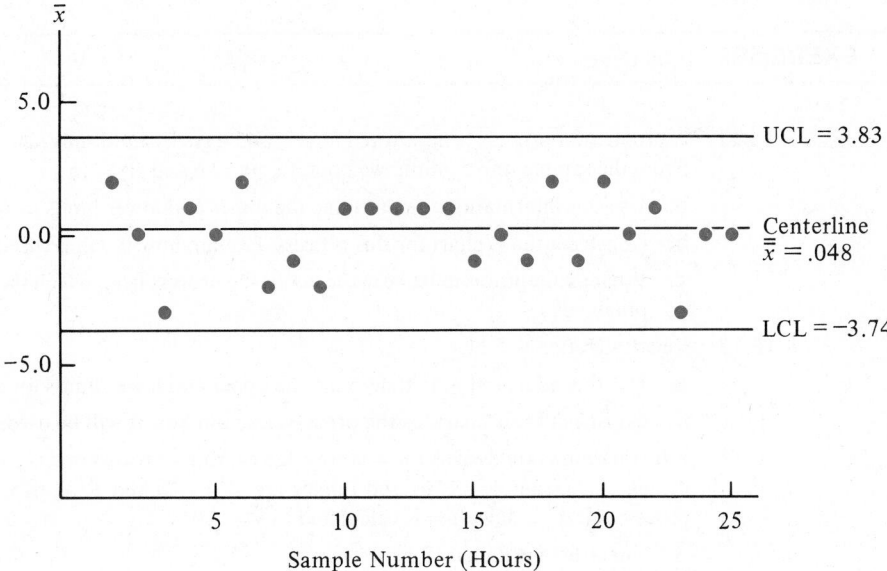

Figure 15.12 $\bar{x}$ Chart of the Data in Table 15.5

Figure 15.12 shows the $\bar{x}$ chart for this data. None of the points fall outside the control limits, so the process mean appears to be in control.

◆ ◆

After the $\bar{x}$ and R charts have been constructed, they are used to monitor future samples from the process. As new samples are taken, their sample means and ranges are plotted and compared with the control limits to check for possible problems or changes in the process. Periodically, after many more samples have been taken, the control limits and centerline should be revised by using the additional data.

EXERCISES   Conceptual Level

15.6   Explain what is meant when a process is said to be in statistical control.

15.7   Discuss the construction of a control chart.
   a.   What types of statistics can be used?
   b.   What is the purpose of the centerline and the control limits?
   c.   How is the control chart used to determine whether or not the process is in control?

15.8   What is the difference between control limits and specification limits? How are they related?

15.9   What would be the purpose of a study using each of the following control charts? Give an example of a quality control problem in which each might be useful.
   a. $\bar{x}$ *chart*   **b.** R chart

15.10   In the construction of an $\bar{x}$ chart, what two estimates are available for σ, the standard deviation of the sampled population? Why might you use one estimate rather than the other?

EXERCISES Skill Level

15.11 A production process is monitored for $k = 45$ days by randomly sampling $n = 10$ items daily. From the sample information, we obtain $\bar{\bar{x}} = 32.6$ and $\bar{R} = 15.1$.

    a. Use this information to determine the upper and lower limits for an R chart.

    b. Construct the R chart for this process. Explain how it will be used.

    c. What assumptions must be made about the process from which the sample information was obtained?

15.12 Refer to Exercise 15.11.

    a. Use this information to determine the upper and lower limits for an $\bar{x}$ chart.

    b. Construct the $\bar{x}$ chart for the process. Explain how it will be used.

15.13 Fifty random samples of size $n = 8$ are selected from a process that is judged to be in control. The means of the sample means and ranges are $\bar{\bar{x}} = .640$ and $\bar{R} = .015$, respectively; the overall process standard deviation is calculated as $s = .003$.

    a. Construct an R chart for the process.

    b. Construct an $\bar{x}$ chart for the process, using s as an estimate of σ in the control limits.

    c. Construct an $\bar{x}$ chart for the process, using the range estimate of σ in the control limits.

    d. Compare the results of parts b and c. Are the control limits approximately the same?

15.14 Refer to Exercise 15.13. A sample of size $n = 8$, selected one week after constructing the control charts, is as follows:

    .630   .641   .643   .650   .629   .636   .643   .644

    Does the process still appear to be in control?

EXERCISES Applications

15.15 A bottle manufacturer has observed that over a period of time when his manufacturing process was assumed to be in control, the average weight of the finished bottles was 5.2 oz with a standard deviation of .3 oz. The observed data was gathered in samples of six bottles selected from the production process at 50 different times. The average range of all the samples was found to be .6 oz, and the standard deviation of ranges was .2. During each of the next five days samples of size $n = 6$ were selected from the manufacturing process, showing the results that follow.

Day	$\bar{x}$	R
1	5.70	.43
2	5.32	.51
3	6.21	1.25
4	6.09	.98
5	5.63	.60

    a. Construct the $\bar{x}$ chart and the R chart, using the data collected during the period for which the process was assumed to be in control. Plot the data for the last five days.

b. Use the control charts constructed in part a to monitor the process for the sample data from the next five days.

c. Does the production of bottles appear to be out of control during any of these five days? Interpret your results.

15.16 Refer to Exercise 15.15. A particular soft drink manufacturer specifies that the bottles she purchases from the manufacturer must weigh at least 4.8 oz but not more than 5.5 oz. Assume the manufacturing process is in control.

a. What is the probability that the manufacturing process is capable of meeting the stated specification limits?

b. How many bottles in a shipment of 10,000 bottles from the manufacturer can be expected *not* to meet the bottler's specification limits?

15.17 The Nuclear Regulatory Commission requires that the amount of radiation present at any industrial site may not exceed .005 microcurie [*The Press Enterprise* (Riverside, Calif.), February 19, 1988). In a check of a process whose by-product is radioactive polonium, random samples of size $n = 4$ were taken twice daily for 12 days at a particular plant and the amount of radiation was measured. The results are given in Table 15.6.

Table 15.6   Sample Data for Exercise 15.17

Sample	Measurements	Sample	Measurements
1	.0041, .0026, .0046, .0041	13	.0017, .0021, .0017, .0020
2	.0033, .0055, .0031, .0036	14	.0030, .0039, .0038, .0041
3	.0032, .0044, .0018, .0039	15	.0043, .0037, .0030, .0033
4	.0021, .0027, .0040, .0001	16	.0046, .0017, .0006, .0034
5	.0043, .0036, .0037, .0035	17	.0026, .0036, .0023, .0022
6	.0014, .0017, .0010, .0022	18	.0024, .0049, .0033, .0056
7	.0015, .0034, .0044, .0013	19	.0039, .0036, .0017, .0032
8	.0038, .0035, .0033, .0039	20	.0041, .0033, .0025, .0030
9	.0047, .0033, .0037, .0041	21	.0044, .0025, .0023, .0030
10	.0030, .0037, .0032, .0030	22	.0033, .0029, .0026, .0034
11	.0028, .0027, .0021, .0040	23	.0019, .0008, .0006, .0024
12	.0045, .0004, .0048, .0027	24	.0030, .0030, .0029, .0053

a. Calculate the sample means and ranges for each of the 24 samples.

b. Construct an $\bar{x}$ chart for the data.

c. Do the control limits lie within the specification limits?

d. Samples 25–30 (Table 15.7) were taken over the next three days after the control chart was constructed. Does the process mean appear to be in statistical control? Is it capable of meeting the specification limits?

Table 15.7   Additional Data for Exercise 15.17

Sample	Measurements	Sample	Measurements
25	.0066, .0048, .0047, .0048	28	.0028, .0054, .0057, .0042
26	.0028, .0041, .0058, .0016	29	.0030, .0034, .0034, .0077
27	.0030, .0042, .0050, .0063	30	.0044, .0040, .0048, .0031

15.18   Refer to Exercise 15.17.

a. Construct an R chart, using samples for the first 24 days.

b. Plot the ranges for samples 25–30 on the R chart. Does the process variation appear to be in statistical control?

15.6

Process Capability

After a process involving variables data has been monitored with control charts and deemed to be in statistical control, the capability of the process can then be determined. *process capability* **Process capability** refers to the ability of a process to stay within its specification limits.

In Section 15.3 histograms were used to evaluate process capability by simply looking at how much of the histogram fell between the upper and lower specification limits. We can go a step further, though, by calculating a numerical measure of process capability in the following way. Assuming that the measurements generated by the process approximately follow a normal distribution, then almost all of the readings *actual process spread* should fall within a range of 6σ. This range is referred to as the **actual process spread**; *allowable process* the distance between the specification limits is called the **allowable process** *spread* **spread**. From these measures the process capability index, denoted C_p, is derived.

Definition

process capability index

> The **process capability index** is defined by
>
> $$C_p = \frac{\text{USL} - \text{LSL}}{6\hat{\sigma}}$$
>
> where $\hat{\sigma}$ is an estimate of the standard deviation of the measurements from the process.

Distribution of measurements
from the process

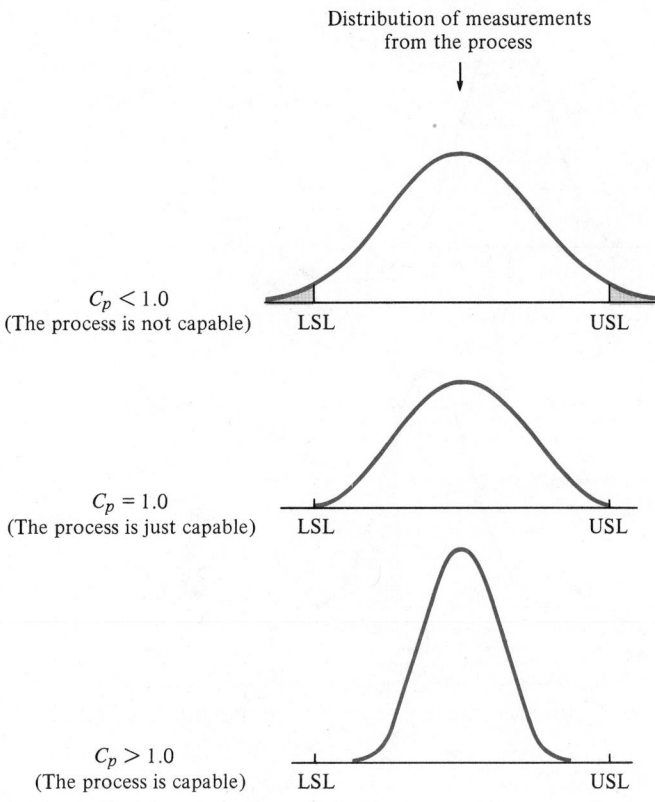

$C_p < 1.0$
(The process is not capable)   LSL   USL

$C_p = 1.0$
(The process is just capable)   LSL   USL

$C_p > 1.0$
(The process is capable)   LSL   USL

Figure 15.13   Interpretation of the Process Capability Index C_p

The index C_p is interpreted as follows: If $C_p = 1.0$, then the process is said to be *capable* (but just barely) of meeting its specification limits. Values of C_p that exceed 1.0 are better, since then the probability is higher that the measurements will be able to stay within the specification limits. A C_p exceeding 1.33 (i.e., an 8σ range fits within the specification limits) is usually considered very good and is commonly used as a target in many applications. On the other hand, a C_p less than 1.0 implies that a process is not capable of meeting its specifications. Figure 15.13 illustrates typical values of the index C_p along with the associated distributions of measurements from the process.

The C_p is one of five measures, originally developed in Japan, that are now used in almost all quality control programs. These indexes are useful because they convey much information about the process being studied in a very simple fashion. Capability indexes are also unitless measures, which allows them to be used to compare two entirely different processes. For example, if copper-plating thicknesses (in inches) from a chemical-plating process have a C_p of .81 and resistance measurements (in ohms) on electronic components have a C_p of 2.30, then one can immediately conclude that the electronic component process is the better of the two, even though their measurement units, inches and ohms, are unrelated.

The index C_p does not take into account the location of the process, only its *potential* for meeting specification. Figure 15.14 illustrates this characteristic by

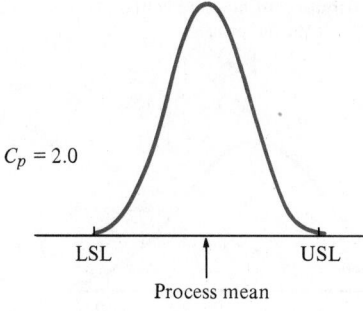

$C_p = 2.0$

LSL   Process mean   USL

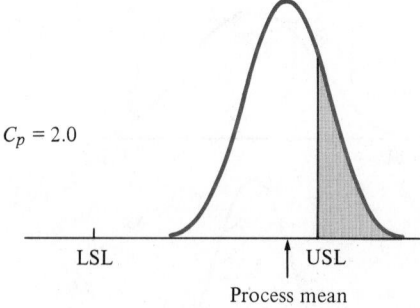

$C_p = 2.0$

LSL   Process mean   USL

Figure 15.14   The Index C_p as a Measure of Process Potential but Not Process Location

showing two processes with C_p's of 2.0, one centered between the specification limits and the other located closer to its upper specification limit. The latter process has the potential to be capable since its C_p is greater than 1.0, but something will have to be done to shift the location of this process (estimated by $\bar{\bar{x}}$) closer to the center of the specification range.

A closely related index that *does* take the process mean into account is the C_{pk} index.

The C_{pk} Index

$$C_{pk} = \min\left[\frac{\text{USL} - \bar{\bar{x}}}{3\hat{\sigma}}, \frac{\bar{\bar{x}} - \text{LSL}}{3\hat{\sigma}}\right]$$

In the definition of C_{pk}, $\bar{\bar{x}}$ is the centerline of the process $\bar{x}$ chart, $\hat{\sigma} = \bar{R}/d_2$ is an estimate of the process variation, and min represents minimum. The k in the subscript of C_{pk} refers to the so-called k factor,

$$k = \frac{|(\text{USL} + \text{LSL})/2 - \bar{\bar{x}}|}{(\text{USL} - \text{LSL})/2}$$

which measures the extent to which the process location ($\bar{\bar{x}}$) differs from its desired value midway between the specification limits. When $\bar{\bar{x}}$ lies between USL and LSL it can be shown that k always lies between 0 and 1 and that the C_p and C_{pk} indexes are related by the formula

$$C_{pk} = C_p(1 - k)$$

Since $0 \leq k \leq 1$, this formula shows that C_{pk} never exceeds C_p and that $C_{pk} = C_p$ precisely when the process is centered midway between its specification limits. Together, C_p and C_{pk} give a clear picture of how well a process is performing when compared with its specification limits.

EXAMPLE 15.7   In Example 15.5 acceptable bearing diameters were required to have diameters of .500 ± .010 in. The control charts in Examples 15.5 and 15.6 showed that the process producing the bearings was in statistical control, so its process capability can now be measured. Calculate C_p and C_{pk} for the bearing diameter data in Table 15.5, and interpret the results.

SOLUTION   The data in Table 15.5 represent differences (in thousandths of an inch) between the measured diameters and their target value of .500 in. Thus, the process specification limits of .510 and .490 translate into limits of USL = 10 and LSL = −10 for the measurements in Table 15.5. Since $\bar{R} = 6.56$ for this data, the variation in the measurements can be estimated by

$$\hat{\sigma} = \frac{\bar{R}}{d_2} = \frac{6.56}{2.326} = 2.820$$

so

$$C_p = \frac{\text{USL} - \text{LSL}}{6\hat{\sigma}} = \frac{10 - (-10)}{6(2.820)} = 1.182$$

In other words, this process is capable of staying within its specification limits.

Whether or not it actually *is* staying within these limits can be determined by calculating the C_{pk} index:

$$C_{pk} = \min\left[\frac{\text{USL} - \bar{\bar{x}}}{3\hat{\sigma}}, \frac{\bar{\bar{x}} - \text{LSL}}{3\hat{\sigma}}\right] = \min\left[\frac{10 - .048}{3(2.820)}, \frac{.048 - (-10)}{3(2.820)}\right]$$
$$= \min[1.176, 1.188] = 1.176$$

Alternatively, we could find the k factor

$$k = \frac{|(\text{USL} + \text{LSL})/2 - \bar{\bar{x}}|}{(\text{USL} - \text{LSL})/2} = \frac{|[10 + (-10)]/2 - .048|}{[10 - (-10)]/2} = .0048$$

and then calculate C_{pk} by

$$C_{pk} = C_p(1 - k) = (1.182)(1 - .0048) = 1.176$$

Since C_{pk} is greater than 1.0, the process appears to be capable of staying within its specification limits. The value of C_p is close to 1.0, however, which means that this process is just capable of meeting specification as long as the process mean is not too far from the center of the specification range. So this process should be monitored closely to make sure that its mean does not vary too much.

EXERCISES   Conceptual Level

15.19   Explain the distinction between the actual process spread and the allowable process spread, which are used to calculate the process capability index C_p.

15.20   Discuss the interpretation of the process capability index C_p.

15.21   Discuss the relationship between the two indexes, C_p and C_{pk}. Explain how they can be used together to determine how well a process is performing.

EXERCISES   Skill Level

15.22   Refer to Exercise 15.11 in which a process was observed with $\bar{\bar{x}} = 32.6$ and $\bar{R} = 15.1$ by selecting random samples of size $n = 10$. If this process is designed to produce measurements in the interval (30 ± 10), is the process capable of meeting its specifications? Calculate the process capability index C_p. Interpret its value.

15.23   Refer to Exercise 15.22. Calculate the C_{pk} index. Use this value to describe the extent to which the location of the process differs from its desired value, 30.

15.24   Refer to Exercise 15.13 in which a process produced $\bar{\bar{x}} = .640$, $\bar{R} = .015$, and $s = .005$.

    a.   Estimate the population standard deviation σ, using two different estimators.

    b.   If the process specification limits are $(.650 \pm .005)$, calculate the C_p index, using the two different estimates of σ from part a. Compare the results.

EXERCISES   Applications

   15.25   In Exercise 15.15 a bottling process was sampled to monitor the average weight of the finished bottles. The average weight of the overall sample was 5.2 oz, with a standard deviation of .3 oz. Is this process capable of meeting the specification limits, 4.8 to 5.5 oz, set by the soft drink manufacturer in Exercise 15.16? Use C_p and C_{pk} to measure this capability.

   15.26   Refer to Exercise 15.17, in which the amount of radiation at an industrial site was measured. The Nuclear Regulatory Commission has set the specification limits not to exceed .005 microcurie. Calculate C_p and C_{pk}. Is the plant capable of meeting the specification limits? Explain.

15.7

The _p_ Chart

The previous two sections dealt with control charts and capability analysis for variables data. We turn now to control chart methods used with attribute data. Recall from Section 15.2 that attribute data refers to measurements that count either the number of defects per item or the number of defective items in samples from a process.

One of the most basic process measurements that can be made is whether or not an item meets its specifications. If it doesn't, the item is said to be defective or nonconforming.* So a ball bearing will be considered defective if its diameter is not within the specification limits set by the design engineers, or an accounting record can be called defective if it is found to contain any errors.

A control chart to monitor defective items is constructed by taking periodic samples of n items from a process and plotting $\hat{p}$, the sample proportion defective. If the process is in statistical control and p is the overall proportion defective, then the sample proportions should fall within three standard deviations $(3\sigma_{\hat{p}})$ of p, where

$$\sigma_{\hat{p}} = \sqrt{\frac{p(1-p)}{n}}$$

For the estimation of p the sample proportions from k samples are averaged. This estimate.

$$\bar{p} = \frac{1}{k} \sum_{i=1}^{k} \hat{p}_i$$

is then used as the centerline of the p chart. When we substitute $\bar{p}$ for p, the estimated variation in the fraction defective becomes

$$\hat{\sigma}_{\hat{p}} = \sqrt{\frac{\bar{p}(1-\bar{p})}{n}}$$

The upper and lower control limits are then

$$\text{UCL} = \bar{p} + 3\sqrt{\frac{\bar{p}(1-\bar{p})}{n}} \quad \text{and} \quad \text{LCL} = \bar{p} - 3\sqrt{\frac{\bar{p}(1-\bar{p})}{n}}$$

Sometimes, because of the small values of $\bar{p}$ that are often encountered in practice, the LCL can be negative. In those cases we then replace LCL by 0.

---

* The American Society for Quality Control (ASQC) recommends that one distinguish between nonconforming units and defective units. Nonconforming units are units that do not meet specifications, whereas defective units are those deemed unsuitable for any intended or foreseeable use requirements. In other words, defective units are worse than nonconforming units. Technically, attribute control charts are used to monitor nonconforming items, but we will also refer to these items as defectives in order to simplify the discussions.

The p Chart

Centerline: $\bar{p} = \dfrac{1}{k} \sum_{i=1}^{k} \hat{p}_i$.

Upper control limit: $\text{UCL} = \bar{p} + 3\sqrt{\dfrac{\bar{p}(1 - \bar{p})}{n}}$.

Lower control limit: $\text{LCL} = \bar{p} - 3\sqrt{\dfrac{\bar{p}(1 - \bar{p})}{n}}$.

Note: If LCL is negative, replace it by 0. The number of samples k should be at least 25.

Unlike sample sizes for $\bar{x}$ and R charts, the sample sizes n are usually much larger for the p charts. For example, a common practice is to use a day's production of items as the sample size and to record the proportion defective produced each day. With attribute data keeping the sample size constant from sample to sample is also very difficult. The number of parts a process makes, for example, is rarely the same from day to day.

Three methods are available to handle the problem of varying sample sizes:

1. If the sample sizes do not vary much from each other, then use their average size, $\bar{n}$, in the control limit formulas.
2. If the sample sizes vary considerably, then they should individually be used when calculating the control limits. These limits will then vary from sample to sample, depending on the particular sample size.
3. In case 2, if you desire to keep the control limits constant, then you can plot the sample z scores,

$$\frac{\hat{p} - \bar{p}}{\sqrt{\bar{p}(1 - \bar{p})/n_i}}$$

instead of the $\hat{p}$'s. The three-sigma control limits will then be UCL $= 3$ and LCL $= -3$, regardless of the size of the n_i's.

EXAMPLE 15.8   In recent years the Department of Defense has increased its quality requirements for weapons systems and other military hardware (F. C. Collins, "Department of Defense Renews Emphasis on Quality," *Quality Progress*, March 1988, pp. 19–21). Aerospace contractors and subcontractors who manufacture these systems must often demonstrate, using control charts, that they are capable of meeting these new requirements. Many such systems use a large number of circuit card assemblies, which consist of printed-circuit boards with various electronic components soldered to them. The components are soldered in place by using a wave solder machine that passes the boards over a surface of liquid solder. Soldered boards are then connected to test stations that test the circuits and classify the boards as either defective or nondefective.

Table 15.8   Circuit Board Data for Example 15.8

Day, i	Rejects	Tested, n_i	Percent, $\hat{p}_i$	Day, i	Rejects	Tested, n_i	Percent, $\hat{p}_i$
1	14	286	0.049	16	15	297	0.051
2	22	281	0.078	17	14	283	0.049
3	9	310	0.029	18	13	321	0.040
4	19	313	0.061	19	10	317	0.032
5	21	293	0.072	20	21	307	0.068
6	18	305	0.059	21	19	317	0.060
7	16	322	0.050	22	23	323	0.071
8	16	316	0.051	23	15	304	0.049
9	21	293	0.072	24	12	304	0.039
10	14	287	0.049	25	19	324	0.059
11	15	307	0.049	26	17	289	0.059
12	16	328	0.049	27	15	299	0.050
13	21	296	0.071	28	13	318	0.041
14	9	296	0.030	29	19	313	0.061
15	25	317	0.079	30	12	289	0.042

$$\bar{p} = \frac{1}{30} \sum_{i=1}^{30} \hat{p}_i = .054 \quad \bar{n} = \frac{1}{30} \sum_{i=1}^{30} n_i = 305.17 \approx 305$$

Table 15.8 contains records of the daily numbers of rejected circuit boards for a 30-day period. Construct a p chart for this data. Since the daily production levels are not constant, use both the average sample size method and the variable–control limit method to construct the chart. What are the advantages of using each of these methods?

SOLUTION   Since $\bar{p} = .054$ and $\bar{n} = 305$ (given in Table 15.8), the average sample size method gives control limits of

$$\text{UCL} = \bar{p} + 3\sqrt{\frac{\bar{p}(1 - \bar{p})}{\bar{n}}} = .054 + 3\sqrt{\frac{(.054)(.946)}{305}} = .093$$

and

$$\text{LCL} = \bar{p} - 3\sqrt{\frac{\bar{p}(1 - \bar{p})}{\bar{n}}} = .054 - 3\sqrt{\frac{(.054)(.946)}{305}} = .015$$

Figure 15.15 shows the p chart with these control limits. From this chart the process appears to be in control, since none of the points exceed the control limits.

In the variable sample size method the control limits for the first sample would be

$$\bar{p} \pm 3\sqrt{\frac{\bar{p}(1 - \bar{p})}{n_1}} \quad \text{or} \quad .054 \pm 3\sqrt{\frac{(.054)(.946)}{286}} \quad \text{or} \quad .054 \pm .040$$

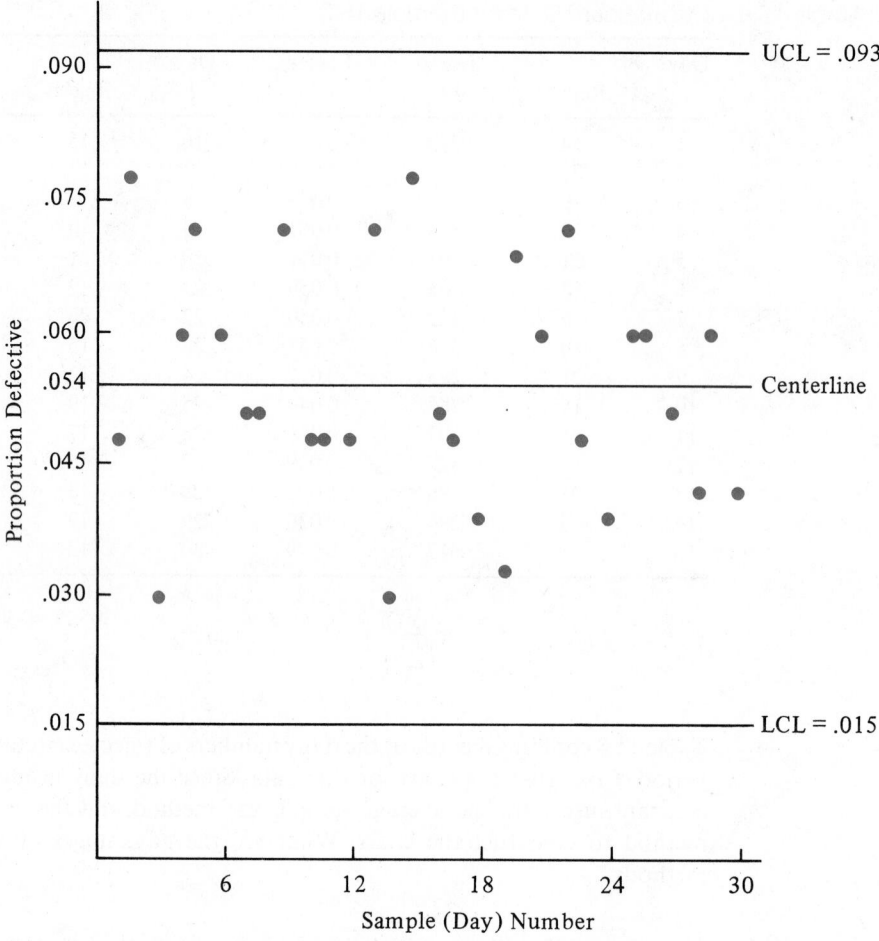

Figure 15.15 *p* Chart for Example 15.8, Using Control Limits Based on Average Sample Size

Similarly, control limits for, say, sample 12 would be

$$\bar{p} \pm 3 \sqrt{\frac{\bar{p}(1 - \bar{p})}{n_{12}}} \quad \text{or} \quad .054 \pm 3 \sqrt{\frac{(.054)(.946)}{328}} \quad \text{or} \quad .054 \pm .037$$

and so on. Figure 15.16 shows the *p* chart using these control limits, which take into account the changing sample sizes. Once again, the process appears to be in control.

Both charts give relatively the same results because the sample sizes n_i do not vary much around their average value of 305. If they had varied, then the chart with variable control limits would have been preferred.

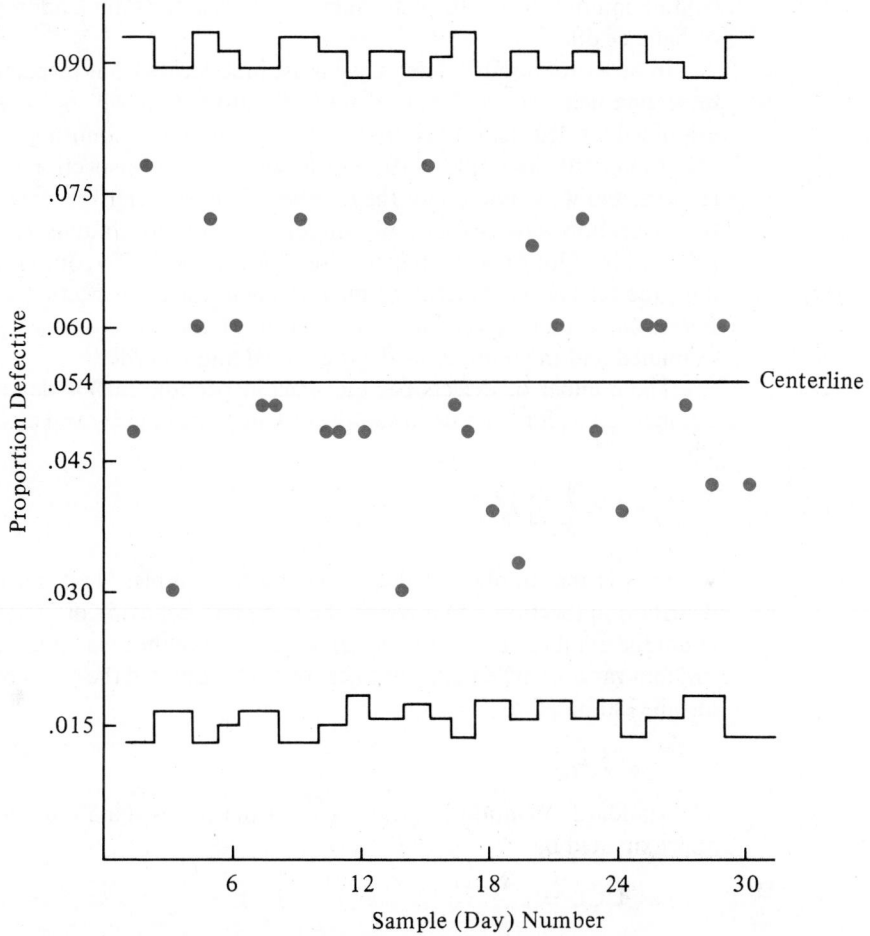

Figure 15.16   *p* Chart for Example 15.8, Using Variable Control Limits Based on Changing
Sample Sizes

15.8

The *c* Chart

defects per item
Besides monitoring the number of defective items made by a process, one can also chart the number of **defects per item**. As you will recall, defects and defectives are two different things. A defect is just a flaw or nonconformity; therefore, a defective item may have more than one defect. For example, an incorrect (defective) accounting record might

contain multiple errors (defects), such as errors in the billing address, account number, or balance due.

inspection unit
In order to monitor defects, we must first decide what inspection unit to use. The **inspection unit** defines the fixed unit of output that will be regularly sampled and examined for defects. For example, when examining accounting records for errors, we may want to use a sample of 10 records each day. The inspection unit would then be 10 records, and we would count the number of errors per 10 records sampled. Choosing the inspection unit is especially important with continuous processes such as the production of long rolls of paper, wire, cloth, or metal. To count the number of surface flaws (defects) in sheet metal, we may decide to use an inspection unit of 1 square foot. Periodically, then, a 1-square-foot section of the metal surface produced would be examined and the number of flaws counted and recorded.

The number of defects per unit (i.e., inspection unit) is denoted by c. To create a control chart for c, we use a sample of k inspection units and calculate the centerline $\bar{c}$ by

$$\bar{c} = \frac{1}{k} \sum_{i=1}^{k} c_i$$

where c_i is the number of defects in the ith sample. Next, recall that the Poisson distribution (Section 5.5) governs the sampling behavior of statistics, such as c, that count the number of events (defects) per unit. Since the mean and variance of a Poisson random variable are equal, and since we have estimated the mean by $\bar{c}$, the variance can also be estimated by

$$\hat{\sigma}_c^2 = \bar{c}$$

The standard deviation is then $\hat{\sigma}_c \sqrt{\bar{c}}$, so the three-sigma ($3\sigma_c$) control limits can be approximated by

$$\text{UCL} = \bar{c} + 3\sqrt{\bar{c}} \quad \text{and} \quad \text{LCL} = \bar{c} - 3\sqrt{\bar{c}}$$

The c Chart

> Centerline: $\bar{c} = \dfrac{1}{k} \sum_{i=1}^{k} c_i.$
>
> Upper control limit: $\text{UCL} = \bar{c} + 3\sqrt{\bar{c}}.$
>
> Lower control limit: $\text{LCL} = \bar{c} - 3\sqrt{\bar{c}}.$
>
> *Note:* If LCL is negative, replace it by 0. This problem can be avoided if the inspection unit is chosen so that $\bar{c}$ exceeds 9.

EXAMPLE 15.9   The defective circuit card assemblies in Example 15.8 could have been defective for a variety of reasons, one of which is defects in the solder connecting the electronic components to the cards. Circuit boards typically have hundreds of solder connections, and finished boards are visually inspected for solder flaws. An inspector records the location of the defects by circling the corresponding connections on a photocopy of the

Table 15.9   Solder Defects Data for Example 15.9

Board, i	Number of Defects, c_i	Board, i	Number of Defects, c_i	Board, i	Number of Defects, c_i
1	14	18	17	35	1
2	17	19	16	36	9
3	15	20	13	37	9
4	12	21	18	38	4
5	10	22	13	39	10
6	16	23	24	40	0
7	16	24	14	41	9
8	20	25	17	42	5
9	18	26	13	43	11
10	13	27	11	44	8
11	15	28	21	45	12
12	14	29	16	46	10
13	13	30	8	47	7
14	17	31	10	48	9
15	15	32	15	49	6
16	16	33	11	50	13
17	13	34	13		

$$\bar{c} = \frac{1}{50} \sum_{i=1}^{50} c_i = 12.54$$

circuit card. With this sheet as a guide, other workers can then resolder the defective connections so that the board will not have to be scrapped.

Table 15.9 contains the numbers of solder defects found on a sample of 50 circuit cards of the same type. Construct a c chart for this data. Suppose that a new wave solder machine was used starting with sample 30. Can you conclude that the new machine has reduced the number of solder defects per board?

SOLUTION   The average number of defects per board is $\bar{c} = 12.54$ (given in Table 15.9), so the control limits are

$$\text{UCL} = \bar{c} + 3\sqrt{\bar{c}} = 12.54 + 3\sqrt{12.54} = 23.16$$

and $$\text{LCL} = \bar{c} + 3\sqrt{\bar{c}} = 12.54 - 3\sqrt{12.54} = 1.92$$

Figure 15.17 shows the c chart, control limits, and centerline for this data. Notice that two of the points on the chart (samples 35 and 40) exceed the lower control limit. Exceeding the lower control limit is, in this case, a good sign since it means that the process is creating significantly fewer defects than normal. This result is fairly strong evidence to conclude that the new wave solder machine has improved the process. Since it appears that the process has improved, the control limits should also be recalculated by using the data after the installation of the new machine. Note that the reduced number of defects will translate into cost savings because there will be correspondingly less rework required.

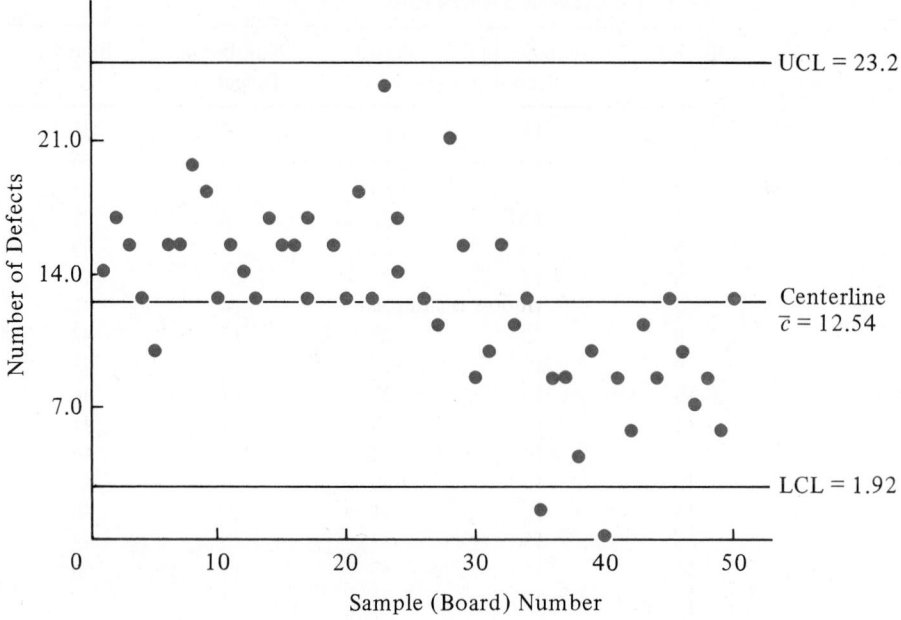

Figure 15.17   *c* Chart for Example 15.9

EXERCISES   Conceptual Level

15.27   Explain the difference between variables data and attribute data in the context of statistical process control (SPC).

15.28   What is the purpose of a study using each of the following control charts? Give an example of a business problem where each chart would be useful in SPC.

   a.   *p* chart   **b.**   *c* chart

EXERCISES   Skill Level

15.29   Random samples of $n = 50$ items were selected twice daily from a process over a one-month period and produced an overall defective rate of $\bar{p} = .036$.

   a.   Construct a *p* chart for the process.

   b.   Explain how the *p* chart constructed in part a will be used.

15.30   Random samples of various sizes were selected hourly from a process judged to be in control. The average proportion of defectives was $\bar{p} = .06$, and the average sample size was $\bar{n} = 325$. If the sample sizes do not vary too much, construct a *p* chart for the process. Explain how it will be used.

15.31   Refer to Exercise 15.30. If the sample sizes selected from the process vary considerably, how would you calculate the control limits for a particular sample of size $n_i = 350$?

15.32   Random samples of 100 inspection units produced an average of 6 defects per unit.

   a.   Construct a *c* chart for the process.

   b.   Explain how the *c* chart constructed in part a will be used.

15.33   Random samples of 75 inspection units produced a *total* of 265 defects.

    a.   Calculate the average number of defects per unit.

    b.   Construct a *c* chart for the process.

EXERCISES   Applications

15.34   In one of the production lines at an automobile assembly plant, side panel and door windows are produced. A window is categorized as defective if it contains any scratches, cracks, bubbles, or other obvious visual imperfections. A quality control engineer samples the output of the production line once every day by randomly selecting 100 items and noting the number of defectives. The results of the past 20 days are shown in Table 15.10. Construct a *p* chart, and check to see whether the process is in control.

Table 15.10   Sample Data for Exercise 15.34

Sample	Number of Defectives	Sample	Number of Defectives
1	8	11	4
2	5	12	3
3	3	13	5
4	9	14	6
5	4	15	2
6	5	16	5
7	8	17	0
8	5	18	3
9	3	19	4
10	6	20	2

15.35   The copy editor for a daily newspaper, in an attempt to control copy and typesetting errors, has decided to employ a quality control study. For each of the past 20 days' newspapers he has randomly selected 10 pages and recorded the number of errors. The results are given in Table 15.11. Construct a *c* chart with which the copy editor can monitor future error rates.

Table 15.11   Errors Data for Exercise 15.35

Day	Number of Errors	Day	Number of Errors
1	11	11	13
2	14	12	7
3	20	13	11
4	13	14	22
5	16	15	4
6	14	16	13
7	18	17	12
8	9	18	10
9	19	19	18
10	12	20	16

15.9

Acceptance Sampling

lots

traceability

After products are manufactured, they are usually grouped together into **lots** before shipping to customers. Lot numbers are assigned to each such group of items to facilitate **traceability**. That is, if any questions or problems should arise concerning the product, the manufacturer can use the lot number to trace these problems back to the particular factors involved when the lot was made.

Many times, lots are inspected to see whether they are of acceptable quality. The inspection can occur in two places. If the manufacturer inspects the lots before shipping them, it is called **outgoing inspection**. If the customer inspects the lots before accepting them, it is called **incoming inspection** or receiving inspection. Figure 15.18 depicts how lots are screened during incoming inspection.

outgoing inspection

incoming inspection

In either case, the inspection proceeds by taking a sample from the lot and either measuring a particular characteristic of the product or simply counting the number of sampled items not meeting specifications. For measured (i.e., variables) data a lot is rejected whenever its sample mean exceeds some specified limits. For attribute data lots are rejected when the number of defectives in the sample is too large. Lots that pass the inspection are then shipped (if the manufacturer does the inspection) or put into stock (if the customer does the inspection). Because of their relative simplicity, acceptance sampling plans for attribute data are used more frequently than those for variables data.

The use of acceptance sampling plans has subsided somewhat in recent years due to increased emphasis on controlling processes rather than inspecting finished products. Some people have argued, for example, that inspecting finished products does not do much to improve product quality; it merely sorts out some of the bad lots. However, acceptance sampling still has a role to play. Although it should not be used in place of control charts, there are many situations where it is economically advisable to use acceptance sampling. New processes, for example, which are replete with unknowns and uncertainties, are not always immediately controllable with charting techniques. Until such processes are brought into statistical control, some amount of inspection is necessary to protect against serious production errors.

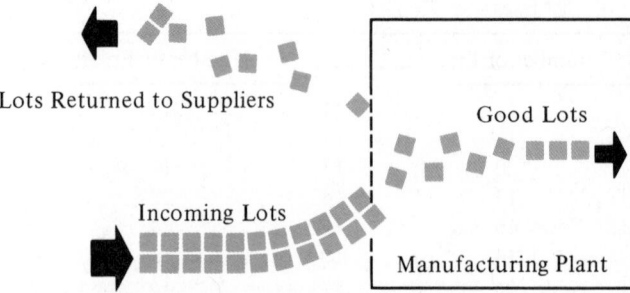

Figure 15.18   Screening Lots During Incoming Inspection

Definition

acceptance sampling
plan for attributes

> An **acceptance sampling plan for attributes** is characterized by a sample size n and an acceptance number a. With the use of these numbers a lot is rejected if the number of defectives in a sample of n items from the lot exceeds the acceptance number a. Otherwise, the lot is accepted.

Changing one or both of these numbers changes the sampling plan and, therefore, changes the ability of the plan to screen out bad lots. For the screen to operate satisfactorily, we would like the probability to be high of accepting lots with a low fraction defective and the probability to be low of accepting lots with a high fraction defective.

EXAMPLE 15.10   Many practitioners use acceptance sampling plans with an acceptance number of $a = 0$. These plans are called **zero acceptance plans** and have become popular because many customers, not wanting any defectives in a lot, intuitively think that zero acceptance plans offer the best protection (see R. C. Baker, "Zero Acceptance Sampling Plans: Expected Cost Increases," *Quality Progress*, January 1988, pp. 43–46).

Calculate the probability of accepting a lot whose fraction defective is $p = .05$ when using an acceptance plan that employs sample sizes of $n = 25$. Repeat this calculation for $p = 0, .1, .2,$ and 1.0.

SOLUTION   If we assume that the sample size, $n = 25$, is small compared with the number of items in the lot, then the number of defectives, x, in a sample should follow a binomial distribution with $n = 25$ and $p = .05$ (see Section 5.3). Therefore,

$$P(\text{accept lot}) = P(x = 0) = C_0^{25} p^0 q^{25} = (1 - p)^{25} = (1 - .05)^{25} = .277$$

Thus, this plan will let 27.7% of all lots that are 5% defective ($p = .05$) pass through the inspection. Repeating the calculation for $p = 0, .1, .2,$ and 1.0, we get

$$P(\text{accept lot when } p = 0) = (1 - 0)^{25} = 1$$
$$P(\text{accept lot when } p = .1) = (1 - .1)^{25} = .072$$
$$P(\text{accept lot when } p = .2) = (1 - .2)^{25} = .004$$
$$P(\text{accept lot when } p = 1.0) = (1 - 1)^{25} = 0$$

◆ ◆

operating
characteristic (OC)
curve

To see how well a sampling plan will work, we plot the probability of lot acceptance against the lot fraction defective, p, for a range of values. The resulting graph is called the **operating characteristic (OC) curve** of the sampling plan. For example, Figure 15.19 shows the OC curve for the $n = 25$, $a = 0$ sampling plan of Example 15.10. This curve was drawn by using the lot acceptance probabilities calculated in Example 15.10. Notice that this OC curve opens upward, which is a characteristic unique to sampling plans with $a = 0$. From the OC curve we can quickly

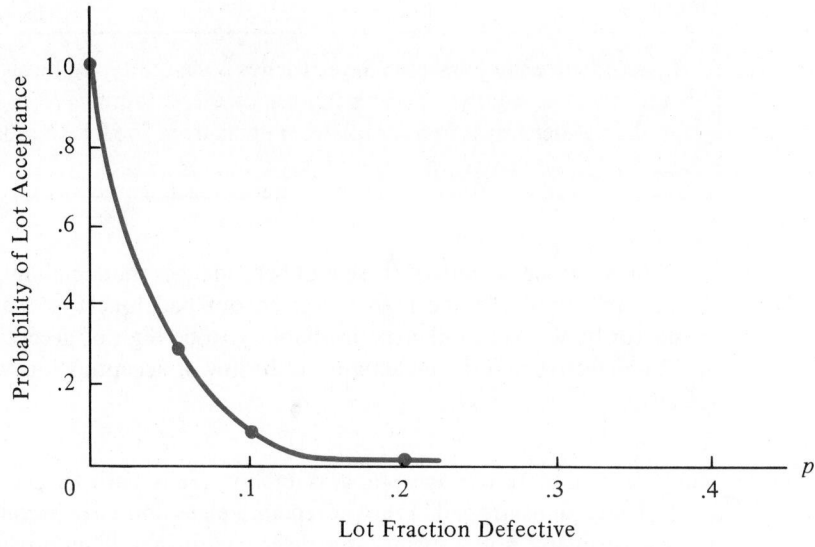

Figure 15.19   Operating Characteristic Curve ($n = 25$, $a = 0$) for Example 15.10

see what level of protection the plan will give for various values of p. Lots that are 5% defective, for example, will slip through the inspection about 27.7% of the time, but almost no lots (only .4%) that are 20% defective will be accepted.

Depending on the industry, producers and consumers are often more specific about how well a sampling plan is expected to perform. Sometimes, they agree on a **acceptable quality level (AQL)**, such that any lot whose fraction defective doesn't exceed the AQL will be accepted by the customer. A sampling plan should then have a high probability of accepting lots whose fraction defective is less than or equal to the AQL and a high probability of rejecting the rest of the lots.

EXAMPLE 15.11   In Example 15.10 suppose that the manufacturer and customer agree to an AQL of .05. What is the impact on both parties of using the zero acceptance plan $n = 25$, $a = 0$ of that example? Next, draw the OC curve for a plan based on $n = 25$, $a = 1$. What are the advantages of using this plan?

SOLUTION   From Example 15.10, 27.7% of all lots that are 5% defective will be accepted by the $n = 25$, $a = 0$ plan. That means 72.3% of such lots will be rejected. From the producer's point of view this plan is not a good one since 5% defective lots are considered acceptable (the AQL is .05), yet this plan will reject the majority of them!

For the $n = 25$, $a = 1$ plan the probability of lot acceptance will be

$$P(\text{accept lot}) = P(x \leq 1) = P(x = 0) + P(x = 1) = C_0^{25}p^0q^{25} + C_1^{25}p^1q^{24}$$

From Table 1 in the Appendix these acceptance probabilities for $p = 0, .05, .1, .2,$ and

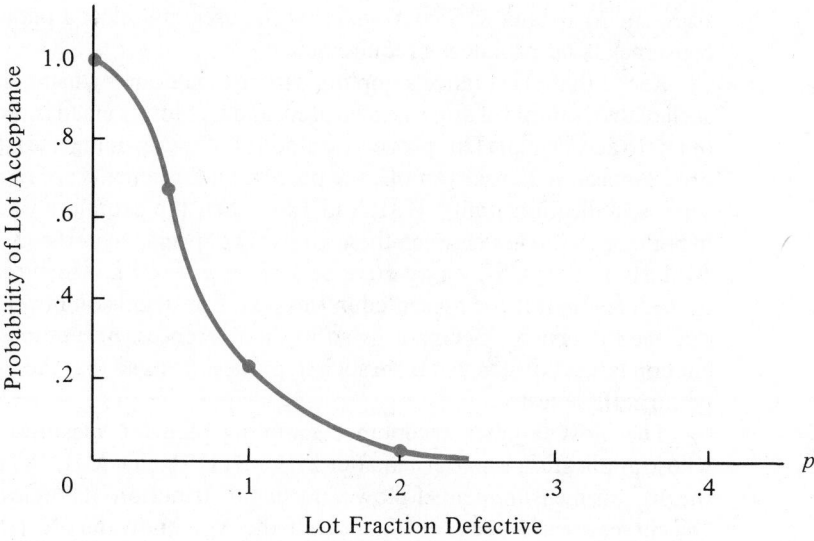

Figure 15.20   Operating Characteristic Curve ($n = 25$, $a = 1$) for Example 15.11

1.0 are found to be

$$P(\text{accept lot when } p = 0) = 1$$
$$P(\text{accept lot when } p = .05) = .642$$
$$P(\text{accept lot when } p = .1) = .271$$
$$P(\text{accept lot when } p = .2) = .027$$
$$P(\text{accept lot when } p = 1.0) = 0$$

The OC curve for this plan is shown in Figure 15.20. Lots that are 5% defective (and considered acceptable since AQL $= .05$) are now accepted by the plan 64.2% of the time, which is somewhat better for the producer than the previous plan was. With the new plan the consumer assumes a slightly greater risk of getting a few lots with defect rates higher than the AQL but otherwise benefits from the increased number of AQL lots that will be shipped by the new plan. Adjustment of the sample size and the acceptance number will eventually produce a plan that is economically acceptable to both parties.

◆ ◆

Many forms of acceptance sampling plans have been developed over the years. An excellent compilation of these plans can be found in *Acceptance Sampling in Quality Control* by E. G. Schilling (listed in the references at the end of this chapter). One of the most widely used plans for attributes data is Military Standard 105D, which is usually abbreviated MIL–STD–105D. It has become a standard in many industries, especially those involved with government contract work. The plan gives tables of AQLs, sample sizes, and associated OC curves so that the user can select a plan that meets both the consumer's and producer's requirements.

Recall that acceptance sampling can also be done by using data for variables. In such plans a sample of size n is measured, and its sample mean is used to decide whether or not to accept a lot. This procedure amounts to performing a test of hypothesis for the process mean μ. If, for example, the process measurements are required to be between some specification limits, USL and LSL, then the sample mean is used to test the hypothesis that μ lies between these limits. The specific hypotheses being tested here are $H_0: \text{LSL} < \mu < \text{USL}$ versus $H_a: \mu < \text{LSL}$ or $\mu > \text{USL}$, which require more complex procedures than those presented in this text. For details on how such tests are carried out, see the text by DeGroot listed in the references. Acceptance plans for variables contain tables constructed from such hypothesis tests so that the user need not actually perform these tests.

The most familiar acceptance sampling plan for variables is MIL–STD–414, whose application is patterned after MIL–STD–105D. MIL–STD–414 assumes that the process measurements follow a normal distribution. Its tables of sample sizes and OC curves are indexed according to whether you know the process standard deviation σ or are estimating it. If σ is unknown, the tables are further subdivided according to how you estimate σ (using either the $\bar{R}/d_2$ method or the sample standard deviation s). As a general rule, acceptance sampling plans for data for variables involve much more effort than attributes plans. For a good overview of the types of plans available and their intended uses, see Schilling's "An Overview of Acceptance Control" listed in the references.

EXERCISES   Conceptual Level

15.36   What is the basic purpose of lot acceptance sampling?

15.37   What information is provided by an operating characteristic curve, and how is it useful?

EXERCISES   Skill Level

15.38   A buyer and seller agree to use a sampling plan with sample size $n = 5$ and acceptance number $a = 0$. What is the probability that the buyer would accept a lot having the following fractions of defectives?

    a. $p = .1$ **b.** $p = .2$ **c.** $p = .4$ **d.** $p = 0$ **e.** $p = 1$

    Construct the operating characteristic curve for this plan.

15.39   Repeat Exercise 15.38 for $n = 5$ and $a = 1$.

15.40   Repeat Exercise 15.38 for $n = 10$ and $a = 0$.

15.41   Repeat Exercise 15.38 for $n = 10$ and $a = 1$.

15.42   Graph the operating characteristic curves for the four plans given in Exercises 15.38 through 15.41 on the same sheet of graph paper. What is the effect of increasing the acceptance number a when n is held constant? What is the effect of increasing the sample size n when a is held constant?

15.43   A buyer and a seller agree to use sampling plan $n = 15$, $a = 0$ or sampling plan $n = 25$, $a = 1$.

   a.   Sketch the operating characteristic curves for the two sampling plans.

   b.   The buyer specifies an acceptable quality level (AQL) of $p = .10$. Which of the two sampling plans would he prefer? Why?

EXERCISES   Applications

15.44   A California winery, which distributes inexpensive table wines through supermarkets, buys 1-liter wine bottles in large lots from a Brazilian supplier. To be acceptable, each bottle must be uniform in color, conform to size and shape requirements, and be free of cracks or chips. Each lot received from the supplier is inspected by selecting $n = 25$ bottles from the lot and counting the number of bottles within the sample that are found to be unacceptable.

   a.   On the same sheet of graph paper, construct the operating characteristic curves for the sampling plans $n = 25$ with $a = 1$, 2, and 3.

   b.   Which sampling plan best protects the Brazilian supplier of wine bottles from having acceptable shipments of bottles rejected and returned by the winery?

   c.   Which sampling plan best protects the winery from accepting shipments of bottles for which the fraction of unacceptable bottles is exceedingly large?

   d.   How might a sampling inspector arrive at an acceptance level that compromises between the risk to the supplier and the risk to the buyer (the winery)?

15.45   With increasing concerns over product recalls and liability suits, manufacturers are employing increased caution in product testing and review. The automobile industry has been especially affected by these requirements, but the cost of product loss in the testing process offers them an interesting dilemma. Suppose the Ford Motor Company routinely selects $n = 25$ new passenger cars from the Ford assembly line and tests each in a collision test to measure the protection offered to simulated passengers during a 50-mile-per-hour (mph) crash. Assume further that Ford is willing to tolerate no more than a 10% chance that its products are not detected as defective if as much as 20% of the entire production line would fail to afford protection to passengers during a 50-mph crash.

   a.   How many of the automobiles involved in the collision tests must fail to provide required protection to passengers before Ford concludes that the production line needs review?

   b.   Since new cars are destroyed in the testing process, Ford wishes to compromise in the size and frequency of production testing. What are the implications of making such compromises?

15.46   In the language of industrial quality control, the probability that a good lot is incorrectly assumed to contain too many defective items is called the *producer's risk*, and the *consumer's risk* is the chance that a lot is accepted when it actually contains a high fraction of defective items. A quality control engineer wishes to study the comparative risks to the producer and consumer of using four different 1-in-5 sampling plans: $n = 5$, $a = 1$; $n = 10$, $a = 2$; $n = 15$, $a = 3$; and $n = 25$, $a = 5$. Assume the true lot fraction defective is $p = .10$.

   a.   Which sampling plan minimizes the consumer's risk?

   b.   Which sampling plan minimizes the producer's risk?

   c.   Since each sampling plan specifies lot acceptance if the sample fraction defective does not exceed 20%, why are the producer's risks not the same for the four 1-in-5 sampling plans?

   d.   Why are the consumer's risks not the same?

15.10

Summary

Quality control is concerned with the design and manufacture of goods and services. Statistical methods, which have been used in the field of quality control since its inception in the 1920s, are used to help in both the design and production phases. Experimental design methods, such as those introduced in Chapter 10, are used in product development, and control charts and acceptance sampling are applied as the products are manufactured.

Several simple graphical techniques are used to pinpoint problem areas and possible solutions. These techniques include Pareto charts, which separate the major problems from the minor ones, and fishbone diagrams, which attempt to locate problem causes. In addition, histograms provide a visual record of the measurements of important quality characteristics.

Control charts monitor the output of a process by simply plotting a relevant sample statistic for many successive samples. Points falling too far from the process average signal serious problems whose causes can often be found and eliminated. One goal of these charts is to detect problems quickly so that the problems may be fixed before product quality suffers. A second goal is to eventually bring a process into a state of statistical control so that the process output is predictable and process capability can be measured.

Acceptance sampling is used in quality control to inspect for serious quality problems before releasing products for use. Samples of finished products are examined and, on the basis of some statistic calculated from the sample, the batch of products is either accepted or rejected. Acceptance sampling does not offer the same degree of control over processes that control charts are designed to have, but such sampling is still important when one is working with new processes that are not yet in a state of statistical control.

Supplementary Exercises

Applications

15.47 A production process at a food-processing plant is designed to fill 1-quart (32 fluid ounces) containers with a mixed fruit juice. An inspector randomly selected four filled containers at each of 25 different times and measured their content. From the sample information he observed the overall sample mean and standard deviation to be $\bar{\bar{x}} = 30.8$ and $s = 1.20$ fluid ounces, respectively.

    a. Use the sample information to obtain control limits for the average content of the containers.

    b. Is there another method for obtaining the control limits? If so, what additional information is necessary?

15.48 A company that publishes classroom newspapers for elementary schools also offers a paperback book club through which classroom teachers place a joint order for all of their students.

The company does a large volume of business, shipping over 165,000 books per day. Since the company is concerned about incorrectly packaged book orders, the shipping cartons are periodically checked at the shipping dock. The number of books in the carton and the number of incorrectly packed books (books that were not ordered by the customer) are recorded in Table 15.12 for 30 random checks.

a. Construct a control chart to monitor the proportion of incorrectly packed books.

b. If the company is willing to tolerate as much as a 5% rate of incorrectly packed books, is the packing process exceeding the upper specification limit?

Table 15.12   Book Data for Exercise 15.48

Carton	Number of Books	Number Incorrectly Packed	Carton	Number of Books	Number Incorrectly Packed
1	52	2	16	45	8
2	13	1	17	71	6
3	26	2	18	99	2
4	67	5	19	36	0
5	12	0	20	43	3
6	33	2	21	17	0
7	42	4	22	60	5
8	19	3	23	29	2
9	86	4	24	95	8
10	73	10	25	87	2
11	35	4	26	45	0
12	65	5	27	74	9
13	32	7	28	41	2
14	64	5	29	51	2
15	112	12	30	80	0

15.49 Augat/Alcoswitch is a miniature-switch company that manufactures switches often used in PC boards, each consisting of from 6 to 12 components. The absence or misalignment of any component will render the switch malfunctional, and since PC boards often contain hundreds of switches, "zero quality control" is a goal set by the company [see Elisabeth BenDaniel, "Using Statistical Process Control with Robotic Testing Improves Quality Level," *Industrial Engineering*, Vol. 20, No. 2 (February 1988)]. The company uses robotic testers called "bakayokes" to test the switches. Suppose the bakayoke tests random samples of 10,000 switches each day for a period of 60 days and obtains an overall defective rate of .0008.

a. Construct a p chart for the process.

b. Suppose that the next three samples produce 12, 15, and 17 defectives, respectively. What can be said about the process? What, if any, action should be taken?

15.50 The following represent the number of imperfections (scratches, chips, cracks, blisters) noted in 25 finished (4 × 8) walnut wall panels:

7 5 4 10 9 5 6 3 8 8 3 5 4 9 3 3 2 4 1 5 7 3 2 6 3

The total number of defects on 75 finished panels previously inspected was 375.

a. Assuming the manufacturing process was in statistical control during the period when the data was gathered, construct a c chart to monitor the process (use the total number of defects for the 100 panels to construct the chart). Plot the number of defects listed above for the 25 panels.

b. If one wall panel is found to have more imperfections than the upper control limit allows, should the quality control engineer assume the manufacturing process is out of control, or should he wait until he finds repeated panels with an excessive number of imperfections before assuming the process is out of control? Explain.

15.51 A quality control engineer wishes to study the alternative sampling plans $n = 5$, $a = 1$ and $n = 25$, $a = 5$. On the same sheet of graph paper, construct the operating characteristic curve for both plans, making use of acceptance probabilities at $p = .05$, $p = .10$, $p = .20$, $p = .30$, and $p = .40$ in each case.

a. If you were a seller producing lots with fraction defective ranging from $p = 0$ to $p = .10$, which of the two sampling plans would you prefer?

b. If you were a buyer wishing to be protected against accepting lots with fraction defective exceeding AQL $= .30$, which of the two sampling plans would you prefer?

15.52 A radio and television manufacturer who buys large lots of transistors from an electronics supplier wishes to accept all lots for which the fraction defective is less than 6%. The manufacturer's sampling inspector selects $n = 25$ transistors from each lot shipped by the supplier and notes the number of defectives.

a. On the same sheet of graph paper, construct the operating characteristic curves for the sampling plans $n = 25$, $a = 1, 2$, and 3.

b. Which sampling plan best protects the supplier from having acceptable lots rejected and returned by the manufacturer?

c. Which sampling plan best protects the manufacturer from accepting lots for which the fraction of defectives exceeds AQL $= 6\%$?

d. How might the sampling inspector arrive at an acceptance level that compromises between the risk to the producer and the risk to the consumer?

15.53 A buyer and a seller agree to use sampling plan $n = 15$, $a = 1$ or sampling plan $n = 10$, $a = 0$. Under each of these plans, determine the probability that the buyer would accept the lot if the fraction defective of the lot is as given. Construct the operating characteristic curve for each of the plans.

a. $p = .05$ **b.** $p = .10$ **c.** $p = .20$ **d.** $p = .40$ **e.** $p = 1.0$

15.54 Refer to Exercise 15.52 and assume the manufacturer wishes the probability to be at least .90 of her accepting lots containing 1% defective and the probability to be about .90 of rejecting any lot with 20% defective. If the manufacturer's sampling inspector samples $n = 25$ items from the supplier's incoming shipments, what is the acceptance level (a) that meets the requirements?

15.55 The accompanying table lists the number of defective 60-watt light bulbs found in samples of 100 light bulbs selected over 25 days from a manufacturing process. Assume that during these 25 days the manufacturing process was not producing an excessively large fraction of defectives.

Day	1	2	3	4	5	6	7	8	9	10	11	12	13	14	15
Defectives	4	2	5	8	3	4	4	5	6	1	2	4	3	4	0

Day	16	17	18	19	20	21	22	23	24	25
Defectives	2	3	1	4	0	2	2	3	5	3

a. Construct a *p* chart to monitor the manufacturing process, and plot the data.

b. How large must the fraction of defective items be in a sample selected from the manufacturing process before the process is assumed to be out of control?

c. During a given day, suppose a sample of 100 items is selected from the manufacturing process and that 15 defective bulbs are found. If a decision is made to shut down the manufacturing process in an attempt to locate the source of the implied controllable variation, explain how this decision might lead to erronous conclusions.

15.56 A hardware store chain purchases large shipments of light bulbs from the manufacturer described in Exercise 15.55 and specifies that each shipment must contain no more than 4% defectives. When the manufacturing process is in control, what is the probability the hardware store's specifications are met?

15.57 Refer to Exercise 15.55. During a given week the number of defective bulbs in each of 5 samples of 100 were found to be 2, 4, 9, 7, and 11. Is there reason to believe that the production process has been producing an excess proportion of defectives at any time during the week?

15.58 A process evaluation study examined the lower glue joint gaps of corrugated boxes, whose known standard is 48 units (measured in 1/128 in.) [see Boris Iglewicz and David C. Hoaglin, "Use of Boxplots for Process Evaluation," *Journal of Quality Technology*, Vol. 19, No. 4 (October 1987), p. 180]. The specification limits for the glue joint gaps are 48 ± 16 units. Measurements were taken on 80 boxes, randomly selected in groups of 5. The sample means and ranges are given in Table 15.13.

a. Construct an $\bar{x}$ chart for the process, graphing both the control limits and the specification limits.

b. Is the process in control?

c. Construct an *R* chart for the process to monitor the variability in the glue joint gaps.

Table 15.13   *Sample Data for Exercise 15.58*

Sample	$\bar{x}$	R	Sample	$\bar{x}$	R
1	19	49	9	25	40
2	16	30	10	32	17
3	21	34	11	30	12
4	51	37	12	31	25
5	46	34	13	28	28
6	42	32	14	27	20
7	24	24	15	28	18
8	14	25	16	21	15

15.59 Refer to Exercise 15.58. The quality control engineer at the manufacturing plant concluded that a machine setting was too low and adjusted the process. The measurements in Table 15.14 were then recorded.

a. Construct the $\bar{x}$ and *R* charts for the process after adjustment.

b. Does the process appear to be in control?

c. Are there any outliers in the process? What might be the cause?

Table 15.14   Additional Sample Data for Exercises 15.58
and 15.59

Sample	$\bar{x}$	R	Sample	$\bar{x}$	R
17	53	40	25	44	20
18	48	13	26	48	17
19	45	16	27	53	7
20	56	22	28	48	15
21	46	19	29	45	19
22	53	22	30	57	49
23	53	23	31	46	26
24	54	45			

Supplementary Exercises

Interpretive

15.60 One way of describing a production process in which smaller units are nested within larger units is to refer to the smaller units as subvessels and the larger units as vessels. Hockman and Lucas describe a case study in which the total production (vessel) is the cumulative sum of many similar machines (subvessels) producing the same product [K. K. Hockman and James M. Lucas, "Variability Reduction Through Subvessel CUSUM Control," *Journal of Quality Technology*, Vol. 19, No. 3 (July 1987)]. An analysis of variance of historical process data was used to determine significant sources of variation in the production process. The ANOVA table in which the important factors given by day, group, subvessels, and subvessel indicator (the number of subvessels/machines operating outside of specification limits) is summarized in the Table 15.15. The sums of squares have been rescaled to the percentage of the total variation; that is, they have been divided by the total sum of squares and expressed as a percentage. The Pareto principle suggests that 80% of the problems can be accounted for by fewer than 20% of the problem

Table 15.15   ANOVA for Exercise 15.60

Source	d.f.	% Total Variation
Day	29	.87
Group	4	2.04
Day × group	29 × 4	.40
Subvessel (within-group)	9 × 5	4.47
Subvessel indicator	p^*	84.31
Day × subvessel	$29 \times 45 - p$	.19
Error	30 × 50 × 2	7.72
Total	$(30 \times 50 \times 3) - 1$	100.00

Source: Data from K. K. Hockman and James M. Lucas, "Variability Reduction Through Subvessel CUSUM Control," *Journal of Quality Technology*, Vol. 19, No. 3 (July 1987).

* p is the number of vessels showing level shifts.

categories. Construct a Pareto chart, using the sources of variation as the problem areas and the percentage of variation values as "relative frequencies." Given the limited information provided here, how would you interpret this resulting Pareto chart?

15.61 What follows is a summary of a study reported by S. Ophir, U. El-Gad, and M. Snyder ["A Case Study of the Use of Experimental Design in Preventing Shorts in Nickel-Cadmium Cells," *Journal of Quality Technology*, Vol. 1, No. 1 (January 1988), pp. 44–50]. SPC and experimental design were both successfully used in arriving at a production process with almost zero defectives.

A high proportion of nickel-cadmium batteries produced by Tadiran's nickel-cadmium battery plant (Givat-Shmuel, Israel) were being scrapped because of shorts in the battery cells. A quality control team organized to investigate the causative factors and suggest control measures, where possible, used the classic SPC approach to isolate and correct problems in the production process. The control procedures initially in place consisted of testing the batteries *after* the battery core had been filled with electrolyte. The team, which began by testing the cores before they were filled, found, first, that shorted cells could be isolated before they were filled, on the basis of their electrical resistance. Second, from a Pareto analysis only *two* further causes of shorts were found; at this point the team suggested a simple solution to the first cause, which was achieved by modifying a welding procedure. The second cause was attacked by using the Ishikawa fishbone diagram shown in Figure 15.21.

Four main causes were charted, and a terse description of the subcauses follows. The workers ("Man") who roll the plates and separators may be insufficiently trained ("Training") or not sufficiently careful ("Caution"). The various "Machines" used in the process consisted of the "Rolling machines" used to roll the plates, the "Guillotines" used to cut the plates, and the "Tab

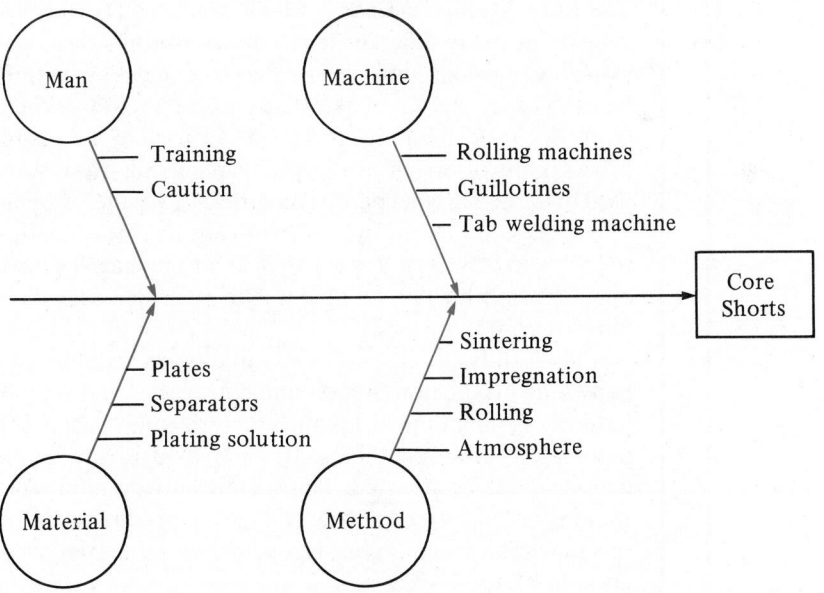

Figure 15.21   Ishikawa Fishbone Diagram for Exercise 15.61
Source: S. Ophir, U. El-Gad, and M. Snyder, "A Case Study of the Use of Experimental Design in Preventing Shorts in Nickel-Cadmium Cells," *Journal of Quality Technology*, Vol. 1, No. 1 (January 1988), pp. 44–50. Reprinted with permission.

welding machines" used to weld a tab to the positive plate. The "Material" relates to the thickness of the plates and their propensity to crumble; the "Separators," used to separate the positive and negative electrodes in the battery, were possibly too thin or too weak; and the "Plating solution" and/or "Method" may be in error. As for the method itself, faulty "Sintering" (the fusing of the nonmetallic material to make the plates), "Impregnation" (an overabsorption during an impregnation process), the order of "Rolling" the plates (negative first or positive first), and moisture on the plates due to atmospheric conditions were listed as possible causes.

A series of experiments involving several factors, each at two levels, indicated that the percentage of batteries with shorts was at a maximum when the settings of these experimental factors were those *in use at the time*, and that *the percent defective was minimized when a new technology was used to accomplish the sintering, when the separators were thicker, and when the negative plate was rolled first*. In summary, the cost of the quality control project was equal to about 1/8 of the savings achieved in the cost of scrapped cells in the first year.

Discuss the various quality control methods used in this study. How did SPC result in cost savings for the plant owners? (You may wish to review the original article in which many more details concerning the process under study are given.)

Case Study

Does Statistical Process Control Really Work?

The Ford Motor Company, which produces the various parts for its vehicles in many different locations and brings these parts to central assembly locations, must ensure that these parts are within specification limits in order that the parts be assembled into a working, nondefective entity with a minimum of defects. As part of its ongoing statistical process control, one of Ford's problem-solving teams selected a process used to harden the fuel pump eccentric of a 3.8-liter, V6 engine camshaft. The process produced inconsistent case hardness depth, causing 12% rework and 9% scrap. Excessive drill bit breakage occurred at the next operation in which an oil hole was to be drilled near the hardened fuel pump eccentric.

The study group decided initially to sample parts from the hardening production process and construct $\bar{x}$ and R charts of the case hardness depth at the fuel pump eccentric lobe "nose." The process was automated; however, the electric coil used in the hardening process (coil A) could be adjusted. Thirty consecutive samples of $n = 5$ pieces were recorded, as well as any changes or repairs to the process during this time. The $\bar{x}$ and R charts, together with the trial control limits, are given in Figure 15.22.

At point A the power on the coil was increased from 8.2 to 9.2; at point B the team discovered and straightened a bent coil; at point C power on the coil was reduced to 8.8. At point D the coil shorted out and needed to be straightened; at this time the team devised a gauge to check the coil spacing to the camshaft. At point *E* the team decreased the spacing between the camshaft and the coil; and at point *F* the first coil (A) was replaced with a second coil (B) of the same type.

1. How would you interpret the results of the actions taken at points *A*, *B*, *C*, *D*, *E*, and *F*?
2. After coil B was installed, the $\bar{x}$ and *R* charts in Figure 15.23 resulted. Do these charts indicate that the process has been stabilized? If specification limits are 3.5 to 10.5 mm, is the process capable of producing *individual* parts that will be within the specification limits?
3. The last chart (Figure 15.24) was plotted after a third redesigned coil was installed. Is the process in control? Is the process capable of producing parts within the specification limits? Has SPC been effective in this study?

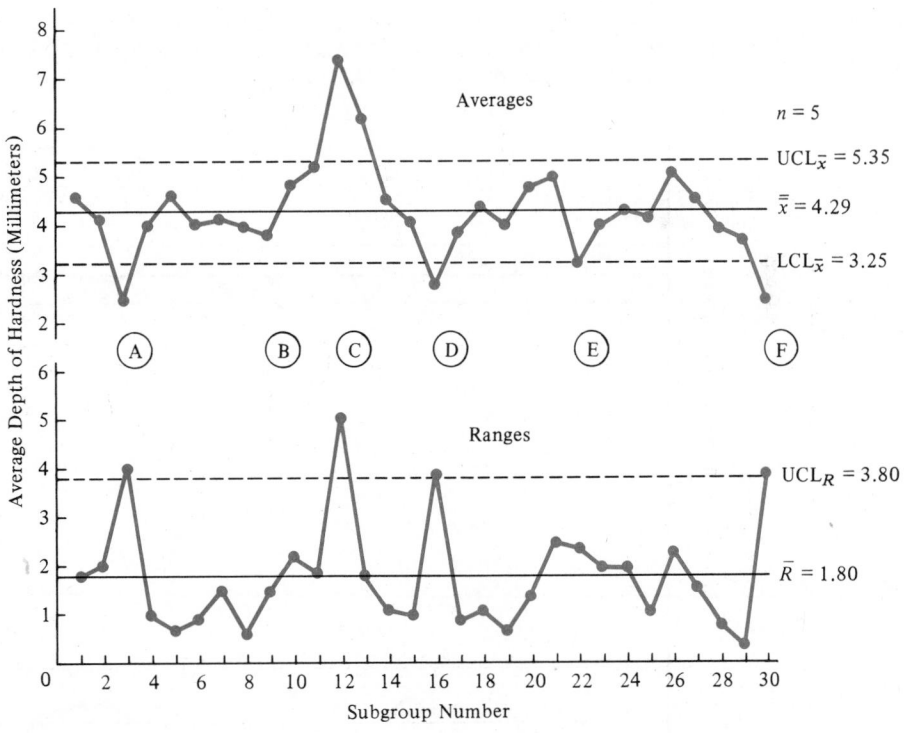

Figure 15.22   Initial Sampling for Coil A: Fuel Pump Eccentric Nose; Case Study
(Redrawn with permission from James C. Seigel, "Managing with Statistical Models," SAE Technical Paper No. 820520, Society for Automotive Engineers, Inc., Warrendale, Pa., 1982)

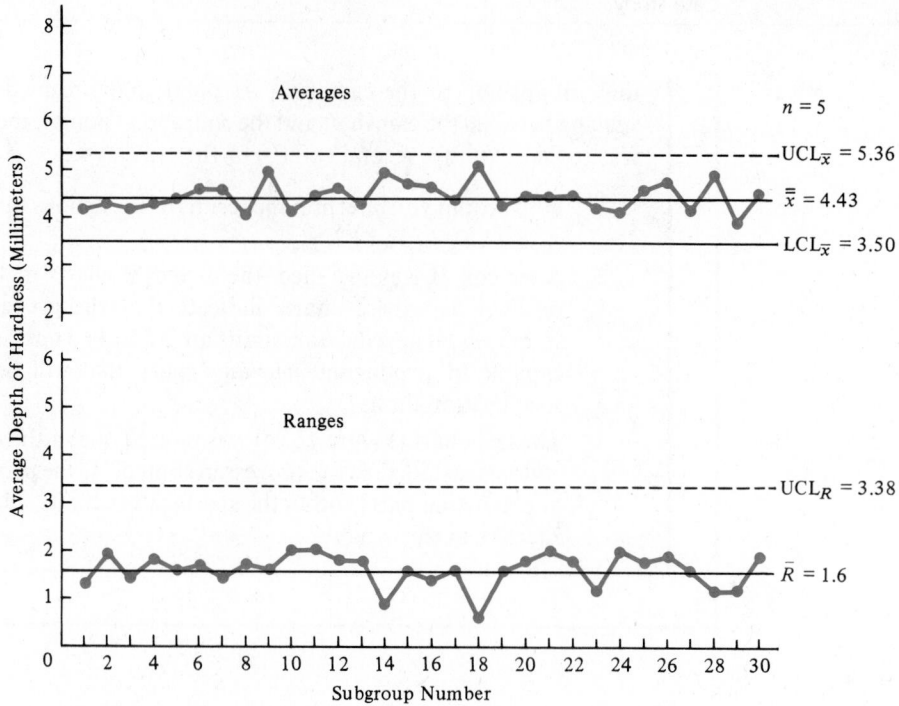

Figure 15.23   Sampling for Coil B: Fuel Pump Eccentric Nose; Case Study
(Redrawn with permission from James C. Seigel, "Managing with Statistical Models," SAE Technical Paper
No. 820520, Society for Automotive Engineers, Inc., Warrendale, Pa., 1982)

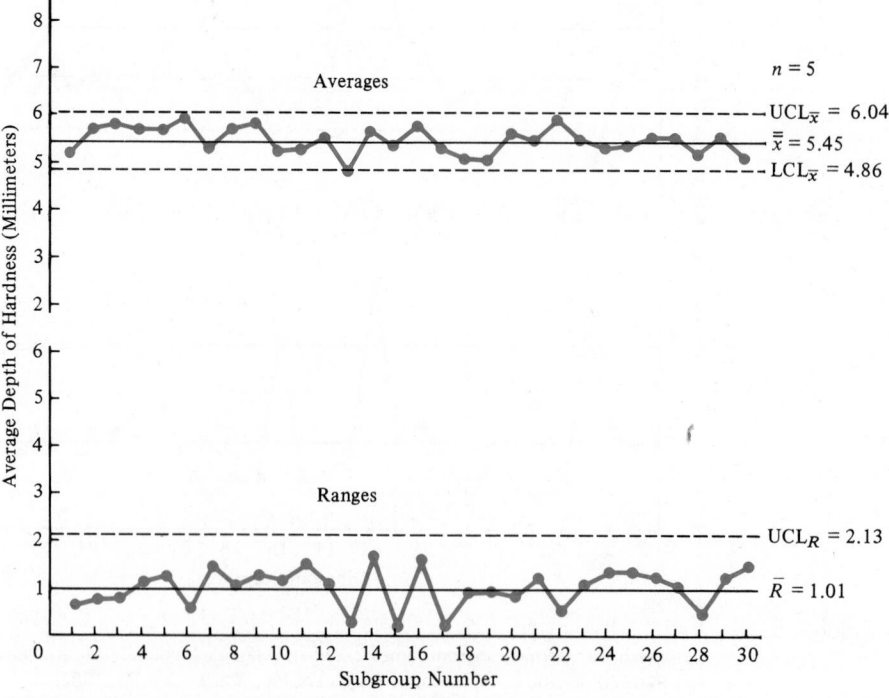

Figure 15.24   Sampling for Coil C: Fuel Pump Eccentric Nose; Case Study
(Redrawn with permission from James C. Seigel, "Managing with Statistical Models," SAE Technical Paper
No. 820520, Society for Automotive Engineers, Inc., Warrendale, Pa., 1982)

References and Suggested Readings

Box, G. E. P. "Signal-to-Noise Ratios, Performance Criteria, and Transformations." *Technometrics*, Vol. 30, No. 1 (1988), pp. 1–38.

Box, G. E. P., and N. R. Draper. *Empirical Model-Building and Repsonse Surfaces*. New York: Wiley, 1987.

Clements, J. A. "Suggestions for SPC Programmers, Part I." *Quality Progress*, February 1988, pp. 60–61.

Degroot, M. H. *Probability and Statistics*. New York: Addison-Wesley, 1975, pp. 406–407.

Dixon, W. J., and F. J. Massey. *Introduction to Statistical Analysis*. 4th ed. New York: McGraw-Hill, 1983, p. 147.

Duncan, A. J. *Industrial Quality Control*. 5th ed. Homewood, Ill.: Irwin, 1986, pp. 445–447.

Duncan, A. J. *Quality Control and Industrial Statistics*. 5th ed. Homewood, Ill.: Irwin, 1986.

Freund, R. A. "Definitions and Basic Quality Concepts." *Journal of Quality Technology*, Vol. 17, No. 1 (1985), pp. 50–56.

Grant, E. L., and R. S. Leavenworth. *Statistical Quality Control*. 6th ed. New York: McGraw-Hill, 1988.

Juran, J. M., and F. M. Gryna, Jr. *Quality Planning and Analysis*. 2d ed. New York: McGraw-Hill, 1980.

Kackar, R. N. "Off-Line Quality Control, Parameter Design, and the Taguchi Method." *Journal of Quality Technology*, Vol. 17, No. 4 (1985), pp. 175–207.

Kane, V. E. "Process Capability Indices." *Journal of Quality Technology*, Vol. 18, No. 1 (1986), pp. 41–52.

Marquardt, D. W. "New Technical and Education Directions for Monitoring Product Quality." *American Statistician*, Vol. 38, No. 1 (1984), pp. 8–14.

Messina, W. S. *Statistical Quality Control for Manufacturing Managers*. New York: Wiley, 1987, p. 159.

Montgomery, D. C. *Introduction to Statistical Quality Control*. New York: Wiley, 1985.

Schilling, E. G. *Acceptance Sampling in Quality Control*. New York: Marcel-Dekker, 1982.

Schilling, E. G. "An Overview of Acceptance Control." *Quality Progress*, April 1984, pp. 22–25.

Shewhart, W. A. *Economic Control of the Quality of Manufactured Product*. New York: Van Nostrand, 1931.

Shewhart, W. A. "Quality Control Charts." *Bell System Technical Journal*, 1926.

Vardeman, S. B. "The Legitimate Role of Inspection in Modern SQC." *American Statistician*, Vol. 40, No. 4 (1986), pp. 325–328.

TO
THE
READER

The inferential procedures for testing, estimation, and prediction presented in preceding chapters were based on sampling distributions that arise under random sampling from populations that are large compared with the sample size. Although random samples can be drawn in many ways, the procedures we have used were mainly based on simple random sampling from large or infinite populations. The paired-difference experiment and the randomized block design are examples of an alternative sampling design that increases precision in testing and estimation by isolating variation due to pairs or blocks.

Survey sampling deals with techniques for sampling human populations or populations consisting of social or economic units. In many applications it is not unusual for the sample size to be large in relation to the size of the population sampled. As a result, calculational forms in survey sampling involve the population size and will differ somewhat from those encountered earlier. In this chapter we introduce some commonly used survey sampling designs together with appropriate methods for estimating population parameters under the various designs.

16

SURVEY SAMPLING

16.1

Introduction

As we have repeated a number of times throughout this text, the main objective of statistics is to make inferences about a large body of data, the population, based on information contained in a sample. In previous chapters we have discussed several statistical procedures that can be used to analyze our available sample data set and to make inferences about certain characteristics of the population (e.g., the population mean μ or the population proportion p). In this chapter we offer a variety of methods for selecting the sample, called **sampling designs**, which can be used to generate our sample data set.

sampling designs

The main objective of any sampling design is to provide guidelines for selecting a sample that is *representative* of its underlying population, thus providing a specified amount of information about the population at a minimum cost. If the underlying population is uniform in the characteristics to be measured, almost any sample provides acceptable results. For example, the Environmental Protection Agency (EPA) bases its diagnosis of the purity of a city's water supply on the analysis of a few pints of water. This small sample is possible because the EPA assumes that one drop of water provides more or less the same proportionate amount of impurities as another drop. Consider, on the other hand, a nationwide survey to determine attitudes of adults toward renting versus buying a home. Just owing to pure chance, it is possible that only residents of New York City are selected as members of the sample. However, such a sample may not accurately reflect the attitudes of all adults, since residents of New York City are much less likely to own their own homes than, say, residents of Los Angeles. A more representative sample could be obtained by randomly selecting individuals from different regions of the country, perhaps by first dividing the population into states and then by randomly selecting a specified number of individuals from each state. The information provided by the random selection of adults chosen from each state would be combined to enable us to make inferences about the rent-versus-buy attitudes of the entire population of adults of the country.

census

It might appear that the only way to *guarantee* that our experimental data set truly represents the population is by conducting a **census**, a recording of every element contained in the population. Why, then, do we typically conduct a sample investigation instead of a complete census? Samples are used because in the majority of cases their advantages outweigh the advantages of a complete census.

Some advantages of sampling, like economy and practicability, are obvious; others are more subtle. For instance, a sample drawn from a continuous production process can give an instantaneous indication of production quality. Another situation in which sampling is more efficient than a complete census is in an experiment that results in the destruction of the elements of the sample. For example, in quality control studies, such as the testing of flashbulbs, testing destroys the product. Thus, sampling must be used because a census would leave nothing to market. Sometimes, sampling may actually provide more accurate results than a census, since the heavy workload of a census may cause researcher fatigue, which, in turn, may be responsible for increasingly careless sampling habits by the researchers. Also, the population itself may be dynamic and never in one state long enough to allow for a complete measurement of its characteristics.

When using sampling, we should take every precaution to ensure that our sampling is conducted in a "random manner" and that our sample is a *random sample*. As you will recall, in Section 4.7 we found that a random sample is selected in such a way that every sample of size *n* drawn from the population has an equal probability of being selected. The primary advantage of using random sampling designs is that when the samples are random, we know the probabilities of including various observations in the sample. Hence, we can make *probabilistic* statements about the underlying population. If the samples are selected in a deterministic (nonrandom) manner, the probabilities of observing various sample measurements are unknown, and we can only make *descriptive* statements about the sample. As you will subsequently see, most survey designs employ some restrictions in selecting the sample, but ultimately, random sampling is employed to obtain the sample observations.

In this chapter we consider the problem of sampling from a *finite population* of measurements; only occasionally will we refer to sampling from an infinite population. Our study will focus on the selection of an appropriate sample design, which, as you will see, involves a balance between the quantity of information obtained and the cost of sampling. The ultimate objective of the sample survey—an inference concerning the underlying population—is based on estimates of the population mean, total, or proportion. For each of these estimates we will give a bound on the error of estimation. Naturally, the estimation formulas vary according to the particular sampling design used in the survey.

Before proceeding with our discussion, we introduce some terms commonly used in survey sampling. These terms are introduced in the order in which they are involved in a sample survey.

Definition

sampling design
survey design

The **sampling design** or **survey design** specifies the method of collecting the sample.

The design does not specify a method of collecting or measuring the actual data. It specifies only a method for collecting the objects that contain the required information. These objects are called *elements*.

Definition

element

An **element** is an object on which a measurement is taken.

The elements may occur individually or in groups in the population. A group of elements, like a household of community residents or a carton of light bulbs, is called a *sampling unit*.

Definition

sampling units

Sampling units are nonoverlapping collections of elements from the population. In some cases a sampling unit is an individual element.

To select a random sample of sampling units, we need a list of all sampling units contained in the population. Such a list is called a *frame*.

Definition

frame

A **frame** is a list of sampling units.

When we conduct a sample survey, our first task is to identify the sampling units and to construct a frame that provides a list of the sampling units. According to our sampling design, a predetermined number of sampling units are then randomly selected from the frame. Our sample consists of the elements contained in the chosen sampling units. We can then use the information obtained from the sample to make inferences about certain characteristics of the population.

EXERCISES Conceptual Level

16.1 What is meant by a random sample? Why is it preferable to use random, as opposed to nonrandom, sampling?

16.2   Why is sampling usually preferable to conducting a census of the population?

16.3   A commonly used sampling technique in public opinion polling is called *quota sampling*. Using this technique, the interviewer selects, according to his or her own discretion, a predetermined number (a quota) of individuals from each of several segments of the population. For instance, he or she may be requested to interview 10 mechanics, 36 housewives, or 7 lawyers.

   a. Under what conditions does quota sampling provide a random sample?

   b. Under what conditions does it provide a nonrandom sample?

16.4   One of the most familiar uses of survey sampling in business is in the practice of auditing. Here, the CPA examines a sample of a client firm's records and tests to see whether they present a fair representation of the firm's operating results and financial position. For each of the following audits, discuss the selection of an appropriate sampling unit and frame.

   a. An IRS representative wishes to conduct an audit to estimate the total value of inventory held by a retail hardware store.

   b. A CPA is interested in conducting an audit of the financial records of a large wholesale machinery company to estimate the firm's average balance of delinquent accounts.

   c. A potential investor in a large commercial office building would like an audit conducted to test the current owner's claim regarding vacancy rates.

   d. An auditor from the Government Accounting Office wishes to examine a government contractor's claim regarding overtime benefits to employees involved in a highway construction project.

16.2

Bias and Error in Sampling

There are two major types of statistical inference: estimation and decision making. And as we have seen, each type has an associated error. Decision-making errors, usually referred to as type I and type II errors, were defined in Chapters 5 and 8. The error of estimation, discussed in Chapters 8 and 9, is defined here.

Definition

error of estimation

> Let $\hat{\theta}$ be a sample estimator of the population parameter θ. The **error of estimation** is the absolute difference $|\hat{\theta} - \theta|$.

bound B on the error

   When choosing a sample size, the experimenter should specify a **bound B on the error** of estimation that he or she is willing to tolerate (see Sections 8.12 and 11.7). That is, the experimenter should specify the value B and then choose the sample size so that $\hat{\theta}$ and θ differ by more than B only a very small fraction of the time. For example, an auditor may wish the chances to be very small that the average account balance obtained from a sample of a firm's accounts receivable differs from the true account balance by more than some specified amount, say $2.00. As you might expect, B and n are inversely related; the smaller the tolerable error of estimation, the greater is the quantity of information required to satisfy the specified bound.

sources of errors
random variation

There are three **sources of errors** in sample surveys. The most common source is **random variation**. For example, suppose a retailing organization is interested in estimating the average income per household in a particular community. In the selection of a random sample of households, by chance all those selected may be in the higher income brackets. Common sense and intuition would suggest the presence of an error of estimation if, for instance, an average household income of $149,000 is obtained for the community. However, when the estimation error is moderate, it may go undetected, leading to erroneous inferences and, perhaps, to faulty decision making. If a slightly inflated, but believable, average household income of $35,000 is obtained, the retailing organization may decide not to market lower-priced economy product lines that are thought to appeal more to those in moderate-to-lower-income communities. On the basis of the inflated average household income, the firm may be erroneously assuming that the community contains very few moderate-to-lower-income families.

misspecification of
the population

Another source of error in sample surveys is a **misspecification of the population**. Such errors are fairly common in public opinion polling for election surveys. The true population of interest in an election survey consists of those who will vote on election day. However, typical election surveys deal with the opinions of registered voters, many of whom will not vote.

Errors due to a misspecification of the population may also arise from sources such as an incorrect list of the population elements, incorrect information recorded on an inventory ledger, erroneous selection of sample elements (such as substituting a neighbor when a respondent is not found at home), question sensitivity, mistakes in the collection of information from the sample due to intentional or unintentional interviewer bias, or errors in processing the sample information. For the most part such causes are controllable; in other cases, such as in the measurement of the dimensions of timber or lumber that swells with an accumulation of moisture, the causes are uncontrollable.

The classic example of a misspecification error in an election survey is the 1936 *Literary Digest* poll. The *Literary Digest* obtained its sample from telephone directories and magazine subscription rolls and, on the basis of the sample responses, predicted that Republican Alfred M. Landon would soundly defeat incumbent Democratic President Franklin D. Roosevelt. As we know, the results were exactly the opposite—Roosevelt defeated Landon by one of the greatest pluralities ever, causing the *Literary Digest* to lose much of its credibility. What went wrong? During the economically depressed times of the thirties, only the affluent, who were typically Republicans, could afford telephones and magazine subscriptions. The sample did not provide a representative cross section of the voting public.

Errors due to misspecification of the population are also common in consumer research surveys, where the sample typically consists entirely of housewives, excluding men, working women, and students because of their relative inaccessibility. In addition, we noticed in Chapter 13 that the U.S. Bureau of Labor Statistics is guilty of a certain amount of misspecification in its published Consumer Price Index. Such surveys cannot expect to reflect the unique buying habits of segments of the population not represented in the sample.

nonresponse

An additional source of error in sample surveys is due to **nonresponse** by some members of the sample. Researchers commonly assume that respondents and non-respondents provide similar cross sections of the population when, in fact, they seldom do. In consumer surveys the nonrespondents are typically working people and the respondents are usually housewives; in public opinion surveys nonrespondents (those registering "no opinion") are usually the very contented members of the sample who generally prefer things as they are.

The researcher can minimize the chances of errors due to random variation only by selecting a proper sampling design. The researcher can have a much more direct effect on errors due to nonresponse. Continued efforts can be made to reach nonrespondents, or in some cases, nonrespondents can be replaced by randomly selected alternates.

To minimize the chances of incurring errors due to misspecification, a researcher can make a very careful statement of the survey objective in advance of the study, thus providing a clear image of the elements that comprise the population. Most important of all, the researcher should take great care in phrasing inferences in terms of the actual population from which the sample information was derived and not in terms of some other, perhaps conceptual, population of greater appeal.

Experience is the best guide to use for controlling sources of error in survey sampling. Individuals or agencies who have designed or conducted numerous surveys of a particular type (e.g., public opinion, market research, account audits, inventory audits) develop a reputation for anticipating certain possible pitfalls in the survey. They are then able to design the sample and the survey methods to avoid most common controllable sources of bias and error while minimizing the effect of uncontrollable sources of error.

EXERCISES   Conceptual Level

 16.5   When conducting an audit of a firm's accounts receivable, the auditor typically will select a random sample of accounts and then verify the account balance with each account holder. An accounting firm has undertaken an audit of the accounts receivable held by a large municipal hospital. Suggest some possible controllable and uncontrollable sources of error in the audit of the hospital's accounts.

16.6   Refer to Exercise 16.5. Suggest some auditing guidelines that may help control the bias and error that may occur in the auditing process. How might the auditor minimize the effect of random variation?

 16.7   Most prime-time television shows in the United States are at the mercy of weekly ratings issued by the A.C. Nielsen organization. Each week the Nielsen organization obtains nationwide samples randomly selected from residential directories (not telephone directories). In a mail survey the occupant of each residence selected is asked to keep a weekly tabulation of the channel selection during the evening prime-time hours and to return the results immediately at the week's end. Suggest some possible sources of error in the Nielsen survey process.

 16.8   You represent an advertising agency that buys considerable network television advertising time during the evening prime-time hours. Would you be willing to rely solely on the Nielsen ratings to select a network program and time slot for your advertisements? Explain. What additional information would you request from the Nielsen organization other than their widely published ratings?

16.3

How to Select a Random Sample

In this section we explain how to implement the most basic sample survey design, a *simple random sample.*

Definition

random sampling

simple random
sample

> A sample of n measurements is selected from a *finite* population of N measurements. If the sampling is conducted in such a way that every possible sample of size n has an equal probability of being selected, the **sampling** is said to be **random** and the result is said to be a **simple random sample**.

As we noted in Section 4.7, perfect random sampling is difficult to achieve in practice. If the population is not too large, each of the N measurements can be written on a slip of paper or on a poker chip and then placed in a bowl. A random sample of n measurements can then be drawn from the bowl.

The best way to ensure that we are employing random sampling is to use a table of random numbers. One such table is Table 11 in the Appendix. This table has been constructed so that the integers from 0 through 9 occur randomly and with equal frequency.

We illustrate the use of the random number table with an example.

EXAMPLE 16.1   Business organizations must manage cash flows effectively for the proper budgeting and control of their present and future resources. When cash flows are high, the firm is in a position to purchase inventories and capital goods on short notice, thereby taking advantage of price cuts offered by suppliers. When the firm is short of cash, it cannot buy ahead and therefore usually ends up paying more for goods and supplies.

One of the best measures of the cash position of a retail merchandise organization is provided by the short-term accounts receivable held by the firm. In an analysis of the cash position of a department store, an accounting firm decides to select a simple random sample of $n = 15$ monthly retail accounts receivable from among the $N = 1,000$ current monthly retail accounts of the department store in order to estimate the total amount due on all outstanding accounts receivable. We know that a simple random sample will be obtained if every possible sample of $n = 15$ accounts has the same chance of being selected. We will determine which accounts are to be included in the sample of size $n = 15$ by using the table of random numbers, Table 11 in the Appendix.

SOLUTION   We can think of the $N = 1,000$ accounts receivable as being numbered $001, 002, \ldots, 999, 000$. That is, we have 1,000 three-digit numbers, where 001 represents the first account, 999 the 999th account, and 000 the 1,000th.

We refer to Table 11 of the Appendix and arbitrarily select a starting point. Suppose our starting point is the first number in the fifth column. If we drop the last two digits of each five-digit number, we see that the first three-digit number formed is 816, the second is 309, the third is 763, and so on. If a random number occurs twice, the second occurrence is omitted and another number is selected as its replacement. Taking a random sample consisting of the first 15 nonrepeated three-digit numbers from column 5, we obtain the following numbers:

816	277	709
309	988	496
763	188	889
078	174	482
061	530	772

If the accounts receivable are numbered, we merely choose the accounts with the corresponding numbers and form our simple random sample of $n = 15$ from $N = 1,000$. If the accounts receivable are not numbered, we can refer to a list of the accounts and select the 61st, the 78th, the 174th, and so on until $n = 15$ accounts have been selected.

In Example 16.1 the population size $N = 1,000$ allowed us to associate each element in the population uniquely with a different three-digit number. What would we do if, say, $N = 964$? Clearly, we can associate the three-digit numbers $001, 002, \ldots, 963, 964$ with the elements of the population. All the remaining three-digit numbers, $965, 966, \ldots, 999, 000$, are simply ignored when selecting our sample of n three-digit numbers from the table of random numbers.

Sometimes, an experimenter uses personal judgment to select a representative sample or applies some intuitive means to "randomly" select the sample. Both procedures are subject to experimenter bias and should be avoided when seeking a simple random sample.

EXERCISES   Conceptual Level

16.9   Define a simple random sample.

16.10   Why is it preferable to use scientific random sampling techniques rather than to use your own judgment to select the sample "at random"?

EXERCISES   Skill Level

16.11   Use the random number table, Table 11 in the Appendix, to select a simple random sample of $n = 10$ six-digit random numbers.

16.12   Refer to Exercise 16.11. You have been asked to select a simple random sample of size 10 from a population containing $N = 1,000,000$ elements. Explain how you could use the $n = 10$ random numbers selected in Exercise 16.11 to achieve your objective.

16.13   Use the random number table, Table 11 in the Appendix, to select a simple random sample of size $n = 5$ from a population containing $N = 75$ elements.

16.14   Use the random number table, Table 11 in the Appendix, to select a simple random sample of size $n = 7$ from a population containing $N = 500$ elements.

16.15   Refer to Exercise 16.14. If you chose to use three-digit random numbers to select the simple random sample, many of the random numbers you selected may have exceeded $N = 500$ and hence were discarded. Can you suggest an alternative technique for associating population elements with random numbers that would eliminate the necessity of discarding so many random numbers?

EXERCISES   Applications

 16.16   The number and size of delinquent accounts are of vital concern to companies offering consumer credit. Excessive delinquency can seriously impair the cash position of the company and possibly entail costly account collection activities. Table 16.1 lists the account receivable balances of the

Table 16.1   Account Receivables Data for Exercise 16.16

Acct. No.	Acct. Bal.[†]	Acct. No.	Acct. Bal.	Acct. No.	Acct. Bal.	Acct. No.	Acct. Bal.
1	$136	26	$152	51	$146	76*	$233
2	216	27*	858	52	173	77	85
3	520	28	130	53	235	78	162
4	312	29	416	54*	220	79	165
5	180	30*	144	55*	600	80	483
6	250	31	320	56	405	81	342
7*	235	32	210	57	135	82	290
8	345	33	275	58	160	83*	95
9	260	34	540	59	325	84*	338
10	185	35	350	60	290	85	710
11	285	36	560	61	302	86	155
12	310	37*	365	62	250	87	115
13*	430	38	125	63*	280	88	200
14	605	39	312	64	235	89	450
15*	310	40	165	65	560	90*	245
16	60	41*	360	66	320	91	260
17	155	42*	450	67	305	92	530
18*	190	43	235	68	160	93	200
19	425	44	190	69	255	94	216
20	75	45	345	70*	204	95*	195
21	315	46	240	71	120	96	395
22	240	47	389	72	403	97	126
23*	209	48	80	73	265	98	250
24	178	49	205	74*	322	99	405
25	313	50*	215	75	602	100	196

* Delinquent accounts.

† All account balances have been rounded to the nearest dollar.

$N = 100$ regular customers of a wholesale hardware company, with delinquent accounts denoted by an asterisk. Suppose you represent an auditing firm that has been hired by the company to examine its accounts receivable. Sufficient time and money are available to investigate only $n = 15$ of the accounts. Using the table of random numbers, Table 11 in the Appendix, select a simple random sample of $n = 15$ accounts from all $N = 100$ accounts receivable of the wholesale hardware company.

16.17   You have been hired as a marketing representative for a clothing manufacturer who has developed a new line of casual clothing designed to appeal to college students. You are requested to select a representative sample of the students at your college (or university) and to survey their attitudes toward the new line of clothing.

   a.   Using your college's student directory as a frame, use the table of random numbers to select a simple random sample of $n = 50$ students.

   b.   How well does your sample of $n = 50$ students reflect the underlying population of all students at your college (or university)? (*Hint:* Using figures provided by the registrar's office, compare your sample distribution with the actual distribution of students according to the male-female ratio, according to academic class, and according to major.)

16.4

Measuring the Goodness of Estimators in Survey Sampling

In Section 2.9 we introduced the Empirical Rule as a rule for describing the variability of symmetrical, mound-shaped distributions. From this rule we saw, for instance, that the interval ($\mu \pm 2\sigma$) contains at least 75% and most likely 95% of the measurements resulting from a mound-shaped distribution. Later, in Chapter 6, we learned that 95% of the measurements from a distribution of *normally distributed* measurements are contained within the interval ($\mu \pm 1.96\sigma$). In our study of the estimation of means in Chapter 8, we used the results based on assumptions of normality for the development of bounds on the error of estimation for μ. This development was possible because the Central Limit Theorem prescribes that the distribution of sample means drawn from a population with finite mean and standard deviation approaches a normal distribution with increasing sample size.

The Central Limit Theorem also applies to sample surveys, provided that the sample size n and the population size N are both large and the sampling fraction n/N is small. There is, however, the occasional limitation on sample size for which the sampling distribution resulting from a sample survey may be skewed and, hence, nonnormal. For this reason, we will use 1.96 × (standard deviation of the estimator) as *approximate* 95% bounds on the error of estimation, or we will construct a confidence interval with confidence coefficient *approximately* .95. If you wish to change the confidence coefficient to ($1 - \alpha$), you simply replace 1.96 by $z_{\alpha/2}$, the appropriate value from the standard normal distribution. However, the accuracy of these confidence coefficients will depend upon how well the sampling distribution of the estimator is approximated by a normal distribution.

16.5

Estimation Based on a Simple Random Sample

The selection of a simple random sample, the most elementary survey design, was presented in Section 16.3. After we gather the sample observations, our next objective is to estimate certain population parameters of interest. Most often, we are interested in estimating the population mean μ or the population total $\tau = N\mu$. For example, the accounting firm in Example 16.1 might be interested in the mean dollar value for the accounts receivable as well as the total dollar amount of these accounts.

population mean μ   The computational formulas for estimating the **population mean μ** and the **popu-**
population total τ   **lation total τ**, for simple random sampling, are shown in the displays. Recall, however, that a point estimate such as $\hat{\mu}$ or $\hat{\tau}$ tells us nothing about the *goodness* of our estimation. Therefore, variance formulas are given so that we can place bounds on the error of estimation of μ and τ.

The estimator for the population mean μ for simple random sampling follows.

Estimation of the Population Mean for a Simple Random Sample

Estimator:

$$\hat{\mu} = \bar{x} = \sum_{i=1}^{n} \frac{x_i}{n}$$

Variance of the estimator:

$$\hat{\sigma}_{\bar{x}}^2 = \left(\frac{s^2}{n}\right)\left(\frac{N-n}{N}\right) \quad \text{where} \quad s^2 = \sum_{i=1}^{n} \frac{(x_i - \bar{x})^2}{n-1}$$

Bounds on the error of estimation:

$$\bar{x} \pm 1.96\hat{\sigma}_{\bar{x}}$$

When the survey objective is to use simple random sampling to estimate the population total τ, we have the following formulas.

Estimation of the Population Total for a Simple Random Sample

Estimator:

$$\hat{\tau} = N\bar{x}$$

Variance of the estimator:

$$\hat{\sigma}_{\hat{\tau}}^2 = N^2\hat{\sigma}_{\bar{x}}^2$$

Bounds on the error of estimation:

$$N\bar{x} \pm 1.96\hat{\sigma}_{\hat{\tau}}$$

finite population
correction (fpc)
factor

Note that the estimated variance of the sample mean, $\hat{\sigma}_{\bar{x}}^2$, is the same as that given in Chapter 7 except that it is multiplied by a **finite population correction (fpc) factor**, $(N - n)/N$, to adjust for sampling from a finite population. When n is small relative to the population size N, the fpc, $(N - n)/N$, is close to 1. Practically speaking, the fpc can be ignored if $n \leq N/20$. In that case $\hat{\sigma}_{\bar{x}}^2$ reduces to the more familiar quantity s^2/n.

EXAMPLE 16.2   Refer to the accounts receivable audit of Example 16.1. The simple random sample of $n = 15$ accounts provided the 15 account balances listed in Table 16.2.

Table 16.2   Account Balances for Example 16.2

$14.50	$23.40	$42.00
30.20	15.50	13.30
17.80	27.50	23.70
10.00	6.90	18.40
8.50	19.50	12.10

a.   Estimate the average μ due for all $N = 1,000$ accounts receivable of the department store, and place a bound on the error of estimation.

b.   Estimate the total τ due on all outstanding accounts receivable, and place a bound on the error of estimation.

SOLUTION   It will help us in our computations to list the sample data as shown in Table 16.3.

Table 16.3   Computations for Data of Table 16.2

x_i	x_i^2
$14.50	210.25
30.20	912.04
17.80	316.84
10.00	100.00
8.50	72.25
23.40	547.56
15.50	240.25
27.50	756.25
6.90	47.61
19.50	380.25
42.00	1,764.00
13.30	176.89
23.70	561.69
18.40	338.56
12.10	146.41
$\displaystyle\sum_{i=1}^{15} x_i = 283.30$	$\displaystyle\sum_{i=1}^{15} x_i^2 = 6{,}570.85$

a. The estimate of the mean account balance μ is

$$\bar{x} = \frac{\sum\limits_{i=1}^{15} x_i}{15} = \frac{283.30}{15} = \$18.89$$

To find a bound on the error of estimation of μ, we first compute

$$s^2 = \frac{\sum\limits_{i=1}^{15} (x_i - \bar{x})^2}{14} = \frac{\sum\limits_{i=1}^{15} x_i^2 - \left(\sum\limits_{i=1}^{15} x_i\right)^2 \bigg/ 15}{14}$$

$$= \frac{1}{14}\left[6{,}570.85 - \frac{(283.30)^2}{15}\right] = \frac{1}{14}(6{,}570.85 - 5{,}350.5927) = 87.1612$$

The estimated variance of $\bar{x}$ is therefore

$$\hat{\sigma}_{\bar{x}}^2 = \left(\frac{s^2}{n}\right)\left(\frac{N - n}{N}\right) = \left(\frac{87.1612}{15}\right)\left(\frac{1{,}000 - 15}{1{,}000}\right) = 5.7236$$

An estimate of the mean account balance μ, with a bound on the error of estimation, is

$$\bar{x} \pm 1.96\hat{\sigma}_{\bar{x}} \quad \text{or} \quad \$18.89 \pm 1.96\sqrt{5.7236} \quad \text{or} \quad \$18.89 \pm \$4.69$$

b. An estimate of the total amount due on outstanding accounts receivable is provided by

$$\hat{\tau} = N(\bar{x}) = (1{,}000)(\$18.89) = \$18{,}890$$

Since the estimated variance of $\hat{\tau}$ is $\hat{\sigma}_{\hat{\tau}}^2 = N^2\hat{\sigma}_{\bar{x}}^2$, an estimate of the total due on all $N = 1{,}000$ accounts, with a bound on the error of estimation, is

$$\hat{\tau} \pm 1.96\hat{\sigma}_{\hat{\tau}} \quad \text{or} \quad N\bar{x} \pm 1.96N\hat{\sigma}_{\bar{x}} \quad \text{or}$$

$$\$18{,}890 \pm (1.96)(1{,}000)\sqrt{5.7236}$$

or

$$\$18{,}890 \pm \$4{,}690$$

◆◆

population
proportion p

In an experimental investigation we may wish to estimate the **proportion p of the population** that possesses a specified characteristic. An auditor may be interested in the proportion of delinquent accounts; a market researcher may be interested in the firm's proportionate share of total market sales; or a corporate executive may be interested in the proportion of shareholders favoring a particular company policy decision.

Recall that we first studied the estimation of the population proportion p in Section 8.10. There, we assumed that our estimation was based on a simple random sample selected from an *infinite* population. When the population size N is *finite*, the population proportion p is estimated as shown in the display.

Estimation of the Population Proportion for a Simple Random Sample

> Estimator:
>
> $$\hat{p} = \frac{x}{n}$$
>
> Variance of the estimator:
>
> $$\hat{\sigma}_{\hat{p}}^2 = \left(\frac{\hat{p}\hat{q}}{n-1}\right)\left(\frac{N-n}{N}\right) \qquad \text{where} \qquad \hat{q} = 1 - \hat{p}$$
>
> Bounds on the error of estimation:
>
> $$\hat{p} \pm 1.96\hat{\sigma}_{\hat{p}}$$

In this case x is the total number of the n sample elements that possess the specified characteristic.

EXAMPLE 16.3 Manufacturing organizations often resort to short-term price discounts to encourage their customers to increase their order size and buy ahead, thus enhancing the manufacturer's cash position. Consistent with this basic intent, a manufacturer and wholesale distributor of frozen food products is considering discounting by 20% the price of frozen dinners to buyers who double their monthly order. Since frozen foods require costly storage, it is not certain that the buyers will take advantage of the discount offer. A random sample of $n = 50$ of the firm's $N = 430$ buyers was contacted, with 15 of the 50 indicating they would accept the price discount offer and double their monthly order. Estimate the proportion p of all $N = 430$ of the firm's buyers who will take advantage of the offer, and place a bound on the error of estimation.

SOLUTION An estimate of the proportion p of all the firm's buyers who will take advantage of the firm's price discount is

$$\hat{p} = \frac{x}{n} = \frac{15}{50} = .30$$

To place bounds on the error of estimation, we must first compute the variance $\hat{\sigma}_{\hat{p}}^2$. We find

$$\hat{\sigma}_{\hat{p}}^2 = \left(\frac{\hat{p}\hat{q}}{n-1}\right)\left(\frac{N-n}{N}\right) = \left[\frac{(.30)(.70)}{49}\right]\left(\frac{430-50}{430}\right)$$

$$= \left(\frac{.21}{49}\right)(.88372) = .003787$$

An estimate of p, with a bound on the error of estimation, is

$$\hat{p} \pm 1.96\hat{\sigma}_{\hat{p}} \quad \text{or} \quad .30 \pm 1.96\sqrt{.003787} \quad \text{or} \quad .30 \pm .12$$

That is, we estimate that the proportion of all the firm's buyers who will take advantage of the price discount is .30, with a bound on the error of estimation of .12.

◆ ◆

EXERCISES Conceptual Level

16.18   Explain what is meant by evaluating the goodness of an estimator. What part does sampling play in that evaluation?

16.19   Why is the finite correction factor (fpc) sometimes needed when evaluating the variance of an estimator? How does one decide if it is needed?

EXERCISES Skill Level

16.20   A simple random sample of size $n = 50$ was selected from a population of size $N = 1,500$. The sample mean and standard deviation were 82.3 and 11.7, respectively.

  a.   Estimate the population mean μ, and place bounds on the error of estimation.

  b.   Estimate the population total τ, and place bounds on the error of estimation.

16.21   A simple random sample of size $n = 36$ was selected from a population of size $N = 2,200$. The sample mean and standard deviation were .065 and .012, respectively.

  a.   Estimate the population mean μ, and place bounds on the error of estimation.

  b.   Estimate the population total τ, and place bounds on the error of estimation.

16.22   A simple random sample of size $n = 100$ is selected from a population of size $N = 2,586$, and the number of "successes" in the sample is $x = 25$. Estimate the proportion p of "successes" in the population, and place bounds on the error of estimation.

16.23   A simple random sample of size $n = 65$ is selected from a population of size $N = 11,650$, and the number of "successes" in the sample is $x = 32$. Estimate the proportion p of "successes" in the population, and place bounds on the error of estimation.

16.24   A simple random sample of size $n = 30$ from a finite population containing 140 elements produced the following observations:

3.2	4.2	2.4	6.1	5.9	5.2
3.9	4.6	8.9	6.0	2.6	9.4
5.1	9.9	5.4	4.1	6.9	2.1
6.6	9.2	3.5	2.7	6.3	3.0
8.5	0.7	4.7	4.0	5.5	7.1

  a.   Estimate the population mean μ, and place bounds on the error of estimation.

  b.   Estimate the population total τ, and place bounds on the error of estimation.

EXERCISES Applications

16.25 In an attempt to gather information about the costs of banning nonreturnable soft drink bottles in supermarkets, the Food Marketing Institute commissioned a study that surveyed $n = 100$ stores in the six states that mandate nickel-back bottles. The results of the study suggest that extra handling costs an average 2.4 cents per bottle with a standard deviation of .9 cent in stores prohibited by state law from using nonreturnable soft drink bottles. Estimate the average increase μ in handling costs per bottle for all stores in states in which nonreturnable soft drink bottles are prohibited, and place a bound on the error of estimation. (Assume that N is sufficiently large to ignore the fpc factor.)

16.26 Refer to Exercise 16.25. Estimate the average increase in handling cost per case of 24 soft drink bottles for all stores in states in which nonreturnable soft drink bottles are prohibited, and place a bound on the error of estimation.

16.27 The Federal Trade Commission (FTC) has developed legislation to allow pharmacies to advertise prices for retail drugs. Such legislation will allow price competition for prescription drugs among retail pharmacies, thus enabling the consumer to obtain the best available bargain for prescriptions. An FTC staff member undertook a study to examine the disparity in prices charged by retail pharmacies in Cincinnati for the drug raudixin, a high blood pressure medication. A table of random numbers was used to select $n = 20$ pharmacies from among the $N = 152$ retail pharmacies of Cincinnati. The price charged for 100 tablets of raudixin by each of the selected pharmacies is shown below.

$3.75	$4.10	$10.40	$7.50	$2.95	$5.75	$7.50	$8.90	$ 4.75	$11.75
5.85	7.65	8.10	6.50	7.50	5.50	8.00	4.50	10.25	4.95

Estimate the average price μ charged for 100 tablets of raudixin by all the 152 retail pharmacies in Cincinnati, and place a bound on the error of estimation.

16.28 Labor unions are organized to represent the interests of member employees in their jobs. Wages and salaries comprise only part of these interests. Fringe benefits, such as retirement and insurance benefits, vacation plans, and working conditions, are often of equal concern. Suppose the Textile Workers Union is interested in determining the proportion of the $N = 352$ employees it represents in a South Carolina textile plant who are satisfied with the company's current retirement and insurance benefits. Using a table of random numbers, a union official selects $n = 50$ of the employees and finds that 27 of them are satisfied with the retirement and insurance benefits currently offered by the company. Estimate the proportion p of all the textile company's employees represented by the union who favor the current benefits offered by the firm. Place a bound on the error of estimation.

16.29 Refer to Exercise 16.16. Use your simple random sample of $n = 15$ accounts drawn from all $N = 100$ accounts listed in Table 16.1.

   a. Estimate the average balance μ for all 100 accounts. Place a bound on the error of estimation.

   b. Estimate the total receivable on all accounts, and place a bound on the error of estimation.

   c. Estimate the proportion of delinquent accounts, and place a bound on the error of estimation. How does your estimate of p compare with the true p for the population of all $N = 100$ accounts receivable?

   d. Estimate the total number of delinquent accounts, and place a bound on the error of estimation. (*Hint:* If p is the population proportion of delinquent accounts, then Np is the *total* number of delinquent accounts. An estimator of Np is $N\hat{p}$, which has an estimated variance of $N^2\hat{\sigma}_{\hat{p}}^2$.)

 16.30 Flextime, the idea of letting workers choose their own time schedules, has been used in many governmental agencies since 1978. To examine the attitude of its employees toward flextime, the Government Accounting Office (GAO) conducted a survey among $n = 50$ auditors randomly chosen from its staff of $N = 5{,}725$ auditors and found that 27 of them prefer a flextime work schedule. Estimate the proportion p of all GAO auditors favoring flextime, and place a bound on the error of estimation.

16.6

Stratified Random Sampling

A second type of sampling design, which often provides a specified amount of information at less cost than simple random sampling, is called *stratified random sampling*. This design is recommended when the population consists of a set of heterogeneous (dissimilar) groups.

Definition

stratified random sample
strata

> A **stratified random sample** is a random sample obtained by separating the population elements into nonoverlapping groups, called **strata**, and then selecting a simple random sample within each stratum.

The stratified random sampling design has three major advantages over simple random sampling. First, the *cost* of collecting and analyzing the data is often reduced by stratifying the population into groups so that the elements possess similar characteristics within the groups but are dissimilar between groups. For example, in a survey of industrial buyers it would cost more to obtain information from those in foreign countries than from domestic buyers. We should, therefore, take small samples from the strata with high sampling costs (foreign buyers) to satisfy our objective of minimizing the total cost of sampling.

The second advantage concerns the *variance* of the estimator of the population mean. This variance is often reduced by using stratified random sampling, because the variation within strata is usually smaller than the overall population variance. For example, the power use for industrial users is likely to be much more variable than the power use for residential users. Therefore, the power company official who wishes to estimate the average use for all company subscribers should select larger samples from the less homogeneous industrial sector in order to obtain good estimates of the population parameters.

The third advantage of stratified random sampling is that separate estimates are provided for the parameters of each stratum, without selecting another sample and incurring additional cost. For instance, it may be more useful to know the mean monthly power use for the residential users and the industrial users of a city than to know only the mean monthly use for all the city's subscribers. Stratified random sampling allows us to analyze stratum differences, so that those groups within the population that deserve special attention can be more easily noted.

proportional allocation procedure

In this chapter we will use a **proportional allocation procedure**, which partitions the sample size among the strata in proportion to the size of the strata. The major advantage of using proportional allocation is that it provides a "self-weighting" sample, since the sampling fraction is the same in each stratum. Sampling cost savings are thus obtained when many estimates have to be made. Optimal allocation, which partitions the sample size according to cost, variability, and stratum size, produces estimates with smaller variances than those obtained by using proportional allocation when the strata sampling costs and variances differ widely. However, Cochran (see the references) notes that the superiority of optimal allocation is exaggerated because of errors obtained when estimating strata sampling costs and variances. Furthermore, he suggests that the simplicity and self-weighting feature of proportional allocation should offset up to a 20% increase in the variance of the derived estimators.

The first step in selecting a stratified random sample is to clearly specify the strata, associating each element of the population with one and only one stratum. In some cases this task may not be so easy. In an opinion poll example, would we classify those living in a city of 1,000 as urban or rural constituents? In the power use example, would the residence of an accountant whose office is in his home be placed in the residential or commercial stratum? Resolution of such conflicts does not affect our results so long as it is always done consistently. For example, we could say that cities of under 2,500 population will always be considered rural, those larger, urban; combination business-residence units will be classified according to which use occupies the greatest floor space of the dwelling.

Once the strata have been specified, we can use the method of Section 16.3 to select a simple random sample within each stratum. The overall sample size n depends on our available budget for sampling and on how precise and accurate we wish the estimate to be (these latter requirements are discussed in Section 16.8). For proportional allocation the sample size is allocated among the L strata so that $n = n_1 + n_2 + \cdots + n_L$, where each n_i is as given by the formula in the display.

Allocation of the Sample Among the Strata

$$n_i = n\left(\frac{N_i}{N}\right) \qquad i = 1, 2, \ldots, L$$

where N_i is the number of elements in stratum i and $N = \sum_{i=1}^{L} N_i$ is the size of the population.

From information obtained from the sample elements, we can compute the estimated mean $\bar{x}_i$ and the variance s_i^2 of the observations within each stratum by using the formulas given in the next display.

Estimation of the Mean and Variance of Each Stratum

$$\bar{x}_i = \frac{\sum\limits_{j=1}^{n_i} x_{ij}}{n_i}$$

$$s_i^2 = \frac{\sum\limits_{j=1}^{n_i} (x_{ij} - \bar{x}_i)^2}{n_i - 1} \qquad i = 1, 2, \ldots, L$$

where x_{ij} is the jth observation in stratum i.

The variance s_i^2 is an estimate of the corresponding true stratum variance σ_i^2.

The estimator $\bar{x}_{st}$ of the population mean μ for stratified random sampling is given in the display.

Estimation of the Population Mean for a Stratified Random Sample

Estimator:

$$\bar{x}_{st} = \frac{1}{N} \sum_{i=1}^{L} N_i \bar{x}_i$$

Variance of the estimator:

$$\hat{\sigma}_{\bar{x}_{st}}^2 = \frac{1}{N^2} \sum_{i=1}^{L} N_i^2 \left(\frac{N_i - n_i}{N_i} \right) \left(\frac{s_i^2}{n_i} \right)$$

Bounds on the error of estimation:

$$\bar{x}_{st} \pm 1.96 \hat{\sigma}_{\bar{x}_{st}}$$

EXAMPLE 16.4   The period from 1978 through 1982 witnessed a rapid decline in the number of new housing starts in the United States. This decline was primarily due to a shortage of available home loan funds from banks and other savings institutions. To increase the supply of funds available for home mortgages, a large manufacturing organization instituted a policy to encourage its employees to regularly invest part of their incomes with local savings institutions. The firm later decided to conduct a survey of its employees' savings habits to judge the effectiveness of the savings campaign. The firm desires to estimate the average amount invested in savings by the employees from their past month's incomes. Suggest a survey design for this problem.

SOLUTION   The employees of the firm can be categorized as clerical workers and laborers, foremen and middle managers, and higher-level executives. A stratified random sample with

$L = 3$ strata appears to be the most appropriate sample survey design. Within each stratum the spending and investment habits of the employees should be reasonably homogeneous. Simple random sampling should be used to select a sample of employees from each stratum for questioning about their savings investment from their last month's incomes.

Suppose the manufacturing organization employs a total of 5,000 people, of which 3,500 are clerical workers or laborers, 1,000 are foremen or middle managers, and 500 are executives. The research department of the firm has enough time and money to interview only $n = 50$ employees. Using proportional allocation, we would partition the sample size $n = 50$ as follows:

$$n_1 = n\left(\frac{N_1}{N}\right) = 50\left(\frac{3,500}{5,000}\right) = 35$$

$$n_2 = 50\left(\frac{1,000}{5,000}\right) = 10 \quad \text{and} \quad n_3 = 5$$

An alphabetical list of employees in each category, available through the payroll office, provides a base from which we can select our sample. Arbitrarily beginning in column 8 of Table 11 in the Appendix, we select the first 35 nonrepeated four-digit random numbers between 0001 and 3500 to identify the 35 clerical workers and laborers to be included in our sample. The results are shown in Table 16.4. Therefore, the first member of the sample should be the clerical worker or laborer who is 84th in alphabetical order, the next, the 118th, and so on. Similarly, three-digit random numbers from 000 to 999 should be used to select a sample of $n_2 = 10$ foremen and middle managers, and three-digit numbers from 001 to 050 used to select $n_3 = 5$ executives.

Having selected the sample elements (employees), we proceed with the interview. From the responses of the employees we compute the mean $\bar{x}_i$ and the variance s_i^2 of the observations within each stratum. The computed stratum means and variances are shown in Table 16.5.

We estimate the average investment $\bar{x}_{st}$ in savings from last month's income, using the data from Table 16.5, as

Table 16.4   Random Numbers to Identify the
Employees to Sample from Stratum 1,
Clerical Workers and Laborers

1419	0473	0151	1798	0691
2483	2638	0118	3159	2096
1873	2872	2349	2084	3068
0585	1539	3409	0827	0084
1761	2533	3208	2635	1411
2985	0815	1505	2217	2191
1360	3001	1256	0664	1842

Table 16.5   Calculations for Example 16.4

Stratum 1	Stratum 2	Stratum 3
$n_1 = 35$	$n_2 = 10$	$n_3 = 5$
$\bar{x}_1 = \$10.16$	$\bar{x}_2 = \$25.50$	$\bar{x}_3 = \$21.80$
$s_1^2 = 16.81$	$s_2^2 = 22.09$	$s_3^2 = 125.44$
$N_1 = 3,500$	$N_2 = 1,000$	$N_3 = 500$

$$\bar{x}_{st} = \frac{1}{N} \sum_{i=1}^{L} N_i \bar{x}_i = \frac{1}{5,000} [(3,500)(10.16) + (1,000)(25.50) + (500)(21.80)]$$

$$= \frac{1}{5,000} (71,960) = \$14.39$$

Therefore, the estimated average amount of last month's income invested in savings by all the employees of the firm is $14.39.

The estimated variance of $\bar{x}_{st}$ is

$$\hat{\sigma}_{\bar{x}_{st}}^2 = \frac{1}{N^2} \sum_{i=1}^{3} N_i^2 \left(\frac{N_i - n_i}{N_i} \right) \left(\frac{s_i^2}{n_i} \right)$$

$$= \frac{1}{(5,000)^2} \left[\frac{(3,500)^2(.99)(16.81)}{35} + \frac{(1,000)^2(.99)(22.09)}{10} \right.$$

$$\left. + \frac{(500)^2(.99)(125.44)}{5} \right] = .5688$$

The estimate of the average savings, with a bound on the error of estimation, is provided by

$$\bar{x}_{st} \pm 1.96 \hat{\sigma}_{\bar{x}_{st}} \qquad \text{or} \qquad \$14.39 \pm 1.96\sqrt{.5688} \qquad \text{or} \qquad \$14.39 \pm \$1.48$$

◆◆

If the survey objective is to use stratified random sampling to estimate the population total τ, then the estimator is as given in the next display.

Estimation of the Population Total for a Stratified Random Sample

Estimator:

$$\hat{\tau} = N\bar{x}_{st}$$

Variance of the estimator:

$$\hat{\sigma}_{\hat{\tau}}^2 = N^2 \hat{\sigma}_{\bar{x}_{st}}^2$$

Bounds on the error of estimation:

$$\hat{\tau} \pm 1.96 \hat{\sigma}_{\hat{\tau}}$$

EXAMPLE 16.5   Refer to Example 16.4. Estimate the total of last month's income invested in savings by the employees of the manufacturing organization. Place a bound on the error of estimation.

SOLUTION   From our previous computations $\bar{x}_{st} = \$14.39$. Therefore, an estimate of the total savings is

$$\hat{\tau} = N\bar{x}_{st} = (5,000)(\$14.39) = \$71,950$$

To find bounds on the error of estimation of τ, we must first compute the estimated variance $\hat{\sigma}_{\hat{\tau}}^2$:

$$\hat{\sigma}_{\hat{\tau}}^2 = N^2\hat{\sigma}_{\bar{x}_{st}}^2 = (5,000)^2(.5688) = 14,220,000$$

The estimate of the total savings, with a bound on the error of estimation, is given by

$$\hat{\tau} \pm 1.96\hat{\sigma}_{\hat{\tau}} \quad \text{or} \quad \$71,950 \pm 1.96\sqrt{14,220,000}$$

or

$$\$71,950 \pm (1.96)(3,771) \quad \text{or} \quad \$71,950 \pm \$7,391$$

Therefore, we are approximately 95% certain that the total investment in savings by the employees is contained in the interval from $64,559 to $79,341.

So far, we have been concerned with the use of stratified random sampling to estimate the population mean and total. In contrast, suppose the manufacturing organization wishes to estimate the proportion of employees who invested part of their last month's income in a savings account. Using the same strata as defined earlier, the researcher can select a simple random sample from each stratum and find the proportion $\hat{p}_i$ of employees in stratum i who invested part of last month's income in savings. The sample proportions obtained from the strata can be combined to provide an estimate of the population proportion.

Estimation of the Population Proportion for a Stratified Random Sample

Estimator:

$$\hat{p}_{st} = \frac{1}{N}\sum_{i=1}^{L} N_i\hat{p}_i$$

Variance of the estimator:

$$\hat{\sigma}_{\hat{p}_{st}}^2 = \frac{1}{N^2}\sum_{i=1}^{L} N_i^2\left(\frac{N_i - n_i}{N_i}\right)\left(\frac{\hat{p}_i\hat{q}_i}{n_i - 1}\right) \qquad \hat{q}_i = 1 - \hat{p}_i$$

Bounds on the error of estimation:

$$\hat{p}_{st} \pm 1.96\hat{\sigma}_{\hat{p}_{st}}$$

EXAMPLE 16.6   Of the $n = 50$ employees interviewed in the savings investment study, the number of those who indicated that they had actually participated is given in Table 16.6. Estimate the proportion of all employees participating in the savings program, and place a bound on the error of estimation.

Table 16.6   Data for Example 16.6

Stratum	Sample Size	Number Participating	$\hat{p}_i$
1	$n_1 = 35$	21	.60
2	$n_2 = 10$	7	.70
3	$n_3 = 5$	4	.80

SOLUTION   The desired estimate is given by $\hat{p}_{st}$, where

$$\hat{p}_{st} = \frac{1}{5,000}[(3,500)(.60) + (1,000)(.70) + (500)(.80)] = .64$$

Bounds on the error of estimation can be found by first computing the variance:

$$\hat{\sigma}^2_{\hat{p}_{st}} = \frac{1}{(5,000)^2}\left\{(3,500^2)\left(\frac{3,500 - 35}{3,500}\right)\left[\frac{(.6)(.4)}{34}\right]\right.$$
$$\left. + (1,000^2)\left(\frac{1,000 - 10}{1,000}\right)\left[\frac{(.7)(.3)}{9}\right] + (500^2)\left(\frac{500 - 5}{500}\right)\left[\frac{(.8)(.2)}{4}\right]\right\}$$
$$= .004744$$

The estimate of the proportion of employees participating in the manufacturing organization's savings program, with a bound on the error of estimation, is given by

$$\hat{p}_{st} \pm 1.96\hat{\sigma}_{\hat{p}_{st}} \quad \text{or} \quad .64 \pm 1.96\sqrt{.004744}$$

or

$$.64 \pm (1.96)(.069) \quad \text{or} \quad .64 \pm .14$$

◆◆

EXERCISES   Conceptual Level

16.31   What are strata? When is it appropriate to use stratified random sampling?

16.32   What are the advantages of using stratified random sampling when it is appropriate?

16.33   What is proportional allocation? How does it differ from optimal allocation?

EXERCISES   Skill Level

16.34   A population can be divided into three strata of sizes $N_1 = N_2 = 2,560$ and $N_3 = 2,675$. Simple random samples of size $n_1 = n_2 = n_3 = 28$ were drawn from these strata, and the means and variances for the three samples are shown in the accompanying table.

Stratum	$\bar{x}$	s^2
1	25.4	12.1
2	45.0	11.0
3	10.4	2.5

   a. Estimate the population mean μ, and place bounds on the error of estimation.

   b. Estimate the population total τ, and place bounds on the error of estimation.

   c. Estimate the average for stratum 1. Place bounds on the error of estimation.

16.35 Determine the proportional allocation of a sample of size $n = 100$ among the strata in the following situations.

   a. $N_1 = N_2 = 600, N_3 = N_4 = 700$

   b. $N_1 = 500, N_2 = 1{,}000, N_3 = 2{,}000$

   c. $N_1 = N_2 = N_3 = 100, N_4 = 300, N_5 = 400$

16.36 A population can be divided into two strata of sizes $N_1 = 1{,}000$ and $N_2 = 2{,}000$. Simple random samples were drawn from these strata by using proportional allocation, with an overall sample size $n = 270$. The means and standard deviations for the two samples are given below.

Statistic	Stratum 1	Stratum 2
$\bar{x}$	.0023	.0015
s	.0020	.0170

   a. Determine the allocation of the overall sample size between the two strata.

   b. Estimate the population mean, and place bounds on the error of estimation. Interpret your results.

   c. Estimate the population total, and place bounds on the error of estimation.

16.37 A finite population can be divided into five strata, each containing 100 elements. Simple random samples of size 50 were drawn from these five strata, and the number of "successes" observed were 26, 34, 19, 44, and 20, respectively. Estimate the proportion p of "successes" in the population, and place bounds on the error of estimation.

EXERCISES   Applications

16.38 An auditor for the GAO has been charged with the responsibility of estimating cost overruns in government defense contracts. For convenience, the auditor has stratified his study according to the specific branch of the military service for which each contract was originally arranged. In an analysis of 270 defense contracts issued in 1988, the auditor randomly selected 45 and found the results shown in Table 16.7.

   a. Estimate the average cost overrun for all 60 defense contracts arranged by the U.S. Naval Department in 1988, and place a bound on the error of estimation.

   b. Estimate the average cost overrun for all 270 defense contracts in 1988, and place a bound on the error of estimation.

   c. Estimate the total of cost overruns for all 270 defense contracts in 1988, and place a bound on the error of estimation.

Table 16.7   Sample Data for Exercise 16.38

Statistic	Air Force	Army	Navy
N_i	120	90	60
n_i	20	15	10
$\bar{x}_i$	$71,468	$68,709	$89,918
s_i	$16,095	$18,452	$21,065

16.39 Job dissatisfaction among employees may result in costs to a firm because of poor quality of workmanship or needless employment costs caused by nonjustified employee absence from the job. In an examination of this latter cost, a corporate personnel manager wished to determine the number of workdays lost owing to sick leave and leave without pay by her firm's 2,700 employees. For administrative convenience the personnel manager decided to use stratified random sampling and proportionately allocate a sample of $n = 27$ among the laborers, technicians, and administrators employed in the company. The data (workdays lost) obtained from sampling 15 laborers, 10 technicians, and 2 administrators are given in Table 16.8.

    a. Estimate the average number μ of days lost owing to sick leave and leave without pay by all 2,700 employees of the firm, and place a bound on the error of estimation.

    b. Estimate the total number of days lost owing to sick leave and leave without pay by the employees of the firm. Place a bound on the error of estimation.

Table 16.8   Workdays Lost Data for Exercise 16.39

Laborers			Technicians		Administrators
8	24	0	4	5	1
0	16	32	0	24	8
6	0	16	8	12	
7	4	4	3	2	
9	5	8	1	8	

16.40 Depreciation allowances offer business the opportunity to create cash reserves to replace outdated equipment and other capital items. An audit is conducted among the three branch plants of a manufacturing firm to estimate the proportion of capital equipment items that have been in use for 10 years or more. From available records the auditor has used proportional allocation to select a stratified random sample of $n = 60$ capital equipment items from among the $N = 7,200$ items listed on the inventory registry. The results in Table 16.9 were obtained. Estimate the proportion p of capital equipment items among the three branches that have been in use for 10 years or more, and place a bound on the error of estimation.

Table 16.9   Sample Data for Exercise 16.40

Statistic	Branch Plant		
	1	2	3
Number of capital equipment items, N_i	3,600	2,400	1,200
Sample size, n_i	30	20	10
Number in use 10 years or more, x_i	7	4	3

16.41   Chain stores and most banking chains process all credit accounts through a central or regional office, not separately through each branch. This technique provides more efficient central control of branch management activities. The credit manager of a chain of four wholesale bakeries is concerned about the volume of delinquent accounts currently outstanding. To reduce the cost of sampling, the manager uses stratified random sampling, with each bakery serving as a separate stratum. From records available in his office the credit manager decides to use proportional allocation to select a stratified random sample of $n = 50$ accounts from all $N = 200$ accounts receivable. The results of the sample survey are given in Table 16.10.

a. Estimate the proportion p of delinquent accounts for the chain of bakeries, and place a bound on the error of estimation.

b. There is reason to believe that the manager of bakery 3 is too lenient in granting credit to his customers. Estimate the proportion p_3 of delinquent accounts for bakery 3, and place a bound on the error of estimation.

Table 16.10   Sample Data for Exercise 16.41

	Bakery			
Statistic	1	2	3	4
Number of accounts receivable	$N_1 = 56$	$N_2 = 68$	$N_3 = 40$	$N_4 = 36$
Sample size	$n_1 = 14$	$n_2 = 17$	$n_3 = 10$	$n_4 = 9$
Number of delinquent accounts	$x_1 = 5$	$x_2 = 7$	$x_3 = 5$	$x_4 = 1$

16.7

Cluster Sampling

Frequently, it is easier to sample *clusters* of elements rather than the individual elements themselves.

Definition

cluster sample
clusters

> **A cluster sample** is obtained by randomly selecting a set of m collections of sample elements, called **clusters**, from the population and then conducting a complete census within each cluster.

Cluster sampling will usually provide a specified amount of information at a minimum cost when either

1.   a frame listing the elements of the population does not exist or would be very costly to obtain, or
2.   the population is large and spread out over a wide area.

As an illustration, consider an economist who wishes to estimate the average weekly expenditures on food per household in a city. To use either simple random sampling or stratified random sampling, the economist must have a list from which the sample elements (households) can be selected. However, a list of all the households in the city may be very costly or even impossible to obtain. Even if such a list is available, the survey costs may still be substantial, because with simple random sampling or stratified random sampling, the households chosen in the sample would probably be scattered over a wide area. As a result, the cost of conducting the survey among the scattered households would be great because of the interviewer's travel times and other related expenses.

Rather than select a sample of households scattered throughout the entire city, the economist could use cluster sampling and divide the city into clusters—political wards, perhaps—and then select a random sampling of clusters. This procedure should be easy to accomplish since a list of wards is readily available. Every household within each chosen ward would then be surveyed. The total survey costs are reduced because the economist has eliminated the need to develop a costly list of all households and because households within each ward are close together geographically, thereby reducing the interviewers' expenses.

While cost savings often result by using cluster sampling, this benefit is not always gained without a concession in return. The cluster design sometimes increases the sampling error, since elements within the same cluster tend to have common characteristics. For example, in surveys of human populations, clusters are often neighborhoods, but neighborhoods typically consist of families of the same age, income, ethnic background, and occupational class. Therefore, when a random selection of clusters is used for the survey, there is a chance that certain socioeconomic classes will not be represented at all if it should happen that their neighborhoods are not included. On the other hand, certain other classes of households may be overrepresented.

We can reduce sampling error by selecting many small clusters instead of a few large clusters. The smaller the cluster sizes (e.g., city blocks instead of school districts), the lesser is the chance that we are excluding certain classes of elements from our sample. Therefore, more information about the population can be gained by selecting a larger number of smaller-sized clusters.

Once the clusters have been specified, a list containing all clusters must be prepared. Simple random sampling is then used to select a random sample of m clusters from among the M clusters in the population.

For cluster sampling the population mean μ is estimated as shown in the display. In the formulas n_i is the number of elements in the ith cluster and t_i is the total of the measurements in cluster i.

Estimation of the Population Mean for a Cluster Sample

Estimator:

$$\hat{\mu} = \bar{x}_{\text{cl}} = \frac{\displaystyle\sum_{i=1}^{m} t_i}{\displaystyle\sum_{i=1}^{m} n_i}$$

Variance of the estimator:

$$\hat{\sigma}^2_{\bar{x}_{cl}} = \left(\frac{M - m}{Mm\bar{n}^2} \right) \left[\frac{\sum_{i=1}^{m} (t_i - \bar{x}_{cl}n_i)^2}{m - 1} \right]$$

Bounds on the error of estimation:

$$\bar{x}_{cl} \pm 1.96\hat{\sigma}_{\bar{x}_{cl}}$$

where

$$\bar{n} = \frac{1}{m} \sum_{i=1}^{m} n_i$$

M = number of clusters in population
m = number of sampled clusters

The estimate of the population total τ is given in the next display.

Estimation of the Population Total for a Cluster Sample

Estimator:

$$\hat{\tau} = \frac{M}{m} \sum_{i=1}^{m} t_i$$

Variance of the estimator:

$$\hat{\sigma}^2_{\hat{\tau}} = M^2 \left(\frac{M - m}{Mm} \right) \left[\frac{\sum_{i=1}^{m} (t_i - \bar{t})^2}{m - 1} \right] \qquad \text{where} \qquad \bar{t} = \frac{1}{m} \sum_{i=1}^{m} t_i$$

Bounds on the error of estimation:

$$\hat{\tau} \pm 1.96\hat{\sigma}_{\hat{\tau}}$$

As we stated earlier, the number of elements in the ith cluster is denoted by n_i, and t_i is the total of the measurements within cluster i. Therefore,

$$t_i = \sum_{j=1}^{n_i} x_{ij}$$

where x_{ij} is the jth observation within cluster i. The terms $\bar{n}$ and $\bar{t}$ denote, respectively, the mean cluster size and mean cluster total based on the sampled clusters.

EXAMPLE 16.7   The objective of product advertising is to increase sales or to create awareness of and interest in the company's products. Therefore, advertising should be placed with the media most likely to reach the buying public. An advertising agent for a firm that sells household products wishes to estimate the monthly expenditures on magazines and newspapers by the households of a certain midwestern city to determine whether such expenditures are sufficient to warrant the use of these media for advertising. Since no list of households is available, and since direct interviewer costs must be controlled, cluster sampling is used, with voting precincts forming the clusters. A simple random sample of 10 precincts is selected from the 50 precincts within the city. Interviewers then survey every household within the 10 precincts and record the total household expenditures on magazines and newspapers during the past month. The data are shown in Table 16.11.

Table 16.11   Magazine and Newspaper Expenditures Data for Example 16.7

Sampled Precinct, i	Number of Households, n_i	Total Expenditures, t_i	Sampled Precinct, i	Number of Households, n_i	Total Expenditures, t_i
1	62	$380	6	69	$403
2	55	517	7	58	555
3	49	480	8	74	486
4	71	613	9	57	450
5	70	540	10	65	395
Sum				$\sum_{i=1}^{10} n_i = 630$	$\sum_{i=1}^{10} t_i = \$4{,}819$

a. Estimate the average monthly household expenditures on magazines and newspapers in the city, and place a bound on the error of estimation.

b. Estimate the total monthly expenditures on magazines and newspapers by all the households in the city, and place a bound on the error of estimation.

SOLUTION

a. The population mean μ per household is estimated by

$$\bar{x}_{c1} = \frac{\sum_{i=1}^{10} t_i}{\sum_{i=1}^{10} n_i} = \frac{\$4{,}819}{630} = \$7.65$$

To calculate $\hat{\sigma}^2_{\bar{x}_{c1}}$, we first evaluate the sum of squares term,

$$\sum_{i=1}^{m} (t_i - \bar{x}_{c1} n_i)^2$$

It can be shown that

$$\sum_{i=1}^{m} (t_i - \bar{x}_{c1} n_i)^2 = \sum_{i=1}^{m} t_i^2 - 2\bar{x}_{c1} \sum_{i=1}^{m} t_i n_i + \bar{x}_{c1}^2 \sum_{i=1}^{m} n_i^2$$

When the terms are taken individually, we have

$$\sum_{i=1}^{10} t_i^2 = (380)^2 + (517)^2 + \cdots + (395)^2 = 2,374,613$$

$$\sum_{i=1}^{10} t_i n_i = (380)(62) + (517)(55) + \cdots + (395)(65) = 304,124$$

$$\sum_{i=1}^{10} n_i^2 = (62)^2 + (55)^2 + \cdots + (65)^2 = 40,286$$

Substituting these values into the sum of squares equation, we find

$$\sum_{i=1}^{10} (t_i - \bar{x}_{cl} n_i)^2 = 2,374,613 - 2(7.65)(304,124) + (7.65)^2(40,286)$$

$$= 79,153.235$$

The average cluster size is

$$\bar{n} = \frac{1}{m} \sum_{i=1}^{m} n_i = \frac{1}{10}(630) = 63$$

Since the total number of clusters in the population is $M = 50$, we find

$$\hat{\sigma}_{\bar{x}_{cl}}^2 = \left(\frac{M-m}{Mm\bar{n}^2}\right)\left[\frac{\sum_{i=1}^{m}(t_i - \bar{x}_{cl}n_i)^2}{m-1}\right]$$

$$= \left[\frac{50-10}{(50)(10)(63)^2}\right]\left(\frac{79,153.235}{9}\right) = .1773$$

Therefore, an estimate of μ, the average monthly household expenditures on magazines and newspapers, with a bound on the error of estimation, is

$$\bar{x}_{cl} \pm 1.96\hat{\sigma}_{\bar{x}_{cl}} \quad \text{or} \quad \$7.65 \pm 1.96\sqrt{.1773} \quad \text{or} \quad \$7.65 \pm \$.83$$

b. An estimate of the total monthly expenditures on magazines and news-papers is

$$\hat{\tau} = \frac{M}{m} \sum_{i=1}^{m} t_i = \frac{50}{10}(\$4,819) = \$24,095$$

which does *not* depend on knowledge of the population size N.

To place a bound on the error of estimation, we first calculate the expression

$$\sum_{i=1}^{m}(t_i - \bar{t})^2 = \sum_{i=1}^{m} t_i^2 - \frac{1}{m}\left(\sum_{i=1}^{m} t_i\right)^2$$

$$= 2,374,613 - (1/10)(4,819)^2$$

$$= 52,336.90$$

The estimated variance is

$$\hat{\sigma}_{\hat{\tau}}^2 = M^2\left(\frac{M-m}{Mm}\right)\left[\frac{\sum\limits_{i=1}^{m}(t_i - \bar{t})^2}{m-1}\right]$$

$$= 50^2\left[\frac{50-10}{(50)(10)}\right]\left(\frac{52{,}336.90}{9}\right)$$

$$= 1{,}163{,}042.222$$

The estimate of total monthly expenditures on magazines and newspapers by all the households in the city, with a bound on the error of estimation, is

$$\hat{\tau} \pm 1.96\hat{\sigma}_{\hat{\tau}} \qquad \text{or} \qquad \$24{,}095 \pm 1.96\sqrt{1{,}163{,}042.222}$$

or

$$\$24{,}095 \pm \$2{,}114$$

◆◆

Often, an experimenter wishes to use cluster sampling to estimate a population proportion p. For instance, in a preelection survey it may be desired to estimate the proportion of residents of a community who favor a particular ballot measure; or it may be desired to estimate the proportion of automobiles in a city that do not pass current emission standards or the proportion of members of a nationwide union favoring a negotiated salary adjustment. To estimate p when using cluster sampling, we first find a_i, the number of elements in cluster i that possess the characteristic of interest, for each cluster $i = 1, 2, \ldots, m$. Then an estimate of the proportion of elements in the population possessing the characteristic is given by the formula in the display.

Estimation of the Population Proportion for a Cluster Sample

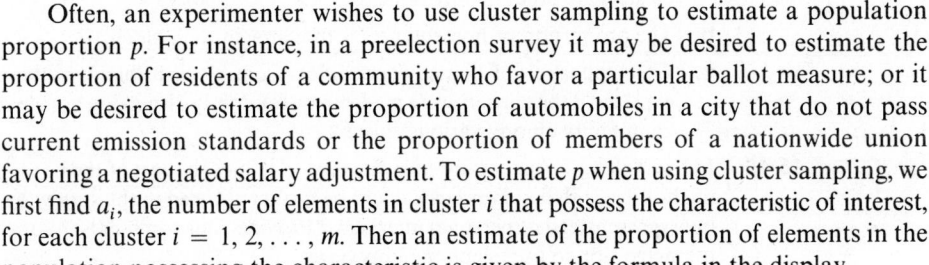

Estimator:

$$\hat{p}_{\text{cl}} = \frac{\sum\limits_{i=1}^{m} a_i}{\sum\limits_{i=1}^{m} n_i}$$

Variance of the estimator:

$$\hat{\sigma}_{\hat{p}_{\text{cl}}}^2 = \left(\frac{M-m}{Mm\bar{n}^2}\right)\left[\frac{\sum\limits_{i=1}^{m}(a_i - \hat{p}_{\text{cl}} n_i)^2}{m-1}\right]$$

Bounds on the error of estimation:

$$\hat{p}_{\text{cl}} \pm 1.96\hat{\sigma}_{\hat{p}_{\text{cl}}}$$

When the cluster sizes $n_1, n_2, \ldots, n_m$ are equal, $\hat{\sigma}_{\bar{p}_{c1}}^2$ is a good estimator of the true variance for any number m of sample clusters. However, experience has shown that when the cluster sizes are not equal, $\hat{\sigma}_{\bar{p}_{c1}}^2$ is a good estimator only when m is large, say, $m \geq 20$.

EXERCISES   Conceptual Level

16.42   What is the distinction between a cluster and a stratum?

16.43   When is it appropriate to use cluster sampling? What are the advantages of using cluster sampling?

16.44   A marketing representative for a firm that sells home products and appliances desires to estimate the average household expenditure on home products and appliances in a certain community. She has decided to use cluster sampling, with city blocks as clusters and households within blocks as elements; but she believes that the item she seeks to measure is highly dependent on whether the residents of the chosen household are homeowners or renters. Explain why the marketing representative should, or should not, use cluster sampling in each of the following situations.

  a.   Most of the households in some clusters are inhabited by homeowners, but those in other clusters are inhabited by renters.

  b.   The proportion of homeowners is the same in each block.

  c.   The proportion of homeowners differs from block to block in the manner that would be expected if the clusters were made up by randomly dispersing homeowners and renters in the community among various clusters.

16.45   In Example 16.7 cluster sampling was used to reduce the cost of the household surveys. Someone might suggest that survey costs could be further reduced in the survey by conducting telephone interviews instead of personal interviews. Provide your comments in support of or opposition to the suggestion to use telephone interviews.

EXERCISES   Skill Level

16.46   A random sample of $m = 10$ clusters is selected from among the $M = 50$ clusters in a population. There are 5 elements in each cluster in the population, and the observed sample produces

$$\sum t_i = 368 \qquad \sum t_i^2 = 14{,}824$$

where t_i is the total of the measurements in the ith cluster.

  a.   Estimate the population mean μ, using the sample information for this cluster sample.

  b.   Calculate $\hat{\sigma}_{\bar{x}_{c1}}^2$, the estimated variance of the statistic used in part a.

  c.   Place bounds on the error of estimation for the estimate calculated in part a.

16.47   Refer to Exercise 16.46.

  a.   Estimate the population total τ, using the sample information given for this cluster sample.

  b.   Calculate $\hat{\sigma}_{\hat{\tau}}^2$, the estimated variance of the statistic used in part a.

  c.   Place bounds on the error of estimation for the estimate calculated in part a.

16.48 A random sample of $m = 8$ clusters is selected from among $M = 100$ clusters in a population. The data in Table 16.12 represent the results of the survey, which involved measuring a quantitative variable within each cluster as well as the presence or absence of a particular characteristic for each element in the cluster. Hence, t_i is the cluster total for the quantitative variable and a_i is the number of elements in the cluster having the particular characteristic of interest to the researcher.

a. Estimate the population mean for the quantitative variable, and place bounds on the error of estimation.

b. Estimate the population proportion p having the particular characteristic of interest to the researcher. Place bounds on the error of estimation.

c. Is the estimate calculated in part b a good estimator of the population proportion? Why or why not?

Table 16.12   Cluster Data for Exercise 16.48

Cluster	n_i	t_i	a_i	Cluster	n_i	t_i	a_i
1	3	18	1	5	2	17	0
2	4	16	2	6	4	15	1
3	8	25	3	7	5	21	4
4	7	19	3	8	5	23	2

16.49 Refer to Exercise 16.48. Estimate the population total τ for the quantitative variable, and place bounds on the error of estimation.

EXERCISES   Applications

16.50 A recent news article noted that the rate of increase of charitable contributions lags far behind the rate of inflation during periods of recession. Concerned about a possible decline in voluntary contributions in her region, a regional director for the American Cancer Society is interested in estimating the average contribution per household and the total contribution from all households in her city. At this stage in the fund-raising campaign, neighborhood volunteers have canvassed a random selection of 12 of the 47 voting precincts in the city and have obtained the data given in Table 16.13 for the amount of donations.

Table 16.13   Contributions Data for Exercise 16.50

Precinct	Number of Households	Total Donations	Precinct	Number of Households	Total Donations
1	36	$117	7	29	$165
2	42	105	8	52	105
3	40	210	9	44	121
4	47	142	10	40	103
5	39	235	11	45	136
6	50	96	12	36	190

a. Estimate the average contribution per household in the city, and place a bound on the error of estimation.

b. Estimate the total contribution from all households in the city, and place a bound on the error of estimation.

16.51 An inspector for a hardware chain wishes to estimate the proportion of defective light bulbs shipped to its warehouse by a manufacturer. The bulbs are shipped in cartons containing 12 boxes, with each box containing 6 bulbs. Design a cluster sampling experiment for the inspector. Should he use cartons of bulbs or boxes of bulbs as clusters? Explain.

16.52 Refer to Exercise 16.51. The inspector decides to use boxes of bulbs as clusters and randomly selects $m = 20$ boxes from among the 100 cartons received in the shipment. The number of defective bulbs found in each of the 20 boxes of bulbs is as follows:

0 2 0 3 1 1 0 1 2 1 0 2 0 1 1 0 3 0 2 1

Estimate the proportion p of defective bulbs in the shipment, and place a bound on the error of estimation.

16.53 An industry is considering revision of its retirement policy and wants to estimate the proportion of employees that favor the new policy. The industry consists of 87 separate plants located throughout the United States. Since results must be obtained quickly and with little cost, the industry decides to use cluster sampling, with each plant as a cluster. A simple random sample of 15 plants is selected, and the opinions of the employees in these plants are obtained by questionnaire. The results are shown in Table 16.14. Estimate the proportion of employees in the industry who favor the new retirement policy, and place a bound on the error of estimation.

Table 16.14   Sample Data for Exercise 16.53

Plant	Number of Employees	Number Favoring New Policy	Plant	Number of Employees	Number Favoring New Policy
1	51	42	9	73	54
2	62	53	10	61	45
3	49	40	11	58	51
4	73	45	12	52	29
5	101	63	13	65	46
6	48	31	14	49	37
7	65	38	15	55	42
8	49	30			

16.8

Finding the Sample Size

One of the first questions asked by someone undertaking a sample survey is, "How many sample elements should I select?" Since sampling is time-consuming and costly, our objective in selecting a sample is to obtain a specified amount of information about a population parameter at a minimum cost. We can accomplish this objective by first deciding on a bound on the error of estimation (which measures our specified information content) and then applying an appropriate sample size estimation formula.

In Section 16.1 we saw that when the population is uniform, a small sample provides the same amount of information as a large sample. Thus, the physician can base a diagnosis on one drop of the patient's blood; the quality control engineer can judge a large lot of transistors by testing only a few. Selecting a large sample in such instances is a waste of time and money. On the other hand, if the population consists of many highly diverse elements, a small sample may provide a poor reflection of the population. In a study to estimate the average height of the male students attending a particular college, a small sample—say $n = 3$ students—may by chance consist entirely of members of the varsity basketball team. A random sampling of $n = 100$ students should, however, provide a much broader coverage of, and hence more information about, the heights of the male students.

The objectives behind the selection of a sampling design and selection of the sample size are the same—to obtain a specified amount of information at a minimum cost. Sampling design decisions are made according to the "lay of the land"—that is, how the elements group themselves together in the population—and according to the cost of recording information contained by those elements. Sample size decisions are made according to the inherent variability in the population of measurements and how accurate the experimenter wishes the estimate to be. These two criteria are, of course, inversely related. To obtain greater accuracy, and hence more information about a population, we must select a larger sample size; the greater the inherent variability in the population, the larger is the sample size required to maintain a fixed degree of accuracy in estimation.

Simple Random Sampling

For simple random sampling the sample size required to estimate the population mean μ, with a bound B on the error of estimation, is as given in the following display.

Sample Size for Estimating μ for a Simple Random Sample

$$n = \frac{N\sigma^2}{(N - 1)D + \sigma^2} \qquad \text{where} \qquad D = \frac{B^2}{z^2}$$

and where σ^2 is the population variance, N is the number of elements in the population, B is the bound on the error of estimation, and z is the value of the standard normal random variable corresponding to a confidence coefficient of $(1 - \alpha)$.

When N is very large, the sample size formula reduces to the more familiar formula given in Chapter 8.

Sample Size for Estimating μ for a Simple Random Sample When N Is Very Large

$$n = \frac{z^2 \sigma^2}{B^2}$$

Notice that if we denote by n_0 the required sample size when N is large, so that

$$n_0 = \frac{z^2 \sigma^2}{B^2}$$

then the required sample size is

$$n = \frac{n_0}{\left(\dfrac{N-1}{N}\right) + \dfrac{n_0}{N}} \approx \frac{n_0}{1 + \dfrac{n_0}{N}}$$

When the objective is to estimate the population total τ, with a bound B on the error of estimation, we must substitute $D = B^2/z^2 N^2$ into the sample size formula in the first display.

Some students may notice a dilemma in the guidelines provided for finding n. To find n, we must know the population variance; but to estimate σ^2, we must have a set of sample measurements from the population. The variance can be estimated by s^2 obtained from a previous sample or by knowledge of the range of measurements, giving the estimate

$$\hat{\sigma}^2 = \left(\frac{\text{range}}{4}\right)^2$$

The range approximation procedure is derived from the Empirical Rule (Section 2.9) and provides a very rough approximation to σ^2.

EXAMPLE 16.8   The manager of the credit division of a commercial bank would like to know the average amount of credit purchases charged to their bankcard each month by the customers who hold bankcards issued by the bank. Since the bank currently has 20,000 open bankcard accounts, time and expense prohibit a complete review of every account. The manager thus proposes selecting a simple random sample of open bankcard accounts to estimate the average monthly account balance μ, with a bound on the error of estimation of $B = \$10$. Although no prior information is available to estimate the variance σ^2 of monthly bankcard account levels, the manager knows that most account levels lie within the range from \$50 to \$450. Find the sample size necessary to achieve the stated bound.

SOLUTION   To estimate the population variance σ^2, we use the rule that says the range is approximately equal to four standard deviations. Therefore,

$$\hat{\sigma}^2 = \left(\frac{\text{range}}{4}\right)^2 = \left(\frac{450 - 50}{4}\right)^2 = 10{,}000$$

Using the formula for finding n when estimating μ, we obtain

$$D = \frac{B^2}{z^2} = \frac{(10)^2}{(1.96)^2} = 26.03$$

$$n = \frac{N\sigma^2}{(N-1)D + \sigma^2} = \frac{(20,000)(10,000)}{(19,999)(26.03) + 10,000} = 376.94$$

Therefore, the credit manager must randomly select approximately 377 accounts to estimate μ, the average monthly bankcard account balance, accurate to within \$10 of the true value.

◆ ◆

Notice that since N was very large in Example 16.8, we could have used the shortcut formula $n = z^2\sigma^2/B^2$. Doing so, we find

$$n = \frac{(1.96)^2(10,000)}{10^2} = 384.16$$

a slightly larger sample size than that specified by the exact formula.

When the purpose is to estimate the population proportion p, the variance $\sigma^2 = pq$ depends on p, the population characteristic to be estimated. We must then use a guessed value for p; or taking a conservative approach by allowing the variance to attain its maximum possible value, we can assume p is near $1/2$ and let $\sigma^2 = .25$. The sample size required to estimate p, with a bound B on the error of estimation, is then the same as that given earlier for the estimation of μ.

Stratified Random Sampling

With stratified random sampling we select a separate simple random sample within each of the L strata. Therefore, we cannot determine n until we know the relationship between n and the sample allocation to the strata $n_1, n_2, \ldots, n_L$. Although there are many ways to allocate n among the strata, in Section 16.6 we used only proportional allocation.

When using stratified random sampling, we must also consider the fact that the variances of the strata, $\sigma_1^2, \sigma_2^2, \ldots, \sigma_L^2$, may not be equal. We will need approximations for each of these variances, which we can obtain from previous samples or by estimating the range of measurements within each stratum. In the latter case the range approximation

$$\hat{\sigma}_i^2 = \left(\frac{\text{range}}{4}\right)^2$$

provides a rough estimate of the variance of the measurements in stratum i based on the *range* of measurements within stratum i.

For stratified random sampling with proportional allocation, the sample size required to estimate the population mean μ, with a bound B on the error of estimation, is as given in the display.

Sample Size for Estimating μ for a Stratified Random Sample

$$n = \frac{\displaystyle\sum_{i=1}^{L} N_i \sigma_i^2}{ND + \dfrac{1}{N}\displaystyle\sum_{i=1}^{L} N_i \sigma_i^2} \qquad \text{and} \qquad D = \frac{B^2}{z^2}$$

where σ_i^2 and N_i are, respectively, the variance and size of the ith stratum.

The sample size required to estimate the population total τ, with a bound B on the error of estimation, is obtained by substituting $D = B^2/z^2N^2$ into the equation in the display.

EXAMPLE 16.9 The dean of a business school is considering canvasing the members of the school's alumni association for the purpose of generating donations to the school's development fund. Currently, there are 3,500 members of the alumni association, 2,100 of whom live in-state while the remainder live out-of-state. The dean has decided to select a stratified sample of alumni (stratified according to current residence) to estimate total donations; using the sample evidence, he will decide whether to contact all remaining alumni. Find the number n of alumni that should be contacted if the dean wishes to estimate the total alumni contributions with a bound on the error of estimation of $10,000. How should this sample size be allocated between in-state and out-of-state alumni? From prior fund-raising drives the standard deviations for donations by in-state and out-of-state alumni were found to be $30 and $20, respectively.

SOLUTION We let the in-state alumni constitute stratum 1 and the out-of-state alumni constitute stratum 2. From prior experience we know that $\sigma_1 = \$30$ and $\sigma_2 = \$20$. Therefore,

$$\sigma_1^2 = (30)^2 = 900 \qquad \text{and} \qquad \sigma_2^2 = (20)^2 = 400$$

Since we are finding the sample size for the estimation of a population total, we compute

$$D = \frac{B^2}{z^2 N^2} = \frac{(10{,}000)^2}{(1.96)^2(3{,}500)^2} = 2.12$$

Before finding n, we compute

$$\sum_{i=1}^{2} N_i \sigma_i^2 = (2{,}100)(900) + (1{,}400)(400) = 2{,}450{,}000$$

The required sample size is then

$$n = \frac{\displaystyle\sum_{i=1}^{2} N_i \sigma_i^2}{ND + \dfrac{1}{N}\displaystyle\sum_{i=1}^{2} N_i \sigma_i^2} = \frac{2{,}450{,}000}{(3{,}500)(2.12) + (1/3{,}500)(2{,}450{,}000)} = 301.08$$

Therefore, the dean should select 302 alumni to estimate the total contributions τ by all alumni, with a bound on the error of estimation of $10,000. With proportional allocation the sample should be partitioned according to the formula

$$n_i = n\left(\frac{N_i}{N}\right)$$

Therefore, he should randomly select

$$n_1 = (302)\left(\frac{2,100}{3,500}\right) = 181.2$$

or 181 in-state alumni and

$$n_2 = (302)\left(\frac{1,400}{3,500}\right) = 120.8$$

or 121 out-of-state alumni.

◆ ◆

When using stratified random sampling to estimate the population proportion p, we must substitute $\sigma_i^2 = p_i q_i$ for the variance of stratum i and let $D = B^2/z^2$. In this expression p_i is the population proportion for stratum i, which can be estimated either by a judicious guess or by a conservative approach of letting $p_i q_i = .25$ for stratum i. With proportional allocation and the conservative approach for assessing each p_i, the formula for the sample size to estimate the population proportion p, with a specified bound B on the error of estimation, is approximately that shown in the display.

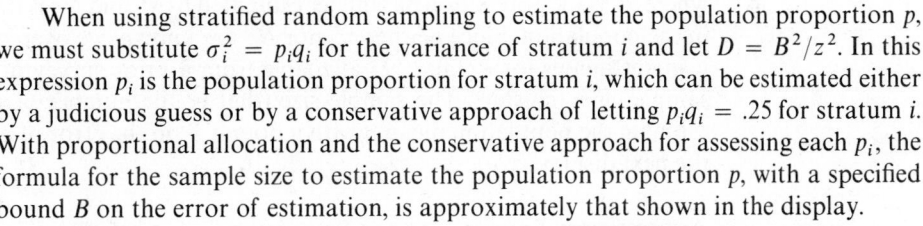

Approximate Sample Size for Estimating p for a Stratified Random Sample When N Is Very Large.

$$n = \frac{N}{NB^2 + 1}$$

Cluster Sampling

With stratified sampling we first partition the population into strata; then we select a random sample from every stratum. The procedure is reversed with cluster sampling. After the population is divided into clusters, a few clusters are randomly selected from the group. Within each chosen cluster every sample element is recorded. The sampling units are the individual elements of the population when using stratified random sampling, but with cluster sampling the sampling units are clusters of elements.

Finding the sample size when using cluster sampling thus amounts to choosing the number m of clusters of elements we will select. The quantity of information provided by a cluster sample is affected not only by m but also by the size of the clusters. For cluster sampling, more information about the population can be gained by selecting a larger number of smaller-sized clusters. The rare exception to this rule is the case in which the population consists of many small homogeneous (similar) segments.

In what follows, we will assume that the relative cluster size has been selected in advance and that the problem is to choose the number of clusters m.

The estimated variance of $\bar{x}_{cl}$ is

$$\hat{\sigma}^2_{\bar{x}_{cl}} = \left(\frac{M - m}{Mm\bar{n}^2} \right) s^2_{cl}$$

with

$$s^2_{cl} = \frac{\sum\limits_{i=1}^{m} (t_i - \bar{x}_{cl} n_i)^2}{m - 1}$$

The actual variance of $\bar{x}_{cl}$ is approximately

$$\sigma^2_{\bar{x}_{cl}} = \left(\frac{M - m}{Mm\bar{N}^2} \right) \sigma^2_{cl}$$

where σ^2_{cl} is the population parameter estimated by s^2_{cl} and $\bar{N}^2$ is the average cluster size in the population. Since neither σ^2_{cl} nor $\bar{N}^2$ are known, we overcome this difficulty by using estimates for σ^2_{cl} and $\bar{N}^2$ available from a prior survey or from a preliminary sample of m' clusters. The sample size (that is, the number of clusters) required to estimate the population mean μ with a bound B on the error of estimation is given in the next display.

Approximate Sample Size for Estimating μ for a Cluster Sample

$$m = \frac{M\sigma^2_{cl}}{MD + \sigma^2_{cl}} \quad \text{with} \quad D = \frac{B^2\bar{N}^2}{z^2}$$

where σ^2_{cl} and $\bar{N}^2$ are estimated by s^2_{cl} and $\bar{n}^2$.

If we denote by m_0 the solution when M is very large, then

$$m_0 = \frac{z^2 \sigma^2_{cl}}{B^2 \bar{N}^2}$$

and the approximate number of clusters is

$$m = \frac{m_0}{1 + (m_0/m)}$$

EXERCISES   Conceptual Level

16.54   Why should one be concerned about sample size? Why not simply use a very large sample size as a general practice?

16.55   Describe the relationship between the bound on error and the sample size.

EXERCISES   Skill Level

16.56   A simple random sample must be chosen in order to estimate a population mean μ to within $B = 10$ with 95% confidence. Find the sample size necessary to estimate μ in the following situations.

    a. $N = 250; \sigma^2 = 1000$ **b.** $N = 5000; \sigma^2 = 500$ **c.** $N = 326; \sigma^2 = 1560$

    d. N is very large; σ^2 is unknown; but the sample range is $R = 1216.5$

16.57   A stratified random sample must be drawn from a population consisting of $L = 3$ strata, using proportional allocation. The strata sizes and variances (estimates based on previous information) are given in the accompanying table. Determine the sample size and proportional allocation necessary to estimate the population mean μ in the following situations.

Strata i	σ_i^2	N_i
1	4,500	2,000
2	5,100	2,500
3	5,240	1,500

    a. $B = 10$; 95% confidence   **b.** $B = 25$; 99% confidence

16.58   Refer to Exercise 16.48. A follow-up survey is planned in an attempt to estimate μ to within 1 unit with 95% confidence. Use the sample information given in Exercise 16.48 to determine the necessary sample size.

EXERCISES   Applications

16.59   Timber cruisers typically use a formula based on basal circumference and topped-off height to estimate the board footage in standing timber. A timber cruiser wishes to estimate the board footage in a stand of timber containing 150 trees that would top off to 30-ft logs, 100 that would top off to 40-ft logs, and 50 that would top off to 50-ft logs. How many trees should he select in each height stratum if he wishes to estimate the total board footage in the stand, with a bound on the error of estimation of 20,000 board feet? From past experience he knows that the standard deviation of the board feet of lumber in 30-ft, 40-ft, and 50-ft logs is 300, 350, and 400, respectively.

 16.60   In the face of a recessed economy a textile firm is considering reducing its work schedule to a four-day workweek. As an alternative, the company would likely choose to close one of its four main manufacturing divisions, terminating the employees in that division. To gain a feeling of employees sentiment, the firm's personnel manager wishes to choose a sample of employees from among the four divisions and estimate the proportion p favoring a reduced workweek, with a bound on the error of estimation of $B = .10$. The firm employs 75 individuals each in divisions 1 and 2, 65 in division 3, and 40 in division 4. It is estimated that about 75% of those in division 4 favor a reduced workweek, and the employees in the other three divisions appear equally divided on the issue. Find the sample size and allocation necessary to achieve the stated bound.

 16.61   Refer to Exercise 16.60. Use the conservative approach for the estimation of the variance σ_i^2 for each division. How does this approach affect the required sample size and allocation for estimating p with a bound on the error of estimation of $B = .10$?

 16.62   A local mass transit commission is interested in exploring the feasibility of expanding bus service within a medium-sized western community. A sample survey will be conducted among the

households of the community to estimate the proportion of adults who would use the expanded bus service at least once a week. The community is divided into four geographic sectors: NW, NE, SW, and SE. Thus, stratified random sampling will be used, with the adult residents within each sector comprising the respective strata. Recent census data suggest that the number of adults residing in each geographic sector of the community is as shown in the accompanying table. Find the approximate sample size and the allocation necessary to estimate the actual proportion of adults who would use the expanded bus service, with a bound of .05 on the error of estimation.

Sector (Strata)	NW	NE	SW	SE
Number of Adults	8,500	7,000	10,000	12,000

16.63   In an effort to reduce gasoline consumption the director of a state motor pool is considering the installation of an experimental carburetor on each of the 412 automobiles operated out of the motor pool. Currently, the motor pool consists of 126 Plymouth Horizons, 69 Chevrolet Celebritys, and 217 Ford Escorts. Before deciding whether to buy a new carburetor for each car, the director wishes to select a sample of cars from the motor pool and to note the gain in gasoline economy in terms of the reduced mileage per gallon. Past evidence indicates that the variance of gasoline mileage for Ford Escorts is about 4.0, for Chevrolet Celebritys is about 4.8, and for Plymouth Horizons is about 6.1. Find the sample size and allocation among the three brands of cars for estimating the average gasoline mileage using the new carburetor, with a bound on the error of estimation of $B = 2$ miles per gallon.

16.9

Other Sampling Designs and Procedures

The sampling designs we have described so far—simple random sampling, stratified random sampling, and cluster sampling—are those most commonly used in business and economics. Other less frequently used sampling designs and methods are briefly introduced in this section. For a complete discussion of these and other sampling procedures, refer to the text by Cochran listed in the references.

Systematic Sampling

systematic sampling   A design that avoids the cumbersome data collection requirements of simple random sampling is **systematic sampling**. A systematic sample is obtained by randomly selecting one element from the first k elements in the frame and then selecting every kth element thereafter. Since it is easier and less time-consuming to perform than simple random sampling, systematic sampling can provide more information per sampling dollar. It is especially useful in auditing, when the relevant information is recorded in an orderly form—say in computer storage or on file cards. Selecting credit accounts, equipment maintenance records, or sales force data from computer-stored company records can be accommodated easily, cheaply, and efficiently by using systematic sampling.

There are some instances in which systematic sampling should not be used. If hidden periodicities exist in the population, the 1-in-k systematic sample may bias results by introducing sampling error resulting from the periodic influence. Sales records and financial data observed over time often have inherent cyclical behavior — restaurant sales are greater on the weekend than during the week, cash levels are highest around the tenth of the month, personal loans are more frequent during the winter months. In addition, production processes often exhibit periodic behavior when certain individuals or machines are responsible for the output of the process at regular intervals. In such instances the 1-in-k systematic sample may either be completely in phase or totally out of phase with the cyclic component, thus misrepresenting its influence in the sample.

Systematic sampling should also be avoided when the population size is unknown. As evidence of this caution, note that the sampling frequency k must be selected to be less than or equal to N/n, where the sample size n is specified in advance. If N is unknown, we cannot accurately specify k. However, this restriction is not a problem if N is assumed to be infinitely large. (See the text by Scheaffer, Mendenhall, and Ott listed in the references.)

The estimation formulas for a systematic sample can be found in Chapter 7 of *Elementary Survey Sampling*, 3rd edition, by Scheaffer, Mendenhall, and Ott. Not surprisingly, these formulas are the same as those offered in Section 16.5 for simple random sampling.

Ratio Estimation

ratio estimation **Ratio estimation** is an estimation procedure based on the relationship between two variables y and x measured on the same set of sampled elements. Ratio estimation, like linear regression (Chapter 11), uses information on a variable x to estimate μ_y or τ_y. For example, an auditor may wish to estimate the actual dollar value of inventory from the inventory recorded on computer accounts; a lumber dealer may wish to estimate the total board feet τ_y in a stand of 50-ft Douglas fir trees based on the total of their basal circumferences τ_x; or a corporate executive may be interested in estimating the total sales τ_y for the firm in the next quarter based on the current quarter's sales τ_x.

A ratio estimate requires the measurement of two variables, y and x, on each element in the sample and a computation of the ratio of their sums,

$$\hat{R} = \frac{\sum\limits_{i=1}^{n} y_i}{\sum\limits_{i=1}^{n} x_i}$$

We assume that

$$R = \frac{\tau_y}{\tau_x}$$

Then $\hat{R}$ is an estimator of R (not unbiased) and $\hat{\tau}_y = \hat{R}\hat{\tau}_x$.

The use of ratio estimation is based on the assumption that the relationship between the variables y and x is stable over the entire population. Peterson and his coauthors (see the references) report on the use of assessment/sales ratios to monitor the assessment performances of public assessment offices. When y is the assessed value of a property and x is its market price, the authors note a much higher ratio for "blighted" neighborhoods than for "upward transitional" neighborhoods. Therefore, if their study reflects assessment practices in most urban communities, we must assume that a single assessment/sales ratio does not adequately portray the quality of assessments for a particular community. In short, ratio estimation is appropriate when the measure of linear correlation between y and x is strong, at least equal to $1/2$ according to Scheaffer, Mendenhall, and Ott (1986). If it is not, the relationship between y and x will be unstable, resulting in estimates for μ_y and τ_y that are less precise than those obtained by using $\bar{y}$ or $N\bar{y}$.

As we have noted, ratio estimation is an estimation procedure; it is not a sampling design. That is, an appropriate sampling design is chosen for the selection of measurements on the variables y and x. Then ratio estimation is used to estimate the ratio R from the sample information. It can be used with any of the survey designs presented earlier in this chapter.

Two-Stage Cluster Sampling

For cluster sampling, convenient groups or clusters of elements are randomly selected from the population. After the clusters are chosen, *all* the elements from each selected cluster are sampled.

two-stage cluster sample

A two-stage cluster sample is obtained by choosing a simple random sample of clusters and selecting a random sample of elements from each cluster. Therefore, when cluster sizes are very large or the elements within each cluster are quite similar, two-stage cluster sampling provides an economical and efficient alternative to cluster sampling.

In recent years commercial banks have become very competitive. Many have sought to attract new depositors and to encourage current depositors to place more money in checking accounts and less in savings accounts by offering free check-writing privileges. Suppose a large branch-banking system wishes to survey the opinions of its current depositors on their reaction to a proposal that would offer them free checking accounts. The bank could consider each branch bank as a separate cluster, select a random sample of branch banks, and then draw a random sample of depositors from each branch. Similarly, a nationwide study to examine the total expenditures by families on medical services could be obtained by first selecting a simple random sampling of states and then choosing a sample of families from each selected state.

area sampling

This latter situation is an example of **area sampling**, in which the initial clusters are geographical areas. Area sampling need not be two-stage but can be a multistage analysis. For example, in the medical expenditures study the states selected in the first stage of the analysis may be partitioned into counties, and then selected counties partitioned into school districts. Then from the selected school districts, the researcher could canvas all families or select a random sampling of families. Public opinion polling agencies tend to rely on area sampling, usually using stratified sampling within selected clusters.

Cost savings are the primary advantage of two-stage or multistage cluster sampling over more conventional sampling procedures. A frame listing all the elements in the population may be very costly, if not impossible, to obtain. It may, however, be easy and inexpensive to obtain a list of clusters. In addition, cluster sampling reduces the cost of sampling by concentrating the sampling effort in smaller areas where the elements are physically close together.

As with cluster sampling, the use of two-stage or multistage cluster sampling may lead to imprecise estimates for μ, τ, and p, since we are intentionally excluding part of the population from our sample. Thus, biases may result when using cluster sampling. It is the responsibility of the problem analyst to decide whether the potential cost savings gained by using a clustering design offset the effect of possible sampling biases.

Randomized Response Sampling

In a sampling of human populations two nonsampling errors that frequently distort the research findings involve a refusal on the part of some of the respondents to answer all the questions or their act of deliberately providing incorrect information. Often, such distortions result when the respondent is afraid of losing prestige or becoming embarrassed by truthful responses to sensitive questions. The bias produced by these nonsampling errors is sometimes large enough to make the sample estimates seriously misleading.

Nevertheless, personal opinions, controversial topics, and intimate behavior are frequently the topics the business researcher is asked to survey. For instance, a retailer may be interested in estimating the incidence of shoplifting; a wholesale distributor of sundries may wish to know the frequency of purchase of birth control products in a certain city; the Internal Revenue Service may be interested in determining the frequency of intentional errors on federal income tax forms; or a police agency may be interested in estimating the frequency of drug use in its jurisdiction. Even such seemingly harmless topics as age, income, or marital status sometimes trigger sensitive reactions in the respondent.

*randomized response
sampling*   A special sampling procedure has been created to handle surveys dealing with potentially sensitive or embarrassing material. This procedure is called **randomized response sampling**, and it requires that a question on the sensitive topic be paired with an innocuous question. The respondent then answers only one of the two questions, which he or she selects at random. For example, the sensitive question

S: "Have you ever willfully shoplifted from Macy's?"

could be paired with the innocuous question

A: "Have you ever visited the state of Florida?"

The interviewer is given an answer but is unaware of which question has been answered by the respondent. Thus, the respondent is saved possible embarrassment or incrimination, since his or her answer is camouflaged by the presence of the innocuous question.

In randomized response sampling a randomization device is created so that the respondent selects the sensitive question with a known probability. Commonly, the randomization device consists of a black bag containing a predetermined mixture of black and white beads. The respondent simply selects a bead, notes its color (without letting the interviewer see the color of the bead selected), and answers the appropriate question.

Note that randomized response sampling does not *compete* with simple random sampling, stratified random sampling, or cluster sampling. To the contrary, the randomized response design is used for collecting data *within* those designs, in cases when survey questions may be somewhat sensitive, in order to reduce the chances of bias and error due to nonresponse or incorrect responses. Formulas for the analysis of experimental results obtained in a randomized response survey can be found in the article by Greenberg and coauthors listed in the references.

EXERCISES   Conceptual Level

16.64   What is systematic sampling? When is it appropriate to use systematic sampling?

16.65   Explain ratio estimation. When is it advantageous to use ratio estimation?

16.66   Describe two-stage cluster sampling. What is gained by applying this technique?

16.67   What is the distinction between area sampling and cluster sampling?

16.68   Describe randomized response sampling. When is it appropriate to use randomized response sampling? What are the advantages of this type of sampling?

16.69   Under what condition is it desirable to use systematic sampling as an alternative to simple random sampling? When should systematic sampling not be used?

EXERCISES   Applications

16.70   For each of the following problem settings, indicate whether a systematic sample, a two-stage cluster sample, or ratio estimation is appropriate.

   a.   A department store chain with a number of branch stores located in four eastern states wishes to estimate the total accounts receivable for all stores. Accounts are kept separately in each store.

   b.   A local electric utility co-op is interested in determining the effect of an energy conservation program begun last month. It wishes to obtain a measure of the average household power use during the current month.

   c.   An economist wishes to conduct a consumer survey to estimate the average monthly food expenditures per household for the households of a certain community. From previous studies the economist believes that the amount spent on food by a household is correlated with total household income.

d. An advertising firm has undertaken a promotional campaign for a new product. It wishes to sample potential customers to determine market acceptability of the product in a certain small community.

    e. The city council of a large southwestern city wishes to know the proportion of residents of the city who favor the introduction of a city income tax whose proceeds are earmarked for the reduction of property taxes.

16.71 Telephone companies and other utilities are examples of regulated monopolies. They cannot arbitrarily adjust their rate structure, as do unregulated companies, but must make an appeal for a rate change to the public utilities commissioner of their state. To keep pace with increasing costs resulting from a high inflation rate, Pacific Northwest Bell has proposed an increase in all long-distance calls for its subscribers in the state of Washington. However, the Washington public utilities commissioner claims that increased rates will simply dampen use of long-distance by Washington subscribers, thus creating no revenue increase for Pacific Northwest Bell. Design a survey by which Pacific Northwest Bell can survey its Washington subscribers to estimate the company's total monthly revenue gain under the proposed rate structure.

16.72 The control of highway traffic and problems of traffic intensity are issues of vital concern to planning and law enforcement agencies, as well as to commercial and private users of public highways. The greater the traffic intensity on any highway, the more frequent are the needed repairs and the need for law enforcement agencies to assist in cases of accidents or traffic congestion. For those using the highway for business purposes, traffic intensity introduces the likelihood of opportunity costs due to delivery delays and other factors. To measure the average traffic intensity per day, the state highway commission has installed a mechanical traffic counter along a busy portion of a state highway. Traffic counters are not always reliable, though, since trucks with multiple axles tend to inflate the count. To accommodate this problem, a researcher obtains an actual traffic count, which is compared with the count provided by the traffic counter. Design a sampling procedure to estimate the average daily traffic on the highway from the count provided by the traffic counter.

16.73 In an audit of several contracts with suppliers to the Department of Defense (DOD), a Government Accounting Office (GAO) auditor has detected the presence of fraud in the form of illegal cash kickbacks by suppliers to DOD officials. To determine the extent of this problem the GAO auditor wishes to estimate the proportion of all DOD contracts issued within the past five years in which an illegal cash kickback was involved. Suggest an appropriate research design for the auditor, write an appropriate questionnaire, develop a randomization device, and explain its use.

16.10

Nonrandom Sampling Procedures

In Section 16.1 we observed that when samples are random, we can make probabilistic statements about the underlying population. On the other hand, when samples are selected in a nonrandom manner, we can make only descriptive statements about the sample. Three common nonrandom sampling procedures are convenience sampling, judgment sampling, and quota sampling. **Convenience sampling** refers to the selection of elements that can be obtained simply and conveniently. Selecting the first *n* customers to enter a store on a particular date would comprise a convenience sample. **Judgment sampling** involves the selection of elements that, according to the judgment and

convenience sampling

judgment sampling

quota
samplingintuition of the sampler, accurately reflect the population. **Quota sampling** requires that the sampler arbitrarily select a predetermined number of individuals from different population sectors—usually different occupational classes. Quotas are usually predetermined in an attempt to characterize the population, but the actual selection is not necessarily performed in a random manner.

Nonrandom sampling procedures should *never* be used when the objective of the sampling exercise is inference—that is, drawing conclusions about the population from information contained in the sample. Nonrandom sampling should be used only if we are willing to restrict our study to descriptive statements about the sample and to forgo inferential statements about the population.

16.11

Summary and Concluding Remarks

The sampling designs discussed in this chapter—simple random sampling, stratified random sampling, cluster and two-stage cluster sampling, and systematic sampling—all are designs that provide random samples. Consequently, when using these designs, we know the probabilities of including various observations in the sample, which

Table 16.15   Summary of Characteristics of Random Sampling Designs

Sampling Design	Basic Characteristics
Simple random sampling	Simple, easy to apply
	No attempt to reduce the cost of obtaining the desired quantity of sample information
Stratified random sampling	Population divided into homogeneous groups, called strata
	Usually provides an estimator with smaller variance, at a reduced cost, than could be obtained by using simple random sampling
	Provides separate estimates of population parameters within the strata
Cluster sampling	Clusters (collections) of elements randomly chosen from the population
	Used when an adequate frame unavailable or travel costs excessive
	Survey costs reduced because of proximity of elements to one another
Systematic sampling	Selects every kth element from a list or file after a random start
	Used when simple or stratified random sampling is costly and impractical
	Cannot be used when periodicities exist within the list

allows us to make probabilistic statements about the population. To summarize, Table 16.15 offers the basic characteristics of the various random sampling designs discussed in this chapter.

Two sampling procedures were introduced in this chapter for handling special types of sampling problems. The randomized response model is useful for collecting survey information when the survey topic is sensitive or potentially embarrassing to the respondent. Ratio estimation provides a model for incorporation of the effects of an auxiliary variable x in the estimation of μ_y or τ_y. But remember that neither randomized response sampling nor ratio estimation by itself constitutes a sampling design. Both are used in conjunction with some traditional sampling design—randomized response sampling is used for collecting sensitive data within some traditional sampling design, and ratio estimation is applied after some traditional design has been employed to select measurements on the variables y and x.

Supplementary Exercises

Applications

16.74 What is the objective of sampling? How does this objective relate to the objective of statistics? (*Hint*: See Section 16.1.)

16.75 Do the following examples represent applications of random sampling or nonrandom sampling? Explain.

   a. The constituents returning a questionnaire on U.S. foreign policy to their congressional representative

   b. The apples in an enclosed 5-lb bag purchased at a local supermarket

   c. A 5-lb bag of apples selected by a housewife from a bin of apples at a local supermarket

   d. Department store queries to every tenth credit account customer about new store hours

   e. Warranty cards providing personal and demographic information received by a manufacturer from those who have recently purchased one of the manufacturer's small kitchen appliances

16.76 Discuss the advantages of conducting a sample survey instead of a census in each of the following instances.

   a. A candidate for governor of the state of Illinois wishes to know the proportion of Illinois voters favoring his candidacy one week prior to the election.

   b. A marketing representative for a breakfast food company is interested in determining the total first-year sales of a new packaged breakfast food the company has developed.

   c. A local newspaper has adopted a more liberal news editorial policy. To obtain reader reaction to this change, an agent for the newspaper randomly selects 10 local subscribers from a subscription list, contacts them by phone, and asks them for their opinion of the change in editorial policy.

   d. To determine the proportion of residents favoring a municipal bond levy, the city manager proposes selecting every 25th individual listed in the city's telephone directory and obtaining his or her opinion by a telephone poll.

e. An oil company executive is interested in determining the average price per gallon of low-lead gasoline charged by its retail stations in the state of Kansas. From a list of stations the executive randomly selects 20 of the 249 retail stations and obtains their retail price by telephone.

16.77 Refer to Exercise 16.76. For each sampling problem, suggest modifications to the sampling procedure that may lessen the chances of incurring bias and error in the survey.

16.78 "Performance risk" concerns whether a product will perform as it should. "Psychological risk" concerns the consumer's perceived self-concept when purchasing the product. Advertising cannot affect the former but can affect the latter. A discount store in a city containing 745 households has adopted a new advertising theme designed to alleviate psychological risk associated with its merchandise. From a residential directory a simple random sample of $n = 50$ households is selected. One month after the new advertising campaign began, the households are contacted, with 13 heads of the households indicating that they preceive the discount store's merchandise as inferior to that offered by competing stores. Estimate the true proportion of households that perceive psychological risk in the merchandise offered by the discount store, and place a bound on the error of estimation.

16.79 A company auditor is interested in estimating the total number of travel vouchers that have been incorrectly filed. In a simple random sample of $n = 50$ vouchers taken from a group of $N = 250$, 20 were found to have been filed incorrectly. Estimate the total number of vouchers from the $N = 250$ that have been filed incorrectly, and place a bound on the error of estimation. (For assistance in computing the variance, see the hint given in Exercise 16.29.)

16.80 The starting point for the development of an understanding of consumer behavior is consumer demography, the descriptive measures characterizing the buying public. From company records the manager of an automobile dealership has obtained a simple random sample of 25 customer files of the 582 customers who have purchased a certain small economy car in the past year. The mean and variance of the ages of the 25 customers are found to be $\bar{x} = 27.5$ and $s^2 = 16.81$. Estimate the average age of purchasers of the economy car from the dealership, and place a bound on the error of estimation.

16.81 Test marketing provides an opportunity to investigate the nature and degree of customer acceptance of a product by marketing the product in certain selected areas. The sales manager for a typewriter manufacturer wishes to know whether sufficient demand exists in a large city to justify adding a new electric portable typewriter to her stock. Currently, she serves four large chains, consisting of 25, 20, 30, and 25 stores. For administrative convenience she decides to use stratified random sampling, with each chain of stores representing a stratum. The sales manager has enough time and money to obtain sales data in only 20 stores. Using proportional allocation, she randomly selects 5 stores from the first chain, 4 from the second, 6 from the third, and 5 from the fourth. After a month the sales figures are as shown in Table 16.16. Estimate the average monthly sales per store, and place a bound on the error of estimation.

Table 16.16   Sales Data for Exercise 16.81

Stratum 1	Stratum 2	Stratum 3	Stratum 4
16	10	5	17
12	17	18	11
10	12	13	12
13	6	15	15
9		20	18
		12	

16.82   Refer to Exercise 16.81. Estimate the total monthly sales if the new electric portable typewriters were sold in all 100 stores, and place a bound on the error of estimation.

16.83   Manufacturing organizations spend millions of dollars each year in the development, promotion, and marketing of new products. Nevertheless, the rate of success of new products is miniscule. One useful market research activity to measure new-product acceptance is to test-market the product in a representative sales region. As a case in point, consider a manufacturer of farm implements who wishes to introduce a new orchard sprayer in three western states, Washington, Oregon, and California. To test the market acceptability of the new sprayer, the manufacturer selects 30 retail outlets in the three states, and the number of sprayers sold over a 12-month period is observed. The 30 retail outlets used in the test marketing are selected by using stratified random sampling and proportionally allocating the sample among the three states. The results are given in Table 16.17.

a.   Estimate the average number μ of sales for all 250 retail outlets in the three western states, and place a bound on the error of estimation.

b.   Estimate the total sales in the three western states if the new sprayer were made available to all 250 retail outlets. Place a bound on the error of estimation.

Table 16.17   Sample Data for Exercise 16.83

Statistic	Washington	Oregon	California
n_i	9	6	15
$\bar{x}_i$	26	23	39
s_i^2	31.2	19.3	38.5

16.84   An insurance executive is concerned that the high inflation rate may leave many clients with insufficient fire insurance coverage for their homes. He has proposed an automatic escalator clause that increases the client's coverage (and annual premiums) according to the annual inflation rate. To gain client reaction to the proposed policy, the insurance executive decides to select a stratified random sample of the clients served by his company in the three counties of his jurisdiction. The results are shown in Table 16.18. Estimate the proportion p of clients in the three counties favoring the escalator policy, and place bounds on the error of estimation.

Table 16.18   Sample Data for Exercise 16.84

Statistic	County A	County B	County C
Total clients	231	407	187
Number surveyed	21	37	17
Number approving escalator policy	8	20	9

16.85   A forester wants to estimate the total number of farm acres planted in trees for a state. Since the number of acres of trees varies considerably with the size of the farm, he decides to stratify on farm sizes. The 240 farms in the state are placed in one of four categories according to size. A stratified random sample of 40 farms, selected by using proportional allocation, yields the results shown in Table 16.19 on number of acres planted in trees. Estimate the total number of acres of trees on farms in the state, and place a bound on the error of estimation.

Table 16.19   Tree Plantings Data for Exercise 16.85

Stratum I, 0–200 Acres		Stratum II, 200–400 Acres		Stratum III, 400–600 Acres		Stratum IV, Over 600 Acres	
$N_1 = 86$		$N_2 = 72$		$N_3 = 52$		$N_4 = 30$	
$n_1 = 14$		$n_2 = 12$		$n_3 = 9$		$n_4 = 5$	
97	67	125	155	142	256	167	655
42	125	67	96	310	440	220	540
25	92	256	47	495	510	780	
105	86	310	236	320	396		
27	43	220	352	196			
45	59	142	190				
53	21						

MKTG   **16.86**   A manufacturer of chain saws has received complaints from buyers about excessive repair costs. To study this problem, the manufacturer wishes to estimate the average repair cost per saw per month for saws he has sold to logging and timber companies. He cannot obtain the repair cost for each saw, but he can find the total repair costs and the number of saws owned by different companies. Thus, he decides to use cluster sampling, with each company as a cluster. From the $M = 87$ logging and timber companies that buy chain saws from the manufacturer, he selects a simple random sample of $m = 12$. The data in Table 16.20 represent the repair costs during the past month for each company. Estimate the average repair cost per saw for the past month, and place bounds on the error of estimation.

Table 16.20   Sample Data for Exercise 16.86

Company	Number of Saws	Repair Costs	Company	Number of Saws	Repair Costs
1	4	$ 55	7	9	$103
2	7	83	8	1	15
3	5	47	9	8	110
4	11	210	10	11	164
5	15	235	11	7	80
6	6	88	12	10	146

16.87   Refer to Exercise 16.86. Estimate the total amount spent by the 87 logging and timber companies on repair of the manufacturer's chain saws during the past month, and place bounds on the error of estimation.

MKTG   **16.88**   Upon checking his sales records, the manufacturer cited in Exercise 16.86 notes that he has sold 703 chain saws to the 87 logging and timber companies. Using this additional information, estimate the total amount spent on repairs by the 87 companies, and place a bound on the error of estimation. (*Hint*: If $\bar{x}_{cl}$ is the mean obtained by cluster sampling and N is the number of elements in the population, then $\hat{\tau} = N\bar{x}_{cl}$ and $\hat{\sigma}_{\hat{\tau}}^2 = N^2\hat{\sigma}_{\bar{x}_{cl}}^2$.) How do these results compare with those obtained in Exercise 16.87?

   16.89   In an effort to budget the coming year's expenses, the manager of a company motor pool wishes to estimate the total annual maintenance costs for the 170 company cars presently operated by company personnel. Currently, those operating the company cars pay for maintenance themselves and are reimbursed at the year's end. How many company employees with company cars should the manager contact if she wishes to estimate the total annual maintenance costs with a bound of $500 on the error of estimation? Assume that past evidence suggests that the standard deviation of annual maintenance costs is about $100 per car.

16.90 Time and motion studies are often used in industry to equate work loads, to determine bases for compensation, and to design machinery for assisting certain manual operations. The production manager of a large assembly plant is interested in knowing the average time spent on a certain assembly operation by the firm's $N = 225$ assembly employees. He feels that men and women probably show a difference in assembly time, so he wants to stratify on sex. In previous studies the assembly times for the 100 male assemblers ranged from 8 to 18 minutes, and for the female assemblers the range was from 5 to 12 minutes. Using proportional allocation, find the approximate sample size necessary to estimate the average assembly time for all employees to within 1 minute. How many men and how many women should be included in the sample?

16.91 In the face of the energy crisis and rising numbers of highway deaths, Congress in 1974 created legislation imposing a maximum 55-mile-per-hour (mph) speed limit on all public highways. Since then, much debate has occurred on the public's acceptance of this law. To study this issue, the California Highway Patrol decided to randomly select $n = 25$ vehicles traveling a certain section of interstate highway and to measure their speeds. The average speed of the 25 vehicles was found to be 57.5 mph with a standard deviation of 9.3 mph. Estimate the average speed μ of all vehicles traveling on the highway, and place a bound on the error of estimation. (Assume that N is sufficiently large so that we can ignore the fpc factor.)

16.92 A timber cruiser for the Bureau of Land Management (BLM) is interested in estimating the total number of Douglas fir trees of harvestable size on 1,200 acres of BLM property. To simplify his study, the timber cruiser plans to use 1-acre plots within the BLM property as sampling units. From past experience the timber cruiser knows that the number of mature Douglas fir trees per acre in the region of interest ranges from 65 to 173. Determine the sample size required (number of acres of BLM land to investigate) in order to estimate the total number of Douglas fir trees of harvestable size on the 1,200 acres, with a bound on the error of estimation of $B = 1,000$ trees.

16.93 The assessment/sales (A/S) ratio is used as a standard measure of assessment performance by public and private assessment offices. In a jurisdiction requiring assessment at full market value, the A/S ratio should be near 1.0. When the A/S ratio exceeds 1.0, property taxation based on the assessments is said to be regressive. In a community with $N = 510$ residential properties a random sample of $n = 10$ properties with recent selling prices is observed. The assessed valuations and selling prices are shown in Table 16.21. Assuming that the 10 sample properties represent a random sample of all residential properties from the community, compute the assessment/sales ratio for the community.

Table 16.21 Sample Data for Exercise 16.93

Property	Assessed Valuation	Selling Price	Property	Assessed Valuation	Selling Price
1	$77,000	$78,000	6	$ 96,000	$93,500
2	75,000	74,500	7	79,000	80,000
3	81,000	83,500	8	87,000	84,500
4	88,000	92,000	9	101,000	99,000
5	68,000	69,500	10	83,000	84,500

16.94 Refer to Exercise 16.93. Use ratio estimation to estimate the total assessed valuation of all residential properties in the community. If all residential properties in the community are taxed at the rate of 2.5% of assessed valuation, estimate the property tax revenue generated by the community from residential properties.

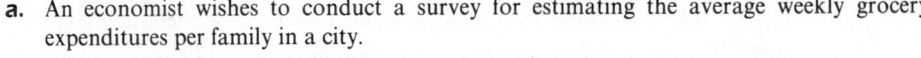

Supplementary Exercises

Interpretive

16.95 For each of the following sample surveys, discuss the selection of an appropriate sampling unit and frame.

   a. An economist wishes to conduct a survey for estimating the average weekly grocery expenditures per family in a city.

   b. An administrative assistant to the governor of a midwestern state wishes to estimate the proportion of the voting public in the state who favor legislation prohibiting strip mining.

   c. A supermarket chain that operates retail grocery stores throughout the United States wishes to survey the opinions of its employees about a company-sponsored health insurance plan.

   d. A marketing representative for a large merchandising chain wishes to survey buyers to determine their attitudes toward a new product line.

16.96 According to a New York management consulting firm, employees in the United States and Canada presently steal more than $10 million a day in cash and merchandise from their employers. The president of a small manufacturing firm has commissioned a group to study employee theft in her firm. She wants to estimate the proportion of employees who have ever willfully stolen anything of value from the firm. You represent the group commissioned by the president. Develop an appropriate survey design to investigate the problem. Include an appropriate questionnaire in your analysis.

16.97 Concerned with untraceable inventory shortages, the manager of a clothing store wishes to estimate the proportion of shoppers entering her store who can be considered shoplifters. The manager figures that she has sufficient resources to interview up to 60 randomly selected shoppers, but she does not know how to set up her sample survey. Place yourself in the position of a consultant to the manager and suggest an appropriate research design for her problem. In particular, write an appropriate questionnaire, develop a randomization device, and explain its use.

16.98 Select an example from business or government that illustrates an effective application of systematic sampling. Select another example that illustrates a case where simple random sampling, but not systematic sampling, should be used to generate the survey information.

16.99 What is meant by area sampling? Give an example from business or government in which area sampling may prove useful in generating survey information.

16.100 To improve typing service, an executive of a large publishing company wishes to estimate the total number of pages typed by secretaries in the company in one day. The company contains eight departments, each generating about the same quantity of typed material per day. Each department employs approximately 25 secretaries, and the work output varies considerably from secretary to secretary. Using a two-stage cluster sampling model, design a survey to answer the executive's question. Carefully identify the clusters and elements in the survey, and specify whether you would suggest large or small numbers of each.

Case Study

Lane County Port District

Lane County is a large western county with a population of approximately 300,000. The economic base of the country is provided by agriculture, wood products, commercial fishing, tourism, and some light industry. The largest community in Lane County is Eugene, the county seat, which has a population of approximately 125,000. Most of the remaining population of the county is located in smaller communities, which vary distinctly in character. The communities in the central-valley area serve as agricultural distribution centers; wood products is the dominant industry in those communities that lie in the mountainous area to the east. Most of the small communities on the coast rely on tourism and the commercial fishing industry. The Eugene economy has elements of all these industries, with additional diversification offered by the state university, the county government offices, and most of the county's nonwood products manufacturing.

Lane County officials are now considering the formation of a Lane County Port District to encourage the location of new industry, which might serve to further diversify the county's economic base. However, before a port district can be formed, the issue requires approval from the county's registered voters. And many residents have expressed concerns that economic diversification might result in uncontrolled population growth and danger to the county's aesthetic qualities. In addition, attitudes appear to be dividing along political party lines, the local Republican organization favoring economic diversification, and the Democrats opposing it. Thus, approval of the port district proposal is uncertain.

Before proceeding with a full-scale promotional campaign in support of a port district, the county manager feels that a survey should be taken to explore voter attitudes and concerns regarding economic diversification. A list of all registered voters in the country is available from the office of the county elections director. A brief description of each of the seven state legislative districts represented in Lane County is given in Table 16.22.

Develop a sampling plan to assist the county manager. In your discussion, consider the following aspects of your sampling design:

1. The definition of the population of interest.
2. The definition of the sampling elements, units, and frame
3. Monetary and time constraints
4. The reason for your particular choice of a sampling design

5. The limitations of the sampling design you have chosen
6. The method to be used in determining an appropriate sample size
7. Potential biases and their effect on interpreting the results of the survey
8. Sampling errors that may result
9. Assumptions you have made in the selection of your chosen sampling design

Table 16.22   Lane County Data for the Case Study

District	Number of Registered Voters	Predominant Economic Base	Predominant Political Affiliation
17	11,650	Fishing, tourism	Republican
18	13,509	Agriculture	Republican
19	9,684	Forest products	Democratic
21	15,235	Light manufacturing	Independent
22	14,476	Forest products, light manufacturing	Republican
27	13,695	Forest products	Democratic
32	19,128	Government, light manufacturing	Democratic

References and Suggested Readings

Cochran, W. G. *Sampling Techniques.* 3d ed. New York: Wiley, 1977.

Greenberg, B. G., R. T. Kuebler, Jr., J. R. Abernathy, and D. G. Horvitz. "Application of Randomized Response Technique in Obtaining Quantitative Data." *Journal of the American Statistical Association,* Vol. 66 (1971) pp. 243–250.

Jessen, R. J. *Statistical Survey Techniques.* New York: Wiley, 1978.

Kish, J. L. *Survey Sampling.* New York: Wiley, 1965.

Peterson, G. E., A. P. Solomon, H. Madjid, and W. C. Apgar. *Property Taxes, Housing, and the Cities.* Lexington, Mass: D. C. Heath, 1973.

Scheaffer, R. L., W. Mendenhall, and L. Ott. *Elementary Survey Sampling.* 3d ed. Boston: PWS-KENT Publishing Co., 1986.

TO
THE
READER

The statistical methods and techniques presented in Chapters 8 through 16 are appropriate for analyzing quantitative data that arise under random sampling from normal populations and in those cases when the sampling distribution of a statistic is approximately normal owing to the Central Limit Theorem.

Many types of business surveys and experiments, however, result in qualitative rather than quantitative response variables, so that the responses can be classified but not quantified. For example, to assess similarities or differences in opinion regarding the best way to reverse a business decline, a sample of business executives, economists, and government officials could be polled and their responses categorized into one of five "opinion" categories, ranging from "increased governmental spending" to "cuts in personal taxes." The resulting data would be the count or number of individuals in each profession-opinion category. Data of this type are usually referred to as *count* or *enumerative* data. Chapter 17 is concerned with methods of analysis appropriate for counts arising in this and similar situations.

17

ANALYSIS OF ENUMERATIVE DATA

17.1

A Description of the Experiment

Many experiments, particularly in the social sciences, result in enumerative (or count) data. For instance, the classification of people into five income brackets results in an enumeration or count corresponding to each of the five income classes. Or we might be interested in studying the reaction of a sales prospect to a particular product-promoting device. If the reactions of a prospect to the promotional device can be classified in one of three ways, and if a large number of prospects were subjected to the device, the experiment would yield three counts, indicating the number of prospects falling in each of the reaction classes. Similarly, a traffic study might require a count and classification of the type of motor vehicles using a section of highway. An industrial process manufactures items that fall into one of three quality classes: acceptable, seconds, and rejects. A student of the arts might classify paintings in one of k categories according to style and period in order to study trends in style over time. We might wish to classify ideas in a philosophical study, or style in the field of literature. The results of an advertising campaign would yield count data that indicate a classification of consumer reaction. Indeed, many observations in the physical sciences are not amenable to measurement on a continuous scale and result in enumerative or classificatory data.

multinomial experiment

The illustrations in the preceding paragraph exhibit, to a reasonable degree of approximation, the characteristics listed in the following display, which define a **multinomial experiment**.

Characteristics of a Multinomial Experiment

1. The experiment consists of n identical trials.
2. The outcome of each trial falls into one of k classes, or cells.

3. The probability that the outcome of a single trial will fall in a particular cell—say cell i—is p_i $(i = 1, 2, \ldots, k)$ and remains the same from trial to trial. Note that

$$0 \le p_i \le 1 \qquad \text{for all } i$$

and

$$p_1 + p_2 + p_3 + \cdots + p_k = 1$$

4. The trials are independent.
5. The experimenter is interested in $n_1, n_2, n_3, \ldots, n_k$, where n_i $(i = 1, 2, \ldots, k)$ is equal to the number of trials in which the outcome falls in cell i. Note that

$$n_1 + n_2 + n_3 + \cdots + n_k = n$$

A multinomial experiment is analogous to tossing n balls at k boxes (cells), where each ball will fall in one of the boxes. The boxes are arranged so that the probability that a ball will fall in a box varies from box to box but remains the same for a particular box in repeated tosses. Finally, the balls are tossed in such a way that the trials are independent. At the conclusion of the experiment we observe n_1 balls in the first box, n_2 in the second, ..., and n_k in the kth. The total number of balls is

$$\sum_{i=1}^{k} n_i = n$$

Note the similarity between the binomial and multinomial experiments, and in particular, note that the binomial experiment represents the special case for the multinomial experiment when $k = 2$. The single parameter p of the binomial experiment is replaced by the k parameters $p_1, p_2, \ldots, p_k$ of the multinomial experiment. The object of this chapter is to make inferences about the cell probabilities $p_1, p_2, \ldots, p_k$. The inferences will be expressed in terms of a statistical test of an hypothesis concerning their specific numerical values or their relationship to one another.

If we proceeded as in Chapter 5, we would derive the probability of the observed sample $(n_1, n_2, \ldots, n_k)$ for use in calculating the probability of the type I and type II errors associated with a statistical test. Fortunately, we have been relieved of this chore by the British statistician Karl Pearson, who proposed a very useful test statistic for testing hypotheses concerning $p_1, p_2, \ldots, p_k$ and gave its approximate probability distribution in repeated sampling. This test statistic is discussed in the next section.

17.2

The Chi-square Test

Suppose that $n = 100$ balls are tossed at the cells (boxes) and that we know that p_1 is equal to .1. How many balls would be expected to fall in the first cell? Referring to

Chapter 5 and utilizing knowledge of the binomial experiment, we calculate

$$E(n_1) = np_1 = 100(.1) = 10$$

Similarly, the expected number falling in the remaining cells may be calculated by using the formula

$$E(n_i) = np_i \qquad i = 1, 2, \ldots, k$$

Now suppose that we hypothesize values for $p_1, p_2, \ldots, p_k$ and calculate the expected value for each cell. Certainly, if our hypothesis is true, the cell counts n_i should not deviate greatly from their expected values np_i $(i = 1, 2, \ldots, k)$. Hence, it seems intuitively reasonable to use a test statistic involving the k deviations

$$(n_i - np_i) \qquad i = 1, 2, \ldots, k$$

In 1900 Karl Pearson proposed the following test statistic, which is a function of the square of the deviations of the observed counts from their expected values, weighted by the reciprocal of their expected value.

Chi-square Test Statistic

$$X^2 = \sum_{i=1}^{k} \frac{[n_i - E(n_i)]^2}{E(n_i)} = \sum_{i=1}^{k} \frac{(n_i - np_i)^2}{np_i}$$

Although the mathematical proof is beyond the scope of this text, it can be shown that when n is large, X^2 possesses, approximately, a chi-square probability distribution in repeated sampling. Experience has shown that the cell counts n_i should not be too small in order that the chi-square distribution provide an adequate approximation to the distribution of X^2. As a rule of thumb, **we will require that all expected cell counts equal or exceed 5,** although Cochran (see the references) has noted that this value can be as low as 1 for some situations.

Recall from Section 9.6 that the chi-square probability distribution is used for testing an hypothesis concerning a population variance σ^2, and that the shape of the chi-square distribution varies according to the number of degrees of freedom associated with s^2. We discussed the use of Table 5 in the Appendix, which presents the critical values of χ^2 corresponding to various tail areas of the distribution for different degrees of freedom. Therefore, we must know which chi-square distribution to use for approximating the distribution of X^2. That is, we must determine the number of degrees of freedom associated with the appropriate χ^2 distribution. Furthermore, we must determine whether the rejection region for the test is one-tailed or two-tailed. The latter problem is easily solved. Since large deviations of the observed cell counts from those expected tend to contradict the null hypothesis concerning the cell probabilities $p_1, p_2, \ldots, p_k$, we reject the null hypothesis when X^2 is large and employ a one-tailed statistical test, using the upper-tail values of χ^2 to locate the rejection region.

The determination of the appropriate number of degrees of freedom to be employed for the test can be rather difficult and therefore will be specified for the practical applications described in the following sections. In addition, we state here the principle involved (which is fundamental to the mathematical proof of the approximation) so that you may understand why the number of degrees of freedom changes with various applications. **This principle states that the appropriate number of degrees of freedom equals the number k of cells less one degree of freedom for each independent linear restriction placed on the observed cell counts.** For example, one linear restriction is always present because the sum of the cell counts must equal

$$n_1 + n_2 + n_3 + \cdots + n_k = n$$

Other restrictions will be introduced for some applications because of the necessity for estimating unknown parameters required in the calculation of the expected cell frequencies or because of the method in which the sample is collected. These restrictions will become apparent as we consider various practical examples.

17.3

A Test of an Hypothesis Concerning Specified Cell Probabilities

The simplest hypothesis concerning the cell probabilities is one that specifies numerical values for each cell. For example, consider a customer preference study in which three different package designs are used to display the same product. We wish to test the hypothesis that the buyer has no preference in the choice of package design; that is, we wish to test

$$H_0 : p_1 = p_2 = p_3 = 1/3$$

against

$$H_a : \text{at least one } p_i \text{ is different from } 1/3$$

where p_i is the probability that a customer will choose package design i ($i = 1, 2,$ or 3).

Suppose that the product to be packaged is a food product and that the three packages are displayed side by side in several supermarkets in a particular city. In one day's time it was noted that $n = 90$ customers purchased the product, of which $n_1 = 23$ purchased package design 1, $n_2 = 36$ purchased package design 2, and $n_3 = 31$ purchased package design 3. Thus, $n_1 = 23$, $n_2 = 36$, and $n_3 = 31$ represent the observed cell frequencies for cells 1, 2, and 3. The expected cell frequencies are the same for each cell, namely,

$$E(n_i) = np_i = 90(1/3) = 30$$

The observed and expected cell frequencies are presented in Table 17.1. Noting the discrepancy between the observed and expected cell frequencies, we wonder whether the data present sufficient evidence to warrant rejection of the hypothesis of no preference.

Table 17.1   Observed and Expected Cell Counts for the
Customer Preference Study

Cell Frequency	Package Design		
	1	2	3
Observed	$n_1 = 23$	$n_2 = 36$	$n_3 = 31$
Expected	30	30	30

The chi-square test statistic for this example possesses $(k - 1) = 2$ degrees of freedom since the only linear restriction on the cell frequencies is that

$$n_1 + n_2 + \cdots + n_k = n$$

or, for our example,

$$n_1 + n_2 + n_3 = 90$$

Therefore, if we choose $\alpha = .05$, we will reject the null hypothesis when $X^2 > 5.991$ (see Table 5 of the Appendix).

Substituting into the formula for X^2, we obtain

$$X^2 = \sum_{i=1}^{k} \frac{[n_i - E(n_i)]^2}{E(n_i)} = \sum_{i=1}^{k} \frac{(n_i - np_i)^2}{np_i}$$

$$= \frac{(23 - 30)^2}{30} + \frac{(36 - 30)^2}{30} + \frac{(31 - 30)^2}{30} = 2.87$$

Since X^2 is less than the tabulated value 5.991 of χ^2, the null hypothesis is not rejected. We conclude that the data do not present sufficient evidence to indicate that the buyers of the area have a preference for a particular package design.

EXERCISES   Conceptual Level

17.1 List the characteristics of a multinomial experiment.

17.2 How is the binomial experiment related to the multinomial experiment?

17.3 Under what conditions is Pearson's chi-square statistic appropriate for multinomial experiments?

17.4 Does the multinomial experiment always lead to a one-tailed statistical test or may it, under certain conditions, require a two-tailed statistical test? Explain.

EXERCISES   Skill Level

17.5 Consider a test of hypothesis concerning specified cell probabilities for a multinomial experiment consisting of k possible outcomes. Determine the appropriate rejection region for the test in the following situations.

a. $k = 8; \alpha = .01$ **b.** $k = 14; \alpha = .05$ **c.** $k = 9; \alpha = .05$

d. $k = 3; \alpha = .10$ **e.** $k = 6; \alpha = .025$

17.6   A multinomial experiment consisting of $k = 5$ possible outcomes was conducted $n = 200$ times, resulting in the following observations.

Outcome	1	2	3	4	5
n_i	36	31	45	42	46

Test the null hypothesis $H_0 : p_1 = p_2 = p_3 = p_4 = p_5 = 1/5$ by using the X^2 test statistic with $\alpha = .05$. Explain your results.

17.7   A multinomial experiment consisting of $k = 6$ possible outcomes was conducted $n = 600$ times, resulting in the following observations.

Outcome	1	2	3	4	5	6
n_i	85	102	96	97	121	99

Do the data provide sufficient evidence to indicate that the six outcomes are not equally likely? Test by using $\alpha = .01$.

17.8   An experiment that can result in one of four outcomes is repeated 250 times. The number of times each of the four outcomes occurred is shown in the accompanying table. Test the hypothesis that $p_1 = .05$, $p_2 = .55$, $p_3 = .35$, and $p_4 = .05$.

Outcome	1	2	3	4
n_i	10	130	89	21

a.   Write the null and alternative hypotheses.

b.   Calculate the expected cell counts.

c.   Calculate the observed value of the test statistic.

d.   Calculate the approximate p-value for the test.

e.   Use the results of part d to make a decision to reject or not reject H_0. Explain your results.

EXERCISES   Applications

 17.9   During a given day the manager of a large supermarket observed the number of shoppers choosing each of the market's six checkout stands. The observed results are given in the accompanying table. Do these data present sufficient evidence to indicate that some checkout stands were preferred over others? Use $\alpha = .05$.

Stand Number	1	2	3	4	5	6
Frequency	84	110	146	152	61	47

 17.10   Over a two-year period the manager of a motel recorded the number of vacant rooms in his motel each evening. The relative frequencies of the occurrence of $0, 1, 2, \ldots$ vacant rooms enabled the manager to calculate the approximate probabilities given in the first table. However, since the manager recorded these data, a new motel has been built near the manager's motel. In the first 100 days since completion of the new motel, the manager recorded the numbers of room vacancies per day. These data are shown in the second table. Do these data present sufficient evidence to indicate that the pattern of vacancies in the manager's motel has changed since the opening of the new motel? Test by using a 5% level of significance.

Number of Vacant Rooms	0	1	2	3	≥ 4
Probability	.10	.25	.35	.20	.10

Number of Vacant Rooms	0	1	2	3	≥ 4
Number of Days	3	16	35	25	21

17.11   The social and economic costs of cardiovascular disease continue, but management can fight back with fitness programs for employees. In fact, many firms are now offering work release for voluntary exercise programs to improve their employees' physical and mental fitness. To measure the effect of such a program, a manufacturing company recorded the time spent per week in vigorous exercise undertaken by 83 of its employees two years after the initiation of a volunteer exercise program. These data were then compared with the level of participation by the firm's employees in vigorous exercise activities prior to the firm-sponsored exercise program. Do the data in Table 17.2 provide sufficient evidence to indicate that the level of participation in vigorous exercise activities has changed since the initiation of the firm-sponsored exercise program? Test with $\alpha = .05$. Has the program proved effective in increasing employee involvement in vigorous exercise?

Table 17.2   Data for Exercise 17.11

Hours per Week in Vigorous Exercise	Percentage Before the Program	Number After the Program
No exercise	52	33
Some, but less than 3 hours	29	23
3 hours to 6 hours	11	14
6 hours to 9 hours	6	9
Over 9 hours	2	4

17.12   Wolgast noted in 1958 that when it is time to buy a family car, in 4% of the cases the wife is the family member who selects the car to be purchased; 31% of the time this decision is shared equally between the husband and the wife; 56% of the time the husband makes the decision; and 9% of the time the decision is made by someone else (E. H. Wolgast, "Do Husbands or Wives Make Purchasing Decisions?" *Journal of Marketing*, October 1958). To see whether Wolgast's findings still hold true in light of the greater awareness and influence of today's woman, a marketing representative randomly selected 200 families that had recently purchased a new automobile. In 18 of the families the wife selected the car to be purchased; in 75 cases the decision was shared equally between the husband and wife; in 92 cases the husband made the decision; and in the remainder of cases the decision was made by someone else. Do these data contradict the findings reported by Wolgast in 1958? Use $\alpha = .05$.

17.4

Contingency Tables

contingency table

A problem frequently encountered in the analysis of count. data concerns the independence of two methods of classifying observed events. For example, suppose we wish to classify defects found on furniture produced in a manufacturing plant—first, according to the type of defect and, second, according to the production shift during which the piece of furniture was produced. If the proportions of the various types of defects are constant from shift to shift, then classification by defects is independent of the classification by production shift. On the other hand, if the proportions of the various defects vary from shift to shift, then the classification by defects is *contingent* upon the shift classification, and the classifications are dependent. In investigating whether one method of classification is contingent upon another, we display the data by using a cross-classification in an array called a **contingency table**.

A total of $n = 309$ furniture defects were recorded and the defects were classified as being one of four types: A, B, C, or D. At the same time, each piece of furniture was identified according to the production shift in which it was manufactured. These counts are presented as a contingency table in Table 17.3. (*Note:* Numbers in parentheses are the expected cell frequencies.)

column probabilities

Let p_A be the unconditional probability that a defect will be of type A. Similarly, define p_B, p_C, and p_D as the probabilities of observing the three other types of defects. Then these probabilities, which we will call the **column probabilities** of Table 17.3, will satisfy the requirement

$$p_A + p_B + p_C + p_D = 1$$

row probability

In like manner, let p_i ($i = 1, 2,$ or 3) be the **row probability** that a defect will have occurred on shift i, where

$$p_1 + p_2 + p_3 = 1$$

Then if the two classifications are independent of each other, a cell probability will equal the product of its respective row and column probabilities in accordance with the Multiplicative Law of Probability.

Table 17.3   Contingency Table Classifying Defects of Furniture According to Type and Production Shift

| | Type of Defect | | | | |
Shift	A	B	C	D	Total
1	15 (22.51)	21 (20.99)	45 (38.94)	13 (11.56)	94
2	26 (22.99)	31 (21.44)	34 (39.77)	5 (11.81)	96
3	33 (28.50)	17 (26.57)	49 (49.29)	20 (14.63)	119
Total	74	69	128	38	309

For example, the probability that a particular defect will occur in shift 1 and be of type A is $(p_1)(p_A)$. Thus, we observe that the numerical values of the cell probabilities are unspecified in the problem under consideration. **The null hypothesis specifies only that each cell probability will equal the product of its respective row and column probabilities, a condition that implies independence of the two classifications. The alternative hypothesis is that this equality does not hold for at least one cell.**

The analysis of the data obtained from a contingency table differs from the problem discussed in Section 17.3, because we must *estimate* the row and column probabilities in order to estimate the expected cell frequencies.

If proper estimates of the cell probabilities are obtained, the estimated expected cell frequencies may be substituted for the $E(n_i)$ in X^2, and X^2 will continue to possess a distribution in repeated sampling that is approximated by the chi-square probability distribution. The proof of this statement, as well as a discussion of the methods for obtaining the estimates, is beyond the scope of this text. Fortunately, the procedures for obtaining the estimates, known as the method of maximum likelihood and the method of minimum chi-square, yield estimates that are intuitively obvious for our relatively simple applications.

It can be shown that the estimator of a column probability equals the column total divided by $n = 309$. If we denote the total of column j as c_j, then we have

$$\hat{p}_A = \frac{c_1}{n} = \frac{74}{309} \qquad \hat{p}_C = \frac{c_3}{n} = \frac{128}{309}$$

$$\hat{p}_B = \frac{c_2}{n} = \frac{69}{309} \qquad \hat{p}_D = \frac{c_4}{n} = \frac{38}{309}$$

Similarly, the row probabilities $p_1, p_2,$ and p_3 may be estimated by using the row totals $r_1, r_2,$ and r_3:

$$\hat{p}_1 = \frac{r_1}{n} = \frac{94}{309} \qquad \hat{p}_2 = \frac{r_2}{n} = \frac{96}{309} \qquad \hat{p}_3 = \frac{r_3}{n} = \frac{119}{309}$$

Denote the observed frequency of the cell in row i and column j of the contingency table by n_{ij}. Then the estimated expected value of n_{11} is

$$\hat{E}(n_{11}) = n(\hat{p}_1 \hat{p}_A) = n\left(\frac{r_1}{n}\right)\left(\frac{c_1}{n}\right) = \frac{r_1 c_1}{n}$$

where $\hat{p}_1 \hat{p}_A$ is the estimated cell probability. Similarly, we may find the estimated expected value for any other cell, say $\hat{E}(n_{23})$:

$$\hat{E}(n_{23}) = n(\hat{p}_2 \hat{p}_C) = n\left(\frac{r_2}{n}\right)\left(\frac{c_3}{n}\right) = \frac{r_2 c_3}{n}$$

Thus, we see that the estimated expected value of the observed cell frequency n_{ij} for a contingency table is equal to the product of its respective row and column totals divided by the total frequency.

Estimated Expected Cell Frequency

$$\hat{E}(n_{ij}) = \frac{r_i c_j}{n}$$

The estimated expected cell frequencies for our example are shown in parentheses in Table 17.3.

We may now use the expected and observed cell frequencies shown in Table 17.3 to calculate the value of the test statistic:

$$X^2 = \sum_{i=1}^{3} \sum_{j=1}^{4} \frac{[n_{ij} - \hat{E}(n_{ij})]^2}{\hat{E}(n_{ij})}$$

$$= \frac{(15 - 22.51)^2}{22.51} + \frac{(26 - 22.99)^2}{22.99} + \cdots + \frac{(20 - 14.63)^2}{14.63} = 19.18$$

The next step is to determine the appropriate number of degrees of freedom associated with the test statistic. We give the rule here, which we will later attempt to justify: **The degrees of freedom associated with a contingency table possessing r rows and c columns will always equal $(r - 1)(c - 1)$.** Thus, for our example we will compare X^2 with the critical value of χ^2 with $(r - 1)(c - 1) = (3 - 1)(4 - 1) = 6$ degrees of freedom.

You will recall that the number of degrees of freedom associated with the X^2 statistic equals the number of cells (in this case, $k = rc$) less one degree of freedom for each independent linear restriction placed on the observed cell frequencies. The total number of cells for the data of Table 17.3 is $k = 12$. From this we subtract one degree of freedom because the sum of the observed cell frequencies must equal n; that is,

$$n_{11} + n_{12} + \cdots + n_{34} = 309$$

In addition, we used the cell frequencies to estimate three of the four column probabilities. Note that the estimate of the fourth column probability will be determined once we have estimated p_A, p_B, and p_C, because

$$p_A + p_B + p_C + p_D = 1$$

Thus, we lose $(c - 1) = 3$ degrees of freedom for estimating the column probabilities. Finally, we used the cell frequencies to estimate $(r - 1) = 2$ row probabilities and therefore we lose $(r - 1) = 2$ additional degrees of freedom. The total number of degrees of freedom remaining is

$$\text{d.f.} = 12 - 1 - 3 - 2 = 6$$

In general, we see that the total number of degrees of freedom associated with an $r \times c$ contingency table is

$$\text{d.f.} = rc - 1 - (c - 1) - (r - 1) = (r - 1)(c - 1)$$

Therefore, if we use $\alpha = .05$, we will reject the null hypothesis that the two classifications are independent if $X^2 > 12.592$. Since the value of the test statistic,

$X^2 = 19.18$, exceeds the critical value of χ^2, we reject the null hypothesis. The data present sufficient evidence to indicate that the proportion of the various types of defects varies from shift to shift. A study of the production operations for the three shifts would probably reveal the cause.

EXAMPLE 17.1 The introduction of unit pricing in food stores makes it easier for shoppers to choose cheaper items. However, Isakson and Maurizi (see the references) have found that lower-income shoppers do not appear to be taking advantage of unit pricing, perhaps because they do not understand the unit-pricing labeling system as well as do middle-income and upper-income shoppers.

In a follow-up study to provide a check on the authors' findings, an economist observed the purchase selection of $n = 1,000$ shoppers in a large supermarket. The supermarkets were located in three different areas of a city where the buyers were, respectively, of lower-income, middle-income, and upper-income families. Packages of the same brand but with different unit prices were placed adjacent to one another on the store shelves. The data for the $n = 1,000$ shoppers, classified according to purchase decision and income class, are shown in Table 17.4. Do these data present sufficient evidence to support the findings of Isakson and Maurizi?

Table 17.4 Sample Data for Example 17.1

Purchase Decision	Income Class			Total
	Lower	Middle	Upper	
Lower unit price	249 (259.60)	494 (490.88)	201 (193.52)	944
Higher unit price	26 (15.40)	26 (29.12)	4 (11.48)	56
Total	275	520	205	1000

SOLUTION The question concerning the validity of the findings of Isakson and Maurizi can be restated in terms of whether the data provide sufficient evidence to indicate a dependence between a buyer's income classification and his or her ability to properly interpret unit-pricing labels by choosing the package with the lower unit price. Therefore, we analyze the data as a contingency table.

The estimated expected cell frequencies may be calculated by using the appropriate row and column totals in the formula

$$\hat{E}(n_{ij}) = \frac{r_i c_j}{n}$$

Thus, we have

$$\hat{E}(n_{11}) = \frac{r_1 c_1}{n} = \frac{(944)(275)}{1,000} = 259.60$$

$$\hat{E}(n_{12}) = \frac{r_1 c_2}{n} = \frac{(944)(520)}{1,000} = 490.88$$

and so on. These estimated values are shown in parentheses in Table 17.4.

The value of the test statistic X^2 can now be computed and compared with the critical value of χ^2 possessing $(r-1)(c-1) = (1)(2) = 2$ degrees of freedom. Then for $\alpha = .05$, we will reject the null hypothesis when $X^2 > 5.991$. Substituting into the formula for X^2, we obtain

$$X^2 = \frac{(249 - 259.60)^2}{259.60} + \frac{(494 - 490.88)^2}{490.88} + \cdots + \frac{(4 - 11.48)^2}{11.48} = 13.25$$

Observing that X^2 falls in the rejection region, we reject the null hypothesis of independence of the two classifications. A comparison of the percentage of shoppers in each income class who properly interpret the unit-pricing labels suggests that the lower-income classes are not taking advantage of unit pricing. Therefore, the Isakson and Maurizi findings have been supported by the economist's survey.

EXERCISES Conceptual Level

17.13 What is the null hypothesis tested when Pearson's chi-square statistic is applied to a contingency table?

17.14 How do we determine the degrees of freedom for Pearson's chi-square statistic when working with a contingency table?

EXERCISES Skill Level

17.15 Determine the appropriate degrees of freedom for the chi-square statistic used for testing independence of classifications for a contingency table possessing r rows and c columns in the following situations.
  a. $r = 5; c = 3$ **b.** $r = 2; c = 4$
  c. $r = 6; c = 7$ **d.** $r = 3; c = 3$

17.16 Determine the appropriate rejection region for the test of independence of classifications for a contingency table with the following values.
  a. $r = 2; c = 4; \alpha = .05$ **b.** $r = 4; c = 4; \alpha = .01$
  c. $r = 6; c = 2; \alpha = .10$ **d.** $r = 3; c = 5; \alpha = .05$

17.17 A test of independence of classifications for an $r \times c$ contingency table results in the following values of X^2. Find the approximate p-value for each test.
  a. $r = 3; c = 3; X^2 = 4.32$ **b.** $r = 4; c = 2; X^2 = 10.55$
  c. $r = 5; c = 3; X^2 = 13.27$ **d.** $r = 2; c = 2; X^2 = 11.08$

17.18 A test of independence of classification was conducted for a 2×3 contingency table. The observed value of the chi-square test statistic is $X^2 = 7.61$.
  a. Calculate the approximate p-value for the test.
  b. Does the observed value of the test statistic provide sufficient evidence to indicate that the two classifications are not independent?

17.19 A total of $n = 287$ items have been classified according to two different criteria, resulting in a 4×3 contingency table (see Table 17.5). The number n_{ij} in row i and column j represents the number of items falling in cell ij.

a. State the null and alternative hypotheses used to investigate whether one method of classification is contingent on the other.

b. Write the formula for the test statistic used to test the hypotheses you stated in part a.

c. Calculate $\hat{E}(n_{ij})$ for each row and column combination.

d. How many degrees of freedom are associated with the test statistic in part b?

e. Find the rejection region for testing the hypotheses in part a, with $\alpha = .05$.

f. Calculate the observed value of the test statistic and use it to draw a conclusion concerning the hypotheses in part a.

Table 17.5 Data for Exercise 17.19

Classification II (Row)	Classification I (Column)		
	1	2	3
1	30	41	15
2	18	22	14
3	34	27	25
4	21	20	20

17.20 A 3×3 contingency table with 3 rows and 3 columns resulted in the observed cell counts given in Table 17.6. Do the data present sufficient evidence to indicate that the two methods of classification are not independent? Use $\alpha = .01$.

Table 17.6 Cell Count Data for Exercise 17.20

Row	Column		
	1	2	3
1	22	26	14
2	18	35	14
3	22	30	25

EXERCISES Applications

17.21 Applied marketing research is intended as a support activity for the marketing manager, since information from research can be a key factor in the marketing decision-making process. The extent to which research influences marketing management, however, is heavily dependent on how marketing managers view marketing research. A recent survey of marketing managers in four different industries provided the data in Table 17.7, which gives the managers' attitudes toward marketing research and its value in marketing decision making. Do these data present

Table 17.7   Attitude Data for Exercise 17.21

Perceived Value of Market Research	Industry Type			
	Consumer Firms	Industrial Organizations	Retail and Wholesale	Finance and Insurance
Little value	9	22	13	9
Moderate value	29	41	6	17
Great value	26	28	6	27
Total	64	91	25	53

sufficient evidence to indicate that the perceived value of marketing research differs among marketing managers in the four industries involved in the study? Use $\alpha = .05$.

17.22 Many insurance companies are beginning to question the policy of offering reduced rates to owners of subcompact cars because the companies claim that their rate of serious and fatal accidents is higher than that of owners of larger-sized cars. To investigate this issue, a researcher made an analysis of accident data to determine the distribution of numbers of cases in which at least one individual was either fatally or critically injured for automobiles of three sizes. The data for 346 accidents are shown in Table 17.8. Do these data indicate that the frequency of fatal and critical injuries in auto accidents depends on the size of automobiles? Use $\alpha = .10$.

Table 17.8   Accident Data for Exercise 17.22

Result	Size of Auto		
	Subcompact	Compact	Full Size
Fatal or critical injury	67	26	16
No fatal or critical injury	128	63	46

17.23 Are there some companies whose securities should be systematically excluded in an investment portfolio for social, political, or moral reasons? Should the investment manager select securities for his clients' portfolios to further such noneconomic goals? Some researchers have noted that reaction to these questions varies considerably from one fund management group to another. To explore this issue, a researcher asked fund managers employed by four of the major fund management agencies whether they approve or disapprove of selecting securities for a portfolio to further social, political, or moral goals. The number of fund managers falling in each of the eight possible categories is shown in Table 17.9. Do these data present sufficient evidence to

Table 17.9   Response Data for Exercise 17.23

Response	Fund Management Agency			
	A	B	C	D
Favor noneconomic goals	7	6	11	13
Do not favor noneconomic goals	23	31	30	41
Total	30	37	41	54

assume that the fraction of fund managers favoring the use of noneconomic goals in portfolio selection differs in the four fund management agencies? Use $\alpha = .10$.

17.24   In 1987 the number of fatalities resulting from automobile accidents decreased substantially, with the decrease being attributed to more stringent seat belt laws and a crackdown on drivers under the influence of alcohol or drugs. Is there a difference in seat belt use between men and women? A survey was conducted in which the interviewee was asked how often he or she wore a seat belt. The results of the survey are shown in Table 17.10.

a.   Does the data provide sufficient evidence to indicate that there is a difference in seat belt use between men and women? Test by using $\alpha = .05$.

b.   Find the approximate *p*-value for the test. Does the *p*-value support the same conclusion as was drawn in part a?

Table 17.10   Seat Belt Data for Exercise 17.24

| Gender | Seat Belt Use | | | |
	Always	Most of the Time	Sometimes	Never
Male	37	60	54	64
Female	39	58	49	39

Source: Adapted from U.S. Department of Commerce, Bureau of the Census, *Statistical Abstract of the United States*, 107th ed. (Washington, D.C. 1987), p. 591.

17.25   To assist advertisers, a magazine subscription service conducted a survey to study the relationship between the number of magazine subscriptions per household and family income. The survey, based on $n = 1,000$ interviews, produced the results shown in Table 17.11. Do these data present sufficient evidence to indicate that the number of magazine subscriptions per household depends on family income? Test at the $\alpha = .10$ level of significance.

Table 17.11   Sample Data for Exercise 17.25

| Number of Subscriptions per Household | Family Income | | | |
	Less Than $15,000	$15,000– $24,999	$25,000– $34,999	At Least $35,000
0	28	54	78	23
1	29	151	301	73
2	10	31	69	57
More than 2	5	19	40	32

17.5

r × *c* Tables with Fixed Row or Column Totals

In the preceding section we described the analysis of an *r* × *c* contingency table by using examples that, for all practical purposes, fit the multinomial experiment described in Section 17.1. While the methods of collecting data in many surveys may obviously satisfy the requirements of a multinomial experiment, other methods do not. For example, we might not wish to randomly sample the population described in Example 17.1 because we might find that, owing to chance, one income category contains a very small number of people (or even might fail to appear in the sample). Thus, we might decide beforehand to interview a specified number of people in each column category, thereby fixing the column totals in advance.

Although these restrictions tend to disturb somewhat our visualization of the experiment in the multinomial context, they have no effect on the analysis of the data. So as long as we wish to test the hypothesis of independence of the two classifications, and either the row or column probabilities are specified in advance, we may analyze the data as an $r \times c$ contingency table. It can be shown that the resulting X^2 possesses a probability distribution in repeated sampling that is approximated by a chi-square distribution with $(r - 1)(c - 1)$ degrees of freedom.

EXAMPLE 17.2   Because of the almost universal presence of TV sets in American households and because of television's unique combination of sight and sound, television offers an ideal medium for advertisers whose products require demonstration. This is especially true with advertisements designed for children, since their attention span and reading comprehension are typically not sufficient so that they can be reached effectively by written sources. But how do children react to commercials? Ward (see the references) conducted a study to determine whether a relationship exists between a child's age and his or her degree of understanding of certain selected TV commercials. Three hundred children were selected in the study, with the children equally divided among three age categories. The results of the study are shown in Table 17.12, with level I of understanding implying practically no understanding of the message of the commercial. Do these data present sufficient evidence to indicate that the level of understanding of TV commercials is related to a child's age?

Table 17.12   Sample Data for Example 17.2

Level of Understanding	Age			Total
	5–7	8–10	11–12	
I	55 (35.67)	37 (35.67)	15 (35.67)	107
II	35 (48.33)	50 (48.33)	60 (48.33)	145
III	10 (16)	13 (16)	25 (16)	48
Total	100	100	100	300

Source: C. Ward, "Children's Reactions to Commercials," Reprinted with permission from the *Journal of Advertising Research.* © Copyright 1972, by the Advertising Research Foundation.

SOLUTION   You will observe that the test of an hypothesis concerning a lack of relationship between a child's age and his or her level of understanding of a TV commercial is identical to the test of an hypothesis implying independence of the row and column classifications. Suppose we denote the fraction of children with level I of understanding by p_1 and the fraction with level II of understanding by p_2. If we hypothesize that p_1

and p_2 are the same for all children of ages 5 to 12, we imply that the fraction of children possessing level III of understanding is $p_3 = 1 - p_1 - p_2$ and that the row probabilities are p_1, p_2, and p_3, respectively. The probability that a child from the sample of $n = 300$ children falls in any one of the three age classes is $1/3$ because equal sample sizes were employed for the three age groups. Then if the two classifications, age and level of understanding, are independent (i.e., the null hypothesis is true), the cell probabilities for the table are obtained by multiplying the appropriate row and column probabilities.

The estimated expected cell frequencies, calculated by using the row and column totals, appear in parentheses in Table 17.12. We then find that

$$X^2 = \sum_{i=1}^{3} \sum_{j=1}^{3} \frac{[n_{ij} - \hat{E}(n_{ij})]^2}{\hat{E}(n_{ij})}$$

$$= \frac{(55 - 35.67)^2}{35.67} + \frac{(37 - 35.67)^2}{35.67} + \cdots + \frac{(25 - 16)^2}{16} = 36.93$$

The critical value of χ^2 for $\alpha = .05$ and $(r - 1)(c - 1) = (2)(2) = 4$ degrees of freedom is 9.488. Since X^2 exceeds this critical value, we reject the null hypothesis and conclude that children's understanding of TV commercials is not the same over the age span 5 to 12. In fact, their understanding appears to increase with age.

◆ ◆

From Example 17.2 we see that if the column totals are fixed in advance, the data then represent the outcomes of c multinomial experiments, each with r cells, and that testing for independence of rows and columns is equivalent to testing for equality of the r cell probabilities across the c multinomials. The same result holds if the row totals rather than the column totals are fixed in advance. Notice that when $r = 2, c = 2$, and the column totals are fixed in advance at n_1 and n_2, a test of independence of row and column classifications reduces to a test of equality of two binomial proportions. It can be shown that for the 2×2 table with fixed columns the value of X^2 is equal to z^2 for

$$z = \frac{\hat{p}_1 - \hat{p}_2}{\sqrt{\hat{p}\hat{q}(1/n_1 + 1/n_2)}}$$

the statistic used for testing the hypothesis of equality of two binomial proportions in Chapter 8.

EXERCISES   Conceptual Level

17.26   Explain the difference between a standard $r \times c$ contingency table and an $r \times c$ table with fixed rows or columns.

17.27   Is there a difference in the method of analysis when testing for independence of classifications for the two types of contingency tables described in Exercise 17.26?

17.28   Consider the special case of a 2×2 contingency table with fixed row totals given by n_1 and n_2. If the columns represent the number of successes and failures, what two statistics are available for testing equality of the two binomial proportions? Are the two statistics equivalent?

EXERCISES   Skill Level

17.29   Random samples of size 100 were selected from each of 3 populations and classified into one of 4 categories. The resulting 4×3 contingency table is shown in Table 17.13. Do the data provide sufficient evidence to indicate that there is a difference in the proportion of items falling into the 4 categories over the 3 populations? Use $\alpha = .05$.

Table 17.13   Sample Data for Exercise 17.29

	Sample		
Category	1	2	3
1	10	12	8
2	41	38	36
3	22	17	24
4	27	33	32

17.30   A 2×2 contingency table with row totals fixed at 200 is shown below.

	Column	
Row	1	2
1	111	89
2	87	113

Use the chi-square statistic to test for the independence of rows and columns. Use $\alpha = .05$.

17.31   Refer to Exercise 17.30. Suppose that column 1 represents the number of successes and column 2 represents the number of failures observed in samples of size 200. Use the z statistic given in Section 8.14 to test for the equality of two binomial proportions with $\alpha = .05$. Does the observed value of z^2 equal the observed value of X^2 found in Exercise 17.30?

17.32   Random samples of size 100, 200, 200, and 300 were selected from each of four binomial populations, and the number of successes observed are shown below for each of the four samples.

Sample Size	100	200	200	300
No. of Successes	21	49	57	72

a. Complete the 2×4 contingency table by computing the number of failures for each sample. The sample sizes become the column totals.

b. Calculate the test statistic for testing whether the proportion of successes varies significantly from sample to sample. Find the approximate p-value for the test, and interpret your results.

EXERCISES   Applications

17.33   By tradition U.S. labor unions have been content to leave the management of the company to the managers and corporate executives. But in Europe worker participation in management decision making is an accepted idea and one that is continually spreading. To study the relationship of worker participation in managerial decision making and worker satisfaction, a researcher interviewed 100 workers in each of two separate West German manufacturing plants. One plant had active worker participation in managerial decision making; the other did not. Each selected worker was asked whether he or she generally approved of the managerial decisions made in the firm. The results of the interviews are shown in Table 17.14.

a. Do the data provide sufficient evidence to indicate that approval or disapproval of management's decisions depends on whether workers participate in decision making? Test by using the X^2 test statistic. Use $\alpha = .05$.

b. Do these data support the hypothesis that workers in a firm with participative decision making more generally approve of the firm's managerial decisions than those employed by firms without participative decision making? Test by using the *z* test presented in Section 8.14. This problem requires a one-tailed test. Why?

Table 17.14   Worker Response Data for Exercise 17.33

Response	Participative Decision Making	No Participative Decision Making
Generally approve of the firm's decisions	73	51
Do not approve of the firm's decisions	27	49

17.34   A common objective of all investors is to select an investment portfolio with minimum possible risk. Typically, risk is measured by the fluctuation of security prices—volatile securities are risky; securities with stable prices are not risky. An investor is interested in studying the risk behavior of security prices from different industrial classes. In the course of her study she randomly selects 50 manufacturing companies, 30 retail chains, and 25 utilities and records their average weekly price change over the past year. The results are shown in Table 17.15. Do these data present sufficient evidence to assume that security price risk varies according to the industrial class of the issuing firm? Test with $\alpha = .05$.

Table 17.15   Investment Data for Exercise 17.34

| Average Price Change | Industrial Class | | |
	Manufacturing	Retailing	Utility
Less than 2 points (low risk)	27	9	5
From 2 to 5 points (moderate risk)	16	13	10
More than 5 points (high risk)	7	8	10
Total	50	30	25

17.35   Refer to Exercise 17.34. A correct analysis of these data shows that security price risk does vary according to the industrial class of the issuing firm. But does security price risk vary uniformly between each pair of industrial classes? That is, where exactly do the differences lie? (*Hint:*

Examine separately the average price changes between each of the three pairs of industrial classes. Use $\alpha = .05$.)

 17.36 Usually, the small investor attempting to develop a personal investment portfolio considers only stocks or bonds. There is, however, a third possibility that is especially appealing in times preceding an anticipated bull market. Convertible bonds—bonds that can be converted into stocks—offer the stable return of a bond when the securities market is bearish and offer the opportunity to convert to stocks should the issuing firm's security prices begin to rise. To properly advise his clients, an investment banker contacted 50 investment analysts from each of three investment firms that are members of the New York Stock Exchange. Each was asked whether he or she currently advises investing in stocks, bonds, or convertible bonds. The responses are shown in Table 17.16.

a. Do these data present sufficient evidence to indicate that the three investment firms differ in their advice for investors? Use $\alpha = .10$.

b. Do these data provide sufficient evidence to indicate that the proportion of investment officers favoring investment in convertible bonds differs significantly for the three investment firms? Use $\alpha = .10$.

Table 17.16   Investment Advice Data for Exercise 17.36

| | Investors Employed by Firm | | |
Advise Investing In	A	B	C
Stocks	13	16	7
Bonds	31	24	35
Convertibles	6	10	8
Total	50	50	50

MKTG **17.37** In a study of alcohol advertising and adolescent drinking, Lieberman and Orlandi found that of $n = 1747$ young adolescents in the study 63% recalled the specific brand of alcohol being advertised and 89% of 1108 adolescents perceived the ads as depicting young adults drinking in social situations, partying and having fun [L. R. Lieberman and M. A. Orlandi, "Alcohol Advertising and Adolescent Drinking," *Alcohol Health and Research World*, Vol. 12, No. 1 (Fall 1987), p. 30]. Suppose that in another study $n_1 = 100$ adolescents and $n_2 = 100$ adults were shown these same advertisements and were asked to identify the ages of the people in alcohol advertisements, with the results shown in Table 17.17. Do the data present sufficient evidence to indicate that there is a difference in the perceived age(s) of people shown in alcohol advertisements between adolescents and adults? Test by using $\alpha = .05$.

Table 17.17   Sample Data for Exercise 17.37

Perceived Age	Adolescents	Adults
Young adults	76	58
Teens/kids	10	22
Mixed ages	7	15
Older adults	7	5

17.6

Other Topics and Applications

The applications of the chi-square test for analyzing frequencies described in Sections 17.3, 17.4, and 17.5 represent only a few of the interesting classification problems that may be approximated by the multinomial experiment and for which our method of analysis is appropriate. By and large, these applications are complicated to a greater or lesser degree because the numerical values of the cell probabilities are unspecified and hence require the estimation of one or more population parameters. Then, as in Sections 17.4 and 17.5, we can estimate the expected cell frequencies and use X^2 as the test statistic. Although we omit the mechanics of the statistical tests, several additional applications of Pearson's chi-square statistic are worth mentioning.

Goodness-of-Fit Tests

Goodness-of-fit tests are used to determine whether observed cell frequencies are consistent with expected cell frequencies that are calculated under the hypothesis that a given model is true. The test described in Section 17.3 is, in effect, a goodness-of-fit test of a model in which three cells are equally likely.

As another example, suppose that we wish to test the hypothesis that a population possesses a normal probability distribution. The cells of a sample frequency histogram correspond to the k cells of the multinomial experiment. The observed cell frequencies are the number of measurements falling in each cell of the histogram. Given the hypothesized normal probability distribution for the population, we could use the areas under the normal curve to calculate the theoretical cell probabilities and, hence, the expected cell frequencies. If μ and σ are unknown, they are estimated from the data with $\bar{x}$ and s, and the degrees of freedom associated with the chi-square statistic are reduced by two. In general, cells corresponding to values in the tail of the normal distribution whose expected frequencies are less than 5 are combined with adjacent cells so that every expected cell frequency is greater than or equal to 5.

Time-Dependent Multinomials

A second and interesting application of our methodology is its use to investigate the rate of change of a multinomial (or binomial) population as a function of time. For example, we might study the decision-making ability of a human (or any animal) being subjected to an educational program and tested over time. If, for instance, the subject is tested at prescribed intervals of time and the test is of the yes-no type, yielding a number x of correct answers that follows a binomial probability distribution, we would be interested in the behavior of the probability p of a correct response as a function of time. If the number of correct responses was recorded for c time periods, the data would fall into a $2 \times c$ table similar to that in Example 17.1 (Section 17.4). We would then be interested in testing the hypothesis that p is constant—that is, that no learning has occurred—and we would then proceed to more interesting hypotheses to determine whether the data present sufficient evidence to indicate a gradual (say, linear) change over time as opposed to an abrupt change at some time. The procedures we have described can be extended to decisions involving more than two alternatives.

You will observe that our learning example is common to business, to industry, and to many other fields, including the social sciences. For example, we might wish to study campaigns as a function of the length of time that a campaign has been in effect. Or we might wish to study the trend in the lot defective in a manufacturing process as a function of time. Both of these examples, as well as many others, require a study of the behavior of a binomial (or multinomial) process as a function of time.

Multidimensional Contingency Tables

The construction of a two-way contingency table to investigate dependence between two classifications can be extended to three or more classifications. For example, if we wish to test the mutual independence of three classifications, we employ a three-way table. The reasoning and methodology associated with the analysis of both the two-way and three-way tables are identical, although the analysis of the three-way table is a bit more complex. Another method of analyzing an $r \times c$ contingency table utilizes *log-linear models* in which the logarithm of the (i, j)th cell probability is modeled as having a parameter associated with the ith row, another parameter associated with the jth column, and an interaction parameter associated with the ith row and jth column. This technique is an interesting and more complicated method of analysis, which, however, easily generalizes to higher-dimensional tables. Details about the method can be found in the book by Bishop, Fienberg, and Holland listed in the references.

Measures of Dependence

In a test for independence of row and column probabilities in an $r \times c$ contingency table, the observed value of X^2 is, in itself, a measure of dependence between the variables represented by the rows and columns of the table. A value of $X^2 = 0$ occurs only if there is exact agreement between observed and expected cell frequencies under the null hypothesis of independence. On the other hand, large values of X^2 are indicative of dependent rows and columns. Several measures of dependence within a contingency table are based on the observed value of X^2.

One very simple measure that lies in the range from 0 to 1 is given by the probability that a chi-square variable is less than the observed value of X^2. Hence, when $X^2 = 0$, this measure is 0; and as X^2 gets larger and larger, this measure gets closer and closer to 1.

A second and quite popular measure of dependence is the ϕ coefficient (phi coefficient), which in a simple 2×2 table corresponds to the correlation between row and column variables whose values are either 0 or 1. In the 2×2 case,

$$\phi = \frac{ad - bc}{\sqrt{r_1 r_2 c_1 c_2}}$$

where the cell entries are a, b, c, and d, with row totals $r_1 = a + b$ and $r_2 = c + d$, and column totals $c_1 = a + c$ and $c_2 = b + d$. In the general $r \times c$ case, ϕ is defined as

$$\phi = \sqrt{\frac{X^2}{N}}$$

where N is the number of observations in the table. The maximum value of X^2 is $N(q - 1)$, where q is the smaller of r and c; therefore, the maximum value of ϕ is $\sqrt{q - 1}$. Hence, ϕ lies between 0 and 1 only for a 2×2 table.

The coefficient of contingency uses this same information in a different way. The coefficient of contingency is defined as

$$\sqrt{\frac{X^2}{X^2 + N}}$$

and has a minimum value of 0 and a maximum value of $\sqrt{(q - 1)/q}$.

Cramer's V uses this information in the form

$$\sqrt{\frac{X^2}{N(q - 1)}}$$

Since the maximum possible value of X^2 is $N(q - 1)$, this measure lies between 0 and 1.

Each of these measures is considered significantly different from 0 if the observed value of X^2 is significant.

The applications and measures of dependence we have just described are intended to suggest the relatively broad application of the chi-square analysis of frequency data, a fact that should be borne in mind by the experimenter concerned with this type of data.

17.7

Analysis of an $r \times c$ Contingency Table Using Computer Packages

Most statistical program packages include a program for analyzing data displayed as an $r \times c$ contingency table. To illustrate and compare their outputs, we present the computer printouts for the analysis of the data in Example 17.1, using the MINITAB, SAS, and SPSS computer packages, in Tables 17.18, 17.19, and 17.20.

Table 17.18 MINITAB Computer Printout for the Chi-square Analysis of the Data in Example 17.1

```
MTB > CHISQUARE C1-C3

EXPECTED COUNTS ARE PRINTED BELOW OBSERVED COUNTS

              C1          C2          C3      TOTAL
    1        249         494         201        944
           259.6       490.9       193.5

    2         26          26           4         56
            15.4        29.1        11.5

TOTAL        275         520         205       1000

CHISQ =    0.43 +      0.02 +      0.29 +
           7.30 +      0.33 +      4.87 = 13.25

DF = 2
```

Table 17.19 SAS Computer Printout for the Chi-square Analysis of the Data in Example 17.1

```
              TABLE OF PURCHASE BY INCOME

PURCHASE                      INCOME

FREQUENCY
EXPECTED        LOWER     MIDDLE     UPPER    TOTAL
--------------------------------------------------
                 249        494        201      944
LOWER PRICE     259.6      490.9      193.5
--------------------------------------------------
                  26         26          4       56
HIGHER PRICE     15.4       29.1       11.5
--------------------------------------------------
TOTAL            275        520        205     1000

      STATISTICS FOR TABLE OF PURCHASE BY INCOME

STATISTIC                          DF    VALUE     PROB
-------------------------------------------------------
CHI-SQUARE                          2    13.246    0.001
LIKELIHOOD RATIO CHI-SQUARE         2    13.650    0.001
MANTEL-HAENSZEL CHI-SQUARE          1    13.002    0.000
PHI                                       0.115
CONTINGENCY COEFFICIENT                   0.114
CRAMER'S V                                0.115

SAMPLE SIZE = 1000
```

Table 17.20   SPSS Computer Printout for the Chi-square Analysis of the Data in Example 17.1

```
------------------------------CROSSTABULATION OF ------------------------
                            PURCHASE BY INCOME
------------------------------------------------------------  PAGE 1 OF 1
                      INCOME
            COUNT   I
            ROW PCT I LOWER     MIDDLE     UPPER        ROW
            COL PCT I                                   TOTAL
            TOT PCT I   1.00 I   2.00 I     3.00 I


PURCHASE------------------------------------------------------
           1.00 I 249     I 494     I 201     I    944
  LOWER PRICE    I  26.4  I  52.3   I  21.3   I    94.4
                 I  90.5  I  95.0   I  98.0   I
                 I  24.9  I  49.4   I  20.1   I
                 ----------------------------
           2.00 I  26     I  26     I   4     I    56
  HIGHER PRICE   I  46.4  I  46.4   I   7.1   I    5.6
                 I   9.5  I   5.0   I   2.0   I
                 I   2.6  I   2.6   I    .4   I
                 ----------------------------
         COLUMN    275       520       205        1000
         TOTAL     27.5      52.0      20.5       100.0


  CHI-SQUARE    D.F.    SIGNIFICANCE    MIN E.F.    CELLS WITH E.F. < 5
  ----------    ----    ------------    --------    -------------------
   13.24589      2        0.0013        11.480            NONE
```

The MINITAB printout in Table 17.18 shows the contingency table in which the columns correspond to income groups and the rows correspond to the purchase decisions (lower price, higher price). The cell entries are the observed and estimated expected frequencies, respectively. The value of X^2 is given as CHISQ = 13.25 with DF = 2. Since X^2 exceeds $\chi^2_{.05} = 5.99$, the hypothesis of independence is rejected, and we conclude that the purchase decision varies across income groups. In fact, since X^2 exceeds $\chi^2_{.005} = 10.60$, the smallest right-tailed critical value listed in Table 5 of the Appendix, the p-value for this test is less than .005 and the results are highly significant.

In the SAS printout in Table 17.19 the observed and estimated expected frequencies are displayed in the 2×3 decision-by-income contingency table. In the list of statistics provided, we find the value of CHI-SQUARE with 2 degrees of freedom (DF) to be 13.246. The observed significance level is given under the column labeled PROB and correct to three-decimal accuracy is .001, indicating that the observed result is very highly significant. The phi coefficient, the coefficient of contingency, and Cramer's V are measures of dependence as given in Section 17.6. The other statistics listed in the printout are not pertinent to our analysis.

The SPSS printout in Table 17.20 differs somewhat from the other two in that four entries are displayed for each cell in the contingency table. From top to bottom, these entries are the observed cell frequency, the cell frequency expressed as a percentage of the row total, the column total, and the grand total. The estimated expected frequencies are not given. For example, the cell in the first row and first column had an observed frequency of 249. This frequency represented 26.4% of 944, the first row total; 90.5% of 275, the first column total; and 24.9% of 1,000, the grand total. The value $X^2 = 13.24589$; the degrees of freedom, D.F. = 2; and the observed significance level of .0013 are found directly below the table. Other statistics usually given in the printout are not relevant to our analysis.

The information as well as reported decimal accuracy varies from one printout to another, but the basic information is the same. All three printouts show the calculated value of X^2 and the corresponding degrees of freedom. These quantities, together with the critical values found in Table 5 of the Appendix, are all that we need to test for independence of the classification variables represented in the contingency table. The value of the observed significance level given in the SAS and SPSS printouts eliminates the need for the chi-square table and directly gives us a measure of the evidence favoring the rejection of the null hypothesis of independence.

17.8

Assumptions

The following assumptions must be satisfied if X^2 is to possess, approximately, a chi-square distribution and, consequently, if the tests described in this chapter are to be valid.

Assumptions

> 1. The cell counts, $n_1, n_2, \ldots, n_k$, satisfy the conditions of a multinomial experiment (or a set of multinomial experiments created by restrictions on row or column totals).
> 2. The expected values of all cell counts should equal or should exceed 5.

Assumption 1 must be satisfied. The chi-square goodness-of-fit tests, of which these tests are special cases, compare observed frequencies with expected frequencies and apply only to data generated by a multinomial experiment.

The larger the sample size n, the more closely the chi-square distribution will approximate the distribution of X^2. We have stated in assumption 2 that n must be large enough so that all the expected cell frequencies will be equal to 5 or more. This is a safe figure. Actually, the expected cell frequencies can be smaller for some tests. For information on the minimum expected cell frequencies for specific goodness-of-fit tests, see the paper by Cochran listed in the references.

17.9

Summary

The preceding material concerned a test of an hypothesis regarding the cell probabilities associated with a multinomial experiment. When the number n of observations is large, the test statistic X^2 can be shown to possess, approximately, a chi-square probability distribution in repeated sampling, the number of degrees of freedom being dependent on the particular application. In general, we assume that n is large and that the minimum expected cell frequency is equal to or greater than 5.

Several words of caution concerning the use of the X^2 statistic as a method of analyzing enumerative data are appropriate. The determination of the correct number of degrees of freedom associated with the X^2 statistic is very important in locating the rejection region. If the number is incorrectly specified, erroneous conclusions might result. Also, note that nonrejection of the null hypothesis does not imply that it should be accepted. We would have difficulty in stating a meaningful alternative hypothesis for many practical applications and, therefore, would lack the knowledge of the probability of making a type II error. For example, we hypothesize that the two classifications of a contingency table are independent. A specific alternative would have to specify some measure of dependence, which may or may not possess practical significance to the experimenter. Finally, if parameters are missing and the expected cell frequencies must be estimated, the estimators of missing parameters should be of a particular type in order that the test be valid. In other words, the application of the chi-square test for other than the simple applications outlined in Sections 17.3, 17.4, and 17.5 requires experience beyond the scope of this introductory presentation of the subject.

Tips on Problem Solving

1. Be certain to assign the correct number of degrees of freedom associated with the chi-square statistic. Check the application to obtain the correct number.

2. Remember that the rejection region is *always* located in the upper tail of the chi-square distribution.

3. Check to make certain that you have included all the cells associated with your multinomial experiment. For example, if you are recording counts on the number of people in each of four social classes who favor a political issue, do not forget to include cells for those who *do not favor* the issue. The check involves testing individual experimental units to make certain that each falls in one and only one cell of your table.

Supplementary Exercises

Applications

17.38 The Securities and Exchange Commission (SEC) has adopted reserve recognition accounting as a requirement for financial statements, thus bringing forecasts into the financial statement. This requirement is especially important to mineral and petroleum companies, who must now publish an estimate of their reserves. To observe compliance with this new regulation, a researcher analyzed financial statements from 40 firms in each of three industries. The results are shown in the accompanying table. Do these data present sufficient evidence to indicate that compliance with the SEC's reserve recognition requirement varies according to industry? Test the hypothesis that $p_1 = p_2 = p_3$, using $\alpha = .05$.

Industry	Oil and gas	Coal	Metals
No. in Compliance	32	24	29

17.39 A city expressway utilizing four lanes in each direction was studied to see whether drivers preferred to drive on the inside lanes. A total of 1,000 automobiles were observed during the heavy early-morning traffic and their respective lanes were recorded. The results are shown in the table. Do the data present sufficient evidence to indicate that some lanes were preferred over others? Test the hypothesis that $p_1 = p_2 = p_3 = p_4 = 1/4$, using $\alpha = .05$.

Lane	1	2	3	4
Observed Count	294	276	238	192

17.40 A manufacturer of buttons wishes to determine whether the fraction of defective buttons produced by three machines varies from machine to machine. Samples of 400 buttons are selected from each of the three machines and the number of defectives counted for each sample. The results are shown in the table. Do these data present sufficient evidence to indicate that the fraction of defective buttons varies from machine to machine? Test with $\alpha = .05$.

Machine Number	1	2	3
Number of Defectives	16	24	9

17.41 A manufacturer of floor polish conducted a consumer preference experiment to see whether a new floor polish, A, was superior to those of four of his competitors. A sample of 100 housewives viewed five patches of flooring that had received the five polishes, and each indicated the patch that she considered superior in appearance. The lighting, background, and so forth were approximately the same for all five patches. The results of the survey are given in the accompanying table. Do these data present sufficient evidence to indicate a preference for one or more of the polished patches of floor over the others? If you were to reject the hypothesis of no preference for this experiment, would this rejection imply that polish A is superior to the others? Can you suggest a better method for conducting the experiment?

Polish	A	B	C	D	E
Frequency	27	17	15	22	19

17.42 A quality control engineer for a factory wishes to examine the operating efficiency of two assembly machine operators. The operators are in charge of the same machine but work during different shifts. During a given week the numbers of good and defective finished items produced by the assembly machine while each operator was on duty were recorded and are as shown in Table 17.21. Do these data present sufficient evidence to indicate that the operators work with about the same efficiency? (Test the hypothesis that the proportion of defective items produced by each operator is the same.)

a. Test by using the X^2 statistic. Use $\alpha = .05$.

b. Use the z test from Section 8.14.

Table 17.21   Efficiency Data for Exercise 17.42

Finished Item	Operator A	Operator B
Good	551	416
Defective	16	17

17.43   To encourage energy conservation, many public utility corporations are now offering reduced rates to their consumers who reduce their energy use by a certain fixed percentage each month. A public utility that serves three communities has offered reduced rates to its consumers who reduce their monthly electrical consumption by at least 15%. After the first month of the energy conservation program, 510 subscribers in the three communities were selected at random for analysis. The observed results are given in Table 17.22. Do these data present sufficient evidence to indicate a difference in the proportion of subscribers participating in the energy conservation program in the three communities? Use $\alpha = .05$.

Table 17.22   Energy Conservation Data for Exercise 17.43

	Community		
Result	Appleton	Beechwood	Cherryville
Reduced consumption by more than 15%	16	63	29
Did not reduce consumption by more than 15%	72	202	118

17.44   A printer is interested in examining the relationship between the number of printing errors and the type size used. He selects three different books recently printed by his company, each printed using a different type size. From each book he records the number of pages with printing errors and the number of error-free pages. The results are shown in Table 17.23. Do the data indicate a dependence between type size and printing errors?

Table 17.23   Printer's Data for Exercise 17.44

	Type Size		
Result	A	B	C
Pages with errors	23	17	41
Pages without errors	241	183	210

17.45 Following the lead of Oregon and Vermont, the Environmental Protection Agency has supported the concept of national legislation to control nonreturnable beverage containers. The enactment of such legislation has been found to result in reduction of roadside litter, and in addition, EPA officials estimate that the equivalent of 92 thousand barrels of oil a day could be saved in reduced energy consumption by mandatory deposit legislation. To examine regional differences in opinion about the EPA proposal, an EPA official randomly chose 100 public officials each from four sectors of the United States. Their responses are recorded in Table 17.24. Do these data present sufficient evidence to indicate a difference in the proportion of public officials in the various regions who favor national legislation to control nonreturnable beverage containers? Test by using a 5% level of significance.

Table 17.24   Response Data for Exercise 17.45

	Region in Which the Public Official Resides			
Opinion	West	South	Midwest	Northeast
Approve of mandatory deposit legislation	54	48	45	39
Disapprove of mandatory deposit legislation	46	52	55	61

17.46 The American consumer is very interested in obtaining useful investment information and in increasing his or her personal knowledge concerning the various financial products and services available in today's market. A survey conducted by Synergistics Research Corporation of Atlanta, Georgia, investigated the various sources used by consumers to obtain investment information and the effect of tax reform on the investment opportunities available to them ("Who Consumers Look to for Financial Advice," *Journal of Accountancy*, January 1988, p. 24). The study was conducted among 400 households on a nationwide basis. In a similar survey the head of each household was asked two questions:

A: Do you find the variety of financial products and services to be confusing?

B: Are you knowledgeable about the new tax advantages available under the Tax Reform Act of 1987?

The results of these two questions are cross-classified with household income in Table 17.25.

Table 17.25   Survey Data for Exercise 17.46

	Household Income		
Response	< $25,000	$25,000–$75,000	> $75,000
Question A			
Knowledgeable	12	79	71
Not knowledgeable	46	151	41
Question B			
Confusing	51	170	59
Not confusing	7	60	53

a. For each of the two questions, determine whether the data present sufficient evidence to indicate that the proportions answering the question in a certain way differ with the income level of the household. Use $\alpha = .01$.

b. Are the two tests in part a independent? What effect will this result have on the validity of your conclusions in part a?

 17.47 A survey of the opinions of the stockholders of a corporation about a proposed merger was taken to determine whether the resulting opinion was independent of the number of shares held. Two hundred stockholders were interviewed, with the results shown in Table 17.26. Do these data present sufficient evidence to indicate that the opinions of the stockholders concerning the merger were dependent on the number of shares held by the stockholder? Test with $\alpha = .05$.

Table 17.26   Opinion Data for Exercise 17.47

	Opinion		
Shares Held	In Favor	Opposed	Undecided
Under 100	37	16	5
100–500	30	22	8
Over 500	32	44	6

 17.48 Suppose the responses for the data in Exercise 17.47 were reclassified according to whether the stockholder was male or female, as shown in Table 17.27. Do these data present sufficient evidence to indicate that stockholder reaction to the proposed merger depended on whether the stockholder was male or female? Test with $\alpha = .05$.

Table 17.27   Opinion Data for Exercise 17.48

	Opinion		
Gender	In Favor	Opposed	Undecided
Female	39	46	9
Male	60	36	10

17.49 Because more people earn a college degree today than did in earlier times, a college degree has lost some of its advantages. Nevertheless, there are still definite advantages to a college degree, such as the potential for greater lifetime earnings and greater job satisfaction. In a study of this issue 400 people were randomly selected from the census data of a large midwestern city, 200 of whom were college graduates. The reported annual incomes of those chosen for the study are shown in Table 17.28.

a. Do these data indicate that annual income depends on whether or not an individual has earned a college degree? Use $\alpha = .10$.

b. If you have rejected the implicit hypothesis of independence in part a, how can you determine whether the data suggest that annual income increases or decreases when earning a college degree?

Table 17.28   Sample Data for Exercise 17.49

Education	Income Less Than $25,000	Income $25,000–$50,000	Income Over $50,000
No college degree	50	99	51
College degree	27	124	49

17.50   A pharmaceutical manufacturer conducted a study to determine the effectiveness of a new drug (serum) for arthritis. The study involved the comparison of two groups, each consisting of 200 arthritic patients. One group was inoculated with the serum; the other received a placebo (an inoculation that appears to contain serum but actually it is nonactive). After a period of time each person in the study was asked to state whether his or her arthritic condition was improved. The observed results are shown in Table 17.29. Do these data present sufficient evidence to indicate that the serum was effective in improving the condition of arthritic patients?

a.   Test by using the X^2 statistic. Use $\alpha = .05$.

b.   Use the z test of Section 8.14.

Table 17.29   Drug Study Data for Exercise 17.50

Condition	Treated	Untreated
Improved	117	74
Not improved	83	126

17.51   A company selling airbrushes is about to mount a new advertising campaign. Before doing so, it wishes to determine whether the campaign should be directed to the male or female market individually, or to both together. A sample of 1,750 users of the airbrush revealed the distribution of regular and occasional users shown in Table 17.30, by gender. Do these data present sufficient evidence to indicate that the use of the airbrush on an occasional or regular basis is related to the gender of the user? Use $\alpha = .01$.

Table 17.30   Airbrush Data for Exercise 17.51

Type of User	Male	Female
Regular	170	465
Occasional	475	640

17.52   A study of the purchase decisions for three stock portfolio managers, A, B, and C, was conducted to compare the rates of stock purchases that resulted in profits over a time period that was less than or equal to one year. One hundred randomly selected purchases obtained for each of the managers showed the results given in Table 17.31. Do these data provide evidence of differences among the rates of successful purchases for the three portfolio managers? Test with $\alpha = .05$.

Table 17.31   Purchase Decision Data for Exercise 17.52

		Manager	
Result	A	B	C
Purchases show profit	63	71	55
Purchases do not show profit	37	29	45
Total	100	100	100

17.53   A carpet company is interested in comparing the fraction of new-home builders favoring carpet over other floor coverings for homes in three different areas of a city. The objective is to decide how to allocate sales effort to the areas. A survey is conducted, and the resulting data are as given in Table 17.32. Do the data indicate a difference in the percentage favoring carpet from one region of the city to another?

Table 17.32   Preference Data for Exercise 17.53

Preferences of Builders	Area 1	Area 2	Area 3
Carpet	69	126	16
Other materials	78	99	27

17.54   A study was conducted to determine whether individuals earning over $30,000 per year use the services of an accountant in the preparation of their income tax returns at the same rate in different regions of the United States. Four states were selected as representative of the four regions (Northeast, South, Midwest, and West). From each state a random selection of individuals with annual incomes in excess of $30,000 was obtained. Each individual was then asked whether he or she uses the services of an accountant in the preparation of income taxes. The results are given in Table 17.33. Do the data indicate a difference in the rate of use of the services of an accountant from one region to another? Test with $\alpha = .05$.

Table 17.33   Sample Data for Exercise 17.54

Classification	Rhode Island	Florida	Iowa	California
Use an accountant	46	121	63	108
Prepares own taxes	149	179	178	192

17.55   A survey was conducted by an auto repairman to determine whether various auto ills depended on the make of the auto. His survey, restricted to this year's model, showed the results given in Table 17.34. Do these data present sufficient evidence to indicate a dependency between auto makes and type of repair for these new-model cars? Note that the repairman did not utilize all the information available when he conducted his survey. In a study of this type, what other factors should be recorded?

Table 17.34   Auto Repair Data for Exercise 17.55

Make	Type of Repair		
	Electrical	Fuel Supply	Other
A	17	19	7
B	14	7	9
C	6	21	12
D	33	44	19
E	7	9	6

17.56   The city manager of a midwestern city conducted a survey to determine whether the incidence of various types of crime varied from one part of the city to another. The city was partitioned into three regions, and the crimes were classified as homicide, auto theft, grand larceny, petty larceny, and other. An analysis of 1,599 cases showed the results given in Table 17.35. Use the MINITAB printout in Table 17.36 to determine whether the occurrence of various types of crime depends on city region.

Table 17.35   City Crime Data for Exercise 17.56

City Region	Homicide	Auto Theft	Grand Larceny (Omitting Auto Theft)	Petty Larceny	Other
1	12	239	191	122	47
2	17	163	278	201	54
3	7	98	109	44	17

Table 17.36   MINITAB Computer Printout for the Data of Exercise 17.56

```
MTB > CHISQUARE C1-C5

EXPECTED COUNTS ARE PRINTED BELOW OBSERVED COUNTS

        HOMICIDE  AUTOTHFT  GRDLRCNY  PTYLRCNY    OTHER    TOTAL
   1         12       239       191       122       47      611
             13.8     191.1     220.9     140.2     45.1

   2         17       163       278       201       54      713
             16.1     223.0     257.7     163.6     52.6

   3          7        98       109        44       17      275
              6.2      86.0      99.4      63.1     20.3

 TOTAL       36       500       578       367      118     1599

 CHISQ =   0.22 +   12.03 +    4.04 +    2.37 +   0.08 +
           0.06 +   16.12 +    1.59 +    8.53 +   0.04 +
           0.11 +    1.68 +    0.93 +    5.79 +   0.53 = 54.11
 DF = 8
```

17.57 To obtain fair treatment and to eliminate discrimination in banking practices against women, female financiers are launching their own savings and lending institutions in many cities across the country. These savings institutions are largely owned and directed by women and intend to provide women with easier access to credit and banking careers. The president of a women's bank in a large eastern city conducted a survey to determine the attitudes of women of different ages toward the concept of a bank owned and operated by women. The results of her survey of 480 women from five different age groups are shown in Table 17.37. Do these data present sufficient evidence to assume that the proportion of women favoring the concept of a bank owned and operated by women differs among women from different age groups? Use $\alpha = .10$.

Table 17.37   Women's Bank Data for Exercise 17.57

	Age Group				
Attitude	21–30	31–40	41–50	51–60	Over 60
Support concept of a woman's bank	52	67	38	34	35
Do not support concept of a woman's bank	43	43	48	71	49
Total	95	110	86	105	84

17.58 During times of business decline and recession, many suggestions are offered to spur the economy into a turnaround. A survey was conducted among 100 business executives, 100 economists, and 100 government officials to find the opinion of each regarding the best way of reversing the trend of a business decline. Their responses are shown in Table 17.38. Do these data present sufficient evidence to assume that opinion regarding the best way to spur the economy into a turnaround during times of recession differs among business executives, economists, and government officials? Test by using a 5% level of significance. Use the MINITAB printout in Table 17.39.

Table 17.38   Response Data for Exercise 17.58

Opinion	Business Executives	Economists	Government Officials
Increase government spending	10	15	39
Cut personal income taxes	37	37	33
Decrease interest rates	24	34	15
Offer tax incentives to business	29	14	13
Total	100	100	100

17.59 Many businesses have encouraged employee ownership of company stock to increase employee commitment to the firm. However, an offer by a company to sell its stock to its employees is sometimes perceived in different ways by workers and management. Hammer and Stern

Table 17.39   MINITAB Computer Printout for the Data of Exercise 17.58

```
MTB > CHISQUARE C1-C3

EXPECTED COUNTS ARE PRINTED BELOW OBSERVED COUNTS

                EXEC        ECON        GOVT       TOTAL
        1         10          15          39          64
                21.3        21.3        21.3

        2         37          37          33         107
                35.7        35.7        35.7

        3         24          34          15          73
                24.3        24.3        24.3

        4         29          14          13          56
                18.7        18.7        18.7

TOTAL            100         100         100         300

CHISQ =       6.02 +      1.88 +     14.63 +
              0.05 +      0.05 +      0.20 +
              0.00 +      3.84 +      3.58 +
              5.72 +      1.17 +      1.72 = 38.86

DF = 6
```

examined this issue by surveying 250 employees of an eastern manufacturing firm who own stock in the company. Each was asked his or her reason for owning company stock. The results are shown in Table 17.40. Do these data suggest that sufficient evidence exists to indicate that the reasons for employee ownership of stock differs among employee groups? Let $\alpha = .05$.

Table 17.40   Stock Ownership Data for Exercise 17.59

Reason	Blue-Collar Workers	White-Collar Workers	Middle Management	Top Management
To save my job	77	25	11	8
As an investment	37	13	8	4
I believe in employee ownership	35	14	7	11

Source: T. Hammer and R. Stern, "Employee Ownership: Implications for the Organizational Distribution of Power," *Academy of Management Journal*, March 1980. Reprinted with permission.

17.60   Portable personal computers, sometimes called "laptops," represent a fast-growing segment of the PC market. According to Market Intelligence Research Company, the use of portable "laptop" computers can be classified in the following user segments ("Laptop's Three Musts for Success in Sales," *Sales and Marketing Management*, February 1988, p. 91):

Business-professional	69%
Government	21%
Education	7%
Home	3%

A sample of $n = 150$ laptop computer owners were surveyed, and the user segments were tabulated as follows:

Business-professional	102
Government	32
Education	12
Home	4

Do the data provide sufficient evidence to indicate that the figures given by Market Intelligence Research Company are not accurate? Calculate the approximate *p*-value for the test, and use it to make your decision.

Supplementary Exercises

Interpretive

17.61   In recent years many managers have begun to realize that there is a positive relationship between worker productivity and the quality of work life. Minor management concessions such as improved lighting, attractive lunchroom facilities, and positive attention to worker grievances have often been met by marked improvements in worker productivity. To determine management attitudes on this subject, a researcher questioned managers and executives from several different types of organizations. Table 17.41 gives the number chosen from each type of

Table 17.41   Survey Data for Exercise 17.61

Type of Organization	Number Surveyed	Percentage Favoring Quality-of-Work-Life Improvement
U.S. Postal Service	170	29.0
Light manufacturing	212	37.8
Heavy manufacturing	87	31.7
Nondurable retailing	568	36.4
Durable goods retailing	342	34.5
Public service agencies	186	27.6

organization and the percentage from that group who favor improvement as a means of increasing productivity. Do these data provide sufficient evidence to indicate that the probability that a manager or executive favors improving the quality of employee work life to increase productivity is dependent on the type of organization?

17.62 A survey of the United States banking industry was conducted to measure the strategic marketing response of different organizational types to recent environmental changes such as inflation, rapid advances in technology, deregulation of depository institutions, and the rapid entry of new competitors into the financial services field. Of the 310 survey responses, 279 of the responding organizations classified themselves into one of three of the Miles and Snow strategic types: defenders (tend to remain within narrow product-market domains), prospectors (tend to search for market opportunities and tend to be creators of change in the industry), and analyzers (a mixture of the first two, operating in possibly one stable and one changing product-market domain). A summary of the survey results is given in Table 17.42, where the 279 banks were cross-classified according to their strategic type and bank type, and according to their strategic type and bank size. What information concerning the relationship of strategic type to bank type and bank size can be gleaned from these data?

Table 17.42   Survey Data for Exercise 17.62

| | Strategic Type | | |
Bank Type or Size	Defender	Prospector	Analyzer
Type			
Independent unit bank	36	18	70
Independent branch bank	2	2	16
Holding company, affiliated unit bank	17	30	47
Holding company, affiliated branch bank	2	17	22
Size (in assets)			
Under $50 million	36	16	66
$50 million to $1 billion	20	33	68
Over $1 billion	1	17	21

Source: Data summarized from S. W. McDaniel and T. W. Kolari, "Marketing Strategy of the Miles and Snow Strategic Typology," *Journal of Marketing*, Vol. 51 (October 1987), pp. 19–30.

17.63 As mass marketers look to overseas markets to maintain growth and expand profit bases, they must decide whether or not to modify their marketing and advertising strategies to reflect the prevalent cultural values. The *global marketing theory* is based on the assumption that people the world over have the same tastes and desires and are remarkably alike in many ways that relate to advertising appeal. To confirm or deny the global marketing theory, Mueller examined a sample of advertisements from Japan and the United States. The advertisements were selected from two magazines from each country, matched according to format, audience, and circulation, one a general newsmagazine and the other a magazine targeted at women. The general results of the study are given in Table 17.43, in which 146 Japanese advertisements and 232 American advertisements were categorized according to the advertisement's appeal. This tabulation covers all products and both types of magazines. Do these data support the global marketing theory? Defend your answer.

Table 17.43   Advertisement Data for Exercise 17.63

	Advertisements	
Appeal	Japanese (%) (N = 146)	American (%) (N = 232)
Group/consensus	6.8	9.0
Individual/independence	6.2	4.7
Soft-sell appeal	21.2	5.2
Hard-sell appeal	1.4	5.6
Elderly/traditional	11.0	1.3
Modernity/youth	2.1	3.9
Status appeal	17.1	9.9
Product merit	28.1	56.0
Nature-oriented	6.2	4.3

Source: B. Mueller, "Reflections of Culture: An Analysis of Japanese and American Advertising Appeals," *Journal of Advertising Research*, June/July 1987, pp. 51–59. Reprinted from the *Journal of Advertising Research* © Copyright (1987), by the Advertising Research Foundation.

Case Study

The Gray Flannel Suit Syndrome

Does a woman's attire affect her climb up the corporate ladder? As late as a decade ago, women in executive positions were attired in somber business suits and blouses with bows at the neck. As women achieved more security in executive positions through their performance within the corporate structure, media reports indicated that successful female executives could now look more feminine. However, interviews with many male and female executives indicated that even with relaxed dress codes in the business workplace, it is still not acceptable for women to appear too feminine in the workplace.

According to the *Wall Street Journal*, this result does not mean that a dark suit is the only acceptable professional dress; tailored dresses and skirts worn with a jacket are also acceptable business attire for women.* Apparently, a woman whose dress is "extremely feminine gives the message that she needs to be taken care of" or that she is not as serious as someone dressing more conservatively. This dual standard concerning

* "Business Women's Broader Latitude in Dress Codes Goes Just So Far," by Kathleen A. Hughes, *The Wall Street Journal*, September 1, 1987. Reprinted by permission of *The Wall Street Journal*. © Dow Jones & Company, Inc. 1987. All rights reserved.

dress codes seems to have little to do with actual performance. Nonetheless, "women whose clothes were described as conservative were *twice* as likely to receive promotions as those whose dress was labeled as frilly, frivolous, or sexy."

As part of a survey conducted by John T. Malloy, 298 women in corporate sales positions were classified according to their perceived job performance and their type of dress. The results are given in the following table:

	Attire			
Job Performance	Very Professional, Very Conservative	Appropriate	Sexy, Frilly, Fashionable	Poorly Dressed
Top performers	12.5%	12.6%	11.5%	1.9%
Consistently above average	60.9%	41.1%	31.0%	30.8%
Average or below	21.9%	44.2%	54.0%	50.0%
Failing	4.7%	2.1%	3.4%	17.3%

Source: John T. Malloy, *Wall Street Journal*, September 1, 1987. Reprinted by permission of *The Wall Street Journal*. © Dow Jones & Company, Inc. 1987. All rights reserved.

1. Suppose that of the 298 women surveyed, 92 were classified as very professionally dressed, 95 dressed appropriately, 59 dressed in a sexy, frilly, or fashionable manner, and 52 were poorly dressed. Does the information provided in Table 17.44 appear to support the contention that job performance evaluations for women in corporate positions depend upon their choice of attire?
2. What is the statistical interpretation of the comment that women whose clothes were described as conservative were twice as likely to receive promotions as women whose dress was labeled as frilly, frivolous, or sexy?

References and Suggested Readings

Anderson, R. L., and T. A. Bancroft. *Statistical Theory in Research*. New York: McGraw-Hill, 1952.

Bishop, Y. M. M., S. E. Fienberg, and P. W. Holland. *Discrete Multivariate Analysis: Theory and Practice*. Cambridge, Mass.: MIT Press, 1975.

Cochran, W. G. "The X^2 Test of Goodness of Fit." *Annals of Mathematical Statistics*, Vol. 23 (1952), pp. 315–345.

Dixon, W. J., and F. J. Massey, Jr. *Introduction to Statistical Analysis.* 4th ed. New York: McGraw-Hill, 1983.

Isakson, H. R., and A. R. Maurizi. "The Consumer Economics of Unit Pricing." *Journal of Marketing*, 1973.

Lefkowitz, J. M. *Introduction to Statistical Computer Packages.* Boston: PWS-KENT Publishing Co., 1985.

Ryan, B. F., B. L. Joiner, and T. A. Ryan. *MINITAB Handbook.* 2d ed. Boston: PWS-KENT Publishing Co., 1985.

Summers, Y. W., W. S. Peters, and C. P. Armstrong. *Basic Statistics in Business and Economics,* 4th ed. Belmont, Calif.: Wadsworth, 1985.

Ward, S. "Children's Reactions to Commercials." *Journal of Advertising Research*, 1972.

TO
THE
READER

Chapters 8 and 9 presented statistical techniques for comparing two populations by comparing their respective population parameters (usually their respective population means). The statistical techniques from Chapters 8 and 9 are applicable to data measured on a continuum and, in Chapter 9, to data possessing normal population relative frequency distributions.

The purpose of Chapter 18 is to present several statistical test procedures for comparing populations for the many types of business data that do not satisfy the assumptions specified in Chapters 8 and 9. For example, product or attribute rankings and measurements on an ordinal scale do not satisfy the assumptions of Chapters 8 and 9 and thus require less demanding statistical techniques for analysis. These techniques are called *nonparametric* statistical methods.

In Chapter 18 we will also examine nonparametric counterparts to the analysis of variance procedures for a completely randomized design and for a randomized block design as described in Chapter 10.

18

NONPARAMETRIC
STATISTICS

18.1

Introduction

Some experiments yield response measurements that defy quantification. That is, they generate response measurements that can be ordered (ranked), but the location of the response on a scale of measurement is arbitrary. For example, the data may admit only pairwise directional comparisons—that is, whether one observation is larger than another or vice versa. Or we may be able to rank all observations in a data set but may not know the exact values of the measurements. To illustrate, suppose that a judge is employed to evaluate and rank the sales abilities of four salespeople, the edibility and taste characteristics of five brands of cornflakes, or the relative appeal of five new automobile designs. Clearly, it is impossible to give an exact measure of sales competence, palatability of food, or design appeal. Thus, the response measurements here differ markedly from those presented in preceding chapters. The judge's scale of measurement may be a Likert scale,* a semantic differential, or an ordinal scale of his or her own design. Although experiments of this type occur in almost all fields of study, they are particularly evident in social science research and in studies of consumer
nonparametric
statistical methods
preference. The data they produce can be analyzed by using **nonparametric statistical methods**.

Nonparametric statistical procedures are also useful in making inferences in situations where serious doubt exists about the assumptions that underlie standard methodology. For example, the t test for comparing a pair of means (Section 9.4) is based on the assumption that both populations are normally distributed with equal variances. The experimenter will never know whether these assumptions hold in a practical situation but will often be reasonably certain that departures from the

---

* See the footnote on page 940.

assumptions will be small enough so that the properties of the statistical procedure will be undisturbed. That is, α and β will be approximately what the experimenter thinks they are. On the other hand, it is not uncommon for the experimenter seriously to question the assumptions and wonder whether he or she is using a valid statistical procedure. This difficulty may be circumvented by using a nonparametric statistical test and thereby avoiding reliance on a very uncertain set of assumptions.

Research has shown that nonparametric statistical tests are almost as capable of detecting differences among populations as the parametric methods of preceding chapters when normality and other assumptions are satisfied. They may be, and often are, more powerful in detecting population differences when the assumptions are not satisfied. For this reason, many statisticians advocate the use of nonparametric statistical procedures in preference to their parametric counterparts.

In this chapter we will discuss only the most common situations in which nonparametric methods are used. For a more complete treatment of the subject, see the texts by Siegel (1956) or Conover (1980) listed in the references.

18.2

The Sign Test for Comparing Two Population Distributions

Suppose that we wish to compare consumer ratings (on a scale of 1 to 10) of two window cleaners. Six homemakers are randomly selected from the consumer group, and each rates one window treated with cleaner A and one with cleaner B. A 10 rating is "best." As you can see, we have used a paired-difference experiment (a randomized block design) to make the comparison. The cleaners A and B are treatments. We have asked each homemaker to examine both cleaners in order to block out homemaker-to-homemaker variation.

The data for the paired-difference experiment are shown in Table 18.1. Do the data present sufficient evidence to indicate a difference in consumer preference for the two cleaners?

Table 18.1   Consumer Ratings for Window Cleaners

Homemaker	Cleaner	
	A	B
1	10	7
2	7	5
3	8	7
4	5	2
5	7	6
6	9	6

No complicated statistical test is needed to answer the question. Indeed, we can use a rough-and-ready nonparametric test procedure, known as the **sign test**, that can almost be performed "by eye." That is, we note that the rating for cleaner A exceeds the rating for B for all six homemakers (thus, the signs of the six differences are all positive). Assuming no difference between the cleaners, this result is equivalent to flipping a balanced coin six times and observing six heads (or tails). The probability of observing such an event, $(1/2)^6 + (1/2)^6 = 1/32$, is quite small. Hence, we would be likely to reject an hypothesis that the distributions of consumer preferences for the two cleaners are identical.

You will recall that we employed the same nonparametric statistical test as an alternative procedure for determining whether evidence existed to indicate a difference in the mean wear for the two types A and B of tires in the paired-difference experiment of Section 9.5. Each pair of responses was compared and x (the number of times A exceeded B) was used as the test statistic. **This nonparametric test is called a sign test because x is the number of positive (or negative) signs associated with the differences. The implied null hypothesis is that the two population distributions are identical,** and the resulting technique is completely independent of the form of the distribution of differences. Thus, regardless of the distribution of differences, **the probability that A exceeds B for a given pair is $p = .5$ when the null hypothesis is true** (that is, when the distributions for A and B are identical). Then x will possess a binomial probability distribution, and a rejection region for x can be obtained by using the binomial probability distribution of Chapter 5. In fact, the sign test uses exactly the same test procedure that was discussed in detail in Section 5.7. Before reading further, you may wish to review Chapter 5 and, in particular, Section 5.7.

The sign test is summarized in the next display.

The Sign Test

Null hypothesis H_0:   The two population distributions are identical and $P(A$ exceeds B for a given pair$) = p = .5$.

Alternative hypothesis H_a: The two population distributions are not identical and $p \neq .5$. Or H_a: The population of A measurements is shifted to the right of the population of B measurements and $p > .5$.*

Test statistic: For n, the number of pairs in which no ties were reported, use x, the number of times that $(A - B)$ was positive, or equivalently, the number of times A exceeded (or was preferred to) B.

Rejection region: For the two-tailed test $H_a : p \neq .5$ and a given value of α, reject H_0 if $x \leq x_L$ or $x \geq x_U$, where x_L and x_U are the lower- and upper-tailed values of a binomial distribution, with $p = .5$ satisfying $P(x \leq x_L) \leq \alpha/2$ and $P(x \geq x_U) \leq \alpha/2$. For the one-tailed test of $H_a : p > .5$ and a given value of α, reject H_0 if $x \geq x_U$, where x_U is the upper-tailed value of a binomial distribution, with $p = .5$ satisfying $P(x \geq x_U) \leq \alpha$.*

---

* For the one-tailed test when the alternative hypothesis is $H_a : p < .5$, simply interchange the letters A and B.

EXAMPLE 18.1 In a market research experiment a food-processing company undertook a study to determine the acceptability of a sugar substitute in their canned orange juice. Eleven families were given a liberal supply of both the current product (A) and the new one with the sugar substitute (B) and asked to use them over a four-week period and to state their preference. The results are shown in Table 18.2.

Table 18.2   Sample Data for Example 18.1

Family	Product A	B	Preferred Product
1	−	+	B
2	−	+	B
3	+	−	A
4	−	+	B
5	0	0	No preference
6	−	+	B
7	−	+	B
8	+	−	A
9	−	+	B
10	−	+	B
11	−	+	B

Assume that the 11 families are representative of potential users of the company's product. Let x denote the number of families preferring the current product (A). Do the data present sufficient evidence to indicate a preference for one of the two orange drinks, the orange juice with a sugar substitute (B) or the current orange juice (A)? State the null hypothesis to be tested, and use x as a test statistic.

SOLUTION Because each pair of preferences A and B corresponds to a particular family, we see that the data were collected by using a paired-difference experiment. Consequently, the observed responses are paired as they appear in the data tabulation. Let x be the number of times that a family indicates a preference for product A. Under the null hypothesis that the two products are equally preferred, the probability p that A is preferred over B for a given family is $p = .5$. Or equivalently, we wish to test an hypothesis that the binomial parameter p equals .5.

What should be done in case of ties between the paired responses? This difficulty is avoided by omitting tied pairs and thereby reducing the number n of pairs. In this example the effective sample size is 10, not 11.

Very large or very small values of x are certainly contradictory to the null hypothesis. Therefore, the rejection region for the test will be located by including the most extreme values of x that also provide an α that is feasible for the test.

Suppose that we wish α to be about .05 or .10. We would begin the selection of the rejection region by including $x = 0$ and $x = 10$ and calculating the α associated with this region, using $p(x)$ (the probability distribution for the binomial random variable; see Chapter 5). With $n = 10$ and $p = .5$, we refer to Table 1 in the Appendix and find

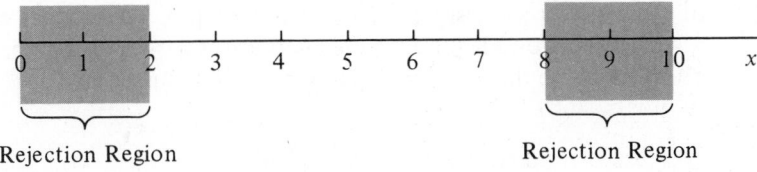

Rejection Region   Rejection Region

Figure 18.1   Rejection Region for Example 18.1

that

$$\alpha = p(0) + p(10) = .002$$

Since this value of α is too small, the rejection region is expanded by including the next pair of x values that are most contradictory to the null hypothesis—namely, $x = 1$ and $x = 9$. The value of α for this rejection region ($x = 0, 1, 9, 10$) is obtained from Table 1 in the Appendix:

$$\alpha = p(0) + p(1) + p(9) + p(10) = .022$$

This α is also small, so we again expand the region. This time we include $x = 0, 1, 2, 8, 9, 10$. You may verify that the corresponding value of α is .11.

At this point we have a choice. We can choose α equal to either .022 or .11, the choice depending on the risk we are willing to take of making a type I error. For this example we choose $\alpha = .11$ and, consequently, employ $x = 0, 1, 2, 8, 9, 10$ as the rejection region for the test. (See Figure 18.1.)

From the data we observe that $x = 2$, and therefore we reject the null hypothesis. We conclude that sufficient evidence exists to indicate that for the population of buyers the two products are not equally preferred. In fact, a greater preference appears to exist for product B, the orange juice with the sugar substitute. The probability of rejecting the null hypothesis when it is true is only $\alpha = .11$, and therefore, we are reasonably confident of our conclusion.   ◆ ◆

In Example 18.1 we used the sign test as a rough tool for detecting a preference between two products. The rather large value of α does not disturb us. We can collect additional data (by allowing more families to conduct the experiment) if we are concerned about making a type I error in reaching our conclusion.

The values of α associated with the sign test can be obtained by using the normal approximation to the binomial probability distribution (as discussed in Section 6.4.) You can verify (by comparing exact probabilities with their approximations) that these approximations will be quite adequate for n as small as 10, owing to the symmetry of the binomial probability distribution for $p = .5$.

For $n \geq 25$ you can conduct the sign test, saving time and effort, by using the z statistic (Chapters 7 and 8) as a test statistic, where

$$z = \frac{x - np}{\sqrt{npq}} = \frac{x - .5n}{.5\sqrt{n}}$$

In using z, you are testing the null hypothesis $p = .5$ against either the alternative $p \neq .5$, for a two-tailed test, or the alternatives $p > .5$ or $p < .5$, for a one-tailed test. The tests utilize the familiar rejection regions discussed in Chapter 8.

Sign Test for Large Samples ($n \geqslant 25$)

Null hypothesis $H_0: p = .5$ (one treatment is not preferred to a second treatment, and vice versa).

Alternative hypothesis $H_a: p \neq .5$, for a two-tailed test. (*Note:* We use the two-tailed test as an example. Many analyses might require a one-tailed test.)

Test statistic: $z = \dfrac{x - .5n}{.5\sqrt{n}}$.

Rejection region: Reject H_0 if $z \geq z_{\alpha/2}$ or if $z \leq -z_{\alpha/2}$, where $z_{\alpha/2}$ is the z value from Table 3 of the Appendix corresponding to an area of $\alpha/2$ in the upper tail of the normal distribution.

EXAMPLE 18.2   A production superintendent claims that there is no difference in the employee accident rates for the day and evening shifts in a large manufacturing plant. The number of accidents per day is recorded for both the day and evening shifts for $n = 100$ days. It is found that the number of accidents x_E per day for the evening shift exceeded the corresponding number of accidents x_D on the day shift on 63 of the 100 days. Do these results provide sufficient evidence to indicate that more accidents tend to occur on one shift than on the other, or equivalently, that $P(x_E > x_D) \neq 1/2$?

SOLUTION   This study is a paired-difference experiment, with $n = 100$ pairs of observations corresponding to the 100 days. To test the null hypothesis that the two distributions of accidents are identical, we use the test statistic

$$z = \frac{x - .5n}{.5\sqrt{n}}$$

where x represents the number of days in which the number of accidents on the evening shift exceeded the number of accidents on the day shift. Then for $\alpha = .05$ we will reject the null hypothesis if $z \geq 1.96$ or $z \leq -1.96$. Substituting into the formula for z, we have

$$z = \frac{x - .5n}{.5\sqrt{n}} = \frac{63 - (.5)(100)}{.5\sqrt{100}} = \frac{13}{5} = 2.60$$

Since the calculated value of z exceeds $z_{\alpha/2} = 1.96$, we reject the null hypothesis. The data provide sufficient evidence to indicate a difference in the accident rate distribution for the day and evening shifts.

◆◆

Note that the data of Tables 18.1 and 18.2 are the result of paired-difference experiments. Suppose that the paired differences for an experiment are normally distributed with a common variance σ^2. Will the sign test detect a shift in location of the two populations as effectively as the Student's t test? Intuitively, we would suspect that the answer is "no," and this answer is correct, because the Student's t test utilizes comparatively more information. In addition to giving the sign of the difference, the t test uses the magnitudes of the observations to obtain a more accurate value of the sample means and variances. Thus, one might say that the sign test is not as "efficient" as the Student's t test, but this statement is meaningful only if the populations conform to the assumption stated above; that is, the differences in paired observations are normally distributed with a variance σ_d^2. The sign test might be more efficient when these assumptions are not satisfied.

The major advantage of the sign test is its ease of application. Furthermore, it is one of the few tests that can be employed when the only information available is that an A observation exceeds a B observation (or vice versa). However, if the data contain more information—specifically, if they can be ranked in order of their magnitudes—better nonparametric tests are available for comparing two population relative frequency distributions. These tests are presented in Sections 18.3 and 18.4.

EXERCISES Conceptual Level

18.1 What is meant by the term *nonparametric statistics*?

18.2 When are nonparametric techniques preferred over the parametric techniques presented in earlier chapters?

18.3 What sampling distribution is appropriate for the sign test when the sample size is small? When the sample size is large?

EXERCISES Skill Level

18.4 Using the tabulated binomial probabilities listed in Table 1 of the Appendix, find the significance levels between $\alpha = .01$ and $\alpha = .15$ that are available when using $n = 10$ paired observations. What are the corresponding rejection regions? Assume that you are using a one-tailed test.

18.5 Repeat the instructions of Exercise 18.4, assuming that you are using a two-tailed test.

18.6 Repeat the instructions of Exercises 18.4 and 18.5 for sample sizes $n = 15, 20$, and 25.

18.7 A random sample of $n = 28$ consumers were given a supply of two different products to be used for ordinary household purposes. Eighteen of the 28 preferred product B to product A, and 3 stated no preference for A or B.

a. If the objective of this study is to determine whether one product is preferred over the other, state H_0 and H_a, and give the test statistic and the appropriate rejection region for $\alpha = .05$.

b. Using the sample observations and the test of part a, what is your conclusion regarding these two products?

18.8 Refer to Exercise 9.41. The data, reproduced next, are from a paired-difference experiment with

$n = 7$ pairs of observations. Use a sign test to determine whether the population distributions for treatments 1 and 2 are different.

	Treatment	
Pair	1	2
1	6.1	4.8
2	9.2	7.4
3	4.1	4.2
4	10.2	8.9
5	9.6	8.6
6	7.6	6.4
7	8.7	7.1

a. State the null and alternative hypotheses for the test.

b. Determine an appropriate rejection region with $\alpha \approx .01$.

c. Calculate the observed value of the test statistic.

d. Does the data present sufficient evidence to indicate that the population distributions for treatments 1 and 2 are different?

e. Compare the results of part d with the results found in Exercise 9.41. Do the results agree? Why or why not?

EXERCISES Applications

18.9 Historically, postmasterships have been considered political patronage positions and were filled by individuals who had demonstrated loyalty to the political party of the current president. In more recent times efforts have been made to ensure that such appointments are filled on the basis of qualification and without regard to political affiliation. During the term of a recent Republican president, 12 postmasterships were filled in the state of Oklahoma. Of those appointed to the posts, 7 were Republicans, 4 were Democrats, and 1 was an independent. Do these appointments suggest that Republicans were more likely to be chosen for vacant postmasterships in Oklahoma than non-Republicans during the term of that president? Test by using a significance level as near as possible to 5%.

18.10 Marketing personnel have usually assumed that a decision by a married couple to purchase a new automobile is primarily motivated by the husband's preferences. Thus, automobile advertising programs generally have been directed to men and not women. Since more women have entered the work force, this assumption has come under question. To investigate this issue, the marketing director for an automobile manufacturer randomly selected the names of 12 married couples and asked each whether the husband or wife was the primary decision maker in their selection of a new automobile. Their responses indicated that in 7 instances the husband was the primary decision maker, in 4 instances the wife was the primary decision maker, and in 1 case the purchase decision was shared by both. Do these results tend to support the traditional belief that new-car decisions by a married couple are more likely to be made by the husband? Test by using a significance level as near as possible to 5%.

18.11 Each of $n = 27$ prominent economists was asked whether he or she believes federal control of domestic oil companies would be in the best interests of the American economy. Fourteen of the economists indicated that they favored controls for oil companies, 11 were opposed to controls, and 2 said they were not certain. Do these responses indicate that economists generally favor federal control of domestic oil companies as a policy in the best interests of the American economy? Test by using a level of significance as near as possible to 5%.

 18.12 The negotiation of an FOB (free-on-board) delivered contract in place of an FOB shipping-point contract may offer cost savings to the user of transportation services since the Uniform Commercial Code (UCC) assumes that FOB contracts are shipping-point contracts unless otherwise specified. In spite of this opportunity for potential cost savings, Decker notes that users often fail to negotiate an FOB delivered contract because of lack of awareness of UCC requirements and policy (R. Decker, "Trim Freight Costs with FOB Awareness," *Purchasing*, March 13, 1980). To examine this issue, a researcher inspected $n = 100$ bills of lading of several interstate haulers; in 42 cases the negotiation of an FOB delivered contract was undertaken by the customers. Do these results suggest that customers generally fail to negotiate an FOB delivered contract? Test by using a level of significance as near as possible to 5%.

 18.13 One effect of a recession is a decline in the nation's industrial production. The data in Table 18.3 show the U.S. production of 10 industrial products during the month of April for two consecutive years. Use a sign test, with a level of significance of 5%, to determine whether these data suggest a decline in national production from year 1 to year 2, and hence, that we had entered a period of economic recession.

Table 18.3   Production Data for Exercise 18.13

	Year	
Production Indicator	1	2
Raw steel (thousands of net tons)	11,200	10,000
Automobiles	493,000	480,400
Trucks	133,000	91,000
Electrical power (millions of kilowatt-hours)	162,000	159,000
Crude oil (daily average, thousands of barrels)	58,000	54,500
Bituminous coal (thousands of net tons)	55,000	61,000
Paperboard (thousands of tons)	2,460	2,550
Paper (thousands of tons)	2,300	2,450
Lumber (millions of board feet)	836	640
Rail freight traffic (billions of ton-miles)	69	65

18.3

The Use of Ranks for Comparing Two Population Distributions: Independent Random Samples

A statistical test for comparing two populations based on independent random samples was proposed by F. Wilcoxon in 1945 (see the references). Suppose that you were to select independent random samples of n_1 and n_2 observations, each from two populations, call them I and II. Wilcoxon's idea was to combine the $n_1 + n_2 = n$ observations and to rank them, in order of magnitude, from 1 (the smallest) to n (the largest). Then if the observations were selected from identical populations, the **rank sums** for the samples should be more or less proportional to the sample sizes n_1 and n_2. For example, if n_1 and n_2 were equal, you would expect the rank sums to be nearly equal. In contrast, if the observations in one population—say population I—tended to be larger than those in population II, then the observations in sample I would tend to receive the highest ranks and would have a larger-than-expected rank sum. Thus (sample sizes being equal), if one rank sum is very large (and correspondingly, the other is very small), you may have evidence to indicate a difference between the two populations.

Mann and Whitney (see the references) proposed a statistical test in 1947 that also used the rank sums of the two samples, and their test can be shown to be equivalent to the Wilcoxon test. Because the **Mann-Whitney U test** and tables of critical values of U occur so often in the literature, we will explain its use in this section.

Assume that you have independent random samples of sizes n_1 and n_2 from two populations—say A and B. The first step in finding the Mann-Whitney U statistic is to rank all $(n_1 + n_2)$ observations in order of magnitude, assigning a 1 to the smallest observation, a 2 to the second smallest, and so on. Ties in the observations can be handled by averaging the ranks that would have been assigned to the tied observations and assigning this average to each. Then calculate the sums of the ranks, T_A and T_B, for the two samples.

For example, suppose that you have random samples of $n_1 = n_2 = 4$ observations selected from each of populations A and B, as shown in Table 18.4. The first step is to assign ranks to the $(n_1 + n_2) = 8$ observations, a rank of (1) to the smallest, 25, a rank of (2) to the next smallest, 27, and so on. The ranks corresponding to these eight measurements are shown in parentheses in Table 18.5. The rank sums for the two samples, $T_A = 12$ and $T_B = 24$, are also shown.

The formula for the Mann-Whitney U statistic can be given in terms of T_A or of T_B; one value of U will be larger than the other, but the sum of the two U values will always equal $n_1 n_2$. Since it is easier to construct a table of probabilities of U for only one tail of the U distribution, we will agree always to use the smaller value of U as a test statistic. The formulas for the two values of U, which we will denote as U_A and U_B are given in the display.

Table 18.4   Independent Random Samples Selected from Populations A and B: $n_1 = n_2 = 4$

A	B
28	33
31	29
27	35
25	30

Table 18.5   Assignment of Ranks to the Data of Table 18.4

A		B	
28	(3)	33	(7)
31	(6)	29	(4)
27	(2)	35	(8)
25	(1)	30	(5)
$T_A = 12$		$T_B = 24$	

Formulas for the Mann-Whitney U Statistic

$$U_A = n_1 n_2 + \frac{n_1(n_1 + 1)}{2} - T_A$$

$$U_B = n_1 n_2 + \frac{n_2(n_2 + 1)}{2} - T_B$$

where $U_A + U_B = n_1 n_2$ and T_A and T_B are the rank sums for samples A and B, respectively.

As you can see from the formulas for U_A and U_B, U_A will be small when T_A is large, a situation that is likely to occur when the population distribution of the A measurements is shifted to the right of the population distribution for the B measurements. Consequently, to conduct a one-tailed test to detect a shift in the A distribution to the right of the B distribution, you will reject the null hypothesis of no difference in the population distributions if U_A is less than some specified value, U_0. That is, you will reject H_0 for small values of U_A. Similarly, to conduct a one-tailed test to detect a shift of the B distribution to the right of the A distribution, you would reject H_0 if U_B is less than some specified value—say U_0. Consequently, the rejection region for the Mann-Whitney U test would appear as shown in Figure 18.2.

Table 7 of the Appendix gives the probability that an observed value of U will be less than some specified value—say U_0. (For more comprehensive tables of the U statistic, see the sources listed in the references.) This value is the value of α for a one-tailed test. To conduct a two-tailed test—that is, to detect a shift in the population distributions for A and B measurements in either direction—we will use U, the smaller of U_A or U_B, as the test statistic and reject H_0 for $U < U_0$ (see Figure 18.2). The value of α for the two-tailed test will be double the tabulated value given in Table 7 of the Appendix.

To see how to locate the rejection region for the Mann-Whitney U test, suppose that $n_1 = 4$ and $n_2 = 5$. Then you would consult the table corresponding to $n_2 = 5$ in Table 7 in the Appendix. The first few lines of Table 7 for $n_2 = 5$ are shown in Table 18.6. Across the top of Table 18.6 you see values of n_1. Values of U_0 are shown down the left side of the table. The entries give the probability that U will assume a small value—namely, the probability $U \le U_0$. Since for this example $n_1 = 4$, you move across the top of the table to $n_1 = 4$. Move to the third row of the table corresponding to $U_0 = 2$ for $n_1 = 4$. Then you see the probability that U will be less than

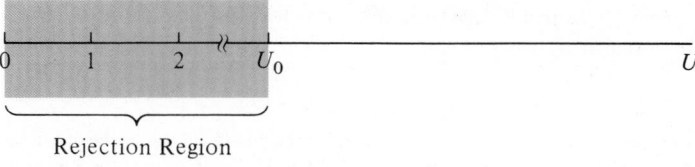

Figure 18.2　Rejection Region for a Mann-Whitney U Test

Table 18.6   An Abbreviated Version of Table 7 in the Appendix;
$P(U \leq U_0)$ for $n_2 = 5$

			n_1		
U_0	1	2	3	4	5
0	.1667	.0476	.0179	.0079	.0040
1	.3333	.0952	.0357	.0159	.0079
2	.5000	.1905	.0714	.0317	.0159
3		.2857	.1250	**.0556**	.0278
4		.4286	.1964	.0952	.0476
5		.5714	.2857	.1429	.0754
⋮			⋮	⋮	⋮

or equal to 2 is .0317. Similarly, moving across the row for $U_0 = 3$, you see that the probability that U is less than or equal to 3 is .0556. (This value is shown in color in Table 18.6.) So if you conduct a one-tailed Mann-Whitney U test with $n_1 = 4$ and $n_2 = 5$ and wish α to be near .05, you would reject the null hypothesis of equality of population relative frequency distributions when $U \leq 3$. The probability of a type I error for the test is $\alpha = .0556$. If you use this same rejection region (that is, $U \leq 3$) for a two-tailed test, α would be double the table value, or $\alpha = 2(.0556) = .1112$.

The procedure for the Mann-Whitney U test is given in the next display.

Mann-Whitney U Test

> Null hypothesis H_0:   The population relative frequency distributions for A and B are identical.
>
> Alternative hypothesis H_a: The two population relative frequency distributions are shifted in respect to their relative locations (a two-tailed test). Or H_a: The population relative frequency distribution for A is shifted to the right of the relative frequency distribution for population B (a one-tailed test).*
>
> Test statistic: For a two-tailed test use U, the smaller of
>
> $$U_A = n_1 n_2 + \frac{n_1(n_1 + 1)}{2} - T_A$$
>
> and
>
> $$U_B = n_1 n_2 + \frac{n_2(n_2 + 1)}{2} - T_B$$
>
> where T_A and T_B are the rank sums for samples A and B, respectively. For a one-tailed test, use U_A.

---

* For the sake of convenience we will describe the one-tailed test as the one designed to detect a shift in the distribution of the A measurements to the right of the distribution of the B measurements. To detect a shift in the B distribution to the right of the A distribution, just interchange the letters A and B in the discussion.

> Rejection region: For the two-tailed test and a given value of α, reject H_0 if $U \leq U_0$, where $P(U \leq U_0) = \alpha/2$. [*Note:* Observe that U_0 is the value such that $P(U \leq U_0)$ is equal to half of α.] For a one-tailed test and a given value of α, reject H_0 if $U_A \leq U_0$, where $P(U_A \leq U_0) = \alpha$.

We illustrate the use of the Mann-Whitney U test with an example.

EXAMPLE 18.3   A large accounting firm with branch offices in several major cities devised two different CPA study programs to assist junior accountants with their preparation for an upcoming CPA examination. To compare the effectiveness of the study programs, the firm selected eight large branch offices, and 50 junior accountants were chosen from each office. The accountants from four offices were trained under program A, the accountants from the other four under program B. Following the completion of the CPA examination, the number of accountants passing the CPA exam was recorded. The results are shown in Table 18.7. Do these data present sufficient evidence to indicate a difference in the population distributions for training programs A and B?

Table 18.7   CPA Exam Data for Example 18.3

A	B
28	33
31	29
27	35
25	30

Table 18.8   Data and Ranks for Example 18.3

A		B	
28	(3)	33	(7)
31	(6)	29	(4)
27	(2)	35	(8)
25	(1)	30	(5)
$T_A = 12$		$T_B = 24$	

SOLUTION   The ranks are shown in Table 18.8 alongside the $(n_1 + n_2) = 8$ measurements. The rank sums for the two samples are also shown. Then

$$U_A = n_1 n_2 + \frac{n_1(n_1 + 1)}{2} - T_A = (4)(4) + \frac{(4)(4 + 1)}{2} - 12 = 14$$

$$U_B = n_1 n_2 - U_A = 16 - 14 = 2$$

Since the question we wish to answer implies a two-tailed test, and since Table 7 gives values of $P(U \leq U_0)$ for specified sample sizes and values of U_0, we must double the tabulated value to find α. Suppose that we desire a value of α near .05. Checking Table 7 for $n_1 = n_2 = 4$, we find $P(U \leq 1) = .0286$. Using $U \leq 1$ as the rejection region, α will equal 2(.0286) = .0572, or rounding to three decimal places, $\alpha = .057$. Because the smaller observed value of U is 2 (calculated above), U does not fall in the rejection region. Hence, there is not sufficient evidence to show a difference in the population distributions for training programs A and B.

◆ ◆

Table 18.9   MINITAB Output for the Data of Example 18.3

```
MTB > SET C1
DATA > 28   31   27   25
DATA > END
MTB > SET C2
DATA > 33   29   35   30
DATA > END
MTB > MANNWHITNEY C1 C2

MANN-WHITNEY CONFIDENCE INTERVAL AND TEST

   C1    N = 4    MEDIAN = 27.500
   C2    N = 4    MEDIAN = 31.500
   POINT ESTIMATE FOR ETA1 - ETA2 IS -4.0006
   97.0 PCT C.I. FOR ETA1 - ETA2 IS (-10.0,    2.0)
   W = 12.0
   TEST OF ETA1 = ETA2 VS. ETA1 N.E. ETA2 IS SIGNIFICANT AT 0.1124

CANNOT REJECT AT ALPHA = 0.05
```

The Mann-Whitney test can be implemented by using the MANN-WHITNEY command in the MINITAB package. As input, the program uses the values from the first and second samples, which are stored in two separate columns. The MINITAB output for the Mann-Whitney test using the data in Example 18.3 is given in Table 18.9. Data summaries appear in lines 1 and 2; point and interval estimates for the difference in population medians are given in lines 3 and 4. The value $W = 12.0$ in line 5 is the sum of the ranks for the values stored in column 1—and in general, for the column number given first in the program command. The value of W can be used to calculate U_A and U_B if desired. The significance level (.1124) of the test is given in line 6. Since the significance level of .1124 is greater than $\alpha = .05$, consistent with our earlier results, we cannot conclude that the underlying population distributions are different.

EXAMPLE 18.4   As a result of the success achieved by the Japanese industry, much concern is now directed toward the relationship between quality control and productivity. To investigate the impact of quality control procedures, a company that manufactures intricate electronic equipment initiated a modern quality control program among the employees of one of its two shifts—call it shift A—involved with assembly operations. The other shift (shift B) used the standard quality control procedures. After one month complete assemblies were examined from the output of each of nine employees randomly chosen from shift A and nine from shift B. The assembly errors for each employee are recorded in Table 18.10. Test the hypothesis of no difference in the location of the distributions of assembly errors that result when using standard control procedures or a modern quality control program.

Table 18.10   Output Data and Ranks
for Example 18.4

Shift A	Shift B
25 (15)	2 (1)
19 (11)	10 (7)
34 (18)	6 (4)
14 (9)	5 (3)
24 (14)	3 (2)
22 (12)	11 (8)
31 (17)	8 (5.5)
29 (16)	17 (10)
23 (13)	8 (5.5)
$T_A = 125$	$T_B = 46$

SOLUTION   We will use the Mann-Whitney U test to test this hypothesis. Thus, for $n_1 = n_2 = 9$ we have

$$U_A = n_1 n_2 + \frac{n_1(n_1 + 1)}{2} - T_A$$

$$= (9)(9) + \frac{(9)(10)}{2} - 125 = 1$$

From Table 7 with $n_1 = n_2 = 9$, the observed level of significance is p-value $= 2\,P[U \leq 1] = 2(.000) = .000$. Hence, we reject the null hypothesis; since p-value $= .000$ we feel very confident in our decision.

◆◆

A simplified large-sample test can be obtained by using the familiar z statistic of Chapter 8. When the population distributions are identical, it can be shown that the U statistic has an expected value and variance of

$$E(U) = \frac{n_1 n_2}{2} \quad \text{and} \quad \sigma_U^2 = \frac{n_1 n_2 (n_1 + n_2 + 1)}{12}$$

The distribution of

$$z = \frac{U - E(U)}{\sigma_U}$$

tends to normality, with mean 0 and variance equal to 1, as n_1 and n_2 become large. This approximation will be adequate when n_1 and n_2 are both larger than, say, 10. Thus, for a two-tailed test with $\alpha = .05$ we would reject the null hypothesis if $|z| \geq 1.96$.

Mann-Whitney U Test for Large Samples ($n_1 > 10$ and $n_2 > 10$)

Null hypothesis H_0:   The population relative frequency distributions for A and B are identical.

Alternative hypothesis H_a: The two population relative frequency distributions are not identical (a two-tailed test). Or H_a: The population relative frequency distribution for A is shifted to the right (or left) of the relative frequency distribution for population B (a one-tailed test).

Test statistic: $z = \dfrac{U - (n_1 n_2/2)}{\sqrt{n_1 n_2 (n_1 + n_2 + 1)/12}}$, and let $U = U_A$.

Rejection region: Reject H_0 if $z \geq z_{\alpha/2}$ or $z \leq -z_{\alpha/2}$ for a two-tailed test. For a one-tailed test place all of α in one tail of the z distribution. To detect a shift in the distribution of the A observations to the right of the distribution of the B observations, let $U = U_A$ and reject H_0 when $z < -z_{\alpha}$. To detect a shift in the opposite direction, let $U = U_A$ and reject H_0 when $z > z_{\alpha}$. Tabulated values of z are given in Table 3 of the Appendix.

What constitutes an "adequate" approximation is a matter of opinion. You will observe that by using the z statistic, you will reach the same conclusion as when using the U test for Example 18.4. Thus,

$$z = \frac{1 - [(9)(9)/2]}{\sqrt{(9)(9)(9 + 9 + 1)/12}}$$

$$= \frac{-39.5}{\sqrt{128.25}} = -3.49$$

This value of z falls in the rejection region ($|z| \geq 1.96$) and hence agrees with the conclusions obtained from the U test.

EXAMPLE 18.5   The U.S. Bureau of the Census reports that all 50 states require taxing jurisdictions to evaluate the effectiveness of property valuations. Valuation is done by an analysis of the ratios of assessed value (A) to sales value (S) for properties contained in the jurisdiction. Ideally, the A/S ratio is constant for all properties in a jurisdiction. When it is not, inequities may result in the property taxation structure.

A tax consultant wishes to compare the ratio of assessed value to sales value for properties in two sections of a city. In the analysis, 24 property sales, $n_1 = 11$ from section 1 and $n_2 = 13$ from section 2, were randomly selected from the two sections of the city. The assessed value for each property was obtained from tax records, and the ratio of assessed value to sales value was computed. The ratios for the two sections of the city are shown in Table 18.11. Ranks of the ratios are shown in parentheses. Do these data provide sufficient evidence to indicate a difference in the distributions of the ratios of assessed value to sales value for properties in the two sections of the city? Test a null hypothesis of no difference, using $\alpha = .05$.

Table 18.11   *A/S Ratio Data for Example 18.5*

Section 1		Section 2	
.55	(9)	.49	(5)
.67	(17.5)	.68	(19)
.43	(1)	.59	(10.5)
.51	(6.5)	.72	(21)
.48	(3.5)	.67	(17.5)
.60	(12)	.75	(22.5)
.71	(20)	.65	(14.5)
.53	(8)	.77	(24)
.44	(2)	.62	(13)
.65	(14.5)	.48	(3.5)
.75	(22.5)	.59	(10.5)
		.51	(6.5)
		.66	(16)
$T_1 = 116.5$		$T_2 = 183.5$	

SOLUTION   Since n_1 and n_2 are both larger than 10, we use the z test. Substituting into the formulas for $E(U)$ and σ_U^2, we obtain

$$E(U) = \frac{n_1 n_2}{2} = \frac{(11)(13)}{2} = 71.5$$

$$\sigma_U^2 = \frac{n_1 n_2 (n_1 + n_2 + 1)}{12} = \frac{(11)(13)(11 + 13 + 1)}{12} = 297.9167$$

$$\sigma_U = \sqrt{297.9167} = 17.26$$

The observed value of U is

$$U_2 = n_1 n_2 + \frac{n_2(n_2 + 1)}{2} - T_2 = (11)(13) + \frac{(13)(14)}{2} - 183.5 = 50.5$$

We will test the null hypothesis that the distributions of ratios of assessed values to sales values for the two sections of the city are identical against the alternative that they differ. Since this alternative implies a two-tailed test, we will reject the null hypothesis if $|z| \geq 1.96$ ($\alpha = .05$).

The observed value of the test statistic is

$$z = \frac{U - E(U)}{\sigma_U} = \frac{50.5 - 71.5}{17.26} = -1.22$$

Since z does not fall in the rejection region, there is not sufficient evidence to indicate a difference in the ratios of assessed value to sales value for properties located in the two sections of the city.

Why use the Mann-Whitney U test for Example 18.5 rather than the parametric test employing the Student's t? Distributions of ratios often tend to be nonnormal, and hence, there is a good possibility that the requirements of the Student's t test are not met. Parametric test procedures, using the z statistic (Section 8.14) or the t statistic (Section 9.4), are appropriate to test the hypothesis of no difference between the two population distributions if the assumptions underlying the tests are satisfied. If you have doubts on this point, the Mann-Whitney U test is a very good alternative.

EXERCISES   Conceptual Level

18.14   The Wilcoxon test and the Mann-Whitney U test are both based on rank sums for samples. Under what circumstances would it be appropriate to use these rank sum tests?

18.15   State the null hypothesis that is tested in using the Mann-Whitney U test. What types of alternative hypotheses are available to the experimenter?

18.16   Why do we choose to use the smaller of U_A and U_B as the test statistic for the Mann-Whitney U test?

EXERCISES   Skill Level

18.17   Two random samples have been independently drawn from populations A and B, whose distributions are known to be nonnormal. A Mann-Whitney U test will be used to determine whether or not population A lies to the right of population B, using U_A as the test statistic. Find the appropriate rejection regions in the following situations.

    a. $n_1 = 7, n_2 = 8, \alpha \approx .05$ **b.** $n_1 = 10, n_2 = 6, \alpha \approx .01$

    c. $n_1 = n_2 = 15, \alpha \approx .05$

18.18   Refer to Exercise 18.17. If you must determine whether or not there is a difference between the distributions for populations A and B, give the appropriate values of α assuming that the rejection regions found in Exercise 18.17 are used for the test.

18.19   Find the value of α when using a Mann-Whitney U test in the following situations.

    a.   Two-tailed test; $n_1 = 6, n_2 = 5$; reject H_0 if the smaller of U_A and U_B is less than or equal to 4.

    b.   One-tailed test; $n_1 = n_2 = 8$; reject H_0 if U_A is less than or equal to 19.

18.20   Refer to Exercise 9.21. Observations from two random and independent samples, drawn from populations 1 and 2, are reproduced below.

Sample 1	Sample 2
1	4
3	7
2	6
3	8
5	6

Use the Mann-Whitney U test to determine whether population 1 is shifted to the left of population 2.

a. State the null and alternative hypotheses to be tested.

b. Find the rejection region for $\alpha \approx .05$.

c. Rank the combined sample from smallest to largest. Calculate T_A, T_B, U_A, and U_B.

d. What is the observed value of the test statistic?

e. Does the data provide sufficient evidence to indicate that population 1 is shifted to the left of population 2?

f. Compare the results of part e with those found in Exercise 9.21. Do the results agree? Why or why not?

18.21 Independent random samples of size $n_1 = 20$ and $n_2 = 25$ are drawn from nonnormal populations A and B. The combined sample is ranked, and $T_A = 252$ and $T_B = 783$. Use the large-sample approximation to the Mann-Whitney U test to determine whether there is a difference in the two population distributions. Calculate the observed significance level, or p-value, for the test.

EXERCISES Applications

18.22 E. Goetz, T. Tyler, and F. Cook ("Promised Incentives in Media Research: A Look at Data Quality, Sample Representativeness, and Response Rate," *Journal of Marketing Research*, May 1984) examined the effects of using promised incentives to increase compliance in media research. To examine viewer perception of a television program, they asked viewers in nine communities to evaluate the program, and these viewers were not offered any incentive. Viewers in nine other communities were each offered $5 to evaluate the program. The percentage of viewers responding to the survey from each community involved in the study is given in the accompanying table.

No Incentive	51	82	47	50	38	61	56	40	45
$5 Incentive	81	80	76	66	70	60	52	88	85

Do these data present sufficient evidence to indicate that the response rate improves when the $5 incentive is used? Use $\alpha = .05$.

a. Use the Mann-Whitney U test.   **b.** Use the Student's t test.

18.23 The sales of new homes are intrinsically tied to interest rates. High interest rates mean that mortgage funds are in short supply and, when available, are very costly to the borrower. A real estate developer is interested in building speculative homes in one of two communities. Knowing that interest rates vary from region to region, the contractor has recorded the interest rate for a conventional $80,000, thirty-year first mortgage from five financial institutions in community A and from seven in community B. These data are shown in the accompanying table. Do these data suggest that the average interest rate for a conventional $80,000, thirty-year first mortgage differs between the two communities? Test by using a 10% level of significance. What do these results suggest regarding which community the contractor should select for building the speculative houses?

Community A	10.6	10.9	10.6	10.7	10.5		
Community B	10.6	10.8	11.0	10.9	11.1	11.2	11.0

18.24 Many studies have suggested that worker productivity can be enhanced when workers feel that they have a stake in the ownership and management of the firm. T. Hammer and R. Stern examined this issue by measuring the perceptions by worker-owners and managers of their role as owners, financial partners, and joint decision makers. Interviews were conducted in several firms among workers and managers who owned at least 100 shares of stock in the firm in which they were employed. Each person was asked to respond on a six-point scale to the following statement: "Management and workers are equal partners in this company." The responses of worker-owners and managers in one firm are shown in the accompanying table, with a response of 1 implying strong disagreement and 6 implying strong agreement.

Worker-Owners	4.5	3.0	3.2	3.7	2.4	1.8	4.0
Managers		5.0	4.0	4.6	5.5	4.4	

Source: T. Hammer and R. Stern, "An Employee Ownership: Implications for the Organizational Distribution of Power," *Academy of Management Journal*; March 1980. Reprinted with permission.

a. Use the value of W given in the MINITAB printout in Table 18.12 to calculate U_A and U_B.

b. Do these data indicate that the perceptions of worker-owners and managers differ regarding the distribution of power within the firm? Use $\alpha \approx .05$.

c. Use the MINITAB printout in Table 18.12 to find the observed level of significance for the test. Do the MINITAB results agree with the results obtained in part b?

Table 18.12 MINITAB Output for the Data of Exercise 18.24

```
MTB > MANNWHITNEY C1 C2

MANN-WHITNEY CONFIDENCE INTERVAL AND TEST

C1          N =   7     MEDIAN =        3.2000
C2          N =   5     MEDIAN =        4.6000
POINT ESTIMATE FOR ETA1-ETA2 IS       -1.4000
96.5  PCT C.I. FOR ETA1-ETA2 IS (    -2.60,     -0.30)
W =        30.5
TEST OF ETA1 = ETA2  VS.  ETA1 N.E. ETA2 IS SIGNIFICANT AT 0.0185
```

18.25 Companies requiring extreme precision in assembly operations, especially those in high-technology industries, often find that the rate of productivity differs considerably between male and female employees involved in assembly work. To examine this issue, an electronics manufacturer observed the time for completion of an intricate electronics component by 13 women and 12 men employed as assemblers. The order in which the 25 employees successfully assembled the component was as follows:

W W W M W M M W W W W M W M M M W M M W W W M M M

(That is, a woman was first to successfully assemble the product, another woman completed the assembly in the second shortest time, and so on.) Using the large-sample Mann-Whitney U test, determine whether these data suggest that there is a difference in assembly times required by men and women to assemble the electronic component. Use $\alpha = .05$. What do these data suggest regarding the comparative productivity of male and female assemblers in this company?

18.4

Comparing $t > 2$ Population Distributions: Independent Random Samples

Kruskal-Wallis test

As an extension of the Mann-Whitney U test for comparing two population distributions, the **Kruskal-Wallis test** provides for a comparison of **more than two** population distributions. In essence, the Kruskal-Wallis test is a one-way analysis of variance procedure based on ranked data and is the nonparametric counterpart to the completely randomized design described in Section 10.3. It is to be used in instances when the sample observations do not satisfy the requirements of the F test, such as those involving ordinal data or instances when the probability distribution of observations is distinctly nonnormal in appearance.

To illustrate the Kruskal-Wallis test, consider, for example, a study conducted prior to the Tax Reform Act of 1986 to compare the use of tax shelters among executives in four major industries: banking, broadcasting, electronics, and retailing. Table 18.13 lists the federal tax paid, as a percentage of the total income, for 21 executives from these four industries. We assume that the high-salaried executives would be in the 50% marginal federal tax bracket without tax shelters and that the federal tax rate for each executive is reduced according to the executive's use of tax shelters.

The question of interest is whether the use of tax shelters varied among the executives in the four major industries under investigation. More specifically, we wish to test whether the differences observed among the independent samples signify genuine population differences, or whether they represent random variation that might be expected if all observations were chosen from a common population. We will use the Kruskal-Wallis test and not the parametric F test since distributions of percentages often tend to appear nonnormal.

For the tax shelter study the Kruskal-Wallis test examines these hypotheses:

H_0: All four populations (industries) have identical probability distributions.

H_a: At least one population has a probability distribution different from the other three.

Table 18.13   Federal Tax as a Percentage of Total Income for 21 Executives

Banking	Broadcasting	Electronics	Retailing
16	41	32	22
37	39	30	25
21	35	38	33
29	46	28	19
	40	47	20
	43		27

Table 18.14   Ranked Observations for the Tax Shelter Study

Banking	Broadcasting	Electronics	Retailing
1	18	11	5
14	16	10	6
4	13	15	12
9	20	8	2
	17	21	3
	19		7

The Kruskal-Wallis test is based upon the *ranks* of n sample observations. All observations from each of the t separate samples are combined and then ranked in a single series. Then the n sample observations are replaced by their respective ranks. Completing this task for the data of Table 18.13, we note that the ranked observations within banking and retailing appear to be smaller than those in broadcasting and electronics (see Table 18.14). But is this difference due to chance, or is it due to a distinct difference among the populations? That is, did a real difference exist in the use of tax shelters for executives among these four industries?

If we denote R_j as the sum of the ranks from the jth sample (column), then the Kruskal-Wallis test statistic is given by

$$H = \frac{12}{n(n+1)} \sum_{j=1}^{t} \frac{R_j^2}{n_j} - 3(n+1)$$

where n_j is the number of observations in the jth sample and n is the total number of observations. When the null hypothesis is true, H is distributed approximately as a chi-square (χ^2) distribution with $(t-1)$ degrees of freedom.

In the tax shelter study we note that $n = 21$, $n_1 = 4$, $n_2 = 6$, $n_3 = 5$, and $n_4 = 6$; the rank sums are $R_1 = 28$, $R_2 = 103$, $R_3 = 65$, and $R_4 = 35$. Therefore, the Kruskal-Wallis test statistic is

$$H = \frac{12}{21(22)} \left[\frac{(28)^2}{4} + \frac{(103)^2}{6} + \frac{(65)^2}{5} + \frac{(35)^2}{6} \right] - 3(22) = 12.268$$

From Table 5 of the Appendix the critical value of χ^2 for $\alpha = .05$ and $(t-1) = 3$ degrees of freedom is 7.81473. Since the test statistic $H = 12.268$ exceeds this critical value, we reject the null hypothesis and assume that tax shelters were not utilized equally among executives from banking, broadcasting, electronics, and retailing. From the data it appears that tax shelters were used more extensively among banking and retailing executives than among executives from broadcasting and electronics.

The Kruskal-Wallis test is an easy-to-apply alternative to the completely randomized design studied in Section 10.3. It should be used in all cases involving the comparison of three or more population distributions when doubt exists about the appropriateness of using the parametric F test or when the sample observations comprise ordinal measurements.

Kruskal-Wallis Test for Comparing $t > 2$ Population Distributions

> Null hypothesis H_0:   The relative frequency distributions of all of the $t > 2$ populations are identical.
>
> Alternative hypothesis H_a: At least two of the t relative frequency distributions differ.
>
> Test statistic: $H = \dfrac{12}{n(n + 1)} \sum_{j=1}^{t} \dfrac{R_j^2}{n_j} - 3(n + 1).$
>
> Rejection region: Reject H_0 if $H > \chi_\alpha^2$ with $(t - 1)$ degrees of freedom.

The only difficulty that may arise when one uses the Kruskal-Wallis test is the occurrence of tied sample observations. In a manner similar to that suggested for the Mann-Whitney U test, ties are handled by averaging the ranks that would have been assigned to the tied observations and, instead, assigning the rank averages.

EXAMPLE 18.6   In the introduction of new consumer products effective packaging is critical to gaining initial product acceptance. Perceptions about packaging often affect the consumer's judgment of product quality. Four different packages for a new soft drink are separately test-marketed by providing the product free to 19 different families. At the end of the test period each family was asked to provide its opinion of product quality on a five-point scale, with a score of 1 implying complete disfavor with the product and 5 complete approval. The data are shown in Table 18.15. Do these data imply a difference in perceived product quality among the users of the four packages? Let $\alpha = .05$.

Table 18.15   Package Design Observations for Example 18.6

A	B	C	D
4.2	3.3	1.9	3.5
4.6	2.4	2.4	3.1
3.9	2.6	2.1	3.7
4.0	3.0	1.7	4.4
	3.8	2.7	4.1

SOLUTION   In this case we have $n = 19$ observations that must be collectively ranked, with the rank of 1 assigned to the smallest observation, 1.7. A ranking of the data is shown in Table 18.16, along with the summed ranks for each column. Applying the Kruskal-Wallis formula for the test statistic H, we find

$$H = \frac{12}{19(20)} \left[\frac{(65)^2}{4} + \frac{(41.5)^2}{5} + \frac{(17.5)^2}{5} + \frac{(66)^2}{5} \right] - 3(20) = 13.678$$

Table 18.16   Ranks of Observations for Example 18.6

A	B	C	D
17	10	2	11
19	4.5	4.5	9
14	6	3	12
15	8	1	18
	13	7	16
$R_1 = 65$	$R_2 = 41.5$	$R_3 = 17.5$	$R_4 = 66$

From Table 5 in the Appendix the critical value of χ^2 for $\alpha = .05$ with $(t - 1) = 3$ degrees of freedom is 7.81473. Since $H = 13.678$ exceeds the critical value from Table 5, sufficient evidence exists to assume that perceived quality based on package design differs among the four designs.

EXERCISES   Conceptual Level

18.26   The Kruskal-Wallis test is the nonparametric analogue of which parametric technique?

18.27   Under what conditions would you elect to use the Kruskal-Wallis procedure rather than its parametric analogue?

EXERCISES   Skill Level

18.28   Determine the appropriate rejection region for a Kruskal-Wallis test under the following conditions.

    a. $t = 5, \alpha = .05$

    b. $t = 3, \alpha = .01$

    c. $t = 6, \alpha = .05$

18.29   Independent random samples of sizes $n_1 = 5, n_2 = 6, n_3 = 6$, and $n_4 = 7$ are drawn from four nonnormal populations and produce $R_1 = 75, R_2 = 27, R_3 = 92$, and $R_4 = 106$. Use the Kruskal-Wallis test to determine whether the four populations are identical.

    a.   State the null and alternative hypotheses for the test.

    b.   Calculate the observed value of the test statistic.

    c.   Calculate the approximate p-value for the test.

    d.   Use the results of part c to determine whether the data present sufficient evidence to conclude that the four populations are different.

18.30   Refer to Exercise 10.9. The data, reproduced next, resulted from an experiment run in a completely randomized design in which each of four treatments was replicated five times.

Sample 1	Sample 2	Sample 3	Sample 4
6.9	8.3	8.0	5.8
5.4	6.8	10.5	3.8
5.8	7.8	8.1	6.1
4.6	9.2	6.9	5.6
4.0	6.5	9.3	6.2

a. Use the Kruskal-Wallis test to determine whether there are significant differences among the four populations. Use $\alpha = .05$.

b. Find the approximate p-value for the test.

c. Compare the results of parts (a) and (b) with the results obtained in Exercise 10.9.

EXERCISES   Applications

18.31   As a result of studies by T. Peters and R. Waterman [*In Search of Excellence* (New York: Harper & Row, 1982)] and other researchers, much attention has been directed to the issue of management style as an important contributor to organizational success. A management consultant sought to compare the degree of authoritarianism among presidents of corporations from four different industries. Table 18.17 lists the measures of authoritarianism for 20 corporate presidents from public accounting, communications, publishing, and transportation, using an authoritarianism scale common among behavioral scientists. Do these data suggest that a difference exists in the level of authoritarianism of corporate presidents for these four industries? Use $\alpha = .05$.

Table 18.17   Authoritarianism Data for Exercise 18.31

Public Accounting	Communications	Publishing	Transportation
88	99	127	77
79	101	119	103
83	82	100	90
106	116	124	85
94	112	121	87

18.32   Much legitimate debate exists about proper training for productive sales personnel. Some claim that sales ability is the result of an education emphasizing business and marketing, and others claim successful sales ability depends more on personal initiative and the drive to compete and succeed. To examine this issue, the sales manager for a national distributor of hardware ranked the 16 sales agents under her supervision according to achievement of sales goal (1 = best) and then categorized them according to educational background. These data are given in Table 18.18. Do these data provide sufficient evidence to assume a difference in sales ability exists among the sales agents with different educational backgrounds? Use $\alpha = .05$.

Table 18.18   Ranked Data for Exercise 18.32

Business Degree	Liberal Arts Degree	No College Degree
1	3	6
2	8	7
4	15	13
5	10	11
9	14	12
		16

 18.33   Refer to Exercise 10.16 and use the Kruskal-Wallis test to determine whether the data provide sufficient evidence to conclude that rates of return differ among the three investment firms. Use $\alpha = .05$. Explain any difference in the conclusions you reach when using the F test of Section 10.3 and the Kruskal-Wallis test.

18.5

The Wilcoxon Signed-Rank Test for a Paired Experiment

A signed-rank test proposed by Wilcoxon can be used to analyze the paired-difference experiment of Section 9.5 by considering the paired differences corresponding to treatments A and B. Under the null hypothesis of no difference in the distributions for A and B, you would expect (on the average) half of the differences in pairs to be negative and half to be positive. That is, the expected number of negative differences would be $n/2$ (where n is the number of pairs). Furthermore, positive and negative differences of equal absolute magnitude should occur with equal probability. For example, the probability of a difference equal to $+2$ is the same as the probability of a difference equal to -2. If you were to order the differences according to their absolute values and rank them from smallest to largest, the expected rank sums for the negative and positive differences should be equal. Sizable differences in the rank sums of the positive and negative differences would provide evidence to indicate a difference between the distributions of responses for the two treatments A and B,

Calculation of the Test Statistic and the Wilcoxon Signed-Rank Test

1. Calculate the differences $(x_A - x_B)$ for each of the n pairs. Differences equal to zero are eliminated and the number of pairs, n, is reduced accordingly.

2. Rank the **absolute values** of the differences, assigning 1 to the smallest, 2 to the second smallest, and so on. Tied observations are assigned the average of the ranks that would have been assigned with no ties.

3. Calculate the **rank sum** for the **negative** differences and label this value T^-. Similarly, calculate T^+, the **rank sum** for the **positive** differences.

> 4. For a **two-tailed test** we use the *smaller* of these two quantities, T, as the test statistic to test the null hypothesis that the two population relative frequency histograms are identical.

In a two-tailed test the smaller the value of T, the greater will be the evidence favoring rejection of the null hypothesis. **Hence, we will reject the null hypothesis if T is less than or equal to some critical value, say T_0.**

To detect the one-sided alternative, that the distribution of the A observations is shifted to the right of the B observations, we use the rank sum T^- of the negative differences and reject the null hypothesis for small values of T^-, say $T^- \le T_0$. If we wish to detect a shift of the distribution of B observations to the right of the distribution of A observations, we use the rank sum T^+ of the positive differences as a test statistic and reject for small values of T^+, say $T^+ \le T_0$.

The probability that T is less than or equal to some value, T_0, has been calculated for various sample sizes and values of T_0. These probabilities, given in Table 8 of the Appendix, can be used to find the rejection region for the T test. An abbreviated version of Table 8 is shown in Table 18.19. Across the top of the table you see the number n of differences (the number of pairs). Values of α (denoted by the symbol P in Wilcoxon's table) for a one-tailed test appear in the first column of the table. The second column gives values of α (P in Wilcoxon's notation) for a two-tailed test. Table entries are the critical values of T. The critical value of a test statistic is the value that locates the upper (or lower) extreme of the rejection region.

For example, suppose you have $n = 7$ pairs and you are conducting a two-tailed test of the null hypothesis that the two population relative frequency distributions are identical. Checking the $n = 7$ column of Table 18.19 and using the second row (corresponding to $P = \alpha = .05$ for a two-tailed test), you see the entry 2 (shown in color in Table 18.19). This value is T_0, the critical value of T. As noted earlier, the smaller the value of T, the greater will be the evidence to reject the null hypothesis. Therefore, you will reject the null hypothesis for all values of T less than or equal to 2. The rejection region is shown in Figure 18.3.

Table 18.19   An Abbreviated Version of Table 8 in the Appendix: Critical Values of T

One-sided	Two-sided	$n = 5$	$n = 6$	$n = 7$	$n = 8$	$n = 9$	$n = 10$
$P = .05$	$P = .10$	1	2	4	6	8	11
$P = .025$	$P = .05$		1	2	4	6	8
$P = .01$	$P = .02$			0	2	3	5
$P = .005$	$P = .01$				0	2	3

One-sided	Two-sided	$n = 11$	$n = 12$	$n = 13$	$n = 14$	$n = 15$	$n = 16$
$P = .05$	$P = .10$	14	17	21	26	30	36
$P = .025$	$P = .05$	11	14	17	21	25	30
$P = .01$	$P = .02$	7	10	13	16	20	24
$P = .005$	$P = .01$	5	7	10	13	16	19

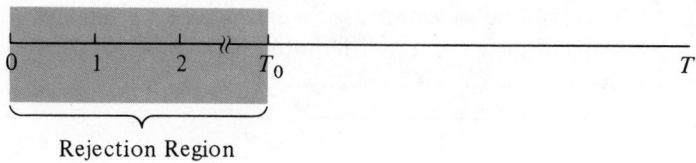

Rejection Region

Figure 18.3   Rejection Region for the Wilcoxon Signed-Rank Test for a Paired Experiment

Wilcoxon Signed-Rank Test for a Paired Experiment

Null hypothesis H_0:   The two population relative frequency distributions are identical.

Alternative hypothesis H_a: The two population relative frequency distributions are not identical (a two-tailed test). Or the relative frequency distribution for population A is shifted to the right (or left) of the relative frequency distribution for population B (a one-tailed test).

Test statistic: T, the smaller of the rank sum for positive differences and the rank sum for negative differences.

Rejection region: Reject H_0 if $T \le T_0$, where T_0 is the critical value given in Table 8 of the Appendix.

EXAMPLE 18.7   Effective marketing requires knowledge of the seller's markets, because all marketing decisions are based to some degree on how wholesalers, distributors, and consumers will react to those decisions. To provide information useful in marketing decision making, a frozen foods manufacturer undertook a study to compare consumer opinion of her line of frozen dinners (A) with those of her most active competitor (B). The study was conducted by having the purchasing managers of each of six nationwide supermarket chains respond on a 10-point Likert scale* according to their evaluation of the two products. In their analysis the respondents were asked to consider price, package design, variety, company promotion, and time before spoilage when scoring each product on a Likert scale. The results are shown in Table 18.20. Do these data indicate that the population distributions of scaled responses differ significantly between product A and product B?

---

* A Likert scale is an instrument that associates ordinal values with qualitative attributes. In this example the attribute is "measure of acceptability," for which the associated Likert scale would appear as follows:

Impression: Very poor   Acceptable   Outstanding

1	2	3	4	5	6	7	8	9	10

The respondent is then asked to mark his or her response on the scale. The mark defines an assumed measure of acceptability for the product in question.

Table 18.20   Response Data for Example 18.7

Respondent	Likert Score		Difference $(A - B)$	Rank
	A	B		
1	6.5	4.0	2.5	3
2	4.0	8.0	-4.0	4
3	5.5	6.5	-1.0	1.5
4	5.5	10.0	-4.5	5
5	7.0	8.0	-1.0	1.5
6	4.0	9.5	-5.5	6

SOLUTION   The differences, $x_A - x_B$, are calculated and ranked according to their absolute value, as shown in Table 18.20. The sum of the positive ranks is $T^+ = 3$ and the sum of the negative ranks is $T^- = 18$. The smaller rank sum is that for positive differences, and hence $T = 3$. From Table 8 the critical value of T for a two-tailed test with $\alpha = .10$ is $T_0 = 2$. Since the observed value of T exceeds the critical value of T, the null hypothesis of identical population relative frequency distributions is not rejected. There is not sufficient evidence to indicate a significant difference in the population distributions of scaled responses associated with the two products.   ◆ ◆

The MINITAB package includes a command WTEST for implementing the Wilcoxon signed-rank test for a paired experiment. The initial input and output for this program using the data in Example 18.7 is given in Table 18.21. The paired scores are

Table 18.21   MINITAB Output for the Data of Example 18.7

```
MTB > SET C1
DATA > 6.5 4 5.5 5.5 7 4
DATA > END
MTB > SET C2
DATA > 4 8 6.5 10 8 9.5
DATA > END
MTB > LET C3 = C1 - C2
MTB > PRINT C3
C3
    2.5    -4.0    -1.0    -4.5    -1.0    -5.5

MTB > WTEST 0 C3

TEST OF MEDIAN = 0 VERSUS MEDIAN N.E. 0

                 N FOR    WILCOXON               ESTIMATED
            N    TEST     STATISTIC    P-VALUE    MEDIAN
C3          6      6         3.0        0.142     -2.500
```

stored in columns 1 and 2, and their differences are set in column 3. The command WTEST 0 C3 specifies that the Wilcoxon signed-rank procedure will be used to test for a zero median using the values in column 3. Notice that none of the six differences was equal to zero, so all six were used in calculating the value of the test statistic given as 3.0. Since the p-value of .142 exceeds $\alpha = .05$, there is not sufficient evidence to conclude that the underlying population distributions differ significantly.

When n is large (say 25 or more), T is approximately normally distributed, with a mean and a variance of

$$E(T) = \frac{n(n+1)}{4} \quad \text{and} \quad \sigma_T^2 = \frac{n(n+1)(2n+1)}{24}$$

Then the z statistic is

$$z = \frac{T - E(T)}{\sigma_T}$$

and can be used as a test statistic. The test is conducted in exactly the same manner as for the large-sample, Mann-Whitney U test. Thus, for a two-tailed test and $\alpha = .05$ we reject the hypothesis of identical population distributions when $|z| \geq 1.96$.

Large-Sample, Wilcoxon Signed-Rank Test for a Paired Experiment ($n \geq 25$)

Null hypothesis H_0:   The population relative frequency distributions for A and B are identical.

Alternative hypothesis H_a: The two population relative frequency distributions differ in location (a two-tailed test). Or H_a: The population relative frequency distribution for A is shifted to the right (or left) of the relative frequency distribution for population B (a one-tailed test).

Test statistic: $z = \dfrac{T - [n(n+1)/4]}{\sqrt{[n(n+1)(2n+1)]/24}}$, and T can be either T^+ or T^-.

Rejection region: Reject H_0 if $z \geq z_{\alpha/2}$ or $z \leq -z_{\alpha/2}$ for a two-tailed test. For a one-tailed test, place all of α in one tail of the z distribution. To detect a shift in the distribution of the A observations to the right of the distribution of the B observations, let $T = T^+$ and reject H_0 when $z > z_\alpha$. To detect a shift in the opposite direction, let $T = T^-$ and reject H_0 if $z < -z_\alpha$. Tabulated values of z are given in Table 3 of the Appendix.

If any doubt exists about the applicability of the use of the t test for a paired experiment, the Wilcoxon test offers a very efficient alternative.

EXERCISES   Conceptual Level

18.34   When is it appropriate to use the Wilcoxon signed-rank test?

18.35   If an analysis is performed in a paired experiment by using both a sign test and the Wilcoxon signed-rank test, is it possible for the test results to be different? (Assume α is the same in each test.) If the results are different, what would you conclude?

EXERCISES	Skill Level

18.36   A paired-difference experiment for $n = 6$ randomly selected pairs of observations produced the differences shown below. Use Wilcoxon's signed-rank test to determine whether the two populations are different.

   1.2   3.2 -3.4 -5.1 2.6   2.7

   a. State the null and alternative hypotheses for the test.

   b. Rank the observed differences according to their absolute value, and calculate T^+ and T^-. What is the observed value of the appropriate test statistic?

   c. Do the data provide sufficient evidence to indicate that there is a difference between the two populations? Use $\alpha = .10$.

18.37   A paired-difference experiment for $n = 10$ randomly selected pairs of observations produced the differences $(A - B)$ shown below. Use Wilcoxon's signed-rank test to determine whether population A is shifted to the right of population B.

   0   1.5   4.2 -3.2 0   2.1   2.4   1.5 -2.1 1.5

   a. State the null and alternative hypotheses for the test.

   b. Rank the observed differences according to their absolute value, and calculate T^+ and T^-. What is the observed value of the appropriate test statistic?

   c. Do the data provide sufficient evidence to indicate that population A is shifted to the right of population B? Use $\alpha = .05$.

18.38   A paired-difference experiment for $n = 35$ randomly selected pairs of observations produced $T^- = 288$ and $T^+ = 342$. Use Wilcoxon's signed-rank test to determine whether the two populations are different. Use $\alpha = .01$.

18.39   Refer to Exercise 9.41. The data reproduced in Table 18.22 are from a paired-difference experiment with $n = 7$ pairs of observations.

   a. Use Wilcoxon's signed-rank test to determine whether there is a significant difference in the two populations.

   b. Compare the results of part a with result obtained in Exercises 18.8 and 9.41. Interpret these results.

Table 18.22   Differences Data for Exercise 18.39

Pair	Treatment	
	1	2
1	6.1	4.8
2	9.2	7.4
3	4.1	4.2
4	10.2	8.9.
5	9.6	8.6
6	7.6	6.4
7	8.7	7.1

EXERCISES   Applications

18.40   Refer to Exercise 18.22. Suppose the media research study is conducted by surveying viewers in only *nine* communities but that within each community viewers are divided into two groups, one receiving a $5 incentive, the other group receiving no incentive. The percentage of viewers responding to the media research survey within each community is given in the accompanying table. Do these data provide sufficient evidence to indicate that an incentive improves the response rate in media research? Let $\alpha = .05$. Explain the difference between the design used in this exercise and the one used in Exercise 18.22.

Community	1	2	3	4	5	6	7	8	9
No Incentive	55	66	50	64	58	55	48	56	66
$5 Incentive	70	60	72	81	66	55	62	66	69

18.41   Local public assessment offices are frequent users of statistical and computer-assisted procedures to maintain property valuations that are fair representations of market value. Such procedures allow the offices to update property valuations rapidly during times of inflation without incurring the expense of drastically increasing the size of the assessment force. A county assessment office has installed a computer-assisted property valuation program and would like to compare the efficiency of the new program with valuations generated by members of the current assessment staff. The assessment-sales ratio, the common measure of assessment accuracy used by all assessment offices, was computed for $n = 10$ recently sold properties from the county. Two assessment-sales ratios were computed for each property, one based on the computer assessment and the other based on the assessment provided by the assessor. These data are shown in Table 18.23.

a.   Do these data suggest that the mean assessment-sales ratio differs according to whether the computer model or the assessor is used to assess values of properties in the county? Use $\alpha = .10$.

b.   If an ideal assessment-sales ratio is 1.0, implying that the assessment is a perfectly accurate prediction of the selling price, do the data provide sufficient evidence to indicate that the assessments generated by one of the two methods come closer to the ideal assessment-sales ratio than assessments generated by the other?

Table 18.23   Assessment Ratio Data for Exercise 18.41

Assessment Method	Property									
	1	2	3	4	5	6	7	8	9	10
Computer model	.92	.98	.90	1.05	1.01	.95	1.10	1.05	.96	1.10
Assessor	.88	.92	.95	.95	.93	.97	.97	.98	.92	.99

18.42   As part of a collective-bargaining agreement, a corporation has agreed to offer a new health insurance plan to its employees. Two large insurance companies have each presented a health insurance plan to management; both are within the guidelines specified in the collective-bargaining agreement and should be equally costly to the corporation. To obtain workers' attitudes toward the two competing plans, the personnel officer randomly selected eight employees from the employee registry. Both insurance plans were explained in detail to each employee in the sample. The employees were then asked to rate each plan on a 10-point Likert scale, with 1 indicating the plan is completely unacceptable and 10 indicating it is perfectly acceptable. The recorded responses are shown in Table 18.24. Do these data suggest that the employees' attitudes differ significantly toward these two insurance plans? Use $\alpha = .10$.

Table 18.24   Response Data for Exercise 18.42

Insurance Plan	Employee							
	1	2	3	4	5	6	7	8
A	8.0	7.5	7.0	8.5	9.5	9.0	8.5	8.0
B	5.0	4.0	9.0	7.5	8.0	7.0	6.0	8.0

18.6

The Friedman Test for a Randomized Block Design

randomized block design
treatments
blocks

As described within Section 10.7, the **randomized block design** is an extension of the paired-difference design that allows for the comparison of t population means. In this design the t populations are called **treatments**. To compare the treatments, we randomly assign experimental units within **blocks** of relatively homogeneous experimental material in an attempt to block out, or isolate, response variation not attributable to treatment differences.

Friedman test

The **Friedman test** provides a nonparametric counterpart to the randomized block design F test described in Section 10.8. With the F test we must assume that the observations taken from each of the t treatment populations are normally distributed and that the treatment variances are equal. **No distributional assumptions are necessary to use the Friedman test to compare the t treatments.**

When the Friedman test is used, all of the t treatments are assigned in a random manner within each of the b blocks (hence, the name randomized block design). After the observations are recorded for each treatment-block combination, the data are displayed within a two-way table in which the rows represent the blocks and the columns represent treatments. Our data table then consists of b rows and t columns. Within each row (block) the data are ranked. The Friedman test then provides an analysis of the column (treatment) rank sums, in a manner similar to that for the Kruskal-Wallis test.

The null and alternative hypotheses for the Friedman test are as follows:

H_0: The probability distributions for the t treatment populations are identical.

H_a: At least two of the treatment populations are different.

The test statistic for the Friedman test is

$$X^2 = \frac{12}{bt(t + 1)} \sum_{j=1}^{t} R_j^2 - 3b(t + 1)$$

where t is the number of treatments, b the number of blocks, and R_j is the sum of the ranks in the jth column (treatment). As with the Kruskal-Wallis test, the Friedman test statistic X^2 is approximated by the chi-square distribution with $(t - 1)$ degrees of freedom.

EXAMPLE 18.8   When considering a site for industrial expansion, business executives weigh such factors as tax policy, quality of educational facilities, construction costs, and labor quality. In the end, however, the decision boils down to a subjective assessment of "overall business climate." The president of a California-based electronics company has asked each of five vice presidents to provide an analysis of four states as prospective sites for the location of a satellite manufacturing facility. The vice presidents conducted their analyses of the four states in random order and then provided their assessment of "overall business climate" on a 10-point scale (1 = totally unacceptable, 10 = outstanding). Their responses are summarized within Table 18.25. Do these data present sufficient evidence to conclude that a difference exists in perceptions of overall business climate among the four states? Test using $\alpha = .05$.

Table 18.25   Response Data and Ranks for Example 18.8

Vice President	State			
	Arkansas	Colorado	Illinois	Iowa
A	8.5 (1)	8.0 (2)	3.5 (4)	6.0 (3)
B	7.5 (2)	8.0 (1)	6.0 (3)	5.5 (4)
C	9.0 (1)	6.0 (3)	4.0 (4)	7.0 (2)
D	8.0 (1)	6.0 (3)	7.0 (2)	4.0 (4)
E	7.0 (2)	5.5 (3)	4.5 (4)	7.5 (1)
R_j	7	12	17	14

SOLUTION   The rank sums for each column (state) are included within Table 18.25. There are $t = 4$ treatments (states) and $b = 5$ blocks (vice presidents) in this example. Thus, the Friedman test statistic is

$$X^2 = \frac{12}{bt(t + 1)} \sum_{j=1}^{t} R_j^2 - 3b(t + 1)$$

$$= \frac{12}{(5)(4)(5)} (7^2 + 12^2 + 17^2 + 14^2) - 3(5)(5) = 81.36 - 75 = 6.36$$

From Table 5 in the Appendix we find $\chi_{.05}^2$ based on $(t - 1) = 3$ degrees of freedom is 7.81473. Since our calculated test statistic $X^2 = 6.36$, we do not reject the hypothesis of equality of perceptions of overall business climate among the four states and cannot assume that, in fact, a difference in perception of business climate does exist.

◆ ◆

When you use the Friedman test, note that the analysis is based on the **ranks** of observations within each block (that is, the ranking of the t treatments within each block) and not the magnitude of difference between observations. In Example 18.8, for example, the analysis could have been performed if the five vice presidents had simply **ranked** the four states according to their respective assessments of business climate. The fact that vice president A provided a margin of preference of only .5 rating point when choosing Arkansas over Colorado (8.5 vs. 8.0) as his first choice while vice president D chose Arkansas over Illinois by 2 rating points is ignored in the Friedman analysis. Thus, although the Friedman test provides an alternative to the F test for the randomized block design of Section 10.8, the Friedman test is based upon less information and is hence less powerful than the F test, assuming the parametric assumptions are met.

The Friedman test for randomized block designs is summarized in the following display.

Friedman Test for Randomized Block Designs

Null hypothesis H_0: The probability distributions for the t treatments (populations) are identical.

Alternative hypothesis H_a: At least two of the t treatments have different probability distributions.

Test statistic: $X^2 = \dfrac{12}{bt(t+1)} \sum_{j=1}^{t} R_j^2 - 3b(t+1).$

Rejection region: Reject H_0 if $X^2 > \chi_\alpha^2$ with $(t-1)$ degrees of freedom.

EXERCISES   Conceptual Level

18.43   For which experimental design can the Friedman test procedure be used?

18.44   Under what conditions would one use the Friedman test rather than its parametric analogue?

EXERCISES   Skill Level

18.45   Determine the appropriate rejection region for the Friedman test for a randomized block design in the following situations.

    a. $t = 5, \alpha = .025$ **b.** $t = 3, \alpha = .10$ **c.** $t = 6, \alpha = .01$

18.46   Refer to Exercise 10.20. The data, reproduced in Table 18.26, resulted from an experiment run in a randomized block design in which each of four treatments were randomly applied to experimental units in each of six blocks. Use the Friedman test for a randomized block design to determine whether there are differences among the four treatments. Use $\alpha = .05$.

Table 18.26   Data for Exercise 18.46

	Treatment			
Block	1	2	3	4
1	9.6	11.3	8.3	11.6
2	9.2	9.0	14.1	15.0
3	9.7	10.3	12.8	15.5
4	11.5	18.3	18.8	15.4
5	12.4	14.8	14.2	12.3
6	13.2	10.4	15.3	15.4

18.47   Data from an experiment run in a randomized block design were ranked within each block, producing the rankings shown in Table 18.27. Use the Friedman test for a randomized block design to determine whether there are significant differences among the three treatments. Calculate the approximate p-value for the test, and use it to draw your conclusions.

Table 18.27   Data for Exercise 18.47

	Treatment		
Block	1	2	3
1	1	3	2
2	2	1	3
3	1	3	2
4	3	1	2
5	1	2	3
6	3	2	1

EXERCISES   Applications

18.48   Refer to Exercise 10.47 and use the Friedman test to determine whether a difference in gasoline mileage exists for the five brands of gasoline. Test by using $\alpha = .01$.

18.49   In an effort to locate new venture investment opportunities for its clients, a major investment banking company asked five of its experienced business analysts to review the business plans of four new ventures seeking investment capital. Business plans for the four new ventures were assigned to the analysts in random order, with the analysts each being asked to rank the new ventures in order of performance. Their responses are provided in Table 18.28. Do these data present sufficient evidence to assume that a difference exists among the preferences of analysts for these four new business ventures? Test by using $\alpha = .05$.

18.50   An article in *Venture* ("Looking for Cash," *Venture*, December 1984) described the various pros and cons of using public stock offerings, equipment leasing, debt, and private placements as means for generating working capital by entrepreneurial companies. An analysis of financing

Table 18.28   Response Data for Exercise 18.49

	New Business Venture			
Analyst	1	2	3	4
A	3	1	4	2
B	1	2	4	3
C	2	1	4	3
D	3	1	4	2
E	1	2	3	4

sources by entrepreneurial companies in five industries during 1988 is recorded in Table 18.29. The recorded data are the number of firms within a particular industry class that gained working capital from each source during 1988. Do these data present sufficient evidence to assume that a difference exists in the preferences expressed by companies in 1988 for sources of working capital? Test by using $\alpha = .05$.

Table 18.29   Financing Data for Exercise 18.50

	Source of Working Capital			
Industry	Public Offerings	Equipment Leasing	Debt	Private Placement
Wholesale and distribution	71	21	39	96
Retailing	69	53	84	76
Manufacturing	196	109	62	98
Natural resources	38	12	5	17
Services	58	3	39	64

18.51   With the necessary investment of time and money in sales force training and development, sales force turnover is a chief concern of industry. In fact, C. Futrell and A. Parasuraman ("The Relationship of Satisfaction and Performance to Salesforce Turnover," *Journal of Marketing,* Fall 1984) report that some companies consider themselves fortunate if they retain 50% of new salespeople after three years and that training and development costs can range as high as $75,000 for each salesperson in some industries. In an attempt to curtail turnover a national conglomerate with seven subsidiary companies explored the use of three different sales training programs. Each of the three training programs was applied in random order within each of the seven subsidiary companies. The data listed in Table 18.30 provide the retention rates (percentages) of employees remaining as active members of the sales force three years following the conclusion of sales training. Do these data provide sufficient evidence to assume that a difference exists in the retention rates resulting from the three sales force training programs? Test by using $\alpha = .05$.

18.52   A common problem in the manufacture of plywood is the development of resins (glues) that resist separation of the plywood veneer when exposed to moisture. An experiment was conducted to examine the separation resistance when using four different resins. The four resins were randomly applied in the manufacture of plywood produced from three types of material: Douglas fir, hemlock, and lodgepole pine. The results of the randomized block design are

Table 18.30   Training Program Data for Exercise 18.51

Subsidiary Company	Training Program		
	A	B	C
Appliance Center	42	36	30
Bi-Less	45	55	35
Caravan	63	71	55
Dutchess	49	60	40
Emporium	55	45	52
First Choice	70	67	50
Garden Mart	50	56	33

included in Table 18.31, with the recorded data representing the degree of separation (as measured on a 1–10 scale) after immersion of the plywood in a vat of water. (In this case 1 indicates no separation and 10 indicates complete separation.) Do these data present sufficient evidence to indicate a difference in the separation of plywood when using the four resins? Test by using $\alpha = .05$.

Table 18.31   Separation Data for Exercise 18.52

Wood Material	Resin			
	A	B	C	D
Douglas fir	3.2	4.6	1.5	2.4
Hemlock	5.5	2.5	2.5	3.0
Lodgepole pine	4.0	3.8	2.5	2.0

18.7

The Runs Test: A Test for Randomness

Consider a production process in which manufactured items emerge in sequence and each is classified as either defective (D) or nondefective (N). We studied how we might compare the fraction defective over two equal time intervals by using the normal z test in Chapter 8. We extended this procedure to a test of an hypothesis about a constant fraction defective over two or more time intervals by using the chi-square test of Chapter 17. The purpose of these tests was to detect a change or trend in the fraction defective p. Evidence to indicate an increasing fraction defective might show the need for a study to locate the source of difficulty. A decreasing value might suggest that a quality control program was having a beneficial effect in reducing the fraction defective.

Trends in a fraction defective (or other quality measures) are not the only indication of a lack of process control. A process may cause periodic runs of defectives, with the average fraction defective remaining constant, for all practical purposes, over long periods of time. For example, flashbulbs are manufactured on a rotating machine with a fixed number of positions for bulbs. A bulb is placed on the machine at a given position, the air is removed, oxygen is pumped into the bulb, and the glass base is flame-sealed. If a machine contains 20 positions, and several adjacent positions are faulty (too much heat in the sealing process), surges of defective bulbs will emerge from the process in a periodic manner. Tests to compare the fraction defective over equal intervals of time will not detect this periodic difficulty in the process. The periodicity, indicated by runs of defectives, is indicative of nonrandomness in the occurrence of defectives over time and can be detected by a **test for randomness**. The statistical test that we present, known as the **runs test**, is discussed in detail in the article by Wald and Wolfowitz (1940) listed in the references. Other practical applications of the runs test will follow.

    As the name implies, the runs test studies a sequence of events in which each element in the sequence may assume one of two outcomes—say success S or failure F. The runs test is thus applied to a sequence of n_1 successes and n_2 failures. If we think of the sequence of items emerging from a manufacturing process as defective F or nondefective S, the observation of 20 items might yield the following:

$$S\ S\ S\ S\ S\ F\ F\ S\ S\ S\ F\ F\ F\ S\ S\ S\ S\ S\ S\ S$$

We notice the groupings of defectives and nondefectives and wonder whether it implies nonrandomness and, consequently, lack of process control.

Definition

> **A run** is a maximal subsequence of like elements.

For example, the first five successes are a subsequence of five like elements and it is maximal in the sense that it includes the maximum number of like elements before encountering an F. (The first four elements form a subsequence of like elements, but it is not maximal because the fifth element could also be included.) Consequently, the 20 elements shown above are arranged in five runs, the first containing 5 S's, the second containing 2 F's, and so on.

    A very small or very large number of runs in a sequence indicates nonrandomness. Therefore, we will use R (the number of runs in a sequence) as a test statistic to test the null hypothesis that the elements are arranged in random sequence. The rejection region when employing the alternative hypothesis that the sequence is nonrandom is $R \leq k_1$ and $R \geq k_2$, as indicated in Figure 18.4. A one-tailed test, using values of R in the upper tail, is employed for an alternative hypothesis that the expected value of R is

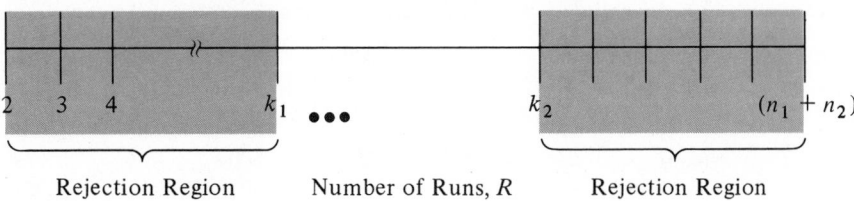

Figure 18.4   *Rejection Region for the Runs Test*

larger than when randomness is present. This alternative would imply an overmixing of the elements of the sequence. In contrast, a one-tailed test, using values of R in the lower tail, is used for an alternative hypothesis that the expected value of R is less than when randomness is present. This alternative would be used if you would not expect overmixing to occur and you could expect only a small number of runs (large groups of like elements). Or you can reject the null hypothesis for small values of R if overmixing is unimportant from a practical point of view.

To illustrate the selection of the alternative hypothesis for the runs test, consider a test of the null hypothesis that the sequence of defective and nondefective flashbulbs emerging from a production line is random. If specific heat-sealing positions on the rotating sealing machine were operating improperly, defectives would be produced in a spaced and systematic manner, a condition that could result in either a large- or a smaller-than-expected number of runs (overmixing or undermixing of defectives and nondefectives). To detect this situation, you would want to use a two-tailed test, rejecting the null hypothesis of randomness for either very large or very small values of R. Then you would place part of α in the upper tail of the R distribution and part in the lower tail and select a two-tailed rejection region, as shown in Figure 18.4.

On the other hand, if the heat supply (natural gas) to all sealing positions were inadvertently reduced, all sealing positions would produce defectives. The machine would produce, alternatively, surges of defectives and nondefectives, depending on whether the gas supply were deficient or adequate. This situation would result in only undermixing, yielding a smaller number of runs than expected. To detect this situation, you would run a one-tailed test, rejecting the null hypothesis of randomness for small values of R.

The rejection region for the runs test can be found by using Table 9 of the Appendix, which gives the probability that the number of runs R is less than or equal to

Table 18.32 An Abbreviated Version of Table 9 in the Appendix for the Runs Test

	a							
(n_1, n_2)	2	3	4	5	6	7	8	9
(2, 3)	.200	.500	.900	1.000				
(2, 4)	.133	.400	.800	1.000				
(2, 5)	.095	.333	.714	1.000				
(2, 6)	.071	.286	.643	1.000				
(2, 7)	.056	.250	.583	1.000				
⋮	⋮	⋮	⋮	⋮				
(3, 7)	.017	.083	.283	.583	.833	1.000		
(3, 8)	.012	.067	.236	.533	.788	1.000		
(3, 9)	.009	.055	.200	.491	.745	1.000		
(3, 10)	.007	.045	.171	.455	.706	1.000		
(4, 4)	.029	.114	.371	.629	.886	.971	1.000	
(4, 5)	.016	**.071**	.262	.500	.786	.929	.992	1.000
(4, 6)	.010	.048	.190	.405	.690	.881	.976	1.000
(4, 7)	.006	.033	.142	.333	.606	.833	.954	1.000
(4, 8)	.004	.024	.109	.279	.533	.788	.929	1.000

some number a. An abbreviated version of Table 9 is shown in Table 18.32. In the table the sample sizes (n_1, n_2) are shown at the left of the table and the number a is recorded across the top of the table. The table entries give the probability that $R \leq a$.

To illustrate the use of the table, suppose you have $n_1 = 4$ elements of one type and $n_2 = 5$ of another. Furthermore, suppose you wish to conduct a one-tailed test of the hypothesis of randomness and reject H_0 if the number of runs is too small. Choosing a rejection region with $\alpha \approx .05$, we choose $R \leq 3$ since the probability of observing $R \leq 3$ is $\alpha = .071$, which is shown in color in Table 18.32.

On the other hand, suppose you wish to detect an overly large number of runs, rejecting the null hypothesis of randomness when R is large. Then for $n_1 = 4$ and $n_2 = 5$, the table gives $P(R \leq 7) = .929$. Therefore, it follows that $P(R \geq 8) = 1 - P(R \leq 7) = 1 - .929 = .071$. Hence, if we choose $R \geq 8$ as the rejection region, $\alpha = .071$ for this one-tailed test.

The two preceding examples explain how to find α if you conduct either a one-tailed test for undermixing or a one-tailed test for overmixing. A two-tailed test of nonrandomness to detect either undermixing or overmixing can be conducted by locating the rejection regions as shown in Figure 18.4 and determining the critical values in the same manner as for the one-tailed tests.

Runs Test

> Null hypothesis H_0:   The sequence of elements, call them S's and F's, has been produced in a random manner.
>
> Alternative hypothesis H_a: The elements have been produced in a nonrandom sequence (a two-tailed test). Or the process is nonrandom owing solely to overmixing (an upper one-tailed test) or solely to undermixing (a lower one-tailed test).
>
> Test statistic: R, the number of runs.
>
> Rejection region: For a given value of α and for a two-tailed test, reject H_0 if $R \leq k_1$ or $R \geq k_2$ (see Figure 18.4), where $P(R \leq k_1) + P(R \geq k_2) = \alpha$ and k_1 and k_2 are obtained from Table 9. For a given value of α and for a lower one-tailed test, reject H_0 if $R \leq k_1$, where $P(R \leq k_1) = \alpha$ and k_1 is obtained from Table 9; for an upper one-tailed test reject H_0 if $R \geq k_2$, where $P(R \geq k_2) = \alpha$ and k_2 is obtained from Table 9.

EXAMPLE 18.9   After a two-week study program to acquaint company personnel with U.S. tariff regulations, the manager of the international business division of a large manufacturing firm constructed a 20-question, true-false examination to test the effectiveness of the study program. The examination was constructed with correct answers running in the following sequence:

T F F T F T F T T F T F F T F T F T T F

Does this sequence indicate a departure for randomness in the arrangement of T and F answers?

SOLUTION   The sequence contains $n_1 = 10$ true and $n_2 = 10$ false answers, with $R = 16$ runs. Nonrandomness can be indicated by either an unusually small or an unusually large number of runs, and consequently, we will use a two-tailed test.

Suppose that we wish to use an α approximately equal to .05, with .025 or less in each tail of the rejection region. Then from Table 9 with $n_1 = n_2 = 10$, we note $P(R \leq 6) = .019$ and $P(R \leq 15) = .981$. Then $P(R \geq 16) = .019$, and we will reject the hypothesis of randomness if $R \leq 6$ or $R \geq 16$. Since $R = 16$ for the observed data, we conclude that evidence exists to indicate nonrandomness in the manager's arrangement of answers. His attempt to mix the answers was overdone.   ◆ ◆

time series   A second application of the runs test is for detecting nonrandomness of a sequence of quantitative measurements over time. These sequences, known as **time series**, occur in many fields. For example, the measurement of a quality characteristic of an industrial product, the blood pressure of a human, and the price of a stock on the stock market all vary over time. Departures in randomness in a series, caused either by trends

deviations   or periodicities, can be detected by examining the **deviations** of the time series measurements from their average. Negative and positive deviations could be denoted by S and F, respectively, and we could then test this time sequence of deviations for nonrandomness. We illustrate this application with an example.

EXAMPLE 18.10   The management of a retailing firm feels that the firm's sales respond to a seasonal influence. If they were, a better monthly sales forecast could be prepared by recognizing the effects of seasonality. The firm's sales over the past year are shown for each month in Figure 18.5. The average sales $\bar{x}$ for the 12 months appears as shown. Note the deviations around $\bar{x}$. Do these sales data indicate a lack of randomness and thereby suggest a periodic relationship between sales and time of the year?

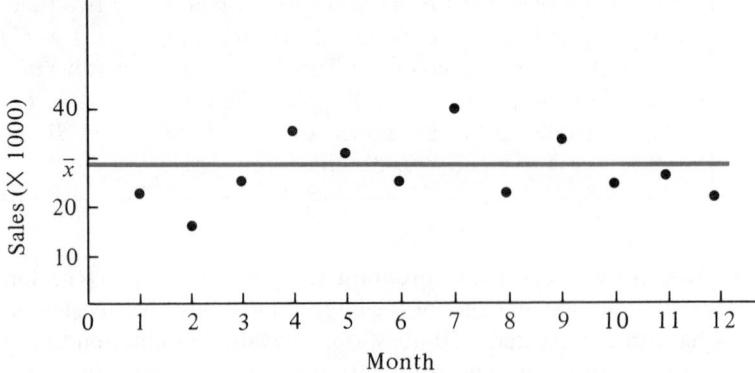

Figure 18.5   *Monthly Sales for the Firm in Example 18.10*

SOLUTION   If there is a seasonal trend in monthly sales, we would expect runs of high (or low) monthly sales corresponding to particular seasons. This result would produce a smaller-than-expected number of runs and lead us to reject the hypothesis of randomness if $R \leq k_1$. That is, we will use a lower one-tailed test, placing α entirely in the lower tail of the R distribution. The sequence of negative (N) and positive (P) deviations, as indicated in Figure 18.5, is

$$N \; N \; N \; P \; P \; N \; P \; N \; P \; N \; N \; N$$

Then $n_1 = 8$, $n_2 = 4$, and $R = 7$. Consulting Table 9, we find p-value $= P(R \leq 7) = .788$. This value of R (i.e., $R = 7$) is not improbable, assuming the hypothesis of randomness to be true. Consequently, there is not sufficient evidence to indicate nonrandomness in this sequence of monthly sales. Therefore, there is insufficient evidence to support the notion of a seasonal (periodic) component in the firm's sales pattern.

◆ ◆

The runs test can also be used to compare two population frequency distributions for a two-sample unpaired experiment. It provides an alternative to the Mann-Whitney U test of Section 18.3. If the measurements for the two samples are arranged in order of magnitude, they will form a sequence. The measurements for samples 1 and 2 can be denoted as S and F, respectively, and we are once again concerned with a test for randomness. If all measurements for sample 1 are smaller than those for sample 2, the sequence will be $S \; S \; S \ldots S \; F \; F \; F \ldots F$, giving $R = 2$ runs. A small value of R provides evidence of a difference in population frequency distributions, and the rejection region is $R \leq a$. This rejection region implies a one-tailed statistical test. An illustration of the application of the runs test to compare two population frequency distributions is left as an exercise.

As in the case of the other nonparametric test statistics studied in preceding sections of this chapter, the probability distribution for R tends to normality as n_1 and n_2 become large. The approximation is good when n_1 and n_2 are both greater than 10. Consequently, we may use the z statistic as a large-sample test statistic, where

$$z = \frac{R - E(R)}{\sigma_R}$$

$$E(R) = \frac{2n_1 n_2}{n_1 + n_2} + 1 \quad \text{and} \quad \sigma_R^2 = \frac{2n_1 n_2 (2n_1 n_2 - n_1 - n_2)}{(n_1 + n_2)^2 (n_1 + n_2 - 1)}$$

Here, $E(R)$ is the expected value of R and σ_R^2 is the variance of R. The rejection region for a two-tailed test with $\alpha = .05$ is $|z| \geq 1.96$.

Large-Sample Runs Test ($n_1 > 10$ and $n_2 > 10$)

Null hypothesis H_0:   The sequence of elements, call them S's and F's, has been produced in a random manner.

Alternative hypothesis H_a: The elements have been produced in a nonrandom sequence (a two-tailed test). Or the process is nonrandom owing solely to overmixing (an upper one-tailed test) or solely to undermixing (a lower one-tailed test).

Test statistic: $$z = \frac{R - \dfrac{2n_1 n_2}{n_1 + n_2} - 1}{\sqrt{\dfrac{2n_1 n_2 (2n_1 n_2 - n_1 - n_2)}{(n_1 + n_2)^2 (n_1 + n_2 - 1)}}}.$$

Rejection region: For a given value of α and for a two-tailed test, reject H_0 if $z \geq z_{\alpha/2}$ or if $z \leq -z_{\alpha/2}$. For a given value of α and for a one-tailed test, place all of α in one tail of the z distribution and reject H_0 if $z \geq z_\alpha$, for an upper one-tailed test, or if $z \leq -z_\alpha$, for a lower one-tailed test. Tabulated values of z are given in Table 3 of the Appendix.

EXERCISES   Conceptual Level

18.53 What is a run? How are the length and the number of runs related to randomness in a sequence of events?

18.54 Explain the reasoning underlying the runs test.

EXERCISES   Skill Level

18.55 A sequence of elements contains n_1 S's and n_2 F's. Find the appropriate rejection region, using the test statistic R, the number of runs, to test for randomness of the sequence in the following situations.

    a. $n_1 = 4, n_2 = 8, \alpha \approx .10$ **b.** $n_1 = 7, n_2 = 9, \alpha \approx .05$

    c. $n_1 = 5, n_2 = 5, \alpha \approx .01$

18.56 A sequence consisting of $n_1 = 8$ S's and $n_2 = 6$ F's produces $R = 10$ runs. Test for randomness of the sequence of elements, using the runs test with $\alpha \approx .05$.

18.57 A sequence consisting of $n_1 = 15$ S's and $n_2 = 20$ F's produces $R = 5$ runs. Test for randomness of the sequence of elements, using the large-sample approximation to the runs test with $\alpha \approx .01$.

EXERCISES   Applications

18.58 The home office of a corporation selects its executives from the staff personnel of its two subsidiary companies, company A and company B. During the past three years nine executives have been selected by the parent company from the subsidiaries—the first selected from B, and the second from A, and so on. The following sequence shows the order in which the executives have been selected and the subsidiary firms from which they came:

$B\ A\ A\ A\ B\ A\ A\ A\ B$

Does this selection sequence provide sufficient evidence to imply nonrandomness in the selection of executives by the parent company from its subsidiaries?

18.59   Paper is produced in a continuous process. A brightness measurement x is made on the paper once every hour and the results are as shown in the table. Compute the average brightness measure $\bar{x}$. Then note the deviations about $\bar{x}$. Do these results indicate a lack of randomness and thereby suggest periodicity in the process and lack of control?

Times (Hours)	1	2	3	4	5	6	7	8	9	10	11	12	13	14	15
Brightness, x	3.1	2.3	2.6	3.2	4.2	3.9	3.2	5.0	4.8	2.8	4.2	3.2	3.4	3.1	2.6

18.60   The so-called random walk hypothesis suggests that the movement of certain security prices is completely random and follows no discernible pattern over time. Listed in Table 18.33 are the daily closing prices (rounded to the nearest dollar) over 20 consecutive market days for a certain security listed on the New York Stock Exchange. Do these data tend to support the notion that the daily closing prices for this security follow a random walk? (*Hint:* Compute the mean security price for these data and note the runs of negative and positive deviations of daily prices from the 20-day average.)

Table 18.33   Stock Price Data for Exercise 18.60

Time (Day)	Price	Time (Day)	Price
1	$21	11	$22
2	22	12	21
3	24	13	20
4	24	14	18
5	23	15	18
6	24	16	19
7	25	17	18
8	23	18	17
9	23	19	16
10	22	20	18

18.61   When a regression analysis is based on time series data, one often wishes to determine whether the residuals (errors of prediction) are random or suggest the presence of some hidden factor. One means of testing the residuals for randomness is to apply a runs test to the residuals arranged in time order, noting the changes in the sign of the residuals over time. Eddy and Saunders examined the impact of the publication of new product announcements on stock returns of 66 firms by using a standard market model and regressing price against the Standard and Poor's 500 Stock Price Index. The purpose of their study was to explore for the effects of isolated factors that may systematically affect stock prices and, as a result, cause the residuals to vary in a nonrandom manner. The market model residuals observed by Eddy and Saunders over a 20-month period are shown in Table 18.34. Do the market model residuals indicate a lack of randomness over time and, hence, the presence of some isolated factor that is systematically affecting stock prices?

18.62   Fifteen experimental batteries were selected at random from a pilot lot at plant A, and 15 standard batteries were selected at random from production at plant B. All 30 batteries were simultaneously placed under an electrical load of the same magnitude. The first battery to fail was an A, the second a B, the third a B, and so forth. The following sequence shows the order of

Table 18.34   Residuals Data for Exercise 18.61

Month	Market Model Residual	Month	Market Model Residual
1	.0032	11	−.0018
2	−.0135	12	−.0080
3	.0022	13	−.0178
4	.0034	14	−.0065
5	−.0062	15	.0130
6	.0101	16	−.0181
7	.0033	17	−.0129
8	.0216	18	.0027
9	.0150	19	−.0033
10	.0059	20	.0068

Source: A. Eddy and G. Saunders, "New Product Announcements and Stock Prices," *Decision Sciences*, January 1980. Reprinted with permission.

failure for the 30 batteries:

A B B B A B A A B B B B A B A B B B B A A B A A A B A A A A

a. Using the large-sample U test, determine whether there is sufficient evidence to conclude that the distribution of battery life for the experimental batteries lies to the right of the distribution for the standard batteries. Use $\alpha = .05$.

b. If, indeed, the experimental batteries have the greater mean life, what would be the effect on the expected number of runs? Using the large-sample runs test, determine whether there is a difference in the distributions of battery life for the two populations. Use $\alpha = .05$.

18.8

Rank Correlation Coefficient

rank correlation

rank correlation coefficient

In preceding sections we have used ranks to indicate the relative magnitude of observations in nonparametric tests for comparison of treatments. We will now employ the same technique in testing for a *monotonic relation* (one that is ever increasing or ever decreasing) between two ranked variables. A **rank correlation** provides a measure of the degree of linearity between the ranking variables or a measure of the degree of monotonicity between the variables being observed. Thus, a **rank correlation coefficient** is frequently referred to as a coefficient of agreement for preference data. Two common rank correlation coefficients are the Spearman r_s and the Kendall τ. We will present the Spearman r_s because its computation is identical to that

Table 18.35   Preference Rankings and Prices of
Television Sets

Manufacturer	Preference	Price
1	7	$449.50
2	4	525.00
3	2	479.95
4	6	499.95
5	1	580.00
6	3	549.95
7	8	469.95
8	5	532.50

Table 18.36   Preference and Price Rankings of
Television Sets

Manufacturer	Preference Ranks, x_i	Price Ranks, y_i
1	7	1
2	4	5
3	2	3
4	6	4
5	1	8
6	3	7
7	8	2
8	5	6

for the sample correlation coefficient r of Chapter 11. Kendall's rank correlation coefficient is discussed in detail in the text by Kendall and Stuart (1979) listed in the references.

Suppose an individual who is in the market for a 24-in. color television set ranks the standard models for each of the eight major television manufacturers. His preferences are listed in Table 18.35 next to the manufacturer's suggested retail price for the set under consideration. Do the data suggest an agreement between the individual's preference rankings and the manufacturers' suggested retail prices? That is, is there a correlation between preferences and prices?

The two variables of interest are the preference rankings and the prices. The former is already in ranked form, and the prices may be ranked as shown in Table 18.36. The ranks for tied observations are obtained by averaging the ranks that the tied observations would have had if no ties were observed. The Spearman rank correlation coefficient r_s is calculated by using the ranks as the paired measurements on the two variables x and y in the formula for r given in Chapter 11.

Spearman's Rank Correlation Coefficient

$$r_s = \frac{S_{xy}}{\sqrt{S_{xx}S_{yy}}}$$

where x_i and y_i represent the *ranks* of the *i*th pair of observations and

$$S_{xy} = \sum_{i=1}^{n}(x_i - \bar{x})(y_i - \bar{y}) = \sum_{i=1}^{n} x_i y_i - \frac{\left(\sum_{i=1}^{n} x_i\right)\left(\sum_{i=1}^{n} y_i\right)}{n}$$

$$S_{xx} = \sum_{i=1}^{n}(x_i - \bar{x})^2 = \sum_{i=1}^{n} x_i^2 - \frac{\left(\sum_{i=1}^{n} x_i\right)^2}{n}$$

$$S_{yy} = \sum_{i=1}^{n}(y_i - \bar{y})^2 = \sum_{i=1}^{n} y_i^2 - \frac{\left(\sum_{i=1}^{n} y_i\right)^2}{n}$$

When there are no ties in either the x observations or the y observations, the expression for r_s algebraically reduces to the simpler expression shown in the next display.

Shortcut Formula for r_s

$$r_s = 1 - \frac{6\sum d_i^2}{n(n^2 - 1)} \qquad \text{where} \qquad d_i = x_i - y_i$$

If the number of ties is small in comparison with the number of data pairs, little error will result in using this shortcut formula for calculating r_s. We illustrate the use of the formula with an example.

EXAMPLE 18.11   Since the association between brand preference and price is an important aspect of pricing, one of the TV manufacturers wishes to know the degree of association, measured by the correlation r_s, between TV preference and price. Calculate r_s for the preference and price data shown in Table 18.36.

SOLUTION   The values of d_i and $d_i^2, i = 1, 2, \ldots, 8$, are shown in Table 18.37. Substituting into the formula for r_s, we have

$$r_s = 1 - \frac{6\sum_{i=1}^{n} d_i^2}{n(n^2 - 1)} = 1 - \frac{6(144)}{8(64 - 1)} = -.714$$

Table 18.37   Data for Example 18.11

Manufacturer	Rank, x_i	Rank, y_i	d_i	d_i^2
1	7	1	6	36
2	4	5	-1	1
3	2	3	-1	1
4	6	4	2	4
5	1	8	-7	49
6	3	7	-4	16
7	8	2	6	36
8	5	6	-1	1
				144

The Spearman rank correlation coefficient may be employed as a test statistic to test an hypothesis of no association between two populations. We assume that the n pairs of observations (x_i, y_i) have been randomly selected and therefore that the hypothesis of no association between the populations implies a random assignment of the n ranks within each sample. Each random assignment (for the two samples) represents a sample point associated with the experiment, and a value of r_s could be calculated for each. Thus, we can calculate the probability that r_s assumes a large positive or negative value owing solely to chance and thereby suggests an association between populations when none exists.

The rejection region for a two-tailed test is shown in Figure 18.6. If the alternative hypothesis is that the correlation between x and y is negative, we reject H_0 for large negative values of r_s (in the lower tail of Figure 18.6). Similarly, if the alternative hypothesis is that the correlation between x and y is positive, we reject H_0 for large positive values of r_s (in the upper tail of Figure 18.6).

The critical values of r_s are given in Table 10 of the Appendix. An abbreviated version of Table 10 is shown in Table 18.38. Across the top of Table 18.38 (and Table 10 of the Appendix) are recorded the values of α that you might wish to use for a one-tailed test of the null hypothesis of no association between x and y. The number of ranked pairs n appears at the left side of the table. The table entries give the critical value r_0 for a one-tailed test. Thus, $P(r_s \geq r_0) = \alpha$.

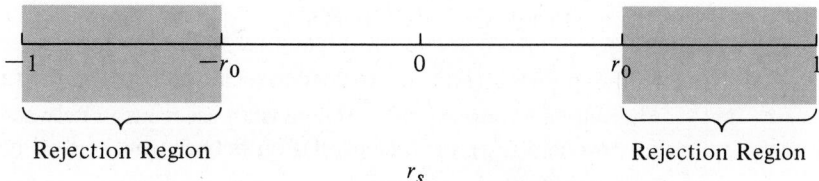

Figure 18.6   Rejection Region for a Two-Tailed Test of the Null Hypothesis of No Association, Using Spearman's Rank Correlation Test

Table 18.38   An Abbreviated Version of Table 10 in the Appendix
for Spearman's Rank Correlation Test

n	$\alpha = .05$	$\alpha = .025$	$\alpha = .01$	$\alpha = .005$
5	0.900	—	—	—
6	0.829	0.886	0.943	—
7	0.714	0.786	0.893	—
8	0.643	0.738	0.833	0.881
9	0.600	0.683	0.783	0.833
10	0.564	0.648	0.745	0.794
11	0.523	0.623	0.736	0.818
12	0.497	0.591	0.703	0.780
13	0.475	0.566	0.673	0.745
14	0.457	0.545		
15	0.441	0.525		
16	0.425			
17	0.412			
18	0.399			
19	0.388			
20	0.377			

For example, suppose you have $n = 8$ ranked pairs and you know that if any correlation exists between the ranks, it must be positive. Then you would reject the null hypothesis of no association only for large positive values of r_s and would use a one-tailed test. Referring to Table 18.38 and using the row corresponding to $n = 8$ and the column for $\alpha = .05$, you read $r_0 = .643$. Therefore, you would reject H_0 for all values of r_s greater than or equal to .643.

The test is conducted in exactly the same manner if you wish to test only for a large negative correlation. The only difference is that you would reject the null hypothesis if $r_s \leq -.643$. That is, you just use the negative of the tabulated value of r_0 as the lower-tail critical value.

To conduct a two-tailed test, you reject the null hypothesis if $r_s \geq r_0$ or if $r_s \leq -r_0$. The value of α for the test will be double the value shown at the top of Table 10. For example, if $n = 8$ and you choose the .025 column, you will reject H_0 if $r_s \geq .738$ or if $r_s \leq -.738$. The α value for the test is $2(.025) = .05$.

Spearman's Rank Correlation Test

Null hypothesis H_0:   There is no association between the ranked pairs.

Alternative hypothesis H_a: There is an association between the ranked pairs (a two-tailed test). Or the correlation between the ranked pairs is positive (or negative) (a one-tailed test).

Test statistic: r_s.

Rejection region: For a given value of α and for a two-tailed test, reject H_0 if $r_s \geq r_0$ or if $r_s \leq -r_0$, where r_0 is given in Table 10 of the Appendix; double the

> table value of α to obtain the value of α for the two-tailed test. For a given value of α and for a one-tailed test, reject H_0 if $r_s \geq r_0$, for an upper-tailed test, or if $r_s \leq -r_0$, for a lower-tailed test; the α value for a one-tailed test is the value shown in Table 10.

EXAMPLE 18.12 Refer to Example 18.11. Management realizes that the significance of a correlation depends on the true association and the sample sizes used to estimate the correlation coefficient. In Example 18.11 the rank order correlation between TV preference and price was found to be $-.714$. Now management wishes to know the significance of this correlation tested against the null hypothesis of no association.

SOLUTION The critical value of r_s for a one-tailed test with $\alpha = .05$ and $n = 8$ is .643. Let us assume that the correlation between the individual's preference rankings and the manufacturers' suggested retail prices could not possibly be positive. (That is, assume that a low preference rank means a television set is highly favorable and should be associated with a high price if the manufacturers' prices are indicators of quality, additional features, guarantee, and so forth.) The alternative hypothesis is that the population correlation coefficient is less than zero, and we are concerned with a one-tailed test. Thus, α for the test is the table value of .05, and we will reject the null hypothesis if $r_s \leq -.643$. The calculated value of the test statistic, $r_s = -.714$, is less than the critical value for $\alpha = .05$. Hence, the null hypothesis of no association is rejected at the $\alpha = .05$ level of significance. It appears that some agreement does exist between preference and price.

◆ ◆

EXERCISES Conceptual Level

18.63 Indicate whether the following statements are true or false. If false, correct the statement.

a. If $r_s = 1$, then r computed from the same data will equal 1.

b. If $r = -1$, then r_s computed from the same data will equal -1.

c. If an analysis of two variables suggests that an association exists between the two variables at the $\alpha = .10$ level of significance, the association would also prove to be significant at the $\alpha = .05$ level of significance.

d. The shortcut formula for r_s can be used whether or not ties exist for the x observations and the y observations.

e. For a two-tailed Spearman rank correlation test at a given level of α, we use as a critical value the tabulated Spearman value under the column headed by $\alpha/2$.

EXERCISES Skill Level

18.64 A random sample of n pairs of observations (x_i, y_i) have been selected, and the x and y observations have been ranked separately. Indicate the appropriate rejection region for testing for rank correlation in the following situations.

a. Test for negative rank correlation; $n = 16$, $\alpha = .01$

b. Test for positive rank correlation; $n = 10$, $\alpha = .05$

c. Test for positive or negative rank correlation; $n = 25$, $\alpha = .10$

18.65 A random sample of $n = 18$ pairs of observations produced $r_s = .482$. Do the data provide sufficient evidence to indicate that the population rank correlation coefficient is greater than zero? Use $\alpha = .01$.

18.66 Refer to Exercise 11.10. The data reproduced in the table display the coordinates for $n = 7$ pairs of observations (x, y).

x	7	8	2	3	5	3	7
y	2	0	5	6	4	9	2

a. Calculate the Spearman rank correlation coefficient, r_s, between x and y.

b. Do the data present sufficient evidence to indicate that there is an association between x and y? Test by using $\alpha = .05$.

EXERCISES   Applications

18.67 The data in Table 18.39 represent the monthly sales and the promotional expenses for a women's apparel store that specializes in sportswear for younger women.

a. Calculate the coefficient of linear correlation r (Chapter 11) between monthly sales and promotional expenses.

b. Do these data present sufficient evidence to indicate that a linear relationship exists between sales and promotional expenses? Test the hypothesis $\rho = 0$, using $\alpha = .01$.

c. Calculate r_s, the Spearman rank correlation between monthly sales and promotional expenses.

d. Do these data suggest there is an association between sales and promotional expenses? Test at the 1% level of significance.

e. Compare your results from part b and part d. What do these results suggest about the consistency of the test for the significance of linearity between two variables (Section 11.8) and the test for the significance of association (Section 18.8)?

Table 18.39   Sales and Expense Data for Exercise 18.67

Month	Sales ($\times$ $1,000)	Promotional Expense ($\times$ $1,000)	Month	Sales ($\times$ $1,000)	Promotional Expense ($\times$ $1,000)
1	$62.4	$3.9	7	$ 98.6	$ 7.9
2	68.5	4.8	8	104.3	9.0
3	70.2	5.5	9	106.5	9.2
4	79.6	6.0	10	107.3	9.7
5	80.1	6.8	11	115.8	10.9
6	88.7	7.7	12	120.1	11.0

18.68 Since marketing activities often pose ethical problems, Hunt, Chonko, and Wilcox examined the attitudes of marketing researchers toward a broad range of ethical issues. In particular, the authors asked a group of in-house marketing researchers and a group of agency researchers to separately rank-order 10 ethical issues common to problems in marketing research. Their responses are presented in Table 18.40. Do these data present sufficient evidence to suggest that a positive association exists between the ranking of importance of ethical issues by in-house marketing researchers and agency researchers? Let $\alpha = .05$.

Table 18.40   Attitude Rankings Data for Exercise 18.68

Issue	In-house Researchers	Agency Researchers
Research integrity	1	1
Treating clients fairly	4	2
Confidentiality	5	3
Social issues	2	5
Personnel issues	6	6
Treating respondents fairly	3	7
Treating others in company fairly	7	9
Interviewer dishonesty	10	4
Gifts, bribes	9	8
Treating suppliers fairly	8	10

Source: S. Hunt, L. Chonko, and J. Wilcox, "Ethical Problems of Marketing Researchers," *Journal of Marketing Research*, August 1984. Reprinted with permission.

 18.69 In a study to describe planning methods used in U.S. commercial banks, Giroux and Rose compared the frequency with which large banks (assets over $500 million) and medium-sized banks (assets of $100 to $500 million) involve certain elements in their bank's planning program. The results of their survey are shown in Table 18.41. Calculate r_s. Do these data indicate that a

Table 18.41   Banking Data for Exercise 18.69

Planning Element	Large Banks (Over $500 Million)	Medium-Sized Banks ($100–$500 Million)
Feasible and measurable goals	66%	68%
Top management commitment	83	84
Management/staff involvement	70	54
Department preparation of plan	80	54
Information accounting systems	43	62
Cooperation between management and planning personnel	77	64
Commitment to improve plans	78	75
Assignment of responsibility to implement plan	53	51
Monitoring of goal achievement	73	75
Critique of planning methods	40	36
Keeping employees informed	35	36

Source: G. Giroux and P. Rose, "An Update of Bank Planning Systems: Results of a Nationwide Survey of Large U.S. Banks," *Journal of Bank Research*, Autumn 1984. Reprinted with permission.

positive association exists between large and medium-sized banks' use of planning elements in their bank's planning programs? Let $\alpha = .01$.

 18.70   The controversy over whether the country can afford the staggering costs of federal safety standards continues. Economists advocate a risk-benefit analysis to calculate whether preventing some deaths is worth the cost; consumer advocates feel no price is too high to save lives. To shed more light on the controversy, a researcher asked three groups to rank 30 products and activities from most to least risky. These rankings were compared with the actual statistics on deaths per year in each category. The results are shown in Table 18.42.

Table 18.42   Risks and Rankings Data for Exercise 18.70

Actual Risk (Based on Deaths per Year)	Ranking of Risk by		
	League of Women Voters	College Students	Business and Professional Club Members
1. Smoking	4	3	4
2. Alcohol	6	7	5
3. Motor vehicles	2	5	3
4. Handguns	3	2	1
5. Electric mowers	18	19	19
6. Motorcycles	5	6	2
7. Swimming	19	30	17
8. Surgery	10	11	9
9. X rays	22	17	24
10. Railroads	24	23	20
11. Aviation, private	7	15	11
12. Large construction	12	14	13
13. Bicycles	16	24	14
14. Hunting	13	18	10
15. Home appliances	29	27	27
16. Fire fighting	11	10	6
17. Police work	8	8	7
18. Contraceptives	20	9	22
19. Aviation, commercial	17	16	18
20. Nuclear power	1	1	8
21. Mountain climbing	15	22	12
22. Power mowers	27	28	25
23. Football	23	26	21
24. Skiing	21	25	16
25. Vaccinations	30	29	29
26. Food coloring	26	20	30
27. Food preservatives	25	12	28
28. Pesticides	9	4	15
29. Antibiotics	28	21	26
30. Spray cans	14	13	23

Source: Data derived from N. Howard and S. Antilla, "What Price Safety? The 'Zero-Risk' Debate." *Dun's Review*, September 1979. Reprinted with permission.

a. Calculate r_s between the actual risk and the level of perceived risk for each of the three groups of respondents.

b. Do these data suggest that an association exists between the actual risk and the level of perceived risk for any of the three groups of respondents? Use $\alpha = .05$.

c. Do these data suggest that the rankings of perceived risk differ between the various pairs of respondent groups? (That is, is there an association between perceived risk rankings for the League of Women Voters and college students; between the League of Women Voters and business and professional club members; between college students and business and professional club members?) Use $\alpha = .05$.

18.9

Summary

The nonparametric statistical tests presented in the preceding pages represent only a few of the many nonparametric statistical measures of inference available. A much larger collection of nonparametric test procedures, along with worked examples, is given in the text by Conover (1980), and an extensive bibliography of publications dealing with nonparametric statistical methods has been presented by Savage (1953). These texts are listed in the references at the end of the chapter.

We have indicated that **nonparametric statistical procedures are particularly useful when the experimental observations are susceptible to ordering but cannot be measured on a quantitative scale.** Since parametric statistical procedures usually cannot be applied to this type of data, you would use nonparametric methods to analyze the data.

A second application of nonparametric statistical methods is in testing hypotheses associated with populations of quantitative data when uncertainty exists about whether the assumptions for the population distributions are satisfied. Just how useful are nonparametric methods for this situation? Many statistical test and estimation procedures are only slightly affected by moderate departures from the assumed form of the population frequency distribution, and parametric test procedures are more efficient than their nonparametric equivalents when the assumptions underlying parametric tests are true. However, the gain in efficiency is often very slight; and when the assumptions are not true, nonparametric methods may be more efficient than their parametric counterparts. For this reason, and because of their ease of application, nonparametric methods are becoming more accepted by statistical practitioners.

Supplementary Exercises

Applications

18.71 The number of defective electrical fuses proceeding from each of two production lines A and B was recorded daily for a period of 10 days, with the results as shown in Table 18.43. Assume that both production lines produced the same daily output. Compare the number of defectives produced by A and B each day, and let x denote the number of days when B exceeded A. Do these data present sufficient evidence to indicate that, on the average, production line B produces more defectives than A?

a. Use the sign test, with a level of significance as near as possible to .01.

b. Use the Wilcoxon signed-rank test, with $\alpha = .01$.

Table 18.43   Defective Fuses Data for Exercise 18.71

	Line			Line	
Day	A	B	Day	A	B
1	172	201	6	142	170
2	165	179	7	190	182
3	206	159	8	169	179
4	184	192	9	161	169
5	174	177	10	200	210

18.72   Much skepticism appears to exist about reported EPA mileage figures for new cars, because the government tests only simulate actual driving conditions while ignoring such factors as wind, road conditions, traffic intensity, and driver differences. A private testing agency compared the EPA mileage rating of 11 new makes of automobiles with the mileage recorded on a test run under actual driving conditions. The results are shown in the table. (*A* indicates that the mileage obtained under actual driving conditions was greater than the EPA figure; *E* indicates that the EPA mileage rating was greater; 0 indicates that the two mileage ratings were equal.) Do the findings of the independent testing agency indicate that the EPA ratings are higher than what you can expect to obtain for gasoline mileage while driving under normal conditions? Use a test with a level of significance as near as possible to 10%.

Car Make	1	2	3	4	5	6	7	8	9	10	11
Test Results	E	0	E	A	E	E	E	E	E	A	E

18.73   A plywood manufacturer wished to compare the durability, under the presence of moisture, of two different brands of exterior veneer glue. The experiment consisted of applying both brands of glue *A* and *B* to each of ten 4 × 8, 3/4-in. plywood panels, with one brand being applied to a different 4 × 4 half of each panel. All 10 panels were then immersed in a vat of water until one end peeled. The data shown in the table indicate which of the glued ends of each panel were noted to peel under the immersion experiment. In the case of panel 7 both glued ends appeared to peel simultaneously. Do these results suggest that there is a difference between glues *A* and *B* in bonding strength under the presence of moisture? Use $\alpha = .05$.

Panel	1	2	3	4	5	6	7	8	9	10
Glue	A	A	B	A	A	A	0	B	A	A

18.74   Variable-rate mortgages have become popular in recent years but offer uncertainty to both lender and borrower during a period when interest rates are unstable. Many borrowers prefer to lock in a high prevailing rate rather than accept a variable-rate mortgage that may result in even higher future rates. To examine this issue, a savings and loan officer surveyed 17 applicants for home loans and found that 6 preferred a variable-rate mortgage, 9 favored a fixed-rate mortgage, and 2 had no opinion. Do these responses indicate that more people generally favor fixed-rate, as opposed to variable-rate, mortgages? Use a test with a level of significance as near as possible to 5%.

18.75   Refer to Exercise 9.82, in which the annual percentage change in manufacturing productivity was reported for the United States, Canada, and the United Kingdom from 1975 through 1986. The data are reproduced in Table 18.44.

   a.   Using Wilcoxon's signed-rank test, determine whether there is a significant difference in the percentage change in manufacturing productivity between the United States and Canada. Between the United States and the United Kingdom? Between Canada and the United Kingdom?

   b.   Test for a significant difference in the percentage change in manufacturing productivity among these three countries, using Friedman's test statistic with $\alpha = .05$. What do you conclude from your analysis?

Table 18.44   Productivity Data for Exercise 18.75

Year	United States	Canada	United Kingdom
1975	2.5	2.9	2.5
1976	4.6	4.5	2.9
1977	3.0	3.2	1.9
1978	1.5	1.2	.5
1979	−.1	−.3	−.2
1980	.0	.4	1.3
1981	2.2	1.4	2.4
1982	2.2	2.9	5.3
1983	5.8	2.9	6.1
1984	5.5	2.3	4.4
1985	5.1	2.5	3.8
1986	3.7	−.2	3.6

Source: United States figures from *The World Almanac and Book of Facts, 1988* (New York: World Almanac, 1987), p. 132; other figures reflect the averages reported by A. Neef and J. Thomas, "Trends in Manufacturing Productivity and Labor Costs in the U.S. and Abroad," *Monthly Labor Review*, Vol. 110, No. 12 (December 1987), pp. 25–30.

18.76   The advertising agent for a mail-order company conducted an experiment to determine the effect of color versus black and white advertisements. To advertise a special sale, the agent selected 1,000 households in each of eight randomly selected cities across the nation. In each city the head of 500 households received a color advertising circular and another 500 received a black and white advertisement. Two months after all advertisements had been mailed, the number of mail orders placed by the recipients of each type of advertisement from each of the eight cities was found to be as shown in Table 18.45. Test the hypothesis that the two types of advertising are equally effective, as measured by the number of orders placed by the recipients of each type of advertisement, against the alternative that the two types of advertisements are not equally effective. Use $\alpha = .05$.

18.77   Two makes of automobile tires were tested on the rear wheels of 12 different automobiles. The number of miles before tire failure was recorded for each tire. The results are shown in Table 18.46.

   a.   Using the sign test, test the hypothesis that the two makes of tires have the same usable life, as measured by the miles before failure, against the alternative that the usable lives are not equally effective. Use $\alpha = .05$.

b. Suppose that make B is a belted tire and make A is a fiberglass-belted tire, suggesting that make B has the capability of withstanding much more wear. Test the hypothesis of equal usable life against the alternative that make B is superior. Use $\alpha = .05$.

Table 18.45   Mail-Order Data for Exercise 18.76

	Type of Advertisement	
City	Color	Black and White
Atlanta	113	87
Boston	126	101
Chicago	89	90
Denver	105	82
Los Angeles	135	80
San Francisco	117	79
Seattle	175	113
St. Louis	71	93

Table 18.46   Tire Wear Data for Exercise 18.77

	Make of Tire			Make of Tire	
Automobile	A	B	Automobile	A	B
1	17,500	22,000	7	27,500	25,000
2	26,450	23,000	8	26,000	37,000
3	25,000	24,000	9	22,400	31,500
4	28,000	32,000	10	28,300	25,900
5	16,500	19,650	11	18,900	31,000
6	24,000	33,000	12	24,500	30,000

18.78 A recent study by C. Duncan and C. Lillis ("Directions for Marketing Knowledge Development: Opinions of Marketing Research Managers," *Journal of the Academy of Marketing Science*, Winter–Spring 1982) reports on a survey of marketing research managers to reveal their perceptions of the importance of marketing decision-making topics. Four marketing research managers were each asked to rank six prominent, marketing, decision-making topics in order of importance. Their responses are indicated in Table 18.47. Do these rankings provide sufficient

Table 18.47   Survey Data for Exercise 18.78

Marketing Research Manager	Marketing Decision-Making Topics					
	Market Potential	Competitive Conditions	Product Positioning	Sales Forecasting	Pricing	Concept Testing
Adams	1	4	3	2	6	5
Baker	2	5	1	3	4	6
Clayton	1	3	2	4	5	6
Davis	2	1	3	5	6	4

evidence to assume that a difference exists in the importance attributed to these six decision-making topics by market research managers? Let $\alpha = .01$.

18.79 Although some contend that pay increases are signs of organizational recognition, others suggest they are primarily valued as inflation adjustments. L. Krefting ("Differences in Orientation Toward Pay Increases," *Industrial Relations*, Winter 1980) contends that these variances in opinion are much more likely to occur between management and workers than within one group or the other. Six managers and eight line employees of an industrial organization were asked to respond to the following seven-point scale regarding their interpretation of the meaning of company pay increases:

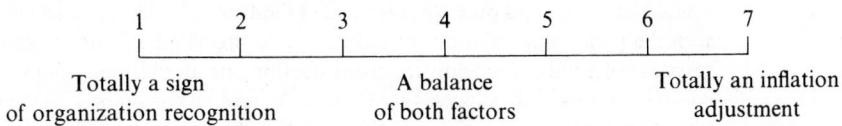

The survey provided the results shown in the table. Do these data present sufficient evidence to indicate a difference between the perceptions of management and line employees of the intent of pay raises? Use $\alpha = .05$.

Managers	2.4	4.3	3.0	3.5	5.4	2.9		
Line Employees	6.1	4.8	5.1	5.5	7.0	6.0	6.6	4.8

18.80 The outputs from two filler machines in a food-processing plant were examined to compare their fill levels. The measurements given in the table represent the fluid ounces of content from five filled containers selected from the output of each machine. Do these data present sufficient evidence to indicate a difference in the population of fill levels for the two machines? Use a level of significance of $\alpha = .10$.

a. Use the Mann-Whitney U test.

b. Use the Student's t test.

Machine A	30.5	30.2	30.0	31.2	30.7
Machine B	30.9	31.0	31.5	31.4	31.3

18.81 The deregulation of the banking industry has raised questions about consistency of service (and rates charged to customers) between large urban and smaller rural commercial banks. Table 18.48 lists the average interest rate paid on most short-term loans by rural, small city, and

Table 18.48   Interest Rate Data for Exercise 18.81

Bank	\multicolumn{10}{c}{Year–Quarter}
	1–1	1–2	1–3	1–4	2–1	2–2	2–3	2–4	3–1	3–2
Rural	17.1	18.2	18.6	17.6	17.4	17.4	15.9	14.6	13.8	13.4
Small city	17.8	19.1	19.6	18.0	17.7	17.5	15.5	14.2	13.7	13.1
Urban	18.3	19.7	20.1	18.1	18.1	17.8	14.8	13.7	12.7	12.7

Source: J. Scott and W. Dunkelberg, "Rural Versus Urban Bank Performance: An Analysis of Market Competition for Small Business Loans," *Journal of Bank Research*, Autumn 1984. Reprinted with permission.

urban banks over 10 consecutive quarters. Do these data present sufficient evidence to indicate a difference in the loan rates charged by rural, small city, and urban banks over the period of time indicated? Test by using $\alpha = .05$.

18.82   In an effort to cut prices, neighborhood groups are forming buying clubs—small, informal cooperatives. The objective of these co-ops is to save money by eliminating part of the markup a retail store must include in its prices to cover overhead and allow for a profit. The disadvantage associated with buying clubs is their limited selection, which eventually causes many co-op members to lose interest in the co-op and to drop their membership. Interviews were conducted among 13 members of a local food co-op to examine this issue. Eight of the co-op members had been members for at least two months and 5 had been members for less than two months. Each was asked to respond on a 10-point scale to indicate the degree to which he or she was satisfied with the prices and selection offered by the co-op. These results are shown in Table 18.49. (A response of 1 indicates complete dissatisfaction and 10 indicates complete satisfaction with the co-op.) Do these data suggest that those who have been members of the food co-op for at least two months are less satisfied with their membership than those who have been members for less than two months? Use $\alpha = .05$.

Table 18.49   Response Data for Exercise 18.82

Time of Membership	Measure of Satisfaction							
At least 2 months	5.0	4.5	6.0	9.5	3.5	5.0	8.5	4.0
Less than 2 months	8.0	9.0	7.5	9.5	10.0			

18.83   Items emerging from a continuous production process were classified as defective or nondefective. A sequence of items observed over time was as follows:

$D\,N\,N\,N\,N\,N\,N\,D\,D\,N\,N\,N\,N\,N\,D\,D$
$D\,N\,N\,N\,N\,N\,D\,N\,N\,N\,D\,D\,N\,N\,N\,D\,D$

  a. Find the probability that $R \le 11$, when $n_1 = 11$ and $n_2 = 23$.

  b. Do these data suggest lack of randomness in the occurrence of defectives D and nondefectives N? Use the large-sample runs test.

18.84   A quality control chart has been maintained for a certain measurable characteristic of items taken from a conveyor belt at a certain point in a production line. The measurements obtained today, in order of time, are listed from left to right as follows:

68.2	71.6	69.3	71.6	70.4	65.0	63.6	64.7
65.3	64.2	67.6	68.6	66.8	68.9	66.8	70.1

  a. Classify the measurements in this time series as above or below the sample mean and determine (use the runs test) whether consecutive observations suggest lack of stability in the production process.

  b. Divide the time period into two equal parts and compare the means, using the Student's t test. Do the data provide evidence of a shift in the mean level of the quality characteristics?

18.85   A large corporation selects college graduates for employment, using both interviews and a psychological achievement test. Interviews conducted at the home office of the company were far more expensive than the tests conducted on the campuses. Consequently, the personnel office was interested in determining whether the test scores were correlated with interview ratings and whether tests could be substituted for interviews. The idea was not to eliminate interviews but to reduce their number. To determine whether correlation was present, a personnel officer ranked 10 prospects during interviews and then tested the prospects. The paired scores are shown in Table 18.50. Calculate the Spearman rank correlation coefficient r_s. Rank 1 is assigned to the candidate judged to be the best. Do these data present sufficient evidence to indicate that the correlation between interview rankings and test scores is less than zero? If this evidence does exist, can we say that test scores could be used to reduce the number of interviews?

Table 18.50   Interview and Test Data for Exercise 18.85

Subject	Interview Rank	Test Score
1	8	74
2	5	81
3	10	66
4	3	83
5	6	66
6	1	94
7	4	96
8	7	70
9	9	61
10	2	86

18.86   Refer to Exercise 9.108. The data from that exercise are reproduced in Table 18.51, giving the average price of a gallon of gasoline (in United States dollars) during 1984 and 1985 for six industrial nations. Use Wilcoxon's signed-rank test procedure to determine whether the average price of gasoline changed significantly from 1984 to 1985. Does your conclusion using Wilcoxon's test procedure agree with your conclusion using the t statistic? If not, how do you account for this discrepancy?

Table 18.51   Gasoline Prices Data for Exercise 18.86

City, Country	1984	1985
United States	$1.37	$1.34
Paris	2.27	2.18
Rome	2.77	2.52
Bonn	1.85	1.66
Mexico City	1.20	1.36
Tokyo	2.42	2.37

Source: Data from U.S. Department of Commerce, Bureau of the Census, Statistical Abstract of the United States, 107th ed. (Washington, D.C., 1987), p. 592.

18.87   According to a survey of $n = 155$ fleet managers, General Motors cars represent the best fleet car deal. However, the GM vehicles were criticized for their quality in comparison with their competitors' quality (*Sales and Marketing Management*, November 1987, p. 34). Suppose that of the $n = 155$ fleet managers surveyed, 96 preferred GM cars over those of Ford or Chrysler as the best fleet vehicle buy. Use the sign test to determine whether this information provides sufficient evidence to conclude that GM cars are preferred to Ford or Chrysler cars as the best fleet buy.

18.88   In a report issued by California's Department of Insurance, the complaints among the state's top 25 sellers of homeowner's insurance ranged from .061% to 1.049% per 1,000 policies during 1986. The data are presented in Table 18.52.

  a.   Use Spearman's r_s to summarize the relationship, if any exists, between the number of homeowner's policies sold in 1986 and the percentage of complaints per 1,000 policies sold.

  b.   Do these data suggest that, on the average, the percentage of complaints per 1,000 policies increases as the number of policies sold increases? Use $\alpha = .05$.

Table 18.52   Homeowners' Complaints Data for Exercise 18.88

	Company	Policies Sold in 1986	Total Complaints	Percent Complaints per 1,000 Policies
1.	USAA	130,795	8	.061
2.	20th Century	61,109	5	.082
3.	Oregon Mutual	44,197	4	.091
4.	State Farm	1,386,404	173	.125
5.	Chubb & Son	55,652	7	.126
6.	Calif. State Auto Assn.	195,147	27	.138
7.	Hartford Fire & Casualty	116,529	17	.146
8.	Crum & Forster (All West)	78,662	12	.153
9.	CalMutual	63,133	10	.158
10.	Farmers	1,215,953	206	.169
11.	Allstate	982,186	177	.180
12.	Ohio Casualty	93,690	19	.203
13.	California Casualty	98,125	20	.204
14.	Prudential	68,166	14	.205
15.	Reliance	72,998	16	.219
16.	Travelers	114,909	26	.226
17.	Safeco	225,477	54	.239
18.	Residence Mutual	87,860	21	.239
19.	Fireman's Fund	201,442	49	.243
20.	Transamerica	81,515	20	.245
21.	Aetna Life & Casualty	110,329	34	.308
22.	Continental Corp.	91,978	33	.359
23.	Royal	31,733	12	.378
24.	Cigna	43,307	22	.508
25.	Republic Financial	157,282	165	1.049

Source: California Department of Insurance, March 1988.

18.89   A management scientist wished to examine the relationship between the buyer's image of a particular manufacturer's product line and the distance between the buyer and the manufacturer's plant. Each of 12 buyers for regional variety stores rated the manufacturer's product line on a scale of 1 to 20. The data are given in Table 18.53. Calculate the Spearman rank correlation coefficient r_s. Do these data provide sufficient evidence to indicate a negative correlation between rating and distance? Interpret your results in the context of this exercise.

Table 18.53   Sample Data for Exercise 18.89

Buyer	Rating	Distance	Buyer	Rating	Distance
1	12	75	7	9	680
2	7	165	8	18	60
3	5	2,750	9	3	1,800
4	19	10	10	8	1,150
5	17	950	11	15	700
6	12	2,230	12	4	510

18.90 According to some people, brand loyalists possess blind loyalty to a brand even though they may agree that a competing brand is more efficient in terms of the primary use of the product. To examine this theory, a researcher contacted nine housewives and asked them to rank the seven leading brands of powdered laundry detergent in terms of the "cleaning power" they perceived the brands to possess. Five of the nine housewives were loyal to brand *A*, but the other four possessed no brand loyalty among the five competing brands. The ranks given to brand *A* by the nine housewives are listed in the table. (A rank of 1 implies greatest cleaning power, 7 implies least cleaning power.) Do these data suggest that loyalists to brand *A* rank it higher in "cleaning power" than those who possess no brand loyalty? Test at a 10% level of significance.

Brand Loyalists	2	1	3	5	2
Brand Switchers	3	4	1	6	

18.91 Claxton and Ritchie report on a study of prepurchase shopping problems in retail stores undertaken to discover differences in shopping difficulties between men and women. Recorded in Table 18.54 are the consensus rankings by men and women of the relative difficulty in shopping for 24 different products and services, as reported by these authors. Do these data provide sufficient evidence to assume that an association exists between the assessments of men and women regarding the relative difficulty of prepurchase shopping problems?

Table 18.54   Shopping Difficulty Data for Exercise 18.91

Rank Order of Shopping Difficulty		
Men	Women	Product/Service
2	1	Auto repairs
4	2	Automobiles
1	3	Home improvement
13	4	Clothing and footwear
5	5	Household furnishings
7	6	Household appliances

(continues)

Table 18.54   (Continued)

Rank Order of Shopping Difficulty		
Men	Women	Product/Service
6	7	Housing and real estate
8	8	Home entertainment
3	9	Life insurance
9	10	Household moving
17	11	Sporting goods
19	12	Groceries
12	13	Photographic equipment
10	14	Auto insurance
15	15	Children's toys
18	16	Home-gardening supplies
11	17	Legal services
23	18	Dry cleaning
20	19	Stationery supplies
16	20	Travel agency services
21	21	Drugs and pharmaceuticals
14	22	Jewelry
22	23	Financial services
24	24	Personal care

Source: J. Claxton and J. Ritchie, "Consumer Prepurchase Shopping Problems: A Focus on the Retailing Component," *Journal of Retailing*, Fall 1979. Reprinted with permission.

Supplementary Exercises

Interpretive

18.92   When should a statistician choose to employ a nonparametric statistical method instead of a parametric test like the t test or the z test?

18.93   For each of the following problem settings, specify an appropriate nonparametric statistical test.

a.   The loan officer of a commercial bank would like to measure the association between loan size and the order in which loan requests are received each month.

b.   An investor is interested in the pattern of U.S. Treasury bill rates and whether past rate changes have been random over time.

c.   An industrial relations officer is interested in comparing management and worker acceptance of a set of personnel policies designed to improve productivity. Individual responses were measured on a five-point scale.

d.   A personnel officer would like to determine the comparative preferences of 25 employees regarding two proposed employee benefit plans.

e. Twenty police cars are run for a period of time with ordinary gasoline and later with a mixture of gasoline and alcohol (gasohol). It is of interest to compare the performance of the police cars using gasoline and gasohol as fuel.

18.94 The biomedical industry in the United States is rigorously regulated by the Food and Drug Administration (FDA) with regard to product quality, product labeling, and advertising standards. In assessing the impact of FDA regulation upon young biomedical firms, Hauptman and Roberts sampled firms formed in Massachusetts during the period from 1968 to 1975 for the purpose of marketing biomedical and/or pharmaceutical products and ultimately compiled information on 26 such firms. As one aspect of their study, they investigated the effect of a 1976 FDA regulation dealing with medical devices and pharmaceuticals on these firms. Table 18.55 summarizes the schedule of launching of the first three new products (the number entering the market) by each of these 26 firms during three marketing periods; 1970–1974, the period prior to the 1976 FDA regulation; 1975–1979, the transition period for the regulation; and 1980–1983, at least four years after the regulation. Not all 26 firms launched three new products during these marketing periods. Calculate the average number of new products launched per year for each of the three time periods. Do the data reflect a decrease in the average number of new products during the transition period? Does the number of new products introduced differ significantly for the three time periods in the study?

Table 18.55   New-Product Data for Exercise 18.94

New Products	Marketing Periods		
	1970–1974	1975–1979	1980–1983
1	17	6	3
2	8	6	6
3	4	3	9

Source: Data from Oscar Hauptman and Edward B. Roberts, "FDA Regulation of Product Risk and Its Impact upon Young Biomedical Firms," *Journal of Product Innovation Management*, Vol. 4 (1987), pp. 138–148.

18.95 An accounting firm is considering the use of a new format for its audit reports of subscribing firms. The new format consists of numerous figures, graphs, and charts instead of numerical and verbal descriptions of the firm's financial records. Twelve subscribing firms were issued audit reports using the new format. Of this group 8 indicated that they prefer the new format, and 4 said they prefer the old format. Decide on a null and alternative hypothesis that might be of interest to the accounting firm, and use the data to test the hypothesis. Interpret your results.

18.96 In a study to determine the acceptability of a new package design, a firm decided to issue the product only in its current package to one store (store *B*) and only in its newly designed package to another store (store *A*). In the past when both stores sold the product in its current package, weekly sales were almost identical. Over a period of 15 weeks sales of the product showed the sales patterns given in the table. (0 implies that the sales in store *A* and store *B* were equal for that week.) The firm must decide whether or not to begin marketing its product with the new package design. Decide on an appropriate null and alternative hypothesis for the firm, and use the data to test the hypothesis. Explain your conclusions.

Week	1	2	3	4	5	6	7	8	9	10	11	12	13	14	15
Store with Greatest Sales	B	0	A	A	B	A	B	B	A	A	A	A	0	A	A

18.97 A filler machine in a food-processing plant is set so that the mean fill is 16 oz. When the process is in control, the actual load per can should vary about this mean in a random manner, with a variance that remains stable over time. Fill readings for the process, collected at one-hour intervals, were as shown in the table. Do these data suggest that the process is out of control?

Time Period	1	2	3	4	5	6	7	8
Fill Reading	15.87	16.14	15.98	16.03	16.05	16.01	16.08	15.94

18.98 Eight corporate advertising executives were asked to evaluate the comparative effectiveness to their firms of using television and radio advertisements versus advertisements in the printed media. Their responses, as listed on a 10-point Likert scale, are shown in Table 18.56. The paired-difference t test, the sign test, and the Wilcoxon signed-rank test are all possible tests that can be used in analyzing these data. Analyze the data by using these three tests. and compare the results. Explain any discrepancies in the test results. Are all the parametric test requirements met?

Table 18.56   Response Data for Exercise 18.98

Executive	Radio and TV	Printed Media
A	5.0	4.0
B	6.2	3.0
C	7.5	5.0
D	8.0	4.0
E	4.5	6.0
F	6.6	5.0
G	6.0	6.0
H	5.0	5.5

Case Study

Market Proxies: How Well Do They Perform?

In the capital asset pricing model (CAPM) for monitoring investment performance, an asset's measured risk factor depends upon the proxy (an observed index or measure of CAPM) selected as well as other security-specific characteristics. One viewpoint is that the choice of different proxies can lead to opposite rankings of portfolios; others contend that errors in the approximating index are unimportant in the long run.

In investigating whether a reasonable estimate of CAPM is likely to lead to invalid conclusions, Keith C. Brown and Gregory D. Brown evaluated the historical performance of publicly traded investment portfolios relative to six different market proxies (or indexes). In formulating these market proxies, Brown and Brown obtained the historical rates of return and the market values of five different classes of market assets covering the 32-year period from 1947 to 1978. These five classes were (1) common stock, (2) fixed-income corporate issues, (3) real estate, (4) U.S. government issues, and (5) municipal bonds. Brown and Brown created five market proxies (indexes), summarized as follows:

Index 1: Common stock
Index 2: Index 1 plus fixed-income corporate issues
Index 3: Index 2 plus real estate
Index 4: Index 3 plus U.S. government issues
Index 5: Index 4 plus municipal bonds

Of those mutual funds in operation during this time period, the final selection of 32 funds shown in Table 18.57 included an equal representation of Weisenberger's (Weisenberger Financial Services, *Investment Companies*, 1987) four investment classifications: growth, income, income and growth, and balanced investments.

Using each of the five indexes, the authors ranked these 32 funds according to measures that reflect the excess return on investments when compared with the risk-free returns on investments. The rankings for each fund by each of the five indexes and a ranking of each fund according to its average rate of return are given in Table 18.58.

Evaluate Spearman's rank correlation coefficient for all pairs of the five indexes, as well as for each index and the mean return for each of the 32 mutual funds. What can you conclude about the value of these indexes as they relate to each other and to the mean return of these 32 mutual funds?

Table 18.57 Mutual Funds Data for the Case Study

Fund Number	Code*	Name	Mean	Standard Deviation
1	GI	Affiliated Fund, Inc.	10.65	14.74
2	I	Axe-Houghton Income Fund, Inc.	9.52	15.90
3	G	Axe-Houghton Stock Fund, Inc.	9.94	19.74
4	I	Century Shares Trust	10.37	20.11
5	G	Chemical Fund, Inc.	10.13	20.54

Annual Returns (%)

(continues)

Table 18.57   (Continued)

Fund Number	Code*	Name	Annual Returns (%) Mean	Annual Returns (%) Standard Deviation
6	GI	The Colonial Fund, Inc.	9.77	16.25
7	B	Composite Bond & Stock Fund, Inc.	7.72	11.52
8	GI	Delaware Fund, Inc.	7.98	17.53
9	B	Dodge & Cox Balanced Fund	8.04	12.16
10	B	Eaton & Howard Balanced Fund	7.53	10.74
11	GI	Fidelity Fund, Inc.	11.46	17.21
12	G	Growth Industry Shares, Inc.	10.68	18.98
13	GI	The Investment Company of America	12.05	17.05
14	GI	Investment Trust of America	10.69	18.05
15	B	Investors Mutual, Inc.	7.30	11.63
16	I	Investors Selective Fund, Inc.	5.25	5.78
17	G	Johnston Mutual Fund, Inc.	9.68	15.22
18	B	Loomis-Sayles Mutual Fund, Inc.	7.37	11.12
19	G	Massachusetts Investors Growth Stock Fund	11.21	19.08
20	I	Mutual Investing Foundation–MIF Fund	9.72	15.82
21	G	National Investors Corporation	11.92	18.58
22	G	National Growth Fund	10.46	20.74
23	B	Nation-Wide Securities Company, Inc.	8.07	10.94
24	I	Puritan Fund, Inc.	11.19	15.79
25	B	The George Putnam Fund of Boston	9.07	13.42
26	G	Scudder Common Stock Fund, Inc.	9.72	17.85
27	I	Scudder Income Fund, Inc.	6.76	12.08
28	GI	Selected American Shares, Inc.	8.89	16.31
29	GI	State Street Investment Corporation	10.80	15.73
30	I	United Income Fund	10.17	16.50
31	B	Wellington Fund	7.13	11.18
32	I	Wisconsin Income Fund, Inc.	8.36	15.04

* Weisenberger classifications: G, growth; GI, growth and income; and B, balanced.
Source: Keith C. Brown and Gregory D. Brown, "Does the Composition of the Market Portfolio *Really Matter?" Journal of Portfolio Management*, Winter 1987, p. 28. Reprinted with permission.

Table 18.58   Performance Rankings Data for the Case Study

Fund Number*	Index 1	Index 2	Index 3	Index 4	Index 5	Index Average	Mean Return
1	3	4	6	10	3	5.2	9
2	8	9	9	4	6	7.2	19
3	11	13	5	6	10	9.0	14
4	7	8	4	8	12	7.8	11
5	12	14	3	5	7	8.2	13
6	16	15	10	12	8	12.2	15

(*continues*)

Table 18.58   (Continued)

Fund Number*	Index 1	Index 2	Index 3	Index 4	Index 5	Index Average	Mean Return
7	15	20	11	16	19	16.2	26
8	32	30	32	32	32	31.6	25
9	25	23	19	21	22	22.0	24
10	21	22	13	15	20	18.2	27
11	5	5	14	11	5	8.0	3
12	20	19	28	26	16	21.8	8
13	1	1	7	2	2	2.6	1
14	18	11	23	13	11	15.2	7
15	27	31	22	25	26	26.2	29
16	14	21	1	9	13	11.6	32
17	10	10	16	19	14	13.8	18
18	22	24	17	20	25	21.6	28
19	13	7	25	7	9	12.2	4
20	17	16	21	22	15	18.2	17
21	4	3	8	3	4	4.4	2
22	23	25	31	31	31	28.2	10
23	9	12	12	14	17	12.8	23
24	2	2	2	1	1	1.6	5
25	19	17	15	18	21	18.0	20
26	28	26	30	30	30	28.8	16
27	31	32	26	27	29	29.0	31
28	30	28	29	29	28	28.8	21
29	6	6	18	17	18	13.0	6
30	24	18	24	24	23	22.6	12
31	26	27	20	23	24	24.0	30
32	29	29	27	28	27	28.0	22

*Fund numbers as shown in Table 18.57.
Source: Keith C. Brown and Gregory D. Brown, "Does the Composition of the Market Portfolio *Really* Matter?" *Journal of Portfolio Management*, Winter 1987, p. 30. Reprinted with permission.

References and Suggested Readings

Beyer, W. H., ed. *Handbook of Tables for Probability and Statistics.* Cleveland, Ohio: The Chemical Rubber Company, 1968.

Conover, W. J. *Practical Nonparametric Statistics.* 2d ed. New York: Wiley, 1980.

Kendall, M. G., and A. Stuart. *The Advanced Theory of Statistics, Vol. 2.* 4th ed. New York: Hafner Press, 1979.

Mann, H. B., and R. Whitney. "On a Test of Whether One of Two Random Variables Is Stochastically Larger Than the Other." *Annals of Mathematical Statistics*, Vol. 18 (1947), pp. 50–60.

Ryan, B. F., B. L. Joiner, and T. A. Ryan. *MINITAB Handbook.* 2d ed. Boston: PWS-KENT Publishing Co., 1985.

Savage, I. R. "Bibliography of Nonparametric Statistics and Related Topics." *Journal of the American Statistical Association*, Vol. 48 (1953), pp. 844–906.

Siegel, S. *Nonparametric Statistics for the Behavioral Sciences*. New York: McGraw-Hill, 1956.

Wald, A., and J. Wolfowitz. "On a Test of Whether Two Samples Are from the Same Population." *Annals of Mathematical Statistics*, Vol. 2 (1940), pp. 147–162.

Wilcoxon, F. "Individual Comparisons by Ranking Methods." *Biometrics*, Vol. 1 (1945), pp. 80–83.

TO
THE
READER

In this chapter we show how decision analysis extends the classical approach to decision making (Chapters 8 and 9) by formally integrating the decision maker's personal preferences and perceptions regarding uncertainty into the decision framework. Decision analysis requires that the decision maker identify available alternatives and payoffs associated with each alternative and assign probabilities to the uncertain states of nature. In this instance decision makers are forced to assume a more active role in the decision process, relying more on logic and intuition than on prescribed sets of rules and formulas.

Decision analysis has broad application in business because many business problems defy the requirements of classical statistical techniques. For example, when contemplating the acquisition of a competing business, an executive cannot rely on a sampling distribution to judge the probable success of the acquisition. His evaluation must instead be based on judgment and experience gained from an understanding of pertinent business and economic factors affecting the acquisition.

19

DECISION ANALYSIS

Introduction

The statistical tests of hypotheses and inferential methods presented in Chapters 8 and 9 are all based on sample information. As we recall, the objective of this **classical, or empirical, approach** to statistics is to make inferences about certain characteristics of a population—usually its mean or proportion of successes—based on information contained in a sample drawn from the population. Certain assumptions are made about the distribution of the population as suggested by the nature of the population or as implied by the Central Limit Theorem. On the basis of these assumptions we can determine a sampling distribution for our statistic of interest and hence construct an appropriate confidence interval or test of our hypothesis.

The classical approach was extended by Robert Schlaifer and others in the early 1960s to enable the decision maker to formally integrate his or her personal preferences and perceptions regarding uncertainty and value into the decision framework. This modern approach to decision making, which has come to be known as **decision analysis**, can be defined as the logical and quantitative analysis of all the factors that influence a decision. Decision makers are thus forced to assume more active roles in the decision-making process. In the end they rely more on rules consistent with their logic and personal behavior and less on the mechanical use of a set of formulas and tabulated probabilities.

decision analysis

One particular area of difference between the classical approach to statistics and decision analysis concerns the treatment of errors that may result from the application of each procedure. The classical approach employs the probabilities α and β of the two types of errors to evaluate the goodness of a test. In application, the classical approach often causes an arbitrary reliance on rules and formulas—as evidenced by the almost religious zeal attached to the significance level $\alpha = .05$. We know that great care should be taken when selecting the error probabilities α and β in a statistical test of an hypothesis. For example, these probabilities should consider losses associated with the errors they define as well as any prior information that may not be formally involved with the sample data used to calculate the test statistic. But quite often, an experimenter *arbitrarily* selects an α value (say $\alpha = .05$). In contrast, the decision analysis approach suggests that the **expected loss is a more practical and realistic criterion for comparing two statistical test procedures.** As we will see, decision analysis utilizes the concept of gain (or loss) associated with every possible decision available to the decision maker and selects the decision that maximizes the expected gain.

The primary purpose of decision analysis is to increase the likelihood of good outcomes by making good decisions. A good outcome is one that we would like to occur, and a good decision is one that is consistent with the information and preferences of the decision maker. Thus, decision analysis attempts to provide a decision-making framework based on any and all information, whether it be sample information, judgmental information, or a combination of both. In this sense, *judgment* refers to the ability of the decision maker to decide which cost and risk considerations are important and should be included in the analysis, which noneconomic factors are important for consideration, and which probability distributions are appropriately assigned to uncertain events. It does not refer to snap judgments, decisions made without careful analysis of all available information.

In summary, decision analysis is not really in conflict with the classical approach to statistical inference. The difference is simply in the degree of formality of the decision-making procedures employed and the extent to which the decision maker plays an active or a passive role in the formulation of the decision.

19.2

Certainty Versus Uncertainty

In most decision-making situations the decision maker must choose one of several possible actions or alternatives. Usually, the alternatives and their associated payoffs are known in advance to the decision maker. An investor choosing one investment from several investment opportunities, a store owner determining how many of a certain type of commodity to stock, and a company executive making capital budgeting decisions are all examples of a business decision maker selecting from a multitude of alternatives. However, the decision maker does not know which alternative will be best in each case unless he or she also knows with certainty the values of the economic variables that affect profit. We call these variables **states of nature** because they represent different events that may occur or states over which the decision maker has no control.

states of nature

Generally, the states of nature in a decision problem are denoted by the symbols $s_1, s_2, \ldots, s_k$. That is, the first state of nature is denoted by s_1, the second by the symbol s_2, and so forth. **It is assumed that the states of nature are mutually exclusive (no two states can be in effect at the same time) and collectively exhaustive (all possible states are included within the analysis).**

actions   If *n* **actions** or alternatives are available to the decision maker, they are generally labeled $a_1, a_2, \ldots, a_n$. **It is also assumed that the alternatives constitute a mutually exclusive, collectively exhaustive set.**

The exact identity of the alternatives and states of nature associated with a decision-making situation is unique to that problem. An investor may wish to invest a certain sum of money in a savings account, a Treasury bond, or a number of shares of a mutual fund. His available alternatives are the investment of his money in one of the three mentioned investment opportunities, and the states of nature that may influence his economic payoffs are the possible states of the securities market in the near future. Another investor may wish to invest none, part, or all of the same or a different sum of money among the same three investment opportunities. Her alternatives include all possible splits of her investment sum among the three investment opportunities and may be practically endless. In such cases alternatives may have to be grouped and others eliminated in order to arrive at an analytical solution to the decision-making problem. Any decision-making problem requires ingenuity and perceptiveness of the decision maker, since he or she must identify the available actions or alternatives and their associated payoffs and must identify the states of nature that may affect the outcome.

When the state of nature s_j is known or, if unknown, has no influence on the outcome of the alternatives, we say that the decision maker is operating under certainty   **certainty**. Otherwise, we say the decision is made under **uncertainty**.

uncertainty   Decision making under certainty is usually the simpler of the two techniques. The decision maker simply evaluates the outcome of each alternative and selects the one that best meets the objective. However, if the number of alternatives is very large, even in the absence of uncertainty, the best decision may be difficult to identify. For example, consider the problem of a delivery agent who must make 100 or so deliveries to different residences scattered over a broad urban area. There may be literally thousands of different alternative routes the agent could select, yet only one supplies a least cost alternative. However, if he had only three stops to make, the agent could easily find the least cost route.

Under uncertainty the decision-making task is always complicated. Probability theory and mathematical expectation offer tools for establishing the logical procedures for selecting the best alternative. Statistics provides the structure to reach the decision. But the decision maker must inject his or her intuition and knowledge of the problem into the decision-making framework to arrive at the decision that is both theoretically justifiable and intuitively appealing. A good theoretical framework and a commonsense approach are both essential ingredients for decision making under uncertainty. One cannot function properly without the other.

To understand these ideas, consider an investor who wishes to invest $1,000 in one of three possible investment opportunities. Investment *A* is a savings plan with a commercial bank returning 6% annual interest, and investment *B* is a government bond returning the equivalent of 4 1/2% annual interest. These two investments involve no risk and hence no uncertainty to the investor. Investment *C* consists of shares of a mutual fund with a wide diversity of available holdings from the securities market. The annual return from an investment in *C* depends on the *uncertain* behavior of the mutual fund under varying economic conditions.

Assuming that the investor either will invest his $1,000 in one of the afore-mentioned investment opportunities or will not invest at all, his available actions are as follows:

a_1:   Do not invest.
a_2:   Select investment A, the 6% commercial bank savings plan.
a_3:   Select investment B, the 4 1/2% government bond.
a_4:   Select investment C, the uncertain mutual fund.

Actions a_1, a_2, and a_3 do not involve uncertainty since the outcomes associated with the selection of any of these actions do not depend on the uncertain market conditions. We could therefore say it is clear that action a_2 *dominates* actions a_1 and a_3.

Definition

dominating action

> One **action** is said to **dominate** another when the second action is never preferred to the first, regardless of the state of nature.

Unless the investor attaches patriotic considerations to the selection of the government bond, investment B would never be selected when investment A is available. Similarly, action a_1, which provides for no growth of the principal amount, is clearly inferior to the risk-free, positive-growth investment alternatives a_2 and a_3.

Investment C, the mutual fund investment, is another matter. This alternative (a_4) is associated with an uncertain outcome that, depending on the state of the economy, may produce either a negative return or a positive return. Thus, there exists no immediately apparent dominance relationship between action a_4 and action a_2 (the best among the actions involving no uncertainty, a_1, a_2, and a_3).

Suppose that the investor believes that if the market is down in the next year, an investment in the mutual fund would lose 10%; if the market stays the same, the investment would stay the same; and if the market is up, the investment would gain 20%. The investor has thus defined the states of nature for his investment decision-making problem as follows:

s_1:   The market is down.
s_2:   The market remains unchanged.
s_3:   The market is up.

A study of the market combined with economic expectations for the coming year may lead the investor to attach subjective probabilities of 1/4, 1/4, and 1/2, respectively, to the states of nature s_1, s_2, and s_3. The question of interest is, then, how can the investor use the information regarding investments A, B, and C and the expected market behavior as an aid in selecting the investment that best satisfies his objectives? We will consider this question in Sections 19.3 and 19.4.

EXERCISES Conceptual Level

19.1 For each of the following business decision-making problems, list the available actions and the states of nature that might result to affect the payoff.

    a. The selection of an electronic computer for the management of the financial records and inventory of a pharmaceutical wholesaling company

    b. The choice of one from among five different television advertising formats prepared by an advertising agency for a domestic airline

    c. The decision of a delicatessen manager about the number of loaves of French bread she should order when she knows the daily demand ranges from 10 to 15 loaves

    d. The decision by an entrepreneur to sell his company to a large conglomerate or retain ownership of his enterprise

    e. The pricing of a newly developed airless spray dispenser for paint

    f. The choice by an investor of one of the five major asset classes (stocks, corporate bonds, real estate, U.S. government securities, municipal bonds) for the investment of $75,000

19.2 Do good decisions always lead to good outcomes? Does the observation of a good outcome imply that the decision maker initially made a good decision? Explain.

19.3 As an individual involved in the construction of a new personal residence, suppose you are confronted with the problem of choosing from among two heating systems: electric or natural gas. The installation costs of each system are the same, but rate charges currently favor a forced-air electric heating system. However, cost projections offered by the local utility company suggest that rate charges for electric heat may in the near future surpass those for natural gas. Identify the actions and the states of nature for this decision problem. What additional criteria might be included in your ultimate decision involving the choice of a heating system? How would your decision be affected if you were building the house for immediate resale instead of building it as a personal residence?

19.4 Select a major decision you have had to make recently, such as the choice of where to go to college, the choice of which job offer to accept upon graduation, the choice of where to invest a certain part of your income, or some other appropriate decision problem. Identify the alternatives available to you and the states of nature associated with your decision problem. Did your decision problem involve decision making under certainty or under uncertainty?

19.3

Analysis of the Decision Problem

In any problem involving the choice of one from many alternatives, we first must identify all the actions we may take and all the states of nature whose occurrence may influence our decision. The action to take none of the listed alternatives whose outcome is known with certainty may also be included within this list. Associated with each action is a list of payoffs to the decision maker. Payoffs are the value consequences to him if he takes a specific action, assuming that each of the states of nature occurs. Of course, if an action does not involve risk, the payoff will be the same no matter which state of nature occurs.

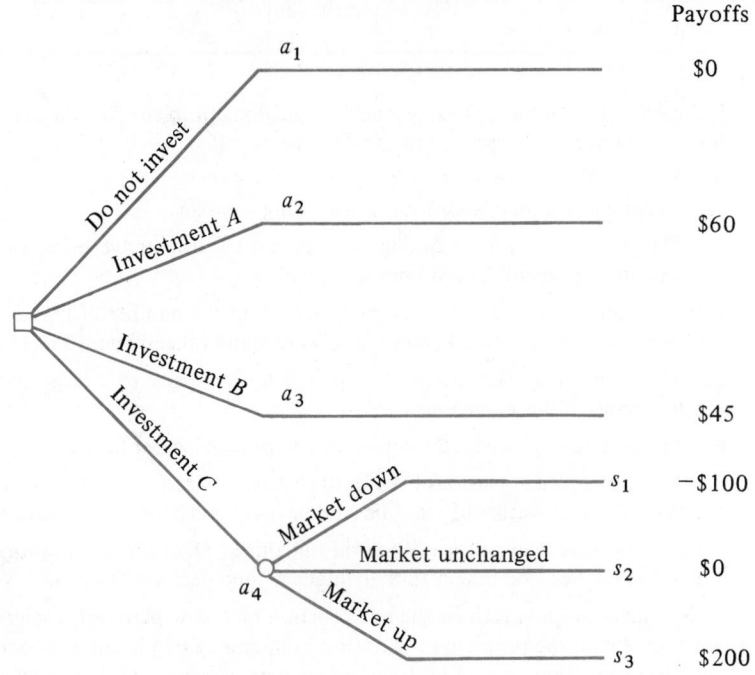

Payoffs

a_1 — $0

Do not invest

Investment A a_2 — $60

Investment B a_3 — $45

Investment C

a_4

Market down s_1 — -$100

Market unchanged s_2 — $0

Market up s_3 — $200

Figure 19.1   Decision Tree for the Investment Problem of Section 19.2

decision tree

One of the most illuminating things an analyst can do to clearly describe a decision problem is to draw its decision tree. A **decision tree** is created by first drawing branches, represented by lines, depicting the various actions $a_1, a_2, \ldots, a_n$ available to us. To each of these primary branches we then attach secondary branches denoting the possible states of nature $s_1, s_2, \ldots, s_k$. Finally, at the end of the last set of branches we list the payoffs associated with these states of nature. For example, the simple investment problem described in Section 19.2 would give rise to the tree shown in Figure 19.1.

The payoffs associated with each possible outcome in a decision problem can also be listed in a *payoff table*, which is defined in the display.

Definition

payoff table

A **payoff table** is a listing in tabular form of the value payoffs associated with all possible actions under every state of nature in a decision problem.

The payoff table is usually displayed in grid form, with the states of nature indicated in the columns and the actions in the rows. If we label the actions $a_1, a_2, \ldots, a_n$ and the states of nature $s_1, s_2, \ldots, s_k$, a payoff table for a decision problem would appear as shown in Table 19.1. A payoff is entered in each of the nk cells of the payoff table, one for the payoff associated with each action under every possible state of nature.

Table 19.1   Payoff Table

Action	State of Nature				
	s_1	s_2	s_3	$\cdots$	s_k
a_1					
a_2					
a_3					
$\vdots$					
a_n					

At this point let us digress for a moment and examine what is meant by the concept of a "payoff." What is the decision maker attempting to accomplish? Assuming that he is a rational decision maker, he will study the list of alternatives and **select the action that best satisfies his objectives**. The payoffs he assigns to each possible outcome must be assessed in value units consistent with his objectives. These value units may, in fact, not be measurable in monetary units at all. Or the value consequences may be represented in a payoff table as some measure combining profit and a subjective measure such as "possible environmental impact."

The following problem settings illustrate the importance of satisfying a goal by other than short-term profit maximization.

A doctor is faced with the problem of treating a group of factory employees, all of whom exhibit a rash on their upper bodies. She does not know the source of the rash, but for similar rashes she knows that an inexpensive medicated lotion works quite well. She also has access to an experimental, but costly, drug that may prove effective. Ten patients are given the lotion and 10 are given the experimental drug. After three days those who received the new drug are all back at work and completely free of the effects of the rash. Nine of the 10 patients who received the lotion are cured within six days, but the lotion does not effectively cure the tenth patient at all. The experimental drug is then administered to all future employees who exhibit the rash. Clearly, the cost of administering the new drug is incidental in this case. Cure rate and time lost from the job were the measures of payoff considered.

A company is considering building a subsidiary plant in one of two locations. The most costly location for building is in an area with chronic unemployment and low per capita income. The more expensive site location is chosen. Thus, the company executives have sought to maximize some social goal, perhaps identified as the firm's public image, instead of minimizing building costs. Payoff has been measured in social value units or public image units and is represented accordingly in the payoff table.

Defense planning is made by using probability-of-damage payoff as a measure to rank alternatives. Timber companies adopt a payoff model combining timber harvesting costs and environmental considerations, such as clear-cutting versus selective harvesting. Scientific expeditions measure payoff on an arbitrary scale in terms of the comparative worth of expected findings to the scientific community and the public.

value model

An example of the construction of a **value model** follows. This example involves the selection of one from among three alternatives under certainty since the effect of each attribute of value is known in advance to the decision maker.

EXAMPLE 19.1   The president of a large manufacturing company is considering three potential locations as sites at which to build a subsidiary plant. To decide which location to select for the subsidiary plant (only one location will be selected), the president will determine the degree to which each location satisfies the company's objective of minimizing transportation costs, minimizing the effect of local taxation, and having access to an ample pool of available semiskilled workers. Construct a payoff model, and select payoff measures that effectively rank each potential location according to the degree to which each satisfies the company's objectives for transportation costs, taxation costs, and the available work force.

SOLUTION   Let us call the three potential locations site A, site B, and site C. To determine a payoff measure to associate with each attribute (company objective) under each alternative, the president subjectively assigns a rating on a 0-to-10 scale to measure the degree to which each location satisfies the company's objectives for transportation costs, taxation costs, and the available work force. (For each objective a rating of 0 indicates complete dissatisfaction and a rating of 10 indicates complete satisfaction.) The results are shown in Table 19.2, and the associated decision tree is shown in Figure 19.2.

Table 19.2   Ratings Data for Example 19.1

Company Objective	Alternative		
	Site A	Site B	Site C
Transportation costs	6	4	10
Taxation costs	6	9	5
Work force pool	7	6	4

To combine the components of payoff, the president asks himself, "What are the relative measures of importance of the three company objectives I have considered as components of payoff?" Suppose he decides that minimizing transportation costs is most important and twice as important as either the minimization of state and local taxation or the size of the work force available. He thus assigns a weight of 2 to the transportation costs and weights of 1 to taxation costs and work force. The following payoff measures result:

$$\text{payoff (site } A) = 6(2) + 6(1) + 7(1) = 25$$
$$\text{payoff (site } B) = 4(2) + 9(1) + 6(1) = 23$$
$$\text{payoff (site } C) = 10(2) + 5(1) + 4(1) = 29$$

Rating

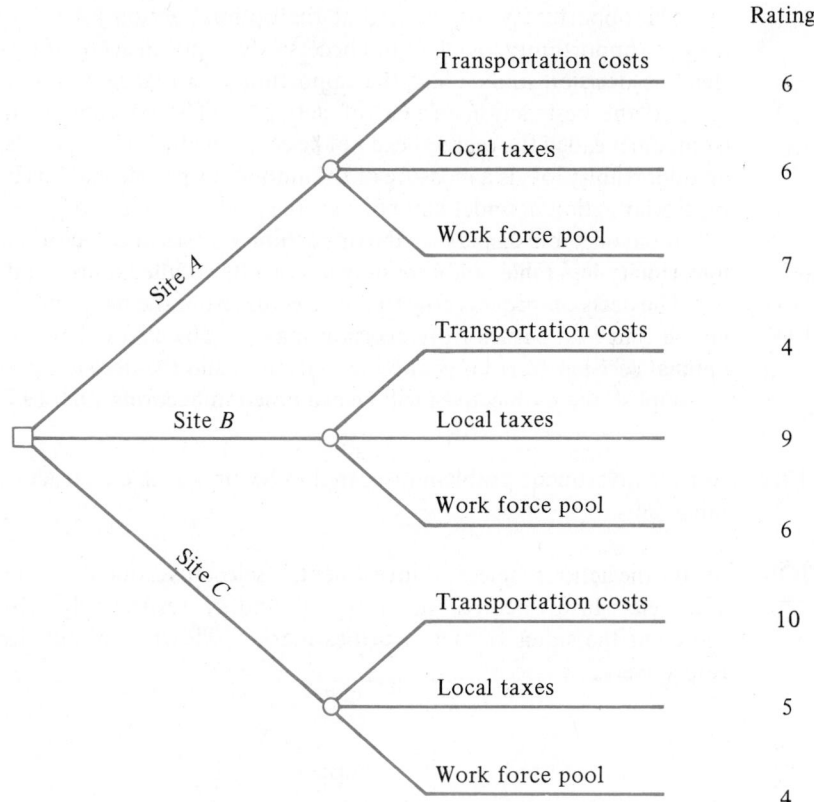

Transportation costs	6
Local taxes	6
Work force pool	7
Transportation costs	4
Local taxes	9
Work force pool	6
Transportation costs	10
Local taxes	5
Work force pool	4

Site A

Site B

Site C

Figure 19.2   Decision Tree for Example 19.1

◆ ◆

Before a decision maker can rationally evaluate a set of alternatives, he must first clearly identify his goals and objectives. Secondly, he must define a payoff measure that can effectively rank the outcomes according to the amount by which they satisfy his goals and objectives. The payoff measure is thus unique to the decision maker and his preferences and certainly unique to each new problem setting.

Advanced methods can and do handle most classes of nonmonetary outcomes in a decision analysis. However, these methods are reserved for a later course. For economy of time and space the examples in this chapter will assume an analysis involving monetary payoff, measurable by the profit or opportunity loss associated with each.

Definition

opportunity loss

The **opportunity loss** $L_{i,j}$ for selecting action a_i, given a state of nature s_j, is the difference between the maximum profit that could be realized if s_j occurs and the profit obtained by selecting action a_i. The opportunity loss is never negative.

The opportunity loss is zero at the optimal action for each state of nature. A nonzero opportunity loss does not necessarily imply an accounting loss. It only implies that the decision maker had the opportunity to realize $L_{i,j}$ more in profit had he selected the best action instead of action a_i. The concept of an opportunity loss **regret** (sometimes called the **regret**) need not be concerned with lost profits. In a general sense, an opportunity loss is a measure of the amount of payoff that has been lost by taking a particular action a_i under state of nature s_j.

A payoff table displaying the opportunity losses in a decision analysis is called an **opportunity loss table**, and one showing profits is called a **profit table**.

The decision process consists of selecting from the list of possible alternatives the action that best satisfies the decision maker's objectives. This decision is called the **optimal decision**. Various economic objectives and the decision processes necessary to accomplish these objectives will be examined in Sections 19.4, 19.7, and 19.8.

EXAMPLE 19.2   For the investment problem discussed in Section 19.2, construct the opportunity loss table.

SOLUTION   Define the actions "select no investment," "select investment A," "select investment B," and "select investment C" as a_1, a_2, a_3, and a_4, respectively. Also, let s_1, s_2, and s_3 represent the states of the securities market, "down," "about the same," and "up," respectively.

Table 19.3   Profit Table for the Investment Problem (Dollar Units)

Action	State of Nature		
	s_1, Down	s_2, the Same	s_3, Up
a_1(none)	0	0	0
$a_2(A)$	60	60	60
$a_3(B)$	45	45	45
$a_4(C)$	−100	0	200

Table 19.4   Opportunity Loss Table for the Investment Problem

Action	State of Nature		
	s_1, Down	s_2, the Same	s_3, Up
a_1(none)	60	60	200
$a_2(A)$	0	0	140
$a_3(B)$	15	15	155
$a_4(C)$	160	60	0

As shown in the decision tree in Figure 19.1, actions a_1, a_2, and a_3 are not affected by the states of nature, but action a_4 is. The resultant profits associated with this problem are listed in Table 19.3. We can see that if state of nature s_1 or s_2 is in effect, the best action is to select a_2, the 6% savings plan. If the securities market is up, the $200 profit for investing in investment C far exceeds what we could realize by selecting any other action.

From the definition of opportunity loss given earlier and from the information in Table 19.3, we can now construct the opportunity loss table. It is given in Table 19.4.

◆ ◆

EXAMPLE 19.3 A retailer must decide each morning how many of a perishable commodity to include in her inventory. The commodities cost her $2 per unit and are sold for $5 per unit, realizing a profit of $3 per unit. Any unsold items at the day's end are a total loss to the retailer. Construct the opportunity loss table.

SOLUTION In this problem the states of nature are the demand levels, and the actions are the possible inventory levels. The profits associated with each action are given in Table 19.5, where inventory levels 1, 2, and 3 are denoted by a_1, a_2, and a_3, respectively. Similarly, the demand levels are labeled s_1, s_2, and s_3. The opportunity losses are as shown in Table 19.6.

Table 19.5   Profit Table for the Demand-Inventory Problem

	State of Nature, Demand		
Action, Inventory	$s_1(1)$	$s_2(2)$	$s_3(3)$
$a_1(1)$	3	3	3
$a_2(2)$	1	6	6
$a_3(3)$	−1	4	9

Table 19.6   Opportunity Loss Table for the Demand-Inventory Problem

	State of Nature, Demand		
Action, Inventory	$s_1(1)$	$s_2(2)$	$s_3(3)$
$a_1(1)$	0	3	6
$a_2(2)$	2	0	3
$a_3(3)$	4	2	0

◆ ◆

In the investment problem of Section 19.2, the states of nature that will be in effect have no bearing on the outcome of investments *A* and *B*. But with investment *C* the unknown state of nature that will be in effect does influence the outcome. We hope that this observation illustrates the importance of an earlier remark: Personal judgment and subjective reasoning are required of the decision maker to identify the relevant states of nature and to place precise probabilities on their occurrence. The intuition, perception, and judgment of the decision maker are thus an integral part of the analysis in modern decision making.

Definition

prior probabilities

> The probabilities representing the chances of occurrence of the identifiable states of nature in a decision problem before gathering any sample information are called **prior probabilities**.

How should the prior probabilities be determined? As indicated in Example 19.2, we may sometimes be able to estimate the probabilities of different states of nature from previous experience. Our probability estimate may then just be the relative frequency of occurrence of that particular state of nature in the past. However, in other decision problems our feeling about which states of nature are more likely to occur cannot be summarized solely by such historical relative frequencies. In those cases some other method of assessing prior probabilities is required. [See Howard (1966) or Winkler (1972) listed in the references for some techniques available in eliciting personal opinion concerning the prior probabilities.]

In short, there is no simple answer to the question of how to select the prior probabilities. Realizing that the optimal action is often quite sensitive to only a slight change in the set of prior probabilities, decision makers seek a precise and meaningful set of prior probabilities. To accomplish this end, they use all available information at their disposal and rely on their judgment and experience to process that information and to identify, as well as possible, the prior probabilities.

We will not expand on the necessary aspects of the subjective elements of our decision framework and will assume now that they have been formulated. The discussion in the remainder of this chapter presents some logical procedures for selecting the best action after all states of nature have been identified and their associated payoffs and probabilities have been computed.

EXERCISES   Conceptual Level

19.5   What information is contained in a payoff table? How should a decision maker use that information?

19.6   Reconstruct your decision to attend the college at which you are currently enrolled. Pick one or two other colleges that you might have chosen, and construct a payoff model for your selection decision like the one discussed in Example 19.1. As the value measures, choose educational quality and social environment of the college and location; then rate each college on a 0-to-10 scale according to each dimension of value. Combine the value measures according to their comparative measures of importance. Was your decision to attend the college at which you are currently enrolled consistent with the outcome of your analysis?

EXERCISES   Skill Level

19.7   A decision problem consists of three possible actions, a_1, a_2, and a_3, and four possible states of nature, s_1, s_2, s_3, and s_4. The profit table for this decision problem is shown in Table 19.7.

Table 19.7   Profit Table for Exercise 19.7

	State of Nature			
Action	s_1	s_2	s_3	s_4
a_1	10	−5	−15	20
a_2	5	0	−10	15
a_3	10	5	−20	10

a.   Construct a decision tree for the problem.

b.   Construct the opportunity loss table.

EXERCISES   Applications

   19.8   In the face of stiff competition from foreign competitors American automobile manufacturers have begun to phase out production of large, gas-guzzling cars in favor of smaller economy cars. But as a service to dealers, manufacturers must still supply essential parts for discontinued models. Suppose Ford Motor Company has contracted with a manufacturer of automobile carburetors to supply Ford dealers with replacement four-barrel carburetors for a discontinued engine style. Carburetors will cost Ford $45 each and can be sold by dealers for $65. If ultimately unsold, four-barrel carburetors for the discontinued engine will be valueless. A Ford market analyst has provided the accompanying distribution for demand for the carburetors.

Demand	2,000	4,000	6,000	8,000	10,000	12,000	14,000
Probability of Demand	.05	.15	.35	.25	.10	.08	.02

a.   Identify the alternatives and the states of nature for Ford's decision.

b.   Construct a decision tree.

c.   Construct Ford's profit table for this decision problem.

d.   Construct Ford's opportunity loss table.

e.   Do the demand distribution data imply that actual demand will only be one of the seven listed values, eliminating, for instance, the likelihood that 6,524 carburetors will ultimately be sold? Explain.

 19.9 One researcher notes that floating currency exchange rates have created a great amount of uncertainty for those dealing in import and export trade (W. D. Serfass, Jr., "You Can't Outguess the Foreign Exchange Market," *Harvard Business Review*, March–April 1976). For example, consider the case of an importer of goods who is considering an offer from a New Zealand manufacturer of sheepskin rugs to distribute the manufacturer's finished goods in the United States. Recent New Zealand currency devaluations have created a favorable pricing situation for New Zealand products in foreign markets, but an upward float of the New Zealand dollar against the U.S. dollar would, of course, have an opposite effect. If the importer accepts the manufacturer's offer, he envisions the revenues shown in Table 19.8 under each of three possible exchange relationships that might occur between U.S. and New Zealand currencies over the next fiscal year. If the import agreement is accepted, the rugs will be transported to the U.S. market from New Zealand by air freight at an annual cost of $9,000.

a. Identify the alternatives and the states of nature for the importer's decision problem.

b. Construct the importer's profit table.

c. Construct the importer's opportunity loss table.

Table 19.8   Data for Exercise 19.9

Exchange Relationship	Net Revenues
Downward float of N.Z. dollar	$30,000
No change	15,000
Upward float of N.Z. dollar	7,500

 19.10 A manufacturing entrepreneur must decide how many tons of plastic shipping pellets to produce each week for sale. His manufacturing manager has determined, with an accountant's assistance, that total cost per ton to produce pellets is $25.00. There is also a cost of $2.00 per one-ton bag for each ton sold to shippers. The bagged pellets sell to shippers for $50.00 per ton. Pellets neither shipped nor bagged at the end of the week loose their spongy property and cannot be sold to shippers. However, a local farmer will pay $5.00 per ton for the older pellets, which he uses as filler in his cattle feed (the EPA and Department of Agriculture have approved this practice). Under the assumption that the weekly demand for plastic shipping pellets will range from 0 to 5 tons, construct the manufacturer's decision tree.

 19.11 Recent increases in oil prices have created greater incentives and opportunities for independent oil drillers, although drilling costs remain very expensive. Consider the case of an independent oil driller who holds a lease to land in a region with good prospects for oil discovery. The driller is considering three options: drill alone, join a syndicate with a group of local investors and share costs and profits on a 50–50 basis, or sell his lease rights to another driller for $100,000. Drilling costs are estimated to run $1,500,000, with an additional $350,000 required for equipment if oil is discovered. Oil can be sold at the wellhead for $15 per barrel. From past experience the driller believes that there are three likely outcomes if drilling is undertaken on his leasehold: a gusher providing 750,000 barrels of recoverable oil, a moderate find providing 150,000 barrels of recoverable oil, or a dry hole. Construct a profit table for the oil driller's decision problem.

19.4

Expected Monetary Value Decisions

One decision-making procedure, which employs both the payoff table and the prior probabilities associated with the states of nature to arrive at a decision, is called the *expected monetary value (EMV) decision procedure.*

Definition

expected monetary value (EMV) decision

An **expected monetary value (EMV) decision** is the selection of an available action based on either the expected opportunity loss or the expected profit of the action.

The optimal expected monetary value decision is the selection of the action associated with the minimum expected opportunity loss or the action associated with the maximum expected profit, depending on the decision maker's objective. We will note later that expected opportunity loss decisions and expected profit decisions are always associated with the same optimal decision.

The concept of expected monetary value is an application of mathematical expectation (Section 4.5), where opportunity loss or profit is the random variable and the prior probabilities represent the probability distribution associated with the random variable. The expected opportunity loss for each action a_i is found by evaluating the formula given in the display.

Computing the Expected Opportunity Loss by Using Prior Probabilities

$$E(L_i) = \sum_{\text{all } j} L_{i,j} P(s_j) \qquad i = 1, 2, \ldots$$

where $L_{i,j}$ is the opportunity loss for selecting action a_i given that state of nature s_j occurs and $P(s_j)$ is the prior probability assigned to state of nature s_j.

If we are interested in computing the expected profits for each action, the analysis would be the same except that we would substitute the profits, denoted by the symbol $G_{i,j}$, for the opportunity losses $L_{i,j}$ in the formula given in the display. (The symbol G, for gain, is used to denote profits to avoid the confusion that might result by using P, which throughout this text is used to represent the probability of an event.)

The following examples should clarify the computational procedures involved with the expected monetary value decision procedure.

EXAMPLE 19.4   Refer to Example 19.2. Find the optimal decision that minimizes the investor's expected opportunity loss.

SOLUTION   The prior probabilities associated with the states of nature s_1, s_2, and s_3 are $P(s_1) = 1/4$, $P(s_2) = 1/4$, and $P(s_3) = 1/2$. Thus, the expected opportunity losses are (refer to Table 19.4)

$$E(L_1) = (\$60)(1/4) + (\$60)(1/4) + (\$200)(1/2) = \$130$$
$$E(L_2) = (\$0)(1/4) + (\$0)(1/4) + (\$140)(1/2) = \$70$$
$$E(L_3) = (\$15)(1/4) + (\$15)(1/4) + (\$155)(1/2) = \$85$$
$$E(L_4) = (\$160)(1/4) + (\$60)(1/4) + (\$0)(1/2) = \$55$$

Action a_4, the mutual fund investment, is the optimal expected monetary value decision, since its expected opportunity loss is less than those for other actions.

Actions a_1 and a_3 could have been eliminated from the analysis at the beginning, since their payoffs are never better than those for a_2. They were retained in our example for illustrative purposes.

As an aside, suppose that we compute the expected profits for each action of Example 19.2. If G_i represents the profit realized by selecting action a_i, then we find (refer to Table 19.3)

$$E(G_1) = (\$0)(1/4) + (\$0)(1/4) + (\$0)(1/2) = \$0$$
$$E(G_2) = (\$60)(1/4) + (\$60)(1/4) + (\$60)(1/2) = \$60$$
$$E(G_3) = (\$45)(1/4) + (\$45)(1/4) + (\$45)(1/2) = \$45$$
$$E(G_4) = (-\$100)(1/4) + (\$0)(1/4) + (\$200)(1/2) = \$75$$

Consistent with our analysis based on opportunity losses, we see that action a_4 is best, since it is associated with the maximum expected profit. This result is no accident but will always happen. A proof of this assertion follows from the definition of opportunity loss. **The difference between the expected opportunity losses of any two actions is equal in magnitude to, but opposite in sign from, the difference between their expected profits.** For instance, in Example 19.4 if we look at actions a_2 and a_4, we find that

$$E(L_4) - E(L_2) = \$55 - \$70 = -\$15$$
$$E(G_4) - E(G_2) = \$75 - \$60 = \$15$$

Under either method action a_4 is \$15 better than a_2.

EXAMPLE 19.5   By recording the daily demand over a period of time for the perishable commodity described in Example 19.3, the retailer was able to construct a probability distribution for the daily demand level(s); the distribution is shown in Table 19.9. Find the inventory level that minimizes the expected opportunity loss for the demand-inventory problem described in Example 19.3.

SOLUTION   The historical frequencies of demand—the prior probabilities for this case—are given

Table 19.9   Probability Distribution
for Daily Demand Levels

s_j	$P(s_j)$
1	.5
2	.3
3	.2
4 or more	.0

in Table 19.9. The expected opportunity losses are found by evaluating

$$E(L_i) = \sum_{j=1}^{3} L_{i,j}P(s_j)$$

for each inventory level, $i = 1, 2, 3$.
    The expected opportunity losses at each inventory level are

$$E(L_1) = (\$0)(.5) + (\$3)(.3) + (\$6)(.2) = \$2.10$$
$$E(L_2) = (\$2)(.5) + (\$0)(.3) + (\$3)(.2) = \$1.60$$
$$E(L_3) = (\$4)(.5) + (\$2)(.3) + (\$0)(.2) = \$2.60$$

Thus, the retailer's optimal decision is to stock two units of the commodity.

Note that in Example 19.5 the prior probabilities assigned to the states of nature were constructed empirically from historical data. However, subjectivity and intuition are also implicit in this case, as evidenced by the fact that historical frequencies are believed to depict the future adequately. In any decision problem the prior probabilities reflect all available information, both empirical and subjective, related to the likelihoods of occurrence of the states of nature.

EXERCISES   Conceptual Level

19.12   What is an expected monetary value decision? What information is needed to make such a decision?

19.13   Explain the concept of expected opportunity loss. How does it relate to expected monetary value decisions?

EXERCISES   Skill Level

19.14   Refer to Exercise 19.7. If the prior probabilities for the four states of nature are $P(s_1) = .1$, $P(s_2) = .4$, $P(s_3) = .2$, and $P(s_4) = .3$, find the optimal decision using the expected monetary value (EMV) decision procedure.

    a. Find the expected profit for each action. Which action maximizes the expected profit?

    b. Find the expected opportunity loss for each action. Which action minimizes the expected opportunity loss?

    c. Do the results of parts a and b agree? Verify that the difference between the expected opportunity loss for any two actions is equal in magnitude to, but opposite in sign from, the difference between their expected profits.

EXERCISES   Applications

19.15 Refer to Exercise 19.9. Suppose that after consultation with an expert on international monetary issues, the importer decides that the prior probabilities are as follows: The probability that the New Zealand dollar will float downward from the U.S. dollar within the next year is .10, the probability that the exchange rate will stay about the same is .50, and the probability that an upward float of the New Zealand dollar will occur is .40. If the importer wishes to maximize his expected profits, should he accept the importing agreement with the New Zealand manufacturer? What is the expected profit associated with the decision to accept the importing agreement?

19.16 A real estate developer owns a parcel of land adjacent to a larger tract that will soon be rezoned for industrial use, for an office park, or for residential purposes. The developer must decide before the zoning decision is made whether to develop her property for a grocery store, restaurant, or gasoline service station. From her experience and judgment the developer constructs the profit table and prior probabilities as representative of her decision problem; see Table 19.10. If the developer seeks to maximize her expected profits, what development decision should she select? What is the expected profit for undertaking the best development alternative?

Table 19.10   Data for Exercise 19.16

Zoning Decision	Prior Probabilities	Development Decisions		
		Grocery Store	Restaurant	Service Station
Industrial	.25	−$20,000	$18,000	$25,000
Office park	.25	−10,000	50,000	15,000
Residential	.50	60,000	−15,000	−20,000

19.17 An investor is considering one of two alternatives for the investment of $25,000. On the one hand, he can accept a risk-free investment by investing the sum at 7% annual interest with a local savings and loan institution. On the other hand, he can invest the sum with a speculative developer who is building an elegant restaurant in a reasonably undeveloped area of the city. If successful, the restaurant investment is seen to return a net present value of $100,000 at the end of one year on the investor's $25,000 investment, with a probability of .25. If unsuccessful, it could be run as a fast-food restaurant with a one-year net present value return of $15,000; or if that proves unsuccessful, it could be turned into a storage warehouse at a return of $7,500 on the original $25,000 investment. The probabilities of these latter two events occurring are seen to be .50 and .25, respectively. If the investor wishes to maximize his expected return over the next year, should he invest in the risk-free savings account or invest in the speculative restaurant venture? What is the investor's expected capital gain for a one-year investment in each of the two investment opportunities?

19.18   Refer to Exercise 19.10. Suppose that after consultation with an experienced market analyst the entrepreneur determines the accompanying probability distribution of weekly demand by shippers for the plastic pellets. For this demand distribution, how many tons of plastic pellets should be produced to meet each week's demand if the entrepreneur wishes to maximize his expected profits?

Demand (Tons of Pellets)	0	1	2	3	4	5
Probability of Demand	.05	.10	.25	.40	.15	.05

19.19   Refer to Exercise 19.11. After a careful analysis of the geological characteristics of his leasehold land, the independent driller has concluded that chances are 80 in 100 that no recoverable oil will be discovered, 15 in 100 that a moderate find would result (150,000 barrels), and 5 in 100 that a drilling would provide a gusher (750,000 barrels). From this information, what is the independent driller's best decision if his objective is to maximize his expected profits?

19.20   A businessman is trying to decide whether to take one of two contracts or neither. He has simplified the situation somewhat and feels it is sufficient to imagine that the contracts provide the alternatives shown in Table 19.11.

a.   Which contract should the businessman select if he wishes to maximize his expected profit?

b.   What is the expected profit associated with the optimal decision?

c.   Give a practical interpretation for the expected profit from part b. Is it actually the profit the businessman can expect if he selects the optimal contract?

d.   Which alternative would *you* select? Explain.

Table 19.11   Profit Data for Exercise 19.20

Contract A		Contract B	
Profit	Probability	Profit	Probability
$100,000	.2		
50,000	.4	$40,000	.3
0	.3	10,000	.4
−30,000	.1	−10,000	.3

19.5

Justification of Expected Monetary Value Decisions

Expected profit or opportunity loss decisions are most easily interpreted if we look at the decision in the long run. If the optimal action is repeated many times, the average of the opportunity losses incurred using one procedure would be less than if the decision maker employed some other decision-making scheme. This idea is easy to see in

Example 19.5. If the retailer stocks two units every day, her average of all daily opportunity losses will be a minimum and will be about $1.60, which is less than the average loss incurred by stocking one or three units each day.

In an experiment that cannot be repeated, such as the investment problem of Section 19.2 or a capital budgeting problem, expected monetary value decisions are not so easily justified. For the investment problem we would expect the average profit for investment C to be $75, or a return of 7 1/2% on the investment. Thus, the expected monetary value decision maker selects investment C, since its *expected* return of 7 1/2% is greater than the *sure* returns of 6% for investment A and 4 1/2% for investment B.

The best way of interpreting nonrepetitive expected monetary value decisions is to examine the expected opportunity losses. The expected opportunity loss associated with the optimal decision represents the loss to be expected if we select the optimal decision. This does not imply that the opportunity loss we will actually realize by selecting the optimal decision will equal this value. Often, it is impossible for this value to be realized. Refer to Example 19.4; the optimal decision was to select action a_4, with an associated expected opportunity loss of $55. Note, however, that if action a_4 is selected, the only possible opportunity losses that can be realized are $160, $60, and $0, none of which equals the expected opportunity loss of $55.

cost of uncertainty   What is implied is that the minimum expected opportunity loss measures the expected **cost of uncertainty** owing to our uncertain knowledge about which state of nature will be in effect. If the opportunity presented itself, we should be willing to pay up to the amount of the cost of uncertainty to obtain information identifying the state of nature that will be in effect. Minimizing our expected cost of uncertainty seems to be a logical objective, and this is what we are doing by selecting the action associated with the maximum expected profit and the minimum expected opportunity loss.

Another way to justify the use of this decision method is to examine the components of the expected monetary value. The profit or loss values are economic outcomes, and their relative frequencies of occurrence are the prior probabilities assigned to the states of nature. Thus, an expected monetary analysis provides a model that combines both real economic data and qualitative or subjective information available to the decision maker related to the outcome of the economic data. It is the only decision-making model available to the decision maker at the time that incorporates all available information into the decision. Also, it makes the decision maker more than just an impartial observer by forcing him or her to construct meaningful prior probabilities to associate with the states of nature. This latter concept is the most important and often the most difficult task in decision making under uncertainty.

A reassessment of the values assigned to the prior probabilities often changes the optimum decision. Suppose that after reconsidering his prior beliefs, the investor confronted with the investment problem of Section 19.2 decides to assign these prior probabilities:

$$P(s_1) = 1/4 \qquad P(s_2) = 1/2 \qquad P(s_3) = 1/4$$

His expected profit for investment C is now

$$E(G_4) = (-\$100)(1/4) + (\$0)(1/2) + (\$200)(1/4) = \$25$$

But the expected profits for investments A and B remain at $60 and $45, respectively. Hence, under the new set of prior probabilities investment C is only third best behind the "new" optimum decision to select investment A.

Expected monetary value decisions, since they employ probabilities in the decision analysis, assume that the prior probabilities assigned to the states of nature are the true probabilities for the problem. The optimal expected monetary value decision is meaningful only in terms of its associated prior probabilities. If the decision maker does not have complete confidence in the prior probabilities that he has assigned to the states of nature, he should try other likely sets of prior probabilities to examine the sensitivity of the decision to the selection of the prior probabilities. The conduct of a sensitivity analysis is demonstrated in the following example.

EXAMPLE 19.6   R. V. Brown ("Do Managers Find Decision Theory Useful?" *Harvard Business Review*, May–June 1970) offers an elementary example of a sensitivity analysis, which we will revise in our discussion here. A New York metal broker has the opportunity to buy 100,000 tons of iron ore from a foreign government, at the advantageous price of $5 a ton. The broker is certain that he can sell the ore in the United States for $8 a ton, but there is a chance that the U.S. government will refuse to grant him an import license. In such an event the import contract would be annulled, at a cost of $1 a ton.

    a.   If the broker seeks to maximize his expected profit, should he buy the ore from the foreign government if he believes the chances are 50–50 that the U.S. government will grant him an import license?

    b.   Because of conflicting reports from a number of advisors, suppose the broker is very uncertain about the chances of the U.S. government granting him an import license. How great does the probability that the U.S. government will grant him a license have to be for the broker to accept the advantageous offer from the foreign government?

SOLUTION   The alternatives available to the broker are that he accept the offer from the foreign government and buy the ore (a_1) or that he not buy the ore (a_2). The states of nature concern the actions of the U.S. government regarding the granting of an import license; either the license is granted (s_1) or it is not granted (s_2). The payoff for each possible outcome is listed in the payoff table (Table 19.12). The payoffs listed in the payoff table are profits, or the difference between revenues and costs for each possible outcome.

Table 19.12   Payoff Table for the Broker's Problem in Example 19.6

	State of Nature	
Action	License Granted, s_1	Not Granted, s_2
Buy ore, a_1	$300,000	−$100,000
Do not buy ore, a_2	0	0

    a.   If the chances of a license being granted are 50–50, then $P(s_1) = 1/2$ and $P(s_2) = 1/2$. We find that the expected profit associated with each

alternative is

$$E(G_1) = (\$300,000)(1/2) - (\$100,000)(1/2) = \$100,000$$
$$E(G_2) = (\$0)(1/2) + (\$0)(1/2) = \$0$$

Therefore, the broker is maximizing his expected profits by buying the ore.

b. If the broker does not buy the ore, his expected profit is $0 (nothing ventured, nothing gained or lost). Therefore, he should buy the ore as long as $E(G_1) > E(G_2)$, or as long as $E(G_1) > 0$. Suppose we let p be the probability that the U.S. government will grant the broker a license. Then we find

$$E(G_1) = (\$300,000)(p) - (\$100,000)(1 - p)$$

Then $E(G_1) > 0$ if and only if $(\$300,000)(p) - (\$100,000)(1 - p) > 0$. Solving for p, we find that $E(G_1) > 0$ if and only if $p > 1/4$. Therefore, the probability that the U.S. government grants him a license has to be at least 1/4 before the broker decides to buy the ore. If the probability of a license being granted is less than 1/4, he will not buy the ore.

EXERCISES Conceptual Level

19.21 How does one justify expected monetary value decisions in one-time-only situations?

19.22 Explain the concept of expected cost of uncertainty.

EXERCISE Skill Level

19.23 Refer to Exercise 19.14. What is the expected cost of uncertainty for this decision problem?

EXERCISES Applications

19.24 The president of a manufacturing plant is faced with the possibility of accommodating a drastic increase in demand for his firm's product. His alternatives are to ignore the increase in demand, since his firm is currently operating at production capacity, or to buy out a small competing firm at a cost of $100,000 and use its resources to generate additional production. The firm stands to gain an additional $600,000 in revenues if the additional demand materializes, but it will not gain in additional revenues if demand continues to be stable. The chance that demand for the firm's product will increase is believed to be .10.

a. Construct an opportunity loss table for the decision problem.

b. If the president wishes to minimize his firm's expected opportunity loss, should he buy the competing firm to allow for additional production capacity?

c. Suppose the president is very undecided about the chance that the demand for his firm's product will show a sizable increase. How great must the chance of a sizable increase in demand be for the president to be justified in buying the competing firm?

19.25 A manufacturer of precision parts for aerospace assemblies is considering the submission of a bid on a Department of Defense (DOD) contract to manufacture 1,000 computerized guidance devices. However, the bid requires the submission of a prototype guidance device in addition to a cost estimate. Management believes that if it submits a bid and wins, the work can be accomplished in one of three ways:

a_1: Current employees can perform the task as an overtime assignment at an average cost of $18 per man-hour.

a_2: Current employees can perform the task during overtime with the assistance of a specially designed assembly machine costing $200,000 (the machine would be valueless after use).

a_3: After the development of a prototype, if the bid is won, the assembly work could be accomplished by subcontracting to another firm at a cost of $300 per device.

Costs for developing a prototype are estimated at $500,000, and 10 hours would be required to assemble each guidance device without the revolutionary assembly machine. Assembly time would be reduced to 3 hours with the machine. From the firm's reputation as a quality supplier in past DOD contracts and knowing that DOD considers supplier quality as well as cost when issuing contracts, the manufacturer believes chances are 60% that she will win the contract if she chooses to submit a bid.

a. If the manufacturer submits a bid and wins the contract, how should she choose to manufacture the guidance devices?

b. What would you recommend to the manufacturer as a minimum total bid if she wishes to achieve a 15% pretax profit margin?

19.26 A building contractor is considering whether to submit a bid on a contract to build a large modern apartment complex. Analysis of the proposed plan and other preliminary planning required by the contractor and his firm before submitting a bid on this contract would cost the building firm $50,000. If they bid, they will submit a bid large enough to realize a $250,000 profit for themselves (including the cost of preliminary planning). If the contractor's objective is to maximize his expected profit, how large must the probability of his winning the contract be before he will decide to submit a bid on the apartment contract?

19.6

The Economic Impact of Uncertainty

Ideally, we would like to know in advance which state of nature will be in effect. In the investment problem, knowing that the market will be down or about the same, the decision maker would select investment *A*, the 6% commercial bank savings account; knowing that the market will be up, he would select investment *C*, the mutual fund investment. In the demand-inventory example, if the retailer knows which demand level will be in effect, she selects an inventory level to meet demand exactly.

Under normal conditions uncertainty prevails, and decision makers must act according to their best information and interests. Not knowing the state of nature that will be in effect, they cannot expect a return as great as the return they would expect if the true state of nature were known.

Using an expected monetary value approach, the decision maker, in the face of uncertainty, selects the alternative that maximizes expected profit and minimizes expected opportunity loss. We will now define a concept that was mentioned earlier only in passing.

Definition

cost of uncertainty

> The expected opportunity loss associated with the optimal decision is called the **cost of uncertainty**.

The cost of uncertainty is the maximum amount the decision maker would pay to know precisely the state of nature that will be in effect. Therefore, it is the value the decision maker would place on perfect information (information removing all uncertainty). The cost of uncertainty is sometimes called the **expected value of perfect information**, or simply EVPI.

expected value of perfect information

In Example 19.4 the optimal decision is to select action a_4, with expected opportunity loss $E(L_4) = \$55$. We saw in Example 19.5 that action a_2, with expected opportunity loss $E(L_2) = \$1.60$, is best. These two values are the costs of uncertainty associated with, respectively, the investment problem and the demand-inventory problem. In each case these values represent the maximum amount the decision maker would be willing to pay to know precisely the state of nature that will be in effect.

To better understand the concept of the cost of uncertainty or EVPI, let us focus on Example 19.4, the investment problem. Suppose that the decision maker knows with *certainty* which state of nature will be in effect. If he knows that the state will be s_1 or s_2, he selects action a_2, since the opportunity losses for selecting a_2 under states s_1 and s_2 are the least. Similarly, if he knows that state of nature s_3 will be in effect, he selects action a_4. In each case the opportunity loss is zero, so the decision maker's expected opportunity loss *under certainty* is

$$E(L_0) = \$0(1/4) + \$0(1/4) + \$0(1/2) = \$0$$

The best the decision maker can expect *under uncertainty* is the expected opportunity loss of \$55 by selecting the optimal action a_4. Therefore, the effect of the uncertainty of market conditions costs the investor, on the average, \$55 each time he faces an investment decision such as the one outlined in Section 19.2.

The cost of uncertainty or EVPI may be more easily interpreted if defined in terms of expected profits. Under certainty of the state of nature the decision maker can expect a profit of

$$E(G_0) = (\$60)(1/4) + (\$60)(1/4) + (\$200)(1/2) = \$130$$

by selecting action a_2 under states s_1 and s_2 and by selecting action a_4 under state of nature s_3. This value represents his expected profit under perfect information.

The best the investor can expect under *uncertainty* is \$75, the expected profit associated with the optimal decision, action a_4. The difference between expected profits under certainty and uncertainty is $130 - 75 = 55$, the EVPI. **The cost of uncertainty, or, equivalently, the EVPI, is then the amount of profits forgone or additional losses incurred owing to uncertain conditions affecting the outcome of a decision problem.**

Suppose that the investor has the opportunity to hire the services of the Securities Consulting Agency (SCA). This group would survey many market experts to determine whether they believe the securities market will fall, stay about the same, or rise within the next year. The SCA would then sell the consensus of the survey to the investor.

How much should the investor pay for the SCA report? If the SCA provided *perfect* market information, the investor would pay up to $55 for the report, but no more. That is, the value of the perfect information supplied by the SCA does not exceed $55. The reasoning behind this limiting value is easy to see. For instance, if the decision maker paid $60 to the SCA for perfect market information, his expected profit would be $130 (the expected profit for perfect information) minus $60 (the cost of buying the perfect information), or $70. But an expected profit of $70 is $5 less than the $75 expected profit by taking action a_4 under *uncertainty* and without auxiliary information.

Naturally, the investor does not expect perfect market information from the SCA. Thus, the amount he should pay for the SCA report is some amount less than $55, an amount that is determined from the measures of *reliability* of the SCA market analysis techniques. The concept of placing a value on *imperfect* sample information will be discussed in the next section.

EXERCISE   Conceptual Level

19.27   Explain the concept of EVPI. How should it be used in decision analysis?

EXERCISE   Skill Level

19.28   A decision problem results in the profit table presented in Table 19.13.

Table 19.13   Profit Table for Exercise 19.28

	State of Nature				
Action	$s_1(.05)$	$s_2(.35)$	$s_3(.25)$	$s_4(.20)$	$s_5(.15)$
a_1	6	8	3	2	0
a_2	3	5	0	1	−2
a_3	5	9	−3	7	8

a.   Is there any one action that dominates another action? If so, which action is dominated? This action can be eliminated from consideration as the optimal decision.

b. Use the profit table to determine the action that maximizes the expected profit, using the prior probabilities given in parentheses in the profit table.

c. What is the EVPI for this decision problem?

EXERCISES Applications

19.29 Businesses and other organizations are constantly seeking new-product or new-service ideas that will help them achieve their product mix and market share objectives. A newspaper publisher, seeking diversifications in his publishing activity, is considering the publication of a monthly entertainment guide for sale in his community. The publication will cost 50¢ to print and distribute, will sell for 85¢, and will feature upcoming sporting events, theater attractions, and special activities as well as a guide to local restaurants. With the assistance of a market analyst the publisher has established the accompanying demand distribution for the monthly publication.

Demand	500	1,000	1,500	2,000	2,500
Probability	.10	.20	.35	.25	.10

a. Construct an opportunity loss table for the publisher's problem.

b. If the publisher wishes to minimize his expected opportunity loss, how many entertainment guides should he publish for distribution during a given month?

c. What is the maximum amount the publisher would be willing to pay to learn in advance the exact level of demand for the entertainment guide?

19.30 Refer to Exercise 19.29. After introduction of the entertainment guide to the marketplace, it is found that when readers cannot buy a particular issue, they become irritated and are likely to avoid buying the publisher's entertainment guide in the future. If the loss of goodwill incurred when demand for the publication exceeds supply is estimated to be 75¢ per unit, what is the publisher's optimal level of production? What is his expected value of perfect information?

19.31 Inflation tends to boost interest rates on loans of all types. As a partial hedge against inflation, some lending institutions offer variable-rate loans to borrowers with high credit ratings. The borrower can then accept either a fixed-rate loan with a certain interest obligation or a variable-rate loan that may result in either interest savings or additional interest costs to the borrower, depending on the future state of the economy. Suppose an individual wishes to borrow $10,000 for one year when the prevailing interest rate is 9%. After consulting with an economist, the individual establishes the accompanying probability distribution for the prevailing lending rate over the next year. The borrower must decide whether to accept the prevailing interest rate (9%) or to undertake a variable-rate loan and pay interest according to the prevailing rate at the end of the borrowing period.

Interest Rate, i	8%	8.5%	9%	9.5%	10%
Probability, $P(i)$	.10	.25	.50	.10	.05

a. Construct an opportunity loss table for the borrower.

b. If the borrower wishes to minimize her expected opportunity loss, which type of loan should she undertake?

c. How much would the borrower be willing to pay to know exactly the interest rate that will be in effect at the end of the year?

19.32 Refer to Exercise 19.16. Suppose the developer has sought out a zoning lawyer to assist her in her planning and development activities. What is the maximum amount the developer should pay the lawyer if the lawyer is able to judge with certainty the ultimate outcome of the zoning affecting the large tract of adjacent land?

19.33 Refer to Exercises 19.10 and 19.18. What is the entrepreneur's expected value of perfect information? In common language, explain the meaning of the EVPI to the entrepreneur.

19.7

Decision Making That Involves Sample Information

In the definition and discussion of the prior probabilities in Sections 19.3 and 19.5, we said that they are acquired either by subjective selection or by computation from historical data. No current information describing the chance of occurrence of the states of nature was assumed to be available.

Perhaps observational information or other evidence may be available to the decision maker either for purchase or at the cost of experimentation. For instance, a retailer whose business depends on the weather may consult a meteorologist before making his decision, or an investor may hire a market consultant before investing. Market research surveys before the release of a new product represent another area in which the decision maker seeks additional information. In each instance the decision maker is acquiring information relative to the occurrence of the states of nature from a source other than that from which the prior probabilities were computed.

If such information is available, Bayes' Law can be employed to revise the prior probabilities to reflect the new information. These revised probabilities are called *posterior probabilities*.

Definition

posterior probability

> The **posterior probability** $P(s_k \mid x)$ represents the chance of occurrence of the state of nature s_k given the experimental information x. This probability is
>
> $$P(s_k \mid x) = \frac{P(x \mid s_k)P(s_k)}{\sum_{\text{all } i} P(x \mid s_i)P(s_i)}$$

The probabilities $P(x \mid s_i)$ are the conditional probabilities of observing the observational information x under the states of nature s_i, and the probabilities $P(s_i)$ are the prior probabilities. The notation used here for Bayes' Law is different from what was presented in Chapter 3, but the forms are equivalent.

The expected monetary value decisions are formulated in the same way as before except that we use the posterior probabilities in place of the prior probabilities. If our objective is to minimize the expected opportunity loss, then this quantity is computed for each action a_i. As before, the optimal decision is to select the action associated with the smallest expected opportunity loss.

Computing the Expected Opportunity Loss by Using Posterior Probabilities

$$E(L_i) = \sum_{\text{all } j} L_{i,j} P(s_j \mid x) \qquad i = 1, 2, \ldots$$

EXAMPLE 19.7   An assembly machine operates at a 5% or a 10% defective rate. When believed to be running at the 10% defective rate, the machine is judged out of control, shut down, and readjusted. From past history the machine is known to run at the 5% defective rate 90% of the time. A sample of $n = 20$ has been selected from the output of the machine and $x = 2$ defectives have been observed. From both the prior and sample information, what is the probability that the assembly machine is in control (5% defective rate)?

SOLUTION   The states of nature in this example relate to the possible assembly machine defective rates. Thus, the states of nature are

$$s_1 = .05 \qquad \text{and} \qquad s_2 = .10$$

with assumed prior probabilities of occurrence of .90 and .10, respectively. We want to use these prior probabilities, in light of the observed sample information, to find the posterior probability associated with state s_1.

In this example the "experimental information x" is the observation of $x = 2$ defectives from a sample of $n = 20$ items selected from the output of the assembly machine. We must find the probability that the experimental information x could arise under each state of nature s_j. We find this probability by referring to the tables for the binomial probability distribution (Table 1 of the Appendix). Under state of nature $s_1 = .05$ we find

$$P(x \mid .05) = P(x = 2 \mid n = 20, p = .05) = .925 - .736 = .189$$

Similarly, under state of nature $s_2 = .10$ we find

$$P(x \mid .10) = .285$$

Now we are ready to employ Bayes' Law to find the posterior probability that the machine is in control (s_1), using both the prior and experimental information.

The application of Bayes' Law is most clearly illustrated by using a columnar approach, as shown in Table 19.14. With a columnar approach the states of nature, their associated prior probabilities, and the probabilities that the experimental information x could arise under each state of nature are listed in the first three columns. In column 4 we compute the product of the corresponding entries from columns 2 and *joint probabilities*   3. These values measure the **joint probabilities** $P(xs_j) = P(s_j)P(x \mid s_j)$. We then sum the entries in column 4. This sum is the term in the denominator of the formula for Bayes'

Table 19.14   Columnar Approach to Use of Bayes' Law in Example 19.7

(1) State of Nature, s_j	(2) Prior, $P(s_j)$	(3) Experimental Information, $P(x \mid s_j)$	(4) Product, $P(s_j)P(x \mid s_j)$	(5) Posterior, $P(s_j \mid x)$
s_1	.05	.9	.189	.1701	.86
s_2	.10	.1	.285	.0285	.14
		1.0		.1986	1.00

marginal probability   Law and measures the **marginal probability** of observing the experimental information x. The posterior probabilities, column 5, are thus obtained by taking each entry in column 4 and dividing it by the sum of the entries in column 4. This manipulation simply rescales the joint probabilities of column 4, causing them to sum to 1.00. For instance,

$$P(s_1 \mid x) = \frac{.1701}{.1986} = .86 \qquad P(s_2 \mid x) = \frac{.0285}{.1986} = .14$$

Even though we found 10% of the sample items to be defective (2 of 20), the posterior probability that the machine is running at the 10% defective rate (out of control) is only .14, only slightly greater than the prior probability that the machine is out of control. Therefore, the probability that the machine is not running out of control is .86. ◆ ◆

In reviewing the solution to Example 19.7, you may have noticed that the columnar approach is nothing but a restructuring of Bayes' Law. It is bound to give the same numerical answers as would be obtained by substituting the appropriate numbers into the formula for Bayes' Law. Those who are not convinced of this fact should repeat Example 19.7 by using the formula substitution approach instead of the columnar approach.

The experimental information x often presents itself other than as sample information. It may be in the form of an expert opinion, the result of an engineering survey or a structural test, or the observance of a sudden, unpredictable economic trend. How can such information best be used to help us in the task of decision making under conditions involving uncertainty? We must first evaluate the reliability of the information and then incorporate these measures of reliability into the decision analysis with the other problem data.

The discussion involving the hypothetical Securities Consulting Agency presented in the preceding section offers a case in point. Notice that the "experimental information" in this example is in the form of an opinion; from its studies the SCA will indicate the market condition (state of nature) that it believes will occur. In this example the experimental information x may then present itself at any of three levels.

That is, the SCA may report one of these conditions:

$$x_1 = \text{the market will be down}$$
$$x_2 = \text{the market will remain at the same level}$$
$$x_3 = \text{the market will be up}$$

The probabilities $P(x \mid s_j)$ should be rewritten as $P(x_i \mid s_j)$ to denote the different possible responses x_i, $i = 1, 2, 3$, that may be given by the SCA. These probabilities measure the *reliability* of the experimental information x_i provided by the SCA. If the information provides *complete certainty*, then

$$P(x_i \mid s_j) = \begin{cases} 1 & \text{if } i = j \\ 0 & \text{if } i \neq j \end{cases}$$

The further removed the $P(x_i \mid s_j)$ values are from these limits, the less reliable, and hence the less valuable, is the experimental information that is available.

EXAMPLE 19.8   Reconsider the investment problem of Example 19.2. Suppose that the investor is considering hiring the services of the Securities Consulting Agency, the group discussed in Section 19.6. In the past when the SCA has conducted a market survey for a client, the results shown in Table 19.15 have been noted. For example, $P(x_1 \mid s_2) = .15$ is the probability that the SCA concludes the market will drop when it actually stays the same. Find the investor's best decision by using the expected opportunity losses associated with each possible response from the SCA.

Table 19.15   Conditional Probabilities Indicating the Reliability of SCA Market Surveys, $P(x_i/s_j)$

	State of Nature		
Response, x	s_1, Down	s_2, the Same	s_3, Up
x_1, down	.60	.15	.05
x_2, the same	.30	.50	.25
x_3, up	.10	.35	.70

SOLUTION   Recall from Table 19.3 that regardless of which state of nature is in effect, action a_2 guarantees a greater profit than either action a_1 or action a_3. Thus, we could ignore actions a_1 and a_3, which are dominated by a_2, reducing the investor's decision to one of choosing either action a_2 or action a_4, depending on the information received from the SCA. An analysis of the investor's problem involves three decisions, one for each response from the SCA: "down," "stay the same," or "up." We first need to compute the revised probabilities—the posterior probabilities—for each response.

If the SCA suggests that the market will drop, we employ Bayes' Law, using the columnar approach, and find (see Table 19.16)

$$P(s_1 \mid x_1) = .706 \qquad P(s_2 \mid x_1) = .176 \qquad P(s_3 \mid x_1) = .118$$

Table 19.16   Columnar Approach for Using Bayes' Law in Example 19.8

(1) s_j	(2) $P(s_j)$	(3) $P(x_1 \mid s_j)$	(4) $P(s_j)P(x_1 \mid s_j)$	(5) $P(s_j \mid x_1)$
s_1, down	1/4	.60	.1500	.706
s_2, the same	1/4	.15	.0375	.176
s_3, up	1/2	.05	.0250	.118
			.2125	1.000

Similarly, under the response from the SCA that the market will stay the same, we find

$$P(s_1 \mid x_2) = .230 \qquad P(s_2 \mid x_2) = .385 \qquad P(s_3 \mid x_2) = .385$$

If the SCA says it will rise, we find

$$P(s_1 \mid x_3) = .054 \qquad P(s_2 \mid x_3) = .189 \qquad P(s_3 \mid x_3) = .757$$

Now suppose that the SCA reports that the market will drop. Then using the opportunity loss values from Table 19.4, we find the expected opportunity losses associated with actions a_2 and a_4 to be

$$E(L_2) = (\$0)(.706) + (\$0)(.176) + (\$140)(.118) = \$16.47$$
$$E(L_4) = (\$160)(.706) + (\$60)(.176) + (\$0)(.118) = \$123.53$$

Thus, if the SCA suggests the market will drop, the investor should choose the 6% savings plan, action a_2. The expected opportunity losses for the other two SCA responses have been computed and are listed in Table 19.17. The investor's optimal decisions are easily read from this table. He should select action a_2 if "down" or "the same" is reported, but he should select a_4 if the SCA reports that the market will be up in the coming year.

Table 19.17   Expected Opportunity Losses for
Example 19.8

	Response from the SCA		
Action	Down	The Same	Up
a_2	$ 16.47	$53.85	$105.95
a_4	123.53	60.00	20.00

◆◆

The value of revising our prior probabilities and arriving at this more complex solution is twofold. First, the independent source of information we are using to compute the posterior probabilities gives us a check that the values we assign to the prior probabilities of the states of nature are near what the true values should be. That is, we are updating our guesses with someone else's guesses or with experimental evidence. This procedure allows us to incorporate all available information into our decision analysis. As a second point, related to the first, the additional information allows us to tailor our decision to the specific outcome of the experimental evidence. Thus, instead of having a single decision, we have a whole array of decisions corresponding to the nature of the experimental evidence. In the investment problem each response from the SCA initiates a decision by the investor. Thus, he is able to "partition" his decision and obtain the most use possible from the additional information.

What is the value of the sample information? Certainly, the investor would not pay as much as $55 for the sample information, but how much should he be willing to pay? We will answer that question in the paragraphs that follow.

Although it was not mentioned earlier, we can find the marginal probabilities by using the formula given in the display.

Computing Marginal Probability $P(x_i)$

$$P(x_i) = \sum_{\text{all } j} P(x_i \,|\, s_j) P(s_j) \qquad i = 1, 2, \ldots$$

This value is the sum of the entries in column 4 when using the columnar approach. Therefore, the probability that the SCA report will indicate that the market will be down (x_1) is

$$P(x_1) = (.6)(1/4) + (.15)(1/4) + (.05)(1/2) = .2125$$

Similarly, the probabilities that its report will indicate that the market will be about the same (x_2) and up (x_3) can be found by using the same formula and are, respectively,

$$P(x_2) = .3250 \qquad \text{and} \qquad P(x_3) = .4625$$

Returning to Example 19.8, we recall that when the SCA reports $x_1 =$ down or $x_2 =$ the same, the investor should select action a_2. However, when the report suggests $x_3 =$ up, action a_4 is best. Referring to Table 19.17, we can see that the investor's expected opportunity loss when using the SCA service is

$$E(L) = \$16.47P(x_1) + \$53.85P(x_2) + \$20.00P(x_3)$$
$$= (\$16.47)(.2125) + (\$53.85)(.3250) + (\$20.00)(.4625) = \$30.25$$

Thus, the experimental information is worth, at most,

$$\$55.00 - \$30.25 = \$24.75$$

since the experimental information has effectively reduced the investor's expected cost of uncertainty by that amount. This value, called the **expected value of sample information** (EVSI), represents the maximum amount the investor would pay for the

expected value of
sample information

SCA market service. Under no conditions would the investor pay more than $24.75. For any amount less than $24.75 that he pays for the SCA report, the investor incurs a net saving by buying the additional information.

EXERCISES   Conceptual Level

19.34   Explain the concept of a posterior probability.

19.35   Define expected opportunity loss, and explain how it relates to posterior probability.

19.36   How is Bayes' Law related to decision analysis?

EXERCISES   Skill Level

19.37   Refer to Exercises 19.7 and 19.14. A research organization provides the following sample information to the decision maker:

$$P(x\,|\,s_1) = .35 \qquad P(x\,|\,s_2) = .50 \qquad P(x\,|\,s_3) = .05 \qquad P(x\,|\,s_4) = .45$$

a.   If the sample indicates that x has occurred, use Bayes' Law to find the posterior probabilities, $P(s_j\,|\,x)$, for $j = 1, 2, 3,$ and 4.

b.   From the sample information, what is the optimal decision?

c.   What is the EVSI for this decision problem?

EXERCISES   Applications

19.38   After observing three years of continued growth of sales and profits, the president of a company that manufactures home computers must address the issue of expanding corporate facilities. After careful consideration of all options, the president has decided that the only feasible alternatives for his company are to continue with present facilities, expand present facilities, or develop a remote manufacturing facility to supplement present facilities. An economic consulting agency has assisted the president in developing Table 19.18 to represent expected profits (in thousands of dollars) for each of three possible economic conditions. The economic consulting agency foresees an impending recession and estimates the likelihood that sales will be low, normal, or high to be .5, .3, and .2, respectively.

Table 19.18   Profit Data for Exercise 19.38

	State of Economy		
Alternative	Low Sales	Normal Sales	High Sales
Maintain present facilities	$1,010	$1,100	$1,200
Expand present facilities	− 500	2,010	2,200
Add a remote facility	− 2,000	− 500	3,010

a. For the given information, what is the best option available to the corporate president?

b. How much would the president be willing to pay to learn with complete reliability the future state of the economy (and the level of sales to expect)?

 19.39 Refer to Exercise 19.38. Suppose that a market research firm has offered to conduct a focused sales study for the company in the coming months for a fee of $100,000. From past experience the market research firm has provided the accuracy shown in Table 19.19 in their estimates of future economic activity.

a. If the market research firm is employed to provide an analysis of future sales potential, what is the president's best option if the market research firm projects future sales to be low? If it projects future sales to be normal? If it projects future sales to be high?

b. What is the expected value of the sales information provided by the market research firm? Is its $100,000 fee warranted?

Table 19.19   Estimate Accuracy Data for Exercise 19.39

	State of Economy		
Firm's Estimate	Low Sales	Normal Sales	High Sales
Low	.80	.10	.05
Normal	.15	.80	.15
High	.05	.10	.80

 19.40 A newspaper publisher with headquarters in Portland, Oregon, is considering the distribution of her newspaper in Eugene, 100 miles to the south of Portland. Her potential profits will depend on which of two methods of transportation she uses for the distribution of the newspaper from Portland to Eugene and the proportion of the 50,000 Eugene households that subscribe to her newspaper. The two options for transporting the newspaper are to contract for daily shipment with an interstate bus service (method *A*) or to sublet the transportation between the two cities to an independent trucker (method *B*). The daily profits associated with each method of distribution under every assumed likely proportion of subscribing households are given in Table 19.20.

a. Using only the given information, find the best method of distribution for the publisher.

b. How much would the publisher be willing to pay to know the exact proportion of households who would subscribe to her newspaper?

Table 19.20   Profit Data for Exercise 19.40

Number of Households Subscribing	Proportion Subscribing, p	$P(p)$	Daily Profits	
			Method A	Method B
20,000	.4	.3	$1,500	$1,800
25,000	.5	.5	2,100	2,250
30,000	.6	.2	2,800	2,700

 19.41 Refer to Exercise 19.40. To reduce her uncertainty, the publisher commissions a group to conduct a door-to-door survey of the Eugene households to estimate the proportion who would subscribe to her newspaper. The results of the survey show that 25 households were randomly selected from among the 50,000 households of Eugene, with 14 indicating an interest in subscribing to the newspaper. What is the best method of distribution in light of this sample information?

19.8

Other Topics in Decision Analysis

Over the past 20 years decision analysis has evolved from a controversial set of rules of thumb into a rigorous discipline involving techniques now generally accepted by both the theoretician and the practitioner. Only the most important topics involved with decision analysis have been included in the chapter up to this point. Other topics and techniques of lesser importance will be introduced in this section. For a complete discussion of the modern concepts of decision analysis, see the texts by Schlaifer (1978), Winkler (1972), Baird (1983), or Holloway (1979) listed in the references.

Decisions Ignoring Prior Information

Occasionally, the decision maker's attitude toward risk suggests that a limit exists on the amount of money he can afford to lose. Such an individual may be typified by an investor with a very small bankroll or a retailer on the verge of bankruptcy. Such an individual would ignore the prior probabilities attached to the states of nature, even if the probabilities associated with unfavorable states are extremely small. He looks only at the magnitudes of the losses for each action, shows virtually no interest in the magnitude of potential profits, and makes his decision without computing an expected monetary value. A decision criterion that satisfies a conservative economic objective like that just described is provided by the *minimax* decision.

Definition

minimax decision

> The **minimax decision** is the selection of the action whose maximum possible opportunity loss is a minimum.

The minimax decision maker says that to remain competitive in his business, he must, at all costs, avoid the large opportunity losses. He is minimizing his maximum opportunity loss and, by so doing, he is usually rejecting those alternatives associated with the greatest possible return. He thus assures himself of, at best, a stable financial position.

EXAMPLE 19.9   Find the minimax decision for the investment problem of Example 19.2.

SOLUTION   The maximum opportunity losses associated with each action are given in the table.

Action	a_1	a_2	a_3	a_4
Maximum Opportunity Loss	\$200	\$140	\$155	\$160

Thus, the minimax decision is to select action a_2, the 6% savings plan.

maximax decision   Another decision criterion is the **maximax decision** method, which selects the action that maximizes the maximum possible profit. The individual who employs this technique has a very small aversion toward risk and might be categorized as a gambler. He stands a great chance of losing everything but is concerned only with the possibility of making a quick fortune.

maximin decision   **Maximin**, maximizing the minimum profit, is similar in nature but much more conservative than maximax.

All these techniques have two common traits. First, the decision maker employing minimax, maximax, or maximin strategies tends totally to ignore prior information relating to the probabilities of occurrence of the states of nature. His attitudes toward risk make some economic payoffs appear attractive and others unattractive, regardless of the occurrence of the states of nature. The second similarity is that minimax, maximax, or maximin decisions are usually one-shot decisions. If they were not, the "averaging effect" of employing minimax, maximax, or maximin continuously could tend to deplete one's profits, since whether accepted or not, the states of nature do occur according to some probability distribution.

The expected monetary value decision and the minimax, maximax, or maximin decisions are often the same. This only occurs by chance and does not imply any similarity between objectives. The basic difference is that expected monetary value decisions incorporate prior information (probabilities) into the decision analysis, but the others ignore this information.

Decision Trees

A tree diagram is a useful device to illustrate single-stage and multistage decision problems. Decision trees are most useful for multistage decision problems, especially decision problems sequenced over time. The decision tree graphically X-rays the decision problem and displays the anatomy of the decision. The decision tree becomes more useful as the decision problem becomes more complex, because it focuses attention on the interrelationship of activities and events over time.

For a clear interpretation of the tree diagram, decision points are represented by squares, and chance points, points over which the decision maker has no control, are represented by circles. At the base of the tree diagram are the available first-stage alternatives. From each alternative the decision tree constructs the chronological path, leading through chance points and other decision points, to each assumed possible terminal outcome. Payoffs associated with each path and probabilities of occurrence associated with the chance events are shown on the tree diagram.

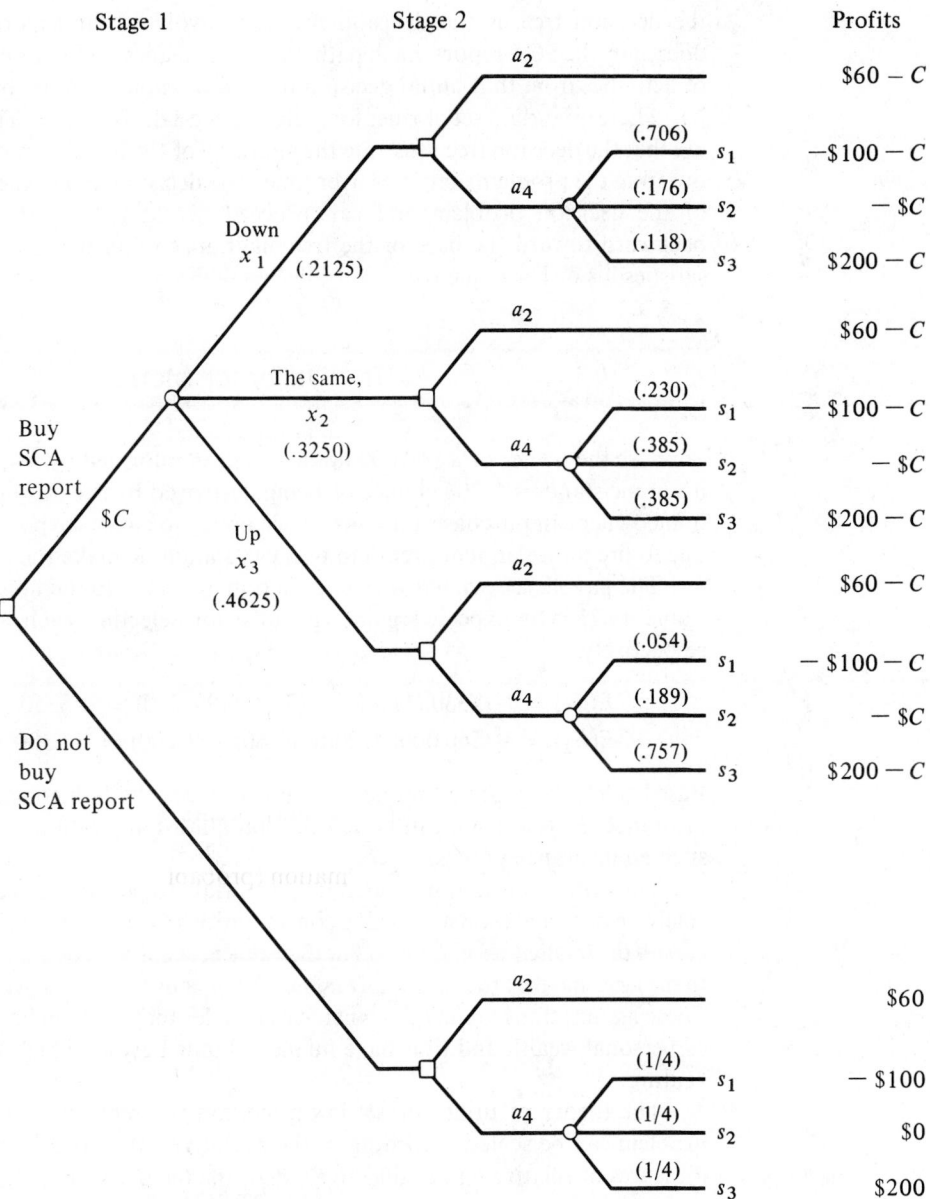

Figure 19.3   Decision Tree for the Investment Problem of Example 19.8

In Figure 19.3 a decision tree is used to illustrate the investment problem. Notice that the anatomy of the problem appears on the tree diagram according to how its activities relate over time. Before the investor decides between actions a_2 and a_4 (a_1 and a_3 have been eliminated because of their dominance by a_2), he must first decide whether or not to buy the SCA report at a cost of $$C$. His stage 1 activity, and hence the base of the decision tree, is the decision problem involving the experimental information offered by the SCA report. Each path on the tree diagram illustrates a possible sequence of activities from that initial decision to each possible terminal outcome.

There is no rigid set of rules for constructing a decision tree. The only requirements are that the decision tree illustrate the anatomy of the decision problem while properly ordering the problem activities over time. The decision maker then has a clear picture of the decision problem and can proceed, working from the terminal outcomes backward toward the base of the tree diagram, to find the set of decisions that best satisfies his or her objectives.

The Utility for Money

Suppose the owner of a $60,000 home has been informed by insurance actuaries that his home stands a 1:200 chance of being destroyed by fire during a given year. If the homeowner can purchase a fire insurance policy to cover the possible loss of his home due to fire for an annual premium of $350, should he make the purchase?

The payoffs associated with each action available to the homeowner are given in Table 19.21. His expected gains (payoffs) for selecting each available action are, respectively,

$$E(G_1) = -(\$350)(1/200) - (\$350)(199/200) = -\$350$$
$$E(G_2) = -(\$60,000)(1/200) + (\$0)(199/200) = -\$300$$

Based solely on expected monetary value criteria, the homeowner *would not buy* the insurance. However, we can be certain that almost any rational individual *would buy* such an insurance policy.

The problem is that it is often inappropriate to perform a decision analysis based solely on expected monetary value considerations. **An expected monetary value decision carries the implicit assumptions that the value of a dollar does not differ from one person to the next and that the value of D dollars is equal to D times the value of a single dollar.** These assumptions are invalid when we consider that we each have different measures of personal wealth and thus have financial limits beyond which we are not willing to venture.

The theory of utility prescribes a method whereby the outcomes of a decision problem can be scaled according to their relative value to the decision maker. These measures of relative value collectively describe the decision maker's **utility curve**. The

utility curve

Table 19.21   Homeowner's Payoffs

Action	Fire, s_1	No Fire, s_2
Buy insurance, a_1	−$350	−$350
No insurance, a_2	−60,000	0

decision analysis is then performed by substituting utility values for monetary values and computing the expected utility associated with each action. Maximization of utility then ensures maximization of *value* to the decision maker in terms of how he perceives the comparative personal measures of value of the possible outcomes associated with his decision problem. The actual assessment of utility for an individual is a complex activity involving the quantification of his or her subjective feelings toward risk, and the topic will not be discussed in this text. For detailed discussions of some modern developments in utility assessment, refer to the texts by Winkler (1972) or Baird (1983) listed in the references.

Utility measures are needed primarily when the decision problem contains some possible outcomes that, if incurred, may place the decision maker in financial or personal jeopardy. As opposed to adopting a minimax decision, the decision maker may choose to rescale the monetary outcomes according to his or her preferences and perform an expected utility value analysis. This analysis allows full utilization of all available resources, including probability measures associated with the uncertain events that the minimax decision ignores, and provides a decision that is guaranteed to maximize the decision maker's personal interpretation of value.

EXERCISES   Conceptual Level

19.42   Define each of the following terms:

   a. Minimax decision   **b.** Maximax decision   **c.** Maximin decision

19.43   Explain the concept of a utility curve. What use might utility curves have for decision analysis?

19.44   Some authors have suggested that a decision maker who consistently adopts a minimax decision criterion is a "pessimistic sore loser" and that under normal conditions he will ultimately deplete his profits. Discuss this viewpoint.

EXERCISES   Skill Level

19.45   Refer to Exercises 19.7 and 19.14.

   a. Find the minimax decision for the decision problem.

   b. Find the maximin decision for the decision problem.

   c. Are the decisions in parts a and b the same? Explain.

19.46   Refer to Exercise 19.28. The utility values given in the accompanying table are assigned to the profits. Use the utility table to choose the action that maximizes the expected utility for this decision problem.

Profit	-3	-2	-1	0	1	2	3	5	6	7	8	9
Utility	0	.1	.1	.2	.3	.3	.4	.4	.5	.8	.9	.9

EXERCISES Applications

19.47 Find the minimax decision for the decision problem described in each of the following exercises.

a. Exercise 19.8   **b.** Exercise 19.9

c. Exercise 19.10   **d.** Exercise 19.16

e. Exercise 19.40

19.48 Refer to Exercises 19.38 and 19.39. Construct a decision tree to graphically portray the decision problem faced by the company president. As shown in Figure 19.3, indicate the chance points by circles and the decision points by squares.

  19.49 Shown in the accompanying table is a set of utility values that have been established for the associated dollar-valued outcomes by a decision maker. If the decision maker wishes to maximize her expected utility, how should she act on each of the following investment problems?

Dollar-Valued Outcome	−$10,000	−$5,000	−$1,000	$0	$5,000	$10,000	$25,000
Utility	0	.45	.50	.55	.70	.80	1.0

a. The investment of $1,000 in an oil-drilling venture returning either a $10,000 profit or nothing. The probability of success in the oil-drilling venture is estimated to be .1.

b. The investment of $10,000 in a new motel-restaurant facility. Depending on the success of the project, the investment is estimated to return a $25,000 profit with a probability of .2, a $5,000 profit with a probability of .3, a $5,000 loss with a probability of .4, or a loss of the entire $10,000 investment with a probability of .1.

c. In both of the preceding investment problems, compare the optimal decision using a maximum expected utility objective with the optimal decision using a maximum expected payoff objective. How do you account for any differences in the selection of an optimal decision between these two objectives?

19.9

Summary

Decision analysis is characterized by a decision maker selecting from among many alternatives, some of which have uncertain outcomes. The decision maker must determine which among the available alternatives best satisfies his or her objective.

Usually, the objective in a decision analysis is involved with an expected monetary value decision. If so, the decision maker proceeds as follows:

1. List all possible actions that may be taken and all the states of nature that may affect the outcomes of those actions.

2. List the payoffs associated with each action under every state of nature.

3. Draw the decision tree associated with the problem.

4. Determine meaningful probabilities to assign to the states of nature to represent their likelihoods of occurrence.

5. Compute the expected profit or expected opportunity loss for each action.
6. Select the action associated with the maximum expected profit or the minimum expected opportunity loss, both of which arrive at the same decision.

If additional information is available to the decision maker, Bayes' Law can be employed to revise the prior probabilities on the basis of experimental information. The posterior probabilities then reflect all available information, both subjective and experimental, relevant to the likelihood of the occurrence of the states of nature.

Supplementary Exercises

Applications

 19.50 An owner of a newsstand buys a certain weekly newsmagazine for $.60 a copy and sells it for $1.00 a copy. Past evidence suggests that the weekly demand for the newsmagazine ranges from 9 to (but not including) 13 copies. If unsold copies cannot be returned for refund at the week's end, construct the owner's weekly profit table for the newsmagazine. Construct an opportunity loss table.

 19.51 Refer to Exercise 19.50. From historical evidence the owner of the newsstand believes the weekly demand for the newsmagazine is as shown in the accompanying table. Find the number of issues of the weekly newsmagazine the newsstand owner should buy each week to maximize his expected weekly profits.

Demand	9	10	11	12
Probability of Demand	.20	.15	.35	.30

 19.52 Decision analysis is often used when deciding whether to introduce new products. For instance, an executive must decide whether or not her company should introduce a new product that would appeal to a limited class of consumers. The executive has determined that the revenues returned to the firm are directly related to the proportion of consumers who buy the product by the equation $revenue $= \$50,000 + \$100,000(p)$, where p is the proportion of consumers who buy. By a careful analysis of the market of potential consumers, the executive has developed the accompanying probability distribution for p. If development, advertising, and promotional costs would run to as much as $75,000 for the product, should the executive recommend that her company introduce it? (*Hint:* Do expected revenues cover costs?)

Proportion of Consumers, p	.10	.20	.30	.40	.50	.60
Probability, $P(p)$	.10	.15	.30	.25	.15	.05

 19.53 Refer to Exercise 19.52. Suppose a market research firm has approached the executive of the firm considering the introduction of the new product with the offer that for $2,000 it can conduct a market survey and determine with virtual certainty the proportion of consumers who will buy the firm's product. Should the executive accept the offer of the market research firm? Explain.

 19.54 Conservation is usually more visible when there are immediate, direct dollar savings to the consumer. But since energy-efficient items usually cost more and their cost-saving benefits are usually not recovered in the short run, many customers are not willing to undertake the initial purchase of such items. A utility company is considering offering clock thermostats for nightly cutbacks in room temperature to their customers. The utility would pay $30 for each thermostat and would offer them to customers for a price of $45. A careful survey of the utility's customers suggests the accompanying probability distribution for the number who would buy a thermostat. Assume that unsold thermostats cannot be returned to the manufacturer.

Demand	50	60	70	80	90
Probability of Demand	.15	.25	.35	.20	.05

a. Construct the profit table.

b. Construct the opportunity loss table.

c. Find the minimax decision.

d. Find the utility company's optimal order quantity if it wishes to maximize its expected profits on the sale of thermostats.

e. If unsold thermostats can be returned to the manufacturer at full refund of cost, how many should the utility company order? Explain your answer.

19.55   Capital budgeting requires a careful periodic review of the operating characteristics and costs of all capital equipment. The operations manager of a manufacturing facility is considering the replacement of a ten-year-old machine that is used to produce special-order molded window frames for the aircraft industry. The manager has just received an order for 50,000 window frames, which can be manufactured by the old machine at a cost of $39 per unit. However, a new molding machine is available from a supplier at a cost of $75,000; because of its greater speed, the new machine could manufacture the window frames at a unit cost of $25. The old machine operates at a 95% level of efficiency, producing a 5% proportion of defectives. If sold, the old machine would bring $5,000. The defective rate of the new machine is uncertain and characterized by the accompanying probability distribution. If the entire cost of the new machine is to be allocated to the current order for 50,000 window frames, should the manufacturer buy the new machine or retain the old machine?

Proportion Defective, p_D	.01	.04	.07	.10	.13
Probability, $P(p_D)$	.10	.20	.50	.15	.05

19.56   Refer to Exercise 19.55. What is the maximum amount the operations manager should be willing to pay to learn the exact defective rate of the new machine?

19.57   Manufacturers seek to design product packages that fit the retailers' available shelf space, are easily stored, and appeal to the consumer. The marketing manager for a food products company is considering a new package design for his company's primary product. Before deciding whether to adopt the new design, the manager must decide whether to hire a market survey group, at a cost of $5,000, to obtain buyer reaction to the new design. The manager has determined that if the new design is introduced and sales are "good," $40,000 will be returned monthly to the company. If sales are "fair," $20,000 will be returned; and if sales are "poor," $6,000 will be returned. Past experience with the market survey group leads the marketing manager to develop the probabilities shown in Table 19.22 regarding the reliability of the market survey group's information, $P(x_i|s_j)$. If a decision is made to not introduce the new design, the old design will be continued, with a monthly return of $16,000 to the company. The monthly advertising cost is $10,000 for the new design and $5,000 for the old design. It is assumed that the probabilities of good, fair, or poor sales resulting after switching to the new design are .425, .25, and .325, respectively.

a. Construct a payoff table for the decision problem. (Use either profit or opportunity loss as the payoff measure.)

b. If the marketing manager seeks to maximize his firm's expected profits, should he recommend that his firm adopt the new package design?

c. What is the cost of uncertainty to the firm associated with the decision problem?

d. If the market survey group is employed to conduct a survey of consumer preferences, should the new design be employed if the survey conclusion is favorable? If it is unfavorable?

e. What is the expected value of the sample information provided by the market survey group? Would the manager then recommend hiring its services for a $2,500 fee?

Table 19.22 Probability Data for Exercise 19.57

	State of Nature, Actual Sales		
Survey Conclusion	Good, s_1	Fair, s_2	Poor, s_3
Favorable, x_1	.75	.40	.10
Unfavorable, x_2	.25	.60	.90

19.58 A retail trade association is considering publishing a quarterly trade journal describing the activities of the firms in the association. The journal is a feasible undertaking only if a sufficient number of members agree to subscribe. The association believes that the accompanying probabilities adequately represent the chance that a given percentage of the trade association membership will subscribe to the journal. Suppose there are 1,000 members in the trade association and that fixed costs of printing the journal will amount to $160 per quarter, with a variable cost of $.25 per journal. How much would the trade association have to charge per journal to break even?

Percentage, p	.30	.40	.50	.60
Probability, $P(p)$	.1	.3	.4	.2

19.59 The Chemical Corporation is interested in determining the potential loss on a binding purchase contract that will be in effect at the end of the fiscal year. The corporation produces a chemical compound that deteriorates and must be discarded if it is not sold by the end of the month during which it was produced. The total variable cost of the manufactured compound is $25 per unit, and it is sold for $40 per unit. The compound can be purchased from a competitor for $40 per unit plus $5 freight per unit. It is estimated that failure to fill orders would result in the complete loss of customers' orders for the compound. The corporation has sold the compound for the past 30 months. Demand has been irregular, and sales per month during this time have been as shown in Table 19.23.

Table 19.23 Sales Data for Exercise 19.59

Units Sold per Month	Number of Months
4,000	6
5,000	15
6,000	9

a. Find the probability of sales of 4,000, 5,000, or 6,000 units in any month.

b. Assuming that all orders will be filled, construct the profit table and the opportunity loss table.

c. What is the expected monthly income of the corporation if 5,000 units are manufactured every month and all orders are filled?

d. What is the inventory level that maximizes the corporation's expected monthly income?

19.60 Factors such as demand elasticity, competitive retaliation, threat of future price weakness, and potential entry of new competitors influence the pricing policy of manufacturers. The marketing manager for a nationwide brewery has suggested that his firm reduce the price of its product by 20% in the hope of inducing greater market penetration. Studies indicate that if the firm's primary competitor does not also reduce its price, the brewery would incur an annual gain in profits of $500,000; but if the competitor does retaliate with a price reduction, annual profits would drop by $200,000. If the objective of the brewery is to maximize expected profits, what is the minimum probability that the competitor will retaliate with a price reduction to justify the suggestion of the marketing manager?

19.61 Refer to Exercise 19.57. Construct a decision tree to graphically portray the decision problem faced by the marketing manager.

19.62 An electronics corporation produces electron tubes for use in transmitting equipment. Currently, the principal material used for the tube casings is made of di-essolan. The management is considering the use of tetra-essolan for the casings, to be used on production runs where a relatively low proportion of the tubes are defective. Tetra-essolan is more expensive than di-essolan but is also more attractive to the consumer. Since there is great difficulty determining in advance of a production run what the proportion of defective tubes will be, it is not certain which would be the best material to use for the casings. After considerable research the corporation has determined that the information shown in Table 19.24 best describes the relationship among proportion defective, their associated probabilities of occurrence (prior probabilities), and the dollar profits per production run by using each of the casing materials.

a. If management's objective is to maximize the expected profits, which casing material should be used?

b. What is the expected profit associated with the optimal decision?

c. How much would the corporation be willing to pay to know in advance of a production run the exact proportion of tubes that will be defective?

Table 19.24   Probabilities for Exercise 19.62

Proportion of Tubes Defective per Production Run	Prior Probability of Proportion Defective	Casing Material	
		Di-essolan	Tetra-essolan
.01	.5	$ 50	$180
.05	.2	80	100
.10	.2	120	70
.20	.1	160	40

19.63 Refer to Exercise 19.62. Suppose that the electronics corporation has decided to select a sample of 25 finished tubes from the production process before deciding whether or not to switch to tetra-essolan for the remainder of a day's production run. Three defective tubes are found in the sample. From the tables for the binomial distribution (Table 1 of the Appendix), the probability of observing three defectives in a sample of size $n = 25$, given that the fraction defective within the population is p_D, has been determined and is given in Table 19.25 for each possible value of p_D.

a. Using Bayes' Law, construct the posterior probabilities associated with the fraction defective (p_D) levels.

b. Incorporating both the prior and the sample information, determine whether the electronics corporation should use di-essolan or tetra-essolan for the casings during that day.

c. How much would the electronics corporation have been willing to pay to obtain the given sample information? Explain.

Table 19.25   Data for Exercise 19.63

p_D	$P(n = 25, 3 \text{ Defectives} \mid p_D)$
.01	.002
.05	.093
.10	.227
.20	.136

19.64   Refer to Exercises 19.62 and 19.63. Suppose that during another day a sample of $n = 25$ finished tubes is selected from the process and that two defectives are found. Use the binomial probability tables (Table 1 of the Appendix) to compute the probability of observing two defectives in a sample of size $n = 25$, given that the fraction defective within the population is .01, .05, .10, and .20. With regard to the sample information of observing two defectives in a sample of size $n = 25$ and their associated probabilities for each p_D level, answer the questions posed in Exercise 19.63.

19.65   A sampling plan is defined as a decision rule by which the decision maker selects a sample of n items and makes one decision if not more than a defectives are observed but makes a complementary decision if more than a defectives are observed in the sample. With regard to the results of Exercises 19.63 and 19.64, suggest a sampling plan for the electronics corporation, using a sample size of $n = 25$.

19.66   A manufacturer of electric power tools is contemplating the purchase of 1/4-horsepower electric motors from a foreign supplier. The motors are currently produced internally and, from experience, it is known that about 97% of all internally produced motors are nondefective. The product from the foreign supplier, although available at a unit cost of $1.00 less than the cost to produce a motor internally, has an uncertain quality. From the best available information the probability distribution shown in Table 19.26 describes the quality (fraction defective) of motors available from the foreign supplier. The cost to the power tool manufacturer of having to

Table 19.26   Probability Distribution for Exercise 19.66

Fraction Defective	Probability
.01	.3
.05	.5
.10	.2

replace or repair a power tool with a defective motor is estimated to be $15 per tool. Assuming that the manufacturer will need 50,000 electric motors during the next year, should he purchase the motors from the foreign supplier or manufacture them internally?

19.67   Refer to Exercise 19.66. Before deciding whether or not to purchase the 50,000 motors from the foreign supplier, suppose that the power tool manufacturer obtains a sample of 25 motors from the foreign supplier, subjects the motors to a battery of rigorous tests, and finds 3 of them to be defective. From the information contained in Exercise 19.66 and this sample information, should the manufacturer buy from the foreign supplier?

19.68   A building contractor must decide how many speculative homes to build. Each home will be of the same design and will cost the contractor $60,000 to build. He plans to sell each for $75,000 but intends to auction off unsold homes at the end of six months to recover construction costs. Auction prices on homes of the design proposed by the contractor have been bringing $45,000. Suppose that the contractor's prior beliefs regarding the likelihood of selling any number of the speculative homes within six months are as shown in Table 19.27. If the contractor's objective is to maximize his expected return, how many of the speculative homes should he build?

Table 19.27   Probabilities for Exercise 19.68

Number of Speculative Homes Sold Within Six Months	Probability
0	.05
1	.15
2	.40
3	.30
4	.10
5 or more	0

19.69   The manager of a manufacturing plant that produces an item for which the demand has been quite variable must decide whether to buy a new assembly machine or to have the current assembly machine repaired, at a cost of $500. The new machine would cost $7,000 and would assemble finished items at a cost of $.60 per unit; the current machine incurs a variable cost of $1.10 per unit. Units are produced as demand occurs so that management will not be left with unsold merchandise. The selling price of each unit of production is $2.00, and the manager believes that demand will occur according to the accompanying probability distribution.

Remaining Demand	5,000	10,000	25,000	50,000
Probability	.2	.4	.3	.1

a.   Construct the profit table, listing the actions and possible states of nature.

b.   If the manager's objective is to maximize the expected profits to the manufacturing plant, what is the optimal decision?

c.   What is the expected profit associated with the optimal decision?

d.   Give a practical interpretation to the expected profit calculated in part c.

19.70   Refer to Exercise 19.69. Suppose that another new machine is available; it would cost $2,000 and would incur a variable cost of $1 per unit in the process of assembling a finished unit. Thus, the manager can choose from among three machines, two new ones and the old one. If her objective is to maximize the manufacturing plant's expected profit, what is the manager's optimal decision?

19.71 A publisher is considering marketing a new monthly magazine with articles and other information of special interest to investors. From past experience and her perceptions of potential demand for a monthly such as the one she is considering, the publisher has established the profit table shown in Table 19.28 to represent her annual profits under three assumed possible levels of buyer response. The publisher estimates the probabilities of occurrence of the three buyer responses to be $P(R_1) = .5$, $P(R_2) = .2$, and $P(R_3) = .3$. Should the publisher go ahead with the project and publish the monthly magazine?

Table 19.28   Profit Table for Exercise 19.71

Buyer Response	Annual Profits	
	Do Not Publish	Publish
Poor (R_1)	$0	$-\$2,500,000$
Fair (R_2)	0	500,000
Good (R_3)	0	3,000,000

19.72 Refer to Exercise 19.71. Suppose the publisher recognizes that competition would be very fierce owing to the presence of many current and successful publications dealing with the same subject. She thus wishes to test-market her proposed monthly and base her decision on information other than her own experience and perceptions. A test market will result in either an unfavorable reaction (O_1) or a favorable reaction (O_2). From historical experience of market tests on other publications, the publisher has established the following conditional probabilities, given buyer response:

$$P(O_1|R_1) = .9 \qquad P(O_1|R_2) = .4 \qquad P(O_1|R_3) = .3$$
$$P(O_2|R_1) = .1 \qquad P(O_2|R_2) = .6 \qquad P(O_2|R_3) = .7$$

a. What is the publisher's best decision if the market test is unfavorable?

b. What is the publisher's best decision if the market test is favorable?

c. Construct a decision tree to illustrate the publisher's decision problem.

Supplementary Exercises

Interpretive

19.73 What is the basic feature that distinguishes decision analysis from the classical approach to statistics discussed in earlier chapters?

19.74 What is the justification for using an expected monetary value objective in decision problems involving uncertainty?

19.75 For each of the following business decision-making problems, list the actions available to the decision maker and the states of nature that might result to affect the payoff.

a. The selection of one from among seven competing suppliers for a fleet of cars for a state agency motor pool.

b. The leasing of a computer by a commercial bank to process checks and handle internal accounting.

c. The expansion of the market of a regional brewery from a two-state market to either a four-state market or a seven-state market.

d. The selection of one from among four grain crops by a farmer for a plot of farmland.

e. The investment of a company pension fund.

f. The daily demand for French bread from customers of a bakery when it is known that demand ranges from 10 to 40 loaves; the baker must decide each morning how many loaves to prepare for sale.

19.76 Refer to Exercise 19.75. Indicate whether the outcomes associated with the decision maker's available actions in parts a through f are known with certainty or involve uncertainty. Do any of the six decision-making problems involve an exercise of decision making under certainty?

19.77 Refer to Exercise 19.75. Identify an appropriate payoff measure for each of the decision-making problems in parts a through f.

19.78 Although the current political climate has ruled out the immediate development of thermal energy generation facilities, a special commission appointed by the governor of the state of Oregon is currently studying the issue of site location for a nuclear power generating plant. The alternatives they are currently considering are (a_1) to locate the plant at the mouth of the Columbia River, 85 miles west of Portland, (a_2) to locate the plant inland on the Willamette River near Salem, and (a_3) not to build the plant and depend on hydroelectric power. Because of the heat generated in cooling operations, a_1 may prove damaging to the commercial salmon fishing industry, vital to the economy of Oregon. However, there may be insufficient water flow available at site a_2 for adequate cooling. If alternative a_3 is chosen, sufficient energy resources may not be available to meet future needs.

a. If you were an advisor to the commission, what would you recommend be used as a measure of payoff for evaluating the three alternatives?

b. Suppose the commission has decided that the important attributes of value in the decision model are (i) the ability of each alternative to meet future commercial power needs, (ii) the ability of each alternative to meet future residential power needs, (iii) the effect on the state's salmon industry, and (iv) other environmental effects. How would you recommend the commission incorporate these four attributes into a model to derive some reasonable measure of payoff associated with each alternative?

c. Using your own judgment, follow the format prescribed in Example 19.1 to assign a payoff measure to each of the three alternatives while incorporating the four attributes of value agreed upon in part b.

Case Study

Using Decision Analysis in Selecting Personnel

F. Winter and K. Rowland ("Personnel Decisions: A Bayesian Approach," *California Management Review*, Spring 1980) note that personnel decisions involving the selection of individuals for employment, promotion, and transfer require an information base, but that typically such decisions ignore the dollar costs and values of the information used. As the authors state:

> The cost and values of such (informational) inputs change over time. If the cost of an employee training program increases, the costs of unsatisfactory performance for those who successfully complete that program increase also.

The decision analysis model provides an appropriate mechanism for evaluating these costs and values in the process of selecting personnel.

In the example provided within their article the authors assign the value of selecting a satisfactory performer for a certain task to be $6,000, while the value of selecting an unsatisfactory performer is $-$2,000$. It is further assumed that the proportion of applicants who will prove to be satisfactory performers is $P(S) = .75$, and the proportion who will prove to be unsatisfactory is $P(U) = .25$. Thus, without any information about each applicant, the personnel officer can reject each applicant with a resulting expected value of $0, or accept each with an expected value of

$$(\$6,000)(.75) - (\$2,000)(.25) = \$4,000$$

Suppose now that the firm represented in the example establishes a contractual arrangement with an assessment center to assess each applicant's performance potential as good, fair, or poor. From past experience with a variety of client firms, the assessment center has developed the probabilities shown in the following table to measure the reliability of its assessments.

	State of Nature (Job Performance)	
Assessment, x	s_1, Satisfactory	s_2, Unsatisfactory
x_1, good	.90	.05
x_2, fair	.08	.20
x_3, poor	.02	.75

1. Construct a decision tree to graphically portray the personnel selection problem, assuming the firm uses the services of the personnel assessment center.
2. What is the maximum amount the firm should be willing to pay the assessment center to conduct a performance assessment of any given job applicant?
3. Should the firm accept or reject for job placement an applicant whose performance potential is judged to be good? Whose performance potential is judged to be fair? Whose performance potential is judged to be poor?

4. What is the expected value of the information provided by the assessment center to the firm? (That is, find the EVSI.) What purpose does this value serve?

References and Suggested Readings

Baird, Bruce F. *The Technical Manager: How to Manage People and Make Decisions.* Belmont, Ca: Lifetime Learning, 1983.

Bell, D. E., R. L. Keeney, and H. Raiffa. *Conflicting Objectives in Decisions.* New York: Wiley, 1977.

Brown, R. V., A. S. Kahr, and C. Peterson. *Decision Analysis for the Manager.* New York: Holt, Rinehart and Winston, 1974.

Holloway, C. A. *Decision Making Under Uncertainty: Models and Choice.* Englewood Cliffs, N. J.: Prentice-Hall, 1979.

Howard, R. "Decision Analysis: Applied Decision Theory," *Proceedings of the Fourth International Conference on Operational Research* (1966).

Raiffa, H. *Decision Analysis: Introductory Lectures on Choices Under Uncertainty.* New York: Random House, 1986.

Raiffa, H., and R. Schlaifer. *Applied Statistical Decision Theory.* New York: Harper and Row, 1961.

Schlaifer, R. *Analysis of Decisions Under Uncertainty.* Melbourne, Fla: Keiger, 1978.

Winkler, R. L. *An Introduction to Bayesian Inference and Decision.* New York: Holt, Rinehart and Winston, 1972.

APPENDIX TABLES

Table 1 Binomial Probability

Tabulated values are $P(x \leq a) = \sum_{x=0}^{a} p(x)$. (Computations are rounded at the third decimal place.)

(a) $n = 5$

a	0.01	0.05	0.10	0.20	0.30	0.40	0.50	0.60	0.70	0.80	0.90	0.95	0.99	a
0	.951	.774	.590	.328	.168	.078	.031	.010	.002	.000	.000	.000	.000	0
1	.999	.977	.919	.737	.528	.337	.188	.087	.031	.007	.000	.000	.000	1
2	1.000	.999	.991	.942	.837	.683	.500	.317	.163	.058	.009	.001	.000	2
3	1.000	1.000	1.000	.993	.969	.913	.812	.663	.472	.263	.081	.023	.001	3
4	1.000	1.000	1.000	1.000	.998	.990	.969	.922	.832	.672	.410	.226	.049	4

(continues)

Table 1   (*Continued*)

(b) *n* = 10

a	0.01	0.05	0.10	0.20	0.30	0.40	0.50	0.60	0.70	0.80	0.90	0.95	0.99	a
0	.904	.599	.349	.107	.028	.006	.001	.000	.000	.000	.000	.000	.000	0
1	.996	.914	.736	.376	.149	.046	.011	.002	.000	.000	.000	.000	.000	1
2	1.000	.988	.930	.678	.383	.167	.055	.012	.002	.000	.000	.000	.000	2
3	1.000	.999	.987	.879	.650	.382	.172	.055	.011	.001	.000	.000	.000	3
4	1.000	1.000	.998	.967	.850	.633	.377	.166	.047	.006	.000	.000	.000	4
5	1.000	1.000	1.000	.994	.953	.834	.623	.367	.150	.033	.002	.000	.000	5
6	1.000	1.000	1.000	.999	.989	.945	.828	.618	.350	.121	.013	.001	.000	6
7	1.000	1.000	1.000	1.000	.998	.988	.945	.833	.617	.322	.070	.012	.000	7
8	1.000	1.000	1.000	1.000	1.000	.998	.989	.954	.851	.624	.264	.086	.004	8
9	1.000	1.000	1.000	1.000	1.000	1.000	.999	.994	.972	.893	.651	.401	.096	9

(c) *n* = 15

a	0.01	0.05	0.10	0.20	0.30	0.40	0.50	0.60	0.70	0.80	0.90	0.95	0.99	a
0	.860	.463	.206	.035	.005	.000	.000	.000	.000	.000	.000	.000	.000	0
1	.990	.829	.549	.167	.035	.005	.000	.000	.000	.000	.000	.000	.000	1
2	1.000	.964	.816	.398	.127	.027	.004	.000	.000	.000	.000	.000	.000	2
3	1.000	.995	.944	.648	.297	.091	.018	.002	.000	.000	.000	.000	.000	3
4	1.000	.999	.987	.836	.515	.217	.059	.009	.001	.000	.000	.000	.000	4
5	1.000	1.000	.998	.939	.722	.403	.151	.034	.004	.000	.000	.000	.000	5
6	1.000	1.000	1.000	.982	.869	.610	.304	.095	.015	.001	.000	.000	.000	6
7	1.000	1.000	1.000	.996	.950	.787	.500	.213	.050	.004	.000	.000	.000	7
8	1.000	1.000	1.000	.999	.985	.905	.696	.390	.131	.018	.000	.000	.000	8
9	1.000	1.000	1.000	1.000	.996	.966	.849	.597	.278	.061	.002	.000	.000	9
10	1.000	1.000	1.000	1.000	.999	.991	.941	.783	.485	.164	.013	.001	.000	10
11	1.000	1.000	1.000	1.000	1.000	.998	.982	.909	.703	.352	.056	.005	.000	11
12	1.000	1.000	1.000	1.000	1.000	1.000	.996	.973	.873	.602	.184	.036	.000	12
13	1.000	1.000	1.000	1.000	1.000	1.000	1.000	.995	.965	.833	.451	.171	.010	13
14	1.000	1.000	1.000	1.000	1.000	1.000	1.000	1.000	.995	.965	.794	.537	.140	14

Table 1 (*Continued*)

(d) $n = 20$

							p							
a	0.01	0.05	0.10	0.20	0.30	0.40	0.50	0.60	0.70	0.80	0.90	0.95	0.99	a
0	.818	.358	.122	.012	.001	.000	.000	.000	.000	.000	.000	.000	.000	0
1	.983	.736	.392	.069	.008	.001	.000	.000	.000	.000	.000	.000	.000	1
2	.999	.925	.677	.206	.035	.004	.000	.000	.000	.000	.000	.000	.000	2
3	1.000	.984	.867	.411	.107	.016	.001	.000	.000	.000	.000	.000	.000	3
4	1.000	.997	.957	.630	.238	.051	.006	.000	.000	.000	.000	.000	.000	4
5	1.000	1.000	.989	.804	.416	.126	.021	.002	.000	.000	.000	.000	.000	5
6	1.000	1.000	.998	.913	.608	.250	.058	.006	.000	.000	.000	.000	.000	6
7	1.000	1.000	1.000	.968	.772	.416	.132	.021	.001	.000	.000	.000	.000	7
8	1.000	1.000	1.000	.990	.887	.596	.252	.057	.005	.000	.000	.000	.000	8
9	1.000	1.000	1.000	.997	.952	.755	.412	.128	.017	.001	.000	.000	.000	9
10	1.000	1.000	1.000	.999	.983	.872	.588	.245	.048	.003	.000	.000	.000	10
11	1.000	1.000	1.000	1.000	.995	.943	.748	.404	.113	.010	.000	.000	.000	11
12	1.000	1.000	1.000	1.000	.999	.979	.868	.584	.228	.032	.000	.000	.000	12
13	1.000	1.000	1.000	1.000	1.000	.994	.942	.750	.392	.087	.002	.000	.000	13
14	1.000	1.000	1.000	1.000	1.000	.998	.979	.874	.584	.196	.011	.000	.000	14
15	1.000	1.000	1.000	1.000	1.000	1.000	.994	.949	.762	.370	.043	.003	.000	15
16	1.000	1.000	1.000	1.000	1.000	1.000	.999	.984	.893	.589	.133	.016	.000	16
17	1.000	1.000	1.000	1.000	1.000	1.000	1.000	.996	.965	.794	.323	.075	.001	17
18	1.000	1.000	1.000	1.000	1.000	1.000	1.000	.999	.992	.931	.608	.264	.017	18
19	1.000	1.000	1.000	1.000	1.000	1.000	1.000	1.000	.999	.988	.878	.642	.182	19

(continues)

Table 1   (*Concluded*)

(e) $n = 25$

a	0.01	0.05	0.10	0.20	0.30	0.40	0.50	0.60	0.70	0.80	0.90	0.95	0.99	a
						p								
0	.778	.277	.072	.004	.000	.000	.000	.000	.000	.000	.000	.000	.000	0
1	.974	.642	.271	.027	.002	.000	.000	.000	.000	.000	.000	.000	.000	1
2	.998	.873	.537	.098	.009	.000	.000	.000	.000	.000	.000	.000	.000	2
3	1.000	.966	.764	.234	.033	.002	.000	.000	.000	.000	.000	.000	.000	3
4	1.000	.993	.902	.421	.090	.009	.000	.000	.000	.000	.000	.000	.000	4
5	1.000	.999	.967	.617	.193	.029	.002	.000	.000	.000	.000	.000	.000	5
6	1.000	1.000	.991	.780	.341	.074	.007	.000	.000	.000	.000	.000	.000	6
7	1.000	1.000	.998	.891	.512	.154	.022	.001	.000	.000	.000	.000	.000	7
8	1.000	1.000	1.000	.953	.677	.274	.054	.004	.000	.000	.000	.000	.000	8
9	1.000	1.000	1.000	.983	.811	.425	.115	.013	.000	.000	.000	.000	.000	9
10	1.000	1.000	1.000	.994	.902	.586	.212	.034	.002	.000	.000	.000	.000	10
11	1.000	1.000	1.000	.998	.956	.732	.345	.078	.006	.000	.000	.000	.000	11
12	1.000	1.000	1.000	1.000	.983	.846	.500	.154	.017	.000	.000	.000	.000	12
13	1.000	1.000	1.000	1.000	.994	.922	.655	.268	.044	.002	.000	.000	.000	13
14	1.000	1.000	1.000	1.000	.998	.966	.788	.414	.098	.006	.000	.000	.000	14
15	1.000	1.000	1.000	1.000	1.000	.987	.885	.575	.189	.017	.000	.000	.000	15
16	1.000	1.000	1.000	1.000	1.000	.996	.946	.726	.323	.047	.000	.000	.000	16
17	1.000	1.000	1.000	1.000	1.000	.999	.978	.846	.488	.109	.002	.000	.000	17
18	1.000	1.000	1.000	1.000	1.000	1.000	.993	.926	.659	.220	.009	.000	.000	18
19	1.000	1.000	1.000	1.000	1.000	1.000	.998	.971	.807	.383	.033	.001	.000	19
20	1.000	1.000	1.000	1.000	1.000	1.000	1.000	.991	.910	.579	.098	.007	.000	20
21	1.000	1.000	1.000	1.000	1.000	1.000	1.000	.998	.967	.766	.236	.034	.000	21
22	1.000	1.000	1.000	1.000	1.000	1.000	1.000	1.000	.991	.902	.463	.127	.002	22
23	1.000	1.000	1.000	1.000	1.000	1.000	1.000	1.000	.998	.973	.729	.358	.026	23
24	1.000	1.000	1.000	1.000	1.000	1.000	1.000	1.000	1.000	.996	.928	.723	.222	24

Table 2   Values of e^{-x}

x	e^{-x}	x	e^{-x}	x	e^{-x}	x	e^{-x}
0.00	1.000000	2.55	0.078082	5.05	0.006409	7.55	0.000526
0.05	0.951229	2.60	0.074274	5.10	0.006097	7.60	0.000500
0.10	0.904837	2.65	0.070651	5.15	0.005799	7.65	0.000476
0.15	0.860708	2.70	0.067206	5.20	0.005517	7.70	0.000453
0.20	0.818731	2.75	0.063928	5.25	0.005248	7.75	0.000431
0.25	0.778801	2.80	0.060810	5.30	0.004992	7.80	0.000410
0.30	0.740818	2.85	0.057844	5.35	0.004748	7.85	0.000390
0.35	0.704688	2.90	0.055023	5.40	0.004517	7.90	0.000371
0.40	0.670320	2.95	0.052340	5.45	0.004296	7.95	0.000353
0.45	0.637628	3.00	0.049787	5.50	0.004087	8.00	0.000335
0.50	0.606531	3.05	0.047359	5.55	0.003887	8.05	0.000319
0.55	0.576950	3.10	0.045049	5.60	0.003698	8.10	0.000304
0.60	0.548812	3.15	0.042852	5.65	0.003518	8.15	0.000289
0.65	0.522046	3.20	0.040762	5.70	0.003346	8.20	0.000275
0.70	0.496585	3.25	0.038774	5.75	0.003183	8.25	0.000261
0.75	0.472367	3.30	0.036883	5.80	0.003028	8.30	0.000249
0.80	0.449329	3.35	0.035084	5.85	0.002880	8.35	0.000236
0.85	0.427415	3.40	0.033373	5.90	0.002739	8.40	0.000225
0.90	0.406570	3.45	0.031746	5.95	0.002606	8.45	0.000214
0.95	0.386741	3.50	0.030197	6.00	0.002479	8.50	0.000203
1.00	0.367879	3.55	0.028725	6.05	0.002358	8.55	0.000194
1.05	0.349938	3.60	0.027324	6.10	0.002243	8.60	0.000184
1.10	0.332871	3.65	0.025991	6.15	0.002133	8.65	0.000175
1.15	0.316637	3.70	0.024724	6.20	0.002029	8.70	0.000167
1.20	0.301194	3.75	0.023518	6.25	0.001930	8.75	0.000158
1.25	0.286505	3.80	0.022371	6.30	0.001836	8.80	0.000151
1.30	0.272532	3.85	0.021280	6.35	0.001747	8.85	0.000143
1.35	0.259240	3.90	0.020242	6.40	0.001662	8.90	0.000136
1.40	0.246597	3.95	0.019255	6.45	0.001581	8.95	0.000130
1.45	0.234570	4.00	0.018316	6.50	0.001503	9.00	0.000123
1.50	0.223130	4.05	0.017422	6.55	0.001430	9.05	0.000117
1.55	0.212248	4.10	0.016573	6.60	0.001360	9.10	0.000112
1.60	0.201897	4.15	0.015764	6.65	0.001294	9.15	0.000106
1.65	0.192050	4.20	0.014996	6.70	0.001231	9.20	0.000101
1.70	0.182684	4.25	0.014264	6.75	0.001171	9.25	0.000096
1.75	0.173774	4.30	0.013569	6.80	0.001114	9.30	0.000091
1.80	0.165299	4.35	0.012907	6.85	0.001059	9.35	0.000087
1.85	0.157237	4.40	0.012277	6.90	0.001008	9.40	0.000083
1.90	0.149569	4.45	0.011679	6.95	0.000959	9.45	0.000079
1.95	0.142274	4.50	0.011109	7.00	0.000912	9.50	0.000075
2.00	0.135335	4.55	0.010567	7.05	0.000867	9.55	0.000071
2.05	0.128735	4.60	0.010052	7.10	0.000825	9.60	0.000068
2.10	0.122456	4.65	0.009562	7.15	0.000785	9.65	0.000064
2.15	0.116484	4.70	0.009095	7.20	0.000747	9.70	0.000061
2.20	0.110803	4.75	0.008652	7.25	0.000710	9.75	0.000058
2.25	0.105399	4.80	0.008230	7.30	0.000676	9.80	0.000055
2.30	0.100259	4.85	0.007828	7.35	0.000643	9.85	0.000053
2.35	0.095369	4.90	0.007447	7.40	0.000611	9.90	0.000050
2.40	0.090718	4.95	0.007083	7.45	0.000581	9.95	0.000048
2.45	0.086294	5.00	0.006738	7.50	0.000553	10.00	0.000045
2.50	0.082085						

Table 3   Normal Curve Areas

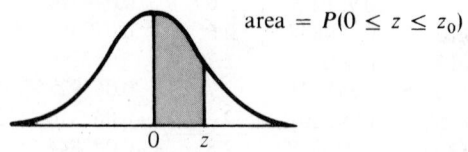

area $= P(0 \le z \le z_0)$

z_0	.00	.01	.02	.03	.04	.05	.06	.07	.08	.09
0.0	.0000	.0040	.0080	.0120	.0160	.0199	.0239	.0279	.0319	.0359
0.1	.0398	.0438	.0478	.0517	.0557	.0596	.0636	.0675	.0714	.0753
0.2	.0793	.0832	.0871	.0910	.0948	.0987	.1026	.1064	.1103	.1141
0.3	.1179	.1217	.1255	.1293	.1331	.1368	.1406	.1443	.1480	.1517
0.4	.1554	.1591	.1628	.1664	.1700	.1736	.1772	.1808	.1844	.1879
0.5	.1915	.1950	.1985	.2019	.2054	.2088	.2123	.2157	.2190	.2224
0.6	.2257	.2291	.2324	.2357	.2389	.2422	.2454	.2486	.2517	.2549
0.7	.2580	.2611	.2642	.2673	.2704	.2734	.2764	.2794	.2823	.2852
0.8	.2881	.2910	.2939	.2967	.2995	.3023	.3051	.3078	.3106	.3133
0.9	.3159	.3186	.3212	.3238	.3264	.3289	.3315	.3340	.3365	.3389
1.0	.3413	.3438	.3461	.3485	.3508	.3531	.3554	.3577	.3599	.3621
1.1	.3643	.3665	.3686	.3708	.3729	.3749	.3770	.3790	.3810	.3830
1.2	.3849	.3869	.3888	.3907	.3925	.3944	.3962	.3980	.3997	.4015
1.3	.4032	.4049	.4066	.4082	.4099	.4115	.4131	.4147	.4162	.4177
1.4	.4192	.4207	.4222	.4236	.4251	.4265	.4279	.4292	.4306	.4319
1.5	.4332	.4345	.4357	.4370	.4382	.4394	.4406	.4418	.4429	.4441
1.6	.4452	.4463	.4474	.4484	.4495	.4505	.4515	.4525	.4535	.4545
1.7	.4554	.4564	.4573	.4582	.4591	.4599	.4608	.4616	.4625	.4633
1.8	.4641	.4649	.4656	.4664	.4671	.4678	.4686	.4693	.4699	.4706
1.9	.4713	.4719	.4726	.4732	.4738	.4744	.4750	.4756	.4761	.4767
2.0	.4772	.4778	.4783	.4788	.4793	.4798	.4803	.4808	.4812	.4817
2.1	.4821	.4826	.4830	.4834	.4838	.4842	.4846	.4850	.4854	.4857
2.2	.4861	.4864	.4868	.4871	.4875	.4878	.4881	.4884	.4887	.4890
2.3	.4893	.4896	.4898	.4901	.4904	.4906	.4909	.4911	.4913	.4916
2.4	.4918	.4920	.4922	.4925	.4927	.4929	.4931	.4932	.4934	.4936
2.5	.4938	.4940	.4941	.4943	.4945	.4946	.4948	.4949	.4951	.4952
2.6	.4953	.4955	.4956	.4957	.4959	.4960	.4961	.4962	.4963	.4964
2.7	.4965	.4966	.4967	.4968	.4969	.4970	.4971	.4972	.4973	.4974
2.8	.4974	.4975	.4976	.4977	.4977	.4978	.4979	.4979	.4980	.4981
2.9	.4981	.4982	.4982	.4983	.4984	.4984	.4985	.4985	.4986	.4986
3.0	.4987	.4987	.4987	.4988	.4988	.4989	.4989	.4989	.4990	.4990

This table is abridged from Table 1 of *Statistical Tables and Formulas*, by A. Hald (New York: John Wiley & Sons, Inc., 1952). Reproduced by permission of A. Hald and the publishers, John Wiley & Sons, Inc.

Table 4   Critical Values of *t*

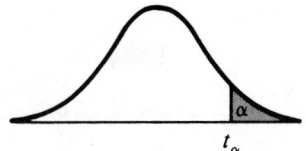

t_α

d.f.	$t_{.100}$	$t_{.050}$	$t_{.025}$	$t_{.010}$	$t_{.005}$	d.f.
1	3.078	6.314	12.706	31.821	63.657	1
2	1.886	2.920	4.303	6.965	9.925	2
3	1.638	2.353	3.182	4.541	5.841	3
4	1.533	2.132	2.776	3.747	4.604	4
5	1.476	2.015	2.571	3.365	4.032	5
6	1.440	1.943	2.447	3.143	3.707	6
7	1.415	1.895	2.365	2.998	3.499	7
8	1.397	1.860	2.306	2.896	3.355	8
9	1.383	1.833	2.262	2.821	3.250	9
10	1.372	1.812	2.228	2.764	3.169	10
11	1.363	1.796	2.201	2.718	3.106	11
12	1.356	1.782	2.179	2.681	3.055	12
13	1.350	1.771	2.160	2.650	3.012	13
14	1.345	1.761	2.145	2.624	2.977	14
15	1.341	1.753	2.131	2.602	2.947	15
16	1.337	1.746	2.120	2.583	2.921	16
17	1.333	1.740	2.110	2.567	2.898	17
18	1.330	1.734	2.101	2.552	2.878	18
19	1.328	1.729	2.093	2.539	2.861	19
20	1.325	1.725	2.086	2.528	2.845	20
21	1.323	1.721	2.080	2.518	2.831	21
22	1.321	1.717	2.074	2.508	2.819	22
23	1.319	1.714	2.069	2.500	2.807	23
24	1.318	1.711	2.064	2.492	2.797	24
25	1.316	1.708	2.060	2.485	2.787	25
26	1.315	1.706	2.056	2.479	2.779	26
27	1.314	1.703	2.052	2.473	2.771	27
28	1.313	1.701	2.048	2.467	2.763	28
29	1.311	1.699	2.045	2.462	2.756	29
inf.	1.282	1.645	1.960	2.326	2.576	inf.

Table 5   Critical Values of Chi-square

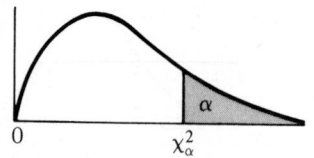

d.f.	$\chi^2_{0.995}$	$\chi^2_{0.990}$	$\chi^2_{0.975}$	$\chi^2_{0.950}$	$\chi^2_{0.900}$
1	0.0000393	0.0001571	0.0009821	0.0039321	0.0157908
2	0.0100251	0.0201007	0.0506356	0.102587	0.210720
3	0.0717212	0.114832	0.215795	0.351846	0.584375
4	0.206990	0.297110	0.484419	0.710721	1.063623
5	0.411740	0.554300	0.831211	1.145476	1.61031
6	0.675727	0.872085	1.237347	1.63539	2.20413
7	0.989265	1.239043	1.68987	2.16735	2.83311
8	1.344419	1.646482	2.17973	2.73264	3.48954
9	1.734926	2.087912	2.70039	3.32511	4.16816
10	2.15585	2.55821	3.24697	3.94030	4.86518
11	2.60321	3.05347	3.81575	4.57481	5.57779
12	3.07382	3.57056	4.40379	5.22603	6.30380
13	3.56503	4.10691	5.00874	5.89186	7.04150
14	4.07468	4.66043	5.62872	6.57063	7.78953
15	4.60094	5.22935	6.26214	7.26094	8.54675
16	5.14224	5.81221	6.90766	7.96164	9.31223
17	5.69724	6.40776	7.56418	8.67176	10.0852
18	6.26481	7.01491	8.23075	9.39046	10.8649
19	6.84398	7.63273	8.90655	10.1170	11.6509
20	7.43386	8.26040	9.59083	10.8508	12.4426
21	8.03366	8.89720	10.28293	11.5913	13.2396
22	8.64272	9.54249	10.9823	12.3380	14.0415
23	9.26042	10.19567	11.6885	13.0905	14.8479
24	9.88623	10.8564	12.4011	13.8484	15.6587
25	10.5197	11.5240	13.1197	14.6114	16.4734
26	11.1603	12.1981	13.8439	15.3791	17.2919
27	11.8076	12.8786	14.5733	16.1513	18.1138
28	12.4613	13.5648	15.3079	16.9279	18.9392
29	13.1211	14.2565	16.0471	17.7083	19.7677
30	13.7867	14.9535	16.7908	18.4926	20.5992
40	20.7065	22.1643	24.4331	26.5093	29.0505
50	27.9907	29.7067	32.3574	34.7642	37.6886
60	35.5346	37.4848	40.4817	43.1879	46.4589
70	43.2752	45.4418	48.7576	51.7393	55.3290
80	51.1720	53.5400	57.1532	60.3915	64.2778
90	59.1963	61.7541	65.6466	69.1260	73.2912
100	67.3276	70.0648	74.2219	77.9295	82.3581

Table 5 (*Concluded*)

$\chi^2_{0.100}$	$\chi^2_{0.050}$	$\chi^2_{0.025}$	$\chi^2_{0.010}$	$\chi^2_{0.005}$	d.f.
2.70554	3.84146	5.02389	6.63490	7.87944	1
4.60517	5.99147	7.37776	9.21034	10.5966	2
6.25139	7.81473	9.34840	11.3449	12.8381	3
7.77944	9.48773	11.1433	13.2767	14.8602	4
9.23635	11.0705	12.8325	15.0863	16.7496	5
10.6446	12.5916	14.4494	16.8119	18.5476	6
12.0170	14.0671	16.0128	18.4753	20.2777	7
13.3616	15.5073	17.5346	20.0902	21.9550	8
14.6837	16.9190	19.0228	21.6660	23.5893	9
15.9871	18.3070	20.4831	23.2093	25.1882	10
17.2750	19.6751	21.9200	24.7250	26.7569	11
18.5494	21.0261	23.3367	26.2170	28.2995	12
19.8119	22.3621	24.7356	27.6883	29.8194	13
21.0642	23.6848	26.1190	29.1413	31.3193	14
22.3072	24.9958	27.4884	30.5779	32.8013	15
23.5418	26.2962	28.8454	31.9999	34.2672	16
24.7690	27.5871	30.1910	33.4087	35.7185	17
25.9894	28.8693	31.5264	34.8053	37.1564	18
27.2036	30.1435	32.8523	36.1908	38.5822	19
28.4120	31.4104	34.1696	37.5662	39.9968	20
29.6151	32.6705	35.4789	38.9321	41.4010	21
30.8133	33.9244	36.7807	40.2894	42.7956	22
32.0069	35.1725	38.0757	41.6384	44.1813	23
33.1963	36.4151	39.3641	42.9798	45.5585	24
34.3816	37.6525	40.6465	44.3141	46.9278	25
35.5631	38.8852	41.9232	45.6417	48.2899	26
36.7412	40.1133	43.1944	46.9630	49.6449	27
37.9159	41.3372	44.4607	48.2782	50.9933	28
39.0875	42.5569	45.7222	49.5879	52.3356	29
40.2560	43.7729	46.9792	50.8922	53.6720	30
51.8050	55.7585	59.3417	63.6907	66.7659	40
63.1671	67.5048	71.4202	76.1539	79.4900	50
74.3970	79.0819	83.2976	88.3794	91.9517	60
85.5271	90.5312	95.0231	100.425	104.215	70
96.5782	101.879	106.629	112.329	116.321	80
107.565	113.145	118.136	124.116	128.299	90
118.498	124.342	129.561	135.807	140.169	100

Table 6 Percentage Points of the F Distribution

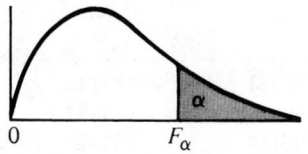

v_2	α	1	2	3	4	5	6	7	8	9
1	.100	39.86	49.50	53.59	55.83	57.24	58.20	58.91	59.44	59.86
	.050	161.4	199.5	215.7	224.6	230.2	234.0	236.8	238.9	240.5
	.025	647.8	799.5	864.2	899.6	921.8	937.1	948.2	956.7	963.3
	.010	4052	4999.5	5403	5625	5764	5859	5928	5982	6022
	.005	16211	20000	21615	22500	23056	23437	23715	23925	24091
2	.100	8.53	9.00	9.16	9.24	9.29	9.33	9.35	9.37	9.38
	.050	18.51	19.00	19.16	19.25	19.30	19.33	19.35	19.37	19.38
	.025	38.51	39.00	39.17	39.25	39.30	39.33	39.36	39.37	39.39
	.010	98.50	99.00	99.17	99.25	99.30	99.33	99.36	99.37	99.39
	.005	198.5	199.0	199.2	199.2	199.3	199.3	199.4	199.4	199.4
3	.100	5.54	5.46	5.39	5.34	5.31	5.28	5.27	5.25	5.24
	.050	10.13	9.55	9.28	9.12	9.01	8.94	8.89	8.85	8.81
	.025	17.44	16.04	15.44	15.10	14.88	14.73	14.62	14.54	14.47
	.010	34.12	30.82	29.46	28.71	28.24	27.91	27.67	27.49	27.35
	.005	55.55	49.80	47.47	46.19	45.39	44.84	44.43	44.13	43.88
4	.100	4.54	4.32	4.19	4.11	4.05	4.01	3.98	3.95	3.94
	.050	7.71	6.94	6.59	6.39	6.26	6.16	6.09	6.04	6.00
	.025	12.22	10.65	9.98	9.60	9.36	9.20	9.07	8.98	8.90
	.010	21.20	18.00	16.69	15.98	15.52	15.21	14.98	14.80	14.66
	.005	31.33	26.28	24.26	23.15	22.46	21.97	21.62	21.35	21.14
5	.100	4.06	3.78	3.62	3.52	3.45	3.40	3.37	3.34	3.32
	.050	6.61	5.79	5.41	5.19	5.05	4.95	4.88	4.82	4.77
	.025	10.01	8.43	7.76	7.39	7.15	6.98	6.85	6.76	6.68
	.010	16.26	13.27	12.06	11.39	10.97	10.67	10.46	10.29	10.16
	.005	22.78	18.31	16.53	15.56	14.94	14.51	14.20	13.96	13.77
6	.100	3.78	3.46	3.29	3.18	3.11	3.05	3.01	2.98	2.96
	.050	5.99	5.14	4.76	4.53	4.39	4.28	4.21	4.15	4.10
	.025	8.81	7.26	6.60	6.23	5.99	5.82	5.70	5.60	5.52
	.010	13.75	10.92	9.78	9.15	8.75	8.47	8.26	8.10	7.98
	.005	18.63	14.54	12.92	12.03	11.46	11.07	10.79	10.57	10.39
7	.100	3.59	3.26	3.07	2.96	2.88	2.83	2.78	2.75	2.72
	.050	5.59	4.74	4.35	4.12	3.97	3.87	3.79	3.73	3.68
	.025	8.07	6.54	5.89	5.52	5.29	5.12	4.99	4.90	4.82
	.010	12.25	9.55	8.45	7.85	7.46	7.19	6.99	6.84	6.72
	.005	16.24	12.40	10.88	10.05	9.52	9.16	8.89	8.68	8.51
8	.100	3.46	3.11	2.92	2.81	2.73	2.67	2.62	2.59	2.56
	.050	5.32	4.46	4.07	3.84	3.69	3.58	3.50	3.44	3.39
	.025	7.57	6.06	5.42	5.05	4.82	4.65	4.53	4.43	4.36
	.010	11.26	8.65	7.59	7.01	6.63	6.37	6.18	6.03	5.91
	.005	14.69	11.04	9.60	8.81	8.30	7.95	7.69	7.50	7.34
9	.100	3.36	3.01	2.81	2.69	2.61	2.55	2.51	2.47	2.44
	.050	5.12	4.26	3.86	3.63	3.48	3.37	3.29	3.23	3.18
	.025	7.21	5.71	5.08	4.72	4.48	4.32	4.20	4.10	4.03
	.010	10.56	8.02	6.99	6.42	6.06	5.80	5.61	5.47	5.35
	.005	13.61	10.11	8.72	7.96	7.47	7.13	6.88	6.69	6.54
10	.100	3.29	2.92	2.73	2.61	2.52	2.46	2.41	2.38	2.35
	.050	4.96	4.10	3.71	3.48	3.33	3.22	3.14	3.07	3.02
	.025	6.94	5.46	4.83	4.47	4.24	4.07	3.95	3.85	3.78
	.010	10.04	7.56	6.55	5.99	5.64	5.39	5.20	5.06	4.94
	.005	12.83	9.43	8.08	7.34	6.87	6.54	6.30	6.12	5.97
11	.100	3.23	2.86	2.66	2.54	2.45	2.39	2.34	2.30	2.27
	.050	4.84	3.98	3.59	3.36	3.20	3.09	3.01	2.95	2.90
	.025	6.72	5.26	4.63	4.28	4.04	3.88	3.76	3.66	3.59
	.010	9.65	7.21	6.22	5.67	5.32	5.07	4.89	4.74	4.63
	.005	12.23	8.91	7.60	6.88	6.42	6.10	5.86	5.68	5.54
12	.100	3.18	2.81	2.61	2.48	2.39	2.33	2.28	2.24	2.21
	.050	4.75	3.89	3.49	3.26	3.11	3.00	2.91	2.85	2.80
	.025	6.55	5.10	4.47	4.12	3.89	3.73	3.61	3.51	3.44
	.010	9.33	6.93	5.95	5.41	5.06	4.82	4.64	4.50	4.39
	.005	11.75	8.51	7.23	6.52	6.07	5.76	5.52	5.35	5.20

					v_1							
10	12	15	20	24	30	40	60	120	∞	α	v_2
60.19	60.71	60.22	61.74	62.00	62.26	62.53	62.79	63.06	63.33	.100	1
241.9	243.9	245.9	248.0	249.1	250.1	251.2	252.2	253.3	254.3	.050	
968.6	976.7	984.9	993.1	997.2	1001	1006	1010	1014	1018	.025	
6056	6106	6157	6209	6235	6261	6287	6313	6339	6366	.010	
24224	24426	24630	24836	24940	25044	25148	25253	25359	25465	.005	
9.39	9.41	9.42	9.44	9.45	9.46	9.47	9.47	9.48	9.49	.100	2
19.40	19.41	19.43	19.45	19.45	19.46	19.47	19.48	19.49	19.50	.050	
39.40	39.41	39.43	39.45	39.46	39.46	39.47	39.48	39.49	39.50	.025	
99.40	99.42	99.43	99.45	99.46	99.47	99.47	99.48	99.49	99.50	.010	
199.4	199.4	199.4	199.4	199.5	199.5	199.5	199.5	199.5	199.5	.005	
5.23	5.22	5.20	5.18	5.18	5.17	5.16	5.15	5.14	5.13	.100	3
8.79	8.74	8.70	8.66	8.64	8.62	8.59	8.57	8.55	8.53	.050	
14.42	14.34	14.25	14.17	14.12	14.08	14.04	13.99	13.95	13.90	.025	
27.23	27.05	26.87	26.69	26.60	26.50	26.41	26.32	26.22	26.13	.010	
43.69	43.39	43.08	42.78	42.62	42.47	42.31	42.15	41.99	41.83	.005	
3.92	3.90	3.87	3.84	3.83	3.82	3.80	3.79	3.78	3.76	.100	4
5.96	5.91	5.86	5.80	5.77	5.75	5.72	5.69	5.66	5.63	.050	
8.84	8.75	8.66	8.56	8.51	8.46	8.41	8.36	8.31	8.26	.025	
14.55	14.37	14.20	14.02	13.93	13.84	13.75	13.65	13.56	13.46	.010	
20.97	20.70	20.44	20.17	20.03	19.89	19.75	19.61	19.47	19.32	.005	
3.30	3.27	3.24	3.21	3.19	3.17	3.16	3.14	3.12	3.10	.100	5
4.74	4.68	4.62	4.56	4.53	4.50	4.46	4.43	4.40	4.36	.050	
6.62	6.52	6.43	6.33	6.28	6.23	6.18	6.12	6.07	6.02	.025	
10.05	9.89	9.72	9.55	9.47	9.38	9.29	9.20	9.11	9.02	.010	
13.62	13.38	13.15	12.90	12.78	12.66	12.53	12.40	12.27	12.14	.005	
2.94	2.90	2.87	2.84	2.82	2.80	2.78	2.76	2.74	2.72	.100	6
4.06	4.00	3.94	3.87	3.84	3.81	3.77	3.74	3.70	3.67	.050	
5.46	5.37	5.27	5.17	5.12	5.07	5.01	4.96	4.90	4.85	.025	
7.87	7.72	7.56	7.40	7.31	7.23	7.14	7.06	6.97	6.88	.010	
10.25	10.03	9.81	9.59	9.47	9.36	9.24	9.12	9.00	8.88	.005	
2.70	2.67	2.63	2.59	2.58	2.56	2.54	2.51	2.49	2.47	.100	7
3.64	3.57	3.51	3.44	3.41	3.38	3.34	3.30	3.27	3.23	.050	
4.76	4.67	4.57	4.47	4.42	4.36	4.31	4.25	4.20	4.14	.025	
6.62	6.47	6.31	6.16	6.07	5.99	5.91	5.82	5.74	5.65	.010	
8.38	8.18	7.97	7.75	7.65	7.53	7.42	7.31	7.19	7.08	.005	
2.54	2.50	2.46	2.42	2.40	2.38	2.36	2.34	2.32	2.29	.100	8
3.35	3.28	3.22	3.15	3.12	3.08	3.04	3.01	2.97	2.93	.050	
4.30	4.20	4.10	4.00	3.95	3.89	3.84	3.78	3.73	3.67	.025	
5.81	5.67	5.52	5.36	5.28	5.20	5.12	5.03	4.95	4.86	.010	
7.21	7.01	6.81	6.61	6.50	6.40	6.29	6.18	6.06	5.95	.005	
2.42	2.38	2.34	2.30	2.28	2.25	2.23	2.21	2.18	2.16	.100	9
3.14	3.07	3.01	2.94	2.90	2.86	2.83	2.79	2.75	2.71	.050	
3.96	3.87	3.77	3.67	3.61	3.56	3.51	3.45	3.39	3.33	.025	
5.26	5.11	4.96	4.81	4.73	4.65	4.57	4.48	4.40	4.31	.010	
6.42	6.23	6.03	5.83	5.73	5.62	5.52	5.41	5.30	5.19	.005	
2.32	2.28	2.24	2.20	2.18	2.16	2.13	2.11	2.08	2.06	.100	10
2.98	2.91	2.85	2.77	2.74	2.70	2.66	2.62	2.58	2.54	.050	
3.72	3.62	3.52	3.42	3.37	3.31	3.26	3.20	3.14	3.08	.025	
4.85	4.71	4.56	4.41	4.33	4.25	4.17	4.08	4.00	3.91	.010	
5.85	5.66	5.47	5.27	5.17	5.07	4.97	4.86	4.75	4.64	.005	
2.25	2.21	2.17	2.12	2.10	2.08	2.05	2.03	2.00	1.97	.100	11
2.85	2.79	2.72	2.65	2.61	2.57	2.53	2.49	2.45	2.40	.050	
3.53	3.43	3.33	3.23	3.17	3.12	3.06	3.00	2.94	2.88	.025	
4.54	4.40	4.25	4.10	4.02	3.94	3.86	3.78	3.69	3.60	.010	
5.42	5.24	5.05	4.86	4.76	4.65	4.55	4.44	4.34	4.23	.005	
2.19	2.15	2.10	2.06	2.04	2.01	1.99	1.96	1.93	1.90	.100	12
2.75	2.69	2.62	2.54	2.51	2.47	2.43	2.38	2.34	2.30	.050	
3.37	3.28	3.18	3.07	3.02	2.96	2.91	2.85	2.79	2.72	.025	
4.30	4.16	4.01	3.86	3.78	3.70	3.62	3.54	3.45	3.36	.010	
5.09	4.91	4.72	4.53	4.43	4.33	4.23	4.12	4.01	3.90	.005	

(continues)

Table 6   (Continued)

						v_1				
v_2	α	1	2	3	4	5	6	7	8	9
13	.100	3.14	2.76	2.56	2.43	2.35	2.28	2.23	2.20	2.16
	.050	4.67	3.81	3.41	3.18	3.03	2.92	2.83	2.77	2.71
	.025	6.41	4.97	4.35	4.00	3.77	3.60	3.48	3.39	3.31
	.010	9.07	6.70	5.74	5.21	4.86	4.62	4.44	4.30	4.19
	.005	11.37	8.19	6.93	6.23	5.79	5.48	5.25	5.08	4.94
14	.100	3.10	2.73	2.52	2.39	2.31	2.24	2.19	2.15	2.12
	.050	4.60	3.74	3.34	3.11	2.96	2.85	2.76	2.70	2.65
	.025	6.30	4.86	4.24	3.89	3.66	3.50	3.38	3.29	3.21
	.010	8.86	6.51	5.56	5.04	4.69	4.46	4.28	4.14	4.03
	.005	11.06	7.92	6.68	6.00	5.56	5.26	5.03	4.86	4.72
15	.100	3.07	2.70	2.49	2.36	2.27	2.21	2.16	2.12	2.09
	.050	4.54	3.68	3.29	3.06	2.90	2.79	2.71	2.64	2.59
	.025	6.20	4.77	4.15	3.80	3.58	3.41	3.29	3.20	3.12
	.010	8.68	6.36	5.42	4.89	4.56	4.32	4.14	4.00	3.89
	.005	10.80	7.70	6.48	5.80	5.37	5.07	4.85	4.67	4.54
16	.100	3.05	2.67	2.46	2.33	2.24	2.18	2.13	2.09	2.06
	.050	4.49	3.63	3.24	3.01	2.85	2.74	2.66	2.59	2.54
	.025	6.12	4.69	4.08	3.73	3.50	3.34	3.22	3.12	3.05
	.010	8.53	6.23	5.29	4.77	4.44	4.20	4.03	3.89	3.78
	.005	10.58	7.51	6.30	5.64	5.21	4.91	4.69	4.52	4.38
17	.100	3.03	2.64	2.44	2.31	2.22	2.15	2.10	2.06	2.03
	.050	4.45	3.59	3.20	2.96	2.81	2.70	2.61	2.55	2.49
	.025	6.04	4.62	4.01	3.66	3.44	3.28	3.16	3.06	2.98
	.010	8.40	6.11	5.18	4.67	4.34	4.10	3.93	3.79	3.68
	.005	10.38	7.35	6.16	5.50	5.07	4.78	4.56	4.39	4.25
18	.100	3.01	2.62	2.42	2.29	2.20	2.13	2.08	2.04	2.00
	.050	4.41	3.55	3.16	2.93	2.77	2.66	2.58	2.51	2.46
	.025	5.98	4.56	3.95	3.61	3.38	3.22	3.10	3.01	2.93
	.010	8.29	6.01	5.09	4.58	4.25	4.01	3.84	3.71	3.60
	.005	10.22	7.21	6.03	5.37	4.96	4.66	4.44	4.28	4.14
19	.100	2.99	2.61	2.40	2.27	2.18	2.11	2.06	2.02	1.98
	.050	4.38	3.52	3.13	2.90	2.74	2.63	2.54	2.48	2.42
	.025	5.92	4.51	3.90	3.56	3.33	3.17	3.05	2.96	2.88
	.010	8.18	5.93	5.01	4.50	4.17	3.94	3.77	3.63	3.52
	.005	10.07	7.09	5.92	5.27	4.85	4.56	4.34	4.18	4.04
20	.100	2.97	2.59	2.38	2.25	2.16	2.09	2.04	2.00	1.96
	.050	4.35	3.49	3.10	2.87	2.71	2.60	2.51	2.45	2.39
	.025	5.87	4.46	3.86	3.51	3.29	3.13	3.01	2.91	2.84
	.010	8.10	5.85	4.94	4.43	4.10	3.87	3.70	3.56	3.46
	.005	9.94	6.99	5.82	5.17	4.76	4.47	4.26	4.09	3.96
21	.100	2.96	2.57	2.36	2.23	2.14	2.08	2.02	1.98	1.95
	.050	4.32	3.47	3.07	2.84	2.68	2.57	2.49	2.42	2.37
	.025	5.83	4.42	3.82	3.48	3.25	3.09	2.97	2.87	2.80
	.010	8.02	5.78	4.87	4.37	4.04	3.81	3.64	3.51	3.40
	.005	9.83	6.89	5.73	5.09	4.68	4.39	4.18	4.01	3.88
22	.100	2.95	2.56	2.35	2.22	2.13	2.06	2.01	1.97	1.93
	.050	4.30	3.44	3.05	2.82	2.66	2.55	2.46	2.40	2.34
	.025	5.79	4.38	3.78	3.44	3.22	3.05	2.93	2.84	2.76
	.010	7.95	5.72	4.82	4.31	3.99	3.76	3.59	3.45	3.35
	.005	9.73	6.81	5.65	5.02	4.61	4.32	4.11	3.94	3.81
23	.100	2.94	2.55	2.34	2.21	2.11	2.05	1.99	1.95	1.92
	.050	4.28	3.42	3.03	2.80	2.64	2.53	2.44	2.37	2.32
	.025	5.75	4.35	3.75	3.41	3.18	3.02	2.90	2.81	2.73
	.010	7.88	5.66	4.76	4.26	3.94	3.71	3.54	3.41	3.30
	.005	9.63	6.73	5.58	4.95	4.54	4.26	4.05	3.88	3.75
24	.100	2.93	2.54	2.33	2.19	2.10	2.04	1.98	1.94	1.91
	.050	4.26	3.40	3.01	2.78	2.62	2.51	2.42	2.36	2.30
	.025	5.72	4.32	3.72	3.38	3.15	2.99	2.87	2.78	2.70
	.010	7.82	5.61	4.72	4.22	3.90	3.67	3.50	3.36	3.26
	.005	9.55	6.66	5.52	4.89	4.49	4.20	3.99	3.83	3.69
25	.100	2.92	2.53	2.32	2.18	2.09	2.02	1.97	1.93	1.89
	.050	4.24	3.39	2.99	2.76	2.60	2.49	2.40	2.34	2.28
	.025	5.69	4.29	3.69	3.35	3.13	2.97	2.85	2.75	2.68
	.010	7.77	5.57	4.68	4.18	3.85	3.63	3.46	3.32	3.22
	.005	9.48	6.60	5.46	4.84	4.43	4.15	3.94	3.78	3.64
26	.100	2.91	2.52	2.31	2.17	2.08	2.01	1.96	1.92	1.88
	.050	4.23	3.37	2.98	2.74	2.59	2.47	2.39	2.32	2.27
	.025	5.66	4.27	3..67	3.33	3.10	2.94	2.82	2.73	2.65
	.010	7.72	5.53	4.64	4.14	3.82	3.59	3.42	3.29	3.18
	.005	9.41	6.54	5.41	4.79	4.38	4.10	3.89	3.73	3.60

					v_1						
10	12	15	20	24	30	40	60	120	∞	α	v_2
2.14	2.10	2.05	2.01	1.98	1.96	1.93	1.90	1.88	1.85	.100	13
2.67	2.60	2.53	2.46	2.42	2.38	2.34	2.30	2.25	2.21	.050	
3.25	3.15	3.05	2.95	2.89	2.84	2.78	2.72	2.66	2.60	.025	
4.10	3.96	3.82	3.66	3.59	3.51	3.43	3.34	3.25	3.17	.010	
4.82	4.64	4.46	4.27	4.17	4.07	3.97	3.87	3.76	3.65	.005	
2.10	2.05	2.01	1.96	1.94	1.91	1.89	1.86	1.83	1.80	.100	14
2.60	2.53	2.46	2.39	2.35	2.31	2.27	2.22	2.18	2.13	.050	
3.15	3.05	2.95	2.84	2.79	2.73	2.67	2.61	2.55	2.49	.025	
3.94	3.80	3.66	3.51	3.43	3.35	3.27	3.18	3.09	3.00	.010	
4.60	4.43	4.25	4.06	3.96	3.86	3.76	3.66	3.55	3.44	.005	
2.06	2.02	1.97	1.92	1.90	1.87	1.85	1.82	1.79	1.76	.100	15
2.54	2.48	2.40	2.33	2.29	2.25	2.20	2.16	2.11	2.07	.050	
3.06	2.96	2.86	2.76	2.70	2.64	2.59	2.52	2.46	2.40	.025	
3.80	3.67	3.52	3.37	3.29	3.21	3.13	3.05	2.96	2.87	.010	
4.42	4.25	4.07	3.88	3.79	3.69	3.58	3.48	3.37	3.26	.005	
2.03	1.99	1.94	1.89	1.87	1.84	1.81	1.78	1.75	1.72	.100	16
2.49	2.42	2.35	2.28	2.24	2.19	2.15	2.11	2.06	2.01	.050	
2.99	2.89	2.79	2.68	2.63	2.57	2.51	2.45	2.38	2.32	.025	
3.69	3.55	3.41	3.26	3.18	3.10	3.02	2.93	2.84	2.75	.010	
4.27	4.10	3.92	3.73	3.64	3.54	3.44	3.33	3.22	3.11	.005	
2.00	1.96	1.91	1.86	1.84	1.81	1.78	1.75	1.72	1.69	.100	17
2.45	2.38	2.31	2.23	2.19	2.15	2.10	2.06	2.01	1.96	.050	
2.92	2.82	2.72	2.62	2.56	2.50	2.44	2.38	2.32	2.25	.025	
3.59	3.46	3.31	3.16	3.08	3.00	2.92	2.83	2.75	2.65	.010	
4.14	3.97	3.79	3.61	3.51	3.41	3.31	3.21	3.10	2.98	.005	
1.98	1.93	1.89	1.84	1.81	1.78	1.75	1.72	1.69	1.66	.100	18
2.41	2.34	2.27	2.19	2.15	2.11	2.06	2.02	1.97	1.92	.050	
2.87	2.77	2.67	2.56	2.50	2.44	2.38	2.32	2.26	2.19	.025	
3.51	3.37	3.23	3.08	3.00	2.92	2.84	2.75	2.66	2.57	.010	
4.03	3.86	3.68	3.50	3.40	3.30	3.20	3.10	2.99	2.87	.005	
1.96	1.91	1.86	1.81	1.79	1.76	1.73	1.70	1.67	1.63	.100	19
2.38	2.31	2.23	2.16	2.11	2.07	2.03	1.98	1.93	1.88	.050	
2.82	2.72	2.62	2.51	2.45	2.39	2.33	2.27	2.20	2.13	.025	
3.43	3.30	3.15	3.00	2.92	2.84	2.76	2.67	2.58	2.49	.010	
3.93	3.76	3.59	3.40	3.31	3.21	3.11	3.00	2.89	2.78	.005	
1.94	1.89	1.84	1.79	1.77	1.74	1.71	1.68	1.64	1.61	.100	20
2.35	2.28	2.20	2.12	2.08	2.04	1.99	1.95	1.90	1.84	.050	
2.77	2.68	2.57	2.46	2.41	2.35	2.29	2.22	2.16	2.09	.025	
3.37	3.23	3.09	2.94	2.86	2.78	2.69	2.61	2.52	2.42	.010	
3.85	3.68	3.50	3.32	3.22	3.12	3.02	2.92	2.81	2.69	.005	
1.92	1.87	1.83	1.78	1.75	1.72	1.69	1.66	1.62	1.59	.100	21
2.32	2.25	2.18	2.10	2.05	2.01	1.96	1.92	1.87	1.81	.050	
2.73	2.64	2.53	2.42	2.37	2.31	2.25	2.18	2.11	2.04	.025	
3.31	3.17	3.03	2.88	2.80	2.72	2.64	2.55	2.46	2.36	.010	
3.77	3.60	3.43	3.24	3.15	3.05	2.95	2.84	2.73	2.61	.005	
1.90	1.86	1.81	1.76	1.73	1.70	1.67	1.64	1.60	1.57	.100	22
2.30	2.23	2.15	2.07	2.03	1.98	1.94	1.89	1.84	1.78	.050	
2.70	2.60	2.50	2.39	2.33	2.27	2.21	2.14	2.08	2.00	.025	
3.26	3.12	2.98	2.83	2.75	2.67	2.58	2.50	2.40	2.31	.010	
3.70	3.54	3.36	3.18	3.08	2.98	2.88	2.77	2.66	2.55	.005	
1.89	1.84	1.80	1.74	1.72	1.69	1.66	1.62	1.59	1.55	.100	23
2.27	2.20	2.13	2.05	2.01	1.96	1.91	1.86	1.81	1.76	.050	
2.67	2.57	2.47	2.36	2.30	2.24	2.18	2.11	2.04	1.97	.025	
3.21	3.07	2.93	2.78	2.70	2.62	2.54	2.45	2.35	2.26	.010	
3.64	3.47	3.30	3.12	3.02	2.92	2.82	2.71	2.60	2.48	.005	
1.88	1.83	1.78	1.73	1.70	1.67	1.64	1.61	1.57	1.53	.100	24
2.25	2.18	2.11	2.03	1.98	1.94	1.89	1.84	1.79	1.73	.050	
2.64	2.54	2.44	2.33	2.27	2.21	2.15	2.08	2.01	1.94	.025	
3.17	3.03	2.89	2.74	2.66	2.58	2.49	2.40	2.31	2.21	.010	
3.59	3.42	3.25	3.06	2.97	2.87	2.77	2.66	2.55	2.43	.005	
1.87	1.82	1.77	1.72	1.69	1.66	1.63	1.59	1.56	1.52	.100	25
2.24	2.16	2.09	2.01	1.96	1.92	1.87	1.82	1.77	1.71	.050	
2.61	2.51	2.41	2.30	2.24	2.18	2.12	2.05	1.98	1.91	.025	
3.13	2.99	2.85	2.70	2.62	2.54	2.45	2.36	2.27	2.17	.010	
3.54	3.37	3.20	3.01	2.92	2.82	2.72	2.61	2.50	2.38	.005	
1.86	1.81	1.76	1.71	1.68	1.65	1.61	1.58	1.54	1.50	.100	26
2.22	2.15	2.07	1.99	1.95	1.90	1.85	1.80	1.75	1.69	.050	
2.59	2.49	2.39	2.28	2.22	2.16	2.09	2.03	1.95	1.88	.025	
3.09	2.96	2.81	2.66	2.58	2.50	2.42	2.33	2.23	2.13	.010	
3.49	3.33	3.15	2.97	2.87	2.77	2.67	2.56	2.45	2.33	.005	

(*continues*)

Table 6   (*Concluded*)

v_2	α	v_1 1	2	3	4	5	6	7	8	9
27	.100	2.90	2.51	2.30	2.17	2.07	2.00	1.95	1.91	1.87
	.050	4.21	3.35	2.96	2.73	2.57	2.46	2.37	2.31	2.25
	.025	5.63	4.24	3.65	3.31	3.08	2.92	2.80	2.71	2.63
	.010	7.68	5.49	4.60	4.11	3.78	3.56	3.39	3.26	3.15
	.005	9.34	6.49	5.36	4.74	4.34	4.06	3.85	3.69	3.56
28	.100	2.89	2.50	2.29	2.16	2.06	2.00	1.94	1.90	1.87
	.050	4.20	3.34	2.95	2.71	2.56	2.45	2.36	2.29	2.24
	.025	5.61	4.22	3.63	3.29	3.06	2.90	2.78	2.69	2.61
	.010	7.64	5.45	4.57	4.07	3.75	3.53	3.36	3.23	3.12
	.005	9.28	6.44	5.32	4.70	4.30	4.02	3.81	3.65	3.52
29	.100	2.89	2.50	2.28	2.15	2.06	1.99	1.93	1.89	1.86
	.050	4.18	3.33	2.93	2.70	2.55	2.43	2.35	2.28	2.22
	.025	5.59	4.20	3.61	3.27	3.04	2.88	2.76	2.67	2.59
	.010	7.60	5.42	4.54	4.04	3.73	3.50	3.33	3.20	3.09
	.005	9.23	6.40	5.28	4.66	4.26	3.98	3.77	3.61	3.48
30	.100	2.88	2.49	2.28	2.14	2.05	1.98	1.93	1.88	1.85
	.050	4.17	3.32	2.92	2.69	2.53	2.42	2.33	2.27	2.21
	.025	5.57	4.18	3.59	3.25	3.03	2.87	2.75	2.65	2.57
	.010	7.56	5.39	4.51	4.02	3.70	3.47	3.30	3.17	3.07
	.005	9.18	6.35	5.24	4.62	4.23	3.95	3.74	3.58	3.45
40	.100	2.84	2.44	2.23	2.09	2.00	1.93	1.87	1.83	1.79
	.050	4.08	3.23	2.84	2.61	2.45	2.34	2.25	2.18	2.12
	.025	5.42	4.05	3.46	3.13	2.90	2.74	2.62	2.53	2.45
	.010	7.31	5.18	4.31	3.83	3.51	3.29	3.12	2.99	2.89
	.005	8.83	6.07	4.98	4.37	3.99	3.71	3.51	3.35	3.22
60	.100	2.79	2.39	2.18	2.04	1.95	1.87	1.82	1.77	1.74
	.050	4.00	3.15	2.76	2.53	2.37	2.25	2.17	2.10	2.04
	.025	5.29	3.93	3.34	3.01	2.79	2.63	2.51	2.41	2.33
	.010	7.08	4.98	4.13	3.65	3.34	3.12	2.95	2.82	2.72
	.005	8.49	5.79	4.73	4.14	3.76	3.49	3.29	3.13	3.01
120	.100	2.75	2.35	2.13	1.99	1.90	1.82	1.77	1.72	1.68
	.050	3.92	3.07	2.68	2.45	2.29	2.17	2.09	2.02	1.96
	.025	5.15	3.80	3.23	2.89	2.67	2.52	2.39	2.30	2.22
	.010	6.85	4.79	3.95	3.48	3.17	2.96	2.79	2.66	2.56
	.005	8.18	5.54	4.50	3.92	3.55	3.28	3.09	2.93	2.81
∞	.100	2.71	2.30	2.08	1.94	1.85	1.77	1.72	1.67	1.63
	.050	3.84	3.00	2.60	2.37	2.21	2.10	2.01	1.94	1.63
	.025	5.02	3.69	3.12	2.79	2.57	2.41	2.29	2.19	2.11
	.010	6.63	4.61	3.78	3.32	3.02	2.80	2.64	2.51	2.41
	.005	7.88	5.30	4.28	3.72	3.35	3.09	2.90	2.74	2.62

				v_1							
10	12	15	20	24	30	40	60	120	∞	α	v_2
1.85	1.80	1.75	1.70	1.67	1.64	1.60	1.57	1.53	1.49	.100	27
2.20	2.13	2.06	1.97	1.93	1.88	1.84	1.79	1.73	1.67	.050	
2.57	2.47	2.36	2.25	2.19	2.13	2.07	2.00	1.93	1.85	.025	
3.06	2.93	2.78	2.63	2.55	2.47	2.38	2.29	2.20	2.10	.010	
3.45	3.28	3.11	2.93	2.83	2.73	2.63	2.52	2.41	2.29	.005	
1.84	1.79	1.74	1.69	1.66	1.63	1.59	1.56	1.52	1.48	.100	28
2.19	2.12	2.04	1.96	1.91	1.87	1.82	1.77	1.71	1.65	.050	
2.55	2.45	2.34	2.23	2.17	2.11	2.05	1.98	1.91	1.83	.025	
3.03	2.90	2.75	2.60	2.52	2.44	2.35	2.26	2.17	2.06	.010	
3.41	3.25	3.07	2.89	2.79	2.69	2.59	2.48	2.37	2.25	.005	
1.83	1.78	1.73	1.68	1.65	1.62	1.58	1.55	1.51	1.47	.100	29
2.18	2.10	2.03	1.94	1.90	1.85	1.81	1.75	1.70	1.64	.050	
2.53	2.43	2.32	2.21	2.15	2.09	2.03	1.96	1.89	1.81	.025	
3.00	2.87	2.73	2.57	2.49	2.41	2.33	2.23	2.14	2.03	.010	
3.38	3.21	3.04	2.86	2.76	2.66	2.56	2.45	2.33	2.21	.005	
1.82	1.77	1.72	1.67	1.64	1.61	1.57	1.54	1.50	1.46	.100	30
2.16	2.09	2.01	1.93	1.89	1.84	1.79	1.74	1.68	1.62	.050	
2.51	2.41	2.31	2.20	2.14	2.07	2.01	1.94	1.87	1.79	.025	
2.98	2.84	2.70	2.55	2.47	2.39	2.30	2.21	2.11	2.01	.010	
3.34	3.18	3.01	2.82	2.73	2.63	2.52	2.42	2.30	2.18	.005	
1.76	1.71	1.66	1.61	1.57	1.54	1.51	1.47	1.42	1.38	.100	40
2.08	2.00	1.92	1.84	1.79	1.74	1.69	1.64	1.58	1.51	.050	
2.39	2.29	2.18	2.07	2.01	1.94	1.88	1.80	1.72	1.64	.025	
2.80	2.66	2.52	2.37	2.29	2.20	2.11	2.02	1.92	1.80	.010	
3.12	2.95	2.78	2.60	2.50	2.40	2.30	2.18	2.06	1.93	.005	
1.71	1.66	1.60	1.54	1.51	1.48	1.44	1.40	1.35	1.29	.100	60
1.99	1.92	1.84	1.75	1.70	1.65	1.59	1.53	1.47	1.39	.050	
2.27	2.17	2.06	1.94	1.88	1.82	1.74	1.67	1.58	1.48	.025	
2.63	2.50	2.35	2.20	2.12	2.03	1.94	1.84	1.73	1.60	.010	
2.90	2.74	2.57	2.39	2.29	2.19	2.08	1.96	1.83	1.69	.005	
1.65	1.60	1.55	1.48	1.45	1.41	1.37	1.32	1.26	1.19	.100	120
1.91	1.83	1.75	1.66	1.61	1.55	1.50	1.43	1.35	1.25	.050	
2.16	2.05	1.94	1.82	1.76	1.69	1.61	1.53	1.43	1.31	.025	
2.47	2.34	2.19	2.03	1.95	1.86	1.76	1.66	1.53	1.38	.010	
2.71	2.54	2.37	2.19	2.09	1.98	1.87	1.75	1.61	1.43	.005	
1.60	1.55	1.49	1.42	1.38	1.34	1.30	1.24	1.17	1.00	.100	α
1.83	1.75	1.67	1.57	1.52	1.46	1.39	1.32	1.22	1.00	.050	
2.05	1.94	1.83	1.71	1.64	1.57	1.48	1.39	1.27	1.00	.025	
2.32	2.18	2.04	1.88	1.79	1.70	1.59	1.47	1.32	1.00	.010	
2.52	2.36	2.19	2.00	1.90	1.79	1.67	1.53	1.36	1.00	.005	

Table 7   Distribution Function of U

$P(U \leq U_0)$; U_0 is the argument; $n_1 \leq n_2$; $3 \leq n_2 \leq 10$.

$n_2 = 3$

U_0	n_1		
	1	2	3
0	.25	.10	.05
1	.50	.20	.10
2		.40	.20
3		.60	.35
4			.50

$n_2 = 4$

U_0	n_1			
	1	2	3	4
0	.2000	.0667	.0286	.0143
1	.4000	.1333	.0571	.0286
2	.6000	.2667	.1143	.0571
3		.4000	.2000	.1000
4		.6000	.3143	.1714
5			.4286	.2429
6			.5714	.3429
7				.4429
8				.5571

Table 7 (*Continued*)

$n_2 = 5$

			n_1		
U_0	1	2	3	4	5
0	.1667	.0476	.0179	.0079	.0040
1	.3333	.0952	.0357	.0159	.0079
2	.5000	.1905	.0714	.0317	.0159
3		.2857	.1250	.0556	.0278
4		.4286	.1964	.0952	.0476
5		.5714	.2857	.1429	.0754
6			.3929	.2063	.1111
7			.5000	.2778	.1548
8				.3651	.2103
9				.4524	.2738
10				.5476	.3452
11					.4206
12					.5000

$n_2 = 6$

			n_2			
U_0	1	2	3	4	5	6
0	.1429	.0357	.0119	.0048	.0022	.0011
1	.2857	.0714	.0238	.0095	.0043	.0022
2	.4286	.1429	.0476	.0190	.0087	.0043
3	.5714	.2143	.0833	.0333	.0152	.0076
4		.3214	.1310	.0571	.0260	.0130
5		.4286	.1905	.0857	.0411	.0206
6		.5714	.2738	.1286	.0628	.0325
7			.3571	.1762	.0887	.0465
8			.4524	.2381	.1234	.0660
9			.5476	.3048	.1645	.0898
10				.3810	.2143	.1201
11				.4571	.2684	.1548
12				.5429	.3312	.1970
13					.3961	.2424
14					.4654	.2944
15					.5346	.3496
16						.4091
17						.4686
18						.5314

(*continues*)

Table 7   (*Continued*)

$n_2 = 7$

U_0	1	2	3	4	5	6	7
				n_1			
0	.1250	.0278	.0083	.0030	.0013	.0006	.0003
1	.2500	.0556	.0167	.0061	.0025	.0012	.0006
2	.3750	.1111	.0333	.0121	.0051	.0023	.0012
3	.5000	.1667	.0583	.0212	.0088	.0041	.0020
4		.2500	.0917	.0364	.0152	.0070	.0035
5		.3333	.1333	.0545	.0240	.0111	.0055
6		.4444	.1917	.0818	.0366	.0175	.0087
7		.5556	.2583	.1152	.0530	.0256	.0131
8			.3333	.1576	.0745	.0367	.0189
9			.4167	.2061	.1010	.0507	.0265
10			.5000	.2636	.1338	.0688	.0364
11				.3242	.1717	.0903	.0487
12				.3939	.2159	.1171	.0641
13				.4636	.2652	.1474	.0825
14				.5364	.3194	.1830	.1043
15					.3775	.2226	.1297
16					.4381	.2669	.1588
17					.5000	.3141	.1914
18						.3654	.2279
19						.4178	.2675
20						.4726	.3100
21						.5274	.3552
22							.4024
23							.4508
24							.5000

Table 7 (*Continued*)

$n_2 = 8$

U_0	n_1							
	1	2	3	4	5	6	7	8
0	.1111	.0222	.0061	.0020	.0008	.0003	.0002	.0001
1	.2222	.0444	.0121	.0040	.0016	.0007	.0003	.0002
2	.3333	.0889	.0242	.0081	.0031	.0013	.0006	.0003
3	.4444	.1333	.0424	.0141	.0054	.0023	.0011	.0005
4	.5556	.2000	.0667	.0242	.0093	.0040	.0019	.0009
5		.2667	.0970	.0364	.0148	.0063	.0030	.0015
6		.3556	.1394	.0545	.0225	.0100	.0047	.0023
7		.4444	.1879	.0768	.0326	.0147	.0070	.0035
8		.5556	.2485	.1071	.0466	.0213	.0103	.0052
9			.3152	.1414	.0637	.0296	.0145	.0074
10			.3879	.1838	.0855	0406	.0200	.0103
11			.4606	.2303	.1111	.0539	.0270	.0141
12			.5394	.2848	.1422	.0709	.0361	0.190
13				.3414	.1772	.0906	.0469	.0249
14				.4040	.2176	.1142	.0603	.0325
15				.4667	.2618	.1412	.0760	0415
16				.5333	.3108	.1725	.0946	.0524
17					.3621	.2068	.1159	.0652
18					.4165	.2454	.1405	.0803
19					.4716	.2864	.1678	0974
20					.5284	.3310	.1984	.1172
21						.3773	.2317	.1393
22						.4259	.2679	.1641
23						.4749	.3063	.1911
24						.5251	.3472	.2209
25							.3894	.2527
26							.4333	.2869
27							.4775	.3227
28							.5225	.3605
29								.3992
30								.4392
31								.4796
32								.5204

(*continues*)

Table 7   (*Continued*)

$n_2 = 9$

U_0	n_1								
	1	2	3	4	5	6	7	8	9
0	.1000	.0182	.0045	.0014	.0005	.0002	.0001	.0000	.0000
1	.2000	.0364	.0091	.0028	.0010	.0004	.0002	.0001	.0000
2	.3000	.0727	.0182	.0056	.0020	.0008	.0003	.0002	.0001
3	.4000	.1091	.0318	.0098	.0035	.0014	.0006	.0003	.0001
4	.5000	.1636	.0500	.0168	.0060	.0024	.0010	.0005	.0002
5		.2182	.0727	.0252	.0095	.0038	.0017	.0008	.0004
6		.2909	.1045	.0378	.0145	.0060	.0026	.0012	.0006
7		.3636	.1409	.0531	.0210	.0088	.0039	.0019	.0009
8		.4545	.1864	.0741	.0300	.0128	.0058	.0028	.0014
9		.5455	.2409	.0993	.0415	.0180	.0082	.0039	.0020
10			.3000	.1301	.0559	.0248	.0115	.0056	.0028
11			.3636	.1650	.0734	.0332	.0156	.0076	.0039
12			.4318	.2070	.0949	.0440	.0209	.0103	.0053
13			.5000	.2517	.1199	.0567	.0274	.0137	.0071
14				.3021	.1489	.0723	.0356	.0180	.0094
15				.3552	.1818	.0905	.0454	.0232	.0122
16				.4126	.2188	.1119	.0571	.0296	.0157
17				.4699	.2592	.1361	.0708	.0372	.0200
18				.5301	.3032	.1638	.0869	.0464	.0252
19					.3497	.1924	.1052	.0570	.0313
20					.3986	.2280	.1261	.0694	.0385
21					.4491	.2643	.1496	.0836	.0470
22					.5000	.3035	.1755	.0998	.0567
23						.3445	.2039	.1179	.0680
24						.3878	.2349	.1383	.0807
25						.4320	.2680	.1606	.0951
26						.4773	.3032	.1852	.1112
27						.5227	.3403	.2117	.1290
28							.3788	.2404	.1487
29							.4185	.2707	.1701
30							.4591	.3029	.1933
31							.5000	.3365	.2181
32								.3715	.2447
33								.4074	.2729
34								.4442	.3024
35								.4813	.3332
36								.5187	.3652
37									.3981
38									.4317
39									.4657
40									.5000

Table 7 (*Continued*)

$n_2 = 10$

U_0	n_1									
	1	2	3	4	5	6	7	8	9	10
0	.0909	.0152	.0035	.0010	.0003	.0001	.0001	.0000	.0000	.0000
1	.1818	.0303	.0070	.0020	.0007	.0002	.0001	.0000	.0000	.0000
2	.2727	.0606	.0140	.0040	.0013	.0005	.0002	.0001	.0000	.0000
3	.3636	.0909	.0245	.0070	.0023	.0009	.0004	.0002	.0001	.0000
4	.4545	.1364	.0385	.0120	.0040	.0015	.0006	.0003	.0001	.0001
5	.5455	.1818	.0559	.0180	.0063	.0024	.0010	.0004	.0002	.0001
6		.2424	.0804	.0270	.0097	.0037	.0015	.0007	.0003	.0002
7		.3030	.1084	.0380	.0140	.0055	.0023	.0010	.0005	.0002
8		.3788	.1434	.0529	.0200	.0080	.0034	.0015	.0007	.0004
9		.4545	.1853	.0709	.0276	.0112	.0048	.0022	.0011	.0005
10		.5455	.2343	.0939	.0376	.0156	.0068	.0031	.0015	.0008
11			.2867	.1199	.0496	.0210	.0093	.0043	.0021	.0010
12			.3462	.1518	.0646	.0280	.0125	.0058	.0028	.0014
13			.4056	.1868	.0823	.0363	.0165	.0078	.0038	.0019
14			.4685	.2268	.1032	.0467	.0215	.0103	.0051	.0026
15			.5315	.2697	.1272	.0589	.0277	.0133	.0066	.0034
16				.3177	.1548	.0736	.0351	.0171	.0086	.0045
17				.3666	.1855	.0903	.0439	.0217	.0110	.0057
18				.4196	.2198	.1099	.0544	.0273	.0140	.0073
19				.4725	.2567	.1317	.0665	.0338	.0175	.0093
20				.5275	.2970	.1566	.0806	.0416	.0217	.0116
21					.3393	.1838	.0966	.0506	.0267	.0144
22					.3839	.2139	.1148	.0610	.0326	.0177
23					.4296	.2461	.1349	.0729	.0394	.0216
24					.4765	.2811	.1574	.0864	.0474	.0262
25					.5235	.3177	.1819	.1015	.0564	.0315
26						.3564	.2087	.1185	.0667	.0376
27						.3962	.2374	.1371	.0782	.0446
28						.4374	.2681	.1577	.0912	.0526
29						.4789	.3004	.1800	.1055	.0615
30						.5211	.3345	.2041	.1214	.0716
31							.3698	.2299	.1388	.0827
32							.4063	.2574	.1577	.0952
33							.4434	.2863	.1781	.1088
34							.4811	.3167	.2001	.1237
35							.5189	.3482	.2235	.1399
36								.3809	.2483	.1575
37								.4143	.2745	.1763
38								.4484	.3019	.1965
39								.4827	.3304	.2179

(*continues*)

Table 7   (*Concluded*)

$n_2 = 10$

U_0	n_1									
	1	2	3	4	5	6	7	8	9	10
40								.5173	.3598	.2406
41									.3901	.2644
42									.4211	.2894
43									.4524	.3153
44									.4841	.3421
45									.5159	.3697
46										.3980
47										.4267
48										.4559
49										.4853
50										.5147

Computed by M. Pagano, Department of Statistics, University of Florida.

Table 8   Critical Values of *T* in the Wilcoxon Signed-Rank Test; $n = 5$ (1) 50

One-sided	Two-sided	$n = 5$	$n = 6$	$n = 7$	$n = 8$	$n = 9$	$n = 10$
$P = .05$	$P = .10$	1	2	4	6	8	11
$P = .025$	$P = .05$		1	2	4	6	8
$P = .01$	$P = .02$			0	2	3	5
$P = .005$	$P = .01$				0	2	3

One-sided	Two-sided	$n = 11$	$n = 12$	$n = 13$	$n = 14$	$n = 15$	$n = 16$
$P = .05$	$P = .10$	14	17	21	26	30	36
$P = .025$	$P = .05$	11	14	17	21	25	30
$P = .01$	$P = .02$	7	10	13	16	20	24
$P = .005$	$P = .01$	5	7	10	13	16	19

One-sided	Two-sided	$n = 17$	$n = 18$	$n = 19$	$n = 20$	$n = 21$	$n = 22$
$P = .05$	$P = .10$	41	47	54	60	68	75
$P = .025$	$P = .05$	35	40	46	52	59	66
$P = .01$	$P = .02$	28	33	38	43	49	56
$P = .005$	$P = .01$	23	28	32	37	43	49

Table 8   (*Concluded*)

One-sided	Two-sided	$n = 23$	$n = 24$	$n = 25$	$n = 26$	$n = 27$	$n = 28$
$P = .05$	$P = .10$	83	92	101	110	120	130
$P = .025$	$P = .05$	73	81	90	98	107	117
$P = .01$	$P = .02$	62	69	77	85	93	102
$P = .005$	$P = .01$	55	68	68	76	84	92

One-sided	Two-sided	$n = 29$	$n = 30$	$n = 31$	$n = 32$	$n = 33$	$n = 34$
$P = .05$	$P = .10$	141	152	163	175	188	201
$P = .025$	$P = .05$	127	137	148	159	171	183
$P = .01$	$P = .02$	111	120	130	141	151	162
$P = .005$	$P = .01$	100	109	118	128	138	149

One-sided	Two-sided	$n = 35$	$n = 36$	$n = 37$	$n = 38$	$n = 39$	
$P = .05$	$P = .10$	214	228	242	256	271	
$P = .025$	$P = .05$	195	208	222	235	250	
$P = .01$	$P = .02$	174	186	198	211	224	
$P = .005$	$P = .01$	160	171	183	195	208	

One-sided	Two-sided	$n = 40$	$n = 41$	$n = 42$	$n = 43$	$n = 44$	$n = 45$
$P = 0.5$	$P = .10$	287	303	319	336	353	371
$P = .025$	$P = .05$	264	279	295	311	327	344
$P = .01$	$P = .02$	238	252	267	281	297	313
$P = .005$	$P = .01$	221	234	248	262	277	292

One-sided	Two-sided	$n = 46$	$n = 47$	$n = 48$	$n = 49$	$n = 50$	
$P = .05$	$P = .10$	389	408	427	446	466	
$P = .025$	$P = .05$	361	379	397	415	434	
$P = .01$	$P = .02$	329	345	362	380	398	
$P = .005$	$P = .01$	307	323	339	356	373	

From "Some Rapid Approximate Statistical Procedures" (1964), 28, F. Wilcoxon and R. A. Wilcox. Reproduced with the kind permission of Lederle Laboratories, a division of American Cyanamid Company.

Table 9   Distribution of the Total Number of Runs R in Samples of Size (n_1, n_2); $P(R \leq a)$

					a				
(n_1, n_2)	2	3	4	5	6	7	8	9	10
(2, 3)	.200	.500	.900	1.000					
(2, 4)	.133	.400	.800	1.000					
(2. 5)	.095	.333	.714	1.000					
(2, 6)	.071	.286	.643	1.000					
(2, 7)	.056	.250	.583	1.000					
(2.8)	.044	.222	.533	1.000					
(2,9)	.036	.200	.491	1.000					
(2, 10)	.030	.182	.455	1.000					
(3,3)	.100	.300	.700	.900	1.000				
(3, 4)	.057	.200	.543	.800	.971	1.000			
(3, 5)	.036	.143	.429	.714	.929	1.000			
(3, 6)	.024	.107	.345	.643	.881	1.000			
(3, 7)	.017	.083	.283	.583	.833	1.000			
(3, 8)	.012	.067	.236	.533	.788	1.000			
(3, 9)	.009	.055	.200	.491	.745	1.000			
(3, 10)	.007	.045	.171	.455	.706	1.000			
(4, 4)	.029	.114	.371	.629	.886	.971	1.000		
(4, 5)	.016	.071	.262	.500	.786	.929	.992	1.000	
(4, 6)	.010	.048	.190	.405	.690	.881	.976	1.000	
(4, 7)	.006	.033	.142	.333	.606	.833	.954	1.000	
(4, 8)	.004	.024	.109	.279	.533	.788	.929	1.000	
(4, 9)	.003	.018	.085	.236	.471	.745	.902	1.000	
(4, 10)	.002	.014	.068	.203	.419	.706	.874	1.000	
(5, 5)	.008	.040	.167	.357	.643	.833	.960	.992	1.000
(5, 6)	.004	.024	.110	.262	.522	.738	.911	.976	.998
(5, 7)	.003	.015	.076	.197	.424	.652	.854	.955	.992
(5, 8)	.002	.010	.054	.152	.347	.576	.793	.929	.984
(5, 9)	.001	.007	.039	.119	.287	.510	.734	.902	.972
(5, 10)	.001	.005	.029	.095	.239	.455	.678	.874	.958
(6, 6)	.002	.013	.067	.175	.392	.608	.825	.933	.987
(6, 7)	.001	.008	.043	.121	.296	.500	.733	.879	.966
(6, 8)	.001	.005	.028	.086	.226	.413	.646	.821	.937
(6, 9)	.000	.003	.019	.063	.175	.343	.566	.762	.902
(6, 10)	.000	.002	.013	.047	.137	.288	.497	.706	.864
(7, 7)	.001	.004	.025	.078	.209	.383	.617	.791	.922
(7, 8)	.000	.002	.015	.051	.149	.296	.514	.704	.867
(7, 9)	.000	.001	.010	.035	.108	.231	.427	.622	.806
(7, 10)	.000	.001	.006	.024	.080	.182	.355	.549	.743
(8, 8)	.000	.001	.009	.032	.100	.214	.405	.595	.786
(8, 9)	.000	.001	.005	.020	.069	.157	.319	.500	.702
(8, 10)	.000	.000	.003	.013	.048	.117	.251	.419	.621
(9, 9)	.000	.000	.003	.012	.044	.109	.238	.399	.601
(9, 10)	.000	.000	.002	.008	.029	.077	.179	.319	.510
(10, 10)	.000	.000	.001	.004	.019	.051	.128	.242	.414

Table 9   (Concluded)

					a					
(n_1, n_2)	11	12	13	14	15	16	17	18	19	20
(2,3)										
(2,4)										
(2,5)										
(2,6)										
(2,7)										
(2,8)										
(2,9)										
(2,10)										
(3,3)										
(3,4)										
(3,5)										
(3,6)										
(3,7)										
(3,8)										
(3,9)										
(3,10)										
(4,4)										
(4,5)										
(4,6)										
(4,7)										
(4,8)										
(4,9)										
(4,10)										
(5,5)										
(5,6)	1.000									
(5,7)	1.000									
(5,8)	1.000									
(5,9)	1.000									
(5,10)	1.000									
(6,6)	.998	1.000								
(6,7)	.992	.999	1.000							
(6,8)	.984	.998	1.000							
(6,9)	.972	.994	1.000							
(6,10)	.958	.990	1.000							
(7,7)	.975	.996	.999	1.000						
(7,8)	.949	.988	.998	1.000	1.000					
(7,9)	.916	.975	.994	.999	1.000					
(7,10)	.879	.957	.990	.998	1.000					
(8,8)	.900	.968	.991	.999	1.000	1.000				
(8,9)	.843	.939	.980	.996	.999	1.000	1.000			
(8,10)	.782	.903	.964	.990	.998	1.000	1.000			
(9,9)	.762	.891	.956	.988	.997	1.000	1.000	1.000		
(9,10)	.681	.834	.923	.974	.992	.999	1.000	1.000	1.000	
(10,10)	.586	.758	.872	.949	.981	.996	.999	1.000	1.000	1.000

From "Tables for Testing Randomness of Grouping in a Sequence of Alternatives," F. Swed and C. Eisenhart, *Annals of Mathematical Statistics*, Vol. 14 (1943). Reproduced with the kind permission of the author and of the editor, *Annals of Mathematical Statistics*.

Table 10   Critical Values of Spearman's Rank Correlation Coefficient

n	$\alpha = .05$	$\alpha = .025$	$\alpha = .01$	$\alpha = .005$
5	0.900	—	—	—
6	0.829	0.886	0.943	—
7	0.714	0.786	0.893	—
8	0.643	0.738	0.833	0.881
9	0.600	0.683	0.783	0.833
10	0.564	0.648	0.745	0.794
11	0.523	0.623	0.736	0.818
12	0.497	0.591	0.703	0.780
13	0.475	0.566	0.673	0.745
14	0.457	0.545	0.646	0.716
15	0.441	0.525	0.623	0.689
16	0.425	0.507	0.601	0.666
17	0.412	0.490	0.582	0.645
18	0.399	0.476	0.564	0.625
19	0.388	0.462	0.549	0.608
20	0.377	0.450	0.534	0.591
21	0.368	0.438	0.521	0.576
22	0.359	0.428	0.508	0.562
23	0.351	0.418	0.496	0.549
24	0.343	0.409	0.485	0.537
25	0.336	0.400	0.475	0.526
26	0.329	0.392	0.465	0.515
27	0.323	0.385	0.456	0.505
28	0.317	0.377	0.448	0.496
29	0.311	0.370	0.440	0.487
30	0.305	0.364	0.432	0.478

From "Distribution of Sums of Squares of Rank Differences for Small Samples," E. G. Olds, *Annals of Mathematical Statistics*, Vol. 9 (1938). Reproduced with the kind permission of the editor, *Annals of Mathematical Statistics*.

Table 11   Random Numbers

Line	1	2	3	4	5	6	7	8	9	10
					Column					
1	10480	15011	01536	02011	81647	91646	69179	14194	62590	36207
2	22368	46573	25595	85393	30995	89198	27982	53402	93965	34095
3	24130	48360	22527	97265	76393	64809	15179	24830	49340	32081
4	42167	93093	06243	61680	07856	16376	39440	53537	71341	57004
5	37570	39975	81837	16656	06121	91782	60468	81305	49684	60672
6	77921	06907	11008	42751	27756	53498	18602	70659	90655	15053
7	99562	72905	56420	69994	98872	31016	71194	18738	44013	48840
8	96301	91977	05463	07972	18876	20922	94595	56869	69014	60045
9	89579	14342	63661	10281	17453	18103	57740	84378	25331	12566
10	85475	36857	53342	53988	53060	59533	38867	62300	08158	17983
11	28918	69578	88231	33276	70997	79936	56865	05859	90106	31595
12	63553	40961	48235	03427	49626	69445	18663	72695	52180	20847
13	09429	93969	52636	92737	88974	33488	36320	17617	30015	08272
14	10365	61129	87529	85689	48237	52267	67689	93394	01511	26358
15	07119	97336	71048	08178	77233	13916	47564	81056	97735	85977
16	51085	12765	51821	51259	77452	16308	60756	92144	49442	53900
17	02368	21382	52404	60268	89368	19885	55322	44819	01188	65255
18	01011	54092	33362	94904	31273	04146	18594	29852	71585	85030
19	52162	53916	46369	58586	23216	14513	83149	98736	23495	64350
20	07056	97628	33787	09998	42698	06691	76988	13602	51851	46104
21	48663	91245	85828	14346	09172	30168	90229	04734	59193	22178
22	54164	58492	22421	74103	47070	25306	76468	26384	58151	06646
23	32639	32363	05597	24200	13363	38005	94342	28728	35806	06912
24	29334	27001	87637	87308	58731	00256	45834	15398	46557	41135
25	02488	33062	28834	07351	19731	92420	60952	61280	50001	67658
26	81525	72295	04839	96423	24878	82651	66566	14778	76797	14780
27	29676	20591	68086	26432	46901	20849	89768	81536	86645	12659
28	00742	57392	39064	66432	84673	40027	32832	61362	98947	96067
29	05366	04213	25669	26422	44407	44048	37937	63904	45766	66134
30	91921	26418	64117	94305	26766	25940	39972	22209	71500	64568
31	00582	04711	87917	77341	42206	35126	74087	99547	81817	42607
32	00725	69884	62797	56170	86324	88072	76222	36086	84637	93161
33	69011	65795	95876	55293	18988	27354	26575	08625	40801	59920
34	25976	57948	29888	88604	67917	48708	18912	82271	65424	69774
35	09763	83473	73577	12908	30883	18317	28290	35797	05998	41688
36	91567	42595	27958	30134	04024	86385	29880	99730	55536	84855
37	17955	56349	90999	49127	20044	59931	06115	20542	18059	02008
38	46503	18584	18845	49618	02304	51038	20655	58727	28168	15475
39	92157	89634	94824	78171	84610	82834	09922	25417	44137	48413
40	14577	62765	35605	81263	39667	47358	56873	56307	61607	49518
41	98427	07523	33362	64270	01638	92477	66969	98420	04880	45585
42	34914	63976	88720	82765	34476	17032	87589	40836	32427	70002
43	70060	28277	39475	46473	23219	53416	94970	25832	69975	94884
44	53976	54914	06990	67245	68350	82948	11398	42878	80287	88267
45	76072	29515	40980	07391	58745	25774	22987	80059	39911	96189
46	90725	52210	83974	29992	65831	38857	50490	83765	55657	14361
47	64364	67412	33339	31926	14883	24413	59744	92351	97473	89286
48	08962	00358	31662	25388	61642	34072	81249	35648	56891	69352
49	95012	68379	93526	70765	10592	04542	76463	54328	02349	17247
50	15664	10493	20492	38391	91132	21999	59516	81652	27195	48223

(*continues*)

Table 11   (Concluded)

Line	Column									
	1	2	3	4	5	6	7	8	9	10
51	16408	81899	04153	53381	79401	21438	83035	92350	36693	31238
52	18629	81953	05520	91962	04739	13092	97662	24822	94730	06496
53	73115	35101	47498	87637	99016	71060	88824	71013	18735	20286
54	57491	16703	23167	49323	45021	33132	12544	41035	80780	45393
55	30405	83946	23792	14422	15059	45799	22716	19792	09983	74353
56	16631	35006	85900	98275	32388	52390	16815	69298	82732	38480
57	96773	20206	42559	78985	05300	22164	24369	54224	35083	19687
58	38935	64202	14349	82674	66523	44133	00697	35552	35970	19124
59	31624	76384	17403	53363	44167	64486	64758	75366	76554	31601
60	78919	19474	23632	27889	47914	02584	37680	20801	72152	39339
61	03931	33309	57047	74211	63445	17361	62825	39908	05607	91284
62	74426	33278	43972	10119	89917	15665	52872	73823	73144	88662
63	09066	00903	20795	95452	92648	45454	09552	88815	16553	51125
64	42238	12426	87025	14267	20979	04508	64535	31355	86064	29472
65	16153	08002	26504	41744	81959	65642	74240	56302	00033	67107
66	21457	40742	29820	96783	29400	21840	15035	34537	33310	06116
67	21581	57802	02050	89728	17937	37621	47075	42080	97403	48626
68	55612	78095	83197	33732	05810	24813	86902	60397	16489	03264
69	44657	66999	99324	51281	84463	60563	79312	93454	68876	25471
70	91340	84979	46949	81973	37949	61023	43997	15263	80644	43942
71	91227	21199	31935	27022	84067	05462	35216	14486	29891	68607
72	50001	38140	66321	19924	72163	09538	12151	06878	91903	18749
73	65390	05224	72958	28609	81406	39147	25549	48542	42627	45233
74	27504	96131	83944	41575	10573	08619	64482	73923	36152	05184
75	37169	94851	39117	89632	00959	16487	65536	49071	39782	17095
76	11508	70225	51111	38351	19444	66499	71945	05422	13442	78675
77	37449	30362	06694	54690	04052	53115	62757	95348	78662	11163
78	46515	70331	85922	38329	57015	15765	97161	17869	45349	61796
79	30986	81223	42416	58353	21532	30502	32305	86482	05174	07901
80	63798	64995	46583	09785	44160	78128	83991	42865	92520	83531
81	82486	84846	99254	67632	43218	50076	21361	64816	51202	88124
82	21885	32906	92431	09060	64297	51674	64126	62570	26123	05155
83	60336	98782	07408	53458	13564	59089	26445	29789	85205	41001
84	43937	46891	24010	25560	86355	33941	25786	54990	71899	15475
85	97656	63175	89303	16275	07100	92063	21942	18611	47348	20203
86	03299	01221	05418	38982	55758	92237	26759	86367	21216	98442
87	79626	06486	03574	17668	07785	76020	79924	25651	83325	88428
88	85636	68335	47539	03129	65651	11977	02510	26113	99447	68645
89	18039	14367	61337	06177	12143	46609	32989	74014	64708	00533
90	08362	15656	60627	36478	65648	16764	53412	09013	07832	41574
91	79556	29068	04142	16268	15387	12856	66227	38358	22478	73373
92	92608	82674	27072	32534	17075	27698	98204	63863	11951	34648
93	23982	25835	40055	67006	12293	02753	14827	23235	35071	99704
94	09915	96306	05908	97901	28395	14186	00821	80703	70426	75647
95	59037	33300	26695	62247	69927	76123	50842	43834	86654	70959
96	42488	78077	69882	61657	34136	79180	97526	43092	04098	73571
97	46764	86273	63003	93017	31204	36692	40202	35275	57306	55543
98	03237	45430	55417	63282	90816	17349	88298	90183	36600	78406
99	86591	81482	52667	61582	14972	90053	89534	76036	49199	43716
100	38534	01715	94964	87288	65680	43772	39560	12918	86537	62738

Abridged from *Handbook of Tables for Probability and Statistics*, 2nd ed. Edited by William H. Beyer (Cleveland: The Chemical Rubber Company, 1968). Reproduced by permission of CRC Press, Inc.

Table 12 Factors Used When Constructing Control Charts

Number of Observations in Sample, n	Chart for Averages — Factors for Control Limits			Chart for Standard Deviations — Factors for Central Line		Chart for Standard Deviations — Factors for Control Limits				Chart for Ranges — Factors for Central Line			Chart for Ranges — Factors for Control Limits			
	A	A_1	A_2	c_2	$1/c_2$	B_1	B_2	B_3	B_4	d_2	$1/d_2$	d_3	D_1	D_2	D_3	D_4
2	2.121	3.760	1.880	0.5642	1.7725	0	1.843	0	3.267	1.128	0.8865	0.853	0	3.686	0	3.276
3	1.732	2.394	1.023	0.7236	1.3820	0	1.858	0	2.568	1.693	0.5907	0.888	0	4.358	0	2.575
4	1.501	1.880	0.729	0.7979	1.2533	0	1.808	0	2.266	2.059	0.4857	0.880	0	4.698	0	2.282
5	1.342	1.596	0.577	0.8407	1.1894	0	1.756	0	2.089	2.326	0.4299	0.864	0	4.918	0	2.115
6	1.225	1.410	0.483	0.8686	1.1512	0.026	1.711	0.030	1.970	2.534	0.3946	0.848	0	5.078	0	2.004
7	1.134	1.277	0.419	0.8882	1.1259	0.105	1.672	0.118	1.882	2.704	0.3698	0.833	0.205	5.203	0.076	1.924
8	1.061	1.175	0.373	0.9027	1.1078	0.167	1.638	0.185	1.815	2.847	0.3512	0.820	0.387	5.307	0.136	1.864
9	1.000	1.094	0.337	0.9139	1.0942	0.219	1.609	0.239	1.761	2.970	0.3367	0.808	0.546	5.394	0.184	1.816
10	0.949	1.028	0.308	0.9227	1.0837	0.262	1.584	0.284	1.716	3.078	0.3249	0.797	0.687	5.469	0.223	1.777
11	0.905	0.973	0.285	0.9300	1.0753	0.299	1.561	0.321	1.679	3.173	0.3152	0.787	0.812	5.534	0.256	1.744
12	0.866	0.925	0.266	0.9359	1.0684	0.331	1.541	0.354	1.646	3.258	0.3069	0.778	0.924	5.592	0.284	1.719
13	0.832	0.884	0.249	0.9410	1.0627	0.359	1.523	0.382	1.618	3.336	0.2998	0.770	1.026	5.646	0.308	1.692
14	0.802	0.848	0.235	0.9453	1.0579	0.384	1.507	0.406	1.594	3.407	0.2935	0.762	1.121	5.693	0.329	1.671
15	0.775	0.816	0.223	0.9490	1.0537	0.406	1.492	0.428	1.572	3.472	0.2880	0.755	1.207	5.737	0.348	1.652
16	0.750	0.788	0.212	0.9523	1.0501	0.427	1.478	0.448	1.552	3.532	0.2831	0.749	1.285	5.779	0.364	1.636
17	0.728	0.762	0.203	0.9551	1.0470	0.445	1.465	0.466	1.534	3.588	0.2787	0.743	1.359	5.817	0.379	1.621
18	0.707	0.738	0.194	0.9576	1.0442	0.461	1.454	0.482	1.518	3.640	0.2747	0.738	1.426	5.854	0.392	1.608
19	0.688	0.717	0.187	0.9599	1.0418	0.477	1.443	0.497	1.503	3.689	0.2711	0.733	1.490	5.888	0.404	1.596
20	0.671	0.697	0.180	0.9619	1.0396	0.491	1.433	0.510	1.490	3.735	0.2677	0.729	1.548	5.922	0.414	1.586
21	0.655	0.679	0.173	0.9638	1.0376	0.504	1.424	0.523	1.477	3.778	0.2647	0.724	1.606	5.950	0.425	1.575
22	0.640	0.662	0.167	0.9655	1.0358	0.516	1.415	0.534	1.466	3.819	0.2618	0.720	1.659	5.979	0.434	1.566
23	0.626	0.647	0.162	0.9670	1.0342	0.527	1.407	0.545	1.455	3.858	0.2592	0.716	1.710	6.006	0.443	1.557
24	0.612	0.632	0.157	0.9684	1.0327	0.538	1.399	0.555	1.445	3.895	0.2567	0.712	1.759	6.031	0.452	1.548
25	0.600	0.619	0.153	0.9696	1.0313	0.548	1.392	0.565	1.435	3.931	0.2544	0.709	1.804	6.058	0.459	1.541
Over 25	$\dfrac{3}{\sqrt{n}}$	$\dfrac{3}{\sqrt{n}}$	—	—	—	*	†	*	†	—	—	—	—	—	—	—

*$1 - \dfrac{3}{\sqrt{2n}}$. †$1 + \dfrac{3}{\sqrt{2n}}$.

Source: Reproduced by permission from ASTM Manual on Quality Control of Materials, American Society for Testing Materials, Philadelphia, Pa., 1951.

GLOSSARY

Acceptance region   (Chapters 5, 8, and 9)

In the theory of hypothesis testing the set of values of the test statistic such that if the test statistic assumes one of these values, the null hypothesis is accepted.

Alternative or research hypothesis   (Chapters 5, 8, 9, and throughout the text)

In the theory of testing hypotheses the hypothesis that will be accepted as true if the null hypothesis, H_0, is proved false.

Analysis of variance   (Chapters 10, 11, and 12)

An analysis in which the total variation displayed by a set of observations (as measured by the sum of squares of deviations from the mean) is separated into components associated with various factors used in classifying the observations.

Area sampling   (Chapter 16)

A method of sampling in which the total area of interest is divided into subareas that are then randomly sampled. Each subarea in the sample is completely enumerated.

ARIMA models   (Chapter 14)

An acronymn for the *autoregressive integrated moving average* models used in the Box-Jenkins forecasting procedure.

Arithmetic mean   (throughout the text)

The arithmetic mean of a set of n measurements $x_1, x_2, x_3, \ldots, x_n$ is equal to the sum of the measurements divided by n.

Attribute data   (Chapter 15)

The phrase used in quality control when referring to counted (discrete) data, especially when counting the number of defective items or the number of defects per item.

Autocorrelation   (Chapters 12 and 14)

The internal correlation between members of a time series separated by a constant interval of time.

Autocorrelation coefficient   (Chapter 14)

In a time series the autocorrelation coefficient of lag k, r_k, measures the amount of correlation between observations that are k periods apart.

Autocorrelation function   (Chapter 14)

A graphical display of the autocorrelation coefficients of a given time series.

Autoregression   (Chapter 14)

The generation of a series of observations whereby the value of each observation is partly dependent on the values of those that have immediately preceded it. A regression structure in which lagged response values assume the role of the independent variables.

Bayes' Law   (Chapters 3 and 19)

If A is some event that occurs if and only if either B or $\bar{B}$ occurs, then

$$P(B \mid A) = \frac{P(A \mid B)P(B)}{P(A \mid B)P(B) + P(A \mid \bar{B})P(\bar{B})} = \frac{P(AB)}{P(A)}$$

Binomial experiment   (Chapter 5)

An experiment consisting of n independent trials in which the outcome at each trial is a "success" with probability p or a "failure" with probability $(1 - p)$. We are interested in x, the number of successes observed during the n trials.

Binomial probability distribution   (Chapter 5)

A probability distribution giving the probability of x, the number of successes observed during the n trials of a binomial experiment:

$$p(x) = C_x^n p^x (1 - p)^{n-x} \qquad x = 0, 1, 2, \ldots, n$$

Block design   (Chapter 10)

See Randomized block design.

Box-Jenkins forecasting procedure   (Chapters 13 and 14)

A four-stage procedure for the development of a forecasting model that combines an autoregressive component involving past observed values with a moving average component involving current and past error terms.

Box plot   (Chapter 2)

A graphical display that allows for the assessment of asymmetry and the detection of outliers in a data set.

Census   (Chapter 16)

A recording of every element contained in a population.

Centered moving average   (Chapter 13)

An average of M terms in a time series, calculated for successive groups of M terms and plotted at the time periods associated with the middle of each particular group.

Centerline   (Chapter 15)

The line midway between the control limits on a control chart. The centerline is the grand average of all the sample statistics plotted on the control chart.

Central Limit Theorem   (Chapters 7, 8, and 9)

If random samples of n observations are drawn from a population with finite mean μ and standard deviation σ, then when n is large, the sample mean $\bar{x}$ will be approximately normally distributed with mean μ and standard deviation $\sigma/\sqrt{n}$. The approximation will become more and more accurate as n becomes larger.

Chi-square statistic   (Chapters 9, 17, and 18)

A statistic having the distribution of a sum of independent squared standard normal variables. The number of independent squared normal variables is called the degrees of freedom.

Cluster sample   (Chapter 16)

A cluster sample is obtained by first randomly selecting a set of m collections of sample elements, called clusters, from the population and then conducting a complete census within each cluster.

Coefficient of correlation   (Chapter 11)

A number between -1 and $+1$ that measures the linear dependence between two random variables. The limiting values -1 and $+1$ indicate perfect negative and perfect positive correlation, respectively; a correlation of zero suggests a complete lack of association between the two variables.

Coefficient of determination   (Chapters 11 and 12)

A measure of the goodness of fit of a regression model that is equal to the square of the coefficient of correlation (simple or multiple, whichever is appropriate).

Combination   (Chapter 3)

The number of combinations of n objects taken r at a time denoted by the symbol C_r^n, where

$$C_r^n = \frac{n!}{r!(n-r)!}$$

Complement of an event   (Chapter 3)

The complement of an event A is the collection of all sample points in the sample space that are not in A. The complement of A is denoted by the symbol $\bar{A}$, and $P(\bar{A}) = 1 - P(A)$.

Completely randomized design   (Chapters 10 and 18)

A simple form of experimental design in which the treatments are randomly allocated to the experimental units.

Compound event   (Chapter 3)

An event composed of two or more simple events.

Conditional probability   (Chapter 3)

The probability of occurrence of an event A given that another event B has occurred is called the conditional probability of A given B and is denoted as $P(A \mid B)$. Computationally, $P(A \mid B) = P(AB)/P(B)$.

Confidence coefficient   (Chapters 8, 9, and throughout the text)

A probability associated with a confidence interval that expresses the probability that the interval will include the parameter value under study.

Confidence interval   (Chapters 8, 9, and throughout the text)

An interval computed from sample values. Intervals so constructed will straddle the estimated parameter $100(1 - \alpha)\%$ of the time in repeated sampling. The quantity $(1 - \alpha)$ is called the confidence coefficient.

Contingency table   (Chapter 17)

A two-way table for classifying the members of a group according to two or more identifying characteristics.

Continuous random variable   (Chapter 4)

A random variable defined over, and assuming the infinitely many values associated with, the points on a line interval.

Control chart   (Chapter 15)

A graph of a sample statistic over many successive samples from a production process. Used in quality control to monitor the product consistency.

Control limits   (Chapter 15)

Values on a control chart that are three standard deviations (of the sample statistic used) on either side of the centerline.

Convenience sampling   (Chapter 16)

The selection of a sample that can be obtained simply and conveniently.

Correlogram   (Chapter 14)

A graph illustrating the autocorrelations between members of a time series (vertical axis) for different separations in time k (horizontal axis).

Cost of uncertainty   (Chapter 19)

The smallest expected opportunity loss in a decision-making problem involving uncertainty.

Critical value   (Chapters 5, 8, and 9)

In a statistical test of an hypothesis the value of the test statistic that separates the rejection and acceptance regions.

Cyclic effect   (Chapter 14)

A periodic movement in a time series that occurs as a result of stimuli from the economy and is generally not predictable.

Decision analysis   (Chapter 19)

The logical and quantitative analysis of all the factors that influence a decision.

Degrees of freedom   (Chapter 9 and throughout the text)

The number of linearly independent observations in a set of n observations. The degrees of freedom are equal to n minus the number of restrictions placed on the entire data set.

Delphi method   (Chapter 14)

A method for forecasting in the absence of a relevant data base in which a sequence of questionnaires is used to identify the factors than can best be of assistance in focusing attention on the resultant forecast.

Dependent variable   (Chapters 11, 12, and 14)

The predictand in a regression equation. The variable of interest in a regression equation, which is said to be functionally related to one or more independent or predictor variables.

Descriptive statistics   (Chapter 1 and throughout the text)

That branch of statistics dealing with procedures for summarizing the information in a set of measurements and for describing the characteristics of the set.

Deseasonalized series   (Chapter 13)

A time series in which the seasonal component has been estimated and then removed from the series.

Design of an experiment   (throughout the text)

The sampling procedure that enables the gathering of a maximum amount of information for a given expenditure.

Deterministic model   (Chapters 11 and 12)

A model that contains no random elements, as opposed to a probabilistic model.

Discrete random variable   (Chapter 4)

A random variable over a finite or a countably infinite number of points.

Dummy variable   (Chapter 12)

An artificial variable used to represent the level of a qualitative variable in a multiple regression analysis.

Econometric model   (Chapter 13)

An econometric model is a probabilistic model consisting of a system of one or more equations that describe the relationship among a number of economic and time series variables.

Empirical Rule   (throughout the text)

Given a distribution of measurements that is approximately bell-shaped, the interval

$(\mu \pm \sigma)$ contains approximately 68% of the measurements

$(\mu \pm 2\sigma)$ contains approximately 95% of the measurements

$(\mu \pm 3\sigma)$ contains almost all the measurements

Error of estimation   (Chapters 8 and 16)

The difference between an estimated value and the true value.

Event   (Chapter 3)
A collection of sample points.

Expected value   (Chapters 4, 11, and 12)
Let x be a discrete random variable with probability distribution $p(x)$ and let $E(x)$ represent the expected value of x. Then

$$E(x) = \sum_x xp(x)$$

where the elements are summed over all values of the random variable x.

Exponential probability distribution   (Chapter 6)
A distribution used to model lifetimes or waiting times in which the right tail of the distribution decreases exponentially. The probability function is given as

$$f(x) = \lambda e^{-\lambda x} \qquad x \geq 0$$

Exponential smoothing   (Chapter 13)
In time series analysis a computational method that averages the first t time series values by increasingly weighting out the contribution of remote values. The exponentially smoothed value at time t is

$$S_t = \alpha \sum_{i=0}^{t-2} (1 - \alpha)^i y_{t-i} + (1 - \alpha)^{t-1} y_1$$

where $y_1, y_2, \ldots, y_t$ are the response values and α is the smoothing constant ($0 \leq \alpha \leq 1$).

***F* distribution**   (Chapters 9, 10, 11, and 12)
The sampling distribution of the ratio of two independent quantities each of which is distributed like a sample variance from a normal distribution.

Factor   (Chapters 10 and 12)
An independent variable under examination in an experiment as a possible cause of variation.

Factorial experiment   (Chapter 10)
An experiment designed to examine the effect of one or more factors, each factor being applied at two or more levels. The experiment utilizes every combination of factor levels as treatments in the experimental design.

First-order difference transformation   (Chapter 14)
A first-order difference transformation (∇y_t) is a series developed by computing the differences between adjacent values ($y_t - y_{t-1}$) in a time series.

Fishbone diagram   (Chapter 15)
A graph used in quality control to identify possible problem causes. Also called an Ishikawa diagram after its inventor.

Frame   (Chapter 16)
A list of sampling units.

Frequency distribution   (Chapter 2)
A specification of the way in which, probabilistically, the relative frequencies of members of a population are distributed according to the values of the variates they exhibit.

Frequency histogram   (Chapter 2)
A specification of the way in which the frequencies of members of a population are distributed according to the values of the variates they exhibit.

Friedman test   (Chapter 18)
A nonparametric test to compare $t > 2$ treatments when the randomized block design has been used. The data are ranked within each block, and the rank sums for each of the t treatments are used in the Friedman test statistic, X^2.

Goodness of fit   (Chapters 12 and 17)

A measure of agreement between the observed values of a random variable and corresponding values derived according to a theoretical model, or an hypothesis under test.

Historical analogy   (Chapter 14)

A method for forecasting the sales of a newly introduced product (or service) that uses the sales history of some previously introduced product as a guide.

Hypergeometric probability distribution   (Chapter 5)

Appropriate when sampling from a finite population of N elements of which k are identified as "successes" and $(N - k)$ are "failures." Then $p(x)$ gives the probability of observing x successes in a sample of n selected from the population.

Independent variable   (Chapters 11 and 12)

A nonrandom variable related to the response in a regression equation. One or more independent variables may be functionally related to the dependent variable. They are used in the regression equation to predict or estimate the value of the dependent variable.

Index number   (Chapter 13)

A quantity that shows the changes over time from some base period to a reference period of a variable process that is not directly observable in practice.

Inferential statistics   (Chapter 1 and throughout the text)

That branch of statistics dealing with procedures used to make inferences about a population from information contained in a sample.

Interaction   (Chapters 10 and 12)

A measure of the extent to which the effect of one factor varies as the levels of the other factors are changed.

Intersection   (Chapter 3)

If A and B are two events in a sample space S, the intersection of A and B is the event composed of all sample points that are in both A and B and is denoted by AB.

Judgment sampling   (Chapter 16)

The selection of a sample that, according to the judgment and intuition of the sampler, accurately reflects the population.

Kruskal-Wallis test   (Chapter 18)

A nonparametric test to compare $t > 2$ population distributions when the completely randomized design has been used. The t samples are jointly ranked and the test statistic H is based on the rank sums of the t individual samples.

Laspeyres index   (Chapter 13)

If the prices of a set of commodities in a base year are $p_{01}, p_{02}, p_{03}, \ldots,$ and $q_{01}, q_{02}, q_{03}, \ldots$ are the quantities sold in the base period, and $p_{n1}, p_{n2}, p_{n3}, \ldots$ are the prices of the same commodities in a given year, the Laspeyres index is

$$L = \frac{\sum p_n q_0}{\sum p_0 q_0}$$

Least squares   (Chapters 11 and 12)

See Method of least squares.

Levels of a factor   (Chapters 10 and 12)

The settings of values of a factor in an experiment.

Linear correlation   (Chapters 11 and 12)

A measure of the strength of the linear relationship between two variables y and x that is independent of their respective scales of measurement. Linear correlation is commonly measured by the coefficient of correlation.

Linear regression analysis   (Chapter 11)

See Simple linear regression.

Linear statistical model   (Chapter 12)

An equation of the form

$$y = \beta_0 + \beta_1 x_1 + \beta_2 x_2 + \cdots + \beta_k x_k + \epsilon$$

that relates a single dependent variable y to a set of independent variables $x_1, x_2, \ldots, x_k$, where $\beta_0, \beta_1, \ldots, \beta_k$ are unknown parameters and ϵ is a random error.

Mann-Whitney U test   (Chapter 18)

A nonparametric test proposed by Wilcoxon for comparing two population distributions by first ordering the combined observations from the samples selected from each population and then summing the ranks of the observations from the samples. The Mann-Whitney U test uses a function of the rank sums as its test statistic.

Mean absolute deviation (MAD)   (Chapter 14)

A measure of accuracy of a forecasting model calculated by averaging the absolute values of the forecast errors:

$$\text{MAD} = \frac{1}{n} \sum_{t=1}^{n} |y_t - \hat{y}_t|$$

Mean absolute percentage error (MAPE)   (Chapter 14)

A measure of forecast accuracy calculated by averaging the absolute values of the percent errors made by the forecasting model:

$$\text{MAPE} = \frac{1}{n} \sum_{t=1}^{n} \left| \frac{y_t - \hat{y}_t}{y_t} \right| (100\%)$$

Mean square error (MSE)   (Chapter 14)

In forecasting the MSE measures forecasting accuracy by averaging the squares of the forecast errors:

$$\text{MSE} = \frac{1}{n} \sum (y_t - \hat{y}_t)^2$$

Median   (Chapter 2)

The median of a set of n measurements $x_1, x_2, x_3, \ldots, x_n$ is the value of x that falls in the middle when the measurements are arranged in order of magnitude.

Method of least squares   (Chapters 11 and 12)

A technique for the estimation of the coefficients in a regression equation that chooses as the regression coefficients the values that minimize the sum of squares of the deviations of the observed values of y from those predicted.

Minimax decision   (Chapter 19)

In a decision analysis the minimax decision is the decision to select the action whose maximum opportunity loss is the smallest.

Mode   (Chapter 2)

The mode of a set of measurements $x_1, x_2, x_3, \ldots, x_n$ is the value of x that occurs with the greatest frequency. The mode is not necessarily unique.

Moving average forecast model   (Chapter 14)

A forecast computed by averaging together the past k values observed from a time series.

Multinomial experiment   (Chapter 17)

An experiment consisting of n identical and independent trials, each of which results in one of k possible outcomes with probabilities $p_1, p_2, \ldots, p_k$, where $\sum p_i = 1$ and $0 \leq p_i \leq 1$. The experimenter is interested in $n_1, n_2, \ldots, n_k$, the number of trials resulting in each of the k outcomes.

Multiple regression analysis   (Chapter 12)

A regression analysis in which the mean value of a dependent variable y is assumed to be related to a set of independent variables $x_1, x_2, \ldots, x_k$ by an expression of the form

$$E(y \mid x_1, x_2, \ldots, x_k) = \beta_0 + \beta_1 x_1 + \beta_2 x_2 + \cdots + \beta_k x_k$$

Mutually exclusive events   (Chapter 3)

Two events A and B are said to be mutually exclusive if the event AB contains no sample points.

Nonparametric hypothesis   (Chapter 18)

A statistical hypothesis that does not involve population parameters but is concerned with the form of the population frequency distribution.

Nonparametric methods   (Chapter 18)

Methods used for testing hypotheses that do not explicitly involve assertions about parameter values.

Nonresponse   (Chapter 16)

In a sample survey the failure to obtain information from a designated element for any reason.

Normal probability distribution   (throughout the text)

A symmetric bell-shaped probability distribution of infinite range represented by the equation

$$f(x) = \frac{1}{\sigma \sqrt{2\pi}} e^{-(x - \mu)^2 / 2\sigma^2} \qquad (-\infty < x < \infty)$$

where μ is the mean and σ^2 is the variance of the distribution.

Null hypothesis   (throughout the text)

In a statistical test of an hypothesis the statement of the hypothesis to be tested.

Observed significance level   (Chapter 8 and throughout the text)

See p-value.

Off-line quality control   (Chapter 15)

Statistical techniques, especially experimental design methods, used in quality control to improve the design of a product.

One-tailed statistical test   (Chapters 8 and 9 and throughout the text)

A statistical test of an hypothesis in which the rejection region is wholly located at one end of the distribution of the test statistic.

On-line quality control   (Chapter 15)

The use of statistical methods, especially control charts, to monitor the consistency of manufactured products.

Operating characteristic curve   (Chapters 5 and 8)

A plot of the probability of accepting the null hypothesis when some alternative is true (the probability of a type II error) against various possible values of the alternative hypothesis.

Opportunity loss   (Chapter 19)

The opportunity loss L_{ij} for selecting action a_i, given that the state of nature s_j is in effect, is the difference between the maximum profit that could be realized if s_j occurs and the profit obtained by selecting action a_i.

Optimal decision   (Chapter 19)

In a decision analysis the optimal decision is a decision to select the action that maximizes the decision maker's objective.

Outlier   (Chapter 2)

One or more observations in a data set that are far in value from the body of the set. Outliers may come from a separate population or may result from sampling or recording errors.

Paired-difference test   (Chapter 9)

A test to compare two populations by using pairs of elements, one from each population, that are matched and hence nearly alike. Thus, the test involves two samples of equal size, where the members of one sample can be paired against members of the other. Comparisons are made within the relatively homogeneous pairs (blocks).

Panel consensus method   (Chapter 14)

A method for forecasting in the absence of a relevant data base in which experts within the organization derive a forecast that represents a consensus of the opinions of the experts.

Parameter   (throughout the text)

A numerical descriptive measure for the population.

Parametric hypothesis   (throughout the text)

A statistical hypothesis about a population parameter.

Pareto chart   (Chapter 15)

A bar chart used in quality control to separate the important problem areas from the less important ones. The bars are drawn in order of decreasing height, from left to right.

Payoff table   (Chapter 19)

In a decision analysis a two-way table displaying the payoffs, either opportunity losses or profits, for selecting a particular action given that a specific state of nature is in effect.

Pearson product-moment coefficient of correlation   (Chapter 11)

See Coefficient of correlation.

Percentiles   (Chapter 2)

The set of values that divide the total frequency of a set of data into 100 equal parts. The $100p$th percentile is a value such that at most $100p\%$ of the measurements are less than the value and at most $100(1 - p)\%$ are greater.

Period of a cyclic or seasonal effect   (Chapters 13 and 14)

The number of time points between identifiable points of recurrence—between peaks and valleys—in a time series. The number of time points for one complete cycle or for a complete seasonal pattern to be exhibited.

Permutation   (Chapter 3)

An ordered arrangement of r distinct objects. The number of n objects selected in groups of size r is denoted by P_r^n, where

$$P_r^n = n(n - 1)(n - 2) \cdots (n - r + 1)$$

Pie chart   (Chapter 2)

A descriptive technique used to exhibit the way in which a single total quantity is apportioned to a group of categories. Each category is assigned a sector of the circle according to the percentage of the total it represents.

Point estimator   (Chapters 8 and 9)

A single number computed from a sample and used as an estimator of a population parameter.

Poisson probability distribution   (Chapter 5)

A model for finding the probability of count data resulting from any experiment, where the count x represents the number of rare events observed in a given unit of time or space.

Population   (throughout the text)

A finite or infinite collection of measurements or individuals that comprises the totality of all possible measurements within the context of a particular statistical study.

Posterior probability   (Chapters 3 and 19)

The probability p_1 of an event at the outset of an experiment might be modified to p_2 in light of experimental evidence. The posterior probability p_2 is usually determined by employing Bayes' Theorem.

Power of a statistical test   (Chapters 8 and 9)

The probability that the statistical test rejects the null hypothesis when some particular alternative is true. Power equals $(1 - \beta)$. The power is greatest when the probability of a type II error is least.

Prediction equation   (Chapters 11 and 12)

In regression the equation used to predict the value of the dependent variable y for specified values of the independent variables $x_1, x_2, \ldots, x_k$. This equation is generally obtained using the method of least squares.

Prior probability   (Chapter 19)

The probability representing the likelihood of occurrence of an event before experimental evidence relevant to the event has been observed. The unconditional probabilities used in Bayes' Theorem are prior probabilities.

Probabilistic model   (Chapters 11 and 12)

A model that incorporates some random element in its formulation as opposed to a deterministic model.

Probability distribution   (Chapter 4)

A formula, table, or graph providing the probability associated with each value of the random variable if the random variable is discrete or providing the fraction of measurements in the population falling in specific intervals if it is continuous.

Probability of an event   (throughout the text)

The probability of an event A is equal to the sum of the probabilities of the sample points in A.

Probability tree   (Chapter 3)

A visual model to display the outcomes of an experiment in which each successive branch of the tree corresponds to a step necessary to generate the possible outcomes of an experiment.

Proportional allocation   (Chapter 16)

An allocation procedure that partitions the sample size among the strata proportional to the size of the strata when using stratified random sampling.

***p*-value**   (Chapter 8 and throughout the text)

The smallest value of α (the probability of making a type I error) for which test results become statistically significant. (*See also* Significance level.)

Qualitative variables   (Chapters 10, 12, and 17)

Variables concerning qualitative or attribute data. The data are not necessarily representable in numerical form.

Quantitative variables   (Chapters 10, 12, and 17)

Variables concerning quantitative or measurement data. The data can be represented on a numerical scale.

Quartiles   (Chapter 2)

Three values that divide the total frequency of a set of data into four equal parts. The central value is called the *median* and the other two are called the *upper* and *lower quartiles*, respectively.

Quota sampling   (Chapter 16)

The selection of a predetermined number of elements from different sectors of the population.

Randomized block design   (Chapter 10)

An experimental design whereby treatments are randomly assigned within a set of blocks to eliminate bias.

Randomized response sampling   (Chapter 16)

A survey sampling procedure for dealing with potentially sensitive or embarrassing material that requires a question on the sensitive topic be paired with an innocuous question. The respondent answers only one of the two questions, which he or she has selected at random.

Random sample   (Chapters 4, 16, and throughout the text)

Suppose that a sample of n measurements is drawn from a population consisting of N total measurements. If the sampling is conducted in such a way that each of the C_n^N samples has an equal probability of being selected, the sampling is said to be random and the result is said to be a random sample.

Random variable   (Chapters 3, 4, and throughout the text)

A numerical-valued function defined over a sample space.

Range   (Chapters 2 and 16)

The range of a set of n measurements $x_1, x_2, x_3, \ldots, x_n$ is the difference between the largest and smallest measurement.

Rank correlation coefficient   (Chapter 18)

A number between -1 and 1 that measures the correspondence between two sets of rankings. Spearman's rank correlation coefficient is calculated exactly like the coefficient of correlation (Chapter 11), using ranks rather than actual observations.

Ratio estimation   (Chapter 16)

An estimation procedure based on the relationship between two variables y and x that have been measured on the same set of sampled elements.

Regression analysis   (Chapters 11 and 12)

The process of fitting a regression equation to a set of data by using the method of least squares. Also includes the various statistical tests and estimates associated with the use of the fitted equation.

Regression equation   (Chapters 11, 12, and 14)

An equation expressing dependence of the mean of a dependent variable y on one or more independent or predictor variables $x_1, x_2, x_3, \ldots$.

Rejection region   (Chapters 5, 8, and 9)

In the theory of hypothesis testing the set of values such that if the test statistic assumes one of these values, the null hypothesis is rejected.

Research hypothesis   (Chapters 5, 8, 9, and throughout the text)

See Alternative hypothesis.

Residual   (Chapters 11, 12, and 14)

The difference between the observed value of y and the value predicted by a model, $(y - \hat{y})$, sometimes referred to as the error of prediction.

Root mean square error (RMSE)   (Chapter 14)

A measure of forecast accuracy calculated by taking the square root of the mean square error (MSE).

Run   (Chapter 18)

In a series of observations of attributes the occurrence of an uninterrupted series of the same attribute is called a run. A run can be of length 1.

Sample   (throughout the text)

Any subset of a population.

Sampling design   (Chapter 16)

A method for selecting a sample.

Sampling distribution   (Chapter 7 and throughout the text)

The distribution of a statistic based on all possible samples that can be drawn according to a specific sampling plan. The term almost always refers to random sampling, and the statistic is usually a function of the n sample observations.

Sampling units   (Chapter 16)

Nonoverlapping collections of elements from the population.

Scatter diagram   (Chapter 11)

A plot of the pairs of values (x, y) on a rectangular coordinate system, used to assess the relationship between x and y.

Seasonal effect   (Chapters 13 and 14)

The rises and falls in a time series, which always occur at a particular time of year because of changes in the seasons.

Seasonal index   (Chapter 13)

An index measuring the seasonal effect of a particular month (when using monthly data) or quarter (when using quarterly data).

Second-order difference transformation   (Chapter 14)

A second-order difference transformation $(\nabla^2 y_t)$ is a series developed by computing the differences between adjacent first differences in a time series. Symbolically, we compute $(y_t - y_{t-1}) - (y_{t-1} - y_{t-2})$ for all pairs of first differences.

Serial correlation   (Chapter 12)

See Autocorrelation.

Signed-rank test   (Chapter 18)

A nonparametric statistical test of the difference between two treatments using paired observations. The differences are ranked according to their absolute magnitude, and each rank is given the sign of the original difference. The sum of the positive or negative ranks provides the test statistic developed by Wilcoxon.

Significance level   (Chapter 7)

The probability α of making a type I error.

Sign test   (Chapter 18)

A nonparametric statistical test of significance depending on the signs of differences between matched or unmatched pairs and not on the magnitudes of the differences.

Simple linear regression   (Chapter 11)

A regression analysis in which the mean value of a dependent variable y is assumed to be related to a single independent variable x by the expression $E(y \mid x) = \beta_0 + \beta_1 x$.

Simple random sample   (Chapters 4, 7, and 16)

A simple random sample results when sampling is conducted in such a way that every possible sample of size n has an equal probability of being selected from a population.

Skewed distribution   (Chapter 2)

A frequency distribution that is not symmetrical about its mean.

Smoothed statistic   (Chapters 13 and 14)

In time series analysis an average of a set of process values, usually a moving average or an exponentially smoothed statistic, used to represent a time series after eliminating random fluctuations.

Smoothing constant   (Chapters 13 and 14)

The constant α $(0 \le \alpha \le 1)$ employed to weight out the contribution of remote response values in an exponential-smoothing scheme.

Spearman's rank correlation coefficient   (Chapter 18)

See Rank correlation coefficient.

Specification limits   (Chapter 15)

The upper and lower specification limits are the largest and smallest values that a quality characteristic can have and still be considered acceptable for use.

Specific seasonal index   (Chapter 14)

The ratio of the time series at time t to a corresponding centered moving average at time t. The ratio measures by how much the seasonal component affects the series at time period t.

Standard deviation (Chapter 2 and throughout the text)
The standard deviation of a set of n measurements $x_1, x_2, x_3, \ldots, x_n$ is equal to the positive square root of the variance of the measurements.

Standard error (throughout the text)
The precision of an estimator is measured by its standard deviation; hence, the standard deviation of an estimator is called its standard error.

Standard normal probability distribution (Chapters 6, 7, and throughout the text)
A normal probability distribution having a mean equal to zero and a variance equal to one.

State of nature (Chapter 19)
In a decision analysis the states of nature are the uncertain events over which the decision maker has no control.

Statistic (throughout the text)
A value computed from sample measurements, usually but not always as an estimator of some population parameter.

Statistical control (Chapter 15)
Refers to the situation when all the points on a control chart appear to be random and fall within the control limits.

Statistical process control (Chapter 15)
The name used to describe the application of certain statistical techniques to the design and manufacture of products.

Stem and leaf display (Chapter 2)
A tabular-graphical method whereby each observation is divided into a stem value and a leaf value. The stems define a set of classes, and the number of leaves within a stem defines the frequency with which the values in that class occur.

Stepwise regression (Chapter 12)
A method of selecting the "best" set of independent variables for a regression equation. Variables are introduced one at a time, with the criterion for accepting a variable based on the correlation of that variable and y in the presence of the other variables already in the model.

Stratified random sample (Chapter 16)
A sample obtained by separating the population elements into nonoverlapping groups, called strata, and then selecting a simple random sample within each stratum.

Student's t distribution (Chapters 9, 11, and 12)
The distribution of $t = (\bar{x} - \mu)/(s/\sqrt{n})$ for samples drawn from a normally distributed population and used for making inferences about a population mean when the population variance σ^2 is unknown and the sample size n is small.

Symmetrical distribution (Chapter 2)
A frequency distribution for which the values of the distribution that are equidistant from the mean occur with equal frequency.

Systematic sampling (Chapter 16)
A method of selecting a sample by a systematic method as opposed to random sampling, such as selecting each tenth name from a list or by sampling every third resident in every every other block in an area sample.

Table of random numbers (Chapter 16)
A table constructed so that each entry is equally likely to be any five-digit number from 00,000 to 99,999. The table of random numbers is used in selecting a random sample.

Tchebysheff's Theorem (Chapter 2)
Given a number k greater than or equal to 1 and a set of n measurements $x_1, x_2, \ldots, x_n$, at least $(1 - 1/k^2)$ of the measurements will lie within k standard deviations of their mean.

Test statistic (Chapters 5, 8, 9, and throughout the text)
A function of a sample of observations that provides a basis for testing a statistical hypothesis.

Time series   (Chapters 13 and 14)

Any sequence of measurements taken on a variable process over time. Usually illustrated as a graph whose vertical coordinate gives a value of the random response plotted against time on the horizontal axis.

Treatment   (Chapter 10)

A stimulus that is applied in order to observe its effect on the experimental situation. A treatment may refer to a physical substance, a procedure, or anything capable of controlled application according to the requirements of the experiment.

Two-stage cluster sample   (Chapter 16)

A two-stage cluster sample is obtained by choosing a simple random sample of clusters and then selecting a random sample of elements from each cluster.

Two-tailed statistical test   (Chapters 8, 9, and throughout the text)

A statistical test of an hypothesis in which the rejection region is separated by the acceptance region and is located in both ends of the distribution of the test statistic.

Type I error   (Chapters 5, 8, and throughout the text)

In the statistical test of an hypothesis the error incurred by rejecting the null hypothesis when the null hypothesis is true.

Type II error   (Chapters 5, 8, and throughout the text)

In the statistical test of an hypothesis the error incurred by accepting the null hypothesis when the null hypothesis is false and some alternative to the null hypothesis is true.

Unbiased estimator   (Chapter 8)

An unbiased estimator $\hat{\theta}$ of a parameter θ is an estimator for which the expected value of $\hat{\theta}$ is equal to θ. That is, $E(\hat{\theta}) = \theta$.

Uniform probability distribution   (Chapter 6)

A distribution having a constant probability over a finite interval from a to b, represented by

$$f(x) = \frac{1}{b - a} \qquad a \leq x \leq b$$

Union   (Chapter 3)

If A and B are two events in a sample space S, the union of A and B is the event containing all sample points in A or B or both.

Variables data   (Chapter 15)

The phrase used in quality control to refer to measured data, that is, data from some continuous random variable.

Variance   (Chapter 2 and throughout the text)

The variance of a set of n measurements $x_1, x_2, \ldots, x_n$ is the average of the squares of the deviations of the measurements about their mean.

Venn diagram   (Chapter 3)

A diagram portraying graphically each sample event as a sample point in a sample space S.

Wilcoxon signed-rank test   (Chapter 18)

See Signed-rank test.

Reference

Kendall, M. G., and W. R. Buckland. *A Dictionary of Statistical Terms.* 4th ed. London: Longman Group, 1982.

ANSWERS TO SELECTED EXERCISES

CHAPTER 2

2.1 from 5 to 20 classes, depending on the amount of data

2.3 (I) overlapping class limits; (II) class intervals not of equal widths; (III) subintervals noninclusive

2.5 **a.** 15.6

2.7 **b.** 7/24

2.9 **b.** 29/50   **c.** no

2.29 mean; mode

2.31 **a.** skewed right   **b.** symmetric
c. symmetric   **d.** skewed left

2.33 $\bar{x} = 4.143$; median $= 4$; mode $= 4$

2.35 **a.** $\bar{x} = 12.9$; median $= 12.5$; mode $= 12$
b. skewed right

2.37 **a.** $\bar{x} = 1.642$; median $= 0$; mode $= 0$
b. skewed right

2.43 $\bar{x} = 3$; $s^2 = 4$; $s = 2$

2.45 **a.** at least 3/4 between 28.0 and 38.0; at least 8/9 between 25.5 and 40.5
b. approximately 68% between 30.5 and 35.5; approximately 95% between 28.0 and 38.0; almost all between 25.5 and 40.5

2.47 **a.** approximately 68% between 9.81 and 10.03; approximately 95% between 9.70 and 10.14; almost all between 9.59 and 10.25
b. almost all

2.49 **a.** no

2.53 $\bar{x} = -1$; $s^2 = 6.4444$; $s = 2.54$

2.55 $\bar{x} = 4.808$; $s^2 = 2.219$; $s = 1.490$

2.57 **a.** $\bar{x} = 1.64$; $s^2 = 85.56$; $s = 9.25$
b. median $= 0$; $Q_1 = -4.87$; $Q_3 = 7.80$

d.

k	$\bar{x} \pm ks$	Interval Boundaries	Fraction in Interval
1	1.64 ± 9.25	-7.61 to 10.89	.62
2	1.64 ± 18.5	-16.86 to 20.14	.98
3	1.64 ± 27.75	-26.11 to 29.39	1.00

2.59 identical

2.61 **b.** $\bar{x}_g = 5.133$; $s_g = 1.756$

2.63 **a.** $\bar{x}_g = 39.1$; $s_g = 11.9$ **c.** $\bar{x} = \$50,500$; $s = \$18,566.02$ **d.** 20%   **e.** 41%

2.67 inner fences: (6, 22); outer fences: (0, 28); 5 and 27, mild outliers; 38, an extreme outlier

2.69 inner fences: $(-23.45, 26.55)$; outer fences: $(-42.20, 45.30)$; no outliers

2.71 **c.** $\bar{x} = 6.70$; $s^2 = 4.0745$; $s = 2.02$ **d.** .64   **e.** 1.00; yes   **f.** yes; yes

2.75 **a.** $\bar{x} = .0871$; $s^2 = .0017668$; $s = .042$ **b.** .159; 3.8; 4

2.79 1.825; 2.05

2.81 **a.** 3806.67   **b.** 4   **c.** 7.208   **d.** 25,887.5

2.83 **a.** .95   **b.** .16

2.85 .16

2.87 \$693.64

2.89 **a.** at least 3/4 between 35 and 195; at least 8/9 between -5 and 235
b. slight difference in location; slightly more variable
c. higher content, but less variable

2.91 $\bar{x}_g = 131.50$; $s_g = 56.700$

CHAPTER 3

3.1 **a.** 0; 1   **b.** 1

3.3 **a.** compound   **b.** simple   **c.** simple   **d.** compound

3.5 **a.** 8   **b.** $P(E_i) = 1/8$ **c.** 1/8; 1/8   **d.** 3/8   **e.** 7/8

3.7 **c.** 1/3; 1/3

3.9 **b.** 35 sample points   **c.** 1/7

3.13 **c.** $P(A) = 1/36$; $P(B) = 1/6$; $P(C) = 15/36$; $P(D) = 1/36$; $P(E) = 0$; $P(F) = 15/36$

3.15 **f.** $P(A) = 1/7$; $P(B) = 3/7$; $P(C) = 5/7$; $P(D) = 5/7$; $P(AB) = 1/7$; $P(A \cup B) = 3/7$

3.17 $P(A) = 5/6$; $P(B) = 5/6$; $P(AB) = 2/3$; $P(A \cup B) = 1$

3.21 **a.** $\bar{A}$: $\{E_2, E_6, E_7, E_8\}$; AB: $\{E_1, E_4\}$; AC: ϕ; BC: $\{E_6\}$ **b.** $P(A) = 1/2$; $P(B) = 1/2$; $P(C) = 3/8$; $P(\bar{A}) = 1/2$; $P(AB) = 1/4$; $P(AC) = 0$; $P(BC) = 1/8$ **c.** no   **d.** yes   **e.** $P(A|B) = 1/2$; $P(A|C) = 0$; yes; no

3.23 $P(A) = .61$; $P(B) = .39$; $P(C) = .37$; $P(D) = .63$; $P(C|A) = .098$; $P(D|A) = .902$; $P(C|B) = .795$

3.25 **a.** 7/15   **b.** 5/8

3.27 **a.** .125   **b.** .875

3.29 **a.** yes; no   **b.** no; yes   **c.** no; yes

3.31 **a.** 2/3   **b.** 1/2   **c.** 1/3

3.33 **a.** 7/10   **b.** 3/10

3.37 .59; .08; .87

3.39 .94

3.41 **a.** .22   **b.** .60   **c.** .79   **d.** .6241   **e.** no

3.43 **a.** 1/2   **b.** 1/8

3.47 **a.** .38   **b.** .7895

3.49 .99

3.51 .80

3.53 .3830

3.55 no; *mn* rule

3.57 **a.** 50   **b.** 80   **c.** 400

3.59 17,576,000

3.61 1,000

3.63 5,040

3.65 53,130

3.67 1,680

3.69 12

3.71 $2,598,960 = C_5^{52}$

3.73 **a.** 25   **b.** 4/25   **c.** .0001

3.75 **d.** $P(A) = 1/8$

3.77 **a.** .59   **b.** .41   **c.** .9914

3.79 1/3

3.81 2/3

3.83 **a.** .36   **b.** .0704   **c.** .2432

3.85 **c.** 3/10   **d.** 1/10   **e.** 7/10

3.87 **a.** .144   **b.** .056   **c.** .944

3.89 **a.** .8725   **b.** .6189; .2435

3.91 **a.** .702   **b.** .06   **c.** .102   **d.** .0084

3.93 **a.** 1/4   **b.** 1/12   **c.** 1/2

3.95 **a.** 1/49   **b.** 13/49

3.97 **a.** .64   **b.** .04   **c.** .96

3.99 **a.** 1/10   **b.** 4/10

3.101   **a.** $P(A) = 30/56$ **b.** $P(B) = 55/56$

3.103   .49

3.105   .8

3.107   **a.** 2/3   **b.** 4/5   **c.** 4/15

3.109   .00039601; .039; .9606; .9996

3.111   **a.** .49   **b.** .51   **c.** .0369

3.113   **a.** 29/52   **b.** 4/52   **c.** 29/52   **d.** 1
           e. 8/23   **f.** 14/29   **g.** 23/52
           h. 38/52   **i.** no; yes

3.115   .00315; not independent

CHAPTER 4

4.1   **a.** discrete   **b.** continuous   **c.** discrete
           d. continuous   **e.** discrete

4.3   **a.** discrete   **b.** discrete   **c.** continuous
           d. continuous   **e.** continuous

4.7   **b.** .70   **c.** .25   **d.** .35

4.9   **b.**

x	$p(x)$
0	1/8
1	3/8
2	3/8
3	1/8

           c. 1/2   **d.** 7/8   **e.** 7/8

4.13 $p(0) = 3/10; p(1) = 6/10; p(2) = 1/10$

4.15   **a.** $p(0) = 1/16; p(1) = 4/16; p(2) = 6/16;$
           $p(3) = 4/16; p(4) = 1/16$
           b. $p(x) = C_x^4(1/2)^4$

4.17   no

4.19   **a.** $\mu = 1.8$

4.21 $E(\text{profit}) = \$500$

4.23 $E(N) = 13,800.39$

4.27 $\sigma^2 = 1$

4.29   **a.** $E(x) = 1.45; \sigma^2 = 1.3475$ **c.** .05; 0

4.31   **a.** $E(x) = \$1666.67$ **b.** \$11,055,555.56
           c. \$8316.65

4.33   **a.** $p(0) = 16/81; p(1) = 32/81; p(2) = 24/81;$
           $p(3) = 8/81; p(4) = 1/81$ **b.** 9/81
           c. $E(x) = 4/3$ **d.** $\sigma^2 = 8/9$

4.35   **a.** 720   **b.** 20   **c.** 1   **d.** 120
           e. 100

4.37 C_{10}^{100}

4.39 C_{15}^{85}

4.41 $p(0) = .3702; p(1) = .4176; p(2) = .1767;$
           $p(3) = .0332; p(4) = .0023$

4.45 $\sigma = .63; 1.00$

4.47 $p(0) = .1056; p(1) = .3185; p(2) = .3604;$
           $p(3) = .1813; p(4) = .0342$

4.49   **b.** .067   **c.** .029

4.51 $\sigma = .9538; 11/12$; yes; no

4.53 $p(1) = .30; p(2) = .60; p(3) = .10; .7$

4.55   \$1.75

4.57   \$12,000

4.59 $p(0) = .064; p(1) = .288; p(2) = .432;$
           $p(3) = .216; P(\text{at least two insured}) = .648$

4.61   **a.** .25   **b.** .1875   **c.** $p(x) = (.75)^{x-1}(.25)$
           for $x = 1, 2, \ldots$ **d.** $P(x > 15) = .0134$

CHAPTER 5

5.1   **a.** binomial
           b. p does not remain constant.
           c. continuous random variable
           d. binomial   **e.** p does not remain constant

5.3   large n

5.5   **a.** $p(0) = 1/16; p(1) = 4/16; p(2) = 6/16;$
           $p(3) = 4/16; p(4) = 1/16$

5.7   **a.** .00007373   **b.** .0886837   **c.** .3125
           d. .387420

5.9   **a.** .965   **b.** .127   **c.** .092   **d.** .035
           e. .873

5.11   **a.** .31744   **b.** .317

5.13   .317; .078

5.15

x	$p(x)$
0	.07776
1	.2592
2	.3456
3	.2304
4	.0768
5	.01024

5.17   **a.** .586   **b.** .034   **c.** suspect $p < .6$

5.19   suspect $p > .01$

5.21   **a.** μ increases and decreases with p.
b. σ^2 maximum at $p = .5$; decreases as the distance between p and .5 increases

5.23   **a.** $\mu = 1.5$; $\sigma = .866$
b. 3/4 of the measurements between .634 and 2.366; all between $-.232$ and 3.232; yes

5.25 $\mu = 2000$; $\sigma^2 = 1600$; at least 3/4 between 1920 and 2080; at least 8/9 between 1880 and 2120

5.27   **a.** $\mu = 40$ **b.** $\sigma = 4.90$ **c.** no

5.29   **a.** Poisson
b. binomial, which can be approximated by Poisson
c. Poisson   **d.** Poisson   **e.** binomial

5.31   **a.** .135335   **b.** .27067   **c.** .593994
d. .036089

5.33   **a.** .677   **b.** .6767

5.35   **a.** .055   **b.** .478

5.37   **a.** .0067   **b.** .1755   **c.** .5595

5.39   .0166

5.43   **a.** .6   **b.** .5143   **c.** .0714

5.45   **a.** $p(0) = 165/455$; $p(1) = 220/455$; $p(2) = 66/455$; $p(3) = 4/455$
c. $\mu = .8$; $\sigma^2 = .50286$
d. .99 fall between $-.618$ and 2.218; .99 fall between -1.327 and 2.927; yes

5.47   **b.** 1/5   **c.** $p(0) = 1/5$; $p(1) = 3/5$; $p(2) = 1/5$

5.49   **a.** .7240   **b.** .2526   **c.** .5041

5.55   **a.** null hypothesis: $p = .7$; alternative hypothesis: $p > .7$
b. x, the number of successes   **c.** $\alpha = .035$
d. do not reject the null hypothesis

5.57   **a.** null: $p = 1/3$; alternative: $p > 1/3$
b. null: $p = 1/2$; alternative: $p \neq 1/2$
c. null: $p = .25$; alternative: $p < .25$

5.59   **a.** null hypothesis: $p = 1/2$; alternative hypothesis: $p \neq 1/2$
b. .03125   **c.** .7378

5.61   **a.** .036   **b.** .352

5.63   yes; reject the null hypothesis with $\alpha = .055$

5.65   **a.** .004   **b.** .891
c. The assumption of independent trials may be violated.

5.67   **a.** .5905   **b.** .9185

5.69   **a.** .5078   **b.** .6564   **c.** .0004   **d.** .0075

5.71 $\mu = 90$; $\sigma^2 = 9$; at least 3/4 between 84 and 96; at least 8/9 between 81 and 99

5.73   **a.** null hypothesis: $p = .1$; alternative hypothesis: $p < .1$ with $p = P(\text{order incorrectly filled})$
b. no; reject the null hypothesis if $x = 0$ ($\alpha = .122$)

5.77   .236

5.79   **a.** .387   **b.** .651   **c.** .264

5.81   **a.** .3277   **b.** .9421

5.83   .3

5.85   null hypothesis: $p = .2$; yes, reject the null hypothesis if $\alpha = .109$; no, do not reject if $\alpha = .047$

5.87   .151

5.89   **a.** .007, larger   **b.** .602, smaller

5.91   **a.** n too small for Poisson approximation to the binomial
b. Poisson   **c.** Poisson
d. p too large for Poisson approximation to the binomial

5.93   **a.** .0067   **b.** .1755   **c.** .2650

5.95   **a.** .1353   **b.** .5940

5.97   .3679; .1353

5.99   **a.** .5457   **b.** $9.9(10^{-6})$ **c.** .4543

5.101   **a.** .9726   **b.** .1608   **c.** yes

5.103   **a.** .0067   **b.** .4405   **c.** .9596   **d.** 5

5.105   1/25

5.107   no; reject $p = .5$ if $x \geq 11$ with $\alpha = .059$

CHAPTER 6

6.3   **a.** .4452   **b.** .1517   **c.** .1548   **d.** .4326
e. .2704   **f.** .6268   **g.** .3125   **h.** .9233

6.5   **a.** .1554   **b.** .2005   **c.** .5328   **d.** .6730
e. .7257   **f.** .0287

6.7   **a.** 1.28   **b.** 1.645   **c.** -1.645
d. 2.575

6.9   **a.** 35th   **b.** 50th   **c.** 71st   **d.** 87th

6.11   .0401

6.13 .0985

6.15 .0192; firm has questionable financial state

6.17 **a.** .5398   **b.** .4286   **c.** .3156; .2063
        d. .2578; .2483

6.19 Poisson approximation

6.21 **a.** .578   **b.** 10; 2.45; .5752

6.23 **a.** .618   **b.** .6156

6.25 **a.** .0228   **b.** .3811

6.27 yes; $z = 2.88$

6.29 963

6.33 **b.** .25   **c.** $\mu = .5$; $\sigma^2 = .0833$

6.35 **a.** 0   **b.** .2212   **c.** .2149   **d.** .6873

6.37 **a.** .25   **b.** .25   **c.** .625
        d. $\mu = 30,000$; $\sigma = 11,547.005$ **e.** .577

6.39 **a.** .135335   **b.** .63212   **c.** .23254

6.41 **a.** .1736   **b.** .2061   **c.** .5318
        d. .1896   **e.** .1896   **f.** .0630

6.43 **a.** .2062; .4092

6.45 85.36 minutes

6.47 293.59

6.49 **a.** .6294   **b.** .8490   **c.** .0446

6.51 **a.** .5557   **b.** .0968   **c.** .5284

6.53 yes; $z = -3.265$

6.55 .9929

6.57 .8980

6.59 263.84 days

6.61 .9474

6.63 .5927

6.65 yes; $z = -2.00$

6.67 **a.** .0537   **b.** \$1162.52

6.69 **a.** 1/2   **b.** 3/4   **c.** 4/10

6.71 .0767 g per gallon

6.73 .025

6.75 **a.** $\mu = 63$; $\sigma = 4.828$ **b.** .9560
        c. .0048

6.77 the serum is effective; $z = 2.53$

6.79 183 books

CHAPTER 7

7.3 **a.** 120   **b.** 1/120

7.5 **d.**

$\bar{x}$	$p(\bar{x})$
4	.1
5	.1
6	.1
7	.2
8	.1
9	.2
10	.1
11	.1

  e. $\mu = 7.6$; 0

7.7 **b.**

s^2	$p(s^2)$
9	.2
21	.3
36	.1
39	.1
57	.2
63	.1

  c. $\sigma^2 = 26.64$; 0

7.11 **a.** $\mu_{\bar{x}} = 50$; $\sigma_{\bar{x}} = 1$ **b.** normal
         c. .0228; .9876

7.15 when $N \geq 20n$

7.17 **a.** .0582   **b.** .0427   **c.** .0033

7.19 **a.** .0571   **b.** .2719   **c.** large or infinite
         population size
         d. normality of sampled population
         e. .1056

7.21 **a.** .0025   **b.** suspect $\mu < 38,000$

7.25 **a.** .0793   **b.** .0082

7.27 **a.** .0985   **b.** .0336

7.29 .0793

7.33 .1212

7.35 **a.** .1314   **b.** .3557

7.37 **a.** .7939   **b.** .7939

7.41 **a.** .1586   **b.** decreases

7.43 .0708

7.45 **a.** 6   **d.** 6.25   **e.** 0

7.47 $\sigma^2 = 12.8$; $E(s^2) = 16$

7.49 **a.** .0951   **b.** .0090

7.51 .0869

7.53　.4681

7.55　approximately 0

7.57　.1922

7.59　.1151; .0228

7.61　.0013

CHAPTER 8

8.7　bound on error increases

8.9　bound on error decreases

8.11　$34.5 \pm .20$

8.13　$2{,}160 \pm 159.38$

8.15　$3{,}200 \pm 98.00$

8.21　**a.** (42.92, 47.08)　**b.** (43.53, 46.47)
　　c. (43.80, 46.20)

8.23　90%: (2.364, 3.042); 95%: (2.299, 3.107)

8.25　(1,099.40, 1,172.60)

8.27　**a.** (16.18, 18.82)
　　b. decrease the confidence coefficient, $1 - \alpha$

8.31　They are not significantly different.

8.33　(−4.60, 2.20)

8.35　$-8.1 \pm .38$

8.37　(2.99, 4.83)

8.39　(−1.01, 4.75)

8.43　$.65 \pm .093$

8.45　**a.** (.409, .491)　**b.** (.401, .499)
　　c. (.386, .514)

8.47　$.3 \pm .06$

8.49　(.54, .72)

8.51　$.72 \pm .038$

8.53　random and independent samples

8.55　**a.** (−.002, .062)　**b.** (−.021, .081)

8.57　**a.** $.284 \pm .122$　**b.** (.182, .386)

8.59　(.07, .15)

8.63　$pq \approx .25$

8.65　**a.** $n = 97$　**b.** $n = 93$

8.67　$n_1 = n_2 = 769$

8.69　$n = 70$

8.73　$n = 115$

8.83　**a.** one-tailed　**b.** $z < -1.645$
　　c. $z = -3.41$　**d.** yes

8.85　**a.** two-tailed　**b.** $|z| > 1.96$
　　c. reject H_0; $z = -3.41$

8.87　do not reject H_0; $z = .154$ (with $\hat{p} = 282/700$)

8.89　yes; $z = -6.52$

8.91　yes; $z = -1.60$

8.93　no; $z = .78$

8.95　yes; $z = 2.86$

8.97　yes; $z = -2.77$

8.101　**a.** .0104　**b.** .001　**c.** .0526
　　d. p-value $< .001$

8.103　p-value $< .001$; reject H_0; yes

8.105　p-value $= .0394$; reject H_0

8.107　.2177

8.109　yes; $z = 1.76$; p-value $= .0392$

8.111　.0104

8.113　.0099

8.115　$11{,}336 \pm 171.01$

8.117　(61.56, 67.44)

8.119　(171.31, 177.59)

8.121　(.801, .999)

8.123　(−55.56, −22.44)

8.125　(.61, .77)

8.127　-17 ± 5.17

8.129　$-.04 \pm .103$

8.131　(1.24, 8.76)

8.133　$n = 385$

8.135　$n = 1{,}068$

8.137　$n = 369$

8.139　$n = 70$

8.141　**a.** $H_0: \mu = 1{,}100$; $H_a: \mu < 1{,}100$
　　b. $z < -1.645$　**c.** yes; $z = -2.69$

8.143　.1342

8.145　yes; $z = -2.40$

8.147　yes; $z = 10.4$

8.149　$z = -.22$; p-value $= .8258$; no significant
　　difference in average yields

8.151　$z = 1.84$; yes

8.153　yes; $z = 7.36$

8.155　p-value $= .0052$; claim is incorrect

8.159　**a.** $\bar{x} = 1.96$; no; 99% confidence interval:
　　(1.82, 2.10)

b. no **c.** reduces width
d. (45,418.08, 52,462.22) **e.** no; $z = 2.89$

CHAPTER 9

9.5 **a.** (8.01, 12.65)
b. do not reject H_0; $t = .321$
c. p-value $> .20$

9.7 yes; $t = -9.01$

9.9 (4.20, 5.00)

9.11 p-value $> .10$

9.13 $t = -1.195$; p-value $> .10$

9.15 $\alpha = .01$

9.21 **a.** $\bar{x}_1 = 2.8$; $\bar{x}_2 = 6.2$; $s_1^2 = s_2^2 = 2.2$;
$s_1 = s_2 = 1.483$
b. $(-5.14, -1.66)$
c. reject H_0; $t = -3.62$
d. p-value $< .005$

9.23 yes; $t = 3.610$

9.25 $.02 < p$-value $< .05$

9.27 **a.** yes; $t = 6.035$
b. p-value $< .005$

9.29 no; $t = -1.788$

9.31 **a.** yes; $t = -2.657$
b. $(-7.588, -1.496)$

9.33 p-value $= .20$

9.35 $(-102.463, -23.537)$

9.39 no

9.41 **b.** $\bar{d} = 1.157$; $s_d^2 = .37619$
c. $H_0: \mu_1 = \mu_2$; $H_a: \mu_1 \neq \mu_2$
d. $t = 4.99$; p-value $< .01$
e. reject H_0

9.43 no; $t = 1.874$

9.45 (3.67, 34.01)

9.47 reduce unexplained variation

9.49 **a.** lower α **b.** higher α

9.53 **a.** 35.1725 **b.** 20.0902 **c.** 18.4926
d. 4.10691 **e.** 45.7222 **f.** 9.59083

9.55 **a.** 19.6751 **b.** no **c.** p-value $> .10$

9.57 no for $\alpha \leq .10$; $\chi^2 = 9.086$

9.59 yes; $\chi^2 = 14.8$

9.63 area in the upper tail of the rejection region is $\alpha/2$

9.65 **b.** $F = 3.33$ **c.** no; $F = 1.881$
d. p-value $> .20$

9.67 yes; $F = 3.28$

9.69 yes; $F = 2.486$

9.71 (2.12, 5.25)

9.73 no; $t = -1.341$

9.75 31, using $t_{.05} \approx z_{.05} = 1.645$

9.77 yes; $t = 7.5$

9.79 p-value $> .10$

9.81 yes; $F = 2.865$; $.01 < p$-value $< .025$

9.83 do not reject H_0; $t = .33$; p-value $> .20$

9.85 $n = 19$

9.87 yes; $t = -1.775$

9.89 **a.** no; $t = -1.71$ **b.** $(-1.52, .02)$

9.91 $(-1.84, .09)$

9.93 $.02 < p$-value $< .05$

9.95 (3515.42, 6118.58)

9.97 **a.** $(-2.013, .493)$ **b.** (8.635, 10.485)

9.99 **a.** (428.08, 447.92) **b.** (554.28, 1457.50)

9.101 (.00896, .05814); (.3786, .9645)

9.103 no; $\chi^2 = 7.008$

9.105 no; $F = 2.39$

9.107 yes; $F = 2.836$

9.109 do not reject H_0; $t = -.10$

CHAPTER 10

10.3 **a.** $\bar{x}_1 = 5.00$; $\bar{x}_2 = 8.62$; $s_1^2 = 3.4822$;
$s_2^2 = 7.9884$
b. reject H_0, $t = -3.38$
c. reject H_0, $F = 11.42$

10.5 reject $H_0: \mu_1 - \mu_2 = 0$; $t = -2.94$, $F = 8.65$

10.9 **a.**

Source	d.f.	SS	MS	F
Treatments	3	38.82	12.94	9.72
Error	16	21.29	1.33	
Total	19	60.11		

b. reject H_0; $F = 9.72$ **c.** p-value $< .005$
d. $\bar{x}_1 = 5.34$, $\bar{x}_2 = 7.72$, $\bar{x}_3 = 8.56$, $\bar{x}_4 = 5.50$;
$(-3.93, -0.83)$

10.11 **a.** $.01 < p\text{-value} < .025$
c. $(29.955, 41.945)$ **d.** $(-3.499, 19.949)$; no

10.13 **a.** 37.00 ± 9.46 **b.** 45.00 ± 9.46
c. -8.00 ± 13.37

10.15 **a.** yes; $F = 36.52$ **b.** -7.62 ± 2.40

10.17 **b.** $(-11.968, 4.368)$

10.21 **a.** 3

b.

Source	d.f.	SS	MS	F
Treatments	2	12.52	6.26	2.14
Blocks	3	32.43	10.81	3.69
Error	6	17.59	2.93167	
Total	11	62.54		

c. no; $F = 2.14$
d. $.05 < p\text{-value} < .10$

10.23 **a.** no; $F = 2.84$ **b.** no; $F = 3.10$
c. 2.91 ± 2.29

10.25 2.875 ± 2.371

10.27

Source	d.f.	SS	MS	F
Premiums	3	406.25	135.4167	21.10
Banks	2	8.17	4.0833	.64
Error	6	38.50	6.4167	
Total	11	452.92		

a. yes; $F = 21.10$ **b.** 11.67 ± 5.06
c. no; $F = .64$

10.31 **a.**

Source	d.f.	SS	MS	F
A	1	31.25	31.25	8.74
B	1	11.25	11.25	3.15
AB	1	1.25	1.25	.35
Error	16	57.20	3.58	
Total	19	100.95		

b. do not reject H_0; $F = .35$
c. factor A is significant, $F = 8.74$; factor B is not significant, $F = 3.15$
e. $(-4.29, -.71)$

10.33 **a.** significant interaction, $F = 10.84$
b. A is not significant, $F = 1.05$; B is significant, $F = 10.10$; no
c. yes   **d.** 11.82

10.35

Source	d.f.	SS	MS	F
Advertising Level	2	40.062	20.031	71.54
Type of Appeal	1	13.230	13.230	47.25
Interaction	2	25.385	12.693	45.33
Error	6	1.680	0.280	
Total	11	80.357		

Since the interaction is significant, we focus our attention on the differences between mean scores at each of the six advertising levels and appeal type combinations.

10.37 **a.** yes
b. There is a significant interaction between price and number of package features.
c. price, $F = 176.35$, $p\text{-value} < .005$; package features, $F = 56.65$, $p\text{-value} < .005$; interaction, $F = 57.90$, $p\text{-value} < .005$

10.39 **a.** 225.00 ± 5.37 **b.** 8.33 ± 9.41

10.41 **a.**

Source	d.f.	SS	MS	F
Makes	2	6.4188	3.20939	5.15
Error	8	4.9867	.62333	
Total	10	11.4055		

b. yes; $F = 5.15$

10.43 **a.** yes; $F = 5.70$ **b.** 5.833 ± 5.411
c. 60.5 ± 3.543 **d.** yes

10.45 **a.** 48.6 ± 4.752 **b.** 56.2 ± 4.752

10.47 **a.** yes; $F = 18.61$ **b.** yes; $F = 7.11$
c. 2.067 ± 1.379

10.49 **a.** randomized block design

b.

Source	d.f.	SS	MS
Treatments	3	198.34375	66.11458
Blocks	7	106.96875	15.28125
Error	21	58.90625	2.80506
Total	31	364.21875	

c. yes; $F = 23.57$

10.51 **b.** yes; $F = 7.50$ **c.** 298.5 ± 172.5

10.53 **a.** difference exists in mean recognition times; $F = 11.67$

b. difference exists between layouts A and D; $t = -2.73$

10.57 Data coded as $x/1000$; F ratios remain invariant:

Source	d.f.	SS	MS	F
Dwellings	4	584.27	146.07	123.79
Models	2	18.90	9.45	8.01
Error	8	9.43	1.18	
Total	14	612.60		

CHAPTER 11

11.1 Probabilistic models have a random error component, while deterministic models do not.

11.3 **a.** 5; -4

11.5 **a.** 6; 3

11.9 **a.** $\hat{y} = 3 - x$

11.11 **a.** $\hat{y} = 6.611 - .292x$

11.13 **b.** $\hat{y} = -1690.400 + .737x$ **d.** $\hat{y} = 520.54$

11.17 SSE $= 1.1$; $s^2 = .367$

11.19 SSE $= 13.7353$; $s^2 = 2.747$

11.21 SSE $= 48.1806$; $s^2 = 6.883$

11.23 SSE $= 67,954.866$; $s^2 = 13,590.973$

11.27 $.7 \pm .609$

11.29 -1 ± 1.192

11.31 **a.** no; $t = -.55$ **b.** yes; $t = 2.95$
c. no; $t = 2.34$

11.33 **a.** $\hat{y} = -.973 + .9543x$ **c.** reject H_0;
$t = 6.33$

11.35 $(.216, 1.148)$

11.39 $(.138, 1.862)$

11.41 $(1.955, 7.135)$

11.43 12.559 ± 1.226

11.45 **a.** $\hat{y} = 1.973 + .207t$
b. The relationship between y and t must continue in the future as it has in the past.
c. $4.250 \pm .487$

11.49 $(-1.111, 3.111)$

11.51 $(.475, 9.703)$

11.53 **a.** $\hat{y} = -3.2328 + 2.20337x$
b. 51.851 ± 1.352 **c.** 51.851 ± 3.856

11.55 520.54 ± 330.34

11.57 **a.** $\hat{y} = 307.917 + 34.583x$
c. $s^2 = 2666.242$ **d.** 515.42 ± 33.21
e. 515.42 ± 119.74

11.59 no

11.63 **a.** $.8836$ **b.** yes; $t = 9.93$

11.65 $r = -.9428$; yes; $t = -4.90$

11.67 **a.** $-.8745$ **b.** yes; $t = -4.416$

11.69 $r^2 = .764$

11.71 **a.** $\hat{y} = 3.4254 - .15249x$ **b.** $-.918$
c. 84.2% **d.** $1.824 \pm .354$

11.73 $80,536.8 \pm 943.5$

11.75 **a.** $\hat{y} = 83.074 - 1.185x$ **c.** $s^2 = 11.5137$
d. yes; $t = -9.42$

11.77 **a.** $\hat{y} = 46.000 - .317x$ **c.** $s^2 = 19.033$

11.79 30.167 ± 2.500

11.81 $r = -.8716$; 75.97%

11.83 $r = .8214$; 67.46%

11.85 79.507 ± 9.932

11.87 **a.** $\hat{y} = 53,088.710 - 131.226x$
b. $20,938.299 \pm 7706.901$
c. a decrease over time; $t = -9.09$

11.89 $\hat{y} = -0.278 + 1.0379x$; reject $H_0: \beta_1 = 0$;
$t = 3.26$ (risky investment)

CHAPTER 12

12.3 **a.** $\hat{y} = 16.38 + 1.42x_1 - 2.85x_2 + .46x_3$
b. $\hat{y} = 13.413$ **c.** no; $t = 1.147$
d. p-value $> .20$ **e.** $(-3.91, -1.79)$

12.5 $\hat{y} = 36.54597 + .80106x_1 - 5.30878x_2 + 2.28814x_3$

12.7 80.106 ± 88.77

12.9 do not reject H_0; $F = 3.734$

12.11 **a.** $\hat{y} = -2.491 + .099x_1 + .029x_2 + .086x_3$
b. x_2 (elevation) and x_3 (slope)

12.13 $(-.002, .200)$

12.17 **a.** 91.5%, $R_a^2 = 89.0\%$
b. regression is significant with p-value $< .005$;
$E = 37.56$

12.19 **a.** $F = 83.7058$ **b.** $.9654$
c. $\hat{y} = -51.034 + 57.600x_1 + 2.956x_2 - .934x_3 - 46.542x_4 + 46.355x_5 - 1.805x_6$

d. x_5 (advertising expenditure); x_3 (utilities index)   **f.** $(-3.149, -.461)$

12.21 (1) $106,438   (2) $78,890   (3) $97,382
(4) $141,936   (5) $108,850

12.23 **a.** yes; $F = 61.991$ **b.** x_1, x_3 and x_5

12.27 $y = \beta_0 + \beta_1 x_1 + \beta_2 x_2 + \beta_3 x_3 + \beta_4 x_4 + \varepsilon$
where

$$x_3 = \begin{cases} 1 & \text{if level 2 of qualitative variable} \\ 0 & \text{if not} \end{cases}$$

$$x_4 = \begin{cases} 1 & \text{if level 3} \\ 0 & \text{if not} \end{cases}$$

12.29 **a.** $y = \beta_0 + \beta_1 x_1 + \beta_2 x_2 + \beta_3 x_3 + \beta_4 x_4 + \beta_5 x_5 + \varepsilon$
b. x_1 = agent's age
x_2 = agent's experience in years

$$x_3 = \begin{cases} 1 & \text{if Southwest} \\ 0 & \text{if not} \end{cases}$$

$$x_4 = \begin{cases} 1 & \text{if Northeast} \\ 0 & \text{if not} \end{cases}$$

$$x_5 = \begin{cases} 1 & \text{if Southeast} \\ 0 & \text{if not} \end{cases}$$

12.35 **a.** $H_0: \beta_4 = 0$ versus $H_a: \beta_4 \neq 0$
b. There is no evidence to indicate a quadratic relationship between time and profits.
c. The director will tend to overestimate or underestimate profits by an amount β_2 if he decides not to include this term in his model.
d. There is a difference in profits between large and small appliances in the presence of x_1, time since initial market introduction.

12.39 **a.** $x_2 = 1: \hat{y} = 1.83 + .83x_1$; $x_2 = 2: \hat{y} = 0.19 + 1.46x_1$; $x_2 = 3: \hat{y} = -1.45 + 2.09x_1$
c. straight lines; yes

12.41 **c.** yes; $F = 15.104$
e. $\hat{y} = 26.66$; $(14.39, 38.93)$

12.43 **a.** -3.3183 **b.** $-.1155; 3.5493$

12.47 If city differences are additive, use the dummy variable

$$x_5 = \begin{cases} 0 & \text{if residence is in Eugene} \\ 1 & \text{if residence is in Springfield} \end{cases}$$

and test $H_0: \beta_5 = 0$ versus $H_a: \beta_5 \neq 0$

12.49 **a.** no significant interaction; $t = 0.28$
b. $F = 0.08$; no interaction

12.51 **b.** $\hat{y} = 16.8166 + 1.0935x_1$; $SSE_1 = 5.7848$
c. $F = 2.380$; do not reject H_0

12.59 **a.** interest rates; largest F value
b. reject H_0; $t = -2.09$ **c.** yes

12.61 yes

12.63 **a.** .6816
b. $\hat{y} = -1.25272 + .42721x_1 + 1.89678x_2 + 2.23333x_3 + 1.11673x_4$
c. x_2 (number of retail outlets); x_1 (promotional budget)   **d.** $1896.78
12.65 $\hat{y} = 1.7125 - .1438x + .0135x^2$; reject $H_0: \beta_1 = 0$ $(t = -3.02)$; reject $H_0: \beta_2 = 0$ $(t = 3.30)$; $R^2 = .5732$

12.67 yes; $F = 7.177$; $R^2 = .6146$

12.69 do not reject $H_0: \beta_3 = \beta_4 = \beta_5 = 0$; $F = .4114$

12.73 $\hat{y} = -10.77 + 1.42x_1 + 15.598x_2 - .028x_1^2 - 3.386x_2^2 + .496x_1 x_2$; $F = 216.213$; $R^2 = .9827$

CHAPTER 13

13.3 **a.** long-term, cyclic, random variation
b. long-term, random variation
c. seasonal, random variation
d. cyclic, seasonal, random variation
e. cyclic, random variation
f. cyclic, seasonal, random variation

13.5 **b.** long-term trend, seasonal variation, random variation

13.7 long-term, cyclic, random variation

13.15 **a.** $S_t = .1y_t + .9S_{t-1}$
b. $S_t = .3y_t + .7S_{t-1}$
c. $S_t = .5y_t + .5S_{t-1}$

13.17 **b.** long-term trend, seasonal effect (peaks in winter and late summer); random variation; 1988 seems to indicate the beginning of a downward cyclical trend

13.23 long-term trend, seasonal component, random variation

13.25 long-term trend, seasonal component, random variation

13.31 **a.**

	I_2	I_3
1	105.4	103.6
2	103.4	97.8
3	102.8	106.7
4	102.3	102.3
5	106.3	105.7

b. $I_2 = 105.3; I_3 = 103.16$

13.33 Percentage gains over 5-year periods are shown in the first table.

Year	Housing	Food	Apparel, Upkeep	Transportation	Health, Recreation
1970–1975	22.9	36.1	6.8	16.3	6.9
1975–1980	59.3	34.9	17.8	48.2	61.2
1980–1985	24.6	25.8	20.0	35.4	18.7

	Estimated Expenditures				
Year	Housing	Food	Apparel, Upkeep	Transportation	Health, Recreation
1975	$4,056	$3,238	$1,004	$1,338	$1,571
1980	6,461	4,369	1,183	1,982	2,532
1985	8,053	5,495	1,419	2,683	3,006

13.35 **a.**

	1960	1965	1970	1975	1980	1985
Milk	112	116	151	177	256	294
Flour	105	115	132	190	243	224
Hamburger	132	139	200	226	446	357
Coffee	80	76	82	131	368	266
Potatoes	143	236	581	436	540	519

b. 100; 99; 103; 135; 171; 368; 295

13.37 **a.** Chicago: 357.19; Detroit: 346.05; LA: 360.19; NY: 361.25; Philadelphia: 358.08
b. 4.39%; .89%   **c.** 4.55%

13.43 depends on base prices in each city in 1967 and the relative importance of each CPI component

13.45 **a.** $I_{80-81} = 123.33; I_{81-82} = 96.53;$ $I_{82-83} = 84.07; I_{83-84} = 87.54;$ $I_{84-85} = 82.00$
b. $I_{84-85} = 81.89$

13.47 **a.** 149, 220, 276   **b.** 149, 221, 279
c. Paasche index

13.49 **a.** long-term, cyclic, seasonal, random variation
b. long-term, cyclic, random variation
c. long-term, random variation
d. seasonal, random variation
e. long-term, seasonal, random variation
f. long-term, cyclic, seasonal, random variation

CHAPTER 14

14.5   **c.** $MSE_a = .008157$; $MSE_b = .008182$; virtually no difference

14.11   **a.** $\hat{y} = 15.7 + .0123t + .00223t^2$
     b. $MAD = .2885$ **c.** independent errors

14.15   **b.** $MSE_{lin} = 16,662.22$; $MSE_{quad} = 3955.43$
     c. linear model: $\hat{y}_9 = 800.03$, $\hat{y}_{10} = 866.70$; quadratic model: $\hat{y}_9 = 1198.03$, $\hat{y}_{10} = 1515.18$

14.17   **b.** $Sales_{96} = 277,478.2$
     c. $Sales_{97} = 515,724.6$

14.21   **c.** $MAD_a = 2.833$, $MAD_b = 4.444$; the one-step forecast is more accurate

14.23   **a.** 76.33   **b.** 90.33   **c.** 76.33

14.25   **b.**

t	y_t	$\hat{y}_t$ (first diff.)	$\hat{y}_t$ (second diff.)
25	24.7	26.67	33.03
26	24.4	25.00	20.03
27	26.0	23.97	23.37
28	23.2	25.90	27.93
29	22.8	22.70	20.00
30	24.3	22.27	22.37
31	27.4	23.73	25.77
32	28.6	28.80	32.47
33	28.8	30.53	30.33
34	25.1	30.30	28.57
35	29.3	24.33	19.13
36	31.4	29.53	34.50
37	28.3	32.27	34.13
38	27.5	29.37	25.40
39	28.8	26.90	25.03
40	22.7	27.93	29.83
41	19.6	20.83	15.60
42	20.3	16.97	15.73
43	20.7	17.47	20.80
44	21.4	20.03	23.27
45	22.6	22.00	23.37
46	28.3	23.37	23.97
47	27.5	30.83	35.77
48	28.1	29.53	26.00

     c. $MSE_1 = 672.75$; $MSE_2 = 691.08$; first differencing provides a more accurate forecast

14.31   **a.** $\hat{y}_{t+T} = .1y_t + .9S_{t-1}$
     b. $\hat{y}_{t+T} = .5y_t + .5S_{t-1}$
     c. $\hat{y}_{t+T} = .8y_t + .2S_{t-1}$

14.33   **a.** $\hat{y}_{t+1} = 3.3457S_t - 3.58025S_t(2) + 1.23457S_t(3)$

     b. $\hat{y}_{t+1} = 4.46939S_t - 5.5102S_t(2) + 2.0408S_t(3)$
     c. $\hat{y}_{t+1} = 9.75S_t - 15.00S_t(2) + 6.25S_t(3)$

14.35   **a.** triple-smoothed forecasts

14.43   when the inherent pattern of the time series is not readily apparent

14.47   **a.** $\nabla \hat{y}_t = .07931 + .1980 (y_{t-1} - y_{t-2})$
     b. $\hat{y}_{23} = 17.15851$ **c.** $\hat{y}_{24} = 17.2692$

14.55   **a.** $\hat{y}_9 = 86$, $\hat{y}_{10} = 107.67$; $MSE = 4613.26$
     b. $\hat{y}_9 = 133.33$, $\hat{y}_{10} = 147.00$, $MSE = 1178.86$
     c. second differencing provides a more accurate forecast

14.57 $MSE = 12.63$

14.61   **a.** $\nabla^2 \hat{y}_t = 2.792 + .624 \nabla^2 y_{t-1}$
     c. 149.67 versus 149.062.

14.69   **a.** EWMA   **b.** multiple linear regression
     c. multiple exponential smoothing model
     d. EWMA   **e.** multiple linear regression or multiple exponential smoothing

14.75 $\bar{p} = .4055$; forecast $N(\bar{p}) = 835.33$

CHAPTER 15

15.11   **a.** $LCL = 3.37$; $UCL = 26.83$
     c. process is in control

15.13   **a.** $LCL = .002$; $UCL = .028$
     b. $LCL = .6347$; $UCL = .6453$
     c. $LCL = .6344$; $UCL = .6456$ **d.** yes

15.15   **a.** $\bar{x}$ chart: $LCL = 4.81$; $UCL = 5.59$;
     R chart: $LCL = 0$; $UCL = 1.6032$ **c.** yes

15.17   **b.** $LCL = .001698$; $UCL = .004462$
     c. yes   **d.** out of control; no

15.23 $C_{pk} = .503$; process is not on center

15.25 $C_p = .389$, $C_{pk} = .333$; no

15.29   **a.** $LCL = 0$; $UCL = .115$

15.31 $LCL = .022$; $UCL = .098$

15.33   **a.** $\bar{c} = 3.53$ **b.** $LCL = 0$; $UCL = 9.17$

15.35 $LCL = 2.54$; $UCL = 24.66$

15.39   **a.** .919   **b.** .737   **c.** .337   **d.** 1.000
     e. .000

15.41   **a.** .736   **b.** .376   **c.** .046   **d.** 1.000
     e. .000

15.43   **b.** $n = 25$, $a = 1$

15.45 **a.** $a = 2$ (review the production line if 3 or more automobiles fail)

15.47 **a.** LCL $= 29.0$; UCL $= 32.6$
b. yes; ranges and $\bar{R}$

15.49 **a.** LCL $= 0$; UCL $= .0016$
b. defectives are increasing; process should be carefully monitored

15.51 **a.** $n = 25; a = 5$ **b.** $n = 25; a = 5$

15.53 **a.** .829; .599   **b.** .549; .349
c. .167; .107   **d.** .005; .006   **e.** 0; 0

15.55 **a.** LCL $= 0$; UCL $= .0848$ **b.** .0848

15.57 yes

15.59 **a.** $\bar{x}$ chart: LCL $= 36.35$, UCL $= 63.51$;
R chart: LCL $= 0$, UCL $= 49.77$
b. yes   **c.** no

CHAPTER 16

16.15 associate two random numbers between 000 and 999 with each population element

16.21 **a.** .065 $\pm$ .004   **b.** 143 $\pm$ 8.55

16.23 .492 $\pm$ .122

16.25 2.4 $\pm$ .176

16.27 6.8075 $\pm$ .970

16.35 **a.** $n_1 = n_2 = 23; n_3 = n_4 = 27$
b. $n_1 = 14; n_2 = 29; n_3 = 57$
c. $n_1 = n_2 = n_3 = 10; n_4 = 30; n_5 = 40$

16.37 .715 $\pm$ .051

16.39 **a.** 7.963 $\pm$ 3.075   **b.** 21,500.1 $\pm$ 8302.37

16.41 **a.** .36 $\pm$ .1183   **b.** .5 $\pm$ .2836

16.47 **a.** 1840   **b.** 28,480   **c.** 1840 $\pm$ 330.77

16.49 1925 $\pm$ 232.295

16.53 .709 $\pm$ .047

16.57 **a.** $n_1 = 61, n_2 = 77; n_3 = 46$
b. $n_1 = 18; n_2 = 22; n_3 = 13$

16.59 74, with $n_1 = 37; n_2 = 25; n_3 = 12$

16.61 70, with $n_1 = n_2 = 21; n_3 = 18; n_4 = 11$

16.63 5 cars: 2 Fords, 2 Chevrolets, 1 Plymouth

16.71 use multistage cluster sampling

16.75 **a.** nonrandom   **b.** random
c. nonrandom   **d.** random   **e.** random

16.79 100 $\pm$ 30.67

16.81 13.05 $\pm$ 1.603

16.83 **a.** 31.9 $\pm$ 1.913   **b.** 7975.0 $\pm$ 478.209

16.85 50,505.593 $\pm$ 8489.862

16.87 9686 $\pm$ 3019.012

16.89 164

16.91 57.5 $\pm$ 3.65

16.93 .9952

CHAPTER 17

17.5 **a.** $X^2 > 18.4753$ **b.** $X^2 > 22.3621$
c. $X^2 > 15.5073$ **d.** $X^2 > 4.60517$
e. $X^2 > 12.8325$

17.7 no; $X^2 = 6.96$

17.9 yes; $X^2 = 95.06$

17.11 yes; $X^2 = 11.58$

17.15 **a.** 8   **b.** 3   **c.** 30   **d.** 4

17.17 **a.** p-value $> .10$ **b.** $.01 < p$-value $< .025$
c. p-value $> .10$ **d.** p-value $< .005$

17.19 **d.** 6   **e.** $X^2 > 12.5916$
f. $X^2 = 7.93$, do not reject H_0

17.21 yes; $X^2 = 20.91$

17.23 no; $X^2 = 1.34$

17.25 yes; $X^2 = 79.98$

17.27 no

17.29 no; $X^2 = 3.04$

17.31 reject H_0, $z = 2.40$; yes

17.33 **a.** yes; $X^2 = 10.27$ **b.** yes; $z = 3.205$

17.35 manufacturing versus retail: $X^2 = 4.67$;
retail versus utility: $X^2 = 1.31$; manufacturing
versus utility: $X^2 = 9.79$; with $\chi^2_{.05} = 5.99$,
the difference lies only between manufacturing
and utility

17.37 yes; $X^2 = 10.16$

17.39 yes; $X^2 = 24.48$

17.41 no; $X^2 = 4.40$

17.43 no; $X^2 = 1.65$

17.45 no; $X^2 = 4.70$

17.47 yes; $X^2 = 11.63$

17.49 **a.** yes; $X^2 = 9.71$ **b.** by observation

17.51 yes; $X^2 = 43.56$

17.53 yes; $X^2 = 6.49$

17.55 no; $X^2 = 12.915$

17.57   yes; $X^2 = 21.06$

17.59   no; $X^2 = 6.70$

17.61   no; $X^2 = 8.61$

17.63   no; $X^2 = 62.59$

CHAPTER 18

18.3   binomial; normal (approximate)

18.5   rejection region (abbreviated RR): $x \le 1$ or $x \ge 9$ with $\alpha = .022$; RR: $x \le 2$ or $x \ge 8$ with $\alpha = .110$

18.7   **a.** $H_0\!:p = .5$, $H_a\!:p \ne .5$; $z = (x - .5n)/.5\sqrt{n}$; reject H_0 if $|z| > 1.96$ **b.** reject H_0, $z = 2.2$

18.9   no; $x = 7$; RR: $x \ge 9$ with $\alpha = .073$

18.11   no; $x = 11$; RR: $x \le 8$ with $\alpha = .054$

18.13   no; $x = 3$; RR: $x \le 2$ with $\alpha = .055$

18.17   **a.** $U_A \le 13$ with $\alpha = .0469$
   b. $U_A \le 9$ with $\alpha = .0112$
   c. use $z = (U_A - \mu_U)/\sigma u$ and rejection region $z < -1.645$

18.19   **a.** .052   **b.** .0974

18.21 $z = -4.75$, p-value $\le .002$; there is a significant difference between the two population distributions.

18.23   yes; $U = 4.5$; RR: $U \le 7$ with $\alpha = .106$; community A has lower borrowing rates.

18.25   There is no difference; $z = -1.47$; RR: $|z| \ge 1.96$ with $\alpha = .05$

18.29   **b.** $H = 10.246$ **c.** $.01 < p$-value $< .025$ **d.** reject H_0

18.31   yes; $H = 9.83$

18.33 $H = 8.06$; the rates of return are significantly different.

18.37   **b.** $T = 7$ **c.** no; RR: $T \le 6$ with $\alpha \approx .05$

18.39   **a.** reject H_0, $T = 1$; RR: $T \le 2$ with $\alpha \approx .05$

18.41   **a.** yes; $T = 5$; RR: $T \le 11$ with $\alpha = .10$
   b. no; $x = 5$; RR: $x \le 2$ or $x \ge 7$ with $\alpha = .1797$

18.45   **a.** $X^2 > 11.1433$ **b.** $X^2 > 4.60517$
   c. $X^2 > 15.0863$

18.47 $X^2 = .333$ with p-value $> .10$; no significant difference

18.49   yes; $X^2 = 9.72$

18.51   yes; $X^2 = 7.714$

18.55   **a.** $R \le 3$ or $R \ge 9$ with $\alpha = .095$
   b. $R \le 5$ or $R \ge 13$ with $\alpha = .060$
   c. $R \le 3$ or $R \ge 10$ with $\alpha = .016$

18.57   reject H_0, $z = -4.61$; sequence is nonrandom

18.59 $\bar{x} = 3.44$; $R = 7$; RR: $R \le 4$ or $R = 11$ with $\alpha = .071$; no evidence of lack of randomness

18.61   no; $R = 11$ and $z = .046$; RR: $|z| > 1.96$ with $\alpha = .05$

18.63   **a.** false   **b.** true   **c.** false   **d.** false
   e. true

18.65   no, $r_s = .482$; RR: $r_s \ge .564$

18.67   **a.** $r = .9892$ **b.** yes; $t = 21.35$; RR: $|t| > 3.169$ with $\alpha = .01$ **c.** $r_s = 1$
   d. yes; $r_s = 1$; RR: $|r_s| \ge .780$

18.69   no; $r_s = .728$

18.71   **a.** no; $x = 2$; RR: $x \le 1$ with $\alpha = .011$
   b. no; $T = 13$; RR: $T \le 5$ with $\alpha = .01$

18.73   no; $x = 2$; RR: $x \le 1$ or $x \ge 8$ with $\alpha = .0391$

18.75   **a.** U.S. versus Canada: $T = 20.5$; U.S. versus U.K.: $T = 26.5$; Canada versus U.K.: $T = 19.5$; none show significant difference
   b. $X^2 = 2.375$; no significant difference

18.77   **a.** equal usable life; $x = 4$; RR: $x \le 2$ or $x \ge 10$ with $\alpha = .0386$
   b. equal usable life; $x = 4$; RR: $x \le 3$ with $\alpha = .073$

18.79   yes; $U = 3$; RR: $U \le 9$ with $\alpha = .0592$

18.81   no; $X^2 = .8$

18.83   **a.** $P(R \le 11) = .0256$ **b.** yes; $z = -1.95$

18.85 $r_s = -.845$; RR: $r_s \le -.564$ with $\alpha = .05$; yes

18.87   reject H_0; $z = (x - .5n)/.5\sqrt{n} = 2.97$; GM cars are preferred

18.89 $r_s = -.504$; yes; RR: $r_s \le -.497$ with $\alpha = .05$

18.91   yes; $r_s = .849$; RR: $r_s \ge .343$ with $\alpha = .05$ (one-tailed)

18.93   **a.** rank correlation coefficient   **b.** runs test
   c. Mann–Whitney U test
   d. Wilcoxon signed rank
   e. Wilcoxon signed rank

18.95   using sign test, do not reject H_0, $x = 8$; RR: $x \ge 9$ with $\alpha = .073$

18.97   using runs test, do not reject H_0; $R = 5$; RR: $R = 2$ or $R = 7$ with $\alpha = .107$

CHAPTER 19

19.7 **b.**

	s_1	s_2	s_3	s_4
a_1	0	10	5	0
a_2	5	5	0	5
a_3	0	0	10	10

19.9 **a.** a_1: agree; a_2: not agree; s_1: downward float; s_2: no change; s_3: upward float

b.

	s_1	s_2	s_3
a_1	21,000	6,000	$-1,500$
a_2	0	0	0

c.

	s_1	s_2	s_3
a_1	0	0	1,500
a_2	21,000	6,000	0

19.11

Action	s_1 (750,000)	s_2 (150,000)	s_3 (0)
a_1: drill alone	$9,400,000	$400,000	$-1,500,000
a_2: syndication	4,700,000	200,000	$-750,000
a_3: sell	100,000	100,000	100,000

19.15 yes; $4,500

19.17 speculative restaurant; $E(\text{gain} \mid \text{restaurant}) = \$9{,}375$; $E(\text{gain} \mid \text{savings account}) = \$1{,}750$

19.19 sell the lease; $E(G) = \$100{,}000$

19.23 4

19.25 **a.** current employees working overtime
b. $782,000

19.29 **a.**

Inventory	Demand				
	500	1,000	1,500	2,000	2,500
500	0	175	350	525	700
1,000	250	0	175	350	525
1,500	500	250	0	175	350
2,000	750	500	250	0	175
2,500	1,000	750	500	250	0

b. publish 1500   **c.** $178.75

19.31 **a.**

	s_1 (8%)	s_2 (8.5%)	s_3 (9%)	s_4 (9.5%)	s_5 (10%)
a_1 (9%)	100	50	0	0	0
a_2 (variable)	0	0	0	50	100

b. variable-rate loan   **c.** up to $10.00

19.33 $17.75; the maximum amount he would be willing to pay to learn the true demand for his pellets

19.37 **a.** $P(s_1 \mid x) = .121$, $P(s_2 \mid x) = .069$, $P(s_3 \mid x) = .345$, $P(s_4 \mid x) = .466$ **b.** choose a_1
c. EVSI $= 1.585$

19.39 **a.** maintain present facilities; expand present facilities; add a remote facility
b. $331,576.90; yes

19.41 method B; posterior probabilities: .121, .611, .268

19.45 **a.** choose a_2 **b.** no decision   **c.** no

19.47 **a.** 6000   **b.** accept offer   **c.** 3 tons
d. grocery store   **e.** method B

19.57 **a.**

	Profits		
	s_1 (Good)	s_2 (Fair)	s_3 (Poor)
a_1 (New)	30,000	10,000	−4,000
a_2 (Old)	11,000	11,000	11,000

b. yes   **c.** $5,125
d. yes, $E(G_1) = \$23,112$; no, $E(G_1) = \$6,418$
e. EVSI = $2,514.78; do not hire survey

19.59 **a.** .2, .5, .3   **c.** $65,500   **d.** 5,000

19.63 **a.** .013, .237, .578, .173   **b.** di-essolan
c. $15.57

19.65 use di-essolan if two or more defectives are found

19.67 foreign supplier; expected savings = $16,775

19.69 **a.**

	Demand			
	s_1 (5,000)	s_2 (10,000)	s_3 (25,000)	s_4 (50,000)
a_1 (Buy New)	0	7,000	28,000	63,000
a_2 (Fix Old)	4,000	8,500	22,000	44,500

b. buy new machine   **c.** $17,500

19.71 no; expected profit: −$250,000

19.49 **a.** do not invest (.53 versus .55)
b. invest (.59 versus .55)

19.51 stock 11 magazines

19.53 yes; EVPI = $2,250

19.55 choose new machine

INDEX

Table of Normal Curve Areas

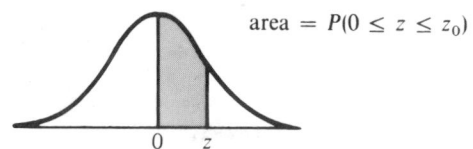

area $= P(0 \le z \le z_0)$

z_0	.00	.01	.02	.03	.04	.05	.06	.07	.08	.09
0.0	.0000	.0040	.0080	.0120	.0160	.0199	.0239	.0279	.0319	.0359
0.1	.0398	.0438	.0478	.0517	.0557	.0596	.0636	.0675	.0714	.0753
0.2	.0793	.0832	.0871	.0910	.0948	.0987	.1026	.1064	.1103	.1141
0.3	.1179	.1217	.1255	.1293	.1331	.1368	.1406	.1443	.1480	.1517
0.4	.1554	.1591	.1628	.1664	.1700	.1736	.1772	.1808	.1844	.1879
0.5	.1915	.1950	.1985	.2019	.2054	.2088	.2123	.2157	.2190	.2224
0.6	.2257	.2291	.2324	.2357	.2389	.2422	.2454	.2486	.2517	.2549
0.7	.2580	.2611	.2642	.2673	.2704	.2734	.2764	.2794	.2823	.2852
0.8	.2881	.2910	.2939	.2967	.2995	.3023	.3051	.3078	.3106	.3133
0.9	.3159	.3186	.3212	.3238	.3264	.3289	.3315	.3340	.3365	.3389
1.0	.3413	.3438	.3461	.3485	.3508	.3531	.3554	.3577	.3599	.3621
1.1	.3643	.3665	.3686	.3708	.3729	.3749	.3770	.3790	.3810	.3830
1.2	.3849	.3869	.3888	.3907	.3925	.3944	.3962	.3980	.3997	.4015
1.3	.4032	.4049	.4066	.4082	.4099	.4115	.4131	.4147	.4162	.4177
1.4	.4192	.4207	.4222	.4236	.4251	.4265	.4279	.4292	.4306	.4319
1.5	.4332	.4345	.4357	.4370	.4382	.4394	.4406	.4418	.4429	.4441
1.6	.4452	.4463	.4474	.4484	.4495	.4505	.4515	.4525	.4535	.4545
1.7	.4554	.4564	.4573	.4582	.4591	.4599	.4608	.4616	.4625	.4633
1.8	.4641	.4649	.4656	.4664	.4671	.4678	.4686	.4693	.4699	.4706
1.9	.4713	.4719	.4726	.4732	.4738	.4744	.4750	.4756	.4761	.4767
2.0	.4772	.4778	.4783	.4788	.4793	.4798	.4803	.4808	.4812	.4817
2.1	.4821	.4826	.4830	.4834	.4838	.4842	.4846	.4850	.4854	.4857
2.2	.4861	.4864	.4868	.4871	.4875	.4878	.4881	.4884	.4887	.4890
2.3	.4893	.4896	.4898	.4901	.4904	.4906	.4909	.4911	.4913	.4916
2.4	.4918	.4920	.4922	.4925	.4927	.4929	.4931	.4932	.4934	.4936
2.5	.4938	.4940	.4941	.4943	.4945	.4946	.4948	.4949	.4951	.4952
2.6	.4953	.4955	.4956	.4957	.4959	.4960	.4961	.4962	.4963	.4964
2.7	.4965	.4966	.4967	.4968	.4969	.4970	.4971	.4972	.4973	.4974
2.8	.4974	.4975	.4976	.4977	.4977	.4978	.4979	.4979	.4980	.4981
2.9	.4981	.4982	.4982	.4983	.4984	.4984	.4985	.4985	.4986	.4986
3.0	.4987	.4987	.4987	.4988	.4988	.4989	.4989	.4989	.4990	.4990

This table is abridged from Table 1 of *Statistical Tables and Formulas*, by A. Hald (New York: John Wiley & Sons, Inc., 1952). Reproduced by permission of A. Hald and the publishers, John Wiley & Sons, Inc.